Solar Technologies for the
21st Century

Solar Technologies for the 21st Century

by Anco S. Blazev

THE FAIRMONT PRESS, INC.

CRC Press
Taylor & Francis Group

Library of Congress Cataloging-in-Publication Data

Blazev, Anco S., 1946-
  Solar technologies for the 21st century / by Anco S. Blazev.
    pages cm
  Includes bibliographical references and index.
    ISBN 0-88173-697-X (alk. paper) -- ISBN 0-88173-698-8 (electronic) -- ISBN 978-1-4665-8291-0
(Taylor & Francis distribution : alk. paper) 1. Solar energy. 2. Solar energy industries. I. Title. II.
Title: Solar technologies for the twenty-first century.

  TJ810.B55 2013
  621.47--dc23

                                                                                2012048663

Published by The Fairmont Press, Inc.
700 Indian Trail
Lilburn, GA 30047
tel: 770-925-9388; fax: 770-381-9865
http://www.fairmontpress.com

Distributed by Taylor & Francis Ltd.
6000 Broken Sound Parkway NW, Suite 300
Boca Raton, FL 33487, USA
E-mail: orders@crcpress.com

Distributed by Taylor & Francis Ltd.
23-25 Blades Court
Deodar Road
London SW15 2NU, UK
E-mail: uk.tandf@thomsonpublishingservices.co.uk

Printed in the United States of America
10 9 8 7 6 5 4 3 2 1

ISBN-0-88173-697-X (The Fairmont Press, Inc.)
ISBN-978-1-4665-8291-0 (Taylor & Francis Ltd.)

While every effort is made to provide dependable information, the publisher, authors, and editors
cannot be held responsible for any errors or omissions.

Contents

Preface

Different methods of converting the sun's energy into electricity have been used since the dark ages (no pun intended). The applications have been slowly evolving from simple sun drying of fish and fruits to using optics to light fires—all the way to today's wide-spread residential and commercial applications of photovoltaics and solar thermal technologies. And this is just the beginning.

The overall potential of solar power generation is astounding. The entire global electricity demand could be provided from just 3% of the world's deserts, while several hundred square miles in the Nevada desert could power all of the US's lower 48 states.

It sounds good and doable, but there are still gaps in solar technologies, energy markets, and global finances, so these must be solved first, and before we can expect serious solar energy generation in the US and abroad.

These gaps caused the latest solar boom and bust cycle (2007-2011) which is now in its last stages. It gave us a number of examples of how to and how not to do things.

Some of the examples from the last solar boom-bust cycle, worth remembering are:

- Government subsidies work, but are not a long-term solution.

- Political maneuvers, in most cases, cause more harm than good.

- Large companies and projects can fail too; the larger they are, harder they fall.

- Quick money-making schemes give the industry a bad name.

- Solar installations in cloudy areas have marginal success at best.

- Large-scale desert solar installations are the way to energy independence.

- Efficient energy storage is needed to make solar power compatible.

- All solar technologies are created equal, but do not perform equally.

- Using cheap materials and labor does not lead to long-term success.

The overwhelming conclusion from the last boom-bust cycle is that a lot of team effort, time and money are (still) needed to bring the solar energy to its rightful place as an efficient, reliable and profitable energy source.

Most people in key technical and administrative decision making positions were not even born during the last energy boom-bust cycles, so the lessons of these are lost on them. Because of that, we've all had to re-live the past and learn those lessons for ourselves. Let's hope that we have learned—and remember—the lessons this time.

Most importantly, we must roll up our sleeves and quickly make the necessary changes needed to ensure full development and implementation of renewable energies in our daily lives during the 21st century. It can be done. It should be done!

Acknowledgments

This book contains a number of complex subjects in many areas of the solar energy field, which require complete understanding of the different principals and subjects at hand. Since neither the author, nor anyone else, can claim absolute expertise in all subject areas, we do appreciate the full and friendly cooperation of our capable co-authors, helpers, assistants and editors.

We would like to thank also all the good people and well-meaning solar professionals and enthusiasts, who gave us permission to use part of their materials in this book. Without these, the book would not have been possible—at least not in its present form.

Even after all this effort, the book is only a glance at the quickly developing solar energy industry. The fast growing body of information—technological, financial, political, regulatory and socio-economic—is constantly changing.

Our thanks go to our many supporters and collaborators for their contributions, hands-on assistance, support and encouragement during the planning, information gathering, sorting, writing and editing processes.

We hope our collaborators and assistants will continue working with us on completing the tasks at hand as outlined herein. The key issues and problems must be put on the table for open, sincere and professional discussion and debate—with you, the reader, as judge and jury.

This is vitally important and absolutely needed to further the goals of the fledgling solar industry, and ensure its progress now and throughout the 21st century.

A. Blazev, 2012

Introduction

This book takes a close look at the solar technologies of today and tomorrow, describes their properties, and evaluates the technological potential of each type. We also take a close look at the logistics of deploying solar power, and analyze related socio-economic benefits of solar energy as a viable and sustainable way to solve urgent energy, environmental, and socio-economic problems.

Large solar power installations (vs. residential or small commercial ones) are clearly the best and fastest way to replace fossil fuels as the main energy source in this century. Implementing such installations is an immense task.

The different technologies have different properties and correspondingly different niche markets, so we need to learn more about each of these to use them efficiently, profitably, and safely.

The future will tell which solar technology belongs where, and we hope that analyses in this book will increase our knowledge base for making proper decisions based on scientific facts and data.

We also hope that the political and regulatory developments and changes, reviewed herein will work in favor of expanding the use of alternative energies—the answer to our energy independence and environmental cleanup—with emphasis on the development of large-scale solar power installations as the quickest way to achieve these goals.

This book, and the issues it discusses, should serve as both a guidebook for beginners, as well as an outline for the debates on solar power technologies, their issues and their proper use by specialists.

Debates must be initiated soon, and escalated proportionally, because we are running out of resources.

Our main objective is to create conditions that will put an end to the fragmentation and confusion in the energy field, stop the half-way solutions, answer the unanswered questions, and correct our myriad mistakes.

One goal of this book is to present examples of major successes and failures of solar companies and projects. While the successes were the clear majority for awhile, in 2011 the world's financial chaos, and poor-quality imports caused a very large number of solar companies and projects to fail.

By mid-2012 we saw an avalanche of failing solar companies and projects—an astounding 10:1 ratio of failures vs. successes.

We documented the key failures to date (fall of 2012) herein, in an attempt to provide a list of how *not* to do things in the future. We hope the reader will not misjudge our intentions, but will instead take a close look at these failures and take from them what is needed to ensure the success of the renewable energies and cleaning the environment in the 21st century.

Chapter 1

Solar Energy

Solar energy has not been developed thus far
only because the oil companies cannot patent sunlight.

—Anco Blazev

We have all heard, time after time, that solar power is badly needed and is here to stay this time around. But is it really? We have gone through the agony of several solar energy boom/bust cycles since the 1970s, and have seen the best and worst these can bring.

Having learned from those, we are now poised to make practical decisions. We must now identify and develop solar power conversion equipment and methods for its manufacturing, installation and operation that are profitable, reliable, sustainable, and safe.

Easier said than done! There are many companies pushing inefficient and unsafe technologies with limited futures which might even negatively affect the environment, further slowing the development of the solar industry as a whole.

Of course, there are also millions of cheaply made PV and wind components flooding world markets. Millions of PV modules installed worldwide were made using questionable materials and procedures, and under-qualified, underpaid labor.

How are we to sort the developments of the past—especially those since 2007—and project them into the 21st century, to ensure a bright future for our children and their children, powered by reliable, efficient and safe renewable energy?

These are the key questions we attempt to answer in this text. These are important questions because the potential of the sun's energy is great, and also because we simply have no other choice. It would be irresponsible to continue relying on the old finite and polluting energy sources.

It would be also stupid of us to not use the progress science and technology have made, which allows us to create the best materials, processes and devices needed to capture and convert sunlight into useful energy.

Figure 1-1 shows the sunlight's potential to provide most, if not all, of our energy needs.

Consider the universities, hi-tech companies, R&D and other organizations around the world, working full-

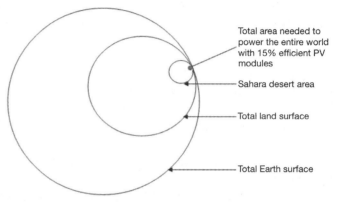

Figure 1-1. How little land is needed to provide power to the world. (Not to scale)

time on developing new solar materials and processes. Today's discoveries and developments are the most prolific ever. Some of the work done lately is on the verge of breaking the established limits of efficiency and performance. And this is just the beginning.

We owe future generations more than exhausted oil wells, empty mineral mines, air, soil, and water contamination; we must work harder at preparing the foundation for new energy technologies that new generations can build on—agreeing on the basics of the most abundant, clean and free energy source—sunlight.

SOLAR POWER GENERATION BASICS

Solar power generation devices are characterized by the type and amount of energy generated by exposing them to the sun's rays and collecting the produced power.

According to their principle of operation and design, solar power generation equipment and installations can be divided into three major categories:

1.   Thermal (T),
2.   Photovoltaic (PV), and

3.   Hybrid (PV-T and CPV-T), which are a combination of the other two.

Thermal solar power generating equipment produces heat of some sort, while photovoltaic power generating equipment generates DC or AC electric power. Hybrid equipment is a combination of T and PV components and could produce either or both types of energy.

Solar power generation equipment is evaluated on the basis of its:

1.   Technical performance, which includes efficiency, reliability, ease of operation and maintenance, and safety, and

2.   The economic results, or how much profit we make at the end of the day, and at the end of the useful life cycle of the equipment or project. This counts most when we compare solar technologies.

We will take very close look at the technical aspects of the different solar technologies, and because the bottom line is so very important, we'll take a close look at it in a special chapter.

So how do we convert sunlight into power? What does it take? The basic prerequisites for creating and operating a successful solar power generating plant are:

1.   Availability of solar energy (preferably unobstructed direct sunlight),

2.   Proper location (favorable climate and available land),

3.   Proper technology selection (PV equipment, inverters etc.),

4.   Availability of interconnection points (transmission lines, substation etc.),

5.   Suitable local, regulatory, and utilities conditions, and

6.   A team of capable and experienced professionals which is absolutely necessary to properly plan, design, set up and operate a PV power plant.

These are the critical factors to tackle when planning a PV power plant. Taking a closer look at some key prerequisites for a successfully designed, installed, and operated solar power plant, we start with the most obvious requirement, sunlight.

SUNLIGHT BASICS AND KEY CONCEPTS

Life on Earth depends on the sun's energy. The sun is a large star at the center of our solar system and is approximately 93 million miles away, but its life-giving energy arrives on Earth every day. This exact distance between the sun and the Earth is crucial to maintaining life. The Earth's orbit and its very appropriate distance from the sun are responsible for providing climate conducive to life which exists here. The other planets are either too close (thus too hot), or too far away (and too cold) to sustain organic life as we know it. The Earth also has an appropriate atmosphere with enough oxygen, carbon dioxide, and water to sustain life.

All life-sustaining elements are in the right proportion, form and shape needed for life to flourish on Earth. These are also in a very delicate balance, so changes or modifications to this balance are detrimental to life on Earth.

Sunlight essentially consists of a range of energy bands, which we generally refer to as the solar spectrum. We can see some of these bands, but most are invisible to the human eye. We can also feel some of these when they impact our skin and are perceived as heat. So, sunlight is radiation of the electromagnetic type, with what we see and feel being only a small part of its entire spectrum.

From our perspective, we think of visible light as the most important component of the solar spectrum, because we can see it. The infrared part of the solar spectrum is also noticeable since we feel it as heat. However, these are just small parts of the entire light (sunlight) spectrum, and they are clearly distinguished by their different wavelengths and respective properties.

As the study of light has advanced, physicists have found that while it is usually best to consider light as energy traveling through space as a wave, in some situations light behaves as if it were made up of tiny "packets" of energy or "particles" that have no mass and always travel at the same speed, which we call the speed of light. When light is discussed in this manner, the packets of energy are called photons, which is the term we use in this text.

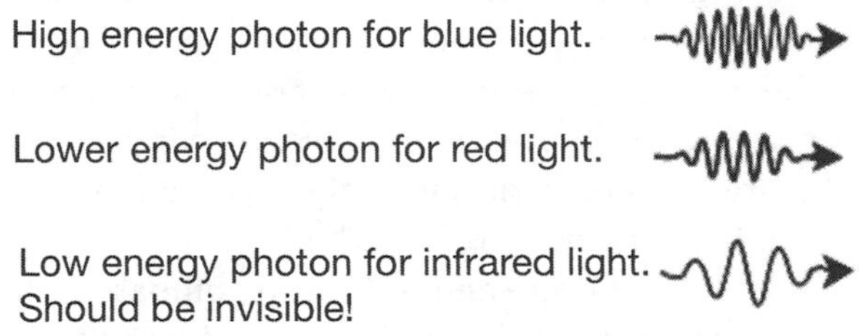

High energy photon for blue light.

Lower energy photon for red light.

Low energy photon for infrared light. Should be invisible!

Figure 1-2. Paths of sunlight energy particles (photons) (1)

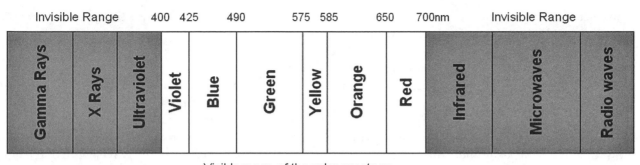

Visible range of the solar spectrum

Figure 1-3. The solar spectrum

The Solar Spectrum

The solar spectrum shows the range of energy coming from the sun. It is subdivided into sections and classifications based on the wavelengths and their particular energies. The key components of the solar spectrum follow.

Ultraviolet (UV) Radiation

This part of the spectrum accounts for less than 10% of the energy from the sun that reaches the Earth's surface. Most ultraviolet energy is filtered out by our atmosphere. Ultraviolet energy is divided into three parts:

a.   Ultraviolet C or (UVC) spans the range of 100 to 280 nm. The term ultraviolet refers to the fact that the radiation is at a higher frequency than violet light (and hence invisible to the human eye). Due to absorption by the atmosphere, very little UVC reaches the Earth's surface. This spectrum of radiation has germicidal properties and its property of killing bacteria and viruses is used in germicidal lamps.

b.   Ultraviolet B or (UVB) range spans 280 to 315 nm. This part of the spectrum is of interest to us because these are the rays that cause our skin to tan and burn, and our bodies to produce vitamin D. It is also absorbed by the atmosphere; along with UVC it is responsible for the photochemical reactions leading to the production of the ozone layer.

c.   Ultraviolet A or (UVA) range spans 315 to 400 nm. This part of the spectrum also has rays that cause our skin to tan, and hence is used in tanning beds and UVA therapy for psoriasis.

Visible Radiation

Most of the sun's energy reaching the Earth' s surface is in the visible (and some in the infrared) part of the spectrum. Visible radiation spans 400 to 700 nm. As the name suggests, it is this range that is visible to the naked eye. This is also the part of the spectrum that is most useful to power generation, since it is readily captured by PV equipment and turned into electricity.

Infrared (IR) Radiation

The infrared range spans 700 nm to 106 nm. It is responsible for an important part of the electromagnetic radiation that reaches the Earth. This is the part of the spectrum that heats water in solar thermal technologies which is then used for heating or to generate electricity. IR radiation is considered "parasitic" energy in most PV technologies, since it heats and overheats the PV cells and modules, rendering them inefficient and even causing them to fail.

The different wavelengths of the solar spectrum have different energies, with the most energetic being in the 400 to 700 nm range. Some wavelengths, like the IR, are even harmful to PV devices because they generate heat within the PV cells and modules.

Note: If the reader is interested in learning more about the nature of light, we recommend an excellent book, *QED: The Strange Theory of Light and Matter*, by Richard Feynman, where QED stands for "quantum electro-dynamics." Feynman gives the basics of light and its interactions with matter in a very clear and entertaining way.

He cautions us that we may not understand what he is saying, not because of technical difficulty, but because we may be unable to believe, or refuse to accept, the simplicity (or absurdity) of the truths he is discussing. "The theory of quantum electrodynamics describes Nature as absurd from the point of view of common sense. And it fully agrees with experiment. So I hope you can accept Nature as She is—absurd."

For all practical purposes, most solar conversion devices capture only certain limited wavelength ranges of the solar spectrum, but some PV technologies, using special non-silicon substrates and multi-junction solar cells are capable of capturing most of the wavelengths, including IR radiation, falling upon them and efficiently converting them into electric energy.

We consider the latter devices as the precursor of the PV technologies of the future, since they would be able to fully use the incoming sunlight, while at the same time withstanding hostile climates during long-term operation.

Factors Influencing Solar Power Reaching the Earth

Now that we have examined the solar spectrum and discussed its properties, let's see how much of that energy can be captured and converted into electric energy by our PV devices. These properties and characteristics of sunlight are very important for calculating and designing solar power generating installations.

Sunlight travels from the sun to Earth in approximately 8 minutes, and while it loses some of its energy during this journey, most of it arrives here safe and sound. Measured at the top of the atmosphere, we find the highest power density reaching up to 1,367 W/m².

That is to say that if we could mount a PV module that was 1 m² in an area 200 miles above the Earth's surface, and if that PV module were 100% efficient (which is not yet possible), we would be able to generate 1,367 W DC electric power from it. The problem of how to transport the produced electricity back to Earth is a separate issue for which we have no answer at this time.

Several factors commonly used to characterize sunlight and measure its properties follow (1)

Spectral Radiance

The spectral irradiance as a function of photon wavelength (or energy), denoted by F, is the most common way of characterizing a light source (sunlight in our case). It gives the power density at a particular wavelength. The units of spectral irradiance are in Wm-2μm-1. The Wm-2 term is the power density at the wavelength $\lambda(\mu m)$. Therefore, the m-2 refers to the surface area of the light emitter and the μm-1 refers to the wavelength of interest.

In the analysis of solar cells, the photon flux is often needed as well as the spectral irradiance. The spectral irradiance can be determined from the photon flux by converting the photon flux at a given wavelength to W/m². The result is then divided by the given wavelength, as shown in the equation below.

$$F = \left(\frac{W}{m^2\mu m}\right) = q\Phi \frac{1.24}{\lambda^2(\mu m)} = q\Phi \frac{E^2(eV)}{1.24}$$

where:

- F is the spectral irradiance in Wm-2μm-1;
- Φ is the photon flux in # photons m-2sec-1;
- E and λ are the energy and wavelength of the photon

in eV and μm respectively; and q, h and c are constants

Photon Energy

Sunlight consists of many photons traveling together as a photon flux. Each photon in the flux is characterized by either a wavelength, denoted by λ or equivalently energy denoted by E. There is an inverse relationship between the energy of a photon (E) and the wavelength of the light (λ) given by the equation:

$$E = \frac{hc}{\lambda}$$

where:

- h is Planck's constant and
- c is the speed of light.

The above inverse relationship means that light consisting of low energy photons (such as "red" light) has a long wavelength. When dealing with "particles" such as photons or electrons, a commonly used unit of energy is the electron-volt (eV) rather than the Joule (J). An electron volt is the energy required to raise an electron through 1 volt, thus 1 eV = 1.602 x 10-19 J.

By expressing the equation for photon energy in terms of eV and μm we arrive at a commonly used expression which relates to the energy and wavelength of a photon, as shown in the following equation:

$$E = \frac{1.24}{\lambda(\mu m)}$$

Note: The exact value of 1 × 106 (hc/q) is 1.2398 but the approximation 1.24 is sufficient for most purposes.

Photon Flux

The photon flux (and its quality and quantity) determine the intensity of the sunlight reaching our PV devices. The photon flux is defined as the number of photons per second per unit area:

$$\Phi = \frac{\text{\# of photons}}{\sec m^2}$$

The photon flux is important in determining the number of electrons which are generated, and hence the current produced from a solar cell. As the photon flux does not give information about the energy (or wavelength) of the photons, the energy or wavelength of the photons in the light source must also be specified. At a given wavelength, the combination of the photon wavelength or energy and the photon flux at that wavelength can be used to calculate the power density for photons at the particular wavelength.

Power Density

The power density is calculated by multiplying the photon flux by the energy of a single photon. Since the photon flux gives the number of photons striking a surface in a given time, multiplying by the energy of the photons comprising the photon flux gives the energy striking a surface per unit time, which is equivalent to a power density. To determine the power density in units of W/m², the energy of the photons must be in Joules. The equation is:

$$H = \left(\frac{W}{m^2}\right) = \Phi \times \frac{hc}{\lambda} (J) = q\Phi \frac{1.24}{\lambda(\mu m)}$$

where:

$\quad \Phi \quad$ is the photon flux.

One implication of the above equations is that the photon flux of high energy (or short wavelength) photons needed to give a certain radiant power density will be lower than the photon flux of low energy (or long wavelength) photons required to give the same radiant power density. In the animation, the radiant power density incident on the surface is the same for both the blue and red light, but fewer blue photons are needed since each one has more energy.

The total power density emitted from a light source (sunlight in this case) can be calculated by integrating the spectral irradiance over all wavelengths or energies of interest. However, a closed form equation for the spectral irradiance for a light source often does not exist. Instead the measured spectral irradiance must be multiplied by a wavelength range over which it was measured, and then calculated over all wavelengths. The following equation can be used to calculate the total power density emitted from a light source.

$$H = \int_0^\infty F(\lambda)\, d\lambda = \sum_{i=0}^\infty F(\lambda)\, \Delta\lambda$$

where:

$\quad H \quad$ is the total power density emitted from the light source in W m-2;

$\quad F(\lambda) \quad$ is the spectral irradiance in units of Wm-2μm-1; and

$\quad d\lambda$ or $\Delta\lambda \quad$ is the wavelength.

As the sunlight travels down and gets close to Earth, it has to travel through the atmosphere where it collides with dust and water vapor particles, thus losing some of its power. So, on a perfectly clear day on the equator we could measure up to 1,110 W/m² (down from 1,367 W/m² measured above the atmosphere).

In the Arizona desert we measure 900-1100 W/m² on a clear summer day. That's not bad for generating useful power from a piece of otherwise "useless" desert land during daylight periods, when we need it most, especially during the summer months.

Clouds, fog and dust will rob some of the power that sunlight is trying to deliver to Earth, but even then most PV modules will be able to convert some of the energy reaching their surface to electric power (10-90%, depending on the cloud cover density). Although that is a greatly reduced amount, it is still enough to produce a lot of usable energy.

Ask the Germans, where under mostly cloudy conditions the number of PV installations has risen at an unbelievably high pace lately, and continues to rise.

The curvature of the Earth also influences the amount of energy that strikes its surface at any given location. The intensity of the sunlight is greater in the area of the equator between the tropics, and it loses intensity towards the South and North Poles. This is because the sunlight, due to the Earth's curvature, has to travel a much greater distance at a sharper angle through the atmosphere before reaching the Earth's surface.

In addition, since the Earth is tilted on its axis, the time of year also influences the amount of energy which strikes its surface at different locations and times.

So, the important characteristics of incident solar energy are:

1. The spectral content of the incident light (visible, UV and IR light content),

2. The average radiant power density of sunlight (W/m²) at the location,

3. The angle at which the incident solar radiation strikes a PV module,

4. The seasonal sunlight energy (summer vs. winter), and the local variations, and

5. The atmospheric and weather conditions (clouds, fog, smog, etc.).

All these properties of sunlight are important for designing PV projects, and are used extensively by design engineers and installers alike, making them integral parts of the proper execution of any solar project. We will take a closer and much more detailed look at all these parameters in the text below.

Practical Characteristics of Solar Energy and Their Use

To produce electricity, sunlight has to be captured and converted into thermal or electric energy suitable for human consumption. As we discussed in the previous section, sun energy races toward the Earth at a very high speed, and if its path were not obstructed by space junk or clouds and dust in the atmosphere, it would arrive here at full power.

We call this "beam" or "direct" radiation. Arriving "directly" from the sun it is most powerful and measures around 1367 W/m² just above our atmosphere. Its power drops after crossing the atmosphere to approximately 900-1,100 W/m² as measured at noon in the deserts during the summer months, and much less than that in other parts of the globe and during different seasons of the year.

When the sunlight hits clouds, dust, or man-made gasses in the air, it gets scattered and we call that "diffused" radiation. Diffused radiation has properties very different from direct radiation. It contains less energy, and thus PV modules will produce less power under diffused radiation—in some cases much less. Particularly, concentrating thermal or PV equipment is affected by this diffusion, causing it to lose its focus and operate well below its maximum efficiency rating, if at all, under extreme conditions.

Sunlight hitting the Earth is reflected from its surface, and we call this effect "albedo." Different materials have different reflecting properties, but most do reflect and some reflect a lot. Take fresh snow, for example. It will reflect almost 80% of the light falling on its surface. Water, on the other hand, absorbs most of the sunlight, instead, and gets heated in the process. Thus, reflected sunlight can be captured by our PV modules installed nearby as well. The albedo always has an effect on PV module performance, so it should be taken into consideration, especially in areas with snow cover or other highly reflective ground surface cover.

As we discussed in the previous section, another factor that affects the amount of energy available for conversion into electricity is the distance that the sunlight travels once it enters the Earth's atmosphere. At certain times of the day and year, the sun seems to be overhead at a 90-degree angle to the Earth's surface, and this is when sunlight travels the shortest path and is the strongest.

We call this Air Mass = 1 (or AM 1). AM 0 is measured above the atmosphere and is much stronger than AM 1. In the early morning and later afternoon, the sun is at a sharper angle and sunlight travels a longer path through the atmosphere, so the angle decreases (approaching 45 degrees), the sunlight has a longer path to travel and the AM number increases: AM 1.15, 1.5, etc. depending on the angle. The sharper the angle of the sun's rays, the larger the AM number and objects' shadows. AM 1.5 is measured at an angle close to 45 degrees.

The air mass number can be determined by the formula:

$$AM = \sqrt{1 + \left(\frac{s}{h}\right)^2}$$

where:
  h   is the object's height, and
  s   is its shadow length

The revolution of the Earth around the sun and its rotation on its axis produces seasonal and daily effects, which vary by location on the globe. This location is measured on the world map in terms of longitude (east-west direction), and latitude (north-south direction). The intersection of these provides us with a precise point on the map.

All these components taken together represent what we call "global radiation," which is an important factor in the proper design, installation and operation of solar energy generating systems. Solar professionals need to be very familiar with it, if a properly designed PV system is their goal.

The Environmental Variables

The solar spectrum components and the quantity of sunlight traveling in space at any moment are constant, with only slight and usually predictable variations. When the sunlight hits the Earth's atmosphere, however, it becomes a variable, depending mostly on the contents of the atmosphere, the seasons, local weather, and time of day. Thus the amount of sunlight we receive on Earth is a variable, and out of our control, but must be considered when working with solar equipment.

Water (moisture) in the atmosphere (clouds), dust, CO_2 and other gasses absorb radiation, thus diminishing the amount and level of energy that reaches Earth. So the large amount of atmospheric water vapor and CO_2 in the sunlight's path determine how much energy reaches the Earth's surface. The influence of these variables is significant and must be well understood and taken into consideration in our solar calculations and designs, if we are to optimize the performance of our PV devices.

There is not much we can do about these variables, for they obey higher orders, and so we can call them "fixed" variables. We can, however, work around them, building and using renewable energy technologies that are best suited for local conditions.

Major variables over which we have no control are local weather conditions—clouds, rain, snow, fog, smog, dust, humidity, etc. These must also be well understood and studied carefully, when designing, installing and operating solar energy generating systems—thermal or photovoltaic.

We cannot control the weather, dust, smog and other natural phenomena, but we can study historical weather data and anticipate their behavior, to compensate for their effects, thus obtaining the highest possible power output at a given location.

Ideally, large-scale solar power plants should be installed in locations with the least cloud, fog and humidity levels. However since there is no place with perfectly clear skies all through the year, we must consider clouds and other effects of weather, estimate their influence, and design systems accordingly.

There are, however, variables which we can and should control, such as the size, and position of trees, buildings, smoggy factories, dusty fields, and other obstructions which will diminish system performance with time. Tall trees and structures near a power field will reduce the output, so they must be avoided, removed, or trimmed.

Air pollution generated by industrial activities can hinder the performance of PV devices, so it must be considered. The effects of smog and fog from nearby populated centers or large bodies of water must also be considered and calculated as accurately as possible.

Air and ground temperatures are other variables worthy of mention, due to their effect on power output of solar power generating devices. Our Earth provides a marvelous thermodynamic balance, where the temperature increase due to incoming sunlight during the day is balanced by the outgoing heat during the cool of night, so that balance is maintained.

During the summer months, however, the balance in some areas, like deserts, is temporarily altered. The air and ground get extremely hot during the day, over heating solar energy generating equipment. Solar power field structures could reach temperatures well over the operating limits of the PV devices in them (temperatures near 180°F have been measured inside PV modules in the Arizona desert), thus causing drastic power output decrease and electromechanical failures.

Prolonged exposure to extreme heat can cause deterioration, damage and even destruction to most present-day photovoltaic devices, so serious consideration must be given to proper use of these devices in extreme climates. Their function and behavior must be well understood and proven, before the design stage and definitely before installing and exposing them to harsh conditions.

Finally, we must also take into consideration the environmental consequences of our efforts. Installing acres upon acres of solar collectors (thermal or PV power fields) in the deserts, for example, could have serious effects on their delicate ecosystems. Unfortunately, there are not enough data to predict these effects, although efforts to understand and control them are ongoing.

There may be some positive effects of large-scale installations on the environment too. Those structures will provide shade which could cool the ground in summer; however the Earth has a fragile environmental balance, and every action people take for their own comfort and convenience, has an effect on the environment—usually not for the better. So, covering a large land mass (thousands of acres) with any solar technology will have an effect, which we must evaluate and consider before acting.

In all cases, all variables must be well understood, thoroughly analyzed, taken into consideration, and their impacts incorporated into system design. Remember that for every action of man, the environment responds with a counter action.

SOLAR ENERGY AVAILABILITY AND USE IN POWER GENERATION

Many things must be considered in the planning and design of an efficient and profitable solar power installation. The most important is figuring how much solar energy is available at a location, and the most efficient ways to use it.

We'll start with figuring how much energy we have to work with, and how it is used in our daily activities.

The total solar energy reaching the Earth's atmosphere is estimated at 175PW (PW is 1,000,000,000MW). 30% is reflected back into space and some is absorbed by the clouds. Whatever sunlight reaches the Earth's surface is absorbed by the oceans and land mass.

A major portion of the sunlight reaching the Earth falls on the world's deserts, because they are mostly void of cloud cover. Most large-scale solar power plants are located in desert areas, and that is also where most solar power will be generated in the future. This is why our emphasis is on the operation and challenges of PV technologies in the world's deserts.

Annual world-wide energy consumption is approximately 1/8000 of the total sunlight energy reaching the Earth' surface, so theoretically only a small part of our deserts could provide the entire world's energy demand.

Putting this in practice, however, is quite complicated, because capturing and converting sunlight into useful energy, and delivering it to the point of use is not easy. So let's go deeper into it to isolate and sort out the problems.

Conversion of Sunlight for Power Generation

We mentioned above that sunlight has to be captured and converted into the two forms most useful for human consumption, thermal and electric energy. To do this right, we need to know the properties of sunlight as well as the properties and behavior of everything related to the process of converting it into useful energy.

We know that full unobstructed sunlight, like what we see in the deserts, is called "beam" or "direct" radiation, while when obstructed by clouds, dust and other particles in the atmosphere it is called "diffused" radiation. The direct and diffused radiations, as well as the albedo (light reflection from ground surfaces) are useful forms of energy.

They have different properties under different atmospheric conditions and seasons, so we do need to know how they affect solar power generation in order to design a most efficient PV power plant.

Another factor to consider is the path that sunlight travels to reach the location of our PV plant. For this we need to take a close look at the Earth' path around the sun, and around its axis, and understand the behavior of sunlight at the location we are considering.

The Earth makes a full circle around the sun every 364.99 days, while at the same time making one full revolution around its axis every 24 hours. The rotation of the Earth around the sun determines the seasons, while the rotation around its axis determines time of day. The combination of these two factors is very important for the proper design and operation of solar generating systems. Calculating these two parameters, we can predict how much sunlight we can get at anytime of year and adjust the solar power collecting devices for maximum performance.

Figure 1-4 shows that on June 21 of every year the northern hemisphere is inclined slightly towards the sun. This slight inclination is enough to provide a shorter path for the sunlight, more intense direct radiation, and generally warmer and sunnier summertime. On December 21 of every year, the Earth is inclined in the opposite direction, taking the northern hemisphere farther from the sun, where the sunlight travels a longer distance to reach us, translating to less direct radiation and shorter, colder, darker days.

Figure 1-5 shows the amount of power produced by a solar power generating system at different seasons and different hours of the day. Basically it tells us that we will get maximum power at noon on June 21st and much less at noon on December 20th, due to the Earth's position.

Anytime in between, we get solar energy between the maximum and minimum levels. Again, it all depends on local weather conditions, because a cloudy or rainy day in December would produce almost the same amount of

sunlight as such a day in June.

We also see from Figure 1-5 that PV modules will generate more energy if they are pointed directly at the sun all day (in effect tracking it with face pointed directly at the sun), instead of being anchored in a fixed position.

The difference is exaggerated during winter, because of the sharper angle of the sunlight falling on fixed-tilt modules. We will expand on these details, including tracking vs. fixed solar power collection in this text, so for now we just need to understand that the most important parameters for efficient solar power generation are geographic location, seasons, local weather, time of day, the technology we use, and whether we use tracking or fixed

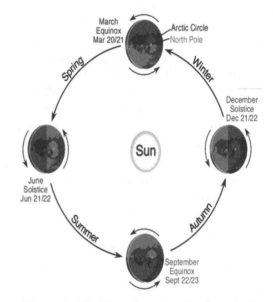

Figure 1-4. Earth path and cycles (seasons)

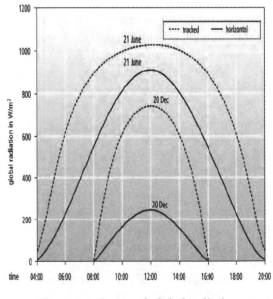

**Figure 1-5. Seasonal global radiation.
(Tracking vs. fixed mounted PV modules.)**

PV modules.

The darker areas of the map in Figure 1-6a show the highest solar radiation areas of the world, which are most suitable for solar power generation. Of course any area will generate some power, but the darker areas will generate many times that of the lighter areas. Again—location, location, location.

Similarly, the darker areas on the US map in Figure 1-6b show where the sunlight is more intense and where the power output of any type of solar generating device will be greatest. Obviously, the deserts of the southwest USA seem to be the most appropriate for this purpose, and that is why there are such a great number of solar (thermal and PV) power plants installed and planned for installation in those areas.

These areas are a focal point of this book as well, because we believe that the energy future is related to the proper and efficient use of PV technologies in US and world deserts. Unfortunately, the inhospitable climatic conditions in the deserts are problematic for most PV technologies, which is also something we are addressing herein, with the hope of shedding more light on those issues and encouraging open discussions about solutions.

Insolation and Climate Effects

We can see now that the weather (clouds, rain, snow, fog and smog) plays a great role in the performance of any PV installation. Inconsistent cloud cover causes variability issues (power fluctuations) which are undesirable and even harmful.

There are some variability problems on mostly sunny days and fully cloudy days, mostly expressed in reduced output, but the variability is greatest during partially cloudy days, when the sun is randomly going in and out of the clouds.

During this time, power output is going up and down uncontrollably, creating serious and harmful fluctuations in power being injected into the power grid as well.

Power output increases gradually from zero in early morning to full (peak) power at noon and then gradually back to zero by late afternoon. This is another, albeit more predictable and controllable, variable. On a clear sunny day that ascent and descent will be a smooth bell curve, but on a cloudy day it would be a jumble of ups and downs of solar intensity and related power fluctuations.

Some fluctuations would be gradual, some fast, and some slow, but basically with no way to predict and compensate for them in advance. See Figure 1-7.

There are a number of effects that influence the behavior of the PV system on partially cloudy days, the majority of which are related to these conditions: spatial-temporary, incident angle, inverter, area affected, and temperature, which follow in more detail.

Spatial-temporary Effects

These are short-time variations in cloud cover—the sun going in and out randomly. To summarize the effect of this variable on the power output of a solar plant, we need to take into consideration all other factors (see below) but basically speaking, the larger the plant, the larger the output power fluctuations will be. Also, the shorter the duration of the sunlight intensity (insolation) fluctuations, the larger the power output effect. In other words, if the sun comes in and out of the clouds every second or two, the PV system (collectors-transmission-inverters) will not have enough time to process the signal and convert it properly

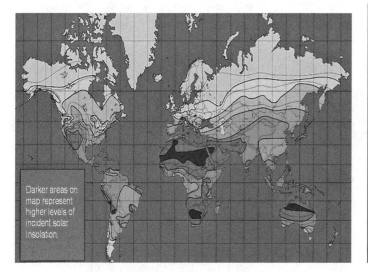

Figure 1-6a. World's solar radiation.

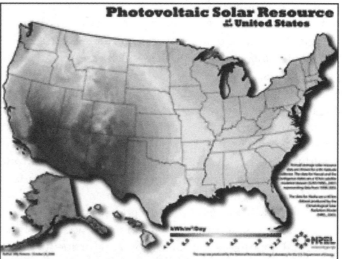

Figure 1-6b. US solar radiation.

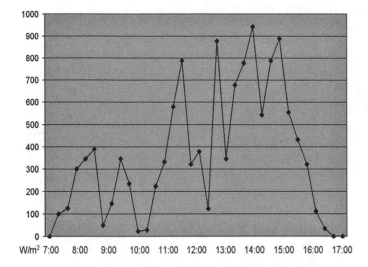

Figure 1-7. Partially cloudy day solar radiation readings

into useable energy. This is one of the major reasons we always recommend that large solar power plants should be installed in areas with high solar insolation and minimal weather fluctuations.

Incident Angle Effects

Incident angle effects are caused by improper positioning of the PV modules in relation to the sun. This anomaly could be caused by a number of factors, the most important being the awkward position of fixed PV modules' systems during early morning and late afternoon. During these times, fixed systems collect only a small percent of the overall PV power reaching their surface.

Inverter Effects

Inverter effects are caused by the inverters' inability to properly follow and adapt to the random ramp-up and ramp-down patterns of generated power during variable conditions. Because of that, inverters are not able to convert 100% of the incoming DC power, which causes additional power losses.

Basically, when a cloud covers the sky partially, the inverters sense the change and start regulating the power-down cycle, in a less-than-efficient manner in most cases. At a certain point—which varies from inverter to inverter—the cloudy sky will reduce the power output from the modules to such a low level that the inverters cannot handle it and simply shut down.

So even if the modules are generating some power, the inverters do not "see" it, and it is wasted. These abnormal conditions are caused by and/or identified as MTTP issues, IEEE 1547 dropouts, inverter "clipping" abnormalities, etc.

Tracking Systems

Trackers are much better at collecting the maximum sun energy, because their x-y drive controllers position the PV modules to face the sun properly at all times. Trackers, however, are affected by partial cloud cover even more than their fixed PV module cousins, because tracker controls might get confused by the sun hiding behind erratic cloud cover, thus not providing the best angle for sunlight collection.

Cloud Cover

The effects of cloud cover vary with the size of the power plant. Large plants cover large areas of land, where partially cloudy days might complicate things even further by throwing a shadow over one side of the field, while the other side is clear and gets full sun. This situation will make attempts to get a complete handle on incoming vs. outgoing power almost impossible. This condition is especially critical in fields using trackers because the shaded side will stop generating power altogether, adding to the complexity of the variability factor.

Temperature Effects

Temperature effects are abnormalities caused by high temperature on solar cells and panels. Silicon and thin film PV modules drop 0.5% efficiency with each degree C temperature increase. Temperature increases are implicated in several failure or degradation modes of PV cells and modules, as elevated temperatures increase the stresses associated with thermal expansion and significantly increase degradation rates.

These factors are hard to keep track of, let alone control as needed for producing consistent maximum power output. Because of that, we need to take them into consideration during the design of the power field and then do whatever is necessary to control their effect during operation.

Derating Factor

Once we have captured solar energy via PV devices (solar cells or modules) we need to convert it into electricity. Some of the energy is lost during the conversion process; i.e., only 15% is converted by a PV module, and the losses continue down the line before the energy can be used. In a PV power plant, the loses are from PV equipment, wires, connectors, controllers, inverters, transformers, etc.—equipment that is needed to capture and convert sunlight into AC power.

The losses from each piece of equipment determine its efficiency, and the corresponding power loss amount is the derating factor. The derating factors of all plant components must be used to calculate the PV system's ability to produce DC and AC power. Some of these factors are detailed as follow:

DC Rating (Nameplate)

The size of a PV system is its nameplate or DC power rating, which is the amount of DC power it is supposed to produce according to the manufacturer at STC (1,000 W/m² solar irradiance and 25°C module determined by adding the total PV modules' power listed on each PV module in watts and then dividing the sum by 1,000 to convert it to kilowatts (kW); our 100 MWp PV system consists of 500,000 PV modules, each measuring 200 Watts at STC, and is expected to generate 100 MWp of power at STC.

This is a theoretical number, simply because PV modules seldom operate at STC in the field, and for sure not in the fields we are most interested in—harsh desert areas with the highest temperature extremes. Nevertheless, the DC rating is important for establishing a baseline of the generation power and is used in many consequent calculations and considerations.

DC Derating Factor

This is basically the difference between the DC rating (nameplate) as provided by the manufacturer and the actual power generated at location. We perform a series of field tests after installation to verify and baseline the performance at this particular location. The tests could be done with each string of modules independently, providing an average DC power produced for each string. The different strings could be added to obtain the actual total DC power produced by the entire installation. When compared with the nameplate, the difference gives us the actual DC derating factor of the entire power plant under full power. Field measurements are usually lower than the nameplate rating.

$$DC\ derating\ factor =$$
$$Actual\ generated\ DC\ power/Nameplate$$

A derate factor of 0.95, for example, indicates that post-install field testing power measurements at the location were 5% less than the manufacturer's nameplate rating measured at STC. Or, our power plant with 100 MWp nameplate produced only 95 MWp during a certain time period under actual operating conditions. In this case:

$$DC\ derating\ factor =$$
$$95\ MWp/100MWp =$$
$$0.95,\ or\ 95\%\ of\ nameplate$$

During this test and after consecutive time-lapse tests, we can find a number of variables and discrepancies which must be taken into consideration when estimating the initial overall plant efficiency, its performance levels and the related DC derating.

DC to AC Derating Factor

Another critical variable of the overall performance estimate of a PV power-generating system is its DC to AC derating factor (or DC to AC conversion efficiency), which is simply the multiple of the derate factors of the different power plant components.

$$DC\ to\ AC\ Derating\ factor =$$
$$multiple\ of\ components'\ derate\ factors$$

Table 1-1. Average derating factors as used in PVWATT (NREL)

Default	*Low*	*High*	
PV Module Nameplate DC Rating	0.95	0.88	0.96
Inverter and Transformer	0.92	0.88	0.96
Mismatch	0.98	0.97	0.99
Diodes and connections	0.995	0.99	0.997
DC Wiring	0.98	0.97	0.99
AC Wiring	0.99	0.98	0.993
Soiling	0.95	0.30	0.995
System Availability	0.980	0.00	0.995
Shading	1.00	0.00	1.00
Sun-tracking	1.0	0.95	1.35
AGE 1.00	0.70	1.00	
Overall DC-to-AC Derating Factor	**0.77**	**0.00**	**0.77**

In this case, DC to AC derating factor = 0.95 x 0.92 x 0.98 x 0.995 x 0.98 x 0.99 x 0.95 x 0.98 x 1.00 x 1.00 x 1.00 = 0.77. Or, 77% of the nameplate DC power will be converted into AC power at this particular location with this particular equipment. In other words, our 100 MWp power plant will generate 77 MWp of AC electricity and send it into the grid.

The Derating Factor Components in More Detail:

• *PV module nameplate DC rating*

This accounts for the accuracy of the manufacturer's nameplate rating. Field measurements of PV modules may show that they are different from their nameplate rating or that they experience light-induced degradation upon exposure. A derate factor of 0.95 indicates that testing yielded power measurements at STC that were 5% less than the manufacturer's nameplate rating.

• *Inverter and transformer*

This reflects the inverter's and transformer's combined efficiency in converting DC power to AC power. A list of HYPERLINK "http://www.gosolarcalifornia. ca.gov/equipment/inverters.php" inverter efficiencies by manufacturer is available from the Consumer Energy Center. The inverter efficiencies include transformer-related losses when a transformer is used or required by the manufacturer.

• *Mismatch*

The derate factor for PV module mismatch accounts for manufacturing tolerances that yield PV modules with slightly different current-voltage characteristics. Consequently, when connected together electrically, they do not operate at their peak efficiencies. The default value of 0.98 represents a loss of 2% because of mismatch.

• *Diodes and connections*

This derate factor accounts for losses from voltage drops across diodes used to block the reverse flow of current and from resistive losses in electrical connections.

• *DC wiring*

The derate factor for DC wiring accounts for resistive losses in the wiring between modules and the wiring connecting the PV array to the inverter.

• *AC wiring*

The derate factor for AC wiring accounts for resistive losses in the wiring between the inverter and the connection to the local utility service.

• *Soiling*

The derate factor for soiling accounts for dirt, snow, and other foreign matter on the surface of the PV module that prevent solar radiation from reaching the solar cells. Dirt accumulation is location- and weather-dependent. There are greater soiling losses (up to 25% for some California locations) in high-traffic, high-pollution areas with infrequent rain.

For northern locations, snow reduces the energy produced, and the severity is a function of the amount of snow and how long it remains on the PV modules. Snow remains longest when sub-freezing temperatures prevail, small PV array tilt angles prevent snow from sliding off, the PV array is closely integrated into the roof, and the roof or another structure in the vicinity facilitates snow drift onto the modules.

For a roof-mounted PV system in Minnesota with a tilt angle of 23°, snow reduced the energy production during winter by 70%; a nearby roof-mounted PV system with a tilt angle of 40° experienced a 40% reduction.

• *System availability*

The derate factor for system availability accounts for times when the system is off because of maintenance or inverter or utility outages. The default value of 0.98 represents the system being off 2% of the year.

• *Shading*

The derate factor for shading accounts for situations in which PV modules are shaded by nearby buildings, objects, or other PV modules and arrays. For the default value of 1.00, the PVWatts calculator assumes the PV modules are not shaded. Tools such as HYPERLINK "http://www. solarpathfinder.com/" Solar Pathfinder can determine a derate factor for shading by buildings and objects. For PV arrays that consist of multiple rows of PV modules and array structures, the shading derate factor should account for losses that occur when one row shades an adjacent row.

• *Sun tracking*

The derate factor for sun-tracking accounts for losses for one- and two-axis tracking systems when the tracking mechanisms do not keep the PV arrays at the optimum orientation. For the default value of 1.00, the PV Watts calculator assumes that the PV arrays of tracking systems are always positioned at their optimum orientation and performance is not adversely affected.

• *Age*

The derate factor for age accounts for performance losses over time because of weathering of the PV modules.

The loss in performance is typically 1% per year. For the default value of 1.00, the PV Watts calculator assumes that the PV system is in its first year of operation. For the eleventh year of operation, a derate factor of 0.90 is appropriate.

Note: Because the PV Watts overall DC-to-AC derate factor is determined for STC, a component derate factor for temperature is not part of its determination. Power corrections for PV module operating temperature are performed for each hour of the year as the PV Watts calculator reads the meteorological data for the location and computes performance. A power correction of –0.5% per degree Celsius for crystalline silicon PV modules is used.

SOLAR CELLS AND MODULES

To generate photo-electricity (electric power from sunlight) using solar cells or devices, we need sunlight first. We know that sunlight can vary significantly with the seasons, from day to day, from hour to hour and even minute-by-minute. Provided that we have good, unobstructed sunlight, we need only well-functioning solar cells and modules.

We see solar cells, usually in solar panels, all around us. They come in different types and shapes, but all do the same work—converting sunlight into useful DC electric energy—which then can be used as is, or be converted into AC electricity for use in home appliances or commercial enterprises.

To understand how solar cells work, however, we need to look deeper into their structure and function, to look closely at the theory behind the solar cell's electric generation from the semiconductor point of view.

A solar cell needs to be an efficient converter of the sun's energy into electricity, even as it is working against a certain set of constraints. These constraints will be defined by the working environment in which solar cells are produced. For example in a commercial environment where the objective is to produce a competitively priced solar cell, the cost of fabricating a particular solar cell structure must be taken into account. However, in a research environment where the objective is to produce a highly efficient laboratory-type cell, maximizing efficiency rather than cost is the main consideration.

Sometimes these two conditions do not meet in real life, so we must use care when reading solar cells' specifications, and news related to their efficiencies.

Silicon Solar Cell Efficiency

The theoretical efficiency for photovoltaic conversion is in excess of 86.8%. However, the 86.8% figure uses detailed balance calculations and does not describe device implementation. For silicon solar cells, a more realistic efficiency under one sun operation is about 29%. (1)

The maximum efficiency measured for a silicon solar cell is currently 24.7% under AM1.5G. The difference between the high theoretical efficiencies and the efficiencies measured from terrestrial solar cells is due mainly to two factors. The first is that the theoretical maximum efficiency predictions assume that energy from each photon is optimally used, that there are no unabsorbed photons, and that each photon is absorbed in a material which has a band gap equal to the photon energy.

This is achieved in theory by modeling an infinite stack of solar cells of different band gap materials, each absorbing only the photons which correspond exactly to its band gap.

The second factor is that the high theoretical efficiency predictions assume a high concentration ratio. Assuming that temperature and resistive effects do not dominate in a concentrator solar cell, increasing the light intensity proportionally increases the short-circuit current. Since the open-circuit voltage (Voc) also depends on the short-circuit current, Voc increases logarithmically with light level. Furthermore, since the maximum fill factor (FF) increases with Voc, the maximum possible FF also increases with concentration. The extra Voc and FF increases with concentration which allows concentrators to achieve higher efficiencies.

In designing single junction solar cells, the principles for maximizing cell efficiency are:

- Increasing the amount of light collected by the cell that is turned into carriers;

- Increasing the collection of light-generated carriers by the p-n junction;

- Minimizing the forward bias dark current;

- Extracting the current from the cell without resistive losses.

Solar Cells' Structure
Silicon solar cells consist of:

Substrate Material, Usually Silicon
Bulk crystalline silicon dominates the current photovoltaic market, in part due to the prominence of silicon in the integrated circuit market. As is also the case for transistors, silicon does not have optimum material parameters.

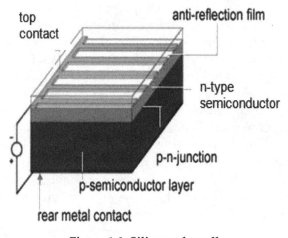

Figure 1-8. Silicon solar cell

In particular, silicon's band gap is slightly too low for an optimum solar cell, and since silicon is an indirect material it has a low absorption co-efficient.

While the low absorption co-efficient can be overcome by light trapping, silicon is also difficult to grow into thin sheets. However, silicon's abundance and its domination of the semiconductor manufacturing industry have made it difficult for other materials to compete.

Cell Thickness, 100-500 µm
An optimum silicon solar cell with light trapping and very good surface passivation is about 100 µm thick. However, thickness between 200 and 500 µm are typically used, partly for practical issues such as making and handling thin wafers, and partly for surface passivation reasons.

Doping of Base, 1 Ω•cm
A higher base doping leads to a higher Voc and lower resistance, but higher levels of doping result in damage to the crystal.

Reflection Control, Textured Front Surface
The front surface is textured to increase the amount of light coupled into the cell.

Emitter Dopant, n-type
N-type silicon has a higher surface quality than p-type silicon, so it is placed at the front of the cell where most of the light is absorbed. Thus the top of the cell is the negative terminal and the rear of the cell is the positive terminal.

Emitter Thickness, <1 µm
A large fraction of light is absorbed close to the front surface. By making the front layer very thin, a large fraction

of the carriers generated by the incoming light are created within a diffusion length of the p-n junction.

Doping Level of Emitter, 100 Ω/cm²
The front junction is doped to a level sufficient to conduct away the generated electricity without resistive loses. However, excessive levels of doping reduce the material's quality to the extent that carriers recombine before reaching the junction.

Grid Pattern, Fingers 20 to 200 µm Width,
Placed 1-5 mm Apart
The resistivity of silicon is too low to conduct away all the current generated, so a lower resistivity metal grid is placed on the surface to conduct away the current. The metal grid shades the cell from the incoming light so there is a compromise between light collection and resistance of the metal grid.

Rear Contact
The rear contact is much less important than the front contact since it is much further away from the junction and does not need to be transparent. The design of the rear contact is becoming increasingly important as overall efficiency increases and the cells become thinner.

We will take a more detailed look at silicon and thin films solar cells' structure and operation in Chapter 2.

PV Modules Function and Properties
PV modules have a number of mechanical, chemical, and electric characteristics with which design engineers, customers, installers, and investors must be familiar.

These specs and characteristics describe the history and the state of the module. They are the first thing to pay close attention to when considering certain types, brands or sizes of PV cells and modules, and when designing, installing and operating a large-scale power plant. (1)

A PV module consists of individual solar cells electrically connected to increase their power output. They are packaged so that they are protected from the environment and so that the user is protected from electrical shock. However, several aspects of PV module design which may reduce either the power output of the module or its lifetime need to be identified.

Following chapters describe how solar cells are encapsulated into PV modules and examine some of the issues which arise as a result of interconnection and encapsulation.

The most important effects in PV modules or arrays are:

- Losses due to the interconnection of mismatched solar cells;
- The temperature of the module; and
- Failure modes of PV modules.

A PV module consists of a number of interconnected solar cells (typically 36 connected in series) encapsulated into a single, long-lasting, stable unit. The key purpose of encapsulating a set of electrically connected solar cells is to protect them and their interconnecting wires from the typically harsh environment in which they are used.

For example, solar cells, being relatively thin, are prone to mechanical damage unless protected. In addition, the metal grid on the top surface of the solar cell and the wires interconnecting the individual solar cells may be corroded by water or water vapor.

The two key functions of encapsulation are to prevent mechanical damage to the solar cells and to prevent water or water vapor from corroding the electrical contacts. Many different types of PV modules exist, and the module structure is often different for different types of solar cells or different applications.

For example, amorphous silicon solar cells are often encapsulated into a flexible array, while bulk silicon solar cells for remote power applications are usually rigid with glass front surfaces.

Module lifetimes and warranties on bulk silicon PV modules are more than 20 years, indicating the robustness of an encapsulated PV module. A typical warranty will guarantee that the module produces 90% of its rated output for the first 10 years and 80% of its rated output up to 25 years.

A third-party insurance company could re-insure these warranties in the (quite likely these days) event that the manufacturer goes bankrupt. The same, or other, insurance company could also insure the operation and performance level of the PV modules and the entire power field, for the duration. This could be an expensive undertaking, but it is one way to avoid financial loss, and/or total failure.

PV Modules Structure

Most PV bulk silicon PV modules consist of a transparent top surface, an encapsulant, a rear layer and a frame around the outer edge. In most modules, the top surface is glass, the encapsulant is EVA (ethyl vinyl acetate) and the rear layer is Tedlar, as shown in Figure 1-9.

Front Surface Materials

The front surface of a PV module must have a high transmission in the wavelengths which can be used by the solar cells in the PV module. For silicon solar cells, the top surface must have high transmission of light in the wavelength range of 350 nm to 1200 nm. In addition, the reflection from the front surface should be low.

While theoretically this reflection could be reduced by applying an anti-reflection coating to the top surface, in practice these coatings are not robust enough to withstand the conditions in which most PV systems are used. An alternative technique to reduce reflection is to "roughen" or texture the surface.

However, in this case the dust and dirt are more likely to deposit on the top surface, and are much less likely to be dislodged by wind or rain. These modules are not therefore "self-cleaning," and the advantages of reduced reflection are quickly outweighed by losses incurred due to increased top surface soiling.

In addition to its reflection and transmission properties, the top surface material should be impervious to water, should have good impact resistance, should be stable under prolonged UV exposure, and should have a low thermal resistivity.

Water or water vapor ingress into a PV module will corrode the metal contacts and interconnects, dramatically reducing the lifetime of the PV module. In most modules the front surface is used to provide mechanical strength and rigidity, therefore either the top surface or the rear surface must be mechanically rigid to support the solar cells and the wiring.

There are several choices for a top surface material including acrylic, polymers, and glass. Tempered, low iron-content glass is most commonly used as it is low cost, strong, stable, highly transparent, impervious to water and gases and has good self-cleaning properties.

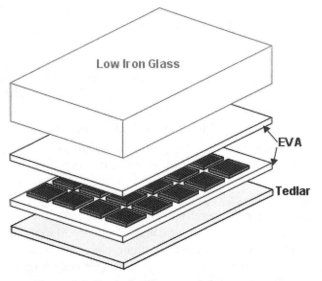

Figure 1-9. Typical silicon module components

Encapsulant

An encapsulant is used to provide adhesion between the solar cells, the top surface and the rear surface of the PV module. The encapsulant should be stable at elevated temperatures and high UV exposure. It should also be optically transparent and should have a low thermal resistance.

EVA (ethyl vinyl acetate) is the most commonly used encapsulant material. EVA comes in thin sheets which are inserted between the solar cells and the top surface and the rear surface. This sandwich is then heated to 150°C to polymerize the EVA and bond the module together.

Rear Surface

The key characteristics of the rear surface of the PV module are that it must have low thermal resistance and that it must prevent the ingress of water or water vapor. In most modules, a thin polymer sheet, typically Tedlar, is used as the rear surface. Some PV modules, known as bifacial modules are designed to accept light from either the front or the rear of the solar cell. In bifacial modules both the front and the rear must be optically transparent.

Frame

A final structural component of the module is its edging or framing. A conventional PV module frame is typically made of aluminum. The frame structure should be free of projections which could result in the lodgment of water, dust or other matter.

Electrical Circuit

A bulk silicon PV module consists of multiple individual solar cells connected, nearly always in series, as needed to increase the voltage above close to what is needed for practical applications.

In a typical small commercial module, 36 cells are connected in series to produce a voltage sufficient to charge a 12V battery. In it, each individual silicon solar cell has a voltage of just under 0.6V under 25°C and AM1.5 illumination. Taking into account an expected reduction in PV module voltage due to temperature and the fact that a battery may require voltages of 15V or more to charge, most modules contain 36 solar cells in series. This gives an open-circuit voltage of about 21V under standard test conditions, and an operating voltage at maximum power and operating temperature of about 17 or 18V.

The remaining excess voltage is included to account for voltage drops caused by other elements of the PV system, including operation away from maximum power point and reductions in light intensity.

Large commercial PV modules contain many (even hundreds) of cells, their combined voltage being much higher—in the range of 50 volts and more.

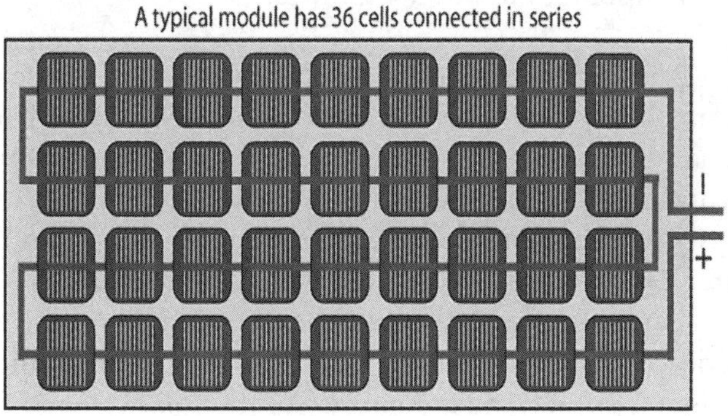

Figure 1-10. PV module top view

While the voltage from the PV module is determined by the number of solar cells, the current from the module depends primarily on the size of the solar cells and also on their efficiency. At AM1.5 and under optimum tilt conditions, the current density from a commercial solar cell is approximately between 30 mA/cm^2 and 36 mA/cm^2. Single crystal solar cells are often 100cm^2, giving a total current of about 3.5 A from a module.

Multicrystalline modules have larger individual solar cells but a lower current density, hence the short-circuit current from these modules is often approximately 4A. However, there is a large variation in the size of multicrystalline silicon solar cells, and therefore this current may vary. The current from a module is not affected by temperature in the same way that the voltage is, but instead depends heavily on the tilt angle of the module.

If all the solar cells in a module have identical electrical characteristics, and they all experience the same insolation and temperature, then all the cells will be operating at exactly the same current and voltage. In this case, the IV curve of the PV module has the same shape as that of the individual cells, except that the voltage and current are increased.

The equation for the module's electrical circuit becomes:

$$I_T = M \cdot I_L \qquad M \cdot I_0 \left[\exp\left(\frac{q\frac{V_T}{N}}{nkT} \right) 1 \right]$$

where:

N   is the number of cells in series;
M   is the number of cells in parallel;
I_T is the total current from the circuit;
V_T is the total voltage from the circuit;

I_0 is the saturation current from a single solar cell;
I_L is the short-circuit current from a single solar cell;
n   is the ideality factor of a single solar cell; and
q, k, and T are constants

Thin film solar cells and modules share some of the above characteristics and properties, but are differentiated by the lack of silicon material, which is replaced by a thin film of different materials.

Solar Power Installations

Having a good understanding of the structure and function of solar cells and PV modules, we could buy a batch or two and install them on a roof, or in a large power field to generate electric power.

Roof installations are the simplest of the bunch. All it takes is a good roof preparation, done by qualified roofers. These and other specialists should verify that the roof can handle the weight of the new installation, then design and install the metal support frame on the roof, using appropriate techniques.

This is not as easy as it sounds, for improper roof preparation and installation can result in serious roof damage and eventually leakage.

Once the frame is up and secured, the PV modules are mounted on it. Each module is usually wired to the next one, with the final module in the string having long wires that lead into the house for use by DC appliances.

Most often, however, the DC power is fed into an inverter, mounted somewhere in the house. The inverter takes the DC power generated by the modules and converts it into AC power of appropriate quality (frequency modulation, etc.).

The inverter then sends the converted power around the house for use by home appliances. Excess power not used by the appliances is sent into the local power grid per a previously agreed upon contract.

Ground-mounted installations (residential or commercial) are different from their roof-mounted cousins in that the metal frames and modules are mounted on the ground. These could be tracking or fixed mounted, and the generated power is used for powering residential or small commercial customers. The excess power could be sold to the local utility.

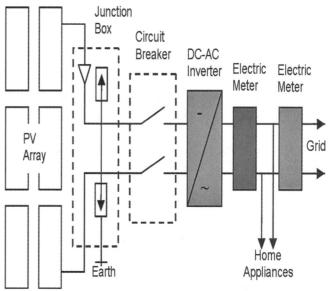

Figure 1-12. Residential solar installation

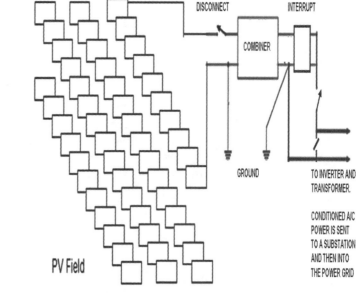

Figure 1-11. PV modules installation

Figure 1-13. Large-scale PV power plant

When a very large number of PV modules are installed in one place, then we have a large-scale power plant. Since utilities normally have say-so in such installations, and usually buy the generated power, we also call these installations utility-type power plants.

There are a number of complications arising at even the best installations of any type and size, so it suffices to say that low quality products, improper installation or operation and maintenance will double and triple the problems.

In the following chapters, we will take a closer look at manufacturing and function of solar cells and PV modules, their design, and the power fields in which these are used.

Since our emphasis is on the types of different solar technologies, we will sort and review these in the following categories:

1. Solar technologies in operation today are discussed in Chapter 3.

2. The most promising solar technologies, which we hope to have ready for use in the 21st century, are reviewed in Chapter 3.

3. Exotic solar technologies, which are still in some stage of development, are reviewed in Chapter 4.

Present and future application of the different technologies in practice (in actual solar installations and markets), as well as the related (financial, political, regulatory, and environmental) issues are discussed in the chapters that follow.

We wish you a fun and informative journey through the pages of this useful text, hoping to meet again on the front lines of solar development and environmental cleanup in the near future.

Notes and References

1. PVCDROM by C.B. Honsberg and S. Bowden, www.pveducation.org, 2010
2. Understanding PV Module Specifications, Justine Sanchez, Jan. 2009.
3. http://solarsystem.nasa.gov/index.cfm
4. http://www.energysavers.gov/renewable_energy/solar/index.cfm/mytopic=50012
5. Photovoltaics for Commercial and Utilities Power Generation, 2011. Anco S. Blazev
6. Planck M., Distribution of energy in the normal spectrum. http://hermes.ffn.ub.es/luisnavarro/nuevo_maletin/Planck%20(1900),%20Distribution%20Law.pdf
7. Einstein A., Generation and transformation of light.
8. Feynman R.P., QED : The Strange Theory of Light and Matter.
9. Backus C.E., Solar Cells.
10. Parrott J.E., Choice of an equivalent black body solar temperature.
11. Rai G.D., Solar Energy Utilization.
12. Hu C., White RM. Solar Cells: From Basic to Advanced Systems.
13. Mack M., Solar Power for Telecommunications.

Chapter 2

Today's Solar Power Generating Technologies

Not everything that is faced can be changed,
but nothing can be changed until it is faced.
—James Baldwin

There are a number of different solar power generating technologies operating in the field today, while others are under development in R&D labs, and many are still in the dreams, or nightmares, of scientists.

For the purpose of this text we sort and review these as follows:

1. Solar technologies in operation today are discussed in some detail in this chapter.

2. The most promising solar technologies which are ready, or which we hope to have ready in the 21st century, are reviewed in Chapter 3.

3. Exotic solar technologies, which are still in some stage of development, are reviewed in Chapter 4.

Subsequent chapters deal with power installations and markets, and related financial, political, regulatory, and environmental issues.

Technologies in operation today are those seen on housetops and in power fields in the US and abroad, while those most promising are under serious consideration for use soon—some having been deployed, albeit in small numbers or as prototypes.

Exotic solar technologies employ those materials, processes and devices which look promising, but are still in the initial design stage, on the lab bench, or in the early planning stage. Some of these look very promising, but lack one or another element needed to make them complete.

Sometimes it is difficult to draw the line between the "most promising" and "exotic," because some of the technologies in these two categories overlap. We are sure, however, that the reader will appreciate our reasoning and draw his/her own conclusions on the matter.

The first category—solar technologies in full operation today—cannot be disputed in most cases, because many of these are seen in operation all around us.

There are cases, however, when we should question the use of some solar technologies—even some of those deployed in large numbers. Some have not been proven efficient and/or safe for long-term use, especially under extreme climate conditions.

Using such technologies is not justifiable, and might lead to serious failures, financial losses, or environmental damage.

On the other hand, the technologies in the other two groups (most promising and exotic) are not clearly defined technically or practically, so their possible utilization cannot be predicted with any accuracy.

This is a living and quickly developing field, so anything can happen. We fully expect a shift of technologies between the categories, where some technologies operating today will be replaced by new ones not yet widely used.

At the same time, the technologies in the most promising and exotic categories could progress quickly to take their place in the energy market.

But most probably, shifts among the categories will be partial and sometimes temporary, where strategies occupy niche markets in special areas for awhile, until replaced by others.

We will attempt to sort all this (somebody has to do the dirt work), hoping that the reader will be patient and understanding of the difficulty surrounding such an important task.

Putting hi-tech concepts and works into categories is like lining up cats in a row; they simply refuse to co-operate. Just as we think we have them lined up, one will jump out of the pack and run away. Similarly, any day a disruptive technology could pop up, mess up our carefully put together lineup and take the lead—pushing others into the background or into oblivion.

Until then, we will line them up according to present-day, state-of-the-art technology and our understanding of the matters at hand.

In this chapter we will cover solar technologies in full use today in the US and around the world. Those that are actually installed and operational in significant numbers, and/or those that we have enough information about, as needed to make an educated decision for their full deployment now and in the near future.

Solar technologies in full use today can be divided into several categories:

1. Solar Thermal Technologies
   a. Flat plate water heater
   b. Stirling engine
   c. Parabolic trough
   d. Power tower

2. Photovoltaic Technologies
   a. Silicon Solar Technology
   b. Thin Film Solar Technology
   c. CPV technologies

3. Hybrid Solar Technologies
   a. PV-T technologies
   b. CPV-T technologies

Each of the above technologies is characterized by the fact that it is designed to absorb the incident sunlight and convert it into:

1. Heat, which then can be used as is for heating (hot water or steam generation), or could be converted into electricity.
   Note: Storage of the heated liquids for nighttime

use is also possible, which is the greatest advantage of this technology today.

2. DC power, which can be used as is to power specialized applications, or (most often) converted into AC power and sent into the grid.
   Note: No reliable large-scale storage is available yet, due to cost, efficiency and reliability restrains. This is a great obstacle for this technology, but efforts continue.

SOLAR THERMAL AND THERMO-ELECTRIC POWER GENERATION

Converting sunlight into heat or electricity is nothing new, but only recently have we seen serious developments in this area. From these developments we can confidently say that solar thermal power generation is the energy of the future, and the fastest way to our energy independence and a cleaner environment.

Solar thermal (heat generation) and thermo-electric (electric power generation) are the most reliable and mature solar energy conversion technologies today.

Solar thermal equipment uses sunlight to convert its energy directly into heat which is then used for heating, or for electric power generation. The heat can be used for heating homes, or as a heat source in commercial processes. Most often, however, the heat is converted into electric power which is then sent into the electric grid.

The conversion of sunlight to heat is a straightforward process, while the conversion of heat to electricity is somewhat more complex and requires expensive equipment and large installations to make it cost effective. This conversion is usually done at the so-called

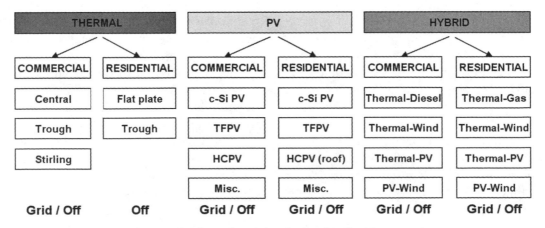

Figure 2-1. Major solar technologies for the 21st century

"utility scale power plants," using concentrated solar power (CSP) equipment which is the technology of choice today.

CSP technologies require direct (clear sky) sunlight for efficient operation, as needed for the optics to reflect and concentrate the reflected light onto a receiver which converts it into heat which can be used for heating or for electricity generation.

Relatively flat land is best for CSP systems, with slopes not exceeding 3 percent being recommended in most cases. The area of land required depends on the type of solar plant, but on average it is about 5-6 acres per installed megawatt (MW) of electric energy.

Cost-effective utility-scale CSP power plants are 100 MW in size or larger, requiring a minimum of 500-600 acres of land for each installed 100 MWp.

This large land base requirement involves significant surface disturbance (digging, land leveling and other modifications) with associated potential impact on a variety of resources on public and private lands.

These types of facilities also require roads, water source, wind protection, security fencing and such for safe and efficient operation.

The generated electricity is usually sold to the local utilities under a power purchase agreement (PPA), or other long-term power sale agreements.

We are taking a closer look at CSP technology later on in this text.

Introduction to CSP

The sun's energy falling on the Earth's surface for just 60 minutes is equivalent to the entire annual global energy consumption. From that fact alone we can easily conclude that the potential for getting free energy from the sun is virtually unlimited. The deployment of concentrating solar power (CSP) technologies is a good example of our attempts to capture that potential.

CSP's capacity is expected to increase exponentially over the next several years. Worldwide installed CSP capacity estimates vary widely, from 20-35 GW by 2025 to 1000-2000 GW by 2050. These estimates might be too optimistic, considering decreased government subsidies, water shortages, and the growing trend of conversion of CSP to PV power plants.

Some restrictions in EU, such as FiT reductions in most countries, will have profound effects on the CSP industry. We do believe that it will grow as the needs and the energy markets of developing countries continue to expand, but the pace of this growth is uncertain.

CSP technology is facing increasing challenges from PV competitors who have leveraged PV's declin-

ing costs and adaptability, to create a large global market. CSP will have trouble competing directly with PV on a cost per kWh basis in the near future but might be able to occupy niche markets with its ability to provide more stable power by providing after-hours energy via on-site thermal storage.

The need for cooling water is a great problem facing the CSP industry. PV doesn't have this problem and is taking full advantage of that.

Though we estimate that the total solar power (CSP and PV) produced around the world will continue to grow, anything can happen at any time to alter the growth pattern. Technology types, proper design, manufacturing, installation and operation have a lot to do with it, but other—even greater and unrelated—forces will be shaping the overall future of solar energy generation. These forces are demand and supply balance, material prices, energy costs, financial conditions, land availability and permitting, transmission and interconnection, socio-economic, political, regulatory, and other factors that contribute to the complexity and the degree of difficulty in deploying solar power generating equipment.

There are a number of thermal solar energy converting technologies, and we will review the major ones, focusing on those most likely to take a major part in, and have the largest impact on, commercial and utilities type power generation development and use in the 21st century.

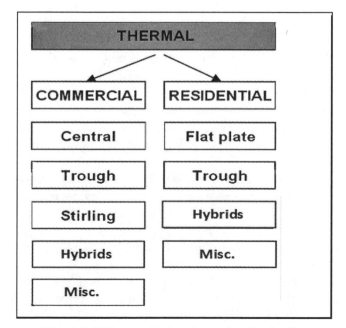

Figure 2-2. Types of solar thermal technologies

Definition of Solar Thermal and Solar Thermo-electric Technologies

Solar thermal equipment uses sunlight to convert the sun's energy into heat used for heating or for electric power generation. The heat can be used for heating homes, or as a heat source in commercial processes. Most often, however, the heat is converted into electric power which is then sent into the electric grid.

The conversion of sunlight to heat is a straightforward process, while the conversion of heat to electricity is more complex and requires expensive equipment and large installations to be cost effective. This conversion is usually done at the so-called "utility scale power plants," using concentrated solar power (CSP) equipment, today's technology of choice. There are five types of solar thermal systems in use:

1. Flat plate water heater
2. Stirling engine (dish)
3. Parabolic trough
4. Power tower
5. Hybrids

The **flat plate water heater** solar thermal energy generator is in its own category, because it generates only heat and is the simplest and cheapest of the bunch. It can be mounted on the roofs of houses and businesses and is used only to heat water (or other liquids) to a moderate temperature. This technology has been successfully used by commercial operations, such as restaurants, laundromats, and canning facilities for several decades.

Note: Lately, hybrids (combination water heaters and PV modules) have been used for generating both hot water and electricity from the same piece of equipment.

Smaller size, roof-mounted, parabolic troughs were also popular in the Southwest USA in the 1980s, and are making a slow comeback, while the major CSP technologies are large, ground-mounted, grid-connected systems.

The **Stirling engine, parabolic trough** and **power tower** are in the category of concentrated solar power (CSP) technologies, because they do capture sunlight and concentrate (focus) it onto a receiver. The heat is then normally used to make electricity by several methods which we will review below.

A number of **hybrids** use some elements, or modifications of the 4 major types in order to improve efficiency, or adapt them to different use. We discuss those in this text as well.

We will review the solar thermal and thermo-electric technologies, focusing on the CSP technologies, since they are most suitable for commercial and utilities power generation.

The solar thermal technologies of today can be described as follow:

Flat plate water heater

Solar water heater (solar thermal) systems operate on the principle of capturing sun's energy to heat water for residential and commercial uses, and/or to generate heat energy for industrial operations.

These heaters can be described as active (as is the case of the solar collectors), or passive (as used in some building design and energy applications).

Usually water (or other liquids) are passed through an insulated, solar collector (by convection if passive, or with a pump if active), and provide buildings with heat and/or hot water. Such systems are quite flexible and can successfully replace PV, or conventional electricity generation, and even gas heating.

Solar water heating systems can also be concentrating or non-concentrating. The concentrating collectors use lenses or mirrors to focus the sunlight on receivers which heat the water. These can be low-, medium-, and high-concentration, which respectively produces low, medium, and high temperatures.

Non-concentrating collectors are the ones we see on house roofs, and are generally shaped as a flat panel with the collector area equal to that of the absorber area. These also can generate low, medium or high tempera-

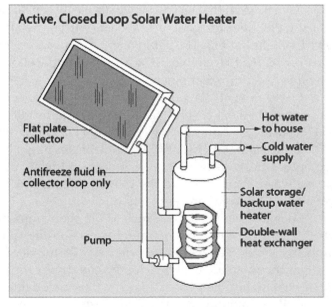

Figure 2-3. Solar water heater

tures according to their design and application.

According to temperature levels these heaters can be divided into:

1. Low temperature, when the operating temperature of the device is up to 180°F above ambient temperature. These units are used for swimming pools and other such low-temperature uses. The cost is also low—approximately $20-40/ft².

2. Medium temperature, when the operating temperature of the device is up to 1300°F above ambient temperature. These devices can be used for generating domestic and light commercial hot water and space heating. The cost is much higher—around $100/ft².

3. High temperature, when the operating temperature of the device is over 1300°F above the ambient temperature. These devices can be used for absorption cooling and/or electricity generation in industrial operations and even utility type installations. The cost of these units is also around $100/ft² retail, or less for large-scale installations.

Solar water heating systems used for domestic and light commercial applications are: a.) Flat-plate collector, b.) Evacuated-tube solar collector, and c.) Direct circulation water heating system.

Proper installation and operation of these units is critical and will determine their efficiency and durability.

These systems are most cost-effective when:

1. Cost of electric or gas fuels is high in the particular location,

2. The location has enough direct solar insolation (especially needed for the concentrating solar devices), and

3. The demand for heated water or space heating is constant throughout the year.

The benefits of using these systems are undeniable, especially when large quantities of water are used, because hot water heating is a significant portion of the operation costs of residential and many commercial operations. This is especially true for processing facilities using lots of hot water (such as bottling and canning operations), restaurants, laundries, hospitals, and many others.

Solar water heaters reduce the consumption of electricity and natural gas. There are already regulations in place that favor such installations and are designed to lower the installation and operation cost of small solar water heating installations.

CSP Technologies

With the flat plate water heater covered as much as needed for our purposes and filed in the category of thermal (heat) generation for residential and small commercial applications, we will now concentrate on the three major types of thermo-electric systems presently used in larger installations: the Stirling dish, the parabolic trough, and the power tower.

These three technologies have one thing in common: they all use trackers and optics of some type to optimize their efficiency. They also require relatively flat land with slopes not exceeding three percent to accommodate the solar collectors.

The area of land required depends on the type of plant, but it is about five acres per installed megawatt (MW), so commercial scale CSP facilities of any type are 100 MW or larger, and will require in excess of 500 acres of land, plus whatever else is needed for installation and support infrastructure.

Unlike solar photovoltaic technologies which use semiconductors to convert sunlight directly into electricity, CSP plants generate electricity by converting sunlight into heat first. Much like a reflective mirror, their reflectors focus sunlight onto a receiver. The heat absorbed by the receiver is used to move an engine piston (Stirling engine), or generate steam that drives a turbine to produce electricity (parabolic troughs and power tower).

Power generation after sunset is possible also by storing excess heat in large, insulated tanks filled with liquids or molten salt during the day and using it at night.

Since CSP plants require high levels of direct solar radiation to operate efficiently, deserts make ideal locations. As a matter of fact, these types of systems cannot operate efficiently in any other environment.

A study by Ausra, a solar energy company based in California, indicates that more than 90% of fossil fuel-generated electricity in the U.S. and the majority of U.S. oil usage for transportation could be eliminated by using solar thermal power plants and will cost less than it would cost to continue importing oil.

The land requirement for the CSP plants would be roughly 15,000 square miles in the SW USA deserts, or the equivalent of 15% of the land area of Nevada. While this may sound like a large tract of land, in the long run

CSP plants use less land per equivalent electrical output than large hydroelectric dams when flooded and wasted land is included and less than coal plants when factoring in the land used for mining and waste disposal.

Another study, published in *Scientific American*, details the possibility of using CSP and PV plants in the deserts to produce 69% of U.S. electricity and 35% of total U.S. energy including transportation by 2050.

Far fetched? Yes… for now. But looking at the next generations, we clearly see that they will use solar as a primary energy source. They'll have few other choices.

For now, we must fully develop and implement the major CSP technologies—the Stirling engine-dish tracker, the parabolic trough tracker, the power tower (central receiver)—and vigorously seek other possibilities.

We will take a close look at and discuss each of these below, focusing on their technological advancements and use in large-scale solar installations.

The Stirling Engine

One of the most elegant and flexible solar thermal power conversion technologies today is the Stirling engine-dish system. It consists of mirrors mounted on a frame which is continuously tracking the sun all through the day. The mirrors focus the reflected sunlight onto the receiver of the Stirling engine mechanism which is activated by the heat and turns on a shaft, connected to the rotor of an electric generator similar to that of the alternator of your car. The generator rotor turns with the engine shaft and generates electric power while the sun is shining and the receiver is hot enough to activate the engine and rotate the shaft.

A Stirling engine system is actually a solar electricity generator because the heat produced by the mirrors attached to it is converted into electric energy on the spot, so small installations of a few units are possible—something that is just not practical with the other CSP technologies. The Stirling engine needs cooling just like a car engine for more efficient operation and to cool the engine walls, bearings and other moving parts.

The mirror, or mirrors, are mounted on a metal frame which is driven by two motor-gear assemblies (x-y drives), programmed to move the frame in such a fashion that it follows the sun's movement precisely all day long, providing accurate focusing of the sunlight onto the heating plate of the Stirling engine. When the plate gets hot enough, the air in one of the cylinders in it is compressed and forces the piston in it to move up. This action forces the piston in the other cylinder (which is simultaneously cooled) to move down. Eventually, the compression in the second cylinder increases to the point that its piston is forced to go back, thus forcing the piston in the first cylinder to assume its initial position. The cycle repeats over and over while there is enough heat to maintain the process.

The Stirling engine function, under ideal operating conditions, can be represented by four cycles, or thermodynamic process segments, of interaction between the working gases, the heat exchanger, pistons and the cylinder walls.

The Stirling engine is an ingenious and very efficient piston engine, without the noise and exhaust of internal combustion engines. As a matter of fact, it can be classified as an "external combustion" engine. The gasses inside the cylinders are not exhausted, so there is no pollution and there are few moving parts with very little noise, so it can be used virtually anywhere.

Since its invention in 1816 by the Scottish inventor Dr. Robert Stirling, the Stirling engine has been considered and proposed for use in many different applications. Presently it is used in some specialized applications, where quiet, clean, no-exhaust operation is required. Some fancy and special purpose submarines and vessels use Stirling engines part of the time under special conditions.

Unfortunately, mass-market application for the Stirling engine is not found as yet, although many scientists and inventors are working on it. Its use in solar power generating equipment might be a good start in that direction. There are a number of installations using this technology, but most of them are smaller, demo-type systems.

No large CSP power plant using Stirling engine technology is in operation, to our knowledge, and as a matter of fact, two of the largest (and only) Stirling en-

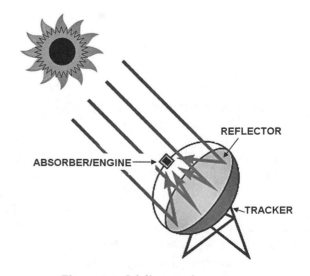

Figure 2-4. Stirling engine system

gine power plants planned for installation in the California deserts were cancelled in 2011, some because of the cost and reliability of the technology as proposed for use in the deserts.

Parabolic Trough Technology

Parabolic trough solar systems consist of a frame in a parabolic trough shape in which glass, metal or plastic reflectors are mounted to focus the sun's energy onto a receiver pipe running above and in parallel to the trough's length. The receiver pipe, or heat collection element (HCE), is centered at the focal point of the reflectors and is heated by the reflected sunlight to very high temperatures. Liquid of some sort is pumped through the receiver pipe and is heated in the process.

The HCE of the parabolic trough units is usually composed of a metal pipe with a glass tube surrounding it, and with the space between these evacuated to provide low thermal losses from the pipe. The pipe is coated with a material that improves the absorption of solar energy. Several improvements have been made or are underway to improve performance, the most significant of which is the seal between the glass and the pipe. This seal has not been as reliable as desired and development of a better configuration is still underway.

Parabolic troughs can focus the sunlight many times its normal intensity on the receiver pipe, where heat transfer fluid (HTF—usually mineral or synthetic oil) flowing through the pipe is heated. This heated fluid is then used to generate steam which powers a turbine that drives an electric generator. The collectors are aligned on an east-west axis and the trough is rotated north-south, following the sun as needed to maximize the sun's energy input to the receiver tube.

Parabolic trough power plants, also called solar electric generating systems (SEGS), represent the most mature CSP technology, with the most installed capacity of all CSP technologies. The first SEGS solar trough plant started operating in 1984, with the last one coming on line in 1991. Altogether, nine such plants were built, SEGS I–VII at Kramer Junction and VIII and IX at Harper Lake and Barstow respectively. In February 2005, all but two (I and II) of the Kramer Junction SEGS plants were acquired by FPL Energy and Carlyle/Riverstone and are still operating.

A natural gas system added to the plant "hybridizes" it and contributes up to 25% of the output. This feature also allows operation later at night or on cloudy days to meet the requirements of the grid. FPL now runs these systems, making it the largest solar power generator in the United States. All of the power generated from

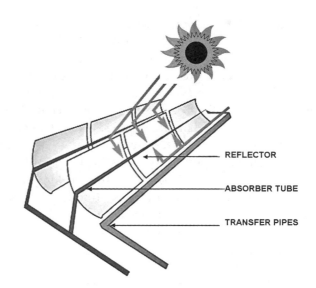

Figure 2-5. Parabolic trough technology

the SEGS projects is sold to Southern California Edison under long-term contracts negotiated by Luz back in the 1980s. There are a number of such plants worldwide, and many others are planned.

One distinct advantage of CSP systems is their ability to generate power after the sun has gone down. Here, the HTF fluid going through the receiver pipe is routed through a thermal storage system which permits the plant to keep operating for several hours after sunset while the electrical demand is still relatively high. The thermal storage system consists of a "hot" storage tank equipped with heat exchanger where HTF circulates and gives up a portion of its heat to heat the storage solution in the tank during the day. At night, the hot storage solution flows through the same heat exchanger heating up HTF which is sent to the steam turbines for generating power. The cooled-down storage solution flows from the heat exchanger to a "cold" storage tank where it stays until daytime when it is reheated and returned to the "hot" storage tank. The cycle is repeated nightly.

The Solar Power Tower

The power tower (or central receiver) power generation uses methods of collection and concentration of solar power based on a large number of sun-tracking mirrors (heliostats) reflecting the incident sunshine to a receiver (boiler) mounted on the top of a high tower, usually in the middle of the collection field. Eighty to ninety-five percent of the reflected energy is absorbed into the working fluid which is pumped up the tower and into the receiver. The heated fluid (or steam) returns down the tower and is fed into a thermal electrical power plant, steam turbine, or an industrial process that uses the heat.

The difference between the central receiver concept of collecting solar energy and the trough or dish collectors discussed previously, is that in this case all of the solar energy to be collected in the entire field is transmitted optically to a relatively small central collection region rather than being piped around a field as hot fluid. Because of this central receiver, systems are characterized by large power levels (100-500 MW) and higher temperatures (540-840°C) of the working fluids, allowing the creation of high-quality superheated steam which is more efficient for electricity generation.

Power tower technology for generating electricity has been demonstrated in the Solar One pilot power plant at Barstow, California, since 1982. This system consists of 1818 heliostats, each with a reflective area of 39.9 m² (430 ft²) covering 291,000 m² (72 acres) of land. The receiver is located at the top of a 90.8 m (298 ft) high tower and produces steam at 516°C (960°F) at a maximum rate of 42 MW (142 MBtu/h).

The reflecting element of a heliostat is typically a thin, back (second) surface, low-iron glass mirror. This heliostat is composed of several mirror module panels rather than a single large mirror. The thin glass mirrors are supported by a substrate backing to form a slightly concave mirror surface. Individual panels on the heliostat are also canted toward a point on the receiver. The heliostat focal length is approximately equal to the distance from the receiver to the farthest heliostat. Subsequent "tuning" and optimization of the closer mirrors is done upon installation.

Another heliostat design concept, not so widely developed, uses a thin reflective plastic membrane stretched over a hoop. This design must be protected from the weather but requires considerably less expenditure in supports and the mechanical drive mechanism because of its light weight. Membrane renewal and cleaning appear to be important considerations with this design. In all cases, the reflective surface is mounted on a pedestal that permits movement about the azimuth and elevation axis. Movement about each axis is provided by a fractional-horsepower motor through a gearbox drive. These motors receive signals from a central control computer that accurately points the reflective surface halfway between the sun and the receiver.

System design and evaluation for a central receiver application is performed in a manner similar to that when other types of collectors are used. Basically, the thermal output of the solar field is found by calculating collection efficiency and multiplying this by the solar irradiance falling on the collector (heliostat) field, minus some optical, transmission and other losses.

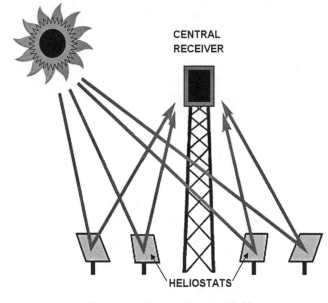

Figure 2-6. Power Tower Field

The components and their actual function in the power tower system are as follow:

Tracking and Positioning

The heliostats must follow the sun all day to focus the sunlight on the tower receiver. This is achieved by two electric motor-drive assemblies on each unit. To keep parasitic energy low, fractional horsepower motors with high gear rations are used to move the heliostat about its azimuth and elevation axes. This produces a powerful, slow, steady and accurate tracking motion. Under emergency conditions, however, rapid movement of the heliostats to a safe or stow position is an important design criterion. A typical minimum speed requirement would be that the entire field defocus to less than 3 percent of the receiver flux in 2 minutes. Higher speed is desired in case of impending disasters, such as high wind, hail and such, to protect the mirrors from mechanical damage.

Since it is currently considered best to stow the heliostats face-down during high wind, hail storms, and at night, an acceptable time to travel to this position from any other position would be a maximum of 15 minutes. The requirement for inverted stow is being questioned since it requires that the bottom half of the mirror surface be designed with an open slot so that it can pass through the pedestal. This space reduces not only the reflective surface area for a given overall heliostat dimension, but also the structural rigidity of the mirror rack. However, face-down stow does keep the mirror surface cleaner and safer.

Receivers

The receiver, placed at the top of a tower, is located at a point where reflected energy from the heliostats can be intercepted most efficiently. The receiver absorbs the energy being reflected from the heliostat field and transfers it into a heat transfer fluid. Taking a closer look at the receivers, we see that there are two basic types of receivers, external and cavity receivers.

a.   External receivers normally consist of panels of many small (20-56 mm) vertical tubes welded side-by-side to approximate a cylinder. The bottoms and tops of the vertical tubes are connected to headers that supply heat transfer fluid to the bottom of each tube and collect the heated fluid from the top of the tubes. The tubes are usually made of Incoloy 800 and are coated on the exterior with high-absorption black paint.

External receivers typically have a height-to-diameter ratio of 1:1 to 2:1. The area of the receiver is kept to a minimum to reduce heat loss. The lower limit is determined by the maximum operating temperature of the tubes and hence the heat removal capability of the heat transfer fluid.

b.   Cavity receivers are an attempt to reduce heat loss from the receiver by placing the flux absorbing surface inside an insulated cavity, thereby reducing the convective heat losses from the absorber. The flux from the heliostat field is reflected through an aperture onto absorbing surfaces forming the walls of the cavity. Typical designs have an aperture area of about one-third to one-half of the internal absorbing surface area.

Cavity receivers are limited to an acceptance angle of 60 to 120 degrees Therefore, either multiple cavities are placed adjacent to each other, or the heliostat field is limited to the view of the cavity aperture. The aperture size is minimized to reduce convection and radiation losses without blocking out too much of the solar flux arriving at the receiver. The aperture is typically sized to about the same dimensions as the sun's reflected image from the farthest heliostat, giving a spillage of 1-4%.

Heat Transfer Fluids

The choice of the heat transfer fluid to be pumped through the receiver is determined by the application. The primary choice criterion is the maximum operating temperature of the system followed closely by the cost-effectiveness of the system and safety considerations. The heat transfer fluids with the lowest operating temperature capabilities are heat transfer oils. Both hydrocarbon and synthetic-based oils may be used, but their maximum temperature is around 425°C (797°F).

However, their vapor pressure is low at these temperatures, thus allowing their use for thermal energy storage. Below temperatures of about –10°C (14°F), heat must be supplied to make most of these oils flow. Oils have the major drawback of being flammable and thus require special safety systems when used at high temperatures. Heat transfer oils cost about $0.77/kg ($0.35/lb).

Water has been studied for many central receiver applications and is the heat transfer fluid used in many power tower plants. Maximum temperature applications are around 540°C (1000°F) where the pressure must be about 10 MPa (1450 psi) to produce a high boiling temperature. Freeze protection must be provided for ambient temperatures less than 0°C (32°F). The water used in the receiver must be highly deionized to prevent scale buildup on the inner walls of the receiver heat transfer surfaces. However, its cost is lower than that of other heat transfer fluids. Use of water as a high-temperature storage medium is difficult because of the high pressures involved.

Nitrate salt mixtures can be used as both a heat transfer fluid and a storage medium at temperatures of up to 565°C (1050°F). However, most mixtures currently being considered freeze at temperatures around 140-220°C (285-430°F) and thus must be heated when the system is shut down. These mixtures have good storage potential because of their high volumetric heat capacity. The cost of nitrate salt mixtures is around $0.33/kg ($0.15/lb), making them an attractive heat transfer fluid candidate.

Liquid sodium can also be used as both a heat transfer fluid and storage medium, with a maximum operating temperature of 600°C (1112°F). Because sodium is liquid at this temperature, its vapor pressure is low. However, it solidifies at 980°C (208°F), thereby requiring heating on shutdown. The cost of sodium-based systems is higher than nitrate salt systems since sodium costs about $0.88/kg ($0.40/lb).

For high-temperature applications such as Brayton cycles, the use of air or helium as the heat transfer fluid and operating temperatures of around 850°C (1560°F) at 12 atm. pressure are being proposed. Although the cost of these gases would be low, they cannot be used for storage and require very large-diameter piping and expensive compressors to transport them through the system.

Linear Fresnel Reflector

The Linear Fresnel reflector (LFR) systems fall in the "miscellaneous" category, because they are a variation of the parabolic trough system. They are similar to the parabolic trough in design and function, but use an array of nearly flat Fresnel reflectors instead of a mirror surface.

These reflectors concentrate solar radiation onto elevated, inverted linear receivers. Water, or other liquid, flows through the receivers and is converted into steam. This system is also line-concentrating with the advantages of low costs for structural support and reflectors, fixed fluid joints, a receiver separated from the reflector system, and long focal lengths that allow the use of flat mirrors.

LFR technology is seen as a potentially lower-cost alternative to trough technology for the production of solar process heat. Planned commercial applications are estimated at a size from 50-200 MW. Linear Fresnel applications are mostly at the experimental stage. Companies working in the field claim higher efficiency and lower costs per kWh than its direct competitor, parabolic trough, due to high density of mirrors.

Fresnel mirrors are available at little more than €7.00 per m². According to Ausra, this technology can generate electricity for €0.10 per kWh now and under €0.08 per kWh within next 3 years.

The Fraunhofer Institute has contributed greatly in making the key components such as the absorber pipe, secondary reflectors, primary reflector array, and their control ready for operation. Based on theoretical investigations and the specific conditions found in sunny climates, Fraunhofer researchers have calculated that electricity production costs will not rise above €0.12 per kWh.

Linear Fresnel CSP technology derives its name from a type of optical system that uses a multiplicity of small flat optical faces, invented by the French physicist Augustin-Jean Fresnel who, while Commissioner for Lighthouses, invented the segmented lighthouse lens. Flat moving reflectors follow the path of the sun and reflect its radiation to the fixed pipe receivers above. Molten salt or other operating liquid powers a steam turbine, or is stored for night use. The technology itself is simple; the biggest challenge is setting mirrors to track the sun and reflect rays effectively. Flat mirrors are much cheaper to produce than parabolic ones, so this is a bonus.

Another advantage of the compact linear Fresnel reflector CLFR is that it allows for a greater density of reflectors in the array. In addition, Fresnel technology is less sensitive to wind loads and allows parallel land use to a large extent.

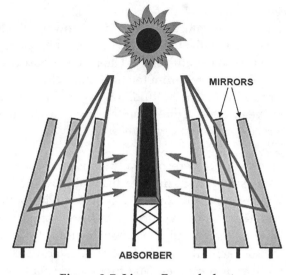

Figure 2-7. Linear Fresnel plant

LFR technology is more competitive economically due to:

— More effective land use than rival technologies;
— Low visual impact on landscape;
— Lower infrastructure costs due to its design;
— Lighter base, less steel used, flat instead of curved mirrors.

TODAY'S PV TECHNOLOGIES

Photovoltaic (PV) technologies include devices and entire fields which generate DC power when exposed to sunlight. The most important PV technologies today—those used widely in residential, commercial and utility type installations—are listed in the order we'll review them:

a. Silicon Solar Technologies
    — Single crystal silicon
    — Polysilicon
    — Other

b. Thin Film Solar Technologies
    — CdTe
    — CIGS
    — Other

c. Misc.
    — a-Silicon
    — Multi-junction

Types of PV Technologies

The most common, commercially available PV products today, which we focus on in this chapter, are a.) crystalline silicon (c-Si) solar cells and modules, b.) thin film (TF) cells and modules, and c.) some branches, related to these technologies.

Figure 2-8 is a graphical representation of the technologies, which we consider "today's solar technologies." Thousands upon thousands of homes and commercial building roofs and hundreds of mega-watt power fields are covered with these products where they operate day in and day out.

We are far from saying that just because they are used today they are the only, or best way to generate solar power in the 21st century. Far from it. As a matter of fact, we do not see most of today's silicon and/or thin film technologies surviving past the mid-century line. New, much more efficient and cheaper solar technologies will surface and be widely used then.

Nevertheless, what we have today is what we have today, and we can only guess about the future. So, let's review what we have today, namely, silicon and thin film materials, solar cells and modules, and some of their relatives.

Silicon based PV technology

Crystalline silicon (c-Si) based photovoltaic (PV) technologies are a major part of energy markets today and promise to be an even greater part in the future. They are the most mature of all PV technologies and compare successfully to the competitors.

c-Si PV technologies also compete successfully with concentrating solar thermal (CSP) and wind energy generators, simply because they are efficient enough and because they can be used in more versatile ways.

Because of that, the future of c-Si PV technologies looks very promising.

This success is reflected in the latest installed capacity, with c-Si solar cells taking 75% of the world market, and c-Si modules 71%, while the other PV technologies are in the single digits. The total quantity of c-Si cells and modules sold in 2009 was remarkable, keeping in mind that there were only a few MW of PV modules made and sold just 4-5 years ago. The ratios are changing somewhat today with thin film modules taking a more prominent role in the world's PV installations, but there is no question that silicon will lead the pack for the foreseeable future.

The bright future of c-Si in the world's PV energy markets comes at a price and has a number of clouds hanging over it in the shape of unresolved technical and financial issues.

Our goal, and sincere intent in this text, is to provide a clear down-to-earth description of the c-Si technologies, their materials, processes, and function, and related issues such as their manufacturing and suitability for, and application in, commercial and utilities power generation projects. We will focus especially on their use in the world's deserts and other extreme climate areas, where most of the large-scale power plants will be located. Since many such installations are presently planned for implementation in the US and abroad, we see this as a timely opportunity to bring the issues for discussion and analysis.

Definitions

For ease of reading, here are some of the basic energy terms and key terminologies related to them, as well as those related to photovoltaic power generation and usage.

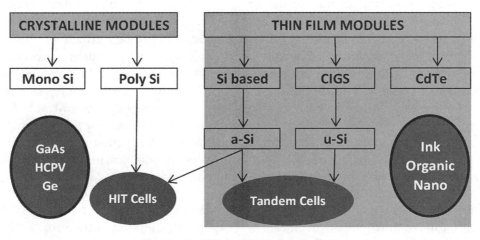

Figure 2-8. Major PV technologies

Solar energy (or power) is the energy that comes from the sun, and which we can collect and convert into thermal or PV power for everyday use.

Solar power generation is the actual process of capturing and converting solar energy into useful heat or electricity.

Photovoltaics, or PV, is the branch of the solar energy power generating industry that deals with direct conversion of solar energy into electricity by one device.

Note: PV electric generation is different from thermal electric power generation, where the solar energy is first converted into heat and then into electricity.

Metallurgical grade (MG) silicon is the material from which solar grade silicon is obtained via special processing and refining.

Solar grade (SG) silicon is the base material from which solar wafers and cells are made.

Crystalline silicon (c-Si) is the general category to which all (mono, poly, ribbon, etc.) crystalline silicon materials, wafers, cells and modules belong.

Single crystal, or mono-crystalline (sc-Si) silicon has the original symmetrical silicon material structure which is obtained by growing (pulling) a seed of the pure elemental silicon into large rods which are then sliced into thin round wafers.

Multi-crystalline (m-Si) silicon is lower quality silicon (from a solar point of view) with asymmetrical structure consisting of different type and shape strands of c-Si mixed into its bulk. It is made by melting SG Si chunks into large square blocks, letting the blocks cool, sawing them into smaller pieces and then slicing square wafers from those.

Poly-crystalline (p-Si) silicon, or poly, is a thin film material that is deposited on semiconductor devices. It is used in the solar industry for depositing thin films on special solar cells, as we will see below.

Note: There is some confusion about the use of "multi" vs. "poly" silicon terminology, terms that are used interchangeably, so we need to clarify the difference at the onset. "Poly-silicon," or "poly" is actually a thin film used in the semiconductor industry, but "poly" has been widely accepted to refer to "multi-crystalline" silicon widely used in the solar industry. We will also use the terms "multi" and "poly" interchangeably in this text to mean one and the same—multicrystalline silicon.

c-Si solar wafers are thin slices of crystalline silicon upon which solar cells are built. They are similar to semiconductor wafers in shape, but are of lower quality in terms of impurities, thus of much lower cost too.

c-Si solar cells are devices made out of silicon wafers, which when exposed to sunlight produce a certain amount of electric energy.

c-Si solar modules are flat plates (trays) onto which c-Si solar cells are arranged and sealed in, so that when exposed to sunlight they produce electric energy.

c-Si solar arrays or systems are groups of c-Si PV modules positioned and interconnected in such a way as to produce a maximum amount of electricity.

c-Si solar power plants are groups (strings) of c-Si PV arrays arranged so as to produce a maximum amount of electricity.

SILICON BASED SOLAR MATERIALS, CELLS AND MODULES

There are a number of different PV technologies which we will take a look later, but in this section we focus on crystalline silicon (c-Si) based materials, wafers, solar cells and modules.

We will do this in some detail, because c-Si technology is the most mature and widely used PV technology today.

POLYSILICON

c-Si cells modules and many related technologies (a-Si etc.) are based on polysilicon material which is made from sand. The sand is dug out and transported to huge melting furnaces where it is melted and, with the help of chemicals and additives, it is converted into polysilicon material. It is also called metallurgical grade (MG) silicon, because it is used mostly in the metallurgical industry in this state, and at this stage of its purification process.

Thus produced polysilicon contains many impurities, and is basically useless for solar cells manufacturing. So, it has to be purified by going through the solar grade polysilicon production process.

Note the great difference between the as-melted, metallurgical-grade silicon, vs. the highly purified semiconductor-grade silicon material. The metallurgical grade is not good for making any devices due to the high level of impurities which simply "kill" the devices' function by not allowing normal semiconductor or photovoltaic processes to take place.

On the other hand, although the semiconductor-grade silicon is perfect for making solar cells, it is too expensive, thus prohibitive for solar cells manufacturing. Because of that a compromise material with purity

Table 2-1. Silicon purity extremes.

(a) Metallurgical grade silicon

Element	Concentration (ppm)	Element	Concentration (ppm)
aluminum	1000-4350	manganese	50-120
boron	40-60	molybdenum	< 20
calcium	245-500	nickel	10-105
chromium	50-200	phosphorus	20-50
copper	15-45	titanium	140-300
iron	1550-6500	vanadium	50-250
magnesium	10-50	zirconium	20

(b) Semiconductor grade silicon

Element	Concentration (ppb)	Element	Concentration (ppb)
arsenic	< 0.001	gold	< 0.00001
antimony	< 0.001	iron	0.1-1.0
boron	≤ 0.1	nickel	0.1-0.5
carbon	100-1000	oxygen	100-400
chromium	< 0.01	phosphorus	≤ 0.3
cobalt	0.001	silver	0.001
copper	0.1	zinc	< 0.1

somewhere between the metallurgical and semiconductor grade materials is usually used. The quality varies, as the impurities vary from manufacturer to manufacturer and even batch to batch.

The purification process (conversion from MG to SG polysilicon) is dominated by two different chemical vapor deposition (CVD) approaches—the well established production approach known as the Siemens process, and a manufacturing scheme based on fluidized bed (FB) reactors.

With some modifications and improvements the different versions of these two basic types of processes are, and will be, the workhorses of the polysilicon production industry for the near future.

1. **The Siemens process** is usually done in the so-called Siemens reactor, which was developed in the late 1950s and has been the dominant production route historically. Until recently about 80% of the total polysilicon manufactured was made through a Siemens type process.

It involves deposition of silicon from a mixture of purified trichlorosilane or silane gas (which is made by processing solid polysilicon into gas or liquid), plus excess hydrogen, onto hairpin-shaped filaments of high-purity polysilicon crystals.

Pure polysilicon growth occurs inside a vacuum "bell jar" into which the gases are pumped. The filaments, assembled as electric circuits in series, are heated to the vapor deposition temperature by an external direct current. As the gases enter the bell jar, the high temperature (1,100-1,175°C) on the surface of the silicon seed filaments, with the help of the hydrogen, causes trichlorosilane to reduce to elemental silicon and deposit as a thin-layer film onto the hot seed filaments. The film layer grows with time until a thick rod is formed.

Thus-produced rods are taken out of the chamber and broken into smaller pieces to be used for making solar cells.

2. **The Fluidized Bed (FB) process** is used by a number of world-class manufacturers. It has its origins in a 1980s-era program sponsored by the U.S. DOE with the goal of devising a less energy-intensive method.

In an FB process, tetrahydrosilane or trichlorosilane and hydrogen gases are continuously introduced into the bottom of the reactor at elevated temperatures and pressures. High-purity silicon particles are inserted from the top of the reactor and are suspended by the upward flow of gases.

At approximately 750°C, the silane gas is reduced to elemental silicon, which deposits on the surface of the seed particles. As the seed crystals grow, they overcome the resistance of the uplifting air flow and fall to the bottom of the reactor, where they are removed continuously.

As the grown silicon granules are removed, fresh seed crystals are injected into the top of the reactor, so the process continues *ad infinitum*.

In addition to consuming less energy, FB processes offer the ability for continuous production, as opposed to the batch production of the Siemens route.

Thus-produced pure (solar grade) polysilicon is then melted (again) to be converted into either:

1. Single (mono) crystal silicon, which is done in vertical furnaces, where the chunks of polysilicon are melted and pulled up as a thick mono-silicon rod that can be sliced into round wafers to make solar cells, or

2. Poly-silicon, which is produced by melting the poly-silicon chunks in crucibles. When the crucibles are cooled down, the polysilicon in them is cut into bars, each of which is sliced into square wafers that can be used for solar cells manufacturing, or

3. Some other silicon form and shape—ribbon, thin film, etc.

The polysilicon production industry is very large, with lots of big players on all levels, and we see the need for polysilicon increasing in the future. At the same time, materials, equipment and processes are getting more sophisticated and the final product cheaper and of better quality.

All of this leads us to believe that polysilicon will remain as the material of choice in the 21st century.

SOLAR CELLS

Solar cells are the foundation of PV industry, so we need to take a closer look at their materials, structure, and function.

Silicon Material Basics

Silicon solar cells are thin slices (wafers) made from polysilcon material which is actually semiconductor material.

The chemical element silicon is the foundation not only of the c-Si PV technology, but also that of its older and richer cousin—the semiconductor technology. In its purest form silicon consists of a large number of atoms, each atom with 4 electrons orbiting its periphery.

Since adjacent atoms share electrons, it appears that each atom is surrounded by 8 electrons. Each individual atom has a nucleus, which consists of positively charged protons and neutral particles called neutrons, with 8 electrons hovering around the nucleus. There are equal numbers of electrons and protons in the atom, so that the overall charge of each atom is zero (neutral). The electrons are arranged so that they occupy different energy bands (levels) around the nucleus, which are determined by a number of forces and circumstances.

The four electrons in a silicon atom are held together by a covalent bond, which simply means that two adjacent atoms have the ability to share an electron and the force that holds them together. The electrons in these covalent bonds are held in place by the energy of the bond and are kept close to each other and the nucleus by a significant force. Since the electrons are kept close together by the covalent bond forces, they cannot move up and down the energy levels, and are therefore not free to move with the electric flow (if any) around them under normal conditions.

So, silicon in its natural and very pure form is inert; it cannot conduct electricity and acts as a good insulator. If, however, the temperature is increased sufficiently (or if special impurities are added), some electrons can gain enough energy to break the bond and escape, at which point they could potentially conduct electricity to an outside electrical circuit.

This ability to shed electrons under certain conditions is what makes silicon a semiconductor material and why it is perfect for manufacturing both semiconductor devices and solar cells.

The energy needed to break from the bond and keep an electron free from the covalent forces, is called band-gap energy, and sunlight provides this additional energy. Silicon is an excellent material for manufacturing solar cells since its indirect energy band-gap is approximately 1.1 eV at room temperature, which is close to the energy of some of the photons in the solar spectrum. This is the energy that allows the creation of free electrons and holes, thus initiating and propagating the photoelectric process in solar cells.

So, the more sunlight that falls on the cell, the more free electrons and current will be created, up to the physical and electronic limits of the solar cell to free electrons from their bonds and conduct electric current through its lattice.

When an electron leaves its place in the covalent bond, it leaves a hole (or a positive charge) behind, giv-

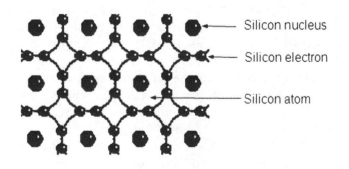

Figure 2-9. The silicon atom

ing the appearance of positive charges moving thought the lattice.

When sunlight hits the surface of the silicon solar cell it impinges enough energy onto the bonded electrons, freeing some of them from their captivity. The higher the number of free electrons (carriers), the more electric current they conduct. The number, or concentration, of free electrons is called intrinsic carrier concentration.

The bonded electrons reside in what is called the valence band (or low energy band), while the free electrons jump into the conduction band (high energy band) where they move freely and could be eventually extracted as electric current. The holes remain mainly in the valence band of the cell.

To optimize this process, we create the so called p-n junction in the silicon material by saturating a narrow surface area between the valence and conduction bands. With this addition, the silicon material (silicon wafers in most cases) is then on its way to becoming a solar cell as we will see below.

At the atomic level, silicon has a number of special qualities that make it a very good solar cell material. One particularly useful effect is the way it limits the energy levels which the electrons can occupy, and the way they move about the silicon crystal, all of which facilitate the initiation and propagation of the photoelectric effect, which is the foundation of the solar cell function.

c-Si Solar Cell Structure

Crystalline silicon (c-Si) solar cells are relatively simple at first glance, but their physical and chemical composition, structure, and electrical properties are quite complex.

We will take a closer look at the c-Si solar cell's structure to get a good understanding of the different aspects of its function. We will also review the major issues and failure mechanisms related to its manufacturing, installation and operation.

Silicon solar cells are made of silicon wafers, which have been processed (as discussed above) in order to achieve the structure needed for capturing and converting sunlight into electricity.

Basically, silicon solar cells (mono- or poly-crystalline) consist of:

1. Top metal contacts, where the generated electrons are collected and sent to the outside electric circuit for use as electric current.

2. Antireflective (AR) coating, which serves a dual purpose; to reduce reflection of sunlight, and to protect the surface from attacks by the elements.

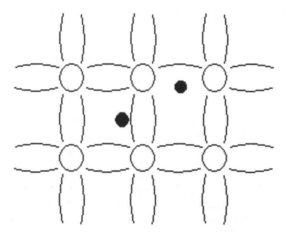

a. Free electrons in the Si lattice

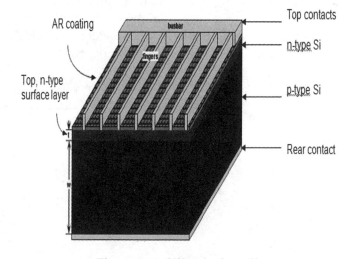

b. An incoming photon frees an electron

Figure 2-10. The solar power generating effect

Figure 2-11. Silicon solar cell

3.   Silicon wafer (bulk, or substrate) onto which all other components are built or attached. The wafer's bulk is doped with only a very thin layer of p-type dopant; the surface is doped with p-type dopant.

4.   p-n junction is the area between the p- and n- type doped areas in the bulk of the wafer.

5.   Rear metal contact is the area on the back of the wafer (solar cell) that is covered with a thin metal layer, which, together with the top metal contacts and the connecting wires, is part of the outside electric circuit through which the electrons flow.

So, when sunlight hits the top surface of the solar cell, the photons free electrons from the cell's bulk at the p-n junction, and send them to the top metal contacts. Wires attached to the top and bottom metal contacts make the outside electric circuit, which can power any device attached to it.

The p-n Junction

The p-n junction is what makes solar cells work. It is an integral part of the solar cell operation, since it is where the photoelectric process starts and where the electric current is generated. Let's take a close look at it:

Look at the cross section of a solar cell in Figure 2-12, with its top surface (N-type region) on the right-hand side and the bottom (P-type region) on the left. The p-n junction is in the middle and represents an electrochemical boundary between the P and N regions. The N region and the resulting p-n junction are created during the diffusion process by doping the N side (usually the top of the wafer) with phosphorous or other such element, thus saturating a very shallow area (less than a micron) below the top surface of the cell with phosphorous atoms. These atoms have excess electrons, which are loosely attached and are readily knocked out to fa-

cilitate the electric power generation, if and when energized by sunlight.

The p-n junction is located at the border line between the phosphorus saturated area and the original silicon bulk (P type region) which was very lightly doped with boron (or other similar chemical atoms) during the final silicon ingot melting process. The P region will provide the holes (+ charge) when the electron-hole pairs are broken by the incoming sunlight and the electrons extracted during the photoelectric effect.

When at rest, the p-n junction and the areas around it are static with no meaningful activities in or across them. When photons from sunlight with proper energy levels impinge onto the solar cell surface, they penetrate into the cell material and impact onto the electrons (in the electron-hole pairs) thus transferring energy to them and breaking the electron-hole pairs. This creates free electrons (–) which start moving around, while the holes (+) from the corresponding electron-hole pairs are mostly left in their original place. Thin areas on both sides of the p-n junction, called the "depletion" region effectively separate the holes from the electrons which are forced towards the N-type region and are finally extracted from the cell as electric current.

The constant movement of electrons back and forth across the p-n junction and through the layers is quite complex, but the final and practical result of it is the creation of DC electric current which flows in an outside circuit attached to the cell. A large part of the electrons do leave the cell through the metal fingers and bars on the N-type side (usually on top of the solar cell), and if we close the circuit, electric current will flow through it.

The electrons in the outside circuit will re-enter the cell via the back side metallization on the P-type side of the cell. Here they will recombine with excess holes in the region and the process repeats indefinitely, or at least while there is enough sunlight to keep it going.

The electron flow provides the current (I) in Amperes and the cell's electric field creates a potential, or

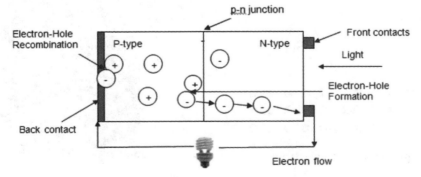

Figure 2-12. The p-n junction function

voltage (V) in Volts. With both current and voltage above zero we have power (P) in Watts, which is a product of the two and which is what we use to define the PV cells and modules power output.

$$P_{Watts} = I_{Amperes} * V_{Volts}$$

When an external load (such as an electric bulb or a battery) is connected between the front and back contacts of the cell, DC electricity flows through the cell and the external circuit, and powers the load connected to the closed external circuit.

Note: Remember that the actual electric current of the external circuit flows in the opposite direction of the electron flow.

c-Si Solar Cells Manufacturing Sequence

The basic c-Si solar processing sequence (mono- or poly-silicon solar cells), as used in the late 1990s by an associate company is outlined below. This process, or a variation of it, is used by most world class c-Si PV cells and modules manufacturers today.

The major steps in solar cells manufacturing are:

Wafers Inspection and Sorting

Wafers are placed on inspection tables and are inspected visually and with optical equipment. Wafers with visible mechanical defects are rejected. Then the wafers are tested with a 4-point probe and are sorted according to their resistivity. Wafers, or wafer samples are sent to an outside lab for metal and organic contamination analysis. Results from these tests determine the level of quality of the finished cells.

Wafers Cleaning and Etch

The wafers that pass all initial inspections and tests go to the wet cleaning line and are chemically processed in special chemicals where they are cleaned and etched to remove damage and oxide formed on the surface. They are then rinsed with de-ionized (DI) water and dried via spin dryer.

Surface Etch (Chemical Etch)

This process is used only on single-crystal silicon with 1-0-0 orientation. (Polycrystalline wafers cannot be textured, because the different strings of silicon have different orientation and the resulting surface is only partially and unevenly textured, if at all.) A controlled chemical solution (composition, concentration, temperature and time) etches the pyramid-like structures in the wafer surface and the surface takes on a dark-gray appearance. The pyramids blend into each other and block excess light reflection. Each pyramid is approximately 4-10 microns high. This step has critical process parameters. The wafers are then rinsed, spin-dried, and stored in special containers for processing.

Note: In a variation of this process, wafers are loaded in a fixture two-by-two with the backs of each pair touching, so that the pyramid structure is formed only on their front surfaces.

Diffusion for P-N Junction Formation

The clean wafers are oven dried, placed in the diffusion furnace at 900-950°C in reactive carrier gasses to impregnate the wafers. POCl3 gas is used for n-type diffusion, which diffuses phosphorous atoms in the wafer's surfaces. This creates a p-n junction in the lightly boron-doped wafers.

Note: This method is easier to control and has more uniform distribution of dopant than using the spray-on diffusion liquid and belt furnace diffusion process used by many companies today. This step has critical process parameters, so any compromise will be reflected in the cells' overall performance and longevity. In a variation of this process, wafers are loaded in a fixture two-by-two with the backs of each pair touching, so that the diffusion layer is formed only, or mostly, on their front surfaces. This facilitates the processing of the back surface later on.

Mass production solar cell operations use a different method in which the wafers are sprayed with a dopant chemical and run through a conveyor belt type furnace, where the dopant is diffused in the wafers' surface. Both methods have advantages and disadvantages.

Plasma Etch for Removal of Edge Layer (diffusion etc.)

The diffusion process implants P dopant in the wafers' side edges, causing an electrical short circuit between the top and bottom (negative and positive) surfaces of the cell, so it is necessary to remove the dopant with a wet chemistry or plasma etch. Wafers are coin-stacked and etched for a brief period in an RF plasma etch reactor; only the edges of the wafer are exposed to the plasma which removes the diffusion coating.

Wafers are then etched gently in a bath of dilute hydrofluoric acid to remove any oxides formed during the plasma etch step, rinsed with DI water, and finally spin dried.

Anti-reflective Coating

AR coating is deposited on the front surface of the wafers. The purpose of the AR coating is to reduce the amount of sunlight reflected from the finished cell surface. The AR coating is deposited via chemical vapor deposition (CVD) or by spraying the chemicals on the wafers and then baking. Both methods achieve similar outcome of enhancing the solar cells' output and giving them the distinctive dark blue color (for poly solar cells).

Note: Different manufacturers deposit and fire the AR coating using different process parameters and sequence order. This is an important process, nevertheless, so its proper design and execution will determine the final, most important aesthetic and performance aspects of the cells.

Printing (Metallization)

Several screen printing steps are used to apply the metallization on the front and back of the wafers. First, silver paste is printed on the top surface which then becomes the front metal pattern (top contacts, or fingers). The paste is dried and the wafers are flipped for printing the back surface with aluminum paste. After drying, silver paste is printed in special slots in the dried aluminum and the wafers are then transported into the firing furnace.

Metal Firing

Thus metalized on both sides, wafers are run slowly through an IR-heated furnace where the metal pastes on the top and bottom sides of the wafers diffuse into the substrate, to make an electric contact with the p-n junction and the back surface. This step has critical process parameters.

Note: The firing of the front contacts is a very delicate process, where time and temperature are controlled to achieve the desired depth of penetration of the metal into the silicon surface. The depth of penetration determines the electro-mechanical properties of the finished cell. Specially designed automated printing-firing equipment is available for more precise and consistent process control.

Inspection and Quality Control

Solar wafers and cells are inspected and tested at several stages of the process sequence. This is done by eye inspection, using magnification and other instrumentation. Electrical tests are also performed at some steps in the process. The final inspection is the most important step and must be performed by well trained and experienced operators.

Cell Flash Testing and Sorting

A certain percent of the completed wafers are placed on a test stand in the solar simulator and are illuminated for a period of time. I/V curve is generated for each cell and the output data are used to sort the cells into groups according to the I/V curve characteristics, prior to soldering and lamination into modules.

Cell Storage

The cells are finally loaded into cassettes, or coin-stacked (with protective material in between) and packed for ease of handling, transportation, or storage prior to laminating into solar modules, or shipping to another location.

Solar Cell Function

A solar cell is basically a semiconductor type p–n diode, for which the current density J (without any illumination—as in a dark room) obeys the foundational p-n diode equation:

$$J(V) = J_0 \left(\exp \left(\frac{1}{n_{id}} \frac{q}{k_B T} V \right) - 1 \right) \qquad (1)$$

Where:

     V   is the applied voltage
     J_0 is the saturation current density
     q   is the elementary charge
     k_B is the Boltzmann constant
     T   is the temperature
     n_{id} is the diode-ideality factor

The other significant variable, the saturation current density J0, is a function of:

$$J_0 = k_B T \left(\frac{\mu_{e,p} n_{i,p}^2}{L_{e,p} N_{A,p}} + \frac{\mu_{h,n} n_{i,n}^2}{L_{h,n} N_{D,n}} \right)$$

$$J(V) = J_0 \left(\exp \left(\frac{1}{n_{id}} \frac{q}{k_B T} V \right) - 1 \right) - J_{ph} \qquad (2)$$

Where:

     p and n are the respective sides of the p-n junction
     e and h are the electrons and holes in and around the junction
     μ is the carrier mobility
     n_i is the intrinsic carrier concentration
     L is the diffusion length and
     N_A and N_D are the acceptor and donor levels respectively

When the diode is illuminated (as when exposed to sunlight), a photon with an energy higher than the band-gap Eg of the semiconductor material (i.e. silicon or CdTe) is absorbed and if all conditions are met, it generates an electron–hole pair (aka generation). The electron–hole pair could undergo a number of transformations: a) it could be lost again if it bumps into barriers and defects in the bulk, or b) it could be lost at an interface (recombination). Here, however, it has some probability to do useful work by being collected if it makes it to the terminal's (electrodes). It then contributes to the light-generated current density Jph in the external electric circuit (which contrary to logic) flows opposite to the normal, conducting direction of the diode.

Ideal Case Conditions

In this case, or theoretically, Jph depends only on the light's intensity, so that the current generated by the solar cell under illumination is simply a superposition of Equation (1) and a constant term:

$$J(V) = J_0 \left(\exp\left(\frac{1}{n_{id}} \frac{q}{k_B T} V \right) - 1 \right) - J_{ph} \qquad (3)$$

Equations (1) and (3) yield current/voltage (I/V) curves, similar to those plotted in a standard I-V curve plot.

The illuminated I/V characteristics of a solar-cell device are summarized by four key parameters:

— The open-circuit voltage, Voc
— The short-circuit current density, Jsc
— The fill factor, FF and
— The efficiency, η

The most important variable, the maximum-power point (MPP) is defined as the point where the electrical power, P, generated by the solar cell is at its maximum level. At this point the fill factor becomes the quotient:

$$FF = \frac{V_{MPP} \cdot J_{MPP}}{V_{OC} \cdot J_{SC}} \qquad (4)$$

Efficiency then depends on:

$$\eta = \frac{P'_{MPP}}{P'_{light}} = \frac{V_{OC} \cdot J_{SC} \cdot FF}{P'_{light}}$$

Where:
P'   is the power density of the light at STC (standard testing conditions)

Real-world Conditions

In this case, there are several loss mechanisms:

1.   Shunts, which allow current to bypass the exter-

nal circuit (causing an increase in slope around the short-circuit point);

2.   Series resistance which brings about a voltage drop under operation (causing the slope around the open-circuit point to decrease), and

3.   Voltage-dependent photo-current collection, which is specific for thin-film solar cells, and which distorts and even invalidates the superposition principle discussed above.

Note: At first glance, the voltage-dependent collection may be mistaken for a shunt, but it is generally easy to distinguish at a closer look, because it only affects the illuminated current-voltage curve.

So basically, provided that everything is equal, the circuit in Figure 2-13a will generate much more electricity, simply because it has no restrictions to speak of. The efficiency and the amount of electrical generation of the circuit in Figure 2-13b, on the other hand, will depend on the type and magnitude of the voltage-dependent light-current source, diode, shunt conductance, and series resistance variables.

These variables are the greatest obstacle to the limited power generation and subsequent usefulness, of

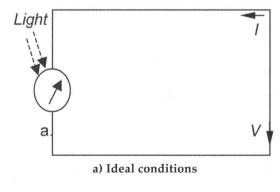

a) Ideal conditions

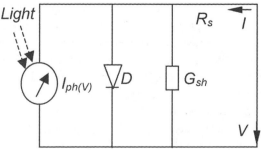

b) Real world conditions, with voltage-dependent light-current source, Iph(V), diode, D, shunt conductance Gsh, and series resistance R

Figure 2-13. Different operating conditions

present-day solar cells and modules. The different technologies have different levels of dependence, depending on their substrate materials and superstructures. Thin film technologies would behave differently from silicon based technologies under different conditions, where the internal obstacles and the external influences affect them differently. We will take a closer look at these dependencies in following chapters.

One critical thing to remember from the above explanation is the fact that the internal variables are just that—varying variables, affected by a number of internal and external factors. In conjunction with the internal factors, we must also remember that the external environmental elements—heat, cold, humidity, etc.—play tremendous roles here as well. Extreme heat, for example, will increase the internal resistance of the solar cells accordingly. The higher the air temperature, the higher the internal temperature, followed closely by a linear increase of internal resistivity, reduction of power and even destruction of the devices in some extreme cases.

The combination of internal and external factors and related variables is very complex, so we need to always keep it in mind. This is the only way to ensure long-term, reliable operation of solar cells and modules under any operating conditions. Few available devices approach ideal quality, though most manufacturers aim for good quality.

Serious violations of the quality standards can result in drastic reduction of power and failures on the test stands and/or during long-term exposure to the elements in the field.

There are a number of characteristics and operational parameters of today's solar cells, that we need to understand in order to properly apply them in our daily lives.

Key Characteristics of Solar Cells

The key elements in the conversion of sunlight into electric energy are, a) the availability and intensity of incoming sunlight, and b) the ability and efficiency of the solar cell to capture and process the sunlight into electric energy via the "photovoltaic" effect which takes place close to the top surface of the solar cell in and around the p-n junction as discussed above.

The key characteristics of, and processes taking place in, the solar cell are:

Band Gap Energy

When sunlight hits the solar cell, its photons with energies greater than the energy band-gap of the semiconductor material (1.1eV for Si) are absorbed, thus pro-

moting electrons from the valence band to the conduction band (P-type to N-type respectively). This action leaves a corresponding number of holes in the valence band which then recombine with the electrons returning at the back of the cell.

The energy "band gap" is the energy difference between the top of the valence band (electrons at rest) and the bottom of the conduction band (electrons with extra energy). When sunlight hits the solar cell, the energy of the electrons in the valence band (where they reside when not activated) is increased, due to internal (parasitic heat) or external (sunlight) energy transfer due to bombardment from photons (light) and phonons (heat) contained in the sunlight.

When the energy of the incoming photons in the sunlight reaches the bang gap energy of the material, the electrons in it are activated and are able to jump from the valence (rest) to conduction (energized) energy band. This is only IF they get enough energy for the transition effort.

Remember that the band gap energy has different values for different materials and that for silicon it is 1.1eV. This means that the incoming photons must have at least as much energy as the band gap (preferably a bit more) to activate the band gap's electrons. This energy and the resulting interactions are the engine of the photoelectric effect.

The silicon energy band gap at 300°K (80.33°F) is calculated as:

$$E_g(300\ K) = E_g(0K) - \frac{\alpha T^2}{T + \beta} = 1.166-$$

$$\frac{0.473 \times 10{-}3 \times (300)^2}{300 + 636} = 1.12\ eV$$

where:

T is the temperature, and

Eg(0), α and β are fitting parameters for different materials

Note: Eg(0) for Si=1.116eV, Ge=0.744e; α for Si=4.73x10-4 eV/K, Ge=4.77x10-4 eV/K, also that β for Si=636K, and for Ge=235.

With the energy exchange initiated by the sunlight falling on the solar cell, its temperature increases and the amplitude of atomic vibrations increases which in turn leads to larger spacing between the atoms in the silicon lattice. The interaction between the lattice phonons and the free electrons and holes affects the band gap too, and its magnitude changes proportionally with temperature increase, so this relationship is reflected by Varshni's empirical expression:

$$E_g(T) = E_g(0) - \frac{\alpha T^2}{T + \beta}$$

where:

T is the temperature

Eg(0), α and β are fitting parameters for different materials

If the electron-hole pairs are generated within the depletion region of the p-n junction, the electric field in the depletion region separates the pairs and drives the electrons through an external load as DC electricity. Maximum power is generated when the solar cell receives the maximum amount of sunlight, and thus generated power is delivered to the load when its impedance matches that of the illuminated device. The total output, however, will start decreasing as the temperature of the cell increases.

Power Generation

Solar cells are characterized by the power they generate, which is at maximum when the current and voltage are at their maximum levels. This can happen only when the cell is fully operational (no defects), and receives maximum solar insolation.

$$P_{max} = I_{max} * V_{max}$$

With these terms at zero, the conditions

$$V = Voc/I = 0 \text{ and } V = 0/I = Isc$$

also represent zero power.

A combination of maximum current and maximum voltage maximizes the generated power and is called the "maximum power point" (MPP).

$$MPP = I_{max} * V_{max}$$

So the MPP of a solar cell with 3.0 A current and 0.5 V voltage would be 1.5 W. A 100 pc. PV modules made of these solar cells connected in series would generate 150 Wp under full solar insolation of 1000 W/m^2 and much less at lower insolation levels—falling proportionally with the reduction of sunlight hitting the modules' surface.

The solar conversion efficiency η of a PV cell or module is used most commonly to express and compare performance. The efficiency is given by:

$$\eta = Voc*Isc*FF/P$$

Where:

Voc is the open circuit voltage (the voltage generated when the load resistance is infinite, or there is no resistance),

Isc is the short circuit current (the current generated when the load resistance is zero,

FF is the fill-factor, calculated as follows:

$$FF\% = \frac{I_{max} * V_{max} \text{ (actual measurements)}}{MPP \text{ (maximum obtainable power)}}$$

This is simply the ratio of the actual measurements (Voc and Isc generated by the cell under the specific testing conditions) divided by the maximum power the cell can generate. In other words, the efficiency is basically the ratio of the amount of power a solar cell or module could produce vs. the total amount of power contained in the incoming sunlight and how efficiently the cell converts it into electric power.

So, if a PV module with 1.0 m^2 active surface area is rated at 15% efficiency which is average for c-Si PV modules, we can quickly deduct that it theoretically could produce 150 Watts DC power under 1000 W/m^2 solar insolation (its maximum power). So if it generates 50 V and 3 A, then its FF will be 1. If it produces 50 V and only 2 A (or 100 W), then its FF will be (50 * 2)/150 W, or FF = 0.67.

It's a mouthful of concepts, but a good simplified description of the major practical effects of solar power generation which are the foundation of solar cells' and modules' ability to convert sunlight into useful DC power.

PV Effects in Solar Cells and Modules

Below is a list of some of the key effects encountered in solar cells and modules (1). These effects and their influence on the operation of solar cells and modules must be thoroughly understood and applied in the successful design, installation and operation of PV components and power generating facilities.

Light Absorption

The intensity of sunlight incident on a solar cell varies with time, which has profound effect on its performance. The light intensity changes the short-circuit current, the open-circuit voltage, the FF, and the efficiency and impact of series and shunt resistances. The level of light intensity on a solar cell is called "number of suns," where 1 sun corresponds to standard illumination of 1.0 kW/m^2 at AM1.5, while 10.0 kW/m^2 incident on the solar cell means that it is operating at 10 suns, or at 10X.

Concentrating PV systems operate at up to 1000 suns, which require special materials (non-silicon) and more sophisticated equipment and processes. Solar cells experience daily variations in light intensity, with the incident power from the sun varying between 0 in the early morning and 1.0 kW/m^2 and higher at noon in some cases. At low light levels, the effect of the shunt resistance becomes increasingly important. As the light intensity decreases, the bias point and current through the solar cell also decrease and the equivalent resistance of the solar cell may begin to approach the shunt resistance.

When these two resistances are similar, the fraction of the total current flowing through the shunt resistance increases, thereby increasing the fractional power loss due to shunt resistance. Consequently, under cloudy conditions, a solar cell with a high shunt resistance retains a greater fraction of its original power than a solar cell with a low shunt resistance. In either case, the efficiency and final output of the solar cell are directly proportional to the amount and quality of sunlight incident on its surface, minus the above mentioned parasitic effects. Thin film cells and modules seem to perform better under cloudy conditions for these and other reasons.

One characteristic of PV cells and modules is that under any sunlight level some photons falling on the solar cell surface will be reflected, some will be absorbed in the material and some will go right through it. But only photons which are absorbed in the substrate will generate power. If the photon is absorbed it will give its energy to the electron from the valence band to transfer it to the conduction band and generate power into the external circuit. More photons are absorbed, more electrons will be transferred, and more power will be generated by the cell.

A key factor in determining if a photon is absorbed or transmitted is its energy. Photons falling onto a semiconductor material can be divided into three groups based on their energy compared to that of the semiconductor band gap:

1. Photons with energy less than the band gap energy, will interact only weakly with the semiconductor material, passing through it as if it were transparent.

2. Photons with just enough energy to create an electron hole pair and which are efficiently absorbed to generate free electrons and produce power as a result.

3. Photons with energy much greater than the band gap that are strongly absorbed will not contribute to freeing electrons and might even be harmful.

The absorbed photons create both majority and minority carriers. In many photovoltaic applications the number of light-generated carriers is on an order of magnitude lower than the number of majority carriers already present in the solar cell due to doping.

Consequently, the number of majority carriers in an illuminated semiconductor does not alter significantly. However, the opposite is true for the number of minority carriers. The number of photo-generated minority carriers outweighs the number of minority carriers existing in the solar cell in the dark, and therefore the number of minority carriers in an illuminated solar cell can be approximated by the number of light generated carriers.

Understanding these effects is critical for the proper design and use of PV cells and modules, for they determine the efficiency and overall behavior of the devices.

Absorption Coefficient

The absorption coefficient is the variable that determines how far into the silicon material the sunlight of a particular wavelength can penetrate before it is absorbed. In a material with a low absorption coefficient, light is only poorly absorbed, and if the material is thin enough, it will appear transparent to that wavelength.

The absorption coefficient depends on the material and also on the wavelength of light which is being absorbed. Semiconductor materials have a sharp edge in their absorption coefficient, since light which has energy below the band gap does not have sufficient energy to raise an electron across the band gap. Consequently this light is not absorbed.

For those photons with energy levels above the band gap, the absorption coefficient is not constant but depends strongly on the prevailing wavelength. The probability of absorbing a photon depends on the likelihood of having a photon and an electron interact in such a way as to move from one energy band to another.

For photons which have energy very close to that of the band gap, the absorption is relatively low since only those electrons directly at the valence band edge can interact with the photon to cause absorption.

As the photon energy increases, there are a larger number of electrons which can interact with it and result in its full energy absorption. In c-Si solar cells, the photon energy greater than the band gap is wasted as electrons quickly thermalize back down to the band edges.

The absorption coefficient, α, is related to the extinction coefficient, k, by the following formula:

$$\alpha = \frac{4\pi k}{\lambda}$$

Where:

    λ is the light wavelength and

    k   is the extinction coefficient (~0.06 for silicon and photon energies less than 3.0 eV)

The dependence of absorption coefficient on wavelength causes different wavelengths to penetrate different distances into a semiconductor before most of the light is absorbed. The absorption depth is given by the inverse of the absorption coefficient, or $\alpha - 1$. The absorption depth is a useful parameter which gives the distance into the material at which the light drops to about 36% of its original intensity, or alternately has dropped by a factor of $1/e$.

Since high energy light has a large absorption coefficient, it is absorbed in a short distance (for silicon solar cells within a few microns) of the surface, while red is absorbed less strongly. Even after traveling several hundred microns, not all red light (IR) is absorbed in silicon.

Light Trapping

The optimum device thickness is not controlled solely by the need to absorb all the light. For example, if the light is not absorbed within a diffusion length of the junction, then the light-generated carriers are lost to recombination. In addition, a thinner solar cell which retains the absorption of the thicker device may have a higher voltage.

Consequently, an optimum solar cell structure will typically have "light trapping" in which the optical path length is several times the actual device thickness, where the optical path length of a device refers to the distance that an unabsorbed photon may travel within the device before it escapes out of the device. This is usually defined in terms of device thickness. For example, a solar cell with no light trapping features may have an optical path length of one device thickness, while a solar cell with good light trapping may

have an optical path length of 50, indicating that light bounces back and forth within the cell many times.

Light trapping is usually achieved by changing the angle at which light travels in the solar cell by having it be incident on an angled surface. A textured surface will not only reduce reflection as previously described but will also couple light obliquely into the silicon, thus giving a longer optical path length than the physical device thickness.

The angle at which light is refracted into the semiconductor material is, according to Snell's Law, as follows:

    n1 sin θ1 = n2 sin θ2

where:

    θ1 and θ2 are the angles for the light incident on the interface relative to the normal plane of the interface within the mediums with refractive indices n1 and n2 respectively.

If light passes from a high refractive index medium to a low refractive index medium, there is the possibility of total internal reflection (TIR). The angle at which this occurs is the critical angle and is found by setting θ2 in the above equation to 0.

$$\theta_1 = \sin^{-1}\left(\frac{n_2}{n_1}\right)$$

Using total internal reflection, light can be trapped inside the cell and make multiple passes through the cell, thus allowing even a thin solar cell to maintain a high optical path length.

Generation Rate

The generation rate gives the number of electrons generated at each point in the device due to the absorption of photons. Neglecting reflection, the amount of light which is absorbed by a material depends on the absorption coefficient (α in cm–1) and the thickness of the absorbing material.

The intensity of light at any point in the device can be calculated according to the equation:

$$I = I_0 e^{-\alpha x}$$

where:

    α is the absorption coefficient typically in cm–1;

    x   is the distance into the material at which the light intensity is being calculated; and

    I_0 is the light intensity at the top surface.

The above equation can be used to calculate the number of electron-hole pairs being generated in a solar cell. Assuming that the loss in light intensity (i.e., the absorption of photons) directly causes the generation of an electron-hole pair, then the generation G in a thin slice of material is determined by finding the change in light intensity across this slice.

Consequently, differentiating the above equation will give the generation at any point in the device, hence:

$$G = \alpha N_0 e^{-\alpha x}$$

where:

N_0 = photon flux at the surface (photons/unit-area/sec.);

α = absorption coefficient; and

x = distance into the material.

The above equations show that the light intensity exponentially decreases throughout the material and further that the generation at the surface is the highest at the surface of the material. For photovoltaic applications, the incident light consists of a combination of many different wavelengths, and therefore the generation rate at each wavelength is different.

Carrier Lifetime

The minority carrier lifetime is a measure of how long a carrier is likely to stay around before recombining. It is often just referred to as the "lifetime" and has nothing to do with the stability of the material. Stating that "a silicon wafer has a long lifetime" usually means minority carriers generated in the bulk of the wafer by light or other means will persist for a long lifetime before recombining.

Depending on the structure, solar cells made from wafers with long minority carrier lifetimes will usually be more efficient than cells made from wafers with short minority carrier lifetimes. The terms "long lifetime" and "high lifetime" are used interchangeably.

With low level injected material (where the number of minority carriers is less than the doping) the lifetime is related to the recombination rate by:

$$\tau = \frac{\Delta n}{R}$$

where:

τ is the minority carrier lifetime,

Δn is the excess minority carriers concentration and

R is the recombination rate.

Recombination Losses

Recombination losses effect both the current collection (and therefore the short-circuit current) as well as the forward bias injection current (and therefore the open-circuit voltage). Recombination is frequently classified according to the region of the cell in which it occurs. Typically, the main areas of recombination are at the surface (surface recombination) or in the bulk of the solar cell (bulk recombination). The depletion region is another area in which recombination can occur (depletion region recombination).

Current Losses

In order for the p-n junction to be able to collect all of the light-generated carriers, both surface and bulk recombination must be minimized. The two conditions commonly required for current collection are:

1. The carrier must be generated within a diffusion length of the junction, so that it will be able to diffuse to the junction before recombining; and

2. In the case of a localized high recombination site (such as at an unpassivated surface or at a grain boundary in multicrystalline devices), the carrier must be generated closer to the junction than to the recombination site. For less severe localized recombination sites (such as a passivated surface), carriers can be generated closer to the recombination site while still being able to diffuse to the junction and be collected without recombining.

The presence of localized recombination sites at both the front and the rear surfaces of a silicon solar cell means that photons of different energy will have different collection probabilities. Since blue light has a high absorption coefficient and is absorbed very close to the front surface, it is not likely to generate minority carriers that can be collected by the junction if the front surface is a site of high recombination. Similarly, a high rear surface recombination will primarily affect carriers generated by infrared light, which can generate carriers deep in the device. The quantum efficiency of a solar cell quantifies the effect of recombination on the light generation current.

Voltage Losses

The open-circuit voltage is the voltage at which the forward bias diffusion current is exactly equal to the short circuit current. The forward bias diffusion current is dependent on the amount of recombination in a p-n junction. Increasing the recombination increases the forward bias current, in turn reducing the open-circuit voltage.

The material parameter which gives the recombination in forward bias is the diode saturation current. The recombination is controlled by the number of minority carriers at the junction edge, how fast they move away from the junction, and how quickly they recombine.

Consequently, the dark forward bias current, and hence the open-circuit voltage are affected by the following parameters:

1. **The number of minority carriers at the junction edge.** The number of minority carriers injected from the other side is simply the number of minority carriers in equilibrium multiplied by an exponential factor which depends on the voltage and the temperature. Therefore, minimizing the equilibrium minority carrier concentration reduces recombination. Minimizing the equilibrium carrier concentration is achieved by increasing the doping.

2. **The diffusion length in the material**. A low diffusion length means that minority carriers disappear from the junction edge quickly due to recombination, thus allowing more carriers to cross and increasing the forward bias current. Consequently, to minimize recombination and achieve a high voltage, a high diffusion length is required. The diffusion length depends on the types of material, the processing history of the wafer, and the doping in the wafer. High doping reduces the diffusion length, introducing a trade-off between maintaining a high diffusion length (which affects both the current and voltage) and achieving a high voltage.

3. **The presence of localized recombination sources within a diffusion length of the junction.** A high recombination source close to the junction (usually a surface or a grain boundary) will allow carriers to move to this recombination source very quickly and recombine, dramatically increasing the recombination current. The impact of surface recombination is reduced by passivating the surfaces.

Surface Recombination

Any defects or impurities within or at the surface of the semiconductor promote recombination. Since the surface of the solar cell represents a severe disruption of the crystal lattice, the surfaces of the solar cell are a site

of particularly high recombination. The high recombination rate in the vicinity of a surface depletes this region of minority carriers. A localized region of low carrier concentration causes carriers to flow into this region from the surrounding, higher concentration regions.

Therefore, the surface recombination rate is limited by the rate at which minority carriers move towards the surface. A parameter called the "surface recombination velocity," in units of cm/sec, is used to specify the recombination at a surface. In a surface with no recombination, the movement of carriers towards the surface is zero, and hence the surface recombination velocity is zero. In a surface with infinitely fast recombination, the movement of carriers towards this surface is limited by the maximum velocity they can attain, and for most semiconductors is on the order of 1×10^7 cm/sec.

The solar cells' surface area, regardless of the surface finish (mechanical polishing or chemical etch), is a most critical area, for it is on or near this surface where the photoelectric effect takes place. The surface area is often ignored, or improperly cleaned and processed, and the resulting solar cells will exhibit low performance. In worst cases, the defective surface area will cause the cells to deteriorate quickly or fail after some time in the field (latent effect).

Note: Latent effects in this context are hidden defects, triggered by a set of conditions, usually long after the manufacturing process has been completed. These are especially pertinent and dangerous in the case of PV cells and modules, because they are exposed to harsh environmental conditions (excessive heat, freezing, chemical attacks, high wind loads, etc.) for many years of non-stop field operation. Any of these conditions could trigger a latent effect with time.

Solar Cell Types

There are a number of different technologies today, the most important of which can be seen in Table 2-2.

We looked at silicon material and wafer types above, so we only need to remind the reader that the

CRYSTALLINE Si	THIN FILMS	OTHER
Mono-Crystalline	Amorphous Si	GaAs
Multi-Crystalline	Epitaxial Si	InP
Micro-Crystalline	CdTe	Other III-V
Super c-Si	CIGS	Germanium
Si Ribbon	Organic/Polymer	CPV Cells

Table 2-2. Different types PV technologies

category "crystalline silicon," c-silicon, or c-Si is the general designation of all types of crystalline silicon-based products, including c-Si wafers, solar cells, and PV modules made from them.

The major types of silicon solar cells and modules in use today, some of which we discuss in more detail in the next chapter, are as follow:

Single Crystal Solar Cells

Single crystal silicon, also called mono-crystalline, mono-silicon, or mono-Si, or sc-Si, is a type of silicon that was grown by the very special and expensive Czochralski (or CZ) method, or via the float zone (FZ) method. Both methods use similar equipment and production methods and produce a superior silicon material for semiconductor devices and solar cells manufacturing. Solar cells and panels made out of CZ or FZ silicon material have the highest efficiency and longevity of all silicon-based PV devices, primarily due to the uniform, stable and predictable nature of the bulk material.

Multi-crystalline Silicon

Multicrystalline, mc-silicon, or mc-Si is the most widely used silicon material. It is most often called "poly," "polysilicon," or "polycrystalline" silicon (which is what we will call it in this text too) because it consists of many (poly) strings instead of a single crystal. Poly is made by melting and casting silicon chunks into large blocks, splitting the blocks into smaller rectangular blocks and slicing these into thin, square-shaped wafers.

Poly-crystalline Silicon Solar Cells

Poly crystalline silicon, also called poly silicon, or poly, or pc-Si is a thin film of silicon, deposited via CVD, or LPCVD processes on semiconductor type wafers, to be used as a gate material in MOSFET transistors and CMOS microchips. The solar industry uses similar equipment and processes to deposit very thin layers of silicon (pc-Si and a-Si) onto polysilicon or other substrates. The resulting devices are of lower efficiency, as compared to sc-Si or mc-Si.

Note: There is a confusion created by the term "poly" as it is used widely to identify PV cells modules made out of multi-crystalline silicon, instead of its actual use in the semiconductor thin film. Since we cannot change the decades-long use of the term "poly" in the solar industry to identify multi-crystalline silicon products, we will continue using it too with a certain degree of caution and with due clarification when needed.

Amorphous Silicon Solar Cells

Amorphous silicon is also thin film silicon, alpha silicon, or a-Si and is used in p-i-n type solar cells. Typical a-Si modules include front side glass, TCO film, thin film silicon, back contact, polyvinyl-butyral (PVB) encapsulant and back side glass. a-Si has been used to power calculators for some time now, mostly because it is easily and cheaply deposited on any substrate.

Silicon Ribbon and Foils

Silicon ribbon, just like the name suggests, is a ribbon made out of silicon material, by using a metallurgical extrusion type of process. The resulting material is much thinner than the usual silicon solar wafers and cells, thus much less material is used (as in bulk ingot growth) and wasted (as in wafers slicing) than in the conventional processes.

The key advantage of this process is using less material and energy for equal production output, so this technology has a promising future of rising materials costs. This advantage, however, comes at the expense of increased fragility of the base material, which is a significant barrier in optimizing the mass production of solar cells and modules made of string ribbon.

Silicon foils are even thinner pieces of silicon, produced via kerf-free wafering technology (less waste of silicon material during slicing ingots into wafers). Some manufacturers have produced very thin films of silicon by using special slicing tools and techniques, and we are aware of a 20μm-thick wafers, used to manufacture solar cells, which due to their thinness are called "foils."

These very thin and even flexible mono-crystalline silicon materials are neither a thin-film nor a wafer, and give the "foils" unique form and physical attributes.

Epitaxial Silicon

Epitaxial thin-film solar cell is made by depositing a very thin film of silicon on a cheap substrate, such as with highly doped sc-Si wafers (e.g., from low-grade silicon or scrap Si material). The epi layer of Si is deposited by chemical vapor deposition (CVD).

The resulting mix of a high quality epi layer and a cheap substrate is a compromise between high cost and efficiency, yet it offers a solution to gradual transition from a wafer-based (heavy material dependent) to a thin-film technology (less material and more sophisticated processing).

This process is easier to implement than most other thin-film technologies today, but it remains to be seen if its efficiency and cost will be able to compete on the energy market.

Micro-crystalline Silicon Solar Cells

Micro crystalline silicon, also called nano-crystalline silicon, uc-Si, or nc-Si, is a form of silicon in its allotropic form, very similar to a-Si. nc-Si has small grains of crystalline silicon in the amorphous phase, and if grown properly can have higher electron mobility due to the presence of silicon crystallites. It also shows increased absorption in red and infrared wavelengths.

Super Monocrystalline Silicon

This is a new purer type of silicon with more perfect crystalline structure which exhibits reduced phonon-phonon and phonon-electron interactions. This phenomenon increases certain transport properties, resulting in 60% better room-temperature thermal conductivity than natural silicon.

In this chapter we focus on mono and polycrystalline silicon solar cells and modules, their structure, function, and the related properties and issues. We will take an even closer look at some of these technologies in the next chapter, because they fall in the category of "most promising PV technologies."

C-SI PV MODULES

A PV module consists of a number of interconnected solar cells (typically 36 to 72 connected in series for battery charging), and many more for large-scale applications. Individual solar cells are soldered in strings and encapsulated into a single, hopefully long-lasting unit simply because PV modules cannot be disassembled for repairs. The main purpose for encapsulating a set of electrically connected solar cells into a module is to protect them from the harsh environment in which they will operate.

Solar cells are relatively thin and fragile and are prone to mechanical damage due to vibration or impact unless well protected. In addition, the metal grid on the top and bottom surfaces of the solar cells, the wires interconnecting the individual solar cells, as well as the soldered junctions can be corroded by moisture or water vapor entering the module, if the protecting materials are damaged or absent. So the encapsulation: a) provides a manageable package that can be installed in the field, b) prevents mechanical damage to the solar cells, and c) prevents water or water vapor from penetrating the module and corroding the electrical contacts and junctions.

Many different types of PV modules exist, and module structure is often different for different types of solar cells or for different applications. For example, amorphous silicon, and other thin film solar cells are often encapsulated in a flexible array, while crystalline silicon solar cells are usually mounted in rigid metal frames with a glass front surface.

Silicon PV modules' lifetime and warranty are often 20-25 years, which assumes robust and durable encapsulation of the PV modules. Pool encapsulation will cause performance degradation and failure with time.

c-Si PV Modules Structure

Most bulk silicon PV modules consist of a transparent top surface, an encapsulant, a string of PV cells, rear encapsulant, rear cover and a frame around the outer edge. In most modules, the top surface is glass, the encapsulant is EVA (ethyl vinyl acetate), and the rear layer is usually Tedlar or a number of similar plastic and thermo-plastic materials.

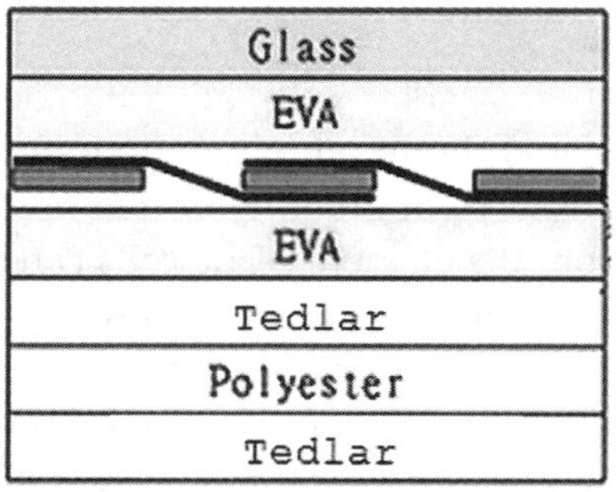

Figure 2-14. PV Module cross section

Typical module components are:

Cover Glass

The cover of the front surface of a PV module is usually glass with high transmission in the wavelengths which can be used by the solar cells, usually in the range of 350 nm to 1200 nm. In addition, the reflection from the front surface should be low. While theoretically the reflection could be reduced by applying an anti-reflection coating to the glass surface, in practice these coatings are not robust enough to withstand some of the conditions in which most PV systems are used.

An alternative technique used to reduce reflection is to "roughen," or texture the top glass surface. In this case dust and dirt are very likely to adhere to the top surface and less likely to be dislodged by wind or rain. These glass surfaces are not "self-cleaning" and the advantages of reduced reflection are quickly outweighed by losses incurred due to increased top surface soiling. Texturing the inside of the glass is also practiced by some manufacturers, but there are some disadvantages in doing this as well, so the proper glass has to selected according to the module type and designation.

In addition, the top surface should have good safety properties and impact resistance, should be stable under prolonged UV exposure, and should have a low thermal resistivity. There are several choices for a top surface material including acrylic, polymers and glass. Tempered, low iron-content glass is most commonly used as it is low cost, strong, stable, highly transparent, mostly impervious to water and gases, and has good self-cleaning properties.

This type of glass is the most stable and trouble-free component of the entire module assembly. It doesn't deteriorate easily regardless of the harshness of the elements, and unless it is broken, it will withstand the test of time for 25-30 years largely unaffected. Once the glass is broken, however, the module must be removed or put on a special maintenance schedule.

Cell Strings

A number of solar cells are interconnected and sealed (laminated) between plastic materials, which insulate them from each other, from the interconnecting wires, and from the elements. The cells can be arranged and wired in a number of ways, as shown in Figure 2-15. Thus-generated DC power is extracted from the module via two wires protruding from the module and routed into a junction (or terminal) box which is fitted with connectors for quick interconnect within the other array components.

Encapsulant (Front Surface)

An encapsulant is used to provide adhesion between the solar cells, the top surface and the rear surface of the PV module. The encapsulant should be stable at elevated temperatures and high UV exposure. It should also be optically transparent and should have a low thermal resistance. EVA (ethyl vinyl acetate) is the most commonly used encapsulant material. EVA comes in thin sheets which are inserted between the solar cells and the top surface and the rear surface. This sandwich is then heated to 150°C to polymerize the EVA and bond the module together.

EVA is responsible for protecting the cells from moisture and reactive species entering the module. Long exposure to UV and IR radiation tends to damage the EVA and it becomes yellow, which reduces its transmittance and module efficiency. Cracks and pores created under long exposure will allow the elements to enter the module and destroy the cells.

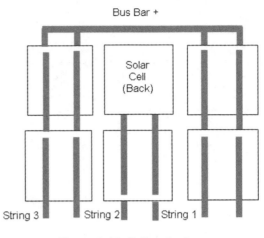

Figure 2-15. Cell stringing

Back Cover

Most c-Si modules have a thin sheet of aluminum as a back cover, which is screwed into the frame, with the terminal box attached to it. A key characteristic of the back cover is that it must have low thermal resistance and that it must prevent the ingress of water or water vapor. In most modules, a thin polymer sheet, typically Tedlar, is used as the rear surface, which provides electrical and environmental protection for the solar cells.

Some PV modules, known as bifacial modules, are designed to accept light from either the front or the rear of the solar cell. In bifacial modules both the front and the rear must be optically transparent, so glass is the preferred material for these.

Note: The c-Si module's standard configuration is solid and strong enough to withstand transportation bumps, handling, and high winds during operation. Nevertheless, it is not intended to be used for support, to be stepped on, or even leaned on, as is the practice of some installers.

Even if the glass doesn't break in such cases of careless handling and installation, the weight and/or impact put enough temporary stress on the cells to cause micro-cracks and other interruptions, which eventually grow into much bigger problems.

Side Frame

A final structural component of the module is the edging or framing, which provides additional mechanical strength and isolation from the elements. Module edges are sealed by the encapsulant layers in the framing and by additional adhesive materials for even greater protection against the elements. An aluminum frame is then fastened around the edges of the module. The frame structure should be free of projections or pockets which could trap water, dust or other foreign matter.

c-Si PV Module Manufacturing

A number of potential issues are encountered during the cell and module manufacturing processes, all of which must be taken into consideration, if we are to have reliably performing PV cells and modules, lasting 25-30 years.

The major issues to keep in mind when designing or planning to use c-Si PV modules are:

1. Quality of silicon material, chemicals and consumables
2. Cell type and design parameters
3. Quality of the cells' manufacturing equipment and process
4. Module type and design parameters
5. Modules' manufacturing equipment and process
6. Possible cell malfunctions within this type and make of module
7. Possible module malfunctions within the particular array

Once the materials—PV cells, laminates, glass, back cover, wiring etc.—have been received and gathered at the module production site, the module is assembled in the following sequence:

Cell Sorting, Arranging and Soldering

Finished solar cells are flash tested and sorted by their I-V characteristics and power output. Cells that pass the test are placed in bins according to their performance and stored or taken to the module assembly area.

Wiring and Assembly

Cells are connected in a series circuit manually, or by a semi-automated soldering machine using solder coated metal ribbon (usually two in parallel) soldered to the top of one cell and to the bottom of the next cell. This process forms a string of cells which could be as long as desired but is usually shaped to fit in the respective PV modules tray. Electrical continuity and resistivity tests are performed on some modules to make sure that the bonds are good. "Pull" tests are sometimes done, to check the mechanical strength of the bonds. Thus, connected cells make a complete circuit (string), which is ready for lamination into a completed module.

Lamination and Framing

A lay-up for lamination is prepared with clean top glass, then EVA film and strings of wired cells are placed

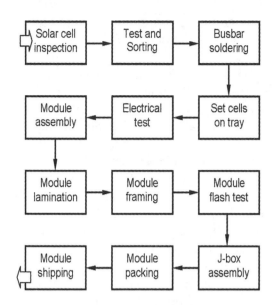

Figure 2-16. PV module assembly sequence

on it. Sometimes the backing materials (Tedlar and back cover) are placed on top too, forming a complete module. Several lay-ups, each consisting of the above components are lined up in a large cabinet called a laminator. Using silicone vacuum blankets, the batch of lay-ups is heated and vacuum laminated at one time. After cooling, the modules are ready for use. This method of laminating is much cheaper than laminating one or two units at a time. The excess lamination is trimmed and terminal wiring is attached. In most cases, an aluminum extruded frame is assembled around the material and the unit is ready for shipping.

Basically, module laminators consist of a large-area heated metal platen mounted in a cabinet-like vacuum chamber. The top of the cabinet opens for loading and unloading modules. A flexible diaphragm is attached to the top of the chamber, and a set of valves allows the space above the diaphragm to be evacuated during the initial pump step and backfilled with room air during the press step. A pin lift mechanism is sometimes used to lift modules above the heated platen during the initial pump step, but most standard modules don't require it.

Temperature uniformities of ±5°C at the lamination point are sufficient for obtaining good laminations with acceptable gel content and adhesion across the module. While more uniform temperatures are available from some laminator suppliers, there is no real benefit to the module manufacturer.

Laminators are available with two types of cover opening systems, clamshell and vertical post. In the clamshell design, the cover is mounted on a hinge at the back of the laminator, which opens like the hood of a car. This leaves the laminator wide open on three sides, making it easy for an operator to load and unload modules manually.

Automated belt-fed laminators, on the other hand, use the vertical post method, which lifts the cover horizontally above the process chamber. Because the cover does not need to travel much for belt loading, the chamber opening and closing times, and resulting process steps (heating and vacuum pump down) are reduced. As a result, most high throughput module lines use belt-fed laminators with vertical cover lifts.

Note: Fully automated cell assembly and lamination lines exist today, but most low- to mid-volume assembly operations, especially those in Asia, still prefer manual lay-up and stringing operations, combined with low-throughput clamshell type laminators. This is due mostly to the availability of cheap labor, which is a source of quality issues.

Automating the labor-intensive processes provides a more consistent and higher quality product, but since there are so many other factors, automation alone does not guarantee high quality.

Modules Flash Testing and Sorting

After completing the assembly process by adding edge sealers, side frame, terminal box, etc., the modules are placed on a test stand in the solar simulator (flasher) and are illuminated with a special type of light that resembles the solar spectrum at STC, for a period of time. The temperature of the modules is kept at 25°C during the test by active or passive cooling. I-V curve is then

generated for each module. The output data—efficiency, voltage etc.—are used to identify, sort and label the modules according to their output. Modules that pass the test are packed and shipped to customers.

PV Modules, Technical Specifications

PV modules are complex electro-mechanical assemblies with a number of components which determine their function. Modules usually come with a set of specifications and installation and operation instructions which describe their type, structure, use and expected behavior. Some of the key concepts of PV modules' function and in-the-field operation are described below (2).

Rated Power at STC (in Watts)

This is the module power rating at STC, or 1,000 watts per square meter of solar irradiance on the module surface at 25°C (77°F) cell temperature. Because module power output depends on environmental conditions, such as irradiance and temperature, each module is tested at STC so they can be compared and rated on a level playing field. When less sunshine hits the module, less power is produced. Likewise, the hotter it gets, the less power modules produce due to the cells' temperature degradation phenomena.

STC references the actual cell temperature, not ambient air temperature. As dark PV cells absorb radiant energy, their temperature increases and will be significantly higher than the ambient air temperature. For example, at an ambient air temperature of about 23°F,

Solar cell material	Cell efficiency η_z (laboratory) (%)	Cell efficiency η_z (production) (%)	Module efficiency η_M (series production) (%)
Monocrystalline silicon	24.7	21.5	16.9
Polycrystalline silicon	20.3	16.5	14.2
Ribbon silicon	19.7	14	13.1
Crystalline thin-film silicon	19.2	9.5	7.9
Amorphous silicon[a]	13.0	10.5	7.5
Micromorphous silicon[a]	12.0	10.7	9.1
CIS	19.5	14.0	11.0
Cadmium telluride	16.5	10.0	9.0
III-V semiconductor	39.0[b]	27.4	27.0
Dye-sensitized cell	12.0	7.0	5.0[c]
Hybrid HIT solar cell	21	18.5	16.8

Table 2-3. Performance of different solar cells and modules

a PV cell's temperature will measure about 77°F—the temperature at which its power is rated.

If the ambient air temperature is 77°F (and irradiance is about 1,000 W/m²), module cell temperature will be about 131°F and power output will be reduced by about 15%. If the air temperature is 120°F, as it is at noon on a clear summer day in the Sonora desert, the power output might drop by another 15-20%.

Rated Power Tolerance (%)

This is the specified range within which a module will either over perform or underperform its rated power at STC. Power tolerance is a much-debated module specification. Depending on the module, this specification can vary greatly, from as much as +10% to –9%. A 100 W module with a –9% power tolerance rating may produce only 91 W straight out of the box. With potential losses from high temperatures, it will likely produce even less.

Because modules are often rated in small increments, it is not uncommon for those that fall under the lower power tolerance of the next model to be rated as a higher wattage module. Case in point: A module with a +/–5% power tolerance rating that produces 181 W during the factory testing process could be classified as a 190 W module, as opposed to a 180 W module. For maximum production, look for modules with a small negative (or positive only) power tolerance.

Rated Power Per Square Foot (Watts)

This is the power output at STC, per square foot of module (not cell) area. It is calculated by dividing module rated power by the module's area in square feet and is also known as "power density." The higher the power density, the less space is needed to produce a certain amount of energy. With some of the newer-generation modules, power density values are higher due to increased cell and module efficiency.

The greatest variation in this specification is in comparing crystalline PV modules to thin-film modules. If space is tight for array placement, consider choosing modules with higher power densities, though more efficient modules can be more expensive. Choose modules with lower power densities, and you'll need more modules for the same amount of energy. That means more infrastructure (module mounts, hardware, etc.) and more installation time.

Module Efficiency (%)

This is the ratio of output power to input power, or how efficiently a PV module uses the photons in sunlight to generate DC electricity. If 1,000 W of sunlight hit 1 square meter of solar module, and that solar module produces 100 W of power from that square meter, then it has an efficiency of 10%. Similar to power density, the higher the efficiency value, the more electricity generated in a given space.

Series Fuse Rating (Amps)

This is the current rating of a series fuse used to protect a module from overcurrent, under fault conditions. Each module is rated to withstand a certain number of amps. Too many amps flowing through the module—perhaps backfed amps from paralleled modules or paralleled strings of modules—could damage the module if it's not protected by an overcurrent device rated at this specification. Backfeeding from other strings is most likely to exist if one series string of modules stops producing power due to shading or a damaged circuit. Because PV modules are current-limited, there are some cases where series fusing may not be needed.

When there is only one module or string, there is nothing that can backfeed, and no series string fuse is needed. In the case of two series strings, if one string stops producing power and the other string backfeeds through it, still no fuse is needed because each module is designed to handle the current from one string. Some PV systems even allow for three or more strings with no series fuses. This is due to 690.9 Exception B of the NEC and is possible when the series fuse specification is substantially higher than the module's short circuit current (Isc). When required, series fuses are located in either a combiner box or in some batteryless inverters.

Connector Type

This is the way the module output terminal or cable/connector is configured. Most modules come with "plug and play" weather-tight connectors to reduce installation time in the field. These are connectors such as Solarlok (manufactured by Tyco Electronics), and MC and MC4 (manufactured by Multi-Contact USA). Solarlok and MC4 are lockable connectors that require a tool for opening.

Because so many PV systems installed today operate at high DC voltages, lockable connectors are being used on modules in accessible locations to prevent untrained persons from "unplugging" the modules, per 2008 NEC Article 690.33(C). Due to this new code requirement, most PV manufacturers are updating their connectors to the locking type. Depending on how fast this change is reflected in the supply chain, connectors on a particular module may be of an older style or lockable—so be sure to check.

Some manufacturers still offer modules with junction boxes (J-boxes). J-boxes allow the use of conduit between modules, as raceways are required for PV source and output circuits (with a maximum system voltage greater than 30 volts) installed in readily accessible locations per 2008 NEC Article 690.31(A). This approach is used to prevent an unqualified person from accessing array wiring.

Materials Warranty (Years)

This is a limited warranty on module materials and quality under normal application, installation, use, and service conditions. For most modules, material warranties vary from 1 to 5 years. Most manufacturers offer full replacement or free servicing of defective modules.

Read the warranty conditions carefully prior to purchasing modules, because they vary from vendor to vendor. The problem that is most often encountered and one that is most disputed is shipping charges and replacement (labor) cost. In other words, the expense of shipping a batch of modules from abroad, uninstalling the defective ones and reinstalling the new batch might be higher than the cost of the modules themselves. A vendor might object to absorbing these charges, unless all these conditions and exemptions are clearly identified and addressed in the warranty.

Power Warranty (Long-term)

This is a limited warranty for module power output based on the minimum peak power rating (STC rating minus power tolerance percentage) of a given module. The manufacturer guarantees that the module will provide a certain level of power for a period of time—at least 20 years. Most warranties are structured as a percentage of minimum peak power output within two different time frames—90% over the first 10 years and 80% for the next 10 years.

For example, a 100 W module with a power tolerance of +/−5% will carry a manufacturer guarantee that the module should produce at least 85.5 W (100 W x 0.95 power tolerance x 0.9) under STC for the first 10 years. For the next 10 years, the module should produce at least 76 W (100 W x 0.95 power tolerance x 0.8).

Module replacement value provided by most power warranties is generally prorated according to how long the module has been in the field. Again, the cost of shipping and replacement of the non-conforming modules could be high, and replacement conditions must be well understood and agreed upon by both parties prior of purchasing the modules.

Cell Type

This describes the type of silicon that comprises a specific cell, based on the cell manufacturing process. There are four basic types of modules for the non-commercial market—monocrystalline, multicrystalline, ribbon, and amorphous silicon (a-Si). Each cell type has pros and cons.

Monocrystalline PV cells are the most expensive and energy intensive to produce but usually yield the highest efficiencies. Though multicrystalline and ribbon silicon cells are slightly less energy intensive and less expensive to produce, these cells are slightly less efficient than monocrystalline cells.

However, because both multi- and ribbon-silicon solar cells leave fewer gaps on the module surface (due to square or rectangular cell shapes), these modules can often offer about the same power density as monocrystalline modules per unit area.

Thin-film modules, such as those made from amorphous silicon cells, are the least expensive to produce and require the least amount of energy and raw materials, but they are the least efficient of the cell types. They require about twice as much space to produce the same power as mono-, multi-, or ribbon-silicon modules.

Thin-film modules do have better shade tolerance and high-temperature performance, but are of lower overall efficiency and often more expensive to install, because their lower power density requires additional framing and BOS components.

Some manufacturers offer cells that are a combination of cell types—Sanyo's "bifacial" HIT modules, for example, are composed of a monocrystalline cell and a thin layer of amorphous silicon material. In addition to generating power from the direct rays of the sun on the module face, this hybrid module can produce power from reflected light on its underside, increasing overall module efficiency.

This approach. although quite clever, is questionable, because of wasted land (active) area and the marginality of power produced by the back surface.

Cells in Series

This is the number of individual PV cells wired in series, which determines the module design voltage. Crystalline PV cells each operate at about 0.5 V. When cells are wired in series, the voltage of each cell is additive. For example, a module that has 36 cells in series has a maximum power voltage (Vmp) of about 18 V.

Why 36? Historically, these modules—known as 12-V modules—were designed to push power into 12-V batteries. But to deliver the 12 V, they needed to have enough excess voltage (electrical pressure) to compensate for the voltage loss due to high temperature conditions. Modules with 36 ("12-V") or 72 ("24-V") cells are designed for battery-charging applications.

Modules with other numbers of cells in series are intended for use in batteryless grid-tied systems. Grid-tied modules now combine a certain number of cells for the goal of maximizing power with grid-tied inverters and their maximum power point tracking (MPPT) capabilities. Due to the increased availability of step-down/ MPPT battery charge controllers, grid-tied PV modules can also be used for battery charging, as long as they stay within the voltage limitations of the charge controller.

Maximum Power Voltage

Maximum power voltage (Vmp) is the voltage generated by a PV module or array when exposed to sunlight and connected to a load—typically an inverter or a charge controller and/or a battery. Batteryless grid-tied inverters and MPPT charge controllers are built to track maximum power throughout the day. The Vmp of each module and array, as well as array operating temperatures, must be considered when sizing an array to a particular inverter or controller.

Increasing temperatures cause voltage to decrease; decreasing temperatures cause voltage to increase. Fortunately, series string-sizing programs for grid-tied inverters will allow you to input both the high and low temperatures at your installation site, and calculate the correct number of modules in series to maximize system performance.

Maximum Power Current (Imp)

This is the maximum amperage produced by a module or array (under STC) when exposed to sunlight and connected to a load. This specification is most commonly used in performing calculations for PV array disconnect labeling required by NEC Article 690.53, as the rated maximum power-point current for the array must be listed. Maximum power current is also used in array and charge controller sizing calculations for battery-based PV systems.

Open-circuit Voltage (Voc)

This is the maximum voltage generated by a PV module or array when exposed to sunlight with no load connected. All major PV system components (modules, wiring, inverters, charge controllers, etc.) are rated to handle a maximum voltage. Maximum system voltage must be calculated in the design process to ensure all

components are designed to handle the highest voltage that may be present. Under certain low-light conditions (dawn/dusk), it's possible for a PV array to operate close to open-circuit voltage.

PV voltage will increase with decreasing air temperature, so Voc is used in conjunction with historic low temperature data to calculate the absolute highest maximum system voltage. Maximum system voltage must be shown on the PV array disconnect label.

Short-circuit Current (Isc)

This is the amperage generated by a PV module or array when exposed to sunlight and with the output terminals shorted. The PV circuit's wire size and over-current protection (fuses and circuit breakers) calculations per NEC Article 690.8 are based on module/array short-circuit current. The PV system disconnect(s) must list array short-circuit current (per NEC 690.53).

Maximum Power Temperature Coefficient (% per degree C)

This is the change in module output power in percent of change per degree Celsius for temperatures other than 25°C (STC temperature rating). This specification allows us to calculate how much module power will be lost or gained due to temperature shifts. In hot climates, cell temperatures can reach in excess of 70°C (158°F). Consider a module maximum power rating of 200 W at STC, with a temperature coefficient of –0.5% per degree C. At 70°C, the actual output of this module would be approximately 155 W.

Modules with lower power temperature coefficients will fare better in higher-temperature conditions. Notice the relatively low values listed for thin film modules. This specification reflects their usually better high-temperature performance.

Open-circuit Voltage Temperature Coefficient (mV per degree C)

This is the change in module open-circuit voltage in millivolts per degree Celsius at temperatures other than 25°C (STC temperature rating). Expressed as millivolts per degree Celsius, it can be shown as percentage per degree Celsius, volts per degree Celsius, or volts per degree Kelvin. If given, this specification is most commonly used in conjunction with open-circuit voltage to calculate maximum system voltage (per NEC Article 690.7) for system design and labeling purposes.

For example, consider a series string of ten 43.6 V (Voc) modules installed at a site with a record low of –10°C. Given a Voc temperature coefficient of –160mV per degree Celsius, the rise in voltage per module will be 5,600 mV [–160 mV per degree Celsius x (–10°C – 25°C)], making an overall maximum system voltage of 492 V [10 x (5.6 V + 43.6 V)]—under the 600 VDC limit for PV system equipment.

Nominal Operating Cell Temperature (NOCT)

This is the temperature of each module, given an irradiance of 800 W/m² and an ambient air temperature of 20°C. This specification can be used with the maximum power temperature coefficient to get a better real-world estimate of power loss due to temperature. The difference in cell temperature and ambient temperature is dependent on sunlight's intensity (W/m²). Less-than-ideal sky conditions are common in many areas, so a standard of 800 W/m² is the basis for this specification, rather than 1,000 W/m², which is considered full sun. The construction and coloring of each module is slightly different, so actual cell temperature under these conditions will vary per module.

For example, if a particular module has an NOCT of 40°C and a maximum power temperature coefficient of –0.5% per degree Celsius, power losses due to temperature can be estimated at about 7.5% [0.5% x (40°C – 25°C)].

Practical Operational Effects

The actual solar cell design and evaluation processes encompass a number of theoretical considerations, some of which were discussed above. More general considerations, related to the cells' functionality are reviewed below.

We agreed above that when sunlight hits a solar cell it breaks apart electron-hole pairs on impact. Each photon with enough energy (band-gap energy of 1.1 eV) frees exactly one electron, which frees a hole in the process. Millions of photons of this energy level will generate millions of electron-hole pairs and millions of electrons released from the pairs' bond will generate current which will flow through the solar cell and into an external load.

Photons with much lower or much higher energy levels do not contribute much to the overall energy generation and might even impact it negatively by interfering with the generation process and/or increasing the cell temperature.

How much sunlight energy does our PV cell absorb that is converted in useful electric power? Unfortunately, it is not much. Regular silicon cells (arranged in PV modules) convert only 15-18% of the incoming sunlight today.

Thin film cells and modules convert 8-9%, while some exotic cells and modules convert 4-6%. On the other end of the efficiency spectrum, some specialized multi-junction cells convert over 40% of the sunlight they see.

Why so much difference? It is due to the special characteristics of solar cells, which we review below, starting with the different effects taking place in the silicon solar cells and modules when exposed to sunlight.

Energy Interactions and Losses

Sunlight can be separated into different wavelengths, which we can see in the form of a rainbow. Visible light is what we see in the rainbow, but it is only part of the sunlight spectrum, while the other sunlight components are made up of a range of different wavelengths with different energy levels and overall behavior.

Energy capture, conversion processes, and related problems are:

a. Not all photons have enough energy to generate electron-hole pairs, so they'll simply pass through the cells as if they were transparent.

b. Other photons have too much energy which could create negative effects. If a photon has more energy than required, some of the extra energy is lost. A photon with twice the required energy can create more than one electron-hole pair, but this effect is negligible in most cases and could even be harmful in larger quantities.

    Note: The above two negative effects account for 70-80% of the loss of incoming energy on the solar cell.

c. Material with lower band gap can use more of the photons (which have lower energy) but the band-gap also determines the potential (the voltage) of the newly generated electric field, and if it's too low it will make up current by absorbing more photons. In this case, the voltage might be too low and not practical for everyday use.

    Note: The optimal band gap, balancing these two effects, is around 1.3-1.5 eV for single junction, single material cells of any type.

d. The solar cell's surface quality is of utmost importance, for it is where major optical interactions and interferences occur.

Inefficiencies are partially due to imperfections in the bulk crystal (there are many more imperfections in polysilicon material). Observe the perfectly symmetrical geometry of the silicon atoms in the mono-crystalline silicon in Figure 2-17.

Surface imperfections (also Figure 2-17), are mostly due to manufacturing process abnormalities (growing, slabbing, sawing, etching), and can also cause serious efficiency problems.

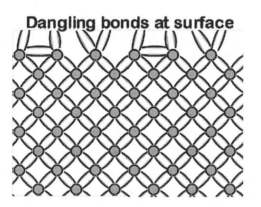

Figure 2-17. Silicon solar cell surface area (mono-crystalline silicon wafer)

However, because both multi- and ribbon-silicon solar cells leave fewer gaps on the module surface (due to square or rectangular cell shapes). These modules can often offer about the same power density as monocrystalline modules per unit area.

Note: Mono-crystalline silicon wafers are usually round in shape, so when assembled as such they leave large gaps in the PV modules' surface. Cutting the round wafers into squares to minimize the gaps is possible, but it is an expensive process that creates a lot of waste.

Temperature Losses

Temperature losses are expressed by the so-called temperature coefficient which is basically the rate of change of the generated power (and other measurable parameters) with respect to increase of temperature, usually measured in degrees C above STC (25°C). Changes can be measured and calculated for: short-circuit current (Isc), maximum power current (Imp), open-circuit voltage (Voc), maximum power voltage (Vmp), and maximum power (Pmp), as well as the fill factor (FF) and efficiency (h).

ASTM standard methods for performance testing of cells and modules addresses only two temperature coefficients, one for current and one for voltage. Actual field characterization of PV modules and arrays performance indicates that four temperature coefficients for Isc, Imp, Voc, and Vmp, are necessary to sufficiently and accurately model electrical performance for a wide range of operating conditions.

ASTM also specifies that temperature coefficients are determined using a standard solar spectral distribution at 1,000 W/m² irradiance, but from a practical standpoint they need to be applied at other irradiance levels as well, in order to get a complete picture of the cell/module behavior; i.e., some PV modules operate under the mostly cloudy skies of central Europe or the northeastern US, while others are exposed daily to the ferociously bright and hot deserts of the Southwest.

Obviously there is a huge difference in those operating conditions, thus different test and measurement parameters must be used to determine which type PV modules are best for which climate and operating condition.

This is extremely important, because solar cells and modules must be designed and manufactured to operate properly under full sunlight exposure, which could bring cell temperature up to 195°F. This is almost 3 times the STC test standard and will have a profound effect on the cells' and modules' performance and longevity.

The actual cell temperature inside the module determines the drop of output and is therefore a most critical part of the temperature degradation. The cell temperature can be calculated using a measured back-surface temperature and a predetermined temperature difference between the back surface and the cell.

$$Tc = Tm + E/Eo \cdot \Delta T$$

where:

Tc	=	Cell temperature inside module, (°C)
Tm	=	Measured back-surface module temperature, (°C)
E	=	Measured solar irradiance on module, (W/m²)
Eo	=	Reference solar irradiance on module, (1000 W/m²)
ΔT	=	Temperature difference between the cell and the module back surface at an irradiance level of 1000W/m².

This temperature difference is typically 2 to 3°C for flat-plate modules in an open-rack mount. For flat-plate modules with a thermally insulated back surface, this temperature difference can be assumed to be zero. For concentrator modules, this temperature difference is typically determined between the cell and the heat sink on the back of the module.

For c-Si solar cells and modules, a 0.5% drop of power output per each degree C increase of temperature above STC (25°C) has been accepted as an average; i.e., a c-Si solar cell in a module operating in the Arizona desert could reach 85°C internal temperature.

Note that this is 60°C above STC, or (60 x 0.5) = 30% drop of output. This is a serious reduction of power must be anticipated during the design and evaluation stages. Those who have no experience with extreme desert climates, and are planning installations in the desert, must rethink and reevaluate their options.

The desert not friendly, and everything left in it is constantly attacked until it is changed destroyed. This daily barrage going on for 30 years can put any material on the defense and even destroy it. Thorough understanding of desert phenomena is the only way to make our proposed technologies work.

Different types of solar cells behave differently under desert conditions but they all show reduced power output under extreme heat, so we need to know how to estimate the power reduction with the seasons and for the duration.

The negative effect of elevated operating temperature is a well known and understood phenomenon. Increasing the temperature reduces the band gap of most semiconductors, thereby affecting most of the key operating parameters and especially the open-circuit voltage.

So when cell temperature goes up (due to internal problems or external excess operating temperatures), the voltage goes down. With that, cell efficiency and output decrease proportionally as well.

The elevated temperatures decrease the voltage of the cell, which translate into loss of power output.

Resistive Effects

Some of the major negative effects caused by defective material properties and/or improper processing are the so-called resistive (parasitic) effects which reduce the efficiency of the solar cell by dissipating power while increasing the internal resistances. The most common and most important of these are series and shunt resistance. The influence of temperature and aging on these, as well as the overall solar cells' performance, needs very close investigation as well.

A closer look at the resistive parasitic effects in c-Si solar cells:

a.   Series resistance, Rs, in a solar cell is expressed first by the movement of current through the emitter and base of the solar cell; second, the contact resistance between the metal contacts and the bulk silicon; and finally the resistance of the top and rear

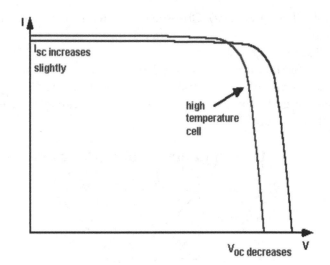

Figure 2-18. Elevated temperature effect

metal contacts themselves.

The main impact of elevated series resistance is reducing the fill factor, although excessively high values may also reduce the short-circuit current and ultimately result in overheating and destruction of the cell and the module housing it.

b.   Significant power losses caused by lower than normal shunt resistance, RSH, are typically due to manufacturing defects, rather than poor solar cell design. Low shunt resistance causes power losses in solar cells by providing an alternate current path for the light-generated current. Such a diversion reduces current flowing through the solar cell junction and reduces voltage from the solar cell.

The high Rs and Rsh reduce the FF and the efficiency of the cells, which results in power output decrease. The effect of shunt resistance is particularly severe at low light levels, since there will be less light-generated current. The loss of this current to the shunt therefore has a larger impact. In addition, at lower voltages, where the effective resistance of the solar cell is high, the impact of a resistance in parallel is large.

Other Losses

The electrons are forced to flow from one side of the cell to the other through an external circuit going to and from the load (battery, motor, lamp, etc.). This means that we need good contacts on the top and bottom of the cell to collect and recapture the electrons flowing to and from the external circuit. We can cover the bottom of the cell with a conducting metal, but we cannot cover the top completely because that will stop the electrons from

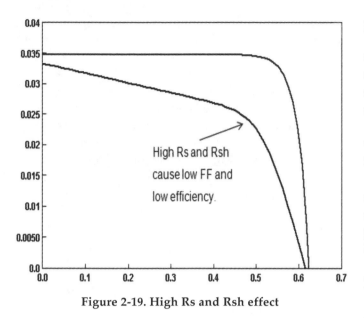

Figure 2-19. High Rs and Rsh effect

reaching the cell surface.

So we deposit appropriately spaced metallic grid on top of the cells to shorten the distance electrons travel from the bulk material to the metal grid, where they are collected and sent into the electrical circuit to provide electrical energy to a load. At the same time, we try to cover only a small part of the top cell surface with this metal grid, because we need a maximum amount of exposed top surface, where the photons can fall onto and generate free electrons (electricity).

The metal grid is just wide and long enough to capture a maximum number of electrons, while not covering (shading) too much of the top active surface. Even so, some photons are still blocked by the grid, while others can't reach it before they recombine with available holes and go back to a state of rest, where they are useless until energized by a photon impact again.

Practical Design Parameters

A lot of experience is needed to design most efficient and reliable solar cells. One of the key elements in this process is the design of the metal contacts which we will cover now.

Cell Thickness

The overall thickness of the solar cell must be kept in mind when designing the cell—the metal contacts in particular—because it is critical for proper cell operation and its mechanical strength.

The indirect energy band-gap results in a low optical absorption coefficient, and this means that the silicon solar cells need to be several hundred microns thick if they are to absorb most of the incident sunlight (and its

photons) in order to prevent greater losses.

Also, for the sake of this discussion, the metal contacts penetrate the wafers' surface to a certain depth, and are also quite heavy, so if the cells are too thin, the metal contacts might exert too much pressure on the surface, causing them to warp and even break.

But the cells cannot be too thick either, because that will increase the distance that charged particles need to travel, and contribute to decreased efficiency.

The amount of light absorbed depends on the optical path length and the absorption coefficient. Thicker cells also cost more, and are heavier, so the PV modules get more expensive and heavier, both of which are highly undesirable factors.

Top Metal Contacts

Metallic top contacts are necessary to collect the current generated by a solar cell. "Busbars" are connected directly to the external leads, while "fingers" are finer areas of metallization which collect current for delivery to the busbars. The key design trade-off in top contact design is the balance between the increased resistive losses associated with a widely spaced grid and the increased reflection caused by a high fraction of metal coverage of the top surface.

Contact resistance losses occur at the interface between the silicon solar cell and the metal contact. To keep top contact losses low, the top n+ doped layer must be as heavily doped as possible. However, a high doping level creates other problems. If a high level of phosphorus is diffused into silicon, the excess phosphorus lies at the surface of the cell, creating a "dead layer," where light-generated carriers have little chance of being collected.

Many commercial cells have a poor "blue" response due to this "dead layer." Therefore, the region under the contacts should be heavily doped, while the doping of the emitter is controlled by the trade-offs between achieving a low saturation current in the emitter and maintaining a high emitter diffusion length.

Combining the effects for resistive losses allows us to determine the total power loss in the top contact grid. For a typical screen printed cell type, the metal resistivity will be fixed, and the finger width is controlled by the screen size. Typical values for the specific resistivity of silver are 3×10^{-8} Ω m. For non-rectangular fingers the width is set to the actual width and an equivalent height is used to get the correct cross sectional area.

To start the design process, we will assume that the depth of the metal contacts into the wafer surface is optimized, so that it is not too deep (to cause electrical shorts, or warp the wafers) and not too shallow (to cause adhesion and high resistivity problems). These are process design and manufacturing issues that must be resolved before any tests and optimizations can be considered.

The design of the top contacts is one of the most important solar cell design parameters. It seems like a simple thing to place contacts on the wafer surface, but doing it right is a science. Minimization of the fingers' and busbars' resistance and the overall reduction of losses associated with the top contacts is a major goal here. Resistive losses in the emitter, resistive losses in the metal top contact, and shading losses are possible.

Critical top contact features which determine the magnitude of these losses are the metal height-to-width ratio, finger and busbar spacing, the minimum metal line width, and the resistivity of the metal.

Aspect Ratio

Aspect ratio is the relation between the height and width of the contacts. The proper selection of this parameter determines the performance characteristics of the solar cell. The aspect ratio is related to the other surface contacts design parameters discussed below:

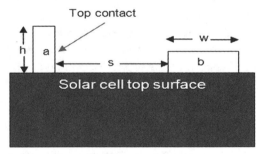

Figure 2-20. Surface contacts parameters

b - Low aspect ratio top contact
s - Contacts spacing
h - Height of contact
w - Width of contact

$$\text{Aspect ratio} = \frac{h}{w}$$

Finger Spacing on Top Surface

A key factor in top contact design is that of controlling the resistive losses in the emitter. We know that the power loss from the emitter depends on the cube of the line spacing; therefore, a short distance between the fingers is desirable for a low emitter resistance. Too short a distance, however, will result in too much shading (coverage) of the front surface, preventing light from reaching the active layer of the cell.

Theoretical calculations, actual tests, and trials are needed to determine the proper spacing of contacts and bus bars for each type and size of solar cell.

Metal Grid Resistance

The grid resistance is determined by the:
— Resistivity of the metal the contact is made of,
— The pattern of the metallization, and
— The aspect ratio of the metallization scheme.

Basically, low resistivity and a high metal height-to-width aspect ratio are desirable in solar cells, but in practice are limited by the capability of the fabrication technology used to make solar cell and the cost of that technology. A tapered contact has lower resistive losses than one of constant width.

Shading Losses

Shading losses are caused by the presence of metal fingers and busbars on the top surface of the solar cell, which prevents light from entering the active layer. Shading losses are also determined by the transparency of the top surface which, for a planar top surface, is defined as the fraction of the top surface covered by metal. The optimum width of the busbar is achieved when its resistive and shadowing losses are equal.

Transparency

The transparency of the front surface is determined by the width of the metal lines on the surface and their spacing. An important practical limitation is the minimum line-width associated with a particular metallization technology. For identical transparencies, a narrow line-width technology can have closer finger spacing, thus reducing emitter resistance losses.

Rear Metal Contacts

The rear contact is much less important than the front contact since it is much further away from the junction and does not need to be transparent. The design of the rear contact, however, is becoming increasingly important as overall efficiency increases and the cells become thinner. The types of metals, their adhesion and thickness are important and are under intense investigation.

AR Coating

AR coating is applied before or after the top metal contacts are deposited and fired. There is a difference in when and how this is done, but we will ignore this factor for now.

There are many antireflective (AR) coatings available today. The type, color, and thickness of the AR coating need to be designed and executed properly to avoid problems later on. The AR coating type will determine how efficient the cell operation is and how it lasts.

The AR color is the visible part of the cell/module surface. It is also one of the major reasons for the success of polysilicon modules, because people like their attractive dark-blue color which is determined by the AR coating type and thickness.

The deposition process parameters determine how well the AR film will perform, and how well adhered to the substrate it will be.

Poor adhesion (due to shallow penetration) will cause it to peel off, thus causing visual and power degradation and premature failure. Deep penetration into the cell surface, on the other hand, will provide good adhesion but might puncture the diffusion layer and shunt the p-n junction.

Depending on the deposition and firing processes of the AR coating, the top metal contacts and the cell performance might be affected. This is a complex issue, so it suffices to say that the resulting effect could be significant, even critical, if manufacturing specs are not followed.

General Material and Process Specs and Characteristics

The type and quality of the materials that make up solar cells and modules are of utmost importance as far as the efficiency, reliability and profitability of the final product are concerned.

Solar cells and modules can be analyzed on the basis of the characteristics of materials and processes, as follow:

Substrate

Silicon solar cells are usually 250 to 450μm thick, as needed for safe handling and manufacturing.

The doping level of the substrate material (solar wafers and cells proper) is kept at an average 1.0 ohm/square.

Doping

N-type silicon is usually the diffusion layer and has higher surface quality than p-type silicon so it is usually on the top of the cell where most of the light is absorbed. The top layer needs to be thin since a large fraction of the carriers generated by the incoming light are created within a diffusion length of the diffusion layer.

100 ohm/square is the average emitter doping level.

Reflection Control

AR coating is used in most cases to reduce the reflection. The top surface of monocrystalline silicon cells is usually textured for the same reason.

Grid Pattern

Low resistivity metal contacts are deposited on the top surface of the cells to conduct away the current, but the contacts shade the cells from incoming light, so a compromise between light collection and resistance of the metal grid has to be reached. Usually fingers 20 to 200 μm width are deposited 1 to 5 mm apart.

Rear Contact

The rear contact is much less important than the front contact since it is much farther away from the junction and does not need to be transparent. The design of the rear contact, however, is becoming increasingly important as overall efficiency increases and the cells become thinner. Full aluminum back with silver fingers (grid) pattern is standard.

Surface Preparation

Both surfaces of as-cut wafers are dirty, rough, and badly damaged by the mechanical action of the wafer saws. The damage is so bad in most cases that if solar cells are made from these wafers, the resulting efficiency and longevity of the cell will be simply unacceptable.

To maximize these parameters, the surface has to be "leveled" and cleaned. This is done by chemically etching the surface. First, the wafers are dipped in cleaning solutions which remove any residue from previous steps, providing a uniformly clean surface for the next steps.

The wafers are then chemically etched to remove some of the damaged top layers. The chemical action dissolves microns from all surfaces, thus removing most of the severe surface damage.

In many cases micron-size pyramids are etched in the surface to increase surface area. This method also provides the cleanest and most damage-free surface possible. The wafers are then rinsed and dried. Now they are ready for processing into solar cells.

These are extremely important processes, because they establish the base upon which the cells' active components will be laid. The surface must be void of blemishes and perfectly clean for the resulting cells to achieve maximum efficiency and longevity.

Temperature Degradation

The influence of temperature is an important factor in solar cells' and modules' performance. As a rule of thumb, the power output from all PV cells and modules decreases with increase of temperature. Depending on materials and structure, the drop could be from 0.3 to 0.6% reduced output per degree Celsius increase of temperature above STC (250°C).

As an example, 85°C (185°F) measured inside PV modules operating in the desert represents approximately 60°C above STC, which comes to about 18-36% power loss.

In practical terms, our 15% efficient module would be only 9.5-12% efficient (depending on material and type of construction) at this temperature.

The combination of the above factors determines to a large extent the level of temperature degradation and its escalation with time. Basically, solar cells made of poor quality materials, using poor processing techniques, are much more likely to drop in efficiency and even fail with time.

Annual Degradation

Finally, the power output from most solar cells and modules decreases with time as well. Studies by independent researchers have revealed that an average of 0.50 to 1.5% of power output is lost every year by cells and PV modules operating under full sunlight.

This degradation is due to a number of phenomena, but part of it is caused by gradual and permanent increase of the internal Rs and Rsh resistances over time.

This is why most PV module manufacturers issue a long-term guaranty of only 80% of the original power to be generated by year 20 of their modules' on-sun operation. What happens after that is anyone's guess, but the manufacturers are basically no longer responsible from that point on—unless special agreements have been reached previously.

Increased annual degradation could cause significant financial losses, because entire strings of PV modules must be shut down for days at a time while new modules are installed. Labor for replacement work will decrease final profits as well.

In summary, silicon solar technologies, in their different variations, are the oldest and most widely used PV technologies. This is especially true for mono- and poly-Si based solar cells and modules, which represent more than half of all PV technologies used today.

Regardless of the problems of late, as well as those we expect in the future, silicon solar cells and modules will no doubt dominate the 21st century's renewable energy markets.

New advances in equipment and processes have brought efficiencies to much higher, levels, while bringing prices to all-time lows.

The combination of these two factors (high efficiency and low price) makes it hard to compete with the silicon PV technologies and, in our opinion, ensures their dominance of the energy markets.

THIN FILM PV TECHNOLOGIES

Thin film PV (TFPV) technologies are a relatively new branch of the solar industry, which has grown much faster in popularity and size than other PV technologies in the last several years. Since the active layers in TFPV cells and modules are deposited in the form of thin films, we refer to them as "thin film" PV products.

Thin films of special photovoltaic materials can produce solar cells or modules with relatively high conversion efficiencies, while at the same time using much less semiconductor material than c-Si cells. In addition, thin film equipment and manufacturing methods allow efficient, cheap, fully automated mass production which is the reason for their success lately.

This, however, comes at the expense of reduced efficiency (average 6-9%), which is not expected to increase much in the future (in mass production mode). On the other end of the efficiency spectrum, multi-layer thin film CPV cells have reached efficiencies over 40% and are getting higher by the day with theoretical efficiency limits in the 80% range.

Due to their versatility, TFPV products have become very popular for use in a number of applications. Recently they have gained a share in large-scale installations as well. Some TFPV modules also show better efficiency under reduced solar radiation than the c-Si competition. This is very useful in many regions with cloudy climates, and could account for their quick rise in European and other world energy markets.

The major types of TFPV technologies considered for commercial and large-scale installations are:

1. Cadmium telluride thin films
2. CIGS thin films
3. Amorphous silicon thin films
4. Silicon ribbon
5. Epitaxial silicon thin films
6. Light absorbing dyes thin films
7. Organic/polymer thin films
8. Ink thin films
9. Nano-crystalline cells
10. Indium phosphide

11. Single-junction III-V cells
12. Multi-junction cells
    a. Gallium arsenide based cells
    b. Germanium based cells
    c. CPV solar cells

Below we see each of these technologies and their specific structure and function, focusing on their use in large-scale PV power generation and related issues.

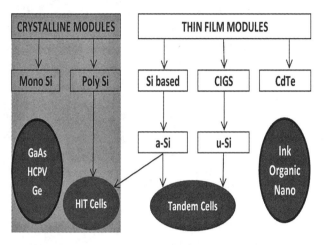

Figure 2-21. Key PV technologies today

Thin Film Manufacturing Process

So let's see how these thin films and the resulting TFPV modules are made today:

Thin Film Deposition Processes

Thin film PV cells and modules manufacturing processes, similar to thin film processes used in the semiconductor industry, are well controlled. The level of process control depends only on the quality requirements and budget restrictions. The actual thin films deposition is usually done on a substrate (glass, plastic or such) which has been thoroughly inspected, cleaned and prepared for the deposition step.

1. Substrate preparation is a key factor in maintaining process control, and determines the overall quality, performance and longevity of the final product. Basically speaking, dirty substrate will not only produce a defective product, but will also contaminate the equipment, forcing lengthy and expensive clean-up procedures.

   Pre-deposition cleaning of large glass substrates (panes) is done in automated washers, where brushes, soap and/or high pressure water solutions remove all particles and organic material from both surfaces. The glass panes are then rinsed

with DI water and dried with forced air and heat applied to both surfaces. This step is also critical, because moisture retained on the large surfaces is a great enemy of vacuum/plasma processes and could seriously affect product quality.

2. The substrate is then placed in a large vacuum chamber (usually on a horizontal or vertical conveyor belt that moves the material along the process path), where powerful vacuum pumps suck out the air, to remove any mechanical (dust) and chemical contaminants (reactive gasses and water). The substrates are then usually heated, to remove any residual moisture and to heat the surface close to the temperature of the deposited species, reducing potential thermal disequilibrium and stress of the thin films to be deposited on it. Argon gas-based DC or RF plasma is ignited and maintained at a proper pressure and power density during the process, facilitating the deposition process and the related reactions.

3. The material to be deposited as thin film is then evaporated (by melting it and directing the resulting vapor clouds onto the deposit surface), or it is sputtered (by dislodging small clusters of it via high voltage-generated ion bombardment) onto the substrate. These processes are also called chemical vapor deposition (CVD) and physical vapor deposition (PVD), respectively.

   In both cases thin film material particles impinge on and adhere to the heated substrate surface on impact. The strength of the adhesion between film and substrate, or between individual films, depends on the design and execution quality of the entire process sequence—quality of all materials, cleanliness of substrate and chamber interior, vacuum integrity, process temperature, forward and reverse plasma and substrate bias power levels, time duration, partial and total gas pressures, speed, and other process variables.

4. Coated substrate with the films deposited on top of it is taken out of the chamber and exposed to a number of additional operations, such as wet chemistry treatment, rinsing, drying, annealing, and wire attachment. The above CVD, PVD and wet chem processes are repeated several times for some devices, following strict process and quality control procedures all through the sequence.

Upon completion of the PV structure creation, the substrate with the deposited thin films is joined to similar substrate (glass usually) with the help of adhesives and encapsulants. Thus, TFPV modules are tested, sorted and packaged for shipment.

Thin Film PV Analysis

Thin film PV (TFPV) technologies are a newer branch of the PV industry, where new sophisticated vacuum/plasma deposition equipment and processes similar to those used in the semiconductor industry are used to deposit very thin (microns in most cases) light-sensitive films onto substrates such as glass, plastics or metals.

TFPV in general are a good alternative for generating power from the sun simply and cheaply. Nevertheless, as with any new, unproven technology, they are faced with challenges.

Background

The physical and chemical properties of TFPV technologies are well understood because they are similar to those of semiconductor devices, which have been around for decades. Thin films have been used in their present configuration in the semiconductor industry since the 1970s, and while the processes are basically the same, the equipment has gone through many changes through the years. The TFPV industry benefits tremendously from using such well understood processes and sophisticated equipment, which lends itself to automation and mass production.

There are, however, several differences in the way semiconductor and TFPV films and products are made and used.

1. The semiconductor industry uses small process chambers, where usually only one or a few wafers are placed and processed as a batch under the strictest of controls. Deviations or exceptions are simply not tolerated, so the quality of raw materials, consumables and procedures is second to none. The product is thoroughly checked for electro-mechanical defects—100% product inspections, several times during the process sequence. Sometimes it is inspected after each step, and even several times after some steps.

   Most high-volume thin film PV deposition processes, on the other hand, are executed in large in-line, conveyor type tools, consisting of a series of deposition chambers. The conveyors move slowly along the line, transferring the glass substrates from chamber to chamber where the different materials of the thin film (layers) are deposited under varying conditions. The emphasis of the process design and execution is on fast deposition times, resulting in high volume production and cheap final product with acceptable, but not superb quality.

2. Finished TFPV products are indeed very cheap—with a market value of ~\$.65/Wp, or ~\$50/m^2. Compared to a market value over \$250,000/m^2 for some finished semiconductor devices (made with similar materials and equipment), this is a huge difference. This also means that TFPV modules must be produced very fast and in very high volume, using the cheapest starting materials to ensure low market value and provide decent profit for module manufacturers.

   Of course there are limits to how fast and how cheaply one can push the production line before starting to produce junk, but TFPV manufacturers have been quite successful in keeping the delicate balance between speed, safety, cost and quality thus far.

3. Both silicon and TFPV modules contain a small quantity of toxic materials in their structure, such as Pb, Sn, Cd, Ga, As, In etc., but this is where the similarities end. The amount of any toxin in silicon modules is extremely low, while the active layers in some TFPV modules are made entirely of toxic materials—such as that of CdTe and some CIGS PV modules. And the hazards are growing, because of the quick entry of these potentially toxic modules into the large-scale energy market. Millions of them, all containing toxic materials (8-9 grams of cadmium equivalent in each module) cover thousands of acres of virgin desert lands.

   This unprecedented massive deployment of a fairly new product that is unproven for long-term exposure to a hostile desert climate elicits a number of questions, the answers to which are unavailable presently, since they are avoided by manufacturers and supporters of the technology.

   We hope these and other unresolved issues will be addressed properly by the US scientific community, manufacturers, regulators and product users in the near future. We should not risk filling the US deserts with poison.

Thin Films Structure

As the name implies, "thin films" are just that—very thin films (layers) of organic or inorganic materials. Thin film PV (TFPV) modules consist of several very thin layers (thin films) of different materials piled on top of each other to form a structure that is suitable for trapping and converting sunlight into electricity.

The thickness of each film is usually several microns (1 micron is 0.001mm, or 0.0004 inch). Visualize the thickness of a human hair (avg. 100 microns) and you'll get a good idea of what 100 microns is—50-100 times the thickness of an individual thin film.

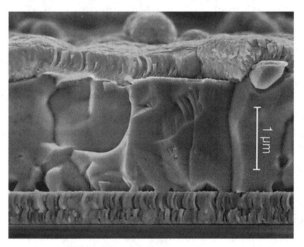

Figure 2-22. TFPV structure (~2 microns thick)

Now visualize layering these super-thin films until there are 8 or 10 of them. This is how TFPV structures are made and it is what they look like. A better visualization might be using a piece of Scotch tape as an example. Standard Scotch tape is ~0.060mm thick, or 60 microns, which is at least 10-20 times thicker than most thin films.

Yes, the entire thin film structure—all different layers combined in your TFPV modules—is many times thinner than a strip of Scotch tape. The various layers (different chemical compounds) are even thinner. They are stacked on top of each other, held together by weak electro-mechanical forces of complex nature and behavior. We'll take a closer look at these forces and the related interactions in this text.

Thin films of any type and size are affected by chemical, mechanical, thermodynamic and electric forces and changes in, between, and around them. TFPV structures also depend heavily on the surrounding materials, glass, laminates, etc., and components in the PV module for protection. It's a complex picture, but well understood by design and process engineers and research scientists, thanks to the broad experience we've gained from the semiconductor industry which is based

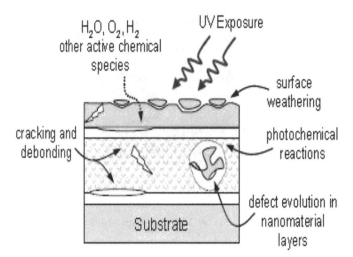

Figure 2-23. Destruction mechanisms

on thin films and processes.

Note: Remember these pictures, because we will revisit them in the next chapters, to explain the behavior of thin films in TFPV modules under different environmental conditions.

THE MAJOR TFPV TECHNOLOGIES

We classify these PV technologies as "major" for the purposes of this text, because they are presently considered for everyday use and/or large-scale PV power generation projects. The major thin film PV technologies we review here are cadmium telluride, CIGS, and amorphous silicon thin films.

Cadmium Telluride (CdTe)

Cadmium telluride (CdTe) is a type of solar cell and module based on thin films of the heavy metal cadmium and its compounds, cadmium telluride (CdTe) and cadmium sulfide (CdS).

CdTe is an efficient light-absorbing material, quite adaptable for the manufacture of thin-film solar cells and modules. Compared to other thin-film materials, CdTe is easier to deposit in mass production environments and more suitable for large-scale production.

CdTe bandgap is 1.48 eV, which makes it almost perfect for PV conversion purposes. At 16.5% demonstrated efficiency in the lab, it is a candidate for a major role in the energy future. Mass production modules are sold with 8-9% efficiency. No significant increase is expected with the present production materials and methods, although manufacturing costs are down—at or below $1.0/Wp.

| Glass Top Cover |
| Conductive Oxide |
| Resistivity Oxide |
| CdS Window Layer |
| CdTe Absorber Layer |
| Metal Contact |
| Glass Bottom Cover |

Figure 2-24. CdTe/CdS thin-film solar cell

With a direct optical energy bandgap of 1.48 eV and high optical absorption coefficient for photons with energies greater than 1.5 eV, only a few microns of CdTe are needed to absorb most of the incident light. Because only very thin layers are needed, material costs are minimized, and because a short minority diffusion length (a few microns) is adequate, expensive materials, processing time and costs can be avoided.

The structure, as shown, consists of a front contact, usually a transparent conductive oxide (TCO), deposited onto a glass substrate. The TCO layer has a high optical transparency in the visible and near-infrared regions and high n-type conductivity. This is followed by the deposition of a resistivity oxide, the CdS window layer, the CdTe absorber layer, and finally the back contact.

For high-volume devices, the CdS layer is usually deposited using either closed-space sublimation (CSS) or chemical bath deposition, although other methods have been used to investigate the fundamental properties of devices in the research laboratory. In all cases, mass production and automation are possible, which is the greatest advantage of this technology.

The CdTe p-type absorber layer, 3-10 µm thick, can be deposited using a variety of techniques including physical vapor deposition (PVD), CSS, electro-deposition, and spray pyrolysis.

To produce the most efficient devices, an activation process is required in the presence of CdCl2 regardless of the deposition technique. This treatment is known to re-crystallize the CdTe layer, passivating grain boundaries in the process, and promoting inter-diffusion of the CdS and CdTe at the interface.

Forming an ohmic contact to CdTe, however, is difficult because the work function of CdTe is higher than all metals. This can be overcome by creating a thin p+ layer by etching the surface in bromine methanol or HNO3/H3PO4 acid solution and depositing Cu-Au alloy or ZnTe:Cu.

This creates a thin, highly doped region that carriers can tunnel through. However, Cu is a strong diffuser in CdTe and causes performance to degrade with time. Another approach is to use a very low bandgap material, like Sb2Te3, followed by Mo or W. This technique does not require a surface etch and the device performance does not degrade with time.

CdTe PV modules manufacturing is a more sophisticated sophisticated process than that of the conventional c-Si modules process which uses simple 1970s manufacturing equipment and materials. CdTe TFPV modules are manufactured with modern, complex, expensive semiconductor type equipment and processes.

Thus, the precision and accuracy of the resulting process steps, and therefore the quality of the final product, are limited only by the quality of the materials and supplies, and the capabilities of the engineers, technicians and operators on the production lines.

CdTe thin-film solar modules are now being mass produced very cheaply, and it is expected with economies of scale that they will achieve the cost reduction needed to compete directly with other forms of energy production in the near future. Since CdTe thin film PV devices still have far to go to achieve maximum efficiencies, it will be interesting to see which materials and methods are most successful.

The most efficient CdTe/CdS solar cells (efficiencies of up to 16.5%) have been produced using a Cd2SnO4 TCO layer which is more transmissive and conductive than the classical SnO2-based TCOs, and include a Zn2SnO4 buffer layer which improves the quality of the device interface.

CdTe PV research, done by manufacturers, universities and R&D labs focuses on some of these challenges.

1. Boosting efficiencies by exploring innovative transparent conducting oxides that allow more light into the cell to be absorbed and at the same time more efficiently collect the electrical current generated by the cell.

2. Studying mechanisms such as grain boundaries that can limit voltage.

3. Understanding the degradation some CdTe devices exhibit at the contacts and redesigning the devices to minimize this phenomenon.

4. Designing module packages that minimize any outdoor exposure to moisture.

5. Engaging aggressively in both indoor and outdoor cell and module stress testing.

Great effort is dedicated today toward addressing the main problems with CdTe PV modules.

1. The relatively low efficiency which contributes to using more land and mounting hardware,

2. Temperature power degradation, which leads to power deterioration and even failures,

3. Annual power degradation, which depends on the above factors, and

4. Other negative long-term effects, such as end-of-life recycling, etc.

Availability of the rare metals used in CdTe TFPV technology, and their increasing cost and potential long-term toxicity are other serious issues which manufacturers and regulators have put on the back burner, as is the fact that the glass-glass design of the frameless CdTe PV modules is simply inadequate for desert applications.

Note: At a meeting of PV specialists in February 2011 (PV Module Reliability Workshop—PVMRW), the degradation and longevity of PV technologies and products were discussed by representatives of several manufacturing companies.

The susceptibility of thin film modules to moisture was addressed as one of the major concerns, and packaging solutions were presented. As steps in the right direction, location-specific reliability tests and evaluations were also topics.

We are glad that such open discussions are underway, since this is the fastest way to resolve the issues and put the promising technologies on the energy market. However, we don't see many changes in the actual PV module designs on the market.

Many modules have serious weaknesses in the edge-sealer area, and millions (thin film mostly) are installed in the desert without a side frame to protect the active films from the elements. The longevity of frameless modules is questionable, simply because the desert will have no mercy on the unprotected plastic edge sealer.

After the edge seal is gone, the fragile thin films in the modules (some of which are toxic) will be destroyed too, causing the modules and the entire power fields to lose efficiency and fail with time.

What will happen to the disintegrated plastics and toxic thin films in the modules is anyone guess, but we would like to err on the side of safety and recommend a review of the frameless design before it is too late.

Let's hope we won't have to wait for a serious accident, before bringing these issues out in the open and discussing possible solutions.

CIGS Technology

Early solar cells of this type were based on the use of CuInSe2 (CIS). However, it was rapidly realized that incorporating Ga to produce Cu(In,Ga)Se2 (CIGS) structure results in widening the energy bandgap to 1.3 eV and an improvement in material quality, producing solar cells with enhanced efficiencies. CIGS have a direct energy bandgap and high optical absorption coefficient for photons with energies greater than the bandgap, such that only a few microns of material are needed to absorb most of the incident light, with consequent reductions in material and production costs.

The best performing CIGS solar cells are deposited on soda lime glass in the sequence—back contact, absorber layer, window layer, buffer layer, TCO, and then the top contact grid. The back contact is a thin film of Mo deposited by magnetron sputtering, typically 500-1000 nm thick.

The CIGS absorber layer is formed mainly by the co-evaporation of the elements either uniformly deposited, or using the so-called three-stage process, or the deposition of the metallic precursor layers followed by selenization and/or sulfidization. Co-evaporation yields devices with the highest performance while the latter deposition process is preferred for large-scale production.

Both techniques require a processing temperature >500°C to enhance grain growth and recrystallization.

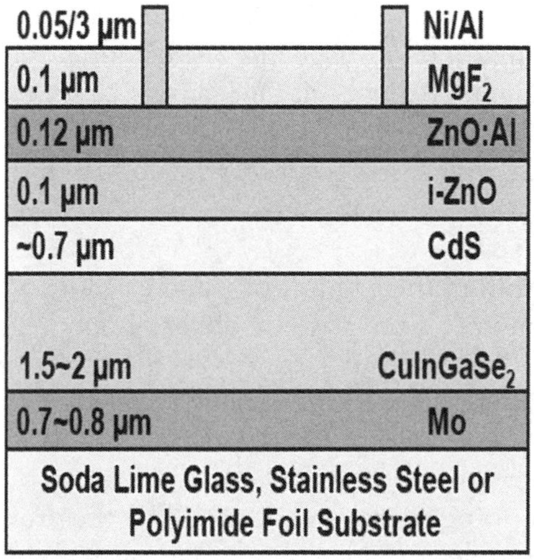

0.05/3 µm	Ni/Al
0.1 µm	MgF$_2$
0.12 µm	ZnO:Al
0.1 µm	i-ZnO
~0.7 µm	CdS
1.5~2 µm	CuInGaSe$_2$
0.7~0.8 µm	Mo
Soda Lime Glass, Stainless Steel or Polyimide Foil Substrate	

Figure 2-25. CIGS cell cross section

Another requirement is the presence of Na, either directly from the glass substrate or introduced chemically by evaporation of a Na compound. The primary effects of Na introduction are grain growth, passivation of grain boundaries, and a decrease in absorber layer resistivity.

The junction is usually formed by the chemical bath deposition of a thin (50-80 nm) window layer. CdS has been found to be the best material, but alternatives such as ZnS, ZnSe, In2S3, (Zn,In)Se, Zn(O,S), and Mg-ZnO can also be used.

The buffer layer can be deposited by chemical bath deposition, sputtering, chemical vapor deposition, or evaporation, but the highest efficiencies have been achieved using a wet process as a result of the presence of Cd^{2+} ions. A 50 nm intrinsic ZnO buffer layer is then deposited and prevents any shunts. The TCO layer is usually ZnO:Al 0.5-1.5 μm. The cell is completed by depositing a metal grid contact Ni/Al for current collection, then encapsulated.

CIGS solar cells have been produced under lab conditions with efficiencies of 19.5%, and lately modules with efficiencies of 15.7% were verified as well. Commercial, mass produced, CIGS PV module efficiency, however, is still lower than CdTe PV modules—and this will have a major impact on their future unless ways to increase their efficiency and reduce their costs are found.

CIGS TFPV modules have similar problems as those plaguing CdTe TFPV technologies. They have low efficiency, require larger mounting infrastructure, exhibit power loss under excess heat and have a significant annual degradation rate. Scarcity of materials and related toxicity issues are, as in the CdTe PV case, on the back burner for now. These issues must be evaluated from the point of view of large-scale installations, where millions of these modules will be installed. In such cases, minute amounts of toxic materials in each module are multiplied many times and become a substantial threat to the environment. Also, special measures must be taken for proper disposal or recycling of these modules.

CIGS research is focused on several of today's challenges in this promising technology:

1. Pushing efficiencies even higher by exploring the chemistry and physics of the junction formation and by examining concepts to allow more of the high-energy part of the solar spectrum to reach the absorber layer.

2. Dropping costs and facilitating the transition to a commercial stage by increasing the yield of CIS modules—which means increasing the percentage of modules and cells that make it intact through the manufacturing process.

3. Decreasing manufacturing complexity and cost, and improving module packaging.

Present SIGS Manufacturing Process Issues

CIGS solar modules are often manufactured by depositing approximately 0.75 μm thick back contact of molybdenum (Mo) layer on float glass. Different metallic foils are used by some manufacturers. This is then followed by the deposition 1.5–2.5 μm thick p-type CIGS absorber layer by means of co-evaporation, selenization/sulfurization of metallic precursors, selenization of metallic precursors, which is done usually by reactive magnetron sputtering. Other non-vacuum technologies such as spraying and printing are used by some manufacturers as well.

The n-type CdS heterojunction layer is grown usually by a special chemical bath process to approximately 75-100 nm thickness, as needed to minimize the absorption of photons with energy greater than the CdS bandgap of 2.4 eV.

Another 500-1000 nm thick layer of n-type TCO is deposited on the CdS heterojunction layer, to minimize the spreading resistance for the passage of the current to the contact.

A 50-100 nm intrinsic ZnO layer is then inserted between the CdS and TCO layer to avoid direct contact between these two in regions not adequately converted by CdS, due to the roughness of the substrate and/or the thin films. This layer can be omitted if the substrate and the thin films are perfectly even.

Some manufacturers and research groups tend to deposit TCO layers of highly n-type aluminum-doped zinc oxide (ZnO:Al) by RF magnetron sputtering, while others deposit a ZnO:B for a TCO layer, via chemical vapor deposition (CVD) process.

Commercial CIGS PV modules on glass or other insulating substrates, narrow and long cell strips are prepared and are monolithically interconnected during the fabrication process. A sequence of laser and mechanical scribing is used to prepare CIGS thin-film solar cell strips and to interconnect them monolithically.

Strips of Mo back-contact layer are scribed by laser immediately after its deposition. CIGS layer is then deposited on the strips of Mo and is also deposited on the portion of glass substrates that is opened by laser scribing. This, however, does not cause significant shunting because the sheet resistance of CIGS is orders of magnitude higher than that of the Mo layer.

The first mechanical scribing is carried out after the deposition of CIGS/CdS/i-ZnO layers to prepare strips of the active layers, followed by the top TCO layer deposition. The top TCO layer fills the vias created by the first mechanical scribes and consequently connects the top TCO contact of one cell strip to the bottom Mo contact of the next cell strip.

And finally the second mechanical scribing of the ZnO:Al, TCO layer is carried out to separate semiconducting layers of cell strips. The separation of cell strips at the bottom Mo contact occurs through the laser scribing of the Mo layer.

Co-evaporation of Cu, In, Ga and Se is used by some manufacturers. The main difficulty of this technique is coating large areas with consistent process stability and homogeneity. Monolithically interconnected CIGS cells on 60 cm×120 cm size sodalime glass substrates have been reported with efficiency of 13%.

Others use coevaporation of Cu, In, Ga and Se for roll-to-roll coating of CIGS absorber, with reported best efficiency of 10.4%. Selenization/sulfurization of DC magnetron-sputtered metallic precursors in a conventional furnace for preparation of CIGS layer is also used by several entities, with reported efficiency of 12.8%.

CuInS2 thin-film PV modules on glass by sulfurization of Cu–In metallic precursor by RTP is another technique that is fast, and has advantages of low thermal budget and thus reduces the energy payback time. Efficiencies of 13.73% have been obtained by this method for CuIn1−xGaxSe2−ySy on glass, and 12% for CuIn1−xGaxS2 (CIGS2) cells on glass.

Ten percent efficient CIGS cells metallic foils have been produced for space applications, and efficiency of 12.78% has been achieved for CuIn1−xGaxSe2−ySy on glass prepared by selenization/sulfurization of metallic precursors in an RTP set-up.

Non-vacuum techniques are also used to deposit a precursor layer and then selenize it to form CIGS. Non-vacuum processes such as spraying or printing use fine particles or nanoparticles of size <100 to several hundred nanometers of compounds such as CuO, InO and (Cu, In)Se in a suitable carrier fluid to deposit the precursor layer.

Non-vacuum processes are predominantly additive, not subtractive. Spraying is fast and simple when used to deposit very thin layers. It can be carried out in a wide variety of ways and has a low material usage, but requires a flat substrate and there are overspray losses. In addition, building thick (0.75–2 μm) layer by deposition of multiple thin layers is quite time consuming and expensive, while printing is fast and cheap. It also has much higher thickness capabilities. For example, printing with paints can provide thickness of up to 100 μm while the CIS thickness in solar cells is maximized at 2 μm.

Printing is used for creating patterns and for layer deposition, by using ink of thoroughly mixed, tiny nanoparticles, in the proper proportions that can be printed on the substrates, thus providing uniform thickness and elemental ratios, even across large-area substrates.

Major innovations in printable semiconductor transparent top electrodes have been achieved, producing low-cost roll-to-roll processing, RTP, fast assembly and fabrication of nanostructured components.

A hybrid process for growing CIGS layer using solid sources has been developed as well, producing 13.5% efficient small-area solar cells and 7.5% efficiency for large-area modules.

The addition of trace amounts of sodium to the CIGS thin films is known to improve the morphology, concentration of p-type doping, and PV characteristics, so some manufacturers use a Mo layer on sodalime glass, counting on the sodium in the glass to diffuse in order to improve the morphology of CIGS layers.

Some deposit a barrier layer on the sodalime glass or on the metal foil substrate. The sodium is then added by deposition of a sodium compound for improving the CIGS properties. If silicon oxide and silicon nitride buffer layers are used to avoid diffusion of sodium from the substrate, sodium can be added by deposition of a sodium-containing compound.

In summary, there are endless combinations and permutations of CIGS and CIGS-like materials, processes and finished devices. There is also a significant effort to develop these technologies as quickly as possible to reach the level of efficiency and reliability required by the energy markets.

Once this is done, CIGS will be the predominant and preferred type of thin film PV power generators. We are confident that this will happen in this century.

Amorphous Silicon

Amorphous silicon (a-Si) is produced via thin film processes, based on depositing thin layers of silicon films on different substrates. Silicon thin-film cells are mainly deposited by chemical vapor deposition (CVD), typically plasma-enhanced (PE-CVD), using silane and hydrogen reactive and carrier gasses for the actual deposition.

Depending on the deposition parameters and the stoichiometry of the process, this reaction can yield different types of thin film structures, such as amorphous silicon (a-Si, or a-Si:H), protocrystalline silicon or nano-crystalline silicon (nc-Si or nc-Si:H), also called micro-crystalline silicon.

These types of silicon feature dangling and twisted bonds, which result in deep defects (energy levels in the bandgap) as well as deformation of the valence and conduction bands (band tails), which lead to reduced efficiency.

Proto-crystalline silicon mixed with nano-crystalline silicon is optimal for high, open-circuit voltage. Solar cells and modules made from these materials tend to have lower energy conversion efficiency than those made from bulk silicon, but have some operating advantages (such as lower temperature degradation). They are also less expensive to produce, although the capital equipment expense is greater, due to equipment complexity.

a-Si has a somewhat higher bandgap (1.7 eV) than crystalline silicon (c-Si) (1.1 eV), which means that it absorbs the visible part of the solar spectrum more efficiently than the infrared portion. nc-Si has about the same bandgap as c-Si, so nc-Si and a-Si can advantageously be combined in thin layers, creating a layered cell called a "tandem cell," where the top a-Si cell absorbs the visible light and leaves the infrared part of the spectrum for the bottom cell in nc-Si.

The material used in these solar cells is actually hydrogenated amorphous Si (α-Si:H), an alloy of Si and hydrogen (5–20 at. % H), in which the hydrogen plays the important role of passivating the dangling bonds that result from the random arrangement of the Si atoms.

The hydrogenated amorphous Si is found to have a direct optical energy bandgap of 1.7 eV and an optical absorption coefficient, α, greater than 105 cm-1 for photons with energies greater than the energy bandgap. This means that only a few microns of material are needed to absorb most of the incident light, reducing materials usage and hence cost.

Most a-Si devices produced today have the p-i-n structure shown here.

a-Si solar cells can be manufactured as single, double and multi-junction devices. This increases the useful range of the technology, since multi-junction solar cells have higher efficiency and could be used in many areas of the energy markets.

a-Si solar cells absorb the solar spectrum more efficiently and exhibit improved cell stability by using multiple p-i-n structures with different energy bandgap i-layers. This produces 'double junction' or 'triple junction' structures as shown above. Narrower energy bandgap layers are produced by alloying the Si with Ge, and

wider energy bandgap layers are produced by alloying the Si with carbon. The highest reported stabilized efficiency of a double-junction is greater than 9.5%, and for a triple-junction module it is greater than 10%.

If the gases used for deposition of amorphous Si are diluted in hydrogen, the deposit consists of regions of crystalline Si immersed in an amorphous matrix. This two-phase material is known as 'microcrystalline Si' or sometimes as 'nanocrystalline Si'.

The physical properties of the material resemble those of crystalline/multicrystalline Si rather than amorphous Si, especially with regard to stability under intense illumination.

Work is currently underway to develop hybrid amorphous Si/microcrystalline Si tandem solar cells and modules, which are referred to as 'micro-morph devices'. Trials indicate that these hybrid devices and modules rival triple-junction amorphous Si in terms of efficiency and stability.

The biggest problem with a-Si TFPV technology, and a barrier to its success, however, is its low efficiency. Today's best cell efficiencies are about 12% in the lab, which is almost 50% lower than other PV technologies. Mass produced a-Si cells and modules are in the 8% efficiency range today.

Another problem of single p-i-n junction a-Si modules is that their efficiency degrades under illumination to less than 5% because of a phenomenon known as the Staebler-Wronski effect. Great effort is underway to better understand and to solve this problem.

Also, a-Si manufacturing cost is quite high, most of which is due to the high initial capital investment, and complex equipment maintenance and operation. This is because very complex thin film deposition equipment is

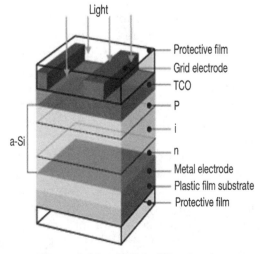

Figure 2-26. a-Si thin film structure

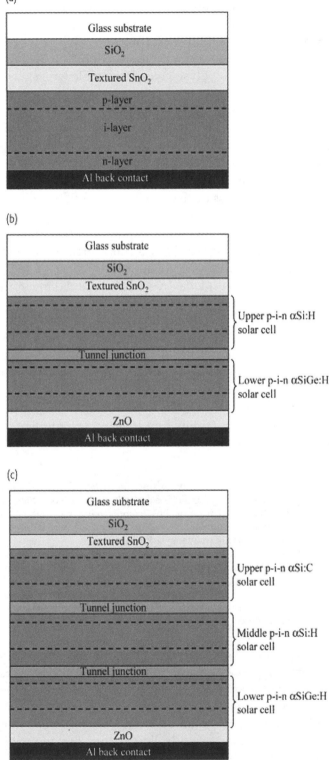

Figure 2-27. A cross-sectional view of
(a) a single junction,
(b) a double junction, and
(c) a triple junction a-Si solar cell.

used, which is expensive and requires much more qualified and expensive labor, as compared with the competing PV technologies.

Two proposed solutions to this problem are higher manufacturing rates and batch (simultaneous) processing of multiple modules. Good progress has been made in rates that are 3-10 times higher than those being used in production, but all this is still on a lab scale and yet to be proven in on a large scale.

On the positive side, while some of the more efficient cells and modules lose 20-30% of their output in the field, due to excess heat exposure, a-Si loses only 5-10%, due to its lower temperature coefficient. Also, the active thin film structure is composed mainly of silicon films which have inert and homogeneous natures that show better chemical and mechanical stability than some of the competing thin films—in case of an encapsulation failure. a-Si modules are also more resistive to the negative effects of shading in the field. Of equal importance, a-Si modules do not contain any hazardous materials, which is paramount where large-scale PV installations are concerned. These qualities put a-Si on the top of the list of PV technologies suitable for large-scale power generation in deserts and other inhospitable areas.

Even with low efficiency (well under 10%), a-Si thin film technology is being successfully developed for building-integrated photovoltaics (BIPV) in the form of semi-transparent solar cells which can be applied as window glazing. These cells function as window tinting while generating electricity. It remains to be seen if the amount of generated electricity covers initial and operating expenses.

A triple-junction a-Si TFPV power system has been operating near Bakersfield, CA, for several years, and is providing proof of the excellent performance of this technology. The 500 kW grid-connects system has been performing well, meeting or exceeding its design goals. Performance data from this larger-scale installation confirm data obtained from smaller a-Si systems and prove that this thin film PV technology can be successfully used in large-scale power plants, if the low efficiency can be justified.

Great research effort is underway at universities and R&D labs around the world, geared towards solving efficiency and cost issues and obtaining a-Si that is truly competitive in the energy market. a-Si manufacturers need to improve their understanding of this technology and focus on:

1. Improving the light-stabilized electronic quality of a-Si and low-gap a-Si:H cells to achieve broader

spectrum conversion, and increased overall efficiency.

2. Increasing the growth rates of a-Si, a-SiGe, etc. layers while maintaining high electronic quality, to obtain increased throughput and reduced capital cost.

3. Developing high-growth-rate methods for nanocrystalline silicon while maintaining high electronic quality for increased efficiency, stability and reduced cost.

4. Understanding and controlling light-induced degradation in a-Si:H as needed for increased efficiency and understanding of the intrinsic limits of the efficiency.

5. Developing *in-situ* in-line process monitoring for increased yield.

6. Improving light-management to obtain maximum efficiency.

7. Improving stability and conversion efficiency of a-Si modules in actual use by addressing the Staebler-Wronski negative effects, where the conversion efficiency of the a-Si module decreases when it is first exposed to sunlight.

8. Reducing capital equipment costs for manufacturing a-Si panels by improved manufacturing processes that include increasing the deposition rates.

9. Improving module-packaging designs to make them more resilient to outdoor environments and less susceptible to glass breakage or moisture ingress.

10. Developing new module designs for building-integrated applications.

The future of a-Si PV products depends the timely resolution of these issues.

Note: The untimely exit of Applied Materials from the a-Si equipment manufacturing business in the summer of 2010 cast a shadow of doubt over a-Si technology. This was a major action that is still on the agenda and which many manufacturers and proponents of a-Si technology are still trying to explain and overcome.

While Applied Materials has left the solar field, a-Si TFPV is still alive—and it is here to stay. It is a promising technology that has already found niche markets, and will become even more popular with time, as it finds its rightful place in the large-scale energy markets.

c-Si vs. Thin Films (Now and Later)

Raw materials for c-Si PV represent 75% of the total cost structure, with polysilicon as the single largest contributor to costs.

Thin film modules, such as CdTe modules (containing cadmium and tellurium heavy metals), and CIGS PV modules (containing In, As, Ga and other metals) account for only 10% of the total cost of the finished PV modules. On the surface this means that thin film PV technologies are less dependent on a potentially volatile commodity metals market.

The problem, however, is that most of these metals (cadmium, tellurium, indium, arsenic, etc.) are exotic, rare, and sometimes very toxic metals. Most of these are also produced under extremely dirty and dangerous working conditions in third-world countries.

A close look reveals the data shown in Table 2-4.

So, put in other words; silicon is a very large part of the silicon PV manufacturing, but it is will understood commodity, and its future quite clearly points towards increased efficiency, reliability and reduced price.

At the same time, thin films PV modules' future (due to metals' availability and toxicity) (in addition to unproven long term reliability), is veiled in uncertainty*.

---

*Author's note: The uncertainty surrounding rare, exotic and toxic metals is due to changing regulations and other conditions, which affect the availability and price of these commodities. We expect that as they become more popular, and more attention is given them, the problems will increase in frequency and intensity.

Also, the materials' availability and pricing discrepancies can be a double-edged sword; the price pressure can provide a large incentive for c-Si manufacturers to reduce silicon material costs and usage, whereas thin film suppliers are locked into dependency on commodity materials that are uncontrollable by virtue of their nature, location, toxicity, availability and price.

Since the availability of raw materials for thin film's PV technologies is uncertain, thin film manufacturers must find other ways to reduce costs, i.e. increasing the efficiency of their modules. Although this makes sense, it will be hard (and expensive) to do, because most thin film technologies are already close to their maximum efficiencies for mass production.

Thus, reducing material costs on a per-watt basis and reducing the balance-of-systems penalty suffered by low-efficiency products is a goal which all thin film manufacturers are chasing, but it is like hitting a moving target.

Condition / Task	c-Si PV	Thin films PV.
Materials Availability	Unlimited	Limited
Process equipment	Simple	Complex
Production process	Simple	Complex
Mfg. plant setup / scale up	Cheap	Expensive
Final PV module price	Decreasing	Going up
Future PV module cost	Decreasing	Leveled
Present PV module efficiency	14-16%	10-12%
Future PV module efficiency	18-20%	12-13%
BOS equipment (quantity)	100%	125-150%
Power field land use	4-6 acres	10-12 acres
Module frame	Wrap-around	None
Esthetics	Normal	Better
PV modules reliability	Proven	Unproven
PV modules toxicity	Slight to none	Considerable
Recycling	Not needed	Mandated (toxic)
Restrictions	None	Increasing
Surprises	None expected	Many expected
Future success	Clear, upward	Complex...???

Table 2-4. Comparing the c-Si and thin film technologies of today and tomorrow

During 2011, record efficiencies were announced, such as CdTe record cell at 17.3%, 17.8% CIGS cell aperture-area efficiency, and flexible module aperture-area efficiency of 13.4 percent. These are lab trials, however, and have little to do with mass production *or* with long-term field performance. For now, commercially available technologies suffer from much lower efficiencies and unproven long-term reliability.

Still, the record lab-test efficiencies are a promising indication that high-efficiency thin film products could be commercially available—eventually.

What can be done to bridge the gap between thin film and c-Si PV manufacturing costs? Each manufacturing process and technology is different, so the method and size of the gap would vary.

Several years ago, a 0.5% increase in module efficiency resulted in approximately 5-6% savings in total module manufacturing costs. Cost savings decrease logarithmically as efficiency increases and gets closer to the maximum possible. In other words, the higher the efficiency, the more thin films cost to achieve quality in mass production.

So this is not the cheapest or fastest way to catch up with c-Si technologies, because their efficiencies can be increased faster and cheaper. One great advantage of thin film manufacturers is the fact that the equipment used in their manufacturing is much more sophisticated, flexible, and readily lends itself to automation. This simply means that thin film manufacturers must focus on increasing process yields, throughput/uptime, and the always-important manufacturing scale.

Scale-up of the production process is another tool that, if properly used, could result in cost reduction. Several years back, doubling of total production capacity of a thin film manufacturing facility could result in a 15-20% drop in production costs, while the same production ramp-up could bring only 10% (or less) decrease in c-Si wafer, cell, and module costs production.

Another key aspect of cost reduction is production capacity utilization; production expansion (especially today) is not a guaranteed-success move. Production utilization rates have been dropping across the board—in all countries and all types of technologies. A number of factors must be kept in mind when considering levels of utilization. Growing too quickly is one of them, and often results in significant capacity underutilization—which is exactly what we saw with a number of Chinese companies who were forced to down-scale or even shut down completely as a result of the untimely manufacturing underutilization.

This can greatly affect the fully loaded costs (which also include depreciation) of PV modules and other components.

Finally, the initial cost of capital equipment and yearly depreciation are of utmost importance to the bottom line as well. Since c-Si capital manufacturing equipment is much simpler, readily available and cheaper, it costs less and its annual depreciation amounts to only 5-8% of the module costs.

In the thin film modules manufacturing process, however, equipment depreciation usually accounts for 20-25% of the operational cost (provided that it is operated at full capacity). So expansion of scale (as needed to reduce costs) may increase costs (via capex), while at the same time c-Si manufacturers continue to successfully produce and sell PV modules at very thin and (recently) even negative margins.

Industry analyses forecast over 20 GW of PV module manufacturing capacity is to come offline by 2015. This is partially due to the global financial slump, accompanied by energy markets' supply-demand imbalance.

Things look even worse the next several years, when considering that the greater PV supply chain, cell and module manufacturers will shut down 60 GW of production capacity by 2015-2016.

Table 2-5. PV manufacturing capacity reductions, 2010-2015 (in GW)

Year	Wafers	Cells	c-Si	Thin film
2010	1.6	0.25	0.5	0.0
2011	1.8	1.0	0.75	0.2
2012	3.2	4.2	4.0	1.25
2013*	9.5	8.0	8.0	1.5
2014*	4.0	4.5	3.5	3.0
2015*	0.5	1.0	1.0	1.0

*Estimates

c-Si module manufacturing costs for Chinese Tier-1 suppliers are projected to fall as low as $0.45 per Watt, with an average selling price as low as $0.61 per Watt by 2015-2016. At the same time, the average c-Si module efficiency is expected to hit 16.4 percent by then, compared to 14.5-15% in 2012.

Most current PV manufacturers—including thin film—will have to make critical decisions in the near future. Their choices are to either exit while they can, or continue fighting in the component markets, while looking for marginal advantages and finding ways to leverage them to compensate for the risks which are unpredictable.

The future is looking bright for PV technologies, but some significant changes are expected and needed to bring them to LCOE.

Some of the disruptive changes will be brought by:

Direct solidification of silicon material provides the cheapest wafers. Direct solidification of molten Si offers true kerf-less wafering (which eliminates losses from sawing). It is the Holy Grail for the solar industry, which has been the goal of many companies since the 1970s. This technology has a potential market size of up to $600 million, and is expected to be the first to reach full commercialization by 2015.

Alternatives to cell efficiency, such as anti-reflective and light-trapping coatings. These are 2nd-tier technologies, also looking at a market size of over $600 million. This is the best way to provide cost-effective efficiency gains. Commercialization is likely during 2013-2015.

New Active Layers

New processes and materials are expected to dominate the 21st century PV technological gains.

1. Copper-zinc-tin-sulfide (CZTS) cell technology will eventually replace some of the competing thin film technologies (possibly CIGS) and gain a significant market share through use of cheaper materials, with the major advantage of eliminating the use of exotic and expensive materials, such as indium and gallium.

2. Epitaxial Si (epi-Si) technology, which is thin monocrystalline silicon, has the potential to replace amorphous silicon (a-Si) infrastructure and reach higher efficiencies than a-Si modules.

In summary, the game changers for the PV modules markets will be emerging PV technologies (or modifications of the existing ones), and especially those that are easy to scale, provide the highest efficiency gains, and, equally as important, those that ensure reduced final costs per generated Watt.

Thin film PV technologies have the potential to break through and dominate the renewable energy markets in the 21st century, but on the basis of what we have seen lately, and the slow progress made by thin film PV technologies, we don't see this happening any time soon.

More importantly, some of the prevailing thin film technologies contain significant amounts of toxic materials (including carcinogenic, heavy metal cadmium.) Regardless of their efficiency or price, we don't see toxic modules as a long-term solution. Worse, we see them as hazardous to the environment and human health.

Presentday thin film modules are not proven safe for 30 years non-stop operation in the extreme heat of deserts, which is where millions are installed today. Until manufacturers prove beyond the shadow of a doubt the long-term safety of potentially toxic thin film PV modules, we will object to covering thousands of acres with large numbers of them.

CONCENTRATING PV (CPV) TECHNOLOGY

Concentrating PV (CPV) technology has been around for 30 years. It was developed by several US companies in the 1970s and 1980s with the help of DOE and its satellite national labs. HCPV is the most efficient PV technology in the world, but its commercialization has been bogged down by a number of factors unrelated to its efficiency.

CPV equipment operation is somewhat different and more complex than other PV technologies. In contrast with 'flat-plate' PV modules, where a large area of photovoltaic material (usually crystalline silicon) is exposed to the maximum sunlight, CPV systems, as the name suggests, use lenses or mirrors to focus (or concentrate) sunlight onto a small amount of non-silicon photovoltaic material.

General Description

CPV systems operate by concentrating sunlight through Fresnel lenses onto specially designed, and very efficient (over 40%) CPV solar cells. Additional components, such as secondary optics, insulators, heat sinks, and trackers are needed to keep the light focused precisely onto the solar cells, to dissipate the generated heat and conduct the generated electricity to the electric grid, or user's site.

CPV solar cells mounted into CPV assemblies are installed in modules, which are mounted on a steel frame—a frame that pivots on a pedestal (Figure 2-28). GPS controllers send a signal to the x-y drive motors which drive actuators to position the frame and modules exactly perpendicular to the sun all day long with a .01% degree margin of error.

Basic Operation

HCPV systems (trackers) consist of several key components, as follow:

1. CPV solar cells
2. Heat sinks
3. Bypass diodes
4. Secondary optical elements
5. CPV assembly (containing the above components)
5. Fresnel lenses
6. CPV modules (housing the above components)
7. Steel frame
8. x-y drive gear-motors
9. GPS based controller

CPV solar cells are made out of GaAs, Ge or other more efficient and more durable semiconductor materials with thin films deposited on them to create a multi-junction device of very high efficiency—over 42.3% presently.

CPV cells are sophisticated semiconductor devices that have followed Moore's law to an extent, as far as increase of efficiency per active area is concerned. At the

Figure 2-28. HCPV tracker

very least, they are much closer to it than any competing PV technology to date.

CPV cells are arranged into a special assembly, called CPV cell assembly, which ensures efficient collection of sunlight and cooling of cells during operation, as well as the proper conduction of electric power into the wiring harness. Each cell is configured with a blocking diode, which protects it from reverse bias currents and surges, and also isolates it from the circuit in case of total failure. The cell and diode are usually mounted on a heat sink, which extracts heat from the cell during operation and dissipates it in the surrounding environment outside the module.

A Fresnel lens is used to capture a large area of sunlight and focus it onto the small CPV cell. The Fresnel lens is basically a flat (or slightly curved) plastic or glass lens that uses a miniature saw-tooth design on its bottom to redirect and focus the incoming light onto the small area cell, several inches away from the lens. When the saw-tooth teeth on the lens are arranged in concentric circles, light is focused at a central point, and the equipment with this design is called the point-focus CPV system. When the teeth run in straight rows, the lenses act as line-focusing concentrators, and the resulting equipment is called the linear-focus CPV system.

The concentration ratio of each CPV device can vary as well, depending on lens and cell design. For example, if sunlight falling onto 100 cm^2 is focused onto 1 cm^2 of PV material, the ratio is considered as 100 suns, or 100 x. If the light from a 1000-cm² Fresnel lens is focused onto a 1-cm² CPV cell, then the ratio is 1000 suns, or 1000 x. If thus concentrated sunlight light falls onto a well designed CPV cell, it will produce 100, or 1000 times the electricity respectively that a normal c-Si PV cell of the same size would produce under the same operating conditions.

Commercial concentration ratios are between 200 and 500 suns, and as much as 1000 suns, but there are theoretical and practical limits of min.-max. sun concentration levels, with the mid-range 400-600 considered the most efficient and practical for now, in the author's opinion.

Most CPV systems use only direct solar radiation, so these installations operate best under direct sunlight (such as in desert areas) and always involve trackers, forcing the modules to rotate and follow the sun all day, which keeps the lenses and CPV cells looking right at the sun at all times, thus generating maximum possible electric energy while the sun is shining.

The cells are placed in modules, which are mounted on a large frame, which rotates around its two axes to follow sun movement all day. Tracking is achieved with one or two gear-motor assemblies, which get a signal from a photo-detector or GPS controller, which knows where the sun is at all times and runs the motors to move the frame accordingly. Precise tracking is needed, to keep the sun at 90 degrees with respect to the Fresnel lens and CPV cell, so that sunlight is focused onto the solar cell at all times. If the sunlight comes at an angle, or is diffused (as on a cloudy or foggy day), some, if not most sunlight will get de-focused, reflected, or otherwise fall away from the cells, resulting in very low light-to-power conversion efficiency.

Tracking systems add cost in terms of motor and controller maintenance, but this cost is relatively small compared with other O&M cost savings that trackers provide. For example, the tracker's motor requires only annual lubrication and a single motor controls more than 50 kWp of PV. Also, tracker bearings require no lubrication and are designed for more than 25 years of use.

The operating and maintenance (O&M) cost of a utility-scale tracking system ends up being less than US$0.001/kWh more than that of a fixed configuration. And this calculation does not factor in the O&M savings from increased energy production for the same power output, which is a different subject all together.

HCPV Technology

As a clarification; HCPV is a variation of the CPV technology, where higher sunlight concentrations—in

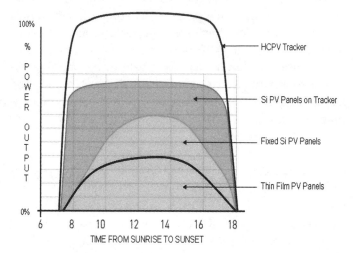

Figure 2-29. Performance of different PV technologies

the 500-1,000 times range—are used.

It is easy to see from the figure that HCPV trackers are twice as efficient as the regular c-Si modules in fixed mount, which is the most widely used technology. The best part is that while c-Si PV technology is hitting the top of its physical limits of 28% efficiency, HCPV cells have a long way to go before reaching the upper limits of their efficiency, which is ~80%.

So, we are looking at 60% efficiency by 2020 and 80% by 2040. Just imagine 80% of the sunlight reaching the CPV cells being converted into useful electricity. With all other factors considered, we are looking at 65-75% efficient HCPV tracking systems in the future. We are fully confident that that future will arrive, and if we are smart enough, even sooner than predicted.

The power balance sheet of an HCPV tracker looks good, but is not perfect.

Grid efficiency = Pt – Ol – Rd – Ci = 29.5%*
Note: Estimate of SolarTech, Inc. USA, 2010

Where:

Pt is total DC power generated by CPV cell	= 41.0%
Ol is Optics loss and misalignment	= 4.0%
Rd is Cell temperature degradation (at 190°F)	= 2.5%
Ci is Inverter and electric circuits loss	= 5.0%

Total DC power converted into AC	= 29.5%

Note: HCPV trackers stop generating power as the air haziness (clouds, fog, etc.) increase, while some PV technologies continue generating proportionally. Because of that, HCPV trackers can only be used in hot desert areas with maximum direct radiation.

So, our 41% efficient CPV cells mounted on a tracker will deliver only part of the generated power into the grid. And according to the above table, it will average ~29.5% efficiency under full power on a bright sunny day. Good, but not spectacular. There are other losses too, related to the electro-mechanical and optical systems' misalignment, and poor maintenance, which must be taken into consideration, in a full and precise power plant design calculation.

Improvements in materials and the various key components (cells, lens materials, optics, heat sinks, GPS controls, drive gears) will account for further increases in overall efficiency in the near future.

Based on R&D work done in the past and combined with that done recently, we are confident that HCPV trackers (point-focus type) will soon be able to obtain 35% grid efficiency with the present state-of-the-art, and up to 45-50% when their efficiency is increased to ~60% by 2015-2020.

The future of this technology looks good, and only hard work separates HCPV technologies companies from broadly deploying their products on the world energy markets.

HCPV Cost Estimates

The cost of the HCPV technology is based on:

1. Location. HCPV operates most efficiently in desert areas with hot and bright sunshine;

2. Price of materials. HCPV trackers consist of 90% steel, aluminum and glass, so the price of these commodities determines the total cost of the installation; and

3. Size of the installation. The larger the installation, the lower the cost, due to bulk orders of materials.

For a large size installation—over 5.0 MWp—installed in the Arizona desert, at 2009 commodity prices, the cost of manufacturing HCPV tracking systems would be ~$1.20/Wp (summer of 2011).

HYBRID TECHNOLOGIES

The combination of photovoltaic and thermal (PV-T) hybrid technologies basically generates both electricity and water at the same time. They can be divided into several types, according to their components and applications.

Presentday hybrid solar technologies involving photovoltaics and heat generation can be divided into:

1. Photovoltaic-thermal hybrid, or PV/T, and

2. Concentrating PV-thermal hybrid, or CPV/T, which can be subdivided into:
    2.1 Concentrating photovoltaic/thermal hybrid, or CPV/T, and
    2.2 High-concentration photovoltaic/thermal hybrid, or HCPV/T.

A Word of Explanation and Clarification

PV/T is a combination of photovoltaic devices, which incorporate both photovoltaics and thermal heating. These can operate at 1x (1 sun) to generate electricity, while running water through the system for use somewhere else.

CPV/T therefore, is a combination (hybrid) of PV concentrators which concentrate the sunlight 10-100 times onto a receiver tube covered with solar cells mounted on a receiver tube, through which water or cooling liquid flow. The trough and receiver track the elevation of the sun all day long (one-axis tracking), and thus the device generates more DC power per active area, and the water/coolant are at much higher temperatures than their PV/T cousin.

HCPV/T is a combination of PV concentrators (mirrors or Fresnel lenses), which operate on higher magnification (100-1000 x) and track the sun in x-y direction (two-axis tracking) all through the day. These devices produce more power per area than their CPV/T cousin, while the water/liquid temperature is controlled as needed to optimize the efficiency of the device.

We'll start with a close look at the mechanics and thermodynamics of the simplest PV-T system—that of a fixed mounted flat panel equipped for generating both PV electricity and hot water capabilities, which makes it a PV-T hybrid.

These systems combine photovoltaic with thermal solar. The advantage is that the thermal solar part carries heat away and cools the photovoltaic cells. Keeping temperature down lowers the resistance and improves cell efficiency. Modified CPV systems have been tested, where the CPV cells are cooled by active flow of liquid, thus generating both heat and electricity.

Below we take a close look at the PV/T device operation and its thermo-dynamics and static behavior, both of which could be used with some translation and extrapolation for use with CPV/T and HCPV/T devices. These calculations are absolutely necessary for the proper design of the devices and the power fields these are supposed to operate in.

The hybrid photovoltaic-thermal (PV/T) system basically consists of an array of PV cells positioned directly on top of the absorber plate of a conventional forced circulation type solar water heater—similar to those we see on roofs of houses.

PV-T Hybrid System

A PV/T solar collector is composed of:

1. Transparent cover allowing sunlight to pass towards the absorber and to create a greenhouse effect. It is composed by one or more glass or plastic panes.
2. PV cells for the production of electricity.
3. Absorber plate for transferring heat to the water or in the tubes built into it.
4. Frame, for protection of the whole of these elements.
5. Insulator, allowing limiting the losses by conduction through the walls, back, and side.

The schematics of the PV/T collector are shown in Figure 2-30. The top cover is represented by a glass sandwich that includes PV cells. The cell area can cover the entire active surface or can be distributed in a grid where the spacing between adjacent columns and rows can allow a direct gain of solar radiation to the absorber plate.

Different configurations of a PV/T collector can be created by changing the cell area density in balance electricity and thermal energy output of the system.

Dynamic Model of a PV/T System

Below, an explicit dynamic model suitable for PV/T system simulation is introduced. (19)

The effectiveness of a PV/T system compared to a photovoltaic panel (PV) is underlined. The model provides the thermal state of various collector components and generates results for hourly and transient performance analysis (thermal/electrical gain).

The dynamic thermal model of the PVT collector is built on the finite-difference control-volume technique. The PV/T collector is composed of four major components which represent the different nodes: a.) Glass cover, b.) solar cell, c.) absorber plate, and d.) water in channels and in storage tank. See below detailed calculations.

The energy and fluid flow equations are developed on the base of the four nodes. All sub-parts in each node are lumped together in proportion to give the average properties of the representing major component, (Figure 2-31).

Where

Tg, Tc, Tp, Tf, Ta are respectively the temperature of glazing, the solar cell, the absorber plate, the water circulation and the ambient temperature.

W   is the wind speed,
G   is the incident solar irradiation,
m   is the mass flow rate of fluid,
P   is the electric power output, and
Q   is the thermal profit.

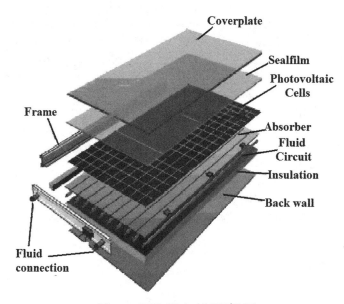

Figure 2-30. Hybrid PV/T (7)

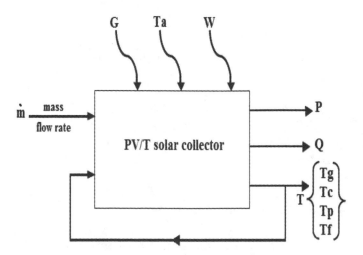

Figure 2-31. Modeling of PV/T system.(7)

Nomenclature

Ac	Area of sensor (m^2).
a	Absorption of the absorber.
αsTc	Temperature coefficient of short-circuit (mA/°C).
Cp	Conductance of the thermal losses (J/kg•K).
Di	Internal diameter of the tube (m).
d	Reflectivity diffuses glazing.
εg	Emittance of glass.
εp	Emittance of absorber.
G	Incident solar irradiation (W/m^2•°C).
G	Flow mass (Kg/s•m^2).
h	Convection coefficient of exchange (W/m•°C).
hf	Thermal coefficient (W/m•°C).
Ipv	Photovoltaic current (A).
Isc,sTc	Current in standard conditions (A).
K	Conductivity of the insulator (W/mk).
Kb	Conductivity of material of the tube (W/mk).
k2	Conductivity of the absorber (W/mk).
L	Length of the tube (m).
m	Mass density (kg/s).
mc	Effective collector capacity (J/K).
λ	Thermal conductivity (W/m•°C).
n1	Number of tubes in the plane sensor
Ng	Number of glass covers.
Pr	Prandtl number.
Qv	Volume through put (m3/s).
σ	Constant of Stefan (W/m^2•K4)
Re	Reynolds number.
S	Solar radiation (W/m^2).
t	Transmissivity of the glazing.
Ta	Ambient temperature (°C).
Tc	Temperature of the cell (°C).
Tfi	Inlet temperature of the fluid (°C).
Tfm	Average temperature of the fluid (°C).
Tfo	Outlet temperature of the fluid (°C).
Tp	Temperature of the absorber (°C).
Tpm	Average temperature of the absorber.
UL	Coefficient of heat loss (W/m^2 C).
Vpv	Photovoltaic voltage (V).
W	Distance between the tubes (m).
w	Wind speed(m•s-1).

Since the PV/T system consists of several different elements working in unison, we take a close look at their separate dynamics by analyzing separately a.) the glass cover, b.) the solar cells, c.) the absorber plate, and d.) the heated water proper. (7)

a. Glass cover sub model

The glass cover temperature is computed by equation 1.

$$m_g C_g \frac{dT_g}{dt} = \alpha_g GA_g + A_g\left(h_{\text{Wind}} + h_{r,g-a}\right)\left(T_a - T_g\right)$$
$$+ A_g\left(h_{c-g} + h_{r-c}\right)\left(T_C - T_g\right) \tag{1}$$

Where mg and Cg are respectively the mass and the specific heat capacity of the glass cover, αg the effective absorptivity of glass. h$_{wind}$, hc-g the convective heat transfer coefficients at the outer and inner glass surfaces respectively and hr,g-a, hr,c-g the radiation heat transfer coefficients at the outer and inner glass surfaces respectively.

The convective heat transfer coefficient is given by Watmuff et. al., 1977:

$$h_{wind} = 2.8 + 3^* v_{wind} \tag{2}$$

The radiation heat transfer coefficient between the front cover and the ambient environment is:

$$h_{r,ga} = \varepsilon_g \sigma \left(T_g^2 + T_a^2\right)\left(T_g + T_a\right) \tag{3}$$

And that between the front cover and the collector plate is:

$$h_{r,c-g} = \frac{\sigma\left(T_g^2 + T_a^2\right) \cdot \left(T_g + T_a\right)}{\dfrac{1}{\varepsilon_g} + \dfrac{1}{\varepsilon_C} - 1} \tag{4}$$

Where εg and εp are the emissivity of the glazing and the collector plate respectively, and σ is the Stefan-Boltzmann constant.

b.) Solar cell sub model

Similarly, the temperature of the PV solar cell depends of the glass cover temperature as:

$$m_c C_c \ \frac{dT_c}{dt} \quad a_c \tau_c GA_c - A_c h_{r,gc}(T_c - Tg) - A_c h_{conv,gc} \tag{5}$$

$$T_c - T_g) - A_s h_{con,cp}(T_c - T_o) - \eta_c \alpha_c \tau_c GA_c$$

where mc, Cc are correspondingly the mass and the specific heat of the solar cell, hconv,gc, hr,gc are respectively the convective and the radiation heat transfer coefficients between glass cover and solar cell, hcon,cp the conduction heat transfer between cell and absorber plate.

The electric power output depends on the instantaneous operating temperature Tc of the PV module, and can be expressed as a function of the electrical current of the PV module:

$$I_{pv} = n_p \left\{ I_{ph} - I_0 \left[\exp\left(\frac{V_{pv} + R_S I_{pv}}{V_T}\right) - 1 \right] \frac{V_{pv} + R_S I_{pv}}{R_{sh}} \right\} \tag{6}$$

Where:

$$V_r = \frac{K_g \cdot T_a \cdot \eta}{q} \tag{7}$$

At the level of the PV panel the solar cell temperature is expressed as:

$$T_c = T_p - \frac{U_t \cdot (T_p - T_a)}{h_{c,p-c} + h_{r,p-c}} \tag{7}$$

The photoelectric current is given by:

$$I_{pd} = [I_{sc,src} + \alpha_{src} (T - T_{a,nf})] \frac{G}{1000} \tag{9}$$

where $K_B = 1.3806 \times 10^{-23}$ (J/K), the Boltzmann constant $q = 1.6 * 10^{-19}$ (C^0), I_0 is the opposite current of saturation of the diode, n is the factor of non-ideality (n = 1,62) of the diode, and V_T is the thermodynamic potential. In the same way, the photovoltaic module voltage with N_S serial is:

$$V_{pv} = N_s V_T \ln\left(\frac{I_{SC,SIC} - I_{pv}}{I_n}\right) \tag{10}$$

The electrical power is given by:

$$P\,pv = V\,pv * I\,pv \tag{11}$$

c.)   Absorber plate sub model:
       The temperature of the absorber plate is:

$$m_p C_p \frac{dT_p}{dt} = A_s h_{con,CP}(T_C - T_p) - A_c h_{con,Pa}(T_p - T_a) \tag{12}$$

Where
m_p, C_p are the mass and the specific heat of absorber,

$h_{cond,cp}$ is the heat transfer coefficient between absorber and solar cells;

$h_{cond,pa}$ the conduction heat transfer coefficient between the absorber and the ambient environment,

$h_{con,Fa}$ the convective heat transfer coefficient of water flow in channel.

The mean plate temperature is used to calculate the useful gain of a collector (Q_u):

$$Q_u = A_c[S - U_L(T_{pm} - T_a)] \tag{13}$$

where S is the absorbed solar radiation. The heat loss coefficient (U_L) is:

$$U_l = U_t = \frac{1}{A} + \frac{B}{D} \tag{14}$$

Where:

$$A = \frac{N_g}{\dfrac{C}{T_P}\left[\dfrac{T_p - T_a}{N_g + f}\right]^{0,33} + \dfrac{1}{h_w}}$$

$$B = \sigma (T_p^2 + T_a^2)(T_p + T_a)$$

$$C = 365.9(1 - 0.00883\beta + 0.0001298\beta^2)$$

$$D = \frac{1}{\varepsilon_p + 0{,}05N_g(1 - \varepsilon_p)} + \frac{2N_g + f - 1}{\varepsilon_p} - N_g$$

$$f = (1 - 0.04h_w + 0.0005h_w^2)(1 + 0.091N_g)$$

The mean plate temperature is given by:

$$T_{pm} = T_{fR} + \left(\frac{Q_u/A_C}{F_R U_L}\right)(1 - F_R) \tag{15}$$

Where the collector heat removal factor is:

$$F_R = \frac{GC_P}{U_L}\left(1 - \exp - \left(\frac{U_L \cdot F'}{GC_P}\right)\right) \tag{16}$$

The collector efficiency factor

$$F' = \cfrac{\cfrac{1}{U_L}}{lai\left(\cfrac{1}{U_L(d + (lai - d))F} + \cfrac{1}{C_b} + \cfrac{1}{\pi d\ ihf}\right)} \qquad (17)$$

The effectiveness coefficient is:

$$F = \frac{Tanh((lai - d)/2)}{m(lai - d)/2} \qquad (18)$$

d.) Water in channels and in storage tank sub model
The fluid temperature is computed by:

$$\dot{m}C_f \frac{dT_f}{dt} = A_f h_{C-f}(T_C - T_f) - C_{fm}\frac{\Delta Tf}{\Delta y} \qquad (19)$$

Where C_f is the specific heat of fluid, A_f the inner surface area of the flow channels per unit surface area of the collector and y is the length of channel.

The convective heat transfer coefficient is given by:

$$h_{C-f} = \frac{Nuf \times k_f}{D_h} \qquad (20)$$

Where D_h is the hydraulic diameter of the channel and k_f is the thermal conductivity of the fluid.

The mean fluid temperature is:

$$T_{fin} = T_{fi} + \left(\frac{Q_u/A_C}{F_R \cdot U_L}\right) \cdot (1 - F'') \qquad (21)$$

Where F'' is the collector flow factor:

$$F'' = \frac{\dot{m}C_f}{A_c U_L F'}\left(1 - exp\left(\frac{-A_C U_L F'}{\dot{m}C_p}\right)\right) \qquad (22)$$

The storage tank temperature is given by:

$$M_R C_R \frac{dT_R}{dt} = \dot{m}C_F(T_{fo}T_{fi}) - A_R H_{R,Ta}(T_R - T_A) \qquad (23)$$

where m_R and A_R are respectively the lumped mass and the outside surface area of the water tank $\cdot$ T_{fo}, T_{fi} water temperature at the inlet and outlet of the tank, h_R, T_a heat loss coefficient at the outside surface of the tank, including the thermal resistance of the tank insulation.

State Equilibrium of a PV/T System

By gathering the system components sub-models, the global energy balance can be written as a state equation:

$$T(t) = AT(t) + B\ m + CE$$
$$y(t) = Dt(t) \qquad (24)$$

Where

T(t) is a vector containing the temperatures at the 3 nodes of the PV/T system,

A is the state matrix which contain the heat exchange coefficients between the system elements,

B is the control matrix which encloses commands applied on the mass flow rate (m) in the PV/T system,

C is the perturbation matrix acting on the perturbation inputs vector (E=[G, Ta]),

y(t) is a vector of exit containing the electrical power, the temperature of exit of fluid and the temperature of storage tank.

The detailed form of the state equation is expressed by:

$$\begin{pmatrix} \dot{T}_c \\ \dot{T}_f \\ \dot{T}_R \end{pmatrix} = \begin{pmatrix} A_{11} & A_{12} & A_{13} \\ A_{21} & A_{22} & A_{23} \\ A_{31} & A_{32} & A_{33} \end{pmatrix}\begin{pmatrix} T_c \\ T_f \\ T_R \end{pmatrix} + \begin{pmatrix} B_{11} \\ B_{21} \\ B_{31} \end{pmatrix}\dot{m} + \begin{pmatrix} C_{11} & C_{12} \\ C_{21} & C_{22} \end{pmatrix}\begin{pmatrix} G \\ T_a \end{pmatrix}$$

$$\begin{pmatrix} P \\ T_{fo} \\ T_{ro} \end{pmatrix} = \begin{pmatrix} D_{11} & D_{12} & D_{13} \\ D_{21} & D_{22} & D_{23} \\ D_{31} & D_{32} & D_{33} \end{pmatrix}\begin{pmatrix} T_c \\ T_f \\ T_R \end{pmatrix} \qquad (25)$$

Simulation Results

Differential equations were converted to numerical format through finite difference scheme. The collector segments were interacted through the glass nodes, the solar cell nodes, the thermal absorber nodes and water nodes. It can simulate the transient performance of the system at a time interval of one minute. Its inputs require total solar irradiance and ambient temperature while its outputs contain the current and voltage generated by the solar cells, the temperatures of the transparent cover, solar cells, absorber plate, water inside the collector and the storage tank.

The output data are analyzed and used to estimate the electrical power output, the amount of heat that can be drawn from the system. A constant wind speed of 5m/s, any hourly total solar irradiance, and a daily ambient temperature are used in the simulation shown in Figure 2-32. The temperature of the photovoltaic modules decreases while water circulates, increasing the electrical output power. Further, heated water may be either used as is or converted into electric energy.

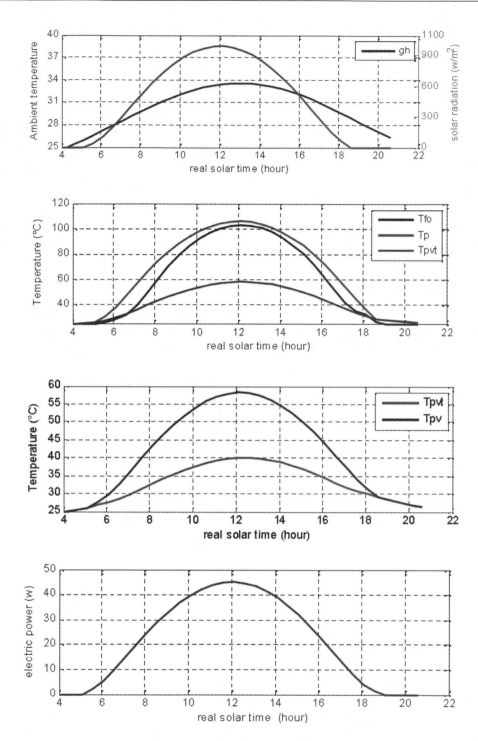

Figure 2-32. Simulation results (7)

Conclusions

The dynamic model data can be used for calculating the working parameters of a functioning PV/T system. Electric and thermal equations, formulated as dynamic and state equations of the system, can be used for different working conditions.

The model simulation evaluates the system performances such as Tfo, Tp, Tpvt and electric power for a considered weather conditions (G and Ta) and a constant operating condition (m).

It is important to note here that combined thermal and electrical performances of the PV/T system are much better when compared to separate, a.) solar thermal panel and/or b.) photovoltaic panel.

This model can be used to establish a sizing algorithm for a PV/T plant, and then to control the mass flow so as to bring optimum function and efficiency for the system.

CPV/T Technology

Concentrating photovoltaic hybrid (CPV/T) technology is a fairly new development, and although a number of companies are working on this concept, full commercialization and large-scale deployment are still far off.

The CPV/T equipment usually consists of a parabolic trough lined with a mirror surface, focused on a bank of solar cells mounted on a heater tube through which flows water or some liquid. When the sun hits the trough, its mirror surface reflects the sunlight directing it to the cells, which generate DP power at 15-20% efficiency. When the cells heat under the sun rays, they transmit the heat to the tube they are mounted on and heat the liquid running thru the tube.

The above dynamic and static calculations for the generic PV/T device could be used here too, because the basic glass-absorber-cells-liquid configuration exists here in almost identical fashion. Their properties and behavior are similar, except for different results, due to much higher solar illumination, DC voltage and temperatures.

Several companies use this technology—by combining concentrated (low concentration) PV electricity production and heat collection to simultaneously deliver electricity and hot water for commercial, industrial and institutional applications.

Mirrors and single-axis trackers are used to focus light on a row of PV cells in a special CPV/T assembly. Heat, which is normally lost in a typical PV-only system is used to heat water that runs through the CPV/T assembly and cools the solar cells.

Cooler solar cells generate more electrical energy and thus generated hot water is used for heating or process water at the same time. Some of these are turnkey systems designed for easy assembly and rooftops or ground installations.

Although the electricity-hot water hybrid cogeneration equipment has a PV element, and thus competes with conventional PV technologies, the dollar-per-watt metric is somewhat less applicable. The metric, in addition to the electricity generated within a certain time, includes gallons of hot water generated during the same time period.

The advantage here is that this type of technology does not compete head-to-head with solar panel makers but rather with natural gas. And since natural gas in the U.S. is cheap (roughly $0.02 to $0.03 per kilowatt-hour),

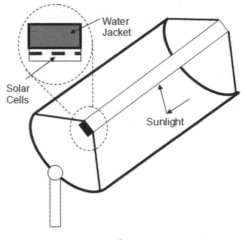

Figure 2-33. CPV/T one-axis tracker

this technology is most suited for locations with higher gas prices such as Japan and Europe.

Of course the climate has to be suitable as well, because for this technology to function efficiently, it needs direct unobstructed (desert-like) sunlight. Appropriate (niche market) customers are needed as well, those with demand for hot water. This includes the food and beverage industry, schools, hospitals, manufacturing facilities, army bases, etc.

Another advantage of the system is the capability to store hot water for later use. The ability to generate energy in off-grid applications is also advantageous, and creates another niche market.

In addition, thus generated heat can also be used for cooling applications, which opens the opportunity for supplying electricity and hot water to buildings with large chilling loads, as well.

These CPV/T systems produces water at 70°C, and require no special permitting or roof modifications to support the additional load. They also come complete with integrated hydronics, controls, and inverters.

Another great advantage is the fact that if the CPV/T system is SRCC (Solar Rating and Certification Corporation) and IEC (International Electrotechnical Commission) certified, it qualifies for both thermal and electric rebates. This includes the California Solar Initiative (CSI), which offers an incentive for solar hot water systems.

Currently, CSI thermal rebates offered in the first-tier level of $12.82 per year for one displaced gas, complement the CSI photovoltaic rebates. This results in faster paybacks than traditional solar hot water or solar electric systems alone.

On top of that, the installations are also eligible for the 30 percent investment tax credit. With the CSI and federal rebates, CPV/T system customers can yield a return on investment in five years, while hedging against future utility price hikes and gas price volatility.

PV-T systems produce 3-5 times more energy (electricity and heat combined) and three times the greenhouse gas reductions of the competing solar technologies. When usage is compared watt-per-watt, solar water heating is 3 to 5 times more efficient than PV, which ensures a much faster payback and essentially doubles the value of a CPV/T system.

Since this is an accepted way of doing things in China, Europe, and Austria, the technology will be widely accepted there.

The value of hot water production is, however, somewhat compromised by the plumbing aspect of hot water generation, placing this technology in niche energy markets which are limited in size.

HCPV/T Technology

High concentration photovoltaic/thermal (HCPV/T) hybrid technology is even newer and more undeveloped than its CPV/T cousin. As a matter of fact we know of only one company (SolarTech of Arizona) that is working on its optimization and commercialization.

The function of the HCPV-T system is similar to that of its older cousin, the CPV-T energy system. It also requires tracking and operates at very high—over 60% overall— combined efficiency.

The dynamic and static calculations for the generic PV/T device could be used here too, because the basic glass-absorber-cells-liquid configuration exists here in almost identical fashion. Their properties and behavior are similar, except for different results, due to the high concentration of sunlight onto the modules, which results in much higher solar illumination, DC voltage and temperatures.

In summary:

- HCPV/T hybrid systems can be used also (with some modifications) without liquid cooling, using air cooling of the heat sinks only, instead.

- The system can be built in different sizes; from 0.5kWp to 50kWp as needed to fit different roofs and applications—including large-scale ground-mounted installations.

- As another option, 6 regular silicon solar cells are added to each HCPV cell assembly in a special arrangement, so that some of the sunlight that is reflected by the HCPV cell illuminates the 6 regu-

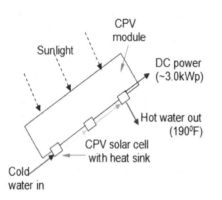

Figure 2-34. HCPV-T module cross section

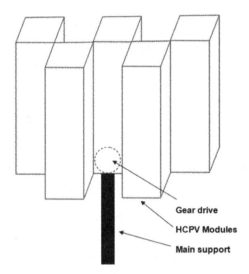

Figure 2-35. HCPV/T two-axis tracker (pole mount)

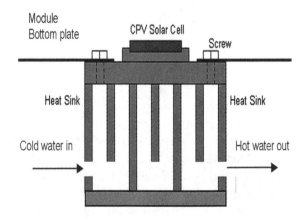

Figure 2-36. Solar cell heat sink and water heating assembly

lar solar cells around it, thus producing additional DC power. This configuration increases a 3.0kWp power output of a standard 7 modules HCPV/T system to 5.5kWp, without significantly reducing the temperature, or amount of hot water produced by the thus modified reference system.

- Different combinations of these options allow unprecedented efficiency, product flexibility, and offer a number of different application choices.

Total Energy Management (via Hybrid Systems)

Figure 2-38 is a graphic representation of a functional, all-purpose combination power generator, consisting of a combo (hybrid) PV power/hot water/gas generator. This is a complete self-contained, self-controlled, hybrid PV-T system for use in remote locations.

This system provides power on a 24/7 basis and covers the best of both worlds—DC/AC power and hot water for everyday needs. Such a system would be convenient for use in field hospitals, refugee camps and other operations, where power and hot water are inaccessible.

This system would be exceptionally efficient in areas with lots of sunlight—desert areas are best, of course. It can be configured with or without batteries, and with or without a gas generator. In all cases, the solar collector(s) will provide much needed DC power and hot water during the day, where the only limiting factor is the amount of sunlight available to the collector(s).

By making adjustments to the system, we could obtain maximum efficiency under different climatic conditions. Such a system is what the future holds for a number of critical applications, so we foresee a bright future for these types of energy systems.

NEW AND DEVELOPING SOLAR TECHNOLOGIES

Obviously, the TFPV industry is in its infancy, and a lot of work remains to be done before it becomes a proven, mature technology. There are many research groups active in the field of photovoltaics in universities and research institutions around the world. This research can be divided into three areas: a) making solar cells and modules cheaper and/or more efficient to effectively compete with other energy sources; b) developing new technologies based on new solar cell architectural

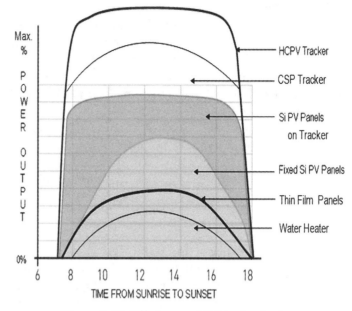

Figure 2-37. Efficiency of PV technologies

designs; and c) developing new materials to serve as light absorbers and charge carriers.

Thin-film photovoltaic cells and modules use less than 1% of the expensive raw material (silicon or other light absorbers) compared to silicon wafer-based solar cells, leading to a significant price drop per Watt peak capacity. Many research groups around the world ac-

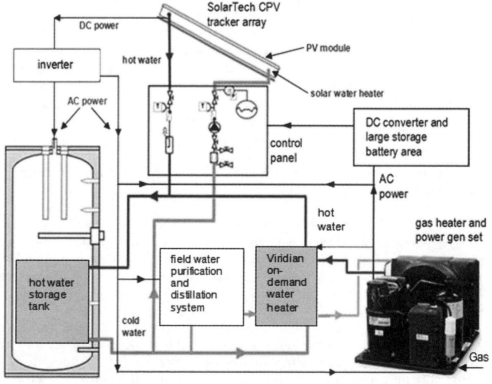

Figure 2-38. Total energy management system.

tively seek different thin-film processes; however, solutions that can achieve a similar market penetration as traditional bulk silicon solar modules remain to be seen.

An interesting aspect of thin-film solar cells is the possibility of depositing cells on all kinds of materials, including flexible substrates (PET for example), which opens a new dimension for applications.

We will take a very close look at some of these budding technologies in the next chapters. Below is a list of some new and exotic thin film materials and processes and a short description of ongoing R&D efforts with a detailed description of each of those in the following chapters.

Metamorphic Multi-junction Solar Cell

This is an ultra-light and flexible multi-junction (MJ) cell, similar to the Germanium-based CPV cell that converts solar energy with record efficiency. It represents a new class of solar cells with clear advantages in performance, engineering design, operation and cost. For decades, conventional cells have featured wafers of semiconducting materials with similar crystalline structure. Their performance and cost effectiveness is constrained by growing the cells in an upright configuration.

Present-day MJ cells are rigid, heavy, and thick with a bottom layer made of Ge semiconductor material. In the new method, the cell is grown upside down. These layers use high-energy materials with extremely high quality crystals, especially in the upper layers where most of the power is produced. Not all of the layers follow the lattice pattern of even atomic spacing. Instead, the cell includes a full range of atomic spacing, which allows for greater absorption and use of sunlight.

The thick, rigid Germanium layer is not used, thus reducing the cell's cost and 94% of its weight. By turning the conventional approach to cells on its head, the result is an ultra-light and flexible cell that also converts solar energy with record efficiency of 40.8% under 326 suns concentration.

Polymer PV

The invention of conductive polymers may lead to the development of much cheaper cells that are based on inexpensive plastics. However, organic solar cells generally suffer from degradation upon exposure to UV light, and hence have lifetimes which are far too short to be viable, at least at this stage of their development.

The bonds in the polymers are always susceptible to breaking up when radiated with shorter wavelengths. Additionally, the conjugated double bond systems in the polymers which carry the charge react more readily with light and oxygen.

So, most conductive polymers, being highly unsaturated and reactive, are highly sensitive to atmospheric moisture and oxidation, making commercial applications difficult.

Nanoparticles PV

Experimental non-silicon solar panels can be made of quantum hetero-structures, e.g. C or quantum dots, embedded in conductive polymers or meso-porous metal oxides. Also, thin films of many of these materials on conventional silicon solar cells can increase the optical coupling efficiency into the silicon cell, boosting overall efficiency.

By varying the size of the quantum dots, the cells can be tuned to absorb different wavelengths. Although the research is still in its infancy, quantum dot-modified photovoltaics may be able to achieve up to 42% energy conversion efficiency due to multiple exciton generation.

Solar cell efficiency could theoretically be raised to more than 60% using quantum dots. Electrons can be transferred from photo-excited crystals to an adjacent electronic conductor. Researchers have demonstrated the effects in quantum dots made of PbSe, but the technique could work for quantum dots made from other materials too.

Photonic Crystals PV

These are nanostructured materials in which repeated variations in the refractive index on the length scale of visible light produces a photonic band gap. This gap affects how photons travel through the material and is akin to the way in which a periodic potential in semiconductors influences electron flow. In the case of photonic crystals, light of certain wavelength ranges passes through the photonic band gap while other wavelength ranges are reflected. The photonic crystal layer could be attached to the back of a solar cell.

Transparent Conductors

Many new solar cells use transparent thin films that are also conductors of electrical charge. The dominant conductive thin films used in research now are transparent conductive oxides (TCO), and include fluorine-doped tin oxide (SnO_2:F, or FTO), doped zinc oxide (e.g., ZnO:Al), and indium tin oxide (ITO). These conductive films are also used in the LCD industry for flat panel displays.

The dual function of a TCO allows light to pass through a substrate window to the active light-absorbing material beneath, and also serves as an ohmic contact to transport the photogenerated charge carriers away from that light-absorbing material. Present TCO materials are effective for research, but perhaps are not yet optimized for large-scale photovoltaic production. They require special deposition conditions at high vacuum, and can sometimes suffer from poor mechanical strength, while most have poor transmittance in the infrared portion of the spectrum (e.g., ITO thin films can also be used as infrared filters in airplane windows). These factors make large-scale manufacturing more costly.

A relatively new area has emerged using carbon nanotube networks as a transparent conductor for organic solar cells. Nanotube networks are flexible and can be deposited on surfaces in a variety of ways. With some treatment, nanotube films can be highly transparent in the infrared, possibly enabling efficient low-bandgap solar cells. Nanotube networks are p-type conductors, whereas traditional transparent conductors are exclusively n-type. The availability of a p-type transparent conductor could lead to new cell designs that simplify manufacturing and improve efficiency.

Infrared Solar Cells

Researchers have devised an inexpensive way to produce plastic sheets containing billions of nano-antennas that collect heat energy generated by the sun and other sources. The technology is the first step toward a solar energy collector that could be mass-produced on flexible materials. While methods to convert the energy into usable electricity still need to be developed, the sheets could one day be manufactured as lightweight "skins" that power everything from hybrid cars to computers and iPods, with higher efficiency than traditional solar cells.

The nano-antennas target mid-infrared rays, which the Earth continuously radiates as heat after absorbing energy from the sun during the day. Also, double-sided nano-antenna sheets can harvest energy from different parts of the Sun's spectrum. In contrast, traditional solar cells can only use visible light, rendering them idle after dark.

UV Solar Cells

Researchers have succeeded in developing a transparent solar cell that uses ultraviolet (UV) light to generate electricity but allows visible light to pass through it. Most conventional solar cells use visible and infrared light to generate electricity. Used to replace conventional window glass, the installation surface area could be large, leading to potential uses that take advantage of the combined functions of power generation, lighting and temperature control Also, easily fabricated

PEDOT:PSS photovoltaic cells are ultraviolet light selective and sensitive.

3-D Solar Cells

These are truly three-dimensional solar cells that capture nearly all the light that strikes them and could boost the efficiency of photovoltaic systems while reducing their size, weight and mechanical complexity. The new 3D solar cells capture photons from sunlight using an array of miniature "tower" structures that resemble high-rise buildings in a city street grid.

Meta-materials

Meta-materials are heterogeneous materials employing the juxtaposition of many microscopic elements, giving rise to properties not seen in ordinary solids. Using these, it is possible to fashion solar cells that are perfect absorbers over a narrow range of wavelengths. This is still pure research.

Photovoltaic Thermal Hybrid

These systems combine photovoltaic with thermal solar. The advantage is that the thermal solar part carries heat away and cools the photovoltaic cells; keeping temperature down lowers the resistance and improves the cell efficiency. Modified CPV systems have been tested, where the CPV cells are cooled by active flow of liquid, thus generating both heat and electricity.

HIT Solar Cell

A novel device developed by Sanyo is the HIT cell. In this device thin film layers of amorphous silicon are deposited onto both faces of a textured wafer of single crystal silicon. This improves the efficiency and reliability of the solar cell by avoiding the usual problems related to doped and fired silicon solar cells.

This type of solar cell would be more expensive to produce, but might provide the much needed solution to the problems c-Si and TFPV modules encounter when exposed to extreme desert temperatures. Such cells would exhibit less heat sensitivity and power degradation, but, we have not seen any proof of that happening as yet.

Organic Solar Cells

Organic materials have been subjects of intense research lately, with the most interesting being the molecular and polymeric semiconductors, Fullerene (C60) and its derivatives. These materials are used in organic LEDs and thin film transistors for use in smart cards. The advantage of using these materials is their manufacturing simplicity and the possibility of use on flexible materials.

The basic structure consists of glass or other such substrate, coated with indium tin oxide (ITO) which is a transparent conductor and acts as the top electrode. Several layers of light-absorbing organic materials could be deposited on the ITO and then completed by depositing a metal back contact onto the organic materials.

Cell efficiencies of 5% have been achieved, but low efficiency and problems with large-scale manufacturing processes must be resolved before we see these technologies on the market.

CTZSS Solar Cells

This is a new type of solar cell, developed by IBM (summer 2010), who claims that it could potentially lower the price of solar power in the future. The new CTZSS solar cell is made from copper, tin, zinc, sulfur and selenium, all of which are somewhat earth-abundant, according to IBM.

The new process does not use vacuum chambers and related processes (the sort of equipment used for chip-making). Instead, it deploys a solution-based approach, which is different and quite interesting, but it brings us back to the days when we used to make c-Si cells via solution-based processes. So we've come a full circle, it seems.

In mass production, that could mean producing cells through printing or dip and spray coating. Bringing such a cell to market, however, won't be easy. Investors have become nervous about championing new types of solar cells. Various CIGS companies also employ printing, and other roll-to-roll manufacturing processes to produce thin film solar cells. It has taken most quite a while to iron out the kinks in the manufacturing process.

So, will IBM make solar cells? We'd guess not, but some other manufacturers might benefit from their research and experience. We will see...

Thermovoltaics and Thermophotovoltaics

Researchers are working on ways to combine "quantum and thermal mechanisms into a single physical process" to generate electricity and make solar power production twice as efficient as existing technologies. It's called PETE photon-enhanced thermionic emission. This process does require some exotic materials (cesium-coated gallium nitride) and works best at high temperatures more likely with concentrators or parabolic dishes than flat solar panels.

Intermediate Band

This is intermediate band nitride thin film semiconductor material, which might be an alternative to the multijunction designs for improving power conversion efficiency of solar cells. The intermediate-band solar cell is a thin-film technology based on highly mismatched alloys. The three-bandgap, one-junction device has the potential of improved solar light absorption and higher power output than III-V triple-junction compound semiconductor devices.

Plasmonics

Plasmonic technology uses engineered (special) metal compounds based structures to guide light at distances less than the scale of the wavelength of light in free space. Plasmonics can improve absorption in photovoltaics and broaden the range of usable absorber materials to include more earth-abundant, non-toxic substances, as well as to reduce the amount of material necessary.

We take a much closer look at some of these technologies in the following chapter.

Notes and References

1. PVCDROM
2. Understanding PV Module Specifications, Justine Sanchez, Jan. 2009.
3. Electroluminescence Investigation on Thin Film Modules, Thomas Weber, Anton Albert, Nicoletta Ferretti, Margarete Roericht, Stefan Krauter and Paul Grunow, Photovoltaik Institut Berlin, Germany
4. Electroluminescence on the TCO Corrosion of Thin Film Modules, Thomas Weber, Elfriede Benfares, Stefan Krauter and Paul Grunow, Photovoltaik Institut, Berlin, Germany
5. Mora Associates, October 2009. http://www.moraassociates.com/publications/0903%20Concentrated%20Solar%20Power.pdf
6. Renewable Energy UK: http://www.reuk.co.uk/First-European-Solar-Power-Tower.htm
7. Environment News Service: http://www.ens-newswire.com/ens/mar2007/2007-03-30-02.asp
8. APS http://www.aps.com/main/green/Solana/About.html
9. Stirling Energy Systems company website: http://www.stirlingenergy.com/
10. National Renewable Energy Laboratory (NREL): dish/stirling report
11. JC Winnie environmental blog: http://www.innovationsreport.com/html/reports/energy_engineering/report-82659.html
12. IEEE Spectrum Online: http://www.spectrum.ieee.org/oct08/6851
13. Global Greenhouse Warming.com: http://www.global-greenhouse-warming.com/solar-parabolictrough.html
14. National Renewable Energy Laboratory (NREL): http://www.nrel.gov/csp/troughnet/
15. Nevada Solar One official website: http://www.nevadasolarone.net/the-plant
16. CSP Today: http://social.csptoday.com/index.php
17. SolarPaces. http://www.solarpaces.org/News/Projects/projects.htm
18. *Photovoltaics for Commercial and Utilities Power Generation*, 2011. Anco S. Blazev
19. A dynamic model of hybrid photovoltaic/thermal panel, by Majed Ben Ammar, Mohsen Ben Ammar and Maher Chaabene

Chapter 3

Most Promising Solar Technologies

Adversity causes some men to break; others to break records.
—William Ward

A number of solar energy generating technologies have proven worthy of high efficiency, reliable, long-term operation under adverse conditions. Many have not. There are many reasons for this, and a lot of ongoing discussions between scientists, engineers, politicians and energy industry specialists. No conclusive decisions have been made on exactly which technology will be most successful in the long run in terms of efficiency and reliability. So the debate continues. We are taking advantage of this situation, and the responsibility that comes with it, to sort the specifics of the "most promising" solar technologies, their usage, and the related advantages and disadvantages for use in the 21st century.

For the purpose of this writing, we will segregate the solar technologies according to their promise (as we see it) to withstand 30 years of non-stop operation under different conditions—which is why we call them "most promising."

In contrast, the "exotic" (for lack of a better word) solar technologies we discuss in the next chapter are not mature enough to be expected to be reliable for most applications. Most of these are still in R&D labs, and belong in different niches of the energy market.

Many of them hold the promise of cheap mass production and cheap installation, but in most cases are quite limited by their low efficiency, operational instability, inability to withstand the elements, and/or difficulty with transfer into mass production.

These are serious issues, so many of these technologies, as good as they may sound on paper, do not necessarily hold the promise of long-term reliable operation—especially in the deserts.

Which technologies will succeed in the 21st century? Without a crystal ball, we must rely on experience, using what we know about solar technologies to make an educated guess.

The energy future depends on the efficiency and reliability of solar devices to operate without major problems for 30+ years, under extreme conditions such as in deserts and humid areas.

Short of that, we are looking at disposable toys, spread over large areas, which will break within several years, leaving vast financial and environmental disasters.

What makes us think there is ANY solar technology ready for this test? As a matter of fact, many lab and long-term field test results show that even the oldest and most reliable technologies, such as c-Si and thin film PV modules, made by the most renown and reliable manufacturers, do deteriorate and fail in the field after several years of non-stop operation.

As some fail faster than others, we believe that the type and quality of materials, and manufacturing labor are of the essence when talking about longevity of PV modules.

Nevertheless, since we have more experience with "established" solar technologies, and since we best understand the risks associated with their use, we will put them in the category of "most promising," until they prove otherwise. We will move on, remembering that there is less risk working with the evil we know well, than with the one of which we have little knowledge.

For the sake of clarity we divide the "most promising" solar technologies into several distinct categories in this text, which are then subdivided into their constituents.

Surprisingly, most people think that all solar technologies can be used anywhere and in any environment. This is not true. That misconception and the variability of solar power constitute the most urgent problem to be addressed.

A good example of this lack of understanding is the deployment of millions of thin film PV modules in large-scale installations. Since most of these modules have not been proven efficient or safe for long-term use in the blistering-hot desert climate, we don't know what will happen to them 5, 10, or 15 years down the road.

Energy storage, or the lack thereof, is another big

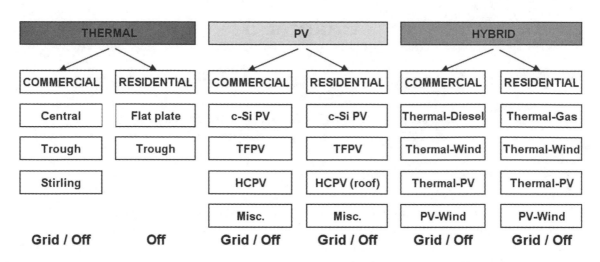

Figure 3-1. Most promising solar energy generating technologies in the 21st century

problem to be resolved. These technologies are also in the "emerging" category, and we will take a close look at them as well.

Nevertheless, solar technologies are getting more efficient, reliable and cost-effective by the day. They are also desperately needed for establishing solar as a competitive energy source, and with the help of storage systems will become a "non-variable" energy source, successfully competing in 21st century energy markets.

SOLAR THERMAL TECHNOLOGIES

We discussed solar thermal technologies in more detail in the previous chapter, so here we will only add

that they will play a significant role in the 21st century. CSP will dominate large-scale power generation for the foreseeable future, while the other technologies will share the energy market by finding niches where they are most practical, efficient and cost-effective.

CSP Technologies' Future

The dark areas in Figure 3-3 show where the European Union is planning to install a large number of wind, CSP, and PV power plants. CSP technologies will take a major part in this effort which will bring the CSP industry to a new level by 2020. This, in addition to plans for many additional GWs of CSP installations in the US and Asia, is an exciting development which paints a bright picture for the future of CSP technologies as a world-class electric power generator.

We have witnessed the quick and successful development of the CSP industry of late and have seen

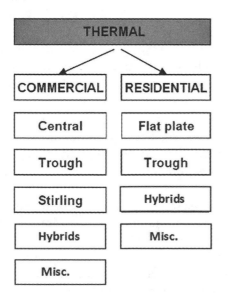

Figure 3-2. Solar thermal technologies in the 21st century

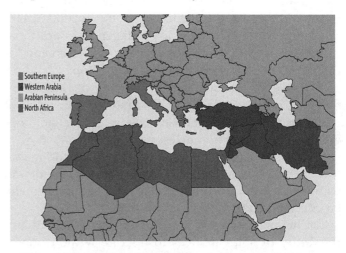

Figure 3-3. CSP technologies: future applications

estimates that place it at the top of the list of large-scale power generation. The fast pace of technological developments today, and the world's ever-changing socio-economic and political climates make it hard to predict what will happen, but we know that CSP is here to stay and that it will grow.

We fully expect CSP technologies will be widely used in the areas of the world with most favorable climate conditions—primarily dry, desert areas. The darkened areas on the map show the approximate locations where we expect full deployment in the 21st century. The power generated in these areas will be carried to adjacent developed countries—mostly in the EU and the Middle East. This will allow these countries to be energy independent, while providing financial prosperity in the underdeveloped countries where CSP technologies are deployed.

Research, technological innovations and improvements, and mass production are making CSP competitive with other energy generating technologies. CSP technologies today are priced between photovoltaics and wind, or approximately $4/Watt. Costs are decreasing constantly due to the need for energy and market expansion.

In the US, CSP is shaping as a promising power provider, although lately there have been some barriers to its wider implementation.

The brightest hope for CSP is the Solar European Industrial Initiative, established by the SET Plan, whose goal is to scale-up the most promising solar technologies from the R&D level to commercial feasibility in the large-scale, multi-MW range of the so-called "lighthouse" projects. Since Europe doesn't have the abundant sun needed to operate CSP power plants, some of the new solar and wind power fields will be located in North-African Mediterranean countries. Fresh water availability is an increasing problem in North Africa and in southern Europe, so some of the new solar installations could be dedicated to a combination of electricity production for the entire EU continent and also for water desalination for local consumption.

The European association of the solar thermal electricity industry (ESTELA) expects to have more than 800 MW connected to the grid by 2020, while the long-term potential for European Mediterranean countries is estimated at 30,000 MW. This, of course, depends on taking the necessary measures to ensure efficient, reliable operation of the power fields for the duration. A much larger contribution could be obtained when the full potential of the North African countries is developed. The annual electricity production could reach 85 TWh per year, covering 4% of the EU electricity consumption by 2020. For 2030, installed capacity is estimated to double and could reach 60 GW for an annual energy production of about 170 TWh.

Increasing the deployment of CSP technologies will require the removal of technical barriers and a number of improvements are needed to increase grid flexibility and enhance storage technologies.

Current electricity transmission and distribution systems are designed to manage more traditional generation technologies and are not ideally suited for introduction of large-scale CSP plants. Also, sunshine is an intermittent power source, which poses a challenge in terms of efficient storage of surplus energy.

Thermal storage and hybrid operation (using biomass or fossil fuel as an alternative heat source) need to be further explored and improved. Also, further research of new applications, like water desalination and hydrogen production, is needed.

Additional critical aspects to be addressed are land use, materials consumption, and conversion efficiency. Removing technical and administrative barriers is only the beginning of the process for reaching the ambitious objectives embraced by the CSP sector for 2020-2030. Ad hoc financial instruments such as long-term and stable feed-in tariffs should be encouraged and the awareness of the potential of the solar sector must be reinforced as well. So the EU energy future looks bright; the power source is there, and human ingenuity can separate Europe from energy dependence.

There is great potential for large-scale CSP power fields in the rest of the world, with China, India and Australia being the most suitable areas, as far as availability of adequate for CSP solar energy is concerned.

China is making steady progress towards implementation of large-scale CSP power plants. The construction of a large concentrated solar power (CSP) project in Gaoshawo town, Yanchi County, Ningxia province in northwest China was started recently. It is the first project of its kind in Asia, which utilizes integrated solar combined cycle (ISCC) hybrid technology, where trough solar thermal generators will operate in tandem with gas turbine generators. The plant will have a capacity of 92.5 MWp, and is scheduled to go online in the fall of 2013 at the cost of $346 million.

This design is an alternative to the traditional trough solar power generation, as it adds energy (thermal) storage facilities and cooperates with combustion turbines for maximizing utilization of solar power resources and optimum power production.

The thermal efficiency of the technology is boosted

by up to 80%, by fully utilizing solar energy as the primary source of high-temperature heating. It also provides a "smoothing" effect via gas-fired generators during cloudy periods or rainy days.

ISCC, therefore, is viewed by specialists as a potential solution to China's major problem with CSP—the inability of CSP to deliver consistent power during erratic changes in output in relation to weather changes and peaks of grid demand across the country.

In addition, China is putting significant investment into developing a nationwide smart grid to ease this problem. To this end, the ability of the ISCC to store solar-generated energy to help prevent fluctuations in supply will prove useful.

If this technology is proven to be a success in "smoothing" the demand and supply ratios, it would become a major factor in the future of the CSP technology in China because it has largely shied away from CSP, believing it to be too unreliable and erratic. In addition, and very important, the CSP sites are located too far away—mostly to the northwest where there is no adequate electrical grid to feed into the power systems into the areas that need it most, southern China and the coastal manufacturing belt.

Again, the problem is cost! ISCC is expensive technology, both to install and operate. It is very complicated and expensive to add a large solar field onto a traditional power plant. The challenges of this undertaking are unprecedented, and it has not been proven efficient and cost-effective thus far. But China has money these days and is energy hungry. Combined with its political goal of increasing energy security and reducing oil-dependency, as well as its comparative advantage of low construction costs, China will pave the way for the technology.

The initial test of this technology would be the delivery of reliable, uninterrupted, cost-effective energy, so the experts in China and around the world are paying close attention. So are the politicians—the launch of construction of the first ISCC project was attended by high level officials of China's ruling party. Beijing seems convinced that ISCC may be one sure way towards energy independence.

India's CSP power projects have faced many hurdles since the highly praised National Solar Mission (NSM) was launched in 2010. Finance, however, is getting in the way of NSM and the seven solar projects planned under Phase I of the program. Major problems are the high capital cost of CSP, and the uncertainty surrounding the rate of return from the solar projects, making banks and financial institutions uneasy with financing such large undertakings.

The uncertainty of the global financial crisis is not helping the situation. The turmoil seen in the debt markets in 2011-2012 has raised further concerns about the prospects of new large-scale solar plants in India, due to the high capital needed to develop CSP.

A number of reports on the financing of Phase I of the NSM have been hazy and even misleading, and only two developers provided financial closure details by the given deadline in July 2011. The other projects continue to show progress, with key components now ordered and project construction beginning. There is still ambiguity surrounding the methods used to finance these plants, but unless financing is provided via project finance or private equity the CSP projects will linger much longer than planned and some might fail to see completion during Phase I. If a large number of plants do not meet their completion deadline in May 2013 and do not raise the needed capital, then the energy market has spoken, and CSP is not bankable in India—at least for now. This will inevitably have a negative effect on the completion of Phase II of the NSM as well.

Political and financial issues are the major holdbacks of full implementation of this promising energy source in these areas of the world. Or, as AREVA Solar CEO Bill Gallo said, "Innovation and commercialization are tied together. The concentrated solar thermal industry must continue to innovate while we commercialize, deploy and scale our technologies. This will drive us to grid parity, which is more than just the price of electricity. Grid parity is about stability, reliability and the true costs of land, water and emissions. These are among the key factors that will determine our success as an industry and our ability to play a larger role in energy markets around the world."

PHOTOVOLTAIC TECHNOLOGIES

There are a number of PV technologies that show great promise for deployment in the 21st century.

These are:
1. Crystalline silicon (c-Si) PV modules
2. CdTe thin film modules,
3. SIGS thin film modules
4. Alpha silicon (a-Si) modules
5. Ribbon silicon modules, and
6. Hybrid technologies

c-Si, CdTe, SIGS and a-Si technologies are dominant, at least for now and until the new technologies prove themselves, and/or until a new disruptive technology comes along. We fully expect this to happen some day, but until then these will remain the most promising and most used in residential, commercial and large-scale installations in the 21st century.

PV Basics

We took a close look at solar cells and modules design, and manufacturing in the previous chapter, so here we present only the very basic practical operation, common to all PV technologies.

The active area in the middle of the device operates as a transistor which generates excess electrons, while the top and bottom contacts are metal wires attached to the cell's anode and cathode to extract and conduct the electrons into an outside electric circuit for practical use. Each interconnect adds the voltage of each successive

device and a constant current flows through the structure.

The encapsulation is typically ethylene vinyl acetate (EVA) and the top of the modules is covered by glass. Plastic cover is used in flexible PV devices.

The equivalent circuit includes the diode junction, the photocurrent source (the two overlapping circles) and a series resistance, due to the solar cell semiconductor material properties and/or that of the contact's material and its properties.

Here we show three silicon solar cells (and their junctions) with interconnects in which the n-type top contact of one device is connected to the p-type back contact of the next via metal wires.

Note: Thin film devices use a modified version of the silicon cell interconnects—the so-called monolithic interconnection.

Sunlight impinging on the solar cell's top surface generates photo-carriers, followed by carrier separa-

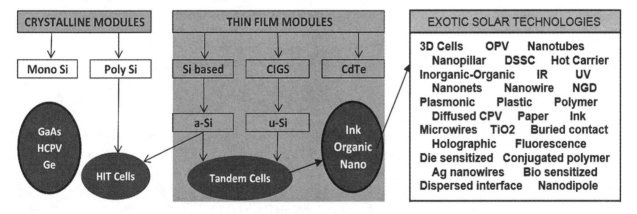

Figure 3-4. Major PV technologies

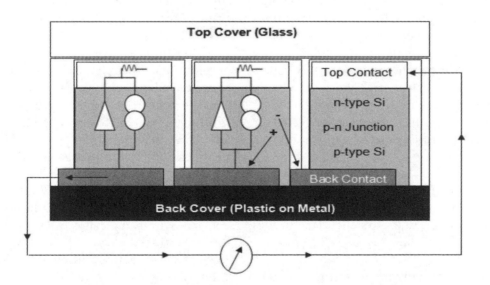

Figure 3-5. A solar module as an electronic device, cross-section.

tion within the device, which produces a photo voltage. The charge (electrons) motion produces a photocurrent, which runs in reverse through the diode junction.

Electrical power is extracted from the device by the attached interconnects (metal wires). If a suitably matched resistive load (battery, light, electric motor, etc.) is attached to the wires, the photo-current flows from the device into the load and thus supplied electric energy is used to charge the batteries, light the light, or turn the motor.

The performance of a solar cell is determined by how efficient the diode function is, how effectively it collects the photocurrent generated, and how much of the photon energy is preserved in the device. Another key factor determining the efficiency is the amount of energy lost in this process, due to defects in the material, intrinsic recombination mechanisms, and parasitic losses (such as series resistances and shunts).

The theoretical limit to efficiency of solar cells is estimated with some reservations, the best method being that described as the Shockley-Queisser (SQ) limit. This method can be summarized as follows:

1. The solar spectrum consists of photons with various energies conforming to a 5600 K blackbody radiation distribution filtered by the Earth's atmosphere. Photons with less energy than the energy gap of the active semiconductor in the device are not absorbed (the device may be transparent to these), so their energy is lost. For a Si solar cell this amounts to about 19% of the energy in the standard "air mass 1.5" solar spectrum.

2. For the photons that are absorbed, the generated electron hole pairs rapidly lose their energy until they reach the energies of the semiconductor band edges. For example, a photon with 2 eV of energy absorbed in Si will produce a 2 eV energy difference between the electron and the hole. This is far above the energy gap of the Si and the excess energy is converted to heat. This results in another 31% energy loss for a Si photovoltaic device.

3. Further energy is lost because the difference in chemical potential of relaxed photogenerated electrons and holes is less than the energy gap. This accounts for another 16% efficiency loss in a Si device.

4. Finally there is the current loss under power generation conditions that leads to losses at the maximum power condition. It varies from device to device and according to the operational conditions.

The Shockley-Queisser limit for Si solar cells is 29%. Considering additional factors, as discussed above, combined with other losses, such as those related to contact schemes, the empirical limit under AM 1.5 sunlight goes down to 26%.

The record-performance of the commercial Si solar cell is currently around 24% efficiency, which is very near the expected (calculated) theoretical limit. The catch here is that the closer to the efficiency limit we get, the more difficult and expensive the processes and devices become. Basically speaking, there is a limit as to how high we can go with the efficiency before breaking the bank.

The most practical approach to avoiding the SQ efficiency limit is to use multijunction solar cells which are basically several cells stacked on top of each other (see Tandem Cells below). In this configuration, some light getting to the top cell is absorbed, but some still passes through to be absorbed in a second lower energy gap cell below it, and so on. At the end, the series connection of the cells results in the addition of the voltage produced by each cell to produce constant current that can be used.

This series connection represents a serious limitation to the overall performance when the solar spectrum is subject to change at different times of day. This is because the maximum output of the device is limited by the maximum output of the weakest link in the system. So when the sun goes down, some cells operate at much reduced power, which proportionally reduces the power output of the entire system, regardless of the higher output of the other cells in the tandem configuration.

This problem can be solved by using tracking concentrators, where lenses or mirrors focus maximum sunlight available on the cells all day long.

The current generated by a solar cell scales with light intensity, while the voltage increases logarithmically with intensity, so the overall efficiency of the cell is expected to improve logarithmically with concentration. This is true, but the cell bulk also increases in temperature with increased light flux, so its efficiency is effectively and proportionally reduced. Series resistance problems within the cell can also increase, and higher carrier density can lead to greater recombination rates, which can also reduce efficiency. There is, therefore, an optimum concentration ratio for each particular device and operating condition which must be matched for best results. Effectively cooling cells might help reduce thermally generated resistive losses, thus optimizing cell performance under high concentration.

Other approaches, such as multi-exciton, hot carrier, and intermediate band solar cells are under evaluation and some of these are reviewed in more detail in Chapter 4. The goal of these technologies is to increase the generated current without necessarily producing a gain in voltage in single junctions without concentration. These devices, while they may not have intrinsic or practical benefits over multi-junctions, are interesting because they can eventually be produced at a lower cost than the more complex crystalline multi-junction devices.

Amorphous Si (a-Si) is one of the most promising technologies, as it can be used in a number of combinations—in single- or multi-junction solar cells—where low-cost devices are easily produced, although the efficiency of most of these is still quite low.

Polycrystalline thin film multi-junctions are under evaluation and development, and some good results have been shown, but their practical application is yet to be demonstrated.

SILICON PV TECHNOLOGIES

Crystalline Silicon (c-Si), with its sub-categories; single-crystal Si and poly Si is the oldest and most understood solar technology. It is also deployed in many countries and has been operating under different conditions for many years. Although it has efficiency and reliability problems, it is still used more than any other solar generating technology—and will be, for the foreseeable future.

c-Si PV technologies also compete successfully with concentrating solar thermal (CSP) and wind energy generators, simply because they are efficient enough and because they can be used in more versatile ways.

Because of that, the c-Si PV technologies future looks promising. As a matter of fact, as confirmation of that statement, a number of large-scale CSP power fields planned for installation in the US have been scrapped, and c-Si will replace the CSP technologies.

Single Crystal Si Solar Cells
Single crystal silicon, also called mono-crystalline (mc-Si), mono-silicon, or mono-Si, or sc-Si, is a type of silicon that was grown by the very special and expensive Czochralski (or CZ) method, or via the float zone (FZ) method. Both methods use similar equipment and production techniques and end up with the best, most efficient, silicon material for semiconductor and solar cells device manufacturing.

Solar cells and panels made out of CZ or FZ silicon material have the highest efficiency and longevity of all silicon-based PV devices, primarily due to the uniform, stable and predictable nature of the bulk material.

This type of cell is the most commonly used, and constitutes about 80% of today's market. It will continue to lead the markets, and until a more efficient and cost effective PV technology is developed.

Monocrystalline silicon manufacturing process consists of a single crystal ingot, which is produced using the Czochralski method. After a Si ingot is manufactured to a diameter between 10 to 15 cm, it is cut into wafers of 02-0.3 mm thick to form a solar cell which generates approximately 35 mA current per cm^2 area, with accompanying voltage of 0.55 V at full illumination. For some other semi-conductor materials the efficiency under different wavelengths can reach 40%.

We took a closer look at the mc-Si materials, processes and devices in the previous chapter and in even greater detail in our first book of the series, *Photovoltaics for Commercial and Utilities Power Generation.*

Attempts to enhance the efficiency of mc-Si solar wafers produced by this process are currently limited by the amount of energy produced by the photons in the silicon wafers, since their energy decreases with increase of wavelengths.

Significant losses are generated during operation due to the combination of silicon material, the metal contact's resistance, and the reflection of sunlight from the cell's surface. Additional losses are observed under radiation with longer wavelengths, which leads to thermal dissipation and performance deterioration. This causes the cell to heat up, thus reducing its efficiency and even shortening its lifetime.

The maximum efficiency of mono-crystalline silicon solar cells today is around 22-23% under STC, but the highest recorded recently in a lab environment was 24.7%. The efficiency of commercially available solar cells and modules is in the 16-18% range.

Solar silicon processing technology has many points in common with the microelectronics industry, and the benefits of the huge improvements in Si wafer processing technologies used in microelectronic applications have been successfully used to improve the performance of c-Si cells, giving this technology a significant advantage and making it the most favorable to manufacturers and customers alike.

We foresee mc-Si solar technology, and variations

thereof, leading solar energy markets during the 21st century, due to land restriction and higher efficiency and quality requirements.

Multi-crystalline Silicon

Multicrystalline, mc-silicon, or mc-Si is the most widely used silicon material today. It is most often (albeit incorrectly) called "poly," "polysilicon," or "polycrystalline" silicon (which is what we will call it in this text too) because it consists of many (poly) strings instead of one single crystal.

Poly is made by melting and casting silicon chunks into large blocks, splitting the blocks into smaller rectangular blocks and slicing these into thin, square-shaped wafers. These wafers are then processed into solar cells.

We looked more closely at the poly materials, processes and devices in the previous chapter and in even greater detail in our first book, *Photovoltaics for Commercial and Utilities Power Generation.*

The solar industry has invested 50 years into reducing costs and increasing production of silicon-based materials and devices, leading to the development of new crystallization techniques. Initially, poly-crystalline was the dominant solar industry, especially while the cost of Si material was very high. But today, even with a silicon price reduction to $50/kg, this technology is still very attractive, mostly because manufacturing cost is lower. This comes at the expense of an efficiency drop in these cells which are less efficient (in the 14-16% range) than mono-crystalline.

Poly-crystalline cells are manufactured by melting silicon and solidifying it in crucibles to orient the crystals in a fixed direction, with the end result of producing rectangular blocks of ingot of multi-crystalline silicon. These blocks are then sliced into smaller blocks and finally into thin wafers. This final wafer-slicing step, and the waste related to it, can be avoided by making thin ribbons of poly-crystalline silicon. See the section on "ribbon silicon" below.

Thus produced wafers are then processed into solar cells by methods similar to those used in the production of mc-Si solar cells.

Due to their ease of manufacturing, relatively high efficiency and reliability, mc-Si solar cells and modules will dominate the world's solar energy markets for the foreseeable future.

Short of an unforeseen and remarkable breakthrough, mc-Si cells and modules will be the most widely used solar energy generating technologies for use under any environmental conditions in residential and commercial applications during the 21st century.

Poly-crystalline Silicon Solar Cells

Poly crystalline silicon, also called poly silicon, or poly, or pc-Si is a thin film of silicon, deposited via CVD, or LPCVD processes on semiconductor type wafers, to be used as a gate material in MOSFET transistors and CMOS microchips. The solar industry uses similar equipment and processes to deposit very thin layers of silicon (pc-Si and a-Si) onto polysilicon or other substrates. The resulting devices are of lower efficiency, as compared to sc-Si or mc-Si.

Note: There is a confusion created by the term "poly" as it is used widely to identify PV cells modules made out of multi-crystalline silicon, instead of its actual use in the semiconductor thin film. Since we cannot change the decades-long use of the term "poly" in the solar industry to identify multi-crystalline silicon products, we will continue using it too with a certain degree of caution and with due clarification when needed.

No doubt, conventional c-Si solar technologies (as described above) will dominate the PV side of the solar energy markets in both residential and commercial applications—including large-scale power fields—for the foreseeable future. These are established and proven technologies, which although not perfect, are performing better than the rest under all environmental conditions. Yes, problems with materials and other issues will hinder the expansion of c-Si technologies, but their development, as far as reliability and cost are concerned, is continuing, and we expect serious improvements in both of these directions. Manufacturing methods are also intensely scrutinized for quality and cost-effectiveness, and this is reflected in recent improvements.

In conclusion, c-Si based PV technologies are here to stay during the 21st century. The question is what and how large a role in our energy future these technologies will play.

3D Solar Cells

These are solar cells that are arranged (externally or internally) in a 3-D fashion—a way that is different from that used today.

Note: As a clarification, "external" arrangement is the actual arrangement (mounting) of solar cells or modules in 3-D configurations and shapes; such as mounting on buildings, monuments and other structures.

The "internal" arrangement is modifying the actual surface of the solar cell or module in a 3-D configuration by, for example, creating shapes into it.

The intent of these designs is to capture most of the light that strikes the setup, to increase the efficiency of photovoltaic systems while reducing their size, weight, complexity and cost.

The new 3D solar cells capture photons in a number of ways, from a different physical arrangement of the cells themselves into arrays (external 3D photons capture), to creating miniature structures within the cells' bulk that resemble high-rise buildings in a city street grid (internal 3D photons capture).

Until now, solar cells and panels have been flat, literally. The flatter, the better, is the common thinking. Whether made out of silicon or thin films, the cells and panels mounted on a house or an industrial solar field, have always been one shape—flat. Over 50 years of complete flatness.

But the world's not flat, and there's no reason why our solar panels should be flat either. Here is where 3-D solar cells, panels and structures come into play.

The 3D cells can add power to some installations, allow for the use of less expensive materials, increase active photovoltaic area while keeping the device footprint constant, orthoganalize the light absorption and carrier extraction axis, solve the thick/thin conundrum, and/or increase the number of interactions between the solar flux and the device surface, thus optimizing output according to incoming sunlight.

This new thrust into three dimensionality fundamentally differs from traditional texturing of solar cells because traditional methods utilize a top down approach to inducing a third dimension (e.g., the use of etchants to form pyramids, or the etching away material by high intensity laser), but recent advances have been achieved via a bottom up approach by utilizing as-grown micro- or nanoscale features, such as silicon nanowires, ZnO nanowires, and carbon nanotubes.

The new flexible (i.e. CIGS based) PV modules offer new possibilities and facilitate the 3-D model by virtue of their easily moldable and changeable shapes.

The 3D technologies have continued to mature such that several US and international patents have been recently issued to cover these additive 3D approaches.

Please note that the following text represents a new, 3rd dimension (no pun intended) technology, in thinking about the function of solar cells and modules. It introduces a totally new concept of capturing incoming sunlight, which forces a different thinking, one that is literally 3 dimensional.

3-D Power Generation

The power per unit area generated by a planar solar cell without front or back texturing under direct solar insolation can be described by: (5)

$$\frac{Pmp}{A} = \eta_{2D} * I * \cos\psi \tag{1}$$

where:

Pmp is the maximum power generated,

A is the active area,

η_{2D} is the absorption efficiency (assuming all other efficiencies are unity),

I is the insolation in power per unit area on a surface orthogonal to the incoming solar flux, and

ψ is the zenith angle of the incoming solar flux.

Many types of devices, both more traditional top-down approaches and newer bottom-up approaches may be approximated by towers with vertical sidewalls and flat tops. For these types of 3D devices, the traditional term for efficiency, η_{2D}, may vary widely for different geometries and testing conditions. Therefore, a new efficiency for this topology, η_{3d}, was proposed.

This treatment for cell efficiency assumes that a model cell is infinite in extent and therefore, all photons, except those traveling parallel to the cell surface, will impinge the surface. It is also assumed that all surfaces are non-rough and that specular reflection occurs at each impingement.

Finally, it is assumed that at each impingement a photon of light is either reflected or absorbed and that the characteristic dimensions of the topological features are significantly larger than both the absorption distance of the material and the wavelength of the impinging photon, so that transmission and refraction through and around features may be ignored.

$$\eta_{3d} = F_o [1 - \eta_{2d} - (1 - \eta_{2d})G'] + \eta_{2d} \tag{2}$$

where

F_o is the area fraction of the cell surface not obstructed by towers,

η_{2d} is the absorbance efficiency of a planar cell made of the same materials, and

G is the average number of interactions between a photon and the tower sidewalls.

The value of G (which should be increased to maximize Pmp) is dependent on the geometry of the system, incident angle of sunlight, and aspect ratio of the towers. Previously, the value of G was found using the Monte Carlo simulations. It was found to vary linearly with towers' height and to the square of the open area fraction.

Derivation of the Value of G

The sun can be approximated as a point source in spherical coordinates having a position determined by the radial vector, δ, the azimuthal angle, ψ, and the zenith angle, ψ. A photon which is emitted from the sun and travels towards the Earth, located at the origin of the coordinate system, has a velocity in vacuum, v, which is collinear with $-\delta$. This velocity can be broken down into x-, y-, and z-component velocities by

$$Vx = -\delta \cos(\omega)\sin(\psi)$$
$$Vy = -\delta \sin(\omega)\sin(\psi)$$
$$Vz = -\delta \cos(\omega) \qquad (3)$$

The maximum dwell time of a photon in the 3D tower system is determined by the velocity in the direction co-linear with the axial direction of the towers, which is denoted as the z-direction. In this direction, a photon enters a space between towers at time zero and travels in the negative z-direction towards the floor.

After a time necessary to traverse a z-distance equal to the height of the towers, h, the photon then reflects off the floor and travels upward. Again after a time necessary to traverse a z-distance equal to h, the photon exits the system. From the initial time zero when the photon enters the system, to the final time (t_f) when the photon exits the system, a z-distance of 2h has been traveled.

$$t_{max} = \frac{z}{v_z} = [\frac{2h}{\delta \cos(\psi)}] \qquad (4)$$

Solving t_{max} in the z-direction makes it possible to derive the x- and y-distances traveled in that time from (eq. 3).

$$X = \{Vx\}t_{max} = \{-\delta \cos(\omega)\sin(\psi)\} * \frac{-2h}{-\delta \cos(\psi)}\}$$

$$= [2h \tan \psi \cos \omega]$$

$$y = \{Vx\}t_{max} = \{2h \tan \psi \cos \omega\} \qquad (5)$$

The value of G_x can be derived by considering a 3D system which is composed of a perfectly reflecting material. The simplest geometry for such a 3D cell would be a single trench, infinite in the y-direction and bounded in the x-direction by two walls of height h and separated by a distance d. A photon is considered which enters the system with a random y- and x-position located x_0

from the rightmost wall and velocities in the x- and y-directions determined by the azimuthal angle, v.

At time zero, the photon has traveled a distance of zero and the value of I'_x is equal to zero at that time. After a time period of t_1, the photon has traveled a distance equal to x_0 to impact the rightmost wall to give I'_x a value of one.

After a second time period, t_2, the photon has traveled a distance of $x_0 + d$ and impacted the leftmost wall to give G_x a value of 2. This reflecting between walls continues until a time of tn +1, which is the last interaction between a photon and a trench wall. At this point the photon has traveled in the x-direction a value of $x_0 +$ nd and reflected off a trench wall n +1 times. After this last interaction with the trench wall, the photon then travels some distance in the x-direction, x_f, before it exits the system at t_{max} to give a value of G_x equal to n + 1.

The total distance traveled by the photon in the x-direction after a time t_{max} is equal to $x_0 + nd + x_f$, which is also equal to (equation 5).

$$x_0 + nd + x_f = x = \{vx\}t = = \{2h \tan \psi \cos \omega\} \qquad (6)$$

Furthermore, since the starting position of the photon can be any value from 0 to d with no preference as to the starting position, as the number of particles

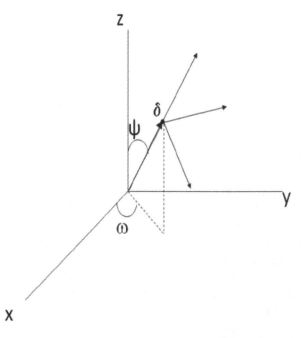

Figure 3-6. The sun (as a point source) can be described relative to the earth (origin) in a spherical coordinate system by the radial vector (δ), the azimuthal angle (ω), and the zenith angle (ψ). A photon traveling towards the earth has a velocity in the direction of $-\delta$ from equation 4. (5)

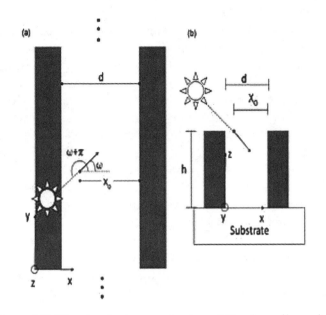

Figure 3-7. The simplest geometry for a 3D solar cell would be a single infinitely long trench, (a) shows this geometry compressed in the z-direction. The gray portions represent smooth vertical sidewalls. (b) shows this geometry compressed in the direction y. (5)

approaches infinity, the average value of x_0 will tend towards a value of $d/2$. The same logic applies to the average value of x_f.

We can then derive a value of G_x this x-trench, by solving (6) for $n + 1$.

$$n + 1 = \frac{2h \tan \psi \cos \omega}{d} = G_x \qquad (7)$$

The overall value of G is equal to $G_x + 1$ due to a reflection off the floor of the trench which is not inherently accounted for:

$$G_{x\text{-trench}} = G x+1 = \left\{ \frac{2h \tan \psi \cos \omega}{d} \right\} +1 \qquad (8)$$

The same logic applies to a y-trench, which is infinite in the x-direction to give

$$G_{y\text{-trench}} = Gy+1 = \left\{ \frac{2h \tan \psi \cos \omega}{d} \right\} +1 \qquad (9)$$

An x- and y-trench which overlap will form a box. In this instance the photon in a box acts as two decoupled oscillators in the x- and y-directions. Since the travel in the y-direction does not affect that in the

x-direction, the total number of reflections will be equal to the number of reflections due to the y-direction plus those due to the x-direction.

$$G_{box} = G_x + Gy+1$$

$$G_{box} = \left\{ \frac{2h \tan \psi}{d} * \cos \omega \right\} + \left\{ \frac{2h \tan \psi}{d} * \sin \omega \right\} + 1 \qquad (10)$$

In practical terms, a team of researchers at UC Berkeley optimized structures with a bounding-volume of area footprint (base area) 10 x 10 m² and height ranging from 2 to 10 m.

Figure 3-8 shows the energy generated in a day as a function of the height of the GA (genetic algorithm)-optimized 3DPV solar cell, compared to that of a flat panel of the same area footprint.

The energy generated by the 3D structures scales linearly with height, thus leading to "volumetric" energy conversion. In addition, the power generated as a function of time during the day (inset, Figure 3-8) shows a much more even distribution for 3DPV, due to the availability of cells with different orientations within the structure.

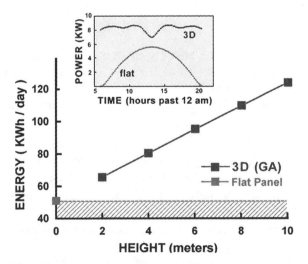

Figure 3-8. Plot of the energy produced in a day by GA optimized 3D PV structures compared to that of a flat panel in the same conditions. The inset shows the power generated during the day for the flat panel compared to the 3DPV at height = 10 m.

The increase in power with height is dominant in the early morning and late afternoon, as expected, although the enhancement is broad in time and remains significant at all times during the day, even at midday. This even supply of power throughout the day can be "built-in" to a 3D structure, in contrast to power generated by a flat panel, which, without dual-axis tracking, decays rapidly around peak-time.

Interestingly, all the GA structures show similar patterns in their shapes, even for different heights. They contain no holes running across the bounding volume, which is necessary to intercept most of the incoming sunlight, and—less intuitively—they all have triangles coinciding with the 12 edges of the bounding box volume, so that they would cast the same shadow on the ground as the open-box. We emphasize that these patterns emerge from randomly generated structures, are not artifacts of the simulations, and are a fingerprint of emergent behavior resulting from the GA calculations.

The primary shape of the GA structure in Figure 3-9 (a) is a box with its five visible faces caved in toward the midpoint.

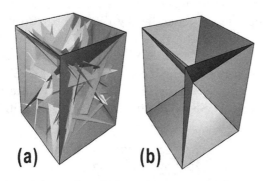

Figure 3-9. Schematics of 3DPV structures: (a) GA-optimized structure shown with all 64 triangles inside the bounding box; (b) funnel, a simplified version of most GA-optimized structures that retains their superior performance over other shapes. (8)

A simplified, symmetric version of this was constructed, as shown in Figure 3-9 (b); this idealized structure, which we refer to as the "funnel," generates only 0.03% less energy in the day than the original GA output, and therefore contains most key ingredients of the complicated GA structures.

We compared the energy generated by simple open-box shapes and the funnel structures through a figure of merit M, defined as the ratio of the energy produced in a day to the total area of active material used, and scaled to one for the flat panel case. The energy of the funnel shape outperforms the open-box at all heights, and while both structures generate more energy than the flat panel case, they use excess material for a given energy, i.e., M<1.

For example, for a height of 10 m the open-box shape generates approximately 2.38 times as much energy as the flat panel but requires 9 times as much active

material (M = 0.26).

The figure of merit for the open box decreases with height indicating that such a shape is not ideal for 3DPV in terms of efficient materials' use. On the other hand, the GA derived funnel shapes maintain a nearly constant figure of merit over this height range, with a cross-over to superior materials performance compared to the open-box at a height of ~5 m, and 30% higher M at 10 m.

The increase in produced energy of the best-performing GA structures is due to a decrease in the total power reflected to the environment and an increase in power generated using light reflected from other cells.

Despite the relatively small increase in energy generation of the GA shapes compared to the open box, these structures shed light on some fascinating aspects of 3DPV and may give significant practical advantages.

In summary, 2-D solar cells generate power by virtue of the energy of the photons impinging on their surface, where each photon has one single chance to generate one single electron. When the photon strikes the solar cell surface, it either bounces off due to the reflectivity of the surface, or is absorbed. The latter process might generate an electron, if the photon hits the right place with the right amount of energy. Or it might not, if it does not meet the requirements, which means that the photon was wasted.

3-D solar cells have the advantage to offer a second or third chance to the bouncing photons to generate an electron. That is, when the electron bounces off of one surface (where it might have, or might not have generated an electron), it is then directed to another surface, to a third one and so on, thus having many more chances to generate an electron, or a number of electrons under special conditions.

This phenomena allows 3-D solar cells to generate more power under most weather conditions. And although this might be true for operation under cloudy sky, we believe that it is even more true for operation under intense direct solar radiation, such as desert sunlight, where the photons have excess energy and would be able to generate more than one electron at the different surfaces they hit.

Although advantageous in many ways, due to the more complex nature of these solar cells and approaches, it still remains to be seen if this technology will reach the reliability, efficiency and cost-effectiveness of the conventional technologies, and if it will have a significant impact on the world's energy markets of the 21st century.

Types of 3-D Solar Cells

According to their design and structure 3-D solar cells can be divided into several categories. For now we'd consider cells with a.) external and b.) internal 3-D photon capture capabilities:

External Capture 3-D Arrays

The external method for 3-D capture of photons and conversion into electricity is quite simple. It consists of arranging solar cells onto 3-D surfaces, such as monuments, building fixtures, etc., and tuning their reflectivity and absorption properties to capture as much light as possible—regardless of the sunlight direction and intensity.

Inspired by the way trees spread their leaves to capture sunlight, researchers are looking into what will happen if the flatness of a solar collector is changed into a three-dimensional shape. Just imagine a large tree covered with solar cells. How efficient and practical would that be? Well, it depends on many factors, but it is surely a new way of thinking and generating solar energy.

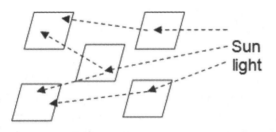

Figure 3-10. External capture by cell arrangement

3-D Surface Capture

A variation of the first external method of 3-D photon capture is building solar cells in a 3-D shape. This won't be that hard using some of today's thin film deposition technologies. It is just a matter of calculating, designing and manufacturing the proper size and shape of the shapes as well as the respective coating processes of the photo-electric thin films.

A non-flat solar cell with some sort of 3-D shape will have several benefits, such as larger surface area and more chances to bounce off and capture photons. As a matter of fact, the major benefit of 3D type solar cells is their ability to capture photons quite efficiently from different angles and intensity, thus making it quite useful, especially under conditions of diffused (cloud cover) weather.

The preliminary results are encouraging. It seems now that this variation of not-so-flat solar cells has the potential to be quite efficient under normal conditions.

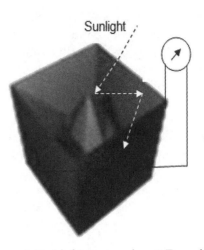

**Figure 3-11. Light capture in a 3-D surface
Note the etched pyramids in the surface.**

Their unique 3-D shape allows them to pick up more direct sunlight and generate more electricity than flat cells and panels using the same amount of ground space. But remarkably, and more importantly, this model is efficient even on overcast, rainy days.

Cloudy skies are the enemy of solar power, but experimental 3-D solar cells and panels can pick up almost as much electricity on a cloudy day as when it's sunny, according to the specialists. This is hard to believe since the sunlight intensity is significantly reduced under cloudy skies, so no matter how efficient the cells are, they cannot produce more energy than what the sunlight allows, which is not much.

The conversion efficiency is created by the dynamic collector shapes in case of the 3-D arrays, which just like tree leaves gobble every photon coming their way, and as is the case of the latter design of 3-D solar cells. Unfortunately, cloudy days do not generate many photons—not even close to full unobstructed sunshine in the desert—so the efficiency is limited by the amount of sunlight falling on the cells.

In all cases, upon its full development this technology must find its own niche market—somewhere between the sunny deserts and totally cloudy skies. The researchers are hoping for the day when 3-D solar panels and structures would be placed all around us, to cover otherwise wasted surfaces, such as statues, monuments, city park features. These would be works of art, instead of simple ugly flat panels hidden on a roof.

A 3-D external capture solar structure using external capture 3-D solar cells would be even more efficient, due to the unique shape of the structure and the cells within it. The concept of three-dimensional photovoltaic cells, modules, arrays and structures can be proven computationally by using a genetic algorithm to optimize the energy produced under certain conditions and time frames by the 3-D shaped solar cells, framed into a given area and volume. Some of this work is underway in several research organizations.

Internal Capture 3-D Solar Cells

The second type of external 3-D capture of photons device is the most complex of the group, because it requires building a totally different solar cell with some complex, miniature, and hard-to-manufacture features.

These 3D solar cells with external photon capture use an array of miniature thin film "tower" structures that resemble high-rise buildings in a city street grid, deposited on the substrate material.

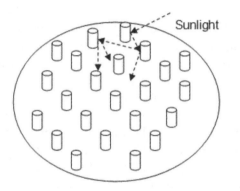

Figure 3-12. Internal capture via thin film towers

This type of photovoltaic cells trap light between their tower structures, which are about 100-200 microns tall, forming arrays of millions of vertically aligned carbon nanotubes.

Manufacturing of these cells consists of preparing a silicon wafer, which serves as the cell's substrate and bottom junction. The wafers are then coated with a thin layer of iron using a photolithography process with the appropriate pattern, from which the foundations of the nanotubes are created via plasma or wet chem etch processes.

The patterned wafers are heated in a high temperature CVD furnace, where in the presence hydrocarbon gases, arrays of multi-walled carbon nanotubes are grown on top of the previously formed iron pattern foundations.

This is followed by molecular beam epitaxy (MBE) which coats the nanotubes with cadmium telluride (CdTe) and cadmium sulfide (CdS) which are the active films, and which will serve as the p-type and n-type photovoltaic layers in the newly manufactured cells.

Then another thin coating of indium tin oxide (ITO), a clear conducting material, is added to serve as the cell's top electrode, while the carbon nanotube arrays serve as both a support for the 3D arrays and as a conductor connecting the photovoltaic materials to the silicon wafer.

Cadmium materials are used in some cases, but a broad range of other photovoltaic materials are readily available and could also be used, which is the goal of future research efforts in addition to finalizing the shape, height and spacing between the towers, as well as determining the optimum angle at which the light hits the structures.

The tower structures can trap and absorb light received from many different angles, so the new cells are efficient even when the sun is not directly overhead, and even in cloudy days with predominantly diffused solar radiation.

This kind of 3D cells absorb many more photons than conventional cells, due to the larger surface area and interaction between the nanotubes. Also, their thin film coatings can be made thinner, which allows the electrons to exit more quickly, thus reducing the likelihood that recombination will take place, which increases their overall "quantum efficiency," or the rate at which absorbed photons are converted to electrons.

There are, however, several serious hurdles before these devices can be commercially produced, in addition to resolving some of the complexities of the manufacturing process. This includes tests of their reliability in long-term commercial applications, especially under adverse climatic conditions, as well as their survival of the launch and operation in space.

And then, mass production procedures and techniques must be developed to replace the current small-scale lab processes and prototype production methods.

A variation of the internal capture solar cells could be produced by etching different geometric patterns into the surface of different types of solar cells.

A great advantage of the 3-D type solar cells is the fact that they could be used on house roofs or on board spacecraft "as is," without the mechanical tracking systems required to maintain a constant orientation to the sun, as is presently required by conventional PV and CPV devices. This is because the 3-D structure has the ability to capture sunlight from different angles and under different weather conditions. The absence of tracking devices will reduce the weight and complexity of the 3-D solar power generating systems, and will improve their reliability.

On a broader scale, these types of cells could also

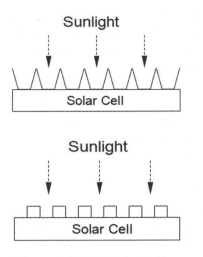

Figure 3-13. 3-D Solar Cells

find near-term applications for enabling efficiency improvements in photovoltaic coating materials, which could also change the way solar cells are designed, manufactured and used in a broad range of applications.

Internal Capture 3-D Micro-cells

Traditional solar cells are 2-dimensional structures, utilizing a single pass sunlight conversion mechanism. The above mentioned 3-D devices with external collection of photons are somewhat more efficient in this respect than their 2-D cousins, but still not as efficient as the internal collection devices we review herein.

There are several ways that the conventional 2-D solar energy conversion devices fail to collect most of the sunlight impinging on them, and waste electrons (electron-hole pairs), which result in a much lower than theoretically possible power conversion efficiency.

The key reasons for these losses are:

1. **Surface Reflection**—Due to fundamental physics, approximately 30% of incident sunlight is reflected off the surface of silicon cells.

2. **Electron Re-absorption**—When a photon strikes the solar cell, an electron is "knocked loose" creating an electron-hole pair that moves through the cell material creating an electrical current. In conventional 2-dimensional solar cell designs, these electron-hole pairs must travel a long distance before reaching a metal contact wire. As a result, they are reabsorbed by the material and do not contribute to the production of electrical current.

The answer to these problems is a variation of the above described method for internal 3-D capture of pho-

tons, including light trapping and electron extraction by means of truly 3-dimensional solar cells, where the photon trapping is done internally (the capture process takes place in their interior).

These cells are designed from the ground up as an integrated opto-electrical device that reduces losses to achieve the highest theoretical efficiency possible by a photo-electric device. By leveraging the scalability of conventional solar and semiconductor processes, these types of 3D solar cells can deliver an unprecedented level of conversion efficiency and possibly reduced cost as well.

In conventional solar cells sunlight passes through the entire cell structure one time, while 3-D solar cells use a mix of 3-D cells and micro-cells that trap sunlight inside photovoltaic structures where photons bounce around and among the cells until most of them are converted into electricity.

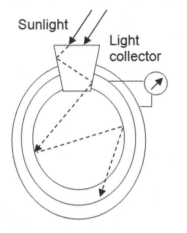

Figure 3-14. Internal capture 3-D solar cells

The key features and benefits of 3-D solar cell design, which determine their practical use, are:

1. Efficient light collection, where instead of reflecting sunlight off the surface, a special array of properly designed light collecting elements is used to guide the incident sunlight photons into an array of highly optimized 3-D PV and micro-PV structures.

2. Specially designed 3-D structure, in contrast to conventional solar cells which have one photon absorbing capacity, use multi-faceted structures, where photons can bounce off many surfaces until all photons that can be absorbed by the material are absorbed.

3. Thin absorbing regions allow PV structures to be fabricated with very thin absorbing regions, designed to enhance charge carrier separation. This will allow electron-hole pairs to travel the shortest possible distances before reaching an outside circuit wire and being extracted to produce current. This approach allows height and material reduction, as compared to conventional crystalline silicon cells.

4. Below surface contacts, which unlike conventional solar cells where electrical contact wires run on the top of the cell, blocking sunlight, are used in a network of contact wires that run below the light collectors. This approach allows 3-D solar cells to trap and utilize nearly 100% of the incident light.

3-D cells and modules as a concept are unbeatable. Their practical implementation in terrestrial or space applications, however, is yet to be proven. We do have great hope in their future as a niche market for some special applications, so we will keep our eyes open for future developments in this area. We do believe that it will be one of the leaders in the energy markets of the 21st century.

Sliver Si Cells

Another version of 3-D cells arrangement is the so-called sliver (note SLIVER, or strip—not SILVER). This is a totally new concept of making solar cells and modules that utilize 3-D principles to increase the surface area of each cell, respectively increasing the conversion efficiency and total power output of the module.

This is done by arranging slivers (strips) of substrate material perpendicular to the module surface and making the entire cell surface area active and ready to generate electrons upon impact by sunlight photons.

The cells could be arranged into a module with a transparent front glass and reflecting bottom surface.

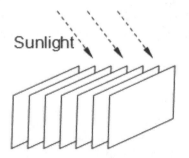

Figure 3-15. Sliver cells arrangement in the module

This way the electrons impact, bouncing between the sides of the slivers where they generate photocurrent. Some reach the bottom surface and bounce back to try again to knock some electrons out of their hiding places.

The active surface area of the new type of module is several times larger than a conventional solar module, and the rear mirror surface helps collect photons which would be wasted otherwise. Both of these improvements contribute significantly to the increase of power output.

The manufacturing process consists of micromachining, or etching, the thin slivers from a silicon wafer (and as wide as the wafer itself), then processing the batch as in conventional solar-cell processes, via diffusion, metallization, AR coating, and other steps.

The final result, according to researchers, is obtaining solar cells and modules that perform at least as well as the same area conventional solar cells and modules, but at 1/10 of material consumption.

This technology could be successfully deployed in niche markets, such as installation on windows, where the slivers would act as electricity-generating blinds on residences and commercial buildings.

The manufacturing process, however, seems too complex and expensive, but with more work and ingenuity it might be simplified and made cheaper. Provided that the power output per surface area is compatible, then higher labor cost might be justifiable.

This technology would be very efficient in some areas, and could be successfully used in applications such as window blinds and other areas that require transparency. Only the manufacturing complexity and higher cost of the finished product separate it from full implementation in 21st century energy markets.

Note: The high costs of silicon used in photovoltaics dictate the price of conventional solar panels. Several new approaches are under consideration now, promising great results with a reduced amount of silicon. The idea is to change the shape and size of the active area of the solar wafers (and the resulting solar cells), or make them much thinner than with conventional technologies.

One such approach, as described above is to make solar cells that use half the silicon in each solar wafer, by slicing the wafer into strips and using only half of these, while still generating 80-90% of the conventional solar cell's power.

This approach would save money because the total cost of the most expensive material—silicon—is cut almost in half, and because silicon is the major expense, the entire device could be cheaper even after adding all other materials and additional manufacturing steps.

The cost can be reduced further by using manufacturing equipment already developed for the semiconductor industry, but this remains to be seen, just as the performance of these types of cells is yet to be proven.

Also, in conventional solar cells, wires for collecting current are placed on top of the cell, where they block some of the incoming sunlight. Space and material can be saved if the wires are placed between the strips of silicon, where they block no light. The wires also don't need to be very thin (as needed to avoid blocking incident light), thus they can be sized accordingly which might improve the collection of electrons and add to the efficient performance of solar cells.

So far so good, but we see some problems here:

1. Slicing the strips from the whole solar cell and arranging them into a new configuration is a delicate operation, which would create waste, contaminate the cell's active surface area, and introduce additional edge damage, negatively affecting the cell's performance and adding expense

2. Depositing the slivers onto a substrate, or etching them from bulk silicon would be too expensive and time consuming a process.

3. Directing light from the areas between the silicon strips onto their surface requires the use of special optics, such as Fresnel lenses, which would be quite expensive. These optics are usually made out of plastic materials, which would lose their transparency and fail prematurely under the blistering desert sun.

4. Proper redirection (focusing) of the sunlight from the in-between areas onto the adjacent silicon strips would also require tracking the sun. Without tracking the redirection would be marginal at best and would waste a large part of the active area, resulting in reduced efficiency. That is, if the cells are fixed mounted, at noon (or soon before or after) they'll be parallel to the sun and very little power would be generated at the best production period.

5. Placing metal wires between the silicon strips might be a good idea, but attaching them securely to the strips is a questionable undertaking.

2012 Update

During the summer of 2012, industry insiders claimed that 3-D solar cells could have delivered $6.0 billion of electricity on top of that generated throughout the world in 2011—IF all installations were made of 3-D solar cells. The analysis is based on the stipulation that a typical 16-17% efficient solar cell is quite inefficient at times, and delivers half or less of the normal output when the sunlight is obstructed by clouds and/or when it is coming at a sharp angle, such as in early morning and late evening.

3-D cells, in contrast, can maintain their high efficiency for a much longer period of time, due to the number of surfaces oriented in different directions, both during cloudy periods and when sunlight arrives at a sharp angle. Over the life-time operation of the devices, this could translate into 100-200% more power than generated by the conventional solar cells, say the estimates.

The system payback period of the 3-D cells, therefore, could be half as short as conventional solar technologies.

All this looks good on paper, and we have no doubt that 3-D cells have added benefits, so making calculations on paper is a good start. However, we see a number of serious complications in making, operating and maintaining 3-D installations of this technology. To begin with, we'd argue that

1. Manufacturing and installation of 3-D cells would be more expensive per active area than any of the conventional PV technologies, and

2. When comparing the total energy generated by equal size active surface areas, 3-D cells cannot produce that much more energy simply because, due to the 3-D configuration, not all cells (or active areas of the cell) are illuminated all the time. This will result in output penalties, depending on the shape of the 3-D device or array, and the angle of the sunlight.

This is also only a supposition, which we are allowed to make because the technology is in its infancy and has no proven record as yet. When the 3-D technology gets more exposure in day-to-day field operations and tests, and puts several thousand sun hours under its belt, the calculations might show different results. Until then, we can only speculate and make estimates.

Nevertheless, we should not be surprised if 3-D cells do show very good results under certain operating conditions and take over some niche markets. We will be looking forward to seeing the field test data and the progress of this technology in the near future.

Emitter Wrap-though Cells

Emitter wrap-through (EWT) cells are unique and very specialized devices, processed by segments of the conventional technologies with the purpose of reducing optical losses from the front surface by eliminating front metal contacts from the front surface. This is achieved by wrapping the metal contacts on the underside of the solar cells.

This new approach has allowed increased efficiency through better cell design, rather than through material savings or process improvements.

Instead of the bulky top metal contacts fused into the front surface (as in conventional solar cells) small, laser-drilled holes in the top surface are used to connect the rear n-type contact with the opposite side emitter.

This approach eliminates the wasted front surface occupied by the top metal contacts in conventional solar cells. In addition, this approach eliminates the problems related to the metal contact's inefficiency and electromechanical failures.

The removal of the top contacts allows the full surface area of the cell to absorb solar radiation because masking by the metal lines is no longer present. Tests show that there are manufacturing gains by putting the contacts on the backs of the cell, thus obtaining a 15-20% increase in overall efficiency.

One suggested manufacturing process includes the deposition of a phosphorous-doped SiOx-layer by means of PECVD with tetramethylcyclotetrasiloxane (TMCTS) and trimethylphosphite (TMPi) as precursor fluids. In the same vacuum chamber, a capping layer of non-doped dioxide based on TMCTS is deposited to protect the strongly hygroscopic layer from ambient influence and from the phosphorus oxychloride atmosphere in the tube furnace.

During the high temperature step, phosphorus is expected to diffuse from the doped silicon dioxide layer into the Si-crystal on the front side, forming a moderately doped emitter independently of the heavy diffusion. To achieve a lower sheet resistance on the front side than on the rear side and in the via-holes, either the layer thickness of the doped silicon dioxide layers or the phosphorus content can be tailored to obtain "finite" dopant sources.

Another process has been described featuring the deposition of a thin PECVD deposited SiOx-layer using silane as precursor gas. The heavy diffusion is mitigated by the silicon dioxide layer, likewise resulting in a shallow diffusion on the front side and a heavy diffusion in the via-holes and on the rear side.

In all cases the EWT cells formation of the via-hole emitter must be unaffected by the PECVD layers deposited on the front side. The via-hole emitter can mitigate or, in case of the capping layer of the previous process, the formation of the via-hole emitter can even be hindered.

For both process sequences an appropriate diversification of the sheet resistances on front and rear surfaces could be reached, resulting in a sheet resistance of around 70 ohm/sq. on the front surface and 35 ohm/sq. on the rear surface.

This type of solar cell can be manufactured, with some modifications, by the conventional PV techniques. It has selective emitter structure, fabricated in a single high-temperature step, with a highly doped emitter at the via-holes and the rear side, allowing for a low via-hole resistivity as well as a low resistivity contact to screen-printed pastes, and a moderately doped front side emitter exhibiting high quantum efficiency in the low wavelength range.

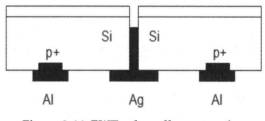

Figure 3-16. EWT solar cell cross section

A major disadvantage of this technology is most evident on large area EWT cells, where the devices exhibit high series resistance, which limits the fill factor and reduces the overall device efficiency. Of course, due to the complexity of the manufacturing process, the cost of EWT solar cells made via any process sequence would be much higher than conventional solar technologies.

Although this technology shows promise with over 18% efficiency, we don't expect it to take over the energy markets anytime soon.

Some elements of its development, however, would be useful in improving operations characteristics of other technologies in the race for higher efficiency and reliable performance in the 21st century.

HIT Solar Cells

HIT cells are a new type of solar cell developed by several manufacturers and based on well-established a-Si thin film technologies for forming high-quality a-Si:H films and a-Si:H solar cells with low-level plasma and low-thermal-damage processes.

HIT solar cells have a structure in which an intrinsic a-Si:H layer is introduced on the silicon surface, followed by a p-type a-Si:H layer, deposited on a randomly textured n-type CZ c-Si wafer, thus forming a p/n heterojunction.

On the back side of the c-Si wafer, intrinsic and n-type a-Si:H layers are deposited to obtain a back surface field (BSF) structure.

Transparent conducting oxide (TCO) layers are formed on both sides of the doped a-Si:H layers, and metal grid electrodes are deposited via screen-printing methods.

The high-quality intrinsic a-Si:H layer mitigates the defects on the c-Si surface by passivating it, thus obtaining higher than normal Voc.

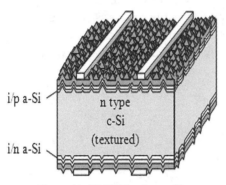

Figure 3-17. HIT solar cell.

Improved manufacturing processes, such as advanced surface cleaning and texturing techniques, as well as use of much more sophisticated thin film deposition equipment and techniques do contribute to the increased efficiency, performance stability, and durability of the new technology.

The HIT structure provides high performance, and conversion efficiencies of 21.8% (Voc: 0.718 V, Isc: 3.852 A, FF: 0.790) were reported previously. Presently, higher efficiency HIT cells are made with efficiency well over 22% and increasing.

In addition, and very importantly, HIT solar cells have much better temperature coefficient compared to conventional p/n junction c-Si solar cells. Therefore, more power is generated by HIT solar cells in actual use, and especially under extreme heat conditions, than the conventional c-Si solar cells.

As can be seen in Figure 3-18, HIT solar cells, which are made with the help of conventional a-Si thin film deposition techniques, are processed at much lower temperatures, and with much shorter process times than most of the competing c-Si technologies. These factors contribute to the performance characteristics, durability,

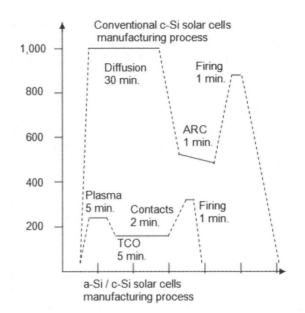

Figure 3-18. Comparing process times of a-Si (as used in HIT mfg. process) and conventional c-Si technologies (not to scale)

and cost of the HIT cells.

The HIT solar cells can also be made much thinner than conventional competitors (less than 100 microns, vs. 200-250 microns for regular c-Si cells), thus saving expensive material. Nevertheless, more expensive materials and processes are used for manufacturing HIT cells—CZ silicon and advanced thin film deposition equipment methods—all of which contribute to the increased cost of these solar cells. Also, the advantage of using a thinner wafer is insignificant when considering the higher breakage and yield loss.

We have seen several installations, successfully using HIT solar cells and modules recently, so we do predict a bright future for the HIT technology in the 21st century.

LDSE Solar Cells

Another new solar cell based on CZ type silicon, which a number of manufacturers are involved in developing and manufacturing, is using the new generation Laser Doped Selective Emitter (LDSE) technology.

These new solar cells, which are built on 6" p-type CZ wafers, have achieved a 20.03% efficiency rate, as verified recently by a world certification lab. This beats the statistics obtained in 2010—19.6% efficiency for LDSE technology.

The new LDSE solar cells are made by proprietary methods, using existing screen printing lines, reducing the cost required for cell processing. Combined with optimized manufacturing techniques used for mass production, the new technology promises to be a strong competitor in the global solar market.

CZ silicon wafers are also mechanically stronger, and have somewhat more stable operating characteristics. This will guarantee higher quality and yield, and will extend the life of the LDSE cells.

However, it remains to be seen if these cells can be successfully mass produced. The 6-inch-diameter LDSE solar cells are larger than the conventional 4- and 5-inch wafers, which will require a new approach to arranging them in PV modules. The gaps between the wafers would be too large if used as is, while cutting such large wafers in squares would generate a lot of expensive scrap, so that might be another challenge to be overcome before successful implementation can be expected. We do see the higher cost of the CZ silicon as the most significant barrier manufacturers must overcome before putting this promising technology in mass quantity on world energy markets.

Still, higher efficiency and more stable operation are extremely important factors in some applications (such as lack of land), which must be taken into consideration when predicting the future of the new solar technologies.

So, we'll keep our fingers crossed for the success of the new LDSE PV technology and will patiently await its introduction into the world energy markets of the 21st century.

Microcrystalline Silicon Solar Cells

Microcrystalline silicon, also called nano-crystalline silicon, nc-Si, is a form of silicon in its allotropic form, very similar to a-Si, which also contains small crystals. It absorbs a broader spectrum of light and is flexible, which is beneficial for some practical applications.

Because of this, and because of its better optical absorption than crystalline silicon, this type of silicon has received great attention during recent years.

nc-Si has fairly small grains—much smaller than the conventional polysilicon—or crystalline silicon within the amorphous phase. It is well proven that the grain boundaries (those of the small and large grains) are important parts of the microstructure of nc-Si and play a great role in its performance.

If grown properly, this material offers higher electron mobility, due to the presence of the small silicon crystallites. As an added advantage, and something that has practical use, it also shows increased absorption in the red and infrared wavelengths.

Use of microcrystalline silicon alone as an active absorber layer in solar cells was not seriously considered until recently. Hydrogenated microcrystalline silicon (nc-Si:H) was originally introduced some time ago, and is now generally obtained by a PECVD process using a mixture of silane and hydrogen.

Due to its efficient doping properties, both for n-type as well as p-type material, nc-Si:H is used as ohmic contact layers in solar cells and in thin-film transistors. nc-Si is a two-phase material in which crystalline regions are embedded in an amorphous matrix. Also, unlike amorphous silicon, it has stable transport properties that are not influenced much by light-induced degradation.

Microcrystalline silicon is much more sensitive to the impurities in its structure, however, so passivation of the impurity is essential for better performance. A possible way to achieve this is to employ low-temperature processing for the intrinsic layer (T < 180°C).

It is also found that low-temperature processing is effective to avoid damage at the p/i interface due to the defect creation accompanying the impurity diffusion into the active area.

Deterioration of the device performance at higher deposition rate can be avoided under higher gaseous pressure in the plasma CVD process. The origin of the deterioration is viewed in terms of the structural and electronic properties.

Thin film microcrystalline silicon (nc-Si:H), prepared by plasma-enhanced chemical vapor deposition (CVD), is a promising material for use as the active absorber layer in low cost solar cells for which efficiencies higher than 8% have been obtained on flexible substrates and over 10% on glass substrates.

The small crystalline structure, however, might be responsible for the relatively lower efficiency of solar cell made out of these materials.

Recently, three different approaches have shown experimentally that microcrystalline silicon, if suitably

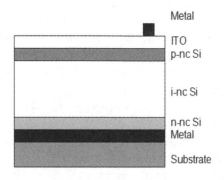

Figure 3-19. nc-Si structure

deposited, can be an active absorber material for photovoltaic solar cells:

1. Very high frequency glow discharge (VHF-GD) technique is most favorable for deposition of hydrogenated microcrystalline silicon at low substrate temperatures, and is thus used for the deposition of entirely microcrystalline cells in p-i-n, and inverted n-i-p structures, achieving 8-9% efficiency.

2. Another successful approach is achieved by using PECVD process at much higher temperatures, obtaining efficiencies of over 10%.

3. Catalytic CVD for nc-Si:H solar cell deposition, or "hot wire" technique has been tried as well, but its success as far as efficiency of thus obtained solar cells is limited and more work is needed.

The open-circuit voltage (Voc) of micro-crystalline silicon solar cells usually decreases with increasing volume fraction of crystallinity Fc > 60%.

One explanation is based partly on broader band tails and higher defect density when Fc is high, since structural relaxation of the μc-Si:H network and passivation of grain boundary defects cannot properly take place due to the low amorphous silicon content in this well-crystallized material.

Another important reason for the decrease in Voc is the assumption of a lower band gap for the more crystallized material—all contributing to lower efficiency.

Higher efficiencies are obtained when used in tandem solar cells, based on a stack of a-Si:H and nc-Si:H cells, as discussed elsewhere in this text.

The use of rough substrates has been demonstrated to increase light trapping efficiency and overall solar cell efficiency. However, the presence of elongated porous regions (cracks and micro-pores) between adjacent agglomerations of nano-crystals is thought to decrease electrical performance, due to shunts between the top and bottom contacts.

This phenomena decreases the open-circuit voltage and the fill-factor of the microcrystalline silicon solar cell.

As-deposited intrinsic microcrystalline silicon tends in general to have an n-type character. The origin of this n-type character was attributed either to the large amounts of defects, and/or to the large amount of oxygen incorporated in these films. By adding small quantities of boron it is possible to compensate this n-type character, and to push the Fermi level to midgap achieving efficiencies of ~5% in the single-junction p-i-n solar cell.

This technique is called "microdoping" and it is extremely delicate in its reproducibility with respect to obtaining the desired midgap character of the material. Adding impurities (boron) improves the efficiency of the solar cell. One can therefore conclude that the midgap character of the base material is of primary importance rather than the concentration of (certain) impurities.

The morphology of the films in terms of crystalline orientation could be modified in a wide range, by varying of deposition parameters.

Single junction microcrystalline-silicon solar cells are stable under elevated light-soaking conditions. For example, amorphous silicon solar cells degrade over 50% under sodium light of 6-sun intensity, while nc-Si:H cells remains completely stable even under 10 suns.

This is one of the most important aspects of nc-Si:H and nc-Si:H-based cells. This is also important for the tandem uc-Si and a-Si solar cells described in this text.

Individual nc-Si:H thin films may show a pronounced post-oxidation as a function of the deposition conditions, which gives rise to an increase in the lateral dark conductivity, which also has not been proven. It can be assumed that the doped layers in the solar cell protect the intrinsic layer of the cell from such post-oxidation.

For example, after one year of field operation, nc-Si:H cells show little or no reduction in cell performance that can be measured, which is very important when considering long-term exposure of solar cells and modules in adverse weather conditions.

Some of the shortcomings of the individual nc-Si behavior can be overcome by alloying it with germanium (Ge). This is done via high-temperature PECVD (RF plasma) process, in the presence of SiH4, GeH4 and H2 gasses. The resulting devices require thinner thin film layers, which is significant from a production and cost-effectiveness point of view. In addition, the reliability and efficiency of these devices is increased, but more work is needed to tune the process and transfer it into production.

Micromorphous Silicon

A variation of the nc-Si technology is the so-called micromorphous silicon technology, which combines two different types of silicon, amorphous and microcrystalline. These are stacked in a top and bottom configuration to form a photovoltaic cell.

Several companies produce these types of cells to

more efficiently capture blue light, thus increasing the efficiency of the cells during cloudy days or in locations with no direct sunlight.

Another variation of this material is the so-called protocrystalline silicon, which could be used to optimize the open circuit voltage of a-Si photovoltaics. This is, however, a more complex technology, which needs more work before being deployed in world' markets.

While microcrystalline silicon, and all its variations, have yet to achieve the efficiency and applicability of their more established cousins, c-Si based materials and devices, these materials and processes are promising because they can be used in some niche applications and also as a foundation in the research for new and improved solar energy technologies.

Molded Silicon Solar Cells

The silicon wafers manufacturing process is a wasteful one. This is true, to some extent, of the numerous attempts to reduce silicon use and the waste related to it.

Any means of reducing the size of the solar wafers today leads to some loss of material as waste. The best, and maybe the only, possible way to utilize 100% of silicon material with no waste, would be to mold (cast) and process each wafer individually. That way, expensive silicon material, and the energy used to produce it, would be saved, the solar manufacturing process would be simplified, and costs would come down.

Examples:

Conventional Si solar cells process:
    Si -> MG Si -> SG Si -> Wafers -> Solar Cells +
        Si waste (up to 20%)

Molded Si solar cells process:
    Si -> Wafers -> Solar Cells (0% Si waste)

Obviously, there is a big difference between these process sequences. While the conventional process has many steps, using expensive chemicals and wasting materials and energy, the new molded Si process bypasses most of the expensive, complex steps and related waste.

This type of technology could be *truly disruptive*, if and when successfully implemented. Thus produced solar wafers would be ready for processing into solar cells, without wasting materials or energy, and without additional steps in the wafer manufacturing process— waste cleanup and disposal, wafer edge finishing, wafer cleaning...

A number of companies and research institutions

are looking into perfecting this amazingly simple way to make cheap polysilicon wafers. Thus far, there's not been much progress, but the work continues with private and DOE funds poured into research. Something good will come out of it soon, we are sure, and when it does, it might revolutionize the entire solar energy business.

Approaches like this are exactly what is needed to achieve the goals ahead of us for 21st century power generation.

Silicon Ribbon

Silicon ribbon, like the name suggests, is a ribbon made of silicon material. It is much thinner than the usual silicon solar wafers and cells, thus less material is used (as in bulk ingot growth) and wasted (as in slicing wafers).

Banking on the key advantage of this process of using less material and energy for equal production output, the Si ribbon technology is looking at a promising future. These advantages, however, come at the expense of increased fragility of the base material, which is a significant barrier in optimizing mass production of string ribbon solar cells and modules.

The conventional silicon wafers manufacturing process calls for casting a solid rod or brick of silicon and then slicing it into thin wafers. Slicing generates waste in the form of dust; almost 50% of the silicon material is wasted this way. It's a dirty, expensive waste, badly contaminated with cutting fluids and metals, so it is no longer pure enough to be recycled.

The string ribbon process reduces both feedstock consumption and wafering slicing cost since the wafers are grown directly in the form of ribbon from a silicon melt. This way they don't have to be cut and shaped, no waste is generated, and the process is much cleaner and cheaper.

There are several techniques for making string ribbons, with the edge-defined film-fed grown (EFG) being the most promising. A graphite dye is immersed into molten silicon, making it rise into the dye by capillary action. It is then pulled as a 8 cm. wide, self-supporting, but very thin sheet of silicon (averaging 200 microns), which hardens in the air above the melt. The ribbon can then be cut in different shapes and sizes for processing into solar cells and modules.

Two ribbons could be pulled simultaneously as well, which makes the process even more efficient.

Another similar process is called "dendritic web" growth process. It consists of two dendrites, which are placed into molten silicon and withdrawn quickly, caus-

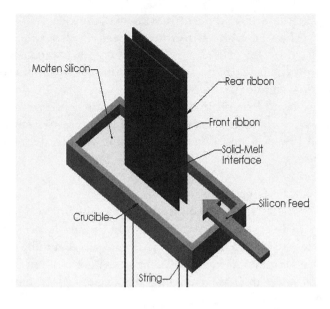

Figure 3-20. Dual ribbon pulling method

ing the silicon to exit and solidify as a thin sheet. A more practical modification of this method now in use is called the "string ribbon method," where two graphite strings are used (instead of the dendrites) to draw the silicon sheet, making process control much easier. Again, the silicon sheet can be cut into different shapes and sizes.

The commercial process is usually continuous, because it is difficult and expensive to start and stop the line. Chunks of silicon enter the furnace and melt, the strings unwind from spools, and the emerging ribbon is cut to lengths, stopping only when needed—for maintenance and such. Each ribbon is then laser-cut into wafers, which go directly onto a belt for the next step in becoming solar cells, and ultimately high-efficiency solar panels.

In all cases, silicon produced by these methods is multi-crystalline with a quality approaching that of the directionally solidified material of the conventional c-Si solar cells. Although lab tests show efficiencies in the 17-18% range, solar modules made using silicon ribbons and produced by these methods, generally have efficiencies in the 10-12% range.

The major issue of this technology, that of its brittleness, has not been successfully resolved, and hinders its wider implementation. Nevertheless, the ribbon technology is a viable way of material conservation and cost cutting, so we should not discount its direct or indirect influence on 21st century energy markets.

Si Ribbon Solar Cells Manufacturing Process

An example of a traditional Si ribbon based solar cells manufacturing sequence is listed below. It is quite similar to that used for processing regular Si wafers, with some slight process modifications and procedural differences:

1. Visual inspection of the Si ribbon strips for shape, color and surface un-uniformity. Ribbon with an average thickness of 300 μm and a resistivity of 3.0 Ωcm and visually even surface area is used.
2. Cutting the ribbon to an optimum size, which depends on the processing equipment size and loaded into cassettes.
3. Cleaned/etch of the wafers in solutions of 2:1:1 H_2O:H_2O_2:H_2SO_4, 15:5:2 HNO_3: CH_3COOH:HF, 2:1:1 H_2O:H_2O_2:HCl.
4. $POCl_3$ diffusion to form ~85 $\Omega/\square$ emitter.
5. HF etch to remove the diffusion glass.
6. Evaporation or printing of Al and alloying at 850°C.
7. Evaporation or printing of Al-Ti-Pd-Ag for back contact formation.
8. Photolithography of the front contacts using evaporation of Ti-Pd-Ag and lift-off followed by Ag plating.
9. Mesa etching for area definition.
10. Forming gas annealing at 400°C.
11. Evaporation of ZnS/MgF_2 for double AR coating.
12. Attach leads for contact soldering.

Automated production processes might vary as follow:

1. Inspection of the Si ribbon strips for shape, color and surface un-uniformity.
2. Cutting the ribbon to an optimum size,
3. Cleaned/etch of the wafers
4. Belt furnace diffusion to form 40-50 $\Omega/\square$ emitter
5. HF etch to remove diffusion glass
6. SiN_x deposition on the front side
7. Screen printing of Al paste on the back.
8. Belt furnace annealing at 700-800°C
9. Screen printing of Ag paste for the front contact,
10. Final belt furnace annealing
11. Attach leads for contact soldering

High efficiency screen-printed silicon ribbon solar cells have been fabricated by optimizing rapid thermal processing (RTP) cycles to improve SiN_x-induced hydrogenation of bulk defects and quality of back surface field. These cells were fabricated by a two-step process where the first step is used to form an effective Al BSF and provide enhanced defect hydrogenation.

The second firing step is performed at a lower temperature (<800°C) to form screen-printed Ag grid on the front. However, the latter step could degrade hydrogen passivation achieved in step one due to evolution of hydrogen from the defects.

While hydrogen is found to be introduced into silicon due to the annealing of SiNx AR coating, it is not clear how much and how fast the hydrogen diffuses into the underlying defective as grown wafer because the diffusion mechanism can be influenced by temperature, defect type and concentration, doping concentration and conductivity type.

Recently, record high efficiency EFG (16.7%) and string ribbon (17.7%) cells were reported, via special processes using photolithography contacts, thermal oxidation for front surface passivation, Al gettering for 30 min, ZnS/MgF2 for double layer anti-reflection (DLAR) coating, and microwave-induced remote hydrogen plasma (MIRHP) treatment for defect passivation.

Disadvantages

String ribbon technology has several disadvantages, as to the shape of the silicon crystals, where the overall thickness of the ribbon varies, thus not every 'silicon strip' can be processed directly into a solar cell without breakage.

In addition to this drawback, the growth process is thermally very inefficient. The radiating area/gram of crystal is extremely high, leading to very high energy expenses during processing which offset the reduced silicon use and expense.

Ribbon wafers generally contain more point defects and higher dislocation density (105-106/cm^2) due to larger temperature gradient during the growth compared to the widely used ingot materials (HEM, CZ). This results in much lower as-grown bulk lifetimes in the range of 1-6 μs in ribbon materials, which are not sufficient for high efficiency cells. Therefore, quality enhancement during cell processing is necessary to compete with other materials.

Nevertheless, and despite the low carrier lifetime, efficiencies of ribbon solar cells are approaching 13-15% in production and 16-18% in the lab, due mostly to bulk lifetime enhancement during cell processing

Future Developments

After the initial hoopla that silicon ribbon technologies would dominate the market, their share is quite small—less than 1% of total sales today—and does not seem to be growing. This is mostly due to the fact that the process is not easy to control, and the wafers' surface is not uniform enough, thus resulting in breakage, pro-

cessing defects, and performance inefficiencies.

The silicon sheets forming process is also complex and uses a lot of energy, which makes it comparably more expensive than some of the other mass produced TFPV technologies.

Improvements in the manufacturing processes and the performance of silicon ribbon cells are expected, with low cost and efficiency reaching 20% in the near future.

Silicon Foil Technology

Silicon foil technology is different from conventional silicon processes, because due to the high cost it strives to reduce its amount in each cell. A number of manufacturers and R&D organizations are looking into using only small amounts of silicon material—as nanocrystals, or thin sheets.

Using kerf-free wafering technology (less waste of silicon material during slicing ingots into wafers), some manufacturers have produced very thin films of silicon by using special slicing tools and techniques.

We are aware of 20μm-thick wafers used to manufacture solar cells, which due to their thinness are called "foils." These very thin and even flexible monocrystalline silicon materials are neither a thin-film nor a wafer, given the foils' unique form factor and physical attributes.

Another way to produce thin silicon foils is by selective etching of silicon material in ethylene-diamine solution of proper concentration. Boron doped silicon is best for this purpose for it is much slower to react. By adjusting the boron doping level and the chemicals' concentration, the etching speed can be reduced 100-1000 times. Additional control is achieved by varying the reaction temperature and turbulence levels. Reactive ion etch can be used to finish the surface, or to modify it as needed for the particular application.

Etching silicon wafers is not the best way to cost efficiency, but this technology has found use in a number of areas where very thin, and/or very light PV devices are required. Thin silicon foils are used in transmission of radiation, such as foils for vacuum windows and as supports for samples or absorption filters. This type of silicon is also a suitable material for soft X-rays, as it can be made thin enough to have a low absorption.

These very thin foils combine the advantages of reduced silicon material utilization (such as thin-film PV) with the high-efficiency potential of mono c-Si to produce efficient and cheap PV devices. The experts believe that Si foil will extend the reach of conventional silicon PV technology well into the future.

The unique properties of the silicon thin materials and the extraordinary flexibility of the foils allows for development of diverse applications in niche markets, such as building integrated PV and flexible PV.

Considering the high cost of specialized slicing, and the production yield loss due to etching and breakage of the fragile very thin silicon substrates, however, we do believe that this technology will remain in the niche markets for some time.

A major breakthrough to push it on the broad energy markets of the 21st century is unlikely, but the materials and techniques used for making these devices are quite useful and we see them applied for the development of other PV technologies.

Thin, Back-mirror Silicon Cells

An old concept of saving silicon base material is under new development, where the solar cells are:

1. Made much thinner, using one of the above described methods (ribbon, foil, thin slicing, etc.),

2. Have optimized front surface texturing, and benefit from

3. Added back surface mirror.

All of these factors increase efficiency. The higher efficiency (in addition to better surface texturing) is achieved by means of trapping sunlight in and keeping the photons inside the active material (via the backside mirror reflection), until their energy can be fully and efficiently used to free electrons and generate an electrical current.

The efficient trapping light in the bulk is greatly assisted by proper texturing of the front surface of the silicon solar cell. Effective texturing of the surface results in facets (such as pyramids and cones) that redirect incoming light, and even refract it.

The photons travel at least partially along the length of the silicon layer, and stay in the material much longer. This way they have a much better chance of being absorbed and of freeing electrons that generate current in the outside circuit.

This phenomena is assisted and amplified by adding a reflective layer at the back of the silicon layer. Thus created "photon mirror" will reflect the passing photons and will keep them moving around the solar cell still longer. This heightens the chance of increasing the number of freed electrons, thus increasing the electric current in the outside circuit.

This design requires thinner silicon material, where the solar cell can be half its ordinary thickness, while absorbing the same amount of light. This results in the use of much less silicon material, reducing final costs.

Also there are less impurities within the thinner material, which improves efficiency. In contrast, conventional silicon solar cells are approximately 200-250 μm thick, so the electrons have to deal with the impurities across the entire width of the material, which is a major reason for their lower efficiency.

The new thinner cells are 100-125 μm thick, so the generated electrons have a shorter distance to travel and are less likely to encounter impurities before escaping to generate electric current.

Cheaper, less pure silicon material could be used with this method as well, reducing costs even further. Let's see how it looks under the lens of our critical analysis:

1. Surface texturing is an established technique, and any improvement is a plus as far as final efficiency is concerned, but it complicates the process and increases cost significantly.

2. Making thinner solar cells is not easy, or cheap, because the resulting material is very fragile. Thus, more expensive equipment and more qualified labor is needed to process and handle the wafers.

3. The final yield is usually drastically reduced due to breakage, so the combination of these factors translates into reduced yields and higher manufacturing costs. It is debatable if the increased cost of materials and labor is justified by reduced material use.

4. The back surface mirror is a good idea, but the added complexity and cost are significant. We need to consider the fact that most electrons generated at the front surface have enough energy to make it to the outside circuit. Also by the time they get to the back mirror they have very little energy left, so it is questionable how much energy can be produced there.

In summary, this very good idea deserves a close look, needs to be proven, adapted for mass production, and deployed in special applications.

The battle for lower cost silicon material—still the preferred material for manufacturing solar cells—continues, so efforts in this area are justified and encouraged. Silicon will be around for a long time, so efforts to optimize solar cells' materials and processes (and hence modules' performance) will continue, to achieve the best and most efficient combinations.

Developments in Japan, where land is scarce and high performance solar cells and modules are preferred, confirm our deductions, so we'll be looking for great developments in these areas in the 21st century.

THIN FILM PV TECHNOLOGIES

We classify the conventional thin film PV technologies as "most promising" for the purposes of this text, because they are presently considered for a number of installations, including large-scale PV power generation projects. Although they will be used for a long time, we reserve the right to be skeptical about their long-term success during operation in large-scale power fields in the desert regions for 30+ years, as attempted today.

In our opinion, most thin film based technologies—and the nature and behavior of their materials—are vulnerable to the extremes of nature. They will not be able to withstand harsh desert environments over 30+ years of non-stop operation, or compete successfully with the c-Si based technologies in these areas. Time will tell.

These commonly used thin film and module designs are not suitable for the torture they will encounter in harsh deserts conditions. Because of the desert climate phenomena, serious power deterioration and total failures are expected, some even within a short time after installation. Several such cases have occurred and are addressed elsewhere in this text.

However, when properly designed and used reasonably in moderate climates, *where it belongs*, thin film technology holds the promise of reducing the cost of PV arrays by lowering material and manufacturing costs without jeopardizing the installations' efficiency or longevity.

Unlike crystalline Si solar cells, where pieces of semiconductors are sandwiched between glass panels to create the modules, thin film panels are created by depositing thin layers of certain materials on the cover glass or stainless steel (SS) substrates, using PVD, CVD and other sophisticated processes employed in the semiconductor industry.

The advantage of this methodology lies in the fact that the thickness of the deposited layers is usually a few microns, as compared to crystalline wafers which tend to be several hundred microns thick. In addition, films

deposited on SS sheets allow the creation of flexible PV modules. Flexibility, however is a misunderstood factor, which needs to be addressed. Thin films, though flexible by nature, do not like to be flexed and bent, and their lifetime is shortened with every flex and bend. There are no exceptions to this rule.

Another advantage of thin film modules manufacturing is lowering the process cost due to high throughput deposition processes, as well as the lower cost of materials. Technically, the fact that the layers are much thinner results in less photovoltaic material to absorb incoming solar radiation.

The efficiencies of thin film solar modules are usually lower than c-Si, although the ability and flexibility to deposit many different materials and alloys has allowed tremendous improvement in efficiencies.

Furthermore, the versatility of thin film PV modules has resulted in a large gain in market share—from non-existent several years ago to about 15-20% today. And thin film based PV installations are increasing in size by the day, promising to take over the US deserts in the very near future.

We do, however, need to caution again that indiscriminate, large-scale use of thin film PV modules in high temperature desert regions has not been proven reliable or safe, and we do foresee a lot of problems in these fields in the near future. Efficiency drops, general failures, and even environmental contamination would not be surprising.

Four kinds of thin film cells have emerged as commercially important:

1. Amorphous silicon (a-Si) cell in single and multi-junction structures,
2. Thin poly-crystalline silicon on a low cost substrate,
3. SIGS hetero-junction cell, and
4. CdTe/CdTe hetero-junction cell.

Note: We took a very detailed look at thin film materials, processes and devices in our first book, *Photovoltaics for Commercial and Utilities Power Generation*, so we'd like to direct the reader to it for additional details.

Silicon-based Thin Film Solar Technologies

A number of thin film technologies are based on the use of silicon material. Although silicon is part of their structure, its shape, size, manufacturing equipment and processes (and most importantly, its field behavior) are drastically different from that of the silicon used in conventional silicon based solar technologies.

In this section we'll review several types and modifications of silicon-based thin films, which we believe will either succeed to penetrate the energy markets of the 21st century, or at the very least some of their components will be used to optimize the head runners.

The key thin film, silicon-based technologies are:

1. Alpha silicon (a-Si), and
2. Epitaxial thin film silicon (epi-Si).

Amorphous (a-Si) Silicon Solar Cells

Amorphous silicon (a-Si) is one of the earliest thin film technologies developed for solar cells applications. This technology is different from crystalline silicon in the way the devices are produced, and in that silicon atoms in a-Si devices are randomly oriented from each other. This randomness in the atomic structure has a major effect on the electronic properties of the material, causing a higher band-gap (1.7 eV) than crystalline silicon (1.1 eV).

The larger band-gap allows a-Si cells to absorb the visible part of the solar spectrum more strongly than the infrared portion of the spectrum, which leads to reduced temperature dependence and better efficiency and reliability under certain conditions.

There are several variations in this technology where substrates can be glass or flexible SS, tandem junction, double and triple junctions, and each one has a different performance.

Amorphous-Si, Double or Triple Junctions

a-Si cells have lower efficiency than their mono- and poly-crystalline silicon counterparts, with maximum efficiency achieved in the laboratory at approximately 10-12%, single junction.

Usually a-Si modules degrade when exposed to sunlight and then stabilize at around 6-8%, due to the so-called Staebler-Wronski effect which causes the changes in the properties of hydrogenated amorphous Si.

To improve efficiency and solve degradation problems, approaches such as developing multiple-junction a-Si devices have been attempted. This improvement is linked to the design structure of such cells where different wavelengths from solar irradiation (from short to long wavelength) are captured. The efficiencies of these technologies is around 6-8%.

Tandem Amorphous-Si and Multi-Crystalline-Si

An alternative method to enhance the efficiency of PV cells and modules is "stacked" junctions, also called micromorph thin films. Here two or more PV junctions are layered, one on top of the other, where the top layer is constructed of an ultra thin layer of a-Si which converts the shorter wavelengths of the visible solar spectrum.

At longer wavelengths, microcrystalline silicon is most effective, which results in higher efficiencies than amorphous Si cells, depending on the cell structure and layer thicknesses. There has been a great push lately to move thin film Si solar panels in this direction.

Amorphous silicon, a.k.a. alpha silicon, or a-Si, is also thin film silicon and is used in p-i-n type solar cells. Typical a-Si modules include front side glass, TCO film, thin film silicon, back contact, polyvinyl butyral (PVB) encapsulant and back side glass.

a-Si has been used to power calculators for some time now, mostly because it is easily and cheaply deposited on any substrate. Some commercial installations have shown promise. Below, we'll take a closer look at these, as well as the properties of a-Si that we believe make it promising for serious applications.

Amorphous silicon (also thin film silicon, alpha silicon, or a-Si) is used in p-i-n type solar cells. Typical a-Si modules include front side glass, TCO film, thin film silicon, back contact, polyvinyl butyral (PVB) encapsulant and back side glass. a-Si has been used to power calculators for some time now, mostly because it is easily and cheaply deposited on any substrate.

a-Si is produced via thin film processes, based on depositing thin layers of silicon films on different substrates. Silicon thin-film cells are mainly deposited by chemical vapor deposition (CVD), typically plasma-enhanced (PE-CVD), using silane and hydrogen reactive and carrier gasses for the actual deposition. Depending on the deposition parameters and the stoichiometry of the process, this reaction can yield different types of thin film structures, such as amorphous silicon (a-Si, or a-

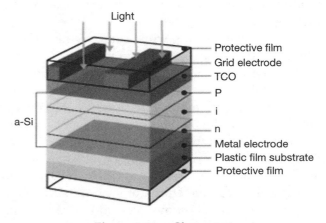

Figure 3-21. a-Si structure

Si:H), protocrystalline silicon or nanocrystalline silicon (nc-Si or nc-Si:H) also called microcrystalline silicon.

These types of silicon feature dangling and twisted bonds, which result in deep defects (energy levels in the bandgap) as well as deformation of the valence and conduction bands (band tails), leading to reduced efficiency. Proto-crystalline silicon mixed with nano-crystalline silicon is optimal for high, open-circuit voltage.

Solar cells and modules made from these materials tend to have lower energy conversion efficiency than those made from bulk silicon, but have some operating advantages (such as lower temperature degradation). They are also less expensive to produce, although the capital equipment expense is greater, due to equipment complexity.

a-Si has a somewhat higher bandgap (1.7 eV) than crystalline silicon (c-Si) (1.1 eV), which means that it absorbs the visible part of the solar spectrum more efficiently than the infrared portion. nc-Si has about the same bandgap as c-Si, so nc-Si and a-Si can advantageously be combined in thin layers, creating a layered cell called a "tandem cell," where the top a-Si cell absorbs the visible light and leaves the infrared part of the spectrum for the bottom cell in nc-Si.

The biggest problem with a-Si TFPV technology and a barrier to its success, however, is its low efficiency. Today's best cell efficiencies are about 12% in the lab, which is almost 50% lower than other PV technologies. Mass produced a-Si cells and modules are in the 8% efficiency range today.

A second problem with a-Si is its manufacturing cost as related to initial capital investment, which is quite high compared to competing PV technologies. Two proposed solutions to this problem are higher manufacturing rates, and batch (simultaneous) processing of multiple modules. Good progress has been made in rates that are 3-10 times higher than those being used in production, but all of this is still on a lab scale and yet to be proven on a large scale.

On the positive side, while some of the more efficient cells and modules lose about 20-30% of their output in the field, due to excess heat exposure, a-Si loses only about 5-10%, due to its lower temperature coefficient. Also, the active thin film structure is composed mainly of silicon films which have inert and homogeneous natures that show better chemical and mechanical stability than some of the competing thin films—in case of an encapsulation failure. a-Si modules are also more resistive to the negative effects of shading in the field. Of equal importance, a-Si modules do not contain any hazardous materials, which is paramount where large-scale PV installations are concerned. These qualities put a-Si on the top of the list of PV technologies suitable for large-scale power generation in deserts and other inhospitable areas.

Even with low efficiency (well under 10%), a-Si thin film technology is being successfully developed for building-integrated photovoltaics (BIPV) in the form of semi-transparent solar cells which can be applied as window glazing. These cells function as window tinting while generating electricity. It remains to be seen if the amount of generated electricity covers initial and operating expenses.

A triple-junction a-Si TFPV power system has been operating near Bakersfield, CA, for several years, and is providing proof of the excellent performance of this technology. The 500 kW grid-connects system has been performing well, meeting or exceeding design goals. Performance data from this larger-scale installation confirm data obtained from smaller a-Si systems and prove that this thin film PV technology can be successfully used in large-scale power plants, if the low efficiency can be justified.

Great research effort is underway at universities and R&D labs around the world, geared towards solving efficiency and cost issues and obtaining a-Si that is truly competitive in the energy market. a-Si manufacturers need to work on understanding the key areas of this technology and the processes, and focus on their optimization by:

a. Improving the light-stabilized electronic quality of a-Si and low-gap a-Si:H cells to achieve broader spectrum conversion, and increased and stable overall efficiency.

b. Increasing the growth rates of a-Si, a-SiGe, etc. layers while maintaining high electronic quality, to obtain increased throughput and reduced capital cost.

c. Developing high-growth-rate methods for nanocrystalline silicon while maintaining high electronic quality as needed for increased efficiency, stability and reduced cost.

d. Understanding and controlling light-induced degradation in a-Si:H as needed for increased efficiency and understanding of the intrinsic limits of the efficiency.

e. Developing *in-situ* in-line process monitoring for increased yield.

f. Improving light management to obtain maximum efficiency.

g. Improving stability and conversion efficiency of a-Si modules in actual use by addressing the Staebler-Wronski negative effects, where the conversion efficiency of the a-Si module decreases when it is first exposed to sunlight.

h. Reducing capital equipment costs for manufacturing a-Si panels by improved manufacturing processes that include increasing the deposition rates.

i. Improving module-packaging designs to make them more resilient to outdoor environments and less susceptible to glass breakage or moisture ingress.

j. Developing new module designs for building-integrated applications. The future of a-Si PV products depends on the timely resolution of these complex issues.

k. Understanding the role of H: (hydrogen) in establishing nanostructures, alloying, doping, and as a basic properties and performance modifier during cell operation.

l. Fundamental understanding and control of the film structure, gas-phase chemistry, and the related reactions on the film surface, as well as their individual and combined effect on the device properties and performance.

Epitaxial Thin Film Silicon

The high cost of silicon material accounts for about half of the production cost of current conventional, industrial-type silicon solar cells. To reduce the amount of consumed silicon, the photovoltaics (PV) industry is counting on a number of options presently being developed. The most obvious is to move to thinner silicon substrates by producing thinner Si wafers, or shaving the thicker wafers, but this is proving hard to do.

A more feasible approach is the so-called epitaxial deposition of a thin film of silicon on a cheap substrate, thus creating efficient but cheap solar cells. Several approaches can be used to create such a thin film cell:

a. **Epitaxial single crystal sc-Si**

    To create an epitaxial thin-film solar cell on a cheap substrate, we start with highly doped sc-Si wafers (e.g., from low-grade silicon or scrap Si material) and deposit an epi layer of Si by chemical vapor deposition (CVD). The resulting mix of a high quality epi layer and a cheap substrate is a compromise between high cost and efficiency, and yet offers a solution to gradual transition from a wafer-based (heavy material dependence) to a thin-film technology (less material and more sophisticated processing).

    This process is easier to implement than most other thin-film technologies today, but it remains to be seen if its efficiency and cost will be able to compete on the energy market.

b. **Epitaxial polysilicon thin film**

    To produce thin-film polysilicon solar cells, a thin layer (only a few microns) of polysilicon Si is deposited on a cheap substrate, such as ceramic, high-temperature glass, or cheap polysilicon wafers, or ribbon Si. The seed layers are then epitaxially thickened into absorber layers several microns thick using high-temperature CVD with a deposition rate exceeding 1 m/min. Polycrystalline silicon films with grain sizes between 1-100 m appear to be particularly good candidates.

    Good polycrystalline silicon solar cells can be obtained using aluminum-induced crystallization of amorphous silicon. This process leads to very thin layers with an average grain size around 5 m. This technology is still in R&D stages, but shows high cost-reduction potential and might become very important, especially in case of silicon shortage, or very high prices in the future.

Epitaxial Thin Film Solar Cells on Silicon Ribbon

We believe this is one of the most promising technologies for the 21st century. High quality thin films, deposited on a low cost silicon ribbon substrate could provide very efficient, and most reliable long-term operation under adverse conditions.

The thin film deposition process, herein referred to as "epi-process," is usually conducted in RF reactors, where high-purity gasses containing the deposition components, such as dichlorosilane are used as a source for the epitaxial growth of the active photo-electric thin films on top of the substrate.

The ribbon silicon slices are cleaned thoroughly in megasonic peroxide-ammonia solutions and their surface is further cleaned by etching with HCl gas just prior to introduction into the reactor. They are then loaded in the reactor (a quartz tube) on a silicon carbide-coated graphite susceptor (holder). RF heater heats up the reactor and the substrates by RF induction into the susceptor, which is sometimes inclined with respect to the horizontal. At the same time the walls of the reactor are cooled by air jets or other means.

The carrier and inert gasses are precisely metered into the system during the process via gas control valves and sensors, which provide automatic gas concentration and timing of the duration of each run. Hydrogen gas is introduced in the system as well, by first purifying it through special filters, and then Pd-Ag diffusion cell or similar device.

The doping gases are usually either arsine or diborane, diluted with hydrogen at approximately 10 to 20 ppm level and further diluted with hydrogen as needed to achieve full stoichiometry. The temperature in the reactor is controlled by an optical pyrometer of the disappearing filament, or similar, type.

The epitaxial structures are grown at elevated temperatures of about 1100-1200°C, and at a growth rate of 5 μm/min. Higher growth rates are possible in mass production, which is desirable and economically advantageous.

Properties of Epitaxial and As-grown EFG Solar Cells

Conventional solar cells are made by diffusing a shallow (~0.3 μm), highly doped n-type junction layer into a lightly doped p-type substrate (~2 x 10^{15} cm^{-3}). The junction layer is kept thin because the minority carrier lifetime is very low (~0.2 nsec) in the heavily doped surface layer. This requires the base to be lightly doped, in the 10^{15} cm^{-3} range, to obtain the very long diffusion lengths (~200 μm) needed for high Isc, which leads to relatively low Voc = 0.55-0.59 V.

The saturation current in such diffused structures is usually dominated by generation recombination in the wide space-charge region. The diffused solar cells can also exhibit large shunt-leakage currents which arise from surface damage during processing, penetrating through to the junction and impeding its function.

The obvious way to increase the Voc is to increase the carrier concentration in the base region. But the results achieved when the base doping is increased to or above 1017 cm^{-3} are controversial because the generation-recombination current that exists in diffused junctions (due to localization of defects within the junction region during the diffusion process) become large compared to the ideal diffusion current component. This prevents a significant increase above that achieved with the epitaxial thin film solar cells described above.

Epitaxial cells have several advantages over the diffused cell described above. Perhaps the most important is that the dopant concentration and width of the n and p layers can be independently adjusted, allowing both regions to contribute significantly to the short-circuit current. By dividing the collection region, the minority-carrier lifetime in the base region can be maintained at a lower level than required in a diffused cell, thus allowing a substantially higher doping level in the base, and consequently, a higher Voc.

Another advantage of the epitaxial structure is that control and grading of the n and p layers during processing give a built-in field which will enhance the collection of photo-generated carriers. The series resistance of the cell is also reduced compared to a typical diffused cell because of the thickness of the n layer.

In addition, the influence of surface damage on the performance of the junction is reduced. The ability to use a more highly doped base region in the epitaxial structure offers another potential advantage in that the cell becomes more suitable than a conventional diffused cell for focused-light collection schemes.

These cells can be used in a CPV (concentrating PV) type system, using a focused-light solar cell array. The cost of the system is reduced in this case by focusing light on the solar cell, thereby reducing the area of high-cost solar cells required per unit power output. The higher intensity of sunlight incident on the cell can result in high current densities and consequently in a plasma carrier concentration at least an order of magnitude higher than the background doping density in a conventional diffused junction solar cell. That can reduce the effective lifetime because the recombination of carriers is dominated by a bimolecular recombination process, instead of the usual monomolecular recombination.

Also, the higher doping level and the smaller thickness of the base layer of the epitaxial cell result in a

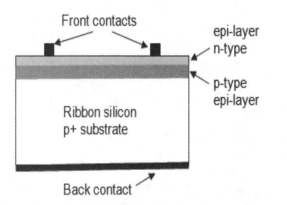

Figure 3-22. Hi-efficiency epitaxial thin film solar cell on silicon ribbon substrate

lower Ohmic voltage drop at the high current densities found in the focused-light solar cells.

Generally speaking, epitaxial p-n junction thin film structures, grown on silicon ribbon substrates, have different electric and photo-electric characteristics than a p-n junction diffused directly into the silicon ribbon substrate. In most cases the latter (as-grown EFG substrate solar cells) have a saturation current density value much in excess of the normal.

The as-grown silicon ribbon substrate has a lot of large defects (inclusions, twin and grain boundaries) and contaminants that degrade the carrier diffusion lengths, which are absolutely necessary for the normal photovoltaic process.

The presence of large-angle grain boundaries negatively impacts current-voltage characteristics of the devices, while twin boundaries are not that harmful, but have a negative effect on the final device efficiency.

In contrast, epitaxial thin films are much superior as far as surface defects and contaminants are concerned, simply because they are grown with superior materials and under precisely controlled process parameters. The superior materials and the resulting well behaved p-n junction of the epitaxial thin films contribute to higher efficiency of the devices. An impurity gradient in the base region provides a drift field, which enhances the carrier transport to the collecting p-n junction at the surface epitaxial thin film layers and adds to their superior performance.

A very thin, about 50 µm, epi-layer thickness is enough to produce approximately 10% efficient solar cells. By additional surface treatments, anti-reflective coatings, and optimized metal pattern, the efficiency of these devices could be significantly increased to approach that of conventional solar cells.

A major problem with this technology is the inconsistency of the silicon ribbon substrate material, which varies in thickness, surface roughness, surface defects, contaminants and impurities. The dislocation-defects density in as-grown silicon ribbon substrate is very high, while that in epitaxial thin films is significantly reduced. Another serious problem is the presence of SiC precipitates in the substrate bulk, which are mostly likely remnants from the die material used in the growing of the silicon ribbons. The relatively high concentration of these contaminants gives rise to the shunt leakage current, reducing the cells' fill factor and open-circuit voltage.

Epitaxial thin films, in contrast, exhibit much lower diffusion of contaminants in the surface layer, leading to improved lifetime and lower saturation current density. They also show considerable improvement of overall performance characteristics and efficiency of the epitaxial thin film solar cells.

Due mainly to the bulk and surface imperfections of as-grown silicon ribbon substrates, the performance (efficiency) of epitaxial thin film solar cells grown on top of these ribbons is approximately 20-30% higher that that of solar cells made directly by diffusing as-grown silicon ribbons.

There are several major technological aspects of the use of epitaxy for solar cells that require close attention to realize their potential. These are: a) better surface passivation is needed to minimize surface recombination, which can be done by proper thermal oxidation of the open surface regions, b) p-n junction properties improvement is needed, and in particular their degree of ideality, contamination and doping level controls, and c) improvement of minority-carrier lifetime in the layers as a function of doping concentration and depth.

Improvements in as-grown silicon ribbon surface preparation (etch and clean), high temperature gettering, and other optimization techniques would eventually lead to the production of stable, high efficiency solar cells, which we believe would be more reliable than solar cells produced by the existing thin film methods such as CdTe and SIGS.

The increased cost of the epitaxial thin film materials, equipment and processing must be weighed against the increased efficiency to determine the benefits of the epitaxial solar cells.

Obviously more work is needed to get to the final epitaxial devices' configuration, and to complete the picture of using these devices on the world's energy markets, but we are confident that this technology will find its place under the sun during the 21st century.

Tandem a-Si/mc-Si Solar Cells

Conventional solar cells consist of a single p-n junction, which is responsible for the entire action and resulting quality of the solar cell. Adding other materials could create an additional junction or a set of junctions that would improve the performance characteristics of thus obtained solar cells. Such a structure is called "tandem," and there are a number of possibilities for creating tandem cells using a-Si solar cells in combination with other materials.

These kinds of material combinations are useful in increasing the overall efficiency of the device, as well as correcting some of the operational problems created by a-Si material deficiencies.

The most successful tandem arrangement to date is microcrystalline and a-Si tandem solar cells.

Microcrystalline and a-Si Tandem Solar Cells

Microcrystalline/alpha silicon tandem solar cells are fairly simple structures, consisting of a microcrystalline silicon (uc-Si) bottom cell and an amorphous silicon (a-Si) top cell. These structures are considered one of the most promising new thin-film silicon solar-cell concepts, in terms of simultaneously achieving high conversion efficiencies and relatively low manufacturing costs.

The concept is based on the VHF-GD (very high frequency glow discharge) deposition method, where the key active element is the hydrogenated microcrystalline silicon bottom cell that opens new perspectives for low-temperature thin-film crystalline silicon technologies.

Microcrystalline silicon is much more complex and very different from an ideal isotropic silicon-based semiconductor. Efficiencies of about 12% have been obtained with these types of cells, but this is only the beginning, because the complexity and variety of tandem microcrystalline silicon offers exiting new developments of this technology in the near future.

The deposition process used for manufacturing these tandem cells is unexploited, so work is still needed to achieve maximum efficiency and cost savings. Still, the results obtained thus far are very encouraging and confirm that this concept has the potential to come close to the required performance criteria concerning price and efficiency.

The combination of an amorphous silicon top cell with a microcrystalline silicon bottom cell to form a stacked tandem cell creates two different gap energy fields. The resulting structure has different characteristics than the well-known double-junction in the a-Si:H=a-Si:H tandem cells.

The concept of superposing two a-Si:H cells is based on the reduction in the Staebler-Wronski effect that can be obtained by keeping each individual i-layer as thin as possible, at the expense of underutilizing the

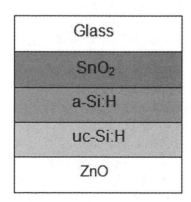

Figure 3-23. uc-Si/a-Si tandem cell

full solar spectrum potential.

The double-junction concept for a-Si:H cells is useful for reducing light-induced degradation, but the uc-Si:a-Si tandem cells offer much better utilization of the sun spectrum, due to the stable mc-Si:H bottom cell, which ensures the better stability of the entire tandem structure. Actually, it is a proven fact that the majority of the light-induced degradation of the tandem cell is due to its effect on the a-Si:H top cell alone.

Making the bottom uc-Si:H active layer thicker, could enhance the photocurrent without adverse stability problems. This change, however, requires increasing the thickness of the top layer as well, due to the required current matching. This brings us back where we started, so this approach needs to be optimized for each cell design. Even though the entire hybrid cell is more stable, the problematic a-Si:H layer remains the weak point and one that determines the final cell performance and longevity.

Hydrogen dilution and hot wire deposition techniques have been used for enhancing the stability of the a-Si:H cell. This provides a relative stability of the layer, but the reduced absorption (higher band gap) of such hydrogen-diluted a-Si:H layers limits the photocurrent of the top cell, and thus the current and stable efficiency of the entire tandem cell.

In assessing new solar-cell technologies we must know their performance characteristics. One of the key questions is how many kWh per installed Wp can be obtained during the 20-30 lifetime of the solar cells. Initial efficiency is important, but field performance and longevity are other factors to consider. Operation in high-temperature areas requires good (lower) temperature coefficient, or the cells will degrade and maybe even disintegrate under the enormous heat load.

Temperature coefficients of c-Si cells, CIGS and a-Si:H cells are higher than those of mc-Si:H and tandem cells, while the Voc and Jsc values are quite similar. This ensures higher efficiency and stable performance of uc-Si:H cells under high temperature regimes and eventually longer lifetime. This makes tandem cells good candidates, for example, for building integration of PV thin-film solar cells, especially under high-temperature conditions and/or in cases where no special aeration of the façade can be provided to keep the temperature of the cell relatively low.

As previously mentioned, tandem cells consist of two absorber materials with two different gap energies. For all tandem solar cells the so-called photocurrent matching between top and bottom cell is imperative, in order to get the maximum power out of the cell. For tandem cells, however, the difference in gap energies between top and bottom cells is pronounced, namely about 1 eV for the mc-Si:H bottom cell and about 1:7 eV for the a-Si:H top cell. Thus, the current-matching problem appears here under a different angle.

Due to the difference in gap energies, the power generation in a tandem cell is by nature shifted to the amorphous top cell because of the different open-circuit voltages of the individual cells.

Note, the power from the top cell is larger by the ratio Voc (top) = Voc (bottom), or (900mV = 500mV) than the power of the bottom cell. This is not an ideal situation because the top cell suffers from the Staebler-Wronski effect, whereas the stable bottom cell shares only one third of the total generated power.

To achieve further improvements of tandem solar cells, the following concepts or combinations of them must be analyzed and their issues solved:

1. Voc value of the bottom cell must be increased as high as possible, the route to achievement of which is not clear.

2. An optical mirror between top and bottom cell can be implemented, which leads to enhanced light-trapping for the top cell. The top cell is thereby kept thinner (leading to a reduced SWE), but it can still absorb enough light, thanks to an optical mirror.

3. a-Si:H=mc-Si:H=mc-Si:H triple-junction cells can be implemented where the current-matching requirements over three individual cells shift the power generation towards the stable microcrystalline cells.

4. Light-trapping is an issue for any thin film solar concept because it helps to keep the total absorption of the cell but at reduced absorber thickness. Light-trapping also allows reduction of the deposition time required, and it improves the ratio between collection length and cell thickness.

5. Basically, a more stable amorphous silicon top cell should be obtained, because it contributes excessively to the degradation, but this is an inherited problem that won't be solved soon, if ever.

6. Apart from, the conversion efficiency of the solar cells, obtaining a high deposition rate in the manufacturing process is another key issue. It will decide whether this technology can be cost-competitive.

Multi-junction a-Si Technology

In addition to the "classical" tandem structures, some of which we review herein, there are other multiple-junction solar cell concepts that are interesting to implement with the help of a-Si silicon based materials. One such concept is the triple-junction combination of a-Si:H/a-Ge:H/mc-Si:H.

This combination shows promise as a more stable and efficient structure, but the materials, equipment, overall manufacturing processes as well as device performance are too complex and underdeveloped to consider here.

Another promising approach is a number of multiple junctions connected electrically in parallel rather than in series. Again, a complex and underdeveloped concept, but with possibilities.

Test and Measurement

The measurement of tandem and triple junction solar cells is loaded with extra uncertainty over single junction cells through their more complex spectral mismatch. (9)

For μ-morph tandem cells the spectral response depends on the thickness of each layer and the collection efficiencies of the top and the bottom cell, respectively. The short circuit current measurement error depends on the spectrum of the sun simulator and the spectral responses of the tandem's top and bottom test cell, but also on the blocking ability of the limiting junction, i.e. the spectral mismatch cannot be calculated according to IEC 60904-7:1998 Ed.2.

Besides the problem of the mismatch correction for the short circuit current, for correcting the standard test condition of the maximum power point of the series, the addition of two IV curves reveals a change in the fill factor as well.

Similar to series connected cells in crystalline cells, the fill factor goes up with an increasing current mismatch, while the short circuit current would decrease and the open circuit remains constant.

In the end, a method for the tolerance evaluation of the power measurement of tandem cells is not developed yet, but we've a good guess from round-robin tests. The calculation of mismatch factors for tandem cells is not straightforward as in the case of single junction cells because of the series interconnection of a top and a bottom cell with inherently different spectral responses.

The uncertainty of the Pmax measurement for μ-morph tandem modules depends on the reverse current characteristic, i.e. the lower the reverse current, the higher the uncertainty. In other words, a less effective blocking of the limiting junction weakens the effect of spectral variations on the measured power of the tandem module.

As long as there is no procedure for the spectral mismatch correction of tandem cells, sun simulators of higher spectral quality or outdoor measurements are the most effective tool to lower the measurement error on μ-morph tandem cells. In the meantime such sun simulators are commercially available even for larger sized modules.

In summary, tandem a-Si based solar cells is a promising technology, because the stability of the final devices is improved (due to the more stable second films), and their efficiency is increased (due to better utilization of the solar spectrum, caused by the ability of the double-junction to reduce light-induced degradation.) Efficiencies of 12-14% have been achieved with some tandem combinations and further potential for improvement is imminent.

Tandem a-Si based solar cell technologies also have the potential to become concept designs for the next generation of thin-film solar cells, in view of their potential for high efficiency, stability, reliability and low cost.

Laboratory research must be coordinated with industrial requirements, where stability and efficiency are key issues, in addition to the demands of mass production simplicity, volume, production costs, and quality. All these requirements must be met before this (or any) technology can find a solid place in the energy markets of the 21st century.

Comparison between c-Si, a-Si, and mc-Si

Accurate determination of power output and energy yield is crucial for profitable operation of PV power plants. While performance of c-Si based modules is well understood and implemented in state-of-the-art simulation programs, thin-film technologies such as a-Si and μ-Si or tandem structures consisting of μ-Si and a-Si based modules provide difficulties in performance determination, even after termination of the degradation process. (10)

Spectral Effect

Due to the different spectral responses of the PV technologies and their different matching to the actual solar spectrum, actual performance may change, as illustrated in Figure 3-24. a-Si is more efficient in the blue part of the solar spectrum, so spectra with low AM (high sun elevation angles) are adequate. Crystalline Si-technologies perform best in the red part of the solar

spectrum, so spectra with high AM (low sun elevation angles) are preferable.

Figure 3-25 shows the relative change of PV conversion efficiency vs. AM 1.5 at standard test conditions (STC).

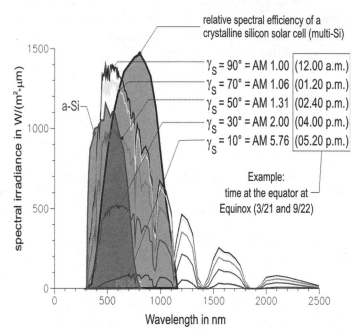

Figure 3-24: Spectral conversion efficiency of a-Si and multi-c-Si solar cells and their match to sun's spectra during a day, corresponding to sun's elevation angles (9)

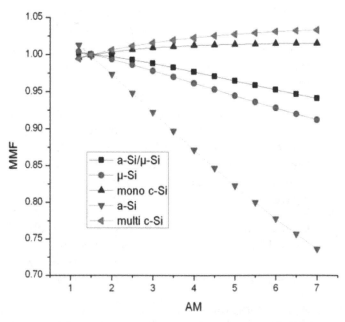

Figure 3-25. Relative change of conversion efficiency (MMF—mismatch factor) vs. STC (AM 1.5) calculated as a function of relative air mass (AM) for different PV technologies using standard spectra given by CIE 85 (10)

Weak Light Performance

Performance evaluation is becoming even more complex due the change of conversion efficiency at different irradiance levels—the so-called "weak light performance." For the different technologies this has been examined indoor via a solar flash simulator (Pasan IIIb) equipped with several neutral density filters (100,200, 400, 800 W/m²) to change the irradiance level and also outdoor measurements. Results are shown in Figure 3-26.

The better performance of μa-Si and in particular of a-Si for low a medium irradiance levels can be attributed to its relatively high series resistance, so best performance cannot be achieved under STC.

The deviation between the simulator results and the outdoor measurements in Figure 3-26 can be explained by additional effects during outdoor measurements, such as spectral effects and changing angle of incidence. Outdoor data have been referenced via Isc at STC to allow calculation of the actual irradiance.

While conversion efficiency is not constant for the range of irradiance levels occurring, the irradiance statistics of location become relevant (instead of just an average irradiance level). An example of irradiance statistics of some locations is given in Figure 3-27.

In the end, the measurement results done by the Photovoltaic Institute in Berlin, Germany, show that amorphous silicon modules show higher energy yields

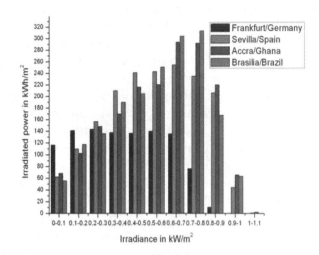

Figure 3-27: Statistics in terms of frequency distribution of global irradiance levels for different locations (module elevation angle = latitude –20°). (10)

compared to crystalline modules for diffuse light irradiation. This was as well calculated with the spectral mismatch given by the sky spectra in CIE 85 and the spectral response for single junction modules and detected from field measurements.

Highly diffuse skies and high sun elevation angles result in the largest current increases for amorphous based modules over crystalline silicon modules. For moderate diffuse skies, c-Si performs better. In total, a slight relative increase of 2% for a-Si and μa-Si has been observed.

Author's note: For detailed measurement data of the different Si technologies, under different conditions, we refer the reader to the Photovoltaic Institute report listed under item #9 in the Notes and References section.

CIGS Solar Cells

The CIGS technology has been used for a long time in a number of specialized applications, but just recently it was considered for deployment in large-scale solar power installations. We do believe that it will find wider applications in the future, as well, thus we consider it a promising solar power generator technology for the 21st century.

CIS devices consist of copper indium diselenide ($CuInSe_2$) thin films, or copper indium selenide CuInSe. CIGS devices, on the other hand, are a mixture of copper, indium selenide and copper gallium selenide, with a chemical formula of $CuIn_xGaSe_2$, where the value of x can vary as needed to obtain pure copper indium selenide, or pure copper gallium selenide.

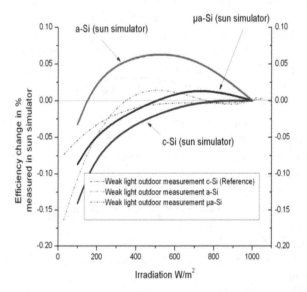

Figure 3-26: Comparing performance at different irradiance levels for different PV technologies (single junction a-Si and c-Si as well as tandem μ-Si/a-Si). Data have been obtained via outdoor measurements and via sun simulator (Pasan IIIb, AM 1.5 plus neutral density filters). (10)

These structures basically represent types of photovoltaic devices that contain semiconductor elements from groups I, III and VI in the periodic table, and are known for their high optical absorption coefficients, and whose electrical characteristics allow precise tuning of the resulting solar devices.

CIS.CIGS are a tetrahedrally bonded semiconductor material, characterized by a specific chalcopyrite crystal structure. Its bandgap varies continuously with x from about 1.0 eV (for copper indium selenide) to about 1.7eV (for copper gallium selenide).

The active layer (the light absorbing material) in CIGS devices can be deposited directly onto molybdenum coated glass sheets or steel band in a polycrystalline form, saving the energy-consuming and expensive step of growing large crystals. In some cases the substrates can be flexible, which allows a number of special applications.

Also, and very importantly, selenide provides excellent uniformity, which diminishes the number of recombination sites in the bulk material, thus improving the quantum and conversion efficiencies of thus manufactured solar cells and modules.

CIGS (which are a variation of CIS structures, where indium is incorporated with gallium for increased band gap), unlike basic p-n junction silicon cells are explained by a multifaced hetero-junction model. This is a complex process, which we don't intend to tackle in this text, but the informed reader can find a lot of references in the literature. A brief analysis follows:

The CIGS process can be basically described as a sequential, multi-step process. It starts with the formation of a Cu-free precursor, $(InGa)_2Se_3$ by the simple reaction of the carrier gasses in the process chamber:

$$2(InGa) + 3(Se) = (InGa)_2Se_3$$

The reaction proceeds when Cu and Se are added (evaporated or sputtered) onto the previously formed $(InGa)_2Se_3$ film, where the CU diffuses to form $Cu(InGa)_2Se_3$. As the Cu molecules and clusters start arriving slowly onto the film, there is a depletion of (not enough) Cu at first. This anomaly creates defect chalcopyrite phases, which will exist in one or another shape, form and quantity as long as Cu is available in the film.

At this stage we can split the reactions into several phases of Cu integration, as follow.

1.   The bulk the Cu reaction is:
     $$3Cu + 5(InGa)_2Se_3 = 3Cu(InGa)_3Se_5 + (InGa)$$

2.   The reaction on the device surface is:
     $$Cu + (3(InGa) + 5Se = Cu(InGa)_3Se_5$$

3.   The final reaction of the CIGS film formation at this step of the process is:
     $$10Cu + 15(InGa)_2Se_3 + 5Se = 10Cu(InGa)_3Se_5$$

     Or:
     $$15(InGa)_2Se_3 + 5Cu_2Se = 10Cu(InGa)_3Se_5$$

Remember, Cu is a easily diffused into any material. This is a major problem in the semiconductor industry, where Cu contamination accounts for device failures and billions of dollars spent on Cu materials and processes segregation, and otherwise preventing "Cu contamination."

We can deduct from above that for every 3 Cu atoms added to the CIGS film, one other (In or Ga) atom has to be removed as needed to keep the system's chemical equilibrium. Here the problem is that the CIGS' surface atoms (whatever type and number available at the moment) have the opportunity to react with the incoming Cu or Se atoms on the surface of the film, where they are free to form new chalcopyrite nests, which increase further the heterogeneous nature of the CIGS film material.

It is also possible to achieve a stoichiometry (or chemical balance) of the film at this point, but this process is hard to control, so we won't spend much time on it. It suffices to say, that in case of reaching complete stoichiometry, all Cu sites are filled and there is no room for additional Cu in the film crystal lattice. This, however, is an unstable state—due to the diffusivity of Cu and the affinity of the surrounding atoms to react with it—and could be ignored at this stage of the CIGS films formation for all practical purposes.

After awhile the reaction (and the resulting films) reach a state of Cu-rich film, where after completing the deposition of Cu_2Se, In and Ga are re-introduced to form the reaction:

$$Cu_2Se + 2(InGa) + 3Se = 2Cu(InGa)Se_2$$

This reaction takes place at the very surface of the thin film structure and constitutes an epitaxial growth of the film on it.

When finally the Cu_2Se layer on the surface is depleted (reacted with the incoming species to form $Cu(InGa)Se_2$, we reach a more stable, Cu-poor (or Cu satisfied) state. The CIGS formation continues in this manner, accompanied by out-diffusion of Cu atoms from the existing CIGS films below. This reduces the Cu content to under its original state of 1, and we arrive to the final CIGS film structure as follows.

$$Cu + (InGa) + 2Se = Cu(InGa)Se_2$$

The stability of this structure depends on a number of factors, such as the process and the related reactions parameters, the type and purity of materials etc. As a rule of thumb, we think of CIGS films as alive and ever "developing" systems. This is due to the instability of the film, mostly caused by the presence of energetic and unstable Cu atoms, which have a high affinity to diffuse in most inorganic materials.

The "stabilization" of the final film in such a system is critical, but hard to do, and because of that we see no good way to ensure that the Cu atoms will remain where we want them during 30 years of non stop bombardment by sunlight photons and the daily attack of other environmental factors like excess heat, humidity etc.

As a matter of fact, excess heat damages all thin films by virtue of the mechanical process within which causes the never ending daily expansion and shrinking process. The presence of humidity (moisture) in a CIGS film, on the other hand, damages the films by destructive chemical reactions and also by activating the Cu atoms to the point to where they will eventually reduce the device's photovoltaic efficiency, and will even destroy it with time. This is why effective encapsulation of the PV modules is such an essential requirement, which is as yet misunderstood and underestimated by some manufacturers.

CIGS Structure and Function

Early solar cells of this type were based on the use of CuInSe2 (CIS). However, it was rapidly realized that incorporating Ga to produce Cu(In,Ga)Se2 (CIGS) structure, results in widening the energy bandgap to 1.3 eV and an improvement in material quality, producing solar cells with enhanced efficiencies.

CIGS have a direct energy bandgap and high optical absorption coefficient for photons with energies greater than the bandgap, such that only a few microns of material are needed to absorb most of the incident light, with consequent reductions in material and production costs.

The best performing CIGS solar cells are deposited on soda lime glass in the sequence—back contact, absorber layer, window layer, buffer layer, TCO, and then the top contact grid. The back contact is a thin film of Mo

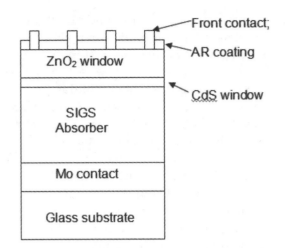

Figure 3-28. CIGS cell cross section

deposited by magnetron sputtering, typically 500-1000 nm thick.

The CIGS absorber layer is formed mainly by the co-evaporation of the elements either uniformly deposited, or using the so-called three-stage process, or the deposition of the metallic precursor layers followed by selenization and/or sulfidization. Co-evaporation yields devices with the highest performance while the latter deposition process is preferred for large-scale production.

Both techniques require a processing temperature >500°C to enhance grain growth and recrystallization. Another requirement is the presence of Na, either directly from the glass substrate or introduced chemically by evaporation of a Na compound. The primary effects of Na introduction are grain growth, passivation of grain boundaries, and a decrease in absorber layer resistivity.

The junction is usually formed by the chemical bath deposition of a thin (50-80 nm) window layer. CdS has been found to be the best material, but alternatives such as ZnS, ZnSe, In2S3, (Zn,In)Se, Zn(O,S), and Mg-ZnO can also be used.

The buffer layer can be deposited by chemical bath deposition, sputtering, chemical vapor deposition, or evaporation, but the highest efficiencies have been achieved using a wet process as a result of the presence of Cd2+ ions. A 50 nm intrinsic ZnO buffer layer is then deposited and prevents any shunts. The TCO layer is usually ZnO:Al 0.5-1.5 μm. The cell is completed by depositing a metal grid contact Ni/Al for current collection, then encapsulated.

CIGS solar cells have been produced under lab conditions with efficiencies of 19.5%, and lately modules with efficiencies of 15.7% were verified as well. Commercial, mass produced, CIGS PV module efficiency, however, is still lower than CdTe PV modules—and this will have a major impact on their future unless ways to increase their efficiency and reduce their costs are found soon.

CIGS TFPV modules have similar problems as those plaguing CdTe TFPV technologies. They have low efficiency, require larger mounting infrastructure, exhibit power loss under excess heat and have a significant annual degradation rate. Scarcity of materials and related toxicity issues are, as in the CdTe PV case, on the back burner for now. These issues must be evaluated from the point of view of large-scale installations, where thousands and millions of these modules will be installed. In such cases, minute amounts of toxic materials in each module are multiplied mega times and become a substantial threat to the environment. Also, special measures must be taken for proper disposal or recycling of these modules.

CIGS research is focused on several of today's challenges of this promising technology:

a.   Pushing efficiencies even higher by exploring the chemistry and physics of the junction formation and by examining concepts to allow more of the high-energy part of the solar spectrum to reach the absorber layer.

b.   Dropping costs and facilitating the transition to a commercial stage by increasing the yield of CIS modules—which means increasing the percentage of modules and cells that make it intact through the manufacturing process.

c.   Decreasing manufacturing complexity and cost, and improving module packaging.

Note: At a meeting of PV specialists in February 2011 (PV Module Reliability Workshop [PVMRW]), the degradation and longevity of PV technologies and products were discussed by representatives of several manufacturing companies. The susceptibility to moisture of SIGS modules was addressed as one of the major concerns, and packaging solutions were presented. Location-specific reliability tests and evaluations were also discussed, which is a step in the right direction.

We are glad that such open discussions are underway, since this is the fastest way to resolve issues and put promising technologies on the energy market.

Fabrication Techniques

There are a number of methods used presently to deposit the precursor materials—Cu, In, and Ga, and Se—onto a substrate, and process these films at high temperatures under proper process conditions. Some of these include electro-deposition, sputtering of metallic layers at low temperatures, CVD at high temperatures, printing of inks containing nano-particles, as well as a special technique used for bonding wafers and components in the semiconductor industry.

We will take look at some of these, hoping they will survive the competition and be successfully used in energy markets of the 21st century.

Electrodeposition Followed by Selenization

Electro-deposition is one of the methods used, although we question its use in large-scale manufacturing operations. Two different methodologies exist presently: deposition of elemental layered structures, and simultaneous deposition of all elements (including Se). Both methods require thermal treatment in a Se atmosphere to make device quality films.

Because electrodeposition requires conductive electrodes, metal foils containing the precursor materials are used in both approaches.

Single layer electrodeposition (sequential deposition of elemental layers) is similar to the process of sputtering elemental layers one by one, except that liquid media is used here, instead of a vacuum chamber. The advantages and disadvantages of this approach are obvious, and currently we are not aware of any commercial scaling up of this process. Nevertheless, some elements of this approach are useful for understanding the properties of CIGS materials and devices and might be used in furthering the CIGS cause in the future.

The method of simultaneous electro-deposition is performed using a working electrode (*cathode*), a counter electrode (*anode*), and a reference electrode. A metal foil substrate is used as the working electrode in industrial processes. An inert material is used for the counter electrode, and the reference electrode exists to measure and control the potential difference between the anode and cathode. The reference electrode allows the process to be performed potentiostatically, meaning the potential of the substrate can be controlled.

Electrodeposition of all elements simultaneously is a difficult processing problem, and to our knowledge there's been no successful large-scale solution to date.

First, the standard reduction potentials of the elements are not the same, causing preferential deposition of a single element. This problem is commonly alleviated by adding different counter ions into the solution for each ion to be deposited (Cu^{2+}, Se^{4+}, In^{3+}, and Ga^{3+}), thus changing the reduction potential for that ion.

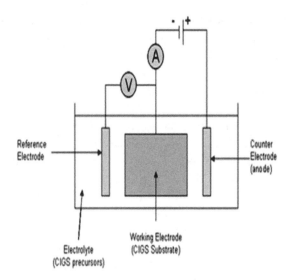

Figure 3-29: CIGS simultaneous electrodeposition

Second, the Cu-Se system has a complicated behavior and the composition of the film depends on the Se4+/Cu2+ ion flux ratio which can vary over the film surface. Because of this behavior, the deposition conditions (specifically precursor concentrations and deposition potential) need to be optimized. Even with optimization, reproducibility is low over large areas due to composition variations and potential drops along the substrate.

The resulting films have small grains, are Cu-rich, and generally contain Cu2-xSex phases along with impurities from the process chemicals. Annealing is always required in order to improve the crystallinity of the resulting devices. To achieve efficiencies higher than 7%, a stoichiometry correction is also required. The correction can be done via high temperature physical vapor deposition (sputtering), but this approach has elements that are impractical for commercial application.

Selenization

Selenization is also required and there are two different techniques for selenization of the CIGS film.

a. Selenization from Se Vapor

The substrate is usually a soda lime glass, coated with a thin film of Mo, which forms the back contact of the solar cell. Cu and In,Ga layers are sequentially deposited on the substrate by sputtering, and the different layers are selenized in an H_2Se atmosphere and then converted into a CIGS thin film via different thermal processes.

The advantage of this process compared with the coevaporation process is that large area depositions of CIGS films are commercially producible. The main disadvantage is that H_2Se is toxic.

b. Annealing of stacked elemental layers

The substrate and deposition technique of Cu and In, Ga are identical to the selenization from vapor processing. The deposition of Se is done by evaporation and the rapid thermal process takes place in an inert atmosphere. With this technique, large-area deposition of CIGS films are commercially producible. The main disadvantage, again, is that H_2Se is toxic.

General Selenization Concerns

The Se supply and selenization environment are extremely important in determining the properties and quality of the film produced from precursor layers. When Se is supplied in the gas phase (for example as H_2Se or elemental Se) at high temperatures, the Se becomes incorporated into the film by absorption and subsequent diffusion. During this step, called chalcogenization, complex interactions occur to form a chalcogenide. See below.

These interactions include formation of Cu-In-Ga *intermetallic* alloys, formation of intermediate metal-selenide binary compounds, and phase separation of various stoichiometric CIGS compounds. Because of the variety and complexity of the reactions taking place, the properties of the CIGS film are difficult to control.[3]

Differences exist between films formed using different Se sources. Using H_2Se yields the fastest Se incorporation into the absorber; 50% Se can be achieved in CIGS films at temperatures as low as 400°C. By comparison, elemental Se only achieves full incorporation with reaction temperatures of 500°C and above. Below 500°C films formed from elemental Se were not only Se deficient, but also had multiple phases including metal selenides and various *alloys*. Use of H_2Se also provides the best compositional uniformity and the largest grain sizes. However, H_2Se is highly toxic and is classified as hazardous to the environment.

Chalcogenization

Here, metal or metal-oxide nanoparticles are used as the precursors for CIGS growth. These nanoparticles are generally suspended in a water-based solution and then applied to large areas by various methods, printing being the most common. The film is then dehydrated and, if the precursors are metal-oxides, reduced in an H2/N2 atmosphere. Following dehydration, the remaining porous film is sintered and selenized at temperatures greater than 400°C.

Several manufacturers are scaling up this process, using oxide particles, or other particles containing inks, usually based on Se content The advantages of this process include uniformity over large areas, non-vacuum or low-vacuum equipment, and adaptability to roll-to-roll manufacturing.

When compared to laminar metal precursor layers, the selenization of sintered nanoparticles is more rapid. The increased rate is a result of the greater surface area associated with porosity. Decreasing high temperature selenization reduces the thermal budget and cost as well.

A major issue is the porosity and a tendency towards rougher absorber surfaces, which create efficiency problems, among others.

Use of particulate precursors allows for printing on a large variety of substrates with high materials utilization—around 90% or more. A major disadvantage is that little is known about the details of this type of deposition.

In some manufacturing processes the printed rolls are cut into cells, which must be binned and integrated in modules similar to how Si devices are made today. The binning process is different from the monolithic integration that many CIGS companies are using, where monolithic integration is far more adaptable to inline production.

Reported cell efficiency to date is 14%, via the coevaporation process, but this has not been officially verified. The verified efficiency is approximately 9%, mostly due to low Voc and low fill factor. This is indicative of a rough surface and/or a high number of defects, which assist the harmful recombination process. Because of that, these CIGS cells have poor transport properties including a low Hall mobility and short carrier lifetime.

These are some issues which the CIGS industry is attempting to solve, before their products become a common element of the 21st century energy markets.

PVD and CVD Methods

In *Photovoltaics for Commercial and Utilities Power Generation*, we described in detail the multi-chamber thin film deposition processes, which we believe are needed to produce quality, cost effective films with defined operational boundaries and little mixing of materials.

We discussed the currently widely used process called co-deposition, or co-evaporation, where all deposition steps are done in the same chamber as the material (substrate) moves along. The obvious advantages of such an approach are increased speed and reduced cost, but as is often the case with any process where speed and cost are the objective, the quality is compromised due to intermixing and other process-related issues.

One doesn't have to be a specialist to understand that when several operations—gasses and material sprays in this case—are conducted in the same confined place, the end result could be a mixture of any type and proportion.

Sputtering of Metallic Layers Followed by Selenization

To avoid the intermixing problems, we use sputtering, which is another way of making CIGS absorbers. In this method, metal films of Cu, In, and Ga are individually sputtered at or near room temperature in a vacuum chamber. The films are sputtered one by one in a precisely controlled environment, and reacted in a Se atmosphere at high temperature. This process has high throughput, close to that of coevaporation, and its compositional uniformity can be more easily controlled and achieved.

Sputtering a stacked multilayer of metal—for example a Cu/In/Ga/Cu/In/Ga structure—produces a smoother surface and better crystallinity in the absorber, when compared to sputtering a simple bilayer (Cu-Ga alloy/In) or trilayer (Cu/In/Ga). These attributes result in higher efficiency devices, but forming the multilayer is a complicated deposition process and probably not worth the cost of extra equipment or process complexity. [19] Additionally, the reaction rates of Cu/Ga and Cu/In layers with Se are different. If the reaction temperature is not high enough, or not held long enough, CIS and CGS form as separate phases. The same considerations outlined in the previous section apply to Se incorporation.

Companies using similar processes include Showa Shell, Avancis (formerly Shell Solar), Miasolé, Honda Soltec, and Energy Photovoltaics (EPV). Showa Shell sputters a Cu-Ga alloy layer and an In layer, followed by selenization in H_2Se and sulfurization in H_2S. The sulfurization step appears to passivate the surface in a way similar to CdS in most other cells. Thus, the buffer layer used is Cd-free which eliminates concerns related to toxicity and environmental impact of Cd. Showa Shell has reported a maximum module efficiency of 13.6% with an average of 11.3% for 3600 cm^2 substrates.

Shell Solar uses the same technique as Showa Shell to create the absorber; however, they use a CdS layer deposited by chemical vapor deposition. Modules sold by Shell Solar have a specification of 9.4% efficiency.

Miasole has had great success in procuring venture capital funds for its process and scale-up. However, little is known about their sputtering/selenization process beyond their stated efficiency of 9 to 10% for modules.

EPV uses a hybrid between coevaporation and

sputtering in which In and Ga are evaporated in a Se atmosphere. This is followed by Cu sputtering and a selenization step. Finally, In and Ga are again evaporated in the presence of Se. Based on Hall measurements, these films have a low carrier concentration and high mobility compared to other devices. EPV films have also been shown to have a low defect concentration.

Coevaporation

Coevaporation, or co deposition, is the most prevalent CIGS fabrication technique, and offers the greatest promise for the future. Some coevaporation processes deposit bi-layers of CIGS with different stoichiometries onto a heated substrate, where they intermix. Another process involves three different deposition steps, some of which are current CIGS efficiency record holders at close to 20%.

The first step in this process is co deposition of In, Ga, and Se. This is followed by Cu and Se deposited at a higher temperature to allow for diffusion and intermixing of the elements. In the final stage, In, Ga, and Se are again deposited to make the overall composition Cu deficient.

Commercial modules have been known to underperform, usually due to low Voc, which is characteristic of high defect density and high recombination velocities.

As most of the CIGS research is in the areas of coevaporation. Companies using this technique will benefit most from the achievements of the scientific community. But the issues are still significant, with device uniformity and related difficulties of inline vacuum coevaporation systems being the most critical.

High film growth temperatures which raise thermal budget and cost is another problem to be tackled, in addition to the notorious material under-utilization (waste of material in the inner chamber walls) of the coevaporation process. The process equipment, its capital cost, and operation and maintenance are expensive as well—all increasing the complexity and cost of the manufacturing process.

One way to enhance the utilization of selenium is to use a thermal or plasma-enhanced selenium-cracking process. This configuration can be coupled with an ion beam source for ion beam-assisted deposition, which enhances the properties of the devices, but complicates further the manufacturing process, making it cost prohibitive.

In the CIGS coevaporation process, thin films are produced by evaporating Cu, In, Ga and Se from elemental sources, as shown below.

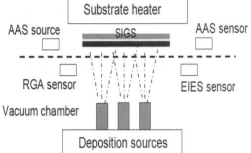

Figure 3-30. Setup including Process Control Units

To achieve the favored film composition, precise control of the evaporation rates is necessary. Therefore an electron impact emission spectrometer (EIES) and an atomic absorption spectrometer (AAS) or a mass spectrometer is used. The process requires a substrate temperature between 300°C and 550°C for a certain time during film growth.

The inverted three-stage process is the most favored coevaporation process. At first, In, Ga and Se are evaporated at different rates and deposited as (In,Ga)2Se3 at 300°C on the substrate.

Afterwards Cu and Se are evaporated and deposited on the substrate at elevated temperatures. Last, In, Ga, and Se are evaporated again.

The inverted three-stage process leads to smoother film morphology and high efficiency solar cells. The inline system in continuous mode can produce large quantities of finished devices, which reduces their final cost, so we see this approach as one long-term solution.

Chemical Vapor Deposition

Chemical vapor deposition (CVD) is a well understood process, which has been used in the semiconductor industry during the last 3-4 decades. It has been implemented in multiple ways for the deposition of CIGS as well. Processes include atmosphere pressure metal organic CVD (AP-MOCVD), plasma-enhanced CVD (PECVD), low-pressure MOCVD (LP-MOCVD), and aerosol assisted MOCVD (AA-MOCVD). Multiple source precursors must be homogeneously mixed and the flow rates of the precursors have to be kept at the proper stoichiometry. Single-source precursor methods do not suffer from these drawbacks and should enable better control of film composition compared to multiple source precursors, but are slower, so a compromise must be reached.

CVD is not yet used commercially for CIGS manufacturing, because CVD-produced films have low efficiency and a low Voc, which is partially a result of a high defect concentration. Also, the film surfaces are generally rough which contributes to further decrease of the Voc.

However, the requisite Cu deficiency can be achieved using AA-MOCVD along with a (112) crystal orientation. Still, if the film quality produced by CVD can be improved, any company using this technique could benefit from knowledge gained in other industries using large area CVD deposition, such as glass coating manufacturers.

CVD deposition temperatures are lower than those used for other processes such as co-evaporation and selenization of metallic precursors. Therefore, CVD is more economical from an energy standpoint. Converting commercial CVD equipment to an inline process as well as the expense of handling volatile precursors is a major problem, which needs to be resolved before this technique is widely used.

Roll-to-roll Deposition

Polyamides or stainless steel foils are used as flexible substrates coated with a thin film of Mo. If stainless steel foil is used as a substrate, an insulator between the steel foil and the Mo layer is necessary for interconnection. Polyamides provide high electric strength and high dimensional stability under heat. Cu, In, Ga and Se are deposited by an ion beam-supported, low-temperature deposition technique. The low cost production and available flexible modules with high power per weight ratio are huge advantages, but the efficiency is still lower compared to the other processes.

Electrospray Deposition

Recently, a new technique for the deposition of CIS films has been introduced as electrospray deposition. This technique involves the electric field-assisted spray of ink-containing CIS nano-particles directly onto the substrate and then sintering in an inert environment. The main advantage is that the process takes place at normal room conditions and it is possible to attach this process to some continuous or mass production system like the roll-to-roll production mechanism.

Breakthroughs and Commercial Applications

Despite CIGS having the advantage over CdTe where both have heavy metal Cadmium usage as well as rare-earth Telluride availability issues, the development of CIGS lags behind CdTe commercially. In the laboratory, 18.7% efficiency on a flexible medium was achieved with CIGS cells, though it remains to be seen how well

mass production values can be achieved with any CIGS cell. Commercial production of flexible CIGS cells has begun in Berlin, Germany; commercial production after initial production runs has begun on an annual 35MW capable facility.

Challenges

The biggest challenge for CIGS modules has been the limited ability to scale up the process for high throughput, high yield and low cost. Several deposition methods are used: sputtering, "ink" printing and electroplating, with each having different throughput and efficiencies. Both glass and stainless steel substrates are used, where the stainless steel substrates could yield flexible solar cells. Another big problem of this technology is indium shortage and rising cost, since it is heavily used in indium tin oxide (ITO), a transparent oxide that is used for flat screen displays such as TVs, computer screens, and many other applications.

Some manufacturing techniques waste only 5% indium in their process, while others lose 50-60% of the material—usually in the vacuum sputtering technology where material gets sprayed inside the chamber walls.

CIGS modules also contain toxic materials, so they need to be recycled, which is easily achieved as demonstrated by the large-scale recycling program in Germany, which supports not only the modules recycling, but also employs industrial regenerative strategies of the thin film materials in the recycled modules.

Conclusion

CIGS as used in photovoltaic cells (CIGS PV cells) are usually in the form of polycrystalline thin films. Unlike the silicon cells based on a single p-n junction formation, CIGS are a more complex hetero-junction system. The best efficiency achieved thus far is in the 20% range. The best efficiencies are obtained by re-arranging the CIGS surface and making it look like CIS. CIGS high efficiency is the highest compared with those achieved by other thin film technologies such as Cadmium Telluride (CdTe) or amorphous silicon (a-Si). As for CIS, and CGS solar cells, world record total area efficiencies are 15.0% and 9.5% respectively.

With record CIGS efficiency at 20% for several years, new trends of CIGS research have shifted to investigation on low cost deposition methods that could be alternatives to expensive vacuum processes.

Non-vacuum solution processes progressed quickly and efficiencies of 10%-15% have been achieved by many parties, such as ISET, Nanosolar and IBM.

CIGS solar cells are not as efficient as crystalline

silicon solar cells, for which the record efficiency is 24.7%, but they are expected to be substantially cheaper due to the much lower material cost and potentially lower fabrication cost.

Being a direct bandgap material, CIGS has very strong light absorption and only 1-2 micron meter of CIGS is enough to absorb the greater portion of the sunlight.

Yet for crystalline silicon, it would require much thicker material to do the same job. Therefore, CIGS belongs to a category of solar cells called thin film solar cell (TFSC).

Other materials in this group include CdTe and amorphous Si. Their record efficiencies are slightly lower than that of CIGS for lab-scale top performance cells. Another advantage of CIGS compared to CdTe PV modules, is the fact that much smaller amount of toxic material cadmium are present in CIGS cells.

CZTS Solar Cells

Copper zinc tin sulfide (CZTS) is fairly to the solar industry. Although it was a well established process for semiconductor devices, its use in solar cells was considered in the early 2000s.

It is different from the rest in that it is a quaternary semiconducting compound of the class of related materials, related to the I2-II-IV-VI4, such as copper zinc tin selenide (CZTSe) and the sulfur-selenium alloy CZTSSe.

CTZS can be obtained from the chalcopyrite CIGS structure by substituting the trivalent In/Ga with a bivalent Zn and IV-valent Sn. CZTS.

Carrier concentrations and absorption coefficient of CZTS are similar to CIGS. Other properties such as carrier lifetime (and related diffusion length) are low (below 9 ns) for CZTS. This low carrier lifetime may be due to high density of active defects or recombination at grain boundaries.

Many secondary phases are possible in quaternary compounds like CZTS, and their presence can affect the solar cell performance. Secondary phases can provide shunting current paths through the solar cell or act as recombination centers, both degrading solar cell performance.

It appears that all secondary phases have a detrimental effect on CZTS performance, and many of them are both hard to detect and commonly present. Common phases include ZnS, SnS, CuS, and Cu2SnS3. Identification of these phases is challenging by traditional methods like X-ray diffraction (XRD) due to the peak overlap of ZnS and Cu2SnS3 with CZTS.

CZTS is made by a variety of semiconductor processing (vacuum and non-vacuum) techniques, similar to those used to make CIGS, but the optimal process parameters are different.

The production methods can be broadly categorized as vacuum deposition vs. non-vacuum, and single-step vs. sulfization/selenization reaction methods. Vacuum-based methods predominate in the CIGS industry today, but there is increasing progress in the less-complex, more flexible possibility to deposit thin films on large area substrates.

A major challenge in the making of CZTS thin films and the related alloys is the volatility of the some of the elements in their structure (Zn and SnS in particular), which are unstable and evaporate easily during processing.

Once CZTS films are formed, element volatility decreases but is still a problem, since CZTS films will decompose into binary and ternary compounds under certain operating conditions—especially when exposed to high temperatures during field operations.

CZTS can be made using wet chemistry methods which allow CZTS formation at low temperatures, thus avoiding decomposition. These processes also avoid volatility problems, which are more evident in vacuum processes.

The optical and electronic properties of CZTS are similar to those of the better known type of CIGS (copper indium gallium selenide) process for making solar cells. CZTC is quite efficient as a thin-film absorber layer in solar cells, and unlike CIGS and CdTe, it is composed of abundant, non-toxic materials. The price and availability (or lack thereof) of indium and tellurium, used in the competing thin film technologies, as well as serious toxicity of cadmium are major disadvantages for their future use.

Recent material improvements for CZTS have proven its clean and green nature, and increased its efficiency to well above 10%. Nevertheless, much more work is needed before we see CZTS solar cells and modules on the energy markets.

Theoretically, the abundance and cleanliness of the materials in CTZS thin films is an attractive alternative to CIGS and CdTe thin films, which are not that abundant—and some are toxic.

These, in addition to ever increasing efficiency, are the greatest benefits of CTZS solar technologies, which is enough to keep them in the list of "most promising" thin films for use in the 21st century.

Cadmium Telluride (CdTe) Solar Cells

Cadmium telluride (CdTe) has long been known to have the ideal band-gap (1.48 eV) with a high direct absorption coefficient for a solar absorber material, and it is recognized as a promising photovoltaic material for thin-film solar cells.

Small-area CdTe cells with efficiencies of greater than 15% (as of the summer of 2012 over 16%) and CdTe modules with efficiencies of greater than 9% have been demonstrated. CdTe, unlike the other thin film technology, is easier to deposit and more apt for large-scale production in addition to its hetero-junction potential.

The toxicity of cadmium (Cd) and the related environmental issues remain somewhat of a problem for this technology, which is why First Solar has introduced a recycling program for decommissioned PV modules.

The CdTe technology has been getting very popular lately, and large installations are popping up in US deserts. This is due mostly to the lower cost of modules and also generous government subsidies and loan guarantees.

Cadmium telluride (CdTe) is a type of solar module based on thin films of the heavy metal cadmium and its compounds, cadmium telluride (CdTe) and cadmium sulfide (CdS). The toxicity of cadmium is well known, so its presence in the CdTe TFPV modules is a serious issue which has been successfully swept under the rug by manufacturers and government officials alike. Sooner or later, however, it must be dealt with because there are thousands of desert acres covered with these unproven potentially toxic CdTe PV modules.

Another potential issue with this technology is the availability of the rare and exotic metal tellurium used in the manufacturing of CdTe modules. Raw-materials constraints could affect the cost of modules.

CdTe is an efficient light-absorbing material, quite adaptable for the manufacture of thin-film solar cells and modules. Compared to other thin-film materials, CdTe is easier to deposit in mass-production environments and more suitable for large-scale production.

CdTe bandgap is 1.48 eV, which makes it almost perfect for PV conversion purposes. At 16.5% demonstrated efficiency in the lab, it is a candidate for a major role in the energy future. Mass production modules are sold with 8-9% efficiency. No significant increase is expected with the present production materials and methods, although manufacturing costs are down—at or below \$1.0/Wp.

With a direct optical energy bandgap of 1.48 eV and high optical absorption coefficient for photons with energies greater than 1.5 eV, only a few microns of CdTe are needed to absorb most of the incident light. Because only very thin layers are needed, material costs are minimized, and because a short minority diffusion length (a

few microns) is adequate, expensive materials processing time and costs can be avoided.

CdTe TFPV cells and modules generally consist of a front contact, usually a transparent conductive oxide (TCO), deposited onto a glass substrate. The TCO layer has a high optical transparency in the visible and near-infrared regions and high n-type conductivity. This is followed by the deposition of a CdS window layer, the CdTe absorber layer, and finally the back contact.

For high-volume devices, the CdS layer is usually deposited using either closed-space sublimation (CSS) or chemical bath deposition, although other methods have been used to investigate the fundamental properties of devices in the research laboratory. In all cases, mass production and automation is possible, which is the greatest advantage of this technology.

Figure 3-31. CdTe cell

The CdTe p-type absorber layer, 3-10 μm thick, can be deposited using a variety of techniques including physical vapor deposition (PVD), CSS, electrodeposition, and spray pyrolysis. To produce the most efficient devices, an activation process is required in the presence of CdCl2 regardless of the deposition technique. This treatment is known to recrystalize the CdTe layer, passivating grain boundaries in the process, and promoting inter-diffusion of the CdS and CdTe at the interface.

Forming an ohmic contact to CdTe, however, is difficult because the work function of CdTe is higher than all metals. This can be overcome by creating a thin p+ layer by etching the surface in bromine methanol or HNO3/H3PO4 acid solution and depositing Cu-Au alloy or ZnTe:Cu. This creates a thin, highly doped region that carriers can tunnel through. However, Cu is a strong diffuser in CdTe and causes performance to degrade with time. Another approach is to use a very low bandgap material, e.g. Sb2Te3, followed by Mo or

W. This technique does not require a surface etch, and the device performance does not degrade with time.

CdTe PV modules manufacturing is a sophisticated process, much more sophisticated than that of conventional c-Si modules, which uses simple 1970s manufacturing equipment and materials. CdTe TFPV modules are manufactured with the help of modern, complex and expensive semiconductor type equipment, so the precision and accuracy of the process steps and the quality of the final product are limited only by the quality of the materials and supplies, and the capabilities of the engineers, technicians and operators on the production lines.

CdTe thin-film solar modules are now being mass produced very cheaply, and it is expected with economies of scale that they will achieve the cost reduction needed to compete directly with other forms of energy production in the near future. Since CdTe thin film PV devices still have far to go to achieve maximum efficiencies, it will be interesting to see which materials and methods are most successful.

The most efficient CdTe/CdS solar cells (efficiencies of up to 16.5%) have been produced using a Cd2SnO4 TCO layer which is more transmissive and conductive than the classical SnO2-based TCOs, and includes a Zn2SnO4 buffer layer which improves the quality of the device interface.

CdTe PV manufacturer research at universities and in R&D labs focuses on some of these challenges:

a. Boosting efficiencies by exploring innovative transparent conducting oxides that allow more light to be absorbed by the cell and at the same time more efficiently collect the electrical current generated by the cell.

b. Studying mechanisms such as grain boundaries that can limit voltage.

c. Understanding the degradation some CdTe devices exhibit at the contacts and redesigning the devices to minimize this phenomenon.

d. Designing module packages that minimize any outdoor exposure to moisture.

e. Engaging aggressively in both indoor and outdoor cell and module stress testing.

These efforts are geared to address the main problems with CdTe PV modules:

a. The relatively low efficiency which contributes to using more land and mounting hardware;

b. Temperature power degradation, c) annual power degradation, and other negative long-term effects;

c. Availability of the rare metals used in CdTe TFPV technology; their toxicity, and other serious issues which are being disregarded by manufacturers.

Let's hope that we won't have to wait for a serious accident before bringing these issues out in the open and discussing possible solutions.

Conclusions

Obviously, the TFPV industry is in its infancy and a lot of work remains to be done before it becomes a proven, mature technology. There are many research groups active in the field of photovoltaics in universities and research institutions around the world. This research can be divided into three areas: a) making solar cells and modules cheaper and/or more efficient to effectively compete with other energy sources; b) developing new technologies based on new solar cell architectural designs; and c) developing new materials to serve as light absorbers and charge carriers.

Thin-film photovoltaic cells and modules use less than 1% of the expensive raw material (silicon or other light absorbers) compared to silicon wafer-based solar cells, leading to a significant price drop per Watt peak capacity. Different thin-film processes are being researched by groups worldwide; however, it remains to be seen if these solutions can achieve a similar market penetration as traditional bulk silicon solar modules.

An interesting aspect of thin-film solar cells is the possibility of depositing cells on all kinds of materials, including flexible substrates (PET for example), which opens a new dimension for applications. The future will surely bring even more good news to the thin film industry.

The existing c-Si and thin film PV technologies are compared in Table 3-1.

In practical terms, c-Si technologies have some advantages that must be considered when designing PV power plants, especially:

1. Large-scale installations in the deserts, where thin film PV is unproven;

2. Areas of limited land availability, where higher efficiency PV is needed; and

Table 3-1. Comparing c-Si and thin film PV technologies

	Description	Thin Film PV	c-Si PV	Therefore c-Si PV is:
	Process			
1	Raw materials	Exotic	Everyday type	No shortages expected
2	Process chemicals	Complex	Simple	Lower cost
3	Equipment	Complex	Simple	Lower capital cost
4	Mfg. process	Complex	Simple	Lower labor cost
5	Process quality	Excellent	Simple	Lower quality expectation
	Electro-mechanical			
1	Power output	4-8 W/ft2	10-15W/ft2	2 times more powerful
2	OC voltage	45V-100V	20V-50V	2 times lower
3	String power	200-750W	1500-3000W	4-6 times higher
4	Stability	Poor	Good	Overall superiority
5	Voltage isolation	Intermittent	Excellent	Overall superiority
6	Design flexibility	Poor	Excellent	Overall superiority
7	Blocking diode	Yes	No	Less maintenance
8	Glass strength	Annealed	Tempered	5-6 times stronger
9	Lamination	Poor*	Good	More reliable
10	Opaque to UV	No	Yes	Better quality
	Field Application			
1	Land use	4-5 acres/MWp	10-12 acres/MWp	2-3 times less land use
2	Cost $/Wp (2011)	$1.80/Wp	$1.80/Wp	Comparable**
3	Cost tendency	Uncertain	Decreasing	Cost decreasing
4	Temp. coefficient	0.3-0.4/C°	0.5-0.6/C°	Somewhat higher
5	Degrading	1.0-1.5%/annum	0.5-1.0%/annum	Lower degrading
6	BOS components	Increased qty.	Standard	Higher cost
7	Test failure rates	Higher	Lower	Less failures***
8	Field failure rates	Higher	Lower	Less failures
9	Hazards	Toxic compounds	Lead, solvent	Less toxicity
10	EOL recycling	100% hazmat	Partial	Lower EOL cost

NOTE: * Modules have no side frames
         ** Cradle-to-grave costs
         *** Damp heat test failures

3. The need for reduced use of toxic materials (such as those used in manufacturing CdTe and some CIGS thin film PV modules).

PROMISING TECHNOLOGIES

Below are descriptions of PV technologies that are already commercially viable, or will become so.

High Efficiency Technologies
Background

A new phenomenon is taking shape around the world—the tendency to use the highest performance solar cells and modules. This is due primarily to the lack or high cost of land available for solar installations.

The other, equally important consideration in preferring high performance PV products is the fact that higher performance solar cells and modules are of relatively higher quality. This means that they will be much more reliable during daily operation, and will last for the duration without problems, even in the harshest climates—something that cannot be said for many solar cell types.

This preferential treatment will become more and more pronounced during the 21st century. Japan announced in the summer of 2012 that it will spend over $10 billion on solar installations—giving priority to cells and modules that are in the above-mentioned "high performance" class.

It happens that most of the high performance solar cells and modules are made by Japanese manufacturers, such as Sony, Sanyo, Sharp, and Mitsubishi, so there might be some bias in this preference.

Nevertheless, the fact that a major player in the solar market is drawing a line based on higher efficiency (and by association higher quality) is a major push toward quality preferences.

The High Performers

There are types of solar cells and modules, which use standard materials and process sequences with some modifications—just enough to give them superior qualities. When evaluating such technologies and devices, keep in mind that changing (even improving) one parameter alone is not enough. If efficiency is increased at the expense of extreme sunlight degradation, then the "improvement" is worthless, and must be abandoned or changed in a way that does not negatively affect the other operating parameters of the final device.

The types of cells we are discussing here fall in the category of "high-efficiency" solar cells, simply because they are more efficient than the standard lot of similar materials and processes.

Some of these technologies are:

HIT Solar Cells

The Hetero-junction with Intrinsic Thin layer (HIT) solar cell is composed of a single thin crystalline silicon wafer, usually 4" or 5" square, sandwiched between two thin films of amorphous silicon (a-Si).

The structure of the HIT cell improves the overall output by reducing recombination loss—the loss of electrical current that occurs when an electron and a hole (carriers) generated by impinging photons within the solar cell combine and disappear.

This effect is achieved by surrounding the energy generation layer of single thin crystalline silicon with high quality ultra-thin amorphous silicon layers.

Several manufacturers have developed special technologies for cleaning the energy generation middle layer and protecting it from damage during construction of the surrounding layer, the result being an increase in the open circuit voltage from 0.718V to 0.722V. This also improves efficiency and reliability of the solar cell by avoiding the usual problems related to doped and fired silicon solar cells.

This type of solar cell would be more expensive to produce, but might provide the much needed solution to the problems c-Si and TFPV modules encounter when exposed to extreme desert temperatures. Such cells theoretically would exhibit less heat sensitivity and power degradation, but we have not yet seen any proof of that happening.

HIT cells have improved efficiency, over 22% in lab tests and around 19% in mass production, mostly due to the good a-Si:H/c-Si/a-Si:H hetero-interface of the HIT structure, which enables a high Voc (well over 0.7 V). This also results in much better temperature coefficient and better heat handling properties (theoretical values).

To reduce manufacturing cost, numerous technologies are in R&D mode presently, with the goal to further improve efficiency, to increase the Voc of HIT solar cells, and to achieve up to 25% efficiency.

Commercial modules have the following key electrical and performance properties:

Cell Efficiency (n%)	19.7%
Module Efficiency (n%)	17.0%
Max. Power Voltage (Vpm)	55.8V
Max. Power Current (Ipm)	3.59A
Open Circuit Voltage (Voc)	68.7V
Short Circuit Current (Isc)	3.83A
Temperature Coefficient (Pmax)	−0.29 %/C
Temperature Coefficient (Voc)	−0.172 V/C
Temperature Coefficient (Isc)	0.88 mA/C

This looks good, but questions related to the major issues surrounding the use of solar cells and modules, remain:

1. Without doubt, the additional a-Si films improve the efficiency of the devices, but do they also improve the device quality and performance characteristics? What if these films misbehave under blistering desert sun? What if with time they cause increased resistance, overheating, and other anomalies, which lead to reduced efficiency or failures?

2. The additional materials and labor add to the final cost of the device, but is the added cost justified by the marginally higher efficiency?

Bifacial Solar Cells

Bifacial solar cells could be a modification of the HIT model, or some other design that basically utilizes both surfaces to collect sunlight and convert it into electricity.

While there is some benefit in using both surfaces, we see it as quite limited, and basically confined to a niche markets such as fences, car-ports and similar structures that have two open sides.

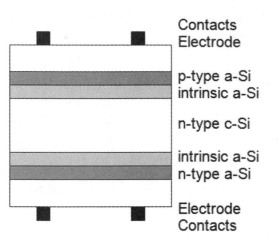

Figure 3-32. Bifacial solar cell

The power that could be generated from the shaded side of a PV module is quite limited. Special modifications and white paint on the surrounding surfaces can be done to increase the reflected sunlight, but that also is expansive, impractical, and more expensive in most cases.

This condition was confirmed by installations done by Solyndra (whose tubular panels also were touted as "bifacial"). Entire roofs were painted white, as needed to reflect more sunlight. While that increased the power output, it was not enough to justify the initial expense and the continuous maintenance. Too, with time the roofs got dirty and their reflectivity decreased, eliminating the slight advantage of the white surface.

So, bifacial technology is a.) too new and not proven reliable as yet, and b.) it comes at a higher price, which in most cases cannot be justified by a marginal increase of electric output.

Although we don't doubt the practicality and flexibility of bifacial solar technologies, we see them filling only small market niches. We do not foresee them as major participants in the large-scale PV power generation of the 21st century.

Buried Contact Solar Cells

The buried contact solar cell is another high efficiency commercial solar cell, this one based on a metal contact formed inside a laser-formed groove in the front surface. The buried contact technology has many advantages and avoids the problems of conventional solar cells related to the old screen-printed contacts method. This allows buried contact solar cells to have performance up to 25% better than commercial screen-printed solar cells.

A key high efficiency feature of the buried contact solar cell is that the metal is buried in a laser-formed groove inside the silicon solar cell, which allows for a large metal height-to-width aspect ratio of the contacts, which improves the device's operational characteristics, since a large metal contact aspect ratio allows a large volume of metal to be used in the contact finger, without having a wide strip of metal on the top surface. This allows a large number of closely spaced metal fingers, while still retaining a high transparency of the front surface.

For example, on a large area device, a screen-printed solar cell may have shading losses as high as 10 to 15%, while in a buried contact structure, the shading losses will be only 2 to 3%. These lower shading losses allow a higher level of photon collection (more photons hit the active surface), and low reflection (less photons hit the metalized front surface areas), which contributes to higher short-circuit currents.

In addition to good reflection properties, the buried contact technology also allows low parasitic resistance losses due to its high metal aspect ratio, its fine finger spacing, and its plated metal for the contacts.

The emitter resistance is reduced in a buried contact solar cell since narrower finger spacing dramatically reduces emitter resistance losses. The metal grid resistance is also low since the finger resistance is reduced by the large volume of metal in the grooves and by the use of copper, which has a lower resistivity than the metal paste used in screen printing.

The contact resistance of a buried contact solar cell is lower than that in screen printed solar cells due to the formation of a nickel silicide at the semiconductor-metal interface and the large metal-silicon contact area. Overall, these reduced resistive losses allow large-area solar cells with high FFs.

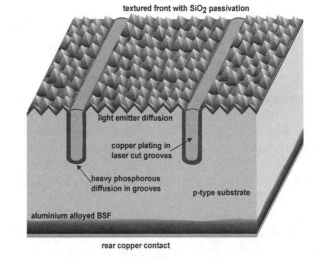

Figure 3-33. Buried contact solar cell (1)

When compared to a screen-printed cell, the metallization scheme of a buried contact solar cell also improves the cell's emitter. To minimize resistive losses, the emitter region of a screen-printed solar cell is heavily doped and results in a "dead" layer at the surface of the solar cell. Since emitter losses are low in a buried contact structure, the emitter doping can be optimized for high open-circuit voltages and short-circuit currents. Furthermore, a buried contact structure includes a self-aligned, selective emitter, which thereby reduces the contact recombination and also contributes to high open-circuit voltages.

Buried contact technology provides significant cost and performance benefits. In terms of $/W, the cost of a buried contact solar cell is the same as a screen-printed solar cell. However, due to the inclusion of certain area-related costs as well as fixed costs in a PV system, a higher efficiency solar cell technology results in lower cost electricity. Also, buried contact technology is can be used for concentrator systems of up to 50x concentration.

PERL Solar Cells

There are several types of solar cells that can be built on silicon substrates, but which require much more expensive materials, sophisticated equipment, and elaborate processing. The passivated emitter with rear locally diffused cell (PERL) is one of these devices.

It is a fairly recent development of a solar cell design that uses micro-electronic techniques to produce solar cells of high efficiency, up to 25% in the PERL case.

Thinner silicon wafers (solar cells) are used in this design, to optimize the efficiency, with 50-μm-thick cells having the highest efficiency. Using such thin wafers in production is impossible due to breakage, so a compromise in the 150-200μm wafer thickness area must be used.

One characteristic of this type of solar cell is the improved texturing of the active front surface area, which is done in the shape of inverted pyramids. If the base size of the inverted pyramid texturing is increased, then the effective surface area increases, which in turn increases absorption in the silicon structure. So a lot of effort is put into the optimized size and shape of the pyramids.

Bigger texture structures (>10 μm) increase the optical absorption and the resulting cell efficiency. Using conventional chemical texturing, the maximum achieved base size of the pyramid is around 9 μm. So, optimizing the pattern size is best done via photolithography methods, similar to those used in the semiconduc-

tor industry. This lithographical step adds an additional processing step, but it has several advantages which outweigh the extra effort and expense.

Placing the metal contacts on top of the pyramidal structure is also critical and requires careful consideration and precise execution. The contacts placement will determine the path the carriers will travel to reach the contact, so the position of the contacts is critical for the overall electricity generation process. Properly executed front metal contact lines could be as thin as 2 microns, which would reduce the front surface shading to about 3% of the surface area. The reduced contact area reduces the total recombination at its proximity, which contributes to higher Voc of the device.

The passivated emitter is the high quality oxide deposited at the front surface of the cell, which significantly lowers the number of carriers recombining at the surface. The rear of the cell is locally diffused only at the metal contacts areas, to minimize recombination at the rear while maintaining good electrical contact.

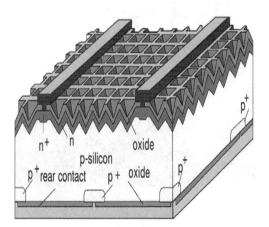

Figure 3-34. PERL solar cell

A major advantage of the PERL solar cell is the improved surface passivation of the front and rear cell areas via thermally grown oxide, and the respective dopants, which significantly reduces the emitter saturation current, thus leading to improved open-circuit voltage (Voc above 0.7V). This also reduces the recombination levels at the contact regions, due to suppressed minority concentration in these areas. PERL cells can reach 25% efficiency, but commercial types operate in the 19-20% range.

PERL solar cells have not been proven reliable for large-scale operation under different environmental conditions. This, combined with their complex manufacturing process and higher cost will have to be tested by time and locations before the technology is fully established as a major competitor in world markets.

PERL solar cells seem to be suitable for some special applications, such as solar powered vehicles and space applications, so we foresee their number increasing as time goes by.

Rear Contacts Solar Cell

Rear contact solar cells eliminate shading losses altogether by putting both contacts on the rear of the cell. By using a thin solar cell made from high quality material, electron-hole pairs generated by light that is absorbed at the front surface can still be collected at the rear of the cell. Such cells are especially useful in concentrator applications where the effect of cell series resistance is greater. An additional benefit is that cells with both contacts on the rear are easier to interconnect and can be placed closer together in the module since there is no need for a space between the cells.

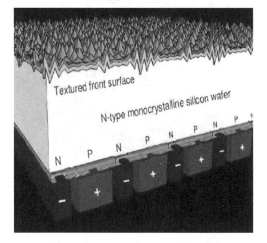

Figure 3-35. Rear contact solar cell

Efficiency of over 23% has been achieved with this cell design. These types of solar cells usually cost considerably more to produce than standard silicon cells, and are typically used in specialized applications such as solar cars and for space exploration.

Passivated Rear Point Contacts Solar Cell

This solar cell is even more specialized, due to its complexity, and although we don't see its use as such in the near-term energy markets, we do believe that some of its features (materials and processes) will be used in the future for developing new technologies. This alone is enough to put it in the category of most promising technologies for the 21st century use.

Efficiency of over 22% has been reached with these cells, but due to their complexity and higher cost, their use is limited.

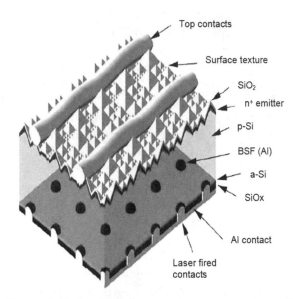

Figure 3-36. Passivated rear point contacts solar cell

Selective Emitter Solar Cells

The "selective" emitter solar cells concept uses laterally different emitter doping which consists basically of:

1.   High doping under the front side metallization for low contact resistance between contact metal and semiconductor interface,

2.   Lower doping between the contact fingers for a better short wavelength response due to less Auger recombination as well as improved emitter passivation.

Unfortunately, most of these designs are very complex as far as the process is concerned, due to a number of masking steps required for achieving proper selective diffusion or emitter etch back.

In particular, such existing concepts are hard to implement into industrial production of solar cells. Nevertheless, several different approaches designed for industrial production are currently under development, and some are already being transferred to the industry.

A number of organizations are developing solar cells with these types of selective emitters, which allows the use of thinner emitters as needed to improve the short-wavelength spectral response.

We foresee the use of this technology limited to niche markets, but its components are valuable instruments in the battle for higher efficiency, reliability and cost effectiveness, which can be successfully used in the development of other solar technologies in the near future.

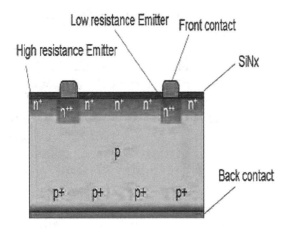

Figure 3-37. Selective emitter solar cells

Intermediate Band

This is intermediate band nitride thin film semiconductor material, which might be an alternative to the multijunction designs for improving power conversion efficiency of solar cells.

The intermediate-band solar cell is a thin-film technology based on highly mismatched alloys. The three-bandgap, one-junction device has the potential of improved solar light absorption and higher power output than III-V triple-junction compound semiconductor devices.

Efficiencies of over 20% have been demonstrated.

Roll-to-roll TFPV

Roll-to-roll process is reviewed in the "High Efficiency Technologies" category not because it produces high efficiency solar devices, but because the manufacturing process is of the highest efficiency possible today.

The process consists of coasting the substrate material from a roll. The process can be conducted at a high speed and relatively low temperatures. Thus it is inexpensive, fast, and less energy intensive than the other PV technologies.

Another advantage is that it can be produced using existing coating and printing equipment, and thus does not require construction of a new facility. This simplifies the manufacturing scale-up and significantly lowers capital and labor costs as compared to the previous generation solar cells.

Compared with the competing PV technologies, the roll-to-roll process:

- Is non-toxic and environmentally friendly,
- Significantly lowers capital cost for new plant construction,
- Uses less energy, and
- Offers low cost retrofit of existing facilities.

Several world-class companies are developing and using this process, which we believe will become a major competitor in several niche energy markets in the 21st century.

III-V PV Cells

There are several technologies that fall in the gap between the most promising and exotic technologies but weigh on the side of most promising only because they are theoretically promising. In most cases, some good results have been shown as well, but generally speaking these technologies have not been fully proven... yet.

Some of these "semi-most promising" solar technologies to watch for in the 21st century are the so-called "III-V PV" devices. These are PV technologies based on the deposition of thin films of the III-V materials. These devices are divided into single- and multi-junction.

Single-junction III-V Devices

Compounds such as gallium arsenide (GaAs), indium phosphide (InP), and gallium antimonide (GaSb) have adequate energy band gaps, high optical absorption coefficients, and good values of minority carrier lifetimes and mobility, making them excellent materials for high efficiency solar cells. These materials are usually produced by the Czockralski or Bridgmann methods, which provide high quality with increased efficiency and reliability, but at a higher price.

After silicon, GaAs and InP (III-V compounds) are the most widely used materials for single-junction (SJ) solar cells manufacturing. These have optimum band gap values (1.4 and 1.3 respectively) for SJ conversion of sunlight. The construction of solar cells made of these materials is similar to the regular single-junction c-Si solar cells we discussed elsewhere.

The major disadvantage of using III-V compounds for PV devices is the high cost of producing their materials and the related manufacturing processes. Also, crystal imperfections, including bulk impurities, severely reduce their efficiencies, so that only very high quality materials could be considered. Too, they are heavier than silicon, which requires the use of thinner cells, but they are weaker mechanically, so their design requires a delicate balance of thickness vs. weight.

The combination of high efficiency, high price, crystal imperfections intolerance, and mechanical weakness makes these devices useful for limited applications, where efficiency and overall behavior is more important than price. Thus they are not widely used in the general PV market, but still can be found in some important niche markets.

Multi-junction III-V Cells

Single-junction PV devices convert only a portion of the sunlight (with photons just above the band-gap level of the semiconductor material). The problem is that photons with lower or higher energy do not generate electron-hole pairs and are lost as heat, which is also detrimental to the cells, reducing their efficiency and deteriorating them over time.

One way to solve this problem and to increase the efficiency of PV devices is to add more junctions, thus creating multi-junction (MJ) solar cells. By selecting materials, properties, number and types of junctions, and manufacturing processes designed to capture the majority of sunlight photons, we can reach very high efficiencies.

Thus far, multi-junction solar cells made primarily using the III-V compounds have clearly proven that by minimizing thermalization and transmission losses, large improvements in efficiency can be made over those of single-junction cells. These devices find use in generating power for space applications and in concentrator systems. They show great promise for high efficiency and reliability under harsh climate conditions, such as those in the deserts.

Future development of multi-junction devices using low-cost thin-film technologies is especially promising for producing more efficient and yet inexpensive devices. Cost reductions will also be significant when thin-film technologies are directly produced on building materials other than glass, because many materials such as tiles and bricks can be substantially cheaper than glass and have much lower energy contents.

The devices in this group are gallium arsenide-based, germanium-based, and CPV solar cells.

Gallium Arsenide Based Multi-junction Cells

High-efficiency multi-junction cells were originally developed for special applications such as satellites and space exploration, but at present, their use in terrestrial concentrators might be the lowest cost alternative in terms of \$/kWh and \$/W. These multi-junction cells consist of multiple thin films produced via metalorganic vapor phase epitaxy. A triple-junction cell, for example, may consist of the semiconductors GaAs, Ge, and GaInP2.

Each type of semiconductor will have a characteristic band gap energy which, loosely speaking, causes it to absorb light most efficiently at a certain color, or more precisely, to absorb electromagnetic radiation over a portion of the spectrum. Semiconductors are carefully chosen to absorb nearly all of the solar spectrum, thus generating electricity from as much of the available solar energy as possible.

GaAs-based multi-junction devices are some of the most efficient solar cells to date, reaching a record high of 40.7% efficiency under "500-sun" solar concentration and laboratory conditions.

This technology is currently being utilized mostly in powering spacecrafts. Demand for tandem solar cells based on monolithic, series-connected, gallium indium phosphide (GaInP), gallium arsenide GaAs, and germanium Ge p-n junctions is rapidly rising. Prices are rising dramatically as well.

Twin-junction cells with indium gallium phosphide and gallium arsenide can be made on gallium arsenide wafers. Alloys of In.5Ga.5P through In.53Ga.47P may be used as the high band gap alloy. This alloy range allows band gaps in the range of 1.92eV to 1.87eV. The lower GaAs junction has a band gap of 1.42eV.

In spacecraft applications, cells have a poor current match due to a greater flux of photons above 1.87eV vs. those between 1.87eV and 1.42eV.

This results in too little current in the GaAs junction, and hampers the overall efficiency since the InGaP junction operates below MPP current and the GaAs junction operates above MPP current. To improve current match, the InGaP layer is intentionally thinned to allow additional photons to penetrate to the lower GaAs layer.

In terrestrial concentrating applications, the scatter of blue light by the atmosphere reduces photon flux above 1.87eV, better balancing junction currents. GaAs was the material of the highest-efficiency solar cell, until recently, when Germanium-based MJ cells capped the world record at 41.4% efficiency.

Indium Phosphide Based Cells

Indium phosphide is used as a substrate to fabricate cells with band gaps between 1.35eV and 0.74eV. Indium phosphide has a band gap of 1.35eV. Indium gallium arsenide (In0.53Ga0.47As) is lattice matched to indium phosphide with a band gap of 0.74eV. A quaternary alloy of indium gallium arsenide phosphide can be lattice matched for any band gap in between the two.

Indium phosphide-based cells are being researched as a possible companion in the manufacturing of gallium arsenide based cells. The two materials, although different, can be either optically connected in series (with the InP cell below the GaAs cell), or through the use of spectra splitting using a dichroic filter.

The efficiencies of these types of cells are in the 15-20% range.

A variation of this technology is the Indium gallium phosphide (InGaP), also called gallium indium phosphide (GaInP), which is a hybrid semiconductor, composed of indium, gallium and phosphorus.

It is used mostly for fabricating high-power, high-frequency electronics and LEDs of different colors. But its use in the manufacturing of high efficiency solar cells used for space applications is growing.

$Ga_{0.5}In_{0.5}P$ is used as the high energy junction on double and triple junction photovoltaic cells grown on GaAs substrates. GaInP/GaAs tandem solar cells have shown efficiency of over 25% under AM 0.

A different alloy of GaInP, lattice matched to the underlying GaInAs, is utilized as the high energy junction GaInP/GaInAs/Ge triple junction photovoltaic cells.

One of the greatest problems of the GaInP grown by epitaxy is its tendency to grow as an ordered material, rather than a truly random alloy, as required in some cases. This changes the bandgap and the electronic and optical properties of the material, which makes the manufacturing process quite complex and expensive.

We see these cells taking place in some niche markets (mostly space, at present), but due to manufacturing complexity and unproven performance characteristics, they will not soon be deployed in large quantities on ground-based utility energy markets.

The presence of varying quantities of toxic materials in these devices must be considered when planning their use in large quantities as well.

Germanium-based Single- and Multi-junction Cells

Germanium (0.86eV band gap) is a semiconductor material, with properties far superior to other substrate materials used for PV cells and modules. It is ~40-50% more efficient than silicon and has a much lower temperature coefficient. It is several times more expensive than silicon, too, but with new superior slicing techniques, it can be cut into very thin wafers, saving a lot of material. This, combined with its higher efficiency and less degradation than silicon, could put it on the competitors' list within the next few years.

Germanium-based solar cells have been used mostly for space applications, but a number of manufacturers have geared up to mass produce them for high concentration HCPV and other high efficiency applications (the record is currently 42.3% efficiency).

III-V Cells Research Trends

Current research on high-performance III-V multijunction thin film devices focuses on several major challenges:

1. Determining high-bandgap alloys based on I-III-VI and II-VI compounds and other novel materials for the top cell.

2. Considering low-bandgap CIS and its alloys, thin-film silicon, and other novel approaches for the bottom cell.

3. Studying the difficult task of integrating the thin-film tunnel junction (interconnect) with the top cell. This work includes understanding the role of defects, how they affect the transport properties of this junction, and the diffusion of impurities into the bulk material.

4. Fabricating a monolithic, two-terminal tandem cell (based on polycrystalline thin-film materials) that requires low-temperature deposition for several layers.

5. Avoiding deterioration of the top cell when fabricating the bottom cell if a low-bandgap cell is fabricated after a high-bandgap cell with a superstrate structure, such as CdZnTe.

6. Avoiding temperatures and processes that could damage the CIS bottom cell if a high-bandgap cell is fabricated on top.

Tandem Cells

One way of increasing the efficiency of a solar cell is to split the solar spectrum into different sections and use a different solar cell that is optimized to capture each section of the spectrum. This can be done with tandem solar cells.

Tandem cells are basically stacks of individual cells with different energy thresholds each absorbing a different band of the solar spectrum. These are usually connected together in series with the ultimate goal to engineer a new type of solar cell that is more efficient than existing technologies. Such material could be 'engineered' using several methods, including the use of quantum dot nanostructures of silicon in a silicon based dielectric matrix.

The confined energy levels in the quantum dots will increase the lowest absorption edge of the material compared to bulk silicon. If the quantum dot density is high enough, the wave functions of the quantum dots will overlap to create true superlattice mini-bands and increase the effective band gap of the material.

Non-silicon materials have been used in tandem

configurations with great success as well. As a matter of fact, some of these have produced the highest conversion efficiency to date.

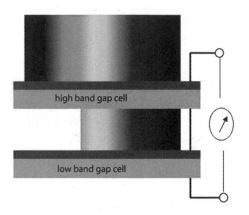

Figure 3-38. Tandem cell structure.

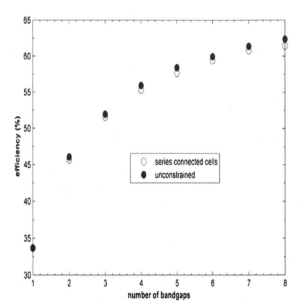

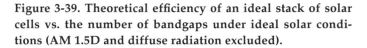

Figure 3-39. Theoretical efficiency of an ideal stack of solar cells vs. the number of bandgaps under ideal solar conditions (AM 1.5D and diffuse radiation excluded).

Basically speaking, tandem cells are two or more solar cells (and two p-n junctions) stacked on top of each other. The top cell absorbs high energy photons, while the bottom cell collects the photons with lower energy. Since these are usually lost in single-cell, single-junction devices, the efficiency of tandem cells is significantly higher under different light conditions.

The cells are connected in series and stacked in the direction of the incident light. The top most solar cell usually has the highest band gap and the bottom most solar cell has the lowest band gap.

Each solar cell (band gap) is designed to target a small portion of the incident solar spectrum for most efficient absorption and conversion to electricity.

The absorption profile for the stack is step-wise, with no overlap or gaps in the absorption between any of the solar cells in the stack. Or, each solar cell absorbs light greater than its own band gap but does not absorb photons with energy greater than the solar cell above it in the stack.

Each solar cell also absorbs all incident photons with energy greater than the cell's band gap and is transparent to photons with energy below the band gap. Each photon produces exactly one electron hole pair upon absorption.

Tandem solar cells can either be individual cells, or connected in series. Series connected cells are simpler to fabricate but the current is the same through each cell, so this contains the band gaps that can be used.

The most common arrangement for tandem cells is to grow them monolithically so that all the cells are grown as layers on a single substrate, and where tunnel junctions connect the individual cells.

"Unconstrained" tandem solar cell stack means that each solar cell is connected to a separate electric circuit and is basically free to operate unrestricted and unhindered at its maximum power point at all times. In the case of "constrained" solar cells, these are connected in series, with any mismatch in the output current of each cell needing to be minimized for optimum operation.

The constrained tandem stack is the preferred configuration for all practical purposes, despite the slight current matching restriction. The advantage of this type of arrangement is that only two electrodes are needed for the whole stack, instead of a pair of electrodes from each cell, which complicates the manufacture and use of the devices and is simply impractical.

It has been proven that the band gap sensitivity of the constrained tandem solar cell stack is greater than that of the unconstrained stack. This is because the constrained case means that the output current of the solar cells in the tandem stack needs to be matched. Or, any change in one of the band gaps can reduce the efficiency of each other cell, and hence the effect on the overall efficiency is large.

The maximum theoretical efficiency for a two-junction tandem under the AM1.5G spectrum and without concentration is 47%. At peak efficiency, the top cell of such a stack has a bandgap of 1.63 eV and the bottom cell has a bandgap of 0.96 eV.

As the number of bandgaps increases, the efficiency of the stack also potentially increases. In reality, increasing the number of cells in the stack will increase its efficiency to a point—depending on the materials properties and process parameter—but eventually the efficiency will start decreasing, due to interferences and because the stack gets so thick that the extraction of electrons is impeded.

Tandem solar cells are complex devices and quite expensive, so their use is restricted to use in specialized applications. One of their most practical applications is in concentrated tracker assemblies. Here, the tandem solar cells follow the sun all day, getting maximum sunlight which is then converted into electricity most efficiently in the different bandgap materials in the cells.

Tandem solar cells made out of different materials have reached the highest conversion efficiency of any commercial solar cell to date. Tandem cells made out of III-V type (non-silicon) materials are over 43% efficient under 500 x concentration. This is a great achievement, and we predict that tandem cell technologies will be widely used in the 21st century.

III-V Type Solar Technologies

The III-V type cells and modules, which efficiency have risen even higher (up to 43.7% in summer of 2012) are used in special applications. While they may become the technology of choice in the future, their use is limited today.

We do expect them—short of another disruptive technology cropping up unannounced—to be the technology of choice by the end of the century. That, however, remains to be seen because in order to achieve such high efficiencies (and to justify their significantly higher cost) the III-V type cells must be used in conjunction with tracking devices, which is a complex and expensive proposition.

Because of that, and because of our inability to make reliable and cheap trackers, their use is quite limited. Let's hope that future generations will find better and cheaper ways to make these cells and trackers, for building efficient solar conversion equipment.

Miniature Solar Cells

Miniature solar cells are a special branch of the solar industry. They are specialized, very small, PV devices based on high-performance semiconductor materials.

They could be shaped as a single cell, or a monolithic string of several solar cells. When exposed to sunlight, or bright artificial light, the energetic photons (optical energy) activate the miniature cell, or array, and generate low output voltage. Some of these cells are ca-

pable of generating floating source voltage in the 2-6V range, and current sufficient to drive and power CMOS ICs, logic gates and other electronic components and devices.

These solar cells have little niche markets in commercial applications for small and portable electronics.

Some of them could be used for providing wireless power and take advantage of the clean power generation without harmful EMI/RFI generation.

Such cells or arrays could also provide power for small LED screens or displays, or "trickle charge" for batteries and other power applications.

There is still some skepticism as far as the efficient and reliable operation of very small solar cells, not to mention the high prices, so it remains to be seen how this branch will develop.

CPV AND HCPV TECHNOLOGIES

Concentrating photovoltaics (CPV) is a branch of the PV industry, using special cells, optics and tracking mechanisms developed in the 1970s by several companies under contracts and financing from the U.S. Departments of Energy.

Early CPV systems used silicon-based CPV cells, which had a problem with elevated temperatures. These were later replaced by GaAs-based multi-junction cells, which have much higher efficiency, but still suffer from the effects of high temperatures.

At first GaAs CPV cells were made by using straight gallium arsenide in the middle junction. Later cells utilize In, Ga and As, due to the better lattice match to Ge, resulting in a lower defect density.

CPV Solar Cells

The result of 30 years of CPV cells development is a complicated stack of crystalline and thin film layers, with different band gaps tailored to absorb most of the solar radiation. These compound semiconductor solar cells have been shown to be more robust when exposed to outer space radiation and extreme climate conditions on earth.

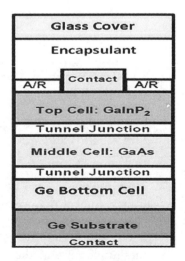

Figure 3-40. CPV cell

These cells take advantage of the successes achieved by the new materials, such as Germanium, GaAs, as well as the development in the processing of III-V semiconductors and multi-layer (tandem) structures in the semiconductor industry.

Since each type of semiconductor has a different characteristic band gap energy which then allows the absorption of light most efficiently, at a certain wavelength, we have absorption of electromagnetic radiation over a portion of the spectrum.

These hetero-junction devices, consisting of layers of various cells with different band-gaps are tuned for utilizing the full spectrum. Initially, light strikes a wide band-gap layer producing a high voltage therefore using high energy photons efficiently enabling lower energy photons to transfer to narrow band-gap sub-devices which absorb the transmitted infrared photons. Gallium arsenide (GaAs)/indium gallium phosphide (InGaP) multi-junction devices have reached the highest efficiency of over 43%.

Originally these cells were fabricated on GaAs substrates; however, to reduce the cost and increase robustness, and because it is reasonably lattice-matched to GaAs, germanium (Ge) substrates are being used more often.

The first cells had a single junction much like the Si p-n junction solar cells; however, because of the ability to introduce ternary and quaternary materials such as InGaP and aluminum indium gallium phosphide (AlInGaP) dual and triple junction devices were grown to capture a larger band of the solar spectrum therefore increasing the efficiency of the cells.

Research Trends

Current research on high-performance multi-junction thin film devices focuses on several major challenges:

1. Determining high-bandgap alloys based on I-III-VI and II-VI compounds and other novel materials for the top cell.

2. Considering low-bandgap CIS and its alloys, thin-film silicon, and other novel approaches for the bottom cell.

3. Studying the difficult task of integrating the thin-film tunnel junction (interconnect) with the top cell. This work includes understanding the role of defects, how they affect the transport properties of this junction, and the diffusion of impurities into the bulk material.

4. Fabricating a monolithic, two-terminal tandem cell based on polycrystalline thin-film materials that requires low-temperature deposition for several layers.

5. Avoiding deterioration of the top cell when fabricating the bottom cell if a low-bandgap cell is fabricated after a high-bandgap cell with a superstrate structure, such as CdZnTe.

6. Avoiding temperatures and processes that could damage the CIS bottom cell if a high-bandgap cell is fabricated on top.

The CPV Technology

The CPV cells described above are mounted under special (Fresnel) lenses, which concentrate and focus sunlight on the cells at 100 to 1000 times its original intensity. This is a lot of sunlight falling on a small area, but it allows high efficiency and reliability, better land utilization, and other benefits.

Cell-lens packages are assembled into large modules, which are mounted on trackers, which track the sun precisely through the day, providing the most power possible.

Efficiencies of 30-32%, measured in the grid, are obtainable with these devices.

HCPV Technology

High concentration PV (HCPV) technology is a variation of the CPV technology, but it uses much higher sun concentration ratios—in the 100 to 1000 times range.

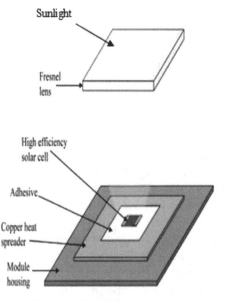

Figure 3-41. CPV assembly

It was developed in the 1980s and tested in small-scale projects during the 1990s. But, it was shelved soon after, so it is still waiting to be introduced and proven in field operations. Several manufacturers have installed capacities, but they seem to be still mostly in the development stages.

HCPV systems are designed and built for operation under extreme desert conditions. They operate at 50-1000 times concentration of sunlight with over 42% efficiency at the cell. Efficiency in the grid is lower, of course, but still at least 2-3 times higher than any of the competing PV technologies. The CPV solar cells are made out of germanium, or GaAs, which are superior semiconductors not affected by temperature extremes. They are also made via sophisticated and precise semiconductor processes, which make them durable and reliable for long-term operation. The optics are made out of glass and silicone, which are also unaffected by sunlight and temperature extremes.

HCPV modules are mounted on trackers which have a number of moving parts, control mechanisms, and electronics that need regular maintenance and tuning for optimum performance. This requires highly trained personnel and increases O&M expenses.

We hope that with its 40+% efficiency, which is increasing with time, HCPV technology will soon become an active participant in the energy markets of the US and the world. HCPV holds great promise of high efficiency (over 60% in the near future), combined with reliability of quality hi-tech components suitable for large-scale generation of electricity in the deserts.

HCPV Tracking Systems

As mentioned previously, HCPV systems use very accurate two-axis tracking, and this combined with other sophisticated components makes HCPV the most efficient, reliable technology to date. Its operation, however, is quite complex for use in today's energy market, as we know it, so it will take a while, but HCPV technology will eventually find its place in energy markets.

HCPV is especially efficient, because in addition to over 42% efficient CPV cells, it tracks the sun with 0.01% accuracy which allows it to capture every single sunbeam from sunrise to sunset. The problem is that if there is any cloud cover or haziness in the sky, HCPV optics cannot handle the diffused light going through them, and the efficiency drops quickly and significantly. Basically, HCPV systems can be used only in bright, hot desert areas, where other technologies have a hard time surviving the harsh elements.

But HCPV is a complex technology and some work

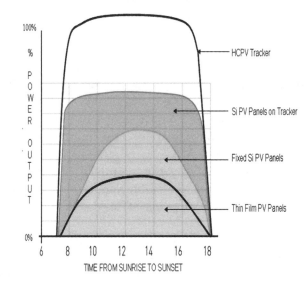

Figure 3-42. Efficiency of the different PV technologies

is still left to be done before we see HCPV trackers installed in large quantities in the world's large-scale PV power plants. At that point the equipment cost can be reduced significantly by placing large orders for materials and components.

In time, technological issues will be resolved, and we clearly see CPV and HCPV technologies as the primary choice for installation and use in large-scale power plants, especially those in desert regions, during the latter part of the 21st century.

ACHIEVEMENTS OF PV TECHNOLOGIES

One of the most encouraging facts today is that there are performance records across the board in all photovoltaic materials types; crystalline silicon, a-Si, CdTe, CIGS, CIS, GaAs, and triple-junction CPV cells.

Many of these were not expected, and some were considered unachievable just several short years back. Nevertheless, the efficiency numbers keep rising.

Here are some recent record-setting results:

Abound Solar (now bankrupt)

Abound Solar was a manufacturer of cadmium telluride PV modules, with planned production of 82.8-watt modules at its Longmont, Colorado, factory. This represents a 12.2% aperture efficiency to be produced on existing production equipment.

The mass production of 82-watt modules was

scheduled to begin in the second half of 2012, but unfortunately Abound Solar shut down and filed for bankruptcy before achieving that goal.

Alta Devices

Alta Devices most recently produced a GaAs-based solar panel with a 23.5% efficiency, as verified by the National Renewable Energy Laboratory (NREL). This seems to be another world record, but no other details are available, so we must assume that this was done in their pilot phase.

Alta's process allows the manufacturing of layers of thin GaAs films that are flexible and measure only one micron in thickness, which has advantage in many niche markets.

First Solar

First Solar holds the world record for CdTe PV module efficiency with a 14.4% total area efficiency. This achievement comes six months after First Solar hit a CdTe solar cell efficiency of 17.3%. Both records were set at the Perrysburg, Ohio, factory.

MiaSolé

MiaSolé is a CIGS thin-film PV manufacturing company, which was third in CIGS panel production in 2011, behind Solar Frontier's 400 MW and Solibro's 66 MW.

MiaSolé announced 17.3% efficient modules, while the manufacturing process for 14 percent efficiency is in production.

Solar Frontier

Solar Frontier is a CIGS/CIS manufacturer with 400 megawatts of modules shipped in 2010. The newly recorded 17.8% aperture area efficiency on a 30-centimeter-square CIS-based PV lab module is a new record for a "fully integrated submodule" manufactured at the Solar Frontier's factories at commercial production scale.

Kunitomi, Japan

Kunitomi recently demonstrated a new CIGS module at 14.5% aperture efficiency, equivalent to a 13.3% module efficiency.

Solar Junction

Solar Junction is a developer of multi-junction cells for high-concentration photovoltaic (HCPV) applications. Working with Semprius it has an agreement to deliver multi-megawatts of epitaxial wafers to the market.

Semprius holds the world-record CPV solar mod-ule efficiency using Solar Junction's III-V multi-junction solar cells based on lattice-matched dilute nitrides. The firm recorded a module efficiency of 43.3%.

SoloPower

SoloPower is a CIGS startup building flexible solar panels in a roll-to-roll electroplating process and has an NREL-measured aperture area efficiency of 13.4%, but the module efficiency is significantly less than that.

The practical application and value added for flexible modules in general is that there is less hardware required to install and the installation is easier—according to the manufacturers, but not yet proven.

SunPower

SunPower has been manufacturing and selling high efficiency cells and modules during the several years. Their back-contact crystalline silicon cells (which move the metal contacts to the back of the wafer, thus maximizing the working cell area, and eliminates redundant wires) have been in production since 2005.

Consistent improvements in efficiency have been demonstrated with each successive generation of commercialized cells, reaching Gen 3 cells efficiencies of over 23%.

Suntech

Suntech's Pluto cell technology achieved a 20.3% efficiency for a production cell using commercial-grade p-type silicon wafers. The new technology is a combination of different elements which combined improve cell efficiency. 21% efficiency is targeted in the near future. The improvements include surface patterning, improved metallization, improved front metal contact dimensions, changes in dopant concentration at the emitter, and improved high-temperature performance.

These are, however, expensive processes, and the new product is not yet in full production. It is, instead, a premium product, offered at premium prices.

MOST PROMISING MATERIALS

The manufacturing process starts with the proper selection of the most efficient and high quality materials. Although silicon and most other materials used in solar cells and modules manufacturing have been around for a long time and are well known, new materials are popping up constantly.

Here we'll take a look at some of the most promising, and hopefully disruptive, materials.

Quasi/Mono Crystalline Silicon Technology

A brand new, proprietary "super large quasi-crystalline silicon ingot casting" technology was developed by a number of manufacturers, and recently approved by a group of experts.

The new method addresses some of the issues of polycrystalline silicon ingot casting waste and cost. When implemented and proven, this technology will have a huge impact on the entire crystalline silicon industry and energy markets. It could be used as a substitute of the more expensive mono-crystalline silicon, while providing similar performance.

Basically, polycrystalline silicon has much lower efficiency than monocrystalline silicon, but it is much cheaper and efficient to produce. At $20/Kg polycrystalline silicon castings per batch, vs. the $60/Kg batch of a Czochralski (CZ) process monocrystalline silicon, not to mention the related complexity of the latter process.

Because of this inequality poly- and mono-crystalline silicon have similar costs per Watt and their share in the energy markets is approximately 50:50.

The intent of the "super large quasi-crystalline silicon ingot casting" is to further reduce the polysilicon casting costs; down to $10/Kg (vs. approximately $25/Kg for making single-crystal silicon presently).

The problem is that no matter what process is used, the CZ single crystal process produces silicon that reaches 18.5% efficiency, while polycrystalline silicon cannot exceed 14-16%. This difference is enough for the CZ to be the material of choice in some applications—regardless of the cost. This difference cannot be changed, because the laws of physics and chemistry, as well as the nature of the semiconductor materials cannot be changed, so that polysilicon used for making solar cells will always be less efficient.

Thus, we must see it to believe it, before buying into the claim that the efficiency of the newly developed quasi-crystalline silicon produced by this new ingot casting technology is comparable to monocrystalline silicon, with efficiency over 18%. If it reaches that efficiency, then the resulting material cannot possibly be poly- or even quasi-silicon; it is rather simply single-crystal silicon, or a variation thereof at best. Nevertheless, the increased efficiency and lower cost are enough to put it on the market in a very short time.

"This technology uses polycrystalline silicon ingot casting technology to produce products with the quality of monocrystalline silicon. It's like producing a product worth $25 for the cost of a $10 product," according to company representatives.

There are at least 45 quasi-crystalline casting furnaces in operation in China, which under full operation represent a production capacity of over 300MW/year. This is no small thing, and if the predictions are correct, it also represents a breakthrough—maybe even disruptive—new technology that could take the markets by a storm.

There are a broad range of commercially available quasi-mono silicon PV cells with efficiencies from 16% to 18.4%, so the new technology seems to be working as predicted. If the cost is low enough, and proves to be reliable in field operations, then we have a successful product which will make a significant difference in the bottom line and help China reach grid parity by 2020.

Still, we must persist with our claim that only single-crystal silicon can be that efficient. This is a simple matter of physics, and there is no way for any other silicon to be that efficient, unless the laws of physics have been changed too...

Because we don't believe that the Chinese can do that, we must insist that no matter what the new product is called, or how it is made, if it reaches 18% efficiency it has mono-crystalline structure. Anything short of that would not be able to achieve such high efficiency.

We'll be watching the developments with the Chinese manufacturers and their new quasi-mono crystalline silicon. It sounds too good to be true, but you never know.

Granular Silicon

This is a new method in the arsenal of polysilicon production techniques. It was developed over the last 15 years and shows much promise, mainly in reducing costs and improving yields, while maintaining the usual quality.

Granular polysilicon is produced by decomposing silane (SiH4) in a fluidized bed reactor. Small seed particles are introduced and a silane/hydrogen mixture is then fed into the reactor. The silane decomposes around the seed particles increasing their average size to approximately 1,000 microns.

The granular polysilicon particles are then withdrawn from the reactor and packaged in drums in a completely enclosed system to maintain purity.

This is a continuous production process, with a number of advantages over the conventional batch methods, based on SiHCl3 chemical vapor deposition (CVD) production process.

The fluidized bed SiH4 CVD reactor benefits from both a lower operating temperature for the decomposition of SiH4 to Si and H_2 as well as a higher throughput from its continuous operation.

The advantages of this process are lower energy consumption (as compared to the Siemens process), continuous production process (as compared with the conventional batch processes), and a very consistent product quality with low variability.

Granular polysilicon can be also transported more easily from a storage container to a crystal puller in an enclosed system. It can be poured into a crucible, eliminating the laborious and tedious tasks of manually stacking silicon chunks of various shapes and sizes with sharp edges and jagged corners.

In 200mm standard hot zone designs, using an external hopper, the poly can be fed into the liquid pool created from melting an initial charge of chunk polysilicon, which facilitates the process further. With this new approach, manufacturers have been able to achieve up to 40% larger charge size than possible with an all-chunk cold charge. It is also possible to reduce the thermal stress on the crucible wall and improve crystal growth yields.

Additionally, optimal combinations of chunk and granular polysilicon fed through a feeder system can be used with great efficiency in the larger, 300 mm hot zone furnaces.

Large charge and multiple recharges of the granular polysilicon production process is a great advantage. This material removes the limitations on charge size in large, complex hot zones and provides for replenishing the melt for another "pull" as the just-completed crystal cools in the removal chamber of the crystal growth furnace.

With a two-crystal recharge process, crystal pulling costs can be reduced by as much as 40% overall, while the throughput can be increased as much as 25% overall.

Manufacturers have commercialized the recharge process for 150 mm, 200 mm, and 300 mm products and intense research and development of granular polysilicon have resulted in robust and proven processes that benefit silicon wafer manufacturers. These benefits will not only help silicon suppliers internally, but should also enable them to provide a higher quality product to solar cells and semiconductor manufacturers.

So, this is another promising—and possibly disruptive—product for 21st century PV manufacturers to use in the battle of optimizing processing techniques and reducing cost.

Super Monocrystalline Silicon

This is a new purer type of silicon with more perfect crystalline structure, and fewer impurities. This type of material is close to the "perfect" silicon, whatever that might be… In the solar cells case, it exhibits reduced phonon-phonon and phonon-electron interactions, which are responsible for the transport of the active species within the silicon material.

The "super" qualities of the purer silicon, and the related phenomenon, increase certain transport properties, resulting in 60% better room temperature thermal conductivity than natural silicon.

The issues are cost of materials and labor, and the durability of the finished devices. Because the title "super" is somewhat all-encompassing, and because material quality is such a broad term, we hesitate to predict the readiness and usability of such superior materials in the 21st century, but the promise is still great.

Certainly "super" often describes good things, but material that is super in one category is not automatically super for all applications.

Finally, remember that the quest for better silicon dates from the mid 20th century, and semiconductor companies have spent a lot of money to make major efforts in this area. These efforts will go on, and we should expect surprises.

HYDROGEN GENERATION

With catalysts created by an MIT chemist, sunlight can turn water into hydrogen. If the process can scale up, it could make solar power a dominant source of energy.

Background

Over the last century technology has developed amazingly fast, and the global population has kept pace, increasing from about 1 billion in 1800 to 3 billion in 1960 and reaching nearly 7 billion today. The demand for energy has increased proportionally in most places and ridiculously disproportionally in others—such as the US—where people use several times more energy than they need… just because it is there.

For the last century we have relied mostly on fossil fuels—coal, oil, and natural gas that are not renewable natural resources. The global demand for energy is expected to double by the middle of this century. By then, the world's oil resources will be exhausted, while gas and coal reserves will be substantially depleted.

We live in a time during which three or four generations (present company included) have consumed natural reserves that took over 100 million years to generate. Some of us are aware of this reality and are doing something about it, while most of us are oblivious to it. What are we doing? What will future generations say about us?

We live in a vicious wealth-depleting cycle, burning fossil fuels equivalent to 9.0 billion gallons of oil per day. Every day, mother Earth contains less coal, oil, and natural gas, while the environment bares the scars of our race to eliminate those natural resources. All of humanity lives on borrowed time, as far as having access to natural resources is concerned.

Just imagine... yes, scientists are allowed to imagine. So imagine life in 2095. Crude oil deposits are gone, and coal and natural gas are near depletion, as are rare and exotic minerals and materials. What does life look like in this world without natural resources? Let's look at two extreme cases:

Best Case Scenario

Technology has replaced oil and other natural resources with new man-made, abundant and cheap materials. There is no danger of shortages, prices are low, and the environment is clean. Life is better than ever. It's hard to imagine that this could happen, but I suppose it could—if the next generations are more capable than we.

One way that could be achieved is for *us* to stop the indiscriminant use of natural resources for producing energy and other, unnecessary, things. The use of solar energy, wind energy, hydro power, ocean wave and tidal power would be maximized. Scarce natural resources would be prioritized and compartmentalized for use in manufacturing only the most essential items such as life-saving medications.

Worst Case Scenario

Wow! We can paint very bleak picture here. But keeping it within reason, we still see a society wrapped in darkness and misery. Things have gone bad during the last 70-80 years. Solar and wind energy sources have not been as good as promised and are not producing enough energy. There is no money for optimization of the existing facilities, or R&D, or implementation of new energy sources. Energy is in high demand, often unavailable, and extremely expensive.

Because of indiscriminant use of natural resources, the environment is badly contaminated and the climate has changed drastically. People live in extreme conditions—boiling summers and freezing winters. Many coastal communities are under water and the deserts have taken thousands of previously agricultural acres.

Plastic bags, medications and cosmetics would be

astronomically high and unavailable to most. This is what we are offering our children and grandchildren.

Solar Hydrogen (Most Possible) Future

Mass hydrogen production is one of the sure ways to reduce the use of natural resources and stop the rampant air, water and soil pollution. Hydrogen is a nontoxic, clean-burning fuel which can be produced from water and sunlight, which can produce "solar hydrogen," which can be then used in every application where the conventional fuels are used today.

Solar hydrogen is hydrogen produced when using solar energy (as a primary energy source) or one of its derivatives—renewable forms (secondary energy source) such as wind, ocean waves, falling water, and biomass. As a bonus, by replacing fossil fuels with solar hydrogen, we can move from an unsustainable wealth-depleting economy to a wealth-enhancing one.

Solar energy, albeit abundant in many areas of the world is inconveniently remote and inaccessible in large quantities. To produce, store, transport and use it requires extensive installations and procedures. Generating hydrogen by splitting water as primary energy source is one of the most promising solutions.

Hydrogen is an efficient fuel, it is carbon-free and is converted back into water during the combustion process. This means that we start with water, use sunlight to split it and generate hydrogen during the day, store it, and then burn the hydrogen at night to generate heat and electricity, which process produces water... again.

It's an almost perfect closed-loop system. A type of a perpetuum mobile, using only the sun's energy to assist in the never ending and very efficient conversion process, where very little energy and water are lost.

The water-to-hydrogen-to-heat-to-electricity process goes something like this:

1. During the day:

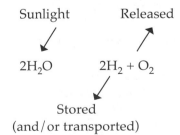

Sunlight   Released

$2H_2O$ $2H_2 + O_2$

Stored
(and/or transported)

2. During the night:

$2H_2 + O_2 \longrightarrow 2H_2O + heat$

$heat \longrightarrow electricity$

The generated water, together with the sunlight used for splitting it, are sources of clean and abundant energy in a carbon-free, natural and renewable cycle. Unlike electricity, hydrogen is a fuel which can be transported over long distances without the presence of expensive transmission lines, and the losses associated with those.

It can be also stored, unlike electricity, and/or transported over long distances and in any imaginary direction, to areas that need it most. Hydrogen also can be stored underground or in containers to be used as needed by industrial enterprises, in households, power stations, motor cars and aviation.

Upon combustion, hydrogen produces heat and clean water. Two pounds of hydrogen release about 120,000 Btu of heat, or as much as one gallon of gasoline. Or, 10 million tons of hydrogen per year can fuel 25 to 30 million hydrogen-fueled cars, or enough to power 6 to 8 million homes. Not bad, and keep in mind that 10 million tons of hydrogen could easily be generated in the Arizona or Nevada deserts, where sunlight is as abundant as sand.

Note: The production of 10 million tons of hydrogen annually would require:

1.   250,000 small neighborhood-based hydrogen electrolysis generators to fuel cars and provide power needs,

2.   16,000 hydrogen vehicle refueling stations, or about one tenth of the current gasoline stations,

3.   35 coal/biomass gasification plants, similar to today's large coal fired plants,

4.   25 large nuclear plants making only hydrogen, and

5.   5 medium-size plants, using oil and natural gas in multi-fuel gasifiers and reformers.

So how do we generate hydrogen using solar power?

Hydrogen is the most abundant element in the universe, for it is in the molecules of most organic and some inorganic materials. However, it does not exist naturally as gas or liquid in large quantities or high concentrations. Hydrogen, therefore, must be produced from other compounds such as water, biomass, or fossil fuels. Various methods of production have unique needs in terms of energy sources (e.g., heat, light, electricity), all of which generate unique by-products and emissions.

Hydrogen Generation Methods

Some of the conventional hydrogen generation methods under consideration and development today are:

1.   Hydrogen generation by steam methane reforming constitutes 95% of the hydrogen produced in the United States today. This is a process of reacting natural gas or other light hydrocarbons catalytically with steam to produce a mixture of hydrogen and carbon dioxide. The mixture is then separated to produce high-purity hydrogen. This method is the most energy-efficient commercialized technology available, and is most cost-effective when applied to large, constant loads. It *does* use expensive, depleting natural resources, and it produces large quantities of carbon dioxide, which must be sequestered.

2.   Partial oxidation of fossil fuels in large gasifiers is another method of thermal hydrogen production. It is done by the reaction of conventional fuels with a limited supply of oxygen to produce a hydrogen mixture, which is then purified. This process can be applied to a wide range of hydrocarbon fuels including natural gas, heavy oils, solid biomass, and coal. Again, large amounts of carbon dioxide are produced.

3.   Hydrogen can also be produced by using electricity in electrolyzers to extract it from water. We call this process "splitting water." This method is not as efficient or cost effective as using fossil fuels in steam methane reforming and partial oxidation, but it is most promising for the near future. It also allows more distributed hydrogen generation and use, and opens possibilities for using electricity made from renewable resources. The primary by-products of this process are oxygen and small amounts of carbon dioxide (as a by-product from electricity generation).

Methods for producing hydrogen without using natural resources are most interesting, due to the obvious avoidance of depleting natural resources and decreased environmental contamination. Some of these methods are our best hope for efficient and safe 21st century power generation. Although components of these systems are still in development phases, *they are our best hope* in the 21st century, because hydrogen generated by solar energy (thermal or electrical) induced water splitting is capable of replacing fossil fuels!

These fossil-free solar hydrogen generation approaches include:

1. Splitting (electrolysis) of water using electricity generated from solar energy or other energy sources (wind, hydro, etc.),

2. Thermochemical reactions that release hydrogen from water or biomass,

3. Thermal dissociation of water using concentrated solar energy; and

4. Microbial activity that releases hydrogen from organic compounds and biomass.

In general, when exposed to special conditions, hydrogen readily reacts with oxygen in a highly exothermic reaction (energy is released—heat in this case), producing pure water as a result of the reaction and as a unique by-product, which could be used in a number of applications, including re-use in the hydrogen generation process:

$$2H_2 (g) + O_2(g) = 2H_2O(l) + 572 \text{ kJ(heat)}$$

In reality, fuel cells use air rather than pure oxygen, and as a consequence small amounts of nitrogen oxides are formed as exhaust products. PEM and SOFC (high temperature) fuel cells, the most efficient H_2 technologies today, emit trace amounts of nitrogen oxide (NO_x) gasses, while others might emit more.

Hydrogen must be generated by extracting it from hydrogen sources, and unfortunately today most of the commercially produced H_2 gas comes from fossils, which process contributes to the overall CO_2 emissions and to global warming.

Once available, however, H_2 is an excellent fuel that can replace hydrocarbons with numerous advantages, including a specific heat capacity 3 times higher than that of our best fossil fuel (natural gas, or methane).

Note: Fossil fuels have high-energy content, and their power densities (the rate of energy production per unit of the earth's area these come from) is approximately 102W per m², while in contrast their energy production is well below 1W per m². At the same time, densities of electricity produced by PV generation are over 40W per m² at peak power when using some highly efficient PV technologies (i.e. HCPV trackers).

One of the big problems of the PV technologies is

their variability; i.e., no sunlight at night, or a rainy day = no electricity generation during that time. Also, when a cloud covers the sky, PV power is drastically reduced. All of this contributes to increased reluctance on part of users to rely on solar power.

The best solution, in our humble opinion, is H_2 energy storage. H_2 is an excellent storage option for storing excess energy during the day for use at nighttime and on cloudy days. There are several methods which rely on the two main solar energy technologies—photovoltaics (PV) and concentrated solar power (CSP).

Water Electrolysis

This is the water electrolysis using electrical current generated by PV of CSP power fields, where:

$$H_2O + 2F = H_2 + \frac{1}{2} O_2$$

where

F is the Faraday constant, measuring 1 mol of electricity.

Once available in gaseous or liquid form, H_2 can be used to generate electricity by the reverse of reaction:

$$H_2 + \frac{1}{2} O_2 = H_2O + 2F \text{ (energy)}$$

This is in fact the process that occurs in an H_2-O_2 fuel cell, which can work as a fuel cell, or as an electrolyzer, depending on the operating conditions. The process is able to create any amount of hydrogen, as needed to power a house or commercial building. Hydrogen can also be used to power cars and boats with internal combustion engines, providing similar power density with added safety.

Note: Although hydrogen is highly flammable, it is 14.4 times lighter than air, rising at 20m/s-1 rate, so if ignited the flame tends to shoot up vertically, while gasoline and diesel are heavy and tend to run and accumulate under vehicles, where they can explode or spread and feed a fire.

The energy needed for the electrolysis can be provided by a number of sources. Presently CSP is the energy of choice for generating large quantities of H_2 from water. The preferred process includes catalytic thermo-chemical processes that make use of the intense heat or electricity generated by the solar radiation to create a large surplus of hydrogen suitable for storage and later use. The entire process is entirely renewable, using abundant energy sources and raw materials: solar energy and water, respectively, without generation of any appreciable CO_2 emissions.

Theoretically speaking, the product yield of the water splitting reaction is 100% efficient since no electrical energy is wasted. And since the cost of water is negligible, it is quite cheap too. The economics of the process are driven mostly by the cost of electricity.

Small amounts of electrolyte (usually KOH) are dissolved in water to enhance conductivity and thus the overall rate of the process of water electrolysis. 30% KOH solution at 80°C (alkaline electrolysis) is used, and the electrolyte can be recovered and re-used time after time. Such electrolysis systems make use of a ceramic microporous separator, whereas the electrodes are usually made of nickel, with Pt coated cathode and MgO_2 coated anode.

The reactions during electrolysis are:

$$2H_2O + 2e = H_2 + 2OH \text{ (cathode)}$$
$$2OH = \tfrac{1}{2}O_2 + H_2O + 2e \text{ (anode)}$$

Typically, a commercial alkaline electrolyzer produces H_2 by consuming 4.49 kW h/m^3 of electricity with current yield and hydrogen purity both close to 100%. The fuel cell makes it possible to use the solar power made available as H_2 when weather conditions are optimal and to store the excess power so that it can be made available for later use as required.

Extracted from water using photovoltaics and electrolysis, H_2 is oxidized in the fuel cell and the only emission is clean water, completing a zero emission energy production cycle.

The necessary investment for the hydrogen infrastructure gains more economic profitability with an increasing number of cells.

Today, when solar panels generate more electricity than a home is using, the excess is simply fed back into the grid, essentially subtracting from the homeowner's utility bill. In an off-grid application, the excess is put into batteries. But fuel cells are more versatile and their price is rapidly declining.

Existing electrolyzers are expensive, hence, the challenge is devising a system that is efficient enough to make energy inexpensively. In general, however, PV electricity should be used as such, for electricity is the highest quality energy available; but, this will require, in turn, the introduction of new generation batteries able to recharge rapidly with large amounts of energy.

Yet, the idea to use PV energy first to crack water molecules into hydrogen and oxygen and later in a fuel cell to make electricity when the sun is not shining is plausible. The concept is a closed-loop system in which

hydrogen oxidized with air in the fuel cell creates water, which is captured and used again.

Thermo-chemical Water Splitting

Achieving energy independence will require large-scale alternative energy generation, where solar-based H_2O splitting could play a deciding role. This process requires practical technology in terms of quantity and cost, of which the thermo-chemical redox-cycle process using simple and robust materials is capable.

This process is a thermal reaction that requires an energy input in a multichannel ceramic honeycomb reactor resembling the familiar catalytic converter of automobiles, coated with active water-splitting materials. The system is heated by concentrated solar radiation using a set of mirrors to concentrate the solar energy, increasing the temperature in the reactor

In the first step of water-splitting, the activated redox reagent (usually the reduced state of a metal oxide) is oxidized by taking oxygen from water and with the help of high temperature heat produces hydrogen, according to the reaction

$$MO_x \text{ (reduced)} + H_2O \longrightarrow MO_x \text{ (oxidized} + H_2$$

During the second step the oxidized state of the reagent is reduced, to be used again (regeneration), delivering some of the oxygen of its lattice according to reaction:

$$MO_x \text{ (oxidized)} \longrightarrow / MO_x \text{ (reduced)} + O_2$$

The metal oxide (such as mixed iron oxide) in this case acts as redox system which is fixed on the surface of a porous absorber. At the beginning, the metal oxide is present in a reduced form. By adding water vapor at 800°C, oxygen is abstracted from the water molecules and hydrogen is produced. When the metal oxide system is saturated, or fully oxidized, it is heated for regeneration at 1100°C to 1200°C in an oxygen-lean atmosphere.

Oxygen is exhausted from the redox system using nitrogen as a flushing gas. The product gas passes through heat exchangers and is cooled down before residual water is separated.

The receiver surface is divided into several square apertures; two apertures make up one receiver pair. One aperture is applied for the dissociation of the water vapor, while the other one is used for the regeneration of the redox system.

Thus, hydrogen can be produced continuously by

alternating the reaction steps.

One advantage here is the production of pure hydrogen and the removal of oxygen in separate steps, avoiding the need for high temperature separation and the chance of explosive mixture formation. The active redox material is in fact capable of water-splitting and regeneration, so that complete operation (water splitting and redox material regeneration) is achieved in a closed solar reactor.

Iodine-sulfur Thermo-chemical Water Splitting

Another promising approach for hydrogen production is the iodine-sulfur (IS) thermo-chemical water splitting cycle. Here, high temperature is needed to start and maintain the reaction, which consists of three sections:

1. Bunsen section,
2. Hydrogen-Iodide (HI) section, and
3. Sulfuric acid section

This complex process is initiated in the Bunsen section under the effects of high heat influx, assisted by the action of sulfur dioxide, SO_2 and iodine, I_2. Excess water and heat are added at this point to keep the exothermic and spontaneous character of the reaction.

Once started, the reaction must be kept exothermic, and when it is stabilized, excess iodine is added to help the mixture split into two immiscible aqueous phases. This is achieved via the so-called spontaneous liquid-liquid (L-L) phase separation process, during which sulfuric acid and poly-hydroiodic acid (HIx) phases are obtained.

The lighter sulfuric acid phase is fed to the sulfuric acid, H_2SO_4 decomposition section, where it is concentrated, while at the same time it is thermally decomposed to sulfur trioxide, SO_3 and water, H_2O.

Increasing the temperature dissociates the sulfur trioxide, which is readily divided into oxygen, O_2, and sulfur dioxide, SO_2.

Thus obtained O_2 can be used as a by-product of the reaction if and when needed, while the SO_2 and H_2 have to be recooled and recycled back to the Bunsen section and stored for further use.

At the same time, the heavier HIx phase is sent to the HI decomposition section, where the hydrogen iodide, HI, is separated from the mixture and exposed to high heat, where it is thermally decomposed into H_2 and I2. The resulting H_2 gas is the final product of this process, and is stored for future use.

The accompanying I_2 and H_2 are also cooled, recycled.

Note: Using the waste heat from solar (CSP) installations to maintain the H_2 generation reaction have been proposed. We do see this as one way to both provide much needed cooling for the CSP plant steam, while at the same time providing heat as needed to maintain the exothermic nature of the H_2 generation reaction.

One problem with this approach is that the volume and heat content of the CSP plant's steam varies with weather conditions. Inconsistent steam volume or heat content would hinder the proper maintenance of the H_2 generating reaction. CSP plants with energy storage, however, might be able to resolve this problem and provide constant quality steam using the stored heat as a source.

The Challenges

Despite the advantages, water electrolysis and hydrogen/oxygen fuel cell technology still face challenges. For instance, the electrodes used in water electrolysis are coated with platinum, which is not a sustainable resource and is quite expensive.

Researchers are investigating the employment of nanomaterials with a large reduction on the amount of precious metal needed. The main advantage of using these materials as catalysts that split water molecules using cobalt phosphate is that they are far cheaper and more abundant compared to metals such as platinum.

Another problem is the intermittent nature of PV electricity (or heat), due to interruptions like nighttime, clouds, etc.

These anomalies create several problems for the H_2 generation process, mostly expressed as:

1. The process activity would decrease in time of cloud activity, and

2. Shutdown of H_2 generation process is unacceptable, with numerous consequences; i.e., Ni dissolution at the cathode is initiated, since it is driven to more positive potentials by short-circuit with the anode.

These shortcomings can be alleviated by using stored heat or electricity as energy sources. As for the Ni cathodes, they could be activated, by coating them with a thin layer of more active, more stable materials.

The Hydrosol Project

The Hydrosol project was designed and proposed by a group of EU scientists and has won a number of prestigious awards as a practical way to generate large amounts of clean energy, including energy storage (for nighttime use) and transport of energy (hydrogen) to remote locations.

The Hydrosol process consists of an innovative solar thermo-chemical reactor for the production of hydrogen from water splitting, resembling the familiar catalytic converter of automobiles.

The reactor contains no moving parts and is constructed from special ceramic multi-channeled monoliths that absorb solar radiation. The monolith channels are coated with active water-splitting nanomaterials capable of splitting water vapor passing through the reactor by trapping its oxygen and leaving as product pure hydrogen in the effluent gas stream.

In a next step, the oxygen trapping material is solar-aided regenerated (releases the oxygen absorbed), and a cyclic operation is established on a single, closed reactor/receiver system. The integration of solar energy concentration systems with systems capable of splitting water will have an immense impact on energy economics worldwide, as it is a promising route to provide affordable, renewable solar hydrogen with virtually zero CO_2 emissions.

The uniqueness of the Hydrosol approach is based on coating nanomaterials with very high water-splitting activity and regenerability (produced by novel routes such as aerosol and combustion synthesis) on special ceramic reactors with high capacity for solar heat absorption. The production of solar hydrogen will offer opportunities to many poor regions of the world which have huge solar potential. Producing solar hydrogen will create new opportunities for countries of southern Europe that can become local producers of energy.

Hydrosol is a new approach, paving the way to a new age in energy generation, storage, transport and use in the 21st century.

With traditional energy sources diminishing, interest in alternative energy supplies is increasing. Hydrogen is considered by many to be the next link in the evolution of energy, after nuclear energy.

Hydrogen Economy

1. Hydrogen promises smooth transition from the existing dirty, inefficient, and unsustainable energy systems of power generation, storage and transport, to clean, efficient, renewable energy.

2. Hydrogen energy generation, storage and transport systems are fully developed now, do not need special materials or processes, and can be implemented immediately.

3. The use of hydrogen as an energy source is ecologically safe, because there is virtually no pollution generated in the process, unlike most of the conventional energy generators and their supply chains.

4. Solar hydrogen generated power could be quite cheap, around 8-10 cents/kWh.

5. Hydrogen is a convenient renewable energy source which can be distributed all over the world, including Third World countries, where such implementation is badly needed.

Hydrogen can be cost competitive with gasoline. This estimate is based on the use of solar thermal energy to generate hydrogen, mass production of the necessary equipment, and subsidies equivalent to those provided to the oil, coal, and utility industries. Large-scale production could deliver hydrogen at under $1 per GGE. Likewise, given the low cost of producing electrical power from existing hydroelectric dams, the use of this electricity to produce hydrogen by electrolysis would cost less than gasoline.

Hydrogen holds the world's record for burning faster than any other fuel and at leaner (lower) fuel-to-air ratios than the ratios for hydrocarbon fuels. These characteristics allow engines that burn hydrogen to operate more efficiently than those that depend on other fuels.

Virtually every existing engine application, from lawnmowers to automobiles and locomotives, can be fueled with hydrogen, and benefits include more power and longer engine life. But perhaps the most astonishing benefit of burning hydrogen in a conventional engine is what we might call minus emissions—that is, the exhaust pipe releases cleaner air than that which enters the engine. Atmospheric levels of carbon monoxide, tire particles, hydrocarbons, pollen, and diesel soot are reduced as air is cleaned by the hydrogen flame. The pollutants are substantially converted into harmless gases.

Kits for high-efficiency combustion of hydrogen can be retrofitted on most engines in the global fleet of motor vehicles. Ordinary engines that have been converted to operate on hydrogen show no sign of metal embrittlement or other degradation after decades of pollution-free service. Engine oil stays clean, spark plugs last much longer, and degradation as measured by corrosion and wear on piston rings and bearings is greatly reduced. These vehicles can therefore last longer, run better, and clean the air. And this transition will facilitate the distribution of hydrogen for the advent of fuel cells on a commercial scale (see "The Fuel Cell Future," *The World & I*, April 1994, p. 192).

Solar hydrogen can make any country energy-independent and pollution-free as far into the future as the sun will shine. Illustratively, just a small portion of North America can produce enough solar hydrogen to supply all the energy needs of the whole continent. Currently, Canada, Germany, Japan, Saudi Arabia, and Russia lead the world in developing plans to employ solar hydrogen. And several major auto manufacturers have experimental fleets of hydrogen-powered vehicles.

In this manner, if solar hydrogen is used to provide energy-intensive goods and services to the world's population, it will facilitate wealth addition as opposed to wealth depletion that results from burning fossil resources. The harder we work in the solar-hydrogen economy, the more goods and security everyone can have, resulting in a lower rate of inflation and less reason for conflicts and strife. These circumstances can create a global incentive to work for higher sustainable living standards, for both present and future generations.

Practical Hydrogen Economy

With most hydrogen today being produced from hydrocarbons, the cost per unit of energy delivered through hydrogen is higher than the cost of the same unit of energy from the hydrocarbon itself.

Duh!

Energy economy powered by solar hydrogen requires complete control of the production, storage, distribution, and utilization of hydrogen and the production of electricity from renewable energy sources. This could happen through:

1. Establishing renewable power fields, where generators convert solar energy and its derivatives, such as wind, falling water, wave motion, and biomass resources into electricity and hydrogen to be used as storage of energy or for transport to be used at remote locations.

2. Storing surplus hydrogen, and that for night power generation, in depleted natural-gas and oil wells and similar geological formations.

3. Using existing infrastructure of electricity grids, natural gas pipelines, highways, and rails to distribute hydrogen and electricity from renewable resources to be used without much modification and expense.

4. Installation of kits that enable the current internal combustion engines of motor vehicles to operate with directly injected hydrogen, landfill gas, or gasoline. A vehicle converted to operate on hydrogen cleans the air through which it travels.

5. Introduction of automobiles that run on hydrogen-based fuel cells, to replace dirty internal combustion engines.

6. Installation of reversible electrolyzers and fuel cells in homes and businesses, where these could produce electricity during off-peak hours, to be converted to storable hydrogen. Thus stored hydrogen can be used to make electricity at peak hours.

DOE Hydrogen Roadmap

The US Department of Energy (DOE) has fully evaluated the need for a new hydrogen-based economy, and has made some attempts to outline and implement the steps. Some of the conclusions and decisions—albeit focusing on the generation, storage and transport of hydrogen for fueling cars—follow herein: (2)

Hydrogen Production, Challenges

Multiple challenges must be overcome to achieve the vision of secure, abundant, inexpensive, and clean hydrogen production with low carbon emissions.

The DOE Roadmap established the following in 2002: (2)

Hydrogen production costs are high relative to conventional fuels. With most hydrogen currently produced from hydrocarbons, the cost per unit of energy delivered through hydrogen is higher than the cost of the same unit of energy from the hydrocarbon itself. This is especially the case when the comparison is made at the point of sale to the customer, as delivery costs for hydrogen are also higher than for hydrocarbons.

The large-scale, well-developed production and delivery infrastructures for natural gas, oil, coal, and electricity keep energy prices low and set a tough price point for hydrogen to meet.

Low demand inhibits development of production capacity. Although there is a healthy, growing market for hydrogen in refineries and chemical plants, there is little demand for hydrogen as an energy carrier. Demand growth will depend on the development and implementation of hydrogen storage and conversion devices, and on a demand pull from products such as hydrogen-powered cars and electric generators. Without demand

Table 3-2. Hydrogen production 2005-2015

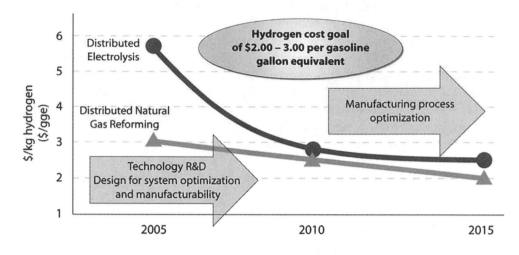

for high-quality hydrogen in the merchant energy carrier market, there is little incentive for industry to completely develop, optimize, and implement existing and new technologies.

Current technologies produce large quantities of carbon dioxide and are not optimized for making hydrogen as an energy carrier. Existing production technologies can produce vast amounts of hydrogen from hydrocarbons but emit large amounts of carbon dioxide into the atmosphere. Existing commercial production methods (such as steam methane reformation, multi-fuel gasification, and electrolysis) require technical improvements to reduce costs, improve efficiencies, and produce inexpensive, high-purity hydrogen with little or no carbon emissions.

Advanced hydrogen production methods need development. While wind, solar, and geothermal resources can produce hydrogen electrolytically, and biomass can produce hydrogen directly, other advanced methods for producing hydrogen from renewable and sustainable energy sources without generating carbon dioxide are still in early research and development phases. Processes such as nuclear thermo-chemical water splitting, photo-electrochemical electrolysis, and biological methods require long-term, focused efforts to move toward commercial readiness. Renewable technologies such as solar, wind, and geothermal need further development for hydrogen production to be more cost-competitive from these sources.

Public-private production demonstrations are essential. Stakeholders need a basic understanding of the different sources of hydrogen production before they will be willing to embrace the concepts. Demonstrations are the best way to gain the needed confidence. The large scale of some production processes, however, makes them particularly difficult and expensive to demonstrate.

Potential of Solar Hydrogen Generation and Use
H_2 Generation

The most developed technologies for alternative electricity generation are photovoltaic, thermal and wind energy conversion. All these technologies are so well developed and field tested that they are becoming household terms. They are also becoming technologically and economically feasible. For example, a 100 MWp photovoltaic power plant installed in a desert area with yearly insolation of 2300 kWh/m, and 15% efficiency of conversion into hydrogen would have a collector surface of about 1/2 km² and would produce yearly 0.5 x 10^9 m³ of hydrogen.

New, more efficient solar technologies, like multi-junction solar cells and efficient CSP receivers, are capable of bringing sunlight-to-AC power efficiencies to 30% now, and over 50% in the near future. This will also reduce land and labor costs, and material and land requirements.

The EU electricity consumption could be met with the European yearly insolation of about 1000 kWh/mZ/y and 15% conversion efficiency, using less than 1% of Europe's land surface. This is how much surface is actually used for the EU roads (2.5 × 10^6 km length). But it doesn't make much sense to use population-dense, cloud-covered Europe for such mass energy production. A better alternative is using the deserts in Africa—and especially those in North Africa—which could produce all the energy Europe needs, and much more.

This is a grand undertaking, yes, but it represents a very smart investment in the future, which will benefit not only the European nations, but will provide the inhabitants of the African countries with opportunities they presently do not have and desperately need. This might bring much needed peace to the region as well.

The yearly insolation in Africa's deserts is approximately 2000 kWh/m^2, so a surface area of less than 0.035% of Africa's land mass would be required by the collector field to meet Europe's entire electricity consumption.

It is easy to see underwater cables crossing the Mediterranean Sea to deliver high voltage DC power to sub-stations in France, Spain, Germany, and Italy for conversion to AC and distribution to the other EU nations.

Some of the energy produced by the solar fields in Africa (approximately 1/4) can be converted to LH$_2$ and delivered in conventional transport vehicles to power stations in the above mentioned EU countries for conversion into electricity and distribution in the EU power network.

Far fetched? Not really. Where are the Europeans of the 21st century going to get their energy? From unreliable Russian gas deliveries? From the almost depleted Saudi Arabian oil wells? The EU nations have very few choices, and the sooner they realize that, the sooner they will be able to start this monumental task. It is the only choice Europe has for long-term energy independence, like it or not.

The continental US has similar problems, although the situation here is somewhat different. We (and Canada) have a lot of oil and gas deposits that could sustainable our wasteful and dirty carbon economy much longer. We would be the most stupid and selfish people on earth to do that. We *must* find smarter and more sustainable ways to meet our energy needs, and solar hydrogen production comes to the rescue as the most efficient, versatile and cost-effective alternative energy technology available today.

Storage and Distribution

Once generated, hydrogen can be stored in its gaseous form, but it is more convenient to convert it into a liquid. In that state, hydrogen can be transported, stored and burned either as free hydrogen or linked to other molecules like ammonia (NH$_3$), methanol (CH$_3$OH), methylcyclohexane (C$_7$H$_{14}$), or in a solid, such as a metal hydride.

Although liquid hydrogen (LH$_2$) has the highest energy density, its disadvantage is the energy required (approximately 30% of its combustion energy) for conversion to liquid and keeping it at a very low temperature of 20 K. Although amassing large quantities of LH$_2$ is a relatively new process, experience from the space sector, and transport of liquid natural gas (LNG) is directly applicable, since the storage techniques of hydrogen as a gas are similar to those of natural gas. It can be stored in storage tanks, or underground in confined aquifers, or in abandoned natural gas reservoirs. The losses from underground reservoirs are in the order of 2-30 per year. Most rock formations are sealed by water in their capillary pores, so that the nature and properties of the contained gas have no effect on the overall leakage.

Transportation/distribution of LH$_2$ can also be accomplished conveniently by pipelines, and in pressurized or cryogenic containers. A gaseous hydrogen pipeline system of some hundred kilometers and a vacuum-jacketed LH$_2$-System of some hundred meters are in operation today, at the Kennedy Space Center and elsewhere in the US.

Another method is to convert (combine) LH$_2$ into Methylcyclohexane, which is liquid at ambient pressure and temperature, so it can be stored and transported in normal oil product carriers in its "as is" state (its consistency and specific weight are similar to diesel fuel). This is important for a smooth energy-system infrastructure transition from crude oil to hydrogen.

Yet another form to which LH$_2$ lends itself for storage and transport is as a solid. Solid, rechargeable metal hydrides, for example, offer favorable volumetric hydrogen packing, although the weight of the non-saturated compound needed for storage of hydrogen is rather high. In a transition period, hydrogen can be mixed with natural gas; i.e., the "town gas" we knew in the past contained up to 60% hydrogen.

End Use of Hydrogen

Hydrogen is a carbon-free fuel with excellent combustion properties. This makes it a clean, universal fuel for use in industry, households, power stations, road vehicles and aviation.

The specific properties of hydrogen induce new end use techniques, such as:

Fuel Cells

The key element of efficient, future electricity generation is the fuel cell, for which hydrogen is the best fuel. Since these are devices which transform chemical energy directly into electric energy, hydrogen generation is not subject to Carnot cycle limitations. It is also done silently without moving parts and with high efficiency.

Efficiencies of future fuel cells like molten carbonate cells operating at 600-800°C with total energy efficiencies of 60-80% with waste heat at 600°C make the fuel cell concept a potential candidate for use in power stations, road and space vehicles.

Advances in catalyst and electrode technology and in cost reductions must be made, however, before fuel cells can seriously be considered for general use. Still, several power plants around the world are using this technology.

Catalytic Burners

Hydrogen, having lower ignition energy than some conventional fuels, including natural gas, burns smokeless when in contact with a suitable catalytic surface. Its burning temperature is quite low too—between 100 and 400°C. This makes it the preferred technology for generating low-temperature heat in residential space heaters, cooking devices and industrial dryers and heaters.

Catalytic combustion has the advantages of higher safety level and very high efficiency of up to 99% with negligible emissions of nitrogen oxides.

Liquid Hydrogen Technologies

Super/hypersonic aviation and direct use in jet engines at super/hypersonic speeds could be powered with liquid hydrogen for these specific benefits:

a.   Higher (three-fold) gravimetric heating value compared with kerosene, and

b.   Cooling capacity, which can be used to cool critical parts (turbine inlet rim, wing leading edges, passenger cabins, etc.), since LH_2 cryogenic storage temperature is approximately –252°C.

As an added benefit, cooling of the wing skin induces laminar flow considerably reducing drag by up to, theoretically, 30% ("laminar flow control").

Automotive Vehicles

The main problem with hydrogen utilization in vehicles is storage due to the low volumetric energy content of hydrogen, which is about one-third that of gasoline. Even liquid hydrogen is far less dense than hydrocarbon fuels. Three storage technologies are available and more or less mature: gaseous hydrogen in pressure vessels, cryogenic liquid hydrogen and hydrogen chemically bonded in metal, and liquid hydrides (methylcyclohexane).

The low ignition energy of hydrogen and its wide ignition range of hydrogen/air mixtures make gas turbines as well as piston engines well adaptable for hydrogen combustion. The ignition within a wide range of nonstoichiometric air/fuel mixture permits combustion

with high amounts of excess air and thus full and load operation with low nitrogen oxides emission.

Many vehicles with hydride storage have been built and operated by Daimler Benz and others through the years, although the operation of the necessary infrastructure, refueling and service is another issue to be solved.

A number of liquid hydrogen cars have been built and operated by BMW and DFVLR (Deutsche Forschungs-und Versuchsanstalt fur Luft-und Raumfahrt). And trucks with the methylcyclohexane technology have been built and operated by the Paul Scherrer Institute in Switzerland.

It should be emphasized that the problems of the necessary infrastructure development are at least as heavy as the development of the different components.

Hydrogen Homesteads

In a transition period, gaseous hydrogen can be utilized in the existing natural gas pipelines of natural gas fueled appliances in homes and in industry, bearing in mind that such a pipe system would not be optimized for use of hydrogen. Hydrogen, having a volumetric heating value of approximately one-third of that of natural gas, requires a 3 times higher flow rate and compressor power at given pipe diameter and energy flow to keep the system under pressure (50-70 bar).

Practical Examples

Enel, Italy, started operation of a 12 MW H_2-powered electricity plant in Venice recently. The plant is fueled by hydrogen by-products from local petrochemical industries. The turbines were specially designed to resist embrittlement from hydrogen, but in any case the only emission of hydrogen combustion is water.

In the last ten years, much hype has been associated with the hydrogen topic, raising questions about the claim that hydrogen is an economically viable fuel for transportation because of the cost and greenhouse gases generated during production, the low energy content per volume and weight of the container, the cost of the fuel cells, and the cost of the infrastructure.

The price of solar electricity in the last 3 years has fallen to such an extent, and the pace of innovation in both fields of solar energy has been so intense, that a number of unexpected practical hydrogen based applications have emerged; i.e. commercial boat engines, such as the commercial electric boat powered by an H_2-O_2 fuel cell.

In 2009 the Austrian companies Fronius, Bitter and Frauscher successfully presented Riviera 600: the first electric boat powered by solar hydrogen fuel cells.

The concept is that of a self-contained energy supply provided by hydrogen simply obtained by photovoltaic electrolysis of water.

The team "Future Project Hydrogen" has created budget calculations for the generation of hydrogen on-site by use of photovoltaics under the premises of 10 boats for commercial use, for example, within a boat rental. With a range of 80 kilometres with a full hydrogen tank and having been awarded a safety certificate by Germany's T€UV, the boat is 6 m long, 2.2 m wide and weighs 1400 kg. Its 4 kW continuous power electric motor has twice the range of conventional battery-powered boats. The 47% efficiency of the noise-free fuel cell engine should be compared to the 18%-20% efficiency of a conventional (steel) internal combustion engine.

The main economic advantage compared with conventional electric boats is the fact that no time must be spent charging the batteries. For conventional electric boats, 6-8 h of charging gives just 4-6 h of use. The hydrogen-powered electric boat requires only the time that it takes to change the cartridge: 5 min. The boat's fueling system consists of a 20 kg cartridge that can be charged with up to 0.7 kg of hydrogen kept at 350 bar. Refueling is done using a standard filler coupling plus a simple exchange of an empty cartridge for a full one.

For example, all three companies involved in the "Future Project Hydrogen" are based in Austria, very close to each other. Scientific and technical advice was provided by the Technical University of Graz, whereas the project was realized with support of the European Union regional programs and further funding from one of Austria's regions. The first 600 Riviera boat is commercialized at 150,000 V, with the first exemplars delivered to customers in 2010.

The energy filling ("Clean Power") station makes use of PV modules integrated in a 250 m^2 flat roof, and further connected to an electrolytic cell. Even at Austria's cold latitudes the station is capable of affording an annual yield of 823 kg hydrogen, equivalent to 1100 cartridges with a 27,200 kW h energy content, namely enough hydrogen to run a boat for 80,000 km. Its installation is simple thanks to the "container construction" design and can be carried out quickly at many different locations. The station is comprised of an electricity power charger, hydrogen and payment units (Figure 3-8). For comparison, storing power in batteries over long periods of time is linked to huge losses due to self-discharge (5-10% per month), while the energy density is a fraction of that for hydrogen, which means that by storing energy in the summer in a battery of the same capacity, one would have no more energy available in winter.

Another example showing that photovoltaic renewable hydrogen is far from being solely a research topic is given by the world's first underground pipeline supplying H_2 to customers in the Italian city of Arezzo.

At present, the pipeline serves 4 companies and the HydroLAb with a main channel of around 600 m where the whole network is around 1 km. Four goldsmith companies use it for industrial and energy needs via four 5 kW fuel cells and two 1 kW fuel cells at the HydroLAb, the laboratory for hydrogen and renewable energies.

The aim was to set up a completely off-grid testing lab for technologies in the renewable energy sector, collecting data to test solar energy technologies linked with hydrogen production and use. Hence, solar panels provide electricity, solar thermal vacuum tube panels provide heat for room heating and feed a 5 kW solar cooling machine in order to get zero emission air conditioning in summer.

Waste water is completely recycled through a special remediation dry technique and rain is collected and stored. The technologies implemented are continuously monitored, with the aim of further optimization in view of widespread commercial application in the building industry.

ENERGY STORAGE

We've seen that variability in PV power plants' output is unavoidable. PV power might be combined with other energy sources (i.e., wind) to smooth the variations and even match the peak load, but even that is not a complete solution. During cloudy or rainy days the output will be limited and the only way to rectify that anomaly is to provide energy storage for supplemental power generation. Stored energy can be used during periods of low energy generation or at night.

There are a number of potential energy storage solutions for use with solar power plants, some of which are applicable for PV power storage too. See below a complete list of presently available energy storage technologies, followed by a discussion of the most promising for use today.

The most promising energy storage technologies under consideration and development for use now and in the future, are:

1. **Thermal storage**
   Steam accumulator
   Molten salt

Cryogenic liquid air or nitrogen
Seasonal thermal store
Solar pond
Hot bricks
Fireless locomotive
Eutectic system
Ice Storage

2.   Electrochemical storage
Batteries
Flow batteries
Fuel cells
Electrical
Capacitor
Supercapacitor
Superconducting magnetic energy storage (SMES)

3.   Mechanical storage
Compressed air energy storage (CAES)
Flywheel energy storage
Hydraulic accumulator
Hydroelectric energy storage
Spring
Gravitational potential energy (device)

4.   Chemical storage
Hydrogen
Biofuels
Liquid nitrogen
Oxyhydrogen and Hydrogen peroxide

5.   Biological storage
Starch
Glycogen

6.   Electric grid storage

We will review here only the applications most suitable for CSP and PV energy storage.

Thermal Energy Storage
Presently, this is the most widely used method of energy storage in thermal solar (CSP) plants. Heat is transferred to a thermal storage medium in an insulated reservoir during the day, and withdrawn for power generation at night. Thermal storage media include pressurized steam, concrete, a variety of phase change materials, and molten salts such as sodium and potassium nitrate.

The most widely used heat transfer liquids today are:

Pressurized Steam Energy Storage
Some thermal solar power plants store heat generated during the day in high-pressure tanks as pressurized steam at 50 bar and 285°C. The steam condenses and flashes back to steam, when pressure is lowered. Storage time is short—maximum one hour. Longer storage time is theoretically possible, but not yet proven.

Molten Salt Energy Storage
A variety of fluids can be used as energy storage vehicles, including water, air, oil, and sodium, but molten salt is considered the best, mostly because it is liquid at atmospheric pressure, it provides an efficient, low-cost medium for thermal energy storage, and its operating temperatures are compatible with today's high-pressure and high-temperature steam turbines. It is also non-flammable and nontoxic, and since it is widely used in other industries, its behavior is well understood and the price is cheap.

Molten salt is a mixture of 60% sodium nitrate and 40% potassium nitrate. The mixture melts at 220°C, and is kept liquid at 290°C (550°F) in insulated storage tanks for several hours. It is used in periods of cloudy weather or at night using the stored thermal energy in the molten salt tank to generate steam and turn a turbine, which in turn generates electricity. These turbines are well established technology and are relatively cheap to install and operate.

Pumped Heat Storage
Pumped heat storage systems are used in CSP power plants and consist of two tanks (hot and cold) connected by transfer pipes with a heat pump in between performing the cold-to-heat conversion and transfer cycles. Electrical energy generated by the PV power plant is used to drive the heat pump with the working gas flowing from the cold to hot tanks. The gas is heated and pumped into the hot tank (+50°C) for storage and use at a later time. The hot tank is filled with solids (heat absorbing materials), where the contained heat energy can be kept at high temperature for long periods of time.

The heat stored in the hot tank can be converted back to electricity by pumping it through the heat pump and storing it back in the cold tank. The heat pump recovers the stored energy by reversing the process.

Some power (20-30%) is wasted for driving the heat pump and during the transfer and conversion cycles, but the technology can be optimized for use in large-scale PV plants.

In all cases, large heat energy losses accompany

the energy storage processes. At least one third of the energy is lost during the conversion of stored heat energy into electricity. More is lost during the storage and following cooling cycles.

Hydrogen Storage

Hydrogen storage is a viable alternative, but it must meet several challenges before hydrogen can become an acceptable energy option for the consumer. The technology must be made transparent to the end user—similar to today's experience with internal combustion gasoline-powered vehicles. (2)

Specific Challenges

Current research and development efforts are insufficient. Hydrogen storage is a critical enabling element in the hydrogen cycle, from production and delivery to energy conversion and applications. Improved storage technologies are needed to satisfy end-user expectations and foster consumer confidence in hydrogen-powered alternatives. Substantial research and development investment in hydrogen storage technologies will be required to achieve the performance and cost targets for an acceptable storage solution.

New media development is needed to provide reversible, low-temperature, high-density storage of hydrogen. These storage characteristics generally describe the technical goals for some of the solid-state materials, including hydrides and carbon adsorption materials. The ultimate hydrogen storage system for meeting manufacturer, consumer, and end-user expectations would be low in cost and energy efficient, provide fast-fill capability, and offer inherent safety.

Energy storage densities are insufficient to gain market acceptance. This barrier directly relates to making hydrogen storage transparent to the consumer and end user. Specifically, transparency would mean a hydrogen storage system that enables a vehicle to travel 300 to 400 miles and fits in an envelope that does not compromise either passenger or storage space. Fundamental limitations on hydrogen density will ultimately limit storage performance. The performance of vehicles, therefore, depends on the overall system performance—the combined vehicle efficiency, energy conversion efficiency, and storage efficiency.

Low demand means high costs. As there are few hydrogen-fueled vehicles on the road today, the more mature compressed and liquid hydrogen storage technologies are quite expensive. High-pressure cylinders will be amenable to high-volume production, once demand warrants it.

Raw material costs could also be reduced substantially if there were sufficient demand. For emerging technologies, manufacturing feasibility and cost reduction measures will play integral roles in the technology development process. The initially low rates at which automakers expect to introduce fuel cell vehicles will present a challenge to the commercialization and cost reduction of hydrogen storage technologies.

Battery Energy Storage

No doubt, this is the most direct and efficient way to store a large amount of electricity generated by PV power plants. The generated DC electric energy is stored as DC power in batteries for later use.

There are several types of batteries, the most commonly used follow.

Lead Acid Batteries

These are the most common types of rechargeable batteries in use today. Each battery consists of several electrolytic cells, where each cell contains electrodes of elemental lead (Pb) and lead oxide (PbO_2) in an electrolyte of approximately 33.5% sulfuric acid (H_2SO_4). In the discharged state both electrodes turn into lead sulfate ($PbSO_4$), while the electrolyte loses its dissolved sulfuric acid and becomes primarily water. During the charging cycle, this process is reversed.

These batteries last a long time, and can go through many charge-discharge cycles, if properly used and maintained. They are affected, however, by high temperatures, when the electrolyte can boil off and destroy the battery. Since there is water in the cells, the electrolyte can freeze during winter weather, which could destroy the battery as well.

Lithium Batteries

These are a mature technology, having been used widely for a long time in consumer electronics. They are actually a family of different batteries, containing many types of cathodes and electrolytes. The most common type of lithium cell used in consumer applications uses metallic lithium as anode, and manganese dioxide as cathode, with a salt of lithium dissolved in an organic solvent. A large model of these can be used to store large amounts of electric power generated by a PV power plant, and due to their highest known power density they could be quite efficient—70-85%.

They are suitable for smaller PV installations, too, because scaling up to large PV plants would be a very expensive proposition.

Sodium Sulphur Batteries

These are high temperature, molten metal, batteries constructed from sodium (Na) and sulfur (S). They have a high energy density, high efficiency of charge/discharge (89-92%) and long cycle life. They are also usually made of inexpensive materials, and due to the high operating temperatures of 300-350°C they are quite suitable for large-scale, grid energy storage.

During the discharge phase, molten elemental sodium at the core serves as the anode, and donates electrons to the external circuit. The sulfur is absorbed in a carbon sponge around the sodium core and Na+ ions migrate to the sulfur container. These electrons drive an electric current through the molten sodium to the contact, through the electric load and back to the sulfur container.

During the charging phase, the reverse process takes place. Once running, the heat produced by charging and discharging cycles is sufficient to maintain operating temperatures and usually no external source is required.

There are, however, safety and corrosion problems, due to the sodium reactivity, which need to be resolved before full implementation of this technology takes place.

Vanadium Redox Batteries

These are liquid energy sources, where different chemicals are stored in two tanks and pumped through electrochemical cells. Depending on the voltage supplied, the energy carriers are electrochemically charged or discharged. Charge controllers and inverters are used to control the process and to interface with the electrical source of energy.

Unlike conventional batteries, the redox-flow cell stores energy in the solutions, so that the capacity of the system is determined by the size of the electrolyte tanks, while the system power is determined by the size of the cell stacks. The redox-flow cell is therefore more like a rechargeable fuel cell than a battery. This makes it suitable as an efficient energy storage for PV installations.

A number of additional types of batteries are under development, and some show great potential for use in larger PV installations in the near future. Most batteries have problems with moisture, high temperature, memory effect, and use of scarce and toxic exotic materials, all of which cause longevity problems and abnormally high prices. If and when all these problems are resolved, the energy storage problems of PV power plants will be resolved as well.

Compressed Air Energy Storage

Compressing air into large high-pressure tanks is one of the most discussed and most promising energy storage methods for use with PV power plants today. It is a simple way of energy storage, using a compressor powered by the electricity produced by the PV plant compressing air into the storage tank. A lot of energy is lost by activating the compressor, and heat is wasted during the compression process, so there are several compressing methods that treat generated heat so as to optimize conversion efficiency. Some of these follow.

Adiabatic Storage

Adiabatic storage retains the heat produced by compression via special heat exchangers, and returns it to the compressed air when the air is expanded to generate power. Its overall efficiency is in the 70-80% range, with the heat stored in a fluid such as hot oil (300°C) or molten salt solutions (600°C).

Diabatic Storage

Here extra heat is dissipated into the atmosphere as waste, thus losing a significant portion of the generated energy. Upon removal from storage, the air must be re-heated prior to expansion in the turbines, which requires extra energy as well. The lost and added heat cycles lower the efficiency, but simplify the approach, so it is the only one implemented commercially these days. The overall efficiency of this method is in the 50-60% range.

Isothermal Compression and Expansion

This method attempts to maintain constant operating temperature by constant heat exchange to the environment. This is only practical for small power plants, which don't require very effective heat exchangers, and although this method is theoretically 100% efficient, this is impossible to achieve in practice, because losses are unavoidable.

There are a large number of other methods using compressed air, such as pumping air into large bags in the depths of lakes and oceans, where the water pressure is used instead of large pressure vessels. Pumping air into large underground caverns is another approach that is receiving a lot of attention lately.

Pumped Hydro Energy Storage

Pumped hydro energy is a variation of the old hydroelectric power generation method used worldwide, and is used quite successfully by some power plants. Energy is stored in the form of water, pumped from a lower elevation reservoir to one at a higher elevation. This way, low-cost off-peak electric power from the PV power plant can be used to run the pumps for elevating the water. Stored water is released through turbines and the generated electric power is sold during periods of high electrical demand. This way the energy losses during the pumping process are recovered by selling more electricity during peak hours at a higher price.

This method provides the largest capacity of grid energy storage—limited only by the available land and size of the storage ponds.

Flywheel Energy Storage

Flywheel energy storage works by using the electricity produced by the PV power plant to power an electric motor, which in turn rotates a flywheel to a high speed, thus converting the electric energy to, and maintaining the energy balance of the system as, rotational energy. Over time, energy is extracted from the system and the flywheel's rotational speed is reduced. In reverse, adding energy to the system results in a corresponding increase in the speed of the flywheel. Most FES systems use electricity to accelerate and decelerate the flywheel, but devices that use mechanical energy directly are being developed as well.

Advanced FES systems have rotors made of high-strength carbon filaments, suspended by magnetic bearings, and spinning at speeds from 20,000 to over 50,000 rpm in a vacuum enclosure. Such flywheels can come up to speed in a matter of minutes—much quicker than some other forms of energy storage.

Flywheels are not affected by temperature changes, nor do they suffer from memory effect. By a simple measurement of the rotation speed, it is possible to know the exact amount of energy stored. One of the problems with flywheels is the tensile strength of the material used for the rotor. When the tensile strength of a flywheel is exceeded, the flywheel will shatter, which is a big safety problem. Energy storage time is another issue, since flywheels using mechanical bearings can lose 20-50% of their energy in 2 hours. Flywheels with magnetic bearings and high vacuum, however, can maintain 97% mechanical efficiency, but their price is correspondingly higher.

Electric Grid Storage

Grid energy storage is large-scale storage of electrical energy, using the resources of the national electric grid, which allows energy producers to send excess electricity over the electricity transmission grid to temporary electricity storage sites that become energy producers when electricity demand is greater. Grid energy storage is a very efficient storage method, playing an important role in leveling and matching electric power supply and demand over a 24-hour period.

There are several variations of this method, one of which is the proposed grid energy storage called vehicle-to-grid energy storage system, where modern electric vehicles that are plugged into the energy grid can release the stored electrical energy in their batteries back into the grid when needed. Far fetched, yes, but the future will demand many such ingenious approaches, if we are to be energy independent.

In conclusion, there are a number of other energy storage methods such as fuel cells, new types of batteries, superconducting devices, supercapacitors, hydrogen production, and more under development, so the future looks bright in this area. In practice, however, there are a number of energy storage installations around the world, totaling over 2,100 MW.

The major technologies used today are:

- Thermal energy storage over 1,140 MW
- Batteries energy storage over 450 MW
- Compressed air energy storage over 440 MW
- Flywheels energy storage over 80 MW

Energy storage has many advantages, in addition to its unique potential to transform the electric utility industry by improving wind and solar power variability, availability and utilization. It can contribute to the overall energy independence and environmental clean-up by avoiding the building of new power plants and transmission and distribution networks. Experts consider energy storage to be the solution to the electric power industry's issues of variability and availability, opening new opportunities for wind and solar power use.

Complexity, safety, price and other restrains must be worked out well before any of the energy storage methods become accepted reality for large-scale PV installations.

Summary

We are witnessing another solar revolution in the making. Will it be successful in changing our ways? Will it bring us energy independence and a clean environment? These are questions for which we have no answers, but must answer them soon.

The amount of solar energy available to our Earth is truly astounding. Several studies show that the entire global electricity demand could be provided from just 3% of the world's deserts, while several hundred square miles in the Nevada desert could power all of the lower 48 states.

We have the technologies to choose from for use in different areas. We have a burning need to secure our energy future and that of the environment. What is stopping us?

Notes and References

1. "PVCDROM," C.B. Honsberg and S. Bowden, www.pveducation.org, 2010.
2. DOE, National Hydrogen Energy Roadmap, 2002
3. Common Failure Modes for Thin-Film Modules and Considerations Toward Hardening CIGS Cells to Moisture A "Suggested" Topic Kent Whitfield, MiaSolè.
4. *Adhesion and Thermomechanical Reliability for PVD and Modules*, by Reinhold H. Dauskardt.
5. Derivation of power gain for three types of three dimensional photovoltaics cells, by Jack Flicker and Jud Ready. *Materials Science and Engineering*, Georgia Institute of Technology, Atlanta, GA.
6. A three dimensional porous silicon p-n diode for betavoltaics and photovoltaics, by Wei Sun and M. Fauchet.
7. A dynamic model of hybrid photovoltaic/thermal panel Majed Ben Ammar, Mohsen Ben Ammar, and Maher Chaabene1.
8. Three-dimensional photovoltaics, Bryan Myers, Marco Bernardi, and Jeffrey C. Grossman. Department of Physics, University of California-Berkeley. Reprinted with permission from the American Institute of Physics.
9. Flasher Tolerances of Power Measurement on Micromorph Tandem Modules, Anja Böttcher, André Prorok, Nicoletta Ferretti, Alexander Preiss, Stefan Krauter* and Paul Grunow, Photovoltaik Institut Berlin AG, Wrangelstr. 100, D-10977 Berlin, Germany.
10. Performance Comparison of A-Si, -a-Si, C-Si as a Function of Air Mass and Turbidity, Stefan Krauter and Alexander Preiss, Photovoltaik Institut Berlin AG, Wrangelstr. 100, D-10977 Berlin, Germany
11. *Photovoltaics for Commercial and Utilities Power Generation*, 2011. Anco S. Blazev.

Chapter 4

Exotic Solar Technologies

One of the greatest discoveries a man makes, and one of his great surprises, is to find that he can do what he was afraid he couldn't do.
— Henry Ford

"Exotic" is what they called Edison's light bulb— what a crazy idea to use something that you cannot see or touch to make light out of!

Or Henry Ford's Model T. What's wrong with our elegant and quiet horse carriages? Isn't it far fetched and crazy to use these noisy, stinky, machines to transport humans, or anything else?

Similarly so, some of the new alternative energy technologies seem exotic, far fetched, and even ridiculous; just think, for example, that organic paste slapped on a piece of paper can produce useful energy!

But, many of these unusual products will be in everyday use in the not-so-distant future.

Technological progress and human needs will see to it that every possibility is exhausted and the best of the best reaches the energy markets—even if it sounds impossible or ridiculous today.

Some of the more exotic PV devices have been classified as "third generation," herein referred to at times as third gen, or 3rd gen, devices. This special designation simply means that these technologies are in line for development and full utilization after, or in parallel with, the development and deployment of their older cousins—first and second generation photovoltaics.

So they all—first, second and third generation solar technologies—are our hope for efficient and cheap energy during the 21st century. Because of that, we need to give all of them the due respect and support of their development.

So what is "exotic" solar technology? These are any and all of the new solar technologies being developed in universities, R&D labs and hi-tech companies in the US and around the world.

We also call them "developing," because they are in the process of evolving from their exotic state to that of *fully developed* solar technologies.

Some seem quite out of this world, and many of them will never see the light of day—no pun intended. But, some of them show potential, as incredible as their function and application might seem to the untrained eye at this point in their development.

One thing we should caution the reader about: we do not see earth-shaking development in this 3rd gen area anytime soon. Instead, we see small, incremental, yet steady, development of the solar technologies, including those we are discussing herein.

No Holy Grail, or Perpetuum Mobile-type device capable of replacing conventional solar technologies is on the horizon. Nor do we foresee such development anytime soon, because the laws of physics and chemistry simply won't allow it, short of a miracle, or some material or force that we are not aware of today.

Because of that, we must look at all solar technologies as only single parts of the complex puzzle—single pieces of the multi-component equation of our energy independence, an equation in which every part, no matter how small, counts. Every part will fill a special need, and have a practical application where it is most efficient and useful, thus reaching and fulfilling its destiny.

More importantly, there is a synergy in the development effort of the technologies in the third gen field, which combined with the optimization efforts of 1st and 2nd gen technologies, will contribute to the development of even better technologies, using new ideas, approaches, materials and techniques.

The benefit of this synergetic action is that some of the materials and processes that prove successful in one type of technology could be used in developing others. By borrowing materials and processes, the entire solar field will be moving steadily forward, contributing one way or another to the goal of achieving acceptable efficiency, reliability and cost-effectiveness as needed to replace fossil fuels and power the 21st century with clean, renewable energy.

Looking at the new technologies, the lines between "virtual reality" and practical applications overlap and fade, so that we might fail to see the obvious, or otherwise see things that are simply not there.

The text below is an attempt to delineate reality from dreams—that which is practical from that which is just wishful thinking. We take a close look at the key solar technologies and spell out their functions, applications, and issues.

Solar technologies of today can be divided as seen in Figure 4-1.

Hybrid organic-inorganic solar cells
Hybrid Polymer-Inorganic Blend Device
Hybrid Solar-Thermoelectric Systems
Hybrid nanotech thermal electric
Hydrogen Generation
Infrared solar cells
Ink PV Cells
Ink Polymer printing

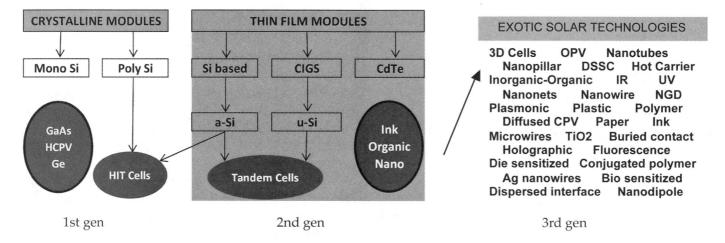

Figure 4-1. The PV technologies

Here is a brief list of some of the 3rd gen technologies, or those we consider exotic, in alphabetical order:

3rd generation solar cells
3D Nanopillar Array
a-Si nanowire
Bio-Sensitized Solar Cell
Carbon Nanotubes
Conjugated Polymers
Crystalline Silver Nanowires
CPV on Flat Glass
Die sensitized Solar Cell
Diffused CPV
Dispersed Interface photovoltaics
Excitonic Solar Cells
Flexible solar cells
Fluorescence concentrators
Infrared solar cells
Holographic Planar Concentrator Technology
Holographic Solar Concentrators
Holographic planar CPV system
Hot carrier cells
Hybrid Solar Cells
Hybrid organic-inorganic solar cells
Hybrid nanotech thermal electric

Inorganic PV
Interband Cascade PV
Inverted Polymer PV
Light absorbing dyes
Luminescent solar concentrator
Meta-materials
Metamorphic multijunction solar cell
Miniature solar cells
Molecular PV
Multiple Energy Levels
Nanocomposite
Nano-crystalline Solar Cells
Nanonets
Nanodipole solar cells
Nanoparticle Lens
Nanopillars
Nanoplasmonic Solar Cells
Nanoporous solar cells
Nanostructured coatings
Nanowire Solar Cells
NGD technology
Organic Photovoltaics
Organic/Polymer Solar Cells
Organic Semitransparent OPV
Organic bladed PV

Optical Guide PV
Photonic Crystals Lens
Photo electrochemical Cells
Plasmonic PV Cells, 3rd Generation
Plastic PV Cells
Polymer Lens
Polymer PV
Printed Paper Solar Cells
Plasmonic Solar Cells
Plastic Solar Cells
Printed Paper Solar Cells
PV H2 Generation
Polymer Solar Cells
Quantum dots solar cells
Quantum Well Solar Cells
Sensitized Solar Cell
Silicon Microwires PV
Single Nanowire PV
Simple single layer devices
Steps and Road Power
Sticker for PV Panels
Tandem Cells
Textile PV
Thermovoltaics
Titanium dioxide solar cell
Transparent Conductors
UV solar cells
Windows PV

We will take a closer look at some of these technologies herein, focusing on the most realistic, feasible and promising, in our opinion.

3rd Generation Solar Technologies

The term "third generation" is thrown around freely these days to denote something new and perhaps superior—a renewed hope for greater things to come in the solar business. How new, useful, and superior it all is and how much of the talk is just hoopla, is debatable, and we're going to debate it herein.

To begin with, we must explain what the "first" and "second" generation solar technologies are, and their relation to the "third" generation:

1. The "first generation" solar technologies are these made out of semiconductor materials, containing one or more p-n junctions. These are mostly the prevailing silicon solar cells and silicon based technologies today, which we consider more stable and more likely to dominate world energy markets for the foreseeable future.

2. The "second generation" solar cells came into existence not long ago, developed on the promises of ease of manufacturing, and lower cost than the "first generation." These are mostly thin film based PV technologies that are finding increased use in all kinds of residential and commercial applications. They are here to stay, for sure, but we see problems related to their use (in their "as is" state) under certain conditions, so it remains to be seen how long and how widely these will be used in the 21st century.

3. The "third generation" solar cells are brand new technologies. Most are still in early stages of development and far from practical implementation. They include some new concepts, like nanotechnology based solar cells, polymer-based cells, bio-mimetics, quantum dot cells, tandem cells, dye-sensitized solar cells, and a large number of different up-conversion technologies. Most of these concepts were not even in the solar vocabulary just a decade or two ago.

With characteristically low efficiency and reliability, most of the newer technologies are struggling to compete, but some are showing promise to vie with conventional technologies soon.

Materials of Construction

Materials used and planned for use in the construction of PV devices can be summarized as follow:

1st Generation:
    Single crystal silicon
    Multicrystalline silicon
    Polycrystalline silicon

2nd Generation
    CuInSe2
    Amorphous silicon
    Thin film crystalline silicon

3rd Generation
    Gallium Indium Arsenide
    Gallium Indium Phosphide
    Germanium

Note: DSSC, OPV, nano-plasmonic, quantum dot, several hybrids, and other technologies discussed below are also part of the 3rd gen "exotic" technologies.

Taking a close look at the major 3rd gen technologies we see that:

1. Dye-sensitized solar cells (DSSC) and devices have a number of advantages, such as higher efficiency (over 10%) than most of their 3rd gen cousins. They also offer the convenience of roll-to-roll processing, which contributes to offering the lowest cost of any commercially available solar cells with energy payback of less than 1 year.

    DSSC devices are light and flexible and could be colored for esthetic variance. Specially important is their uniform (steady) output under different levels of lighting—and they provide decent output even under cloudy skies, unlike the conventional technologies, which are heavily dependent on oncoming light levels. DSSC are easy and cheap to dispose of via conventional waste disposal methods.

    Some types of DSSC devices are already commercially available and we expect to see more of them in different areas of the energy markets soon.

    The major issue DSSC devices must deal with is the rapid degradation under excess heat and UV light, which makes them unsuitable for most long-term applications. The package is also difficult to seal properly, because of the presence of solvents, which evaporate and cause problems in the structure, as well as contribute to corrosion.

    This is also the major reason for listing the DSSC technology as exotic, or "developing": The key issues must be solved for the DSSC technologies to be able to compete on the energy markets.

2. Organic photovoltaic (OPV) devices have the advantage of low cost materials and processing in large area production operations. OPV devices can be processed in fast roll-to-roll fashion, which allows low cost processing. They are lightweight and flexible, but even more importantly; they are tunable which allows them to be efficient under low lighting levels. OPV are also easy and cheap to dispose of via conventional waste disposal methods.

    OPV technologies have a number of issues, as well, the most important of which is their low efficiency, around 6% for single layer, and about 6.5% for multi-layer solar cells. This is in addition to the instability of the materials under excess heat and UV applications. These issues must be solved for the OPV technologies to be able to compete on the large-scale energy markets.

3. Nano-technologies (including quantum dots and plasmonics) are also advantageous as far as cost

and availability of materials, but suffer from the same problems of low efficiency and questionable reliability.

4. On the extreme end of the spectrum are the 3rd gen technologies that are hoping to eventually reach and even overcome the Shockley-Queisser theoretical limit of ~33% power efficiency, which is considered impossible to be achieved by today's the state-of-the-art materials and processes.

    Note: Shockley-Queisser limit is the maximum theoretical efficiency obtainable by solar cells with a single p-n junction and a bangap of 1.1eV (silicon), and is set at 33.7%. This limit does not apply to cells with multiple junctions, which are all time high performers with efficiency limits in the 80% range, and some already approaching 45% efficiency.

5. All 3rd gen technologies offer shelter from the ongoing silicon price wars, and the rising cost of the other materials used in conventional PV technologies (Ag, Cd, Te, etc.), which in addition to the potential of achieving 12% efficiency and the other advantages (flexibility, light weight, ease of manufacturing, etc.) puts them in a different category. These special qualities assure the new technologies a special narrow niche market, which will grow.

We need to clarify from the onset that the term "third generation" is a step away from the conventional wisdom of dealing with solar energy technologies, and a step towards a different and more flexible (no pun intended) way to make flexible, lighter and cheaper PV cells and modules. While 3rd gen devices offer countless possibilities of developing new materials, devices and processes, most of the work on them is still confined to the R&D labs. They are far from popping up in the power fields.

Most of the so-called "exotic" technologies fall in the "third generation" category, so they still struggle with low efficiency, poor reliability, and difficulties in the transition from lab to mass production.

There is a considerable effort of late to develop low-cost materials that would allow tuning their energy bandgap for optimum performance, thus permitting control of the absorptive and power generating properties of the solar cell.

Unfortunately, as yet, it is difficult to envision the budding solar technologies sitting 30 years in a dusty, blistering hot desert, or in the middle of a Louisiana swamp. This is not a fair test, to be sure, but this picture gives the reader a good idea as to what is expected from these devices—when it comes to long-term reliability, no compromises would be tolerated, no matter how sexy and attractive the technology might be.

So at this time, the "third generation" technologies are somewhat like a newborn whose parents are determined to make into a "football star." Regardless of the parents' wishes and efforts, the new baby may or not become a football star.

Similarly, the *solar technologies* under development today may or may not answer the call to become the new disruptive technologies of the future.

Because of the myriad complexities which make it difficult to predict the future, we will review each "exotic" technology in alphabetical order and within the type of technology to which it belongs.

While we see some components of these technologies as very promising, we do not see any single-junction technology becoming disruptive anytime soon. By themselves, the development steps of new, exotic technologies can be useful in creating other technologies, which will bring the overall development of the entire solar sector up a notch.

Some of the new and different directions taken in the development of the new "third gen" solar technologies are:

1. Spectrum conversion cells, which are quite different in their approach to photo-conversion, because they convert the incoming polychromatic sunlight (the sun's spectrum) into a narrower distribution of photons whose energies are adjusted to the bandgap level of the particular solar cell.

2. Intermediate-band cells use one special and complex junction, which contains and supports multiple bandgap levels to increase efficiencies.

3. Hot carrier cells, which in addition to converting the bandgap energy (as normal cells do) also convert the excess energy of above-bandgap photons (which normal solar cells cannot do), into electrical energy.

4. PV-thermal solar technologies, which generate power and hot water simultaneously, and whose popularity is increasing. In some cases the incoming water is used to cool the PV cells, a great benefit during operation under extreme heat conditions.

5. High performers, multi-junction cells, which have multiple junctions built within the solar cell bulk that selectively absorb different regions of the solar spectrum, thus absorbing more photons and producing more electrons and electric power.

The new 3rd gen technologies consist of mostly non-semiconductor (in the classic sense of the word) materials, devices and processes. In the best cases, we are looking at a major modification of these, with the goal of designing and manufacturing different—efficient, reliable and cheap types of photocurrent generating devices.

Thus far, however, organic and plastic solar cells are the only two types of the 3rd gen solar technologies that have reached significant mass production volumes. Nevertheless, present demand is so low, and the market is saturated with these and other PV technologies, so it is hard to see further production volume increases anytime soon.

Third Generation BOS

Balance of system (BOS) components, such as supporting frames, trackers, combiner boxes, inverters, switching networks, transformers, wiring connectors, etc. have also gone through a lot of changes in the past. Comparing the analog inverters of the 1970s with the super-inverters of today we see an almost vertical line of progress. The same is true for the other components, so it suffices to say that BOS components are the most reliable, or at least the most predictable, part of any solar installation today.

This is not the case with PV modules and other solar equipment. Most of these technologies are black boxes, whose long-term operation is unpredictable, and over which we have no control. We don't know what exactly is in the PV modules coming from China. We don't know the quality of materials and labor in those, and have no sure way of making fool-proof qualitative analyses prior to installing them in the field.

Though inverters break and do not last forever, we have enough test/control instrumentation to track and eliminate a malfunction, and even predict a premature failure. With regular maintenance most state-of-the-art BOS can withstand any elements and perform satisfactorily for a long time. Proper regular maintenance costs money, so that's an issue we will tackle in this text as well.

Now back to our new, upcoming solar technologies.

Applications of Third Gen Devices in the 21st Century

Many 3rd gen technologies can be used for the manufacturing of such devices as low-cost solar chargers, thus providing a convenient and affordable power for recharging mobile electronics. This market is especially important in rural areas and developing countries, which opens a huge market, with expected growth of up to two billion handsets by 2015.

Many mobile devices and installations are presently powered by silicon solar panels, and it seems only a matter of time until 3rd gen devices will replace them.

A number of manufacturers and developers are working to take advantage of their ability to operate under different lighting levels, to provide power to non-grid-based lighting applications. There are already products on the market, such as lanterns, that prove the concept design for using DSSCs as a power source.

A variety of other, low-power consumer applications are beginning to emerge. Solar powered sunglasses that use miniature DSSC solar panels cleverly integrated into the lenses have been designed and are available to power up portable electronic devices. Solar calculators, clocks, watches, toys, and greeting cards using these technologies are also expected to enter the market soon.

In the high-tech area, transparent conducting oxides are important electrode materials for third gen, and many other power generating devices. These are usually made of doped metal oxide compounds, especially tin-doped indium oxide (ITO), which suffers from high cost, limited supply, and fragility. So here is a great opportunity for other materials to be used as an electrode substitute.

There are a number of choices today, such as Aluminum doped Zinc oxide (AZO), ATO, Gallium Zinc Oxide (GZO), and doped ZnO.

Conductive polymers are another important class of conducting materials that are expected to enhance the properties of OPV cells in the future.

One of the best-known polythiophene polymers is PEDOT (supplied by a number of companies), which is also utilized for transparent conductive electrodes, given its conductivity and transparency.

Encapsulation of PV devices is essential for their performance and longevity. It usually requires the use of a rigid, heavy and expensive glass cover plate, or plastic sheet-based material. To achieve flexibility and longer device lifetimes, water vapor transmission of the cover at rates of $\sim 10\text{-}3 \text{g}/\text{m}^2/\text{day}$ is considered necessary, and so there is now great interest in developing transparent barrier materials with much lower levels of permeability. Several manufacturers have demonstrated a barrier layer of $0.05 \text{g}/\text{m}^2/\text{day}$ with newly developed transparent, flexible materials, which have an estimated lifetime of over three years.

Painting cars with photosensitive, photo-electricity generating, paint would keep the cars light while providing electric energy to power it. Solar paint is actually one of the most promising pathways for the new technologies to take over a large part of the energy markets. The military is quite interested in this new approach too, for it could provide obvious advantages for vehicle camouflage and solar power.

Roofs of industrial buildings covered with steel sheets could be generating solar electricity using solar paint. Given the fact that there are in excess of one billion square meters of coated steel roofs erected each year around the world, one could foresee the opportunity for using these new solar paints.

Flexible thin glass coating (DSSC based) and transparent conductive electrodes for generating power from the sun are being developed by the US Air Force for providing power to airplanes instrumentation, sensors, etc.

DSSC based powered generating cells and modules are used to provide a "Power Shade" for powering commercial and U.S. Army tents, where they can generate 2-3 kW of power for electronics use and battery recharging.

The same technology could be also used for civilian residential and commercial installations, such as umbrellas, exhibition and car sales tents.

Powering different types of remote sensors is another application that looks promising. These new sensors will be used for environmental monitoring and hazardous gas detection, as well as other critical measurements such as keeping track of the carbon footprint of different installations.

Smart fabrics, where textile applications require highly flexible, low-cost solar power appear to be a good match-up with third-generation thin-film solar cells. Solar bags is one such application, and has been catching on, given the heightened environmental awareness of consumers. Several companies are capitalizing on these emotions and offer a range of biodegradable leather laptop solar bags which have the ability to recharge a phone battery within three hours using direct sunlight.

Solar clothing is a huge area of interest for these technologies as well. Clothing articles such as blouses, jackets, ties and swimwear are being developed to power mobile devices like phones and MP3 players, thus slowly replacing the conventional silicon-based materials presently used for these applications.

Power-to-the-uniform is an application that the

military has been looking into for awhile now as a means to help lighten the soldier's fieldwear and add power generation to it. Detachable patches can be worn to prevent friendly fire or to alert soldiers to the presence of chemical or biological contamination. "Power Shade" that fits over U.S. Army tent structures is under evaluation, where different configurations can generate several kW of power for sheltered electronics or battery recharging.

The U.S. Army's goal is to reduce the weight of the soldier's uniform by using third gen devices to charge vision goggles, laptops, communication devices, and GPS units, keeping them operational in the field, thus providing increased mobility and durability for soldiers, while helping save lives by providing emergency backup power for all kinds of devices.

Conventional and embedded sensors are another direction that third gen designers are looking into. These sensors can be used for environmental monitoring and hazardous gas detection, as well as to provide a more accurate picture of the carbon footprint. The key to creating cost-effective sensors is to use power train components that can be optimized specifically for the application.

Signs and advertising displays powered by third gen technologies are another large market niche. Printed color-changing displays are now being utilized on smart cards, novel packaging solutions, point-of-purchasing advertisements, and a host of other applications. By using these technologies for printed displays, it is possible to achieve cost-effective, self-powering solutions for many applications.

Umbrellas and sunshades for tents and such, outfitted with third gen solar cells, are a unique application that is being pursued by several developers. The sunshades can be used on car lots, shopping centers, parking lot sale events and other public venues. While providing shade for the inventory and people, these structures can also generate electricity which can be stored in batteries for lighting at night.

Such applications can also provide carbon offsets.

When stability and cost issues have been fully resolved, third gen technologies will find even broader applications, and will replace conventional solar technologies in many more areas of the 21st century energy markets.

3rd gen systems are most convenient for use in building integrated PV (BIPV) designs, whereby they can be fully integrated into the building envelope and in many cases replace conventional building materials. Here is where the largest opportunities for these tech-

nologies will likely emerge, because these products can be applied to a variety of lighting levels, and in different settings including roofs tiles, walls, facades, and windows. Several manufacturers are developing products for use in such applications.

Many industrial buildings with steel-sheet roofs will be generating solar electricity using these types of solar cells. There is an estimate that there are in excess of one billion square meters of coated steel roofs erected each year worldwide, and many more billions presently exist, so this is another huge potential market. So a number of companies are making solar tiles and other roof products which incorporate third gen technologies to provide solar power. This field is growing fast and we see it as one of the largest applications of these technologies in the short term.

Solar windows and skylights are another exciting BIPV market opportunity, since there is huge window area available to collect solar energy. Third gen technologies are especially practical in this area due to their aesthetic appeal. Some of these technologies can change the transparency of windows, which may be beneficial to workers' comfort, and could reduce glare and heat at the sunniest times of day.

Solar materials are being designed to replace conventional building materials for integration into semi-transparent glass and other areas of commercial BIPV applications. These materials are more flexible, lightweight and transparent, making them more aesthetically suitable for BIPV applications.

Above we listed some key applications of 3rd gen technologies, but their use doesn't stop there. There are limitless other possibilities for integrating these flexible, lightweight and cheap devices into our daily lives.

The new materials and processes used in their development will be used in optimizing some of the existing technologies, and/or in the design and manufacturing of new such. The sky is the limit...

CPV TECHNOLOGIES

Third generation concentrating PV (CPV) technologies are quite different from 1st and 2nd gen CPV technologies which are efficient and are used in a number of large installations today.

Here again, we need to clarify the differences between the different generations:

1. The 1st gen CPV technology was developed with the help of US Department of Energy in the 1980s.

It consisted of high-efficiency silicon solar cells assemblies equipped with Fresnel lenses, mounted on tracking frames, driven by gear-motor drives. The x-y tracking was controlled by sensors made out of silicon solar cells, using the changing level of shade falling on one side of the cell to detect the sun's position and drive the unit. This type of CPV technology was 18-20% efficient, and had a number of problems related to the different components, ranging from poor materials performance and longevity, to tracking accuracy issues.

2.   The 2nd gen CPV technologies were developed in the late 1990s, with major improvements in the CPV cells, which are now multi-junction cells built on super-efficient GaAs or germanium substrates. The other materials and components underwent serious improvements too, so now the CPV trackers are much more efficient and reliable. The efficiency of the new multi-junction cells is above 40%, and the units are reliable and cheap.

The concept, materials, processes and applications of 3rd gen CPV are totally different, using different materials, processes and operation. Most of them use organic and other "exotic" thin films and most do not track the sun. This is a totally new approach. Take a look:

CPV on Flat Glass

In the search for cheaper and more practical solar panels, researchers have created sheets of glass (or plastic) coated with advanced, light-channeling, organic dyes. The twist here is that these coatings can efficiently concentrate and transport sunlight onto the glass edges.

This is a revolutionary step in the battle for concentrating sunlight onto PV cells and modules that do not require tracking. Because of that, thus coated glass sheets could eventually make solar power as cheap as, or cheaper than, electricity from fossil fuels.

This is actually not a new concept, since it was developed by Ahmed Zewail at Caltech in the late 1970s. He used fluorescing dyes, incorporated in plastic sheets, where the broad visible solar spectrum was absorbed by the dyes at lower frequencies. At the same time the radiation was reflected to the edges of the pane of plastic and collected by leads attached to them.

This system, therefore, acts as a spectral and optical concentrator. Jet Propulsion Laboratory developed the concept further for application in practical architectural designs and PV cells. A number of companies are working on similar designs.

Today's solar concentrators use flat or concave mirrors, or Fresnel lenses mounted on trackers, which follow the sun to focus the light properly on the solar cells or whatever the target might be. In all cases these are heavy, complex and expensive structures, while the new flat glass sheets are lighter and can easily be incorporated into stationary solar panels on roofs or building facades and yet concentrate the incoming sunlight.

The ordinary coated flat glass sheets have a number of advantages over solar concentrators, as we know them today, for they reduce the amount of expensive semiconducting material needed, thus providing a cheap way to make solar cells and panels. An added benefit is that this device can extract more energy from high-energy photons, such as those at the blue end of the spectrum.

Another advantage is that the resulting PV devices are transparent, thus they could also be used as windows connected to solar cells which could generate electricity. This avoids the mechanical systems for tracking the sun to keep the light focused on a small solar cell, so the overall cost of the system could be reduced significantly, since the flat glass concentrators don't require a tracking system.

Instead of using optics and tracking, the glass sheets concentrate light using combinations of specially designed organic dyes. Light is absorbed by the organic dye coating on the glass sheet and emitted into the glass, where it is channeled to the edges of the glass, similar to the way that fiber-optic cables channel light over long distances. Narrow solar cells are laminated to the edges of the glass, where they collect the light and convert it into electricity.

The amount of light, and the level of concentration depend on the size of the glass sheet, and specifically on the ratio between the size of the surface of the glass and the glass edges. The rule of thumb is that the greater the light concentration, the cheaper the generated solar power. Reducing the use of semiconductor materials (used to collect the light at the edges) is an added cost benefit to this technology.

The remaining questions are:

1.   What happens to the output under cloudy and diffused sunlight? Concentrators don't work well, if at all, under clouds...

2.   Organic dyes tend to reabsorb much of the light, and well before it has a chance to reach the edges of the glass and be collected.

3. Organic dyes are also unstable, and do change under IR and UV radiation with time, so how many cycles of light concentration can be expected?

4. Materials and manufacturing must be cheap for the system to justify the added expense.

We see the excitement building at this unique opportunity to achieve "concentration" of several hundred times without the need of tracking. But remember that the sunlight available to each m^2 of this dye-coated glass is no more than what's falling on it. Then, the light goes through "waveguide" channels in the dye coating, before reaching the edges, and finally it is converted into electricity and sent away for use.

So, let's take a $1m^2$ piece of glass coated with the "waveguide" dyes, and attach wires to the glass edges. We then attach c-Si cells to the other end of the wires, and the glass-made-into-PV module is then mounted on the south side of a building in Phoenix, Arizona.

Here is what happens next morning:

The glass-dye PV module will start generating power in early morning, but a significant portion of the incoming light will be lost to reflection, due to the sharp impingement angle.

Author's note: Yes, we are aware that researchers claim these devices can operate efficiently under any angle of incidence, but any glass surface experiences reflection losses that cannot be avoided under a sharp angle of attack.

So we will get a small portion of the 900 W/m^2 converted into electricity first thing in the morning, with the output increasing as time goes on.

At noon, the south-oriented glass PV module sees full 900 W/m^2 insolation, but only 90% of the sunlight enters the organic dye-created waveguide channels, due to 10% reflection from the glass surface. The light that enters the glass is directed toward the edges via the waveguide channels of the organic dye. Assuming additional 20% optical losses in the waveguide channels, only 660 W/m^2 will reach the c-Si solar cells attached to the edges.

At 15% conversion efficiency of the c-Si solar cells, we will get approximately 44 W/m^2 total output at noon, or 4.4% efficiency. Not bad for this setup. But is it enough for any practical purposes? It seems like we will need very large glass areas to produce significant amounts of energy.

The problem is that 44 W/m^2 is the very maximum produced by the system at peak hour. Power output is reduced significantly during early morning and late

evening hours, during cloudy days, and by any shade falling onto the glass from trees, and/or adjacent buildings. Dust from dust storm will complete the output downfall.

The worst part is that all organic dyes, and anything organic, does not last very long under the heat and UV radiation, especially blistering desert sun. So, after two or three summers, power output will be significantly reduced, due to mechanical fatigue and chemical changes, b brought on by the never ending expansion and shrinking, accompanied by ferocious IR and UV bombardment. Too, the coating might decompose and become semi-transparent, changing the appearance of our windows. It's not a rosy picture, but let's err on the side of caution. Let's prove the author wrong before we replace all the windows in the city with this new technology.

This technology is still quite interesting because it offers something that was once thought very difficult and even impossible. Due to it's uniqueness, the impact of using this type of energy generation in niche markets might be surprisingly substantial in the 21st century.

Fluorescence Concentrators

By definition, fluorescence is the emission of light by a substance that has absorbed light, or some other form of electromagnetic radiation. Light is what we refer to here. It is a form of luminescence.

Crystalline silicon and thin-film photovoltaics are the dominant types of photo-energy generation today, but they have some inherited problems which a new type of photovoltaics technology might solve.

It is known as a fluorescent solar concentrator, and uses significantly less photovoltaic material, though still generating lots of useful electricity. While in a normal photovoltaics cell, the active material faces the sun; in a fluorescent solar concentrator the sun-facing component is a polymer sheet about 5mm thick doped with a material that fluoresces when hit by sunlight.

75% of this light emitted by the fluorescing material bounces around, similar to a light beam bounces inside an optical fiber transmission line. The light is directed toward the edge of the fluorescent material sheet, where it is absorbed by thin strips of photovoltaics cells. These are the same width as the thickness of the sheet, mounted perpendicular to the sun-facing surface, so that no space is wasted in this design.

Photovoltaics cells are most efficient at certain wavelengths, but they generate a common wavelength by fluorescence, so they can be tuned to produce the optimum wavelength for the particular cells used in the device.

The fluorescent material can be made out of a dye, similar to that used when this technology was first developed in the late 1970s.

A major problem is that this material tends to degrade when exposed to UV light, so a sophisticated solution, using quantum dots, has been tried. Quantum dots are much more stable, they re-emit virtually all the light they absorb and can be easily tuned for different applications.

Fluorescent concentrators are (theoretically) very practical for use in building-integrated photovoltaics (BIPV), since they are much lighter and cheaper than conventional materials used in conventional solar modules.

A great advantage is that the polymer sheets are transparent and can be used as windows, with the solar cells integrated into the frame. The sheet (windows) can be tinted and the color can be tuned together with the cell for best appearance and performance. This technology is best for use in cloudy weather because they concentrate any type of light—including diffused light and that on a cloudy days just as well as direct sunlight.

Another great advantage is that the angle of the incident light impinging on the solar cells does not matter that much, so they'd generate significant amounts of electric energy regardless of the sun angle.

This and other characteristics of fluorescent concentrators give them a great advantage over conventional technologies. Availability of materials, light weight, flexibility, low cost and their uncanny ability to generate light under any incident light level, angle and weather conditions, are factors that matter a lot. These advantages will assure this technology a place in the energy markets of the 21st century.

Holographic Solar Concentrators

Holographic solar concentrators (SC) open new possibilities of converting solar energy into electricity. Using systems based on holographic concentrators is unique and revolutionary. A hologram can be used both for spectral separation of the light and to focus it into a thin line perpendicular to the plane of the hologram.

Two or more different solar cells are placed along this line in such a way that each of them absorbs only the wavelengths that it can convert into electricity with the greatest efficiency.

Currently, a great deal of effort is being placed on flat-panel modules that are fully populated with photovoltaic (PV) cells and also on high-concentration ratio systems with 100-1000X concentration ratios. Flat-panel modules can accept light over a 180-deg acceptance angle and are typically used at a fixed angle relative to the position of the sun. However, it has been shown that the energy yield of flat-panel systems can be improved by a significant factor (15-40%) with single-axis tracking.

This becomes beneficial especially for higher cost high efficiency crystalline silicon cells. High-concentration photovoltaic (CPV) systems require two-axis tracking because of the relatively narrow acceptance angle dictated by the brightness theorem or power conservation principles. The higher cost of the tracking system can be offset by the higher power that can be extracted from multijunction PV cells.

Another operating domain is low (2X) to medium (50X) concentration ratio concentrator systems. There are several motivations for this group of systems. First, higher output power and energy yields are possible from the same area because higher efficiency cells can be used. This is an important consideration when deployment areas are limited, as in residential and commercial rooftop applications.

The cost of the system can be lower than conventional flat-panel systems by minimizing the use of expensive PV converters.

Finally, the tracking and PV cell cooling requirements are less complex than high-concentration ratio systems that can potentially provide system cost and reliability benefits.

The light concentration is achieved by diffracting incident sunlight onto nearby PV cells. Transmission holograms are designed to provide the required diffraction. The diffracted light can reach the PV cell directly or after total internal reflection (TIR) at the substrate-air interface.

The performance of the HSC is evaluated with respect to acceptance angle, geometric concentration ratio, optical efficiency, and energy collection efficiency. Critical parameters to optimize the performance of the HSC include the hologram's Bragg angles, peak diffraction efficiency (DE), and the spectral and angular bandwidth that is diffracted.

When illuminated on axis, a holographic stack can diffract over a large spectral range corresponding to the spectral response range of many types of PV cells. The spectral performance of the hologram primarily depends on the in-plane angle of incidence.

At ±6 deg from the hologram surface normal along the in-plane axis, the overall optical efficiency is 90% of the value at normal incidence. At an in-plane incident angle of ±16.5 deg, the one-axis tracking HPC system still has 80% of the maximum optical efficiency.

Simulation of an optimized one-axis tracking HPC

design showed that 80% optical efficiency at 2X geometric concentration ratio can be achieved with good irradiance uniformity (<20%).

The acceptance angle for an optimum HPC configuration is ±65 deg in the nontracking direction and ±6 deg in the tracking direction. An error of ±16 deg in the tracking direction is sufficient to maintain 80% of the maximum optical efficiency.

Daily and annual energy yield is modeled for HPC modules and fully PV-cell-populated modules, with different tracking modes, such as fixed tilt nontracking, horizontal N-S one-axis tracking and polar one-axis tracking. Results of polar one-axis tracking HPC indicate 43.8% increase in annual energy yield per unit area over non-tracking HPCs.

A number of manufacturers and research entities are working on developing a proof of concept solar module that uses holograms to concentrate light, possibly cutting the cost of solar modules by as much as 75%, making them competitive with electricity generated from fossil fuels.

The new technology—if and when fully developed—would replace some of the unsightly solar concentrators with sleek flat panels laminated with holograms. Since it concentrates in the 10 x range, it won't replace, or even compete with, the concentrators using very high concentration in the 100-1000 x range.

Instead, it will fill a market niche that requires these types of devices, that are a more elegant solution to traditional concentrators, and can be installed on rooftops, or even incorporated into windows and glass doors.

Some holographic concentrators utilize a two-stage optical concentrator and silicon PV cells. The holograms spectrally select sunlight and collect and concentrate it. The solar cells are interspersed between the holograms as depicted in the cross-section schematic. A layer of holograms (laser-created patterns that diffract light) reflects selected frequencies of light to the mirror-like inside surface of the upper transparent glass where it continues to be reflected until it reaches a solar cell.

This type of system needs 85% less silicon than a crystalline silicon panel of comparable wattage, because the photovoltaic material need not cover the entire surface of a solar panel. Instead, the PV material is arranged in several rows. Reducing the amount of PV material needed could bring the costs down significantly.

Spectrally selecting the desired portion of sunlight allows for "cooler" solar cell operation, requiring no external cooling, while maintaining an increased power output by concentrating specific solar wavelengths unto the cells. The modules also use passive tracking which requires no moving parts.

Different holograms in a concentrator module can be designed to focus light from different angles, so they don't need moving parts to track the sun. This gives greater output in the early morning and late afternoon.

Holograms have advantages that make up for their relatively weak concentration power. They can select certain frequencies and focus them on solar cells that work best at those frequencies, converting the maximum possible light into electricity. They also can be made to direct heat-generating frequencies away from the cells, so the system does not need to be cooled.

Yes, this is a model, which needs to be put in practice and properly tested, to verify the calculations, suggestions and claims.

We don't know if or when this will happen, but this work has great value in applying to other solar technologies (such as CPV) as well. Furthering the development of this and related tracking solar technologies is key to their deployment as the most efficient solar power generators in the 21st century.

Low Concentration CPV

Who better than IBM to show us the way in this new field? With a lot of experience in cooling high-performance computer chips, IBM is developing a way to make concentrating PV (CPV) cells, which are much more efficient and practical than the conventional PV technologies.

Concentrating sunlight onto any solar cells will cause them to lose efficiency and high enough concentration could even melt and destroy the cells. According to the IBM research team working on this project, the concentration of light on photovoltaic cells can be increased by a factor of ten without damaging the cells and increase the amount of usable electrical energy produced by up to five times at the same time.

We assume that they are talking about regular c-Si solar cells, because CPV cells are illuminated with 100 to 1000 suns, which only aggravates the already complicated dynamics of the heat transfer process.

Basically speaking, the principle behind CPV cells is to use a large lens to focus sunlight onto a relatively small piece of semiconductor material, which has PV properties. The benefit is using only a fraction of the expensive semiconductor materials, thus reducing material costs.

A number of companies are making and marketing CPV technologies, but one of the main challenges they all have is coping with the vast amounts of heat produced by the focused sunlight, heating the solar cells to very high temperatures.

As the temperature rises, the cells' efficiency drops and they can be damaged as well. The objective, therefore, is to keep cell temperature as low as possible.

There are only two ways to do this. First, we can use static heat sinks, which are metal blocks onto which the cells are mounted and can transfer excess heat. (Placing the solar cell on a copper heat sink (passive cooling) does not provide adequate interfacial heat transfer.)

Second, we can use dynamic cooling, where cooling liquid or water is pumped through the heat sinks to draw even more of the thermal energy away from the cells. This method is much more efficient, but more expensive, of course.

In both cases the biggest problem is the presence of microscopic indentations, in the form of scratches or "as is" surface imperfections, on the cells and heat sink surfaces. These imperfections interfere in the heat transfer process by decreasing the surface contact between the faces, thus impeding and reducing the heat transfer rate.

The usual remedy for this problem is filling the surface gaps with metallic and organic pastes which act as bonding, and additional thermal, interfaces. These filling and bonding materials, however, are not efficient enough at transferring heat.

IBM has perfected this method by using a specially designed (but secret) material to improve the heat transfer between the solar cell and a water-cooled heat sink. If successful, it will be a major step ahead in CPV technology, with IBM debuting in the most recent solar revolution by introducing a product that 30 years of CPV technology development have failed to find.

We will keep our fingers crossed for IBM, for only time will show how efficient and resilient the new materials are. Our experience with CPV technology and processes, as well the physics and chemistry of the heat transfer process, show that many materials do increase the efficiency of the heat transfer process at first. Stuck between the very hot cell and the cool heat sink, however, these materials suffer serious mechanical degradation and chemical decomposition with time.

Luminescent Solar Concentrator

The definition of luminescence is, "Emission of light by a substance not resulting from heat." This, therefore, is a form of "cold body" (but far from "ideal black body") radiation, caused by chemical reactions, electrical energy, subatomic motions, or stress in a crystal structure.

Note: "Cold" body radiation is different from "warm" body radiation, which is the process of incandescence, as in incandescent light bulbs, which are associated with the emission of heat.

Luminescent solar concentrators are similar to what we discussed in the fluorescence concentrators section, but with a twist. Here, plastics concentrate sunlight into a small spot, where the concentrated solar energy can then be converted into electricity by a multi-junction photovoltaic cell solar cell. This not only increases efficiency, but also decreases cost, as luminescent solar concentrator panels can be made cheaply from plastics, while PV cells need to be constructed from expensive materials such as silicon.

The best part of this type of system is that it concentrates the sunlight and directs it onto a very specific, smaller location where it can be used to generate electricity via germanium multi-junction CPV solar cells with efficiency over 40%.

Unlike a solar tracker, a luminescent solar concentrator is stationary. This is a great advantage over conventional trackers, because there are no moving parts and related expense involved.

The main components here are plastic substrate, dye molecules and common solar cells. Special dye molecules are sprayed onto a sheet of plastic, at the outer edges of which are affixed the solar cells.

The combination plastic/dyes works as a waveguide, which traps the incoming sunlight and directs it along a proper path to a particular destination. In this case, however, when the light hits the plastic, it is absorbed instead of transmitted.

The sun's energy is transferred to the dye, where it creates electrons in their higher excitation state. When the electrons return to their original lower energy level, the dye molecules release the excess energy into the plastic sheet. The energy is trapped in a process called total internal reflection, during which process the light can escape the plastic. The light bounces around in the material, eventually making its way to the outer surface, where solar cells are attached. These solar cells absorb the light that reaches them and generate electricity.

Since the above described (light-to-electron-to-energy-to-electricity) process is non-linear, the efficiency of the system would be the same regardless of the angle of the arriving sunlight rays. This is a great advantage, as the module can be oriented in any which way and doesn't have to move (track the sun) to produce its maximum amount of electricity.

This type of solar concentrator doesn't require a cooling system either, which is another great advantage that, in combination with the lack of moving parts, makes it much less expensive than a solar tracker.

The major drawback is that the light energy bounces around in the plastic in a somewhat uncontrolled manner, so it sometimes gets reabsorbed into the dye molecules and ends up emitted as heat. This energy, then, never makes it to the solar cells, and even worse, it can overheat the device.

Another potential issue is the fact that plastics (and organic dyes) deteriorate when exposed to direct sunlight for a long time. The more intense the light, the faster the deterioration and inevitable failure.

Research today is focused on converting window panes into solar devices that concentrate sunlight and convert it into electricity. A mixture of dyes can be painted onto a pane of glass or plastic, where the dyes absorb sunlight and then re-emit it within the glass at a different wavelength, which then reflects off the interior surfaces of the glass. As the light reflects within the glass pane, it is channeled along the length of the glass to its edges, where it is emitted or converted into electricity by solar cells lined up along the edges.

The sunlight can be concentrated by a factor of 30-50, allowing special solar cells that are optimized for such concentrated sunlight to produce much higher power levels than conventional—including sun tracking—solar technologies.

The unique optics of the approach lead to the creation of an efficient solar concentrator, which doesn't need to be pointed toward the sun, nor does it need to track it throughout the day. These features make it cost-effective as well, at least theoretically. The promise of this technology for use in buildings and other niche markets is great, so provided that the major hurdles (described above) are overcome soon, we will be looking at and through it, all around us.

Quantum Well CPV Cells

Single junction quantum well (quantum dot related) technologies have been used lately for CPV device manufacturing, with demonstrated efficiency of 27.3%. This is the highest value for any nano-structured solar cell to date, and close to the highest efficiency recorded for the conventional single junction cell (27.8%).

The quantum well PV device is made of a p-i-n structure with a gallium arsenide phosphide (GaAsP) indium gallium arsenide (InGaAs) multi-quantum well stack grown in the i-region.

The lower band-gap InGaAs layer is compressively strained, while the GaAsP barrier layer is under tensile strain. Therefore, a judicious choice of composition and layer thickness results in a stack of quantum wells where each GaAsP/InGaAs bilayer exerts no net force on neighboring layers.

By incorporating strained semiconductors into a solar cell without introducing structural defects, the team was able to adjust the absorption threshold for the solar cell. The single junction cell has an absorption edge at 1.33eV, which is fundamentally better matched to the solar spectrum than a 1.42eV gallium arsenide cell. The low defect density allows cells to become increasingly radiatively efficient at concentrator intensities, exhibiting photon recycling effects that further boost efficiency.

Adjusting the absorption threshold becomes particularly important when fabricating highly efficient, multi-junction solar cells. Here the broad solar spectrum is absorbed using a series connected stack of sub-cells with different bandgaps.

While this structure can lead to very high efficiencies, it requires careful control of the absorption threshold of each sub-cell. The series connection ensures that the lowest photocurrent in the stack will limit the photocurrent in the entire cell.

By growing defect-free, strain-balanced stacks of material, we can adjust the absorption threshold of the junctions without relaxing the semiconductor lattice or growing optically thin junctions. This strain-balanced approach makes double junction efficiencies greater than 34% and triple junction efficiencies up to 42% under solar concentration feasible, and work is underway to achieve these goals.

Proponents of CPV say that the cost of focusing optics, steering system, heat sinking and exotic semiconductor materials in a CPV system is still less per kWh than conventional flat panel photovoltaic systems. The argument is that, although conventional PV systems use silicon which is far cheaper than GaAs, they need at least 500 times more semiconductor area to catch and convert the same amount of sunlight. And as silicon solar cells are only 15-16% efficient, 1cm^2 of GaAs in a CPV system should actually be compared with 1,000cm of flat plate silicon.

So, QW CPV is looking into getting its place in the sun (no pun intended). Thus far the results show, at least on paper, that this is a distinct possibility, it will be some time before we see these devices in the solar power fields. Some combinations and permutations, however, will be used in the near future in developing this and similar technologies.

In summary, CPV technology is one of the most promising PV conversion technologies today. The great expectations have not been justified to date, due to the complexity of the system—exotic materials, moving parts, tracking accuracy, etc.

Nevertheless, CPV technologies will continue to grow as the need for solar energy increases, so only time and effort separate us from having efficient and reliable CPV—tracking and stationary (as described above)—as a standard on the world's energy markets.

EXCITONIC PV TECHNOLOGY

If we take a close look at the actual mechanism of electric generation in different PV materials, we see that there are two distinct, and very different, classes.

1. The conventional solar cells (i.e. c-Si, thin films and others), where photons bombard the PV material surface and create free electrons and holes, which eventually become part of the electric current in and outside the device, and

2. The new, excitonic (note EXCITONIC, not EXOTIC) solar cells, or XSCs, where—in contrast with the above mentioned direct photon-electron generation mechanism—the incoming photons create excitons* instead, which assist the electron-hole pairs generation.

The new XSC technology branch includes a number of "exotic" 3rd gen technologies, such as DSSC devices, nano-technology devices, and organic-materials based solar cells.

The exciton generation is a fundamentally and very importantly different process, and since we already took a close look at the electron-hole generation in previous chapters, we need to take a close look at the excitons, and the entire current generating process in XSC devices as well.

Excitons

Excitons are major components in excitonic devices, so they deserve a detailed description and explanation. The exciton concept was discovered by Y. Frenkel in 1931, when working on the properties of insulators. This is actually quite remarkable, because it was before the advent of semiconductors and related concepts, some of which clarify the exciton concept.

Frankel attempted to analyze and describe the excitation of atoms in the insulator's lattice, proposing that the excited state is chargeless and can travel in a *particle-like* fashion through the lattice without any charge exchange.

Basically speaking, an *exciton* is a state (a bound state, to be exact) of an electron and hole which are attracted to each other by the electrostatic Coulomb force. It is considered an electrically neutral (no charge) state, or a quasi-particle that exists in insulators, semiconductors and in some special liquids.

Another way to think of the exciton is as an elementary excitation of condensed matter that can transport energy without transporting net electric charge.

Note: Some of the above concepts are somewhat hard to swallow, so for ease of understanding, let's envision the exciton as an actual neutral particle that can move around the matter, obeying the laws of physics.

An exciton (particle) is formed when a photon impacts upon, and is absorbed by, a semiconductor material. The exciton creation excites an electron from the valence band and propels it into the conduction band. This action leaves behind a localized, positively charged "hole," similar to that of a hole from the electron-hope pair in c-Si semiconductor material.

The electron in the conduction band is attracted to this localized hole by the Coulomb force. This attraction provides a stabilizing energy balance. At this point, the exciton has slightly less energy (think of mechanical energy, not charge) than the unbound electron and hole.

If given a chance, the electron and hole created by the exciton will eventually recombine, which marks the decay of the exciton, which is limited by resonance stabilization due to the overlap of the electron and hole wave functions, resulting in an extended lifetime for the exciton. The exciton decay is sometimes accompanied by emission of light.

The wave function of the bound state is hydrogenic, or similar to the state of a hydrogen atom. However, the binding energy is much smaller and the particle's size much larger than a hydrogen atom.

At this time—and before the electron and hole have a chance to recombine—the electron is extracted from the device and sent into wires that comprise the outside electric circuit. Current from that circuit can be used as needed on the spot, or sent into the grid.

The exciton decays (loses its energy) and is no longer part of the process. Then another photon hits the semiconductor material and is absorbed by it, creating another exciton, which triggers an electron to be excited and ejected into the conduction band. And the process continues until there is no more light with sufficient energy to trigger the excitonic process.

In XSC cells, excitons are generated directly at the hetero-interface upon light absorption, where they photo-generate the charge carriers. The excitons, however, are easily diffused if conditions are not just right. This makes the entire current generation process less defined and somewhat "touchy," as compared with the electron-hole process of conventional PV technologies.

A characteristic of XSCs is that charge carriers are generated and separated simultaneously across the hetero-interface, while in conventional cells the photogeneration of free electron-hole pairs occurs throughout the bulk semiconductor, and carrier separation occurs upon their arrival at the junction in a subsequent process. This constitutes a fundamental difference in the photovoltaic behavior, which must be well understood and properly utilized.

The basic structure and function of an excitonic device follows.

Excitons created by light absorption in organic semiconductors 1 and 2 in Figure 4-2 do not possess enough energy to dissociate in the bulk (except at trap sites), but the band offset at the interface between OSC1 and OSC2 provides an exothermic pathway for dissociation of excitons in both phases, producing electrons in OSC1 and holes in OSC2. The band offset must be greater than the exciton binding energy for dissociation to occur.

Only one side (OSC2) is photoactive and it consists of just one monolayer of sensitizing dye, while OSC1 consists of a film of nanoporous TiO_2 with such a high surface area that the adsorbed monolayer of dye is optically thick. To make intimate electrical contact to such a convoluted film usually requires a liquid electrolyte solution. To minimize the rate of the recombination reaction, the kinetically ultraslow redox couple iodine/tri-iodide is commonly employed.

Note that in the conventional cell, the exponential absorption coefficient is $\alpha = 105$ cm-1 throughout the cell. In the excitonic cell, however, light is absorbed only within 1 nm of the interface (the "excitonic layer") with $\alpha = 107$ cm-1.

The calculated J-V curves under 50 mW/cm² illumination (31.6 mW/cm² is absorbed) at $\lambda = 2.1$ eV.) The only difference, therefore, between the conventional and excitonic cells is the spatial distribution of the photogenerated charge carriers, while the dark current is identical for both devices.

Figure 4-3b shows also that major differences in PV behavior occur in otherwise identical devices that differ only in the spatial distribution of photogenerated carriers. The conventional cell is less efficient than the XSC. There are at least two reasons for this:

1. Recombination throughout the bulk is more efficient than recombination only at the interface (excitonic layer), given the same rate constants for recombination, and

2. The XSC has the photogenerated force, $\Delta\mu h\nu$, operating in concert with Øbi separating the two carriers and driving them toward their respective electrodes; while in the conventional cell, $\Delta\mu h\nu$, opposes Øbi.

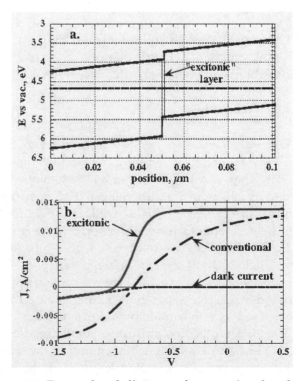

Figure 4-3. Energy band diagram of conventional and excitonic solar cells at equilibrium.

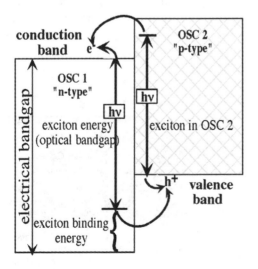

Figure 4-2. Energy level diagram for an excitonic heterojunction solar cell. (4)

Thus, the conventional cell produces Voc = 0.83 V while the XSC produces Voc = 1.00 V. (Both of these are greater than Øbi = 0.64 eV because the band offset at the heterojunction also serves to rectify the photocurrent.) The relative short circuit photocurrents (11.1 mA/cm² versus 13.7 mA/cm²) and fill factors (34% vs. 59%) strongly favor the XSC.

Overall, the simulated XSCs have theoretical conversion efficiency more than 2.5 fold higher than the conventional cell.

So basically, XSC devices exhibit differences, some of which are summarized above, in addition to:

1. Charge generation and separation in XSC devices are simultaneous and this occurs via exciton dissociation at a hetero-interface.

2. Electrons are photogenerated on one side of the interface and holes on the other. Similar to the p-n junction formation and operation, and

3. A dominant difference in XSCs is the great importance of photo-induced chemical potential energy gradient, $\Delta\mu hv_{-}$, whereas $\Delta\mu hv$ is unimportant, and therefore neglected, in theoretical descriptions of conventional PV cells.

Now let's take a look at several types of XSCs devices, starting with those of the DSSC type.

1. Dye-sensitized (Grätzel) solar cell (DSSC).
2. Two-layer, planar (Tang) solar cell.
3. Bulk heterojunction (Sariciftci, Heeger) solar cell, and
4. Related Hybrids

Dye-sensitized solar cells (DSSC) are a new addition to the solar energy generating class of materials and processes. They successfully separate the two functions provided by silicon in a traditional cell design; acting as the source of photoelectrons, as well as providing the electric field to separate the charges and create a current.

In the dye-sensitized solar cell, the bulk of the semiconductor is used solely for charge transport, while the photoelectrons are provided from a separate photosensitive dye. Charge separation occurs at the surfaces between the dye, semiconductor and electrolyte.

The dye molecules are quite small (nanometer sized), so to capture a reasonable amount of the incoming light the layer of dye molecules needs to be made fairly thick, much thicker than the molecules themselves. To address this problem, a nanomaterial is used as a scaffold to hold large numbers of the dye molecules in a 3-D matrix, increasing the number of molecules for any given surface area of cell. In existing designs, this scaffolding is provided by the semiconductor material, which serves double duty.

The following is a review of the major types DSSC cells.

Grätzel Dye-sensitized Solar Cell (DSSC).

The original Grätzel design, which started the entire DSSC industry, consists of three primary parts:

1. On the top is a transparent anode made of fluorine-doped tin dioxide (SnO_2:F) deposited on the back of a glass plate.

2. On the back of the conductive plate is a thin layer of titanium dioxide (TiO_2), which forms into a highly porous structure with an extremely high surface area. TiO_2 only absorbs a small fraction of the solar photons (those in the UV).

3. The plate is immersed in a mixture of a photosensitive ruthenium-polypyridine dye (also called molecular sensitizers) and a solvent.

After soaking the film in the dye solution, a thin layer of the dye is left covalently bonded to the surface of the TiO_2. A separate backing is made with a thin layer of the iodide electrolyte spread over a conductive sheet, typically platinum metal. The front and back parts are then joined and sealed together to prevent the electrolyte from leaking.

The funny thing here is that the "advanced" materials used in these cells are inexpensive compared to the silicon material needed for normal cells, because they require no expensive manufacturing steps. TiO_2, for instance, is already widely used as a paint base.

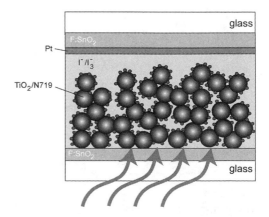

Figure 4-4. Dye sensitized TiO_2 solar cell (4)

Function

Sunlight enters the cell through the transparent Glass/SnO₂:F top contact, striking the dye on the surface of the TiO₂. Photons striking the dye with enough energy to be absorbed will create an excited state of the dye, from which an electron can be "injected" directly into the conduction band of the TiO₂, and from there it moves by diffusion (as a result of an electron concentration gradient) to the clear anode on top.

The dye molecule has lost an electron now, and will decompose if another electron is not provided. The dye strips one from iodide in electrolyte below the TiO₂, oxidizing it into tri-iodide. This reaction occurs quite quickly compared to the time it takes for the injected electron to recombine with the oxidized dye molecule, preventing this recombination reaction that would effectively short-circuit the solar cell.

The electrolyte then recovers its missing electron by mechanically diffusing to the bottom of the cell, where the counter electrode re-introduces the electrons after flowing through the external circuit.

General Characteristics

The Grätzel dye sensitized TiO₂ solar cell system is characterized by:

1.   Large titania surface area to increase absorption of adsorbed dye,

2.   Ultrafast and efficient electron injection from dye to titania, and

3.   Redox couple to transport holes.

Two-layer Planar (Tang) Solar Cell

General characteristics of the two-layer, planar solar cell:

1.   Planar and simple.
2.   Vacuum deposited molecular films.
3.   Absorption by both layers.

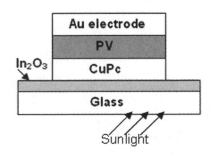

Figure 4-5. Two-layer, planar solar cell. (4)

The cell is constructed on thin film indium tin oxide (ITO) coated glass, which provides the transparent conducting properties of the glass substrate. On top of the ITO is a thin film of copper phthalocyanine (CuPc), 200-300 A thick, deposited by PVD methods.

On the CuPc film is another thin film of organic material, such as perylene tetracarboxylic derivative (PV film), 400-500 A thick. To complete the cell, that film of opaque Ag or Au layer is deposited on the previously formed structure, where the Ag electrode area represents the active area of the photovoltaic cell.

Both CuPc and PV films are thermally very stable, which allows their deposition by high temperature vacuum PVD methods in the 500 and 600°C range. The substrate is kept at room temperature during deposition by external cooling.

Function

The operation of the two-layer cell when illuminated by sunlight can be described as follows:

Light shining on the cell is absorbed by both CuPc and PV layers and creates excitons, which diffuse in the bulk of the films. The interface between the CuPc and the PV layer is the active sites (equivalent to the p-n junction of conventional PV devices). This interface is responsible for the dissociation of the excitons, where the holes are preferentially transported in the CuPc layer and are then collected by the ITO electrode.

At the same time, the electrons are transported in the PV layer towards the Ag electrode and are extracted into the outside electrical circuit to be used as needed.

The effectiveness of the exciton dissociation at the CuPc/PV interface is mostly due to the high built-in field (described as a dipole field of trapped charges) at the interface.

At sufficiently high strength of the field, the carrier generation efficiency would be virtually saturated and could be then determined primarily by the diffusion of excitons into this interface region. If small external voltage is superimposed, then the carrier collection could be affected, which could be interpreted as a weak bias dependence of the collection efficiency of the device.

The polarity of the CuPc layer is always positive with respect to the PV layer regardless of the electrode materials. This is indicative of the conduction types in the two layers. This means that the CuPc film is responsible for transporting the holes, which is consistent with observed CuPc behavior. On the other hand, the PV layer is responsible for electron transport.

Usually the photo ionization potential of CuPc is lower than the PV layer potential by about 0.4-0.5 eV.

When the cell is exposed to continuous illumination under short-circuit condition, both Voc and Isc show little degradation, but the fill factor (ff) decreases significantly. After several days it could go down 25-30%, mostly due to degradation of the Ag electrode, which increases the series resistance of the cell.

Typical cell parameters, under simulated Am^2 illumination (75 mW/cm^2), are

Voc = 400-475 mV
Isc = 2.2-25 mA/cm^2, and
FF = 0.65 ± 0.03.

The power conversion efficiency obtained under such illumination, and without correcting for reflection or electrode absorption losses of these devices, is high as compared with other SXC devices. Their fill factor is about 0.65, which represents a substantial improvement for these types of cell designs as well.

The I-V characteristics show that the photo-generated current has only a weak dependence on the reverse bias voltage. The short-circuit photocurrent and the photocurrent at 0.5 V reverse bias differ by only about 5%. In the forward direction, the dynamic resistance is about 100 Ohm, which represents an upper limit of the series resistance of the two-layer cell.

So, although this two-layer, thin-film configuration is not the ultimate solar cell, it offers a different, and quite efficient, approach to the construction of photovoltaic cells based on organic materials. A key advantage of this cell over the more conventional single-layer organic cells is that it does not suffer from the usual dependence of charge generation on the electric field. This independence leads to large enhancement in both the fill factor and the photovoltaic efficiency of the CuPc/PV two-layer system.

More work is to be done in this area, but the results to date are encouraging and we see this technology being used in some special applications in the near future.

Bulk Heterojunction Hybrid (Sariciftci, Heeger) Solar Cell

This device is solution processable, so that both the PEDOT:PSS and active layer can be deposited from solution (e.g. Spin, Spray, Ultrasonic, Ink-jet, Doctor blade).

Its short exciton diffusion length requires two components of the active layer, the donor and acceptor to phase separate, and also excitons must not have to travel more than 10 nm.

And ultrafast and efficient exciton dissociation

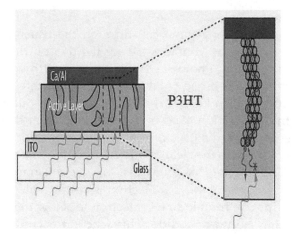

Figure 4-6. Bulk heterojunction device (4)

of the interface between donor and acceptor must efficiently dissociate exciton and inhibit recombination.

Function
P3HT absorbs photon.

Exciton diffuses to interface.

Exciton dissociates.

Carriers diffuse, driven by chemical potential, to electrodes.
— Electrons through PCBM network flow to cathode.
— Holes through P3HT phase flow to anode.

P3HT absorbs further to the red, with optical bandgap at < 2 eV.
— But Short Exciton lifetime, τ, is ~ 300-500 ps.

Exciton migrates to interface
— But Diffusion length < 10 nm. Hence blend.

Exciton dissociates to electron and hole at interface with PCBM.
— Ultrafast, 100% yield, minimal recombination.

Carrier recombination is minimal.
— Numerable reasons

Holes transport through P3HT.
— Hole mobility, μh, > 10-3 cm^2/Vs

Electrons transport (hopping) through PCBM.
— Electron mobility, μe, > 10-3 cm cm^2/Vs

The overall operational efficiency of these devices is presently around 3-4.0%, but is expected to rise to around 6-7.0% in the near future.

Ru Light-absorbing Solar Cell

These are special types of dye-sensitized solar cells, where a ruthenium metalorganic dye (Ru-centered) is used as a monolayer of light-absorbing material. This type of dye-sensitized solar cell depends on a mesoporous layer of nanoparticulate titanium dioxide to greatly amplify the surface area (200-300 m^2/g TiO_2, as compared to approximately 10 m^2/g of flat single crystal).

Photogenerated electrons from the light-absorbing dye are passed on to the n-type TiO_2, and the holes are passed to an electrolyte on the other side of the dye. The circuit is completed by a redox couple in the electrolyte, which can be liquid or solid.

The Ru light absorbing DSSC type of cell allows a more flexible use of materials, and is typically manufactured by screen printing and/or use of ultrasonic nozzles, with the potential for lower processing costs than those used for bulk solar cells.

However, the dyes in these cells also suffer from degradation under heat and UV light, and the cell casing is difficult to seal due to the solvents used in assembly. In spite of these problems, this is a popular emerging technology with special applications and significant commercial impact forecast within this decade.

Solid State DSSC

One possible solution to some of the problems encountered thus far is to replace the liquid electrolyte with a solid charge-transport (solid-state) material, called mesoporous; i.e., mesoporous TiO_2 is called "mesoporous titania."

These materials contain pores with diameters between 2 and 50 nm. A dye-sensitized mesoporous film of TiO_2 can be used for making photovoltaic cells and this solar cell is called a 'solid-state dye sensitized solar cell'. The pores in mesoporous TiO_2 thin film are filled with a solid hole-conducting material (such as p-type semiconductors) or organic hole conducting material.

The process of electron-hole generation and recombination is the same as in the other DSSC cells. Electrons are injected from photoexcited dye into the conduction band of titania and holes are transported by a solid charge transport electrolyte to an electrode.

For solid-state dye-sensitized solar cells, the first challenge originates from disordered titania mesoporous structures. Mesoporous titania structures should be fabricated with well-ordered titania structures in uniform size (~ 10 nm). The second challenge comes from solid electrolyte.

To develop the solid electrolyte, several requirements should be considered. The solid electrolyte is required to have special properties, as follow:

1. The electrolyte should be transparent to the visible spectrum (wide band gap)

2. Fabrication should be possible for depositing the solid electrolyte without degrading the dye molecule layer on titania.

3. LUMO level of dye molecule should be higher than the conduction band of titania.

4. Several p-type semiconductors tend to crystallize inside the mesoporous titania films, destroying the dye molecule-titania contact. Therefore, the solid electrolyte needs to be stable during operation.

In a solid state dye-sensitized solar cell the DSSC electrolyte is replaced with a p-type semiconductor or organic hole conductor materials avoiding problems such as leakage of liquid electrolytes. A solid state dye-sensitized solar cell is schematically shown herein.

The mesoporous metal oxide electrode, commonly, TiO_2 is placed in contact with a solid state hole conductor. Attached to the surface of the nanocrystalline electrode film is a monolayer of the sensitizing dye. After the excitation of the dye an electron is injected into the conduction band of the semiconductor oxide electrode. The sensitizer dye is regenerated by the electron donation from the hole conductor. In the solid state cell, the charge transport is electronic whereas when using liquid or polymer electrolyte, ionic transportation takes place. The hole conductor must be able to transfer holes from the sensitizing dye after the dye has injected electrons into the TiO_2; that is, the upper edge of the valence band of p type semiconductors must be located above the ground state level of the dye.

Furthermore, hole conductors have to be deposited within the porous nanocrystalline layer penetrating

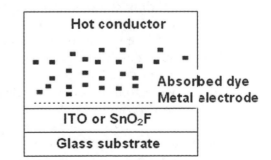

Figure 4-7. Solid state dye-sensitized solar cell (2)

into the pores of the nanoparticle and finally it must be transparent in the visible spectrum, or, if it absorbs light, it must be as efficient in electron injection as the dye. CuI, CuBr or CuSCN were found to be the successful candidates to replace the liquid electrolyte. The energy conversion efficiency of the fully solid state solar cell of nanoporous n-TiO_2/cyanidin/p-CuI was found to be 1%.

The efficiency of the solid state device was further improved by employing CuI as hole transporter and ruthenium bipyridyl dye complex as a sensitizer instead of cyanidin. The cells based on this structure gave a maximum power conversion efficiency of 6% corresponding to a fill factor of about 45%.

A solid-state dye-sensitized solar cell reported lately with 6.8% energy conversion efficiency—one of the highest yet reported for this type—is based on enhanced light harvesting from the increased transmittance of an organized mesoporous TiO_2 interfacial layer and the good hole conductivity of the solid-state-polymerized material.

There is a problem with this design, however, where light intensity of higher than 100 mW/cm^2 reduces the overall cell efficiency to about 4.5%. This is something that needs to be corrected for this type of cell to find a place in the energy markets.

Large Area DSSC

Dye-sensitized solar cells are usually made of TiO_2 and have been extensively studied and developed as a cheap alternative to conventional photovoltaic cells. There are, however, other materials (albeit not so stable) which have similar band gap and other properties that are suitable for dye sensitization. One of these is SnO_2.

Cells based only on SnO_2 suffer severe recombination losses, so an alternative method of using SnO_2/MgO films seems much more sound. Here the SnO_2 crystallite surface covered with an ultra-thin layer of MgO, and the resulting combination shows reasonably high efficiencies. And amazingly enough, the SnO_2/MgO cells show better corrosion resistance to the cell chemicals, thus withstanding the dye and electrolyte degradation well, and even better than TiO_2 cells.

Another important factor here is that the ultra-thin barrier of MgO on SnO_2 remains intact during prolonged usage or storage of the cell. On a practical level, the SnO_2/MgO cells were found to have efficiencies ranging from 6.5-7% at AM 1.5 and 1000 Wm^{-2} illumination.

During long-term tests of continuous exposure to artificial light, the cells did not show signs of deterioration. However, similar cells fail quickly when exposed

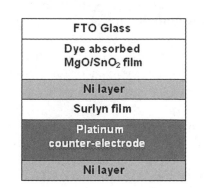

Figure 4-8. Large area dye-sensitized solar cell.

to direct sunlight and as the temperature of the cell increases to 45-55°C.

At the same time, similar cells made from TiO_2 deteriorate in the same manner and as quickly as the SnO_2/MgO cells.

Bio-sensitized Solar Cells

Bio-sensitized solar cells (BSSC) are, in our opinion, the most sophisticated solar conversion devices, the hardest to manufacture, and are difficult to use. The use of novel, or living, matter is not usually associated with reliable solar electricity generation, although solar energy is used by all living matter. We see photosynthesis conducted on a large scale all around us, since all living plants and trees use it in their daily life, but the complexity and the potential of this technology is extraordinary.

Even though we don't expect to see dye- or bio-sensitized solar cells and modules on our roofs, and even less so in large-scale power fields in the deserts, anytime soon, we do predict that these technologies will be important contributors to the advancement of solar energy generation in the 21st century.

Dye-sensitized solar cells (DSSC) are a close relative to the BSSC cells and just like the BSSC devices consist of a photo-active dye, a mesoporous nanocrystalline semiconductor layer coated on a transparent conducting oxide (TCO) substrate (anode), a liquid electrolyte (I^-/I_3^- in acetonitrile), and a Pt-coated TCO substrate (cathode).

Upon excitation, the excited dye molecules (or biomolecules in BSSC devices) inject electrons into the conduction band of the nanocrystalline semiconductor. The electrons are collected by the TCO anode and flow through the outer circuit to the Pt-coated cathode. I_3^- becomes reduced to three I^- by the excess electrons at the cathode, and the three I^- become oxidized back to I_3^- to regenerate the dye molecules at the anode.

In a DSSC, molecules of an appropriately chosen dye play an important role in determining the overall cell performance; many dyes such as ruthenium polypyridine complexes, phthalocyanine, xanthenes, and coumarins have been tested as potential sensitizers.

In this text we take a close look at the dye-sensitized solar cell (DSSC) device, which is very similar to that of its BSSC cousin, except that some organic components are replaced with biomaterials. So we will refer the reader to the more detailed description of DSSC technologies instead.

The main, if not only, difference between the DSSC and BSSC cells is their use of different electrolytes (organic and inorganic vs. biomaterials).

Many biomaterials such as natural dyes extracted from plants, cyanobacteria cells, living whole-cell photosynthetic micro-organisms and protein with solid-state electronics have been tested in replacing the toxic organic dyes (as in DSSC cells) and making safer biosensitized solar cells (BSSC) instead.

Some researchers have used the whole cells of Synechococcus sp.PCC7942 as a biocatalyst and 2,6-dimethyl-1,4-benzoquinone (DMBQ) as a mediator to produce a photosynthetic bioelectrochemical cell with a photoconversion efficiency of 2%.

Others have demonstrated a micro-electromechanical system (MEMS) photosynthetic electrochemical cell (mPEC) that harnesses the subcellular thylakoid photosystems isolated from spinach cells to convert radiant energy into electricity; the device produced Voc = 470mV and Isc=1.1mA cm^{-2} with a photoconversion efficiency 0.01%.

Yet another research group entrapped the bacterial photosynthetic reaction center (RC) from Rb. Sphaeroides strain RS601 on three-dimensional worm-like mesoporous WO_3-TiO_2 films to develop versatile bio-photoelectric devices; the ISC was detected near 30 mA cm^{-2} with a maximal incident conversion efficiency of photons to current (IPCE) of 11% (about 800 nm).

Among all those bio-sensitizers, the most promising are derivatives of metalloporphyrins, of which the porphyrins are integrated with metals of various types. The capability of porphyrins to adsorb visible light in regions 400-450 nm (B band) and 500-700 nm (Q bands) makes it an effective candidate as a sensitizer.

Through chemical modification, 7.1% efficiency has been achieved with a DSSC based on Zn porphyrin but a serious limitation arises from the tendency of the porphyrin to aggregate, which competes with electron injection and yields a decreased overall efficiency of power conversion in the device.

One type of BSSC uses Myoglobin (Mb), reconstituted zinc protoporphyrin-apomyoglobin (ZnMb), and eosin-modified ZnMb (EoZnMb), as photo-sensitizers to functionalize TiO_2 nanocrystalline films for biosensitized solar-cell applications. The Mb-sensitized solar exhibits poor cell performance because of the internal electron transfer that reduces Fe(III) to Fe(II) so that a small photocurrent is observed.

For both ZnMb- and EoZnMb-sensitized solar cells, the efficiencies of power conversion are enhanced ten times those of their un-sensitized counterparts (blank TiO_2 solar cells), through efficient electron injection from the proteins to TiO_2 that improves the charge separation between TiO_2 and the sensitizer, and decreased current leakage between TiO_2 and the electrolyte.

The efficiencies of power conversion of both ZnMb and EoZnMb-sensitized solar cells is enhanced over ten times due to superior charge separation between TiO_2 and the protein, and due to smaller current leakage between TiO_2 and the electrolyte.

Nevertheless, compared with a typical DSSC device, the power conversion efficiency of a BSSC is presently still too small, mostly because the immobilization of the proteins on the surface of TiO_2 is too low.

Mb-sensitized solar cells are easy to manufacture, but have poor performance, which is due to the reduction reaction Fe(III) $\rightarrow$ Fe(II), which produces a photocurrent density of the device smaller than its unsensitized counterpart.

The first practical example of BSSC devices made with artificial proteins as potential photosensitizers have been produced with some success and show the potential of this technology to introduce new materials and parameters in solar energy generation.

A more complex approach demonstrates the photogeneration ability of BSSCs based on Bacteriorhodopsin (bR), which is a natural light activated protein found in Halobacterium salinarum vegetation. It holds high promise for solar energy conversion because of its remarkable functional stability against a high concentration of salt (up to 5 M NaCl), and thermal stability up to 140°C, in dry state, and with its high quantum efficiency. In addition, bR functionally tolerates a broad range of pH (5-11) and is easy and inexpensive to clone and produce.

The wild bR of the type of the triple mutant bR, 3Glu [E9Q/E194Q/E204Q], in combination with wide gap semiconductor TiO_2 can be used to collect sunlight.

Differential scanning calorimetry data show thermal robustness of bR wild type and 3Glu mutant, which make them good candidates as photosensitizers in solar cells. Molecular modeling shows that binding of bR to the exposed oxygen atoms of anatase TiO_2 favors electron transfer and is directed by local, small-distance interactions.

Solar cells, based on bR wild type and bR triple mutant immobilized on nanocrystalline TiO_2 film have been successfully constructed by a number of research groups. The photocurrent density-photo voltage (J-V) characteristics of bio-sensitized solar cells (BSSC), based on the wild type bR and 3Glu mutant adsorbed on nanocrystalline TiO_2 film electrode are confirmed.

They show that the 3Glu mutant displays better photoelectric performance compared to the wild type bR, giving a short-circuit photocurrent density (Jsc) of 0.09 mA/cm^2 and the open-circuit photovoltage (Voc) 0.35 V, which is not much, but is a step in the right direction.

Work is in progress to improve the cell performance of these and other types of BSSCs. The investigations are focused on increasing the adsorption of the proteins on TiO_2 films and seeking more appropriate bio-related electrolytes.

Advances in these efforts will bring more positive practical results and new discoveries in the BSSC field, which can be used in other areas of the solar energy markets.

There are results to date with three different bio-sensitized solar cells (BSSCs), using photosensitizers of these types:

1.   Myoglobin (Mb),

2.   Reconstituted zinc protoporphyrin-apomyoglobin (ZnMb), and

3.   Eosin-modified reconstituted zinc protoporphyrin-apomyoglobin (EoZnMb).

The first, Mb-sensitized SCs, exhibit poor cell performance because of the internal electron transfer that reduces Fe(III) to Fe(II), so that a very small photocurrent is observed.

For both ZnMb- and EoZnMb-sensitized SCs, the efficiencies of power conversion are ten times those of their unsensitized counterparts (blank TiO_2 SC), through efficient electron injection from the proteins to TiO_2 that improves the charge separation between TiO_2 and the sensitizer and decreased current leakage between TiO_2 and the electrolyte.

These are only some of the preliminary trials and practical examples of BSSC devices made with artificial proteins as potential photosensitizers. Relative to a typical DSSC, the power conversion efficiency of a BSSC is still too small because the extent of immobilization of the proteins on the surface of TiO_2 is too slight.

Work is in progress to improve the cell performance of BSSC by increasing the adsorption of the proteins on TiO_2 films and by seeking more appropriate bio-related electrolytes.

We will not get much deeper into this technology in this text, because it is far too complex. One needs a chemistry background to even comprehend the basics, and it only gets more complicated after that.

More importantly, BSSC technology is still in its very early stages of development, and although some good results have been demonstrated, it will not be seen on the energy market anytime soon.

There are practical attempts to produce hydrogen from bio-mass (i.e., using anaerobic sludge at the anode. The process is conducted in a light-assisted microbial electrolysis cell (MEC) with a dye-sensitized solar cell (DSSC), which can be optimized by connecting multiple MECs to a single dye (N719) sensitized solar cell.

Hydrogen production can be expected simultaneously in all the connected MECs when the solar cell is exposed to sunlight. The amount of hydrogen produced in each MEC depends on the activity of the microbial catalyst on their anode and the amount of sunlight falling on the device. Substrate (acetate) to hydrogen conversion efficiencies ranging from 42% to 65% have been obtained in experimental setups.

A moderate light intensity of approximately 400 W/m^2 was sufficient to initiate and maintain hydrogen production in the coupled MEC-DSSC cells. This is an exciting, and very promising, new way of generating hydrogen using bio-mass or other bio materials.

For the time being, we hope that the BSSC processes will be developed sufficiently soon, so that this technology can find a place in the markets. At the very least, materials and processes used in the development work will prove useful in the development of other technologies, so we will be keeping track of the BSSC progress.

Advantages of DSSC Technologies

One area, where DSSCs are particularly attractive is the process of injecting an electron directly into the TiO_2. This is qualitatively different to that occurring in a traditional solar cell, where the electron is "promoted" within the original crystal. In theory, given low rates of production, the high-energy electron in the silicon could re-combine with its own hole, giving off a photon, which in the end results in no current being generated. Although this particular case may not be common, it is also fairly easy for an electron generated in another molecule to hit a hole left behind in a previous photo-excitation.

In comparison, the injection process used in the DSSC does not introduce a hole in the TiO_2, only an extra electron. Although it is energetically possible for the electron to recombine back into the dye, the rate at which this occurs is quite slow compared to the rate that the dye regains an electron from the surrounding electrolyte. Recombination directly from the TiO_2 to species in the electrolyte is also possible although, again, for optimized devices this reaction is rather slow.

To the contrary, electron transfer from the platinum-coated electrode to species in the electrolyte is necessarily very fast. As a result of these favorable "differential kinetics," DSSCs work even in low-light conditions. DSSCs are therefore able to work under cloudy skies and non-direct sunlight, whereas traditional designs would suffer a "cutout" at some lower limit of illumination, when charge carrier mobility is low and recombination becomes a major issue. The cutoff is so low they are even being proposed for indoor use, collecting energy for small devices from the lights in the house.

— The 11% efficiency demonstrated by DSSCs make them attractive as a replacement for existing and more expensive technologies. This is especially true in "low density" applications, like rooftop solar installations, where the mechanical robustness and light weight of the glassless DSSC collector could be a major advantage.

— A practical advantage, one DSSCs share with most thin film technologies, is that the cell's mechanical robustness indirectly leads to higher efficiencies in higher temperatures. In any semiconductor, increasing temperature will promote some electrons into the conduction band "mechanically."

The fragility of traditional silicon cells requires them to be protected from the elements, typically by encasing them in a glass box similar to a greenhouse, with a metal backing for strength. Such systems suffer noticeable decreases in efficiency as the cells heat up internally. DSSCs are normally built with only a thin layer of conductive plastic on the front layer, allowing them to radiate away heat much easier, and therefore operate at lower internal temperatures.

Disadvantages of DSSC Technologies

— The major disadvantage to the DSSC design is the use of the liquid electrolyte in most cases, which has temperature and time driven stability problems. At low temperatures the electrolyte can freeze, ending power production and potentially leading to physical damage. Higher temperatures cause the liquid to expand, making sealing the panels a serious problem.

— Another major drawback is the electrolyte solution, which contains volatile organic solvents and must be carefully sealed. This, along with the fact that the solvents permeate plastics, has precluded large-scale outdoor application and integration into flexible structure.

— Replacing the liquid electrolyte with a solid has been a major ongoing field of research. Recent experiments using solidified melted salts have shown some promise but currently suffer from higher degradation during continued operation and are not flexible. The dyes used in early experimental cells (circa 1995) were sensitive only in the high-frequency end of the solar spectrum, in the UV and blue. Newer versions were quickly introduced (circa 1999) that had much wider frequency response, notably "triscarboxyruthenium terpyridine" [Ru(4,4',4"-(COOH)3-terpy)(NCS)3], which is efficient right into the low-frequency range of red and IR light.

The wide spectral response results in the dye having a deep brown-black color, and is referred to simply as "black dye." The dyes have an excellent chance of converting a photon into an electron, originally around 80% but improving to almost perfect conversion in more recent dyes, the overall efficiency is about 90%, with the "lost" 10% being largely accounted for by the optical losses in top electrode.

— A solar cell must be capable of producing electricity for at least 20 years, without significant decrease in efficiency. DSSC are not proven capable of this yet, but the "black dye" system was subjected to 50 million cycles, which is equivalent to ten years' exposure to the sun in Europe. No discernible decrease in performance was observed.

The dye is subject to breakdown in high light situations, such as exposure to desert sunlight, which is why it cannot be used for large-scale installations—most of which are in desert regions.

—   Because of these problems, and at least for now, DSSC are not suitable for large-scale deployments where higher-cost, higher-efficiency cells are more practical and cost-effective. In the future, however, increasing DSSC conversion efficiency and robustness, and correcting their problems (see below) might make them suitable for these applications as well.

Summary

The "quantum efficiency" of regular solar cells is determined by the chance that one incoming photon will create one or more electrons. In quantum efficiency terms, DSSCs are extremely efficient because, due to the "depth" of their nanostructure, there is a very high chance that a photon will be absorbed, and the dyes are very effective at converting absorbed photons to electrons.

Most of the losses that exist in DSSCs are conduction losses in the TiO_2 and the clear electrode, or optical losses in the front electrode. The overall quantum efficiency for green light is about 90%, with the "lost" 10% being largely accounted for by the optical losses in top electrode.

The quantum efficiency of traditional solar cell designs varies, depending on their thickness, but is (theoretically) about the same as the DSSC. The maximum voltage generated by such a cell, in theory, is simply the difference between the (quasi-)Fermi level of the TiO_2 and the redox potential of the electrolyte. So the open circuit voltage (Voc) of a DSSC is about 0.7 V under normal solar illumination conditions.

That is, if an illuminated DSSC is connected to a voltmeter in an "open circuit," it would read about 0.7 V (Voc), which is close to or better than the Voc of normal solar cells. So, in terms of voltage, DSSCs offer slightly higher Voc than silicon, about 0.7 V compared to 0.6 V. This is a fairly small difference, so the real-world difference is determined by the total current production (Jsc).

The problem is that although the dye is highly efficient at turning absorbed photons into free electrons in the TiO_2, it is only those photons which are absorbed by the dye that ultimately result in current being produced. The rate of photon absorption depends on the absorption spectrum of the sensitized TiO_2 layer and upon the solar flux spectrum.

The overlap between these two spectra determines the maximum possible photocurrent. Typically, used dye molecules generally have poorer absorption in the red part of the spectrum compared to silicon, which means that fewer of the photons in sunlight are usable for current generation.

These factors limit the current generated by a DSSC. For comparison, a traditional silicon solar cell generates about 35 mA/cm², whereas current DSSCs offer about 20 mA/cm²—or—almost 1/3 less power is produced by the DSSC under the same conditions. Combined with a fill factor of about 45%, overall peak power production efficiency for current DSSCs is about 11%.

Another, maybe bigger and more important, problem is that DSSCs degrade when exposed to UV radiation. To correct that, the barrier layer may include UV stabilizers and/or UV absorbing luminescent chromophores (which emit at longer wavelengths) and antioxidants to protect and improve the efficiency of the cell.

Basically speaking, DSSCs are currently the most efficient and most promising of the so-called "third-generation" solar technologies. Thin film technologies are typically around 8%, while traditional low-cost commercial silicon panels operate at approximately 15%, so DSSCs are filling a gap—as far as efficiency and location-specific considerations are concerned.

So, the future of DSSCs is bright, but full of challenges. Due to their originality and practicality (in some areas of their applications) research efforts will continue, and we expect great things in the 21st century from this challenging and promising technology.

DSSC Historical Development
2006

The first successful solid-hybrid dye-sensitized solar cells were reported. To improve electron transport in these solar cells, while maintaining the high surface area needed for dye adsorption, two researchers designed alternate semiconductor morphologies, such as arrays of nanowires and a combination of nanowires and nanoparticles, to provide a direct path to the electrode via the semiconductor conduction band.

Such structures may provide a means to improve the quantum efficiency of DSSCs in the red region of the spectrum, where their performance is currently limited.

In August 2006, to prove the chemical and thermal robustness of the 1-ethyl-3 methylimidazolium tetracyanoborate solar cell, the researchers subjected the devices to heating at 80°C in the dark for 1000 hours, followed by light soaking at 60°C for 1000 hours. After dark heating and light soaking, 90% of the initial photovoltaic efficiency was maintained, which is the first time such excellent thermal stability has been observed for a liquid electrolyte that exhibits such a high conversion efficiency. Contrary to silicon solar cells, whose performance declines with increasing temperature, the dye-sensitized solar-cell devices were only negligibly influenced when increasing the operating temperature from ambient to 60°C.

2007

Wayne Campbell at Massey University, New Zealand, experimented with a wide variety of organic dyes based on porphyrin. In nature, porphyrin is the basic building block of the hemoproteins, which include chlorophyll in plants and hemoglobin in animals. He reports efficiency on the order of 5.6% using these low-cost dyes.

2008

In a joint article published in *Nature Materials* in June, Michael Grätzel and colleagues at the Chinese Academy of Sciences demonstrated cell efficiencies of 8.2% using a new solvent-free liquid redox electrolyte consisting of a melt of three salts, as an alternative to using organic solvents as an electrolyte solution. Although the efficiency with this electrolyte is less than the 11% being delivered using the existing iodine-based solutions, the team is confident the efficiency can be improved.

2009

A group of researchers at Georgia Tech made dye-sensitized solar cells with a higher effective surface area by wrapping the cells around a quartz optical fiber. The researchers removed the cladding from optical fibers, grew zinc oxide nanowires along the surface, treated them with dye molecules, and surrounded the fibers with an electrolyte and a metal film that carries electrons off the fiber. The cells are six times more efficient than a zinc oxide cell with the same surface area. Photons bounce inside the fiber as they travel, so there are more chances to interact with the solar cell and produce more current.

These devices only collect light at the tips, but future fiber cells could be made to absorb light along the entire length of the fiber, which would require a coating that is conductive as well as transparent. Max Shtein of the University of Michigan said a sun-tracking system would not be necessary for such cells, and would work on cloudy days when light is diffused.

2010

Researchers lead by Professor Benoît Marsan at the Université du Québec à Montréal claim to have overcome two of the DSC's major issues. According to Benoît Marsan, "new molecules" have been created for the electrolyte, resulting in a liquid or gel that is transparent and noncorrosive, which can increase the photovoltage and improve the cell's output and stability. For the cathode, the researchers found a way to replace the platinum by cobalt sulphide, which is far less expensive, more efficient, more stable and easier to produce in the laboratory.

Today and in the Future

DSSCs are still in the midst of their development cycle along the lines described above and in a number of variations. Efficiency and reliability improvements are needed, and the key is continuous efforts in these areas, which are actually widespread today. Developments include the use of quantum dots for conversion of higher-energy, higher-frequency light into multiple electrons, using solid-state electrolytes for better temperature response, and changing the doping of the TiO_2 to better match it with the electrolyte being used—just to mention a few.

DSSC technology shows promise, especially in short-term use in low-light, low-temperature niche market applications where other technologies do not function efficiently. Their use in large-scale desert field applications is still off, but it is not to be discounted.

DSSC technology is the first of the third gen technologies to offer commercial products. This is mostly due to the development efforts of EPFL and a host of R&D organizations, in addition to global economic and environmental drivers.

Dyesol and Solaronix are the main material developers supplying G24i (battery chargers for Africa in 2009) and Corus (metal solar roof panels in 2011).

Other developers likely to follow in the next few years include 3G Solar, Fraunhofer ISE/ColorSol, Fujikura, and Pecell Technologies. In addition, there are many DSSC developments going on at the corporate and SME levels in Japan, but to date, there has been little publicity covering these.

Dyesol is one of the pioneers in the field of DSSC technology, with more than 20 patents. It is well funded and is commissioning DSSC plants around the world, to bring the technology close to solar developers, such as in Korea and Taiwan. Dyesol recently entered the U.S. market, to take advantage of the Obama Administration's funding opportunities for the clean tech energy sector. It is building its business infrastructure to design novel commercial DSSC applications with industrial partners such as the Corus Group, RG/Permasteelisa (Italy), and the Australian Defense Science Technology Organization. Dyesol and EPFL are at the forefront of

DSC developments, regarding both technology and materials. Designed using standard industrial cells and off-the-shelf DSC materials, Dyesol's champion cell efficiency runs at 11.3 percent, showing eight months stability.

Solaronix is a Swiss company engaged in the business of manufacturing DSSCs for commercial applications. The main product is DSSC based on nanocrystalline titanium dioxide (nc-TiO$_2$) deposited on a transparent electrically conducting substrate. This product is flexible in many ways, and opens new market niches. The efficiency of the finished product is 10%.

3G Solar plans to commercialize its efficient DSSCs. Material development is now focused on the use of high performance enhanced stability dyes such as C101. Much like Dyesol, 3G Solar plans to license out its technology; however, 3G Solar will do this by establishing small DSSC plants in developing countries, to supply home systems to run lights and refrigerators.

Fujikura is at the forefront of technological development, using novel materials for its 5 cm^2 DSC, which incorporates N719 dye and an ionic liquid electrolyte that shows minimal decomposition over 1000 hours. Its three percent cell efficiency was found to decrease minimally, which corresponds to about 10 years stability on an equivalent amorphous silicon system. Fujikura is in the process of optimizing its DSC prior to scaling up for a commercialization process that will use screen-printing, and is taking advantage of its extensive experience with flexible printed circuits and membrane switches.

G24i is one of the first commercial suppliers to offer a range of consumer electronic products. Its initial market is Africa, where mobile phone usage is set to explode despite the fact that the continent suffers from a poor electricity supply infrastructure. Solar phone chargers will provide a more reliable primary/back-up power source than electricity grids, and current models are able to charge a phone battery in less than three hours.

Peccell Technologies, in collaboration with Fujimori Kogyo and Showa Denko KK, demonstrated a large flexible DSC module (2.1m x 0.8m) on a plastic substrate at PV 2008, which is one of the lightest and most stable solar cells available in the world, with an efficiency of about 3 percent. When installed indoors, its output is more than 110 volts, and stability tests have shown that it can last for more than six months. Pecell has also collaborated with Dyesol, and plans to start its module production shortly.

SolarPrint was set up in 2007 to produce DSSCs technology. The facility will initially use screen-printing,

and at a later date will develop a sprayable technology (similar to Dyesol/Corus) that will enable the solar material to be applied like paint. SolarPrint is currently in discussion with application partners, including mobile phone developers and lithium-cell laptop battery suppliers, since it will initially focus on the consumer electronics market in the developing world. In addition, the company is in negotiations with the German Color-Sol consortium, whose novel printing technology may well remove a few steps in the commercial production processes used for Colorsol's large-area DSSCs for BIPV applications and help it get to market sooner.

Update

The first commercial shipment of DSSC solar modules was recorded in July 2009 from G24i Innovations.

A four-year, €14.2 million European research project under the European Commission's Seventh Framework Program (FP7) aims to develop better and cheaper solar panels. The "SUNFLOWER" project ("SUstainable Novel FLexible Organic Watts Efficiently Reliable"), led by the Swiss Center for Electronics and Microtechnology with more than a dozen partners, started work in October, 2011, to increase the cells' efficiency and lifetime, and decrease production costs of excitonic solar devices.

Goals for an initial prototype include a "tandem" multilayer structure to increase efficiency, better-performing barrier layers and getters, and creation in a roll-to-roll atmospheric printing process. This is a chance to develop a technology that is ideally suited to manufacturing in the EU due to its high level of automation, need for highly trained personnel, low energy consumption, and close proximity to suppliers and markets.

Still, while commercial-scale power generation from these technologies may be years away (if ever), there is a low-hanging fruit: powering consumer electronics devices, where size and weight matter more than efficiency, and where long-term stability is much less of a concern.

Assuming the technology adopts improved packaging and maintains its simplified manufacturability, it should be attractive enough to power all sorts of personal gadgetry—just like CIS technologies captured the solar calculators niche market in the 1980s.

Few digital gadgets last longer than a few years before being upgraded out of necessity, or more likely due to the must-have-it syndrome, so no 20-year warranty or PPA is required.

A proper, good size, niche market... or better yet, several good size markets, would do just fine, thank you!

NANO-PV TECHNOLOGY

Generally speaking, nano-technology and nano-materials are a recent, but quickly developing sector of the high-tech industry. Their propagation in the solar field has been stealthy, due to their increased importance.

During the last several years (a decade or less), nanomaterials have emerged as the new building blocks of some light energy harvesting assemblies. Organic and inorganic structures, as well as hybrids, exhibiting improved selectivity and producing efficient catalytic processes have proven successful. Their tunable, size-dependent properties, such as size quantization effects in semiconductor nanoparticles and quantized charging effects in metal nanoparticles, provide the basis for developing new and effective photoelectric conversion systems.

Some of the nanostructures provide innovative approaches for designing the next generation PV devices. Recently the synthesis of nanostructures, using different shapes, such as spheres, prisms, rods, and wires, as well as their use in 3-dimensional assemblies has piqued the interest of scientists and proven that these are viable alternatives for developing new PV devices.

There are basically three major ways that we can use nanoparticles and nanostructures in practical PV power generation:

1. Photosynthesis with donor-acceptor molecular systems.

    The devices and processes in this category are based on the principles of photosynthesis of living vegetation. A large variety of donor-acceptor systems have been synthesized as engines for sunlight conversion, but donor-acceptor systems containing Chla and porphyrins (which are able to mimic the photo-induced electron-transfer process of natural photosynthesis) are the most interesting.

    Nevertheless, even with some laboratory successes of achieving charge separation, the use of this approach for PV power generation is quite limited.

    To continue this work, we need a better understanding of light absorption, energy transfer, radiative and non-radiative excited-state decay, electron transfer, proton-coupled electron transfer, catalysis processes and such. These are important in designing molecular assemblies for energy conversion, and a good understanding of their nature and behavior will provide insight into the new approaches that are needed for efficient, practical utilization of the donor-acceptor molecular systems.

2. Semiconductor based photo-catalysis.

    Subjecting semiconductor nanoparticles to excitation at the band gap level, forces charge separation, but because of the small size of particles and high recombination rates, only a small fraction of thus generated charges can be utilized to induce redox processes at the interface. Using TiO_2 and other semiconductors has shown the need to overcome the limitations in achieving higher photo-conversion efficiencies in photo-catalytic processes.

    The use of semiconductor nano-structures has been applied for solar hydrogen production by the splitting of water molecules. This process is of great practical interest, since it adds a new useful dimension to energy generation and storage processes.

    Author's note: We review the photo-catalysis of water for hydrogen generation in a different section of this text, so we will only mention that the addition of semiconductor-semiconductor or semiconductor-metal composite nanoparticles to the photo-catalytic processes (including hydrogen generation) offers new possibilities, such as facilitating the charge rectification in the semiconductor nanostructures, which would improve the charge separation, and overall device efficiency.

3. Semiconductor, or semiconductor-organic, nano-structure devices.

    Recently, different nano-technologies and organized nano-structures of organic and inorganic (and the related hybrids) have been used in the making of a number of different PV devices. Most of these systems and strategies are used in the development of 3rd gen solar cells, some of the most promising of which are:
    a. Donor-acceptor based molecular clusters,
    b. Dye sensitization of semiconductor nanostructures,
    c. Quantum dot solar cells, and
    d. Carbon nanostructure based solar cells.

We do review these different categories in different sections of this text, and so we invite the reader to take a look at them in their corresponding areas.

To provide a clearer view on the subject, we will further divide the nano-devices and materials used in the manufacturing of solar cells and modules into two major categories;

a.)   Those containing nano-particles (as used in different technologies), and
b.)   Nano-structured PV devices.

Because we discuss the use of nanoparticles as units in different PV technologies and in other sections of this text, we will only briefly discuss their basics here, while taking a closer look at different nano-structures below.

Nano-particles

What are nano-particles?

"Nano-particle" is a fairly new term, coined in the 1990s to segregate and fully describe these from other types of particles in use.

A nano-particle can be described as a very small piece (one complete unit) of matter (metals, semiconductors, or insulators), usually of sub-micron size. A nano-particle as a unit is bigger than a single molecule, and usually measures 1 to 100 nanometers in size.

Nanoparticles in the context of this text can also be viewed as "metamaterials"—special and sometimes artificially made materials engineered to provide unique properties which may not be readily available in nature.

Most metamaterials rely on their structural properties (shape and size) and location, rather than on their type or composition, as needed to create small interferences and inhomogeneities, to evoke an effective behavior, or misbehavior, on a macroscopic scale.

For example, some metamaterials with negative refractive index allow the creation of "super-lenses" which can have a spatial resolution below that of the wavelength. Others can create a type of 'invisibility' over a narrow wave band.

In solar cells, the presence of nanoparticles in key locations in the bulk or surface can modify significantly the scattering and absorption of light, which in turn changes the performance characteristics of the devices.

For example, it has been found that metal nanoparticles help to successfully scatter incoming light across the surface of the silicon substrate in a controlled manner.

Several-fold photocurrent enhancement at certain wavelength has been achieved by silver nanoparticles used for the specific scattering and absorption of light; silver particles deposited on SOI have shown over 30% photocurrent increase.

Gold nanoparticles have been used for scattering and absorption of light on doped silicon, thus obtaining 80% enhancements at certain wavelengths. Gold nanoparticles on thin film silicon have been used to increase the overall conversion efficiency by 8%;

Nanoparticles dispersed in plastic film form efficient solar cells that are flexible and light, and can be shaped to be incorporated into cases for devices such as mobile phones and laptop computers.

Proper use of nanoparticles in the manufacturing of different types of solar cells can contribute to the following benefits:

1.   Materials and manufacturing costs are reduced as a result of using a low-temperature process similar to printing instead of the high-temperature vacuum deposition process typically used to produce conventional solar cells.

2.   Reduced installation costs are achieved by producing flexible rolls instead of rigid crystalline panels. Some cells made from semiconductor thin films have similar characteristics.

Currently available nano-technology solar cells are not as efficient as traditional ones; however, their lower cost offsets this in some practical applications. In the long term, nano-technology PV devices should be of much lower cost.

The introduction of quantum dots in these structures should make them as efficient as conventional PV technologies.

PV Nano-structures

Here we review a number PV technologies based on nano-structures, which are basically miniature structures within the solar cells structure.

a-Si nano-wires

a-Si thin-film silicon technology is one of the most promising in terms of achieving efficient, low-cost energy production. The a-Si solar cell is generally made of p+-doped a-Si/intrinsic a-Si/n−-doped a-Si to form a p-type/intrinsic/n-type (p-i-n) structure.

The intrinsic layer is depleted at thermal equilibrium. Therefore, a large electric field is present in the intrinsic layer. After light absorption, the internal electrical field forces the photo-generated carriers to drift to the n- and p-contacts, contributing to the photocurrent. The carrier's transport velocity is greater in the drift motion than in the diffusion motion.

As a consequence, the absorption layer thickness in the radial a-Si p-i-n solar cell is smaller than that in the planar p-i-n solar cell. However, the material quality of a-Si is not as good as that of crystalline Si, so the thickness of the absorption layer is limited and is too thin to absorb most of the sunlight. A thick i-layer will reduce the electrical field, and large amounts of carriers might recombine in the a-Si film before they arrive at the contact because of traps in a-Si. This indicates that the absorption layer thickness is limited by a compromise between light absorption and photo-generated carrier transport.

To improve some of these properties, a-Si nanowires have been developed with a-Si:H nanostructures which display greatly enhanced absorption over a large range of wavelengths and angles of incidence, due to suppressed reflection. The enhancement effect is particularly strong for a-Si:H nano-cones arrays, which are a variation of the wires in the form of cones.

The nano-cones structures provide nearly perfect impedance matching between a-Si:H and air through a gradual reduction of the effective refractive index. More than 90% of light is absorbed at angles of incidence up to 60° for a-Si:H NC arrays, which is significantly better than nanowires arrays (70%) and thin films (45%). In addition, the absorption of nano-cones arrays is 88% at the band gap edge of a-Si:H, which is much higher than nanowire arrays (70%) and thin films (53%). Our experimental data agree very well with simulation.

The a-Si:H nanopillars function as both absorber and antireflection layers, which offer a promising approach to enhance solar cell energy conversion efficiency.

Figure 4-9 shows a side view of a vertical cross-section of a single nanowire/nanopillar (a), and a top view of a horizontal cross section (b). This structure has photon absorption and carrier transport that are perpendicular to each other, and which could overcome the efficiency limit of the 2-D a-Si solar cell.

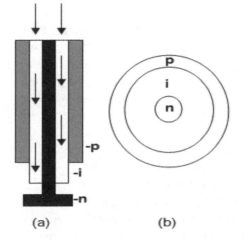

Figure 4-9. a-Si nanowire structure

The nanowire structure has an n-type a-Si nanowire array in which the i-layer and the p-layer a-Si are sequentially grown along the surface of the nanowire. Incoming light is absorbed along the axial direction of the nanowire, and carrier transport is along the radial direction.

Photocurrent of a-Si solar cell with 4000-nm-long nanowires is nearly 40% more than that of a regular a-Si solar cell. Conversion efficiency of 11.6% is obtained, which is around 32% increase over existing a-Si technologies.

Thus far p-i-n s-Si nanowire solar cells are having 3.4% quantum efficiency. However, the axis of the nanowire is not along the light-propagation direction. Therefore, further effort and detailed study of the characteristics of a-Si-based nanowire-array solar cells are needed to optimize the efficiency of this technology.

Hydrogenated s-Si Nanowires

Hydrogenated amorphous silicon (a-Si:H) is a new material based on its unique production technique, and which has advantages over its crystalline form. The a-Si:H thin films can be deposited via conventional PECVD techniques at a very low temperature (100-250°C), which allows deposition not only on glass but also on plastic and stainless steel sheets.

This added capability makes it suitable for roll-to-roll processing, which guarantees low-cost manufacturing, and which puts it in the category of one of the most promising candidates for the new generation solar cells, in addition to many other important applications, such as thin film transistors (TFTs), RFIDs, etc.

The typical film thickness of the a-Si:H films is 1 μm, which guarantees effective light absorption in a-Si:H thin film solar cells, which is two orders of magnitude thinner than that in single crystalline Si.

The minority carrier diffusion length, however, is typically only 300 nm, so the mismatch of light absorption depth and minority diffusion length can cause insufficient absorption, or carrier collection loss.

This mismatch of light absorption and minority carrier diffusion length can be overcome by using nanowire structures. But this is still a theory because, of the large variety of semiconductor nanowires obtained thus far, very few results of successful a-Si:H nanowires have been obtained.

Also, due to the high refractive index of a-Si:H, a large portion of incident light is reflected back from the surface and is lost. 1/4-wavelength transparent thin films are now the industrial standard for antireflection coatings (ARCs) for thin film solar cells like the a-Si:H solar cell. However, this quarter-wavelength ARC is typically designed to suppress reflection at a specific wavelength and at specific angles of incidence only and does not resolve the issue.

Although broadband reflection suppression using nanostructures is achievable, only very few technologies can be applied to the new generation thin film solar cells, due to either restrictions on certain materials, or complexity of the fabrication process.

Because of these issues, we don't expect to see the s-Si nanowire/nanopillar technology taking over the solar markets anytime soon, but it promises to open new horizons to making solar cells—be it as nanowires, nano-cones or AR coatings.

Carbon Nanotubes

Carbon nanotubes (CNTs) are fairly new arrivals in the solar industry. They have very interesting structures with unique electronic and extraordinary mechanical properties, showing great promise for future application in PV devices. Because of that, CNTs have been heavily investigated for use in a variety of applications. The interest in CNTs and related research activities on them have accelerated in the last decade, and the results have been encouraging.

CNTs are basically elongated fullerenes (polymers) with diameters as small as 0.7 nm and several microns long. They have a characteristically large surface area per volume, as well as high electron and thermal conductivities. Their use for field emission displays (FED), strain and other sensors, field effect transistor (FET), future electronics and computing, electrodes, high strength composites, and storage of hydrogen, lithium and other metals is possible and is under active investigation. Their use as photoelectric generators is part of this effort as well.

Photovoltaic applications are focused on CNTs' utilization as either

a)   Photo-induced exciton carrier transport medium impurity within a polymer-based PV layer, or

b)   As the photoactive (photon-electron conversion) layer.

Metallic CNTs are preferred for the application mentioned in a) above, while semiconducting CNTs are preferred for that in b).

To increase the photovoltaic conversion efficiency of organic PV (OPV) devices, special electron-accepting impurities must be added within the photoactive region. Thus, by incorporating CNTs in the polymer, a dissocia-

tion of the exciton pair can be accomplished by the thus formed CNT matrix.

The high CNT surface area (~ 1600 m^2/g), which is one of their main advantages, is a good media for exciton dissociation. The separated carriers within the polymer/CNT matrix are transported via percolation pathways of adjacent CNTs, thus providing excellent means for high carrier mobility and efficient charge transfer. The problem here is that the overall efficiency of the thus formed photovoltaic devices is low (averaging 3-5%), compared to the established PV technologies with efficiencies in the 15-20% range.

CNTs were first discovered by Dr. Iijima of NEC Corporation in 1991 almost by accident. He found them scattered in the process debris, while cleaning his arc discharge apparatus. In the early days of CNT production, laser ablation and arc discharge methods were the dominating approaches, both able to produce single-walled and multi-walled carbon nanotubes (SWCNTs and MWCNTs).

These processes are still used by researchers today, but laser ablation is not amenable for scaleup, so it is used exclusively for R&D, while the arc discharge process is preferred for large-scale production of CNTs.

The purity of the arc process is rather modest, due to the large content of amorphous carbon co-product, while laser ablation has been able to produce SWCNTs with a purity as high as 90%. Chemical vapor deposition (CVD) has been widely used to grow CNTs in recent years. In CVD process, a feedstock of CO or some type of hydrocarbon is heated to 800-1000°C in the presence of a transition metal catalyst, which promotes nanotube growth. CVD is amenable for nanotube growth on patterned surfaces, for fabrication of electronic devices, sensors, field emitters and other applications where controlled growth over masked areas is needed for further processing.

More recently, plasma enhanced CVD (PECVD) has been investigated for its ability to produce vertically aligned nanotubes. A variety of plasma sources and widely varying results have been reported in the literature with variable results and potential for practical applications.

Types of CNT Materials and
Devices Available Commercially

C60 and other Fullerenes

Carbon Nanotube Masterbatches

Carbon Nanotubes Arrays-CNT Arrays

Conductive Nanotubes Composite for Li Ion
Battery Applications

COOH Functionalized CNTs

COOH Functionalized Nanotubes-COOH-CNTs

COOH Functionalized Industrial Grade Nanotubes-COOH-IGMWNTs

Double Walled Nanotubes-DWNTs

Few Layered Graphene Oxide

Fullerenes

Functionalized CNTs

Graphene

Graphene Coatings

Graphene Nanoplatelets

Graphitized Nanotubes

Graphitized Nanotubes-GMWNTs

HDPlas Nanomaterials

Helical MWNTs

Industrial Grade CNTs

Industrial Grade Nanotubes-IGCNTs

Multi Walled Nanotubes-MWNTs

MWNT Arrays

MWNTs

NH2 Functionalized CNTs

OH Functionalized CNTs

OH Functionalized Industrial Grade Nanotubes-OH-IGCNTs

OH Functionalized Nanotubes-OH-CNTs

Short CNTs

Short COOH CNTs

Short COOH Functionalized Nanotubes 0.5-2.0um long

Short OH CNTs

Short OH Functionalized Nanotubes 0.5-2.0um long-Short OH CNTs

Single Layer Graphene Oxide

Single Walled Nanotubes-SWNTs

SWNTs

TWNTS

We obviously cannot cover all items in this extensive list, plus some of them are in the very initial stages of development, so we will focus on the key types as applied in the photovoltaic field. We would direct the reader to search for more details on the web, instead.

CNTs are very versatile, and have some special properties, such as:

- High electrical conductivity
- High tensile strength
- Highly flexible, they can be bent considerably without damage
- Very elastic ~18% elongation to failure
- High thermal conductivity
- Low thermal expansion coefficient
- Good field emission of electrons
- Highly absorbent
- High aspect ratio (length = ~1000 x diameter)

Because of these special properties, CNTs have found applications in a number of commercial operations, such as:

- Conductive plastics
- Energy storage
- Conductive adhesives
- Molecular electronics
- Thermal materials
- Structural composites
- Fibers & fabrics
- Catalyst supports
- Specialized ceramics
- Biomedical applications
- Filtration
- Photovoltaics

CNTs in Photovoltaics

The science behind CNT based PV devices is based on combining the specific physical and chemical characteristics of conjugated polymers with the high conductivity along the CNTs' tube axis. Dispersing CNTs into the photoactive layer to obtain more efficient OPV devices is the main method and its execution is the focus of researchers.

Metal nanoparticles applied to the exterior of CNT, for example, increase the exciton separation efficiency, for the metal provides a high electric field at the CNT-polymer interface, thus accelerating the exciton carriers, transferred to the CNT matrix. Such devices measure open circuit voltage Voc of ~0.35 V, and short circuit current (Isc) of ~6.0 mA/cm^2. The fill factor is ~0.4%, with the white light conversion factor around 0.8%. Single wall nanoparticles (SWNT) in P3OT semiconductor polymer have measured Voc of ~0.94V, while their Isc is ~0.12 mA/cm^2.

CNTs may be used not only as photovoltaic power generators, or add-in material as needed to increase carrier transport (as discussed above), but also as the photoactive layer proper. The semiconducting single walled CNT (SWCNT) is a potentially viable material for PV applications, due to its unique structural and electrical properties.

SWCNTs have high electric conductivity (100 times greater than that of copper) and show great carrier transport, thus greatly decreasing carrier recombination. SWCNTs' bandgap level is inversely proportional to the tube diameter, so this single material may show multiple direct bandgaps matching the solar spectrum, which will allow it to be used in many different ways and applications.

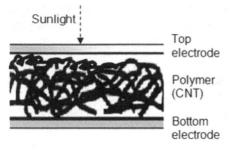

Figure 4-10. CNTs as a carrier transport

The donor-acceptor heterojunction in thus created PV devices can achieve charge separation and collection because of the existence of a bicontinuous network. Along this network, electrons and holes can travel toward their respective contacts through the electron acceptor and the polymer hole donor.

PV conversion efficiency enhancement is due to the introduction of internal polymer/nanotube junctions within the polymer matrix, so that the high electric field at these junctions can split up the excitons, while the single-walled carbon nanotube (SWCNT) acts as a pathway for the electrons.

The dispersion, or growing, of CNTs in a solution of an electron donating conjugated polymer is perhaps the most common strategy to implement CNT materials into OPVs. Usually poly(3-hexylthiophene) (P3HT), or poly(3-octylthiophene) (P3OT) are used for this purpose. These blends are then spin coated onto a transparent conductive electrode with thicknesses that vary from 60 to 120 nm. The conductive electrodes are usually glass covered with indium tin oxide (ITO) and a thin layer of poly(3,4-ethylenedioxythiophene) (PEDOT) and poly(styrenesulfonate) (PSS). PEDOT and PSS help to smooth the ITO surface, decreasing the density of pinholes and stifling current leakage that occurs along shunting paths. Thin layers of aluminum (with intermediate layers of lithium fluoride) are then applied onto the photoactive material via thermal evaporation or sputter coating.

Enhancements of more than two orders of magnitude have been observed in the photocurrent from adding SWCNTs to the P3OT matrix. Improvements were speculated to be due to charge separation at polymer-SWCNT connections and more efficient electron transport through the SWCNTs. However, a rather low

power conversion efficiency of 0.04% under 100 mW/cm^2 white illumination was observed for the device suggesting incomplete exciton dissociation at low CNT concentrations of 1.0% wt.

Because the lengths of the SWCNTs are similar to the thickness of photovoltaic films, doping a higher percentage of SWCNTs into the polymer matrix was believed to cause short circuits. To supply additional dissociation sites, other researchers have physically blended functionalized MWCNTs into P3HT polymer to create a P3OT-MWCNT with fullerene C60 double-layered device. However, the power efficiency was still relatively low at 0.01% under 100 mW/cm^2 white illumination.

Weak exciton diffusion toward the donor-acceptor interface in the bilayer structure may have been the cause in addition to the fullerene C60 layer possibly experiencing poor electron transport.

More recently, a polymer photovoltaic device from C60-modified SWCNTs and P3HT has been fabricated. Microwave irradiating a mixture of aqueous SWCNT solution and C60 solution in toluene was the first step in making these polymer-SWCNT composites. Conjugated polymer P3HT was then added resulting in a power conversion efficiency of 0.57% under simulated solar irradiation (95 mW/cm^2). It was concluded that improved short circuit current density was a direct result of the addition of SWCNTs into the composite causing faster electron transport via the network of SWCNTs.

It was also concluded that the morphology change led to an improved fill factor. Overall, the main result was improved power conversion efficiency with the addition of SWCNTs, compared to cells without SWCNTs; however, further optimization was thought to be possible.

Additionally, it has been found that heating to the point beyond the glass transition temperature of either P3HT or P3OT after construction can be beneficial for manipulating the phase separation of the blend. This heating also affects the ordering of the polymeric chains because the polymers are microcrystalline systems and it improves charge transfer, charge transport, and charge collection throughout the OPV device. The hole mobility and power efficiency of the polymer-CNT device also increased significantly as a result of this ordering.

Emerging as another valuable approach for deposition, the use of tetra-octyl-ammonium bromide in tetra-hydrofuran has also been the subject of investigation to assist in suspension by exposing SWCNTs to an electrophoretic field. In fact, photo-conversion efficiencies of 1.5% and 1.3% were achieved when SWCNTs were deposited in combination with light harvesting cadmium sulfide (CdS) quantum dots and porphyrins, respectively.

The best power conversions achieved to date using CNTs were obtained by depositing a SWCNT layer between the ITO and the PEDOT: PSS or between the PEDOT: PSS and the photoactive blend in a modified ITO/PEDOT: PSS/P3HT: (6,6)-phenyl-C61-butyric acid methyl ester (PCBM)/Al solar cell. By dip-coating from a hydrophilic suspension, SWCNT were deposited after initially exposing the surface to an argon plasma to achieve a power conversion efficiency of 4.9%, compared to 4% without CNTs.

However, even though CNTs have shown potential in the photoactive layer, they have not resulted in a solar cell with a power conversion efficiency greater than the best tandem organic cells (6.5% efficiency).

It has been shown that in most of the previous investigations the control over a uniform blending of the electron-donating conjugated polymer and the electron-accepting CNT is one of the most difficult as well as crucial aspects in creating efficient photocurrent collection in CNT-based OPV devices.

Therefore, using CNTs in the photoactive layer of OPV devices is still in the initial research stages and there is room for more work on existing, new and novel methods to better take advantage of the beneficial properties of CNTs as solar cell materials.

CNT Growth by PECVD

Recently, plasma-enhanced chemical vapor deposition (PECVD) has emerged as a key growth technique to produce vertically aligned nanotubes preferred in some applications. There are various methods for CNT growth and catalyst preparation, as we saw above, but we believe that PECVD has the greatest potential for large-scale implementation in the 21st century for use in the manufacturing of PV devices, thus we'll review it in some detail here.

A CNT is basically a rolled-up tubular shell of graphene sheet which is made up of benzene-type hexagonal rings of carbon atoms. Its diameter is as small as 1 nm. The length can be from a few nanometers to several microns, made of only carbon atoms.

To understand the CNT's structure, it helps to imagine folding a two-dimensional and very thin graphene sheet. Depending on the dimensions of the sheet and how it is folded, several variations of nanotubes can arise. Also, like the single or multilayer nature of graphene sheets, the resulting tubes may be a single- or multi- wall types. The tube's orientation is denoted by

a roll-up vector (see below). Along this vector, the graphene sheet is rolled into a tubular form.

The structure is conveniently expressed in terms of a one-dimensional unit cell. See Figure 4-11. (3)

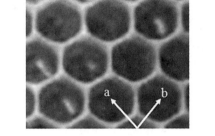

Figure 4-11. Two-dimensional graphene sheet (CNP) with its chiral vector (3).

The circumference of a SWCNT is given by the chiral vector

$$C = na + mb$$

where
n and m are integers, and
a and b are unit vectors of the hexagonal lattice.

The diameter of the nanotube is given by

$$\sqrt{3}a_{c-c}(m^2 + mn + n2)^{0.5}$$

where
a_{c-c} is the C-C bond length.

SWCNTs exhibit unique electronic properties in that they can be metallic or semiconducting depending on their chirality. This allows formation of semiconductor-semiconductor, and semiconductor-metal junctions useful in device fabrication. SWCNTs also possess extraordinary mechanical properties. The Young's modulus of individual SWCNTs has been estimated to be around 1 TPa and the yield strength can be as large as 120 GPa.

A MWCNT is a stack of graphene sheets rolled up into concentric cylinders, see Figure 4-12. The walls of each layer of the MWCNT, i.e. the graphite basal planes, are parallel to the central axis ($\theta = 0$). In contrast, a stacked-cone arrangement (also known as Chevron, bamboo, ice cream cone, or piled cone structures) is also seen where the angle between the graphite basal planes and the tube axis is nonzero.

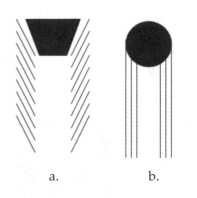

a.   b.

Figure 4-12. MWCNT structure (3)

Figure 4-14. MWCNT grown in an inductive plasma reactor (3)

where:

MWCNF $\theta > 0$ (left), and

MWCNT $\theta = 0$ (right)

A MWCNT, in contrast, has no graphite edges and therefore there is no need for valence-satisfying species such as hydrogen. Since the stacked-cone structures exhibit only small θ values and are not solid cylinders but are mostly hollow, they can be called multi-walled carbon nanofibres (MWCNFs). Note that the terminologies graphitic carbon fibers (GCFs) and vapor-grown carbon fibers (VGCFs) have long been used to denote solid cylinders.

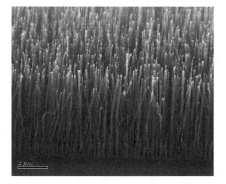

Figure 4-13. High magnification SEM images of MWCNF grown a dc-HFCVD reactor (3)

Challenges of CNT Manufacturing and Applications

As with all emerging technologies, CNTs researchers battle handicaps and challenges, which must be resolved before CNTs can be used in photovoltaic applications. Basically:

1.   CNTs degrade overtime when exposed to oxygen and excess humidity. Special passivation layers and other oxidation-preventing precautions are required to eliminate premature oxidation, which re-

duces the optical transparency of the top electrode, which in turn lowers the overall PV conversion efficiency.

2.   The dispersion of CNT within the polymer photoactive layer presents another serious challenge. CNT material must be equally dispersed within the polymer matrix to form efficient pathways between the occurrence of the excitons and the electrode. If CNT is inefficiently dispersed, with areas of low or very high concentration, the process will result in inefficient charge transfer and respective low efficiency.

3.   The photoactive matrix layer of CNT presents challenges in the lack of capability to form a p-n junction, which results in an internal built-in electric potential, which in turn provides pathways for efficient carrier separation within the PV device. Since it is extremely difficult to dope certain segments of a CNT, the formation of an efficient p-n junction has been difficult to achieve. To overcome this difficulty, energy band bending has been done by the use of two electrodes having different work functions. A strong built-in electric field covering the whole SWNT channel is formed for high-efficiency carrier separation.

Also, oxidation issue of CNT is more critical for this application. Oxidized CNT has a tendency to become more metallic, which increases disproportionally its conductivity, contributing to loss of semiconductor properties, and with that the PV conversion efficiency decreases proportionally.

CNT use in DSSCs

Due to the simple fabrication process, low production cost, and high efficiency, there is significant interest in dye-sensitized solar cells (DSSCs). Thus, improving DSSC efficiency has been the subject of a variety of research investigations because it has the potential to be manufactured economically enough to compete with other solar cell technologies. Titanium dioxide nanoparticles have been widely used as a working electrode for DSSCs because they provide high efficiency, more than any other metal oxide semiconductor investigated.

Yet the highest conversion efficiency under air mass (AM) 1.5 (100 mW/cm^2) irradiation reported for this device is about 11%. Despite this initial success, the effort to further enhance efficiency has not produced major results. The transport of electrons across the particle network has been a key problem in achieving higher photoconversion efficiency in nanostructured electrodes. Because electrons encounter many grain boundaries during transit and experience a random path, the probability of their recombination with oxidized sensitizer is increased.

Therefore, it is not adequate to enlarge the oxide electrode surface area to increase efficiency because photo-generated charge recombination should be prevented. Promoting electron transfer through film electrodes and blocking interface states lying below the edge of the conduction band are some of the non-CNT based strategies to enhance efficiency that have been employed.

With recent progress in CNT development and fabrication, there is promise to use various CNT based nanocomposites and nanostructures to direct the flow of photogenerated electrons and assist in charge injection and extraction. To assist the electron transport to the collecting electrode surface in a DSSC, a popular concept is to utilize CNT networks as support to anchor light harvesting semiconductor particles. Research efforts along these lines include organizing CdS quantum dots on SWCNTs.

Charge injection from excited CdS into SWCNTs was documented upon excitation of CdS nanoparticles. Other varieties of semiconductor particles including CdSe and CdTe can induce charge-transfer processes under visible light irradiation when attached to CNTs. Including porphyrin and C60 fullerene, organization of photoactive donor polymer and acceptor fullerene on electrode surfaces has also been shown to offer considerable improvement in the photoconversion efficiency of solar cells. Therefore, there is an opportunity to facilitate electron transport and increase the photoconversion efficiency of DSSCs utilizing the electron-accepting ability of semiconducting SWCNTs.

Other researchers fabricated DSSCs using the sol-gel method to obtain titanium dioxide coated MWCNTs for use as an electrode. Because pristine MWCNTs have a hydrophobic surface and poor dispersion stability, pretreatment was necessary for this application. As a relatively low-destruction method for removing impurities, H_2O_2 treatment was used to generate carboxylic acid groups by oxidation of MWCNTs. Another positive aspect was the fact that the reaction gases including CO_2 and H_2O were non-toxic and could be released safely during the oxidation process.

As a result of treatment, H_2O_2 exposed MWCNTs have a hydrophilic surface and the carboxylic acid groups on the surface have polar covalent bonding. Also, the negatively charged surface of the MWCNTs improved the stability of dispersion. Then, by entirely surrounding the MWCNTs with titanium dioxide nanoparticles using the sol-gel method, an increase in the conversion efficiency of about 50% compared to a conventional titanium dioxide cell was achieved.

The enhanced interconnectivity between the titanium dioxide particles and the MWCNTs in the porous titanium dioxide film was concluded to be the cause of the improvement in short circuit current density. Here again, the addition of MWCNTs was thought to provide more efficient electron transfer through film in the DSSC.

CNTs Used in OPV Cells

A fundamental issue of organic solar cells made of bulk heterojunctions is the lack of congruence in achieving simultaneous high efficiencies for light absorption, exciton dissociation, and charge carrier collection. The blending of donor and acceptor significantly increases their contact interface and therefore increases the exciton dissociation efficiency. However, free carriers must be transported through the phase-separated domains (either hoping or tunneling) inside the disordered blend.

To minimize the recombination loss, the blend thickness has to be designed with a tradeoff between the light absorption and the carrier collection such that enough photons can be absorbed while the photoexcited charge carriers travel a small distance before extraction.

Another tradeoff is on the nanoscale morphology of the blend, which has to be tuned so that it creates enough interconnected pathways for efficient extraction of charge carriers while still maintaining enough interfaces for dissociation of excitons. Even with finely tuned nanoscale morphology, power conversion efficiency is still limited because of the insufficient light absorption due to the thin active layer (less than 100 nm).

To overcome the incongruence of bulk heterojunctions, ideal interdigitated heterojunctions are considered to be the best solution for high-performance organic solar cells. The two phases of donor and acceptor are interdigitated in percolated highways to ensure high mobility of charge carrier transport with reduced recombination in the bicontinuous pathways.

The two phases are interspaced with an average length scale of around or less than the exciton diffusion length (10-20 nm). A pure donor phase at the hole-collecting electrode and a pure acceptor phase at the electron-collecting electrode are placed to act as diffusion barriers for the wrong sign charge carriers at the respective electrodes. However, such a well-organized nanostructure is a great challenge to implement.

In recent years, several groups have attempted to mimic the ideal interdigitated heterojunctions based on porous titania, zinc oxide nanorod arrays, TiO_2 nanotubes arrays, as well as $ZnO-TiO_2$ core-shell nanorod arrays. Unfortunately, none has reported high power conversion efficiency on these devices.

Vertically aligned (VA) organic polymer nanowires may be a good solution to make real interdigitated heterojunctions for organic solar cells, but these organic nanowires suffer from low carrier mobility.

Another critical issue of organic solar cells with bulk heterojunctions is the short lifetime. Traditionally, the electrode made of aluminum is very much prone to oxidation, and the aluminum atoms can easily diffuse into the active layer to quickly degrade the polymer.

The degradation of the interface between ITO and PEDOT:PSS also occurs frequently. A practical way to minimize device degradation is an inverted structure that incorporates metal oxides such as TiOx and MoO_3 as a diffusion barrier and the use of high work function metals such as Ag and Au as an anode electrode.

A novel inverted structure of organic photovoltaic cells with interdigitated bulk heterojunctions is proposed, which can be implemented by VA-CNTs. In this structure, a layer of semiconducting polymer (polythiophene) is electrochemically polymerized on VA-CNTs to block electrons. The donor-acceptor blend is infiltrated into the gaps between these polymercoated VA-CNTs. A layer of TiOx on top of the blend is deposited to block holes and absorb ultraviolet light to protect the active layer.

The advantages of devices based on such interdigitated heterojunctions include but are not limited to the following.

1.   The proposed interdigitated structure is more practically realistic to be implemented than the ideal interdigitated heterojunctions and other interdigitated structures.

2.   The interdigitated structure effectively decouples the light absorption length and the charge collection length such that both can be tuned independently for obtaining the best device performance. For example, by maintaining the spatial distance between VA-CNTs, the effective collection length can be fixed, while the effective absorption length can be increased by simply increasing the length of the CNT (device thickness). On the other hand, the effective collection length can be tuned by varying the space among VA-CNTs without significantly affecting absorption.

3.   The inverted structure will enable a longer lifetime of the device by avoiding the problem of degradation. Studies suggest that the power conversion efficiency of organic solar cells with the proposed interdigitated bulk heterojunctions can exceed 12% when a small band gap polymer donor, PCP-DTBT, is used. However, such structure requires VA-CNTs to be well controlled over spacing and length.

These efforts and the results thereof call for further study of the growth of VA-CNTs for their practical application as solar cell materials.

Single Nanowire PV

Nanowire is basically a semiconductor type of material, containing elements of the In and Ga type, that is shaped as a long, upstanding strand. Single nanowire, as the name suggests, is different from multi- and other shapes and types of nano-devices. It consists of a single strand (wire) of nano-material, which is solely responsible for the PV conversion effect.

Many millions and billions of these wires are usually deposited on a substrate to make a solar cell capable of converting sunlight into electricity.

The properties and potential of semiconductor nanowires as building blocks for photovoltaic devices based on a single nanowire model is a novel theory which deserves a close look due to its originality and potential practical applications.

Two central nanowire motifs involving p-i-n dopant modulation in axial and coaxial geometries are reviewed herein as platforms for fundamental studies, in addition to the challenges and opportunities for improving efficiency enabled by controlled synthesis of more complex nanowire structures. The potential applications as power sources for emerging nanoelectronic devices are of particular interest as well. (13)

As a prototypical example, semiconductor nanowires are a broad class of materials which, through controlled growth and organization, have led to a number of novel nanoscale photonic and electronic devices. There is a need to elucidate the intrinsic characteristics and limits of nano-enabled solar cells to evaluate their potential for next generation, large-scale, high-efficiency and low-cost solar cells, as well as integrated power solutions for emerging nanoelectronic devices. Toward meeting that need, we have initiated studies exploiting single nanowire heterostructures as stand-alone and active photovoltaic elements.

The use of single nanowires as photovoltaic elements presents several key advantages which may be leveraged to produce high-efficiency, robust, integrated nanoscale PV power sources.

1. The principle of bottom-up design allows the rational control of key nanomaterial parameters, which will determine PV performance, including chemical/dopant composition, diode junction structure, size, and morphology. Importantly, this principle has been demonstrated previously in a wide variety of nanoscale structures and devices.

2. Single or interconnected nanowire PV elements could be seamlessly integrated with conventional electronics and/or future nanoscale electronics to provide energy for low-power applications.

3. Studies of PV properties at the single nanowire level will permit determination of the intrinsic limits, areas of improvement, and potential benefits of nanoenabled PV.

Axial and Radial Nanowires as Key PV Elements

Two unique structural motifs that can yield functional PV devices at the single nanowire level include p-type/intrinsic/n-type (p-i-n) dopant modulation in axial and radial geometries.

A prototypical single p-i-n axially modulated nanowire diode is shown in Figure 4-15.

In this structure, electron-hole pairs are generated throughout the device upon absorption of photons whose energies are equal to or greater than the band-gap of silicon (Eg = 1.12 eV for single-crystal silicon). Carrier generation and separation are most efficient within the depletion region due to the built-in field established

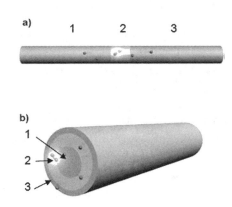

Figure 4-15. Schematic of carrier generation and separation

(a) axial, and
(b) radial p-i-n nanowires.
The different (1, 2 and 3) regions denote the p-type, i-, and n-type diode segments, respectively.
Note: The spheres in #2 sections denote the holes and electrons.

across the p-i-n junction.

Once swept in the direction of the electric field, the photogenerated holes (electrons) traverse through the p-type (n-type) regions and are collected as a photocurrent by ohmic metal contacts. In this axial configuration, the p-type and n-type regions can be made arbitrarily short since their main purpose is to provide contact to the junction embedded within the nanowire.

Therefore, the active device area can be kept very small so as to enhance integration. A closely related structure is the p-i-n radially modulated nanowire diode (Figure 4-15b). The overall device physics are identical to those of the axially modulated motif, with added benefits being that the p-i-n interface extends along the length of the nanowire and that carrier separation takes place in the radial versus the longer axial direction.

Since the latter yields a carrier collection distance smaller or comparable to the minority carrier diffusion length,[1,18] photogenerated carriers can reach the p-i-n junction with high efficiency without substantial bulk recombination. Indeed, recent theoretical studies have suggested that coaxial nanowire structures could improve carrier collection and overall efficiency with respect to comparable single-crystal bulk semiconductors, and especially when relatively low-quality materials are used as absorber materials.

Large numbers of vertically aligned arrays of coaxial nanowires would enable substantial light absorption along the long-axis of the nanowires, and also afford the benefit of short range and efficient radial carrier separation. Together, these advantages would orthogonalize the pathways for light absorption and carrier collection, eliminating a key limitation of conventional planar solar cells, and reducing device surface reflectance, which is not the focus of our work and the current tutorial review.

Nano-crystalline Solar Cells

The nano-crystalline structures based solar devices make use of some of the usual thin-film light absorbing materials, but are deposited as a very thin absorber on a substrate (supporting matrix) of conductive polymer or mesoporous metal oxide having a very high surface area to increase internal reflections. Hence, the probability of light absorption increases.

Using nanocrystals allows one to design architectures on the length scale of nanometers, the typical exciton diffusion length. In particular, single-nanocrystal ('channel') devices (arrays of single p-n junctions between the electrodes, such as TiO_2, and separated by a period of about a diffusion length), represent a new architecture for solar cells and potentially high efficiency.

We envision the development of this type of photoconversion to be in the R&D labs for awhile yet, but it opens new possibilities in areas where other technologies simply cannot compete.

Nanopillars

Nanotechnology researchers have created different types and shapes of micro (or nano) materials such as nano-wires, micro-wires, nano-cones and nano-pillars lately that are excellent at trapping light and reducing the amount of semiconductor material usually needed to make solar cells.

Nanowires and nanopillars use one half to one third as much semiconductor material required by thin-film solar cells made of materials such as cadmium telluride, and as little as 1 percent of the material used in crystalline silicon cells.

Nanopillars make it much easier to extract electric charge from many materials, which could change solar dynamics by making solar cells and modules much cheaper. And because reducing material costs while achieving the same amount of light absorption, we see this technology progressing quickly.

Nanopillars are a recent development, consisting of nanostructures still in R&D mode, with the goal of lowering the cost of solar cells and detectors. Nanowire and nanopillar devices developed for solar energy conversion absorb light just as well as commercial thin-film solar cells, while using much less semiconductor material.

Computer simulations show that, compared to flat surfaces, nanopillar semiconductor arrays have many times the surface area and are therefore much more efficient collectors of sunlight. This can be illustrated by looking at a single solar module lying on the ground, vs. several modules arranged in a columnar array high above the ground. The surface area of the array is several times larger than the single module, therefore it is exposed to much more sunlight.

Of particular interest to cost-effective solar cells the use of novel device structures and materials processing for enabling acceptable efficiencies. One approach is a direct growth of highly regular, single-crystalline nanopillar arrays of optically active semiconductors on aluminium substrates which are then configured as solar-cell modules.

Such a device is a photovoltaic structure that incorporates three-dimensional, single-crystalline n-CdS nanopillars, embedded in polycrystalline thin films of p-CdTe, to enable high absorption of light and efficient collection of the carriers.

Through experiments and modeling, we have the ability to produce highly versatile solar modules on both rigid and flexible substrates with enhanced carrier collection efficiency arising from the geometric configuration of the nanopillars.

The ability to deposit single-crystalline semiconductors on support substrates is of profound interest for high-performance solar-cell applications. The most common approach involves epitaxial growth of thin films by using single-crystalline substrates as the template.

Owing to their single-crystalline nature, nanopillars have the potential to produce high-performance solar modules. Although nanowires can be grown non-epitaxially on amorphous substrates, their random orientation on the growth substrates could limit the explored device structures.

The template assisted, VLS growth of highly ordered, single crystalline nanopillars on aluminum substrates as a highly versatile approach for fabricating solar cell modules is another possibility that explores the potential of simplifying the solar cells manufacturing process, while enabling the development of new materials and structures, and especially those of the 3D structures, which do enhance the optical absorption efficiency of the materials they are made of.

This is true specifically of the development of CdSe and CdTe nanopillar arrays, which have shown that their photoelectrodes exhibit enhanced collection of low energy photons. These photons are absorbed far below the surface, as compared with planar (2-D) photoelectrodes, and the results demonstrate the potential advantage of these non planar cell structures, especially for material systems where the bulk recombination rate of carriers is larger than the surface recombination rate.

This method for making 3-D nanopillar solar cells and modules based on CdTe/CdS consists of forming pillars of 3-D single crystalline CdS nanopillars, which are embedded into polycrystalline thin films of p-CdTe. This structure enables the high absorption of light and allows very efficient collection of the resulting carriers.

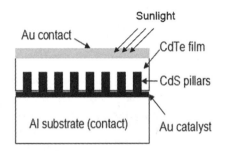

Figure 4-16. 3D CdS/CdTe nanopillars solar cell

These cells' function is similar to that of 2D Cte/CdS solar cells. The space charge and carrier collection region of the nanopillar solar cells, however, is enhanced for H=900 nm, which reduces the total volumetric recombination of photogenerated carriers. This enhances the performance of nanopillar cells as compared with conventional planar CdS/CdTe solar cells, especially for devices with short minority carrier diffusion lengths, ignoring enhanced optical absorption (that is, reduced reflectance) due to the 3D geometric configuration of the nanopillars.

The nanopillar structure may be disadvantageous as compared with conventional planar-structured photovoltaics when interface recombination is the limiting factor for cell performance (for instance, when the bulk minority carrier lifetimes are long).

As an alternative, flexible solar cells can be made from this structure by removing (etching) the aluminum substrate, and substituting it with a flexible indium bottom electrode. Then the entire 3D assembly can be embedded in clear plastic and used as needed as a flexible power generation source. The device can be bent at will, with only marginal effect on performance and not much degradation of performance after repeated bending cycles.

Although conventional equipment and procedures are used, this is still a complex process to execute properly, and is in need of further development and optimization before being introduced in mass production.

Unfortunately, the efficiencies of thus obtained 3-D devices have been very low to date with a maximum of 6% achieved on a lab scale. The inefficiency is due mostly to the un-optimized dimensions, poor density and misalignment of the nanopillars, as well as their low p-n junction interface quality.

On the other hand, single-nanowire devices have much better theoretical efficiencies, approaching the physical limits of the materials. However, controlling the cost and the quality of the manufacturing process for mass production of large-scale single-crystal nanopillar modules, using highly dense and ordered arrays of single crystalline nanopillar arrays, has not been possible to date.

A variation of a new nanopillar material consists of an array of nanopillars that are narrow at the top and thicker at the bottom. The narrow tops allow light to penetrate the array without reflecting off. The thicker bottom absorbs light so that it can be converted into electricity. The design absorbs 99% of visible light, compared to the 85% absorbed by conventional cylindrical nanopillars with the same thickness along their entire length. At the same time, an ordinary flat thin film of the material would absorb only 15 percent of the incoming light. This difference alone ensures the great future of this technology in the 21st century.

The nanopillars are grown to approximately two micrometers in height, but only slightly over 100 nanometers in diameter at the base. The tops of the nanopillars are approximately 50-60 nanometers in diameter, or almost 2 times smaller diameter.

The manufacturing process is much simpler, since it uses molding instead of the complicated process steps and layer-by-layer deposition of materials. Some even require complex and expensive materials that include wires with metal nanoparticles.

As with most newly developed exotic technologies, this method could be cheap, but translating it into a large-scale manufacturing process is not easy and will present technological challenges. Nevertheless, the simple fact that there is proof that nano-structures can dramatically increase absorption is an exciting start on the long road to commercialization.

In time, the pillars' type, shape and size will be optimized to the point where it will be possible to make devices that absorb a wider portion of the solar spectrum, including infrared wavelengths of light. This will enhance the usefulness of the devices in different areas of solar energy generation.

In summary, 3-D nanopillars technology holds a theoretical promise, but its mass production and practical application—due to low efficiency and high cost—are still in the future. However, the concepts and procedures of this technology are quite promising and could be useful for developing other solar processes in the 21st century.

Nanoplasmonic Solar Cells

Plasmonics, which we review in more detail below, have become a focus of recent research in photovoltaic applications, mostly due to their effects in enhancing the absorption performance of solar cells.

Different approaches have been proposed to integrate plasmonics technologies into solar cells, where a range of metallic nanostructures that show plasmon resonance wavelength in the visible and near-infrared regime can be utilized to increase the coupling of light into the solar cell.

This phenomena is widely used to increase the coupling of light that can be trapped in thin layers of active regions as in thin film technologies. In this review, more attention is given to the techniques of fabricating the metallic nanoparticles and the ways to control their plasmon resonance wavelengths.

The shape, size, and dielectric permittivity of the host and the type of the metallic nanoparticles on tuning the resonance wavelength must be optimized to obtain a functional, efficient and reliable device. The cluster of nanoparticles also gives different resonance wave-

length from the individual nanoparticles due to dipolar coupling among the nanoparticles. Nevertheless, the plasmon resonance can be engineered to increase the absorption performance of conventional and specialized solar cells.

A number of researchers are working on the next generation solar cells, the so-called nano-plasmonic solar cells, which promise to be twice as efficient as, and cheaper than, conventional PV technologies.

The new technology allows for the most efficient collection of solar energy in a wider light spectrum range than the currently developed PV technologies in other laboratories around the world.

Nano-tubes PEC

The nano-tube based photo-electric cell (PEC) is similar to the devices we review under photoelectric (PEC) solar cells used for solar hydrogen generation in this text, except for the added nanotube component, which would surely make it more efficient, but also more complex to manufacture, thus more expensive.

Looking at Figure 4-19, in this cell there are organic dyes dissolved in the electrolyte which fills the cell. The electrons at the valence band of the dyes (the S/S mark) are excited by the impinging sunlight and are transferred to the conduction band and injected into the TiO_2 nanotube anode (a).

Those electrons are then collected by the cathode (c) via the external current loop and are sent back to the electrolyte solution through a reduction reaction in it. Electrons are then carried by redox ions in the electrolyte and transported through it into the electrode through a diffusion mechanism. They can then be collected and transported again through an external electrical circuit to the cathode.

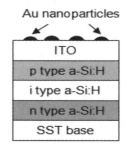

Figure 4-17. Schematic diagram of a-Si:H p-i-n solar cell structure with Au nanoparticles

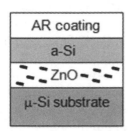

Figure 4-18. Schematic diagram of a solar cell with metallic nanoparticles in between the layers as reflector

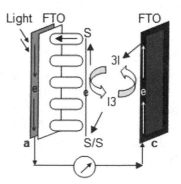

Figure 4-19. Nanotube PEC, operating principle.

Summary

Over the last several years new synthetic materials, technologies and strategies have been developed, geared towards the design of nano-materials and structures. The list of semiconductors, metals, polymers and light harvesting assemblies used in this effort is endless.

We have reviewed herein a number of practical systems, as well as the optical, photo-catalytic, and photo-electrochemical properties of various nano-materials and structures that could be used in the development of the next generation PV devices.

Nevertheless, the optical and light processing properties of the nano-materials and the resulting devices are still poorly understood and somewhat blindly used. Serious efforts are needed to screen potentially useful systems and find ways to design more efficient and reliable nano-based solar energy conversion devices.

New strategies are needed to organize ordered assemblies of two or more components on electrode surfaces, and will be the key to improving the performance of 3rd gen PV solar cells. And as importantly, new sensitizers, and/or special semiconductor systems that can collect and convert IR light need to be developed and optimized, to broaden the photo-response of the new devices.

Some 3rd gen technologies, as Q-dots and carbon nanostructure based solar cells, are still in their infancy. As an example, harvesting multiple charge carriers generated in semiconductor Q-dots will be a major challenge for researchers working in this area.

Hybrids of solar and conventional devices may provide some short-term benefits in the manufacturing and use of economically viable devices, but commercialization of large-scale solar cells based on nanostructure architecture is far from reality.

Advances in basic research and some sectors of the nano-technologies, coupled with the increased demand for clean energy, is gradually changing our perspective of solar energy. Materials and processes developed during the search for new nano-materials and systems are bringing new ways of thinking about solar cells. With that, we believe, these technologies and their elements will be major contributors to the advances in the solar energy sector in the 21st century.

QUANTUM PARTICLES PV TECHNOLOGY

Generally speaking, a quantum dot (QD) is a nanoscale piece of matter or, more precisely, a very small particle of semiconductor material, whose excitons are confined in all three spatial dimensions, so that their electronic characteristics are closely related to the size and shape of the individual quantum dot crystal.

Note: Basically, nano-particles and quantum dots have a lot in common, with quantum dots having some added benefits such as allowing precise tuning.

So, the rule of thumb here is that size matters, and in fact it determines the performance characteristics of QD devices. Another rule stipulates that the smaller the size of the crystal, the larger the band gap, creating a greater difference in energy between the highest valence band and the lowest conduction band. So, more energy is needed to excite the dot and, concurrently, more energy is released when the crystal returns to its resting state.

In fluorescent PV dye applications, this means that higher frequencies of light are emitted after excitation of the dot as the crystal size grows smaller, which results in a color shift from red to blue of the emitted light.

It follows that a key advantage in using QD materials and devices is the high level of control possible over the size of the crystals produced, in turn allowing precise control over the conductive properties of the final material or device.

Below we will review some of the major QD technologies.

Quantum Dots

As outlined above, a key advantage of QDs is that they can be tuned to absorb different parts of the solar spectrum by carefully varying their size, which is seen as a promising approach to capturing solar power to produce electricity. This is in part because they can essentially be spray painted onto a substrate.

So far, however, this approach has been hampered by low efficiencies, in the range of 4-6%, while 10-12% efficiency is theoretically possible by adding more and more layers of quantum dots that are tuned to capture various wavelengths as sunlight travels through them.

QD solar cells have the potential to increase the maximum attainable thermodynamic conversion efficiency of solar photon conversion up to about 66% by utilizing, in addition to the multi-layer structure, hot photogenerated carriers to produce higher photo-voltages or higher photo-currents.

Multi-layer QD cells work best because the electrical resistance between the layers is reduced. Introducing a transition layer made up of four films of different transparent metal oxides will keep resistance between the layers very low, while allowing light to pass through to the bottom layer.

Now we'll take a close look at this promising technology.

QD Basics

Because conversion efficiency is one of the most important parameters to optimize for implementing photovoltaic and photochemical cells on a truly large scale, several schemes for exceeding the Shockley-Queissar (S-Q) limit have been proposed and are under active investigation.

These approaches include tandem cells, hot carrier solar cells, solar cells producing multiple electron-hole pairs per photon through impact ionization, multiband and impurity solar cells, and thermo-photovoltaic/thermo-photonic cells.

The most important characteristics of these are:

1.   Hot carrier and impact ionization effects, and

2.   The effects of size quantization on the carrier dynamics that control the probability of these processes. (6)

The solar spectrum contains photons with energies ranging from about 0.5 to 3.5 eV. Photons with energies below the semiconductor band gap are not absorbed, while those with energies above the band gap create electrons and holes with a total excess kinetic energy equal to the difference between the photon energy and the band gap. This excess kinetic energy creates an effective temperature for the carriers that is much higher than the lattice temperature. Such carriers are called "hot electrons and hot holes," and their initial temperature upon photon absorption can be as high as 3000°K with the lattice temperature at 300°K. The division of this kinetic energy between electrons and holes is determined by their effective masses, with the carrier having the lower effective mass receiving more of the excess energy.

A major factor limiting the conversion efficiency in single band gap cells to 31% is that the absorbed photon energy above the semiconductor band gap is lost as heat through electron-phonon scattering and subsequent phonon emission, as the carriers relax to their respective band edges (bottom of conduction band for electrons and top of valence for holes).

The main approach to reduce this loss in efficiency has been to use a stack of cascaded multiple p-n junctions with band gaps better matched to the solar spectrum; in this way higher-energy photons are absorbed in the higher-band gap semiconductors and lower-energy photons in the lower-band gap semiconductors, thus reducing the overall heat loss due to carrier relaxation via phonon emission. In the limit of an infinite stack of

band gaps perfectly matched to the solar spectrum, the ultimate conversion efficiency at one sun intensity can increase to about 66%.

Another approach to increasing the conversion efficiency of photovoltaic cells by reducing the loss caused by the thermal relaxation of photogenerated hot electrons and holes is to utilize the hot carriers before they relax to the band edge via phonon emission.

There are two fundamental ways to utilize the hot carriers for enhancing the efficiency of photon conversion. One way produces an enhanced photo-voltage, and the other way produces an enhanced photocurrent.

The former requires that the carriers be extracted from the photo-converter before they cool, while the latter requires the energetic hot carriers to produce a second (or more) electron-hole pair through impact ionization—a process that is the inverse of an Auger process whereby two electron-hole pairs recombine to produce a single highly energetic electron-hole pair.

To achieve the former, the rates of photogenerated carrier separation, transport, and interfacial transfer across the contacts to the semiconductor must all be fast compared to the rate of carrier cooling. The latter requires that the rate of impact ionization (i.e. inverse Auger effect) be greater than the rate of carrier cooling and other relaxation processes for hot carriers.

Hot electrons and hot holes generally cool at different rates because they generally have different effective masses; for most inorganic semiconductors, electrons have effective masses that are significantly lighter than holes and consequently cool more slowly. Another important factor is that hot carrier cooling rates are depen-

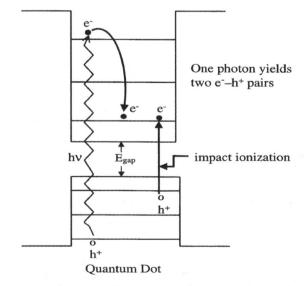

Figure 4-20. Enhanced PV efficiency in QD solar cells by impact ionization (inverse Auger effect). (6)

dent upon the density of the photogenerated hot carriers (viz., the absorbed light intensity).

Here we review briefly several of these technologies:

Quantum Dot-sensitized Nanocrystalline TiO₂ Solar Cells

This configuration is a variation of a recent promising new type of photovoltaic cell that is based on dye-sensitization of nanocrystalline TiO_2 layers. In this PV cell, dye molecules are chemisorbed onto the surface of 10-0 nm size TiO_2 particles that have been sintered into a highly porous nanocrystalline 10-20 nm TiO_2 film.

Upon photo excitation of the dye molecules, electrons are efficiently injected from the excited state of the dye into the conduction band of the TiO_2, affecting charge separation and producing a photovoltaic effect.

For the QD-sensitized cell, QDs are substituted for the dye molecules; they can be adsorbed from a colloidal QD solution or produced *in situ*. Successful PV effects in such cells have been reported for several semiconductor QDs including InP, CdSe, CdS, and PbS.

Possible advantages of QDs over dye molecules are the tunability of optical properties with size and better heterojunction formation with solid hole conductors. Also, a unique potential capability of the QD-sensitized solar cell is the production of quantum yields >1 by impact ionization (inverse Auger effect). Dye molecules cannot undergo this process.

Efficient inverse Auger effects in QD-sensitized solar cells could produce much higher conversion efficiencies than are possible with dye-sensitized solar cells.

Quantum Dots Dispersed in Organic Semiconductor Polymer Matrices

Photovoltaic effects have been reported in structures consisting of QDs forming junctions with organic semiconductor polymers. In one configuration, a disordered array of CdSe QDs is formed in a hole-conducting polymer—MEH-PPV (poly(2-methoxy, 5-(2t-ethyl)-hexyloxy-p-phenylenevinylene).

Upon photoexcitation of the QDs, the photogenerated holes are injected into the MEH-PPV polymer phase, and are collected via an electrical contact to the polymer phase. The electrons remain in the CdSe QDs and are collected through diffusion and percolation in the nanocrystalline phase to an electrical contact to the QD network.

Initial results show relatively low conversion efficiencies but improvements have been reported with rod-like CdSe QD shapes embedded in poly(3-hexyl-thiophene) (the rod-like shape enhances electron transport through the nanocrystalline QD phase). In another configuration, a polycrystalline TiO_2 layer is used as the electron conducting phase, and MEH-PPV is used to conduct the holes; the electron and holes are injected into their respective transport mediums upon photoexcitation of the QDs.

A variation of these configurations is to disperse the QDs into a blend of electron and hole-conducting polymers. This scheme is the inverse of light-emitting diode structures based on QDs.

In the PV cell, each type of carrier-transporting polymer would have a selective electrical contact to remove the respective charge carriers. A critical factor for success is to prevent electron-hole recombination at the interfaces of the two-polymer blends; prevention of electron-hole recombination is also critical for the other QD configurations mentioned above.

All of the possible QD-organic polymer photovoltaic cell configurations would benefit greatly if the QDs can be coaxed into producing multiple electron-hole pairs by the inverse Auger impact ionization process. This is also true for all the QD solar cell systems described above. The most important process in all the QD solar cells for reaching very high conversion efficiency is the multiple electron-hole pair production in the photoexcited QDs, and the various cell configurations simply represent different modes of collecting and transporting the photogenerated carriers produced in the QDs.

Silicon Quantum Dot/Crystalline Silicon Solar Cells

Silicon (Si) quantum dot (QD) materials are a promising PV structure that have been used for 'all-silicon' tandem solar cells. Si nanocrystals embedded in a dielectric matrix are of interest in the field of silicon optoelectronics, and are part of third-generation photovoltaics.

Successful fabrication of (n-type) Si QD/(p-type) c-Si photovoltaic devices is an encouraging step towards the realization of all-silicon tandem solar cells based on Si QD materials.

For a terrestrial solar spectrum (AM1.5G, 1000 W m−2) the optimal bandgap of the top cell required to maximize energy conversion efficiency is 1.7-1.8 eV for a two-cell tandem with a crystalline Si (c-Si) bottom cell. By stacking solar cells of different bandgaps on top of one another, with the highest bandgap cell uppermost, light is automatically filtered as it passes through the stack. Each cell absorbs a slice of the solar spectrum, with photons below the bandgap passing through to underlying cells.

To date, considerable material fabrication work has been done on the growth and characterization of Si QDs embedded in dielectric matrices such as oxide and nitride and in a SiC wide bandgap material matrix.

When very small Si nanocrystals are made (<7 nm in diameter), they behave as quantum dots (QDs) due to the three-dimensional quantum confinement of the carriers.

In indirect bandgap semiconductors optical transitions are allowed only if phonons are absorbed or emitted to conserve the crystal momentum. The localization of electrons and holes inside a QD leads to reflections or folding of phonons in k-space. This relaxes the k-conservation requirement and creates a quasi-direct bandgap. Measurements of photoluminescence (PL) from Si QDs in silicon oxide (SiO_2) have shown a 1.7 eV lowest energy transition from 2 nm QDs: effectively a 1.7 eV bandgap.

Measurements also show that there is a large increase in PL intensity as the QD size decreases, which is consistent with the increase in radiative efficiency with the onset of pseudo-direct bandgap behavior.

The photoluminescence peaks from Si QDs in nitride are more blue-shifted than that of Si QDs in oxide and range from 1.5 to 3.0 eV. The confined energy levels of Si QDs in a carbide matrix are quite similar to those of Si QDs in an oxide matrix.

Solar cells consisting of phosphorus-doped Si QDs in a SiO_2 matrix deposited on p-type crystalline Si substrates (c-Si) have been fabricated, where Si QDs are formed by alternate deposition of SiO_2 and silicon-rich SiOx. The deposition is done by RF magnetron co-sputtering, followed by high-temperature annealing. To get the required current densities through the devices, the dot spacing in the SiO_2 matrix is set at 2 nm or less.

Current tunneling through the QD layer is observed from the solar cells with a dot spacing of 2 nm or less. Voc increases proportionally with reductions in QD size, which may relate to a bandgap widening effect in Si QDs or an improved heterojunction field allowing a greater split of the Fermi levels in the Si substrate.

Although Si QD cells have very high theoretical efficiency, the finished devices do exhibit low efficiency, which must be overcome before they become widely used in 21st century markets.

Nevertheless, the work done with this technology is leading to improvement of materials and processes. Some segments of this work are used in the development of other technologies as well.

Quantum Dot PECs

The novel QD PEC devices are made of a thin layer of CdSe quantum dots deposited on a conductive layer on top of glass substrate that can be used as a photo-anode in photo-electrochemical solar cells. This configuration can yield photovoltage of approximately 675 mV and about 2 mA cm² in short circuit photocurrent.

The response of the QD based photoactive electrodes correlates with the absorption spectra of the QDs, which is easily tuned by size modification. Similar geometry based on acid treated CdS quantum dots, provides a photo-cathode with complementary photovoltage.

A tandem photo-electrochemical cell consisting of the two photoactive electrodes exhibits Voc of 816 mV. The tandem configuration enables achieving the highest photovoltage reported to date for QD based photo-electrochemical cells. The current matching requires optimization of the spectral response and the charge collection efficiencies of the two photoactive electrodes.

This is simply a new approach to the utilization of

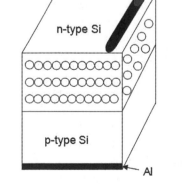

Figure 4-21. n-type Si QD on p-type c-Si

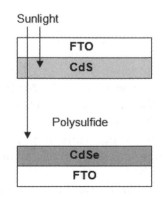

Figure 4-22. Schematic representation of quantum dot (QD) based tandem PEC, illuminated through the photo-cathode.

QDs in photo-electrochemical cells, on the path towards efficient and low cost photovoltaics.

Quantum Well Solar Cells

Energy conversion efficiency is the key parameter for all photovoltaic technologies since it directly impacts device and deployment costs. The efficiency of standard wafer processed solar cells is still below the theoretical limit of photovoltaic energy conversion 86%. One possibility to improve the conversion efficiency of Si based solar cells is the introduction of multiple absorbers consisting of Si/SiO_2 quantum wells, QWs.

Quantum well solar cells (QW) use nanometer-width layers (or quantum wells) comprised of different photovoltaic materials. By carefully controlling which photovoltaic materials are used and the widths of the quantum wells, the absorptive properties of the solar cell can be tailored to make more efficient use of the incident sunlight.

These materials are crystalline in structure, meaning that they are comprised of ordered, repeating sequences of atoms (often referred to as a crystal lattice) and are commonly characterized by the distance between atoms (or lattice constant).

To produce solar cells of maximum possible efficiency, both of the photovoltaic materials used must be made to conform to the same lattice constant, a challenge that has been the subject of intensive research by a number of institutions. The result of this work is the strain-balanced solar cell, in which layers of photovoltaic material under compressive strain (atoms forced closer together than in the original crystal structure) are alternated with layers of material that are under tensile strain (atoms pulled further apart than in the original structure)—effectively giving all the layers in the solar cell the same lattice constant.

The quantum well solar cell represents the most efficient nanostructured solar cell achieved to date. A series of quantum wells are incorporated into a p-i-n diode and serve to extend the absorption of the cell. This can be useful when engineering multi-junction solar cells, as the quantum wells allow the absorption profile of each sub-cell to be adjusted independently to the lattice parameter. Recent work has mainly focused on optimizing the quantum well growth conditions and incorporating the cell into a multi-junction device.

The quantum well solar cell bears some similarity to the intermediate band solar cell, but with some important differences regarding the density of confined states and in the extent to which the carrier population in the confined well states are in equilibrium with those in the barrier material.

In a QW, an intrinsic region is inserted into a conventional p-n solar cell to extend the field-bearing region. The quantum wells extend the absorption below the bulk band-gap Eg to threshold Ea. The carriers produced by the extra photons absorbed in the well "thermalize up" to the bulk band-edge and contribute extra current with quantum efficiency close to unity at room temperature.

A number of companies are working to bring these high efficient devices into commercial production. 28.3% efficient quantum well solar cells were reported recently—a world record for solar cells of its class. With this achievement, the technology is undergoing rapid growth and variations of it are already seen on the energy market.

In the Si/SiO_2 QW absorber material, charge carrier confinement results in effective band gap energies well above the bulk band gap of 1.1 eV, tunable by the Si QW thickness. The Si based tandem cell has a theoretical efficiency limit of 60%.

Several approaches have been undertaken to fabricate high quality Si quantum structures such as Si/SiO_2 QWs and Si quantum dots in a SiO_2 matrix. The band gaps in these structures are in the range of 1.1 eV (bulk Si) to 1.6 eV for QWs and 1.7 eV for crystalline quantum dots, respectively. Current research also focuses on Si quantum structures embedded in alternative host matrices such as Si3N4 or SiC.

Quantum confinement in Si nanostructures, however, raises the fundamental dilemma of sufficient charge extraction necessary for efficient photovoltaic devices. Here, we explore a possible solution by investigating one-dimensional (1D) confinement in recrystallized Si layers and the extraction of charges within the nonquantized dimensions. This means that the carrier extraction takes place parallel to the Si/SiO_2 interfaces of two-dimensional QWs while 1D confinement is sustained in the vertical direction.

It is shown that the developed lateral contact scheme is able to provide four orders of magnitude-enhanced conductivity compared to Si/SiO_2 QWs with standard vertical contacts where the charge transport is limited by insulating SiO_2 barriers.

To analyze the optical properties, Si/SiO_2 QWs with varying layer thicknesses were deposited by remote plasma-enhanced chemical vapor deposition on transparent and insulating sapphire substrates and were subsequently annealed at 1100°C for 30 s by using rapid thermal annealing to achieve partial recrystallization.

Lateral contacts to the deposited Si/SiO_2 QW

layers were fabricated by photolithography and subsequent reactive ion etching of 100-microns-wide mesas and 80-microns-square contact holes into the QW material with a pitch of 10 microns.

The holes were filled with metal followed by a lift-off process, thus defining the contact areas. For the generation of an internal electric field, adjacent contact holes were filled with Al and Pt, respectively, resulting in a Schottky barrier induced band bending.

Finally, a post metallization annealing at 400°C for 30 min was performed, which improves the contact resistances. Dark and illuminated current voltage (I-V) measurements were carried out by using an HP 4156A semiconductor parameter analyzer. Photovoltaic measurements were conducted utilizing a solar simulator providing a spectrum close to the standard AM 1.5 solar irradiation.

To evaluate the band gap energy of the QW material further, the optical absorption in the fabricated QWs was characterized by photothermal deflection spectroscopy (PDS).

All of the spectra show two distinct regions of different exponential slopes separated by the band gap energy. Below the band gap, defect-induced tail states within the band gap (Urbach tail) account for the absorption signal below the band gap energy. For energies above the band gap, e.g., 1.14 eV for the 60 nm, 1.25 eV for the 5 nm, 1.3 eV for the 3 nm, and 1.55 eV for the 2 nm QW samples, the slope of the absorption curve increases due to the higher density of states above the energy gap. The distinct blue shift of the absorption spectra with decreasing QW thicknesses can be attributed to the quantum confinement effect.

In addition, photoluminescence (PL) measurements have also shown the same behavior of an increase in the band gap-related PL peak energy with decreasing QW thickness. A comparison of the room temperature PL spectra with the well known effective mass approximation (EMA) reveals a good correlation between the measured PL peak energies and the EMA model if the excitonic nature of radiative recombination is considered.

For the calculation of the effective band gap and the excitonic binding energy, the effective masses for electrons (me=1.1mO) and holes (mh =0.5mO) known for bulk Si were used. Both absorption and PL measurements have proven the ability to engineer the band gap in the Si/SiO$_2$ QW system which is needed for the implementation of an all Si tandem solar cell.

The obtained I-V curves are linear, i.e., the current voltage characteristics are dominated by the series resistances RS/RS values determined from the slope of the I-V characteristic. The increase in RS in the QW samples compared to the single crystalline SOI layer is attributed to amorphous grain boundaries limiting the current transport in the recrystallized QW material.

The further increase in RS with reduced QW thickness between 50 nm QW and 4 nm QW is related to the decrease in crystallinity of the Si layers, i.e., 90% in the 50 nm Si layers and only 10% in the 4 nm QWs, as determined by Raman measurements.

Due to the high series resistances, the short circuit currents are limited to 2 x10^{-2} mA/cm^2 for the SOI reference and to the 10^{-4} mA/cm^2 range for the QW samples.

Furthermore, the open circuit voltages Voc raise, if the Si QW thickness is reduced. Voc increases from 20 mV in the SOI sample and 40 mV in the 50 nm QW to 191 mV in the 4 nm QW sample.

The enhancement can be directly attributed to the increase in the band gap energy in ultrathin QWs. With increasing band gap, the difference between the Fermi level and conduction band EC-EF as well as between the Fermi level and valence band EV-EF grows.

As a consequence, the Schottky barrier height at both contacts increases, reducing the reverse currents JR in the Schottky contact QW cells according to the thermionic emission formula,

$$J_R = A^* T^2 \exp\left(-\frac{q\Phi_B}{kT}\right) \tag{4-1}$$

with the effective Richardson constant A*, temperature T, and the Boltzmann constant k.

The relation between reverse current JR and open circuit voltage Voc is then

$$V_{oc} \cong \frac{kT}{q} \ln\left(-\frac{J_{ph}}{J_R}\right) = \Phi_B - \ln\left(\frac{A^* T^2}{J_{ph}}\right) \tag{4-2}$$

with the photocurrent density Jph.

Assuming every photon generates an electron hole

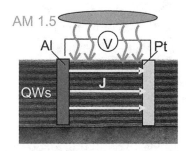

Figure 4-23. Cross section of an Al/QW/Pt Schottky cell structure. (14)

pair contributing to the photocurrent, the theoretical dependence of Voc from the band gap using Eq. 1 and 2 has been calculated by using literature values for the Richardson constant for electrons A* = 120 A cm−2 K−2) and holes (A* = 40 A cm−2 K−2).

For the SOI and the 50 nm Schottky cell structure, a barrier height of 0.56 eV for the band offsets at the Al/ Si and the Pt/Si interface has been found in accordance with literature values.

For the samples with QW thicknesses below 50 nm, the band gap shift extracted from the PDS measurements was added to the barrier height values found for the metal/bulk-Si interface.

For a first estimate, we considered the Schottky barriers to increase equally for the valence and conduction bands leading to an overall growth of the barrier height at both contacts by 0.5 dEG.

In addition, the ideal photocurrent density Jph for all samples is calculated by convoluting the measured QW samples or known bulk S SOI sample, absorption spectra and the AM 1.5 solar spectrum.

The excellent agreement between measured and calculated Voc values supports the conclusion that Voc rises due to a quantum confinement-induced shift of the band gap. Furthermore, the apparently low Voc values in the samples investigated are a direct consequence of high reverse currents in the Schottky contact solar cells as well as the low absorption limiting the theoretically obtainable photocurrent density Jph.

As mentioned above, Jsc is mainly determined by the high series resistances. According to the one diode model, the short circuit current V=0 is given by:

$$J_{sc} = J_R \left[\exp\left(-\frac{J_{sc}R_S}{kT} \right) - 1 \right] - J_{ph} \qquad (4\text{-}3)$$

Applying the same parameters (JR, RS, and Jph) already used for the calculation of Voc. There is a convincing agreement between the measured and calculated Jsc values supporting the validity of a Schottky barrier-induced electric field as a driving force in this solar cell structure.

The data demonstrate that the high series resistances along with the insufficient absorption limiting Jph are the main obstacles for a more pronounced benefit of quantum size effects in the fabricated QW based solar cells so far. The principle idea of separating the directions of confinement and extraction seems to be confirmed.

In conclusion, Si QW solar cells were developed by using a lateral transport concept along with an internal field provided by Schottky contacts. It is shown that the open circuit voltage, and thus conversion efficiency, can be enhanced due to confinement effects in a Si based QW absorber.

Although in this preliminary work the lateral conductivity is still too low for efficient photovoltaic energy conversion, the path to an improved QW solar cell is clear. An optimized lateral device should be comprised of fully recrystallized and weakly doped QWs to minimize series resistance and highly doped polycrystalline silicon instead of Schottky contacts resulting in a p-i-n structure to minimize reverse currents which should yield Voc values potentially above those of standard Si solar cells.

For optimum efficiency of the desired QW/bulk Si tandem cell, the number of Si layers needs to be adjusted in such a way that identical photocurrents Jph are obtained for the QW absorber and the underlying bulk Si.

With these optimizations, it should be possible to benefit from the quantum confinement in fully recrystallized Si layers, which represents a major development toward highly efficient, all-silicon tandem solar cells.

Advantages of QWSCs

Thermo-photovoltaics (TPV) is the generation of electricity in low-band-gap solar cells from the radiant energy emitted by conventional sources of heat. Gas, coal, wood, nuclear fuel and petrol all burn at temperatures in the range 500±2500 K, radiating energy over a relatively broad spectrum like the sun, but at longer wavelengths. This can be converted into electricity by low band gap cells. Recent interest has been stimulated by the possibility of surrounding the source by an "emitter" which re-radiates in a narrower spectrum rather like an old-fashioned gas mantle.

The band-gap of the solar cell should be chosen to be just below this narrow emission band for maximum efficiency of electricity generation. QW solar cells offer a significant advantage in this field, since the band-gap can be tuned simply by changing the well width. It is also important for TPV applications because QW enhance voltage compares to single bandgap cells made from the well material. The improved temperature dependence of the QW devices compared to conventional cells made from either the well or barrier material is also important for TPV applications, as the cells are close to the source and must operate at high temperature.

Lattice-matched quantum wells have been shown to enhance the efficiency of AlGaAs (Aluminum Gallium Arsenide), InP (Indium phosphide) and GaInP (Gallium Indium Phosphide) bulk cells and to enhance the output voltage compared to homogeneous cells made from the well material, by more than expected from confinement.

Using QW technology for Concentrator Solar Cells

As part of a CO_2 reduction strategy, the International Energy Agency recently set a goal of attaining 1TW of peak power capacity from solar electricity by 2050. All photovoltaic technology sectors must grow rapidly to meet this goal. Yet solar panel manufacturing already uses more silicon than the entire microelectronics industry. In addition, though the technology can scale to terawatt levels, expanding the manufacturing infrastructure is very capital-intensive.

Concentrator photovoltaic systems may provide an alternative solution. By using mirrors or lenses, they focus sunlight onto small, highly efficient solar cells. This shifts the manufacturing burden from semiconductors to metal and glass—materials that have more established manufacturing industries.

Moreover, recent improvements in concentrator cell efficiencies suggest that this approach may be cost-effective and rapidly scalable. Single junction quantum well solar cells with an efficiency of 27.3%, the highest value for any nanostructured solar cell to date, were recently demonstrated.

It is also very close to the highest efficiency recorded for a single junction cell (27.8%). The device is made of a p-i-n structure with a gallium arsenide phosphide (GaAsP) indium gallium arsenide (InGaAs) multi-quantum well stack grown in the i-region.

The lower band-gap InGaAs layer is compressively strained, while the GaAsP barrier layer is under tensile strain. Therefore, a judicious choice of composition and layer thickness results in a stack of quantum wells where each GaAsP/InGaAs bilayer exerts no net force on neighboring layers.

By incorporating strained semiconductors into a solar cell without introducing structural defects, the team was able to adjust the absorption threshold for the solar cell. The single junction cell has an absorption edge at 1.33eV, which is fundamentally better matched to the solar spectrum than a 1.42eV gallium arsenide cell. The low defect density allows cells to become increasingly radiatively efficient at concentrator intensities, exhibiting photon recycling effects that further boost efficiency.

Adjusting the absorption threshold becomes particularly important when fabricating highly efficient, multi-junction solar cells. Here the broad solar spectrum is absorbed using a series-connected stack of sub-cells with different bandgaps.

While this structure can lead to very high efficiencies, it requires careful control of the absorption threshold of each sub-cell. The series connection ensures that the lowest photocurrent in the stack will limit the photocurrent in the entire cell.

By growing defect-free, strain-balanced stacks of material, we can adjust the absorption threshold of the junctions without relaxing the semiconductor lattice or growing optically thin junctions. This strain-balanced approach makes feasible double junction efficiencies greater than 34% and triple junction efficiencies up to 42% under solar concentration. Work is underway to achieve these goals.

Summary

The electronic properties of QDs currently limit device performance. To exceed power-conversion efficiencies of 10% in a single-junction planar cell, a material's electron and hole mobility should exceed 10–1 cm² V–1 s–1. Another weak point is their bandgap, which should be as trap-free as possible.

A lot of progress has been made towards this objective through advances in the packing and passivation of QDs in thin solid films and many new device architectures have much to offer the field.

Nanostructured interfaces at the electron-extracting electrode would enable improved electron extraction by allowing a greater volume of light-absorbing colloidal quantum dot films to be incorporated into a device without compromising the internal quantum efficiency. This will eventually lead to greater overall absorption and performance efficiency.

Much more work needs to be done, through systematically engineering high-electron-mobility electrodes such as nanopillars, nanowires and nanopores.

Enhancing photon absorption from a given amount of colloidal quantum dot film offers significant prospects for increasing the available photocurrent. Enhancing photon absorption from a given amount of QD film offers significant prospects for increasing the available photocurrent.

New nanostructured metals exhibiting particularly strong optical scattering, which increases the interaction of light inside the semiconductor relative to today's specular, planar double-pass metals will enhance the processes further.

Plasmonic enhancements, which involve concentrating the optical field in the light absorber, have been exploited in silicon and organic solar cells, and will also provide significant benefits in the quantum-tuned field once suitable-aspect-ratio nanoparticles are tailored to the spectral requirements and the dielectric environment of colloidal quantum dot films.

The urgent need for cost-effective, efficient solar-harvesting solutions creates a powerful motivation for the quantum-tuned photovoltaics community to continue pursuing rapid advances. The scientific disciplines and concepts on which researchers draw are remarkably broad; spanning over the spectrum of organic and inorganic materials chemistry and processing, electronic materials transport, trap spectroscopy, and photonic-electronic device engineering.

It suffices to say that QDs are an exciting, albeit complex, technology with many variations and countless possibilities, and that the field's progress and the further advancement all rely on the interdisciplinary perspective of its growing community of active researchers.

Unfortunately, persistently low efficiencies keep the QD technology off the energy markets for now, but we hope that some of the components of the ongoing effort will contribute to the development of other solar technologies in the 21st century.

PLASMONIC PV TECHNOLOGY

Plasmonic Solar Cells (PSC) are a fairly new development with a great potential which has generated excitement in the scientific community. Their potential as photovoltaic devices has been investigated and many universities and R&D institutions are dedicating a lot of time and effort to their efficiency improvement and potential commercialization. This is why we have included them in this chapter and consider them one of the promising (albeit still under development) solar technologies which will play a large role in our energy future.

For a PV technology to be a viable energy source and to be able to compete with fossil fuels, the price needs to be reduced several-fold. Presently, over 60% of the solar market uses silicon based solar cells built on mono- or poly-crystalline silicon wafers. These wafers are typically 200-300μm thick and represent over 40% of the cost of the resulting PV cells, thus the industry is fully dependent on the cost of silicon raw material.

PSC have great potential in driving the cost of solar power down, because they use 1-2μm-thick thin films. The substrates these films are deposited on (glass or plastics) are much cheaper than silicon and most other solar materials. Present-day PSC, however, suffer from the problem common to all thin films-based PV devices—they just don't absorb as much sunlight as conventional PV technologies. Therefore, discovering

and implementing new, more efficient methods for trapping light on the surface or in the bulk is critical and will determine the success of PSCs.

Plasmonic technology uses engineered metal structures (nano-particles) to guide light at distances less than the scale of the wavelength of light in free space. Plasmonics can improve absorption in photovoltaics and broaden the range of usable absorber materials to include more earth-abundant, non-toxic substances, as well as to reduce the amount of material necessary.

Metals support surface plasmons that are the collective oscillation of excited free electrons, which are characterized by a resonant frequency.

A number of methods for optimizing the light trapping efficiency of PV cells have been developed. Some involve creation of pyramids on the surface which are actually larger in size than most thin films used for making PV cells. Other methods roughen the surface of the substrate by growing SnO_2, ZnO and such on the active surface, the depth of which is on the order of the incoming wavelengths. Although this method increases the generated photo-current, the thin films are of poor material quality and present latent problems, resulting in power deterioration or field failures.

A novel approach receiving a lot of attention lately is to scatter the incoming light using metal nanoparticles, which are excited at their surface plasmon resonance. See Figure 4-24.

Author's note: Surface plasmon resonance is the collective resonant oscillation of the valence electrons in a solid material, when stimulated by sunlight.

The plasmon resonance effect is initiated if and when the frequency of impinging sunlight photons is close to the natural frequency of the surface electrons, which usually oscillate against the restoring force of the positive nuclei.

Electrons, or visible or IR light beam can be used to

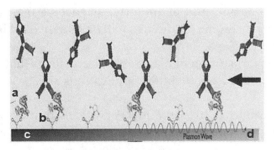

Figure 4-24. Surface plasmon resonance
a. Conjugated ligand
b. Dextran
c. Metal surface
d. Plasmon wave

initiate the surface plasmon resonance effect. The only condition is that the momentum of the incoming light beam must match that of the plasmon.

By manipulating the geometry of the metallic structures, the surface plasmon resonance or plasmon propagating properties can be tuned depending on the applications. The resonances of noble metals are mostly in the visible or infrared region of the electromagnetic spectrum, which is the range of interest for photovoltaic applications.

The surface plasmon resonance is affected by the size, shape and dielectric properties of the surrounding medium. Silver and gold have dominated experimental research in this area although other metals, such as copper, titanium, and chromium also support surface plasmons.

There are several types of plasmonic PV cells, and we'll review those which consist basically of metal (or other) nanoparticles scattered on the active surface. This increases the Raman scattering ability to generate additional photons, which then excite the surface plasmons. This excitation causes electrons to be excited too, and by virtue of the acquired energy to travel through the thin film material thus generating electric current, which is the goal of this process sequence.

The design of PSC PV devices varies depending on the method used to trap and scatter light across the active surface and through the bulk to generate and capture maximum photons and generate the greatest amount of electric energy per unit active area and solar influx level. Maximizing the number and energy levels of the power-generating electrons is the emphasis of R&D efforts today, seeking to make them a viable competitor to conventional technologies.

Other important goals of PSC PV development and commercialization are lowering the amount and cost of materials used in their manufacturing, while increasing the energy production per unit area. These factors are interconnected and as material sciences advance, we believe plasmonic PV devices will be widely used in 21st century markets.

Several different plasmonic mechanisms that could be utilized for photovoltaic applications are:

1. Scattering from the metal particles that act as dipoles (far-field effect),

2. Near field enhancement, and

3. Direct generation of charge carriers in the semiconductor substrate.

Most significant photocurrent enhancements which result in inorganic devices are explained by the first mechanism of scattering, and for organic devices by near field enhancement.

Plasmonic Device Function

When a photon is excited enough in the substrate of a PV device, an electron and hole pair is created and the electron is freed. If more photons are available, more electrons will be created, increasing efficiency of the PV device. Note how the photons in the nanoparticles plasmonic cell in Figure 4-24 bounce around and have more chance to create more electron-hole pairs.

After the separation, however, the electrons and holes have a tendency to recombine, due to their opposite charge. For the device to work, electrons must be collected prior to the recombination—which is very fast—and if the extraction process is fast enough, then the PV device is efficient. One way to speed up the collection process is to make the conducting material as thin as possible. Thinner material ensures less sunlight is absorbed by the bulk of the device, because more light is absorbed by thicker substrate. Also electrons have less bulk to travel to escape the area and be collected in an outside circuit.

The two most important mechanisms that have been proposed to explain photocurrent enhancement by metal particles incorporated into or on solar cells are *light scattering* and *near-field concentration of light*.

The contribution of each mechanism depends mostly on particle size, how strongly the semiconductor absorbs, and the electrical design of the solar cell. Here we focus on scattering by metal particles as a means of enhancing light trapping into thin-film solar cells, because it is the mechanism behind the enhancement in most of the recent experimental work.

Metal nanoparticles are strong scatterers of light at wavelengths near the plasmon resonance, which is due to a collective oscillation of the conduction electrons in the metal. For particles with diameters well below the wavelength of light, a point dipole model describes the absorption and scattering of light well.

The scattering and absorption cross-sections respectively are given by (5):

$$C_{scat} = \frac{1}{6\pi} \left(\frac{2\pi}{\lambda}\right)^4 |\alpha|^2, \; C_{abs} = \frac{2\pi}{\lambda} \, \mathrm{Im}[\alpha] \qquad (4\text{-}4)$$

where

$$\alpha = 3V \left[\frac{\varepsilon_p/\varepsilon_m - 1}{\varepsilon_p/\varepsilon_m + 2}\right] \qquad (4\text{-}5)$$

is the polarizability of the particle. Here V is the particle volume, εp is the dielectric function of the particle and εm is the dielectric function of the embedding medium. We can see that when εp = −2εm the particle polarizability will become very large. This is known as the surface plasmon resonance. At the surface plasmon resonance, the scattering cross-section can well exceed the geometrical cross section of the particle.

For example, at resonance a small silver nanoparticle in air has a scattering cross-section that is around ten times the cross-sectional area of the particle. In such a case (first order), a substrate covered with a 10% areal density of particles could fully absorb and scatter the incident light. For light trapping it is important that scattering is more efficient than absorption, a condition that is met for larger particles, as follows from Equation (4-4).

Typically, an Ag particle with a diameter of 100 nm has an albedo (scattering cross section over sum of scattering and absorption cross sections) that exceeds 0.9. At these larger sizes, dynamic depolarization and radiation damping become important corrections to the quasi-static expressions given in Equations (4-4) and (4-5). Furthermore, for larger size the excitation higher-order plasmon modes (quadrupole, octupole) must be taken into account.

Dynamic depolarization occurs because as the particle size increases, conduction electrons across the particle no longer move in phase. This leads to a reduction in the depolarization field (which is generated by the surrounding polarized matter) at the centre of the particle. As a result, there is a reduced restoring force and hence a red-shift in the particle resonance.

For particle sizes where scattering is significant, this re-radiation leads to a radiative damping correction to the quasi-static polarizability, the effect of which is to significantly broaden the plasmon resonance. The red-shift and broadening of the resonance with increased particle size would generally be expected to be an advantage for solar cell applications, since light-trapping should occur over a relatively broad wavelength range and at wavelengths that are long compared with the quasi-static values of the surface plasmon resonance wavelengths of noble metal particles.

While an increased size leads to a larger absolute scattering cross section, these effects do lead to a reduced cross section when normalized by size. Inclusion of dynamic depolarization and radiative damping effects can give reasonably accurate predictions of many features of the extinction spectra for larger particles in cases where the contribution of higher-order multipoles can be neglected.

Particle shape also plays an important role in the effect of metal nanoparticles on a photovoltaic device. Particle shapes such as disks that have a large fraction of their volume close to the semiconductor can lead to a very high fraction of light scattered into the substrate. Conversely, Sundararajan et al. have shown that nanoparticle aggregates can lead to a reduction of photocurrent, a point that must be considered in colloidal fabrication of nanoparticle assemblies. They have also shown that nanoshells can lead to optical vortexing, which resulted in a reduction in photo-generated current.

For metals with low interband absorption, the dielectric function can be described by the Drude model, which describes the response of damped, free electrons to an applied electromagnetic field of angular frequency ω:

$$\varepsilon = 1 - \frac{\omega_p}{\omega^2 + i\gamma\omega} \tag{4-6}$$

Here ωp is the bulk plasmon frequency, given by $\omega^2\,p = Ne^2/m\varepsilon o$ where N is the density of free electrons, e is the electronic charge, m is the effective mass of an electron and ε_0 is the free-space dielectric constant.

Inserting Eq. (4-6) in Eq. (4-7) leads (in free space) to:

$$\alpha = 3V \frac{\omega_p^2}{\omega_p^2 - 3\,\omega^2 - i\gamma\omega} \tag{4-7}$$

Thus the surface plasmon resonance frequency for a sphere in free space occurs at $\omega_{sp} = \sqrt{3}\,\omega_p$ and mainly depends on the density of free electrons in the particle. The density of free electrons is highest for aluminum and silver, leading to surface plasmon resonances in the ultra-violet, and lower for gold and copper, leading to surface plasmon resonances in the visible. The resonance frequency can be tuned by varying the dielectric constant of the embedding medium: a higher index leads to a red-shift of the resonance.

As an example, at the surface plasmon resonance for a silver nanoparticle, the scattering cross-section is about 10x the cross-section of the nanoparticle. The goal of the nanoparticles is to trap light on the surface of the PV cell. The absorption of light is not important for the nanoparticle, rather, it is important for the PV device. One would think that if the nanoparticle is increased in size, then the scattering cross-section becomes larger. This is true, however, when compared with the size of the nanoparticle, the ratio $CS_{scat}/CS_{particle}$ is reduced.

Particles with a large scattering cross section tend to have a broader plasmon resonance range.

Wavelength Dependence

Surface plasmon resonance depends mainly on the density of free electrons in the particle. The order of densities of electrons for different metals is shown below along with the type of light which corresponds to the resonance. For example:

- Aluminum—ultraviolet
- Silver—ultraviolet
- Gold—visible
- Copper—visible

If the dielectric constant for the embedding medium is varied, the resonant frequency can be shifted. Higher indexes of refraction will lead to a longer wavelength frequency.

Light Trapping

The metal nanoparticles are deposited at a distance from the substrate to trap the light between the substrate and the particles. The particles are embedded in a material on top of the substrate. The material is typically a dielectric, such as silicon or silicon nitride. When performing experiments and simulations on the amount of light scattered into the substrate due to the distance between the particle and substrate, air is used as the embedding material as a reference.

It has been found that the amount of light radiated into the substrate decreases with distance from the substrate. This means that nanoparticles on the surface are desirable for radiating light into the substrate, but if there is no distance between the particle and substrate, then the light is not trapped and more light escapes.

Incident light excites surface plasmons, so surface plasmon frequency is specific for the material. Through the use of gratings on the surface of the film, different frequencies can be obtained. Surface plasmons are also preserved through the use of waveguides which make the surface plasmons easier to travel on the surface, minimizing losses due to resistance and radiation.

The electric field generated by surface plasmons influences electrons to travel toward the collecting substrate. As more electrons are collected, the plasmonic PV device becomes more efficient.

Developments in Plasmonic Solar Cells

The ability to concentrate and scatter light across the surface of the plasmonic solar cell is its key advantage, which enables it to operate at increased efficiencies. Use of metal wires on top of the substrate to scatter the light might be an added benefit, for it utilizes a larger area of the active surface for light scattering and absorption. The problem here is that the lines (which are used instead of dots) might increase the reflectivity, in turn decreasing the efficiency of the cells.

Researchers have discovered a photonic waveguide which collects light at a certain wavelength and traps it within the structure, which can contain 95% of the light impinging on the surface (vs. about 30% for other traditional waveguides.) This new waveguide can also control light within a single wavelength ten times better than traditional waveguides, by selecting the wavelength the device captures. By changing the structure of the lattice, and keeping the light until full absorption, the efficiency of the solar cell would be increased dramatically.

Improving the absorption of light in plasmonic solar cells is one of the major challenges and goals of the research today.

Applications of Plasmonics in Solar Cells

Plasmonic solar cells are useful devices, with potentially wide areas of applications, such as powering solar aircraft and space exploration vehicles. Their use in these and in terrestrial applications is most beneficial where reduced vehicle or structure weight is required.

Metal Nanoparticle Plasmonic Solar Cell

A common design for PSC manufacturing is depositing some type of metal nanoparticles on the top active surface of the thin film PV device, using the special (plasmonic) properties of the nano-particles to enhance the device performance.

The function of these devices is quite simple: when light hits the metal nanoparticles at their surface plasmon resonance, the light is scattered in many directions. This allows light to travel along the bulk and bounce back and forth between the substrate and the nanoparticles, thus enabling the PV device to absorb more light,

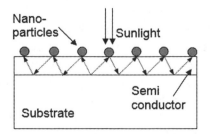

Figure 4-25. Nanoparticles plasmonic cell

increase the energy levels and generate more electricity.

Note: Surface plasmon resonance is the resonant oscillation of the valence electrons in a solid matter, when stimulated by incident light, if and when the frequency of light photons matches the natural frequency of surface electrons, which are oscillating against the restoring force of positive nuclei. In nano-structures it is actually called localized surface plasmon resonance.

Localized surface plasmon resonance in nano-structures has been shown to provide substantial efficiency enhancement in solar cells with a number of semiconducting materials and particle geometries.

Thin Film Plasmonic PV Device

One efficient method of using plasmons for generating solar energy is depositing a thin metal film on a substrate (glass or plastic) with a thin film of silicon or another semiconductor material on top. Incoming sunlight will travel through the thin layer of silicon and generate surface plasmons on the interface (Si-semiconductor). Here an electric field is generated in the silicon thin film, and electrons can move through the silicon and be collected in the outside circuit. If the thin film has nanometer size grooves in it, they will act as waveguides, thus exciting as many photons in the silicon thin film as possible.

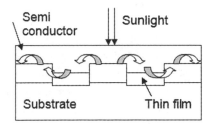

Figure 4-26. Thin metal film plasmonic cell

Thin film based on nanocrystalline silicon, amorphous hydrogenated silicon (aSi:H) and gallium arsenide (GaAs) are the most engineered solar cells to utilize plasmonics effects to enhance the absorption spectra. The use of nanoparticles at sub-wavelength scale in thin film helps to make ultra thin film solar cells. By applying a concept of nanoparticles for a p-n diode in aSi:H an increase in short circuit current density and overall energy conversion of over 8% has been achieved in an aSi:H cell.

An improvement of 8% for GaAs solar cells with 110 nm Ag nanoparticles fabricated using mask deposition through anodic aluminium oxide on glass has also been reported.

Plasmonics in Tandem PV

The plasmonics effect of metallic nanostructures and nanoparticles has also been reported for tandem solar cell. Ag nanoparticles have been engineered as a reflector for efficient carrier recombination and absorption enhancement.

By incorporating Au nanoparticles into Poly (3,4-ethylenedioxythiophene) polystyrenesulfonate (PEDOT:PSS) a short circuit current density enhancement of over 20% was also achieved in a polymer tandem solar cell in comparison with the reference cell 39.

Plasmonics in Organic PV

Works to enhance the performance of organic PV (OPV) using metallic nanostructures and nanoparticles have been reported. The works are dated as early as the 1990s using Al, copper phthalocynine and Ag thin layers, copper metallic cluster and silver island film.

In 2008 the OPV cell was fabricated using 30 nm thick silver anodes on a glass substrate by focused ion beam. By patterning only 16% of the surface an increase in overall power conversion by a factor of 3.2 was observed compared to an unpatterned surface. In other words, surface plasmon polariton of Ag films improved the photocurrent of an organic photodiode by a factor of 3 at resonance wavelength.

Numerous other works have shown that using colloidal nanoparticles made of Ag, Au or Cu shows an improvement in the absorption of organic solar cells. The effect of plasmonics has also been investigated in dye sensitized solar cells using Au nanoparticles and Ag nanoparticles 35. By pulse-current electrode position Ag nanoparticles have been integrated into polymer bulk heterojunction solar cells. The Ag-incorporated solar cell showed a conversion efficiency of 3.69% compared to the reference device's efficiency of 3.00%.

Summary

Surface plasmons provide substantial efficiency enhancement in solar cells, using a range of semiconducting materials and different particle geometries. The basic concepts behind light scattering from metal nanoparticles on a dielectric substrate are quite clear, but the actual manufacturing and practical application of these devices is still in the future.

Much more work remains to be done to determine the optimum particle distribution for solar cell applications, taking into account interactions between the particles as well as realistically including the interaction with the absorbing substrate.

The technologies using simple and inexpensive

methods for fabricating such nanostructured metal patterns over large areas already exist. Soft lithography manufacturing methods are well established, so it won't be long before we see plasmonic devices in broad use for electric power generation. Before then, however, they need to go through mass-production optimization and field testing, to determine the type and scale of use of the different approaches.

In spite of present drawbacks, the field of plasmonics PV brings a new dimension to the old solar power generation area, so we expect to see some types, or at least elements, of this sophisticated technology as active contributors to progress in the solar field in the 21st century.

And since we took a brief look at organic PV above, we will now take a much closer look.

ORGANIC PV TECHNOLOGY

History

Over the past two decades, the science and engineering of organic semiconducting materials have advanced rapidly, leading to a range of organic materials-based solid-state devices, including organic light-emitting diodes (OLEDs), field-effect transistors, photodiodes, and photovoltaic cells.

It began in the 1960s with fundamental studies on the optical and electronic properties of model organic molecules such as acenes—molecules based on up to five fused benzene rings. This area of research gained significant momentum in the late 1970s and in the 1980s, with the development of high-purity small organic molecules, with tailored structure and properties, which were synthesized and processed at room temperature into thin films by means of physical vapor deposition (PVD) techniques.

Tang et. al. developed single heterojunction organic photovoltaic cells in the mid-80s with power conversion efficiency of just over 1%, well over the first report of a device with similar geometry by Kearns and Calvin in 1958. In the 1990s the development of high-purity conjugated polymers allowed the fabrication of organic photovoltaic cells with materials processed from simple solution. (1)

In the 1990s, Grätzel and co-workers proved in practice that the use of nano-structured metal oxide electrodes allowed the fabrication of dye-sensitized photo electrochemical solar cells. Although the efficiencies of these devices were quite low from a commercial point of view, they were considered promising for use in some niche markets.

The development of high-purity conjugated polymers in the following decade allowed the fabrication of organic photovoltaic cells from chemical solutions. These developments have led to the rapid growth of a new field of "organic" or "excitonic" solar cells. The function of these cells is based on light absorption by their molecules to generate relatively localized excited states.

Some of these materials and processes are already considered for commercial use. Interest in these different technologies is enabling further advances in their materials and device manufacturing, and serious attempts are being made to improve and optimize their performance.

Low-temperature processing of organic small molecules from vapor phase, or from polymers from simple solutions provides organic semiconductors with a critical advantage over their inorganic counterparts, where high-temperature processing limits the range of substrates on which they can be deposited.

Background

The organic semiconductors can be manufactured on flexible plastic substrates that can lead to applications and consumer products with lower cost, highly flexible form factors, and light weight. Low-temperature processing reduces the amount of energy used during the manufacturing process, which reduces energy cost and payback time.

These qualities, combined with the ability to tune the physical properties of organic (macro) molecules by means of fine tuning their chemical structure, are another benefit and a main driver of research and industrial interest in organic photovoltaics.

A key difference in the physics of organic semiconductors compared to their inorganic counterparts is the nature of the optically excited states. The absorption of a photon in organic materials leads to the formation of an exciton, i.e., a bound electron-hole pair. The exciton binding energy is typically large, on the order of or larger than 500 meV (such binding energies represent 20 times or more the thermal energy at room temperature, as compared with that of inorganic semiconductors.

Consequently, optical absorption in organic materials does not lead directly to free electron and hole carriers that could readily generate an electrical current. Instead, to generate a current, the excitons must first dissociate. The excitonic character of their optical properties is a signature feature of organic semiconductors, which has impacted the design and geometry of organic photovoltaic devices for decades.

While the physics of conventional p-n junctions is reasonably well understood and solar cell properties can be derived from materials parameters and the nature of the electrical contacts with electrodes, the understanding of the underlying science of organic solar cells is far less advanced and remains an intense subject of research.

For instance, a key difference in the physics of organic semiconductors, compared to their inorganic counterparts, is the nature of the optically excited states. The absorption of a photon in organic materials leads to the formation of an exciton, i.e., a bound electron-hole pair. The exciton binding energy is typically large, on the order of or larger than 500 meV (such binding energies represent 20 times or more the thermal energy at room temperature, kT(300 K)¼ meV, to be compared with a few meV in the case of inorganic semiconductors.

Consequently, optical absorption in organic materials does not lead directly to free electron and hole carriers that could readily generate an electrical current. Instead, to generate a current, the excitons must first dissociate.

The excitonic character of their optical properties is a signature feature of organic semiconductors, which has impacted the design and geometry of organic photovoltaic devices for the past decades.

Organic materials have been subjects of intense research lately, with the most interesting being the molecular and polymeric semiconductors, Fullerene (C60) and its derivatives. These materials are used in organic LEDs and thin film transistors for use in smart cards. The advantage of using these materials is their manufacturing simplicity and the possibility of use on flexible materials.

These (liquid) organic solar devices consist of glass or other such substrate, coated with indium tin oxide (ITO) which is a transparent conductor and acts as the top electrode. Several layers of light-absorbing organic materials could be deposited on the ITO and then completed by depositing a metal back contact onto the organic materials.

Cell efficiencies of 5% have been achieved, but low efficiency and problems with large-scale manufacturing processes must be resolved before we see these technologies on the market.

Organic molecular and polymeric semiconductors can form films with complex morphologies and varying degrees of order and packing modes through the interplay of a variety of non-covalent interactions. Their molecular structure consistently presents a backbone along which the carbon (or nitrogen, oxygen, sulfur) atoms are sp2-hybridized and thus possess a p-atomic orbital.

The conjugation (overlap) of these p orbitals along the backbone results in the formation of delocalized p molecular orbitals, which define the frontier (HOMO and LUMO) electronic levels and determine the optical and electrical properties of the (macro) molecules. The overlap of the frontier p molecular orbitals between adjacent molecules or polymer chains characterizes the strength of the intermolecular electronic couplings (also called transfer integrals or tunnelling matrix elements), which represent the key parameter governing charge carrier mobilities.

In crystalline inorganic semiconductors, the three-dimensional character and rigidity of the lattice ensure wide valence and conduction bands and large charge carrier mobilities (typically on the order of several 10^2 to 10^3 cm^2 V^{-1} s^{-1}).

With inorganic semiconductors: a) the weakness of the electronic couplings, due to their intermolecular character, b) the large electron-vibration couplings, leading to pronounced geometry relaxations), and c) the disorder effects all conspire to produce more modest carrier mobilities.

This is due to charge-carrier localization and formation of polarons, so the transport mechanism then relies on polarons hopping from molecule to polymers' molecule. Thus, the charge carrier mobilities strongly depend on the morphology of the material and vary over several orders of magnitude when going from highly disordered amorphous films to highly ordered crystalline materials.

Function

In general, the process of conversion of sunlight into electricity by an organic solar cell is quite different from that of the conventional solar cells. It can be described by: a) absorption of a photon leading to the formation of an excited state (the bound electron-hole pair (exciton*) creation); b) exciton diffusion to a region where exciton dissociation (charge separation) occurs; c) charge transport within the organic semiconductor to the respective electrodes; and d) extraction into the outside circuit to be used as needed.

Note: *Exciton* is a state (a bound state, to be exact) of an electron and hole which are attracted to each other by the electrostatic Coulomb force. It is considered an electrically neutral (no charge) state, or a quasi-particle that exists in insulators, semiconductors and in some special liquids. See more details in the Excitonic PV chapter of this text.

The bad news is that organic materials have a large band gap (over 2 eV, while 1.1 eV is the ideal band gap), so only a small portion (about 30% of the incident solar light is absorbed.

The good news is that the absorption coefficients of organic materials are as high as 105 cm-1, so only 100 nm thick film is enough to absorb most of the photons, provided that a reflective back contact is used. This is why optimized "spectral" harvesting of photons is needed, which is achieved by lower band gap polymers and/or using energy-transfer cascades.

The primary photo-excitations in organic materials do not generate many free charge carriers (vs. 100% free carrier generation in the case of conventional solar cells). Instead, coulombically bound electron-hole pairs, called excitons, are created first.

So, very strong electric fields are needed for efficient dissociation of excitons, whose fields can be supplied via externally applied electrical fields, as well as through interfaces where, due to abrupt changes of the potential energy, strong local electrical fields are possible (E) -grad U).

Photo-induced charge transfer can occur when the exciton has reached such an interface within its lifetime. Therefore, exciton diffusion length limits the thicknesses of the bi-layers. Exciton diffusion length should be at the same order of magnitude as the donor acceptor phase separation length. Otherwise, excitons decay via radiative or non-radiative pathways before reaching the interface, and their energy for power conversion is lost.

OPV Cell Performance

Here, we discuss the electrical characteristics of organic solar cells and their performance in some detail. (1)

In the dark the solar cell works as a diode. As for conventional p-n solar cells, an organic photovoltaic device can be approximated by an equivalent circuit, comprised of: (i) a diode with reverse saturation current density Jo (current density in the dark at reverse bias) and ideality factor n; (ii) a current source (Jph), which corresponds to the photocurrent upon illumination; (iii) a series resistance (Rs), which must be minimized and takes account of the finite conductivity of the semiconducting material, the contact resistance between the semiconductors and the adjacent electrodes, and the resistance associated with electrodes and interconnections; and (iv) a shunt (Rp) resistance, which needs to be maximized and takes into account the loss of carriers via possible leakage paths. The latter include structural defects such as pinholes in the film, or recombination centers introduced by impurities.

(a)　Spectral photon flux density in the standardized AM 1.5 G illumination conditions and corresponding integrated current that would be produced if each photon contributes to current with unity efficiency.

(b)　Current-density voltage characteristics of a solar cell in the dark and under illumination.

(c)　Semi-logarithmic plot of the same electrical characteristics, illustrating the effects of the parasitic resistances RS and RP in forward and reverse bias.

(d)　Equivalent circuit used to model solar cells.

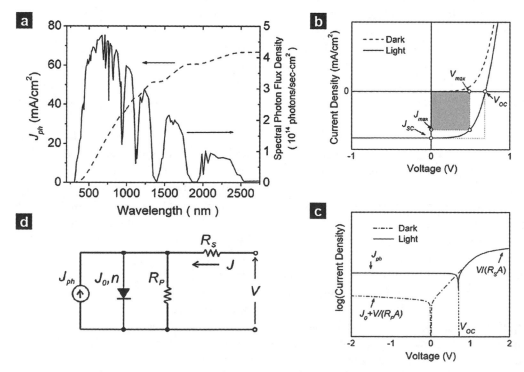

Figure 4-27. Optical and electrical properties of solar cells. (1)

Solving for this simple circuit provides the following analytical expression for the current-voltage characteristics, referred to as the Shockley equation:

$$J = \frac{1}{1 + R_S/R_P} \left[J_0 \left\{ \exp\left(\frac{V - JR_S A}{nkT/e}\right) - 1 \right\} - \left(J_{ph} - \frac{V}{R_p A}\right) \right]$$

(4-8)

where

e denotes the elementary charge,
kT is the thermal energy, and
A is the area of the cell.

Analysis of Eqn (4-8) in various regimes of photocurrent shows that the series resistance is the critical factor, especially in regimes of high photocurrents, Jph. From Eqn (4-8), equations for the open-circuit voltage VOC and the short circuit current density JSC can be derived:

$$V_{OC} = n\frac{kT}{e}\ln\left\{1 + \frac{J_{ph}}{J_0}\left(1 - \frac{V_{OC}}{J_{ph}R_p A}\right)\right\} \approx n\frac{kT}{e}\ln\left\{1 + \frac{J_{ph}}{J_0}\right\}$$

(4-9)

$$J_{SC} = -\frac{1}{1 + R_S/R_P}\left\{J_{ph} - J_0\left(\exp\left(\frac{J_{SC}R_S A}{nkT/e}\right) - 1\right)\right\} \approx -J_{ph}$$

(4-10)

Eqns (4-8)-(4-10) usually need to be solved numerically except for cases where Rs is very small and/or Rp sufficiently large so that the effect of Rs or Rp can be ignored; in such instances, the approximate expressions on the right-hand side of Eqns (4-9)-(4-10) apply.

When the device is under illumination, two quantities can be easily determined experimentally: the intersects of the electrical characteristics with the vertical and horizontal axes, which correspond to Jsc and Voc, respectively, see Figure 4-27b.

At any point on the electrical characteristic in the fourth quadrant (Jsc negative and Voc positive), the solar cell produces an electrical power density given by the product of voltage and current density. This product is maximized at a point that corresponds to voltage V_{max} and current density J_{max}. The power conversion efficiency h, which represents the most important metric for a photovoltaic cell, is then defined as:

$$\eta = \frac{J_{max}V_{max}}{P_{inc}} = FF\frac{J_{sc}V_{OC}}{P_{inc}}$$

(4-11)

where

P_{inc} is the incident power density, and
FF denotes the fill factor.

These parameters are illustrated in Figure 4-27b. The effects of the parasitic resistances on the shape of the current density/voltage characteristics are illustrated in Figure 4-27c. A finite value of the series resistance Rs limits the current density in forward bias, while a finite shunt resistance Rp is responsible for a dark current increase in reverse bias.

To characterize quantitatively the performance of solar cells for terrestrial applications, standardized illumination conditions are used in which the spectrum of the source simulates the solar spectrum (AM 1.5 G, see Figure 4-27a) and has an intensity on the order of $100mWcm^{-2}$; this corresponds to the average intensity of sunlight with an angle of incidence = 48 relative to the normal to the earth's surface (AM denotes the air mass; G stands for global and refers to a small contribution of diffuse light to the direct incident light). Importantly, according to Eqn (4-11), the power conversion efficiency of a solar cell is determined by three parameters: the fill factor, the short-circuit current density, and the open-circuit voltage.

We now examine the limits of these three parameters in organic solar cells and discuss how they relate to materials and contacts properties.

Fill Factor

The maximum value for the fill factor is a function of the open circuit voltage Voc and the ideality factor of the diode n (optimally equal to 1). As for inorganic solar cells based on p-n junctions, its maximum value can be described by the empirical expression:

$$FF_0 - \frac{V_{OC} - \ln\left(V_{OC} + 0.72\right)}{V_{OC} + 1}$$

(4-12)

where

Voc is a normalized voltage defined as Voc=eVoc/nkT. Eqn (4-12) is a good approximation for Voc values > 10.

For organic solar cells with Voc=0.5–1 V, and ideality factors in the range of n ¼ 1.5–2, this condition is satisfied. We note that in p-n junction-based cells, the various recombination mechanisms are taken into account by drawing equivalent circuits with two diodes, where the first diode (with n ¼ 1) describes radiative band-to-band recombinations while the second (with n ¼ 2) describes recombination via impurities with energy states within the band gap.

In organic solar cells, little is known at this stage about the specific recombination processes of excitons and carriers. In view of the higher disorder and impurity levels present in amorphous organic semiconductors compared to the pure grades of silicon wafers that can be fabricated, ideality factors that deviate from unity are expected; this negatively impacts both the maximum fill factor and the maximum open circuit voltage (for instance, for Voc=0.5 V, FFo drops from 0.80 to 0.69 when n goes from 1 to 2, to be compared with FFo=0.85 in monocrystalline silicon solar cells). It is worth mentioning that the impact of the parasitic resistances Rs and Rp in reducing the fill factor varies with cell performance, cell area, and operation conditions.

As a rule of thumb, the value of Rs must be small compared to the characteristic resistance defined as Rch=Voc/JscA (where A is the cell area) while Rp must be large compared to Rch.

Short-circuit Current Density

The maximum Jsc is given by:

$$J_{SC} = \int_{AM1.5} e\eta_{EQE}(\lambda)N_{ph}(\lambda)\,d\lambda \qquad (4\text{-}13)$$

Here, Nph(λ) is the photon flux density in the incident AM 1.5 G spectrum (see Figure 4-27a) at wavelength λ and a total intensity of 100 mWcm^{-2} (integrated over the full spectrum). The external quantum efficiency $\eta_{EQE}(\lambda)$ is defined by how efficiently an incident photon gives rise to an electron flowing in the external circuit. In organic cells, η_{EQE} can be broken down into the product of efficiencies associated with each of the steps discussed previously: absorption, exciton diffusion, exciton dissociation into free carriers, charge transport, and charge collection.

Upper limits on short-circuit current density are obtained according to Eqn (4-13) by integrating from the high photon energy side (short wavelength) of the spectrum to the wavelength corresponding to the optical band gap of the material (as shown in Figure 4-27a). Hence, the smaller the optical band gap, the larger the maximum short circuit current.

In the case of silicon, the band gap is 1.1 eV (~1130 nm), which yields a maximum value Jsc=43.6 mA cm^{-2} for AM 1.5 G. It is worth mentioning that it is the fraction of photons that are harvested in the AM 1.5 G spectrum that matters rather than the fraction of the intensity contained in that same portion of the spectrum. Since the high-energy side of the spectrum gets harvested, the average

photon energy in the absorbed portion of the spectrum is larger than the band gap energy.

In silicon, the average photon energy is 1.8 eV, to be compared with the 1.1 eV band gap. As mentioned earlier, this results in a significant loss mechanism since energy goes down during thermalization of the electron-hole pairs. These losses can be somewhat minimized by using tandem cell geometries in which materials with different optical band gaps are stacked on top of one another and absorb different parts of the spectrum.

Tandem cell geometries have recently been demonstrated with organic cells.

Open-circuit Voltage

As in the case of conventional solar cells, maximization of the short-circuit current density by using organic semiconductors with decreasing optical absorption gap is not an overall effective strategy since the maximum open-circuit voltage presents an opposite trend with optical absorption gap. It follows that the determination of the optimum light-harvesting conditions to maximize the efficiency of organic solar cells depends largely on the understanding of the origin of VOC and its dependence on materials properties, in particular the relative energies of the relevant energy levels at the organic heterojunction.

Neglecting the effects of the parasitic resistances, according to Eqn (4-9), Voc is a logarithmic function of the ratio of the short circuit current and the reverse saturation current. In the dark, in the absence of carriers, the device is in equilibrium and no photovoltage or Voc is observed since the dark J-V characteristics cross the origin. Upon illumination, absorbed photons generate charge carriers, whose distributions can be described by non-equilibrium quasi-Fermi levels (see Figure 4-27a). For illumination levels such that the produced photocurrent is larger than the reverse saturation current (which holds true under average illumination conditions), Voc is observed experimentally to increase logarithmically with intensity.

Since the maximum short-circuit current can be estimated using Eqn (4-13), reaching the maximum VOC value requires that the reverse saturation current density J_0 be kept at a minimum. Recent studies by Scharber and co-workers on bulk heterojunction cells and Rand and co-workers on multilayer solar cells, indicate that Voc in organic solar cells depends on the energy difference between the ionization potential of the D component and the electron affinity of the A component forming the heterojunction.

Studies of the temperature dependence of the reverse saturation current Jo by Waldauf et al. in bulk heterojunctions and Rand in multilayer structures of small molecules show that the reverse saturation current can be approximated by:

$$J_0 = B \exp\left(\frac{-E_{final}}{n'kT}\right) \qquad (4\text{-}14)$$

where

B is a coefficient with a value in the range of
1000A cm^{-2}, and
E_{final}=IP(D) + EA(A).

Eqn (4-14) shows that the reverse saturation current is thermally activated with a barrier height equal to E_{final}/n' where n' is an ideality factor that corrects for effects such as vacuum level misalignments at the heterojunction caused by energy level bending and interfaces dipoles and the formation of charge-transfer states. By combining Eqns. (4-9), (4-10), and (4-14), the maximum open-circuit voltage VOC for an organic cell can then be written as:

$$V_{OC} = \frac{1}{e}\left[\frac{n}{n'}E_{final} - nkT \ln\left(\frac{B}{J_{SC}}\right)\right] \qquad (4\text{-}15)$$

From the energy diagram in Figure 4-27b, E_{final}=IP(D) + EA (A) is increased as the energy offsets Δ between the D and A molecular states are decreased. However, too strong a reduction in Δ compromises the efficiency of exciton dissociation at the heterojunction and thus decreases the photocurrent; it is generally accepted that Δ has to be on the order of ~0.3-0.5 eV to overcome the exciton binding energy. In this case, optimized conditions to maximize the power conversion efficiency in an organic single heterojunction cell are obtained for a band gap in the range Eg=1.6-1.9 eV (775-650 nm).

It is worth mentioning that molecules with EA energies larger than 4.2 eV are less sensitive to oxidation in the presence of oxygen and moisture. Hence, optimized material combinations should not only satisfy conditions associated with relative state energies but also absolute energies referenced to the vacuum level.

Based on this analysis, a limit of the maximum power efficiency of an idealized single junction organic solar cell can be estimated. With an optimized optical band gap of Eg=1.6 eV (775 nm) and a hypothetical average external quantum efficiency of $\eta_{EQE} = 0.8$, a maximum short-circuit current of Jsc, max=20.2 mA cm^{-2} is calculated from Eqn (4-13) (see also Figure 4-27a).

A required energy offset of $\Delta = 0.5$ eV would translate into E_{final}=IP(D) + EA(A) = Eg − Δ = 1.1 eV. From Eqn (4-15) and assuming that the solar cell has an ideality factor of n'=1.5 and with n' ~ n, the maximum open-circuit voltage is estimated to be Voc,max= 0.68 V. Such a Voc,max leads to a normalized voltage Voc=eVoc/nkT=17.4 V, from which, according to Eqn (4-12), an ideal maximum fill factor FF0=0.79 is obtained. Following Eqn (4-11), these values translate into a power conversion efficiency η=10.8%.

OPV Types

Organic PV (OPV) materials can be subdivided into two major categories:

a.) Single molecule materials, which are simple compounds of carbon (C) in which the molecules are kept separate and act individually, forming usually simple compounds, and

b.) Polymers, which are also based on carbon, but their molecules tend to stick together and are usually mechanically and chemically intertwined into much more complex 2-dimensional chains and/or 3-dimensional structures (chains of organic molecules).

The organic photovoltaic (OPV) technologies and the resulting devices (regardless of what types of organic compounds are used) are further subdivided into several groups, depending on their structure and junction (active layer) type, as follow:

1. Single active layer OPV devices
2. Bi-layer OPV devices, and
3. Multi-layer or heterojunction OPV devices, and
4. Hybrids

Another classification would be as follow:
1. Dye sensitized nanocrystalline TiO$_2$ solar cells,
2. Molecular organic solar cells, and
3. Polymer solar cells

The highest efficiencies among these solar cells (11%) have been reported for the dye sensitized nanocrystalline TiO$_2$ solar cells, which are based on photo-electrochemical principles. It was originally invented by Gerischer and Tributsch, but it is named after M Grätzel, who reported the highest efficiency of 11% using nanocrystalline TiO$_2$ to achieve high interface area electrodes.

Molecular organic solar cells use organic dyes. The active layers are cast by vacuum evaporation techniques. The organic dyes show high absorption coefficients and a good match of the solar spectrum. But their efficiencies are limited by the low exciton diffusion length and the low charge mobility of the materials. Efficiencies of 1% were already achieved in 1986 by Tang, using copper-phtalocyanine and a perylene-tetracarboxylic derivative.

Doping with C60 could slightly improve the power efficiencies. Recently, the use of doped pentacene, which shows much higher mobilities, pushed the efficiencies up to 2% for thin film and 4.5% for single crystalline devices.

The third type of organic solar cells, the polymer solar cells (pristine polymer cells, sandwiched between asymmetric contacts), show low efficiencies because of the inefficient charge generation in the polymer layer. The discovery of the photoinduced charge transfer form π-conjugated polymer to fullerene opened a new way for solar cells and photodiodes.

The efficiency for this process is near unity; i.e., this process is much faster than any competing radiative and non-radiative relaxation pathways. Bilayer devices from conjugated polymers and C60 fullerene shows improved efficiencies. But as for molecular cells, only the light absorbed within the distance of the diffusion range of excitons to the heterojunction contributes to the current.

A breakthrough for polymer solar cells was the introduction of the bulk heterojunction. Mixing of the conjugated polymer and C60, respectively a better soluble derivative of fullerenes, leads to a three-dimensional heterojunction and therefore to efficient charge generation within the whole bulk. Very soon, power efficiencies up to 1% could be reached.

Morphology and improving the electrical contact interfaces have been shown to be crucial parameters for device efficiencies. Intense engineering and improvement of the contacts pushed the power efficiency over 3%.

A closer look at some of the different OPV devices follows.

Single-layer OPV

Single-layer organic photovoltaic cells are the simplest of all OPV devices. They consist of a layer of organic electronic materials, sandwiched between two metallic conductors (electrodes). In most cases it is a transparent conductor, such as indium tin oxide (ITO), which has a high work function. The other layer (electrode) is of low work function, usually metal such as Al, or Mg.

The photo-conversion work is initiated by the difference of work function between the two electrodes. This difference sets up an electric field in the organic layer between the electrodes, so that when sunlight falls onto the organic layer it absorbs light, some of its electrons get excited to the Lowest Unoccupied Molecular Orbital (LUMO) and leave holes in the Highest Occupied Molecular Orbital (HOMO). Excitons are formed during this

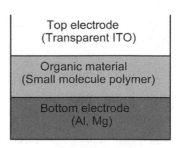

Figure 4-28. Single layer OPV device.

process and help to maintain and propagate it.

The potential difference created by the different work functions separates the exciton pairs, pulling electrons to the positive electrode and holes to the negative electrode. The electrons are then collected in an outside circuit where they create electric energy for use locally, or to be sent into the grid.

The problem with this type of device is that electric fields are not an effective way to break up excitons, since they have no charge, thus charged fields do not affect them. In practice, the efficiency of these devices is very low, less than 1%, so their use is limited to the lab bench for now.

Instead heterojunctions, such as in bi-layer OPV cells, are the preferred and much more effective way to use excitons in the photo-generation and separation process. The resulting efficiencies are much higher as well.

Bi-layer OPV

The general structure of an organic bi-layer photovoltaic device, similar to that reported by Tang in 1986 is shown above. It consists of a transparent electrode, typically a conducting oxide such as indium-tin oxide (ITO), two organic light-absorbing layers, and a second electrode. The two organic layers are made of different organic semiconductors, one with an electron-donor character and the other with an electron-acceptor character.

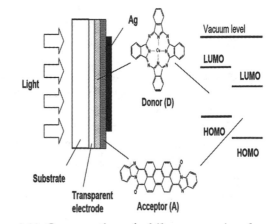

Figure 4-29. Cross-section of a bilayer organic solar cell. (1)

Electron-donor molecules (D) exhibit a low ionization potential (and thus a high-lying HOMO energy), while electron-acceptor molecules (A) possess a high electron affinity (and thus a low-lying LUMO energy). The D and A layers must provide for efficient hole and electron transport, respectively.

For the organic solar cell shown here, a parallel can be drawn between the two organic light-absorbing layers and the n and p-doped semiconductors in an inorganic solar cell. However, unlike their inorganic counterparts, organic semiconductors in the device structure shown herein are in most cases essentially intrinsic (although some amounts of uncontrolled impurities are likely to be present).

The interface between the two layers—the D/A heterojunction—is responsible for efficient exciton dissociation. Therefore, it plays a role for excitons similar to that played by the junction for minority carriers in inorganic cells. The electrons and holes created at the interface can be transported through the A and D layers, respectively, and collected at the two electrodes, thereby contributing to an electrical current in the external circuit.

Heterojunction OPV Cells

Heterojunction (or bulk heterojunction) OPV cells are bi-layer photovoltaic cells, where the electron donor and acceptor are mixed together. So instead of having two separate layers (donor and acceptor) the organic material is a mixture of these. The mix forms a polymer blend of certain polymers and proportions.

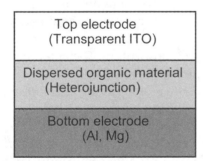

Figure 4-30. Heterojunction layer OPV device.

If the length scale of the blend is similar to the exciton diffusion length, as it should be in a well designed device, most of the excitons generated in either material may reach the interface, where excitons break efficiently. In this case, the free electrons move to the acceptor domains and are then carried through the bulk of the device to be collected by one electrode. The free holes, on the other hand, are pulled in the opposite direction through the bulk, and collected at the other side of the device.

Leads attached to both sides of the device will close an electrical circuit through which current will flow for local use, or be sent into the grid.

Modified bi-layer OPV Cell

This is an example of one of many possible variations of the bi-layer OPV cell design. These devices are still on the research benches—including the one in Figure 4-31. In these devices one or more intermediate layers are squeezed between the polymer layers. These interlayers are introduced with the goal of assisting with, and enhancing, the photo-generation process.

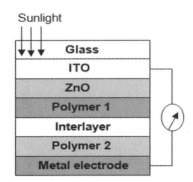

Figure 31. Modified bi-layer OPV cell with intermediate layers.

The function of the device is similar to that of the bi-layer one, except for the added function of the interlayers, which help to initiate and maintain the entire electric power generation process.

The Advantages

The semiconductor industry is working intensively and in parallel with other industries on the development of organic-type devices. Many companies put much effort and many resources into developing technologies for making electronic circuits with organic materials, using non-vacuum (conventional semiconductor) processes, on flexible substrates. And there is a lot of progress in this area.

For example, organic LEDs (OLED) displays, which use similar materials and processes as OPVs, are on the forefront with estimated market growth from US$4.0 billion in 2011 to estimated $20.0 billion by 2015, and as high as $35.0 billion by 2020. OLED displays are only the beginning of this new type of microelectronics, which are expected to dominate the markets soon.

Technical developments have also been made in fully printed non-volatile, rewritable memory for applications in toys and games, logistics, sensor, and ID systems—some of which will be used in organic PV products manufacturing operations.

All these technologies are similar to OPV, and the experience gained from them will be easily transferred to OPV technologies. Although in their early stages, they represent significant large-scale opportunities for high-volume manufacturing of optical and PV devices in the near future.

The fact that their development is enthusiastically supported by SEMI members—the largest semiconductor manufacturers in the world—means that only time separates us from their full implementation in the world markets.

SEMI is planning to support these markets' acceleration through international standards development. While development is still in the early stages, many materials manufacturers are incurring high-costs having to develop unique test methods and characterization specifications for each customer. Technology roadmap activities, public funding of critical R&D, and technical education and promotion of new technologies are other activities that SEMI will take on with member support and guidance.

No doubt, the semiconductor giants are interested in solar energy as a side product, and some have been entering the field with their own products. Looking at the examples from the past, and the speed the semiconductor giants developed the key technologies, we do foresee organic PV growing faster than many of the competitors in the 21st century.

The Disadvantages

As NREL's roadmap for the OPV technology points out, the main challenges holding back the large-scale commercialization of this technology are low cell efficiency (which needs to achieve a rate of least 10 percent), and short lifetime (which must be extended to at least three years). Along with these improvements should come lower-cost electricity generation, with a target rate of $1 per watt considered a viable reality for 2010.

OPV cells are expected to become commercially available in the near future from principal developers and suppliers such as Heliatek, Konarka Technologies, Mitsubishi, Plextronics, Solarmer Energy, Mitsubishi Cooperation.

On the module front, NREL certified what Plextronics claims to be the largest OPV module (232 cm^2 total area), realizing 2.3% active area efficiency, with over 7% efficiency in a lab setting.

There is still far to go before OPV devices can get anywhere near the major competitors efficiency-wise. Reliability issues need to be addressed and resolved too,

before we can discuss wide market acceptance.

Author's note: We've noticed lately that the issue of longevity (or lack thereof) is not given due notice, as if it were something that could be addressed later. The industry should keep in mind the deficiencies of this technology and not rush it to market, for the negative consequences of doing so could be grave. Sunlight is a great enemy to many things, especially organic things.

Summary

Organic semiconductors have large extinction co-efficients compared to crystalline Si leading to efficient light harvesting in layers that are relatively thin with thicknesses in the range of 100-200 nm. Since the organic layers are sandwiched between electrodes with different work functions, a built-in potential appears and results in an electric field that assists the transport of charges. Thus, the electron and hole currents in the device under illumination after exciton dissociation are governed by an interplay of drift and diffusion.

While a critical step in inorganic solar cells is to collect minority carriers before they recombine by having them migrate to the opposite side of the junction, the challenge in organic cells is to dissociate the excitons before they decay to the ground state; this requires that they rapidly diffuse to the D/A heterojunctions, which is the only location where dissociation is efficient in pure materials. Hence, the thickness of the organic layers has to be comparable to the exciton diffusion length L (L¼(Dt)1/2 where D is the diffusion coefficient and t the lifetime of the exciton). An optimal compromise regarding the thickness of the organic layers has to be found between allowing for efficient exciton diffusion to the heterojunction and efficient sunlight absorption. The latter requires that the total absorbance of the film be preferably in the 2-3 range (corresponding to 86-95% absorption efficiency).

However, if the layers of organic semiconductors must be too thin as a result of small L values (e.g., d ¼ 10-20 nm), the incident light does not get absorbed efficiently. For an organic semiconductor such as the copper-phthalo-cyanine compound used by Tang, the peak extinction coefficient at 620 nm is k ¼ 0.74, which leads to an absorption coefficient (a ¼ 4pk/l where l is the wavelength) of a ¼ 0.015 nm_1; for a 10 nm-thick film, this translates into an absorbance ad [absorption efficiency] of only 0.15 [14%] for a single pass and 0.3 [26%] for a double pass assuming the transmitted light gets reflected by the back electrode.

Although devices based on pentacene have recently been shown to exhibit larger exciton diffusion lengths (70 nm),44 this issue remains a limitation—often referred to as the "exciton bottleneck"—and has triggered the design of devices that are based on bulk heterojunctions in which the A and D components are mixed together and form an interpenetrating, phase-separated network D:A with a nanoscale morphology. This layer is then sandwiched between hole and electron transport (undoped or doped) layers.

What Will the Future Bring?

The strongest argument for third-generation organic PV (OPV) technologies is certainly their promise for ultralow costs, with an added bonus of light weight and flexible substrates. The vision of PV elements based on thin organic (plastic) carriers, manufactured by printing and coating techniques from reel to reel, and packaged by lamination techniques is not only intriguing, but highly attractive from an economic and esthetics standpoint.

To achieve such goals, high-volume production technologies for large area coating must be applied to a low-cost material class. Solution-processable organic and inorganic semiconductors have a high potential to fulfill these requirements. Flexible chemical tailoring allows the design of organic semiconductors with the desired properties, and printing or coating techniques like screen, inkjet, offset or flexo-printing are being established for semiconducting polymers today, driven by display or general electronic demands.

OPVs are characterized by a number of different and attractive features, such as:

1. Some are flexible and/or semitransparent,
2. Can be manufactured in a continuous printing (roll-to-roll) process,
3. Can be manufactured in a large-area-coating process
4. Lend to easy integration in different devices,
5. Offer significant cost reduction, compared to traditional solutions,
6. Possess substantial ecological and economic advantages.

Though these features are beneficial for commercialization, OPVs must fulfill the basic requirements for renewable energy production as well, just like its classical counterpart. In the energy market the competitive position of each solar technology is mainly determined by the three factors: efficiency, reliability, and cost (usually calculated per Wp produced).

The potential of OPVs must be judged by these key factors, so successful commercialization can be realized only if all three technology-driving aspects are fulfilled simultaneously.

A product development succeeding in only one or two aspects, e.g., competitive costs and reasonable efficiency, will only be able to address small niche markets unless the third parameter, in that case reliability (lifetime), is also optimized.

Despite the problems and failures of late in the solar field, we're seeing record-setting PV performance in all areas of interest; CdTe, c-Si, CPV, CIGS, GaAs. etc., Ge MJ, etc.

Organic solar cells in general are lightweight, nontoxic, and semi-transparent. They are cheap, due to cheap materials and roll-to-roll, low-cost manufacturing processes, but their efficiencies tend to be very low. More importantly, their long-term reliability, due to the instability of organic materials, especially under extreme climates is questionable.

We estimate worldwide capacity of OPV in 2012 to be just over 5 megawatts.

Some examples of OPV companies and technologies:

Heliatek is one of the companies making OPV cells, and has documented a new record for organic solar cells—a champion cell on a small area that has measured 10.7% efficiency under normal illumination. The efficiency drops to about 9.0% when deposited on a flexible substrate, which is still quite good and useful in many applications.

Note: Based on the materials used, OPV cells can be divided into two major categories:

a.) Polymer-based, or those based on large organic molecules and chains, and

b.) Oligomer-based, or those based on small organic molecules.

While many manufacturers use the polymer technologies, Heliatek uses small molecule organic compounds. These are deposited via low-temperature vacuum technology in very homogenous thin layers at a ratio of 1 gram of organic compound per square meter of substrate.

The dry roll-to-roll process doesn't use solvents as the competing organic ink printing processes, and can be thought of as an OLED in reverse. This unique process, and the innovative materials and the results thereof, give this technology a significant edge over the rest. It only remains to be seen if the level of market acceptance matches the enthusiasm of the proponents of this technology.

Konarka's (printing large molecule polymers) main activity for the last decade has been raising funding. With over $190 million raised thus far, and with precious little to show as far as commercial product is concerned, they are the capital-raising champions of the world. Failure to secure further funding finally forced Konarka Technologies into bankruptcy via a Massachusetts Bankruptcy Court. This means full liquidation and asset sale to pay creditors.

"Konarka has been unable to obtain additional financing, and given its current financial condition, it is unable to continue operations," noted Howard Berke, chairman, president and CEO of Konarka in a statement. "This is a tragedy for Konarka's shareholders and employees and for the development of alternative energy in the United States."

Or is it? Doesn't this rather confirm once again that putting the cart before the horse doesn't work most of the time—regardless of the reasons and good will surrounding the enterprise? Or in Konarka's case, spending millions of dollars without producing even a glimpse of a marketable product proves again that money is not the solution. On the contrary; it might be the reason so many companies are failing.

Konarka's case also confirms that OPV technology is far from full commercial implementation.

Dyesol is another virtual, albeit publicly traded firm (which comes to show that anything can be traded these days). Dyesol builds equipment to manufacture DSSCs. We are not sure exactly what equipment this is, how it is used, and what kind of product it produces, but the firm is very successful in issuing glamorous press releases.

In 2011 Dyesol produced the world's largest DSSC PV module. So if size was the measure of success, then Dyesol is the winner, no doubt. The question they have not answered as yet, however, is how efficient this module is, and how long would it last under a blistering desert sun?

Eight19 Limited raised $7 million from the Carbon Trust and Rhodia to develop plastic organic solar cells. It is unclear what the "Eight19" designation refers to, but we are sure that it could be used anywhere where the numbers 8 and 19 can be squeezed.

In 2011 Eight19 built a brand new production facility in Cambridge Science Park in the UK. It is the largest in Europe for the development of printed organic solar cells, and includes a multi-station roll-to-roll fabrication machine designed to create solar modules at a peak linear speed of over 3.6 kilometres per hour.

Here again, if speed of production is all that matters, then Eight19 would be a clear winner. Unfortunately, it is NOT. What counts is what efficiency we can expect from the PV rolls, and how long they would be able to produce power before the organics in them disintegrate under intense sunlight. This question has not yet been answered.

SolarPrint has eliminated the liquid part of DSSC and replaced it with nanomaterials in a roll-to-roll printing of all components. The first product line to hit the market is the SP5848 Energy Harvesting module. It targets the Wireless Sensor market with estimated 3.3V output and 10 years lifetime.

The lifetime estimate, we assume, is for operation under shade and artificial lighting, because this product would not last nearly that long under direct sunshine.

Intel is (or was) also working on OPV cells, but that effort remains veiled in secrecy... if it is still on.

Heliatek is focused on the promised-land market of building-integrated photovoltaics (BIPV). This includes windows and facades, as well as concrete and other building materials. The BIPV market today, however, is not the true BIPV, but rather a designation for anywhere PV products can be squeezed in a building. Even then, the market is very new, unstable and jittery...

So Heliatek invested $18 million in the construction of a pilot plant and is now looking for $60 million to build a mass production line. This won't be easy, and that might be a good thing. The delay would give Heliatek more time to use the pilot plant to develop a good product, and thoroughly test it before rushing to market.

So what can we expect in the future from OPV materials, cells and modules? Most likely we will see niche markets being developed for the major OPV types. Some of these will take a more prominent place in the energy markets, while others will be put on the shelf.

The most important contribution, in our expert opinion, is the fact that all these technologies are using new materials, equipment and approaches, which can be used in the development of other technologies, and which would accelerate their development and commercialization.

Polymer PV Cells

Polymer solar cells are a branch of the organic PV (OPV) technology, which is under intense development. These cells usually consist of thin films (typically 100 nm or less) of organic semiconductors such as small-molecule compounds like poly-phenylene vinylene, copper phthalo-cyanine (a blue or green organic pigment), and carbon fullerenes and fullerene derivatives, such as PCBM.

Energy conversion efficiencies achieved to date using conductive polymers are low compared to inorganic materials. However, they were improved in the last few years and the highest NREL certified efficiency has reached 6.77%. In addition, these cells could be beneficial for some applications where mechanical flexibility and disposability are important.

These devices differ from inorganic semiconductor solar cells in that they do not rely on the large built-in electric field of a p-n junction to separate the electrons and holes created when photons are absorbed. Instead, the active region of an organic device consists of two materials, one which acts as an electron donor and the other as an acceptor.

When a photon is converted into an electron hole pair, typically in the donor material, the charges tend to remain bound in the form of an exciton, and are separated when the exciton diffuses to the donor-acceptor interface. The short exciton diffusion lengths of most polymer systems tend to limit the efficiency of such devices. Nanostructured interfaces, sometimes in the form of bulk heterojunctions, can improve performance.

Instability of the films, especially under harsh environmental effects is a major problem, which needs to be resolved, before full-scale implementation. Even with its advantages, this technology still has far to go to full market acceptance and serious deployment.

Heterojunction OPV

Today, most polymer solar cells destined for commercial applications are based on the bulk heterojunction (BHJ) concept described in 1995 by Yu et al. In this type of solar cells, the donor material is typically a polymer, which is mixed with an acceptor in the form of a soluble fullerene in an organic solvent. The mix is then spin-coated, or cast, on a substrate of indium-tin oxide (ITO) on glass.

During evaporation of the solvent and latter treatments, a microphase separation takes place with the formation of an interpenetrating network. A hole-blocking layer of, for example, lithium fluoride may be added, and in a last step a metal electrode, usually aluminum thin film, is evaporated on top of the structure.

The bulk heterojunction is important because a large interfacial area between the donor and acceptor materials is created where charge separation can take place.

The process starts, and is maintained, by the exciton generation during excitation of the mixture, whose life time is very short; the diffusion length is in the order of 10 nm or less. The size of the individual domains is critical, and the domains must be interconnected to create continuous paths for both electrons and holes, as needed for transportation to the external electrodes.

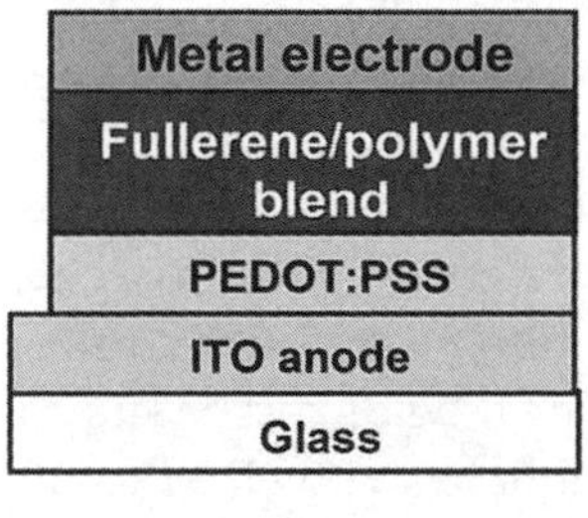

The efficiency of the BHJ-type cell has increased recently, which is a good step in the right direction. Unfortunately, the optimal structure for generating the most efficient device is thermodynamically unstable. A lot of money and effort have been spent on the creation of processes for optimizing the efficiency and reliability of these devices, and the results are good enough for some niche markets.

Many different polymers and small molecules organics have been tested as active materials in organic solar cells, but only very few of these have been tested by more than one research group and most of them failed to produce devices with efficiencies of even 1%. As a consequence, nearly all physical studies of polymer solar cells have been conducted with only a few popular polymers. Experience shows that the efficiency can be increased by adjusting fabrication parameters. This requires additional time and effort to develop the existing materials and processes.

New materials are also under investigation, with the recent trends being,

a)   the low bandgap polymers that extend the harvesting of photons above 600 nm (vs 10 nm of the presently used materials), and

b)   the advent of thermo-cleavable polymers, which are special materials that combine ease of processing with high stability of the finished device.

Author's note: A €14.2 million European research project under the European Commission's Seventh Framework Program (FP7) aims to develop better flexible plastic solar panels. The so-called "SUNFLOWER" project ("SUstainable Novel FLexible Organic Watts Efficiently Reliable"), led by the Swiss Center for Electronics and Microtechnology with more than a dozen partners, started work in Oct. 2011 to increase the cells' efficiency and lifetime, and decrease production costs.

Goals for an initial prototype include a "tandem" multilayer structure to increase efficiency, better-performing barrier layers and getters, and creation on a roll-to-roll atmospheric printing process.

"We have the chance to develop a technology that is ideally suited to manufacturing in the EU due to its high level of automation, need for highly trained personnel,

Figure 4-32. Organic hetero-junction solar cell (9)

low energy consumption, and close proximity to suppliers and markets," says the project coordinator.

The future looks bright for polymer (plastic) solar cells. There is a well defined (albeit small and limited) niche market for their initial deployment.

The key to success is to take small steps, and to not ignore the issues at hand, namely low efficiency, fast degradation (especially under intense heat), and short lifetime. These are organic materials, after all, and we know well that organics don't mix well with IR and UV radiation—no matter how we package and use them. So let's be careful and not put the cart in front of the horse.

Conjugated Polymer PV Cells

The past few years are marked by continuous race for development of novel conjugated polymers for use as electron donor materials in organic solar cells. (9)

This competition of worldwide research groups was very successful and brought the efficiency of plastic solar cells from the level of 4% documented as state-of-the-art three years ago up to the level of 7-8% certified recently.

This is quite a breakthrough in the field, proving that organic photovoltaic devices are indeed promising for mass energy production. Highly efficient solution-based technologies applied to the fullerene/polymer blends should enable industrial production of cheap organic solar cell modules. Grid-connected OPV modules have been demonstrated recently. In addition, some examples of commercially viable devices utilizing small area organic solar cells have been designed.

Conjugated polymers are some of the most promising solar energy generating systems for use in some specialized niche markets. A conjugated system is formed when carbon atoms covalently bond with alternating single and double bonds in hydrocarbons. The hydrocarbons' electrons pz orbitals delocalize and form a delocalized bonding π orbital with a π^* antibonding orbital.

Thus delocalized π orbital is the highest occupied molecular orbital (HOMO), and the π^* orbital is the lowest unoccupied molecular orbital (LUMO). The separation between HOMO and LUMO is defined as the band gap of organic electronic materials. The band gap is typically in the range of 1-4 eV.

The process of conversion of light into electricity by an organic solar cell can be schematically described by the following steps: (17)

1. Absorption of a photon, leading to the formation of an excited state, that is, the bound electron-hole pair (exciton) creation;

2. Exciton diffusion to a region where exciton dissociation, that is, charge separation occurs; and

3. Charge transport within the organic semiconductor to the respective electrodes.

Because of the large band gap in organic materials, only a small portion of the incident solar light is absorbed (see Figure 4-33).

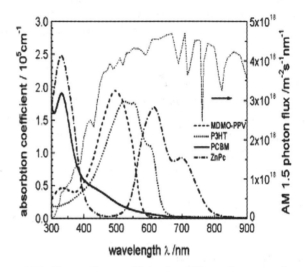

Figure 4-33. Absorption coefficients of films of commonly used materials are depicted in comparison with the standard AM 1.5 terrestrial solar spectrum. (17)

A band gap of 1.1 eV (1100 nm) is capable of absorbing 77% of the solar irradiation on earth. However, the majority of semiconducting polymers have band gaps higher than 2 eV (620 nm), which limits the possible harvesting of solar photons to about 30%.

On the other hand, because the absorption coefficients of organic materials are as high as 105 cm-1, only 100 nm thickness is enough to absorb most of the photons when a reflective back contact is used.

Now we must admit, we need better "spectral" harvesting of solar photons via lower band gap polymers and/or using energy-transfer cascades.

Film thickness is not the bottleneck. The primary photoexcitations in organic materials do not directly and quantitatively lead to free charge carriers but to coulombically bound electron-hole pairs, called excitons.

It is estimated that only 10% of the photoexcitations lead to free charge carriers in conjugated polymers. For efficient dissociation of excitons, strong electric fields are necessary. Such local fields can be supplied via externally applied electrical fields as well as via interfaces.

Blending conjugated polymers with electron acceptors, such as fullerenes, is a very efficient way to break apart photoexcited excitons into free charge carriers.

Ultrafast photophysical studies showed that the photo-induced charge transfer in such blends happens on a time scale of 45 fs. This is much faster than other competing relaxation processes (photoluminescence usually occurs around 1 ns). Furthermore, the separated charges in such blends are metastable at low temperatures (see Figure 4-34).

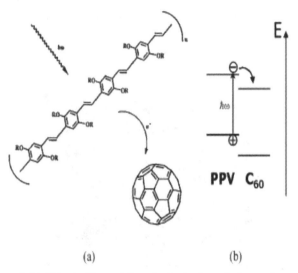

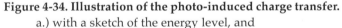

Figure 4-34. Illustration of the photo-induced charge transfer.
a.) with a sketch of the energy level, and
b.) after excitation in the PPV polymer, the electron is transferred to the C60. (17)

For efficient photovoltaic devices, the created charges need to be transported to the appropriate electrodes within their lifetime. The charge carriers need a driving force to reach the electrodes.

A gradient in the chemical potentials of electrons and holes (quasi Fermi levels of the doped phases) is built up in a donor-acceptor junction. This gradient is determined by the difference between the highest occupied molecular (HOMO) level of the donor (quasi Fermi level of the holes) and the lowest unoccupied molecular orbital (LUMO) level of the acceptor (quasi Fermi level of the electrons).

This internal electrical field determines the maximum open circuit voltage (Voc) and contributes to a field-induced drift of charge carriers.

Also, using asymmetrical contacts (one low work-function metal for the collection of electrons and one high work-function metal for the collection of the holes) is proposed to lead to an external field in short circuit condition within a metal-insulator-metal (MIM) picture.

Another driving force can be the concentration gradients of the respective charges, which lead to a diffusion current. The transport of charges is affected by recombination during the journey to the electrodes, particularly if the same material serves as transport medium for both electrons and holes.

As a last step, charge carriers are extracted from the device through two selective contacts. A transparent indium tin oxide (ITO) matches the HOMO levels of most of the conjugated polymers (hole contact). An evaporated aluminum metal contact with a work function of around 4.3 eV matches the LUMO of acceptor PCBM (electron contact) on the other side.

Bi-layer Heterojunction Devices

In a bilayer heterojunction device, p-type and n-type semiconductors are sequentially stacked on top of each other. Such bilayer devices using organic semiconductors were realized for many different material combinations.

In such devices, only excitons created within the distance of 10-20 nm from the interface can reach the heterojunction interface. This leads to the loss of absorbed photons further away from the interface and results in low quantum efficiencies.

The efficiency of bilayer solar cells is limited by the charge generation 10-20 nm around the donor-acceptor interface (Figure 4-35).

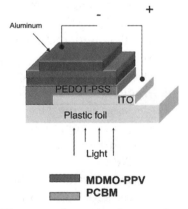

Figure 4-35. Bilayer configuration in organic solar cells. (17)

Using thicker films creates optical filter effects of the absorbing material before the light gets to the interface, resulting in a minimum photocurrent at the maximum of the optical absorption spectrum. Also, the film thicknesses have to be optimized for the interference effects in the multiple stacked thin film structure.

Bulk Heterojunction Devices

Bulk heterojunction is a blend of the donor and acceptor components in a bulk volume (Figure 4-36). It exhibits a donor-acceptor phase separation in a 10-20 nm length scale. In such a nanoscale interpenetrating network, each interface is within a distance less than the exciton diffusion length from the absorbing site.

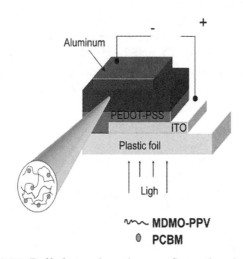

Figure 4-36. Bulk heterojunction configuration in organic solar cells. (17)

The bulk heterojunction concept has heavily increased (orders of magnitude) the interfacial area between the donor and acceptor phases and resulted in improved efficiency solar cells. While in the bilayer heterojunction, the donor and acceptor phases are completely separated from each other and can selectively contact the anode and cathode. In the bulk heterojunction, both phases are intimately intermixed.

This mixture is designed for no symmetry breaking in the volume. There is no preferred direction for the internal fields of separated charges; that is, the electrons and holes created within the volume have no net resulting direction they should move. Therefore, a symmetry breaking condition (like using different work-function electrodes) is essential in bulk heterojunctions. Otherwise, only concentration gradient (diffusion) can act as the driving force.

Furthermore, separated charges require percolated pathways for the hole and electron transporting phases to the contacts. In other words, the donor and acceptor phases have to form a nanoscale, bicontinuous, and interpenetrating network. Therefore, the bulk heterojunction devices are much more sensitive to the nanoscale morphology in the blend.

Bulk heterojunctions can be achieved by co-deposition of donor and acceptor pigments, or solution casting of either polymer/polymer, polymer/molecule, or molecule/molecule donor-acceptor blends.

PPV:PCBM Bulk Heterojunction Solar Cells

2.5% efficient solar cells can be obtained from soluble derivatives of phenylene-vinylenes, for example, poly[2-methoxy-5-(3,7-dimethyloctyloxy)-1,4-phenylene vinylene) (MDMO-PPV) mixed with soluble derivatives of fullerenes, for example, 1-(3-methoxycarbonyl)propyl-1-phenyl-[6,6]-methanofullerene (PCBM). Shaheen et al. showed that a power conversion efficiency of 2.5% under AM 1.5 conditions can be obtained by using chlorobenzene as a solvent for spincasting in the weight ratio of 1:4 for MDMO-PPV:PCBM. Changing the solvent from toluene to chlorobenzene increases the efficiency by nearly a factor of 3, which was assigned to originate from the changes in the nano-morphology.

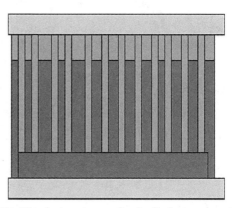

Figure 4-37. Ideal structure of a bulk heterojunction solar cell. (17)

Such bulk heterojunction solar cells have 80 wt% PCBM. However, the polymer MDMO-PPV is supposed to be the main light absorber in these solar cells, because PCBM has almost no absorption in the visible-near-infrared region. Therefore, it is better to increase the volume concentration of MDMO-PPV for better absorption of solar light.

Poly(3-alkylthiophene):PCBM Bulk Heterojunction Solar Cells

Poly(3-alkylthiophenes) (P3ATs) are conjugated polymers with good solubility, processability, and environmental stability. Regioregular poly(3-alkylthiophenes) (RRP3AT) (P3HT:poly(3-hexylthiophene), P3OT:poly(3-octylthiophene), and P3DDT:poly(3-dodecylthiophene) are used as electron donors in polymer:fullerene bulk heterojunction solar cells with record power conversion efficiencies up to 5%.

Energy levels for P3OT, for P3HT, and for P3DDT were almost the same with the optical band gap energy around 1.9 eV. With longer side chain length, their electrochemical band gaps were slightly increased. The absorption coefficient undergoes a systematic decrease by longer side chain polythiophenes due to chromophore dilution (i.e., conjugated segments in ratio to nonconjugated segments decrease upon increasing the side chain length).

Polymer/Polymer Solar Cells

Inspired by the developments of conjugated polymer/fullerene bulk heterojunctions, similar systems are also reported in the literature. Polymer/polymer bulk heterojunction solar cells achieved considerably less efficiencies and attracted less attention, although they might have the potential to be implemented in inexpensive, large-area photovoltaic systems, as well. Bulk heterojunctions of two conjugated polymers have several advantages.

In a conjugated polymer blend, both components can exhibit a high optical absorption coefficient and cover complementary parts of the solar spectrum. It is relatively easy to tune both components individually to optimize optical properties, charge transfer, and charge collection processes.

On the other hand, polymer blends have an intrinsic tendency to phase separate. These phase-separated domains usually have dimensions of several micrometers and are thus too large, compared to exciton diffusion length limitations for polymeric cells. The biggest challenge for the polymer/polymer bulk heterojunction concept is to identify suitable n-type polymers with acceptor properties and good stability.

Donor-acceptor "Double Cable" Polymers

The concept of "double cable" polymers has been introduced to have a control on the morphology at the molecular level. Chemically attaching the electron acceptor moieties directly to the donor polymer backbone prevents the phase separation. Electrons created by photo-induced electron transfer are transported by hopping between the pendent acceptor moieties, leaving the remaining hole on the conjugated chain transporting the positive charge.

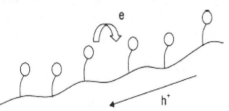

Figure 4-38. Schematic representation of "double cable" polymers.

The charge carriers generated by photo-induced charge transfer can be in principle transported within one molecule, therefore termed as a "molecular heterojunc-

tion". (17)

Several double cable materials have been explored in polymer solar cells. The efficiency of such devices is low, probably due to fast recombination or inefficient inter-chain transport.

PPE-PPV Copolymers

There are two most promising ways that are currently used to improve photovoltaic properties of conjugated polymers. In the first route, conjugated polymers are modified with some structural blocks that reduce their band gaps, therefore enhancing their light harvesting properties. (9) These materials give high short-circuit current densities (Jsc) in solar cells.

According to another approach, the backbones of conjugated polymers are modified to lower their HOMO energy levels and increase open circuit voltages (VOC) of photovoltaic cells. Both concepts are equally promising and should be combined to develop the most efficient polymer materials.

It was demonstrated some years ago that conjugated polymers comprising both phenylene-vinylene and phenylene-ethynylene structural blocks have low lying HOMO energy levels and, therefore, give an appreciably high VOC of 900-1000 mV in organic solar cells. Therefore, this is a promising group of conjugated polymers that might bring state-of-the-art power conversion efficiencies in solar cells providing their band gaps are lowered sufficiently to reach short circuit currents of 15 mA cm^2 and higher. The design of such polymers is currently in progress.

Intermolecular interaction of PPE-PPV copolymers with fullerene-based counterparts in the active layers of organic solar cells is also an important issue. It was shown that variation of the side chains attached to the polymer backbone changes physical and electronic properties of these materials and significantly influences their photovoltaic performance when they are combined with PCBM.

This problem can be approached from an opposite direction by fixing side chains on the polymer backbone and varying organic addends attached to the fullerene cage. Indeed. Dramatic effect of the fullerene derivatives structure and solubility on the photovoltaic performance of their blends with PPE-PPV copolymers can be observed.

It has been demonstrated that photovoltaic performance of two PPE-PPV copolymers depends strongly on the solubility and molecular structure of fullerene derivatives applied as electron acceptor counterparts in the active layer of organic solar cells. The obtained results clearly show that conventional material PCBM gives just moderate efficiencies in combination with some types of copolymers.

The best photovoltaic performances were reached when each copolymer was combined with specific fullerene derivatives with better-suiting molecular structures and solubilities in organic solvents.

Conjugated Polymer PV Cells

An important difference between inorganic and organic solar cells lies in the nature of the primary photoexcited state. In the former, the absorption of photons leads directly to the creation of free electrons and holes at room temperature. The charge carriers can then diffuse and/or drift to their respective collective electrodes.

In organic semiconductors the situation is somewhat different and still the object of strong debate within the scientific community. It is generally accepted that the absorption of a photon induces mainly excitons with binding energies ranging from 0.05 eV up to >1 eV.

According to the Onsager theory, which can be invoked as a first approximation in organic semiconductors, photoexcited electrons and holes are coulombically bound (excitons) and perform a Brownian random walk.

Once excitons have been created by the absorption of photons, they can diffuse over a length of approximately 5-15 nm. Then they decay either radiatively or nonradiatively. The former route gives rise to luminescence that occurs within 500-800 ps for singlet excitons (fluorescence) and several hundreds of nanoseconds for triplet excitons (phosphorescence) at room temperature.

For PV purposes, excitons have to be separated into free charge carriers before they decay. That can be achieved by several different ways:

a. Dissociation by trap sites in the bulk of material,

b. Dissociation by an externally applied electric field,

c. The "hot exciton" dissociation: the excess energy of the photons can quickly be distributed over the conjugation segment of the polymer leading to local temperatures (thermal bath) high enough for a very short period of time (in the femto-second range) to provide the activation energy for exciton dissociation.

d. Exciton dissociation at the discontinuous potential drops at the interfaces between donors and acceptors as well as between semiconductors and metals.

Independently from each other, Dupont's Santa Barbara and Osaka groups reported studies on the photophysics of mixture and bilayers of conjugated polymers with fullerenes. The experiments clearly evidenced that when an exciton reaches the interface between a conjugated polymer (donor) and a C60-based material (acceptor) an ultrafast electron transfer occurs from the lowest unoccupied molecular orbital (LUMO) of the donor to the lower lying LUMO of the acceptor, leaving a hole on the highest occupied molecular orbital (HOMO) of the donor.

Thus, electrons and holes are separated, and the free charge carriers produced can diffuse in their respective environments during their lifetime (up to a millisecond).

The forward electron transfer was observed to happen within 45 fs; that is much faster than any competing relaxation processes. Therefore its efficiency is about 100%.

As mentioned above, photogenerated excitons in a conjugated polymer have to be dissociated into free charge carriers to participate to the extracted current. In the case of pristine polymers, the only efficient way to ensure this dissociation is the action of an electric field. In the simple MIM picture, the number of charge carriers present in the device is small enough to ensure rigid, straight HOMO and LUMO bands.

On the other hand, if the organic semiconductor is doped, band bending can appear, inducing Schottky contact at the metal-electrode interface. In any case, an electric field does appear in the device under the SC condition, if the work functions of the electrodes are asymmetric.

In the former case, the electric field present in the device is constant throughout the entire thickness of the device. In the latter case, the field drops in the depletion zone, but is larger over a smaller thickness. Since large electric fields are mandatory to dissociate excitons, Schottky devices are expected to enhance the charge generation in pristine-conjugated polymer.

The very first efforts to achieve PV devices using conjugated polymers were performed on polyacetylene and some polythiophenes. However, from the conjugated polymers of the first generation, PPV appeared to be the most successful candidate for single-layer polymer PV devices.

The radiative recombination channels of the injected electrons and holes within PPV and its derivatives which resulted in light-emitting diodes (LEDs) suggested this class of materials for PV devices. Interestingly it was found that the same devices exhibit excellent sensitivity as photodiodes under reverse bias, with quantum yields above 20% (electron/photon) at –10 V reverse bias.

This observation opened a route to elaborating monolithic device that can act as light emitter and detector.

Moreover, devices based on derivatives of polythiophene exhibited an even better photoresponse (80% electron/photons at –15 V), competitive with UV-sensitized Si photodiodes, but in both cases, externally applied voltage is necessary to extract significant amount of charges.

For PV effect, it appeared that the intrinsic field induced by the asymmetry of the work function of the electrodes could not be large enough to generate significant power.

While the very first devices based on sublimated merocyanines and phthalocyanines developed in the late 1980s showed efficiency of 1%, recent reports have announced record efficiencies of over 5%. This rapid progress is predominantly based on achievements in the material science.

Novel generations of conjugated polymers reaching purities comparable to some inorganic semiconductors can no longer be compared to the first organic semiconductors, such as polyacetylene.

Beyond the enhancement of the semiconducting and optical properties of the photoactive materials, however, other important developments have to be carried out, like gaining improved control over the solid-state morphology of donor-acceptor composites. The near future will show exciting developments, some of which are already anticipated, like self-assembled materials, nanosized device geometries to release the requirements on the proper morphology, metallic nanoparticles and plasmon absorption and highly complex molecules which fulfill all functions in distinct parts of the molecule.

Simultaneously, further developments of the solution-based techniques such as inkjet, screen, or flexographic printing (a form of rotary web letterpress using flexible relief plates) for the production of organic thin-film solar cells will occur.

Moreover important economic factors addressing the balance-of-system costs will have to be addressed, like novel semiconducting electrodes to replace ITO, low-cost barriers against water and oxygen permeation as well as flexible electronic components to be combined with flexible PV power supply.

Thus, it is much likely that efficiencies close to 10% will be achieved in the relatively short term. The future of this technology appears bright and we expect to see it (or elements of it) in the 21st century markets.

Low Bandgap Polymers

Low bandgap polymers have the ability to absorb sunlight with wavelengths above 600 nm. Traditional polymers used in organic photovoltaics, such as MEH-PPV, have an absorption that extends to maximum wavelengths of 550 nm. Most commonly employed P3HT has an absorption that extends out to 650 nm, and comparing the absorption spectra of this polymer with the solar spectrum, we see a strong mismatch between the absorption spectrum and the emission spectrum of the sun.

Such mismatches could be alleviated by using low bandgap polymers, since the absorption spectrum of the low bandgap polymer overlaps better the sun emission spectra. This fact has increased the interest in low bandgap polymers

The bandgap of a polymer is expressed in terms of the difference between the highest (occupied) molecular orbital (HOMO) and the lowest (unoccupied) molecular orbital (LUMO). The bandgap can be determined from an optical absorption spectrum (UV-vis), which gives the optical bandgap. It can be also determined by means of cyclic voltammetry (CV), which gives the actual electrical bandgap.

The bandgap is usually used to determine the energy generated by a solar cell, but neither the optical nor the electrical bandgaps give this information directly. This is because in the optical bandgap, the binding energy of the exciton is not accounted for, while for CV measurements the energy of solvation of the electrochemical species is usually unknown.

For a close look at the low bandgap phenomena of polymers, we need to take a closer look at the solar spectrum. The solar spectrum generates a number of photons as a function of the different wavelengths, with the maximum number of photons shown at wavelengths of up to 900 nm.

When we look at the integrated number of photons and integrated current as a function of the wavelength, however, we see some important results. If we compare two examples of polymers, one absorbing light with wavelengths up to 500 nm (A) and one absorbing light with wavelengths up to 1000 nm (B), we see that polymer A will absorb 9.4% of the photons in the solar spectrum, while B will absorb 55.1%. By converting to the maximum theoretical current, this corresponds to $Jsc = 5.1$ mAcm^{-2} for A and $Jsc = 33.9$ mAcm^{-2} for polymer B, that is, lowering the bandgap of the polymer results in a higher theoretical current.

The value given for the maximum theoretical current is calculated from the assumption that the polymer absorbs all the photons from 280 to the wavelength given. The current measured depends on the absorption of the device, which includes absorption in the material, but also reflection losses from the window and interfaces.

The actual current measured for an OPV depends on several factors, such as morphology, so to obtain a more precise calculation, the incident photon to current efficiency (IPCE) should be considered, since this is a device measurement and factors like morphology, thickness, carrier mobility, carrier lifetime, and reflection losses have thus been taken into account.

Another important factor to keep in mind when calculating the efficiency of the device is the open-circuit voltage. Usually the maximum voltage obtainable decreases as a function of wavelength when calculated from the optical bandgap, which usually has an optimal value in the region of 0.9-1.2 eV.

Today's low bandgap polymers are based on fused-ring systems or are copolymers with alternating donor and acceptor groups. Fused ring systems enhance the quinoid resonance structure, which in turn reduces the bond alternation.

Some examples demonstrate the effect of additional electron donating groups (EDGs) and electron withdrawing groups (EWGs) to the polymer backbone.

For example, in a bithiophene repeating unit, one thiophene has an EDG and one has a EWG. This results in a donor/acceptor-based copolymer.

These polymers can be divided into two categories:

1. A fused-ring HOMO polymer, for example, poly(isothianaphthene), PITN, and

2. Copolymers based on donor and acceptor majorities, for example, the polymer based on benzo-thiadiazole and thiophene, PBT.

Several copolymers based on fluorene have been developed, where the fluorene and thiophene unit functions as the electron-donating unit, and these are coupled with varying electron accepting groups such as benzothiadiazole, thienopyrazine, and thiadiazolequinoxaline.

Some of the low bandgap polymers mentioned above have been applied in solar cell devices and the efficiency of the devices with the low bandgap polymers is small compared with the efficiency of P3HT based devices.

The measured Isc is several times lower than the maximum theoretical current. The choice of PCBM as an electron acceptor in these devices might explain the low Voc caused by a poor overlap between the HOMO of the polymer and the LUMO of the acceptor.

Inverted Polymer Solar Cells

The inverted polymer solar cell is a very special organic bulk heterojunction system developed recently. It consists of the donor material regioregular poly(3-hexylthiophene) (P3HT), and the fullerene derivative [6,6] phenyl C61 butyric acid methyl ester (PCBM). This system has produced single-junction cell efficiencies over 4%. The most common device of this type is fabricated using the superstrate cell structure: glass/indium tin oxide (ITO)/poly(4,3-ethylene dioxythiophene) (PEDOT):poly(styrene sulfonate) (PSS)/P3HT:PCBM/Al.

Often, a thin evaporated layer of LiF or another alkali metal containing salt is added between the polymer and Al cathode to lower the barrier to electron transport at that interface.

More recently, however, polymer bulk heterojunction devices have been developed that are electrically inverted. The ITO-coated glass or plastic is still the superstrate, but in these devices, the ITO functions as the cathode.

To change the work function of the ITO, various strategies have been used, such as treatment with Cs_2CO_3 via vacuum deposition and solution processing of solution-processed amorphous TiOx and ZnO in self-assembled monolayers.

These devices have reached power conversion efficiencies over 4%. However, there remain problems with this approach. Indium is an expensive metal, and ITO is deposited in a time-consuming sputtering process, which would damage the underlying organic if used as a top electrode.

In addition, the brittle nature of ITO makes it unattractive for use in flexible organic solar cells. Various solution-processable alternatives to ITO have been proposed and used in solar cells, including carbon nanotubes (CNTs) and graphene sheets, and more recently combinations thereof.

Carbon nanotubes (CNTs) have been used recently instead of ITO and even though reasonable device efficiencies have been obtained in organic PV cells, when the ITO was replaced with CNTs in the standard device configuration, the properties of CNT films do not compare favorably to ITO. The sheet resistances are around 200 ohm/sq. at the 85-90% transmissivity, which is required for efficient device performance.

The only polymer bulk heterojunction device reported in which a CNT film was used as a top electrode in an inverted organic PV cell on ITO yielded a poor power conversion efficiency of 0.3%.

Graphene sheets show similar performance to CNT films and in addition require very high temperature annealing steps to obtain conductivities that are still too low for efficient devices to be fabricated.

Combining the two carbon nanostructures into one electrode still does not result in sheet resistivities and transmissivities comparable to ITO.

Using specially prepared Ag nanowire films, as shown in the figure solves the above described problems and inefficiencies. So for now, Ag nanowire films remain the only solution-deposited ITO alternative that meets the performance requirements for photovoltaics, at 10 ohm/sq. with 85% transmissivity over the wavelength from 400 to 800 nm. In addition, their use as a top electrode is highly desirable and it serves to improve the device performance.

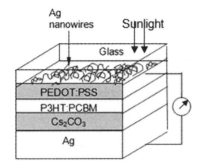

Figure 4-39. Inverted polymer solar cell structure.

Thus fabricated solar cells have reported power conversion efficiency of approximately 2.5% under 100 mW/cm^2 of AM 1.5G illumination, while similar single junction devices of the same type have achieved efficiency of over 4%.

Because they are deposited onto metal, they have unique advantages, some of which are that metals are exceptional barriers to moisture and oxygen, and that they can be textured or shaped into different shapes as needed to enhance light trapping.

Hermetic cell packaging can be manufactured by laminating the top of the cells with glass or plastic materials. Such an approach retains the throughput advantages of roll-to-roll deposition, while still using glass or plastic substrates as a barrier and being convenient for panel installation.

These cells are also potentially useful as components in tandem cells that are not constrained by current matching. Most importantly, because they are ITO-free and fully processed from cheap chemical solution, their cost could be potentially very low.

Efficiency is still a problem, of course, but we expect it to increase significantly in the near future. Thus produced solar cells can be used in a number of applications, where cost is a factor.

Ink Print Polymer Solar Cells

Highly efficient and cost-effective polymer solar cells based on poly(3-hexylthiophene) (P3HT) and 1-(3-methoxycarbonyl)-propyl-1-phenyl-(6,6)C61 (PCBM) have been fabricated by inkjet-printing. Several additives have been added to increase the cells' efficiency.

All solution-processed sequences of creating the different layers in the solar cells—a poly(3,4-ethylenedioxythiophene):poly(styrenesulfonate) (PEDOT:PSS) layer and a photoactive layer based on P3HT and PCBM with additives—were applied via simple and inexpensive inkjet-printing.

Device performance is strongly influenced by the addition of high boiling point additives in the photoactive ink with chlorobenzene (CB) solvent. The morphology, optoelectronic properties, and overall solar cell performance of the devices were dramatically affected by additives, such as 1,8 octanedithiol (ODT), o-dichlorobenzene (ODCB) and chloronaphthalene (Cl-naph). A device fabricated from ink formulated with ODT exhibited the best overall performance with power conversion efficiency (PCE) of 3.71%.

For an optimum ink formulation, different high boiling point additives, such as ODT, ODCB, and Cl-naph, can be added. The addition of these additives greatly and positively influences the morphology, the optical and electrical properties, and the overall performance of the devices.

The special types of devices fabricated from inks formulated with high boiling point additives have exhibited much improved performance, compared to devices prepared without additives. This is due to optimum film morphology, improved light absorption properties, and reduced recombination losses due to an improved crystallinity and charge transport.

The inkjet-printed device with an active layer of P3HT:PCBM containing the high boiling point additive ODT and a modified PEDOT:PSS layer has demonstrated the best solar cell performance with a PCE of ~3.7% and an IPCE of ~55%.

This improvement, especially in the IPCE with the addition of different additives is believed to increase both the optical absorption and the charge transport, as well as reduce recombination losses due to improvements in crystallinity and surface morphology.

Therefore, it can be concluded that this type of inkjet-printing of polymer compounds provides a twofold improvement in the overall solar cell performance:

1. Efficient light harvesting by improved P3HT crystallinity, and

2. Enhanced charge transport and better surface morphology for the top metal contact.

Much higher efficiencies are possible with further improvement of the organic materials, the additives and the respective manufacturing processes.

Conductive Polymers

Conductive polymers, also known as intrinsically conducting polymers (ICPs) are organic polymers that are capable of conducting electricity. They may have metallic or semiconductor properties, but in all cases they have the great advantage of being easily processable by means of a dispersion process.

Conductive polymers are organic materials, but are not thermoplastics so they cannot be thermo-formed. Their electrical conductivity is usually quite high, nevertheless, their properties are different from the mechanical properties of other commercially available polymers.

Their main advantage of conductive polymer materials is that their electrical properties can be fine-tuned by means of organic synthesis and/or by advanced dispersion techniques.

The main types of conductive polymers are the "polymer blacks" (polyacetylene, polypyrrole, and polyaniline) and their copolymers. Historically, these are known as melanins. Poly(p-phenylene vinylene) (PPV) and its soluble derivatives have emerged as the prototypical electroluminescent semiconducting polymers. Today, poly(3-alkylthiophenes) are the archetypical materials for solar cells and transistors.

One extensively studied class of quasi-one-dimensional polymers encompasses species with the abridged formulation MX and MMX where Mis a metal cation, commonly, Ni, Pd or Pt and X is a halogen, like Cl, Br or I, acting as a bridge between sequentially spaced metallic centers.

These two akin families are of great interest, not only for their potential electrical transport behavior, but also for their wide scope of physical properties. Prime examples are luminescence spectra with large Stokes shifts, large third-order nonlinear optical properties, gigantic third-order optical nonlinear susceptibility or progressive resonance Raman scattering.

Also, these materials exhibit the characteristic features of quasi-one-dimensional polymers such as spin density wave (SDW), charge density wave (CDW), solitons, polarons, and bipolarons.

One of the major problems with conductive polymers is that most of them require oxidative doping, so the properties of the resulting state are crucial. Such materials are salt-like (polymer salt), which diminishes their solubility in organic solvents and water and hence their processability. The charged organic backbone is also often unstable towards atmospheric moisture.

Compared to metals, organic conductors can be expensive requiring multi-step synthesis. The poor processability for many polymers requires the introduction of solubilizing or substituents, which can further complicate the synthesis.

Experimental and theoretical thermo-dynamic evidence suggests that conductive polymers may even be completely and principally insoluble and can only be processed by dispersion, complicating the process and limiting their practical application.

Summary

The performance of organic solar cells has recently been greatly improved, opening the path towards an alternative renewable energy source. To realize the main advantages of these cells, low cost and flexible devices, the development of electrodes is of great importance. The high cost, brittleness, and high temperature processing of indium tin oxide (ITO), commonly used in photovoltaic applications, drive a search for alternative electrode technologies.

So far, conductive polymers, silver nanowires, thin metal layers, carbon nanotubes, and graphenes have been investigated as alternatives to ITO for organic solar cells. The conductive polymer poly(3,4-ethylenedioxythiophene):poly(styrenesulfonate) (PEDOT:PSS) is in particular regarded as a promising electrode material because of its good optoelectronic properties and mechanical flexibility.

Several studies have investigated PEDOT:PSS electrodes for application in OSCs and organic light emitting diodes (OLEDs). In a previous study, we reported ITO-free OSCs fabricated on optimized PEDOT:PSS electrodes, using a solvent post-treatment process, resulting in very conductive electrodes.

The organic solar cells with post-treated PEDOT:PSS electrodes showed comparable efficiencies to ITO electrodes. In addition, it was shown that the solvent post-treatment removed insulating and hygroscopic PSS from PEDOT:PSS layers, improving the conductivity and air stability of the films.

While the research for ITO-free OSCs with PEDOT:PSS electrodes has mainly been focused on device performance, the stability of devices has been studied far less, despite of the acidic and hygroscopic nature of PEDOT:PSS, which are potential stability issues.

Thermally treating PEDOT:PSS electrodes has a positive effect on the lifetime of ZnPc:fullerene C60 bulk heterojunction small molecule OSCs and the overall performance of the devices. It is observed that the post-treatment not only improves the conductivity of PEDOT:PSS but significantly enhances the lifetime of devices as well as the short circuit current density and efficiency of OSCs.

We attribute these effects to the removal of insulating and hygroscopic PSS by the solvent post-treatment. These PEDOT_OSCs show an improvement of JSC and PCE by the solvent post-treatment almost independent of the HTL materials used. In addition, the thermal post-treatment considerably improves the lifetime of devices with a strong reduction in FF decay.

These results indicate that the solvent post-treatment for PEDOT:PSS electrodes is a very promising method for the development of low cost and stable ITO-free solar cells.

This is a truly fascinating field, which will have an ever-increasing impact on new and developing solar energy technologies, as well as on their more mature cousins. Nevertheless, we must always keep in mind that the final purpose of these technologies is to provide efficient and cheap power during many long years of non-stop on-sun operation.

So, manufacturers should focus on the basics—conduct thorough R&D, make a good product, and test it completely—before going to market with it. Efficient, reliable product is the key. Skipping steps in the development and testing process and rushing to market is not the right approach.

HYBRID SOLAR TECHNOLOGIES

The term "hybrid", as used in this section of "exotic" technologies, refers to solar cells, which either:

1. Are composed of a number of different types of solar cells, or

2. Contain components of different solar materials, or device combinations.

These could be classified as Hybrid organic-inorganic solar cells, hybrid nanotech thermal electric, polymer-nanoparticle composite, nanostructured inorganic-small molecules solar cells, hybrid solar-thermoelectric systems, and more.

We will take a look at what we see as some of the most promising hybrid technologies under development, and those that show promise for implementation. Some we will look at in their entirety, and others only in part.

Hybrid Organic-inorganic Solar Cells

A hybrid organic-inorganic solar cell, as the name suggests, consists of a combination of both organic and inorganic (some semiconducting) materials. It combines the unique properties of inorganic semiconductors with the film-forming properties of the conjugated polymers.

Organic materials usually are inexpensive, easily processable, and chemically synthesized. On the other hand, inorganic semiconductors can also be manufactured as processable nanoparticulate colloids. By varying the size of the nanoparticles, their band gap can be tuned and their absorption/emission spectra can be tailored.

Such hybrid solar cells have been demonstrated in conjugated polymer blends containing $CdSe$, $CuInS_2$, CdS, or PbS nanocrystals. The use of inorganic semiconductor nanoparticles embedded into semiconducting polymer blends is promising for several reasons:

1. Inorganic semiconductor nanoparticles can have high absorption coefficients and higher photoconductivity as compared to many organic semiconductor materials.

2. The n- or p-type character of the nanocrystals can be varied by synthetic routes, and

3. Band gap of inorganic nanoparticles is a function of nanoparticle size.

If the inorganic nanoparticles become smaller than the size of the exciton in the bulk semiconductor (typically about 10 nm), the electronic structure of such small particles is more like those of giant molecules than an extended solid.

High surface tension of very small inorganic nanocrystals makes them unstable, so they have a tendency to grow to larger particles by a process called "Ostwald ripening". Therefore, nanoparticles are synthesized commonly shielded by an organic ligand.

These ligands prevent the aggregation and oxidation of the nanoparticles and can alter the solution/dispersion characteristics of the particles into the polymer matrices. This organic ligand, on the other hand, is a barrier for transport of charges from nanoparticle to nanoparticle. Therefore, in the hybrid solar cells, such ligands must be removed to ensure intimate electrical contact between the nanoparticles.

Background

Conventional PV cells are made from inorganic materials such as silicon, germanium, gallium, etc. Their efficiency is relatively high, they use very expensive and sometimes rare materials in addition to energy intensive processes. To compensate for the lost energy, some variations of photo-electrochemical (PET) or dye sensitized solar cells (DSSC) have been developed as cheap alternatives for conventional silicon solar cells.

A compilation of materials and techniques used lately allowed the creation of the so-called "hybrid" solar cells, a combination of both organic and inorganic materials. That allows the devices to use the best and unique properties of both the inorganic semiconductors and the active (organic) film forming properties of the conjugated polymers.

Organic materials are usually widely available and inexpensive. They are easy to manufacture and their functionality can be manipulated by molecular design and chemical synthesis.

At the same time, inorganic semiconductor components can be manufactured as nano-particles, which are cheaper to produce than their large cousins—Si wafers and ribbons. Inorganic semiconductor nanoparticles offer the advantage of having high absorption coefficients and can be made in different sizes to fit different requirements, which offers the devices greater "tune-ability". This simply means that by varying the size and distribution of nanoparticles on the surface, the device bandgap can be tuned, thus adjusting the absorption range for greatest efficiency.

According to scientists at LIOS (2), hybrid solar cells are a mix of nanostructures of both organic and inorganic materials. Therefore, they combine the unique properties of inorganic semiconductor nanoparticles with properties of organic/polymeric materials.

In addition to this, low cost synthesis, processability and versatile manufacturing of thin film devices make them attractive. Also, inorganic semiconductor nanoparticles may have high absorption coefficients and particle size induced tunability of the optical band-gap. Thus, the organic/inorganic hybrid concept for photovoltaic solar cells is getting interesting and attractive in recent years.

Theoretically, hybrid solar cells are manufactured using different concepts such as solid state dye-sensitized solar cells and hybrid solar cells using the bulk heterojunction concept with different nanoparticles such as TiOx, ZnO, CdSe, CdS, PbS, $CuInS_2$, and others.

Some DSSC devices could be considered hybrids, because they could contain both organic and inorganic materials in their structure, and because of that we'll dedicate more time to their structure and function from the hybrid system operation point of view.

A dye-sensitized solar cell of Gräetzel type is comprised of several different materials such as nanoporous TiO_2 electrodes, organic or inorganic dyes, inorganic salts and metallic catalysts. After absorption of a photon,

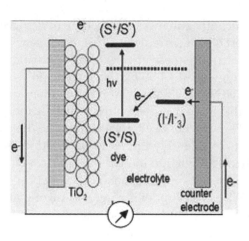

Figure 4-40. Operation principle of a dye-sensitized solar cell.(2)

the excited electron within the sensitizer molecule is transferred to the conduction band of TiO_2, and diffuses through the porous TiOx network to the contact. The oxidized sensitizer molecule is reduced to the original state by supply of electrons through a liquid electrolyte redox couple within the pores.

This photovoltaic conversion system is based on light harvesting by a molecular absorber attached to a wide band-gap semiconductor surface.

A monolayer of dye on a flat surface can only harvest a negligibly small fraction of incoming light. In this case it is useful to enlarge the interface between the semiconductor oxide and the dye. This is achieved by introducing a nanoparticle-based electrode construction which enhances the photoactive interface by orders of magnitude.

The dye sensitization of the large band-gap semiconductor electrodes is achieved by covering the internal surfaces of porous TiO_2 electrode with special dye molecules which absorb the incoming photons.

Sensitization effect can be seen in Figure 4-40 as a shift of the "incident photon to current efficiency (IPCE)" to higher wavelengths when coated with the dye.

The ideal sensitizer dye for a single junction solar cell converting global AM 1.5 sunlight to electricity should attach to the semiconductor oxide surface, absorb all light below a threshold wavelength and inject photoexcited electrons into the conduction band of the oxide.

Many different compounds have been investigated for semiconductor sensitization, such as porphyrins phthalocyanines, transition metal complexes, and coumarin. Metal complex sensitizers usually have anchoring (carboxylated) ligands for adsorption onto the semiconductor surface.

The dyes having the general structure of ML2(X)2,

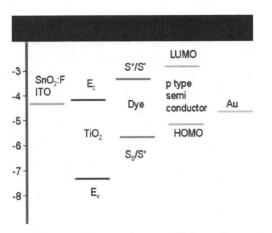

Figure 4-41. Energy diagram for an efficient charge transfer between solid state dye-sensitized solar cell components.(2)

where L stands for 2,20-bipyridyl-4-40-dicarboxylic acid, M for ruthenium or osmium and X for halide, cyanide, thiocynate, or water have been found promising. The excitation of Ru complexes via photon absorption is of metal to ligand charge transfer (MLCT) type.

This means that the highest occupied molecular orbital (HOMO) of the dye is localized near the metal atom, Ru in this case, whereas the lowest unoccupied molecular orbital (LUMO) is localized at the ligand species, in this case at the bipyridyl rings. At the excitation, an electron is lifted from the HOMO level to the LUMO level.

Furthermore, the LUMO level, extending even to the COOH anchoring groups, is spatially close to the TiO_2 surface, which means that there is significant overlap between electron wavefunctions of the LUMO level of the dye and the conduction band of TiO_2. This directionality of the excitation is proposed as one of the reasons for the fast electron transfer process at the dye-TiO_2 interface. Cells based on this concept show energy conversion efficiencies up to 11% on small-area cells, with module efficiencies between 5% and 7%.

In mass-production, organic material is mixed with a high electron transport material to form the photoactive layer. The two materials are combined in a heterojunction type photoactive layer, where one of the materials acts as the photon absorber and exciton donor, and the other facilitates exciton dissociation at the junction by charge transfer.

Both materials are in contact with each other, and this combination produces higher power conversion efficiency than a single material.

The different steps for the charge transfer (conversion) are:

1. Excitation of donor
2. Excitation delocalized of a donor/acceptor complex
3. Charge transfer initiated
4. Charge separation

The acceptor material needs a suitable energy offset to the binding energy of the exciton to the absorber. Charge transfer is favorable if the following condition is satisfied:

$$E_A^A - E_A^D > U_D$$

Where
E_A is the electron affinity, and
U is the coulombic binding energy of the exciton on the donor,
Superscript A is the acceptor, and
Superscript D is the donor.

In commonly used photovoltaic polymers such as MEH – PPV, the exciton binding energy ranges from 0.3 eV to 1.4 eV.

The energy required to separate the exciton is provided by the energy offset between the LUMOs or conduction band of the donor and acceptor. After dissociation, the carriers are transported to the respective electrodes through a percolation network.

The average distance an exciton can diffuse through a material, before annihilation by recombination happens, is the exciton diffusion length. This is short in polymers, on the order of 5-10 nanometers.

The time scale for radiative and non-radiative decay is from 1 picosecond to 1 nanosecond. Excitons generated within this length close to an acceptor would contribute to the photo current.

To avoid the problem of the short exciton diffusion length (vs. a phase-separated bi-layer), a bulk heterojunction structure is used. Dispersing the particles throughout the polymer matrix creates a larger interfacial area for charge transfer to occur.

To justify large-scale manufacturing of hybrid solar cells, however, the efficiency of the product must be much higher. There are three factors that must be considered for this to happen:

1) Bandgap
Current organic photovoltaics have shown 70% of quantum efficiency for blue photons. However, the bandgap should be reduced to absorb red photons, which contain a significant fraction of the energy in the solar spectrum.

2) Interfaces
Contact resistance between each layer in the device

should be minimized to offer higher fill factor and power conversion efficiency.

3)   Charge Transport

Higher charge carrier mobility allows the photovoltaics to have thicker active layers while photovoltaics minimize carrier recombination and keep the series resistance of the device low.

At the same time, nanoparticles are a class of semiconductor materials whose size in at least one dimension ranges from 1 to 100 nanometers. The size control creates quantum confinement and allows for the tuning of optoelectronic properties, such as band gap and electron affinity. Nanoparticles also have a large surface area-to-volume ratio, which presents more area for charge transfer to occur.

The photoactive layer can be created by mixing nanoparticles into a polymer matrix.

For polymers used in these hybrid devices, the hole mobilities are greater than electron mobilities, so the polymer phase is used to transport holes. The nanoparticles transport electrons to the electrode.

The interfacial area between the polymer phase and the nanoparticles needs to be large, and this is achieved by dispersing the particles throughout the polymer matrix. However, the nanoparticles need to be interconnected to form percolation networks for electron transport, which occurs by hopping events.

Aspect ratio, geometry, and volume fraction of the nanoparticles are factors in their efficiency. The structure of the nanoparticles can take a shape of a nanocrystal, nanorods, hyperbranched, and others. Implementing different structures changes the conversion efficiency in terms of nanoparticle dispersion in the polymer and providing pathways for electron transport.

The nanoparticle phase is required to provide a pathway for the electrons to reach the electrode. By using nanorods instead of nanocrystals, the hopping event from one crystal to another can be avoided.

Fabrication methods for these materials include mixing the two in a solution and spin-coating the mixture onto a substrate, followed by solvent evaporation (aka sol-gel). Most of the fabrication methods do not involve high temperature processing.

Annealing increases order in the polymer phase, increasing conductivity. However, annealing for too long would cause the polymer domain size to increase, eventually becoming larger than the exciton diffusion length, and possibly allowing some of the metal from the contact to diffuse into the photoactive layer, reducing the efficiency of the device.

Inorganic semiconductor nanoparticles used in hybrid cells include CdSe (size ranges from 6-20 nm), ZnO, TiO, and PbS. The most common polymers used are P3HT (poly (3-hexylthiophene)), and M3H - PPV (poly[2-methoxy, 5-(2-ethyl-hexyloxy)-p- phenylenevinylene)]). P3HT has a bandgap of 2.1 eV and M3H-PPV has a bandgap of around 2.4 eV. The polymers used as photo materials have extensive conjugation and also happen to be hydrophobic. Their efficiency as a photo material is affected by the HOMO level position and the ionization potential, which directly affects the open circuit voltage and the stability in air.

The electron affinity of CdSe ranges from 4.4 to 4.7 eV. If the polymer used is MEH-PPV which has an electron affinity of 3.0 eV, the difference between the two is large enough to drive electron transfer from the CdSe to the polymer. CdSe also has a high electron mobility of $600 \text{ cm}^2/(\text{V} \cdot \text{s})$.

Values demonstrated for a cell with a PPV derivate as the polymer and CdSe tetropods as the nanoparticle phase has an open circuit voltage of 0.76 V, a short circuit current of $6.42 \text{ mA}/\text{cm}^2$, a fill factor of 0.44, and a power conversion efficiency of 2.4%.

Problems with these systems include controlling the amount of nanoparticle aggregation as the photolayer forms. The particles need to be dispersed in order to maximize interface area, but need to aggregate to form networks for electron transport. The network formation is sensitive to the fabrication conditions. Deadend pathways can impede flow.

A possible solution is implementing ordered heterojunctions, where the structure is well controlled. The structures can undergo morphological changes over time, namely, phase separation. Eventually, the polymer domain size will be greater than the carrier diffusion length, which lowers performance.

Even though the nanoparticle bandgap can be tuned, it needs to be matched with the corresponding polymer, which is not an easy task. The 2.0 eV bandgap of CdSe is larger than an ideal bandgap of 1.4 for absorbance of light, which makes the tuning process even harder.

Overall, compared to conventional semiconductors and solar cells, the properties of these structures, including efficiency and durability, are still lacking. The carrier mobilities are much smaller than that of silicon, which limits their performance and their ability to improve. So different materials and processing techniques must be found and developed for this technology to be deployed broadly.

Several hybrid photovoltaic devices have a significant growth potential as low cost materials with the added benefit of flexibility, which is needed in many applications. They also lend themselves to the most efficient and cost-effective roll-to-roll processing, which allows varying the scale of the solar power conversion and reduces manufacturing costs.

Some of these new and potentially useful hybrid solar technologies follow.

Hybrid PbSe Tandem Solar Cell

A variation of the technology, a tandem photovoltaic device structure, consisting of a PbSe nanocrystal film (inorganic) and a P3HT/PCBM (organic) bulk heterojunction film, was fabricated by research scientists. The PbSe film (top layer) serves as a photocurrent generator as well as a UV protector for the underlying polymer cell. The P3HT/PCBM photovoltaic cell (bottom layer) provides the necessary electric field to the top photoconducting layer to extract the photogenerated charge from that layer.

The charge extraction from the PbSe layer is demonstrated by using light-biased spectral response measurements. In addition, device lifetime measurements were performed under AM 1.5 and UV-enhanced illumination on the tandem cell and on a control P3HT/PCBM device.

These measurements demonstrated that, although the hybrid tandem cell is of lower efficiency, it is significantly more durable than conventional technologies, due to the preferential UV absorption in the upper inorganic PbSe nanocrystal film.

Hybrid Bulk Heterojunction Cells

Another strategy for hybrid solar cells manufacturing is to use blends of inorganic nanocrystals, such as $CuInS_2$, with semiconductive polymers as a photovoltaic layer. The basis of this is the bulk heterojunction concept. Bulk heterojunction concept in inorganic/organic hybrid solar cells is similar to that used in organic/organic solar cells. Excitons created upon photoexcitation are separated into free charge carriers at interfaces between two semiconductors in a composite thin film such as a conjugated polymer and fullerene mixtures. Electrons will then be accepted by the material with the higher electron affinity (electron acceptor, usually fullerene or a derivative), and the hole by the material with the lower ionization potential, which also acts as the electron donor. The solubility of the n-type and p-type components is an important parameter of the construction of hybrid solar cells processed from solutions.

Bulk heterojunction hybrid solar cells have been demonstrated in various semiconducting polymer blends containing CdSe, $CuInS_2$, CdS, or PbS nanocrystals.

This strategy is promising for several reasons:

1. Inorganic semiconductor materials can have high absorption coefficients and photoconductivity as many organic semiconductor materials.

2. The n- or p-type doping level of the nanocrystalline materials can easily be varied by synthetic routes so that charge transfer in composites of n- or p-type organic semiconducting materials with corresponding inorganic counterparts can be studied.

3. If the inorganic nanoparticles become smaller than the size of the exciton in the bulk semiconductor (typically about 10 nm), their electronic structure changes. The electronic structure of such small particles are more like those of a giant molecule than an extended solid. The electronic and optical properties of such small particles depend not only on the material, of which they are composed but also on their size. Band-gap tuning in inorganic nanoparticles with different nanoparticle sizes can be used for realization of device architectures, such as tandem solar cells in which the different bandgaps can be obtained by modifying only one chemical compound. A substantial interfacial area for charge separation is provided by nanocrystals, which have high surface area to volume ratios.

Photovoltaic devices from a composite of 8-13 nm elongated CdSe nanocrystals and regioregular poly(3-hexylthiophene) (P3HT) have been reported. Under 4.8 W/m^2 monochromatic illumination at 514 nm such devices with 80% (vol) CdSe had an Isc of 0.031 mA/cm^2 and a Voc of 0.57 V. For a similar device, others have achieved a power conversion efficiency of 1.7% under simulated AM 1.5 illumination with CdSe nanocrystals of 7-60 nm size.

Hybrid solar cells based on nanoparticles of $CuInS_2$ in organic matrices have been reported by several researchers. Also nanocrystalline $CuInS_2$ is used with fullerene derivatives to form interpenetrating interface donor-acceptor heterojunction solar cells.

Bulk heterojunctions blending of $CuInS_2$ and a p-type polymer (PEDOT:PSS); poly(3,4-ethylenedioxythiophene):poly (styrene sulfonic acid) in the same cell configuration showed better photovoltaic response, with external quantum efficiencies of up to 20%.

Hybrid Solar-thermoelectric

Typical DSSC devices absorb the visible and infrared light of the solar spectrum only, so photons with lower energy do not participate in the photo-conversion process. These photons are basically wasted, thus limiting the photoelectric conversion efficiency of the DSSC.

Not only that, but the wasted solar energy is then absorbed by non-active parts of the device and is then converted to heat energy. This process is harmful, because it increases the internal temperature and degrades the performance of the devices.

If we find a way to utilize the heat energy we would do two things: a) increase the absorption efficiency, and b) decrease the overall device temperature. Both of these processes benefit the conversion efficiency and improve the device performance.

Hybrid thermoelectric-BSSC devices have been proposed that can directly convert heat into electricity by using the Seebeck effect using TiO_2 nanocrystalline DSSC combined with a thermoelectric cell as the upper and lower compartments, respectively.

The DSSC device function depends on incoming photons that are absorbed by the optically and non-optically active materials, where their energy is converted to electrical and heat energies accordingly. In the regular DSC device the heat energy is wasted and that hurts the device performance. In contrast, in the DSSC-TC hybrid the energy produced by the low-energy photons heat is transmitted to the TC part of the device for additional thermoelectric conversion.

The first thermoelectric DSSC based hybrid cell for utilizing the heat energy produced by a solar cell was developed in 2010. This was done by replacing a liquid electrolyte with organic solid material. These solar cells consisted of an organic solid DSSC and nanogenerator which was tuned to harvest mechanical energy.

Typical DSSC dyes only absorb the visible and infrared light of the solar spectrum, and thus photons with lower energy cannot take part in the solar conversion process, limiting the photoelectric conversion efficiency of DSSC. The wasted solar energy is mainly absorbed by non-active parts of a device and is converted to heat energy, increasing the internal temperature and degrading the performance.

It is, however, possible to utilize the heat energy. Thermoelectric (TE) devices can directly convert heat into electricity due to the Seebeck effect, with the advantages of environmental compatibility, vibration-less and noise-less features, small size, long lifetime, etc. To achieve that, TEs can be combined with DSSCs to utilize the heat produced. This should improve significantly the overall device efficiency.

A hybrid tandem cell (HTC) using a TiO_2 nanocrystalline DSSC and a thermoelectric cell (TC) as the upper and lower compartments, respectively has been evalu-

ated. The DSSC absorbes the photons by means of optically and non-optically active materials, where the light energy is converted to electrical and heat energies. The heat produced is transmitted to the TC for thermoelectric conversion.

The typical two-wire HTC device is constructed by connecting DSSC and TC units in series, with wires attached to each unit for transport of the generated electricity.

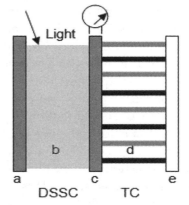

Figure 4-42. DSSC-TC hybrid PV cell

Where:
    a - Electrode with dye and TiO_2 coating
    b - Electrolyte
    c - Electrode with carbon coating
    d - p- and n- type semiconductors
    e - Base

Function

When light hits the top side of the DSSC, the electrons in the dye molecules are excited to a high energy state and then transferred to the conduction band of the TiO_2 electrode. Later, the dye molecules are restored to their original state after acquiring electrons from an organic solvent electrolyte containing an iodide/triiodide couple.

After diffusing in the TiO_2 film, electrons are exported through the cathode of DSSC and then captured by holes at the anode of the TC. The DSSC not only generates the photoelectric conversion, but also produces the heat transferred to the TC. The heat transfer process causes temperature gradient between the two sides of the TC, and thus the balance of carrier density in the semiconductor material is destroyed and then restored through a diffusion process, which changes the shape of the Fermi level of the TC.

At the cathode of the TC, electrons are exported with higher energy, so that the overall maximum output voltage Voc of the HTC device is 0.911V, while the short circuit current, Isc, is 119.4 mA.

The maximum output power P_{max} of this device is 38.02 mW, which compared with a single DSSC, the overall conversion efficiency is enhanced by 10%. This is done by optimizing the DSSC module with a photocurrent matching design of the two compartments of the HTC integrated in series.

The DSSC and TC components can be connected in series, and wires attached to their ends can transmit the generated power to an outside circuit.

Overall, the hybrid basically does not affect the DSSC microstructure or performance, so the efficiency of the DSSC itself is not affected. Compared to a single DSSC, however, the overall conversion efficiency of the hybrid device is enhanced by 10% when the two components are connected in series.

Flat Panel PV-T

Hybrid solar-thermoelectric technology is a combination between electric power and heat generation. This is not a new concept, and has been developed into a number of products, which always consist of two different parts—PV and thermal.

Researchers have now developed a device that can generate electric power and heat (hot water) from ordinary-looking flat PV modules, consisting of one single part. This is done by the use of high-performance nanotech materials, deposited onto the flat module surface with efficiency demonstrated seven to eight times higher than competing thermoelectric generators. This opens up the field of solar-thermal electric power conversion to a broad range of residential and industrial applications.

The two technologies that have dominated harnessing the power of the sun's energy are photovoltaics (PV), which convert sunlight into electric current, and solar thermal power generation, which uses sunlight to heat water and produce thermal energy (T).

PV cells have been deployed widely as flat panels, while solar-thermal power generation employs three-dimensional heat-absorbing surfaces for use in residential and large-scale industrial settings.

Solar thermal devices have failed the expectations to generate enough electric power at cost, so better light-absorbing surfaces have been developed lately, where enhanced nanostructured thermoelectric materials are placed within an energy-trapping, vacuum-sealed flat panel structure.

The combined mechanisms (PV-T) enhance the electricity-generating capacity while at the same time acting as solar-thermal power technology.

Electric power conversion technology based on the Seebeck effect and high thermal concentration enable wider applications, and similar STEGs generators have achieved a peak efficiency of nearly 5% under AM1.5G (at 1 kW m−2) conditions. The new PV-T hybrid flat modules generate hot water and electricity simultaneously at a minimal cost, thus making this type of power generation efficient and cost-effective.

The convenience of generating both types of energy from the same device at the same time, and its lower cost, which promises a 30% quicker payback on initial investment, opens a new energy-market niche. When fully developed, this device could become a major driver in the energy field.

Summary

Hybrids of the third generation type (or made of "exotic" materials and components) are exciting and very promising technologies, pending full understanding and development. Some of them hold the promise for deployment in a number of niche markets in the 21st century—after resolving present-day issues and getting out of the R&D labs. We are looking forward to it!

NEW PV MATERIALS

The real battle for PV efficiency and long-term reliability is just beginning. We will soon see a lot of action in the new and exotic materials area, where we expect major innovations and progress to be achieved. Materials in general determine the function and quality of the finished devices, so their role is undisputed.

We will look at some new materials, with the understanding that this is just the tip of the iceberg, and that we are barely scratching the surface.

3-D Porous Silicon

3-D porous silicon is a variation of the existing silicon technology, where the internal photon capture method discussed in 3-D solar cells in the previous chapter is fully utilized.

This so-called "porous silicon" solar cell employs a clever concept of increased surface area. It consists of a specially designed and manufactured silicon wafer made of mono-silicon, p-doped substrate, which surface is etched into, or otherwise dotted with, microscopic cavities (pores), which process can be done via special electrochemical anodization and other invasive processes. The wafer surface (including the inside of the pores) is coated with n-type silicon thin film.

The pores are actually very small, and can be broken

in: 1-2 nm size (microporous), 2-50 nm (mesoporous) and over 50 nm (macroporous). Due to the miniaturization, the cell fabrication—pore creation and thin film deposition—is difficult and must be extremely precise, which can make it quite expensive.

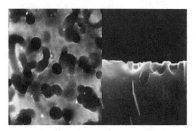

(a) Top view   (b) Side view

Figure 4-43. 3-D porous silicon substrate.

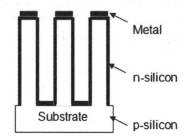

Figure 4-44. 3-D porous silicon solar cell

This porous structure has a cavernous (on the micro-level) internal surface, which on the overall retains the properties of a 2-dimensional solar cell structure. The overall surface area, however, is expanded several fold, which if used properly would provide much higher efficiency of the device.

Since the active area (inner surface) is increased many times, so is the p-n junction which runs along the entire internal surface area, thus the possibility of capturing photons and generating photocurrent is proportionally increased as well. This, of course, increases the efficiency of the device, which usually is limited by the surface area. The potential in this case depends on how many photons reach the p-n junction hidden in the caverns.

The actual measured efficiency of these devices is in the 2-3% range to date, but extrapolating for the entire porous area used properly gives us a theoretical 8% efficiency.

To illuminate the entire surface area, including the sides and the bottom of the pores, however, would require the device (and the pores) to be oriented in parallel to the sun's rays, and respectively to the photons impinging on them, at all times.

For maximum efficiency, however, this would require a tracking mechanism, as needed for maximum illumination of the insides of the pores, which would be too expensive and impractical for this type of product.

With more work, this type of device might find its place in 21st century energy markets, but for now it is in the R&D labs.

Ultrahigh Efficiency (UHE) Porous Silicon

Porous silicon, which is also a type of 3-D material, has not been utilized in the production of solar PV cells. R&D efforts are at a relatively early stage and many researchers believe that it is theoretically possible to achieve efficiencies on the order of 40% using this new technology.

Computer simulations incorporating a number of proposed alterations to the material at the nanometer scale indicate that such efficiencies are indeed possible.

Making porous silicon involves manipulating the cell material at a much higher degree of precision (in the micron and nano-meter level) relative to today's PV manufacturing techniques. Through the use of nano-structures called quantum dots (or quantum wells), a greater portion of the sun's energy can be absorbed and converted into electricity.

The end result is a new silicon material with greatly improved energy conversion efficiencies estimated to be in the 40% range, called ultrahigh efficiency porous silicon (UHE).

Currently there is no viable low-cost material or process to raise the efficiency enough for commercially viable products. Still, a number of manufacturers are working on the new UHE porous silicon technology, attracted by the promise of efficiency more than double the level of today's solar cells.

Porous Silicon for ARC Coating

Porous silicon has not achieved acceptable efficiency thus far, mostly because the active layers are not fully optimized. But porous silicon films can be used as antireflection coating (ARC), due to the extremely low reflection coefficients of ARCs based on porous silicon, which have been obtained by researchers recently.

These were comparable with the most efficient double layer ARCs of MgF_2/ZnS deposited on silicon surface preliminary textured with KOH. However, due to the high thickness of such ARCs they could not be effectively used in the silicon solar cells, because that would lead to increased series resistance of the structure and corresponding ohmic losses of photocurrent.

Using porous silicon layers of 200-nm thick for structures with normal depth p-n junctions does not considerably influence the series resistance and the fill factor, while at the same time decreasing the reflection losses.

In summary, the optical reflection of the surface of thin layers of PS is similar to values of reflection coefficients obtained by applying two-layer antireflective systems that are presently used in silicon SC technology. At the same time, antireflective coatings based on porous silicon are simple and cheap to make using conventional equipment.

However, to implement these AR coatings in large-scale manufacturing operations requires the solution of problems such as improvement and stabilization of the passivating processes and the resulting properties of the porous silicon films.

Comparison of the output characteristics of mono- and multi-crystalline solar cells with and without the porous layer shows that high quality passivation of silicon surface by the porous layer is needed. Only then it is possible to attain increase in the cells' efficiency. The improvement is due not only to minimization of the overall optical losses, but also to decrease in recombination losses on the active surface.

To eliminate resistive losses and corresponding decrease in fill factor, optimization of the solar cell structure is necessary, and especially the depth and profile of the emitter region doping.

It is possible to eliminate the increase of the emitter layer resistance by having optimized parameters of the diffusion process, which will reduce changes in the thickness and subsurface concentration of doping impurity.

This process seems promising, and though it is still in the R&D stage, it might prove to be one a key factor in optimizing the efficiency of solar cells and modules in the near future.

All-carbon Material for Solar Cells

There is another new—potentially disruptive—material (at least on paper), not using any of the established PV materials, where the solar cells are neither c-Si, nor thin films related. Instead, they are made entirely of a new carbon material, thus the name "all-carbon" cells, which (the cells and the name) were recently developed by US researchers. These new cells are made and tuned in a way that is capable of capturing usually unused infrared IR and near-IR energy in the sunlight spectrum.

Conventional solar cells are incapable of harnessing this type of energy, and as a matter of fact it (IR) contributes to overheating the cells, making them lose efficiency. It even damages them to the point of destruction in extreme cases.

The new all-carbon material is fully transparent to visible light, so it would be easy to make a tandem device with it by overlaying conventional cells. This combination of silicon solar cells and all-carbon cells can capture the entire range of the light spectrum, making it very efficient and avoiding overheating, efficiency loss and damage associated with IR energy effects.

The new cell technology is actually a combination of two exotic forms of carbon active layers: carbon nanotubes and carbon C60 which is also known as bucky-balls.

These are brand new technologies, which only recently have been demonstrated in lab mode, with no viable mass production or commercial applications on the horizon. The complications lie in the fact that for the new solar cells to work, the nanotubes must be very pure, uniformly single-walled, and of the same symmetrical configurations (several interchangeable combinations are possible.)

Carbon nanotubes and C60 materials combinations open new opportunities and possibilities for increasing the efficiency of solar cells, since they are "like PV polymers on steroids," as the researchers say. This is an exciting development because it demonstrates the possibility of achieving high efficiency PV power conversion using an active layer that is entirely made from cheap carbon materials.

The new all-carbon cells, however, need a lot of additional work to be fully optimized by precise control over the exact shape and thickness of the layers of active materials in the cell structure. It will be a long time before they find a place on our roofs for large-scale commercial applications.

Still, they open a window for a glimpse into the future of photovoltaic devices, where clumsy c-Si and inefficient thin films will be replaced, or made more efficient, by new much more sophisticated PV technologies of the 21st century.

Conductive Polymers

Presently the inorganic material indium-tin oxide (ITO) is used as a conductive, transparent electrode in most thin film PV devices. While it has good electrical properties, ITO creates a number of problems in OPV devices. The problems range from complex process, to rigid structure, which does not allow roll-to-roll processing. Since these are the major advantages of OPV technology, the search for a more suitable (cheaper and flexible) transparent conducting electrode is underway.

Conductive polymers, or intrinsically conducting polymers (ICPs) are organic materials capable of conducting electricity. They may have metallic or semiconductor properties, but in all cases they have the great advantage of being easily processable by means of a dispersion process; thus, they are cheap and flexible.

The main types of conductive polymers are the "polymer blacks" (polyacetylene, polypyrrole, and polyaniline) and their copolymers. Historically, these are known as melanins. Poly(p-phenylene vinylene) (PPV) and its soluble derivatives have emerged as the prototypical electroluminescent semiconducting polymers. Today, poly(3-alkylthiophenes) are the archetypical materials for solar cells and transistors.

One of the major problems with conductive polymers is that most of them require oxidative doping, so the properties of the resulting state are crucial. Such materials are salt-like (polymer salt), which diminishes their solubility in organic solvents and water and hence their processability. The charged organic backbone is also often unstable towards atmospheric moisture.

At this stage of their development, however, organic conductors can be expensive, requiring multi-step synthesis. The poor processability for many polymers requires the introduction of solubilizing or substituents, which can further complicate the synthesis.

Experimental and theoretical thermo-dynamical evidence suggests that conductive polymers may even be completely and principally insoluble so that they can only be processed by dispersion, which complicates the process and limits their practical application.

Nevertheless, the effort continues, because the future of OPV is directly affected by the type of available conductive electrodes. So conductive polymers will remain on the research bench until a suitable combination is found.

Efficiency Sticker for PV Panels

In the battle for increased efficiency, this approach using specialized materials is a bit different. The claim here is that the power output of solar panels can be boosted by 5-15% by simply applying a large (made of special materials) transparent sticker to the front.

The sticker is actually a polymer film embossed with microstructures (Fresnel lens-like type) that bend incoming sunlight. This results in additional light absorption by the active materials (solar cells) in the panels. This additional light is converted into additional electricity.

The technology is cheap and could potentially lower the cost per watt of solar power. Also, unlike other technologies developed to improve solar panel performance, this one can be added to panels that have already been installed. The polymer film does three main things:

1. Prevents light from reflecting off the surface of solar panels,

2. Traps light inside the semiconductor materials that absorb light and convert it to electricity, and

3. Redirects incoming light so that rather than passing through the thin semiconductor material, it travels along its surface, thus increasing the chances it will be absorbed.

Researchers have designed the stickers' microstructures that accomplish this function by using algorithms that model how rays of light behave as they enter the film and encounter various surfaces within the solar panel, including the protective glass cover, the semiconductor material, and the back surface of the panel—throughout the day.

The key to accomplishing this is bending the light through the optimal paths and right amounts, just enough to enter the solar panel at an angle, but not so much of an angle that the light reflects off and is lost. If any light is reflected off either the glass or semiconductor surfaces, the film redirects much of it back into the solar panel.

Tests at the National Renewable Energy Laboratory showed that the film increases power output on average between 10 and 12.5%. But the best news of this type of improvement is the fact that the performance is best (as compared to conventional solar technologies) under cloudy conditions, when incoming light is diffuse.

Adding the sticker film either in the factory for best performance, or in the field—on solar panels already in use—increases the overall cost of solar panels by 1-10%, but the panels would produce enough additional electricity to justify the price and account for significant long-term savings.

Increasing the power output of a solar panel also decreases other costs—such as shipping and installation—because fewer solar panels are required at each installation.

But here is the catch: the overall benefit and potential savings due to adding the sticker film depend on how long the polymer film lasts.

The cost per kilowatt hour of solar power is figured by estimating the total power output of the solar panel during its lifetime (20 to 25-years) and the related warranty conditions.

If the film is scratched, if it attracts excessive dust, or becomes discolored (non-transparent) after years in the sun, it could lower power output.

Based on our 30 years of experience, we must assure the researchers and manufacturers that any polymer exposed non-stop to desert sunlight will become semi-transparent within 2-3 years (if it doesn't get sand-blasted first), and things will go down hill fast from there.

So, like with all other types and components of this nature, durability is the big, unresolved issue. The materials used in solar panels must last for decades, and the durability of this sticker film hasn't been tested or verified.

Graphene vs. ITO

Indium tin oxide (ITO) and fluorine tin oxide (FTO) have been widely used as window electrodes in opto-electronic devices. Typically indium-tin-oxide (ITO) has been used for one electrode and aluminum, calcium, or magnesium for the other.

The big problem is that the ITO film, which appears to be increasingly problematic due to a) the limited availability of the element indium, b) its instability in the presence of acid or base, c) its susceptibility to ion diffusion into polymer layers, d) its limited transparency in the near-infrared region, and e) its brittle nature.

Note: FTO devices exhibit excess current leakage caused by FTO structure defects, so their use is limited to some special applications and devices.

The search for novel electrode materials with good stability, high transparency and excellent conductivity is therefore a crucial goal for optoelectronics. Graphene, two-dimensional graphite, as a rising star in material science, exhibits remarkable electronic properties that qualify it for applications in future optoelectronic devices.

Recently, transparent and conductive graphene based composites have been prepared by incorporation of graphene sheets into polystyrene or silica. However, the conductivity of such transparent composites is low, typically ranging from 10-3 to 1 S/cm depending on the graphene sheet loading level, which makes the composites incapable of serving as window electrodes in opto-electronic devices.

Fabrication of conductive, transparent, and ultrathin graphene films from exfoliated graphite oxide, followed by thermal reduction is one of the preferred methods. The obtained graphene films with a thickness of ca. 10 nm exhibit a high conductivity of 550 S/cm, which is comparable to that of polycrystalline graphite (1250 S/cm), and a transparency of more than 70% over 1000-3000 nm. The application of graphene films as window electrodes in solid-state dye-sensitized solar cells has been demonstrated and proven.

Graphene sheets have been produced either by mechanical exfoliation via repeated peeling of highly ordered pyrolytic graphite (HOPG) or by chemical oxidation of graphite. Considering the facile solution processing, the oxidation of graphite was preferred for this study. Oxygen-containing functional groups render the graphite oxide (GO) hydrophilic and dispersible in water.

Graphene in the form of chicken wire mat, made of carbon atoms is cheap and readily available. It is an adequate replacement for ITO; but, as with every new technology, there are problems which have barred its use for this application. One problems is handling; graphene sheets tend to stick together, and because they are a single atom thick, separating them is not easy. Another issue is affinity. While graphene clings to itself, it does not have the same affinity for the polymers or small molecules used in OPV. The sheets do not adhere well to the PV material.

Lately, graphene sheets doped with impurities have been made that improve the adhesion to polymers of OPV cells, while at the same time improving their electrical conductivity.

Graphene is flexible too—another benefit. Overcoming the barriers could bring the technology one step closer to cheap, robust, flexible, PV cells that can be manufactured and transported in rolls. They could then be simply unfurled like tar-paper on roofs to form solar power installations.

Ink for PV Cells

There are many types of "inks" used for photovoltaic conversion, some of which we review in different sections in this text, so now we will take only a cursory look at some of the new types.

Si Quantum Dots Inks

A fairly new development, this light-activated power generating product is based on a unique and patented solvent-based silicon nanomaterial platform that can be applied like ink on any substrate. Developers claim that this approach, in addition to its unique optical advantages, has significant cost savings over traditional silicon products by using less (and/or less expensive) semiconductor materials, as well as by having a more efficient manufacturing process, which lends itself to high speed (roll-to-roll) production.

This new technology consists of processing the quantum dots in the silicon "ink" in a way that makes it possible to use the old "roll-to-roll" printing technology used for printing on paper or film. Applying ink directly on any substrate (including a flexible one) allows applications such as tagless printing for clothing labels and portable chargers for consumer and military customers.

By controlling the sizes of the dots from 2 to 10 nm,

the absorption or emission spectra of the resulting film can be controlled. This allows capture of everything from infrared to ultraviolet and the visible spectrum in between which is not possible with conventional technology.

The technology is also used as an efficient light source. By controlling particle size, you can produce light of any color or a combination of particle sizes that will give off white light. This application might provide additional, and possibly larger markets for this technology in the near future—at least in some specialized areas.

CZTS Ink

Researchers at IBM have increased the efficiency of a novel type of solar cell made largely from cheap and abundant materials by over 40 percent. According to industry sources, the efficiency is now 9.6%, which is up from the previous record of 6.7% for this type of solar cell, and near the level needed for commercial solar panels.

Equally important, the IBM solar cells also have the advantage of being made with an inexpensive ink-based process. They convert light into electricity using a semiconductor material made of copper, zinc, tin, and sulfur—all abundant elements—as well as the relatively rare element selenium, under the abbreviation CZTS.

The IBM cells use indium tin oxide (ITO) as a conductive material, which is limited by the availability of indium. But several other conductors could work as well, and IBM is looking into it. In addition, they are also investigating the possibility of completely replacing the selenium in the CZTS-inked solar cells with sulfur. For the record-efficiency cell, researchers replaced half of the selenium used in a previous experimental cell, so if all of the selenium could be replaced, the cells could, in theory, supply all of the electricity needs of the world. This, of course is provided there are suitable means for storing and redistributing power for use at night or on cloudy days.

The IBM solar cells could be an alternative to existing "thin film" solar cells, which use materials that are particularly good at absorbing light, but some are rare and exotic, while others are toxic.

The new cells could also have advantages compared to cells made of copper indium gallium and selenium (CIGS), which are just starting to come to market. That's because the indium and gallium in these cells is expensive, and while the selenium used in the IBM cell is rarer than indium or gallium, its cost is a tenth of those.

The new ink-based manufacturing process solves some of the key challenges to making efficient CZTS cells too. A common approach to making any type of high-quality solar material is to dissolve a precursor substance in a solvent. This isn't possible with the CZTS cells because the zinc compounds required in the new cells aren't soluble.

To get around this, researchers use a combination of dissolved materials and suspended particles, creating a slurry-like ink that could then be spread over a surface that's been heat-treated to produce the final materials. The particles prevent the material from cracking and peeling as the solvent evaporates.

IBM researchers are also looking into ways to improve the efficiency of the new solar cells, with the goal of reaching about 12%—high enough to give manufacturers confidence that they could be mass produced and still have efficiency levels of around 10 percent. The goal is a 15% efficiency improvement and should be possible by improving other parts of the solar cell besides the main CZTS material, or by doping the semiconductor with other trace elements, which is easy with the ink-based process.

Metamaterials

Metamaterials are artificial materials engineered to have properties that may not be found in nature. These materials, and their derivations and combinations, offer the most promise for exceptional discoveries and exciting new developments in the future.

Their special properties are usually due to their unique structure and physical arrangement, rather than from special composition. The usual approach is to use small in-homogeneities in some materials, and arrange them in such a way as to create some sort of macroscopic and unusual behavior, or misbehavior.

These are usually heterogeneous materials employing the juxtaposition of many microscopic elements, giving rise to properties not seen in ordinary materials. Using these properties in appropriate fashion, it is possible to make materials—including solar cells—that are perfect absorbers over a narrow range of wavelengths.

Nevertheless, most of this is still expressed in theoretical calculations and ongoing research, some of which is focused on materials with negative refractive index, which appears to permit the creation of super-lenses. These super-lenses can have a special spatial resolution, which is somewhat below the wavelength—a sort of 'invisibility' that has been demonstrated at least over a narrow wave band with gradient-index materials. It's a "back to the future" thing, but exciting and promising.

In more practical terms, these materials can 'bend' visible light at unusual but precise angles, regardless of its polarization, which is a step toward perfectly transparent solar-cell coatings that could also direct the sun's rays into the active area to improve solar power output.

One version of metamaterials is basically a metal film several hundred nanometers thick. The metal films are patterned with deep cavities shaped as circles or similar symmetrical geometric shapes. Each shape is built around a thin column made of the same, or similar, material.

There is another material (metal usually) filling the space between the wire and the cavity wall. The size and shape of the patterns determines the material function, whether it bends or refracts light of different colors, and to what degree.

Each of the shapes and single-layer films can be made using common lithography, etching and coating techniques similar to those used in making semiconductor devices.

The laws of science stipulate that when light moves from one medium to another it scatters, or gets split into many light bands. This is why a subject dipped in a glass of water appears to be distorted, This is due to the mismatch between the refractive index of water and air.

A solar cell coated with a material whose refractive index is identical to that of air would allow the light to go thru unimpeded, without any bending or reflection. So, in such media the above-mentioned subject in the glass of water would no longer look distorted.

Similarly so, the goal of researchers now is to produce films that have a refractive index exactly equal to that of air. A material with these extraordinary properties (none of which is found in nature) would not bend light, but instead it would transmit it perfectly, without any reflection and/or losses.

In this case, the major challenge in engineering metamaterials is a serious energy loss. As the metal structures interact with light, they lose energy, which is converted to heat, which we well know is detrimental to the proper function of solar cells. This is no exception, and the heat loss measured in these types of devices is so great that just less than half of the incident light can pass through them.

If metamaterials are to become a reliable and efficient source of energy, they must be tuned to pass at least 90 percent of the light. One way to do this is by amplifying the incoming sunlight as it passes through metamaterials. This can be done via optical amplifiers, similar to those used in lasers and in telecommunications. Incorporating the optical amplifiers within the solar cells' thin film structures will enable metamaterials to be used as efficient solar energy conversion devices.

The research in metamaterials is interdisciplinary and involves such fields as electrical engineering, electromagnetics, solid state physics, microwave and antennae engineering, optoelectronics, classic optics, material sciences, semiconductor engineering, nanoscience and others.

The increasing popularity of metamaterials and their potential impact on a number of practical fields—including use as solar cells—makes them very interesting to the scientific community, so the work in this area continues and we expect great results in the near future.

Nanonets

Nanonets are another application under the umbrella of nano-technology for use in solar energy generation. This is a somewhat different but promising approach to use the sun's energy more efficiently. Nanonets have great potential in practical applications such as splitting water into hydrogen gas, which can be stored and used anytime—a very important requirement of the solar industry.

A cheap new nanostructured material could prove an efficient catalyst for performing this critical reaction. This material is called "nanonet" because it has a two-dimensional branching structure, made up of branching wires of titanium and silicon, thus comprising a compound that has been demonstrated to enable and sustain the water-splitting reaction.

Titanium disilicide, which absorbs a broad spectrum of visible light, has been shown to split water into hydrogen and oxygen. Amazingly enough, it can also store the hydrogen, which it absorbs or releases by changing the temperature. Many other semiconducting materials have been tested as water-splitting catalysts as well, but most have proved inefficient, or unstable.

The nanonets are very complex compounds, so making these long, thin structures requires limiting growth in all but one dimension—up. Limiting growth in one dimension (width) while promoting the growth of complex structures in height is very challenging. Under the right conditions, however, the process happens spontaneously.

Nanonets, made of flexible wires about 15 nanometers thick, grow spontaneously from titanium and silicon in a reaction chamber at high temperatures. The high surface area of the compound helps the nanonet to enhance the reaction, and the nanonets material is 10 times more electrically conductive than its bulk form.

Conductivity is an important property for water-splitting catalysts. and the nanostructured version of the material performs about 100 times better than bulk titanium disilicide.

The nanonets' large surface area, conductivity and other properties might be useful as an electrode for water splitting, provided they prove efficient and cheap enough. However, due to their complexity, the fabrication methods used presently may limit their implementation in the near term.

More importantly, when they do get out of the R&D labs and enter the crowded energy market, will they be able to compete?

Nanostructured AR Coatings

A number of researchers are looking into special coatings that mimic structures found in nature to increase the use of solar energy. Nanostructured coatings have special properties, which, if properly understood and applied, could help increase the efficiency of most solar cells.

One way is by reducing the reflectivity of their surface. Most types of solar cells lose about 20-30% of the available sunlight energy because it is simply reflected away by the reflectivity of their surface materials (glass, plastics, etc.).

There are, however, new engineered coatings which can by mimic the way a moth's eye absorbs light! This approach (whenever implemented) could reduce unwanted reflection from 20-30% to less than 2% on a typical solar cell.

The materials are kept secret at this point, but durability testing is underway at National Laboratories, where a number of nanostructured coatings are applied to various solar cell surfaces, as needed to determine the feasibility of use in harsh environments, including heat, humidity, and the radiation encountered in outer space.

This is an important project, and if successful, the result would be a significant increase in overall efficiency of most solar cells.

Nanotube Conductors

Many new solar cells use transparent thin films that are also conductors of electrical charge. The dominant conductive thin films used in research now are transparent conductive oxides (TCO), and include fluorine-doped tin oxide (SnO_2:F, or FTO), doped zinc oxide (like ZnO:Al), and indium tin oxide (ITO). These conductive films are also used in the LCD industry for flat panel displays.

The dual function of a TCO allows light to pass through a substrate window to the active light-absorbing material beneath, and also serves as an ohmic contact to transport the photogenerated charge carriers away from that light-absorbing material. Present TCO materials are effective for research, but perhaps are not yet optimized for large-scale photovoltaic production. They require special deposition conditions at high vacuum, and can sometimes suffer from poor mechanical strength, while most have poor transmittance in the infrared portion of the spectrum (e.g., ITO thin films can also be used as infrared filters in airplane windows). These factors make large-scale manufacturing more costly.

A relatively new area has emerged using carbon nanotube networks as a transparent conductor for organic solar cells. Nanotube networks are flexible and can be deposited on surfaces in a variety of ways. With some treatment, nanotube films can be highly transparent in the infrared, possibly enabling efficient low-bandgap solar cells. Nanotube networks are p-type conductors, whereas traditional transparent conductors are exclusively n-type. The availability of a p-type transparent conductor could lead to new cell designs that simplify manufacturing and improve efficiency.

PV Paper

The major disadvantages of all present approaches for photovoltaics are high production and material costs. Organic photovoltaics based on bulk heterojunctions (BHJs) of a polymer/fullerene blend have the potential to overcome these disadvantages.

In a conventional BHJ photovoltaic cell, a composite of a conjugated polymer (donor) and a fullerene (acceptor) is sandwiched between an opaque metallic and a transparent electrode. The conjugated polymer harvests light and the generated excitons are dissociated at the donor-acceptor interface.

The built-in electric field, which is due to the difference in work functions of asymmetric electrodes, causes electrons and holes to drift towards the cathode and anode, respectively.

The regular device layout of a BHJ photovoltaic cell is substrate/ITO/PEDOT:PSS/photoactive layer/buffer layer/Al, where PEDOT:PSS is poly(3,4-ethylened ioxyt hiophene):poly(styrenesulfonate) and ITO is indium-tin oxide. The most commonly used photoactive layer is a blend of poly(3-hexylthiophene-2,5-diyl) (P3HT) and [6,6]-phenyl-C 61 butyric acid methyl ester (PCBM).

A thin buffer layer of LiF or Ca is commonly used to raise the work function of aluminum (Al) for better electron collection. PEDOT:PSS is used as the hole-conducting and electron-blocking layer for efficient hole collection by ITO.

To improve device stability, inverted layer structures are preferred, with a typical device layout of substrate/ITO/ZnO, TiO x/photoactive layer/PEDOT:PSS/Ag, or Au. In this case ITO is used for electron collection and Ag or Au is used for hole collection. In both structures, regular and inverted expensive electrodes (ITO, Ag or Au) are applied. In fact, expensive ITO has become an integral part of organic photovoltaics.

To simplify the manufacturing process, and bring it to cost-effective levels, roll-to-roll coated polymer/ fullerene based photovoltaics are used. Most of these processes, however, use expensive plastic substrates, and expensive ITO and Ag electrodes, which do not allow free patterning of the layers.

Instead paper substrate could be used in a special process now under development.

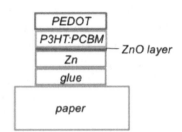

Figure 4-45. Printed paper PV

The manufacturing process consists of:

1. Cold foil transfer printing of a structured Zn bottom cathode on the glue,

2. Printing of P3HT:PCBM photoactive layer on top of oxidized zinc layer, and

3. Printing of PEDOT:PSS as transparent and conducting top anode.

The function of this (paper) cell is pretty simple too; sunlight hits the top transparent layer, and the electrons and holes are collected by the Zn and PEDOT:PSS layers respectively. Current can be extracted by attaching electrodes to the anode and cathode.

The advantage of free patterning of all the layers (using flexible and inexpensive materials) is that it can eliminate any extra process required to interconnect solar cells into a solar module. This type of roll-to-roll printing technique also offers the opportunity to realize more complex and cheaper device structures than the evaporation processes or coatings.

Of course, the applications of this type of solar cells (printed on paper) are limited. Technology will find niche markets where it could be used as a temporary energy source or replacement of PV power. Elements of the manufacturing process are still in development and could be used to optimize other similar processes as well.

Silicon Microwires as Photocathodes

Solar hydrogen generation, like other photo-enhanced processes, use photo-electrolysis of water and other liquids. These processes require efficient and durable photo-cathodes with high surface area. Presently all cathodes in use today are too expensive and not sufficiently effective.

Cathodes made of silicon microwires (arrays of these) meet the requirements by allowing for greater light penetration and offering larger surface area. These qualities will help achieve the goals, if the carrier mobilities are sufficiently high for surface reactions to occur before charges recombine (which is one of the problems).

Researchers report excellent electronic properties on positively doped silicon microwire arrays, grown with copper catalysts and used in a methyl viologen redox system. Although equivalent efficiencies for normal solar fluxes were only 2 to 3%, the high internal efficiencies and low use of the available optical flux suggest that further improvements are possible.

Silicon microwires appear to be a promising, high efficiency material for manufacturing photocathodes for hydrogen generation and other purposes. The work on this novel material is underway, and we are confident that it will find success in the near future.

PV Textiles

Smart, photocurrent-generating textiles are built on the basis of active photovoltaic fibers consisting of nano-layers of polymer-based organic compounds. A flexible solar cell, including a polymer-based anode, two different nano-materials in bulk heterojunction blends as the light absorbing materials, and a semi-transparent cathode to collect the electrons, are formed by coating these materials onto flexible polypropylene (PP) fibers layer-by-layer, respectively, to produce electricity. (12)

Both conventional and technical textiles are indispensable products for daily life. Research and development activities in the field of textiles are running parallel to the advances in smart materials, which sense all relevant environmental stimuli (electrical, chemical, mechanical, magnetic, optical, etc.) and evaluate, react, or sometimes adapt to those conditions.

Smart materials may be in the form of phase-changing materials, chromic materials, shape-memory polymers and alloys, piezo materials, and light-emitting diodes, as well as photovoltaic materials.

For example, smart photovoltaic textiles can produce power for electronic devices such as mobile phones, iPods, and pocket computers by collecting sunlight with nano-based materials. There are limited scientific studies and few commercial applications of wearable solar cells based on inorganic materials.

In fact, the patching process, which generally prevails in the development of wearable photovoltaic materials, may not always meet consumer demands such as flexibility, comfort, and ease of cleaning.

Although, there are some studies about flat textiles integrated with organic solar cells, photovoltaic fibers may form energy-harvesting textile structures in any shape and structure. Therefore, some research has been conducted to develop fiber-based solar cells using inorganic materials, photochemical reactions, etc. In the scientific literature, there are also a few patents, projects, and research papers about fiber-shaped organic solar cells.

To obtain photovoltaic fibers, both polymer and small molecule-based light-absorbing layers were used in previous studies. In one of these studies, the optical fibers, which are not flexible, were coated with poly(3-hexylthiophene) (P3HT): phenyl-C61-butyric acid methyl ester (P3HT:PCBM)-based photoactive materials.

While the light was travelling through the optical fiber and generating hole-electron pairs, the 100 nm top metal electrode (which does not let the light transmit from outside) was used to collect the electrons.

In another study that used small molecule-based materials in an organic active layer of the fiber-shaped solar cell, all layers were deposited onto polyimide-coated silica fibers using the thermal evaporation technique in a vacuum. A semitransparent top electrode that let the light enter the device was used, and the fibers were rotated during the process.

In organic solar cells, the most widely used transparent hole-collecting electrode material is ITO. However, besides being an expensive material due to the low availability of indium, ITO requires expensive vacuum deposition techniques and high temperatures to guarantee highly conductive transparent layers. The advantages of the application of transparent flexible plastic substrates are restricted due to the thermal and mechanical damages of the ITO deposition process.

There are some ITO-free alternative approaches, such as using carbon nanotube (CNT) layers or different kinds of PEDOT:PSS and its mixtures [23-27], or using a metallic layer [28] to perform as a hole-collecting electrode.

Therefore, in order to realize polymer-based solar cells, which are completely flexible, and to substitute the ITO layer, like the highly conductive PEDOT:PSS solution, used as a polymer anode that is more convenient for textile substrates in terms of flexibility, material cost,

and fabrication processes compared with ITO material.

The sun's rays entered into the photoactive layer of photovoltaic fiber by passing through a semi-transparent cathode which is very thin outer electrode layers consisting of ca. 10 nm of lithium fluoride/aluminum (LiF/Al).

The maximum short-circuit current density was obtained as 0.27 mA/cm^2, which is quite low, but it does prove that the fibers can be used as a power-generating source.

This is futuristic, no doubt, but so was the electric bulb less than a dozen decades ago. We would not be surprised if 100 years from now PV textiles are common items used by everyone to power personal electronics and many things that we've not yet imagined.

To enhance the power conversion efficiency of the photovoltaic fiber, existing materials and techniques must be improved. In particular, the optical band gap of the polymers used as the active layer in organic solar cells is

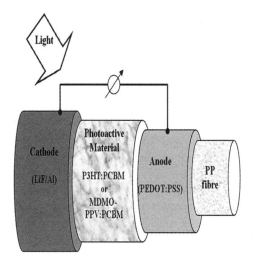

Figure 4-46. Schematic drawing of a photovoltaic fiber with the active layers exposed. (12)

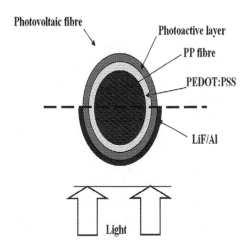

Figure 4-47. Cross section of a photovoltaic fiber.(12)

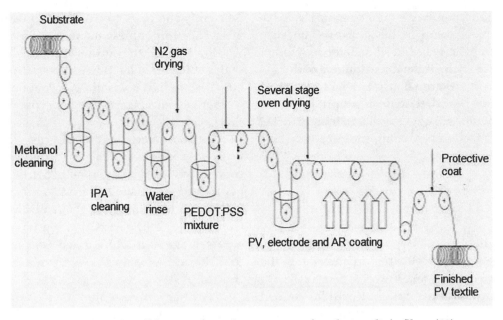

Figure 4-48. Possible manufacturing processes for photovoltaic fiber. (12)

very important.

Generally, the best bulk heterojunction devices based on widely studied P3HT:PCBM materials are active for wavelengths between 350 and 650 nm. Polymers with narrow band gaps can absorb more light at longer wavelengths, such as infra-red or near-infra-red, and consequently enhance device efficiency. Low band gap polymers (<1.8 eV) can be an alternative for better power efficiency in the future, if they are sufficiently flexible and efficient for textile applications.

For further optimization of the commercial-scale production that enhances the optimized power conversion efficiency of photovoltaic fibers, some alternatives are to reduce the thickness of the light-absorbing layer and the PEDOT:PSS layer, to use low band-gap materials, infrared light activated materials which would generate power at night, ultraviolet (UV) light selective materials, and to apply anti-reflective and protective nano-coatings.

After optimization, this photovoltaic fiber design can be produced in the industry with techniques similar to the textile manufacturing steps and may be used to manufacture smart textiles.

More work is needed for all this to happen, but we assure the esteemed reader that there is a lot of interest in this technology (and its products) from important customers like the US Army, so dozens if not hundreds of researchers are working on this as we speak. And they will not stop until all avenues are explored and the best materials and processes are used to get the most out of this promising and useful technology for its full deployment in the 21st century. Money talks.

PV Windows

As discussed above (in CPV Technologies) simple, flat sheets of glass can be used to generate power like solar concentrators, by gathering sunlight over a large area and focusing it onto a small solar cell mounted at the edge to convert the light into electricity.

The new PV glass sheets are lighter and flat, so they can easily be incorporated into solar panels on roofs or building facades. They could also be used as windows which, connected to solar cells, could generate electricity. Flat glass concentrators don't require a tracking system, so the benefit of concentrating the sunlight comes as a free bonus.

Instead of using optics, the glass sheets concentrate light using combinations of organic dyes specially designed to absorb the incoming light falling on the sheets. The dyes then emit the light into the glass, which channels it by the dye film on it to the edges of the glass, similar to the way fiber-optic cables channel light over long distances.

Narrow solar cells laminated to the edges of the glass collect the light and convert it into electricity. The amount of light concentration depends on the size of the sheet—specifically, the ratio between the size of the surface of the glass and the edges. To a point, the greater the concentration, the less semiconductor material is needed, and the cheaper the solar power.

The challenge of using organic dyes as solar concentrators has been that the dyes tend to reabsorb much of the light before it can reach the edges of the glass. Researchers have overcame that problem by using dyes that don't absorb the light that they emit. For example, a dye might absorb a range of colors in the light spectrum, such as ultraviolet through green, but emit light in another color, such as orange, which the dye cannot absorb.

The fact that there are such masses of glass and window usage in the world, that are appropriate for this type of power generation, suggests a large market niche. So we foresee this technology coming to age very successfully in the near future.

NEW PV TECHNOLOGIES

Bi-facial Solar Cells

Bifacial cells are solar cells that, in addition to the regular front surface activity, use the cells' back surface to collect reflected and diffused sunlight to produce added electricity. These cells use p-type wafers to manufacture solar cells with standard c-Si processing equipment and production processes.

High-efficiency, bifacial PV crystalline silicon solar cells are a reality, albeit with somewhat questionable practical applications. The cells manufacturers claim that their product produces 10-30% more power (kWh per kWp installed) when used in standard applications, and up to 50% more when installed in vertical installations—although vertical installations are not proven efficient yet, and there are a lot of unresolved issues with this type of arrangement.

The overall result is equivalent to an efficiency of 21-24% when considering standard installations.

The bi-facial cells can be used in all kinds of applications, such as flat rooftop and ground installations, as well as certain BIPV applications such as solar sound barriers, carports, and greenhouses. The use of the bi-facial cells in such non-standard applications, expands their niche market significantly, but the efficiency must be considered and calculated in each particular application.

If done properly, this disruptive method of manufacturing and use of bifacial cells could significantly improve the financial returns on solar projects, according to manufacturers' sources. One bifacial cells manufacturer is operating in an industrial plant in Heilbronn, Germany, and their cells have been used in bifacial modules worldwide.

This is just a new twist to an old story, keeping in mind that solar cells and panels must be positioned properly to utilize the minimum possible area needed to capture the maximum sunlight energy available.

Although the bi-facial solar cells and panels might offer new possibilities, they may not be able to do this as efficiently and cost-effectively as conventional technologies. Therefore, unless manufacturers know something we don't know, the future of this technology in large-scale, efficient, solar power generation is uncertain at best. Use of it in specialized locations and applications is a better choice, but here again, cost and reliability will determine the winner.

Looking back… Solyndra also thought that having bi-facial (actually 360-degree) sunlight collecting capability would offer some benefits. And it does… in very special and usually hard to reproduce and maintain cases. The fact that sunlight is allowed to bypass the front of the cells means that we have already wasted a percentage of it. How much we can collect from the back depends on many factors, some of which we cannot control.

Author's note: Solyndra recommended painting the roof area white, so it can reflect more sunlight which would then be absorbed by the backside of their PV collectors. Good idea, until the first sand storm leaves mud on the roof, changing its color to brown. We can only guess how much reduction of reflectivity and loss of power follows such an event.

Nevertheless, the new bi-facial technology offers a new avenue toward the use of solar energy. It also expands window of opportunity for solar cells in different locations and applications.

Black Silicon Solar Cells

Black silicon technology is based on an IP NREL which was developed to provide cheap, efficient solar cell material. Black silicon is not a new material, but rather a new process that refers to the surface color of silicon wafers. The black color is obtained by a special etching process, which is accurately controlled to produce a multitude of nano-scale size pores.

Such a surface absorbs incoming light so that the virtual absence of reflected light from the porous wafer surface makes it appear black.

NREL scientists claim that the black surface reflects only a small portion (approximately 0.3%) of the visible and near-infrared region of the solar spectrum, increasing the overall conversion efficiency of the new cells.

Since the original design required the use of expensive noble metals like gold and silver, NREL scientists have changed it to other, cheaper metals. Copper nanoparticles are presently used in the etching process, which eliminates the use of expensive metals.

This will allow the commercialization of the new cells at reduced prices, which is another big step toward reducing costs and increasing the efficiency of solar cells.

CTZSS Solar Cells

The CTZSS solar cell is a new technology, developed by IBM in 2010. The claim, as yet not officially proven, is that this design could potentially and significantly lower the price of solar power. The new CTZSS solar cell—as the initials suggest—is made from copper, tin, zinc, sulfur and selenium, all of which are somewhat earth-abundant.

The new process does not use vacuum chambers and related processes used in making semiconductor devices. Instead, it deploys a simple wet chemistry solution-based approach, which is a different and quite interesting twist, and which brings us back to the days when we used to make Si solar cells via solution-based processes on 2- and 3-inch silicon wafers. So it appears that we've come full circle.

In mass production, the simpler process could mean producing cells on a wet-chemistry bench by simple dipping, spray coating and printing steps. Various CIGS companies also employ similar methods in roll-to-roll manufacturing processes to produce thin film solar cells, but there are still kinks in these processes.

Bringing another cell of this type (unproven commercially) to market won't be easy, especially now, when investors are skittish about pouring more money into new types of solar cells. But this is IBM, and it has enough of its own resources, so anything is possible.

Extremely Thin Absorber Solar Cells

Extremely thin absorber (ETA) solar cells are conceptually close to solid state dye-sensitized solar cells. In the ETA solar cells, an extremely thin layer of a semiconductor such as $CuInS_2$ or CdTe or CuSCN replaces the dye in TiO_2 based (semi-DSSC) devices.

The major purpose of the ETA concept is to allow relatively low-quality materials to be employed, due to the short distance through which photogenerated carriers need to be transported before being separated into their respective electron- and hole-conducting phases.

Until now, most of ETA cells have used relatively good quality absorbers (i.e., with relatively low recombination rates), such as CdS, CdSe, PbS and In2S3, which are hard to obtain and work with, and which make the devices very expensive.

One manufacturing technique for making ETA cells starts with transparent conducting oxide (fluorine-doped tin oxide (FTO) which is coated with a dense layer of TiO_2 onto which a mesoporous titania film (25 nm particle size) is deposited. The resulting structure is sintered for stabilizing the layers.

CdS is formed on the TiO_2 surface by aqueous chemical bath deposition. For Cu_2_xS cells, the CdS-coated film is immersed in a hot bath of CuI in 1M aqueous KI, typically for several seconds. This results in a solution exchange process converting CdS to Cu_2_xS. This exchange process is thought to be topotactic and gives a copper sulfide close to Cu_2S if Cu_2 + ions can be substantially excluded from the Cu+ exchange bath. This can be done by adding a few drops of hydroxylamine to the CuI-KI bath.

The cell is completed by dipping in aqueous 0.5M LiSCN. The wet structure is then dried by directly placing it on a hot surface and then depositing CuSCN from propyl sulfide solution by evaporation of the solvent. CuSCN fills pores in the TiO_2, and coats the electrode with a layer of CuSCN, preventing shorting of the back contact to the TiO_2.

Finally thin gold back contact is added by either electron-beam evaporation or sputtering.

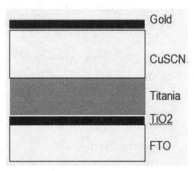

Figure 4-49. ETA solar cell

The ETA solar cell has the advantage of enhanced light harvesting due to the surface enlargement and multiple scattering. Similar to the solid state dye sensitized solar cells, the operation of the ETA solar cell is also based on a heterojunction with an extremely large interface.

Copper sulfides are in many ways ideal absorber materials. They are non-toxic, cheap and, and most importantly they are abundant, and have very good absorption characteristics.

Cu_2S in particular has a bandgap of 1.2 eV, which is nearly ideal for solar photovoltaic applications. An ETA cell is one of the few configurations with potential to exploit the favorable qualities of Cu-S in a low-cost solar cell.

Additionally, Cu-S can be considered a model electronically 'poor' semiconductor, and construction of a good solar cell with it would be clear proof of the above-stated ETA principle.

In the case of the Cu_2xS absorber, a buffer layer a thin layer of material that limits back electron transport, reducing charge carrier recombination of Inx(OH)ySz, (In-OH-S), is required for good cell performance. Note that In-OH-S, whether or not treated in the Cu-exchange solution, does not give any appreciable photocurrent, and this buffer layer is discounted as a possible source of the photoresponse. The In-OH-S layer is also deposited by CBD on the bare TiO_2 surface, followed by CdS and Cu-exchange.

These types of ETA solar cells have fairly low energy conversion efficiency, so lots of work is still needed to bring them close to the performance level of conventional technologies.

Another ETA technique is used for making $TiO_2/CuInS_2$ ETA solar cells via atomic layer chemical vapor deposition technique of the species. A 2nm of Al_2O_3 tunnel barrier and a 10 nm thick In_2S_3 buffer layer can be inserted between the TiO_2 and $CuInS_2$ layers to overcome the interfacial recombination problem.

In yet another ETA manufacturing process, the final structure consists of: top glass-ITO-TiO_2-PBS-PEDOT:PSS-Au. The efficiency of these cells, however, is quite low—less than 1%.

The major contribution of these exotic cells would be in developing and introducing new materials and processes in the ever growing exotic solar energy generation R&D effort.

Footsteps Power Generation

A number of companies and research institutions are looking into an innovative way to convert the power of human foot-steps (or vehicles) impacting upon special devices that can convert the gravitational energy into useful electric energy.

Several approaches have been tried, with some being considered practical when used in flooring (tiles, parquet panels, etc.), incorporated into the appropriate flooring systems. These then convert the kinetic energy of human weight (or vehicles) falling onto the floor (pavement or road) with every step (and vehicle passing) into useful electricity. A basic idea of converting one type of energy into another.

In 2010, the first energy-harvesting floor tiles were installed in a busy hall of a UK university. The floor tiles are connected to mechanisms that convert people's steps into electricity. Simple conversion of mechanical energy into electricity.

Currently, there are paving tiles that contain low-energy light-emitting diodes (LED), which light up in response to the mechanical-to-electric energy transfer process, which is instigated by the steps of the people contacting the floor tiles.

This is a low energy system, consuming approximately 5% of the energy from each footstep to light LEDs

built into the tiles (if necessary). The rest of the generated energy by each footstep is stored in a battery stack, from where the power can be sent to an inverter and/or, if so desired, into the grid. Or it could simply be used to power any device in the vicinity.

In the future, much broader applications will include charging electric cars, personal devices (smart phones, iPads and such) and for any other local power need.

Another interesting, and maybe more plausible application for even larger electric power generation, is using special electro-mechanical devices built into the pavement of city streets and highways. These devices will harvest and convert the gravitational energy impacted upon them by cars and trucks to power street lights, signal lights, advertisement signs and other traffic-related, power-hungry appliances.

Is this possible? Yes, of course—if the efficiency, the cost, and the reliability of the devices matches the needs and/or capabilities of the customers. It's a big IF, and we believe this technology will be used mostly to amuse people—until it is fully developed and capable of competing with conventional energy sources.

Hot Carrier Cells

One of the major problems with solar cells is that a portion of the high energy photons' energy that hits the surface is converted to heat. Even in the most efficient photovoltaic devices, at least 40% of the incident power is lost as heat dissipation. This is a loss of efficiency because the energy of the incoming photons is not fully used. The heat also causes the devices to overheat and further lose efficiency.

The idea behind the hot carrier cell is to utilize some of that incoming energy and convert it into electricity, instead of waste heat. If the electrons and holes can be collected while hot, a higher voltage can be obtained from the cell. The problem with doing this is that the contacts which collect the electrons and holes will cool the material. Thus far, keeping the contacts from cooling the cell has been addressed and achieved theoretically, and lots of work has been done in practice.

Another way of improving the efficiency of a solar cell using the heat generated is to have a cell which allows lower energy photons to excite electron and hole pairs. This requires a small bandgap. Using a selective contact, the lower energy electrons and holes can be collected while allowing the higher energy ones to continue moving through the cell. The selective contacts are made using a double barrier resonant tunneling structure and if the material has a large bandgap of phonons, then the carriers will carry more of the heat to the contact and it won't be lost in the lattice structure.

The hot carrier solar cells (HCSC) concept has been proposed to enable a conversion efficiency exceeding the Shockley-Queisser limit. It provides an attractive solution, thanks to a relatively simple structure, see Figure 4-50, combined with potential efficiency close to the thermo-dynamical limit.

Hot carrier solar cells allow efficiency enhancement by converting the excess carrier kinetic energy, which is normally lost as heat because of carrier thermalization, into electrical work. Two loss processes are at stake: a heat transfer to lattice due to (essentially LO) phonon emission and a heat leakage to the contacts carried by charge carriers as they are extracted from the absorber, which can be minimized when the contacts are energy selective, i.e., allow carrier transmission at a single energy level.

The use of nanostructured semiconductors was found to reduce the electron cooling rate related to phonon emission. On the contact side, the solution proposed so far showed a technological difficulty in achieving a good selectivity and high current densities.

Electron-hole pairs are photogenerated in the absorber where they are kept "hot" (at temperature T_H). They are extracted through selective contacts having a small transmission range δE towards a cold distribution in the electrodes (at temperature T_C).

So, HCSC is an attractive concept allowing a potential efficiency close to the thermo-dynamic limit, with a relatively simple structure, see Figure 4-50.

In such a cell, electron-hole pairs are photogenerated in the absorber where they are kept "hot". They are then extracted through selective contacts having a small transmission range towards a cold distribution in the electrodes.

The carriers are kept in a thermal disequilibrium with the lattice, thanks to a reduced electron-phonon

interaction. They are collected through energy selective contacts, allowing only carriers having one specific energy to be collected, to prevent a heat flow upon carrier extraction towards a cold population.

Hot photogenerated carriers can be incorporated in conventional GaAs and related structures for further increasing their efficiency by efficient use of the generated waste heat energy. The hot carriers can also reduce the radiative recombination in a p-i-n device structure with an intrinsic superlattice absorption region. The reduced radiative recombination leads to higher photovoltage and conversion efficiency.

Models of ideal HCSC were proposed showing a potential efficiency up to 86% under a fully concentrated (~46000 times) black body spectrum.

However, the practical feasibility of such a device has not been clearly stated so far. In particular, achieving the required carrier thermalization rate in the absorber may be challenging. Also, theoretical and experimental study on selective contacts have shown the difficulty to obtain good selectivity and high conductivity.

It is worthwhile noting that HCSC have a thermalization factor of $1W/K/cm^2$, which is 10 times lower than GaAs devices, which are some of the most efficient (with over 40% efficiency of the multi-junction GaAs CPV cells).

This, albeit theoretical, conclusion would give HCSC devices a potential efficiency around 60%.

Even if this high efficiency number is only seen in the R&D lab log books, it and some components of the process are quite intriguing, attractive, and promising. We are sure that this concept and/or its key components will be major contributing factors in the development of efficient and reliable PV technologies in the 21st century.

Photo-electrical Cells

Photo-electrical cells (PECs) have been around for a long time—at least in concept and on paper—but heir practical application is just now becoming important, so we include them in this section, hoping they will soon find wider commercial application.

PECS, just like their conventional cousins are energy generating devices, but with an added wet (liquid) twist. Photo-electro chemistry is a branch of the electrochemistry sciences, which has been receiving extraordinary attention from scientists worldwide, due to its potential to convert light energy into electricity. Amazingly, the efficiencies of some of the related technologies are near those of the competing conventional PV technologies.

All PECs use sunlight to initiate and maintain the process, but vary widely in their materials and construction details, as well as their application.

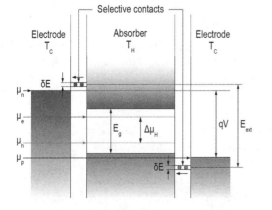

Figure 4-50. Schematic of a hot carrier solar cell (7)

Usually, each cell consists of a semiconducting working electrode (anode) and a metal counter electrode (cathode), usually immersed in an electrolyte. The anode and cathode can be made of many different materials and can be configured in numerous ways.

Some PECs simply produce electrical energy, while others produce hydrogen in a process similar to the electrolysis of water. The hydrogen can then be used to generate electricity on the spot, or stored for later use.

Types of Photo-electrochemical Cells

There are several types of photo-electrochemical cells (systems):

1. Electrochemical photo-voltaic cells,
2. Electrochemical photo-electrolytic cells, and
3. Electrochemical photo-catalytic cells.

By definition, electrochemical photo-voltaic cells are based on a narrow bandgap semiconductor and a redox reactant, so light energy is converted into electrical energy without change of the free energy of the reaction.

Photo-electrolytic cells are those cells in which radiant energy causes a net chemical conversion in the cell, so as to produce hydrogen as a useful fuel. These cells can be classified also as photosynthetic or photo-catalytic.

Photo-catalytic cells, on the other hand, are characterized by the photon absorption, which promotes a reaction without any net storage of chemical energy, and where the radiant energy speeds up a slow reaction.

In the photo-synthetic cells, radiant energy provides a Gibbs energy to drive a reaction such as the dissociation of water; $H_2O = H_2 + 1/2O_2$, during which process electrical and/or thermal energy may be recovered by allowing the reverse, spontaneous reaction to proceed.

We will review some key aspects of these devices, focusing our attention on the electrochemical photo-photovoltaic cells and the related processes.

Electrochemical PV Solar Cell

In the electrochemical PV cell, which is based on a narrow bandgap semiconductor and a redox reactant, light energy is converted into electrical energy without change of the free energy of the reactions within the redox electrolyte. The electrochemical reaction occurring at the cathode is opposite to the photo-assisted reaction occurring at the semiconductor electrode.

Thus, they are also called regenerative photo-electrochemical solar cells. If the photogenerated energy is converted to chemical energy, the free energy of the electrolyte will change too.

Depending on the relative location of the potentials of the redox couples and the type of redox chemicals, the photo-cells containing two redox couples can be further classified as:

a. Photo-catalytic cell, where light merely serves to accelerate the reaction rate, and

b. Photo-electrolytic cell, where the cell reaction is driven by light in the contra-thermodynamic direction.

Compared with electrochemical photo-voltaic cells, anodic and cathodic compartments need to be physically separated to prevent the mixing of the two redox couples in these types of cells.

Titanium dioxide (TiO_2) photoanode and a Pt cathode without an external source have been favored electrode materials for the construction of these devices.

Photo-electrochemical H2 Gas Generation

Hydrogen generation is a big business, which will increase in importance in the near future.

Presently hydrogen is generated by several commercial methods:

1. *Steam reforming* uses natural gas, or methane mixed with steam, where the mixture is exposed to very high temperatures (700-1100°C) to react and release syngas (mixture of hydrogen and carbon monoxide).

2. *Partial oxidation* is a process that occurs when a sub-stoichiometric fuel-air mixture is partially combusted in a reformer, creating a hydrogen-rich syngas. Thermal partial oxidation and catalytic partial oxidation are different types of processes, but which produce the same syngas mixture.

3. *Plasma reforming* is a high power plasma method conducted in a plasma reactor. It is used for the production of hydrogen and carbon black from crude oil.

4. *Coal gasification and carbonization* are processes that convert coal into syngas and methane via low and high temperature processes.

One of the most practical—and for now most useful—ways to use photoelectrochemical devices is the generation of hydrogen gas which can be used for energy production, or stored for later use. The method we describe below uses special photo-electro-chemical cells (PECs) to convert water into hydrogen, using solar energy for a power source.

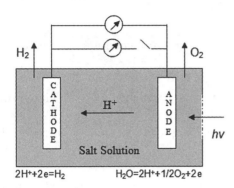

Figure 4-51. Photo-electrochemical cell

The photoelectrochemical (PEC) principle of power generation (involving water decomposition) is based on the conversion of sunlight energy into electricity by a cell consisting of two electrodes immersed in a liquid (usually water) electrolyte.

One of these electrodes is usually a semiconductor material, which when exposed to sunlight is able to absorb part of it to generate electricity.

In theory, there are three options for the arrangement of photo-electrodes in the PEC assembly:

1. Photo-anode made of n-type semiconductor and cathode made of metal;

2. Photo-anode made of n-type semiconductor and photo-cathode made of p-type semiconductor, and

3. Photo-cathode made of p-type semiconductor and anode made of metal.

PEC reactions involve several processes within photo-electrodes and at the photo-electrode/electrolyte interface, including:

1. Semiconductor material (photo-anode) participates into a light-induced intrinsic ionization, forming electronic charge carriers (quasi-free electrons and electron holes);

2. Water at the photo-anode by electron holes is oxidized,

3. Transport of H+ ions from the photo-anode to the cathode through the electrolyte and transport of electrons from photo-anode to the cathode through the external circuit;

4. Reduction of hydrogen ions at the cathode by electrons. Light results in intrinsic ionization of n-type semiconducting materials over the band gap, leading to the formation of electrons in the conduction band and electron holes in the valence band:

The reactions can be summarized and described as follow:

$$2hv \circledR 2e + 2h_0$$

where:
his the Planck's constant,
vis the frequency,
eis the electrons, and
$h_0$is the electron hole.

This reaction may take place when the energy of photons (hv) is equal to or larger than the band gap. An electric field at the electrode/electrolyte interface is required to avoid recombination of these charge carriers. This may be achieved through modification of the potential at the electrode/electrolyte interface.

The light-induced electron holes result in the splitting of water molecules into gaseous oxygen and hydrogen ions:

$$2h_0 + H_2O(liquid) \circledR 1/2O_2(gas) + 2H^+$$

This process takes place at the photo-anode/electrolyte interface. Gaseous oxygen evolves at the photo-anode and the hydrogen ions migrate to the cathode through the internal circuit (aqueous electrolyte).

Simultaneously, the electrons, generated as a result of the initial reaction at the photo-anode, are transferred over the external circuit to the cathode, resulting in the reduction of hydrogen ions into gaseous hydrogen:

$$2H^+ 2e \circledR H_2 (gas)$$

Accordingly, the overall reaction of the PEC may be expressed in the form:

$$2hv + H_2O(liquid) \circledR 1/2O_2(gas) + H_2(gas)$$

The above reaction takes place when the energy of the photons absorbed by the photo-anode is equal to or larger than Ei, the threshold energy:

$Ei = G^o(H_2O)/2N_a = hv = 1.23eV$

where:

$G^o(H_2O)$ is the standard free enthalpy of water, and N_a is Avogadro's number = $6:022 \times 1023$ mol^{-1}

Hence, the electrochemical decomposition of water is possible when the electromotive force of the cell (EMF) is equal to or larger than 1.23 V. The most frequently studied material for the photo-anode is TiO_2.

Despite its high band gap of 3 eV, it is the favored material owing to its high corrosion resistance. The maximal value obtained for the photo-voltage of a PEC equipped with a photo-anode of TiO_2 is ~ 0.7-0.9 V.

The application of TiO_2 as a photo-electrode requires an electrical bias to decompose water through external or internal bias voltage (through the use of different concentrations of hydrogen ions, or a hybrid electrode.

A typical cell involves both a photo-anode (made of an oxide material) and cathode (made of Pt) immersed in an aqueous solution of a salt (electrolyte). The process results in oxygen and hydrogen evolution at the photo-anode and cathode, respectively.

The related charge transport involves the migration of hydrogen ions in the electrolyte and the transport of electrons in the external circuit.

Photo-catalytic Water Decomposition

The principle of photo-catalytic water decomposition is similar to that of photo-electrochemical water decomposition, except for the difference in the location of the sites of their respective reactions—at the photo-anode and cathode in case of photo-electrochemical process, and on the surface of the photo-catalyst in the photo-catalytic process.

Thus, the practical difference between these processes is that the photo-electrochemical water decomposition results in the generation of separate oxygen and hydrogen gas streams, while in the photo-catalytic process the gasses are mixed.

Progress in R&D

The intensity of research on materials for photo-assisted water decomposition, aiming at the development of PEC technology, is increasing rapidly. To achieve success, the focal points are likely to remain as materials selection, types of electrolytes, and the structure of the PEC units.

The end-products must address the following requirements:

1. The photo-electrodes must exhibit high energy conversion efficiencies;

2. The PEC units must be durable and of low maintenance;

3. The costs of manufacturing of the PEC units must be low.

PEC cells have reached 10% efficiency in the lab lately, with the main problem being the corrosion of the materials which are in direct contact with liquids, which leads to premature failure of the devices.

In addition, the liquids tend to evaporate and/or crystallize under long-term exposure to extreme temperatures, which means that the location of their use must be carefully selected.

A minimum service life of 10,000 hours is needed to meet DOE goals, and to provide reliability and cost efficiency to this technology, so the effort is just now starting to address the critical area of reliability.

PAN-PEO Solar Cells

PAN-PEO solar cells are a different type of photo-electrochemical (PEC) solar cells that would be of interest in the future.

Their structure (stack) is as follows:

1. FTO/TiO_2/dye/PAN, EC, PC, $Pr_4N^+I^-$, I_2/Pt/FTO, and

2. FTO/TiO_2/dye/PEO, EC, PC, KI/I_2/Pt/FTO

These cells have been fabricated using a PAN-based gel polymer electrolyte and a PEO-based plasticized polymer electrolyte. The PAN-based gel electrolyte, made of poly-acrylonitrile (PAN), ethylene carbonate (EC), propylene carbonate (PC) and tetra-propylammoniumiodide ($Pr_4N^+I^-$) as the complexing salt exhibits a room temperature conductivity of 2.9×10^{-1} S m$_{-1}$ for the composition, PAN (13%):EC (31%):PC (45%):Pr4N$^+$I$^-$ (7%):I2 (4%) by weight ratio.

The PEO-based polymer electrolyte has a conductivity of 2.2×10^{-3} S cm^{-1} for the composition PEO (37.5%):EC (37.5%):PC (20.7%):KI (3.9%):I_2 (0.4%).

These solar cells have been characterized using current-voltage characteristics and action spectra. The PAN-based solar cells have an overall quantum efficiency of 2.3%. However, the PEO-based solar cells have an overall quantum efficiency of only 0.6%.

The importance of these cells lies in avoiding the

fundamental weakness of liquid based PECs. It is the necessity to replace the liquid electrolyte in dye-sensitized PEC cells with a solid material. Precast PAN-based polymer electrolyte membrane, which is used in this case, offers durability without loss of efficiency.

The plasticizers in the medium provide an efficient structure for the transport of ions in the redox couple. Another advantage of the polymer film is that it can be pre-fabricated on a large scale and even hot press-formed between electrodes to provide effective electric and electronic contact.

Cells made by this method are still of low efficiency (2.3%) under diffused sunlight, but further improvement of the efficiency is possible by optimizing the solar cell structure and fabrication techniques.

Intermediate Band PV

Intermediate band (IB) solar cells are a modified solar technology, which incorporates an additional energy band (simply squeezed between the main energy bands). It is partially filled with electrons, usually residing within the forbidden bandgap of the semiconductor bulk.

The IB solar cell consists of IB material, which is sandwiched between the n- and p-type areas of ordinary semiconductors (i.e. silicon, GaAs, etc.). These act as selective contacts to the conduction band (CB) and valence band (VB), respectively.

In an IB material, sub-bandgap energy photons are usually absorbed through transitions from the VB to the IB and from the IB to the CB bands. When summed, these activities add up to the current of conventional photons absorbed through the VB-CB transition stages.

Researchers have also used hypotheses similar to those adopted by Shockley and Queisser to derive a detailed balance-limiting efficiency, nearing 63% for the IB solar cell, at isotropic sunlight illumination.

Note: "Isotropic illumination" is solar concentration of 46,050 suns, assuming Sun and Earth temperatures to be 6,000°K and 300°K, respectively.

One of the main advantages of this device is the phenomena of capturing photons with insufficient energy to generate electrons from the valence band to the conduction band. These lower-energy photons can use the intermediate band as a "spring" or a stepping stone by gaining some energy, thus completing their goal of generating electron-hole pair.

This concept was proposed by solar cell researchers some time ago, but the idea was initially rejected because it was believed that the additional bands would cause non-radiative recombination, which could counteract the benefits of the current gain. There is, however, a way to inhibit the non-radiative recombination introduced by these centers, so this is no longer an issue.

In addition, earlier researchers did not realize the importance of splitting the Fermi level into three separate quasi-Fermi levels (QFLs), as needed to preserve the energy balance and maintain high output voltage of the cell.

Subsequent modifications and refinements of the initial balance analysis of the intermediate band solar cell have confirmed its potential for high efficiency. And it is the high value of the IB solar cell's limiting efficiency that has attracted many scientists to work in this field.

The limiting (theoretical) efficiency of the IB solar cell is similar to that of a triple-junction solar cell connected in series. However, the IB based cell can be seen as a set of two cells connected in series with; 1. VB-IB, and 2. IB-CB transitions. There is also an additional one connected in parallel, which corresponds to the VB-CB transition. This provides the IB solar cell with additional tolerance to changes in the solar spectrum, thus increasing its efficiency.

An optimal IB solar cell has a bandgap of about 1.95 eV, which is split by the IB into two sub-bandgaps of approximately 0.71 eV and 1.24 eV respectively. The quasi-Fermi levels (QFLs) or electrochemical potentials of the electrons in the different bands are usually close to the edges of the bands.

In the end, the solar cell voltage is comprised of the difference between the CB QFL at the electrode in contact with the n-type side, and the VB QFL at the electrode in contact with the p-type side. This limits the maximum photovoltage of the IB solar cell to just over 1.95 eV. It is, nevertheless, still capable of absorbing photons of energy above 0.71 eV, which helps it to maintain high voltage output under different lighting intensity levels.

In contrast, single-gap solar cells cannot supply a voltage greater than the lowest photon energy they can absorb, which limits their efficiency, especially under lower levels of illumination. Theoretically, IB solar cells operate on the principle of absorbing two sub-bandgap photons, which produce one high energy electron, and still deliver very high photovoltage.

The IB solar cell is a thin-film technology based on highly mismatched alloys. As a three-bandgap and one-junction device, it has the potential of improved solar light absorption and higher power output—even higher than the III-V triple-junction compound semiconductor devices (again theoretically).

Nanostructured materials and certain alloys have been employed in the practical implementation of intermediate-band solar cells, and some of them promise very high efficiency, although a number of challenges still remain for realizing practical, mass produced, intermediate band devices.

All this, of course, is mostly theory, with some plausible lab experimentation to support its main points. Work remains to be done, before these devices become reality—efficient, durable and cheap enough to compete with conventional PV technologies, let alone with III-V multi-junction solar cells (the most efficient solar energy conversion devices at present).

But the promise of reaching 50% efficiency with this seemingly simple device is tempting, and the work will surely continue. We have no doubt that this technology, and/or parts of it, will soon take a prominent place in developing new and more efficient technologies.

Interband Cascade PV

Conventional PV devices are based on semiconductor p-n junctions. Multistage PV devices based on a single material system that allows coverage of different wavelengths might be a more efficient way to use sunlight. Also, devices using InAs/GaSb/AlSb structures that operate in the infrared are also useful as a base technology.

Using the interband cascade "IC" architecture, similar to that used to produce mid-IR lasers is one such technique.

These PV devices comprise multiple cascade stages where each stage is divided into three regions according to the three base processes:

1. Photon absorption,
2. Intraband carrier transport, and
3. Interband tunneling transport,

These absorbers are composed of a type-II quantum well (QW) or superlattice (SL) structure and are connected by both; asymmetric intraband transport and interband tunneling regions. The effective band gap in each absorber region is determined by the layer thicknesses in the SL and can be designed to be either the same or different from the other stages. Several stages with these and different, properly designed band gaps can be stacked, similar to the way different p-n junctions are stacked in a multiple-junction cell to more efficiently make use of the source spectrum.

Several such stages can be designed and stacked, each with a different band gap, to more efficiently divide and use the solar spectrum. The transport regions are constructed either with compositionally graded semiconductor alloys or with digitally graded multiple QWs to form a tilted-band profile.

This way, the conduction band of one end of the structure is near to the conduction band of the adjoining absorber, while the conduction band at the other end of the structure is near the valence band of the absorber adjoining it.

The transport region plays a role similar to the depletion region in a conventional p-n junction to direct current in one direction. However, only the electrons move through the intraband transport region because holes are confined in different regions.

When light hits the PV device, photogeneration of electrons and holes results in a separation of the electron and hole quasi-Fermi levels, and the electrons move to one side due to the asymmetry of the transport region, while at steady state, an internal electric field will be built up to balance this movement of charge carriers, resulting in a sequential potential drop in each cascade stage.

The potential drops in every stage add and contribute to a total forward bias voltage similar to that in a multiple p-n junction cell. In contrast to the p-n junction structure, where heavily doped p- and n- regions are required, doping is not necessary in these structures. This eliminates the detrimental effects of high carrier concentrations such as free-carrier absorption and reduced minority-carrier diffusion length.

The cascade multistage PV absorber architecture is particularly desirable for high-intensity illumination (i.e. with tracking concentrators), where the high-intensity radiation may not be fully absorbed if using single p-n junction of a conventional cell, whose thickness is limited to the photogenerated carrier diffusion length.

For the cascade PV device, multiple stages with the same energy gap can be used to absorb all the photons in a particular portion of the source spectrum while increasing the open-circuit voltage, where the thickness of each stage is shorter than the diffusion length and chosen so that each stage generates the same photocurrent (the stages are current matched). Losses associated with high-current operation are also minimized by increasing the open circuit voltage.

Thus far, this concept has been proven in labs only, but we expect that when properly designed, manufactured, and optimized for PV operation, the interband PV devices could provide high energy conversion efficiency for solar and/or thermal PV systems.

The nature of Sb-based materials will allow these devices to be suitable only and exclusively for IR solar and thermal photon energy conversion. The extension of the interband PV concept to the short-wavelength portion of the solar spectrum depends on the integration of wide band gap materials with Sb-based materials.

This interesting, very promising concept is still in the R&D labs, without practical development and use. The promise of interband conversion, however, is quite tempting and we are sure that the work (or key components of it) will continue until fully explored and implemented—as efficient interband PV devices, or as parts integrated into other PV technologies.

Infrared Solar Cells

Infrared (IR) radiation in the sunlight spectrum is usually not captured by the conventional solar devices, so it is wasted as heat radiated in and from the device. Conventional PV modules also do not generate much power under cloudy skies, because the visible portion of the sunlight spectrum is filtered by the clouds, and the photons reaching the solar cell surface do not have enough energy to generate electrons.

Sunlight under cloudy skies, however, emits a lot of IR radiation. Capturing the IR and even using it to generate electricity is a great benefit to any PV device, and this is what the IRSC devices are good at—collecting IR energy on cloudy days and convert it into electricity.

IR Solar Cells

Infrared solar cells (IRSC) and modules are a new type of technology, which is basically designed to capture the IR and near-IR spectrum of the sunlight and convert it into useful electricity—something that conventional PV modules cannot do.

When the sun is shining directly onto a solar panel, both infrared and non-infrared panels will absorb the same amount of the sun's energy and generate almost the same amount of power. However, when the sun goes behind a cloud, the only light (or most of the light) which can be turned into energy is infrared light, which non-infrared panels cannot capture and therefore stop producing energy at this point.

IRSC have another advantage in the fact that the "low iron tempered glass" used normally in module fabrication, has an emission of 3% in comparison to 88% for "Normal iron tempered glass". The lower the emission the greater the energy conversion efficiency of the solar panel. The emission is the percentage of energy the solar panel can release back into the atmosphere.

To use a building example, double glazed windows are designed to have a very low emission rating because they are supposed to keep the heat inside the home and not let it escape through the glass. A solar panel with "low iron tempered glass", which is used for the IR solar panels, works the same way.

So, this type of infrared panel is basically a standard solar panel whose glass has been treated with a special coating which enables it to harvest infrared sunlight. Infrared light has a wavelength of 800 to 1200nm and is the only light that can be converted into energy if the sun goes behind a cloud.

Therefore, since these types of infrared solar panels are more expensive, it makes sense to install them in a climate where there is often a lot of cloud cover, or if special conditions are to be met.

IR Plastics

This is another less expensive way to produce PV modules on plastic sheets which contain billions of "nano-antennas" collecting heat (IR) energy generated by the sun and/or from other sources.

Nano-antennas target the mid-IR range rays of the sun spectrum, some of which the earth continuously radiates as reflected heat after absorbing energy from the sun during the day. This then opens a new dimension in heat-to-PV generation—that of double-sided nano-antenna sheets that can harvest energy from both the top (sun rays) and bottom (Earth's heat).

One great advantage of IR solar cells is that they can be tuned to capture IR radiation most effectively under cloud cover, where traditional solar cells (which can only use visible light) are rendered idle under cloudy skies.

While methods to convert the energy into usable electricity, including the manufacturing of the devices, still need to be further developed, IR coatings and energy capturing devices and plastic sheets, could one day be manufactured as lightweight "skins" that power everything from hybrid cars to computers and iPods with higher efficiency than traditional solar cells.

This technology is a major step toward a solar energy collector with these unique properties. It could be mass-produced on flexible materials for a very large niche market.

Lift-off Silicon Solar Cells

Thinner silicon is the goal of the industry, and all manufacturers are looking into ways to have that. When we say "thin" we think of 100-150 microns—the minimum required for acceptable yield and performance.

A US manufacturer is gearing to mass produce 35-micron-thick silicon solar cells. The claim is that these will be very high-performance, low-cost mono-crystalline solar cells, made by a very different type of technology called "lift-off", which is based on reusable templates and porous silicon substrates.

The catch here is the extremely thin silicon wafers

are supported at all times during processing (or they'd break).

Another innovation is the replacement of the silver metal usually deposited on the back surface as a metal contact, and uses aluminum metal instead.

This process is all together different, and as a matter of fact there are no wet chemistry (washing, etching, etc.) steps. It is based on CVD deposition of trichlorosilene gas at atmospheric pressure, during which silicon layers are deposited in the template at a rate of 2.5 microns per minute.

This way the silicon bulk is almost 10 times less than regular c-Si solar cells, which use on average 1/2 gram silicon per Watt.

Thus deposited 156 mm diameter solar cells are the largest back-contact cell in the industry. The thinnest and largest cell sounds exceptionally good (to be true). On top of that, according to insiders, the cells are flexible and will be supported (most likely encapsulated) in a type of a resin and fiber mix, similar to that used in circuit board material. The resin-fiber mix will create a cradle that would support the thin cell.

Insiders assure us that the cells can be used without a support backing and/or frame, and are flexible enough to be used as is.

OK, we have to draw a line in the sand right here, because thinnest (most fragile) and largest (and most difficult to process properly) is difficult enough to envision, but using these paper thin, large and fragile pieces of silicon without a frame or any support is a bit too much to accept as a practical possibility.

Silicon cells of any type and thickness—including thin films—have limited bending and flexing abilities. Flexing a very thin piece of silicon even slightly creates fractures. One more flex will crack the thin wafers in many small pieces, even if encapsulated in plastics.

Drawing a parallel with the hasty comments about "strategic customers lined up to buy the cells" (while they are still in the R&D lab), we have to conclude that the manufacturer is way ahead of itself with the technical and marketing claims and plans—as well as their projected 20% efficient modules at 42¢ per Watt.

Note: Current conventional c-Si modules are 15-16% efficient, at a cost of $0.83 per Watt. CdTe modules, which are considered top of the line are 12% efficient, at a cost of $0.73 per Watt. This was achieved at great expense of time, effort and money, not to mention years of testing and optimization. So we wish all the best to the new wanna-be-champion of very thin silicon companies, but we are absolutely sure that they have a long way to go, and many sleepless nights, in producing an acceptable

and practical product, while proving the high efficiencies and low costs at mass production.

Too, we wonder how they would market this new and unproven product today, with cheap PV modules flooding the markets.

Still, if the principals know something we don't, and if even only half of the above claims come true, then the extra thin silicon solar cells and modules might become the disrupting technology we have been waiting for. Anyone who succeeds in making this technology work, and putting it on the market, will become a world-class leader overnight. Promise!

Metamorphic Multijunction Solar Cell

Metamorphic multijunction solar cells are ultra-light and flexible multi-junction (MJ) cells, similar to the germanium based concentrating PV (CPV) cell that converts solar energy with record efficiency. It represents a new class of solar cells with clear advantages in performance, engineering design, operation and cost.

For decades, conventional cells have featured wafers of semiconducting materials with similar crystalline structure. Their performance and cost effectiveness is constrained by growing the cells in an upright configuration. Present-day MJ cells are rigid, heavy, and thick with a bottom layer made of Ge semiconductor material. In the new method, the cell is grown upside down. The active layers use high-energy materials with extremely high quality crystals, especially in the upper layers where most of the power is produced. Not all of the layers follow the lattice pattern of even atomic spacing. Instead, the cell includes a full range of atomic spacing, which allows for greater absorption and use of sunlight.

The thick, rigid germanium layer is not used here,

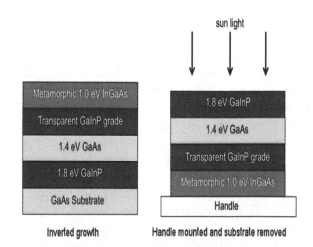

Figure 4-52. Schematic of growth and processing of inverted triple junction solar cell.

thus reducing the cell's cost and 94% of its weight. By turning the conventional approach to cells on its head (no pun intended), the result is an ultra-light and flexible cell that also converts solar energy with record efficiency of 40.8% under 326 suns concentration. This cell was developed by NREL scientists and is produced commercially.

The manufacturing procedure is similar to that used to produce germanium based MJ CPV cells, except that it is reversed to avoid the thick germanium substrate.

The solar cells are grown in a single process step by atmospheric-pressure organo-metallic vapor phase epitaxy (OMVPE) on a (001) GaAs substrate mis-cut 2° toward (111)B. The direction of growth is inverted relative to more conventional multijunction devices, thus the top (lattice-matched) junction is grown first and the bottom (most highly lattice mismatched) junction is grown last. This direction of growth helps prevent the threading dislocations that inevitably originate during mismatched growth from degrading the higher band gap (and more power producing) junctions.

To minimize the dislocations in the metamorphic junctions, a step-graded (Al)GaInP buffer layer, which is transparent to the light intended for the junction, is grown. The grade is engineered to achieve a nearly strain-free metamorphic junction.

The $In_xGa_{1-x}As$ junctions are clad with corresponding $Ga_xIn_{1-x}P$ window and back-surface-field layers with the same lattice constant.

After OMVPE growth of the inverted semiconductor structure, gold is electroplated onto the exposed "back" surface of the inverted structure. The sample is then mounted with a low viscosity epoxy to a structural "handle," in this case silicon wafers or glass slides, however a variety of materials could be used as a handle to optimize cost, heat management, and weight considerations.

The GaAs substrate is finally removed using a selective chemical etch that stops at a GaInP layer. The substrate can be reused if needed.

The inverted growth and processing is shown schematically in Figure 4-52.

This is not an easy or cheap way to make solar cells, but the end justifies the means. Still, these types of cells will not be seen on our roofs anytime soon, but will instead be used in very special applications.

It is worth mentioning here that metamorphic MJ cells have measured efficiency above the Shockley-Queisser limit efficiency of 30% for a single-band gap device at 1 sun. Concentrator cells of this type also measure above the theoretical limit of 37% for single-band gap cells at 1000 suns.

These uncharacteristic properties are due to the special multijunction structure of the cells. Because of that, multijunction cell architectures, using different approaches, such as the addition of metamorphic materials, and four or more junctions, as well as other design modifications have the potential to increase these cells efficiencies over 45%, or even to 50%, when used in tracking concentrators.

Metamorphic, or lattice-mismatched, semiconductors provide an unprecedented degree of freedom in solar cell design, by providing flexibility in band gap selection, unconstrained by the lattice constant of common substrates such as Ge, GaAs, Si, InP, etc.

Even better, and regardless of the success of this technology, other devices will certainly benefit from the development of these materials, such as heterojunction bipolar transistors, 1.3 and 1.55 microns photo-detectors and lasers, and InP-on-GaAs and GaAs-on-Si devices, just to mention a few.

Miniature Solar Cells

Miniature solar cells are essentially different types of solar cells—many of them reviewed herein—that are used exclusively to power small devices such as consumer electronics and microelectronics. The only difference between them and the regular size cells made with the same materials is their very-small size.

They are typically of the thin film solar cells type, based on CIGS, CdTe, and a-Si technologies. Organic solar cells are being developed to join the lineup of technologies synonymous to miniature solar cells.

Some miniature solar cells made of silicon are also gaining popularity. In one such effort, miniature silicon solar cells are used in conjunction with six-junction tandem solar cells, where individual solar cells can be arranged so that each solar cell absorbs the appropriate area of the solar spectrum.

The goal of this six-junction tandem solar cell structure is to achieve the combined efficiency of >50% under 20 suns illumination. One of the cells in the tandem structure is made of silicon and absorbs energy of 1.42 – 1.1 eV. The role of the silicon cell is to convert 7% of the light incident on the tandem structure into electricity. Other cells in the stack contribute the balance of the generated electricity.

A key design parameter for these very small silicon cells, which is dictated by their application (in small devices, where lots of power is needed) is that they should be very efficient.

While the general use of miniature solar cells is still limited to powering small devices, it is an important area that is attracting unprecedented attention from high places. Some areas of specialized use include powering radio sensors and some delicate, small-size devices used by the US Army, where size and weight is of utmost importance.

We see the importance and usage of highly efficient miniature solar cells growing quickly, as the size of electronic devices decreases together with their power needs. Mini solar cells will never be used to power air conditioners, or in large-scale applications, but we'll see them all around us in the 21st century.

Multiple Energy Level Cells

The idea for multiple energy level solar cells is to basically stack thin film solar cells on top of each other. Each thin film solar cell would have a different band gap which means that if part of the solar spectrum were not absorbed by the first cell then the one just below would be able to absorb part of the spectrum. These can be stacked, and an optimal band gap can be used for each cell, to produce the maximum amount of power.

Options for how each cell is connected are available, such as serial or parallel. The serial connection is desired because the output of the solar cell would be just two leads. The lattice structure in each of the thin film cells needs to be the same. If it is not, then there will be losses.

The processes used for depositing the layers are complex. They include Molecular Beam Epitaxy and Metal Organic Vapor Phase Epitaxy. The current efficiency record is made with this process but doesn't have exact matching lattice constants.

The losses due to this are not as effective because the differences in lattices allows for more optimal band gap material for the first two cells. This type of cell is expected (theoretically) to be 50% efficient.

Lower quality materials that use cheaper deposition processes are being researched as well. These devices are not as efficient, but the price, size and power combined allow them to be just as cost effective. Since the processes are simpler and the materials are more readily available, mass production of these devices is more economical.

Great concepts. Great start. We envision this approach gaining momentum as the limits of the single-layer technologies are reached. Due to complexity and cost, however, this might take a while; but, all good things take time.

NGD Technology

Traditional PV designs are limited by a number of key factors.

1. *Expense.* Photovoltaic cells that use abundant materials (Si) are expensive to manufacture (high efficiency but high cost).

2. *Low efficiency.* Less expensive PV (thin-film) is unable to achieve the higher efficiencies of Silicon PV (low cost but low efficiency).

3. *Materials.* The use of rare and toxic materials in the design of many PV designs, preventing them from ever achieving terawatt (global) scale implementation.

The new generation device (NGD) technology is a new approach that uses novel methods, consisting of the addition of a barrier oxide layer, combined with a patent-pending absorber layer, which basically replaces the traditional semiconductor layers used in crystalline silicon and thin-film photovoltaics.

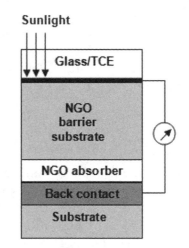

Figure 4-53. Quantum NGD solar cell

The new NGD technology avoids using a semiconductor absorber, thus significantly reducing production costs and eliminating reliance on expensive silicon components. The unique patent-pending NGD™ charge separation system provides a broadband response (full spectrum of light) while avoiding the use of Gallium, Tellurium or Indium—elements that limit large-scale deployment of traditional PV.

Research teams are taking a radical new approach to the development of photovoltaic design and the NGD device has the potential to obtain higher efficiencies at a lower cost per kilowatt/hour than coal without using any rare elements.

Note: Another twist here is the idea of rationalizing the levelized cost of electricity. The industry standard for measuring electrical grid energy costs is referred to as the Levelized Unit Electricity Cost (LUEC) or simply the Levelized Electricity Cost (LCE). These are usually displayed on a megawatt basis to analyze the real cost for electrical plants, however in the case of solar we are showing the cost per kilowatt hour as a way of showing the end-consumer cost comparison.

There are several contributing factors when calculating the LCE but in this case we are removing waste removal or decommissioning expenses and focusing entirely on the cost-to-consumer equation.

The solar industry looks at costs per square meter because solar modules must be linked to achieve appropriate voltage requirements. These linked modules form panels that form the networks for industrial grid-based solar electrical plants. Manufacturing costs don't represent other associated expenses such as construction or land usage.

Another important consideration in determining the cost effectiveness of solar modules is their power conversion efficiency or the amount of sunlight they can effectively convert to electric energy.   The final piece of the LCE equation is the cost per watt peak or $/Wp, as solar modules (representing around 50% of the installed cost) are sold in Watt Peak representing the power rating of the module under standard conditions.

The current cost per kilowatt hour of coal-burning grid electricity globally averages out at $0.10. The NGD technology could match coal's LCE cost. When you factor in the issues of mining, waste removal, pollution and CO_2 emissions, NGD solar power becomes the clear choice for a cleaner, more affordable and endlessly abundant energy future.

Again, theories abound, experimentation continues, with the practical results to follow. But this is how every good thing starts, so we have no doubt that something good is going to come out of this new approach to solar power generation.

Once developed, the new NGD technologies may have the correct combination of attributes to become the "holy grail" technology the solar industry needs to overcome fossil fuel dependencies.

Thermo-photovoltaics

Researchers are working on ways to combine "quantum and thermal" mechanisms into a single physical process to generate electricity, as needed to make solar power production twice as efficient as existing technologies. The process is called PETE, which stands for "photon-enhanced thermionic emission."

This process does require some exotic materials

(cesium-coated gallium nitride, for example), and works best at very high temperatures, so it is most likely suitable for use with concentrators or parabolic dishes, rather than with flat solar panels.

Photon-enhanced thermionic emission (PETE) is a combination of photovoltaic and thermionic (TEC)* effects into a single physical process to take advantage of both the high per-quanta energy of photons, and the available thermal energy due to thermalization and absorption losses. Basically, the device uses all energy coming to it (including thermal) to generate electricity.

A PETE device has the same vacuum-gap parallel-plate architecture as a TEC, except with a p-type semiconductor as the cathode. PETE consists of a simple three-step process. First, electrons in the PETE cathode are excited by solar radiation into the conduction band. Second, they rapidly thermalize within the conduction band to the equilibrium thermal distribution (according to the material's temperature) and diffuse throughout the cathode. Finally, electrons that encounter the surface with energies greater than the electron affinity can emit directly into vacuum and are collected at the anode, generating current.

Each emitted electron harvests photon energy to overcome the material bandgap, and also thermal energy to overcome the material's electron affinity. The total voltage produced can therefore be higher than for a photovoltaic of the same bandgap owing to this 'thermal boost', thus more completely using the solar spectrum.

*Note: Thermionic energy converters (TECs) act as energy engines that convert heat into electricity via direct conversion process. Thermionic energy converters are made out of a hot cathode and cooler anode made of different materials, and separated at a distance by vacuum.

TECs were developed some time ago for deep-space exploration and other applications that require autonomous generators. Due to a number of problems—including the excess temperatures and current densities required for efficient operation—as well as its limited applications, the technology was never fully commercialized.

The function of the PETE devices is fairly well understood, and we know that at low temperatures, the active carriers in the conduction band of the material cannot overcome the electron-affinity barrier, so the PETE current is negligible under that condition. High-energy photons can create direct photoemission as a side benefit as well, but this is not the main function of these devices.

The PETE process gets more efficient as the temperature increases, which leads to overall increase of current as well. At a certain point the produced current reaches a plateau, where every activated electron has been emitted and the process levels or stops.

A pure thermionic emission state dominates at the highest temperatures. This is so because the thermal processes overshadow the effect of photoexcitation at this point, so that the emission current is not constrained by the number of photoexcited electrons.

PETE devices can be made by different techniques, some of which involve GaN films on Mg doped sapphire substrate, and Cs depositions at high temperatures. By changing the temperature levels the device is exposed to, and by modulating the light intensity on the structure (where at the cathode some electrons have significant thermal energy for composing materials and to escape into the vacuum space), current is generated between the two electrodes. Different amounts of photocurrent can be measured under different working conditions.

Reported efficiencies of PETE devices are too low for practical purposes, less than 1%, but the materials and processes used in making this structure are of great interest, and could be successfully used in the development of other technologies.

CSP and CPV technologies are the main candidates for participating in PETE assisted power generation, and we would not be surprised if we see PETE devices mounted on the focal points or other active areas of trackers in the near future.

UV Solar Cells

Researchers have developed a transparent solar cell that uses ultraviolet (UV) light to generate electricity, but allows visible light to pass through it. UV solar cells and modules can be used to replace conventional window glass on large surface areas. This could lead to potential uses that take advantage of the combined functions of power generation, lighting and temperature control.

The transparent, UV-absorbing system is obtained by using an organic-inorganic heterostructure made of the p-type semiconducting polymer PEDOT:PSS film deposited on an Nb-doped strontium titanate substrate.

PEDOT:PSS is easily fabricated into thin films due to its stability in air and its solubility in water. These solar cells are only activated in the UV region and result in a relatively high quantum yield of 16% electron/photon.

Future work in this technology involves replacing the strontium titanate substrate with a strontium titanate film deposited on a glass substrate to achieve a low-cost, large-area manufacture.

Other methods can be used to include UV wavelengths in solar cell power generation. Some companies report using nano-phosphors as a transparent coating to turn UV light into visible light. Others have reported extending the absorption range of single-junction photovoltaic cells by doping a wide band gap transparent semiconductor such as GaN with a transition metal such as manganese.

Web (space) Solar Power

Researchers are working on components of a multi-national project to build space-based solar energy systems that can beam power back to the earth—day or night.

The idea is to launch a power station into space that will concentrate the sun onto solar arrays and then use microwaves or lasers to send the energy anywhere in the world. Initially, smaller satellites would be used to generate enough energy for a small village.

This application would be ideally suited for sending energy to remote regions or to disaster areas. The aim is to eventually put a large enough structure in space that could gather energy capable of powering a large city. The project, which is part of a NASA Institute for Advanced Concepts study, is still in its early phases, but progress has been made, notably in space construction and design.

One of the primary challenges of such space-based solar power generators is getting a large structure up and operating it in space. Researchers are tackling that very problem, with the development of a super lightweight spinning net that could be the foundation of a solar satellite, instead of a heavy metal structure.

The so-called "space web" experiment has been tried on a rocket from the Arctic Circle to the edge of space and demonstrated that larger structures could be built on top of this lightweight net.

The next stage of this project, called SAM (Self-inflating Adaptable Membrane), will develop and test an ultra-light cellular structure that can change shape once deployed in space. The structure is made of cells that self-inflate in a vacuum and can change their volume independently through nano-pumps. These vacuum-sensitive cells are the key component that could transform the structure into a solar concentrator, where thousands of them will collect the sunlight and project it onto solar arrays lower in space or on earth.

In theory, assembling thousands of small individual cell units could be used to build large space systems... and although the entire project, and some of its elements, sound very exciting, we see a number of unresolved issues that will take a lot of time and money to put in practical application.

Nevertheless, we will be tracking the developments of this multi-discipline project carefully, for it might be the beginning of a disruptive solar technology we have been waiting for.

Notes and References

1. Organic Photovoltaics, by Bernard Kippelen and Jean-Luc Bredas. Reproduced by permission of The Royal Society of Chemistry.
2. Review, Hybrid solar cells. Serap Gunes, Niyazi Serdar Sariciftci, Linz Institute for Organic Solar Cells (LIOS).
3. Carbon nanotube growth by PECVD: a review. By M Meyyappan, Lance Delzeit, Alan Cassell and David Hash. NASA Ames Research Center, Moffett Field, CA 94035, USA.
4. Excitonic Solar Cells, Garry Rumble, NREL
5. Plasmonic solar cells. K.R. Catchpole and A. Polman. FOM Institute for Atomic and Molecular Physics, Kruislaan 407, Amsterdam, The Netherlands Centre for Sustainable Energy Systems, Australian National University, Canberra.
6. *Quantum dot solar cells*, by A.J. Nozik. With permission by Elsevier Press.
7. Hot carrier solar cells: in the making, A. Le Bris, L. Lombez, JF. Guillemoles, R. Esteban, M. Laroche, J.J. Greffet, G. Boissier, P. Christol, S. Collin, J.L. Pelouard, P. Aschehoug, F. Pellé.
8. Three-dimensional nanopillar-array photovoltaics on low-cost and flexible substrates Zhiyong Fan, Haleh Razavi, Jae-won Do, Aimee Moriwaki, Onur Ergen, Yu-Lun Chueh, Paul W. Leu, Johnny C. Ho, Toshitake Takahashi, Lothar A. Reichertz, Steven Neale, Kyoungsik Yu, Ming Wu, Joel W. Ager, and Ali Javey
9. Photovoltaic performance of PPE-PPV copolymers: effect of the fullerene component. Diana K. Susarova,a Ekaterina A. Khakina,a Pavel A. Troshin, Andrey E. Goryachev, N. Serdar Sariciftci, Vladimir F. Razumova and Daniel A.M. Egbeb
10. Photo-electrochemical hydrogen generation using band-gap modified nanotubular titanium oxide in solar light. K.S. Raja, M. Misra, V.K. Mahajan, T. Gandhi, P. Pillai, S.K. Mohapatra
11. Optical Absorption Enhancement in Amorphous Silicon Nanowire and Nanocone Arrays, Jia Zhu, Zongfu Yu, George F. Burkhard, Ching-Mei Hsu, Stephen T. Connor, Yueqin Xu, Qi Wang, Michael McGehee, Shanhui Fan, and Yi Cui.
12. A Photovoltaic Fiber Design for Smart Textiles, Ayse (Celik) Bedeloglu, Ali Demir, Yalcin Bozkurt, and Niyazi Serdar Sariciftci.
13. Single nanowire photovoltaics, Bozhi Tian, Thomas J. Kempa and Charles M. Lieber
14. Lateral Si/SiO$_2$ quantum well solar cells, R. Rölver, B. Berghoff, D.L. Bätzner, B. Spangenberg, and H. Kurz
15. One-axis tracking holographic planar concentrator systems, Deming Zhang, Jose M. Castro, and Raymond K. Kostuka,
16. Quantum dot solar cells, V. Aroutiounian, S. Petrosyan,a) and A. Khachatryan
17. Conjugated Polymer-Based Organic Solar Cells, Serap Gunes, Helmut Neugebauer, and Niyazi Serdar Sariciftci.
18. *Photovoltaics for Commercial and Utilities Power Generation*, 2011. Anco S. Blazev

Chapter 5

Solar Power
Successes and Failures

Success is not final, failure is not fatal:
it is the courage to continue that counts.
—Winston Churchill

This chapter shows some of the successes of solar power and highlights some of the recent failures. The key failures during the 2007-2012 solar boom-bust period are listed with very short explanations of events. The bad times are not over yet, but the successes of the recent boom-bust cycle will go down in history as great achievements, while the failures will be analyzed and filed as lessons learned.

We hope this text will serve as both a written record of recent events, and as a living reminder of what to do, and what not to do in the future.

We cannot possibly discuss the successes and failures of the modern solar energy generating industry without mentioning its humble beginnings, and its relative successes and failures in the recent past.

THE U.S. SOLAR ENERGY (R)EVOLUTION

October 17, 1973, was a beautiful day in California, weather holding at 70°F, surfers gliding on the ocean waves, and millions cruising the highways in their 8-cylinder gas-guzzlers. Then, around 3:00PM, LA commuters heard that OPEC was contemplating raising the tax rates and significantly jacking up crude oil prices.

Within two weeks, Arab oil producers dropped output by 25%, beginning the infamous gas shortage—the first in US history. Long gas lines in Los Angeles became a daily event. Then rationing began.

This was Lesson Number One for the US: economic power and military might are insignificant without fuel to drive our personal, commercial and military machines.

What would we do without plastics, medicines, lubricants, cosmetics? What about the environment that we are leaving for those who'll come after us? What can we do to solve some of the problems now, or at least find solutions for the next generations?

As is our goal here, we will delve into the familiar area of alternative energy, learning from the past, and seeking solutions for our future. Some good results of the 1973 oil embargo were the births of the US solar and wind energy industries. In the early to mid-1970s, the US government and businesses were totally preoccupied with energy issues. Seeking solutions, the government established several energy programs, and work was undertaken by private enterprise as well.

The Department of Energy (DOE) was activated on Oct. 1, 1977, providing the framework for a comprehensive national energy plan by coordinating federal energy functions. The new department was responsible for long-term, high-risk research and development of energy technology, federal power marketing, energy conservation, energy regulatory programs, a central energy data collection and analysis program, and nuclear weapons research, development and production.

Billions of dollars were allocated to nuclear research and implementation, coal, gas, and hydro power, and some for development of new alternative technologies.

In parallel, several National R&D institutions were established—the National Renewable Energy Laboratory (NREL), Sandia National Laboratory, Brookhaven National Laboratory, Lawrence Livermore National Laboratory, etc. were also established during that time and/or charged with a task to develop—among other things—alternative energy sources.

The push for new energy sources lasted only until oil prices went down, but continues to be resurrected as oil prices rise from time to time.

In 1979, President Carter unveiled a solar panel installment on the roof of the White House. "No one can ever embargo the sun or interrupt its delivery to us," Carter told the press, "A generation from now this solar heater can either be a curiosity, a museum piece, an example of a road not taken, or it can be a small part of one of the greatest and most exciting adventures ever undertaken by the American people; harnessing the power of the Sun to enrich our lives as we move away from our crippling dependence on foreign oil."

Carter outlined a path to a solar-powered future attempting to make energy independence plans. He asked Congress to approve a $100 million "solar energy bank" with the goal of generating 20% of U.S. power from alternative energy sources by 2000. To fund this ambitious plan, Carter would urge Congress to pass a "windfall profits" tax on the domestic oil industry and subsidies to encourage contractors to install solar panels on new and existing buildings.

Four months later, Islamist militants stormed the American embassy in Iran taking 60 hostages. Domestic issues and Carter's plan for energy independence were put aside. In the 1980s gasoline prices began a steady decline, effectively ending the energy crisis.

All interest in solar and wind energy died down, starting the well known trend of on and off solar. The renewable energy boom-bust cycle of 2007-2012 is its last phase.

In 1986, Reagan had the solar panels removed during routine White House roof maintenance and given away to a Maine state college to heat water for the cafeteria. One of the Carter's solar panels is now on display at the Carter Presidential Library and Museum in Atlanta, GA, so his solar heater *is* truly an example of the "road not taken," which he warned us about so long ago.

But solar energy didn't die completely. There were efforts in the 80s and 90s to resurrect it, and several serious attempts were made to resurrect Carter's plans.

The latest effort to renew solar development started in 2007, but is now ebbing. With hundreds of companies filing for bankruptcy and subsidies going away… again, we are in the midst of a full blown renewable energy bust.

Renewable energy is like any business. There are reasons businesses thrive, and there are reasons they fail. It's important to understand those issues and to think in business terms, if we are going to put renewable energy "out there" and expect success.

Following we look at some great business successes, successful solar companies, successful solar projects, failed renewable energy companies, failed renewable energy projects, and failed thin film solar companies.

We then take a close look at thin films PV products and processes, in order to understand some of the is-

sues, and to explain the high rate of failure in this sector of the solar industry.

We continue the section with a closer look at different types of failure encountered on the manufacturing floor, in the test labs, on the markets, and during field operation.

GREAT BUSINESS SUCCESSES

Some of the world's greatest business successes are Apple Inc., Google, Microsoft, Intel, Nike, McDonalds, Facebook, Southwest Airlines, and Target.

There are a number of common threads among these companies:

1. Their brands became symbolic as their cultural promise endures.
2. Their logos are symbols of reliability, trust and quality.
3. Consumers associate these brands as contributors to their lifestyles.
4. The logos are so powerful that the brand names are not even needed, nor used in most cases.
5. These companies are more resilient, take greater risks and are not afraid to fail during times of uncertainty and change.
6. They push the envelope in pursuit of innovation with their entrepreneurial spirit.
7. They are passionate about what they stand for to create new possibilities.
8. They create a family environment that is focused on giving, sharing and making those around them feel good.
9. The strength of these organization's cultural promise is rooted in their business models, decision-making and strategies, with no deviations or compromises.
10. They overcome temptations to sway from the core values and beliefs. They stay true to their cultural promise which is the key to success and endurance.
11. They are true capitalist enterprises.
12. They invent new rules and cash on them ahead of the competition.

Successful brands stand for something greater than themselves, beyond their core business, that symbolizes their relentless dedication for the advancement of society and the consumers they serve: the 21st century successful capitalist enterprise drivers are trust, consistency and loyalty.

SUCCESSFUL SOLAR COMPANIES

Some of the core values and attributes of successful companies are seen in companies we consider successful in the solar field.

A subsidiary of Abengoa, **Abengoa Solar's** primary activities include the designing, financing, construction and operation of solar power stations that use photovoltaics, concentrated photovoltaics, or concentrated solar thermal technologies. They have specialized in the design and development of solar thermal, or concentrating solar (CSP) projects around the world, and in the US.

Located just outside of Sanlúcar la Mayor, Seville, their Solucar Complex is the largest solar complex in Europe and currently has 183 megawatts (MW) in operation.

The Ecija Solar Complex is composed of two 50-megawatt (MW) parabolic trough plants. Both plants are in commercial operation.

The El Carpio Solar Complex is composed of two 50-megawatt (MW) parabolic trough plants. Both plants began commercial operation in 2012.

Their integrated solar combined-cycle (ISCC) plant in Algeria has a total power output of 150 megawatts (MW), 20 MW of which are obtained from a parabolic trough field composed of 224 parabolic trough collectors. The plant has been in operation since July 2011.

Several CSP plants are under construction:

The Extremadura Solar Complex in Spain is composed of four 50-megawatt (MW) parabolic trough plants. All four plants are under construction. Two of them, 70% owned by Abengoa Solar and 30% by Japanese Itochu, are scheduled to go into commercial operation in 2012.

Solana, one of the largest solar power plants in the world, is a 280 megawatt (MW) parabolic trough plant with six hours of thermal storage. The plant will be located 70 miles southwest of Phoenix, near Gila Bend, AZ. Construction began at the end of 2010 and Solana will begin operation in 2013.

Mojave Solar Project (MSP) is a 280 megawatt (MW) gross parabolic trough plant. The plant will be located 100 miles northeast of Los Angeles, near Barstow, CA. Construction has begun and the Mojave Solar Project will come online in 2014.

Abengoa Solar also has a number of PV power generating plants, in Spain. These are much smaller facilities from 1.0 to 6.0 MWp.

Acciona Energy is a subsidiary of Acciona Group based in Madrid, a Spanish company developing renewable energy projects, including small hydro, biomass, solar energy and thermal energy, and the marketing of bio-fuels.

Acciona is also involved in cogeneration and wind turbine manufacturing, and is carrying out research projects to produce hydrogen from wind power and to manufacture more efficient photovoltaic cells.

Acciona Energy has 164 wind farms in nine countries representing over 4,500 MW of wind power installed or under construction. Acciona Energy is also the developer, owner and operator of Nevada Solar One, the world's first solar thermal plant to be built in more than 16 years, and the third largest facility of its kind in the world.

In 2009, the 100.5 MW EcoGrove Wind Farm in Lena, IL, became operational on 7,000 acres. It comprises 67 Acciona Windpower 1.5 MW turbines, and produces enough power for 25,000 homes, offsetting 176,000 tons of carbon annually.

Acciona Energy has a manufacturing facilities in Europe and in West Branch, IA, which manufactures wind turbines.

Alta Devices was founded in 2007, and is focused on making solar power cost competitive with fossil fuels and accessible anywhere the sun shines. Using a variety of proprietary manufacturing techniques and a unique approach to device design, Alta has invented the world's thinnest and highest efficiency solar cells.

Alta's flexible cells hold world records for cell and module efficiency, and are ideal for incorporating into roofing and building materials, providing power for air and ground transportation, enabling cost effective distributed and off-grid power generation, and for deployment in utility-scale solar farms.

Alta is a development-stage company and has received venture capital funding from August Capital, Kleiner Perkins Caufield & Byers, Crosslink Capital, AIMCo and others. The company is based in Sunnyvale, CA.

In June 2012 Alta Devices™ disclosed details of key technologies that were used to achieve its latest record module efficiency. Alta's thin film solar cells are interconnected into flexible sheets, creating a new and more efficient class of solar material and processes, based on Gallium Arsenide (GaAs).

Traditional solar cells are interconnected to form a module, and the resulting conversion efficiency is compromised because the surface active solar material is covered with metal busbars and wires, preventing some of the light from entering the cells. In addition, there are gaps between cells, which create areas of the module that are not able to convert incident light to electrical energy. More importantly, the failure of one cell can affect the performance of the entire module and string of modules.

Alta's key to improving solar performance at the module level is avoiding these problems, by enabling cell flexibility and using self-interconnected technology that eliminates the wires, thus maximizing the cells' surface which is exposed to incident light.

The self-interconnected cells form a flexible sheet with no gaps, and the sheet can be manufactured to any size or aspect ratio. The efficiency breakthrough allows the use of flexible solar material that can be formed into different shapes and sizes, making possible new and unique solar applications and the penetration of new solar energy niche markets.

This has the potential to change both the applications and economics of solar in the near future, because these new solar materials can be used differently from those that have been available.

The major benefits are mass-scale power production and use almost anywhere the sun is shining. The high efficiency of Alta's material, combined with its thin and flexible nature can be deployed in many new kinds of solar applications including electric vehicles, aircraft and unmanned drones, portable power, roof tiles and other building-integrated uses, and more.

The new solar technology has the potential to reduce the presently high cost of solar modules and entire installations by using solar in nearly any form which broadens the applications. The reduced system complexity and cost are significant, which translates in both economic returns and human benefits.

The construction of Alta's pilot manufacturing facility is underway, with the expectations of putting solar modules on the market in 2013.

Sounds promising, but thin film solar cells and modules have been around for awhile, and have a number of issues related to cost and reliability that should be solved before going to market.

Still, we foresee thin film solar cells and modules having a brilliant future in the 21st century' niche markets.

Amonix

Amonix's technology is superb and represents the ultimate large-scale technology of this 21st century, mostly due to its exceptionally high efficiency. However, this leader in the CPV industry announced layoffs in the summer of 2012.

Can the fledgling CPV industry survive the new bust cycle? There are a number of reports about positive developments in the world of CPV technology (efficiency increases up to 43.5%, new VC funding influxes, and the deployment of good size CPV projects of late). This is all good, but still did not spare this leader of the CPV industry.

Amonix laid off almost 80 employees in May 2012, which is a significant portion of its labor force. And this news comes only a short time (several weeks) after the good news of the completion of the largest CPV power plant in the world; the 30 MWp Alamosa CPV power plant on 225 acres in Colorado.

The Alamosa CPV plant uses more than 500 pieces of 60-kilowatt CPV tracker assemblies, each of which is quite large; 70' wide and 50' tall. The site is interconnected at the 115 kV transmission level and uses a cutting edge PV plant controller, allowing the operator to choose between a variety of smart grid operating modes. The plant's power controls are quite sophisticated and flexible, which allows closed loop power factor control, AC voltage regulation, and also throttling the site's solar energy output as needed to adjust grid frequency. A truly "smart" power plant, you might say. One of the first to join the "smart grid" movement of late.

Alamosa is a great achievement for the solar industry, and the CPV branch in particular, because CPV has a great role to play in the future of alternative energy. This success, despite the obstacles during its design and installation, as well as the consequences from the untimely and tragic death of Amonix' CEO, Brian Robertson in December, 2011 goes to show that CPV has a place in the energy market.

Amonix has plants in Southern California and Las Vegas NV, so the layoffs affect the California plant most.

Since its inception in the late 1990s, Amonix has raised approximately $140 million in venture capital which allowed completion of their new generation CPV tracking systems. In an effort to enter the large-scale energy market, in 2011 Amonix opened the largest CPV trackers production plant in the world in Las Vegas, where the CPV equipment for the Alamosa power plant was manufactured and partially assembled.

Amonix founder, Vahan Garboushian, should be able to reorganize and redirect the company and keep Amonix afloat through the hard times, thus ensuring its successful participation in the world energy markets of the 21st century.

Other CPV companies are trying to beat the Amonix record and install the world's largest CPV power plants. The HCPV technology is over 43% efficient—more than twice as efficient as the most efficient conventional PV modules. It is a serious truly "industrial" type of PV technology that we believe will dominate the energy markets in the 21st century.

Avancis

Avancis was founded in 2006 as a joint venture between Shell and Saint-Gobain. Their goal is the development of efficient and cost-effective CIS technology. Production of high-performance CIS solar modules is based in Torgau, Germany.

In 2011 Avancis opened its 2nd factory at its Torgau, Saxony headquarters. The 25,000 m² factory brings Avancis' annual production capacity to total annual production capacity to 120MWp.

In 2009, Saint-Gobain bought Shell's shares, and is now working with Hyundai on the setup and operation of a third production facility in South Korea with 100MW annual production.

Singulus Technology, AG supplied 3 selenization machines for the CIS PV modules production, which were qualified and ready for production in November, 2011. Avancis' CEO called the production facility state-of-the-art, with modern, German-engineered machines, and superb proprietary factory planning and execution. Avancis appreciated Singulus' on-time delivery of the selenization tools, as well as their quick commissioning.

Now Avancis is ready to go to market. It recently hit 15.8% efficiency of its production modules. All set, with production at 100% capacity and fairly high efficiency. If efficiency, reliability and cost of its PV modules are compatible and affordable enough, success is imminent.

bSolar

bSolar is a manufacturer of a special technology, which might be the answer to many problems. Its bifacial solar cells are just that—they can generate power from both sides of the PV modules. While this might not be a great advantage in some areas, it is surely a great advantage in others.

For example, most PV modules produce very little power under cloudy conditions. Bifacial PV modules would produce 2-3 times more power under these special conditions, simply because both surfaces get some light and are able to generate electricity.

Through its strategic partner and distributor TSBM, bSolar will construct a 730 kW ground-mounted solar project in Nasukarasuyama city, Tochigi, Japan.

The project is expected to go online in 2013 and will feature bSolar's disruptive bifacial PV cells. bSolar also announced that its bifacial cells will be incorporated into several new solar modules by leading manufactures, including Aleo solar, Asola Solarpower and Solar-Fabrik.

Bifacial solar cells and modules sound like a thing of the future, and we are sure to see more of them in some specialized and niche markets. Getting power from both sides of the PV module is a dream come true.

Calyxo

In the summer of 2012, Calyxo (Germany's #2 CdTe modules manufacturer) announced advances in product design that have resulted in independently confirmed peak aperture-area efficiencies of 13.4% for modules and 16.2% for cells.

The even greater news here, which is something we have been talking and concerned about, is that Calyxo has addressed the safe operation of CdTe modules in desert regions. They have developed a suitable product design for achieving the highest reliability possible, that is able to withstand the extreme climates in desert regions.

The new and redesigned CdTe modules installed in Australia's desert show no signs of degradation during the first three years of deployment in the field.

This is a great news, but we need to see 10 or 15 successful years before we can say that the toxic cadmium heavy metal containing PV modules won't harm the desert environment.

These technical advances and the midterm production-cost target of $0.50/Wp also allow a forecast of levelized cost of electricity (LCOE) of under $0.10/KWh, especially in sunny regions of the world.

We applaud Calyxo for taking this very important step in an area that was totally ignored until now.

Infinia

Infinia Corporation develops and manufactures free piston Stirling generators and applications that convert solar, biogas and natural gas heat sources into reliable electrical energy. Infinia is the producer of the PowerDish™—the world's first Stirling-based solar power generation system suitable for automotive-scale manufacturing and deployment from small distribution-scale arrays to multi-megawatt, utility-scale solar power plants. In other words, Infinia is diversified, with solar as only one alternative.

Infinia raised $32 million in debt and options in 2011, and another $50 million in equity before that from investors; but, like its cousin, Stirling Energy Systems,

Infinia has hit a few bumps, recently laying off employees at its previous headquarters in Washington state, in order to move to Utah to "accelerate ...into a world-class solar generator company with geographically consolidated operations."

Infinia has several installations and new customers in India, and completed the first commercial installation of its technology earlier in 2012. Recently they sold 10 MW of engines to solar developers in India. They also installed a 9 KW system on the rooftop of Belen's City Hall in New Mexico.

In May 2012 Infinia won a contract to supply equipment to a $143 million solar project in Cyprus, and has plans to supply a 50-MWp venture in Jordan.

So, Infinia is a potential winner in the Stirling Engine game. There are some useful lessons to learn from Infinia: 1) never give up, regardless of the failures around you, and 2) diversification (technological and geographical) is one of the keys to success.

First Solar

First Solar has experienced both success and failure. Until recently, it was one of the largest and most financially successful solar companies in the world. In 2010, it was considered the second-largest maker of PV modules in the world, ranking sixth in the Fast Company's list of the world's 50 most innovative companies. In 2011, it ranked first on Forbes's list of America's 25 fastest-growing technology companies.

Then it became a victim of the falling PV module prices and was forced to close manufacturing facilities and lay off 2,000 workers in 2012. The FSLR stock price plunged from $325 in 2008 down to $14 in the summer of 2012.

There are also questions related to the excess temperature degradation, as well as the use of exotic and toxic materials.

First Solar manufactures PV modules, consisting of CdTe active layers encapsulated between two glass sheets. Since 2010, its business plan was revised to include the development of large PV power plants and supporting services that include finance, construction, maintenance and end-of-life recycling services.

The CdTe PV modules are generally less expensive than those manufactured from crystalline silicon.

In 2009, First Solar became the first solar panel manufacturing company to lower its manufacturing cost to below $1 per watt, but drastically falling prices of foreign-made c-Si PV modules have thrown a large shadow on that achievement. $0.40 per Watt—the price of c-Si PV modules today—is hard to beat; thin film or no thin film.

First Solar did not give up, but shifted away from existing (mostly rooftop) markets that are heavily dependent on government subsidies, and moved toward providing utility-scale PV systems in sustainable markets with immediate need. This shift, however, successful with the help of large in US government loan guarantees.

Regardless, First Solar showed us a glimpse of how solar companies will operate later on in the 21st century—without the cadmium, of course. Their efficient, mass production equipment and procedures, and complete materials and process quality control are key factors needed to reduce price and maintain quality—those factors distinguish First Solar from the rest.

We believe that First Solar's technology—automated equipment and efficient processes—will become a standard for PV modules manufacturing later in the 21st century.

Sky Fuel

Sky Fuel is a designer and provider of thermal concentrating solar power (CSP) equipment for utility-scale power generation, and also a provider of versatile industrial applications of steam power. Sky Fuel is also working on storing the energy from thermal CSP plants for later use as electricity, thus matching closer the load profile of the utility company.

SkyFuel has two commercially available products—a parabolic trough concentrating solar collector (the SkyTrough) and a high reflectance silverized polymer film (ReflecTech). The SkyTrough is the first utility-scale solar concentrator to use polymer reflectors rather than glass mirrors.

The use of a polymer mirror allows the reflector to be made as one continuous large panel, which is lighter than glass mirrors and therefore the support frame can also be lighter. The SkyTrough system can be deployed in a stand-alone configuration for utility-scale electricity generation. It can also be integrated with fossil fuel power plants, such as natural gas-fired combined cycle power plants, to augment or displace some of the host plant's steam production or, in the case of traditional coal plants, to provide feed-water heating.

SkyTrough systems are also ideally suited for industrial process heat applications and sea-water desalination. SkyFuel is also developing next-generation, high-temperature parabolic trough and linear Fresnel systems.

The U.S. Department of Energy has awarded SkyFuel a $435,000 grant to develop its Linear Power

Tower (LPT) system for utility-scale solar thermal power plants. The LPT is a high-temperature linear Fresnel system.

The versatility and usefulness of Sky Fuel products and services might propel it into the forefront of the solar industry in the near future.

Solel

Solel is an Israeli solar thermal equipment designer and manufacturer based in Beit Shemesh, Israel. Its parabolic trough technology has been around for decades and has been used in test plants for 20 years. Solel, with 400 employees in Israel and 100 in Spain, obtained a $105 million investment from Ecofin and started plans for a $140 million facility in Spain. The company also had an agreement with PG&E for 553 MW at Solel's Mojave Solar Park, which uses parabolic trough technology, in addition to a 150 MW of plants in Spain.

Siemens AG signed a $418-million contract to buy Solel Solar Systems. Siemens will leave the company headquarters and some of its manufacturing facilities in Israel for at least five years. To date, the majority stake has been held by Ecofin Ltd., a London-based investment firm, and another major shareholder.

With the acquisition of Solel, Siemens is counting on strengthening its market position in the promising business of solar thermal power plants, thus being able to expand in the area of green energy.

Solel Solar Systems has a workforce of over 500 and is one of the world's leading suppliers of solar receivers, which are key components of the design of parabolic trough power plants. This company posted revenue of almost $90 million in the first six months of 2009, and was also a leader in the planning and construction of solar fields.

Since 2006, Solel has also been present on the Spanish market, supplying key components for 15 solar thermal power plants with a combined capacity of 750 megawatts. In addition, the company is also active on the important U.S. market.

The market for solar thermal power plants is slowing down, so the estimated double-digit growth rates and increased volume of over €20 billion is now in question. Nevertheless, Solel is primarily focused on the world's growth regions in the US, South Africa, Australia, Spain, India, North Africa and the Middle East, and hopes that its skills in the core competencies will further optimize the water/steam cycle, boosting the efficiency of solar thermal power plants.

Unfortunately, in October, 2012, Siemens decided to exit the solar field all together, without clear plans for the future of Solel. With or without Siemens, however, Solel has proven technology and services, which will ensure their success in the world solar markets.

SolarReserve

SolarReserve, LLC, headquartered in Santa Monica, CA, develops large-scale solar energy projects in the US and worldwide. It holds the exclusive worldwide license to the molten salt, solar power tower technology developed by Pratt & Whitney Rocketdyne, a subsidiary of United Technologies Corporation. This technology uses molten salt to transfer and store energy, which is then used to produce power round the clock.

A spinoff from aerospace giant Hamilton Sundstrand, itself a subsidiary of United Technologies Corp., SolarReserve is one of the most recent entrants in the field. Since 2007, it has amassed a CSP plants development portfolio of more than 25 projects, which use its licensed solar power technology. The potential output of these projects is more than 3,000 MW in the US and Europe. A number of early-stage activities are developed in other international markets including the Middle East, North and South Africa, Australia, China, India and Latin America.

SolarReserve is also developing 1,100 MW of photovoltaic projects across the Western United States, and is actively acquiring new sites to add to the pipeline in the US and overseas. The total of previously developed and financed renewable and conventional energy projects in more than a dozen countries around the world is over $15 billion.

SolarReserve completed a 540-foot solar power tower for its 110 megawatt (MW) Crescent Dunes Solar Energy Plant located near Tonopah, NV. Utilizing the most advanced solar thermal technology worldwide, the Crescent Dunes Plant will be the nation's first commercial-scale solar power facility with fully integrated energy storage and the largest power plant of its kind in the world.

The Crescent Dunes project has secured a 25-year power purchase agreement with NV Energy and will provide clean power to approximately 75,000 homes when complete.

Once operational, the project will expend more than $10 million per year in salaries and operating costs, and is forecasted to generate $47 million in total tax revenues through the first 10 years of operation—contributing to workers' paychecks, service businesses, local school systems and police and fire departments.

The project is being constructed on federal land operated by the Bureau of Land Management. In November 2010, Interior Secretary Salazar signed the project's thirty-year right-of-way and approval to construct. The plant is expected to be operational by the end of 2013.

CSP and energy storage technologies are the key to success and progress of solar and wind power plants, so the success of SolarResrve is important to all sectors of the renewable energy field.

SUCCESSFUL SOLAR PROJECTS

The case studies provided here illustrate the combination of renewable technologies, project scale, financing structures, programs, and unique attributes used to finance and execute successful renewable energy projects.

Ivanpah

After a number of false starts, incidents and controversies, BrightSource Energy's Ivanpah solar thermal (CSP) plant is thought of as the world's largest and most efficient solar farm. It is installed in the deserts of the Mojave National Preserve, and it is expected to operate at 18% efficiency at a capacity factor of over 30%. This alone is a record.

This 392 MWp facility is therefore more efficient than most alternative energy power plants, including c-Si, thin-film cells, or parabolic mirrors. It is installed on three adjacent tracts covering over 3,500 acres of desert land and is expected to run at full capacity 10 to 11 hours a day with the help of a back-up natural gas system which will ensure high performance during cloudy days and late nights, thus reducing unwanted power fluctuations.

At the top of the power towers, boilers absorb sunlight reflected from 7-square-meter ground-mounted mirrors and heat water to more than 1,000 degrees Fahrenheit, the highest temperature in the industry. The super-heated steam drives turbines which generate electric power. A back-up natural gas system permits the long operating hours and the ability to run most of the day at full capacity. The gas is used to warm boilers in the morning and augment solar power on cloudy days to keep output high.

Thus achieved power supply consistency should bring the electricity costs on par with natural-gas plants, something photovoltaic plants simply cannot do now.

BrightSource has raised more than $300 million in financing, expects the plant's efficiency to rise as the company moves beyond its first-generation technology.

Higher-efficiency turbines are already on the market, and additional mirrors, or heliostats, can be deployed. Water temperatures also will rise to above 1,100 degrees.

The real competition is coming from falling solar-panels prices, unless the trade wars slow the trend. BrightSource, however, insists that plant efficiency is only one measure of performance, and not necessarily the best or most significant. Capacity factor, a calculation of a farm's ability to deliver full power over time, may be more important to the customers (the utilities).

Ivanpah's capacity factor (including the use of natural gas) is in the 30% range, matched only by wind farms in an ideal location, where they can have a capacity factor of 40%. Photovoltaic plants generally are lower, and even those in the best locations (in the deserts) have around 20% capacity factor.

The total delivered cost of electricity is of great importance too, and BrightSource claims that this measure makes it "extremely competitive," but we have not seen any data to verify that. California Public Utilities Commission shows that expected delivered costs are to be less than 12.5 cents per kWh, but we don't know how much less. Industry specialists claim that the cost of Ivanpah's electricity will be lower than photovoltaic power and about the same as that of generated by natural gas.

These, however, are only theoretical calculations mixed with speculations, so it remains to be seen, when the plant is in full operation.

The problems Ivanpah faces are numerous:

1. The desert tortoise at the Ivanpah has a mortality rate as high as 98 percent. Ivanpah has to implement Tortoise care programs for hatchlings and improving survival rates. Project owners—NRG, Google and BrightSource Energy—are going to ensure minimal impact to the desert tortoise population at and near the project site. Approximately $22 million were spent on caring for desert tortoises found on or near the site, and more will be spent to protect them. Another $34 million are to be spent to meet the project's federal and state mitigation obligations, which include tortoise habitat restoration, the installation of an additional 50 miles of protective tortoise fencing and the minimum purchase of 7,164 acres of conservation habitat. This is the equivalent of 2 acres for every 1 acre of development.

2. When the $2.2 billion Ivanpah CSP plant (with $1.6 billion of US DOE loan guarantee) is in operation, nearly 350,000 mirrors on 3,600 acres will reflect light onto boilers. Steam will power turbines,

which will generate electricity to power California homes. This creates a number of problems:

a. Great numbers of mirrors installed on the desert floor (with their cement foundation and daily shadow) could change the environment to a degree.

b. Bright sunlight reflected by the mirrors and the towers could cause damage to wild life and even human activities (blinding airplane pilots, etc.).

c. The steam generation and post-process steam cooling require millions of gallons of fresh water daily. The deserts are not the best place to find such a great quantity of water, so it is not clear how this will play out.

We can't predict if the mirrors or the lack of water will become major stumbling blocks for Ivanpah and the other CSP plants. We do know that using such large quantities of technology in the deserts, albeit promising and needed, is a fairly new phenomena, so we must approach it with respect, taking into consideration all abnormalities before and during operation.

Mojave Solar Project

In 2009, Abengoa Solar Inc., sole member of Mojave Solar LLC, filed an Application For Certification (AFC) for its Mojave Solar Project. The proposed project is a 250 MWp solar electric generating facility to be located near Harper Dry Lake in an unincorporated area of San Bernardino County, near Los Angeles, California.

The plant will use well-established parabolic trough technology to convert solar energy to heat via a heat transfer fluid (HTF). The hot HTF generates steam in solar steam generators, which will expand through a steam turbine generator to produce electrical power from twin, independently operable solar fields, each feeding a 125-MW power island.

There are no plans for a supplemental energy source (like natural gas) or energy storage to assist the electrical power production during cloudy days or at night.

The California Energy Commission (CEC), which is in charge of the permitting of the site is exempt from having to prepare an environmental impact report. CEC's certified program, however, does require environmental analysis of the project, including an analysis of alternatives and mitigation measures to minimize any significant adverse effect the project may have on the environment.

In the fall of 2011 Abengoa upped the plant's output design to 280 MWp, and reached agreements with different local firms to promote the economic development of the region. The project is expected to create around 150 jobs during construction, but only a handful during operation.

Solana

Solana is a CSP power plant with total power generation of 250 megawatts, designed and operated by Abengoa Solar, Spain. Operating at full capacity it will provide power to 70,000 Arizona homes. Solana was the largest solar plant in the world in 2011 (on paper), providing the Arizona utility APS with more solar electricity per customer than any utility in the US. It is, and will be for a long time, APS' largest source of renewable energy.

Located 70 miles southwest of Phoenix, near Gila Bend, AZ, the plant's siting and permitting began in 2008, and the plant is to be operational sometime in 2013. During construction, the project employed about 1,500 people, but only a handful of people will remain as permanent employees.

The plant uses concentrating solar power (CSP) technology with thermal energy storage. Solana's parabolic mirrors focus the sun's heat on a tube in which heat transfer fluid is running during operation. The fluid reaches a temperature of 735°F at noon, but gradually drops before and after noon. The hot fluid transfers its heat energy to water, creating steam, which is then used to run conventional steam turbines.

The heat can be stored and used during cloudy days, and/or at night to generate electricity. This is a great plus for this type of solar energy generation, which PV technologies cannot match for now.

The Solana Generating Station covers 3 square miles of abandoned agricultural land, and contains 2,700 parabolic trough collectors.

APS selected Abengoa Solar as a partner in the project after years of investigation of the technology and a thorough consideration of prospective developers. Abengoa Solar will construct, own and operate the Solana Generating Station.

The plant designers claim that Solana uses 75% less water than the property used in the past, but this is a misnomer, because the land was flood irrigated, which is quite wasteful. The lack of water was the reason the land was taken out of operation by BLM, and now the plant will use a million gallons of fresh water daily, which is an incredibly large amount for any desert. Because water is a commodity in the desert, we must be careful how we use it.

Sierra Sun Tower

eSolar's Sierra SunTower is a 5 MW commercial concentrating solar power (CSP) plant built and operated by eSolar. The plant is located in Lancaster, CA, as the only CSP tower facility in North America.

The Sierra SunTower facility is based on power tower CSP technology. The plant features an array of heliostats which reflect solar radiation to a tower-mounted thermal receiver. The concentrated solar energy boils water in the receiver to produce steam. The steam is piped to a steam turbine generator which converts the energy to electricity. The steam out of the turbine is condensed and pressurized back into the receiver. The power is interconnected to the Southern California Edison (SCE) grid and, and in the spring of 2010, it was the only functioning commercial CSP tower facility in North America.

The project site occupies approximately 20 acres in an arid valley in the western corner of the Mojave Desert. It includes two eSolar modules—24,000 heliostats divided between four sub-fields. The heliostats track the sun and focus its energy onto two tower-mounted receivers. The focused sunbeam heats and converts feedwater piped to the receivers into super-heated steam that drives a reconditioned 1947 GE turbine generator to produce electricity. The steam passes through a steam condenser, reverts back to water through cooling, and the process repeats.

Sierra Suntower has been certified by the California Energy Commission as a renewable energy facility. Power from the facility is sold under a Power Purchase Agreement (PPA) with SCE, providing clean, renewable energy for up to 4,000 homes.

Sierra SunTower was designed to validate eSolar's technology at full scale, effectively eliminating scale-up risks. The solar thermal equipment operating at Sierra SunTower forms a blueprint from which future plants will be built.

With $130 million investment from Google, Idealab and Oak, eSolar has secured land rights in the southwestern United States to produce over 1 GW of power using its central power tower receiver surrounded by smaller-than-usual heliostats, in 25 MW clusters.

The only operating solar thermal power tower plant in the US, Southern California Homes' Sierra SunTower will produce 5 MW of electricity powering up to 4,000 homes. It created over 250 construction jobs and 21 permanent jobs, with the clean solar power generation offsetting more than 7,000 tons of CO_2 each year.

County of Yolo

Yolo County installed a 1 MW solar PV project to supply renewable power to a jail and juvenile center. This project represents the first known combined use of QECBs and CREBs in the nation. QECBs and CREBs, known as qualified tax credit bonds, are an inexpensive approach for state and local government to finance renewable energy installation.

The county chose to own the solar PV system and did not select a PPA provider because the PPA's financial benefits were not substantial. The county used a variety of funding sources to help finance the project including CREBs, QECBs, an Energy Commission Energy Conservation Assistance Act loan, and a Tax Exempt Lease Program (TELP) loan. Yolo County was also eligible for a $2.5 million incentive from California Solar Initiative (CSI) and a $1.9 million Pacific Gas and Electric Company rebate.

The total cost of the project with interest payments is $9.4 million (or $9.4 per installed Watt), with total utility bill savings estimated at $18.1 million over 25 years. Without these low-cost finance options, the county might not have been able to finance the project.

The county negotiated a lag time of six months between the system going on-line and when initial payments were due. This allowed it to generate enough funds from the utility savings that were used to make the first payment and schedule its future financial obligations.

Many companies have been able to install solar power lately. With government subsidies and incentives, the US now has more solar power than ever, but this model of operation is only as sustainable as the subsidies and incentives, so we need to find a new model.

Jefferson Union High School District

Perpetual Energy Systems (PES) financed a 1.5 MW solar energy project at the Jefferson Union High School District located south of San Francisco. The project includes more than 8,500 solar PV panels and generates more than 2.3 million kWh of solar energy during the year.

The project required no upfront costs from the school district because it was funded by PES. The district will buy power from PES at a lower rate than the PG&E rate. In return for the financing, the company will get tax and investment credit. The school expects to save 3 percent on annual energy costs over the next 25 years.

This operation, too, appears unsustainable. A

quick calculation shows that PES would not be able to repeat this installation without tax and investment credits. We estimate $230,000 gross ROI annually, which after long-term bad weather losses, and extra O&M expenses would come down to $150,000 net profit yearly. If $100,000 is paid against the estimated $6.0 million investment (without subsidies, etc.) for the installation, this means that PES would need 60 years to repay the capital investment at energy cost of $0.10/kW, 45 years at $0.15/kW, or 30 years at $0.20/kW. None of these variants is acceptable.

Hudson Ranch Project

Hudson Ranch I is a 49.9 MW geothermal generating power facility in the Salton Sea area in California, with EnergySource as the project developer. The electricity produced from the geothermal facility will be sold under a 30-year PPA with local utilities.

The local water board granted the project 800 acre feet of water annually for use at the facility. In addition, the facility took advantage of ARRA tax incentives worth more than $100 million. This $399 million project (or nearly $8.00 per installed Watt) is expected to create 200 jobs during construction, but only a handful during operation. It will provide approximately $3.5 million in annual property taxes.

$100 million tax incentives helps a lot, but $399 million capital expenditure still must be accounted for. So, we estimate $30 million gross income yearly (at $0.10 energy cost). After all expenses are paid, $20 million would be net annual profit, so, barring any big disasters, in 20 years the facility will be paid for. Is this enough?

Hybrid HCPV/T System

In the summer of 2011 a group of technologists and investors considered the use of a hybrid system, consisting of high concentration PV (HCPV), with thermal (hot water) generation capability. This hybrid energy system generates DC electricity and hot water at the same time at efficiency of approximately 40% and 20% respectively, thus delivering combined 60% efficiency at the source. Losses during the conversion and transport would reduce the overall efficiency to the 40-45% range, which is still the highest on the market.

The hybrid HCPV-T unit consists of several HCPV modules mounted on a 6 x 6' frame that tracks the sun constantly. In each module there are several HCPV cells mounted on the bottom with a Fresnel lens on top. Sunlight is captured and focused onto the HCPV cells

by the Fresnel lens at 500 times magnification.

The small (1 cm^2) HCPV cells produce 20 Watts of power under these conditions, but get very hot, and for that reason they are mounted onto heat sinks. In the hybrid design, cooling water is pumped through the heat sinks, which a) cools the HCPV cells down, and help keep their efficiency up, and b) heats the water, which then can be used for residential or commercial applications.

The amount of energy produced depends on the size of the system, its actual setup and operation, and—very importantly—the amount of sunlight available. Since CPV and HCPV technologies require direct beam sunlight, they are best suited for use in the U.S. Southwest and a number of sunny locations around the globe.

For the performance analysis we picked a location in Arizona with 320 sunny days annually, average 900 W/m^2 full solar insolation 8 hrs. per day.

Note: Summertime full solar insolation for CPV use is ~10 hrs, wintertime solar insolation for CPV use is ~6 hrs.

So, under ideal conditions 900 W/m^2 x 8 hrs = 7.2 kW/h energy is received daily per m^2.

Each CPV/T unit is 40% efficient and occupies ~6.0 m^2 area, OR 7.2 kW/h x 6.0m^2 x 40% = ~17.3kW/h daily generation (vs. 5.5 kW/h for PV panels*).

300 CPV/T units (1800 m^2) would generate 1.0MWp or ~8.0MW/h averaged daily.

At $0.20 avg. PPA (including carbon credits, federal and local subsidies and incentives) this represents $1,600/day, or $512,000 annual income from DC/AC power generation.

Also, each HCPV/T unit generates hot water at ~20% efficiency (180W/m^2 = 614 Btu/m^2). Thus, 614 Btu x 8 hrs x 6.0 m^2 x 320 days = 9.43 million Btus are generated annually. Or, our 1.0 MWp plant would generate additional ~$180,000 annually from hot water use.

Useful power generation (considering DC/AC and other conversion losses and O&M inefficiencies) would be ~20% less. Therefore, we can expect gross income during the first year of operation from our 1.0MWp HCPV/T plant to be ~$550,000.

Total capital expense for building the HCPV/T power plant is estimated at $1.30/Wp for equipment and $2.50W/p for BOS, land and administrative expenses, or a total of $3.8 million to install and start operation of the 1.0 MWp plant.

Additional expenses of about 15% of gross income must be assumed and allocated for annual O&M operations, which brings total net income to ~468,000 annually.

Therefore, our 1.0 MWp CPV/T plant will be paid for in about 8 years, providing clean, green energy and reducing global CO_2 generation by over 1 million lbs. annually.

Note: CPV is several times more efficient than fixed-mounted PV panels. It also tracks the sun all day, getting the maximum power possible from morning until evening.

Advanced Residential Installations

Residential installations are the heart of the solar industry, for they are where most solar companies, the utilities and the government place emphasis at present. We believe that residential solar installations help the overall development of the worldwide solar energy industry by providing work for thousands, and contributing to new developments in the area. The amount of generated power, however, is a drop in the ocean. We will never get even close to energy independence relying on solar energy generated on rooftops.

Too, there are serious problems with this fragmented, distributed, way of generating power. Variability of the power output, due to changing weather conditions, inefficiencies and breakdowns of the roof systems over large areas, and related issues contribute to temporary and unpredictable power grid under- and over-loads, which create problems with grid control.

Roof damage is, and will become even more so with time, a serious problem for home owners. Poor craftsmanship of roof installations is causing roof damage, which could create a significant financial burden and nullify the benefits of the installations.

Nevertheless, roof-mounted solar systems are growing daily, and we must account for the benefits they bring to the development of the solar industry.

OTHER ENERGY PROJECTS

BelAir, Maryland

An advanced energy generation and conservation system by Bob Ward Companies started with a basement foundation of precast high-density concrete walls with an interior layer of rigid foam insulation plus R-19 fiberglass bat.

He also included advanced framed upper walls and an R50 blown fiberglass ceiling insulation, plus foam sealing of all band joists, top and bottom plates, and wall penetrations.

A house was kept open for a year to showcase the 3-kW photovoltaic (PV) system, solar hot water system, and super-efficient lighting and appliances.

Pleasanton, California

Builder of more than 350,000 homes, Centex's Avignon development is the first all-solar, energy efficient community built in the county. The 3.5-kW PV systems use SunPower integrated SunTile, which matches the dimensions of the cement roof tiles it replaced to blend in seamlessly with the roofing.

Other energy-saving measures like R-49 attic insulation and R-15 wall insulation, high-efficiency windows and appliances, caulking, sealing, and independent air leakage testing help ensure energy savings of up to 70%.

Okefenokee, Georgia

Passive solar and cooling design elements cut energy use by 43% in this Georgia Department of Natural Resources cottage.

By maximizing cross ventilation using sun-blocking overhangs like long screened porches and adding the heat-shielding power of structural insulated panels, the home's 4.1-kW PV system is able to more than meet all of the home's heating and cooling load in all but the two coldest months of the year.

Wheat Ridge, Colorado

This Habitat for Humanity home is well insulated with extra thick double stud walls sandwiching three layers of R-13 fiberglass, R-60 in the attic and R-30 in the floors. An energy recovery ventilation system retains heat while providing fresh air.

The home is equipped with a 4-kW PV system plus three solar thermal water collectors on a drain-back system with a 200-gallon tank that provide nearly all of the family's electric and hot water needs.

Tucson, Arizona

A southwestern style development by John Wesley Miller Companies includes rooftop solar panels for an integrated solar water heater and a 1.5-kW PV system standard on each of its 99 units.

Steel-reinforced masonry walls with rigid insulation under a three-coat stucco finish, R-38 ceiling insulation, and low-emissivity dual-pane windows ensure a tight thermal envelope and high-efficiency appliances and ducts in conditioned space help reduce energy usage.

Pulte Homes selected a workable combination of active closed-loop solar thermal water heaters to meet the 5% solar requirement and an efficiency package to meet the energy savings requirement of 50% over local code.

The package includes "cathedralized" application of blown insulation along the roof line of the attic to provide conditioned space for the ducts and air handler as well as air leakage testing, high-performance closed-combustion gas appliances, use of a drainage plane, flashing, and other water management details for 1200 to 1500 homes.

Olympia, Washington

It may be hard to believe solar can work in the cloudy, damp Pacific Northwest, but homeowner Sam Garst teamed with Building America and a science-minded builder and architect to do just that.

The house sports a 4.5-kW PV system, radiant heat and hot water from a ground source heat pump, advanced framing, a foam-insulated slab foundation, and Icynene spray wall and ceiling insulation, together with other green features like a greenhouse for passive solar heat, a rainwater cistern, construction recycling, fly-ash concrete, low VOC finishes, and bamboo and recycled tire flooring.

Columbus, New Jersey

Builder Mark Bergman pioneered the first all-solar development of market-rate homes in New Jersey, with 39 homes built to be at least 60% more efficient than code.

The heavily insulated, tightly sealed homes use energy-efficient appliances and lighting to cut energy use. To ensure maximum solar gain for his 2.64-kW photovoltaic systems, he located the panels on detached garden sheds, located anywhere on the lots, for ideal orientation to the sun.

SUCCESSFUL INITIATIVES

Utility Solar Water Heating Initiative

The Utility Solar Water Heating Initiative (USH2O) provides services and support to help interested utilities learn about existing programs and develop their own efforts. USH2O provides information about utility water heating programs and offers services to utility companies and energy service providers considering implementation.

The overall mission of USH2O is to facilitate the successful implementation of utility solar water heating programs and to educate stakeholders about the potential of solar thermal technologies.

USH2O utility partners manage the nation's most successful solar water heating programs. Solar thermal industry partners provide equipment for virtually all of the solar installations across the United States.

USH2O partners include solar thermal program managers from investor-owned utilities, municipal and other publicly owned utilities, as well as manufacturers, distributors, and installers of solar water heating systems.

The U.S. Department of Energy (DOE) and the National Renewable Energy Laboratory provide funding and technical guidance to the initiative. Other stakeholders include the Solar Rating and Certification Corporation and the Solar Energy Industries Association.

SunShot Vision Study

The SunShot Vision Study provides an in-depth assessment of the potential for solar technologies to meet a significant share of electricity demand in the US during the next several decades. The DOE study explores a future in which the cost of solar technologies decreases by about 75% between 2010 and 2020 in line with the SunShot Initiative's cost targets.

The SunShot Initiative is a collaborative national effort to make the US a leader in the global clean energy race by fueling solar energy technology development. SunShot will enable widespread, large-scale adoption of solar across America by making solar energy systems cost-competitive with other forms of energy by the end of the decade.

The SunShot Initiative vision is to make the total cost of solar energy economically viable for everyday use.

Reducing the total installed cost of solar energy systems by 75% can be accomplished by reducing solar technology costs and grid integration costs, and by accelerating solar deployment nationwide.

Some strategies include increasing PV solar cell efficiency, reducing production costs, and opening new markets for solar energy; shortening the time to move new solar technologies from development to commercialization and strengthening the U.S. supply chain for solar manufacturing and commercialization of cutting-edge photovoltaic technologies; reducing the cost of concentrating solar power (CSP), fostering collaboration for utility-scale solutions, and integrating solar into the electric grid; investing in education, policy analysis, and technical assistance to remove barriers and speed penetration; developing a well-trained workforce to foster U.S. job creation in the solar industry.

A number of programs today are geared to finance and accelerate the development of solar power programs and initiatives. SunShot Initiative issues grants to different companies and institutions, working in the solar field.

The scope of the SunShot program is in the areas of Advanced Manufacturing Partnerships (AMP); Extreme Balance of System; Foundational Program to Advance Cell Efficiency; Grid Integration-advanced Concepts; Incubators; Next Generation PV; Non-hardware Balance of System; PV Supply Chain and Cross-cutting Technologies; Rooftop Solar Challenge; SUNPATH.

An example of the goals and benefits of one of these programs is given below.

Next Generation Photovoltaics II Projects

Twenty-three solar projects were investigating transformational photovoltaic (PV) technologies with the potential to meet SunShot cost targets. The projects' goals were to increase efficiency, reduce costs, improve reliability, and create more secure and sustainable supply chains.

On Sept. 1, 2011, the U.S. Department of Energy (DOE) announced $24.5 million to fund the Next Generation Photovoltaics II projects over a performance period of either two years or four years. This early-stage applied research investment seeks to not only demonstrate new photovoltaic concepts, but also to train the next generation of graduate students and post-doctoral fellows who will ultimately lead the development and commercialization of PV technologies in future years.

A list of awardees, and brief description of their projects follows.

Bandgap Engineering ($750,000), Woburn, MA

In this project, silicon (Si) nanowire arrays are being used to engineer an intermediate band solar cell (IBSC). The IBSC has a theoretical efficiency of up to 60%; however, the goal is to engineer a 36% efficient solar cell made only with Si. Bandgap Engineering is seeking the early phase demonstration of an IBSC material produced by growing the Si nanowires epitaxially on the surface of an oriented Si wafer to achieve accurate control over crystallographic orientation and faceting of the nanowires, which will selectively increase coupling between specific electronic states.

California Institute of Technology ($750,000), Pasadena, CA

The goal here is to develop a waferless, flexible, low-cost, tandem, multijunction, wire-array solar cell that combines the efficiency of wafered crystalline silicon (c-Si) technologies with the cost and simplicity of thin-film technologies. The approach synthesizes tandem solar cells by conformal epitaxial growth of III-V compound,

semiconductor, wide-bandgap absorber layers to form dual-junction and triple-junction wire array tandem solar cells. Such high-efficiency multijunction wire arrays represent a transformational, and as-yet unrealized, opportunity for low-cost, high-efficiency photovoltaics.

Colorado School of Mines ($1,484,364), Golden, CO

Researchers are developing a new approach to the synthesis of hydrogenated nanocrystalline silicon (nc-Si:H), which exploits hot-carrier collection as a way of boosting conversion. By using a novel gas-phase plasma process, the team is creating engineered films that incorporate quantum-confined Si nanocrystals with tailored surface termination. These composites of amorphous and nanocrystalline Si have the potential to dramatically increase the efficiency of single junction and multijunction thin-film Si solar cells by mitigating photo-induced degradation, allowing increased absorption, and offering the realistic possibility of hot-carrier devices.

Massachusetts Institute of Technology ($750,000), Cambridge, MA

MIT researchers are developing c-Si thin-film solar cells with a thickness of less than 10 microns at efficiencies greater than 20%. Typical c-Si wafers are about 180–250 micrometers thick and account for approximately 30%-40% of the total module cost. By dramatically reducing the size through nanostructuring surfaces, developing high-performance transparent conductors, and identifying low-cost manufacturing processes, this effort aims to open a new pathway for meeting cost targets.

In a second award of $1,500,000 MIT is using systematic defect engineering to advance thin-film PV cells based on tin sulfide, which offers high optical absorption, high carrier mobilities, and long minority lifetimes. Because tin and sulfur are earth-abundant and require processing temperatures below 400°C, use of these materials in thin-film PV cells has the potential to lead to low-cost fabrication. A rapid ramp-up of efficiency is also possible by leveraging the decades of development of similar thin-film materials.

National Renewable Energy Laboratory ($750,000), Golden, CO

NREL, together with MIT, is developing a novel class of earth-abundant materials for single-junction, tandem-junction, or multijunction thin-film PV applications that can be synthesized with low-cost, scalable methods. A team of NREL researchers has completed proof-of-concept synthesis and characterization of these novel materials. Preliminary results have demonstrated facile synthesis of materials with independent tunability of key material properties, which has the potential to reduce costs. The team is performing exploratory research on these promising materials with a goal to fabricate baseline PV devices by the end of the project.

The project's team (awarded $750,000) includes partners from Colorado School of Mines and Cornell University working to establish a new solar cell paradigm of ternary copper nitride absorbers (Cu-M-N). These absorbers are expected to have favorable properties because of the large valence-band dispersion that results from a nearly perfect energy match of Cu and N energy levels. This match may lead to a defect immunity similar to that exhibited by copper indium gallium diselenide (CIGS) devices. The primary objectives of this project are to identify earth-abundant, thermodynamically stable, and nonreactive Cu-M-N materials, determine their exact chemical stoichiometry and crystallographic structure, and study their physical properties related to photovoltaic applications.

In a third project, awarded $750,000, NREL aims to develop a novel (Zn,Mg)Cu oxysulfide solar absorber material with the potential to reach and exceed 20% energy conversion efficiency. Researchers are substantially modifying the Cu2O base material by alloying with sulfur, zinc, and magnesium. This effectually tailors band-structure properties to match the solar spectrum. In developing a novel optimized solar absorber material, the team is employing both theoretical modeling with electronic structure methods and combinatorial thin-film synthesis and characterization.

PLANT PV ($750,000), Berkeley, CA

PLANT PV is studying the feasibility of using cadmium selenide (CdSe) as the wide band-gap top cell and Si as the bottom cell in a monolithically integrated tandem architecture. The greatest challenge in developing efficient tandem solar cells is achieving a high open circuit voltage (Voc) with the top cell. To achieve tandem power conversion efficiencies greater than 25%, the CdSe top cell must have a Voc greater than 1.1V. Through this project, PLANT PV seeks to determine whether it is possible to epitaxially grow CdSe films with sufficient minority carrier lifetimes and with p-type doping levels necessary to produce an open-circuit voltage greater than 1.1V using close-space sublimation.

Princeton University ($1,476,609), Princeton, NJ

Researchers here are developing silicon/organic heterojunctions (SOH) as a new class of high-efficiency, low-cost photovoltaic technology. In SOH cells, the light is absorbed in silicon just like in conventional crystalline and multi-crystalline silicon photovoltaics, but there is no p-n junction. Instead, the carriers are separated by the field in the silicon created by a silicon/organic heterojunction. These devices are fabricated by spin-coating or spraying a thin layer of an organic semiconductor on silicon. This low-cost room-temperature process eliminates the need for expensive high-temperature diffusion steps required to fabricate p-n junctions.

Purdue University ($750,000), West Lafayette, IN

This project combines earth-abundant copper zinc tin sulfide (CZTS) semiconductor technology with the low-cost and scalable nanocrystal ink technique. This approach has several inherent benefits that contribute to lower module cost, including the ability to uniformly coat large area substrates, automate manufacturing, and reduce labor with faster throughput. The resulting thin-film solar cells are expected to offer high optical absorption coefficients with significantly reduced material and processing costs.

Sandia National Laboratories ($749,853), Livermore, CA

The research aims to introduce a new photovoltaic material—crystalline nanoporous framework (CNF)—that allows detailed control of key interactions at the nanoscale level. This approach can overcome the disorder and limited synthetic control inherent in conventional bulk heterojunction photovoltaic materials. The research team is designing and synthesizing semiconducting CNFs—infiltrating their pores with a complimentary donor or acceptor—and fabricating prototype photovoltaic cells using CNF-composite active layers. This research is reducing the distance that excitons travel before meeting a charge-separating heterojunction, creating a tunable donor-acceptor offset, and maximizing exciton splitting and carrier mobility by eliminating disorder and defects that inhibit charge transport.

Stanford University ($1,380,470), Stanford, CA

This effort aims to develop an efficient up-converting medium capable of converting low-energy transmitted photons to higher-energy photons, which can then be absorbed by any type of commercial solar cell. The research team is using electrodynamic simulations and ab-initio quantum computations to optimize existing up-conversion processes and design new molecular complexes for high-efficiency photovoltaic up-conversion. This technology promises broadband up-conversion at low incident power in a solution-processable, scalable platform.

University of California, Berkeley ($1,500,000), Berkeley, CA

Researchers here are developing a unique method to grow defect-free, III-V compound micro-pillar structures on single- and poly-crystalline silicon substrates. This approach combines the high conversion efficiencies of compound semiconductor materials with the low costs and scalability of silicon-based materials. The dense forest of micron-sized indium gallium arsenide/indium phosphide (InGaAs/InP) pillars is excellent for omni-directional, broadband light trapping and for reducing the amount of rare-earth materials required.

University of California, Irvine ($1,422,130), Irvine, CA

The goal of this project is to build a prototype solar cell made from nontoxic, inexpensive, and earth-abundant iron pyrite (FeS2), also known as fool's gold, with an efficiency of 10% or greater. The research team is developing a stable p-n heterojunction using innovative solution-phase pyrite growth and defect passivation techniques. A pyrite-based device offers a clear pathway to meeting SunShot cost targets, 20% module efficiency, and terawatt scalability using a proven, manufacturable geometry that is suitable for rapid scale-up by a U.S. thin-film photovoltaic industrial partner.

University of California, Los Angeles ($1,500,000), Los Angeles, CA

The primary objective of this project is to identify and develop an appropriate III-Sb quantum dot absorbing medium for intermediate band solar cells (IBSC) via thorough experimental analysis supported by sophisticated band structure modeling. The limiting efficiency of IBSC is on par with three solar cells operating in tandem, though it may have reduced complexity and cost. Supported by a team of internationally recognized experts, both graduate and undergraduate students are working on band structure calculations, quantum dot solar cell device design, materials development, and in-depth experimental analysis.

University of Chicago ($1,500,000), Chicago, IL

Researchers on this project are developing solution-processed, all-inorganic photovoltaic absorber layers composed of colloidal nanocrystals, such as cadmium telluride (CdTe) and lead sulfide (PbS). These are being electronically coupled through novel molecular metal chalcogenide ligands, which provide band-like carrier transport while preserving advantageous quantum confinement effects. The research team anticipates delivering an inexpensive tandem cell using nanocrys-tals of CdTe for the top and PbS for bottom junctions with 20% efficiency.

University of Delaware ($1,278,110), Newark, DE

This university research team is addressing the efficiency limit and high fabrication cost of current light-trapping methods by developing novel low-symmetry gratings (LSG) for next-generation thin crystalline silicon (c-Si) and copper indium gallium selenide (CuInGaSe2 or CIGS) photovoltaic solar cells. The LSG design achieves light-trapping enhancement exceeding the 4n2 Lambertian limit within a specified range of photon wavelengths and can be fabricated using a low-cost, single-step nano-imprint/molding technique. The researchers are also using deposited high-refractive-index glass materials for low-temperature LSG processing, which enables direct imprint/molding sculpting of even complex grating geometries without requiring an additional pattern transfer step.

University of Michigan ($1,500,000), Ann Arbor, MI

This research addresses efficiency, reliability, and scalability (cost) issues that must be resolved to transform organic photovoltaics into a competitively viable solution. The methods for accomplishing this are based primarily on small molecular-weight organic nanocrystalline cells that are stacked to form a high-efficiency tandem architecture. The research team, which includes doctoral candidates as well as undergraduates, is using low-cost light in-coupling schemes to enhance efficiency, exploring deposition by the scalable and manufacturing-ready technologies of liquid phase and organic vapor phase deposition, and subjecting prototype devices to realistic reliability testing.

University of Minnesota ($1,500,000), Minneapolis, MN

Researchers here aim to demonstrate the first functional copper indium aluminum gallium diselenide/copper indium gallium diselenide (CIAGS/CIGS) tandem solar cell through the use of novel materials and processes. They are combining aluminum with both gallium and indium to form a wide bandgap absorber. They are also developing a novel tunnel junction using thermal and air-stable oxides. Finally, they are introducing a graded CdxZn1-xS layer using a novel continuous flow chemical bath deposition system. This system can better control process conditions, reduce particle loading, and allow well-controlled graded films.

University of Washington ($492,865), Seattle, WA

This research team is employing the solution-phase

chemistry methods used for CZTS Se device fabrication to conduct high-throughput experiments with a novel combinatorial deposition platform. This approach allows for the discovery of alloying and doping strategies that produce a back surface field and defect passivation strategies that can dramatically decrease recombination and increase the minority carrier lifetime. By using photoluminescence, current-voltage, capacitance-voltage, and external quantum efficiency analysis, the researchers hope to rapidly converge on practical routes to high-efficiency CZTSSe-based solar cells.

University of Wisconsin–Madison ($462,508),
Madison, WI

This research team is developing nanostructures of pyrite semiconductor to overcome material bottlenecks and allow for application in high-performance solar PV devices. This effort is aimed at developing effective doping methods and improved surface passivation strategies, as well as suitable nanoscale heterostructures of pyrite with other semiconductors. This exploratory research is demonstrating the proof-of-concept of a novel earth-abundant solar material while developing the understanding, materials, and processes needed for its deployment.

Conclusions

The lessons we can draw from the above examples are that all successful companies and projects have one thing in common—a well defined goal and a predetermined trajectory for its achievement.

The most successful companies are also remarkably persistent. Even when circumstances and conditions change unfavorably, they will find the way and will push forward.

FAILED RENEWABLE ENERGY
COMPANIES, 2007-2012

We must remember that there are several types of failures, and several reasons for failing no matter what business we are in. Translating this to the alternative energy sector, minor and major failures can be attributed to technical, financial, administrative, and political problems.

There are accounts of 1,700 companies worldwide failing to date, with more coming and the number doubling by 2016. Here is a list, in alphabetical order and with brief comments, of the most notable failures of renewable energy, and related companies and projects during the solar energy boom-bust cycle from 2007 until the fall of 2012:

3S Soluciones

3S Soluciones is a Spanish solar company, with offices in several Spanish cities and staff of 50-60 technicians and engineers. It was formed in 2005, and has been active in the solar market ever since. It has several small PV installations in Navarra, Malaga, Valencia, Jaén and Almería, which together exceed 4.0 MWp. It also has some larger industrial PV projects in Murcia and Castilla-Leon, consisting of over 3.5 MWp of installed and operating power plants.

Due to the changes of late, including the unprecedented 19% tax placed on solar projects by the Spanish government, 3S Soluciones was not able to continue operations, and has filed for bankruptcy.

3S Soluciones was a fairly small player in the solar game, but it gained a good reputation during its work on smaller PV power plants around the country. It fell victim to Spain's fluctuating solar policies and regulations, which it did not plan for, or expect.

A123 Systems

This is a company failure with a twist.

A123 calls itself the U.S. leader in advanced batteries, and used funds from the DOE to build the Livonia, MI, factory that made the flawed Fisker packs.

A123 designs and manufactures lithium-ion batteries for electrical cars, but had major technical problems, and one of its major batteries deals is on recall. And A123 was forced to spend $55 million for recalling the Fisker Karma battery packs that have misaligned hose clamps and other issues.

A123 was failing and looking for financial rescue, so decided to sell its technology to a Chinese auto-parts maker. After years of hard work, and with the help of $249.1 million in US federal grants, A123's advanced technologies would be owned by a company in China.

However, under the terms of the grant agreement, any changes to the scope of the grant would have to be approved by DOE, and DOE would not approve changes that allowed grant money to be used for investments other than in manufacturing facilities here in the U.S. or for U.S. jobs.

In the fall of 2012, A123 changed its plans and filed for bankruptcy, agreeing to sell their automotive business to Johnson Controls for $125 million, or a return of 20 cents on the dollar. They would be able to repay some of the government and other loans.

AE Polysilicon

Polysilicon prices of $450/kg in 2008 spurred an unprecedented growth of polysilicon production, with many new companies jumping in it, and many existing companies expanding production. AE Polysilicon, in Fairless Hills, PA, subsidiary of Motech, Taiwan, was one of them.

AE Polysilicon produced a special type of granular polysilicon material, via proprietary fluidized bed reactor technology. The new process produced higher yields at a higher throughput, according to the company spokesmen.

But then the polysilicon market prices collapsed and the polysilicon world was turned upside down.

When the dust settled, polysilicon prices hit a low $20/kg range—well below small polysilicon producers' production cost levels. So, Motech Industries cut their losses altogether.

The intention of this new polysilicon manufacturing technology was to provide lower cost polysilicon, but it happened to be the wrong time and place for AE.

Motech, Taiwan, had a $41 million financial exposure to AE Polysilicon, which consisted of investments totaling approximately $26 million and prepayments of $15 million. In order to cut its losses, the company initiated a workforce reduction at the US polysilicon operations, while proceeding with a shutdown of the manufacturing facilities.

In August 2012, AE Polysilicon held an auction of its assets and intellectual property at its in Fairless Hills, PA, production plant. AE is no more, and the polysilicon industry lost another of its viable members.

Aero-Sharp

Aero-Sharp is a Chinese-based manufacturer of power inverters for residential and utility-scale power plants. In 2010, the Australian distributor, Beyond Building Energy, issued a recall of thousands of grid-connected inverters, made by Aero-Sharp, due to some systems displaying slow start-up times and involuntary shut downs.

This is a reminder that all new products should be approached with caution. Testing and track records are paramount. Fortunately for the customers in this case, the problems were discovered soon after installation, so not much damage was done.

There is another issue involved here: the rebranding of solar products. Rebranding means that the name of the original manufacturer was changed on the label, so it becomes very difficult to trace the origin of a product.

Aleo Solar

Aleo is a multi-phase, multi-national failure. This company's is a typical story of poor planning and even worse execution of an imaginary marketing plan that was cut short by a reality that was not expected.

Germany's Aleo solar AG was a silicon solar modules manufacturer, with production plants in Germany, Spain and China. In June, 2012, Aleo closed its production plant in Santa Maria de Palautordera, Spain, with its 20 MWp production capacity. Ninety-two jobs were lost.

Next, Aleo Solar's joint venture based in Gaomi, China, fell victim to cost cutting. It had 246 employees and reached 90 MWp production output before the shutdown at the end of September, 2012.

Now Aleo Solar is focusing on the survival of its Prenzlau production plant in Germany.

Amonix

Amonix is a designer and manufacturer of concentrated photovoltaic (CPV) solar power systems with plants in California and Nevada. They started operations in the mid-1990s with the help of US DOE, and they developed one of the first commercial size HCPV trackers in the world.

Amonix had several functioning HCPV trackers in the APS' Star Test facility in Phoenix, AZ, but these had some problems through the years, so they were decommissioned in 2010-2011.

Amonix also has several other HCPV trackers installed in other locations, but basically they had no experience with mass manufacturing and large-scale installations.

After another loan from US DOE and investors, Amonix built a new mass production plant in North Las Vegas, to supply trackers for the Alamosa power plant, which was also financed by the DOE.

The 214,000-square-foot production plant, complete with 700 employees, started operation in 2011 and did produce the needed trackers. The new plant construction and setup continued throughout the effort, and were almost complete, when the plant closed down, about a year after it opened. There was talk of quality issues, not enough training, and product being returned because of functionality issues.

While Amonix seems to have the most efficient and reliable technology on the market and the most experienced engineers and techs in the field, their HCPV technology is still quite young (as far as world markets are concerned).

Building new facilities and starting mass production before having a proven product is not a good idea.

Amtech Systems

Amtech is a global supplier of production and automation systems and related supplies for the manufacture of solar cells, semiconductors, and sapphire and silicon wafers. It showed a 13% increase in sales early in 2012, but looking at very slim times ahead, it implemented serious cost-cutting measures to tackle the downturn that proved much longer and deeper than expected.

The cost reductions were too late, so now $6-$7 million will be cut in fiscal 2013, though the company would continue to invest in key R&D programs, including its ion implant technology, and would not impact customers.

New orders for the solar segment were $0.7 million, down from orders of $7.2 million in the previous quarter—a 10-fold decrease in revenue.

So, Amtech's solar division is floating up and down the treacherous waters of the solar market, with little guarantee that things will improve. Amtech as a unit will survive, however, because it is diversified into other industries. The solar segment is only a small part of the overall operation.

This is a classical example of a successful diversification. Yes, the solar division will most likely go into hibernation, but the company as a whole will continue unhindered.

Figure 5-1. CPV trackers

Arise Technologies

Based in Waterloo, Ontario, Arise Technologies is divided into three units: the PV silicon division producing high purity silicon for PV cell applications, a PV systems business for rooftop and ground-mounted PV solutions, and the PV cell manufacturing division in Germany.

The company's stock was down 91% in 2011, and as a result Arise Technologies filed for bankruptcy.

In March, 2012, Salamon Group Inc. of Las Vegas, NV, announced that its Sunlogics Power Fund Management Inc. subsidiary has completed the court-approved acquisition of the assets of Arise Technologies, which will be used for its own silicon manufacturing needs.

According to the report, Arise' patented technology will produce solar grade silicon at $13.50/kg, which is 50% below the current cost of manufacturing silicon. This will help Salamon to reduce its own solar technologies costs as well.

AstroPower

AstroPower was headquartered in Newark, DE, where it manufactured solar electric power products, and provided solar electric power systems for the mainstream residential market.

It was in business almost 20 years, starting as a spin-off from the semiconductor labs of the University of Delaware. The focus was on developing a process to manufacture thin-crystalline silicon solar cells on a low-cost substrate. The process showed potential for substantial cost advantages over polycrystalline silicon, without losing its high performance and stability.

The new "Silicon-Film" cells could be manufactured as 'large area' cells, thus overcoming the size limitations of conventional silicon wafers. AstroPower successfully developed another important technology, which was used to process "reject" silicon wafers from the semiconductor electronics industry into useful solar cells.

AstroPower was responsible for many other industry innovations on the solar technology and markets fronts, including the SunChoice pre-packaged systems for the global marketplace.

Arguably though, AstroPower's greatest impact and legacy for the solar PV industry was the company's famous sales and marketing projects. AstroPower boosted the bulk sales of its SunLine, SunChoice and SunUPS line of grid-connected systems to high levels. It also established long-term relationships with the building industry in the US and Europe. Some of the partners in this area were big names such as Shea Homes in California, and the Home Depot chain.

AstroPower experienced a steady decline since 2003, triggered by a failure to disclose financial statements, which resulted in a Nasdaq delisting. They filed for bankruptcy in 2004, selling some of their U.S. business assets to GE Energy.

GE Energy operated the DE facility until 2009, then sold the company to a Taiwanese PV manufacturer who started producing PV modules again in 2010, under the name Motech. This Delaware plant is still in operation, but now under Chinese management.

Ausra

Ausra was formed in 2002 as Solar Heat and Power Pty Ltd in Sydney, Australia. Ausra constructed a 1.0 MW pilot solar plant in 2005 and then began work on a 5.0 MW extension.

Solar Heat and Power relocated to the US in 2007 and was renamed Ausra. It began operation of its 130,000-square-foot (12,000 m^2) solar thermal power systems plant in Las Vegas, NE, in 2008—the first comprehensive solar thermal manufacturing plant in the US.

Ausra attracted big name investors to their main technology, cheap, flat mirrors to concentrate sunlight, and heat tubes full of water. It acquired the Carrizo Energy Solar Farm project, a proposed 177 MW which was under development, but then sold it to First Solar.

The company raised over $115 million with the idea to replace the expensive parabolic mirrors at solar thermal power plants with cheaper flat mirrors. But it never got a chance to even test that idea on a large scale. Big California utilities awarded their largest solar thermal projects to competitors.

Though PG&E actually gave Ausra a contract to build a 177-megawatt plant, and Ausra seemed headed in the right direction, but instead the company gave up its plans and sold the land.

In February of 2009, the company abandoned its plans to build and operate solar thermal power plants using its own technology. Instead, it would just sell equipment to large companies that would operate the power plants. It would focus on supplying equipment and technology, rather than spreading thin in developing massive solar plants.

The French nuclear energy giant Areva bought Ausra in 2010 and renamed it Areva Solar, but Ausra's flat mirror technology is not yet widespread.

Azure Dynamics

Azure Dynamics Corporation was a leader in the development and production of hybrid electric and electric components, and power train systems for com-mercial vehicles. It targeted commercial delivery vehicle and shuttle bus markets, working internationally with a variety of partners and customers.

Azure Dynamics was looking to capitalize on an increase in demand for battery-electric and plug-in hybrid-electric vehicles. Through the first nine months of 2011, the company's revenue almost tripled, but its net loss widened 45% from a year earlier. Their cash balance as of September, 2011, was $1.29 million, down from $11.7 million at the end of 2010.

Unable to continue normal operations, in March 2012 the company finally filed CCAA (Canadian version of Chapter 11). In April, 2012 the Court approved a DIP financing of up to $4,000,000 in funding during the proceedings to support the company's mission to move forward. Azure has been able to re-establish its vehicle warranty obligations, which were temporarily suspended during the filing process. Azure Dynamics is able to cover all warranty claims on vehicles where work was performed from 27 March forward.

Azure is selling its patents to Ottawa-based Mosaid, an intellectual property firm that specializes in the licensing and development of semiconductor and communications technologies. As another victim of the promising but problematic fluid electric vehicle industry, Asure joins other EV manufacturers experiencing similar problems.

Beacon Power

Beacon Power, a flywheel energy storage company with a 20-megawatt energy storage plant in New York, was a spinoff of SatCon's Energy Systems Division, to develop advanced flywheel-based energy storage technology.

In 1998 it became a separate operating entity, and went public in 2000. The first and second generations of the flywheel storage technology were deployed in North America for telecommunications backup power applications.

As an alternate approach, the company focused in 2004 on R&D to develop a system that could "recycle" electricity from the grid, which could be absorbed when demand dropped and injected back into the grid when demand rose. The concept was successfully applied in 2005-2007, which led to the completion of a grid-scale Smart Energy 25 flywheel system.

In 2009 Beacon received a loan guarantee from the US DOE) for $43 million to build a 20 MW flywheel power plant in Stephentown, NY., Unable to pay the bills, Beacon Power was threatened with delisting from Nasdaq and filed for bankruptcy, agreeing to sell its Stephentown facility to repay the DOE loan.

In 2012, Rockland Capital acquired the assets and renamed the company Beacon Power, LLC. Rockland intends to keep the part of the business that's already been built, and maintain the ongoing operations of Beacon's 20-megawatt grid frequency regulation facility in Stephentown, NY, which has been in operation, delivering frequency regulation services since 2011.

Blue Chip Energy GmbH

Blue Chip was a manufacturer of silicon solar cells, operating from its headquarters in Gussing, Austria, since 2008. In July, 2011, the management of Blue Chip Energy GmbH filed for insolvency proceedings and laid off its 110 workers.

Solon Solar had acquired a significant participation in the Austrian cell manufacturer in 2006, to ensure the supply of solar cells for its projects, which was critical.

Solon held 18.28% of the shares in the company, but had troubles of its own and couldn't help its partner. Following the insolvency of Blue Chip Energy, Solon made a valuation allowance in the aggregate amount of €18 million, resulting from shareholders loans in the past. This complicated Solon's already intricate situation, and it filed for bankruptcy shortly thereafter too.

Due to the difficult market situation and the persistent slack demand, the financial situation of Blue Chip Energy became increasingly tense. Despite intensive talks with the financing banks, it was not possible to find a solution. Solon couldn't help either, so, consequently, there was no longer a basis for the continued existence of the company.

BP Solar

BP Solar is one of the oldest and most respected solar-industry pioneers, responsible in great part for advancing solar energy technology and markets in the US and abroad. BP finally halted all of its solar operations, to pursue core technologies, and to focus on renewable-energy technologies such as wind and biofuels.

BP is also credited with over 1.6 GW in photovoltaic installations, with which BP will no longer be involved. The move to withdraw from solar was completed by BP's decision (in 2011) to pull out totally from the Australian government's Solar Flagships projects.

The increasing "commodification" of the energy markets forced the company to attempt to refocus on supplying power-plant scale projects in mid 2011, and abandoning the rooftop market in which it specialized.

BP continues to stand behind its products and projects, and it plans to honor all valid warranty claims arising from its current and past projects and products.

In the fall of 2011 BP also sold its stake in the Tata BP Solar joint venture in India. BP Solar is no more.

This is another confirmation of the volatility and immaturity of the solar energy business, where even large companies cannot compete and stay afloat.

Bright Automotive

Bright Automotive was founded in 2007 and planned to make a vehicle that would be able to go about 40 miles on electric power alone before a gas-powered generator kicked in to give the vehicle (a commercial van) a long range similar to that of the Chevrolet Volt.

The company planned to use an old Hummer automotive plant in Indiana. The business model and the product had merit, and the prototype was almost three years old, but because there was a need to protect the intellectual property, Bright could not disclose more details.

Bright Automotive applied for $400 million in loans from DOE in 2008, and was told for more than 18 months that the loan was close to funding, but the application had not moved forward by March, 2012, and Bright Automotive had to shut down.

The technology and engineering in the electric vehicle remain assets of the company, but that is all.

Bright Automotive was counting on loans, planning to hire hundreds of people, and designing production, before having a viable product that was accepted by the market.

Day4

Day4 Energy of British Columbia, was developing and patented the "stay-powerful™" technology that interconnects solar electric cells and collects the power they generate—an innovation that is a direct replacement of the conventional, high-temperature solar cell soldering process.

The "stay-powerful™" technology is comprised of a polymer film embedded with a number of copper wires specially coated with a proprietary, low-temperature melting point alloy. This combination establishes a low-resistance electrical contact with the surface of the PV cell creating over 2,100 independent electrical contacts.

These innovations were considered important and marketable assets in the company's portfolio. The plan was to license and use, or sell, them for profit, but this was not easy in the midst of a PV frenzy with several hundred competitors jumping at each opportunity.

After an abysmal first quarter, Day4 Energy sold its assets and operations to another company.

Again, we see a company focused on the technical aspects of the operation without fully incorporating risk management planning and plausible exit strategy.

We see this time after time, and markets do what they do best—reward those who are in the right place at the right time, and punish those who are not well prepared. Still, Day4's technology sounds promising.

Energy Innovations

Energy Innovations (EI) was the hope of the CPV industry during 2008-2010. A pioneering CPV solar startup, EI was expected to cover many acres with their highly efficient CPV trackers.

We've long maintained that concentrated photovoltaic (CPV) solar is a niche product, albeit with some potentially large niches, notably, high DNI areas with limited water resources. But scaling up and raising money for solar startups has become particularly challenging.

EI's CPV design had unique qualities. The firm provided CPV product options, including a ground mount, a roof mount, and a carport configuration for its 2-axis CPV assemblies. The trackers were of the micro rather than the macro variety, but still the most efficient in the field.

Unable to sell its product at significant profit, the company auctioned all its assets—another hopeful whose potential was overestimated by investors and developers, and whose timing couldn't have been worse.

Evergreen Solar

A few years ago Evergreen Solar was one of Massachusetts' hottest new sources of "green collar" manufacturing jobs and the nation's third-largest maker of solar panels; but, despite millions in state funding and tax breaks, Evergreen finally shipped the bulk of its operations to China.

This case escaped the national spotlight only because the focus at the time was on Solyndra. So, the less-politicized tale of Evergreen ended up with a quiet bankruptcy and assets sale.

Evergreen's technology includes a new way to make solar wafers at a much lower cost. Unfortunately for Evergreen, solar products prices tumbled and the company had an impossible time repositioning itself in the industry because of its singular technology.

Evergreen's product was reasonably good, but it wasn't cheap, and did not get on the market on time, which made the big difference.

GreenVolts

GreenVolts is a Fremont, CA, based manufacturer of concentrating (CPV) tracking solar systems. According to their website, Greenvolts has established a new paradigm for the solar industry through the design and manufacture of a complete and fully integrated photovoltaic system, which includes advanced concentrating photovoltaic (CPV) modules, high-precision tracking system, inverters that automatically seek the maximum power point for the solar modules, CPUs that continuously monitor performance, and an Intelligent Solar Information System.

The entire system is seamlessly interconnected, making a virtual solar information network that includes GPS and wi-fi for intrasite status, and internet communications via secure IP protocols. Each site even has its own weather and solar monitoring station.

In December, 2011, ABB (a global power and automation technology group) bought it. The promise was great, but GreenVolts and the entire CPV industry went into a free fall in 2012, and in October laid off the majority of its work force.

The advantages of having a strategic investor are numerous, but the risks are great too. The investor can pull out at any time, so the solar company must have a well designed exit plan. Strategic investors must know the technology very well, and realize that when investing in solar they must be willing to take a short- to medium-term loss for long-term gains.

GT Advanced Technologies (GTAT)

Until recently GTAT, based in Nashua, NH, was one of the largest PV equipment suppliers and undisputed leaders in the silicon ingot growth systems market. In October, 2012, however, GTAT announced "major restructuring" due to unfavorable solar market conditions. It would reduce its workforce by approximately 25% and consolidate existing business units into a single 'Crystal Growth Systems' (CGS) group. This, in effect, eliminated its solar equipment, Directional Solidification System (DSS), manufacturing division altogether.

The struggle for survival began. In December, 2012, GTAT decided to remove the majority of its DSS furnace equipment from its backlog and take a $140 million total writedown of its DSS operations. Only existing customers will be supported in the future.

GTAT total expected revenue in 2013 is in the $500-600 million range, but PV segment sales are expected to account for only 1%, vs. 42% of total sales as projected.

This is a serious blow to the viability of the quasi-mono wafers technology, which was seen as a corner stone in reducing silicon prices. It also marks the end of GTAT as a key driver of silicon furnace sales and upgrades for the struggling solar silicon wafering sector.

Hoku

Hoku Corporation, a Hawaii solar energy company, consists primarily of two divisions, Hoku Materials, which provides silicon material for solar wafers and cells, and Hoku Solar, which manufactures solar cells and modules, and provides the related installation services—at least, that was the plan.

The deteriorating solar energy markets, lack of risk management and exit strategy forced Hoku to reduce construction at the facility in Pocatello, ID, in December 2011. By April 2012, all construction contractors ceased.

Having begun as a fuel cell company in 2005, then going into the solar market ill prepared, Hoku Materials terminated approximately 100 of the Pocatello plant employees, and ceased all US business activities. Simultaneously it terminated all its staff at Tianwei Solar, USA, as well as all deals in the US.

Hoku's products and services quickly became overpriced and obsolete—another case of hasty planning and execution.

Interestingly, since 2008 Hoku has received over $280 million in pre-payments from larger PV module manufacturers for polysilicon deliveries; however, Hoku never produced any polysilicon because the plant construction was not finished when the company crashed.

Polysilicon was very expensive ($450/kg in 2008), and investors hoped it would go higher. Today, it is a very depressed $20 per kilogram, and another good lesson in how *not* to.

LDK Solar

LDK Solar, located in Xinyu City, China, manufactures multicrystalline solar ingots and wafers, and provides variety of wafering services for its customers. LDK is one of the world's largest providers of solar materials, with over 3.0 GW production capacity.

It was doing very well during the initial stages of the 2007-2012 cycle, but the total net loss was estimated at $588.7 million, with revenue reaching $2.15 billion in 2011, compared to $2.5 billion in 2010. This discrepancy is blamed on the solar industry's tremendous supply and demand imbalance throughout the value chain during the fourth quarter of 2011 and beyond.

According to company officials, 2012 is dominated by excess capacity and further policy uncertainties in Europe and the US, which result in continued intense competition within the solar industry. Results for 2012 are discouraging.

These are significant losses, even for a large company like LDK Solar. The numbers over-emphasize the discrepancy in the industry, and mainly the large supply and demand imbalances, reflecting unstable and ever changing prices, abnormalities that have plagued the solar industry since 2008, making it hard to price modules and confirm sales agreements.

These discrepancies also show the level of immaturity of the solar industry, with very few standards and only a few regulations in place.

Amazingly, there is much competition among Chinese manufacturers too, with the customers stuck in the middle. Again, too much money thrown too quickly at an enfant industry can disrupt growth.

In addition to, or as a consequence of, the recent losses, in October 2012 LDK sold a 20% stake to Heng Rui Xin (HRX) Energy, a Chinese state-run entity (another form of China government subsidy). The sale price was approximately $23 million, so Heng Rui Xin is paying a 21% premium to LDK's stock closing price of 71 cents at the time of the transaction.

This is another confirmation of the turmoil in the Asian solar industry, and a validation of the claims of unfair government subsidies there.

LSP Energy

LSP Energy, Batesville, MO, is an energy company, engaged in the development, construction, ownership, and operation of gas-fired and other electric generating facilities since 1996. It owns and operates three gas-fired combined cycle electric generating units with a total capacity of 837 megawatts that connect with the Entergy Corp. and Tennessee Valley Authority transmission systems. The electricity is sold through two long-term power purchase agreements with J. Aron & Co. and the South Mississippi Electric Power Association.

Problems with a combustion turbine, and other issues, led the company to seek bankruptcy protection and resulted in a sale of its assets. The mechanical failure of a combustion turbine in May, 2011, evoked a substantial repair bill, as well as a reduction in company revenue. The turbine was rebuilt, but the outage of the unit continued into 2012 with restarts, inspections and parts replacement. In December LSP determined that the replacement rotor installed in October 2011 was faulty. The rotor was replaced again, but the damage was done, and an additional $19 million in costs were not covered by insurance.

LSP was unable to continue funding its operations. In September, 2012 a Delaware bankruptcy judge allowed LSP Energy LP to sell its assets (including the gas-fired power plant) to South Mississippi Electric Power Authority (SMEPA) for $286 million. SMEPA is large enough to afford mechanical problems with the gas-furnace.

If there is a lesson here, it is only a confirmation that large-scale power generation should be left to the large power companies.

Meier Solar Solutions

Meier Solar Solutions of Bocholt and Rossla, Germany, designs and produces encapsulating equipment for PV modules—also called laminators. With a staff of 125 in R&D, design, manufacturing and customer support, the company had the largest number of laminating units installed and operating around the world.

In September, 2010, orders fell to a dismal low. Meier Solar Solutions could not pay its bills and was forced to file for insolvency.

Two days later, the Japanese solar machine builder, NPC Inc., paid $11 million to acquire the operation, and is now optimizing production and increasing sales efforts, though we have no details.

So, another German company falls under foreign domination.

OCI, Korea

Other Asian solar companies are not exempt from the present flood of failures. Excessive overcapacity and continued sharp price declines in the polysilicon sector have forced a number of them—including Korea's largest polysilicon producer, OCI, Korea—to delay previously announced plans to build new plants and even to fold operations.

Due to falling polysilicon prices and deteriorating solar markets in the US and Europe, OCI revenues have fallen by over 50%, which puts the operations at a loss, so planned expansions are now on hold, or cancelled.

OCI is surviving, but has a difficult road ahead. It is in the same shoes as its largest competitor—the Hong Kong based GCL-Poly—who ended up with $330 million operating loss in the fall of 2012, as compared to $3.55 billion profit the year before.

Chinese polysilicon manufacturers are also reporting heavy reduction in volume and prices, so they are in the same situation as all their competitors.

These failures are due mostly to the falling price of polysilicon, down to just above $20/kg, vs. over $500/kg in 2008. Poor planning, however, is also to blame for the overcapacity, overexposure, and lack of exit strategies most of these companies experienced during this period. The race to make a quick buck, instead of following sound business practices, is behind the majority of recent failures among polysilicon manufacturers. Easy access to government subsidies can be blinding to corporate heads, and speed up company demise.

Phoenix Solar

Phoenix Solar was a child of the German "Cooperative of Energy Consumers" initiative, which is responsible for dozens of companies and many MW of solar installations. Phoenix Solar was a German systems integrator, specialized in the designs, construction and operation of large-scale solar power plants. It also wholesales photovoltaic systems, solar modules and related equipment, and with over 1.0 GW of installations under its belt, it benefited from the abnormally high FiTs until they lasted.

Now the FiT and many other perks are gone, and the German solar sector is suffering. Phoenix Solar is no exception. It has been in a restructuring and refinancing mode for awhile, trying to make the company more flexible and operate sustainably during these hard times.

The company reported a 57.1% decline from a year ago—a loss of €13.0 million—while refinancing over €100 million was secured to enable the company to restructure. The stock is down 99% from a year ago.

Personnel reductions of 50% and shaving operating expenses, or around €30 million per annum, are planned for full implementation sometime in 2013.

At the same time, Phoenix Solar has cancelled all previous deals, breaking contracts with suppliers, and has switched to a more "flexible" model to keep expenses in check.

A key part of the restructuring efforts included significantly reducing operating costs, to achieve operational breakeven at around €300 million in sales annually—but sales of what and to whom?

Photovoltech

Photovoltech, a Belgium-based silicon solar cell manufacturer, spun out of the R&D facility, IMEC, in 2001. This old-timer, with 267 workers on payroll, started contemplating restructuring, and maybe even closing down in 2011. A series of announcements at a company meeting in June 2012 confirmed the sharp change of direction the company was taking in these hard times.

Photovoltech's business model was focused on supplying solar cells to European markets, which are seriously affected by the impact of overcapacity, plummeting components prices, module manufacturing facilities closures, company failures and bankruptcies.

Like many other victims of recent developments,

Photovoltech experienced mounting losses and failed to find a buyer for the business or new investors. A total of 267 workers lost their jobs.

Again, risk management and exit strategy plans in the initial proposals and contracts were missing or not well thought out.

Photowatt

Located near Lyon, France, Photowatt with its 20-year-old facilities and 400 employees was one of Europe's oldest and largest solar cells, PV modules, and complete solar systems manufacturers and vendors.

Like most other EU solar manufacturers, Photowatt began a downslide, so the mother company—ATS Automation Tooling Systems—looked to unload it. Photowatt's production was slashed by two thirds, and ATS looked into bankruptcy proceedings, which are difficult in France, due to complex labor laws.

The French government, seeing that the situation at Photowatt was critical, offered Photowatt France an 11th-hour reprieve by investing as much as needed to keep the company afloat.

Environment Minister Nathalie Kosciusko-Morizet said the government was ready to rescue the country's leading PV manufacturer. Photowatt was being hurt by lower-priced competitors and the government "would like to invest" in the company.

French power utility Electricite de France SA (EDF) was approved, to take over Photowatt, France. Further news of ATS is not forthcoming.

Ralos New Energies AG

Ralos New Energies, based in Michelstadt, Germany, was a leader in the design, development, planning, building and operation of photovoltaic plants; anything from private solar electrical systems to high-output solar power stations. With over 100 employees, and branches and subsidiaries across the world, Ralos was among the top 10 comparable photovoltaics companies in Europe.

Unable to pay the bills, in February, 2012 Ralos filed for insolvency. The company's subsidiaries Ralos Projects GmbH and Ralos Solar GmbH were also negatively affected, and their future is uncertain.

The company's problems have been aggravated by the negative impact of the drastic regulatory changes in Germany on its sales strategy.

Ralos' insolvency comes in the midst of a number of bankruptcies of many other German companies, all of whom have suffered by significant decline in module prices during 2011, due to cheap imports and serious FiT

reductions in the region.

The problem doesn't stop at Ralos. The German Solar Industry Association estimates that the proposal of Germany's economic and environment ministries for 30-50% reductions in the solar feed-in tariffs during 2012 will force many companies out of business, will put additional tens of thousands of jobs in the German solar industry in danger, and will make the transition towards renewable energy in the country much more difficult, if not impossible.

Renewable Energy Corp ASA

Renewable Energy Corporation (REC) ASA, Norway, was a leading vertically integrated solar energy company, producing polysilicon, wafers, cells and PV modules for the solar industry. It also produced silicon materials for the semiconductor industry and engaged in project development in selected PV segments around the world.

Founded in Norway in 1996, REC employed around 3,100 people globally. In 2009, REC was involved in a general recall, to repair or change 420,000 solar panels. This recall-replace exercise cost REC over $55 million.

As part of a restructuring plan, REC ceased further funding of its Norway subsidiary, REC Wafer Norway AS, and by summer of 2012, the REC ASA Norway plant was shut down.

Temporary poor quality, resulting in the above mentioned recall, could be identified as the leading cause of the REC wafers' failure. But there are other factors, such as the drop in prices of solar products at the same time, that contributed to its demise.

What's next for REC's US and Singapore production plants is unclear. The mother company as a unit is well diversified, so its long-term survival, barring extraordinary developments, appears guaranteed.

Schott Solar

Schott Solar's main product was manufacturing of c-Si solar cells and modules, with global production capacity of 450 MW per year of both cells and modules in 2010. In January, 2012, it suddenly closed its multicrystalline wafer operations in Jena, Germany, and laid off 290 workers, but insisted that its monocrystalline wafer production in Jena would continue.

Later in 2012, increasing market uncertainties and barriers forced Schott Solar's management to withdraw from c-Si PV manufacturing completely.

The company's thin-film and CSP activities are unaffected by these developments, while the exit from the c-Si sector will affect around 870 employees in Germany, the Czech Republic, and New Mexico.

From the official Schott Solar website, " SCHOTT Solar PV, Inc. has withdrawn from the crystalline photovoltaic business... now focusing on those sectors of the energy industry where we can profitably grow our business. ...to focus on Concentrated Solar Power and Thin Film Solar, businesses that are core to SCHOTT's capabilities."

Schott's remaining in the thin-film, CPV, CSP and BIPV applications market is a mystery, because those sectors are not doing well either. But Schott is trying different strategies to reposition itself in the solar market and has shaved up to 50% off its cost base. Industry analysts claim that some competitors' cost bases, and especially those based in Asia, have seen closer to 60% reductions over the same time period.

So Schott is looking to keep its foot in the solar door, and with its diverse product line, it should survive.

Sharp and Recurrent Energy Divorce

In 2009-2010, Osaka, Japan-based Sharp purchased the successful solar project developer, Recurrent Energy, for over $300 million and 100% stake in the company's business. The move was interpreted as positive news and new hope for developing large-scale solar projects in Japan and worldwide.

Sharp has been a solar cell producer for over 50 years, and not very long ago it was the world's largest. Recurrent Energy is a solar project developer. Sharp saw the synergy of working with Recurrent and the acquisition was thought part of a trend by upstream solar manufacturers to move downstream. Such virtually integrated entities would have more chance to capture some of the value in installation and project development. This move represented the MDV's first exit from the solar business, and maybe the best one they ever made, retreating just before the bottom fell out.

Initially the acquisition seemed to bring together the individual strengths the two companies. Joining forces (and taking control of the operation) Sharp had a complete solution that could bid on any type and size solar power projects and build them using Sharp' own modules.

Recurrent already had a 2-gigawatt product pipeline, which was set to use a lot of Sharp's PV modules, which were piling up in warehouses.

Amazingly, while Sharp seemed happy to make the acquisition, they never made a real effort to put this joint venture on the market. So, Recurrent remained in-dependent from Sharp's solar business, with the ability to build its pipeline with external, and much cheaper PV modules.

With 700 megawatts of contracted projects and over 2.0-gigawatt project pipeline, Recurrent feels secure and is conducting a profitable and growing business, with or without Sharp.

Sharp Corporation, on the other hand, is reorganizing to face the new solar developments. It is selling its largest solar acquisition—Recurrent Energy. The asking price is almost $15 million more than what they paid for the acquisition just two years ago.

The two giants are splitting; the joint venture and its huge potential are gone. We are not sure how to interpret this development, and we are sorry to see such a great opportunity abandoned.

Siemens

The German conglomerate Siemens, one of the world's oldest, largest and most respected hi-tech companies has been in solar since day one. This includes Siemens' solar divisions with nearly 1,000 people around the world. But Siemens is selling its solar business activities as part of a reorganization of its energy division. With the solar and hydro divisions gone, the company plans to streamline its renewable energy activities to wind and hydro power only.

Siemens will also focus on cost reduction; go-to-market; simplified governance; optimized infrastructure and strengthened core activities—including wind and hydro power. The aim of the newly launched program is to enable Siemens to meet its own ambitious goals and to underscore the targets defined in the Siemens framework.

Low growth and strong price pressure in the solar markets have hindered the company's expectations for its solar energy activities and are the major reasons for the divestment. "...Due to the changed framework conditions, lower growth and strong price pressure in the solar markets, the company's expectations for its solar energy activities have not been met." according to company spokesman.

Approximately 680 employees will be affected by the decision, and will be hopefully transferred to the new buyer. There were total of 800 employees in the division.

Consequently, Siemens will not renew its Desertec Initiative membership, exiting the project, losing a major German investor.

In 2009, Siemens took over Israel-based solar thermal manufacturer Solel Solar Systems, which is one of the founding members of Desertec. With Siemens exiting the solar business, Solel becomes part of the divesture and its participation in Desertec, as well as its long-term survival are uncertain.

Siemens also holds a 40% stake in Arava Power Company, in which it invested $15 million in 2009. Some of these groups belong to Siemens' Project Ventures, and not to its solar and hydro division, so their futures are uncertain as well.

Despite all of this, Siemens believes that the Desertec initiative is a technically feasible enterprise, and will support it as a technology partner in the future.

Silicor Materials

Silicor Materials is a solar start-up founded in 2006 in Redwood City, CA, under the name Calisolar. It started building multi-crystalline silicon solar cells, using lower-purity polysilicon for cost reduction. When silicon prices dropped, it switched to producing lower cost polysilicon materials, and was renamed Silicor Materials.

Now Silicor is developing new, cheaper, polysilicon materials, attempting to purify metallurgical silicon (MG Si) into "upgraded metallurgical" silicon, or UMG Si. According to their engineers, this will create a much cheaper alternative. The goal is to develop a process and final product which is cheap enough and pure enough to be made into cells that can compete with conventional solar cells in conversion efficiency and price.

The UMG value proposition might have made sense in 2008 when polysilicon cost over $500 per kilogram, but not so much at its current price of $20 per kilogram.

In January, 2012, Silicor laid off more than 30 employees and halted expansion plans at its plant. The risk for Silicor and its UMG product is great, especially during this time of uncertainty in the solar industry. In any event, UMG Si is lower quality than SG Si, and is still in R&D mode, so its reliability is unproven, and it lacks market approval. Too, the risk of using UMG doesn't go unnoticed by potential customers.

Solar Array Ventures

Solar Array Ventures' goal was to build a sustainable and earth-friendly company that delivers a comprehensive array of technology products and services to their customers. The vision was backed by core competencies in managing solar businesses, large-scale manufacturing, technology and product development, as well as systems and market development.

They were one of many, worldwide, who wanted to establish themselves as solar a manufacturer/developer *and* full-service company.

The original plan was to build a 225,000-square-foot solar plant in New Mexico. The company needed to raise $200 million for the project, and amazingly got as much as $110 million, but the project never got off the ground, and we were unable to verify the numbers or get additional information on the company.

This is an excellent example of the thousands of failed attempts recently, to raise money for a product and service that did not exist; and, this case gives a good insight into the amazing developments during the initial—boom—stages of the boom-bust cycle of 2007-2012.

SolarDay

SolarDay is one of the oldest and largest Italian PV modules manufacturers. Founded in 2004, it specialized in the production of multicrystalline silicon PV modules. At its heyday, the company employed over 140 people, with over €60 million sales in 2009-2010. In 2011 MX Group acquired majority ownership in SolarDay, with the idea to use SolarDay's PV modules in its projects.

SolarDay also promoted itself as a building-integrated PV specialist, and was producing PV modules in various shapes and colors for use in high-end buildings. This is a very unique market segment that a number of Italian manufacturers have targeted, but which didn't last long. SolarDay also boasted its expertise in the design, installation and operation of solar power plants, but did not make much progress in that area either.

Alas, SolarDay went officially into liquidation in the summer of 2012, and from there into the history books. SOLARDAY SpA IN LIQUIDAZIONE, says the company website…frozen in time. Italy's oldest PV modules manufacturer and power field expert is no more.

This failure is particularly surprising, and we really aren't sure why it happened.

Solarhybrid

Solarhybrid, based in Germany, was a large-scale solar plants developer and financier. Large-scale solar projects require a lot of money and effort, so the ROI must be well defined. Many such companies' plans were shuttered under the imbalance of finite investment vs. uncertainty, which was increasing during 2011-2012. Solarhybrid started bending under the pressure of the German solar subsidy cuts.

Especially detrimental to the entire solar sector was the proposed removal of tariffs for plants over 10 megawatts in size, which were Solarhybrid' specialty.

On the bright side, Solarhybrid acquired a 2.25 GWp PV plant's pipeline and a 1.0 GWp system in the US, as well as other large assets.

Still, Solarhybrid started an insolvency process in June, 2012. A cloud of dust and piles of court papers are all that is left from endless meetings and negotiations, plans and designs for muli-billion-dollar solar projects all over the world. Solarhybrid sold its pipeline of projects in the US and Israel to pay some of the bills.

The total lack of contingency planning (or any sound business planning) here is reflected in the sale of the 201 MW solar project rights in Israel, which were acquired just a month and a half before the bankruptcy proceedings were filed.

Solar Milenium

Founded in 1998 in Erlangen, Solar Millennium was a German global company, specializing in the renewable energy sector, in the design, construction and operation of solar thermal (parabolic trough) power plants.

Solar Millennium was building one of the largest solar projects in the world, a 7,000-acre complex near the Riverside County community of Blythe, CA. The plants in the Blythe project are capable of delivering 1,000 megawatts of electricity, enough power for 300,000 homes, according to court papers.

Solar Millennium and affiliated companies listed combined debts of up to $500 million and combined assets below $100 million, according to papers filed with the U.S. Bankruptcy Court in Delaware. The company filed for insolvency in Europe in December, 2011. Solar Millennium, Solar Trust of America and a group of affiliates filed for bankruptcy in April, 2012. The pipeline of projects were sold to other companies.

Solar Millennium had many large and extra-large projects lined up—each of which required an army of specialists and a lot of money—none of which was readily available. And the timing was just so bad, as solar was approaching the final stages of the bust cycle. Financing and operating under these conditions would've been extremely difficult, if not impossible.

Did Solar Millennium have a sound business plan; did they understand the history of solar energy generation and the peripherals, such as government subsidies, variable markets, technical problems? How did they acquire $500 million in debt in such a short time? These are important issues—vital issues.

Solar Systems

Solar Systems, an Australian company, was established in 1997 and specialized in the design and manu-facturing of concentrating PV (CPV) power generating systems. They also managed the design and construction of solar power plants using their CPV technology.

The company's humble beginnings are marked by the installation of 14 small concentrator dishes, providing power to a local grid. Later, Solar Systems was involved in the setup and construction of three concentrator dish power stations in the Northern Territory of Australia, a total of 720 kWp, generating 1,555 MWh of electric power per year.

In 2003 Solar Systems completed the first concentrator dish power station at Umuwa in South Australia, and in 2005 it won the 2005 Engineering Excellence Awards. In 2006 it started building the Mildura super plant, to provide green power to 45,000 homes and create 1,000 jobs during construction.

This $420 million project had federal funds of $75 million, state funds of $50 million, and taxpayer funds of $125 million. The new plant was to be run by Melbourne-based Solar Systems and aviation giant Boeing when operation started by 2013.

In the fall of 2009, however, Solar Systems—for reasons unknown—was placed under voluntary administration, which put its projects at risk of failure. The Mildura Solar Power Station project was put on hold, and two-thirds of the workforce was on furlough.

In March 2010, Silex Systems bailed the projects by purchasing the assets of Solar Systems, while local electrical utilities purchased the operational power plants' stations, since they had agreed to buy the power from them for the duration.

The failure could've been caused partially by the immature technology, which operated at extremely high temperatures, and which—just like Stirling dishes—is very complex and requires a lot of maintenance. Internal issues, lack of sufficient knowledge and understanding of the design, financing, manufacturing, installation and operation of CPV power plants are another possibility.

The failure of Solar Systems signaled the beginning of the world-wide solar bust. Hopefully, Australia's solar power future will be bright, as it is one of the greatest solar potentials in the world.

Solar Trust

Solar Trust of America LLC, which holds the development rights for the world's largest solar power project, filed for bankruptcy protection in April, 2012, after its majority owner began insolvency proceedings in Germany.

Solar Trust holds the rights for the 1,000-megawatt Blythe Solar Power Project in the Southern California desert. Yes, this is the same project, which was awarded $2.1 billion of conditional loan guarantees from the U.S. DOE shortly before that. The bankruptcy will affect the project, but it is unclear what the final tally might be.

Solar Trust ran short of liquidity after Solar Millennium AG (70% stake holder), sought court protection in December, 2011. Solar Millennium then tried to sell that stake to Solarhybrid AG, but that transaction collapsed when Solarhybrid also sought court protection in Germany.

After missing several quarterly rent payments on the Blythe project, Solar Trust could make no other payments, so NextEra Energy Resources LLC stepped in to help with financing and bid for some company assets. The project's future is uncertain.

Solar Trust of America and several affiliates filed for protection from creditors with the U.S. bankruptcy court in Delaware.

The Blythe project they were tackling is way too big for a single company. The US government $2.1 billion loan guarantee is under question as well, but without it the project cannot proceed. Biting off more than one can chew is haunting this project too. This failure is throwing a long dark shadow on the project and the future of the US solar industry.

SolarWatt

SolarWatt, a Dresden, Germany, monocrystalline and polycrystalline c-Si PV modules manufacturer, was also involved in the design and construction of turnkey solar power plants. In operation since 1993, SolarWatt first started manufacturing small modules for powering vending machines and emergency telephones with DC electricity.

The company then expanded into the photovoltaic sector with larger, shock-resistant, vandal-proof solar modules.

In 2010, SolarWatt set up and started production on one of the most modern module manufacturing lines in the world, but the collapsing solar market and other changes forced the company to submit its reorganization plan to the local district court in July, 2012. If the plan is approved by the creditors it will still need to be approved by the local district court. 480 employees are to be laid-off.

In September, 2012, BMW heir, Stefan Quandt was approved by the local court to take over Solarwatt, and a corresponding transaction was registered with the federal cartel authority, but there are no results as yet.

"After the restructuring process is complete, SO-LARWATT AG will be in a position to wade out the current tide of market consolidation from high ground and become a strong player in the growing market for innovative solar energy systems," according to the court documents.

Solarwatt is planning to change its strategy from that of a PV modules manufacturer to a "provider of solar energy systems." This means that they will be jumping into large-scale project design, installation and operation.

Should court and creditors agree with the final reorganization plan, Solarwatt could continue business as a newly reorganized company. But even if the best happens to SolarWatt, and we hope it does, the overall solar market situation in Germany and around the world is not conducive to large-scale solar installations.

SolarWorld

SolarWorld is a worldwide leader in offering brand-name, high-quality, crystalline silicon solar-power technology. The company's strength is its fully integrated solar production systems. Starting with SG silicon as the raw material, making wafers, cells, and panels, all the way to turn-key solar systems of all sizes, SolarWorld controls all stages of the solar value chain.

The main business is selling high-quality PV panels for small and large installations. This is complimented by selling wafers to the international solar cell industry.

Group headquarters are located in Bonn, Germany, with sales sites in Singapore, South Africa, Spain, France and California. The group's largest production facilities operate in Freiberg, Germany, and Hillsboro, Oregon.

Sustainability is the basis of the group strategy. Under the name SolarWorld, the group supports community aid projects using off-grid solar-power solutions in developing countries, exemplifying sustainable economic development. This company employs about 3,300 people worldwide.

Sadly, SolarWorld's Camarillo, CA, manufacturing plant—one of the oldest PV modules manufacturing plants in the world—was recently shut down. This plant had more than 35 years of leadership in the US solar industry, serving as an example of diversity and flexibility, surviving over three decades of solar boom-bust cycles, several owners, and many local and world disasters.

Though SolarWorld invested tens of millions of dollars automating the Camarillo plant after purchasing it in 2006, the company determined it needed to consolidate its U.S. manufacturing in Hillsboro, OR, where it operates the western hemisphere's largest solar plant.

The closure of the Camarillo factory was one of

12 plant shutdowns, layoffs or bankruptcies within the U.S. crystalline silicon solar manufacturing industry since 2007.

The U.S. Department of Labor has determined that all 186 manufacturing employees laid off from SolarWorld Industries America Inc. in Camarillo, as a result of the company's shutdown are eligible for federal trade-adjustment assistance, including grants for education to retrain them for new work.

SolarWorld AG is expected to shed 10-20% of its 3,300-strong workforce in Germany as well, with the ultra-low Asian prices cited as a key reason.

Solon Energy

Solon Energy GmbH is a leading provider of utility-scale and large commercial PV modules and installations to the European, American and world energy markets. SOLON delivers complete and cost-effective turnkey PV systems with a streamlined approach from project development, design and construction to financing and operation.

Founded in 2007, SOLON Corporation is a subsidiary of the SOLON Group, a leading international provider of solar solutions for residential, commercial and utility-scale applications. The SOLON Group operates subsidiaries in Germany, Italy, and the U.S. with more than 600 employees worldwide. With headquarters in Tucson, AZ, and offices in Phoenix and San Francisco, SOLON provides quality custom solar solutions to the US energy market.

In 1998 Solon was the country's first PV manufacturer to list on the stock exchange, and did very well for a decade or so. Then the solar market collapsed and Solon became the highest-profile German casualty.

A victim of shrinking markets, plunging demand and module prices, as well as unsustainable profit margins, Solon was forced to file for insolvency.

With a large €400 million debt, Solon entered into unfruitful negotiations about restructuring its finances, but those failed, so Solon filed for restructuring within the insolvency proceedings. Individual applications for insolvency were filed for its subsidiaries as well.

Finally, after all avenues were exhausted, in August, 2011, Solon closed the doors of its PV modules manufacturing plant in Tucson.

Then, in March, 2012, UAE based Microsol offered assistance, and Solon Energy GmbH emerged under new management—Germany had lost another piece of her solar industry.

Sovello

Germany-based, Sovello, was established in 2005 as EverQ, GmbH and developed into one of the world's largest, fully integrated PV modules manufacturers. They specialized in making silicon solar cells and PV modules, using specialized manufacturing technology.

Sovello produced thin string-ribbon wafers and cells (very thin Si cells made by direct melting and forming of thin films, which are then cut into cells). The cells were assembled into PV modules, with the help of approximately 1,200 employees.

Sovello was at the center of the German solar industry in the border triangle Saxony, Saxony-Anhalt and Thuringia, also called the German "Solar Valley."

Sadly, the thinnest solar cells in the world, making the lightest PV modules, that showed fastest amortization period ever, made by a company awarded Top Brand PV seal had a hard time navigating the rough solar waters of 2011-2012. Good ideas, good product, great results, plausible achievements, but no luck on the energy markets.

Running out of cash and time, the company filed for insolvency In February, 2012. The failure has been attributed to continued plummeting sales. After 6 fruitless months looking for new investors, Sovello GmbH completely shut down terminating the workforce of 1,000 in August, 2012.

The best products with the best engineers and technicians cannot succeed without management that understands their market.

SpectraWatt

Intel joined the solar game in 2008 by spinning off SpectraWatt solar manufacturing division with a $50 million loan, and additional support from Goldman Sachs' Cogentrix Energy, PCG Clean Energy and Technology Fund, and Solon. Solon was supposed to be a customer of the yet-to-come product.

Under Intel's financial umbrella and with the help of its enormous technical expertise, SpectraWatt was to develop and manufacture super-solar cells for sale to large integrators and developers worldwide.

After 5 years in development and close to $100 million spent, SpectraWatt shut down and filed for bankruptcy protection in 2011.

Lesson #1. It is all about the markets. Knowing your market is paramount.

Lesson #2. Proper planning is the key to success in this game. The original plans called for building a factory in Oregon and making only solar cells. Then, the plans changed to putting the production facility in New York...

Lesson #3. Moving a production facility in the midst of a chaotic market can be fatal. The move was executed hastily due to a more generous incentive package awaiting the new plant.

Nearly two years after the inception, and over $90 million spent, not even one module was on the market.

Almost a year later, SpectraWatt started production ramp-up to begin shipment in the second quarter of 2010. It promised to deliver the most efficient solar cells. However, by the time all these things were lined up, the market had shifted, and SpectraWatt wasn't able to compete. Large rivals could supply silicon cells at much cheaper prices.

In December 2010 SpectraWatt started laying off workers and shutting down its factory. The company filed for Chapter 11 in 2011 and put its assets on the auction block.

Lesson #4. Timing is everything.

Stirling Energy Systems

Stirling Energy Systems (SES) was a specialized solar energy equipment manufacturer, based in Scottsdale, AZ. Its technology consisted of a Stirling engine mounted on a dish-shaped frame, lined up with mirrors, tracking the sun all through the day. The sun's heat makes the engine work and generate electricity.

The entire setup looked and performed well in the test beds, so Stirling jumped into mass production, and then into large-scale power fields development—neither of which it had any measurable experience with.

SES had plans to design, install and operate several very large projects. Solar One in the Mojave Desert with 500 MW capacity, with the option to expand to 850 MW. Solar Two, sited in the Imperial Valley near El Centro, CA, would have 300 MW capacity, with the option to expand up to 900 MW. There were couple of smaller projects too.

Figure 5-2. Stirling engine dish

Each site was to be contracted for a 20-year power purchase agreement with Southern California Edison and San Diego Gas & Electric. While still in the planning stages, Stirling also raised $100 million from NTR, so work was under way full blast.

The projects called for using mirrored dishes to focus the sun's heat on the Stirling engines. The heat would heat the trapped gas in the engine' cylinders, making it expand and churn pistons. Over 2000°C of heat was developed at the engines, which would fatigue the materials. The engines would require extensive maintenance and would malfunction frequently.

To provide as much electricity as it promised, Stirling Energy Systems, needed to build over 12,000 dishes. But the company's entire experience was only six working prototypes in a federal laboratory, and those were still in development mode.

Lesson #1. The horse comes before the cart.

A federal assessment, and a number of experts, concluded that the technology was not ready for large-scale deployment. The dishes haven't proven reliable to operate for thousands of hours without breakdown, according to DOE officials and other experts.

Lesson #2. Show your stuff.

Stirling's project ran into several challenges, and after months of negotiations, the project was sold to a third party in February 2010.

SES did make some progress later on, expanding to 10 dishes at the federal laboratory and opening a 60-dish array in Arizona. But the technology was still in development and won't be ready for large-scale projects for another decade or more. In the meantime, SES and the solar industry's reputation took a big hit.

Lesson #3. This is another cautionary tale about how some renewable projects are often more marketing than reality.

See Tessera Solar in the section Failed Projects for a more detailed description of the SES projects listed above.

SunConcept Group

SunConcept Group was a German solar products manufacturing and projects development company, established in 2004 in Elz, and until recently had offices in Italy, the U.K. and South Africa.

The company's main products were poly and mono-silicon PV modules, for the manufacturing of which it employed over 100 full-time professionals and consulting personnel. These people also designed, and developed solar power systems and plants in Germany and other EU countries.

With significant loss of revenues across the board, due to the European economic situation, the flood of cheap PV modules, the overall slowdown in the sector, and falling FiTs, management started working on a restructuring plan. It could not make the changes on time, however, and was forced to shut down its operations.

SunConcept filed for insolvency in February, 2012; seven of its subsidiaries went bankrupt prior to that. The management team continues to work on a continuation plan to maintain profitability and as many jobs as possible, but, SunConcept is badly hurt.

Navigating the murky waters of the solar markets after the subsidies dry up is what separates the winners from the losers. Those with viable business plans and plausible exit strategies have far greater chances of success.

SunPower

SunPower is a PV modules and systems manufacturer, and project developer, based in San Jose, CA. In September, 2011, SunPower got a $1.2 billion loan guarantee to build the California Valley Solar Ranch Project, a 250-megawatt solar plant in San Luis Obispo County.

At the same time, SunPower reported an operating loss of $534 million in 2010.

Then SunPower started consolidating and closing facilities in Europe and Asia in response to reductions in European government incentives.

Before any federal funds were released, however, SunPower sold the California Valley Solar Ranch Project to NRG Energy. NRG is now the owner of the project, the loan guarantees, and the company responsible for repaying them. SunPower is still the lead contractor on the project.

There were other US solar companies who sold their projects built with federal loan guarantees. Some sold them to third parties who were not qualified to receive loan guarantees for the projects. Profiting by selling projects guaranteed by taxpayer money is a new phenomena, but the size of the solar projects that have been "flipped" lately adds a new dimension to the puzzle and is something we want to watch carefully.

SunPower made serious production reductions in the summer of 2012 to control inventory surplus. Over 125 MW production capacity was reduced by cutting production capacities in Fab 1 and 2 in the Philippines and Fab 3 in Malaysia. Fab 1 was shut down permanently in April, 2012, and will be rented to a third party.

Is SunPower a failed company? Only time will tell, but reducing production volume and shutting down major production facilities is not a sign of healthy business.

Suntech

Suntech is a Chinese solar company, which manufactures, develops, and delivers solar energy solutions and is one of the world's largest producers of silicon PV modules for off-grid systems, homes, and large-scale solar power plants. It has offices in 13 countries and projects in over 80, with the key locations in San Francisco, CA; Schaffhausen, Switzerland; and Wuxi, China.

Suntech is only one of hundreds of Chinese companies receiving aid from federal and local governments, which have invested heavily in solar manufacturers and now are facing failure, so much effort is focused on preventing failures of the large solar companies.

Beijing is asking provincial leaders to make plans for deployment of distributed photovoltaic generation (using house and business roofs) to boost the domestic solar industry…at least until the trading wars with the US and the EU are resolved.

So there is still hope for Suntech and many other Chinese companies relying on government subsidies, it's probably an unsustainable situation.

In the US, in late September of 2012, Suntech was notified by the NYSE that it was in danger of being delisted after its share price fell to an average of less than $1/day over a 30-day trading period. Suntech announced a restructuring, laid off 1,500 people, and temporarily closed part of its production in Wuxi.

The other approach taken by Suntech and other Chinese manufacturers is to open production plants in the US. An example of such a move is Suntech's PV modules manufacturing plant in Goodyear AZ, which started production in 2010.

Suntech of Arizona, however, is only an assembly plant, with a small number of low-paying jobs. It builds PV panels with solar cells shipped from Suntech's manufacturing facilities in China. In the past, we exported key components to China for cheap assembly—now we are assembling Chinese key components in the US.

Thomson River Power LLC

Thompson River Power LLC, based in Minnesota, designed and developed advanced products and services to support stable, reliable and efficient electricity grid operation.

The company was incorporated in 2007 and owns and operates a 14 MW power plant in Montana which provides low cost steam and power. Thompson River was an old coal-fired power plant on which a new ownership group spent more than $20 million to bring into compliance with emissions rules to burn "clean coal."

After finishing the work, management announced

that the plant would burn only wood—making it eligible for the Recovery Act money as long as the plant was technologically capable of producing power. But its owners found they couldn't operate the plant profitably by just burning wood.

They filed for and received $6.5 million grant in 2010 from the 2009 economic stimulus package (Section 1603) which paid developers of renewable projects 30% of their costs.

The grant to Thompson River, majority-owned by a Minnesota private-equity firm, was to convert the coal-fired plant to burn wood, which is considered a "renewable" power source. But the plant never operated—either as a coal- or wood-burning plant, before or after receiving the 1603 grant, according to Montana regulators. The plant has not produced power or new jobs.

We don't know how many jobs the firm promised to create, or if anyone is currently employed at the plant, but in July, 2012, Thompson River Power, LLC filed a voluntary petition for liquidation under Chapter 7 in the US Bankruptcy Court for the District of Delaware.

Under bankruptcy procedures, the company is to liquidate its assets to pay creditors. It has listed $26 million in assets and $6.6 million in liabilities—most of which is the $6.5 million 1603 grant. The Treasury has claimed it, but Thomson River did not list the government as a creditor, and is disputing the claim.

Because this money was given as a grant, it will be difficult to recover it through Chapter 11 proceedings.

Again—the lesson is easy money is easy to spend.

Timminco Ltd.

Timminco was involved in silicon material marketing, with production and marketing through Becancour Silicon, a subsidiary. Becancour Silicon purchases materials from Quebec Silicon, of which Becancour owns 51%.

After all is said and done, the solar grade silicon operations were carried on through Timminco Solar, a division of Becancour Silicon. Prior to ceasing operations, Timminco Solar produced solar grade silicon, using its proprietary technology for purifying silicon metal, for customers in the solar photovoltaic energy industry.

Hard times and fluid markets took their toll on the silicon business, and the Ontario Securities Commission ordered a temporary cease trade in Timminco's common shares.

In 2012, the company was unable to proceed on its own and filed for bankruptcy. On the auction block, QSI

partners acquired the silicon metal business and the assets of Becancour Silicon, and 51% ownership in Quebec Silicon for $31.87 million.

The solar grade silicon business and assets of Timminco Solar were sold to Grupo Ferro Atlantica for $2.7 million.

It took 30 long years to build the business, but only a few months for it to die, demonstrating how unstable, fragile and immature this business is.

Vestas

Vestas is a Danish wind energy company—one of many going through extremely hard times. More than 1,500 workers at Vestas' facilities and projects in the US face an uncertain future. This is in addition to the 2,300 layoffs in January 2012. The company's goal is to save $150 million in 2012, with even more job cuts planned if a US federal wind-energy tax credit is not extended at year's end.

In the fall of 2012, wind industry companies (manufacturers and developers) laid off thousands of people, with more slotted to be axed in the near future. Workers from Clipper Wind Power facility in Cedar Rapids, IA, and Vestas Wind Systems in Brighton, CO, and Denver-based Walker Component Group were among the dismissed.

The cuts are caused primarily from uncertainty surrounding the economic slowdown and the Production Tax Credit, which is set to expire at the end of 2012 (or 2013…), and most likely won't be extended by Congress much longer. The credit provides a tax break for wind farm developers, without which they would not be able to make a profit. Thus, if the tax break goes, the wind industry goes with it.

Xtreme Power

Xtreme Power site was the largest energy storage system integrated with a wind farm in the U.S. It was also the first renewable energy project to be completed under the Department of Energy's loan guarantee program, which supplied $117 million for its construction in July 2010.

But the problems with battery energy storage continue, so the only reliable energy storage is that of pumped hydro.

As a confirmation of its immaturity, and illustration of the growing pains of the electrochemical battery storage market, a smoke alarm went off one morning in August 2012 on the north shore of Oahu, Hawaii. This is the site of the 12-turbine, 30-megawatt Kahuku wind farm, where a 15-megawatt battery energy storage from

VC-funded Xtreme Power is/was installed.

The alarm indicated fire in a stack of electrochemical battery arrays, containing 12,000 large size batteries. It was a chemical fire emitting poisons and gasses that can kill. Firefighters couldn't enter the building for seven hours because of toxic fumes coming from the burning batteries.

Similar fires occurred in the same building some time ago, but burned out quickly, and both were attributed to faulty capacitors in the inverters.

Xtreme Power is blaming the inverter manufacturer for the fires, and is suing the company, but the damage is done. The battery technology got another black eye.

The flames destroyed a main building, shutting down the energy storage operation and raising questions about Kahuku Wind's entire project future.

Even when applied successfully for integrating intermittent wind and solar power into the grid, battery storage technology still cannot perform properly, or safely.

Summary

Above are lessons in how to do things and how not to. The US solar industry has gone through a number of boom-bust cycles since the 1970s. The difference this time was that the solar industry had risen higher than ever, having farther to fall.

Indeed, failure rates this time were several magnitudes higher in number and value than all others put together.

This is a good time to review what we've learned:

- Government subsidies work, but are not a long-term solution. Private enterprise is!
- Government support with political favoritism causes more harm than good.
- Large companies and projects can fail too.
- Quick money-making schemes cause failures, and give the industry a bad name.
- New technologies must be tested before mass production and marketing.
- Thin film PV modules manufacturers are especially vulnerable now that we see so many thin film PV companies and projects failing.
- Solar installations in cloudy areas have marginal success, regardless of anything else.
- Large-scale desert solar installations are the fastest way to energy independence.
- Efficient energy storage is needed to make solar power compatible.

- All solar technologies are do not perform equally.
- Cheap materials do not lead to long-term success.
- Sound business planning, complete with risk assessment and exit strategies are absolutely needed in any successful business enterprise that does not rely on subsidies.

The overwhelming lesson of the latest solar boom-bust cycle is that NOT following the basic rules of successful business development in a capitalist society leads to failure.

FAILED RENEWABLE ENERGY PROJECTS, 2007-2012

Following is a list of solar, and solar-related, projects that failed, or are in the midst of a major controversy, which might cause their failure. Our analysis of these projects is geared to provide some insights in the technologies and procedures involved. These are needed as reminders, so we won't keep repeating past mistakes.

Alamosa

Alamosa Solar plant is a 30 MWp installed on 225 acres in Colorado, powered by concentrating PV (CPV) technology manufactured by one of the pioneers in the CPV field, Amonix—backed by financing from DOE, Kleiner Perkins and Westly Group.

The 30 MWp power plant is now the largest CPV plant in the US, taking the place of the much smaller, 5 MWp Hatch CPV installation in New Mexico, which was built and operated by NextEra Energy Resources, and which held the world title until now.

Cogentrix, a subsidiary of The Goldman Sachs Group, is the plant developer and recipient of a $90.6 million loan guarantee from the US DOE. The plant began commercial operation early in 2012 and will sell its electricity output to Public Service Company of Colorado, an Xcel Energy subsidiary.

Cogentrix is known for development of fossil fuel plants so we don't know how it qualified to install CPV power plants. Because of that, we see the role of Amonix in the future O&M of the plant as crucial. And as Amonix is suffering financially, we are not sure that Alamosa will be able to operate successfully during the next 30 years.

Nevertheless, the plant is a sizeable achievement. It consists of over 500 units, each consisting of 60-kilowatt CPV tracker assemblies. Each tracker is approximately 70 feet wide and 50 feet tall. Each tracker unit of the last Amonix CPV generation of trackers, produces approximately 60 kWp DC power.

Solectria inverters are used to convert generated DC to AC, which is then delivered to the local 115 kV power grid. The inverters are capable of operating in a variety of smart grid modes and can also control the site's solar energy output to help regulate grid frequency, as needed.

While we believe that CPV technology, which is over 30% efficient, is the technology of the future, we see this plant as a large-scale prototype which is ahead of its time. As confirmation, Amonix closed its production facility in Las Vegas—the one that produced the 500+ large trackers in use at the Alamosa power plant.

The problem is that the CPV technology, which has been in development over 20 years, is too expensive, compared to conventional and more established renewable power sources.

The CPV trackers, due to a number of moving parts and the need for extreme accuracy of the tracking, are too complex and expensive to operate, thus they have not proven reliable for long-term operation.

A number of other CPV companies are in the race too—SolFocus has a project in Mexico that if completed and successful might eclipse the Amonix CPV farm. Soitec also has large CPV plants in the making, including over 150 MWp of PPAs with SDG&E, and plans for a 50 MWp plant in South Africa.

We have listed the Alamosa power plant among failed projects because

a. the manufacturer and technology principal, Amonix, shut down its operations and laid-off 700 employees even before the Alamosa plant was online;

b. the Alamosa trackers will need a lot of maintenance through the years, so with Amonix gearing down and financially weak, we wonder how the maintenance will be done; and

c. Alamosa power plant cost over $90 million to build, and there will be no additional money to bail it out, if something happens, which means that the 30 MWp plant must perform well in order to repay the loans.

According to official estimates, the Alamosa CPV power plant is expected to generate 87.5 GWh of electric power every year. This is somewhat higher than the numbers we find in quotes for CPV trackers operating in the deserts. Keep in mind that CPV equipment stops working at the slightest cloud formation, and must be stowed at moderate wind gusts, and sand or hail storm warnings. It loses power slowly with time, as the lenses get dirty, which can be a significant loss, especially in the monsoon months. The entire power field also must be stopped completely, or partially, from time to time for lens cleaning, and periodic and emergency maintenance.

So, in our estimate, considering all optical, mechanical, electrical, and other expected and unexpected losses and problems during CPV trackers use, the average yearly electricity production at Alamosa would be: 30 MWp x 6 hrs/day x 300 day/year = 54 GWh annually, or almost 1/3 less than the official estimates. Of course, this might be considered a worst-case scenario, but these are the numbers we use in real life situations of CPV equipment use, and they have proven true.

At an average of $0.10 per kWh payment by the utility, this is approximately $5.4 million gross income annually. After paying the daily O&M costs, and subtracting losses and other expenses (such as taxes, insurance, major replacements of hardware and electronics, etc.) the net income from the generated electricity would average slightly over 50% of the gross.

So, assuming that Alamosa could pay $3.0 million annually against the loans, it will take it over 30 years (depending on the interest rates and other conditions) to repay its debt. And this is if everything goes well, and no major disasters are encountered for the duration.

With Amonix in a downswing, after the Las Vegas production plant shut down, we don't know if Alamosa will get the attention (maintenance, etc.) it needs during 30 years of desert use.

Beacon Power

Beacon Power's predecessor, Beacon Power Corporation, was founded in 1997 as a spin-off of SatCon's Energy Systems Division. Its goal was to develop advanced flywheel-based energy storage technology. In 1998 Beacon Power became a separate operating entity, and went public shortly thereafter.

The first flywheel systems—the 1st and 2nd generations energy storage technologies—were deployed in North America and used mostly for telecommunications backup power applications.

The flywheel technology consists of a carbon-fiber composite rim, supported by a metal hub and shaft and with a motor/generator mounted on the shaft. Together the rim, hub, shaft and motor/generator assembly form the rotor.

When charging (or absorbing) energy, the flywheel's motor acts like a load and draws power from the grid to accelerate the rotor to a higher speed. When discharging, the motor is switched into generator mode, and the inertial energy of the rotor drives the generator which, in turn, creates electricity that is then injected back into the grid. Multiple flywheels may be connected together to provide various megawatt-level power capacities.

The US DOE gave Beacon Power over $25 million in grants, to be used mostly for research on this promising energy storing technology. However, the largest investment came when DOE announced a conditional $43 million loan guarantee to Beacon Power in July, 2009. It was intended to build and operate a 20 megawatt flywheel energy storage plant in Stephentown, NY.

Beacon used $39 million of the $43 million loan guarantee to build the $69 million energy storage plant, which was designed to support 20 MW of power fluctuations. This is the regulation needed to keep the grid stable and free of fluctuations caused by solar (clouds) and wind (lack of wind) power output variations.

It remains to be seen if this a profitable business. There are a lot of spinning wheels involved here, which require a lot of energy to keep them spinning. And the cost benefits are even harder to estimate when operating under the cloudy skies of New York.

Imagine a bank of spinning rotors, which requires significant amount of electric power, which follows the grid fluctuations, and kick in output mode to supply electricity when the solar and wind power level is reduced. This means that the rotors are turning all the time, driven by electricity, and produce some electricity from time to time.

Although this might provide stability to the grid power, the power consumed by the rotors during the day might be greater than that produced. The question becomes one of finances. Who pays for the imbalance in power use? Who pays for the maintenance, which, with so many moving parts, would be quite high.

The company warned investors in 2011 that it might not be able to remain a "going concern," and further reported that that it had received a delisting warning from the Nasdaq exchange for failure to keep its share price above a minimum of $1 through the fall of 2011.

Beacon did not last long enough to take advantage of the new federal ruling to increase the payments made to fast-reacting sources of generation, designed to keep grid frequency within operating boundaries. The new rule could have nearly doubled the revenues Beacon was earning for its services.

We don't know if that would have been enough to keep the company from filing bankruptcy, but it is one more indication of the immaturity of the technology and the uncoordinated character of the entire energy sector.

Beacon had $72 million in assets and $47 million in debts, according to its bankruptcy filing in Delaware. The company had also raised about $125 million through stock sales, and had a $5 million Pennsylvania state grant to build a flywheel plant there, though the odds of that plant being built were low.

But Beacon's bankruptcy doesn't come as a complete surprise, since the company has been losing money for years. The major reason given was that the flywheel energy storage technology was in early stages of development, and was breaking down consistently under continuous operation. There was also money mismanagement.

Lesson #1. Again… easy money, easy go.

With $68 million from DOE, $125 million from stock and $5 million from the state, this is $198 million spent on R&D and buildup of a 20 MW energy storage facility. Almost $10/Watt was spent getting to the operational stages of the 20 MW plant, which Beacon did remarkably well.

The O&M stage, however, is where they stumbled. They were not able to keep the technology operational.

Lesson #2: Don't try to run before you can walk. Beacon was in a hurry to put this new technology on the market, but new technologies should not be rushed.

Texas renewable-energy investor Rockland Capital, paid $30 million to buy Beacon Power and its operational plant. Most of that money went to repay the DOE loan guarantee and other debt, with only just over $5.5 million remaining after all the dust settled, as a reminder that things didn't go as planned.

Lesson #3. Grid energy storage remains a very difficult, albeit very needed, component for solar and wind energy integration with the national grid. Slow, deliberate developments will eventually bring this technology to market.

Author's note: Without energy storage systems the solar and wind technologies, with their variable and unpredictable output, will always be looked at as incomplete and incompatible with the overall energy goals of the country and world. The large variations are also seen as damaging to the proper grid power control. Until the variability is under control, they will not be able to compete with the traditional fossil fuel-fired sources of power generation.

So, grid-scale energy storage remains a very important technology, and we should give due credit to Beacon's New York power storage facility, which was one of the first to bid its services into the real-world power market. Sadly, it was just not ready.

Beacon's flywheel technology is very different, and more promising for large-scale operation, than batteries, fuel cells and other next-generation energy storage technologies. It is also more advanced than these, but the bankruptcy was a big blow to the energy storage market as a whole, so we hope that we will hear good news soon from the new owners of the 20 MW Beacon energy storage plant in NY.

Con Edison's Hixville Historic District Solar Plant

Con Edison's Hixville Historic District Solar Plant is another example of the immaturity (this time of the permitting process) of the solar energy industry.

Con Edison Development had detailed plans for about 9,500 solar panels to be erected on the leased site, where the panels will generate about two megawatts of electricity a year. But residents criticized the company for cutting hundreds of trees, blasting rock ledge, and grading the 14 acres of land where the panels will be erected without notifying neighbors or securing a building permit for the project.

"It could have been done better," company spokesman admitted after being criticized by board members for quiet-hours noise from heavy machinery, and failing to meet with neighbors to explain the project, or even warn them of planned blasting at the site.

According to the rules, however, the regulators' authority doesn't begin until construction starts (?), while land clearing activity is allowed before a building permit is issued by the town. They had no grounds to stop the work because it is an allowed use in any district.

The company submitted the landscaping plan for vegetated screening along property lines of the 14-acre site, a continuous vegetated buffer 30 feet deep that will grow 10 feet high all around the property so that it won't be seen from the road. This plan was the last item required for issuing a permit, and the process cannot be reversed now.

None of the explanations satisfied neighbors shocked to see the unexpected work done on the land and to then find out that more than 9,500 solar panels will soon be going up in their quiet neighborhood.

In a scheduled meeting of the Select Board in February 2012, more than 50 local residents tried to convince town officials to put on hold the 14-acre development.

Select Board members, told the residents they were powerless to halt the project, or prevent construction, due to last year's passage of a large-scale photovoltaic energy farm bylaw at Town Meeting. That bylaw allows such developments "by right" in any zoning district, and without a public hearing or a site plan review.

The Select Board was presented with a petition signed by more than 100 residents, who want to stop the construction, declare a moratorium on such projects, and immediately call a special Town Meeting to reconsider the energy farm zoning bylaw.

Local residents' concerns include the potential negative impact on property values, and possible health impacts from electromagnetic waves, fueled by frustrations with the town's failure to secure an insurance bond before work started.

District fire chiefs had not been consulted on required emergency response plans for dealing with the long list of hazardous materials contained in solar panels.

But stopping the project once it has started "would be very difficult" according to the authorities, while issuing a moratorium was "a Town Meeting action" that could not be initiated by the Select Board; which points to a confusion of who is who, and what they can and cannot do.

Frustrated by town officials' inability to stop the project, the residents are investigating the possibility of taking legal action to delay construction and force Con Edison Development to meet with neighbors about screening plans.

The end result of this case is almost certain; Con Edison gets to complete the power plant, while the residents are left steaming in their anger. Such conflicts can and should be avoided, but due to the newness of solar markets and the inexperience of developers, this case is not the first, nor the last.

Maricopa Solar

In 2009 Stirling Energy Systems (SES) built and installed six of its so-called SunCatcher trackers at Sandia in New Mexico, and later constructed the larger demonstration plant Maricopa Solar Project in Peoria, AZ, where 60 of the large dish-trackers were installed. The total power produced by the facility was rated at 1.5 MWp.

Stirling dishes are large structures—over 40 feet tall and almost that wide—and use a set of complex technologies, where gear-motors drive the tracker frame to follow the sun all day long, while large mirrors focus the sun's rays on an engine mounted in the middle of each dish. The engines convert the sun's heat energy into motion, and the motion is used to make electricity.

The 1.5-megawatt Maricopa Solar Project in Peoria

was intended to demonstrate that the dishes worked and that the company had a supply chain that would manufacture the dishes for the big California projects and another in Texas, which by this time were failing or had already failed.

SES had plans to build tens of thousands of Sun-Catchers in California and Texas but that never happened. See Tessera Solar in this section for more details.

In September, 2011 Phoenix, Stirling Energy Systems Inc. (SES) of Scottsdale, AZ, filed bankruptcy, and the 1.5-megawatt demonstration project in Peoria, which was built a year or two earlier was shut down.

The company that was running the power plant with a license to the technology from SES, Maricopa Solar LLC, also declared bankruptcy and now is liquidating the power plant. But they will keep the intellectual property that is needed to build and run the large Stirling engine dish-trackers.

Maricopa Solar brought in $162,555 in revenue in 2011, before it shut down in September of that year. It earned $273,502 from December 2009 through September 2010, or approximately $27,000 monthly. This breaks down to less than $1,000 per day, which at $0.10 per kWh boils down to approximately 6 hrs. average daily operation for the 1.5MWp facility. Not exceptionally bad, if there were no great O&M expenses for the duration.

One of the great problems with dish-Stirling technology is that the equipment heats up to temperatures greater than 1,000°F, which is an enormous load for a continuous, long-term operation with any type of materials and equipment.

Every day, the sun heats the engine and keeps it at this extreme temperature all day. It cools off at night and soon this huge thermal cycle causes thermal-fatigue. This causes wear and tear that makes it impossible to ensure reliable long-term operation without a lot of

maintenance. We believe that the participants saw these problems and the failure of SES to solve them caused the cancellation of the projects.

When SES ran into a variety of problems in 2008, it got a $100 million investment from Dublin-based NTR plc. But by early 2011, it had sold the California projects to other developers who would use more traditional solar panels for the projects, and NTR wrote off its investment in Stirling.

In April, 2012, online auction (thought to be one of the first of its kind) where large Stirling dishes were on sale, offered the bankrupt Maricopa Solar power plant in Peoria, and its unproven technology, for sale. That was a hard sell, however, because, as part of the deal, the large trackers had to be disassembled and transported to a different location. We don't know the results of the sale, but the 1.5 MWp Maricopa Solar plant is no more.

Moree Solar Farms

Frustrated by the market dominance of the big energy companies, and the confusion they bring to the solar game, Australian's renewable energy generator Pacific Hydro is working on its own retail license, as it tries to save the large solar project, the $923 million large-scale Moree Solar Farms PV facility—a major part of Australia's Solar Flagships initiative.

The proposed 150-megawatt Moree Solar PV Farm was initially backed by Pacific Hydro, BP Solar and Spain's Fotowatio Renewable Ventures, but it suffered a setback in December, 2011 when it lost the rights to a $306 million grant from the federal government's troubled solar flagships program.

Then, BP Solar who was supposed to play a major part in the Moree project, following up on its decision to get out of the solar business altogether, gave a final notice to Moree Solar Farms, and declined to be part of the new bidding process. BP then exited the business, and it its further involvement in this project is unclear.

Instead, Acciona, SA from Spain will fill BP's shoes in this project, which is surprising, because Acciona *was* part of the project in the very beginning, but pulled out in 2010, stating the increased uncertainty in the Australian energy market, as well as the absence of an offtake contact. So, evidently, now Acciona thinks that the Australian energy market has improved, and that offtake contact is easily available.

The northern NSW, Australia, solar farm did not meet its funding agreement obligations, prompting the Energy Minister, Martin Ferguson, to re-open the grant to four shortlisted photovoltaic farm candidates.

All competing bidders—AGL, TRUenergy, and In-

Figure 5-3. Maricopa Solar project, Peoria, Arizona.

figen Suntech—have re-submitted their bids before the deadline, so they are qualified and will be considered for the final decision. In February 2012, the Moree Solar Farm consortium also submitted a revised bid for Solar Flagships funding to the federal government, that fully addresses issues previously raised by the Commonwealth.

In July, 2012 the money in the first round of the federal government's "Solar Flagships" program which was originally awarded to the Moree Solar Farm was revoked after the consortium behind the Moree farm proposed major changes to its project.

There are many reasons Moree is still in trouble. Would it be able to complete the financing? Australia is not exempt from, nor is it a stranger to, the ongoing battle for solar energy advantage, and Moree Solar Farms would be an indicator as of how successful it could be now and in the future.

Sempra US Gas & Power

Sempra US Gas & Power has made a u-turn on a decision announced in the previous year by revealing that the company will not go ahead with its plans to build a 300 MW solar farm on Navy land near Hawaii's Pearl Harbour, according to *Pacific Business News*.

The project was originally announced last year as an unsolicited proposal, made in response to a posting on the Naval Facilities Engineering Command Pacific Enhanced Use Lease website. Sempra wanted the Navy to realize what potentially could be built on its land.

The solar farm could provide the Navy with electricity free of charge, so in order to make a return on its investment, Sempra would sell power to Hawaiian Electric Company (HECO).

If the project had gone ahead, the solar farm would have been one of the biggest solar farms in the country. With HECO's FiT program now in place, Hawaii has become an attractive market for solar energy players. Hawaii has the largest Renewable Portfolio Standard in the US, requiring 40% of the state's energy to be supplied by renewable energy by 2030. HECO's FiT program is designed to encourage the addition of more renewable energy projects in Hawaii.

After announcing the u-turn, Sempra has no plans for new projects in Hawaii; however, Hawaii remains a key market for the company's renewables portfolio, according to company reps. Dropping a 300 MW project is not a u-turn—more like a flat tire.

Solar Dawn

Solar Dawn is a planned 250 MW solar-thermal plant in the Australian state of Queensland, to be de-

veloped and managed by a consortium of companies, including AREVA Solar and Wind Prospect CWP.

AREVA Solar is a subsidiary of AREVA Renewables, which is a global manufacturer of low-carbon power generation equipment. AREVA Solar specializes in solar-generated, superheated steam for stand-alone solar thermal plants, power augmentation for coal- and gas-fired power plants, and for industrial steam processes.

The proposed Solar Dawn power plant will use AREVA Solar's Australian-developed Compact Linear Fresnel Reflector (CLFR) technology, which uses a number of narrow, modular, slightly curved mirrors to focus the sun's heat onto "receivers" mounted above the mirrors. The receivers are in the focal point of the mirrors and get all the sunlight falling on them, thus heating water which circulates in the closed-loop system. The sunlight boils the water in the tubes, directly generating high-pressure, superheated steam. The steam powers a steam turbine generator, producing electricity.

The federal government originally committed A\$464 million to the A\$1.2 billion Solar Dawn solar project under its Solar Flagships support program. One of the conditions was that the state of Queensland put up its share, but the state is saying, no.

After prolonged negotiations, the state government terminated the funding deed to the project, scheduled for installation in 2012. The state withdrew A\$68 million funding after it became apparent that PPA with the local utility was not yet obtained. The AREVA-led consortium was also unable to meet financial close, so now the "fate of the project hangs in the balance," according to insiders.

"The Australian government remains committed to the deployment of large-scale solar energy technology in Australia…" but for now, all the planning and financial investment have no payoff. The promise of 300-450 construction jobs and 30 permanent jobs during operation is gone too.

The plans for the largest solar plant in Australia, the energy it was supposed to generate, and CO_2 it would've saved are now on the back burner. More negotiations will follow. More plans will be made, but under the newly developed economic situation, and the changes it has brought up, we just don't see a project of this nature coming to life any time soon.

San Diego School District

Dozens of schools in the San Diego Unified School District got a facelift in 2005-2006 by installing brand new, shiny, solar panels on their roofs. It was the thing to do in those days of unlimited solar subsidies and unending euphoria.

The panels cost the district nearly nothing to install, and were expected to last at least 20 years. In return, the district agreed on a set price for the energy supplied over the 20-year duration of the PPA contract.

After 6 years under the sun, the panels from 24 San Diego school campuses were taken down in the summer of 2012, when many of them were found to have defects, including premature corrosion, which could cause failure or even roof fires.

Note: The now infamous UNI-SOLAR PV-technology was used for the solar panels on the school district roofs. UNI-SOLAR is actually thin film technology that has been meandering in the making for a long time, and whose inventor and manufacturer went bankrupt.

Under the terms of the contract, the solar panels were installed under a 20-year arrangement with General Electric, which is their legal owner, and Solar Integrated Technologies, which leased the panels from GE. The district purchased the power from GE, which came to about $750,000 per annum, hoping to save $7.0 million during the next 20 years.

Michigan-based Solar Integrated Technologies—who was responsible for the installation and maintenance of the solar panels—filed for Chapter 7 bankruptcy protection in the spring of 2012 and could not be held responsible for the damages.

With Solar Integrated out of business, GE was forced to take over. GE engineers inspected the sites, but declined to take over the maintenance of the panels, due to their poor condition.

GE then offered the district the opportunity to keep the solar panels and maintain them at its cost, but the district declined, citing the dangers involved with continuing to operate the defective panels.

Finally, all panels were removed from the 24 school roofs at GE's expense. The solarless operation will now cost the district about $400,000 annually in additional fees for energy it uses in the coming years.

Having learned the lesson, the school district is more careful now, and is scrutinizing its other solar contracts in an attempt to determine what solar panels would be used on the roofs of other schools, as well as who will do the work and the maintenance.

Note: Solar Integrated Technologies was a subsidiary of Energy Conversion Devices, which is another of the high-profile firms to go under as part of the latest cycle.

The poor quality of the thin film modules is another confirmation of the immaturity of this technology and the pitfalls of hasty marketing.

Tessera Solar

Tessera Solar was created in 2009 by Stirling Energy Systems (SES), a solar energy equipment manufacturer and developer wanna-be. The idea was for newly created Tessera to focus on, and hire specialists for, developing different areas of large-scale projects design, construction and operation, while the mother company, SES, would concentrate on its SunCatcher technology development and marketing, thus increasing equipment sales.

Tessera had ambitious plans. Several huge projects were in the pipeline, and help from willing utilities and government subsidies SES started the planning stages. Here is where the whole thing evaporated into thin air.

Figure 5-4. SunCatcher dish

The technology they would use was the SES proprietary SunCatcher dish-trackers. The SunCatchers had a catchy look to them—large, shiny mirrors blasting sunlight onto a large, impressive looking Stirling engine mounted at the center of the dish, producing 25 kW solar power.

The dish mirrors track the sun, and concentrate its rays on the engine. The Stirling engine converts the sun's thermal energy to grid quality electricity, and sends it into the grid. Some water requirement (4.5 gallons/MWh) was needed to keep the engine running efficiently. The SES dish trackers had the highest efficiency at the time, averaging 25%, with recorded maximum at 31%. The dish systems were both modular and scalable, and their installation did not require extensive ground modification; <5% grade variation was acceptable.

Tessera was busy developing several power plants for use of the Stirling dish trackers during its heyday, some of which were:

1. Maricopa Solar One, 60-dish SunCatchers was located in Peoria, AZ with 1.5 MWp capacity. It was grid

connected and the generated power was purchased by the local Salt River Project power company. The plant started operation on January 22, 2010

2. Calico (SES Solar One) was to generate 850 MW solar power using dish-stirling technology developed, designed and manufactured by Stirling Energy Systems. Tessera Solar was the project developer. A power purchase agreement was signed with SCE. A DOE loan guarantee application was filed and approved to move to Phase 2 of the DOE loan authorization process.

The plant was to be located on 8,230 acres of BLM land in the Mojave Desert, 37 miles East of Barstow, CA. It was to be constructed on previously disturbed land selected in consultation with BLM in 2005.

Transmission lines and interconnects were available. The existing transmission for the 275-MWp Phase 1 was done with only minor substation upgrades required. The upgrade Pisgah-Lugo 230kV line had to be upgraded to 500kV as needed to handle the 575-MWp, Phase 2.

The project was supposed to create close to 700 construction jobs over 3-4 years and would provide approximately $60,000,000 in construction payroll with an average monthly construction workforce of approximately 360. Approximately 160 permanent jobs in the supervisory, administrative, created during the next 30 years for the plant's O&M activities.

3. Imperial Valley Solar (SES Solar Two), was to generate 750 MWp solar power using Stirling dish technology, to be manufactured SES in Arizona. Tessera Solar was the project developer.

The plant was reduced from an original 900 MW proposal to avoid cultural resources at the site, and a power purchase agreement was signed with the local power company, SDG&E. Then, SES filed a loan guarantee application with the US DOE and was approved to move to Phase 2, but that's as far as things got.

The project was to be located on 6,251 acres of BLM land, plus an additional 320 acres of private land in the Imperial Valley, in California. Transmission lines and interconnections were available via the existing transmission for 300 MWp, Phase 1. The additional Sunrise Powerlink was approved by BLM and CPUC for the remaining 450 MWp, Phase 2.

The project was supposed to create close to 700 construction jobs over 3-4 years and would provide approximately $60,000,000 in construction payroll with an average monthly construction workforce of approximately 360. Approximately 160 permanent jobs in the supervisory, administrative, created during the next 30 years for the plant's O&M activities.

All plants boasted lowest-water use CSP technology, which uses water for washing mirrors only. SES was also an innovator in the manufacturing chain. It was to use the US automotive supply chain to make most of the dish components, thus creating new jobs in the US.

Then hard times hit.

In December, 2010—a year after its creation—Tessera Solar announced the sale of its 850 MWp Calico Solar Power project, after it obtained permits for construction, interconnects, PPA, etc., but had trouble lining up financing and lost the power sales contract.

Later, Tessera sold the 709-MW Imperial Valley Solar Project, which was geared to sell the generated electricity to San Diego Gas & Electric. A power purchase agreement (PPA) was already signed and everything was ready to go…but did not go anywhere.

These sales did not come as a surprise considering the fact that the mother company (SES) has been meandering with its technology development and experiencing periodic technical and financial problems.

The combination of complex issues made it difficult for Tessera to line up financing for the projects, which were expected to cost over $3.5 billion to build.

On top of that Tessera was faced with lawsuits for the Calico project and a federal lawsuit from an Indian tribe for the Imperial Valley project. And Southern California Edison sealed the fate of the first project by canceling the PPA.

The sad part is that the Calico Solar Project was under development for 5 years before the sale date and had PPA and all other permits. The project just sat there while the mother company experienced problems with its technology and didn't have enough to continue its development.

An Irish firm, NTR, came to the rescue in 2008 by pumping $100 million into SES, which revived the company and its projects, but didn't go far enough. The new owner was planning to start construction of the project immediately, but had to go through a new permitting process.

In the end, Tessera lost both huge projects. The Maricopa Solar Project One survived for awhile, but was shut down in 2011. See Maricopa Solar Project in this section.

SES filed for bankruptcy in the fall of 2011, showing assets of $1-$10 million, and liabilities of $50-$100 million. The Irish investor ended up with 52% of nothing, after paying $100 million for something that just wasn't there. At least US DOE got a whiff of the problems in time, and did not spend any money on these projects.

Yosemite Solar Power

The photovoltaic system at the El Portal complex at Yosemite National Park is the largest NPS-managed photovoltaic project within the Pacific West Region and the largest grid-connected photovoltaic system owned and operated by the NPS, but it may never pay off.

The 672 kW system consists of 2,800 solar panels and is supposed to produce 800,000 kWh per year. The park estimates saving approximately $50,000 per year on electricity purchased off the grid and is expecting to receive a $700,000 energy rebate from Pacific Gas & Electric Co. (PG&E) over the next five years. This represents an approximate 12 percent reduction in electricity purchased off the grid.

"The collaborative effort to design and build this system has come to fruition and we are extremely proud of the results," Stated Neubacher. "We are committed to being a leader in renewable energy and this project exemplifies our efforts."

The project was funded through the American Recovery and Reinvestment act and cost $5.8 million (plus $700,000 from the local utility). Construction of the system was completed in February 2011 and the interconnection agreement with PG&E was signed in late June 2011.

This looks good on the surface, but

1. The total price of $6.5 million for the 672 kW PV system is almost $10 per Watt installed—at least twice the going commercial rate at the time.

2. How long would it take to pay back the initial $6.5 million investment at $50,000 annual savings? Answer: $6.5M invested/50k saved per year = 130 years ROI, provided no major maintenance or replacement work is done.

Considering 1% annual degradation, we see that during year 10 the park will save only $45,000, during year 20 only 40,000, and down to $35,000 during year 30. Then the installation must be disassembled and hauled away—another $150,000 minimum.

As an educated guess, taxpayers won't get much more than $1.5 million back on the $6.5 million investment, and only a hand-full of temporary jobs were created.

This is another good example of how not to use solar in the US.

Conclusions

2011 left the solar industry confused and damaged. In 2012 an avalanche of renewable energy companies and projects worldwide went down in history as failed. Many larger companies and projects are on the brink

of failure, where smaller and weaker players are most likely to shut down or be snapped up by stronger rivals.

Solar subsidy cuts in the top European and US markets prompted over 50% drop in the price of solar panels, bringing the fast-growing solar industry to a critical tipping point. Many of the stock market heroes find themselves struggling to cut costs in a market awash with solar panels and lacking new projects.

In the long run, the PV products' price declines are healthy for an industry that grew on the promise of government subsidies to make its products competitive. Unfortunately, the subsidies proved unsustainable and only delayed the inevitable.

Some European companies that initiated and helped drive the solar boom of the 2000s were among the hardest hit. Most of these companies' costs are far higher than those of their Asian competitors.

All in all, as of the fall of 2012 around 50 manufacturing companies in the PV industry have failed totally, half since January, 2012. Many of these companies and projects received financial help from their respective governments, which might have accelerated their failure by forcing them into hasty decisions.

Many of the affected are US companies and projects, so as a consequence, US lawmakers are working on a new way to provide grants and loan guarantees from Title XVII of the Energy Policy Act of 2005.

FAILED THIN FILM SOLAR COMPANIES

After compiling the list of failed solar companies, we observed that they exceed the number of failed c-Si solar companies. Because of that, we decided to dedicate a special place for the failed thin film companies, followed by a discussion and analysis of the reasons for their failure.

Thin film solar products are a fairly new development in the worldwide solar industry, and due to their uniqueness and great promise, they deserve special attention. So, here are some of the key thin film solar company failures between 2007-2012 in alphabetical order.

Abound Solar

Abound Solar Inc. was a CdTe manufacturing company, which received $400 million in government loan guarantees to build two factories for mass producing PV modules in Colorado. The work was underway, and things looked good for awhile, but when the 2011-2012 bust cycle storm hit, Abound was basically caught ill prepared.

This was a critical time, when the markets were flooded with cheap PV modules. Abound had no working business model, and wasn't even close to compete in the new situation. They had over 100,000 finished and unsold CdTe PV modules in the warehouse when they went down.

They were sitting on 100,000 modules at $100 each—$10 million in inventory—when they were forced to close the doors, stop production, and lay off most of their people.

This is another case of building factories, instead of making product—at the worst possible time—when large manufacturers flooded the market, and prices were going down daily.

Abound Solar, like so many would-be manufacturers of PV modules, didn't even see what hit it. Risk management? Not much. Exit strategy? None.

Author's note: Several thin film companies fit this Solyndra-like model of building new factories instead of producing and selling product from their old facility.

Many thin film companies were especially affected by bad planning, and/or unable to compete due to high prices and/or low efficiency product. Many failed very quickly, during the last stages of the recent solar boom-bust cycle. Although some of them had a promising product, they will most likely will never be back.

Advent Solar

Advent Solar, based in Albuquerque, NM, was another hi-tech thin film wanna-be manufacturer, quietly developing a new, revolutionary, technology in the midst of the chaotic solar developments.

Advent was hoping to take the market by storm with its superior technology. They started work on a new a-Si based emitter wrap-through technology in 2003, with results that were not well understood. Their technology is quite different from what we know, or what is on the market. It is a specialized approach that uses thin film depositions to form the active layer and the interconnects, which helps to increase the efficiency of PV cells and modules.

Sounds good, but it was a project that led to nowhere. The company had no viable product ready for market, when the solar bust started.

Advent Solar received nearly $17 million through the New Mexico State Investment Council (SIC) and its private equity arms. The state recovered $1.5 million after Advent's assets were sold, but the state lost $15.5 million in the deal.

In 2009 Advent was acquired by Applied Materials, which also exited the solar business soon thereafter.

We know that Abound Solar is no more, but the status of its facilities and equipment is not clear now, and information is unavailable. We also wonder what Applied Materials did, or is doing with the Advent equipment. They most likely offered it for sale to another would-be thin film solar manufacturer.

Applied Materials

Speaking of Applied Materials (AMAT) (one of the largest, most successful and reputable, semiconductor processing equipment manufacturing firms in the world), they decided to get into the solar production equipment business in 2006.

Their timing was perfect. Nobody made better equipment for mass production, depositing high quality silicon thin films—or any type of thin films—on large glass substrates, or any substrates.

AMAT also has extensive experience in the production processes of the related thin film technology (TFT) LCDs, and makes the highest performance, highest quality, and highest efficiency deposition and plasma etch tools (PVD, CVD, PECVD, etc.) available. AMAT has enjoyed an 80% market share of the TFT LCD market since its beginning.

The silicon shortage of 2005-2006, during which silicon spot market prices jumped to an amazing $500/kg and more, provided a large niche market for AMAT to offer an alternative. AMAT's process uses 1/100 of the silicon required to make regular solar cells, so imagine the savings.

Supplies of silicon solar cells were growing tight and expensive then, while AMAT consistently returned 20% net margins and added capacity as fast as possible.

AMAT came up with the newest, most versatile design for a turn-key thin-film PV modules manufacturing line, offering what seemed like a perfect solution to new companies trying to get into the PV cells and modules manufacturing business.

The new production line named SunFab was not tied to expensive Si materials with their volatile prices, it was very fast, and it could produce the largest solar modules in the world. The key advantage was that it promised large quantities and low cost PV modules for the large-utility market. Too, the SunFab lines came with technical support and a production guarantee.

In 2009 AMAT also ventured into acquisition of solar-related companies with investments over $1.0 billion, including $330 million for Italy's Baccini—maker of metallization and testing equipment, $483 million for Swiss solar wafer equipment company HCT Shaping Systems—maker of wafer slicing equipment, and the purchase of Advent.

Unfortunately, the engineers and managers working on the new SunFab production line and the marketing people handling the AMAT solar affairs, knew nothing about solar energy products and even less about the world solar markets.

Most of their customers had little knowledge of solar products and market whims. The technical aspects of the AMAT operation were easy, but they did not understand product logistics or the world energy market's history. Also, AMAT chose to work with new customers, or lower-tier manufacturers, who also lacked experience in the field.

Top tier manufacturers have their own production lines and don't buy off-the-shelf turnkey products, and established manufacturers choose individual tools for specific needs, so AMAT's turn-key and very specialized SunFab production line was locked out of the top level.

Another issue was that Sunfab took the early customers up to 17-18 months to get from construction to production, and almost that long to optimize the line and certify the new products. Semiconductor equipment has a reputation of fast install and production start-up within 4-7 months... By the second half of 2008, while most of the SunFab lines were still in construction, or in planning stages, the window of opportunity started closing. AMAT bailed out of the thin film solar production equipment business in the summer of 2010 and concentrated on making silicon solar wafers slicing and printing equipment, and expanding its LED lighting equipment.

This led to a major restructuring of its "Energy and Environmental Solutions" division, the elimination of nearly 500 jobs, and a 2010 third-quarter charge/loss of around $400 million.

To AMAT' credit, in 2009 management saw what was coming and exited the solar market at the right time, keeping one foot in the door. They continued to lead the solar equipment market (in revenue), but not by thin film production equipment sales.

In a twist of faith, instead of supporting the thin film market (its core expertise), AMAT is supporting the c-Si solar market with the c-Si wafers slicing and printing tools it designs, manufactures and sells today.

In mid-2010, Applied Materials announced a restructuring plan for its Energy and Environmental Solutions segment (200 positions were lost). In the fall of 2012, Applied started shifting production of its PV wafering equipment operations to China, and wafering operations will lose most of the US workers.

These events are confirmation that large produc-tion lines, large companies, or entire governments, cannot dictate the solar energy markets. Those markets will evolve according to their internal abilities, and in response to the world's energy and environmental needs.

AQT Solar

AQT's production facility in Sunnyvale, CA, housed a 15-megawatt manufacturing line based on automated dry sputtering equipment from Intevac. According to insiders, the intention was to manufacture "low-cost CIGS thin film cells and modules that are drop-in replacements for conventional crystalline silicon cells."

AQT was also working on R&D of the reportedly more efficient copper zinc tin sulfide (CZTS) cells—a variation of the SIGS technology, but that work did not produce a marketable product.

AQT raised about $32 million since its founding. The company was building CIGS solar cells on glass, but aspired to build panels as well as to form partnerships in the solar project development business.

The strong technical staff had good initial results, but AQT is no more. They had a good strategy, but they had no product to put on the market. As in many other cases, AQT was able to raise a lot of money for something they never produced.

In August, 2012, AQT was selling its assets and the intellectual property of its CIGS and CZTS materials and modules manufacturing systems, together with all production plant equipment.

Note: A major lesson here is that the promise of CIGS that was planted in the minds of entrepreneurs and investors 10 years ago did not take into account the weak and unstable performance of the CIGS modules, and the sub-$1.00/Watt prices of the competing c-Si solar panels in the quickly changing, shakily subsidized, over-capacity market.

EPV Solar

EPV Solar Inc. was a Robbinsville, NJ-based solar company, with thin films modules manufacturing plants in New Jersey and Senftenberg, Germany. EVP opened for business in 1991, as one of the earliest to start in this business, and one of the first to go under.

A designer, developer, manufacturer and marketer of low-cost amorphous silicon (a-Si) thin-film PV modules, the company had great hopes to be the first developer of large-scale power fields, using its particular technology. But first they decided to expand, to facilitate the manufacture and sale of large quantities a-Si modules. They built factories for the future production of their PV modules.

As many others before them, they thought there was an enormous market for low efficiency, unproven a-Si PV modules. All they had to do was manufacture and sell them, so they financed their endeavor by issuing $77.5 million in senior secured convertible notes maturing in June 2010.

Sales never materialized, since very few large solar farms wanted to use a-Si, EVP did not yet *have* the modules, and 350 people were laid off. They negotiated another $3.6 million in financing before agreeing to sell out to a Turkish company. That sale fell through, and EPV ceased manufacturing operations completely in 2009.

EPV Solar then, with its back to the wall, voluntarily filed for chapter 11 bankruptcy protection shortly thereafter. All 400 employees were laid off by the spring of 2009—some only six months after they were hired—and EPV was gone. The EPV facilities and IP were acquired later on by Sunlogics, which was then acquired by Salamon Group. No further information about the destiny of the solar operations is available.

Author's note: This is an exceptionally unusual case, because the failure occurred at the height of the solar boom. According to insiders, the problem was management's inability of manage money. Building buildings instead of PV modules at the wrong time, is another problem we saw in this case.

Many thin films companies have gone bankrupt since these events took place 2-3 years ago. Bad planning and management, inefficient and immature technology, rushing to market, uncooperative markets…

Did we learn anything? We *have* been here before.

First Solar

This is the greatest (albeit short) solar success story of the century, with First Solar taking the world by storm in just a few years. After securing $4.5 billion in US DOE loans, they planned to take over the US deserts.

The trouble appeared in 2011 and 2012 with the exodus of key managers. Why did they leave? There must be an explanation events reflected in the 4Q11 report which showed net loss of $485.3 million.

There were also nearly $254 million in warranty-related charges which included recalling and replacement of failed modules, and remediation and compensation to customers for failed modules and negative effects on installations.

Additional $35.6 million was "set aside" to cover similar claims in the future. The shock was compounded by news that the entire 6/2008-6/2009 production was recalled, due to sudden loss of power after short field operation in high temperature regions.

First Solar has no proven record of large-scale operations in "hotter climates," since most of its installations are in Germany and France, and are 5-6 years old at best. The installations in high-temperature areas are much newer and fewer.

Since CdTe PV technology is a relative newcomer, their CdTe, frameless, PV modules have not been properly tested, nor proven reliable and safe for long-term use under extreme desert conditions.

The fragile thin films, while reliable enough in computers (where they see minimum mechanical, thermal and chemical exposure and are well protected from the elements), are not sturdy and reliable enough for 30 years direct exposure to desert extremes.

For large-scale customers, lost electrical production, contract non-compliance, service charges, and the consequences of failed modules are serious issues. The failure of even one PV module causes great damage, which can be expressed in financial loss. The reputations of solar power generators hang in the balance of their products.

Manufacturing "excursion" is the new way of identifying rejects, which were formerly listed in the "we screwed up" file. Table 5-1 illustrates First Solar's manufacturing excursions.

Table 5-1. First Solar's rejects and recalls.

Issue / Action	4Q'11	To date
Remove/replace modules	$23.9M	$99.7M
Module cost	70.1	70.1
Power loss compensation	31.8	45.9
Warranty adjustment	37.8	37.8
Total warranty charges	$163.6M	$253.5M

Longevity of the PV modules is the most significant metric for solar panels, and doubly so for thin-film panels, whose performance track record lags behind crystalline silicon modules. Most of these modules are going in desert regions where they have not been proven efficient or safe to operate 30+ years.

Field reliability of thin-film PV modules (any thin film modules—not only CdTe) is un-proven. High temperature degradation of cadmium-telluride (CdTe) panels is not well understood, but it is a most important metric for a company's long-term survivability, so this may not be the last time we hear of warranty-related issues with this technology.

Lower-cost benefits for CdTe TFPV modules do not necessarily outweigh efficiency and performance benefits and potential toxicity issues. Figure 5-5 is a stark reminder of the pitfalls of throwing money into unproven technology.

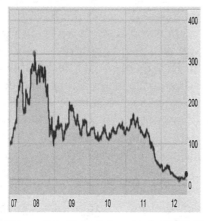

Figure 5-5. First Solar stock, 2007-12

Over $4 billion in US government loan guarantees allowed First Solar to develop several large PV projects in the US deserts, which will take several years to build. First Solar will survive the hard times, but their projects may not.

The fragility of the frameless modules and their carcinogenic heavy metal cadmium are two aspects which make them questionable, and now they've been shown to have uncertified junction boxes. These 232,000 modules are in the field. An electrical malfunction at this point could unleash their toxins on a large scale.

Gadir Solar

Gadir Solar factory in Puerto Real, Spain, started operation in October, 2009, manufacturing a-Si thin film PV modules with production technology supplied by Oerlikon. The plant was established by an investment of approximately €110 million, later funded by additional grants of €13 million, subsidized loans of €10 million, and more than €80 million from shareholders.

There was great hope that a-Si would be the solution to all solar problems, including efficiency, reliability and cost—technology which Gadir was going to introduce. The company website stated: "The photovoltaic amorphous silicon modules are competitive in this market thanks to a number of features… their behavior [in] high temperatures… performance [in] diffuse light and shadows… installation not oriented to the Sun's direct radiation generates high yields…"

The company didn't have a chance to sell many of its modules, although they were certified and ready for market in 2010. The market did not share Gadir's views nor accept the a-Si PV modules as a ready substitute for the well established, more efficient, cheaper c-Si PV modules. Bending under increasing pressure towards the end of the solar boom-bust, Gadir laid off all employees in May, 2012 and closed operations.

Here is another solar manufacturer would-be, who raised a lot of money and built a factory, but picked the wrong time and place. When they had a marketable product at the end of 2010, it was too late. Spain started sliding into the financial abyss, subsidies were withdrawn, FiT was slashed to the bone, and the market was flooded with cheap Asian PV modules. The situation in the other European countries wasn't any better and Gadir could not compete in the new markets.

GE Solar

General Electric (GE) is one of the world's most advanced (in all respects of the word) companies. They decided to get into the solar business by manufacturing CdTe PV modules, and they bought the technology of the fledgling PrimeStar Solar, which was still in R&D mode. Many solar industry specialists considered this a hasty decision of world-class proportions.

First, CdTe is unproven technology, so it would've taken a lot of time to fully develop, certify and market it. Second, First Solar already dominated the CdTe market, so GE ignored one of the key laws of marketing, "Don't start a new business in a market space dominated by a large competitor." Third, the end of the solar bust-boom cycle was clearly visible, so any new business would be hard-pressed to survive in deteriorating market conditions. Finally, they were still moving and building production plants, when most other thin film companies were shutting down operations, and Asia flooded the market with cheap, more efficient modules.

GE's first CdTe thin-film PV plant, was announced in October 2011, with projected capacity of 400 MW and commercial shipments by 2013, but in the summer of 2012 officials began rehearsing exit strategies. Finally, GE announced a change of plans, delaying for at least 18 months construction of the new Colorado CdTe panel manufacturing plant—now halted indefinitely.

GE, however, is technically and financially strong—diversified and well balanced, so there is no danger of serious repercussions from the soar fiasco.

Global Solar Energy

Global Solar Energy, with production plants in the US and Germany (the leader in flexible solar power, according to their website) produces flexible solar technology based on CIGS cells at "...record efficiency and we continue to set the benchmark for integration of solar into innovative applications. Our products range from portable solar chargers and BOP solutions to military applications and emerging innovations" continues the website.

However, the German branch of Global Solar Energy ceased production of its 35 MWp facility in Berlin in June 2012. Global Solar Energy Deutschland started insolvency proceedings at a German court shortly thereafter, but the German executive team's decision will not influence its Tucson, AZ, production line.

The company issued a statement, "The EU renewables market is financially challenging due to high inventories, collapsing prices and significant reductions to European feed-in-tariffs. As a result of this difficult operating environment, a strategic decision has been made to plan and execute an EU capacity reduction and focus investment on the products and technology necessary to meet our customer's needs and fulfill our business plan.

"Global Solar will continue to honor all of its warranty obligations and service European customers, but from its Tucson, USA facilities. While unfortunate, the Berlin facility shutdown provides an opportunity to address financial structure issues, appropriately scale production capabilities and align with growing markets in Asia, the Middle East and North America.

We continue to provide our customers with industry leading products, superior service and competitive pricing, while also ensuring the long-term success of Global Solar," company representatives said.

We wish best to Global Solar, USA, with the warning that flexible solar panels are still in their infancy.

Helianthos Solar

Helianthos, a subsidiary of the Dutch Nuon Corporation, opened a $100 million pilot plant in Amhem, Netherlands, in June, 2009 to develop highly specialized solar cells by a thin film deposition on foil. The plans included a new, larger production facility to be opened in the near future for mass production of their thin film PV cells and modules.

Helianthos' specialized process consists of an innovative approach using a temporary, reusable, substrate onto which flexible thin-film solar materials are deposited, thus forming solar cells and modules. Using the temporary substrate allows relatively high processing temperatures while using a cost efficient, semi-continuous roll-to-roll production processes and abundantly available PV materials.

The active layers of their PV modules comprised a transparent conductive oxide layer (TCO), an active absorbent layer (e.g. thin-film silicon), and a back contact layer (e.g. a reflective metal layer). Using this process, flexible light-weight, and fairly cheap PV modules can be fabricated.

After $100 million investment and only 11 months later, in May 2010, the mother company, Nuon, announced that it was planning to cut its loses. Seeking a strategic investment partner to help launch their flexible solar cell foil, they made an unsuccessful international search over a period of 18 months.

In September 2011, Nuon announced that Helianthos was to stop production all together, and an online auction would give all interested parties an equal opportunity to take over all assets or parts of the business of Helianthos, according to the management.

Helianthos was sold as an entity to HyET Solar, a new company, established by a group of technological companies, which is planning to introduce the flexible solar film on the market shortly. So we now have re-born Helianthos, but "introduce" and "shortly" might be the wrong words to use here. Flexible PV foil is not a hot commodity at the moment.

HelioVolt

One of the oldest CIGS process developing companies in the US, HelioVolt is based in Austin, Texas. During their 10-year trials and over $150 million in venture funding, HelioVolt has not shipped any commercial product to speak of.

The only thing of consequence we've seen come out of HelioVolt are endless sales PowerPoint presentations heralding the potential terawatt energy economy that HelioVolt's technology represents. Instead of looking for markets for this incredible product, HelioVolt's investors decided they "really needed to find a partner that could bring a lot more capital to bear."

In September, 2011, HelioVolt received $50 million infusion (part of an $85 million round) from South Korean giant, SK Innovations. This marks $200 million in investments spent on CIGS R&D during 11 years of trials, the only tangible results of which is a new record of 13.5% efficient CIGS PV modules in a lab-scale trial.

Inventux

The Inventux case reflects the overall situation in Germany, and is linked to the broader question whether Germany can sustain the development of PV technologies such as silicon-based thin-film modules under the barrage of Asian manufacturers pumping out large quantities of low-cost products.

In early 2012 PV thin-film adopters were failing in large numbers at the same time the Berlin-based Inventux filed for bankruptcy. Although Inventux' production technology was supplied by the then leader of thin film solar industry equipment manufacturing, Oerlikon Solar, and was one of the early adopters of its advanced micromorph silicon turnkey technology, it was not able to make a go of it.

Low efficiency, high production costs (compared to the inexpensive Asian products) forced Inventux to rethink its strategy. In the summer of 2012 a temporary insolvency administrator was appointed by the local court in Berlin-Charlottenburg, while the company looked for new investors.

In August, 2012, Inventux's representatives stated that two companies (one from Chile and one from Argentina) have taken over the German photovoltaic manufacturer. The investors are energy-intensive businesses, which also rely on renewable energy for their own operations.

Konarka

Konarka, with its roll-to-roll thin film PV modules production facility in New Bedford, was the hope of the flexible organic thin film PV industry. Initially, it raised over $190 million to develop and market its long-awaited and very promising roll-to-roll OPV (organic photovoltaic) technology.

Konarka received venture capital from a host of big names, including oil companies, Chevron and Total, VC firms, Draper Fisher Jurvetson and New Enterprise Associates as well as Konica Minolta and Good Energies. It also received $24 million in US DOE grants.

Much effort went into the enterprise, but there were not enough sales, so in the summer of 2012, Konarka was forced into Chapter 7 bankruptcy.

Looking soberly at the OPV technology, one cannot help but notice that it is several times less efficient and as many times more expensive to produce than the competing c-Si products. There are niche markets, but they are so small.

Konarka is not alone in the solar boom-bust frenzy of 2011-2012. Many solar companies and especially those in the BIPV and BAPV markets, with their new and partially developed products, are most vulnerable.

We do wonder about the future of Konarka, and all thin film PV technologies in general. We have exposed some of the technical problems—most of which have not been resolved. This, combined with fixed cost of materials and labor, as well as low efficiencies and toxicity, are issues that need to be addressed and resolved.

Masdar PV

Masdar, Abu Dhabi, built a thin film PV modules manufacturing plant in Ichtershausen, Germany. With $230 million spent for the plant setup, it was supposed to be the jewel in Masdar's crown. Using superior German technology, and building another thin film PV plant later on in Abu Dhabi, was suppose to give Masdar the best solar technology ever, allowing it to put its foot in the proverbial solar door, and dominate the solar market in the area.

Masdar put a lot of hopes on its thin film PV modules manufacturing factory. It was furnished with the best equipment money can buy—the famous Applied Materials SunFab turn-key a-Si manufacturing line; it has a state-of-the-art mass production line, and was making a good product that... was going nowhere. Literally.

The new Masdar PV thin film a-Si PV modules manufacturing plant was in the news at first with the dismissal of its top executives. At the same time, Applied Material, which sold the SunFab line to a number of would-be manufacturers, left the solar field amidst the onslaught of the cheaper, more efficient c-Si competition. In doing so, Applied Materials reduced support to its "SunFab" a-Si thin film technology at customers' plants.

Simultaneously, the 90% drop in poly silicon prices made a-Si technology uncompetitive, so most of the large Applied Material's SunFab production line customers had already written off their investment. Most of them don't need support any more, because they are out of business.

Sure enough, the owner and major customer of the Masdar, Germany thin film production turned away from thin films all together and is going to build its power plants with solar thermal technologies instead.

Masdar, Abu Dhabi, partnered with the Spanish solar firm Abengoa in the fall of 2012, to build a new solar thermal plant called "Shams 1" near Abu Dhabi in the near future using the Spanish CSP technology. The new CSP plant will be the largest in the MENA region, at a cost $700 million and 100 MWp power generation.

Masdar PV is fading, as the Abu Dhabi solar power fields are being installed with a reliable CSP technology. We hope there is enough water in the deserts to keep the CSP technology running. Millions of gallons will be needed. If that project fails, there are more efficient technologies that and do not use water.

MiaSolé

MiaSole is a California PV modules manufacturing company. Its main product line is based on CIGS thin film solar panel technology. Similar to other companies that have raised billions of dollars to create this technology, it is now suffering layoffs, Solyndra being one of them.

MiaSole raised over $500 million during the good times of the solar boom from investors. In the spring of 2012 the company raised $55 million to help it enter new markets and boost its sales staff.

According to a representative, the company looked forward to aligning with a partner and collectively executing a flexible product launch and additional capacity to fulfill the 1GW+ commercial pipeline… But in January, 2012 MiaSolé announced layoffs, and by the summer of 2012 over 200 permanent jobs were terminated.

For now MiaSole will concentrate on cost reductions, to enable an internal restructuring (survival mode) and a strategic partnership (takeover).

The company proudly stands behind its track record of not using any government funding or tax brakes, and we do applaud MiaSole for sparing the taxpayers at this critical time.

Instead, MiaSolé decided to rely exclusively on the potential of its flexible product line, which most likely signaled a switch away from the glass module segment of the business and focusing on the flexible product lines.

In a last ditch effort to survive, MiaSole was sold to Hanergy Holding Group, a China-based renewable power provider, for $30 million, and will operate as a wholly owned subsidiary of Hanergy, China.

It took MiaSole 6 years and $555 million to create a production line that is worth only $30 million on today's crowded energy marketplace. Almost 1 to 20 return on investment…not a good sign of the health of the CIGS PV sector.

The Hanergy deal was the best MiaSole investors could expect, short of closing the doors. It will restart operations with an unspecified number of employees in the U.S., according to the media, but the likelihood of the operations ending up in China is quite probable too.

It is also hard to see where the CIGS PV market is going amidst the new market conditions, so there are a number of obstacles on MiaSole's path.

NovaSolar

NovaSolar, is another a-Si manufacturing company. It is based in Hong Kong, and it's sole shareholder is a Chinese company called Portcullis TrustNet. Nova-Solar developed and specialized in the manufacturing of special multi-junction a-Si PV modules.

NovaSolar' thin-film silicon solar modules were made by a thin film deposition on a single sheet of glass. A number of layers of silicon, aluminum or zinc, comprising the active layers, while the back of the panels were covered with a polymer back-coat, which would make them the cheapest thin film panel ever.

NovaSolar was created in April 2009 through an asset and management acquisition of OptiSolar equipment and IP. The new company's headquarters were set up in Hong Kong in early 2010. That year, NovaSolar also acquired a manufacturing facility in Yangzhou, China.

Advertising its amorphous silicon thin film technology as the lowest cost per kilowatt-hour, NovaSolar was negotiating 1,000 megawatt solar farm contracts, and planning to employ 300 people worldwide in 2011-2012.

NovaSolar became involved in construction of fancy buildings—a Fremont R&D facility and a 500,000-square-foot manufacturing site in Yangzhou, China, but both have been halted. The company filed for Chapter 11 bankruptcy, owing $3.5 million in unpaid wages, claiming $6 million in assets and $14 million in liabilities. Its present status is unclear.

Odersun

Odersun, operating in Berlin and Frankfurt, Germany, was a thin-film PV modules manufacturer, using CIGS thin films deposited on copper sheets, or tape, which can be used to generate DC power when exposed to sunlight. The flexible sheets or tape could be cut to any size or length, allowing architects and designers to custom-fit the "modules" in buildings and other applications.

While under development since the mid-1990s, Odersun did not commercialize its technology until 2007, when it opened its first 5 MWp production facility in Germany. With government financing and subsidies of $17.5 million, and the help of a number of other investors, it later opened a second $63 million facility in the German state of Brandenburg.

But the PV sector was already on the cusp of its current crisis of oversupply, so the timing could not have been worse. Like its competitor, Inventux, Odersun had interesting, high-potential technology, but with the waning demand for PV in European markets, these products cannot find markets.

In March, 2012, Odersun filed for bankruptcy. The firm was able to protect the wages and salaries of its

260 employees from the insolvency payments until late May, but could not do it forever. Since the negotiations with foreign investors fell through, the company went into limbo. An attempted joint venture in Beijing, China, was unsuccessful, and Odersun closed down its doors in June 2012.

Oerlikon

Oerlikon Solar, subsidiary of the Swiss Oerlikon Group, was headquartered in Trubbach, Switzerland, employing 675 people in 8 locations worldwide. It offered turn-key production equipment and turnkey manufacturing lines for the mass production of thin film silicon solar modules, primarily tandem-junction a-Si type. It had 870 MWp of contracted capacity and 15 customers in full production in seven countries, and more than 5 million modules produced through its equipment.

It was competing with the US giant equipment maker, Applied Materials, and did well until the solar industry hit the wall. It was sold to Japan's leading semiconductor equipment supplier, Tokyo Electron (TEL) for $275 million.

Oerlikon Solar's major competitor in the a-Si thin-film market had been Applied Materials, which pulled out of the turnkey solar production equipment market several years earlier. As polysilicon prices declined and capacity expansions passed the gigawatt scale, c-Si technologies became highly competitive again. This, and the market dominance of First Solar using CdTe thin-film technology forced many potential new entrants to defer market entry.

Moreover, many of the existing thin film turnkey production line users went bankrupt, while others curtailed further capacity expansions. This significantly limited revenue streams for equipment suppliers, and caused embarrassment and chaos among them—which is what caused Applied Materials to withdraw from the solar equipment market, and contributed to Oerlikon's failure.

Now TEL/Oerlikon is one of the last standing a-Si equipment manufacturers, attempting to resurrect the equipment sales in a thin film PV market that is shrinking quickly. TEL's ingenuity will determine the success or failure of the new enterprise.

Optisolar

OptiSolar was "the great hope" of California and the US solar industry at the beginning of the latest cycle. Unfortunately, they were one of the first in the chain of badly failed solar companies and projects.

Working ambitiously and feverishly on more than 1 GW of solar projects during 2008-2009, OptiSolar wanted to be a one-stop shop—solar panel manufacturer, installer, and project developer, but could do neither well, efficiently and cheaply enough.

OptiSolar seemed destined for success until the very end. It had raised $322 million to complete a vision of futuristic solar manufacturing conceived some time ago. According to the managers, the cost of solar could be radically dropped by building "solar city factory" (city in the factory) complexes, capable of churning out 2.1 gigawatts to 3.6 gigawatts of solar cells a year.

These factories would cost $500 million to $600 million each, and would be composed of factories-within-factories units, focused on different tasks, such as an onsite glass making outfit capable of cranking out 30 million square meters of glass a year, a solar cell unit with 100 identical manufacturing lines, a full-fledged packaging facility, and a recycling facility.

Manufacturing know-how would be imported from the television and computer industries, and even from household construction. Instead of using traditional frames, aluminum frames from regular windows and backing the panels with inexpensive silicon or plastic, would be used.

Modules would cost 0.60 cents to 0.52 cents per watt, and fully installed solar power would cost $1.00 to 0.88 cents a watt.

These ideas—never tried before—became the company's business plan, and it is what the investors paid for. The first factory was built in California, and was supposed to be capable of producing 30-50 megawatts. It landed $20 million in tax breaks in 2007.

However, instead of concentrating on manufacturing solar panels, the company cut costs by installing them itself, and selling the power to a utility. This was to be done through a subsidiary, thus creating a vertically integrated company performing several key functions.

OptiSolar won contracts to install over 200 megawatts in Ontario, Canada, a deal to build a 550-megawatt solar farm near San Luis Obispo for PG&E.

When the crisis hit in 2008, the company was outfitting the new factory and needed another $200 million in funding, which never came.

Ultimately, the size and scope of the entire program was simply too great for a new enterprise. Negotiating land prices, permitting and environmental issues, while planning factories, buying expensive equipment, and hiring engineers is unmanageable and unsustainable.

Another problem for OptiSolar was timing. OptiSolar was planning to sell solar panels to its projects, but these would take years to put in operation—years without revenue.

So, the rise was fast, and the fall was even faster. A few weeks after the factory opening, the company cut 300 (nearly half) of its employees.

Finally OptiSolar shrunk into a smaller company, NovaSolar, which was sold to a Chinese company and managed by the same people. It also filed for bankruptcy within a year or so, in a very similar fashion to the OptiSolar's demise.

Pramac

Another a-Si thin-film PV modules manufacturing firm, the Swiss Pramac, filed for insolvency when shareholders rejected management proposals over restructuring the company after posting net losses of over €94 million in 2011.

The diversified firm had been a customer of Oerlikon Solar, the defunct thin film solar equipment manufacturer. Pramac had a 30MW end-to-end turnkey line using Oerlikon's 'Micromorph' technology. The line was operational, and significant sales were reported.

Pramac had shipped more than 40 megawatts of a-Si solar panels built with equipment from Oerlikon, prior to filing for insolvency in May, 2012.

Other Oerlikon Solar customers, Inventux and Auria Solar also filed for bankruptcy in May, 2012. Also in May, Taiwan-based Green Energy Technology (GET) said it would allocate staff and other resources away from its a-Si thin-film operations. GET had been a customer of Applied Materials 'SunFab' turnkey thin film technology—a competitor of Oerlicon, who exited the solar market in 2009.

Obviously the first wave of turnkey thin film technology adopters from both Oerlikon Solar and Applied Materials were rapidly diminishing either by attrition or by merger and acquisition.

What do these events say about the health of the world thin film PV industry? Is the technology viable? Or are the companies unsuitable for this type of business? We believe that a combination of factors were in play, including lack of sound business development procedures, absence of risk management and/or viable exit strategy.

Q-Cells

Until recently Q-Cells was Germany's (and the world's) largest solar products manufacturer. Unfortunately, it lost hundreds of millions of dollars just during the last 2 years, due to the many effects of the latest cycle. So Q-Cells filed for insolvency proceedings in April, 2012.

Q-Cells had some positive results on the CIGS front with its Solibro thin-film CIGS unit for awhile, and shipped 66 megawatts of CIGS solar modules in 2011, which at that time was more than any other CIGS manufacturer, except Solar Frontier.

The company had a relatively long history and a "fairly mature" co-evaporation technology, so it was likely that the CIGS unit might be eligible for a decent acquisition.

The Q-Cells Malaysia plant's future also looked shaky. While the Germany plant was to be almost certainly closed, the Malaysian facility with its newer equipment and processes was considered as in a better positioned to compete with its much better cost structure. Industry specialists have been warning for awhile about the serious difficulties ahead for the high-cost European PV cell and module makers.

In August 2012, the German based Q Cells SE was acquired by the Korean Hanwha Group. Q Cells, North America is a quasi-independent, solvent unit, so there is a good chance that it will weather the storm.

Scheuten Solar

Scheuten Solar is another German manufacturer of glass-glass, thin film (CIGS) photovoltaic modules and building integrated photovoltaic (BIPV) solutions. It also filed for insolvency in early 2012, after it was unable to maintain its business activities in the face of high inventories and tight profit margins.

Three months later, the company announced that Guangdong Aiko Solar Energy Technology Co., Ltd (Aikosolar), which also owns Chinese project developer, Powerway Renewable Energy, has taken over the "essential components" of Scheuten Solar.

The two companies are said to complement each other, due to Scheuten's "robust" brand and "innovative" product portfolio, Aikosolar's financial strength and Powerway's EPC experience. "The combination sees Scheuten Solar engaged in a synergy with its natural ideal partner, giving rise to a powerful solar solution provider for distribution as well as project solutions," stated Perry Verberne, CCO of Scheuten Solar.

Powerway revealed that the new partnership intends to develop both ground-mounted and rooftop projects in Europe. Furthermore, Scheuten Solar will aim to sell around 200 megawatts of modules annually, "which will bring …[it] back in the position as a top brand in the European PV market."

This is confusing, considering the current state of the market. Who will buy those modules?

Schuco International

Schuco, another German solar technology leader, and one of the largest thin film solar products manufacturers and providers of efficient energy solutions in Europe, experienced heavy operational losses due to weak demand in 2012. They were forced to close both production facilities and their R&D lab, losing about 270 jobs in Germany.

This loss was another heavy blow to the flagging amorphous silicon (a-Si) thin-film sector, and further underscores the challenges facing German PV manufacturers.

Schüco was offering PV and solar-thermal products in the late 1990s as part of its broader sustainable construction materials business. Later, it focused on thin film PV as winning technology, and in 2009 opened its 40MW "Malibu" PV module factory near Magdeburg, Saxony-Anhalt in conjunction with E.ON.

The production lines were setup with thin film production equipment supplied by the large semiconductor and solar production equipment manufacturer Applied Materials, in Santa Clara, CA.

In 2010 Schuco acquired a factory owned by the bankrupt a-Si producer Sunfilm, in Grossröhrsdorf, near Dresden. The Grossröhrsdorf, Dresden, facility was permanently closed, however, at the end August, 2012, and the plant in Osterweddingen, Magdeburg, was closed in September. The R&D centre in Bielefeld, will be shut down in 2013.

Sales at Schuco's New Energy division peaked at 1.0 billion, but declined significantly, to €850 million in 2011. Sales in the first-half of 2012 declined another "double-digit," and the company tried restructuring, but as unsuccessful, and shut down its plants and labs.

The exit of Schüco from the thin-film PV market is another blow to the commercial viability of a-Si thin-film technology. It also gives a black eye to the thin film equipment suppliers, Oerlikon Solar, and Applied Materials. Although both companies are already out of the business, they cannot remove themselves completely from the failures, because their equipment was used, so we must take this into account when we talk about the recent heavy losses and failures of thin film PV manufacturers.

According to Oerlikon Solar, Schüco had only been their customer during the very early R&D evaluations of a-Si technology, and had purchased individual tools for R&D, but not production equipment or complete turnkey lines from them. Later on, with the establishment of the Malibu line, Schüco opted to choose Applied Materials 'SunFab' technology at both its production sites.

The unfortunate exit of so many thin film PV products manufacturers from the solar energy field is more obvious than the exodus from the other sectors, because there was so much money thrown at it at the time. It was newer, more flexible, much cheaper, and basically hailed as the great hope of energy advancement.

Its adaptability to mass production and potential lower cost were advertised as the "new era" in solar power products manufacturing—the solution to all our energy problems—the shortest route to energy independence. Now it all seems like hype gone astray—bunch of technologically impossible dreams that never materialized.

Sharp Solar

Sharp Solar is one of the oldest solar manufacturers in the world, reaching back to the 1960s, with the commercial development of one of the world's first solar cells. Sharp Solar was also one of the world's major solar panel manufacturers, with plants in Japan, the US and the UK, where mono-, poly- and thin film cells and PV modules were made.

Still, in August 2012 Sharp posted a $58 million loss from its solar division Sharp Solar, and their Katsuragi, Japan, thin-film plant was scheduled to ramp down. Then, there was talk about shutting down the production plants in the US and UK too. Total losses were $1.76 billion.

PV module prices continue to decline as the competition intensified on weak demand, but sales in the residential markets remained strong, so management is looking closely at the Japanese market, planning to increase module sales with emphasis on the small commercial and residential markets.

Sharp insists that it will remain a key player also in PV power plant projects within Japan, as well as continuing the development of solar energy peripheral systems such as HEMS, storage batteries and power conditioners.

Significant losses in Sharp's core LCD business forced the company management to undertake major restructuring. For example, the Solar Systems Group merged with the Health and Environmental Systems Group, which was renamed the Health, Environment and Energy Solutions Group—a new name for the old businesses at a cheaper price.

The management of the Solar System Group, which currently operates out of its Katsuragi solar plant, were relocated to the advanced LCD and solar production site in Sakai, Osaka. The production in the 160 MW a-Si thin-film-based solar plant in Katsuragi, Japan, which started production in 2005, together with two other advanced plants in Japan and Sicily, was scaled down significantly as well.

Sharp is implementing overall corporate workforce restructuring and reduction as well; approximately 5,000 of their 57,000 employees, will lose their jobs—most of them indefinitely.

Signet Solar

Signet Solar was founded in 2006 in Menlo, CA. They specialized in the production of a-Si thin film PV modules, from manufacturing plants in India and Germany. The Gen 8.5 modules were made by depositing the active thin films on large area (61 ft²) glass substrates. This is one big piece of glass—10 x 6 feet in size.

The Signet Solar, India, production plant near the city of Chennai was started in 2007, with planned capacity of 300 MWp/year.

Signet Solar', Germany production plant in Mochau, began production in 2008 with planned capacity of 120 MWp/year. These were the good days of the recent solar boom.

In 2008, Signet Solar announced that it would build another factory in Belen, NM, to provide solar panels for a 600 MWp solar plant, providing the majority of the power for local communities.

In April of 2010 the company cancelled plans to construct the $840 million facility after failing to get a US DOE loan for the undertaking. There will be no New Mexico plant.

In the spring of 2011, the company declared its German production plant insolvent too, the operation was shut down, and in June 2012 a petition for liquidation under Chapter 7 was filed.

Signet Solar was also a faithful customer of the acclaimed Applied Materials SunFab production equipment, and as others, it fell victim to the drive to hastily enter the market with slick production equipment from the famous equipment maker. All they needed to do was find the customers, and that was the problem. Most of the customers preferred the cheaper and more efficient c-Si PV modules.

Lesson learned: Good production equipment alone is not enough to ensure successful product sales—regardless of the maker of the equipment or the end product. Thorough planning and execution, detailed risk management and plausible exit strategy must be central parts of any business plan.

Solibro

Solibro was another prominent CIGS PV modules manufacturer, subsidiary of Q-Cells, with humble beginnings as a Swedish solar R&D firm. It was acquired by Q-Cells in 2006 and helped Q-Cells with developing and manufacturing thin film PV products.

Solibro shipped 66 MWp of CIGS solar modules in 2011, more than any other CIGS manufacturer except Solar Frontier, using its fairly mature co-evaporation PV manufacturing technology. The manufacturing cost of the CIGS thin film modules in 2011 was approximately $1.00 per watt at commercial efficiency of 12.2%. That was good, but not good enough. Too, the PV modules had no track record of long-term reliability, which became the norm about that time.

In 2011 Solibro, together with Q-Cells, started seeking help from outside parties. Q-Cells was sold to the Korean Hanwha Group, while Hanergy, a privately held Chinese power generator with gigawatts of hydropower assets, acquired the modestly successful Solibro thin film unit. Hanergy also has wind and solar interests and has spoken of investing $15 billion into the solar industry, so Solibro might be in good hands.

The results: Moving in the European energy markets, as of the fall of 2012, Chinese interests have bought a number of significant German and other solar companies—mostly manufacturers of different solar products.

Soltecture

Soltecture was another Berlin, Germany, based CIGS photovoltaic modules manufacturer working on achieving a commercial efficiency of its PV modules of over 16%. Its high-performance 100 Wp Lionion module with a respectable aperture efficiency of 13.4% percent and a module efficiency of 12.3% was the company's flagship, and its best hopes for success in the energy markets.

Soltecture also reached an aperture efficiency of 13.9% at module efficiency of 12.8% on a 103.9 Wp CIGS PV module, which is another sizeable achievement—representing over 0.5% module output increase in just one year's development work. These successes contributed to attracting over $25 million in investments in early 2011 and building a 35 MWp production line, which made commercially viable product, for awhile.

The collapsing PV markets, and the deteriorating political and regulatory changes forced the company to look into alternatives—one of which was a worldwide search for an investor, a buyer or a partner, and/or do whatever else is needed to endure the hard times.

The company made news in 2011 with a TÜV safety certification and announced the production of its first commercial 100 MW thin-film modules, later partnering with altPOWER for future US-based BIPV projects.

But going quickly through the cash reserves, executive management decided to consider insolvency proceedings which were filed in May, 2012. Provisional liquidators started liquidation procedures, still hoping to find an investor, or a partner.

Assuming that the relatively high efficiency of Soltecture's CIGS modules will attract a new investor, and that lower-than-the-competition efficiency is somehow justified, all these thin film PV modules need is a track record. Until they are proven, they must stay in the R&D labs, or be sold for smaller projects in cooler and shady areas.

Solyndra

Solyndra is the largest solar failure ever. It became the symbol of what failure in the solar industry looks like.

Solyndra's biggest problem was its tubular thin film technology which had a number of serious flaws.

1. The thin films in the glass tubes were close to the bottom of the competitors' efficiencies table.

2. The tubes were spaced so that only 50% of the module's top was struck by sunlight—50% more inefficient in most cases.

3. The bottom part of the tubes was not accessible, would not allow proper cleaning, would get dirty, reduce total output, and eventually stop working.

4. The glass tubes and their fused glass/metal connections (weldment) were quite fragile. Shifting of the frame during transport, installation and operation caused them to stress and crack, thus losing integrity (vacuum) and allowing moisture and the elements to penetrate the tube.

5. The modules could not be used in general purpose and large-scale power installations because they required white background to reflect some of the sunlight as needed to activate the bottom of the modules. This alone limited the market to a very narrow niche, contributing to slow sales and an unsustainable situation.

6. The glass encapsulated thin films tubes required specially designed, custom made, and expensive to purchase, modify and operate processing equipment, engineering and labor.

7. The materials costs were very high as well. The product was more expensive than the competition.

Solyndra got $535 million for its experiment with this spanking new, untested and unproven technology, which is unprecedented in the solar energy market. They used part of the money to build a new production plant, though they had a good plant at the ready. Moving into a new facility took a chunk of the precious little time they had before the doors of opportunity closed, so while they was busy with construction, the markets collapsed.

This was a common problem with the failed solar companies of late. The other problem was rushing to market. And as in so many cases, a number of essential business principles were bypassed only because there was so much money available. The money had to be spent, but the product wasn't competitive.

In November, 2011, the Labor Department approved Trade Adjustment Assistance (TAA) for Solyndra's former employees. All 1,100 ex-employees got federal aid packages, including job retraining and income assistance, valued at $13,000 a head, which means that additional $14.3 million will be spent on this failed venture.

Solyndra's court-ordered assets auction conducted in the summer of 2012 generated sales of $3.81 million, or less than 1% of the federal financing. The final tally, as of the summer of 2012 is: $535.0 in direct federal grants, plus $14.3 million in TAA Federal Aid, minus $3.81 from the equipment auction, totaling $545.49 million in losses.

SunFilm

Another German company, SunFilm was a thin film PV modules manufacturing start-up, once considered the pillar of Q-Cells' push into thin film technologies. Falling victim to changing FiTs and low priced Chinese imports, however, it filed for insolvency in the spring of 2010. 300 workers at its production lines in Grossroehrsdorf and Thalheim were laid off permanently, and the factories closed indefinitely.

SunFilm officials claim the company's plans have been crippled by Germany's sharply reduced solar FiT, as well as by the emerging dominance of US-based cadmium telluride thin-film maker First Solar and the influx of cheap Asian-made PV modules in Europe.

During the past several years the thin-film industry has lost its competitive edge to crystalline silicon modules, which have higher efficiencies, and have seen huge price reductions due to the financial crisis and Chinese government subsidies.

Until its demise, SunFilm manufactured a-Si thin film modules, and was briefly in the headlines after merging with Sontor. This partnership created one of the world's largest thin-film manufacturers at the time, with a combined production capacity of over 145 MW. They had the capacity to manufacture a lot of a-Si PV modules, but sales in their last year or so are uncertain.

By filing for insolvency, SunFilm was hoping for a strategic realignment of the company. Their decision came at the same time the US-based solar equipment manufacturer Applied Materials caved in. Sunfilm used the famous Applied Materials SunFab technology at its production line in Grossroehrsdorf. The exit of both companies from the solar market signals a shift in direction, or at the very least a major realignment of the forces driving it.

Author's note: It appears that during the solar boom stages Applied Materials was able to sell thin film manufacturing equipment (mostly for a-Si production) to at least a dozen companies—most of which are German. Most are now out of business.

Once Applied Materials put SunFab on the market, many companies were attracted by the reputable manufacturer's offer and wanted to believe that the equipment was the key to success.

The problems were: a) They did not consider the different stages a product must go through before achieving market acceptance, and b) They did not have a viable risk management and plausible exit strategy in place.

Trony Solar

Trony Solar is a China based a-Si thin film PV modules manufacturer, with more than 100 patents in solar production and application technologies. It raised $223 million in 2010 to expand its production capacity.

In 2009 Trony claimed to have 115 MW production capacity, which would be expanded to 145 MW, but then they were talking about expanding a 35 MW facility, so it is not clear what their production capacity was. In any case, the company's products (a-Si PV modules) were sold to the off-grid market as well as in the BIPV market—all of which have shrunk significantly recently.

At the end of 2009 Trony had annual sales of $24.34 million, but posted losses for the second-half of 2011, and showed 33% revenue reduction in 2012. In early 2012, Trony Solar warned investors that due to further reductions in demand and continued price erosion, due to increasing competition, it expects further losses.

Trony started restructuring to reduce costs and

minimize the impact of market conditions. At the same time, it was forced to stop trading its shares on the Hong Kong stock exchange (which it joined only little more than a year earlier) due to accounting anomalies.

Even if the company survives, with its reputation it would be hard for it to become bankable again. Being a thin film manufacturer today is not a ticket for an exit from the mess.

Unisolar

Unisolar (aka United Solar Ovonic LLC) is/was a subsidiary of Energy Conversion Devices, and manufactures flexible thin film PV modules from two production plants in Auburn Hills and Greenville, MI. Another ECD company, Solar Integrated, provides the expertise to install Unisolar products, primarily on roofs. The thin film modules were used instead of roof shingles—a clever and successful business venture—for awhile.

Unisolar has benefitted from millions of dollars in state economic development tax breaks and Community Development Block Grant funds, but did not receive a Department of Energy-backed loan as was expected and needed to expand operations.

Unisolar's motto was, "Low-impact solar roof solutions, and lowest cost energy provider." It was not successful. Unisolar filed for Chapter 11 in February 2012, but in May the company discontinued the court-approved assets sale process after not being able to find an "acceptable, qualified bid" by the court deadline.

The company then planned to sell off the assets, and laid off 300 employees in the process. At the same time about 300 workers were on furlough in Greenville.

Then Detroit bankruptcy court determined earlier that even if the auction were successful, it wasn't likely to bring in enough proceeds to pay off the company's $249 million debt and pay creditors and shareholders.

After a single offer of $2.5 million, the buyer Salamon was interested in buying the assets primarily as a "tax write-off" and no indication of future intentions to operate the company as a business.

Amazingly, DOE and The Administration pronounced Uni-Solar a winner, touted for its potential job creation and its new and innovative business model. So what went wrong?

ECD manufactured nickel-metal-hydride batteries for hybrid and electric vehicles when it got into solar through Uni-Solar. It very quickly became a leader in the global solar industry—which was dormant at the time—by making and selling its low efficiency thin film modules for industrial and commercial applications.

It even expanded its operations by adding four

new Michigan-based plants. A clear example, again, of rushing to market. Solyndra and many others were expanding their operations when they hit that bump in the road and never recovered. Over 500 employees were laid off.

Unisolar's partner, the ECD company Solar Integrated, also filed for Chapter 7 protection in the spring of 2012. Twenty-four solar installations in San Diego county, installed by Solar Integrated, using Unisolar's UNI-SOLAR roof solar panels were found defective and removed after 6 years on-sun operation.

Excessive corrosion in many of the solar panels and their connectors made them unsafe for further operation, so the maintenance company refused to continue supporting those installations.

Unisolar never tested their thin film product for long-term, on-sun reliability, and instead rushed to market with disastrous results.

This is a recurring problem with thin film companies, as we can clearly see in this text. The only question now is how many more will follow Unisolar's destiny?

VHF Technologies

VHF-Technologies SA, is a Swiss company also known under the brand name "Flexcell." It was founded in 2000 with the aim of creating an industrial process for a technique for applying amorphous silicon by very high frequency (VHF) plasma deposition, developed at the Institute of Microtechnology (IMT) of the University of Neuchâtel.

This unique technology makes it possible to use a thin, flexible plastic substrate instead of glass. Thus the final product is flexible and can be used in a number of applications, where conventional PV modules cannot be used. In comparison with the production of traditional crystalline modules, the technology requires a very small amount (micron thin silicon film), or 300 times less semiconductor raw materials, resulting in high production volumes and low energy consumption.

Flexcell received a $9 million loan from the now insolvent Q-Cells in 2006, and in 2007 it became part of Q-Cells AG group, which manufactured highly efficient solar cells made of mono- and multicrystalline silicon, but which had its own problems, resulting in bankruptcy and outright sale to a French company in 2012.

Flexcell opened its 25 MW roll-to-roll plant in Yverdon-les-Bains, Switzerland, in 2008. The company focused on the flexible substrate format for BIPV applications. Within the space of a few years, it developed into a high-tech photovoltaics company.

When the hard times hit, the company underwent several years of near collapse, propped up from time to time by owners and other well-wishers. The last handout came in 2012 when Capricorn Capital provided approximately $9 million to keep the company alive, and its 50 employees working until another solution is found.

In September 2012, VHF Technologies announced that it has not been able to find an investor or a buyer, and put its assets for sale, hoping to get over $20 million, to offset its debt.

This thin film company's predicament (as with most other new thin film companies) is not unique, and is much deeper and long lasting than its cash flow problems.

Low efficiency, uncertain long-term reliability, and small-scale production, make the company non-bankable. It stands very little chance of survival in the current solar market environment.

Willard & Kelsey Solar Group

Willard & Kelsey was a CdTe thin film PV modules manufacturer, aiming to be the lowest-cost producer of solar panels in the world, which were manufactured in the company's production plant in Perrysburg, OH. They received several million dollars from Ohio state agencies, despite showing only $500,000 in revenue in 2009 (which actually was also a grant from the state), and a whopping loss of $4.2 million at the same time.

The company received $6 million from the U.S. DOE's "green energy" loan program, shortly thereafter filed for bankruptcy and laid off almost all of its 80 employees.

Assembly-line changes needed to improve the efficiency of the CdTe thin-film solar panels were cited as the cause of the layoffs. Those layoffs were the latest development in a series of delays and financial issues that have created questions about Willard & Kelsey and the government money it has received since moving into new headquarters in 2008.

State officials identified money troubles at the company as early as 2009, yet approved a second $5 million loan for Willard & Kelsey. The state also agreed to defer loan repayments and delay deadlines for company financial reports.

In August, 2012, however, the Ohio Department of Development called a $5 million loan to Willard & Kelsey due because of partial or delinquent payments. If it fails to pay, the money will be collected by the Office of the Ohio Attorney General through a collection process.

This is another example of the effects of free money given to companies that are not yet stable. This is not isolated case, but rather an almost normal behavior, running all through the last solar period.

CONCLUSIONS

This is a good time to review, again, the key lessons we learned during the 2007-2012 solar boom-bust cycle:

- Government subsidies work, but are not a long-term solution. Private enterprise is!
- Government should support development, but should not choose and support winners.
- Political favoritism, in most cases, causes more harm than good.
- Large companies and projects can fail too.
- Quick money-making schemes cause failures, and give the industry a bad name.
- New technologies must be tested before going into mass production and/or being marketed.
- Solar installations in cloudy areas have marginal success, regardless of anything else.
- Large-scale desert solar installations are the fastest way to energy independence.
- Using safe solar technologies in large-scale desert installations is the right thing to do.
- Efficient energy storage is needed to make solar power compatible.
- All solar technologies do not perform equally.
- Using cheap materials and labor do not lead to long-term success.
- Sound business plan, complete with risk assessment and exit strategy are absolutely needed in any successful business enterprise that does not rely on subsidies and hand-outs.

The fact that so many thin film companies went out of business after such a short period of time, and so much money spent, means that there is a problem with the thin film technology and the entire industry. It is a diversified and very complex technology with many different facets. There are many ways to manufacture and use it, and many ways to succeed and fail.

Note: Although Asian module manufacturers' anti-competitive practices had a profound effect on the solar markets, other key reasons for crystalline silicon module price declines were: a) the rapid decrease in polysilicon prices, and b) the ability of some manufacturers to scale production facilities to the gigawatt level, while continuing to improve conversion efficiencies faster than a-Si thin film producers.

These dynamics do not exist in the thin film industry, and will take a long time to develop in the sector.

Since there is no simple, single answer to why thin film companies and their technologies failed recently, we will try to clarify the technical aspects, their manufacturing and use processes and the related issues.

THIN FILM PROCESS AND DEVICES ISSUES

Looking at the above list of failed companies, we cannot help but notice the disproportionately large number of thin film PV manufacturers that failed during this last cycle. Over 30 companies (in our list alone)—most of them new entrants in the thin film solar energy field (some with only 2-3 years existence) are gone, while some have changed hands, and someone else is now struggling with them now. We believe this is not the end of the story. A number of thin film companies are struggling to stay alive, and we just don't know how many of them will survive the ongoing bust cycle. How can we explain all this? What is so special about, or wrong with, thin films PV technologies?

Let's s take a closer look at thin films PV products and processes, since they have a lot to do with what happened.

Thin films have great advantages, but they also have even greater disadvantages. This new and immature technology was in the development labs until very recently, but was yanked out and forced to operate in the field—mostly lured by the free money of subsidies. But it had no experience and no track record of reliable long-term field operation, so when thin film PV manufacturers cut development efforts short and started mass producing, they drew a bad card.

Note how many of the thin film companies failed during expansion, moving or building new facilities.

This is the greatest and most discouraging news of the latest solar cycle. The most elegant, flexible, and promising of our solar technologies, thin film PV, now has a black eye. It will take a long, long time to heal.

This great number of failures should not be left without analysis, so we'll look closely at thin films PV structures and devices—what makes them tick, and what makes them fail. Because, make no mistake about it, the excessive failures of thin film PV manufacturers were partially due to their properties. Their low efficiency, poor reliability, and higher cost of manufacturing were some of the reasons.

And we also foresee a great number of new and serious failures in the power fields, where thin film PV modules were installed, or will be installed. The lack of proven track records means that the problems have not been solved, which will lead to serious field failures.

No manufacturing process is perfect, and a defect could be generated at any step, including the tightly controlled thin film deposition processes in the semiconductor fabs. Starting with the basics, materials selection is paramount for obtaining quality product. "Garbage in, garbage out." So the quality of the substrates, process chemicals and gasses, and all consumables must meet or exceed specifications—not a simple task.

The different pieces of process equipment must be well taken care of, tuned and qualified periodically, to make sure they are in good shape and operating within spec. The labor force must be thoroughly trained.

The thin film PV modules manufacturing process is a complex undertaking, and deserves a closer look. Most thin film technologies use one or more metals in their active layers. So these rare, exotic, and sometimes toxic materials are usually dug out of underground mines. They are then transported to refining facilities, where they are crushed, dissolved in deadly chemical concoctions, and then boiled and baked until finally pure metals are obtained.

The metals are shaped in "targets," which are used for sputtering of thin films, or shaped in different shapes for thermal evaporation. Sometimes they are converted into special gasses by reacting them with other chemicals. The gasses are used in CVD processes, as a deposition source to form solid metal films onto the substrates.

The other major components of the PV modules, the glass panes, are also obtained in a similar fashion. Mountains of sand are dug out of the ground, sifted, melted, refined, and formed into the desired shape. A number of plastic components, produced in large chemical factories, are used for encapsulating the thin films and as an edge-seal.

Some metal and electronic components are also used peripherally for clips, blocking diodes, junction boxes, wiring etc. These components are usually made by third-party vendors, using some of the above described methods.

These materials are then delivered to the thin film production plant, and the actual PV modules manufacturing process starts:

1. Step one of the thin film process deposition (active layer formation) is visual inspection and cleaning of the substrate materials. Contaminated cleaning materials or equipment or improper procedures will render the glass surface unsuitable for proper adhesion of deposited films.

2. The next step, number of steps, is the actual thin film deposition. This process is different at different facilities, but it usually consists of placing the top glass pane on a conveyor belt of sorts, and running it through a number of chambers. Each chamber uses different deposition material and/or methods, as needed to deposit the active PV thin films; i.e., CdTe PV modules are coated with CdTe, CdS and other thin films, while CIGS get copper, indium, gallium and other thin films.

3. Thus coated glass substrate, with the active thin films deposited on it, is then taken for additional processing and final assembly. The top glass can be treated with some chemicals and then covered with a plastic resin, which serves as a protective film for the fragile thin films on the glass. The resin also plays the role of a "glue" which keeps the back cover (another glass pane or metal) fixed steadily onto the front glass.

Generally speaking, there are a number of known, unknown, controlled and uncontrolled factors and variables in most thin film PV modules manufacturing process sequences:

a. Human error is the #1 variable in high-volume thin film manufacturing. People tend to cut corners, improvise, push the wrong buttons, rush through operation, maintenance and inspection steps, etc. These kinds of workplace behaviors cause defects of unpredictable proportions and consequences. Handling substrate materials and consumables, as well as operating the complex process equipment requires highly trained engineers, technicians and operators, who in many cases have bad habits that are hard to break, even with the best of training.

One good thing about thin film PV modules processing is that the majority of the processing (done at reputable companies) is automated. Automation is the best way to reduce human error. Even then, human error comes into play: dirty gloves or wrong process sequences could reduce an otherwise good batch of modules to junk.

Worse, they can introduce latent problems that will not be discovered until the product is installed and operational. We have seen many examples of this anomaly.

b. Equipment quality and malfunctions are #2 on the list of process-related variables in sophisticated thin film manufacturing operations. Thin film process equipment is complex in its design. It requires precise operation and immaculate maintenance. Most production equipment made today is of good quality, but even the best equipment can malfunction.

Key components and process control instrumentation slowly drift out of spec over time, and product quality could vary from batch to batch. One serious drift in the multi-step process could cause serious defects in a batch. Anomalies during processing also occur unexpectedly and quickly, and must be handled promptly.

c. Poor quality, or out-of-spec, supplies, materials and chemicals purchased from third-party manufacturers are another serious problem encountered in the thin film process; i.e., poor quality substrates, contaminated glass panes and chemicals (overlooked during incoming quality control procedures) could cause slight or great defects in the finished product.

Since high volume operations use a lot of materials from third-party vendors, the incoming quality is hard to verify 100% of the time, so any problem at the third-party vendor's plant will have negative effects on final PV product quality.

d. Substrate surface problems are most often associated with thin film malfunction and failures.

Cleaning the top glass substrate with poorly selected chemicals, or shoddy procedures, might introduce fatal impurities, such as Cu, K, Na, Fe metal ions, and phosphates. These impurities would eventually start their own demolition processes from within the thin film structure enclosed in the PV module, reducing its efficiency and longevity.

Dirt particles, fingerprints, residual moisture, and/or strong electrostatic charge on the glass surface entering the vacuum chamber will have a profound, usually negative, effect on process integrity and final product quality. These are only a few of the things related to surface contamination to watch for at all stages of the manufacturing process.

e. Most often, however, it is the process itself that creates problems in thin film manufacturing. The plasma deposition process could introduce serious impurities and defects in the film structures if improper process parameters, or out-of-spec equipment pieces, are used.

Impurities in carrier gasses, contaminated vacuum chamber walls, back-streaming vacuum pumps, unplanned drops or increases of total or partial pressure, air leaks, process time or conveyor belt speed discrepancies, are all process abnormalities which could introduce other impurities or create defects with immediate or latent problems.

A number of additional factors and variables affect the quality of thin film PV modules as well, but it suffices to say that the TFPV manufacturing process consists of dozens of different materials and complex process steps, each of which must be immaculately designed, planned, executed and controlled. Basically, the lowest quality of any process step (or material) in the process sequence determines the highest quality of the final product.

One mishandled step by a careless operator, or one bottle of contaminated chemical or gas, could lead to rejection of a large batch of modules, or worse—cause their failure in the field, where things get very expensive...and embarrassing. And we have a number of thin film PV (TFPV) functional considerations.

Even though the electro-mechanical and chemical properties of PV thin films are thoroughly studied and well understood, improper design and manufacturing procedures are still encountered at times

We already discussed the TFPV manufacturing process and the related issues, so now we'll take a closer look at the behavior of thin films in TFPV modules under on-sun operation.

On-sun Operation

Remember, each of the thin films in the active layer is about 1 micron thin, or thinner—less than 1/100 the thickness of human hair. Recall, too, that the entire thin film structure in TFPV modules is several times thinner than a piece of Scotch tape.

Some of these thin films are so thin in places that we could actually count the molecules across the film thickness, if we had a way to do that. And as we look at thin film very closely, we see all kinds of imperfections as well—interruptions, distortions, breaks, cracks, splits, slips, pits, voids, bubbles, loose particles, and other sub-structures in/on the thin film structure. In other words, the films, although they look perfectly smooth and uniform to the naked eye, are anything but.

Looking closely, we can see a clear delineation between the films—the boundary (interface) between each pair. These are very special areas, which play a huge role in the performance of the PV devices made out of these films. They are also the weakest points in most thin film structures, because this is where the electrical resistivity, mechanical stress and chemical degradation are usually initiated, stored, executed and amplified.

These boundaries represent the time and place where the process was interrupted and switched from one material deposition to another; and/or from one deposition step to another.

This involves moving the substrate, which could be accompanied by longer than specified delays, switching gasses, changing vacuum and plasma power levels, and a number of additional inter-step process modifications.

The inter-step changes do cause momentary out-of-control conditions which could cause a number of process and product abnormalities, such as contamination of the surface by carrier gases or pump oil, and temporary process destabilization (total and partial pressures, gas mixing, power fluctuations, temperature imbalance, etc.).

All combinations and permutations of possible process abnormalities and extraordinary scenarios are too broad to discuss here, but you get the idea—this is a most complex set of parameters and conditions that must be executed perfectly all the time, at all process steps, to obtain the best quality final product.

For the purpose of this writing, however, we'll just agree that the inter-layer boundaries (interfaces) are extremely critical areas, which have different properties than the parent materials. These areas are more fragile, so there is a limit as to how much abuse they can handle during long-term field operation.

Each material pair and its interface has well defined mechanical, chemical, electric and thermal stress limits, depending on the critical de-bonding energy and chemical inertness levels specific for each interface between two different materials.

How much mechanical stress, heat, electrical charge, chemical reactivity or combination of these will be needed to cause delamination, adhesion problems, mechanical, electrical or chemical changes and/or disin-

tegration of the bond between the materials, and eventually the materials themselves?

Amazingly enough, the energy levels and forces acting in these areas are very small, relative to the mass of the materials and strength of the films; i.e., Van der Wall (VDW) $=/\sim 0.001$-0.4 eV/bond, H-bonds $=/\sim 0.11$-0.44 eV/bond, SiO_2 cracking in water $E=/\sim 1.39$ eV/bond.

So, without getting into much detail, just looking at such extremely small numbers one can deduct that it doesn't take much effort to damage an interface which will affect the entire thin film structure. Additionally, in most cases, many of these forces are acting together—on and off many times daily for the duration (30 years or more) of the module's on-sun operation. Most importantly, once destructive processes have started, they cannot be stopped. As a matter of fact, things usually only get worse with every daily cycle, and can accelerate quickly.

Now we'll take a closer look at the major destructive processes acting upon thin film PV structures.

Major Destructive Processes

During lab tests or under field operating conditions, thin films act and react to a number of factors: mechanical stress, chemical attacks, temperature variations, etc. They should be able to withstand all these external forces and internal reactions and counter-reactions for 30 long years. Some will, some won't.

Mechanical Stress: Thin Films Structure and Boundaries Disintegration

Thin film modules are exposed to mechanical stress even before they leave the production line. There are factors and effects that are built into them during processing, due to temperature changes, chemical reactions, handling, etc.

Then, from the moment they leave the production line, they are constantly under mechanical stress—during storage, transport, installation and operation (wind loads etc.). Never ending vibration, hitting, rubbing, pushing, pressing and squeezing of the modules, affects the layers in the thin film structure. Serious mechanical stress is induced during handling, packing, transport (long truck and train rides are the worst), installation (careless handling during installation is a major problem) and operation (high winds, hail, rain and sand storms are some of the worst enemies). Each of these actions and interactions increases the mechanical stress in the film structure.

Because of the films' non-uniformity, and the different coefficients of expansion of the different adjacent films, there is a lot of stress and friction within each film, among the adjacent films, and with the substrate. As with other materials stuck together, they will be slipping and sliding, expanding and contracting, creating never ending tension against each other and the materials around them, be it thin films or encapsulants, glass, etc.

In most cases these activities will produce some usually unpredictable changes. The question is what and how bad will these changes be? Stress, cracks, voids and general weakening of the thin film system is expected with time, followed at times by partial or full disintegration of the films, depending on operating conditions. Results from these changes would be expressed as gradual loss of power, intermittent power, and finally complete failure.

Electrical and Heat Stress

Generation of electric energy is the primary purpose of any PV module. The photoelectric process and the accompanying extraction and transmission of electrons (electric current) through the different thin film layers, their interfaces and contacts is usually accompanied by heat generation.

Parasitic (excess) heat is generated when the resistivity of the materials increases, which inevitably happens when the internal temperature of the cell/module goes up. And it starts going up from the second the sun hits the module. The higher the sun, the more electricity is generated, and the higher the resistivity goes.

On a bright sunny day in the desert, the module interior could see temperatures as high as 180°F, or more. The heat build-up is a combination of the simultaneous increase of air temperatures and internal resistivity of the thin films and their interfaces.

This heat is not enough to damage or destroy the films, since each one can withstand much higher temperatures. The excess heat, however, forces the materials to expand and shift in one direction. They then shrink at night and move in the opposite direction, creating friction at the interfaces and stressing the materials in each layer. The more temperature differential the layers are exposed to, the more they stress and shift.

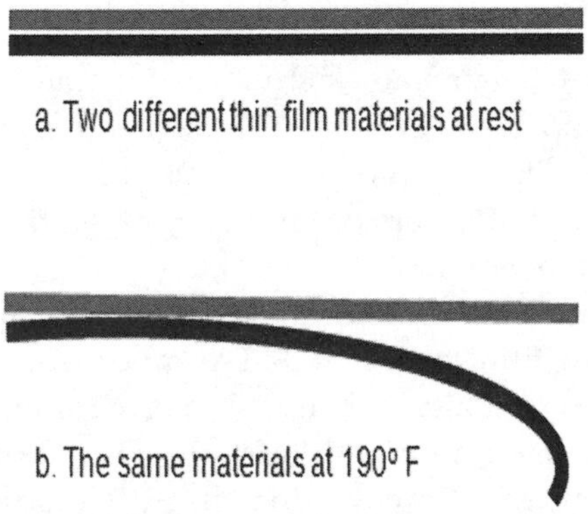

Now imagine these films, packed tightly in the module, getting very hot during the day in the desert (measured inside module temperatures exceed 190°F in Arizona deserts), and freezing at below −20°F at night in winter. This process goes non-stop, 365 days a year. This translates into ~3.650 min-max temperature cycles in 10 years, and close to 11,000 cycles in 30 years. Can the thin films withstand this constant push and pull? Would one of them give up and break down mechanically?

Would that cause an untimely power drop, or complete failure? The results from one, two, or even hundreds, of these cycles would go unnoticed, but the non-stop (30 years/10,000 cycles) effect of the never ending push and pull of the layers will fatigue thin film structure, shifting it into a different performance mode.

As the module gets older, the resistivity of the films and their interfaces increases proportionally, according to the internal heat build-up and dissipation within the module. Electrical output will drop by ~1.0% every year, partially due to the above effects, or combinations of them.

There is no getting away from the harmful effects of excess heat and moisture in field operations. They are variables in the PV generation equation which cannot be ignored, but are hard to control.

Chemical Reactions and Thin Films Decomposition

All unprotected thin films—without exception—react with many chemicals. It is usually the interface (the boundary between the films) that is affected first, and is where problems can be observed. Some films and interfaces will disintegrate instantly with a simple touch (human sweat contains salts), while others will withstand weak chemicals, and some will require strong acids or bases in order to dissolve or decompose them or their interface.

Nevertheless, all thin films are subject to mechanical and chemical changes under certain conditions.

CdTe, CIGS, a-Si and other thin film structures, as well as interfaces between layers, are affected in a similar fashion. Some of them are more chemically resistive than others, but prolonged exposure to weak acids (brought on by moist air, or contained in rain water) penetrating the module lamination (or entering through cracked glass) will eventually cause the films to react,

a. Two different thin film materials at rest

b. The same materials at 190° F

Figure 5-6. Thin film behavior under elevated temperature and humidity

delaminate and decompose.

The type and speed of the decomposition process, and the newly created chemical species, will be determined by the types of thin films, the active chemicals species and the types of reactions these invoke.

Again, we count on the encapsulants and glass or metal frame to keep moisture, air and related chemicals out of the module for 30 years. How many modules will survive the test of time? What would be the total failure rate? These are questions to which we simply have no answers, because there are no precedents.

Moisture ingress is the culprit of pronounced power output degradation of CIS and CdTe modules and reduces their output significantly within months of exposure. This example also leads to the conclusion that good encapsulation is paramount, but since it is never perfect, harsh climate conditions will force moisture to penetrate thin film structure and change its composition and behavior, resulting in power loss and eventual failure.

Environmental effects initiate and accelerate the evolution of defects in TFPV modules. Damaged encapsulation allows moisture to penetrate modules and attack layers, resulting in their delamination or separation from the substrate.

Figure 5-7 depicts the change in Pmax from initial state (in %) after number of hours of damp heat treatment. It is obvious that any and all thin films will degrade and eventually fail with exposure to moisture. This confirms the fact that thin films are chemically reactive and that they will start decomposing soon after being attacked. The speed of reaction, and the degree of degradation is proportional to the moisture content, its

temperature, and the length of exposure.

Imagine what 30 years of this treatment would do to a PV module that does not provide a full protection to the thin films inside it. Hence our fear that the two-glass plates, frameless PV modules will not be able to provide adequate protection for the duration. This is because there is only a very thin plastic strip at the edge of the modules that is stopping the moisture from penetrating inside the modules. This thin strip will be attacked first by the merciless IR and UV radiation, and will be destroyed within a certain amount of time. After that, the fragile thin films will be attacked by anything coming down from the sky and up from the earth, and their lifetime will be shortened significantly.

Just as in tooth decay, once moisture or environmental acids and gasses find a gap in the encapsulation they start the decomposition process which cannot be stopped without intervention. A dentist can drill decay out of a tooth's surface, but drilling decomposition damage out of a TFPV module is impossible. This means that modules will start losing power at a rate proportional to the inflicted damage, usually much quicker than the standard 1.0% power loss per annum.

Eventually—when the internal decay has affected large parts of the critical areas of the thin film structure—the affected modules will fail completely and must be replaced. The process is accelerated when affected modules are exposed to extreme heat and humidity (see Figure 5-8).

The effects depicted in the Figure 5-8 play a significant role in compromising thin film structure integrity. Some of these are vicious and fast acting, while some are slow and cause less damage. The actions of each effect are hard to predict, because different operating conditions have their own peculiarities. It is even harder to predict the combined effect of these processes during non-stop, 30-year, on-sun operation.

The inevitable exposure to excess UV and IR radiation, thermal cycling, mechanical stress, moisture ingress, and chemically active environmental species leads to unpredictable degradation, and unreliable kinetics and reliability models, because of the ever changing types and numbers of forces acting on the materials in inhospitable regions.

In summary, we don't know what to expect of fragile thin film layers in TFPV modules exposed to unending mechanical, chemical, electrical and thermal action and on-sun operation for 30 years. Still, there is enough evidence to conclude that PV modules made with quality materials with proper processing would have a greater chance to survive than those made of

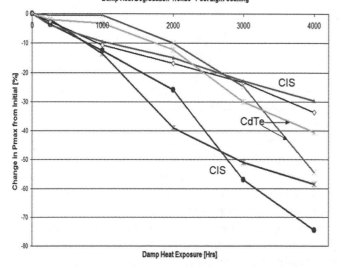

Figure 5-7. Moisture degradation of TFPV modules (17)

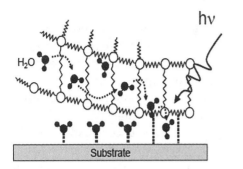

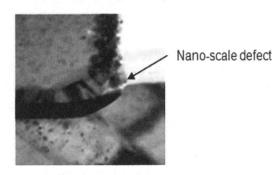

Figure 5-8. The chemical invasion process on molecular level (18)

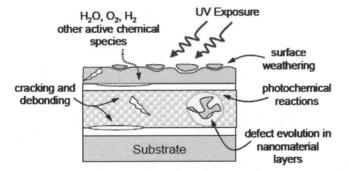

Figure 5-9. The invasion process has started (18)

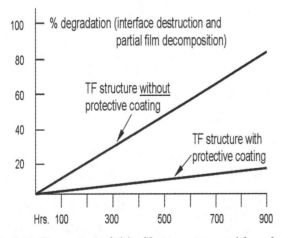

Figure 5-11. Damp test of thin film structures with and without protection.

low-quality materials, using poor design and flawed manufacturing processes.

Tests run by the author and others have determined that most PV thin films do not last very long when exposed to the elements. Any contact with outside elements (air, moisture, rain water, environmental gasses, and any other materials that get access to the thin films) will have an immediate negative impact on their structure, performance and/or longevity.

There are no exceptions to this rule, so utmost attention must be paid to the proper encapsulation and protection of PV thin films in the modules. And here is a good place to emphasize again that most modern thin film PVB modules do not offer sufficient protection against the elements. Especially these modules that consist of two glass panes glued together with a plastic material.

The exposed edges of the module offer very little and very temporary protection, because the elements (especially in hot desert and high humidity regions) will destroy the thin plastic layer at the edges, which will open channels for unobstructed access to the thin films in the modules. The rest is just a matter of time…wasted money and failed modules and projects.

TCO Corrosion

Most thin film PV modules use a thin layer of transparent conductive oxides (TCO) as a metal contact. The TCO is deposited on the back of the top glass. The rest of the thin films (the active layers) are deposited on top of the TCO, thus forming a sandwich the foundation of which is the TCO layer.

The corrosion of the TCO thin film in thin film PV modules is a serious problem, which could disable the entire device. It is, unfortunately, a known reliability problem, which occurs when modules are biased electrically—negative towards ground in warm and humid areas. (4)

Modules that are critical in respect to TCO corrosion are currently restricted to certain inverter topologies and to dry climates. Therefore TCO corrosion tests are under consideration as an additional stress test for the product qualification of thin film modules.

At the moment there is only limited information on how certain modules will behave at high electrical potential against ground in humid areas.

The effect is explained by the drift of positive sodium ions and diffusion of water vapor to the glass/TCO interface. In the subsequent electrochemical reaction the TCO starts to dissolve from the glass at the module edges. In the final stage, the electrical circuit gets inter-

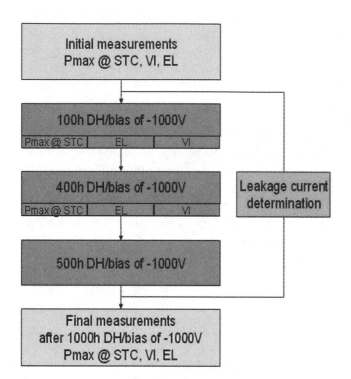

Figure 5-12. Testing scheme for TCO corrosion test incl. leakage current determination. After 100, 400, and 500 hours (total: 1,000 hours) visual, electroluminescent and electrical results are determined. (4)

rupted and the module fails completely. In this study the effect is monitored via electroluminescence for various semiconductor technologies, module encapsulation schemes and fixtures.

Additionally, the power loss and the leakage currents of the biased modules are recorded. The feasibility of this testing sequence for thin film module product qualification is discussed to secure real field conditions. Finally, design rules for thin film modules are suggested.

Results (4)

Module Type A is a commercial amorphous Si module which is framed. During the experiment the module was mounted using clamps.

The second commercial module under evaluation has been module Type B, a micromorph Silicon tandem module, consisting of an amorphous and a microcrystalline layer. This unframed module was fixed with four clamps. The distance between μ-Si semiconductor to module edge is 10 mm.

Module Type C is another commercial micromorph Silicon module. In the experiment the module was fixed with a special mounting scheme to the backside and not to the open edges. The distance from the edge of the active area to module edge was 15 mm.

Comparison

In electroluminescence all modules show inactive areas near the edges. But these areas are not necessarily visible as corrosion; i.e., as a delamination of the TCO layer from the front glass. Corrosion of the thin film silicon modules appears on a later stage after preceding deactivation of the thin film cell.

We did not observe corrosion after damp heat treatment only, as requested in IEC 61646. Although visible corrosion starts at the module edges, thin film Silicon modules show darkening in the area of the junction box, due to damp penetration at the opening in the back glass through the interface of the junction box and the backside glass.

Module fixtures at the backside led to more shunts in the region of their fixation. Nevertheless, the innovative mounting fixtures reduce field intensities compared to mounting clamps at the "open edges." Leakage current pathways are longer and the corrosion of the TCO is avoided.

At the same time, the distance from semiconductor to module edge is important. The module Type C with 15 mm distance shows no corrosion effects compared to the module Type B with 10 mm distance. It seems that for Silicon modules an additional sealing can be omitted if these design conclusions will be taken into account.

Besides the visual and EL inspection we investigated the modules in power determination under standard test conditions (STC) and recorded the leakage currents for Type A, B and C.

Table 5-2 summarizes the data obtained directly after 1000 h in the BDH. Power losses on this stage are

Type	A	B	C	D
Technology	a-Si	a-Si/μ-Si	a-Si/μ-Si	CdTe
Param. Module fixture	frame	clamps	innovative	clamps
ΔP_{max}	-13%	-57%	-3%	-26%
ΔR_s	30%	57%	15%	-5%
ΔR_{sh}	-49%	-90%	-22%	-78%
Corroded area in cm²	9	888	0	-
Share on total area	<0.01	6.5	0	-
Non-active EL	-	22%	4%	-
Charge Q per perimeter in C/cm after 1000 h	0.035	9.263	0.029	-
Charge Q per perimeter in C/cm at ΔP_{max}= -10%	0.031	1.931	0.050	-

Table 5-2. Results after 1,000 hours of damp heat treatment and –1,000 V applied to the modules. (4)

only assignable to modules mounted with frame or clamps by increasing series resistances and decreasing shunt resistances. Type C is very resistant against TCO corrosion. The power determined under standard test conditions (STC) shows a decline of –9% after 2000 h of BDH compared to –3% after 1000 h.

For Type A, B and C we show the transferred charge per perimeter after 100, 500 and 1000 hours and for Type C after 2000 hours. Modules with poor damp tightness show a high charge flow (Type B).

From the charge which has flown, when the power loss reaches –10%, one can conclude that the μ-Si semiconductor of Type B is very inert and on the other hand the a-Si film of Type A is very sensitive to corrosion (Figure 5-13). This comparison at a fixed total charge is independent from the exposure time, mounting structure, applied voltage and module damp tightness and reveals the stability of the thin film cell in the module.

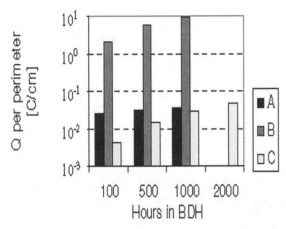

Figure 5-13. Charge Q per perimeter for biased damp heat for a-Si thin film modules. (4)

Conclusions (4)

All modules investigated show corrosion after electrically biased damp heat treatment. Silicon thin film modules show TCO corrosion and degradation depending strongly on the module design. CdTe modules degrade as well, but show different effects regarding TCO corrosion, dissolving and shunting. Results for CIGS modules are not presented here but corrode as well, though maybe independent from the sodium diffusion.

The corrosion affects the maximum power of the modules under STC conditions by an increase of series resistance and reduction of shunt resistance. Electroluminescence reveals different failure modes than visual inspection and enables detection of corrosion at an earlier stage.

From the total charge flow after 1000 h, one can

evaluate the damp tightness of the encapsulant. Corrosion is further influenced by the mounting scheme, dimensions of the modules edges, and the applied system voltage with its polarity. These are the key parameters for the corrosion on the module and system side.

The value of the total charge flow at the point of 10% power drop, allows benchmarking of the thin film cells towards their inertness against corrosion inside the module.

Both aspects are important development targets for stable modules, which do not need any restrictions to a certain inverter topology or even only dry locations. We propose to allow a power drop of less than 20% as an acceptance criterion after 1000 hours of electrically biased damp heat treatment for both polarities.

Because the power drop from the corrosion is nearly linear with the exposure time, a pass criteria not greater than 5% for Pmax after 250 hours appears also to be reasonable, thus allowing a reduction of testing time and related costs.

This corresponds to a lifetime of 45 years for modules passing that criterion on the basis of the total charge flow per perimeter as estimated by Gossla.

PERFORMANCE CHARACTERISTICS

TFPV modules have several serious inherited issues and exhibit a number of abnormal behaviors (discussed above) that need to be brought out. This is even more important now, since there is a lot of misunderstanding, misinterpretation and general confusion on the subject of efficiency vs. cost vs. longevity of TFPV modules.

First and most importantly, we need to keep in mind that TFPV module manufacturers are relative newcomers to the solar field. Most have been in existence only a few years, with even shorter field experience and insufficient on-sun exposure test data. Therefore, they have little experience with long-term field behavior of their TFPV modules, especially under extreme climate conditions.

The other obvious drawback of TFPV modules is their low efficiency, which is even lower under the hot desert sun (due to temperature degradation). Also, they demand use of larger than normal structures and land areas, and there is an annual 1% or more power loss, which further decreases power output over time. Most of the available test data are from "standard" tests done under "normal" conditions, with no mention of performance issues under the extreme conditions of the US deserts where many large TFPV plants are installed, or where future installations are planned.

Efficiency

A major concern with commercially available TFPV modules is their relatively low efficiency, in the range of 8-9% measured at the module level. When the modules are connected in an array, plugged into the grid, and all related losses are taken into consideration, we end up with 4-6% efficiency.

We have seen record low performance of TFPV power plants operating at 2-3% efficiency in the grid. Just think…2% efficiency. It takes ~10 acres to generate 1.0 MWp power with present-day 8% efficiency TFPV modules. At 2% in-the-grid conversion efficiency, we will need more than 40 acres to produce 1.0 MW AC power—not an exciting proposition.

Test Failures

Below is reflected the experience of, and the test results obtained by, researchers at Photovoltaic Institute in Berlin, Germany. (3)

Experimental Setup

In our accredited test laboratory (ISO 17025) we are performing tests according to IEC 61646, among others. Beyond IEC we are performing damp heat tests with bias voltages to simulate real field conditions.

After these stress tests, the modules are observed via electroluminescence (EL). Electroluminescence has been performed using a Nikon D 700 camera with the IR-filter removed from the high-sensitivity 12 Mpixel CCD-chip. A 50 mm f/1.4 high precision IR-optimized optic from Zeiss has been applied.

To get an EL-signal, a current has been applied in forward bias to the module. The brightness of the electroluminescence signal is influenced by the quantum efficiency, dark current losses and band gap shifts of the cell.

The spectra of the electroluminescence signals are measured with the use of a calibrated Tec-5 multichannel diode spectrometer. A MCS-CCD sensor with sensitivity in the wavelength range from 200 nm to 980 nm and an InGaAs sensor with a range between 900 nm and 1,700 nm have been used.

EL Failure Catalogue

In the following, we distinguish between features in the EL image which are related to power losses (failures) and features without power losses (flaw). We observe also the reversible effects, which show recovering

of power losses after the biasing, similar to the polarization effect on crystalline silicon modules (now called PID = potential induced degradation).

Table 5-3 lists the features with power losses as failures and gives the possible root cause.

Table 5-3. Thin-film electroluminescence failure catalogue.(3)

Failure/flaw	*root cause* *(test, where the failure can occur)*
Lightning	Bad or missing edge isolation, bad edge delete (DH, BDH)
Inhomogeneities	Fluctuations in thickness or stoichiometry results in fluctuation of the band gap. Impurities and grain boundaries reacts as recombination centers (initial, BDH)
R_{sh} decrease (shunting)	Impurities, bad or missing laser scribes or isolation trenches (DH, BDH)
R_S increase	Water ingress, corrosion (DH, BDH)
Reversible effects	Charge coupling (BDH)
Glass breakage	(ML, Hot-spot, Reverse Current Overload)
Dots from Hot-Spot test	Locally high temperatures (Hot-spot)
Light stabilization	After UV and LS, EL signal decreases stronger than P

DH = damp heat at 85°C and 85% relative humidity
BDH = DH with bias, ML = mechanical load test

Thin film modules are vulnerable at the edges and must be well water tightened. If damage of the edge isolation (delamination of encapsulant or sealant) occurs, water ingress might destroy the circuit of the surrounding cells.

A module showing brightening at the module edges due to strong shunt resistance decrease after 168 h DH, exhibits lower local photo-voltage, which leads to a decrease in the electroluminescence signal of the cells. Mainly radiative heat is emitted in this area, which can be also detected via IR-imaging.

Another reason for brightening might be an incorrect edge-deletion, which results in high leakage currents giving the same effect in the EL image.

Inhomogeneities (CIS, CIGS, CdTe)

Inhomogeneities in CIGS thin film modules are explained by band gap fluctuations in limiting the cell efficiency. If these lead to lower collection efficiencies and/or higher losses in the IV curve, this becomes visible in the electroluminescence signal as is well established for crystalline silicon.

Additionally, fluctuations in the band gap will shift the luminescence spectra, which results in intensity variations, because of the limited sensitivity of the CCD camera in the infrared. We compared the luminescence spectra of the "normal" emitting areas to the weak emitting areas and we could observe an intensity drop and a spectral shift.

Rsh Decrease (Shunting)

The shunt detection and differentiation after production are already well known. DH and BDH are known to increase the shunt content of CIGS modules. More information to the shunt detection can be found in the literature.

Brightness Evaluation

The investigated modules have been exposed to 100 h and 500 h damp heat atmosphere (DH) at 85°C and 85% relative humidity at negative and positive bias of ±1000V vs. ground.

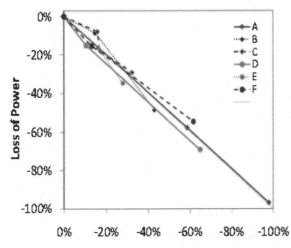

Figure 5-14. Brightness evaluation of biased DH-modules.(3)

The loss of brightness is visualized over the loss of power. Besides the fact that all thin film modules under investigation degrade strongly under BDH conditions (both plus and minus 1000V), we could find a linear dependency between loss of brightness and loss of power.

When damp heat short circuit current and open circuit voltage remained constant, the power losses amount to fill factor losses only. This indicates dark current and series resistance losses.

A huge potential is seen in this measurement method for process improvements in the production because power losses can be spatially resolved in the EL images.

Moreover, module degradation experiments and their result interpretation can profit from the relation of power determination and electroluminescence.

In conclusion, electroluminescence investigation is a powerful tool to reveal failures of thin-film photovoltaic modules during the production stage and testing. Spectral electroluminescence analysis shows need of InGaAs sensors with higher resolution for a better detection of thin-film technologies, except CdTe.

The brightness evaluation presented here shows direct relation between power loss and brightness loss. Finally, a high density of shunts on thin-film modules shows a strong yield impact due to weak light efficiency drop.

SAFETY ISSUES

Recycling

Many thin film modules contain toxic materials and are required by law to be removed, packed, picked up, transported and recycled as hazardous waste at the end of their useful life. This requires the manufacturers of such products to include, in the initial price, the cost of removal, packing, transport, recycling and disposal of the product; or, as an alternative, to have a special fund or insurance that will ensure proper and efficient execution of the end-of-life steps outlined above.

40,000 MT PV components have been forecast for decommissioning and recycling by 2020. This number will double or triple during the following decade. This includes silicon and thin film based PV components recycling. The thin film component will be ~20% by then. 8,000 MT of CdTe thin film PV modules will be the majority of the recycled product in this category.

This number will grow exponentially with the unprecedented growth of the large-scale TFPV power plants.

EU already has directives for voluntary and extended manufacturer responsibility, where the decommissioning, transport, storage, processing and disposal of the modules is the ultimate responsibility of the original manufacturer. The directive has been integrated into the legal system of several member states, and we hope that a similar initiative will be undertaken in the US, before we end up with a huge Hazmat nightmare in the California and Arizona deserts.

Production Cycle Safety

Thin film PV modules are usually manufactured in semiconductor type equipment, using specifications and procedures similar to those used in semiconductor fabs. These procedures—especially in the US and western countries—are quite strict and include health and safety clauses and precautions which have been proven to ensure safe environment under all working conditions and with any materials and components.

This is one of the best points, when comparing thin film PV modules with other technologies, including c-Si. Thin film manufacturing cannot be done in a dirty place, as is sometimes the case with c-Si manufacturing facilities. Nor can it be done sloppily, because the equipment and its sophisticated controls simply will not allow it. c-Si cells and modules could be, and often are, manufactured sloppily, since many of the operations are done by hand, by workers who might be ill-trained or asleep on the job.

Human error can be found anywhere, but it is much less frequent in thin film manufacturing facilities than in c-Si facilities.

Field Operation Safety

All PV modules have a propensity for causing environmental damages. Some of the most dangerous, however, are TFPV modules which contain significant amounts of toxic materials (CdTe, CdS, GaAs etc.). The dangers during manufacturing these products have been well researched and understood.

Their long-term operational safety in the field and especially in hot desert and humid areas, however, have not been thoroughly investigated, and we know practically nothing about their behavior during 30 years under extreme heat or elevated humidity in some of the proposed mega-installations.

The manufacturers and proponents of toxic thin film products have been successful in offering incomplete explanations and solutions until now, which, in light of the impressive success of some TFPV manufacturers, have been accepted by the general public.

These issues have not been reviewed in detail by the scientific community and the regulators, so we hope that an open discussion will follow soon.

In summary, we are supportive of the responsible expansion of all TFPV technologies which can be used in residential and small commercial installations and even bigger installations in moderate climate areas. However, we caution against massive expansion of some TFPV technologies—especially those containing toxic materials—into large-scale power fields in US deserts.

Efficiency, performance, longevity and safety issues must first be addressed and resolved to give us reasonable expectations for long-term success of those PV products yet unproven for use in harsh environments.

FIELD OPERATION ISSUES

Most of the problems of the failed companies and projects described above are of administrative and financial nature. This, however, is only because the underlying technical problems and the related potential failures of their products were not thoroughly addressed, and/or have not had a chance to appear as yet.

Time, combined with harsh environment action, has amazing properties and can do some unexpected—mostly very bad—things to shiny PV modules sitting helplessly under the blistering Arizona desert sun, cooking in their 180°F skin, or bathing in the 100% humidity of the Louisiana swamps.

Are all PV modules and other components installed worldwide ready for this torture? Are they going to successfully and uninterruptedly generate power for 30 years under these conditions?

We do believe that the second wave of failures in the world's solar power fields is coming soon—as a result of hasty implementation of so many untested and unproven PV modules and BOS components recently. We should get ready for it.

Please read on...

Why would PV modules fail? Aren't they made with durable materials, using sophisticated equipment and well trained labor? Yes...partially. Some are, some are not. Even those made by reputable companies might exhibit serious problems in the field.

There are many technical reasons why our shiny PV modules would fail in the field. These are the same reasons that other solar products fail, but we need to drive home the fact that less-than-perfect simply will not cut muster.

Garbage In, Garbage Out

The garbage in, garbage out principle is in full force where high quality products are concerned, so finding the weak links in the sequence is of utmost importance in any manufacturing process. The quality of the *supply chain* (materials purchased from third parties) is where the quality control cycle starts. Immaculate quality control of all incoming materials is paramount for final product quality.

Quality of *production equipment* is very important

as well. Without proven qualified and precisely controllable equipment, even high-quality materials could lose their value. Process *design and execution* are critical as well. One misstep in the solar cell manufacturing process could convert a batch of cells into paper weights.

Finally, and very importantly, are the *people!* Qualified design, equipment, process and quality control engineers, well trained technicians and operators, as well as experienced managers are all part of the team that must function like a well lubricated machine. One single glitch—be it equipment or people related—can introduce enough havoc in the process to reduce the quality of the final product.

The basic rule of thumb is, "The highest quality of the final PV product is determined by the lowest quality of : a) one of the materials, and b) one of the critical steps of the process."

We cannot afford any quality violations through the entire process sequence, regardless of how many steps it consists of—not when this product is expected to operate 30 years under hostile climate conditions.

While total quality control as followed by many well established US companies is needed in all solar products manufacturing operations, the PV industry is not yet well organized and standardized, so we do see a number of variations and deviations in manufacturing operations.

What should we be looking for to ensure the efficiency and reliability of the c-Si solar products in the 21st century? First and foremost, we need to understand the issues related to PV modules and BOS operation, as well as their failure modes.

Customers have no control over the quality control of the manufacturing process. Once the PV modules are installed and operational, they must deal with them, to ensure efficient operation.

The most important thing to watch for in field operations is maximizing the power generation. There are many things that reduce output, so consumers must understand and deal with those.

Dealing with Power Losses at the Module Level

A number of predictable and unpredictable, controllable and uncontrollable losses must be well understood and included in all estimates and design and financial calculations.

Air Pollution

The solar resource can be reduced significantly in some locations due to air pollution from industry and agriculture.

Soiling

Losses due to soiling (dust and bird droppings) depend on the environmental conditions, rainfall frequency, and the cleaning strategy as defined in the O&M contract. This loss can be relatively large compared to other loss factors but is usually less than 4%, unless there is unusually high soiling or problems from snow settling on the modules for long periods of time. Soiling loss may be expected to be lower for modules at a high tilt angle as inclined modules will benefit more from the natural cleaning effect of rainwater.

Incident Angle

The incidence angle loss accounts for radiation reflected from the front glass when the light striking it is not perpendicular. For tilted PV modules, these losses may be expected to be larger than the losses experienced with dual axis tracking systems, for example.

Low Irradiance

The conversion efficiency of a PV module generally reduces at low light intensities. This causes a loss in the output of a module compared with the standard conditions at which the modules are tested ($1,000 W/m^2$). This low-irradiance loss depends on the characteristics of the module and the intensity of the incident radiation.

Shading

Due to mountains or buildings on the far horizon, mutual shading between rows of modules and near shading due to trees, buildings or overhead cabling causes losses.

Module Temperature

The characteristics of a PV module are determined at standard temperature conditions of 25°C. For every degree rise in Celsius temperature above this standard, crystalline silicon modules reduce in efficiency, generally by around 0.5%. In high ambient temperatures under strong irradiance, module temperatures can rise appreciably. Wind can provide some cooling effect which can also be modeled.

Annual Losses

PV modules lose 1% or more power output every year. This fact must be verified, agreed upon by all parties and included in the design and business plans.

Losses in the Field

During the long-term operation of solar power fields we could expect a number of losses that are related to materials quality and workmanship. Some of the issues to be expected in solar power fields are:

Module Quality

Most PV modules do not match exactly the manufacturer's nominal specifications. Modules are sold with a nominal peak power and a guarantee of actual power within a given tolerance range. The module quality loss quantifies the impact on the energy yield due to divergences in actual module characteristics from the specifications.

Module Mismatch

Losses due to "mismatch" are related to the fact that the modules in a string do not all present exactly the same current/voltage profiles; there is a statistical variation between them which gives rise to a power loss.

DC Cable Resistance

Electrical resistance in the cable between the modules and the input terminals of the inverter give rise to ohmic losses (I2R). This loss increases with temperature. If the cable is correctly sized, this loss should be less than 3% annually.

Inverter Performance

Inverters convert from DC into AC with an efficiency that varies with inverter load.

AC Losses

This includes transformer performance and ohmic losses in the cable leading to the substation.

Downtime

Downtime is a period when the plant does not generate due to failure. The number/length of downtime periods will depend on the quality of the plant components, design, environmental conditions, diagnostic response time and repair response time.

Grid Availability and Disruption

The ability of a PV power plant to export power is dependent on the availability of the distribution or transmission network. Typically, the owner of the PV power plant will not own the distribution network. S/he, therefore, relies on the distribution network operator to maintain service at high levels of availability. Unless detailed information is available, this loss is typically based on an assumption that the local grid will not be operational for a given number of hours/days in any one year, and that it will occur during periods of average production.

Degradation

The performance of a PV module decreases with time. If no independent testing has been conducted on the modules being used, then a generic degradation rate depending on the module technology may be assumed. Alternately, a maximum degradation rate that conforms to the module performance warranty may be considered.

MPP Tracking

The inverters are constantly seeking the maximum power point (MPP) of the array by shifting inverter voltage to the MPP voltage. Different inverters do this with varying efficiency.

Curtailment of Tracking

Yield loss due to high winds enforcing the stow mode of tracking systems.

Auxiliary Power

Power may be required for electrical equipment within the plant. This may include security systems, tracking motors, monitoring equipment and lighting. It is usually recommended to meter this auxiliary power requirement separately.

Grid Compliance Loss

This parameter has been included to draw attention to the risk of a PV power plant losing energy through complying with grid code requirements. These requirements vary on a country-by-country basis.

Natural Disasters

Anything can happen during 30 years. Whether the field is in the Arizona desert, the Colorado mountains, or Louisiana swamps, Mother Nature is there and doing what she does best—bringing the good and the bad. All possible natural disasters must be investigated and considered during plant design, and kept in mind during its operation.

Speaking of what can happen during long-term operation, we are presenting herein a couple of studies that reflect typical problems experienced by PV modules during non-stop field operation.

FIELD TEST RESULTS

As justification of our predictions for a second wave of field failures in the near future, resulting from failures of solar products installed around the world during the last solar cycle, we are presenting the following test data, collected by reputable institutions.

Most PV modules go through internal lab tests, followed by certification tests—all of which are done in a lab, with very little actual on-sun exposure and testing. And this is a big problem, because there is a big difference between results from laboratory tests, and what happens during actual long-term field operation. The results depend on the types of modules used and the prevailing weather conditions in the test field, so keep in mind that:

1. The results from tests under the extreme heat in the Arizona desert sun or the high humidity in Southern Florida will be drastically different from those under the cloudy skies of Portland, ME, or Berlin, Germany, and

2. In all cases, failures and degradations increase after 15-30 years of operation.

The field tests in Table 5-4 show the average annual degradation of different PV modules from different manufacturers. Most of the results from c-Si PV modules are actually quite encouraging, but notice also that most of the manufacturers whose products show good results are reputable, well-established companies who have

been in the solar business for a while now and have figured out how to make a high-quality product. In contrast, most new manufacturers have not had a chance to work out the process bugs and/or conduct long-term field tests, to prove the reliability of their products.

Table 5-4 test results prove that c-Si technology, if properly made and used, can be a viable and reliable energy source and that some of the major and well-established manufacturers have achieved reliability.

Keep in mind, however, that most of the above tests (except one) have been conducted only in the average of 5-10 years. The conclusions, therefore, reflect only a fraction of the time, stress and abuse all PV modules will experience during 30 years of operation and exposure to the elements.

Extrapolating for the actual, long-term life of the modules in the desert (30+ years), we could easily come up with numbers several-fold larger than the failures described above.

Long-term Desert Field Tests

As substantiation of the facts and issues discussed in the previous chapters, a research study published in the fall of 2010 confirms the presence of major issues in PV modules operating under harsh desert conditions. The study describes the results from long-term scientific tests done with PV modules operating in an official desert test field for a number of years (ranging from 10 to 17 years).

Note: To our knowledge, this study is the largest-ever evaluation of the long-term performance of production-grade PV modules under desert conditions.

The study was done at one of the most prestigious PV test sites in the world; APS's STAR solar test facility in Phoenix, AZ, where most world-class PV modules manufacturers send their modules for long-term testing, so we would not be surprised if the tests include some famous names.

The work was supervised by one of the world's most reputable PV specialists, Dr. Govindasamy Tamizhmani (Mani) from the Arizona College of Technology and Innovation, and President of TUV Rheinland, Tempe, AZ.

Dr. Mani's team carefully and thoroughly visually inspected, tested electrically, and IR measured each module in the nearly 1,900 PV modules batch.

The final results from the long-term tests in the Arizona desert are shown in Table 5-5.

Amazingly, almost 100% of the nearly 1,900 PV modules tested in that study showed some sort of visual deterioration. Some were more impacted than others,

Table 5-4. Field test results (15)

Manufacturer	Module Type	Exposure Years	Degradation (% per year)
ARCO Solar	ASI 16-2300 (x-Si)	23	-0.4
ARCO Solar	M-75 (x-Si)	11	-0.4
Mfgr. X	X-number (a-Si)	4	-1.5
Eurosolar	M-Si 36 MS (poly-Si)	11	-0.4
AEG	PQ-40 (poly-Si)	12	-5.0
BP Solar	BP-555 (x-Si)	1	+0.2
Siemens Solar	SM 50H (x-Si)	1	+0.2
Atersa	A60 (x-Si)	1	-0.8
Isofoton	I110 (x-Si)	1	-0.8
Kyocera	KC-70 (poly-Si)	1	-0.2
Atersa	APX-90 (poly-Si)	1	-0.3
Photowatt	PW-750 (poly-Si)	1	-1.1
BP Solar	MSX-64 (poly-Si)	1	0.0
Shell Solar	RSM-70 (poly-Si)	1	-0.3
Wurth Solar	WS-11007 (CIS)	1	-2.9
USSC	SHR-17 (a-Si)	6	-1.0
Siemens Solar	M-55 (x-Si)	10	-1.2
Mfgr. Y	Y-number (CdTe)	8	-1.3
Siemens Solar	M-10 (x-Si)	5	-0.9
Siemens Solar	Pro 1 JF (x-Si)	5	-0.8
Solarex	MSX-10 (poly-Si)	5	-0.7
Solarex	MSX-20 (poly-Si)	5	-0.5

Table 5-5. Long term desert field test results (11)

Module Types	Module Count	Modules Affected
A. Mono-Si (17 years; glass/polymer)	**384**	
- Browning in cell center	384	100.0%
- Frame edge seal deterioration	384	100.0%
- Hot spot (IR scan)	4	1.0%
B. Mono-Si (12.3 years; glass/polymer)	**1092**	
- Encapsulant delamination	2	0.2%
- Browning in cell center	1092	100.0%
- Hot Spot (IR scan)	2	0.6%
C. Poly-Si (10.7 years; glass/glass)	**171**	
- Broken cells	47	27.5%
- Encapsulant delamination	55	32.2%
- Hot spot (IR scan)	26	15.2%
- White material near edge cells	2	1.2%
- White material browning	56	32.8%
D. Pol-Si (10.7 years; glass/polymer)	**48**	
- Browning in cell center	37	77.1%
- Hot spot (IR scan)	4	8.3%
E. Mono-Si (10.7 years; glass/polymer)	**50**	
- Bubbling substrate	33	66.0%
- Browned substrate near J-box	50	100.0%
F. Pol-Si (10.7 years; glass/polymer)	**120**	
- Discolored cell patches	6	5.0%
- Backsheet bubbling	1	0.8%
- Browned spots on backsheet	2	1.7%
- Metallization discoloration	22	18.3%
- Frame edge seal deterioration	15	12.5%
- Hot spot (IR scan)	4	3.3%
Total modules tested	**1,865**	

but the overall conclusion was that a large percent of PV modules exposed to constant sunlight and heat in the deserts *deteriorate beyond the manufacturer's warranty* within 10-17 years of operation. Many of the tested modules show average 1.3 to 1.9% annual power degradation, which translates into 23-33% total power loss for the duration (some only 10 years). This, extrapolated for the extended time of use, is well above the manufacturer's warranty of maximum 20% degradation in 20 years.

Interestingly, significant EVA encapsulation discoloration and delamination was observed on many modules, even those that were exposed to the sun only 10 years. Solar cell discoloring was observed on most cells too. Since the oldest modules in this study were made in the mid-1990s we must assume that encapsulation and cell manufacturing processes have improved some during the last 10-15 years to reduce the defect ratio. However, we have no proof of such progress, so this becomes another unknown that needs our attention when dealing with PV modules destined for desert applications.

In summary, browning in the cell center was almost 100%, encapsulation issues followed as a distant second, followed by frame seal deterioration, hot spots, broken

cells and glass. Many of these defects are serious enough to cause power loss and eventually lead to total failure, due to optical interference, electrical resistance, overheating, mechanical disintegration, or any combination of these. This is simply not supposed to happen—not in such large numbers at such an important test site, so we must ask, Was it the materials quality (silicon, metals, EVA, glass), or the manufacturing process (diffusion, metallization, encapsulation), or some combination of these that was primarily responsible for such a large number of defects and failures?

In any case, this study confirms the unfortunate reality that PV modules of any type and size are prone to rapid deterioration and failure under harsh desert conditions. Although we cannot predict future performance and overall condition of modules during years 20, 25 and 30, the logical assumption is that the deterioration processes will continue at the same rate, and even faster in some cases.

Because of that, we must conclude that each module type and size must be very well designed and properly manufactured with high quality materials and proper processes, to have a chance to survive desert elements.

Note: The reasons for the above described defects and the related explanations and justifications are too numerous to launch into at this point, but references and explanation to many of them can be found in the previous chapters, and in our first book, *Photovoltaics for Commercial and Utilities Power Generation*.

Catastrophic Failures

Those of us who live in the desert are well aware of its destructive nature and the forces that are in play nonstop. We know that the desert has no friends or favorites and shows no mercy. Anything left in it for a long period of time will be damaged or destroyed. There are few exceptions to this reality, and unfortunately PV modules are not one of them.

Because of that, we must be prepared for the worst when installing PV fields in the desert. Stifling heat, freezes, highly destructive UV radiation, severe sand storms, strong winds, golf ball-size hail, and fires just to name a few must be expected at any time.

These conditions, added to the high voltage current flowing through the panels create scenarios on the extreme end of most materials' tolerance limits. Plastics crack and disintegrate with time, the active structures inside the modules experience excessive thermal and mechanical abuse, shrinking and expanding with the temperature changes, creating defects in the structure and leading to failure.

The defects caused by mechanical, electric or chemical abnormalities in the modules could lead to partial or total failures. Partial failures result in power degradation resulting in power loss, but the modules are still operational. This is the case with the modules in the ASU field test; they show damage, defects and power loss, but are still functioning. Partial failure could also be caused by sand storms that temporarily disable the power production by soiling the modules, or damaging their support structures. After proper maintenance, the field could be up and running within hours.

Total failures result in permanent damage, be it mechanical damage to the modules or electrical failure causing open circuits, shorts, or chemical disintegration of the active structure. These abnormalities could result in total shutdown of modules and strings of modules, and in some cases even cause fires. These are then classified as catastrophic failures.

Catastrophic failures are not an everyday occurrence in PV power plants, but when they happen they cause a lot of damage. The most dangerous catastrophic module failure, DC arcing, occurs when the modules lose their dielectric properties after long-time operation under the elements (due to lamination break-down) and start leaking electric current across the module elements. DC arcing is an example of premature module failure and is the worst catastrophic field failure.

DC arcing cannot be easily foreseen or prevented, and once started it cannot be stopped. The arc will burn out only when there is no more material to burn, and in worst cases it could ignite the module, burning it and the entire array. This is a catastrophic failure, which could cause damage to the entire project. Proper O&M procedures (including using IR sensors to detect hot spots and overheated modules), albeit expensive and time-consuming, might be able to detect the precursors of DC arc creation, so they are highly recommended.

There is simply no way to predict, let alone prevent, the desert elements and related surprises, but properly designed and manufactured modules, using highest quality materials and processes are a good start in that direction. Proper installation, again using highest quality materials and procedures, is the next best thing we can do to extend the useful life of the modules. Using proper O&M procedures for the duration will close the circle of our responsibilities.

In addition, an adequate manufacturers' warranty, including long-term performance warranty, is needed to provide protection against product defects. Finally, insurance against natural disasters, complete with long-term performance insurance, will bridge the gaps between the manufacturers' warranties and what Mother Nature has in store for our modules during their 30 years. This way, nothing is left to chance, and only now are we ready for anything the desert is going to send our way—for the duration.

Practical Applications

Taking a careful look at the ASU-APS long-term desert field test study makes us pause, take a deep breath and take another, closer look at the results. An almost 100% defect rate and excessive annual deterioration (much higher than expected) after only 10-17 years is something to think about. How would that reflect on the bottom line of our 100 MWp PV project in the Arizona desert? How would we justify the annual power losses and make up for the power expected by, and contracted with, the utility?

This study is further confirmation that many of the unresolved and unaddressed issues presented in this text are real, measurable, and quite significant. They must be resolved before we enter into an irreversible large-scale PV nightmare. These test results are shocking even to those of us with years of experience in the field.

Our 35+ years experience with engineering issues and quality control matters cannot process the near 100% defect ratio suggested in the study, let alone put it into perspective. The practical advice we have for prospective PV customers or investors is to make sure they know EXACTLY what they are dealing with and getting into:

1. Don't let new, shiny PV modules mislead you into thinking that they will perform as well as they look for the duration. These are not disposable toys, but serious high-tech devices, which will be exposed to the most ferocious climate on Earth and endless torture for many years, so they'd better be made right.

2. Don't let the manufacturers' promises mislead you into believing that they know what their modules will do under excess desert heat, unless they have long-term test data to show.

3. Don't let financial gains shown on paper for the first 3-5 years blind you to the fact that what comes during years 10-15 might not be as wonderful, and

4. Do your homework; leave nothing to chance or to unconfirmed promises:

a. Know who the module manufacturer is, and how long they have been making these modules (check if these particular modules are a newly designed product without any desert operation history).

b. Learn the production process details: equipment, process steps, materials, labor conditions and everything else that goes into the manufacturing process.

c. Get as much quality-control documentation as you can: initial inspection, in-process QC and test procedures, final test procedures and results, and long-term test procedures and results.

d. Get information on the supply chain (the quality of materials and chemicals made by a third party).

e. Learn all you can about the modules: where they were made, where and how they were tested and certified, any long-term test results done—when, by whom, and how.

f. Become familiar with the details of the technical information available, describing the modules' structure, function and performance.

g. Discuss and negotiate warranty conditions, in detail.

h. Parts and labor warranty conditions: What is considered a failure? By whom, when, and how will the failed modules be replaced? How would the lost time be compensated?

i. Long-term performance conditions: what is a failure for every year, after year one? Over 1.0% annual power loss, or >20% total power loss should be considered failure. How would the replacement be done and lost time compensated?

j. Temperature coefficient: discuss and agree on the modules' performance under desert conditions, where >180°F in-module temperatures have been measured in the desert. If 0.5% per °C is the maximum accepted under the warranty, then what could be done if it gets higher with time? How would the replacement be done, and lost time compensated?

This is a list of only the key basics. There are numerous additional issues to be addressed and questions to be answered, to get a complete picture and reduce the risks to a manageable level. Since some of the issues are quite complex, and no one person can resolve them all, the intervention of a number of specialists in the different areas of the project is a must.

Challenges for Utility-scale Renewable Projects in the USA

Large-scale (or utility scale) solar projects are the future of solar energy generation in the US and around the world. We can clearly see massive solar projects popping up in the deserts during the 21st century. Planning and permitting, however, are the greatest challenge for utility-scale solar projects today.

Utility-scale solar projects are massive industrial-scale developments on lands that are natural habitat or rural, requiring considerable mitigation of biological resource impacts. In addition, local residents usually feel overwhelmed by multi square mile projects and view these industrial developments as an invasion into the solitude and isolation they sought by living there.

The planning and permitting process itself is also a challenge. In California, this process involves many federal, state, and local agencies that provide regulatory review and ultimately issue permits for electricity generation projects depending on their agency mandate or jurisdiction.

Different agencies, regulators and governing bodies have varying codes, standards, and fees, and each agency is involved at varying degrees, depending on the projects' technology, size, location, and potential for environmental and system reliability impacts. If improperly coordinated, potentially overlapping levels of jurisdictional review can cause delays, add to uncertainty and even cause projects to fail.

Many large solar energy projects are being proposed in California's desert area on U.S. Bureau of Land Management (BLM) land. The BLM has received requests for rights of way encompassing more than 300,000 acres for the development of 34 utility-scale solar thermal power plants totaling about 24,000 MW.

Complying with federal, state, and local agency environmental and permitting requirements related to using the natural resources (land, water etc.) is a challenge for solar energy projects developers. Large utility-scale solar facilities, especially those located in the deserts, face complicated environmental review and mitigation activities. Resolving these issues can lead to legal challenges, causing delays and higher costs and potentially jeopardizing project approval. A number of examples of late come just in time to confirm that.

Large-scale renewable power plants in desert locations face environmental issues to a greater degree since desert locations often provide habitat for sensitive species, have limited water supplies, and are often on federal lands.

The different conditions and requirements across

jurisdictions, multiple permit requirements, lack of zoning and permitting ordinances, and reduced staffing complicate and delay project review and approval.

REAT

To facilitate the development of solar projects, several efforts presently underway are geared to improve permitting processes for utility-scale solar plants. One of these is the Renewable Energy Action Team's (REAT), whose goal is to promote interagency coordination and ease the permitting process to avoid delays, reduce permitting costs, and improve the certainty of permitting decisions.

REAT also aims to develop the Desert Renewable Energy Conservation Plan (DRECP), to identify areas suitable for renewable energy project development and help conserve sensitive species and natural communities in the Mojave and Colorado Desert regions.

BLM

In the fall of 2012 BLM came up with a list of areas amounting to over 250,000 acres of desert lands in the states of Arizona, California, Colorado, New Mexico and Utah that can and should be used for large-scale solar power generation. This is good step forward, and shows the long-term commitment of the US government to achieving energy independence.

But this is only the first step of many needed to complete the chain of events, before we actually see many megawatts of solar power streaming out of the SW USA deserts. Still, we are confident that, if not this generation, the next will take full advantage of the opportunity to generate power from the deserts.

The Desert Phenomena

To wrap up this discussion and put it in the right perspective, we need to take a long, serious look at the actual exposure of PV modules to long-term desert operation. Since deserts offer the best conditions for continuous solar power generation, they are the preferred locations for most solar installations. This leads us to emphasize again and again the importance of well designed and constructed modules and BOS, always keeping in mind that they are not disposable toys; they are high-tech devices destined to experience 30 years of non-stop torture in the world's harshest environments.

We wonder how many module manufacturers are aware of the extreme conditions their modules are expected to endure in the deserts. Have they have taken these conditions into full consideration in their design process and manufacturing operations?

In many cases, by the time the damage is notice-able, by visual inspection or electrical measurements (as in the ASU-APS field test study mentioned above) it means that a) the modules have deteriorated significantly, b) their power output is reduced above the specified values, c) there is a good chance of further power reduction, and d) there is increased chance of total failure with time.

No lab testing can accurately reproduce the different types of field failures. There is no test today that can reproduce 30 years of desert exposure, but underestimating the power of the desert will surely lead to field failure.

On top of that, the modules in our large-scale power field are connected together in long strings that are continuously exposed to the internal torture of several hundred DC volts of electric current running through them. The electric current also generates significant amounts of heat in the cells and modules in its path to the grid. The sun heats the modules from the outside, while the produced electricity heats them from inside.

Any slight flaw in the cells' or modules' structure will eventually create even more heat due to increased resistance or sunlight (IR and UV) absorption. Again, due to the complexity of the external and internal factors at play, no type of lab testing can accurately predict the effects of internal heat generation under desert conditions.

During many years of PV cells and module testing we have seen different variations of field behaviors, but one thing remains constant in the desert: internal overheating of the modules that peaks around the noon hour. Internal mid-summer temperatures of up to 185°F have been measured in the Arizona and Nevada deserts. At these temperatures the abnormalities caused by any small defect in the solar cell/module design or construction are amplified and could lead to a disaster.

In conclusion, the proper PV module design and choice for large-scale desert operation requires:

1. Understanding and respect of the harsh desert conditions and related behavioral abnormalities of PV materials and structures.

   Remember: The desert consumes anything left there for a long time, and PV modules (especially the plastic materials in them) are no exception.

2. Understanding the long-term behavior of solar cells and active structures, and choosing the best and most appropriate materials and manufacturing processes to fit the extreme operating conditions.

3. Understanding the long-term behavior of PV modules operating in the desert.

    Remember: Every new module is expected to last a long time, so it must be made properly and with high quality materials. Any mistake or compromise can bring doom to it, the whole batch of modules, and the entire PV project.

4. Designing and executing module construction that is suitable for desert operation by
    a. Providing best possible edge sealing to protect the active layers
    b. Providing well designed side frame to cover the exposed module edges
    c. Avoiding glass/glass module construction

5. Proper field installation to avoid corrosion and overheating.

6. Proper operation and thorough maintenance, carefully watching for signs of early mortality and power fluctuations. This is needed to identify failing modules or strings and take immediate action, hopefully before the warranty period has expired.

7. Understanding that there is no way to predict with any certainty what will happen, when and where in the desert, so we need to have contingency plans for all possible circumstances:
    a. Manufacturer's and long-term performance warranties must be well designed to account for any materials, labor and performance deviations,
    b. Performance insurance must be well designed to take care of excess power drop, module failures, and any accidents and incidents—manmade or caused by Mother Nature.

In summary, we'd like to use an allegorical illustration of today's reality, which describes most accurately the present-day situation in the world's desert solar fields. The only animals that can live and work in the desert for a long time without any negative consequences are desert camels. These camels are especially designed for long hauls under extreme heat, without a drop of water. Their bodies, skin, brain, lungs, bladder, stomach, legs and feet are especially designed to handle the uniqueness, the harshness and the adversities of the desert elements. No other animal even comes close to the practical use of camels in the deserts.

The type of PV modules we need to install in the deserts must be specially designed and tested for exclu-

sive long-term operation in the deserts. But, instead of camel-like PV modules, what we see today are elegant-looking, exotic creations which can do many tricks to amuse the public, but which have not been designed for a long life in the desert.

Most of today's PV modules designers have never seen a desert, and have no idea what the desert is capable of doing to their product.

So, the lesson is simple: put only desert-designed and desert-tested modules in the deserts. It is very simple task, but we don't see many efforts in this area.

CASE STUDY

In addition to the technical issues surrounding the proper implementation and efficient, long-term operation of solar power fields, we do encounter serious problems of an administrative nature. Problems related to workers, engineers and managers' ignorance or negligence.

These problems are hard to detect during the production process, and most often become obvious during the operation stages, when defects show up in different ways, some resulting in premature power loss and even failure.

Note: The following Case Study—a real case—is published with the permission of the International Finance Corporation, World Bank Group. It reflects some of the major issues that are commonly encountered during the solar plants' design, construction and O&M stages.

This case study highlights a wide range of issues and lessons learned from the development and construction of the 5 MW plant in Tamil Nadu, India. It should be noted that many of these issues (e.g. the losses to be applied in an energy yield prediction or the importance of a degree of adjustability in a supporting structure) come down to the same fundamental point: it is essential to get suitable expertise in the project team.

This not only applies to technical expertise but also to financial, legal and other relevant fields. It can also be achieved in a variety of ways—hiring staff, using consultants or partnering with other organizations.

Issues and lessons described in these case studies will inform the actions of other developers and help promote good practice in the industry to ensure that best practices can be followed to support project financing in this sector.

Energy Yield Prediction

The developer of a the 5MW plant in Tamil Nadu required a solar energy yield prediction to confirm project feasibility and assess likely revenues. The developer did not consider the range of available input data or conduct a long-term yield prediction over the life of the project; both of these would have been useful to derive a more accurate yield figure, particularly for potential project financiers.

The developer sourced global horizontal irradiation data for the site location. Commercially available software was used to simulate the complex interactions of temperature and irradiance impacting the energy yield. This software took the plant specifications as input and modeled the output in hourly time steps for a typical year. Losses and gains were calculated within the software.

These included:
- Gain due to tilting the module at 10°.
- Reflection losses (3.3%).
- Losses due to a lower module efficiency at low irradiance levels (4.2%).
- Losses due to temperatures above 25°C (6%).
- Soiling losses (1.1%).
- Losses due to modules deviating from their nominal power (3.3%).
- Mismatch losses (2.2%).
- DC Ohmic losses (1.8%).
- Inverter losses (3.6%).

The software gave an annual sum of electrical energy expected at the inverter output in the first year of operation.

Although this is a useful indicative figure, an improved energy yield prediction would also consider:
- Inter-row shading losses (by setting up a 3D model).
- Horizon shading, if any.
- Near shading from nearby obstructions including poles, control rooms and switch yard equipment.
- AC losses.
- Downtime and grid availability.
- Degradation of the modules and plant components over the lifetime of the plant.

The results will ideally show the expected output during the design life of the plant and assess the confidence in the energy yield predictions given by analyzing:
- The uncertainty in the solar resource data used.
- The uncertainty in the modeling process.
- The inter-annual variation in the solar resource.

The energy yield prediction for the 5MW plant was provided as a first year P50 value (the value expected with 50% probability in the first year) excluding degradation. The confidence that can be placed in the prediction would typically be expressed by the P90 value, the annual energy yield prediction that will be exceeded with 90% probability.

Projects typically, have a financing structure that requires them to service debt once or twice a year. The year on year uncertainty in the resource is therefore taken into account by expressing a "one year P90." A "ten year P90" includes the uncertainty in the resource as it varies over a ten-year period. The exact requirement will depend on the financial structure of the project and the requirements of the financing institution.

Design

It is vital to ensure that suitable technical expertise is brought to bear on every aspect of the plant design through in-house or acquired technical expertise.

In case of the 5 MW project, the most significant of the design flaws were:

Foundations
- The foundations for the supporting structures consisted of concrete pillars, cast *in situ*, with steel reinforcing bars and threaded steel rod for fixing the support structure base plates. This type of foundation is not recommended due to the inherent difficulty in accurately aligning numerous small foundations.
- Mild steel was specified for the fixing rods. As mild steel is prone to corrosion, stainless steel rods would have been preferable.

Supporting Structures
- The supporting structures were under-engineered for the loads they were intended to carry; in particular, the purlins sagged significantly under the load of the modules. As supporting structures should be designed to withstand wind loading and other dynamic loads over the life of the project, this was a major problem. Extensive remedial work was required to retrofit additional supporting struts.
- The supporting structure was not adjustable (i.e., no mechanism was included to allow adjustment in the positioning of modules). This is a basic mistake which compounded the flaw in the choice of foundation type; the combination of these two mistakes led to extensive problems when it came to attempting to align the solar modules.

Electrical

- String diodes were used for circuit protection instead of string fuses. Current best practice is to use string fuses, as diodes cause a voltage drop and power loss, as well as a higher failure rate.

- No protection was provided at the sub-main or main junction boxes. This means that for any fault occurring between the array junction boxes and the DC distribution boards (DBs), the DBs will trip and take far more of the plant offline than is necessary.

- No load break switches were included on junction box before the DBs. This means it is not possible to isolate the plant at the array, sub-main or main junction box levels for installation or maintenance.

- The junction boxes did not allow for string monitoring. This reduces fault diagnosis capability. The design flaws listed above cover a wide range of issues. However, the underlying lesson is that is it vital to ensure that suitable technical expertise is brought to bear on every aspect of the plant design. Should the developer not have all the required expertise in-house, then a suitably experienced technical advisor should be engaged. It is also recommended that, regardless of the level of expertise in-house, a full independent technical due diligence is carried out on the design before construction commences.

Bear in mind that it is far cheaper and quicker to rectify flaws at the design stage than during or after construction.

Plant Design Conclusions

The performance of a PV power plant may be optimized by reducing the system losses. Reducing the total loss increases the annual energy yield and hence the revenue, though in some cases it may increase the cost of the plant. In addition, efforts to reduce one type of loss may conflict with efforts to reduce losses of a different type. It is the skill of the plant designer to make compromises that result in a plant with a high performance at a reasonable cost.

For plant design, there are some general rules of thumb. But specifics of project locations—such as irradiation conditions, temperature, sun angles and shading—should be taken into account in order to achieve the optimum balance between annual energy yield and economic return.

It may be beneficial to use simulation software to compare the impact of different module or inverter technologies and different plant layouts on the predicted energy yield and plant revenue.

The solar PV modules are typically the most valuable and portable components of a PV power plant. Safety precautions may include anti-theft bolts, anti-theft synthetic resins, CCTV cameras with alarms and security fencing.

The risk of technical performance issues may be mitigated by carrying out a thorough technical due diligence exercise in which the final design documentation from the EPC contractor is scrutinized.

Permits and Licensing

There are many permits required for a multi-megawatt PV power plant in India. In addition to the permits, there will naturally be permits and licenses required as a result of simply operating a business in India (e.g., human resource requirement). These need to be given equal attention in the development phase.

A lesson learnt in the case of the Tamil Nadu plant was that comprehensive legal advice on the permits is required as well as a stringent management and follow up of the application processes. It must be noted that some permit requirements were not relevant to the Tamil Nadu plant. Permission from the Ministry of Defense, for example, was not required as the site was not in a militarily sensitive zone.

The majority of the permits were applied for and in place prior to the start of construction. This is deemed best practice and sets a good example for other developers. One permission, involving land access rights, was overlooked.

The main access route to the plant was through land owned by another party. Until rights are obtained, the project remains vulnerable to the risk of goodwill being withdrawn.

Some permits were issued on the condition that the plant was to be completed before a certain date. This caused problems when the project was delayed. As a result, a re-application or extension was required. This illustrates the importance of effective planning of projects and scheduling of construction.

Construction and Project Management

The 5MW PV power plant in Tamil Nadu, India, was constructed during 2010. At commencement, construction was projected to be completed within 38 weeks; due to various factors, many of which are covered in the case studies in this book, construction took approximately 52 weeks—a significant and costly delay.

In addition, the constructed plant suffered from

serious quality issues. The construction schedule should be carefully thought through by suitably experienced personnel. It is also recommended that project management software tools are used as these enable the developer to track the progress of a project, identify resource constraints and understand the impact of uncompleted tasks.

The main causes for the construction delays were:

- Design flaws. Poor design of components, such as the support structures, lead to costly and time-consuming remedial measures.

- Poorly planned construction schedule. The illogical sequencing of construction tasks caused a number of delays.

- Monsoon rains restricted access to the site as the access road had not been sealed. The access route should have been sealed well before the arrival of the monsoon.

- Modules were damaged (and were at risk of theft) as they were stored unprotected on-site for long periods of time. Modules and other valuable components should not be delivered to a site until shortly before they are required. If they must be delivered earlier, then they should be stored in a controlled and secure environment.

The constructed plant suffered from a range of serious quality issues. These included:

- Foundations in incorrect locations.
- Poor alignment of foundations.
- Cracked and damaged foundations.
- Elements of the supporting structure left unattached.
- Poorly aligned solar modules.
- Damaged solar modules.
- Poor attention to detail in finishing of substation buildings.

While a wide range of factors undoubtedly contributed to each of these issues, the following factors are considered to have been particularly significant:

- Design quality. Certain basic aspects of the design led directly to construction quality issues. The clearest example of this is the design of the foundations and substructures leading to misalignment

of the modules. These problems could have been avoided if suitable expertise had been used in the design stage.

- Design documentation and document control. It is preferable to have a full set of "for construction" design drawings before construction commences. Throughout the construction process, it is vital that document management is thoroughly carried out; in particular, design changes and revision of drawings should be rigorously controlled. The failure to do this led to basic mistakes such as foundations being constructed in the wrong locations.

- Contractor capability. It is fundamentally important to select a suitably experienced and capable construction contractor. Ideally, a contractor with demonstrable experience of similar projects should be selected. In any case, the potential contractors' proposed approach to quality management should be thoroughly scrutinized during the contractor selection process.

- Project Management—supervision/monitoring. Regardless of the capability of the selected contractor, the developer must monitor construction progress closely. Suitably experienced personnel should regularly inspect the progress and quality of the works (and the completeness of the quality records) as they progress. If the developer does not have suitable resources in-house to carry out construction supervision then they should engage a competent party to do it on their behalf.

Quality Management

Controlling construction quality is essential for the success of the project. The required level of quality should be defined clearly and in detail in the contract specifications.

A quality plan is an overview document (generally in a tabular form), which details all works, deliveries and tests to be completed within the project. This allows work to be signed off by the contractor and enables the developer to confirm if the required quality procedures are being met.

A quality plan will generally include:

- Tasks (broken into sections if required).
- Contractor completing each task or accepting equipment.
- Acceptance criteria.

- Completion date.
- Details of any records to be kept (for example, photographs or test results).
- Signature or confirmation of contractor completing tasks or accepting delivery.
- Signature of person who is confirming tasks or tests on behalf of the developer.

Quality audits should be completed regularly. These will help developers verify if contractors are completing their works in line with their quality plans. Audits also highlight quality issues that need to be addressed at an early stage. Suitably experienced personnel should undertake these audits.

Operations and Maintenance Conclusions

It is important to define the parameters for the operation and maintenance of a PV project during its life. These conditions must, as a minimum, cover the maintenance requirements to ensure compliance with the individual component warranties and EPC contract warranties. If an O&M contractor is being employed to undertake these tasks, it is important that the requirements are clearly stated in the contract along with when and how often the tasks need to be conducted.

It is normal for an O&M contractor to provide a warranty guaranteeing the availability of the PV plant. In some cases when the O&M contractor is also the EPC contractor, it is possible for the warranty to include targets for the PR or energy yield. The agreed availability limits are often based on the independently verified energy yield report, but with some leeway.

In general, the O&M activities for a solar PV power plant are less demanding than those related to other forms of electricity generation. This is mainly due to the fact that there are no moving parts in a solar PV system (unless it is a tracking system). However, maintenance is still an important factor in maximizing both the performance and lifetime of the plant components.

Author's note: Although the key factors and issues of solar generation fields' design and implementation vary from project to project and from country to country, the underlying issues of materials quality and labor/management deficiencies prevail.

This particular case is just one of many projects that have experienced problems like those described above. It is a reminder that special consideration, with emphasis on thorough understanding, good training, and utmost attention to detail, must be exercised from the very beginning to the very end of every project.

Experienced engineers, efficient management and highly trained labor—in addition to high-quality materials—are the key to the success of any and every project. Solar power fields are no exception.

Notes and References

1. A Manual for the Economic Evaluation of Energy Efficiency and Renewable Energy Technologies, http://www.nrel.gov/docs/legosti/old/5173.pdf
2. INACCURACIES OF INPUT DATA RELEVANT FOR PV YIELD PREDICTION, Stefan Krauter, Paul Grunow, Alexander Preiss, Soeren Rindert, Nicoletta Ferretti, Photovoltaik Institut Berlin AG, Einsteinufer 25, D-10587 Berlin, Germany
3. PV module characterization, Stefan Krauter & Paul Grunow, Photovoltaik Institut Berlin AG, TU-Berlin, Germany
4. PV MODULE LAMINATION DURABILITY, Stefan Krauter, Romain Pénidon, Benjamin Lippke, Matthias Hanusch, Paul Grunow, Photovoltaic Institute Berlin, Berlin, Germany
5. PV module testing—how to ensure quality after PV module certification. Photovoltaic Institute, Berlin. August, 2011
6. Basic Understanding of IEC Standard Testing for Photovoltaic Panels, by Regan Arndt and Dr. Ing. Robert Puto, TÜV SÜD Product Service
7. Topaz power field http://www.sloplanning.org/EIRs/topaz/FEIR/FEIR/Vol1/C%20files/C09_Hazards_.pdf
8. PV Module Reliability Workshop Cultivates Innovation, Progress, March 15, 2011
9. A.H. Fanney, et al., "Short-Term Characterization of Building Integrated Photovoltaic Modules," Proceedings of Solar Forum 2002, Reno, NV, June 15-19, 2002.
10. Utility Scale Solar Power Plants. A Guide For Developers and Investors, 2012. International Finance Corporation, World Bank Group.
11. Performance Degradation of Grid-Tied Photovoltaic Modules in a Desert Climatic Condition by Adam Alfred Suleske, ASU, Tempe, Arizona
12. Hot Spot Evaluation of Photovoltaic Modules," by Govindasamy (Mani) TamizhMani and Samir Sharma
13. Photovoltaic array performance model, D.L. King, W.E. Boyson, J.A. Kratochvil Sandia National Laboratories, Albuquerque, New Mexico 87185-0752
14. Photovoltaics for Commercial and Utilities Power Generation, 2011. Anco S. Blazev
15. Comparison of Degradation Rates of Individual Modules Held at Maximum Power, C.R. Osterwald, J. Adelstein, J.A. del Cueto, B. Kroposki, D. Trudell, and T. Moriarty
16. B. Kroposki, W. Marion, D. King, W. Boyson, and J. Kratochvil, "Comparison of Module Performance Characterization Methods," 28th IEEE PV Specialists Conference, 2000, pp. 1407-1411.
17. Economic Impacts from the Promotion of Renewable Energy Technologies. The German Experience. Manuel Frondel, Nolan Ritter, Christoph M. Schmidt, Colin Vance http://repec.rwi-essen.de/files/REP_09_156.pdf
17. Common Failure Modes for Thin-Film Modules and Considerations Toward Hardening CIGS Cells to Moisture A "Suggested" Topic Kent Whitfield, MiaSole.
18. Adhesion and Thermomechanical Reliability for PVD and Moddules, Reinhold H. Dauskardt.
19. Photovoltaics International, August, 2012. "Performance characterization and superior energy yield of First Solar PV power plants in high-temperature conditions." by First Solar staff.
20. Photovoltaics for Commercial and Utilities Power Generation by Anco S. Blazev. Fairmont Press, 2011.

UPDATES

SolFocus

SolFocus designs, manufactures and sells concentrated photovoltaic (CPV) tracking systems. This is the most efficient PV technology, and SolFocus was developed to the state of the art. SolFocus has 18 MWp of trackers deployed, and a strong pipeline of future projects. In addition, there was quarter-to-quarter growth in revenue for six consecutive quarters, and the product in the field outperformed expectations, with no product issues.

The future looked brilliant for the young company—until November 2012, when SolFocus announced that it was "restructuring" the company OM and gearing for an immediate sale. The reason given was "cash flow problems not due to product performance or lack of sales," according to company reps.

After all the success stories, the darling of the CPV industry determined that finding a buyer for the technology who can continue work on the project pipeline was the way to go. It appears that the major investors were tired of the solar game, and decided to sell out.

The firm's largest investor, NEA, has $60 million invested in the company since 2006, with the total investment estimated at $230 million from the likes of NEA, NGEN, Apex, and others.

Negotiations with investment bank Advanced Equities to raise a Round E of VC funding collapsed after Advanced Equities went under recently too.

SolFocus' most promising project was the planned 450 MWp CPV project in northern Mexico. Starting with 50 MWp, it had the potential of expanding to 450 megawatts. This would have made it the largest CPV development on earth by far, since currently the world's largest operating CPV power plant is the 30-megawatt Alamosa in Colorado, built by the failing Amonix whose future (the company and the project) is uncertain.

SolFocus is the last in the line of CPV companies, like GreenVolts, Energy Innovations, Skyline Solar, and Amonix that have had to restructure, fold, or sell the business in the recent solar race.

Twin Creeks Technologies

Twin Creeks Technologies is an equipment manufacturer that tried to introduce a revolutionary piece of equipment, used widely in the semiconductor industry, into the solar cells making process. The equipment is an ion implant machine, that can be used for proton-induced exfoliation (or PIE) processing. Here, highly charged hydrogen ions (or rather a proton) generated in the equipment are shot into a crystal substrate. Upon impact, the protons could do some damage or restructuring of the material, depending on the use.

In the solar case, the substrate is silicon and the purpose is using the proton energy to slice very thin wafers from its surface. The hydrogen protons penetrate at a predetermined depth, and line up under the surface. The substrate is heated, and the temperature gradient cleaves the upper part of the substrate, along the crystalline plane at the depth of penetration of the protons. By changing the proton energy, the depth of the protons' penetration can be controlled, and that determines the thickness of the resulting wafer.

The thick silicon substrate can be reused until totally depleted. Amazing, no? Instead of using diamond or a wire saw, which produce a lot of waste and very thick wafers, this new method has no waste, and the wafers are paper thin, flexible and resistant to cracking.

Unfortunately, like many other brilliant ideas, the energy market is not very interested in new solar technologies. So, in November, 2012, the company sold its assets for approximately $10 million to GT Advanced Technologies. This is a 1-to-8 investment ratio, since the firm had raised more than $80 million from a number of investors like Crosslink Capital, Benchmark Capital, Artis, and DAG Ventures.

GT actually has a much better chance of selling the equipment to other industries, since it plans to pursue substrates for power semiconductors, solar wafers, and thin sapphire laminates for touch screens. We doubt that anyone would be able to use this equipment for making solar cells in the present state of the art, because the super-thin wafers would be hard to process into solar cells. So, a good idea goes on the shelf, and GT will use the equipment in other areas until the thin wafers can be successfully processed into solar cells.

SOLAR TRENDS

In November, 2012, one of the most active solar financiers, VC firm Mohr Davidow, announced that it is "narrowing" its cleantech practice. Until recently, Mohr Davidow was the leading cleantech investment firm, but now this practice is under revision and some changes were made in the company's MO.

Now, Mohr Davidow will no longer offer full-spectrum support for greentech companies and technologies. It will, instead, focus on the "convergence of greentech with IT," since IT is the firm's traditional sector of competence and interest. Company experts think that cleantech IT is now a near vertical within the broader IT spread, so that developing business of this nature needs both the greentech and the IT sides of the practice.

Mohr Davidow is still committed to its portfolio of greentech companies and strongly believes that there are winners among them, so it will continue supporting them, according to the reps. However, few, if any, of the companies in the firm's greentech portfolio meet the new IT-green convergence requirement, so the degree of support they will get in the near future seems to be conditional.

The shift at Mohr Davidow points to the changing dynamics in, and the general direction of, the cleantech sector in the present economic and energy situations. The top five VC firms don't admit openly that they invest in cleantech, as they did in the past. Similarly, the US DOE and the White House are silent on the renewables future, putting natural gas, coal and nuclear on the energy agenda instead.

Green VC investment was at $1.6 billion in the third quarter of 2012, which is similar to the second quarter, but it represents a 30% decline from the $2.23 billion invested in the same quarter in 2011. The number of deals fell to 148, compared to 169 in the same quarter of 2011. The projections are that 2012 will see only $6.8 billion in green VC investments, a 28% decline from the $9.4 billion invested in 2011. So, the easy money is gone—no more free lunches for wind and solar—no more Solyndras. A new era is shaping up for the energy sources, which is especially noticeable in the US and China.

The promise of a "semi-green" future offered by large natural gas deposits on one hand, and the dropping prices of solar and wind products on the other, combined with the reluctance of VCs and banks to finance greentech companies and projects is creating a situation we have not seen before.

With all these changes in mind, we see an increase in solar installations in the U.S. and the trend is growing.

According to an Ernst & Young report, California continues to lead in their All Renewables Index, taking the top spot in all technologies across the board, while Colorado, Texas, New Mexico and Massachusetts took the top five places in the Index. Hawaii renewables are also climbing due to the attractive small-scale solar market.

China is quadrupling their estimates for solar energy implementation by 2020 to 50 GWp.

Nevertheless, the fact that solar can survive, and even prosper, in this unfavorable economic situation with limited government and private financial support, and under the increasing pressure of the major competitors—coal, oil and gas fossils—is a clear indicator that solar is finally standing on its own feet. This is what we have been waiting for, for the last 40 years, and it is the best news of the latest solar cycle. The solar baby seems to have grown enough to be able to walk.

In the meanwhile, we will be digging and pumping large quantities of fossils from the ground. It won't be too long before we realize fully that the fossils are badly depleted, and that they are the reason for major environmental and climate changes. Then, and only then, will the renewables be treated as equals and given full support.

Unfortunately, that day is still 2 or 3 decades away, but it might be just enough time for the solar toddler to learn to run—maybe even compete with its older cousins, the fossils. Let's hope so! Future generations depend on it.

Figure 5-15.
The solar baby learns to walk

Chapter 6

Solar Power Fields

*Things turn out best for the people who make the best
out of the way things turn out.*

Art Linkletter

Solar power is a way of life today... albeit not a very important part of the everyday lives of most people. While solar installations are popping up daily around the US and the Western world, mostly geared to reduce power costs, millions on the African and Asian continents still live in darkness, and don't have power for even the most essential needs.

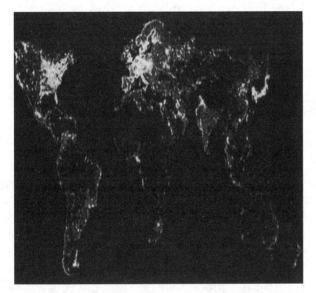

Figure 6-1. World's energy use at night

Looking at the night time picture of the world, guess which are the richest countries in the world today. And guess where solar power has been, is, and will be installed in the future. This is inequality at its best and capitalism at its worse.

Although everyone recognizes the potential of sunlight to power our world, its past is quite shaky, and its future is somewhat uncertain. Every day we hear of new technologies of attempts to expand the use of solar energy into different niche markets, some of which are actually quite successful. But these are very small steps in the marathon to full energy independence and a clean environment.

Solar won't be a major player is this marathon until it is deployed on very, very large-scale power fields in desert areas around the world. This is the most reliable if not the only way to generate Giga watts of power for our ever-increasing population, and to compete with the increasing wattage from conventional (coal, gas and oil) energy generators.

Installing solar panels under the cloudy skies of Northern Germany, or in the state of Maine, is fine to an extent, but it is an iffy situation for the long-term prospects of solar energy. It is also a most inadequate use of solar energy and our resources.

With all due respect to the people involved, such installations—no matter how large and efficient—won't solve our energy problems. This is simply because "Physics 101" tells us there is not enough power in 80-90% cloudy, rainy, and snowy skies to produce reliable solar energy. Period!

For residents in these areas, unsustainable federal and local government subsidies during 2008-2011 have obscured the issues and given false hope to that solar is the way for them, but we need to be realistic. You simply cannot generate wind power without wind, nor can you generate solar power under cloudy, drizzly skies.

No matter how much we want it to happen, or how much money we are willing to throw at it, we cannot change the laws of physics and the basic principles of the photovoltaic process. So the inconsistency and extreme variability of sunlight in these areas will prove too much to handle, and not worth the hassle and expense.

Germany is a good example of billions of euros spent on solar installations with marginal, even controversial, results at best. Solar energy in most of Germany is going sideways, and its future is uncertain. One thing we know: solar power generated under Germany's cloudy skies will not solve their energy problems, but there are other solutions. Using the deserts of North Africa, Australia and the US are the most likely.

Desert areas have unique properties that are:

a.) suitable for generating maximum amounts of reliable energy, and

b.) usually have very little other value for human use.

The combination of these two factors makes the world's deserts the perfect place for large solar installations. We cannot even imagine a better combination, and are sincerely hoping that more effort will be directed into this area in the near future, if we are serious about replacing conventional energy sources with solar energy.

Large-scale power fields of the 21st century—and especially those that deployed towards the end of the century—will look nothing like what we see today. The next generations will certainly use solar technologies and procedures that are much more sophisticated, more efficient and more reliable, and they will be situated in the world's deserts. The laws of physics and common sense dictate that scenario.

In words, today's solar technology is the Model T of the 21st century, slotted to develop into a full-blown Cadillac in the distant future. And while today's solar Model T is meandering under cloudy skies and on inefficient, leaky rooftops, by the end of the century the solar Cadillac will be speeding on paved and unobstructed desert roads, bathed in unlimited amounts of direct, bright sunlight. Then, and only then, can we hope for true and full solar power generation.

Just as the motor car became indispensible in the 20th century, so will solar become part of daily life in the 21st, for we have no other choice. Future generations will be forced to rely heavily, and later on exclusively, on solar and wind energy simply because

a.) they won't have a choice, after we have burned all the fossil fuels,

b.) they will be fighting for survival after we have contaminated everything in sight, and

c.) it is the smart thing to do.

Alternative energies will be the only solution to becoming energy independent by the end of the 21st century.

The only responsible thing for us to do (now, not later) is to lay down a foundation we can build on. We need to build the first "Model T" solar technologies as best as we can, and let future generations take it from there.

To get a glimpse into the future, we need to see what we have available today, how it is used, how to make it better, and how to use it more efficiently. So we will take an unbiased, and at times painful, look at today's solar technologies, projecting them into the future by reviewing the requirements for, and obstacles to, effective and safe performance affecting what we already have.

We divide today's solar technologies into these major categories:

1. Solar thermal power generators,
2. Photovoltaic solar power generators, and
3. Misc. (3rd generation and other technologies)

SOLAR THERMAL TECHNOLOGIES

Solar thermal, mostly concentrated solar power (CSP), technologies have been around awhile now, so we have a lot of experience with them—including their use in large-scale power fields in deserts.

Author's note: See the previous chapters for more details on each of the different CSP technologies.

One typical characteristic of CSP technologies is that they are not as negatively affected by temperature and environmental extremes as PV technologies, so they are well suited for use in inhospitable desert areas. As a matter of fact, they must be used in such areas, because—unlike most of the PV technologies—they need direct sunlight and heat, and do not work well, or at all, under cloudy skies.

Like any other energy technology, they have problems that are difficult to resolve—the major one being the fact that most of them need lots of cooling water to maintain the process efficiency. Deserts are deserts because there is not much water there, and human use consumes what there is. This factor has caused the closure of several CSP projects.

So CSP technologies must find another way to cool the process steam. Using dry cooling (AC type of cooling) is one way, but it uses a lot of energy, so we must find a compromise, to keep the technology compatible.

Too, CSP equipment, as any electro-mechanical device, is prone to degradation when exposed to harsh elements. Here, high pressure, high temperature and high humidity, used daily in the process, are expressed in above-normal wear and tear, material fatigue, corrosion etc., which add to significant maintenance costs.

Dust collected on the receivers, damage from sand, rain and hail become a significant detriments during long-term desert operation.

On the surface, CSP looks like the technology of choice, due to its simplicity, but when we add all problems, it becomes apparent that many of them must be solved first.

Since we cannot claim expertise in solving the electro-mechanical and other problems of CSP systems, we will not spend much time on these technologies, leaving that task to the researchers and CSP specialists. However, we know that CSP is a viable solar energy generating technology, which must be improved if we are to have efficient and cost-effective CSP power fields.

Major CSP technologies expected to compete successfully in the 21st century are:

1. Stirling engine dish trackers,
2. Parabolic trough trackers, and
3. Power towers.

These different solar power generating systems have been used for awhile and some have proven more suited for large-scale use than others. Their advantages and disadvantages are shown in Tables 6-1 through 6-3.

With the CSP technologies covered as much as needed for our purposes here, we will now take a closer look at today's PV technologies, which have much more complex processes, and device designs, functions, and wider applications than their CSP cousins.

The diversity of materials and configurations of these technologies contributes to the complexity of their operation under different conditions, making performance analysis difficult to express in a few words. Because we feel that these technologies will be the technologies of choice in the 21st century, we are dedicating much more space to analyzing the different PV technologies.

Table 6-1. Advantages and disadvantages of Stirling Dish systems

Stirling Engine CSP systems:

Advantages
+ Stirling engine systems have relatively high efficiency. Theoretically, up to 40% of the total energy input could be converted into electricity, which is usually reduced due to thermodynamic and electro-mechanical losses.
+ They use few bearings and seals, which require less lubricant and last longer than steam turbines,
+ Waste heat can be captured, and reused as heat source in some applications
+ They use an external heat source, which could be anything that generates heat: gasoline, diesel, coal, biofuel, or preferably solar energy. Because of that they are well suited for hybrid power generation by using the external heat generated by the different primary heat sources.
+ Stirling engines are well suited for installation on not perfectly leveled terrains, where most other solar technologies require more expensive land leveling and other such modifications.

Disadvantages
- The major disadvantage is the need for cooling.
Efficient cooling is needed, especially in hot desert regions, where fortunately or unfortunately the Stirling engine technology operates best
- They need a lengthy warm up period when starting, so operation in a partially cloudy day, when the engine must stop and start frequently might be challenging
- The working gas will eventually leak out and will have to be replaced, which means lost time, resources and additional expense
- They have a number of moving parts, which require lubrication and maintenance. Periodic maintenance operations increase the cost of operation (COO) by virtue of additional salaries and lost power generation time.
- Hot and wet parts rust with time, which requires expensive maintenance and replacement procedures.
- Mirrors used in these systems have to be of very high quality, because the optical properties of cheap mirrors reduce the efficiency and deteriorate quickly with time.
- Mirrors could get damaged and/or become non-reflective enough with time, thus requiring expensive replacement procedures.
- Basically, like with any engine that operates on and off all day long, the parts will wear down and have to be replaced, or the engine has to be rebuilt, which is an expensive operation.

Table 6-2. Advantages and disadvantages of parabolic trough systems

Parabolic Trough Systems

Advantages

+ Parabolic troughs' components (trough frame, mirrors, receiver, turbine etc.) are well established commercial products, thus risk and uncertainty factors are limited and well controlled

+ This technology is also very well supported by well financed companies and governments (including the US government and its agencies, such as DOE, NREL, SNL etc.)

+ They have relatively high efficiency. Theoretically, up to 40% of the total energy input could be converted into electricity, but thermodynamic and mechanical losses usually decrease the total efficiency

+ Waste heat can be captured and can be used as secondary heat source, and for other thermal and hybrid applications as well.

Disadvantages

- Heat transfer fluids heat is limited to 400°C, which offers only moderate steam qualities
- Steam turbines need cooling, usually lots of water, which is simply unavailable in the desert. The cooling source could be replaced by an AC (cryogenic) system but they reduce the energy output of the systems
- They have a lengthy warm up period at start, so operation in partially cloudy days, when the system must stop and start frequently, might be quite challenging and at reduced efficiency
- The troughs and the steam turbines have large number of moving parts, which require lubrication and periodic maintenance, which require additional parts, labor and lost production time
- **Hot and wet parts rust with time, which requires expensive maintenance and replacement** procedures, and causes increase in lost power generation.
- Miles of hydraulic and liquid transferring pipes, hoses, connectors, expanders etc. contribute to leaks and malfunctions, which increases the O&M cost as well.
- Mirrors used in these systems have to be of very high quality, and need to be kept clean, because the optical properties of cheap and dirty mirrors affect the efficiency and longevity
- Mirrors could get damaged and/or become non-reflective enough with time, thus requiring expensive replacement procedures

PV TECHNOLOGIES

There are a great number of PV technologies today, but most of the newer ones are still in the R&D labs. See Chapter 4 for details on the different "exotic" solar technologies—those most likely to increase in use.

Solar technologies used in everyday electric generation today, and which we consider most efficient and reliable at this time, are:

1. Single crystal silicon PV modules,

2. Thin film PV modules,

3. CPV and HCPV trackers, and

4. Hybrids PV and thermal (DC power and hot water generation)

We covered the structure and function of the different types of PV technologies in previous chapters, so here we focus on the general aspects of their use, emphasizing the actual operating parameters and the related factors and issues.

Performance Factors

PV modules are (supposedly) designed for long-term operation under adverse climate conditions. Although none of them is 100% protected against the adverse effects of nature, the goal is to ensure their proper operation during 25-30 years of non-stop use anywhere in the world.

The factors that influence proper field operation of PV modules are many and of different nature. The key factors which affect the efficiency and thermal behavior

Table 6-3. Advantages and disadvantages of power tower systems

---

Power Tower System

Advantages

+ Power tower components (frame, mirrors, drives, receiver, turbine etc.) are well established commercial products, thus risk and uncertainty factors are limited and well controlled.

+ The technology is supported by large companies and governments (including the US government and its agencies, such as DOE, NREL, SNL etc.)

+ They have relatively high efficiency. Theoretically, up to 40% of the total energy input could be converted into electricity, some of which is reduced by thermodynamic and mechanical losses,

+ Waste heat can be captured also and can be used as secondary heat source and / or for thermal power generation as well.

+ Mirrors do not require much maintenance once installed, however they must be cleaned periodically at a small cost (water and workforce). Production of mirrors when compared to PV cells is less energy-intensive and more environmentally friendly.

+ The noisiest part of the system is the steam turbines, but the plants as a whole are quiet.

Disadvantages

- Power towers need cooling, but water in the desert is scarce and this is a big problem. AC (cryogenic) systems could be used, but they use a lot of power which decreases energy output.

- They use many tracking mirrors, so the terrain before installation needs to be leveled.

- Significant part of light beams from the mirrors is reflected by the collector, which might be dangerous to the safety of traffic and other activities in the vicinity.

- They need a lengthy warm up period when starting, so operation in a partially cloudy day, when the system must stop and start frequently might be challenging.

- They have a large number of moving parts, which require tune-up, lubrication and maintenance. These procedures increase the cost of operation (COO) by adding labor costs and lost power.

- Hot and wet parts rust with time, which requires expensive maintenance procedures.

- Hydraulic, steam and liquid handling apparatus contribute to leaks and malfunctions which also add to the O&M expenses

- Mirrors used in these systems have to be of very high quality, because the optical properties of cheap mirrors cause decease of efficiency and their quality deteriorates quickly with time.

- Mirrors could get damaged and/or become non-reflective enough with time, thus requiring expensive replacement procedures

- The towers and mirrors have a high profile and can be perceived as a visual and esthetic pollution.

---

of PV modules are:

a. The degree of reflection from the top surface of the module;

b. The electrical operating point of the module components;

c. Absorption of sunlight by the regions which are not covered by solar cells;

d. Absorption of low energy (infrared) light in the module or solar cells;

e. The packing density of the solar cells; and

f. The overall quality of the solar cells and modules.

Each of these factors is seemingly independent in their origin and function, but in fact they are closely inter-related in their combined final effect and long-term influence on PV modules' and power fields' efficiency and longevity. Following are the key factors.

Front Surface Reflection

Light reflected from the front surface of the module does not contribute to the electrical power generated. It is considered a loss which needs to be minimized. Reflected light does not contribute to heating the PV module. The maximum temperature rise of the module is therefore calculated as the incident power multiplied by the reflection. For typical PV modules with glass top cover, the reflected light contains about 4% of the incident energy.

Operating Point and Efficiency of the Module

The operating point and efficiency of the solar cell determine the fraction of light absorbed by the cell that is converted into electricity. If the solar cell is operating at short-circuit current or at open-circuit voltage, then it is generating no electricity and hence all the power absorbed by it is converted into heat.

Absorption of Light by the PV Module

The amount of light absorbed by the parts of the module, including the solar cells, will also contribute to the heating of the module and does not contribute to generating power. How much light is absorbed and how much is reflected by the non-solar cell areas is determined by the color and materials of the rear backing layer of the module.

Absorption of Infra-red Light

Light which has an energy level below that of the band gap of the solar cells cannot contribute to generating free electrons and electrical power. On the other hand, if it is absorbed by the solar cells or by the module, this light will contribute to heating. The aluminum metallization at the rear of the solar cell tends to absorb this infrared light. In modules without rear aluminum cover, the IR light may pass through the solar cell and exit from the back of the module.

Packing Factor

Solar cells are specifically designed to be efficient absorbers of solar radiation. The cells will generate significant amounts of heat, usually much higher than the module encapsulation and rear backing layer. Therefore, a higher packing factor of solar cells, in addition to increasing the generated power, will also increase the generated (parasitic) heat per unit area. This increase is harmful for long-term operation of modules, especially in areas of extreme heat, so the packing density must always be considered and adjusted accordingly to balance power output and module heating.

Front Surface Reflection

Light reflected from the front surface of the module does not contribute to the electrical power generated. It is considered a loss which needs to be minimized. Reflected light does not contribute to heating the PV module. The maximum temperature rise of the module is therefore calculated as the incident power multiplied

by the reflection. For typical PV modules with glass top cover, the reflected light contains about 4% of the incident energy.

Operating Point and Efficiency of the Module

The operating point and efficiency of the solar cell determine the fraction of light absorbed by the cell that is converted into electricity. If the solar cell is operating at short-circuit current or at open-circuit voltage, then it is generating no electricity and hence all the power absorbed by it is converted into heat.

Absorption of Light by the PV Module

The amount of light absorbed by the parts of the module other than the solar cells will also contribute to the heating of the module and does not contribute to generating power. How much light is absorbed and how much is reflected by the non-solar cell areas is determined by the color and materials of the rear backing layer of the module.

Absorption of Infra-red Light

Light which has an energy level below that of the band gap of the solar cells cannot contribute to generating free electrons and electrical power. On the other hand, if it is absorbed by the solar cells or by the module, this light will contribute to heating. The aluminum metallization at the rear of the solar cell tends to absorb this infrared light. In modules without rear aluminum cover, the IR light may pass through the solar cell and exit from the back of the module.

Packing Factor

Solar cells are specifically designed to be efficient absorbers of solar radiation. The cells will generate significant amounts of heat, usually much higher than the module encapsulation and rear backing layer. Therefore, a higher packing factor of solar cells, in addition to increasing the generated power, will also increase the generated (parasitic) heat per unit area. This increase is harmful for long-term operation of modules, especially in areas of extreme heat, so the packing density must always be considered and adjusted accordingly to balance power output and module heating.

Mechanical Effects

There are a number of materials in the PV module that act and interact differently with adjacent materials, depending on temperature, pressure and other factors. The major mechanical forces determining the interaction between the different layers in the module are:

a. Coefficient of friction is the force that resists the lateral motion of solid surfaces in contact with each other and moving in different directions. This includes the types of materials (substrates and layers of materials stacked on top of each other) of which PV modules are made.

b. Coefficient of thermal expansion (linear and volumetric) is the way different materials expand and shrink under different temperature gradients. In other words, all materials tend to change shape and size along the surface and across their volume, with changes in temperature, which ultimately results in mechanical friction if in contact with other materials.

These forces also determine a number of events and phenomena that occur in the modules over time. Hot and cold days, windy conditions, hail, sand and storms effect modules, and all these must be taken into consideration in the design and operation of PV power plants. The relations and interactions between the module's components could be described as intimate, ongoing, and relentless.

In Arizona deserts, these materials have to go through blistering 180°F temperatures during the day in summer and sub-freezing temperatures in winter, not to mention the effects of UV light. Also, moisture, air, environmental chemicals, and gasses penetrating the module will cause quick deterioration and even failure.

Because the coefficient of expansion of these materials is very different, they go through never ending shifting, sliding and slipping, expanding and shrinking, constantly rubbing against each other in a soup of dissimilar molecules.

This is where any poor design, or deviation from the materials or manufacturing specs will be revealed, most likely as decreased performance or other failure.

Cracks, chips, pits, crumbling, voids, adhesion failures, delamination, chemical decomposition and reactions within the cells and the modules are expected during the 25-30 years of operation—due to natural elements. Some of these effects, such as yellowing of the EVA, or change of front surface color due to degrading of the AR coating, become visible to the naked eye.

Most other internal and some external processes, however, start and continue undetected visually as time goes by.

Moisture and Chemical Ingress into the Module

This is another serious condition which could reduce the modules' efficiency and life. If it occurs soon after the modules are installed, it would very likely be traced to a manufacturing defect, or handling problems during the transport or installation steps. At times, however, moisture in the modules is found months or years after they have been operating in the field.

The reasons are usually defects in the laminate layers, caused by defective materials or improper handling or processing. Desert sunlight is especially hard on the organic materials in the modules too, so it is usually only a matter of time before they start breaking down mechanically and chemically, at which point moisture and air could penetrate the inside of the modules.

Moisture ingress is a diffusion process (liquid diffusion that is). Its diffusivity depends on the module type, its frame construction and laminate materials composition and application. Moisture could penetrate into PV modules through the edge seals and from front or back cover defects (cracks, pores and such). Once in the module, moisture could easily penetrate the encapsulants, causing them to delaminate, discolor, and mechanically disintegrate. In parallel with that, moisture could oxidize and otherwise attack the solar cells' metal contacts, causing them to degrade or fail completely.

Encapsulants and sealants have changed little during the last 30 years, so the best defense we have in assuring the quality is to make sure that only top-quality materials are used, and that they are applied according to established manufacturing procedures and QC specs.

Quality Considerations

Quality is measurable entity in most cases, which can be given a number or quantity of some sort in all measurable cases. This, however, is not so simple during the manufacturing or operation of solar cells and modules. There are too many internal structural parameters and operational discrepancies to be considered when looking at quality, with most issues either hidden, or revealing themselves very slowly over time.

Quality is also something that must be built into the product. What comes to mind here is the establishment of the so-called, "Guarantee Group" during the design, building, and even the ill-fated voyage of the *Titanic*.

The Guarantee Group functioned as the ultimate authority during all stages of the project, but the most remarkable part of its function is the formation of a group of nine individuals selected to attend a ship's maiden voyage. Their job was to locate and record the performance of all ship components, and fix any prob-

358

Solar Technologies for the 21st Century

lems that arose.

The members in the Guarantee Group were deemed the best in their respected fields of expertise at that time. This elite group of individuals was suppose to change with every consecutive voyage.

The possibility of being part of the Guarantee Group for the *Titanic* was motivation for employees at Harland and Wolff shipyard to work hard to prove their abilities to the company. All nine individuals perished in the sinking of the *Titanic*, so we will never know what they found during the several days at sea, nor what they did during the last several hours of *Titanic's* ordeal.

One thing we know for sure, however, is that this dedication to quality—from the beginning to the end (literally speaking) of the project—is something to be admired and reciprocated. This total dedication to quality—from the beginning to the end of each module manufacturing and project design, setup and operation—is something we need to incorporate in our efforts to make solar a recognizable competitor to, and equal partner with, conventional technologies.

Due to the high number of issues we observe when testing and operating PV cells and modules, we do believe that quality is lacking in many instances. High numbers of modules from many manufacturers fail during reliability and certification tests.

30% failure rate, which is what we see lately, is a frightening number; yet, we see it time after time. But even 20, 15, 10 and 5% are huge numbers. 5% of a 100MWp field is 5MWp of modules going bad for one or another reason, while 30% field failure is unthinkable.

One of the major issues with PV modules (all PV modules with no exception) is the negative effect of high temperature on their performance and longevity. Desert areas—where most of these technologies tend to be installed and operated today—with their harsh climate are the great enemy. As a rule of thumb, anything left in the desert will be destroyed. Including our shiny PV modules.

Any organics in the modules, such as encapsulants, edge sealers, glues, plastic frames, retainers, screws and such will be first to go and within months, in some cases.

They will be followed by anything that gets exposed to the elements. After the destruction of the organic materials, it is the solar cells and/or thin films that are left for the desert to deal with. During 30 years of daily on-sun exposure, the desert elements will have plenty of time to disintegrate and decompose the fragile materials left exposed to its blistering rays, freezing nights, and storms.

So what are we to do? First we must make sure

that we understand that heat in our shiny PV modules is enemy number one. Dealing with the heat is our next objective, so we'll go into more detail now.

Heat Dynamics in PV Modules

We will take a closer look at the temperature effects below, so it suffices to remind the reader at this point that any PV cell or module will lose ~0.5-0.6% of its total output per degree C increase of its internal temperature. This is a significant number and causes a lot of concern about the adaptability of some PV modules for long-term operation in extreme heat conditions.

PV modules operate best in sunny areas, and deserts are best for this purpose. Deserts, however, are home to some of the harshest climates on earth. With an expected 30 years of continuous operation in sand storms, dust storms, rain storms, extreme heat and freezing temperatures, PV modules and other equipment in solar power fields, undergo serious torture. The thermal impact of their environment is the most serious element of their operation, and has the greatest negative effect on them. We must know what to expect and prepare for it.

Understanding heat behavior of a PV module, and the complex processes and interactions within it, is paramount for proper PV module design, manufacturing, installation and operation. With mega-fields planned in US deserts, where the temperatures (measured on cells in the modules) exceed 180°F, heat becomes the primary enemy, which we need to overcome, if we are to be successful there.

The operating temperature of a PV module results from an equilibrium between the heat generated by the PV module and the heat loss to the surrounding environment.

Heat conduction, heat convection, and heat radiation are the three main mechanisms that we should be aware of and fully understand, as they are major contributors to heat generation and loss in PV modules. Understanding and controlling heat will help maintain efficiency and extend the life of our PV modules.

Temperature Effects

A PV module exposed to sunlight generates heat as well as electricity. For a typical commercial PV module operating at its maximum power point, only 10 to 15% of the incident sunlight is converted into electricity, with much of the remainder being converted into heat. Some of the heat is radiated back into the environment, while a significant portion is retained in the module.

Excess heat, such as in deserts, can overheat the

modules, thus reducing their output, accelerating their deterioration and even causing them to fail. A more detailed analysis of the temperature-related factors and issues follows (14).

Heat Conduction

Conductive heat losses are due to thermal gradients between the PV module and other materials (including the surrounding air) with which the PV module is in contact. The ability of the PV module to transfer heat to its surroundings is characterized by the thermal resistance and configuration of the materials used to encapsulate the solar cells.

Conductive heat flow is analogous to conductive current flow in an electrical circuit. In conductive heat flow, the temperature differential is the driving force behind the conductive flow of heat in a material with a given thermal resistance, while in an electric circuit the voltage differential causes a current flow in a material with a particular electrical resistance. Therefore, the relationship between temperature and heat (or power) is given by an equation similar to that relating voltage and current across a resistor.

Assuming that a material is uniform and in a steady state, the equation between heat transfer and temperature is given by:

$$\Delta T = \Phi P_{heat}$$

where

P_{heat} is the heat (power) generated by the PV module;

Φ is the thermal resistance of the emitting surface in °C W-1; and

ΔT is the temperature difference between the two materials in °C.

The thermal resistance of the module depends on the thickness of the material and its thermal resistivity (or conductivity). Thermal resistance is similar to electrical resistance and the equation for thermal resistance is

$$\Phi = \frac{1 * l}{k * A}$$

where:

A   is the area of the surface conducting heat;

l   is the length of the material through which heat must travel; and

k   is the thermal conductivity in units of W m-1/°C-1.

To find the thermal resistance of a more complicated structure, the individual thermal resistances may be added in series or in parallel. For example, since both the front and the rear surfaces conduct heat from the module to the ambient, these two mechanisms operate in parallel with one another and the thermal resistance of the front and rear accumulate in parallel. Alternately, in a module, the thermal resistance of the encapsulant and that of the front glass would add in series.

Heat Convection

Convective heat transfer arises from the transport of heat away from a surface as the result of one material moving across the surface of another. In PV modules, convective heat transfer is due to wind blowing across the surface of the module. The heat which is transferred by this process is given by the equation

$$P_{heat} = hA\Delta T$$

where

A   is the area of contact between the two materials;

h   is the convection heat transfer co-efficient in units of W m-2 °C-1; and

ΔT is the temperature difference between the two materials in °C.

Unlike the thermal resistance, h is complicated to calculate directly and is often an experimentally determined parameter for a particular system and conditions.

Heat Radiation

A final way in which the PV module may transfer heat to the surrounding environment is through heat radiation. As discussed before, any object will emit radiation based on its temperature. The power density emitted by a blackbody (a perfect radiation absorber) is given by the equation

$$P = \sigma T^4$$

where

P   is the power generated as heat by the PV module;

σ is the Stafan-Boltzmann constant, and

T   is the temperature of the solar cell in K.

However, a PV module is not an ideal blackbody, and to account for non-ideal blackbodies, the blackbody equation is modified by including a parameter called the emissivity, ε, of the material or object. A blackbody, which is a perfect emitter (and absorber) of energy has an emissivity of 1.

An emissivity of an object can often be gauged by its absorption properties, as the two will often be very similar. For example, metals (which tend to have reduced absorption) also have a lower emissivity, usually in the range of 0.03. Including the emissivity in the equation for emitted power density from a surface gives

$$P = \varepsilon \sigma T^4$$

where

ε is the emissivity of the surface; and
the remainder of the parameters are as above.

The net heat or power lost from the module due to radiation is the difference between the heat emitted from the surroundings to the module and the heat emitted from the PV module to the surroundings, or in mathematical format

$$P = \varepsilon \sigma \, T^4 sc - T^4_{amb}$$

where

Tsc   is the temperature of the solar cell;
T_{amb} is the temperature of the ambient surrounding the solar cell; and the remainder of the parameters are as above.

NOCT

Since we are still on the very important subject of temperature effects on PV module performance and longevity, we need to discuss the concept of *Normal Operating Cell Temperature* (NOCT). A PV module will typically be rated at 25°C under 1 kW/m². However, when operating in the field, they typically operate at higher temperatures and at somewhat lower insolation conditions. To determine the power output, it is important to determine the expected operating temperature of the PV module.

The nominal operating cell temperature (NOCT) is defined as the temperature reached by open circuited cells in a module under the conditions listed below:

- Irradiance on cell surface = 800 W/m²
- Air temperature = 20°C
- Wind velocity = 1 m/s
- Mounting = open back side.

The equations for solar radiation and temperature difference between the module and air show that both conduction and convective losses are linear with incident solar insolation for a given wind speed, provided that the thermal resistance and heat transfer coefficient do not vary strongly with temperature.

The NOCTs for best case, worst case, and average PV modules are shown below. The best case includes aluminum fins at the rear of the module for cooling to reduce the thermal resistance and increase the surface area for convection.

The best module operated at a NOCT of 33°C, the worst at 58°C and the typical module at 48°C. An approximate expression for calculating the cell temperature is given by

$$T_{cell} = \frac{T_{air} + NOCT - 20}{80} \; S$$

where

S = insolation in mW/cm².

Module temperature will be lower than this when wind velocity is high, but higher under still conditions.

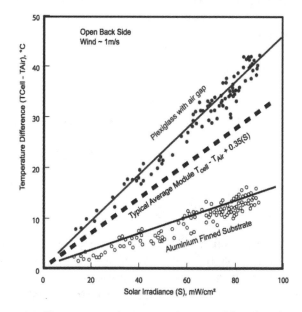

Figure 6-2. Temperature increases above ambient levels, with increasing solar irradiance for different module types

Impact of Module Design on NOCT

Module design, including module materials and packing density, have a major impact on the NOCT. For example, a rear surface with a lower packing density and reduced thermal resistance may make a temperature difference of 5°C or more.

Impact of Mounting Conditions

Both conductive and convective heat transfer are significantly affected by the mounting conditions of the PV module. A rear surface which cannot exchange heat with the ambient (i.e., a covered rear surface such as that directly mounted on a roof with no air gap) will effectively have an infinite rear thermal resistance. Similarly, convection in these conditions is limited to the convection from the front of the module. Roof-integrated mounting thus causes higher operating temperature, often increasing the temperature of the modules by 10°C.

Thermal Expansion and Thermal Stress

Thermal expansion is another important temperature effect which must be taken into account when designing modules. The spacing between cells tries to increase by an amount δ shown as

$$\times / = (\alpha GC - \alpha cD)\Delta T$$

where

αG is the expansion coefficients of the glass;
αC is the expansion coefficients of the cell;
D is the cell width; and
C is the cell center-to-center distance.

Typically, interconnections between cells are looped to minimize cyclic stress. Double interconnects are used to protect against the probability of fatigue failure caused by such stress. In addition to interconnect stresses, all module interfaces are subject to temperature-related cyclic stress which could lead to delamination.

Thermal Stress Prediction

Here we discuss thermal stress (an old concept) by adding a new twist—that of predicting the type and amount of stress in the different module components under different operating conditions. Imagine that! The different aspects of heat stress phenomena, especially their effects on PV modules and other field components operating long-term in very hot areas, could fill volumes.

But, we will squeeze this key subject into several paragraphs, emphasizing the fact that it has been largely ignored until recently. Bringing heat stress to the forefront is something that had to be done a *long* time ago—especially by the large companies who installed mega-fields of PV modules in deserts without having any idea what would happen in the long run.

Some of our technologies, like CdTe and CIGS, have been in mass production only 4 or 5 years, yet millions upon millions of these modules are installed in US deserts. New, untested and unproven for long-term desert operation, many are paid for with taxpayer dollars.

Not only does this constitute blind ignorance, but it borders on blatant, if not criminal, negligence. Some of these millions of PV modules installed in US deserts are made by manufacturers who have little knowledge of deserts. The fast buck calls for selling as many modules as one can, until someone realizes there is something wrong, or missing, and stops the flow. We don't see this happening soon.

One notable exception is the effort of a group of researchers at the University of Santa Cruz, California, who have come up with a detailed ***thermal stress prediction model***. See May 2012 issue of *Photovoltaics International* for details.

This thermal stress model is too complex for entry here, so we would encourage the reader to look it up. It is impressive in its detail, and, we believe, indispensible and urgent in its usefulness.

The model actually takes into account the low-temperature thermal stresses in a c-Si PV module during its design stage—and well BEFORE the solar cells are encapsulated into the PV module forever—which is a key premise for successful quality assurance.

The assessment is done via simple, easy-to-use and physically meaningful analytical (mathematical) predictive models of the key module element: front glass, EVA encapsulant, and the laminate backsheet.

The key thermal stresses analyzed include the normal stresses that are present in the key materials and determine their short- and long-term reliability, the interfacial (shearing and peeling) stresses that affect the assembly's ability to withstand delaminations, and the interfacial stresses, which determine also the cohesive strength of the encapsulant.

The calculated data indicate that the induced stresses under thermal load can be rather high, especially the peeling stress at the encapsulant-glass interface, so that the structural integrity of the module might be compromised, unless the appropriate design-for-reliability (DfR) measures are taken. This includes thorough thermal stress evaluation and prediction and accelerated stress testing.

The authors are convinced, and we overwhelmingly concur, that reliability assurance of PV modules cannot be delayed until after they are manufactured. Instead, the only way to provide long-term reliability starts with materials selection, goes through product and manufacturing stages, and culminates with a properly executed product.

After that, we need to worry about transportation, accidents, installation, operation and maintenance procedures. The most important of these are the initial stages of PV cells and modules design and manufacturing.

Electrical and Mechanical Insulation

The encapsulation system must be able to withstand voltage differences at least as large as the system voltages. Metal frames must also be grounded, as internal and terminal potentials can be well above the earth potential. Any leakage currents to earth must be very low to prevent interference with earth leakage safety devices.

Solar modules must also have adequate strength and rigidity to allow normal handling before and during installation. If glass is used for the top surface, it must be tempered, since the central areas of the module become hotter than areas near the frame. This places tension at the edges, and can cause cracking. In an array, the modules must be able to accommodate some degree of twisting in the mounting structure, as well as to withstand wind-induced vibrations and the loads imposed by heavy snow and ice.

Additionally, some installers may habitually, or be forced to, walk on the modules during installation. This practice induces internal stress to the modules' frame and the cells proper, which might have detrimental results, so it must be avoided altogether.

PV MODULES CERTIFICATION

A PV module consists of a number of interconnected solar cells (typically 36 to 72 connected in series for battery charging), and more in large size PV module designs. Individual solar cells are soldered in strings and encapsulated into a single, hopefully long-lasting unit simply because PV modules cannot be disassembled for repairs. The main purpose of encapsulating a set of electrically connected solar cells into a module is to protect them from the harsh environment in which they will operate.

Solar cells are relatively thin and fragile and are prone to mechanical damage due to vibration or impact unless well protected. In addition, the metal grid on the top and bottom surfaces of the solar cells, the wires interconnecting the individual solar cells, as well as the soldered junctions can be corroded by moisture or water vapor entering the module, if the protecting materials are damaged or absent.

Testing PV modules before installation in the field for long-term operation is a necessary step, but the results are not always indicative of actual lifetime field performance. Many modules qualified and certified as required, and operating for several years in the field, have been found malfunctioning after the abuse of Mother Nature.

Once the solar modules are manufactured, one of the best batches goes to US or European labs for testing and final certification. This is a very important test, which determines the future of this set of modules, and possibly that of the company that manufactures them.

Thus, selected modules undergo standard test procedures. Some pass the testing program, but start deteriorating prematurely and at times even fail completely. Sometimes they fail at the pre-test and pre-certification stages—before the testing procedure has started.

Test and Certification Standards

Since the PV modules are the most critical part of the PV power plant installation, we must make sure that they will perform well. For that we rely on the certification procedures to which all PV modules manufacturers must submit their products. Each locality and technology requires different certification methods(Water mark for Australia, Solar Keymark and CE for Europe, BAFA for Germany, UL and SRCC for USA & Canada, MCS for UK etc.).

All these methods are simply electro-mechanical tests, using similar parameters to identify weak points in the modules' structure or performance characteristics. In all cases, the proper execution and interpretation of the results would tell us what we can expect when the modules are exposed to sunshine.

The results, however, will not be able to tell us what to expect after 15-20 years of non-stop exposure to excess heat and humidity. We can make some educated guesses, but that's all. The rest is up to the quality of the PV products, and of course, Mother Nature has the last word.

A number of standards (methods) for testing and certification of PV modules have been established and accepted for use in different countries, as follow:

a.   Most relevant standards for the U.S. and Canada.
   - UL 1703: Standard for flat-plate PV modules and panels
   - UL 790: Standard for standard test methods for fire test of roof coverage
   - AC 365: Standard for building-integrated PV modules (BIPV)

b. Individual U.S. State Requirements
    - California: California Energy Commission (CEC)
    - Florida: Florida State Energy Center (FSEC)

c. Most relevant standards for Europe and parts of Asia
    - IEC 61730: Standard for PV module safety
      — Part 1: requirements for construction
      — Part 2: requirements for testing

    - IEC 61215: Standard for crystalline silicon terrestrial PV modules

    - IEC 61646: Standard for thin-film terrestrial PV modules

    - IEC 60904-X: Standard for PV devices (measurement procedures and requirements)

Note: A full list of international standards accepted and used by the solar industry around the world can be found at http://www.astm.org/COMMIT/SUBCOMMIT/E4409.htm.

Update: In the summer of 2012, First Solar introduced a new standard for installing PV modules in US deserts without official UL certification. When they were caught by a lowly LA County safety inspector installing millions of CdTe PV modules without proper UL certification in their Antelope Valley Solar project, they invented a new certification standard.

Instead of certifying the millions of uncertified modules as required by law, they went to the negotiating table with LA County officials, where they had several months of talks on the matter. Finally the LA County officials agreed to not "inconvenience the manufacturer excessively," and they waved the UL requirement altogether.

This $1.5 billion project was funded by taxpayers. On their behalf, one must question the ethics of these negotiations. Reputations hung in the balance, and interrupting or delaying the project significantly would have brought the "Solyndra" syndrome back out of the closet where it had been tucked earlier in the year.

So, what we have now is over 4.0 million toxic, cadmium-containing modules installed and operating in the US desert—without UL certification. Actually, there are many more such modules out there, but we don't know how the "alternative certification" was done in those cases. The obvious question is, "What if the unthinkable happens?"

Sadly, the new "standard" in the US—instead of going through the lengthy and expensive UL certification process—can consist of engaging the responsible officials in lengthy negotiations. With 2-3 months of negotiations (much shorter than actual certification timing), an installer can get permission to proceed without UL certification.

Pre-certification Issues

Amazingly, and as confirmation of the immaturity of the solar industry (and some manufacturers in particular), a great number of PV modules sent for certification tests fail even before being tested, and/or shortly after the test is initiated.

We wonder what these manufacturers' engineers and leaders are thinking. Or are they just rushing towards the bottom line without thinking?

Pre-test Failures

Let's look at an unusual trend of failures which were discovered even before the actual testing was begun. These failures tell us a lot about the manufacturers involved, and don't give us too much confidence in their products. See for yourself.

According to Intertek, a certified test facility in California, a number of test modules fail this pre-test screening. Here are five typical reasons why PV modules coming for testing at their facility fail initial inspection—even before the testing can be started.

The reasons are shocking, for a supposedly hi-tech business, such as PV cells and module manufacturing operations:

1. Inappropriate/incomplete installation instructions;

2. Models provided for testing do not accurately represent the entire production model scheme being listed (largest module must be submitted for test);

3. Testing requested without prior construction evaluation being performed;

4. Complete bill of materials with ratings and certification information not provided prior to start of the project;

5. Lack of back-sheet panel RTI rating.

Author's note: According to the author's extensive experience with engineering and quality control operations in the world's semiconductor and solar industries, the pre-certification failure is a phenomena that is quite unusual. It points to weaknesses in the overall manufacturing, QC/QA, and management systems of the manufacturers in question.

Keep in mind that the above listed failures are among the very few that can be readily recognized *before* modules are packed, shipped and installed in the field. How, then, do we catch the possible failures of the thousands of modules we need for our 100MWp power plant, during and after installation in the field? Follow these simple rules:

1.   Take your time to thoroughly check all engineering and quality control procedures used during the processing of your batch of PV modules. The manufacturer must provide some of this information. If not, then go to someone who will.

2.   Find out as much as you can about the "sand-to-module" history of the product, before placing a large order.

3.   Even better, have a team of qualified engineers visit the manufacturer and inspect the process line your modules go through. This is a difficult task, but short of that, you are working with a black box... The choice is yours.

PV Modules Certification Tests

All PV modules must be properly tested and certified before they are allowed on the energy market. For this purpose they are sent to test laboratories for official certification. Then they can be used in the destination country.

The testing process consists of a series of visual and electro-mechanical tests as follow:

Standard Test Conditions (STC)

These are the conditions (light characteristics, operating temperature and time) all PV modules are exposed to, to establish and certify their nominal performance parameters. PV modules are tested using test conditions accepted as "standard," or "standard test conditions," or STC by the solar industry.

The STC specs are as follow:

a.   Vertical irradiance E of 1000 W/m^2;

b.   Temperature T of 25°C with a tolerance of ± 2°C;

c.   Defined spectral distribution of the solar irradiance at air mass AM = 1.5.

I-V Curve

I-V, or current vs. voltage, curve is usually generated and documented during and after these tests, and the efficiency of the solar module is calculated from the test data. Basically, the I-V curve is characterized by the following three points:

a.   The maximum power point (MPP). For this point, the power Pmpp, the current Impp and voltage Vmpp are specified. This MPP power is given in units of watt peak, Wp.

b.   The short-circuit current Isc is approximately 5-15% higher than the MPP current. With crystalline standard cells (10cm x 10cm) under STC, the short-circuit current Isc is around 3A. Some modules will have higher Isc, and some lower.

c.   The open-circuit voltage Voc in silicon cells registers ~0.5V to 0.6V, and for amorphous silicon cells it is ~.6V to 0.9V.
      Note: Wiring the cells in different configurations will produce different voltage and current combinations, as required by the particular type of module or its application.

Efficiency

The efficiency is the most commonly used parameter to compare the performance of one solar module to another. It is the property that identifies each module type and determines its performance in the field.

Efficiency is defined as the ratio of energy output from the solar cell to input energy from the sun. The more efficient a module is, the more expensive it seems to be.

The efficiency of a solar module is determined as the fraction of incident power which is converted to electricity by the module. It is defined as:

$$\eta = \frac{Voc * Isc * FF}{P_{in}}$$

where:
   Voc   is the open-circuit voltage;
   Isc   is the short-circuit current;
   FF   is the fill factor
   P$_{in}$ is the incident power

Environmental and Mechanical Tests

A number of tests, designed to test the resistance of the modules to environmental effects (rain, snow, heat, freeze, etc.), are usually part of the manufacturers' procedures and the product certification procedure.

These tests include:

a. Thermal stress: cold, hot, wet cycles; humidity tests
b. Electrical rigidity, resistance, temperature
c. Hail launching
d. Mechanical stress tests
e. Salt, fog and dampness tests
f. UV light tests
e. Outdoor tests

How much each of these tests reveals about the overall performance of the tested modules cannot be quantified. Our experience shows that modules made by the best world class manufacturers, which pass all these tests, fail in the field. Failure could be instantaneous, or slow under harsh climate conditions.

Either way, these failures shows that the certification process is *not* a 100% guarantee that PV modules will function properly and efficiently for a long time in the field. How can we test better to ensure, or better yet to predict, modules' behavior? Keep reading.

LAB TESTS OF PV MODULES

All PV modules sold on US and EU energy markets are certified, yet many of them (even those made by most reputable manufacturers) underperform and even fail during certification testing and/or in the field.

Why is the certification process not catching the defective modules? a) because not ALL modules are made with the highest quality materials and processes, and b) most modules do NOT go through adequate testing. As a matter of fact, most of them don't get any testing beyond the cursory flash test.

So, more thorough tests with large numbers of modules are needed at manufacturers' sites or in specialized labs. Thorough tests should go well beyond the conventional flash test.

Some of these would be:

Reliability Tests

The experience and conclusions of the researchers at the Photovoltaic Institute in Berlin on the properties and testing procedures to ensuring reliability of PV modules are as follow: (2, 3, 4, 5, 26, 27)

Mechanical Safety

While tests for mechanical safety are relatively easy to perform (2,400 Pa for the mechanical load test accord-

ing to IEC 61215, 61646 and 61730) and should not pose severe problems to the manufacturers, some modules fail these tests, possibly due to enlarging module size without taking into account the mechanical properties.

This issue can be overcome using enhanced mounting clamps with rubber inlays, extra support on the backside, frames with additional cross bars, thicker glass, and smaller formats or stiffer back materials.

Electrical Isolation

Initial electrical isolation problems are typically due to an insufficient distance of the electrically active areas from the metallic frame. Later in the operation phase they are due to moisture ingress from the edge. Electrical isolation is tested using four different methods:

- Application of a high voltage between the terminals and a wrap of conductive foil around of the module. The test voltage for the different tests is: for IEC 61215 & 61646 – 1kV plus twice the maximum system voltage for 1 minute; for IEC 61730-2 class A requirements—2kV plus four times the maximum system voltage; for class B requirements—1kV plus two times the maximum system voltage. If the measured insulation resistance times the area of the module is less than $40M\Omega/m^2$, the module has failed.

- Applying an impulse voltage (MST14 at IEC 61730-2) of up to 8kV at a rise time of $1.2\mu s$ and a fall time of $50\mu s$.

- Measurement of the wet leakage current (module drowned) at 500V or the maximum system voltage (10.15 at IEC 61215 & 61646 and MST17 at IEC 61730-2). If the measured insulation resistance times the area of the module is less than $40M\Omega/m^2$, the module has failed.

- Using the ground continuity test (MST 13 at IEC 61730-2) for modules with a metal frame or a metallic junction box to demonstrate that there is a conductive path between all exposed conductive surfaces of the module and that they can be adequately grounded in a PV system. The resistance between each conductive component of the module shall be less than 0.1Ω for a current of 2.5 times the maximum overcurrent protection rating.

To simulate several years of use, a damp heat test (1000 hours at +85°C and 85% of relative humidity), a thermal cycling test (200 times between -40°C and +85°C) and a humidity freeze test (10 fast drops from 85°C to -40°C at 85% humidity) are applied, at which point the isolation test and the wet leakage current test are repeated.

Several encapsulation and seal materials tested show that while PVB is more susceptible to moisture ingress than EVA, EVA is more commonly used. However, PVB would tend to provide a better fit to the building code requirements. Several different strategies are used to inhibit the moisture ingress, including wrap sealant in the module frame, metal tape around the edge, glass bonding, and in-laminate sealant.

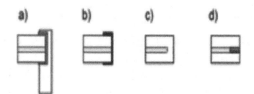

Figure 6-3. Strategies used to counter moisture ingress
(Showing the cross-section at the edge of a PV module)
    a) wrap sealant in the module frame
    b) metal tape around the edges
    c) glass bonding
    d) in-laminate sealant

While glass bonding offers the most secure sealant for moisture in glass-to-glass modules, it is quite costly and not widely used. Manufacturers are researching using 'breathable' membranes in the sealant in different configurations, to provide efficient moisture barrier while avoiding the negative effects in glass-sealed modules.

Polarization Effects

Polarization or the potential induced degradation effect (PID) on crystalline silicon modules creates an alarming failure rate on modern crystalline silicon cells meeting high system voltages. It can cause power losses in the vicinity of 30% and more. Known parameters for PID on the cell level are high electrical resistivity of the SiNx anti-reflective coating via low refractive index and increased thickness and low emitter depth. (26)

Until recently it was commonly believed that module defects caused by voltage stress, and the resulting leakage currents were expected to indicate PID. It was also believed that these effects only occur for special cell and module designs but not affecting the whole silicon solar industry.

In 2010, Solon SE showed that there is also a polarization effect on standard mono and poly crystalline cells.

The influence of different encapsulants and various configurations of the silicon nitride layers could be verified, and a test procedure for crystalline modules was proposed.

Since mid-2010 more and more returns from the field arrived at PI-Berlin's lab showing power degradations caused by polarization. During our research we even found polarization effects on CIGS modules. Cells became more prone to potential induced failures during the last five years. Although the effect is not yet clear on a microscopic level, one can say that it is a consequence of improvements in cell and module design.

On the cell level we tested different encapsulates and the PID behavior of the latest upcoming cell generation—cells with selective emitters. The role of the sodium ions in the polarization effect is still unclear, but during the investigations on the cell level it soon became clear that it is necessary to have a look at the whole module, or more precisely the cell interaction with the module materials and the environment.

Therefore, stress sequences with different working conditions have been evaluated with the aim of creating a reproducible test method which includes all relevant parameters. Finally, a quick polarization test sequence has been developed to help module and cell manufactures check their products.

Author's note: In the summer of 2012 Sharp Solar's polycrystalline, high-performance PV modules with overall module efficiencies of 15.2% were certified as "potential-induced degradation (PID) resistant." Several modules from different manufacturers were tested for PID by Fraunhofer CSP at 50°C (122°F) and a relative humidity of 50%.

Only 1/3 of the tested modules passed the test and exhibited no signs of degradation. The remaining modules showed serious collapses in power outputs of up to 98% compared with their initial power outputs.

Nevertheless, the US National Renewable Energy Laboratory (NREL) recommends these tests be conducted during all certification procedures—for at least 96 hours at 60°C (140°F), 85% humidity at 1 kV load.

What does this tell us? Only that the industry is moving, slowly but surely, towards meeting higher standards. We are confident that soon all PV modules undergoing certification tests will be submerged in boiling water for 3 days under full load. Crazy? Maybe, but have you ever been in the desert during a monsoon? If you haven't, and are planning to sell us PV modules for use in the desert, then you should reconsider your sales strategy and the quality of your modules.

Reliability

Hot-spot Susceptibility

While the photovoltaic conversion process itself is very reliable, the interconnection of the cells in series may cause problems. As with all series connections, the element with the lowest current defines the total current. The current of a single cell may be reduced by local shadowing (due, for example, to dirt on the surface of the module), which limits the total current and power output.

If the string is large enough, the (reverse) voltage at the shadowed cell can surpass the negative breakthrough voltage and could lead to a local power dissipation that could even destroy parts of the cell (hot spot). The hot spot problem can be avoided by reduced voltage or a reduced cell area (limitation of current) or via appropriate bypass diodes or cells with a low reverse breakthrough voltage—interestingly, usually 'bad' (low efficiency) cells.

Electroluminescence

Failure diagnostics are essential to discovering issues of failure susceptibility. Electroluminescence (EL) is a suitable process for checking that the entire module area is incorporated in the photovoltaic energy conversion process. Electroluminescence is the use of a solar cell in a reverse manner to how it was intended to be used; instead of converting irradiance into electricity, electricity (supplied via the cell's electrical contacts) is converted into radiation in the near infra-red and is emitted via the cell's surface.

The intensity of the radiation emission is an indicator for the local efficiency and quality of the photovoltaic conversion process.

Failures in the Lamination Process

Failures in the lamination process can be caused by:
- Old and oxidized EVA
- Insufficient glass-washing
- Wrong temperature
- Insufficient duration and pressure of the lamination process
- Lack of curing of EVA due to shortened process

These failures can be detected by a gel content test of the cured EVA or by a backsheet peel-off test (forcing the peel off the backsheet from the module).

TCO Corrosion

Some technologies that use a transparent conductive oxide (TCO) for the front contacts frequently experience problems if a high negative voltage is applied to the TCO. The effect can be explained by the sodium ions' electrochemical corrosion with water at the TCO/glass interface, causing delamination of the TCO.

Major drivers of this process are:
- Negative cell polarity vs. ground
- Moisture ingression
- High operation temperature
- Na (sodium) content in glass

Therefore, module manufacturers tend to recommend inverters that allow for a positive voltage of the module against ground.

Energy Rating

An electrical energy rating can be carried out from knowledge based on experience of long-term outdoor tests or simulation—or a combination of both—to achieve validation.

Energy Rating Based on Laboratory Measurements

Parameters that influence the energy yield have to be measured in detail as input data for energy yield simulations and comparison of technologies, including efficiency at different irradiance levels (weak light performance), temperature coefficients, spectral efficiency and optical parameters (performance at flat incidence angles, refractive indices).

Energy Rating via Simulation

The correct simulation of direct and diffuse irradiance via their spectral spatial appearance allows for an accurate representation of the module reaching irradiance. After passing the different layers of the encapsulation and being reflected according to the Fresnel laws—considering actual incidence angles and refractive indices—this irradiance forms the cell-reaching spectrum. The photoelectric conversion efficiency depends on matching the cell-reaching spectrum with the cell's spectral response and the actual operating cell temperature (which is derived from a balance of energy flow of absorbed irradiance, electricity generation and heat dissipation).

A further analysis of the different parameters (e.g., performance vs. module inclination angle) can be carried out. An interesting effect is that the inclination angle of the module does not influence the irradiance on the plane of the module, but has a significant effect on the convective heat transfer of the module.

For horizontal mounting (module elevation angle: 0°), the convection capability and convective heat transfer at the module are reduced, causing high operating cell temperatures and a considerable dip in conversion efficiency around noon. This dip is drastically reduced for more inclined modules, allowing an effective flow of air and convection along the module.

The minima of conversion efficiencies 20 minutes after sunrise at 6 a.m. and 20 minutes before sunset at 6 p.m. can be explained by the extremely flat angle of incidence of the direct irradiance during those times of the day.

From these examples, it is clear that the quality of yield prediction depends rather on the comprehension of the entire optical thermal-electrical composition of the installed PV panel than on the knowledge of an isolated PV module.

Energy Rating Using Outdoor Data

Collection and study of outdoor, real-world data is the most accurate, but also the most time-consuming and expensive method of collecting data on energy yield. It is usually done by research institutions, where manufacturers participate if they wish. Presently, mandatory field tests are in the design stages, with no official, uniform and accepted-worldwide test sequence in effect.

Degradation

While the power output of crystalline technologies showed only a little degradation, a-Si modules degrade considerably. To accelerate the process of degradation, so-called 'light soaking' at high irradiance levels (600 (800) – 1,000W/m²) and at constant temperatures (50°C ± 10°C) is applied according to IEC 61646.

Current-induced light soaking was tested to facilitate light soaking. For a-Si, degradation was similar, but current-induced light soaking did not reach the degradation level achieved via light soaking (6% difference.

A PV module based on a combination of amorphous and microcrystalline silicon has shown almost no degradation at all by current soaking, while conventional light soaking has shown a similar level of degradation on an a-Si module.

Conclusion and Outlook

The experience of PI shows that energy rating is most critical for thin-film technologies, while

- Degradation is still the most important factor on energy yields for a-Si and µ-Si/a-Si

- TCO corrosion is mostly solved by in-laminate frames and adequate inverter technology

- Degradation and spectral effects in silicon thin film modules require new modeling in future simulation

- The tandem-junction structure of µ-Si/a-Si is complicating energy yield prediction due to the interdependence of degradation and spectral effects.

Additional Reliability Tests

A number of other, and more vigorous, tests might be needed to verify the quality and performance of PV modules. Most of these tests are not part of the certification or in-house lab testing done by manufacturers. They, however, would allow better understanding and correction of some of the problems which have not been understood, and/or addressed thus far. A closer look at the modules' characteristics and performance would help ensure their quality and reliability for long-term operation under extreme climate conditions.

Some of these tests would be:

Moisture and Chemical Ingress into the Module

This is a serious condition which could reduce the modules' efficiency and life. If it occurs soon after the modules are installed, it would very likely be traced to a manufacturing defect, or handling problems during transport or installation. At times, however, moisture in the modules is found months or years after they have been operating in the field. The reasons are usually defects in the laminate layers, caused by defective materials or improper handling or processing. Desert sunlight is especially hard on organic materials in the modules too, so it is usually only a matter of time before they start breaking down mechanically and chemically, at which point moisture and air could penetrate the modules.

Moisture ingress is a diffusion process (liquid diffusion that is). Its diffusivity depends on the module type, its frame construction and laminate materials composition and application. Moisture could penetrate into PV modules through the edge seals and from front or back cover defects (cracks, pores and such). Once in the module, moisture could easily penetrate through the encapsulants, causing them to delaminate, discolor, and mechanically disintegrate. In parallel with that, moisture could oxidize and otherwise attack the solar cells' metal contacts, causing them to degrade or fail completely.

The encapsulants and sealants have not changed much during the last 30 years, so the best defense we have in assuring quality is to make sure that only quality materials are used, and that they are applied according to established manufacturing and QC specs.

We would also suggest a series of accelerated tests to be conducted with test modules during lab or certification testing.

Accelerated Tests

Accelerated tests can differentiate the performance of various c-Si module designs, pointing to module weaknesses, and outlining failure mechanisms, especially those in high-voltage arrays. c-Si modules in general are robust to thermo-mechanical fatigue, but damp heat with system bias applied to the active layer with respect to the grounded module frame significantly accelerates electrolytic corrosion, and the observed silicon nitride degradation points to areas with higher moisture ingress.

The results of dark I-V curve fitting show that the properties of the cell p-n junction are affected, with positive bias to the active layer improving the p-n junction characteristics, but degrading the series resistance and the saturation current around the maximum power point.

Negative bias catastrophically shunts the modules, except in cases where the module packaging blocks leakage current effectively in the 85°C/85% RH condition.

The damp heat with bias test has proven to be significantly more severe than the thermal cycling test in most test sequences.

Failure of c-Si modules during the thermal cycling phase of qualification testing increases, and it was the single largest failure in qualification testing between 2007 and 2009 at one laboratory.

The 85°C/85% RH test itself is a challenge for modules, and addition of system bias adds significant additional stress. Silicon nitride degradation, however, is not typically seen in modules in the field because the chemical activity of water to hydrolyze silicon nitride is much higher in the damp heat test in the lab.

The damp heat test appears to be a good indicator of the extent of water diffusing into the module and its potential to cause internal damage. The extent to which silicon nitride damage occurs varies greatly according to type of modules, but all modules exhibit some degree of silicon nitride degradation.

Field testing of modules to quantify the degradation associated with system bias has been well reported for thin-film modules, but less so for c-Si modules. Modules under constant system bias display maximum leakage current of 2 µA at the National Renewable Energy Laboratory, Golden, CO, and around 8 µA at the Florida Solar Energy Center, Cocoa, FL.

Modules under accelerated tests show leakage currents in the range of 10 to 100 µA, though some fall out of this range at either extreme, depending on module design. So there is anecdotal evidence that arrays disconnected from a peak power tracking inverter may exhibit degradation, and it is likely that negative bias strings will experience module degradation by the shunting mechanism at some high system voltages.

SunPower has reported degradation of PV modules because of bias, attributed to a static charge that develops at the cell surface. This static charge was also reported to be quickly and completely reversible under special conditions.

The differences between what is observed under accelerated tests with conventional cell modules and those of the high-efficiency SunPower back-contact cells are as follows:

1. Reversing the polarity to reverse the degradation is slow and incomplete in conventional cell modules, unlike what is reported for the SunPower module.

2. The polarity of the bias on the active layer that degrades multicrystalline modules is negative—which differs from the positive bias that degrades the SunPower high-efficiency module. The negative bias to active layer would tend to accumulate a positive charge over the cell surface, which would enhance the front surface field of the conventional cell.

3. The degradation mechanism by a negative charge in the high-efficiency cell is reduction in minority-carrier lifetime, whereas it is shunting in the conventional cells tested herein. By the dark I-V analysis, junction properties are improved by positive bias to the active layer of the conventional cell modules.

4. n+ emitter surfaces of conventional cells need to be of much lower sheet resistivity, making it comparatively difficult to overcome their associated built-in surface field and significantly degrade cell properties.

This does not point to a surface charge created by the damp heat stress, but to some ion drift in the silicon cell that can move to and shunt the front junction, typically 0.20–0.35 µm in depth into the silicon.

While the ion species is not yet identified, elements

exist that primarily act on junction properties, such as copper. The evidence obtained points to the leakage current through glass and the encapsulant as the enablers.

This charge-induced degradation under bias has been reported to be a function of encapsulant resistivity, whereby more resistive encapsulants exhibit less degradation.

The enhancement in performance is consistent with the use of oxide barrier layers in thin-film modules that are used to minimize Na migration that causes degradation.

Other proposed methods of reducing leakage current or its effects and the resulting degradation of thin-film modules include:

a. Building arrays such that the active layer is at positive voltage potential with respect to the frame.

b. Reducing frame area and use of mounting points bonded to the backplane.

c. Reducing humidity in and around the module packaging, which increases conductivity.

d. Reducing electrical potential between the external packaging and the active layer.

Extended Damp Heat Test

Damp heat test, in our expert opinion, is the most relevant in determining and even predicting PV modules' quality and long-term integrity. Heat and moisture are the greatest enemies of anything placed in the desert, where most large-scale installations of the 21st century are and will be.

Acting together, excess heat and moisture—as with summer monsoon storms in the southwestern US deserts—are PV modules' worst enemy. This climatic condition seeks to destroy the modules defenses, penetrate inside and demolish the active elements.

Excess desert heat and UV radiation start their destructive work from day one by attacking the plastic components of the modules—the encapsulation and edge seal. These organic components are no match for the elements, and small cracks and voids start appearing at the edges and sometimes on the top or bottom surfaces of the modules. With time, the cracks and voids grow and get deeper into the materials.

At the same time, temperature cycling weakens the active materials inside the modules (silicon cells, or thin films). In time, these become partially disintegrated and full of mechanical defects—cracks, pits, and similar abnormalities.

Then moisture seeps into the modules through the cracks and voids in the encapsulation and edge seals. Once inside, the moisture and its chemical content (acids and dissolved gasses) begin to decompose the active components.

Now, the silver and aluminum metallization of the silicon solar cells encapsulated into the PV modules starts oxidizing and changing its electrical properties, which results in higher resistance. Elevated resistance reduces the modules' output, but also generates more heat, so the combined effect of outside desert sunlight heating and high resistance overheating of the cells from the inside accelerates the dangerous mechanical and chemical damage of the active components. This results in gradually decreased power output and eventually in total failure.

Similarly, the active layers in thin film PV modules are equally vulnerable to the chemical attacks of the moisture that seeps into them. Most thin film modules today are made without a frame, which means that only a thin layer—several millimeters—of plastic encapsulation material separates the moisture and its destructive forces from the fragile thin films inside.

How long does it take the desert's heat, cold, and elevated UV radiation to damage and disintegrate the thin plastic layer at the edges of the modules? 1 year, 5 years, 10 years? Maybe not, but exposure is expected to be for 20-30 years, so there are plenty of opportunities for the elements to disrupt the composition of the fragile modules.

When the active layers are attacked by moisture and its acidic content, the active thin films—which are already mechanically fatigued by the daily expansion and shrinking—start changing chemically and electrically. Their output decreases and failure follows.

Even worse than their silicon cousins, millions upon millions of thin film PV modules installed in the deserts contain toxic and carcinogenic materials, spread over thousands of acres. We don't know what will happen to the poisons inside those modules when they decompose and start outgassing in the air, or leaking into the soil and water table.

There is no precedent, nor have frame-less thin film modules been proven safe for such long-term exposure to desert elements, so we'd like to remind all involved—from the US administration (which funds some of these projects) to the manufacturers and operators—that they are responsible for environmental damages.

Back to our damp heat test, this is an environmental test, with most relevance to the above-described phenomena and issues.

According to TUV Reinhold, a solar cells and modules testing lab (6), "... the purpose of the test is to determine the ability of the module to withstand long-term exposure to penetration of humidity by applying 85°C ± 2°C with a relative humidity of 85% ± 5% for 1000 hours."

DH1000 is the most "malign" and on the top-list of failure rates in some laboratories accounting for up to 40-50% of total failures for c-Si modules. Similar failures rates can be observed for DH1000 with thin film modules.

The severity of this test particularly challenges the lamination process and the edge sealing from humidity. Important delamination and corrosion of cell parts can be observed as a result of humidity penetration.

Even with no major defects detected after DH1000, the module has been stressed to the point that it becomes "fragile" for the subsequent mechanical load test.

"Fragile" is the key word we think is most applicable to the indiscriminant use of cheap PV modules, which are poorly designed.

And just think: 40-50% of failures happen after 1000 hrs. testing, which is equivalent to less than three months exposure in the desert. What would happen after 2000 hrs.? How about during the 110,000 hours of life expectancy?

There are no scientifically based answers here, just random guesses—some of which are biased in favor of the profitability of a manufacturing operation or project. So why don't we do extended damp tests to get a better idea? Or if we do, then why aren't the results released? Let's guess...

Having been involved in design, manufacturing, testing, installation and operation of many different solar devices and systems, the author is convinced that the mega-watt PV projects popping-up in the world's deserts will experience serious failures and even cause environmental contamination, during the next 10-15 years. This is due to poor design, cheap construction, and little or no testing.

We also predict significant negative environmental impact by some of the PV technologies which contain significant amounts of Cd, As, Pb etc.—especially those installed in mega-fields in the deserts—because *most* PV modules cannot withstand long-term desert conditions.

We say "most" because we believe that it is possible to make PV modules that could withstand desert conditions, but we have not seen many efforts in that area. Instead we see the above issues being dismissed by manufacturers and specialists as insignificant and im-

material.

Now is the time to take a closer look at the damp heat test failures in labs, and figure out why such a high percentage of new modules fail. Now is also the time to fix the problems, instead of waiting for the failures to force us.

By the end of the 21st century we will be surely on the right track, and not surprisingly, many of today's solar technologies will still be in use, but in much more efficient and safe manners.

Dynamic Stress Tests

Today's PV modules' stress test consists mostly of the short-term static mechanical load test, as prescribed by the weak and outdated IEC 61215 and IEC 61646 standards. This test is a mild form of fatigue stress, exerted onto the module for a short period of time and its meaning is hard to translate in real life.

There are no requirements for any elevated stress tests on cells, cell connectors and/or the rigid components, such as glass and framing, to ensure reliable long-term performance; however, TUV Rheinland now offers such dynamic load testing. Such tests are essential, and should be required in future work on modules certification, and in standardizing the test and certification processes.

Extreme Testing

UL and other test and certification labs perform testing and certification for modules to be used in US power fields, while EU customers use slightly different standards, primarily to account for differences in housing construction and weather conditions.

According to UL test engineers, the majority of module "failures" occur with the "humidity freeze" test, when the modules go through a 10-day testing cycle where they are exposed to 85% humidity, which simulates the environment in the tropics and/or desert monsoon season.

Wet modules are then brought down to frigid temperatures of -40^0C where the moisture freezes and expands on and into the modules. UL technicians then scour the panels for any defects or inconsistencies in construction and observe failure modes.

Another test replicates the natural "shading" that occurs when trees or nearby buildings partially block sunlight on a solar roof. A hot, bright light is shined on some panels, basically overtaxing them, while others are kept shaded or totally dark. Some modules get charred during this test, with the glass totally shattering, according to UL engineers. Failures are documented, and it is up to the manufacturer to correct the problems.

In more extreme conditions, a version of the above test is needed to identify some problems and to ensure performance and longevity of modules. The so-called "reverse bias" test is this vehicle, and the author has been using it successfully since the 1980s. We review it below.

Reverse Bias Testing

When exposed to normal illumination, a solar cell in good operating condition is basically a diode of larger area p-n junction operating in forward bias mode. The diode generates photovoltage, which is created by the excess energy of the incident electrons impinging onto the p-n junction.

Abnormal disturbance of the electron flow through the junction is usually related to defects in the base material, such as grain boundaries and grain dislocations.

If a cell in a PV module is shadowed or partially shadowed by tree shade, falling leaves, etc., the solar cells located in the shadowed areas usually enter what we call "reverse bias" state. This is where current from the higher voltage areas flows into the lower voltage (shaded) areas—usually in the opposite direction of the normal flow of current in the device. If the voltage difference is high enough, and if there is nothing to stop the flow of current, it will continue to flow through the cell until either the condition changes or the cell's performance is reduced, shut down or destroyed.

Reverse biasing of solar cells is a common, albeit unwanted and dangerous, effect that influences the I-V characteristics, reduces the total output of the solar cell and contributes to early failures.

Reverse biasing creates high resistance and harmful overheating. If the difference in levels of illumination is high enough, or if the condition continues for a long time, the combined effect of increased resistance and overheating will reduce the performance of the device and even contribute to early failure.

Grain boundaries, crystal dislocations, or other imperfections and in-homogeneity in the cell's microstructure will emphasize and aggravate the condition and will be responsible for further decrease in performance.

Accelerated electrical and thermal effects in the defective areas would affect the dark current voltage characteristics of the p-n junction, which would bring cell efficiency down even further. This is usually accompanied by additional increase of resistance, which contributes to increase of temperature, which increases the resistance further and so on until the cell is damaged by the combination of excess electrical and heat loads.

Defects in the grain boundaries can be seen with some scientific instrumentation as dark areas, because the light-emitting efficiency is much lower in these areas, compared with areas free of defects.

Affected solar cells cannot produce electricity when illuminated, and are rendered useless. In such an event, charge carriers are accelerated in the strong electrical field produced within the solar cell, which generates additional pairs of charge carriers, which leads to increased and even uncontrollable current flow that can destroy the solar cell and the entire module. Fires have been observed under some extreme reverse biased conditions.

Uncontrolled reverse current can overwhelm a PV module, causing electrical abnormalities that negatively affect the individual solar cells. This can cause dramatic changes in the shunt resistance, especially during extended duration of the abnormal reverse current application.

To prevent such an occurrence, fuses and disconnect switches can be installed on the DC side as needed to protect the modules and wiring from overheating. Additional switches are also usually installed between the PV array and the AC grid, to protect the BOS components, such as inverters, wires, over-current and surge protection components.

Since involuntary or accidental reverse biasing of solar installations is such an important factor, we need to make sure that the modules we use are protected against it. For that, extensive analysis of the dark I-V characteristic of solar cells and modules is needed, including analysis of the effects of artificially introduced enhanced reverse current against the normal electron flow in the cell. This is needed to simulate the different aspects and behaviors of reverse biased modules.

Due to the importance of this potential problem, and because it is not given due priority, we suggest that all modules (new batches and configurations) be subjected to increased levels of induced reverse current through the p-n junction. For instance, a reverse current of 50 mA* applied to a new module for a certain period of time would give us enough information about the quality of its materials and construction. It might even tell us how long this module would last under certain operating conditions—such as electric and thermal loads.

Note: Applying higher current to some modules in our tests sometimes resulted in complete shutdown of the electricity generation of the tested modules. Material defects were thought to be the reason for this inconsistent behavior, but there are other factors that are too many and versatile to qualify and document.

The reverse biasing mechanism and its effects are complex, and we won't get into it in this text, so it suffices to say that the artificially induced reverse biasing technique for checking PV cells and modules can be used as a warning and prevention of sudden breakdown of the module by the negative effects of avalanche overcurrent and over-heating.

In our tests, the normal I-V curve usually changes with reverse biasing time, which leads to a decrease in the diffusion voltage, which is also time dependant. The declination of the curve continues for a short time, and then it usually stabilizes for awhile. It sometimes starts declining again until cells or modules are damaged after prolonged exposure to reverse current. This scheme varies with modules from different suppliers and with current densities and time periods used in the test.

The damaging effects of prolonged reverse current application are attributed mainly to crystal imperfections and defects, but also to increased surface leakage along the edge of the cell over time. The time frame of the start-to-failure is short; usually it takes less than 30 minutes to see the entire effect under a constant current regime.

Ideally, the shunt parallel resistance should be infinite and the series resistance 0 ohms under normal operating conditions. Under ideal operating conditions of equilibrium, the net current measured at the PV cell or module is zero, because of the balancing effect of the diffusion, drift and the recombination currents. Upon applying a current in the reverse direction of electron flow, this balance is disturbed, and overheating takes place at the p-n junction, which in turn expands the junction width.

The diffusion current is caused by the diffusion of carriers across the junction of the PV cell, and at reverse bias, the diffusion and the recombination currents will start to increase and grow with time as well.

The shunt resistance drop also causes serious damage, which can be observed on the affected solar cell surface. Sometimes the severity of the effects of the reverse bias are demonstrated as black burn marks, caused by extensive overheating in the p-n junction region of the cell bulk.

The cell overheats because some electric charges get excited and gain extra energy, which allows them to pass through the p-n junction. The damage starts unnoticed and progresses rapidly, with only several minutes needed to initiate and propagate the damage.

The time duration of this effect is too short for anyone to notice—unless they are presently measuring the cell or module output. The cell or module will be destroyed within minutes, and may even burst into flames, unless some type of protection against reverse currents is present in the DC line between the module and the inverter.

In conclusion, reverse current induces stress in the solar cells, the effects of which are decreased performance and damage to the affected structures, which can be measured across the entire module.

Reverse bias is a phenomena which can be controlled by using high quality materials, proper processing techniques, and proper safety devices in field installations. The combination of these measures would ensure efficient and trouble-free performance of PV modules during long-term operation.

Reverse bias tests by manufacturers and customers alike are very useful and extremely important in determining the performance and efficiency of PV solar cells. Even more importantly, the enhanced induced reverse bias test is a good indicator of the operation of affected PV modules under stressful conditions (heat and increased reverse bias current) in the field. It also can be used as a good judge of the longevity of cells and modules in field operations.

It is obvious from the many examples in this text, that we are experiencing significant field failures, especially of PV modules operating under extreme conditions (deserts and swamps). Some of these problems can be explained by the effects of reverse biasing due to mechanical damage (crystal imperfections) or external factors (shading). A number of fires resulting from reverse biasing have been reported as well.

We do predict that enhanced reverse bias tests will become obligatory QC/QA routine in the manufacturing process and during field operations in the 21st century. Though not easy or cheap, it is one of the best tests available to help determine the quality of the materials and processes used in making solar cells, and to predict their long-term behavior.

The bottom line is: *we need a new standard* to ensure proper implementation of this test worldwide.

Conclusions and Recommendations

Some of the conclusions and recommendations of the TUV Rheinland test center team, regarding their lab and field tests and their not-so-good results follow:

- Considering 10-25 years of warranty provided for PV modules, failure rates in qualification testing are still unacceptably high.

- The top 4 failure rates for c-Si modules were related to damp heat, thermal cycling, humidity freeze and diode tests.

- The top 4 failure rates for thin-film modules were related to damp heat, thermal cycling, humidity freeze and static load tests.

- New manufacturers have higher influence on the failure rates of c-Si modules, whereas they have lower influence on the failure rates of thin-film modules. Encouragingly, overall failure rates for both technologies have decreased for the 2007-2009 period as compared to the 2005-2007 period.

- To pass full qualification testing, these top 4 tests are recommended as a minimum, before initiating full qualification testing by the module manufacturers.

- To differentiate among manufacturers who all have qualification certificates, these top 4 tests may be recommended as a minimum, before purchase decisions are made by consumers/system integrators.

- To initiate long-term lifetime reliability or test-to-failure testing, these top 4 tests may be considered as minimum key tests by researchers.

We need to always remember that the above tests, and the conclusions thereof, reflect only a fraction of the time and stress these modules will experience during 20-30 years of non-stop exposure to the elements. The most critical, but not so stable component of the modules—the EVA encapsulant—is easily affected by heat, freeze, UV radiation, etc. Once it is damaged, cells are in danger of degradation and premature failure.

Dr. Dauskardt, Stanford University concludes: (12)

1. Delamination can occur between EVA and the front surface of the solar cells.

2. More frequent in hot and humid climates.

3. Exposure to atmospheric water and/or ultraviolet radiation leads to EVA decomposition to produce acetic acid, lowering the pH and increasing corrosion.

4. EVA, $T_g \sim -15°C$ or lower temperatures may result in "ductile-to-brittle" transition in adhesive/cohesive properties.

This is self-explanatory. Poor performance and failures cannot be avoided, but need to be calculated, and plans for their elimination (or reduction) need to be made in any plant design. It is not easy to bring these issues to the attention of manufacturers, where large-scale PV installations are concerned.

A partnership between manufacturers, installers, investors and customers is the best way to ensure long-term efficiency, longevity and profitability of large PV projects.

c-Si PV MODULES DETAILED FAILURE ANALYSIS

Keeping in mind all the possible behavior conditions and problems we discussed above, we will now look at the possible problems at each process step of PV cells and modules manufacturing. These issues can be traced to materials, equipment, processes and labor-related problems, and they are usually demonstrated during testing or long-term exposure to the elements.

As we examine possible defects and failures which can occur during final module tests (or more importantly, during long-term exposure), note Figure 6-4 showing the key issues and all possible effects of long-term on-sun exposure for PV module elements.

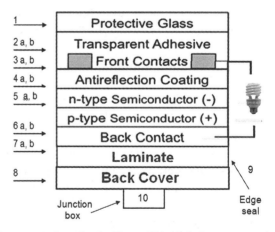

Figure 6-4. Standard silicon PV module cross section

ONE—Glass Cover (Top Cover)

The cover glass is basically designed to provide protection to the fragile solar cells and module components from mechanical damage, such as vibration, impact, etc., as well as to prevent the elements (rain, moisture, dust) from entering the module. If the glass itself is stressed, cracked and broken during processing, handling, transport, installation and operation stages, then the insides of the module might already be damaged. Subsequent damage could occur in such cases, by means of mechanical stress and chemical destruction of the cells and other active components.

This is even more important for thin film PV modules because the active layers are deposited directly on the front or back covers, thus any stress or breakage of the glass will directly, and immediately, affect the performance and longevity of the damaged modules. The advantage of c-Si PV modules in this area is that the EVA layer is enveloping the active components (solar cells), so that even if the top or bottom covers are damaged, the EVA envelope is given a chance to protect the cells from an invasion of the elements and chemical attack.

Modules with damaged cover glass—be it cracked or with hazy appearance—should not be installed and must be removed from service immediately. Repairing damaged glass is not an option in most cases, so the entire module must be properly disposed of, or sent back to the manufacturer for credit or for replacement. Periodic testing of the voltage/current output of modules with partially cracked cover glass, if acceptable at all, should be part of the PM schedule.

Note: In all cases damaged modules should be handled with utmost care. Cover glass surface soiling is another serious issue during normal operation.

Rain and dust will deposit layer upon layer of contaminants and water spots, preventing sunlight from reaching the solar cells underneath. Over time, the layers might grow so thick that very little light goes through them, significantly reducing the output of the modules. Washing the top surface with a soapy water solution, followed by proper rinsing with DI water will restore their efficiency, so it should be part of the PM schedule and carefully executed. This, however, is an expensive undertaking, especially in deserts. Periodic use of water and chemicals also causes concerns with water table contamination, which must be kept in mind when deciding on a PV power field location and O&M procedures.

TWO—EVA Encapsulant
Deterioration and Delamination

The layer of plastic encapsulation (EVA usually) and the cover glass are in intimate contact and partially bonded. These two, however, have different elasticity, and coefficients of expansion and friction. With long-time exposure to the elements (heating, freezing, excess UV radiation) or mechanical stress and fatigue, the EVA plastic will eventually change. Its physical and chemical properties change slowly, and we can see it turning a yellowish color. Yellowing of the EVA results in reduc-

tion of its transmissivity and causes a decrease of power output.

Yellowing is one of the few changes occurring in modules that can be observed with the naked eye. As changes continue, air bubbles, cracks, pits and cavities could form in the EVA material, causing it to separate from the top glass. This further increases optical losses due to reflection and poor transmission, ultimately resulting in serious performance deterioration. Once the EVA material damage and the accompanying delamination processes have started they cannot be stopped. At this point damaged modules are suspect and must be put under periodic observation as part of the O&M schedule.

EVA Delamination from the AR Layer

Some of the above-mentioned damage and defects in the EVA material will also affect the adhesion of the EVA material to the AR coating. EVA delamination from the glass and AR coating surface is caused by the never-ending UV bombardment, expansion and contraction of the layers, and the resulting friction between them. This action might result in an air gap which would cause a second optical barrier, further decreasing power output from the affected cells.

The newly created gaps and cracks might allow moisture, air and reactive gasses to enter the module, possibly attacking other modules components. Changes at the EVA-AR film interphase (air gap, color change, cracking, etc.) are another of the few visually observable alterations in PV modules. Depending on the type of cell (mono or poly crystalline) and type of AR coating (many different formulations are available); the visual effects might change the cell's surface color from black to faded black, and from dark blue to light blue (respectively), as well as a number of shades in between.

THREE—Top Contacts and Interfaces

The top contacts (fingers) consist of silver metal with metal ribbons soldered on top of the contacts to provide connection to the other cells and to the outside circuit. This creates several separate contact points (interfaces) as follow:

a. Silicon to silver metal of the top contact,
b. Silver metal to the soldered metal strip, and
c. Soldered metal strip to the EVA encapsulant.

The different interfaces provide good adhesion to their components and are quite stable under normal operating conditions. However, in abnormal situations, such as excess heat, overcurrent conditions, mechanical or thermal stress, and/or exposure to the elements (water vapor, chemicals and reactive gasses), the metals undergo serious changes which are usually initiated in the interfaces.

The damage could increase with time and cause significant changes in the contact's structure. In all cases these events will result in electrical changes or failures (high resistivity or open circuit). Interrupting the contact with the adjacent cells will lead to a failure of the module.

In some cases, one or more of the affected contact layers might burn, chemically disintegrate, or undergo other destructive processes, which could cause the device to fail partially, or completely, in which case the entire solar module might stop generating power. In all cases, there will be noticeable reduction in power from the module, due to mismatch caused by the affected solar cell(s).

Top Contact Adhesion into the Silicon Substrate

Under normal operating conditions, the adhesion between the front contacts (silver metal) and the silicon material which they are fused into is good enough to keep them together and properly conducting electricity for the life of the module. Abnormally, however, mechanical stress, excess heat, and moisture, and air penetrating into the module could affect the front contacts. Mechanical forces could induce stress in the materials and their interfaces, while ingress of water and chemicals might corrode and separate them from the substrate.

These processes would negatively affect the bond at the interface between these key components (silicon substrate and silver contact). In such cases:

a. The diffusion area might be affected and undergo internal changes which might affect the cell's performance, and/or

b. The resistivity at the affected area might increase, which could result in overheating, and eventually a breakdown (open circuit), caused by partial or complete separation of the contact from the substrate.

These phenomena could bring partial or total failure of the solar cell and/or the entire module.

FOUR—AR Coating Surface Damage

A number of distractive mechanisms on the cell's surface, caused by mechanical stress and chemical attacks could affect the AR layer. These changes could occur under different operating conditions and cause the AR coating top surface to change mechanically or chemically. This could cause a change in the AR layer's optical properties, decreasing the cell's efficiency.

In addition, blistering of large areas of the AR coating might also stress the cell and the top contacts, causing further increase of resistance and heat generation. Overheating could damage the contacts and/or their interface. The warped AR surface might also reflect some of the incoming sunlight at a greater rate than designed. In all cases, the final result is reduction in affected cells' efficiency.

These effects (AR layer delamination, blistering etc.) also result in color change of the AR coating which represents another of the few visually observable changes in the solar cell and module. Depending on the type of cell (mono or poly crystalline) and type of AR coating, the visual effects might change the cell's surface appearance to different variations of the original colors.

AR Coating Adhesion to the Substrate

The antireflection coating (AR) is a fairly thin film of inorganic material, i.e., TiO_2, Si3N4 and such, which is very thin, semi-transparent and fragile under certain conditions. Very thin and lightly bonded to the cell surface, it is easily damaged by mechanical and chemical means. There are several mechanisms that can contribute to changes of the AR layer and its adhesion to the substrate. One is contamination of the silicon substrate surface prior to AR coating, which would result in poor adhesion. Surface contamination could be caused by insufficient cleaning and rinsing, improper handling, or an out-of-spec process during the AR coating deposition. All of these inadequacies would compromise the adhesion integrity of the thin AR film to the silicon substrate.

If the AR layer is not fully adhered to the silicon substrate, there would be physical and optical gaps which would cause excess reflection and/or obstruct sunlight transmission into the cell. Combinations of these conditions greatly reduce a cell's efficiency.

More seriously, changes in the AR film properties could occur more quickly if the damaged area is exposed to the elements via moisture and air leaks. The adhesion forces in the interface between the AR film and the substrate in such cases would weaken even further, and the AR film might disintegrate or separate from the surface.

Most changes in the AR coating's adhesion and other properties are also visible on the module's surface, but equally impossible to correct. When these changes occur, future cell/module behavior is impossible to predict without destructive tests, so the module must be checked periodically as part of the O&M schedule.

Note: All visually observable changes mentioned above start, and might even continue, as microscopic imperfections which eventually grow bigger and more visible. A trained eye is needed to observe the changes in many cases.

FIVE—P-N Junction and Diffusion Process Issues

The p-n junction is the most critical area of solar cells' function. It has the greatest impact on cell/module efficiency, performance and longevity. The p-n junction is formed by diffusing (injecting) a specified concentration of foreign atoms in the silicon bulk, which then initiate and drive the photoelectric effect. Unacceptable amounts of impurities in the bulk silicon, such as Fe, Cu, K, Na and any other contaminants in the process chemicals and gasses will affect the diffusion concentration and depth, ultimately changing the behavior of the p-n junction. Changes in the concentration of the foreign atoms in the junction, due to parasitic effects during its field operation would produce similar results.

If the wafer surface was not properly prepared, cleaned and rinsed, or if the diffusion process was not properly executed (time, gas concentration and temperature), the p-n junction might not be stable enough. Thus affected solar cells might operate well in the beginning, but eventually the diffusion layer will start changing (decreased concentration and further diffusion into the substrate) causing a drop in output. These effects, sometimes occurring long after the module has been in operation, are called "latent" effects. They are most dangerous for the large-scale power plant's long-term success, because after the warranty has expired, failures can cause serious technical and financial difficulties.

The most common negative field effect of the p-n junction is change in the diffusion depth (and the resulting dopant concentration reduction at the junction) over time. While this phenomenon is well understood and the p-n formation parameters are controlled during the manufacturing sequence, certain material and process conditions force greater changes, over time, than allowed by design. In such cases, the cells start losing power quicker than expected and eventually fail altogether.

Diffusion processes techniques vary from manufacturer to manufacturer. Some use the old-fashioned, but most reliable, diffusion furnace process, where the wafers are placed into a glass tube and heated to over 1000°C in a controlled environment. Doping gases are passed through the tube and by precise control of temperature, process time and gas volume, precise deposition concentration and depth are obtained. A high-volume, but less precise, diffusion process has been widely used lately. It consists of spraying or printing the dopant liquid on the wafer surface and then baking it in a conveyor belt type of furnace. This process has more uncontrollable variables and is basically inferior to process control and product quality points of view than the old diffusion furnace method.

Nevertheless, it is cheaper and faster, so we will see it used more and more, and due to its less-than-precise, hard-to-control process parameters, we expect to see more diffusion and p-n junction changes issues in the field as well.

Diffusion layer depth and concentration changes cannot be visually detected, nor can they be measured under field conditions, so at this point we are looking at a black box and relying on the manufacturer's experience and quality control procedures.

P-N Junction Field Performance

What if the diffusion process were less than perfect? First, and most important, if the materials and process specs were even slightly out of control, the solar cells and modules would show low efficiently during the final tests. After sorting, the less efficient would go into the group of "cheaper" cells and modules to be shipped to a customer who hopes they will work well. But if the diffusion process were somewhat out of spec (ergo, creating some lower quality cells), then there would be a good chance that the affected cells and modules might deteriorate much quicker than the rest.

Exposure to extreme temperature is especially testy, and is the reason for many malfunctions and failures in c-Si modules. It is possible for the species in the diffusion layer to start migrating into the bulk under abnormal conditions brought on by extreme temperatures and/or imperfections in the silicon material close to the p-n junction area. These will accelerate the changes in the electrical characteristics of the cells, causing additional efficiency decrease. In summary, as soon as the newly processed solar cells are flash-tested they are ready for encapsulation into a module.

Provided that the materials are of high quality and all process steps have been properly executed, the p-n junction and the cells will operate properly for a long time. But if we had materials quality problems or the diffusion process were not properly executed, then we might be looking for a surprise in our power field, over time.

Note: Here we need to mention the great effect of silicon bulk material on the efficiency and other properties of the solar cells that were made from it. We now know that the quality of the solar wafers determines the quality of the solar cells made from them. The silicon wafers are made from 99.9999% pure solar grade silicon material. Even at that purity, however, a slight increase in one of the harmful contaminants (usually metals) might have serious effects on the solar cells' field performance and longevity.

Over time, and under stress, parasitic metals could start diffusing through the cell and alter p-n junction properties, with the diffusion speed increasing.

SIX—Rear Contact Damage

The rear contact also consists of several interfaces:

a. Silicon bulk to aluminum BSF
b. Silicon to silver back contact, and
c. Silver to soldered metal band

This metal structure and its interfaces are exposed to the same electro-mechanical, thermal and chemical attacks and changes as the front contacts we reviewed above. Although these are somewhat more protected from and less affected by the UV and IR radiation than the front contacts, their long-term quality is as critical. Any elemental impurities, chemical contaminants, mechanical damage, such as cracks, pits and other imperfections in the structure will have profound effects on the cell's efficiency and longevity.

The quality of the metal deposition is of great importance here too, because if it is defective in any way (high resistivity, cracks and pores, poor adhesion, oxidation at the interface, etc.), it will cause overheating, delamination and ultimately reduced efficiency of the cell, and even total failure. Nevertheless, the rear contacts are much less problematic and affect the cells' function less than the front contacts, but failure of the encapsulants and edge sealers could damage them seriously enough to cause performance degradation and failures.

Rear Contact Adhesion

In case of poor contact quality due to process control inadequacies (time, temperature, or metal quality) the metal film can partially separate from the substrate, eventually leading to decreased efficiency due to higher resistivity and overheating.

Excess overheating at the back surface might contribute to further erosion and delamination of the metal film and total failure of the cell and module.

SEVEN—Laminate Issues

Laminate materials (EVA, PET, PVB, Tedlar, etc.) are organic (plastic) materials, which have been in use and remain basically unchanged in composition for the last 30+ years. Also unchanged is their vulnerability to the elements, where EVA and other organic compounds in the module do undergo mechanical and chemical changes and degradation over time.

Continuous expansion during hot days, and shrinking during freezing nights, extreme IR and UV bombardment, and chemicals' ingress will affect the EVA and other organic (plastic) materials. This will eventually result in mechanical changes—discoloration, mechanical stress and disintegration of the lamination materials—creating cracks, bubbles, pits, voids, etc. Ingress of moisture and gasses into the module via such cracks and voids contributes further to the degradation process by decomposing the module materials (solar cells, wiring, contacts etc.).

Rear Laminate

The laminate structure on the back of the module usually consists of the EVA envelope with thin Tedlar and/or other plastic materials backing, laid onto the rear metal cover. The rear EVA-on-Tedlar structure is protected from the elements (UV and IR radiation) so it lasts much longer without damage. Its yellowing and even delamination from the back cover don't have such dramatic effects on module performance.

The adhesion of the EVA envelope to the back of the cells, however, can be affected with time. Excess heating, freezing, and moisture penetrating from the sides might eventually debilitate the adhesion to the solar cells. In that case, moisture and air penetrating the laminate material and reaching the solar cells will cause rapid oxidation of the back surface metallization (aluminum and silver metals), which will also affect the interface between these metals and the back surface of the solar cell.

This deterioration of the cells' adhesion at interface might cause increased resistivity and overheating, and eventually delamination of the metal contacts from the substrate, which will cause performance issues and eventually failure of the affected cells and modules.

EIGHT—Rear Cover and Frame

The purpose of the frame and back cover (aluminum sheet or glass plate) is to protect the module's insides from attack by the elements. It is highly unlikely for the frame and back cover to degrade significantly with time under normal use, but a severe mechanical or chemical attack could compromise its integrity, forming cracks and voids which might allow moisture and air to penetrate the module.

At that point, any of the above-described events might cause the cells and module to decrease in efficiency or fail.

NINE—Edge Seal

c-Si PV modules usually have side protection, as an extension of their edge protection. Ingress of harmful elements into the module is usually initiated through the sides (the edge seal) and could be slowed by a well-installed and sealed metal frame around the edges. A lot of effort is dedicated to finding better sealing materials, and assembly processes have improved significantly lately as a result. Edge sealing—its effects and weaknesses—are well understood. No matter how good the sealing materials are, however, they are made of plastic (organic) materials...

Edge seals are prone to accelerated changes—voids, cracks, pores, and bubbles. Therefore, new types of frames and edge seal protection should be developed to provide additional protection.

Note: Edge seal quality and protection issues are even more important in the case of some thin film PV modules which are frameless. These modules consist of two glass plates with no sides, so that the edge seal is exposed. This makes the thin films inside the modules more vulnerable.

TEN—Junction Box

The junction box is just that—a small box where the wiring "junction" or connection is made. It is a metal container intended for easy, safe and reliable electrical connections. It is also intended to conceal them from the elements and prevent tampering. The box is attached to the back cover of the module by means of screws and/or glue, and contains connectors for wires coming from the module and those connecting it to the external circuitry.

Corrosion of the contacts in the boxes is the most frequent problem encountered during long-term operation in harsh climates. This might cause increased resistance, reducing output power and eventually resulting in fire or an open circuit. This problem, however, is the only one that can be fixed by replacing the defective parts without tearing the module apart.

Summary

If we represent the actual efficiency and longevity of solar cells modules, using the above described conditions and issues, we get

$$\eta a = \eta t - (1 + 2a + 2b... + 7a + 7b + 8 + 9 + 10)$$

and

$$La = Lt - (1 + 2a + 2b... + 7a + 7b + 8 + 9 + 10)$$

Where

ηa is the actual field efficiency of the module, and
ηt is the theoretical (optimum) efficiency of the module,

while

La   is the actual longevity of the module, and
Lt   is the theoretical (optimum) longevity of the module.

Note: Numbers 1 through 10 are the conditions and issues discussed above.

It is impossible to predict, let alone put a numerical value on most of the conditions and issues in 1 through 10 above, so these formulas are good only to show roughly the qualitative dependence of the efficiency and longevity of the modules to the members of this long chain of events and how they could affect (usually in a negative way) the cells and modules during their long-term, on-sun operation.

As we clearly see, any discrepancy in the chain of ten events can reduce the efficiency and/or longevity of the cells and modules, according to the seriousness of the deviation. Or, as mentioned earlier, the basic quality control rule of thumb for any manufactured product is "The highest quality of the finished device is determined by the lowest quality of any process step [or event]."

Applying the related manufacturing process steps, sub-steps and procedures to the each of the members (1-10) of the above formula will result in a long string of variables (literally hundreds of them). Each of these additional sub-routines and variables can have an equally negative effect on the modules' performance and longevity, if not properly executed.

This is why quality of materials, quality of design, and process control, combined with know-how and experience, are so important in ensuring acceptable quality of the final product. We fear that few people are fully aware of these complexities, and we hope that now, with all the issues clearly identified, there will be open discussions about them. Until then, we must use great care when designing, manufacturing, evaluating, and/or using solar cells and modules for large-scale power generation, especially in harsh climates.

Quality Control

The proverbial, "garbage in, garbage out" principle is in full force where high quality products are intended, so looking for and finding weak links in the process sequence is of utmost importance in any manufacturing process. The quality of the supply chain (materials purchased from third parties) is where the quality control cycle starts. Immaculate quality control of all incoming materials is paramount for final product quality.

Quality of production equipment is very important as well. Without proven and precisely controllable equipment, even high-quality materials could lose their value. Process design and execution are without a doubt extremely critical as well. One misstep in the solar cell manufacturing process could convert a batch of cells into paper weights.

Finally, and very importantly, are the people! Qualified design, equipment, process and quality control engineers, well trained technicians and operators, and experienced managers are all part of the team. The basic rule of thumb remains the same. One process misstep can ruin an entire batch of product.

The PV industry, however, is still weak and unstandardized, as is quality in the PV field. We've witnessed the birth (and death) of *many* companies who jumped into the solar business because it looked inviting. They flooded world energy markets with products that were hastily designed and manufactured. Even the best of the best are experiencing gaps in their quality control programs, reflected in yield loss and field failures.

Basically, the major requirements for efficient QC/QA PV programs are

1. Ensuring that equipment, process and product integrity are maintained throughout planning, design, manufacturing, transportation, and installation steps, and

2. Guaranteeing that the final product meets customer needs, that it is safe and that it performs as expected for the duration of its 30-year life. This is usually done during the qualification/certification tests at specialized laboratories.

Qualification/certification tests are designed to address testing of the final product performance. Some of these procedures and tests are the International Electrotechnical Commission (IEC) qualification standards 61215, 61646, and 62108, the safety standards IEC 61730, and Underwriters Laboratory (UL) 1703.

They are good tests, yes, but these were developed in the 1970s and 80s, and simply do not meet the requirements of today's PV materials, processes and products. In the best of cases, these tests are designed to identify "after-the-fact" problems in finished modules.

Initial Inspection

Today, we have better equipment and enough expertise, but most solar cells and modules manufacturers (especially the low-cost ones in Asia) are not aware of, or rush, proper production procedures, skipping some key inspections, including those of incoming Si materials and wafers.

Processing untested and unsorted Si wafers usually increases the rejection rates of the finished product. Most importantly, cells and modules made of lower quality Si wafers will have reduced efficiency and be much more likely to fail before fulfilling their life expectation in field operation.

With millions of PV modules coming from all over the world, we face serious large-scale problems which will become more pronounced—especially after 10-15 years of field operation when the original manufacturers have closed shop.

In his article, "An Eye on Quality" in July 2011, Ian Latchford of Intevac (US equipment manufacturer) addresses this problem and suggests strict quality control, especially at the beginning of the manufacturing process, by using 100% photoluminescence (PL) testing of the wafers' bulk material to identify low-quality or contaminated Si wafers. PL is one of the most effective and reliable non-destructive processes suitable for mass production operations.

Sounds like a no brainer, but not nearly enough PV cells manufacturers are doing PL, or any other 100% initial bulk wafer material quality testing. The reality is that a 10% visual and surface resistivity check is the accepted maximum, and even that is cut significantly during rush production periods.

The industry needs *total* quality control, including PL or similar non-destructive qualitative testing of the integrity and quality of wafers and cells after key steps of the manufacturing process. The key tests, in addition to the initial PL bulk material tests, must include:

a. Initial and in-between-steps wafers surface cleanliness test, where both wafer surfaces must be visually inspected and then scanned or otherwise tested for organic and inorganic contamination and mechanical defects. Building a solar cell on a chemically contaminated or mechanically damaged surface is like building a house on a sandy hill—it will collapse after the first rain.

b. Diffusion length and concentration testing is absolutely necessary to determine any process inadequacies or material imperfections. The diffusion layer is the engine of PV power generation, and not checking it is like entering a race without checking your car engine.

c. Metal fingers depth of penetration into the wafer surface and quality of the alloyed area. Imperfections in the metal-silicon boundaries will result in latent problems which will affect the cells' field performance and longevity.

Not easy or cheap, this is the only way to ensure the quality of each cell and module. Again, the lowest quality solar cell will determine the highest quality of the entire module (and string). So if one cell in a 100-cells module is underperforming, the entire module will respond in kind. If a cell fails, so will the module. What if this failure occurs in our large-scale power field after only 10 years of operation? Can we afford this? Can we afford to not know the *exact* quality of the product we are using?

This issue is complicated to the point that even many production-line engineers are not be able to tell us what quality to expect. Product technical specifications sometimes show only a good final product—for the

short term; but, latent problems not detected during the manufacturing process cause accelerated degradation, higher temperature coefficient and/or premature field failures.

Note: Thin film PV modules manufacturing, on the other hand, is quite different from a process/quality control point of view, because the process is quite sophisticated and quality control procedures are built in. Automated production equipment and clean-room type facilities are state-of-the-art in most cases, and the production process is run by highly qualified engineers and technicians.

Elaborate QC/QA procedures are built into the process too, ensuring superior quality and performance. Mistakes can happen, yes, but they are the exception, unlike the case with c-Si production where quality gaps can be found at every step of the unsophisticated, labor-intensive process.

QC/QA Procedures

The weakest link in the PV modules' "cradle-to-grave" sequence is the lack of efficient QC/QA procedures in every step of the manufacturing, installation and operation process.

There are no formal standards for maintaining the quality of the materials (including that of the supply chain), manufacturing, internal tests, transport, installation and operation. Whatever standards exist are fragmented and unenforceable.

So, manufacturers, developers, installers and operators improvise their parts as they wish. Most of them are more interested in buying the cheapest PV modules and BOS components available, making their projects more bankable, and opening easy access to financing. Assessing the quality of their PV modules is just a matter of guessing.

Project owners tend to go for the near-term solution, but they should be more interested in long-term performance, while investors are mostly interested in long-term success and must be well informed of possible problems, such as temperature degradation, and module failure rates.

How can owners and/or investors know the quality of their new, shiny PV modules? How would they know what QC/QA procedures the manufacturer followed, if any? How would they know what the installer is doing, and if it is done right?

Generally speaking, with a quality product we know the reputation of the manufacturer and his process, the complete and verifiable product specs, and the short- and long-term warranty details. It is of utmost

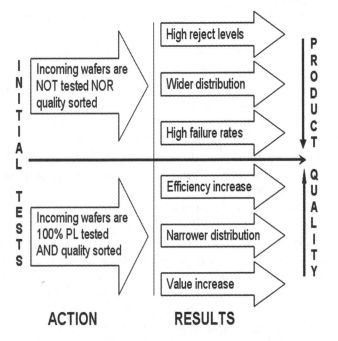

Figure 6-5. Wafers test and sorting procedures

importance also to make sure that the manufacturer (or his representative) has enough funds for potential long-term warranty claims.

Since we don't have much information on QC/QA procedures and results, we must make every effort to ensure that the PV modules and installation are done properly and will operate as specified during their 30-year life.

It's a good idea to add extra protection, such as hiring an outside specialist or commissioner, whose job is to verify the quality of the product and certify the entire job. Such a person will act on behalf of the owners and investors and will dig into the specifics as much as possible, to ensure product and project integrity.

This effort needs to be balanced with additional cost and time required, but it is one of the few controls we have over the quality of our investments.

New standards are badly needed. A number of them follow.

Lack of Standardized Manufacturing QC/QA Procedures

There are no set QC/QA procedures for testing, sorting and using PV materials, equipment, and processes. Some of the basic procedures are borrowed from the semiconductor industry, but even in the best of cases these are fragmented and used in an on-and-off manner. Generally, there is no complete "sand to module" QC/QA system available—even on paper, let alone being implemented by a company or worldwide.

Lack of Standardized System Level Testing of Components

Some combinations of modules and BOS work well in one environment but not in another. There is simply not enough data, covering different combinations (components types and manufacturers), to configure for the best performance under different operating conditions.

Lack of Reliable Climate and Weather Data

The data available in the existing software packages are just not enough for a proper field design. Data from (and the ability to simulate) extreme environments are critical for safe and efficient operation there.

Need for Reliable Simulation of Degradation

All PV products degrade with time. Is the simulation process linear, and would the modules and BOS behave as predicted during 30 years in the field? Proper simulation would more accurately reflect the long-term behavior of components, which is important because abnormal degradation has a non-linear, and some times drastic, impact on a system's performance.

Need for Standardized Warranty Conditions and Insurance Policies

PV modules and BOS manufacturers improvise and negotiate different warranty conditions and insurance policies with the customers and investors. Since in most cases the latter are not specialists in the field, the manufacturers and their associates are free to make adjustments in their favor. Standardized procedures are needed here as well, to level the field.

We are sure this will happen in the 21st century, just as it happened in the semiconductor industry during the last century. Yes, it took a long time, and there are still companies and even entire countries that prefer to improvise instead of standardize. Nevertheless, the foundation is solidly laid and those who build on it are successful.

On-site PV Modules Qualification

On-site qualification is a sign of the times, and of the slow maturing of the solar industry, brought to us by Solar Buyer LLC. This is a fresh and amazingly simple (though hard to implement) idea for ensuring the quality of PV modules **_by direct, on-site_** observations, inspections and tests of the manufacturing process.

We explored this option in our previous book, *Photovoltaics for Commercial and Utilities Power Generation*, but were not successful in finding parties willing to try it. But now we know it can be done... as shown by SolarBuyer LLC, which has established the procedures and is acting as an intermediary between the manufacturers and customers.(8)

SolarBuyer brings together a unique set of experts to design and implement a brand new qualification program based on a broader and more useful set of requirements and metrics than has ever been used before.

The goal is to bring only fully qualified, quality-assured and reputable PV modules manufacturers to the installers and projects developers.

A significant portion of the world's PV module manufacturing is now done in low cost regions—mostly China—so the need for independently ensured product quality has increased.

The requirements are based on the factors that solar installers and project developers regard as important for their business—not just product quality and performance, but also shipping reliability, after-sales service and technical support.

SolarBuyer has offices in China, so their representatives simply camp out on the PV modules manufacturer's production floor and observe the process. Then they report their conclusions to the customers.

This is a simple, reliable way of ensuring product quality. Unfortunately, when we tried this 4-5 years ago we heard loud laughter in the room and were escorted to the door like criminals. So evidently, attitudes are changing fast, which is a good news.

SolarBuyer also offers solar installers and project developers the opportunity to outsource their solar modules procurement to third-party experts.

SolarBuyer helps customers to buy direct from reputable and approved manufacturers by offering the following basic services:

- Product needs assessment
- Procurement strategy and planning
- Aggregated buying power
- Supplier selection and qualification
- Supply negotiation and purchasing
- Supplier auditing and quality assurance
- After sales service

Because SolarBuyer offers access to shared procurement expertise, the service costs are competitive and the resources are flexible. The work is done primarily on the basis of service agreements, retainers and success-based fees.

SolarBuyer is not paid by the manufacturers, nor are they incentivized to offer any particular products. Their focus is on representing the customer, the actual buyer and user of the PV modules.

This new service offers a shared procurement recourse which can fundamentally reduce the costs of procurement. And because SolarBuyer represents the actual PV modules buyer (not the manufacturers), their only goal is to find the best products for the customers' needs.

This service opens the doors to:

Lower purchase prices—
- Purchase direct from manufacturers
- Buying power beyond your own
- Expert, experienced negotiators

Lower operating costs—
- Flexibility through outsourcing
- No hiring of your own procurement experts
- Focus on your own selling
- More efficient after-sales service

Better Product—
- Identify the best, most competitive products
- Access the newest products first

- Easily change or add manufacturers
- Freedom from restrictions to products offered by your distributor

Author's note: This, we believe is the way of the future. It might sound strange and even unorthodox, to sit on the manufacturing floor and watch product being made—it's certainly unique—but keep in mind the product that is under scrutiny. PV modules are perhaps *the most serious of all electric devices*, and they must operate in extreme climates for decades. Do you know any other electronic product that is destined to endure such torture? Let us know, if you think of one.

So this service is not only justifiable, but absolutely necessary, especially when large numbers of modules are ordered and the destiny of a multi-million dollar project is at stake.

We don't know if SolarBuyer will be successful as a business model, but we believe this is how business will be done routinely, as it is the only way to ensure the best quality possible for solar projects.

Bulk Silicon and Wafers Quality

The quality of the bulk silicon material, as well as that of the consequently produced wafers (and of course the resulting solar cells), is one of the most critical aspects of solar cells and modules manufacturing.

Special attention and extreme caution must be exercised when choosing bulk materials and during their processing, because their quality is locked in, difficult to check, and impossible to change.

Silicon wafers are as thick as needed to give them the mechanical strength to withstand handling and other stresses during processing, as well as stresses they might encounter during transport, installation and long-term field operation.

Some of these stresses can be significant, so wafers' thickness must be well understood and calculated prior to processing thousands of them. The natural desire to reduce the wafer's thickness without increasing its strength could lead to a high breakage rate.

Cracking of solar cells is one of the major sources of solar module failure and rejection during processing and operation. So, it is critical to investigate the electrical properties and the mechanical properties of the wafers, and especially their mechanical strength.

The wafer strength and the related damage mechanisms have to be well understood and incorporated into the manufacturing process, to minimize the fracture rate. The mechanical properties must also be considered in the consecutive steps of the modules' use, and to the very end of their useful life. It's not an easy task, but one of utmost importance, and one that manufacturers sometimes ignore.

The fracture strength of silicon wafers can be measured by a 4-point bending method, statistically calculated by Weibull analysis, and interpreted in terms of defects (flaws) distribution in the test wafer.

Results from such tests can be used to determine the nature and source of the defects (flaws) as needed to understand and control fractures and bending (flexing) strength of the silicon wafers and the finished solar cells.

The data can be used to enhance production yields, improve the wafer and cell stability, and to establish a criteria based on mechanical parameters that could lead to a reduction in cell costs (by reducing the number of broken cells.)

Mechanical stress and damage can occur at any step of the manufacturing process and afterwards, but the most important mechanical damage issues are given below. Their origins, creation and behavior must be thoroughly understood, analyzed and observed.

Silicon Solar Cells Quality

The silicon material, which is to be used for melting and slicing into wafers, is the foundation of the quality cycle and determines the quality of the final product. Period! The wafers and solar cells produced from it will be of the same or lower quality, and CANNOT under any circumstances be of any higher quality than the starting materials.

Silicon material is mined and processed in dirty mass-production facilities, where quality is not a priority. This dusty and contaminated (metallurgical grade silicon) is transported to processing plants, where it is washed and melted into ingots, and supposedly its chemical and mechanical properties are improved to a level acceptable for manufacturing solar cells.

While the chemical quality can easily be determined by analyses, the mechanical quality of the material is hard to analyze and document. Silicon blocks (bricks) have different properties from batch to batch, and differ from one end to other of the ingot. Too, the inside of the ingot is different from the sides, and even more so from the ends.

Checking the mechanical quality of the silicon ingots requires sophisticated and expensive equipment, which most manufacturers do not have, so they must rely on the expertise of the suppliers.

In our opinion, this is *problem number one* with the quality of silicon solar cells and is responsible for their poor performance and even failures.

Standardization of the manufacturing procedures, which is lacking today, will solve this issue by implementing equipment and procedures for thorough analysis of incoming silicon material and for documenting the quality of every batch prior to melting and processing into ingots.

Once formed, ingots are cut into blocks and the blocks are sliced into wafers, where another set of mechanical problems are encountered and must be solved.

Crystallinity

The type, size and number of crystals that each wafer consists of are of great importance to its mechanical stability. Just like the strength of a single (monolith) cement column is much greater than that of a similar column made of numerous stacked cement pieces, mono-crystalline silicon wafers are much more stable and stronger mechanically than poly-silicon wafers.

At the same time, poly-silicon wafers consisting of one or more larger crystals, are stronger than similar wafers containing many small crystals.

Basically, as with any brittle material, the mechanical and fracture strength of a silicon wafer depend on both its material-dependent intrinsic properties (grain size, grain boundaries, and crystalline orientation), and on its material-dependent extrinsic variables (surface flaws, pits, and micro-cracks).

Strength reduction caused by the presence of a larger number of small grains is related (and maybe directly proportionate) to the number of grain boundaries in the wafer, since these are proportionate to the number of grains in the wafer.

The level of surface roughness also varies for different crystallinity types, with single crystal silicon producing the smoothest surface, while poly-crystalline silicon with many small grains produces the most uneven surface.

It is difficult to predict the exact influence of the extrinsic and intrinsic variables on the mechanical stress, and/or which has the most significant effect on the mechanical strength of the wafer. As a consequence, we are half blind when it comes to assessing the actual surface and bulk quality of solar wafers, let alone predicting their behavior during the manufacturing process and subsequent long-term use.

Wafer Sawing

The blocks or bricks of ingots are mounted on a wafering saw, which uses either a diamond saw blade or a stainless steel wire. Both processes are high-speed cutting (sawing) operations using large amounts of slurry and cooling liquids. Imagine the noise and the mess one machine makes, then multiply that by tens or hundreds, to get an idea of a large-scale manufacturing operation.

Wafer saws damage the surface of wafers, creating cracks, cavities, pores and other stress-related irregularities. The damage is done by the friction of the saw (blade or wire) on the wafer surface. These surface and bulk flaws cannot be avoided, and we are painfully aware of the fact that wafer strength is related to the density, size and distribution of micro-cracks on its surface and bulk material.

Cut wafers that are not treated after sawing have a much lower strength, due to the presence of the above mentioned micro-cracks. If the density of micro-cracks is high enough, the chance those irregularities increasing the overall stress level is also high. This explains the lower strength in silicon wafers that get no additional chemical treatment.

Micro-cracks induced during the sawing process can be reduced by chemically treating (etching) the wafer surface with acids, such as HF, HNO_3, CH_3COOH and others.

In addition, buffered hot KOH and such chemicals are often used to create pyramids on the wafers' surface (usually used on mono-crystalline silicon wafers), which can further reduce micro-cracks and other defects from the front surface, which is the most critical.

Etching processes reduce the depth of most surface micro-cracks and, in the worst cases, many sharp crack tips are blunted, which reduces their ability to grow and propagate.

These effects reduce the risk of macro-crack initiation, and propagation, making the material less susceptible to breakage and or other types of failure. We have measured increase in the mechanical strength of silicon wafers by a factor of 2 after proper post-saw etching.

Metallization

Metallization is a high-temperature process, which induces another set of stress factors in the already mistreated solar cells-to-be. The wafers are "screen-printed" with a special type of paste containing metal dust, like silver or aluminum, which will become top and bottom contacts.

The printed wafers are placed in a high-temperature furnace where the paste is dried and the metal dust is fused into the wafer surface. The wafer is stressed again and significantly by the high temperature and by the metal "belts" that are now solidly tied to both its surfaces.

The type of metal paste and paste drying and baking process determines the additional stress induced on the newly made silicon solar cells, and defines their ultimate mechanical strength.

Now we have two distinct layers, the bulk silicon wafer, and metal layers on top (silver bands) and bottom (solid aluminum cover) surfaces. Via bending tests it is possible to determine the maximum stress in each layer at the moment of specimen fracture. Unfortunately, the strength of the silicon wafer and the aluminum (Al) layer (i.e. composite beam) cannot be determined precisely because it cannot be determined in which layer a fracture originates.

The different Al metallization paste types have a significant effect on the induced stress and the overall mechanical strength of wafers. When both layers (i.e. the silicon wafer and the Al layer) are loaded in tension the silicon wafer experiences the highest tensile stresses. When reverse loaded, the effect on the mechanical strength of the wafer is much less pronounced.

Usually wafers with an Al layer show an increase in bending strength as compared to un-metalized silicon wafers, which is most likely due to the formation of a eutectic layer (referring to the direct transformation from liquid to solid or vice versa) and a BSF layer (the back surface field).

In all cases the maximum tensile stress in the silicon wafer is located at the interface between the silicon wafer surface and Al layer (in the eutectic and the BSF, and sometimes both, layers). These phenomena are due to the properties, behavior and interaction of the micro-structures in each layer.

The Al layer with its composite-like micro-structure, consists of spherical hyper-eutectic Al–Si particles, a bismuth silicon glass on the surface and a level of porosity within the layer.

The Al layer is un-uniform from top to bottom and from end to end, so it does not fully cover the eutectic layer area. The eutectic layer represents a uniform Al–Si bulk alloy, which is 100% in contact with the BSF layer, and the silicon wafer proper, so it (the eutectic layer) and its interface at the porous layer have a significant effect on the mechanical properties and behavior of the silicon wafer.

Since silicon is a very brittle material that exhibits linear elastic behavior, the presence of a 2nd ductile phase (i.e. the eutectic layer) could induce some plasticity in the system, altering the stress distribution within it. This could affect crack initiation and propagation in a way that is hard to predict. The ductile eutectic layer can also serve as a bridge for some cracks, which would reduce the stress and improve the strength of the silicon solar cells.

Different Al pastes have different effects on the mechanical strength of silicon solar cells, part of which can be explained by the differences in the bulk and surface micro structures. Some of the micro-structure features that affect the mechanical strength of wafers are the Al layer thickness, its porosity, the bismuth glass properties, and the thickness of the eutectic layer.

It is imperative, then, that we have a full understanding of the materials and processes involved in the manufacturing sequence, and control these as much as possible. Proper and timely measurement of stress factors in wafer and solar cell batches is critical.

PV Modules Handling, Packing and Transport Issues

During their assembly, PV modules go through many processing stations where they are exposed to numerous manual transfers, loading, unloading and other operations. During any of these steps the modules could be dropped, hit, scraped, or suffer physical contact which could crack, stress or even break them.

The modules are sold as independent, self-contained "packages." That is, the solar cells are enclosed (encapsulated) in a sandwich of materials, selected, manufactured and installed in the module in such a way as to provide maximum power and protection of the solar cells from impact and from contact with the elements. Any impact, pressure, or similarly careless contact with the modules could be detrimental to the performance and lifetime of the solar cells in them. The resulting damage might be noticeable immediately, but most likely it will show up as loss of performance or failure.

So, if a module is badly damaged—broken glass, bent rear cover or side frame, etc.—it would be detected and separated prior to shipping, or certainly before installation. Some of the stressed and cracked modules will not be detected, however, and might even test OK at first, only to fail a short while later. A manufacturer's warranty usually covers modules that fail during the warranty period, but there are cases when the damage does not manifest itself for months or years. In many such cases, the customer has no recourse and will end up "eating" the damaged product, unless there is special insurance or warranty to cover these exceptions.

PV "packages" are promoted and sold as "hermetic" or nearly hermetic vessels, which means that no water will penetrate through the materials at any time to attack and damage the solar cells. From a scientific point of view, most materials—including all PV modules components (glass and encapsulating materials)—are permeable to a certain degree. Given time and the right con-

ditions, air and moisture *will* penetrate these materials and, yes, even glass. So the packaged modules should be kept away from water and certainly submersion into water must be avoided at all costs. Any signs of excess moisture must be reported for warranty purposes.

In all cases, each individual module must be carefully unpacked, visually inspected and electrically tested prior to installation. Any modules with unusual appearances or performance abnormalities must be separated and held for inspection by the manufacturer's rep. Any unusual drop in power or other abnormal behavior of the modules or strings of modules after installation must be documented and investigated for warranty purposes as well.

Electrical Components

PV modules have other components which also need to be taken into consideration prior to installation. Some of these are wiring, external bypass diodes, junction boxes, and terminals. These are integral parts of the modules' structure and critical to their overall performance, efficiency and longevity.

Any modules with broken or bent wires or parts must be separated for investigation. Abnormalities such as the wrong type and size wiring and poorly designed and assembled junction boxes will cause poor performance or failure if not detected and corrected. Thorough inspection must be performed on each module—especially if these are coming from new manufacturers.

Materials Supply

Although wind and solar technologies are called "renewable," some of their component materials are far from renewable. Some are expensive, exotic and rare materials, in short supply, and some are even on the "endangered species" list.

On top of that, a major part of these exotic materials are found and mined in Third World countries, where quality is often far from being a priority. In all cases, we must know where the materials come from and assess the associated risks, which we've shown elsewhere in this text.

Safety

Generally, all PV modules, including c-Si modules, contain some toxic materials, and must be handled properly. Small amounts of Pb or Sn, for example, could be found inside the modules and sometimes even on their outside surfaces. Some thin film PV modules are especially dangerous, because they contain toxic and carcinogenic compounds of elements like cadmium, tellurium, indium, and arsenic. Trace amounts of these poisons might have remained on the modules' surfaces, which upon contact could trigger allergic reactions in some people.

In all cases, due to the toxicology problems and the fact that PV modules have sharp edges, they must be handled with gloves and with utmost caution. All incidents, accidents, or work-related ailments must be immediately reported and investigated.

Module Encapsulation Issues

Proper encapsulation of PV modules is a well known issue, which has been ignored by some manufacturers in their drive to keep costs low. This has led to the installation of millions of PV modules in the world deserts without proper encapsulation and sealing of the active layers. This short-cut might cause decrease of power output and even massive failures.

Here is a review of the experience and results on the subject, as obtained by researchers at the Photovoltaic Institute in Berlin. (4)

Typically solar cells are developed to show best performance under laboratory conditions for a standard AM 1.5 spectrum, perpendicular irradiance at a level of 1,000 W/m² and room temperature of (25°C). Performance measurements are usually carried out using a solar cell-to-air interface.

These conditions do not reflect the reality. Solar cells are regularly encapsulated within two additional optical slabs, typically EVA and glass, irradiance is hardly ever perpendicular, spectrum is rarely AM 1.5, and operation temperature for relevant irradiance levels is well above room temperature (15–30 K higher) for non-alpine or non-arctic applications. Irradiance level infrequently reaches 1,000 W/m², even for locations with good conditions.

This situation leads to considerable uncertainties for the accumulated electrical energy yield of PV modules. Therefore, a simulation toolbox was created to increase yield accuracy considering optical, thermal and electrical parameters.

Cell to Module Interface

The module manufacturer usually acquires solar cells probed and optimized using an optical interface consisting of the solar cell and the surrounding air. Usually the integration of the cell into a module encapsulation boosts the electrical power output by 1–2% due to the better matching of the optical interfaces involved: antireflection coating (nAR = 1.9–2.3) to EVA (nEVA = 1.45) to glass (nGlass = 1.54) to air (nAir = 1.0) instead

of antireflection coating to air only. Therefore, accumulated reflection losses are reduced, the photon flux is increased, as is the photo current and power output.

Usually this effect is neglected by the manufacturers due to a lack of treatment measures, but also due to the fact that the electrical interconnections between the cells in the module tend to recompense that effect.

Nevertheless, for an accurate modeling and simulation, this effect should be taken into consideration. Therefore the simulation program is designed to trace light through all optical interfaces on the way from the real-world environment into the semiconductor material of the cell. This enables the treatment of cells and modules within.

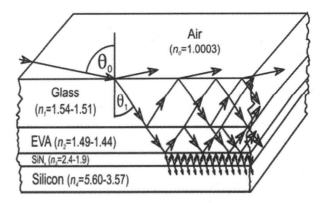

Figure 6-6. Cross section through a typical PV module together with the traces of a single ray (at the incidence angle θ0).

Module to Real World Interface

Real world also considers actual spectra, irradiance levels and angle of incidence for direct solar irradiance throughout a day, together with the elevation angle and the azimuth angle of the sun (plotted for equinox, module and elevation angle)

Modeling of Transmittance
Single Interface

Each optical interface divides an incidencing ray into a transmitted (T) and a reflected (R) component according to the Law of Fresnel. For perpendicular incidence (θin = 0) the reflected part is given by:

$$R = \frac{(n_1 - n_0)^2}{(n_0 + n_1)^2} \qquad T = 1 - R \qquad (6\text{-}1)$$

For non-perpendicular incidence the reflected and transmitted components depend on the state of polarization of the ray towards to the plane of incidence, parallel ($\parallel$) or perpendicular ($\perp$):

Optical Slab

Figure 6-7 shows the situation for an optical slab (or layer) with two optical interfaces and attenuation (or absorption) within the slab.

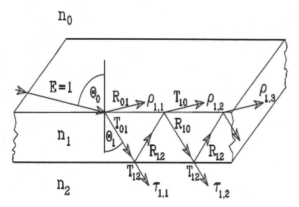

Figure 6-7. Transmitted and reflected components at an optical slab.

Within the optical slab a ray is bounced at its interfaces infinitely, but with decreasing irradiance levels. If transformed into a so-called "geometrical series" the mathematical expression can be simplified.

Multiple Slabs

Considering all possible bounces within the different slabs, among slabs as well as within and among systems of slabs the total transmission equation of a series of slabs is more complex.

Lamination Durability

One of the critical issues facing the photovoltaic industry is the durability of the modules during long-term use under extreme conditions.

Adhesion of the lamination is one of the major issues that needs to be fully addressed and resolved, for the customers to have complete confidence in the long-term field durability of PV modules.

Adhesion of a standard PV module lamination consists of the sequence glass-EVA-cell-EVA-glass or glass-EVA-cell-EVA-backsheet. The adhesion quality depends mainly on: (5)

1. the cleanliness of the glass sheets

2. the condition of the EVA (prior to lamination)

3. the lamination process (process temperature, profile and duration, pressure, homogeneity)

To test lamination quality, three different tests (on the module center as well as close the module edges) have been applied:

1. Chemical analysis of samples to determinate the state of "curing" or cross-linking of the co-polymer EVA after lamination;

2. Peel tests to determine the force to peel off the layer of the laminate;

3. Damp-heat treatment (1000 h at a temperature of 85°C and 85% relative humidity). It was found that chemical analysis of the EVA curing state is helps ascertain the accuracy of the lamination process, following the recommended temperature profile and homogeneity, and the initial condition of the EVA used.

However, this test method may be misleading to allow a statement on the overall quality of the lamination: the surface glass sheet may be treated with oil (e.g., to prevent adhesion of the individual glass sheets during storage) which has not been removed properly before lamination, thus causing low laminate adhesion (especially after the damp heat treatment, early moisture ingression and a reduced module lifetime. Therefore peel tests are essential to determine the overall quality of the lamination.

To avoid (or at least to limit) moisture ingression, an adequate sealing at the module edges is very helpful, but also a good adhesion of the laminate layers is necessary.

Adhesion of a standard PV module lamination consists of the sequence glass-EVA-cell-EVA-glass or glass-EVA-cell-EVA-backsheet. The package of a PV module is a very complex system with many surfaces.

Careful selection of the components is required to achieve an optimum result. Just the backsheet itself consists of 2-4 layers (e.g., Tedlar®-Polyester-Tedlar®) which are bonded to each other.

The standards IEC 61215 (for modules with crystalline cells) and IEC 61646 (for modules consisting of thin film cells) have been established to verify overall quality of PV modules and to ensure performance and lifetime.

These similar standards, IEC 61730 and UL 1703, are both focused on module safety. In addition to the standards, some banks and other financing institutions demand additional tests and proofs—for example, proof of sufficient curing of EVA.

Storing the Material

In most photovoltaic modules an encapsulate material consisting of ethylene-vinyl-acetate (EVA) is used. This encapsulate is produced as a film and is stored and delivered on rolls. The length and the width of the rolls may vary. Most EVA manufacturers recommended a storage temperature below 30°C (optimum at 22°C) and tight wrapping with the original packaging material. A film cut in sheets should be stacked and used within 8 hours.

The company STR gives in an overview in the technical manual of Photocap® products: reduction of adhesion of the EVA to other materials by the duration of storage.

Via customer reply and their own investigations, PI Berlin compiled a list of possible reasons for partial or complete failure of the laminate. Some of the problems are a direct result of not following the storing parameters mentioned above (PI Berlin, 2011):

1. *Unclean glass or other parts*
    Dirt or process chemicals might stick on a surface of one the components of the compound. This prevents adhesion of the EVA or the backsheet or the glass.

2. *Evaporation of the curing agent before curing*
    Due to wrong storage or mistakes on the producer side, there is not enough curing agent left for the curing process.

3. *Curing time too short*
    To increase production numbers, insufficient curing time may be applied by the process engineers. The result might be inhomogeneous curing within a module, or only partial curing.

4. *Wrong curing parameters*
    The temperature for curing is selected too high or too low. In combination (or not) with inappropriate curing duration, partial curing might result or the material can be irreversibly damaged.

5. *Not uniformly cured*
    Since the development of ultra-fast, fast, or similar curing formulas in combination with increasing module sizes, the curing process might not be completed across the whole range of the module. We found that the curing level could deviate up to 20% between the positions. The problem is the time difference when curing temperature is reached at the central part of the module and at the edge or corner of the module. In addition, thermal stress might result in a bending of the module, which lifts parts of the module from the surface of the lamina-

tor and reduces heat flow considerably. During the testing at PI Berlin, gel-content rates for different positions were obtained which deviated more than 30% from each other.

6. *Error by the supplier*
    It could happen that the supplier does not put enough peroxide in the EVA, stores it too long, or just mixes good EVA with bad EVA. In contact with customers PI Berlin was told that it was a single batch of EVA, yet in the lab it proved to have different curing properties.

7. *Process problems on the supplier side*
    Non-uniformly distributed curing agents and other chemicals in the films cause local different properties. In highly optimized process it might lead to localized curing. (Schulze, 2011) has shown that for a standard EVA there is already a noticeable local variation of the peroxide in the sheet.

8. *Chemicals in the EVA are damaging the back-sheet*
    In contact with manufacturers of back-sheets, we were told that a too-high remaining peroxide level can damage bonding. Results at PI-Berlin seem to prove that, yet this issue is still under investigation.

Chemical analysis of the EVA

System of Cross-linking

The film is composed of a standard co-polymer EVA which is produced for example by DuPont™ under the name of Elvax®. The bulk material of the EVA is a thermoplastic. The manufacturer of the film adds a curing agent and other chemicals.

The curing agent is a peroxide (Klemchuka, 1997) which decomposes with the increase of the temperature and starts the reaction in the film. That should be the curing process in the laminator. When the curing process is finished the former thermoplastic becomes an elastomer and cannot be melted anymore. The material is irreversibly cured.

Soxhlet Extraction at the PI Berlin

The extraction method at the PI Berlin is based on an extraction according to soxhlet. The apparatus used by the PI Berlin has proven to give the most reliable results due to good controlling of the extraction cycle and an active flushing of the specimen at the end of the cycle. Further information can be found in the information sheet about the extraction at the PI Berlin.

Removing of the EVA

The removing of the EVA is a difficult part of the analysis. At defined or from customer requested positions, the backsheet is removed and the EVA taken out. This takes a lot of time and practice. Many modules have completely different adhesion qualities.

After taking the EVA, the samples must be cleaned. In this process it is also possible to determine if a filled EVA (fiberglass or otherwise supported) is used which results in correctional parameters for the extraction. It is possible to see, in some cases, the different adhesion qualities of the different parts of the module; a weak connection to the cell and metallization paste, and a good connection to the busbar and glass.

Extraction Process

PI Berlin elaborated on a useful set of parameters to achieve satisfying results. Different EVAs with different curing times are behaving very different in the extraction process. Typically, 45 and 75 cycles are run. If the deviation of 45 and 75 is too high, the measurements are repeated doing 150 cycles.

Extended extraction may be required to reach the stability on the gel content determination. It is essential to provide the most accurate result. Stability is reached after specific extraction times or extraction cycles—depending on the tested module (or lamination process and material). For different scenarios different sets of filters have been used: cellulose (low level) and stainless steel (medium- and high-level gel content). For solving EVA, PI Berlin uses stabilized tetrahydrofuran (THF). This solvent avoids further curing during the extraction process and minimizes the melting and softening of the polymer due to a high extraction temperature.

Problems

In addition to the above problems, extraction adds even more, the primary one being the lack of standardization which allows on the market a large number of EVA films with differing properties.

When the customer receives the datasheet, he must check if test results are within acceptable limits to guarantee the properties he has been promised by the manufacturer. The results show only if the curing happened and give a fairly good approximation of the cross-linking level.

Peel Test

Similar to extraction there is no definitive standard for the peel-off test in the PV industry. The peel test originates from the field of adhesives and is described internationally in many standards as EN 1895, or EN 28510. The principle of the peel test is the pulling of a thin, flexible film from a rigid substrate. The angle for pulling at PI Berlin is 90°.

The peel-off test also includes the surface condition in the test procedure. In some cases the curing sate of EVA may be satisfying, but the adhesion (and consequently the lifetime) of the module may be poor, due to the low adhesion caused by unclean surfaces at the glass, backsheet or cells.

Therefore, the peel-off test is giving more comprehensive information about the durability of the lamination, but the EVA gel-content test helps to find details.

The peel-off procedure is carried out by cutting off stripes partly from the backsheet laminate and applying an increasing pull force to them. For the test, the speed of the peel is constant and the force is recorded. At a specific force the rest of the stripe starts to dismantle from the module laminate. That specific pull-off force is recorded. The test is carried out for 20 times at different locations of the module (center, corner, edge, busbar, cell and glass). In addition, when possible, the different layers (backsheet, EVA or parts of the backsheet) are separated and pulled off.

During the peeling, the module moves in the opposite direction to the peel to keep the peeling angle at 90°.

Damp-heat Test

The damp-heat test is a standardized procedure as described in IEC 61215/61646 (85°C, 85% relative humidity for 1000 h). This test is quite helpful to determine module durability, but while it takes at least 41.5 days to be performed, it is not the main focus herein.

Other Methods

In any other test, stress applied on the encapsulation system, e.g., ammonia chamber (for modules installed on farm roofs with livestock), UV (especially for modules made without glass), and humidity freeze test, increase the possibility of failure of the encapsulation.

None of the tests can be applied as the only measure to describe module durability. For example, it is possible to have an almost perfectly cured EVA (extraction test) and still poor adhesion. On the other hand it also possible that the EVA sticks quite well to back-sheet (peel test) but most of the peroxide has been remaining in the EVA, thus reducing long-term durability. In addition, many non-durable encapsulations may be detected via the standard tests of IEC 61215/61646.

Close attention must be paid to the results which

can indicate discoloration, delaminating or moisture ingression (visual inspections during the IEC test sequence). Due to the lack of standardization and the complex system of the encapsulation, problems might be indicated by a small indicator rather than the total failure within the standard test sequence of IEC 61215/61646.

For the manufacturer of the module, the extraction test and the peel-off test can be easily applied in their production facilities to constantly monitor quality of their production process.

The suppliers of components like EVA will usually provide them with the necessary information, and different test equipment is available on the market.

For testing labs, the challenge is to address the constantly changing properties of the EVA and backsheet materials, and in some cases the unwillingness of the encapsulation manufacturers to pass technical details to third parties.

Many of the tests are made for financing institutions and resellers. Before testing, the only information these groups have is the information on the number plate on the module. Sometimes, the encapsulation production process changes without giving notice to the customer, the testing lab, or the certification body.

Steady control and supervision of the production process seems to be necessary in all cases.

In Summary

The module's encapsulation

a)   provides a convenient package that can be installed and used in the field,

b)   prevents mechanical damage to the solar cells, and

c)   prevents water or water vapor from penetrating the module and corroding the electrical contacts and junctions.

Many different types of PV modules exist, and module structure often differs for different types of solar cells or for different applications. For example, amorphous silicon, and other thin film solar cells are often encapsulated in a flexible array, while crystalline silicon solar cells are usually mounted in rigid metal frames with a glass front surface.

Module construction—materials and type of construction—is of utmost importance and can mean the difference between success and failure, in long-term desert tests.

One useful observation here is that many thin film modules are of the glass/glass type construction, which is basically two glass panes joined together with encapsulation and the thin film structure in between. This configuration is preferred by many manufacturers today (TFPV modules especially) to reduce costs. We insist, however, that these initial cost savings are going to cost dearly in the long run.

An interesting conclusion in several test studies was that glass/glass construction of some PV modules contributes to heat retention which causes additional increase of temperature in the modules and could result in excess delamination and performance deterioration. "It is immediately obvious that Model C has the highest percentage of hot spots. It is believed that the thermal stress of the glass/glass construction and large size could be the primary causes for this failure type. Again, seeing as how this study was conducted in a hot and dry climate, it is not to say that the glass/glass module construction would not succeed elsewhere."

In addition, since most of the modules in the study (65%) were frameless (no metal side frame to cover the exposed edges), the high defect rate must be re-evaluated from that point of view as well. Intense IR and UV radiation could quickly degrade the plastic edge seal, which would cause failure of the module.

So many tests confirm that due to the above issues, frameless glass/glass PV module construction is incompatible with desert applications. This should be a warning to all manufacturers that frameless glass/glass construction has its place in energy markets, but that deserts are not the best choice.

Also, it was noted in the studies, that some manufacturers have replaced the metal back covers of their modules with plastic sheets, which would allow moisture penetration with time. Plastic materials have some advantages (mainly lighter weight and breathability) so these plusses and minuses must be considered in all cases.

Our recommendation would be to use modules with metal back covers for long-term operation in the desert, because plastics just don't last too long there—regardless of their other benefits.

Connecting many modules in strings causes additional over-current, overheating and other issues, or as one study concludes, "The higher degradation rates of grid-tied PV modules as compared to individually exposed modules are attributed to the system voltage related corrosion, module mismatch and insolation levels."

This also confirms that the modules' temperature coefficients in the field are very different from those measured in the lab, so the study recommends using field measurements instead of taking them from the manufacturers' spec sheets. This is also in line with our conclusions and recommendations.

In addition to being disappointing, the excess encapsulation damage and delamination of frameless modules (some due to glass/glass construction overheating) is also quite dangerous because some all-glass, frameless TFPV modules contain toxic materials.

The IR and UV accelerated destructive processes would easily damage the encapsulation in the open edges, where cracks and voids would allow moisture and environmental gases to enter the modules. The unwanted penetrants would then attack and slowly overstress and decompose the thin film structure, which has the potential of contaminating the surrounding area.

This is especially dangerous when considering the huge number of toxic materials containing TFPV modules planned to be installed on thousands of acres of desert in the near future.

So the question here is how to predict and prevent quick delamination, moisture penetration, premature failures of protective encapsulation layers, and the subsequent destruction of the active module components. What are the variables that we should keep in mind, and what should be done to ensure proper protection in preserving optimum performance of PV cells and modules in large-scale PV plants? What kind of quality control is needed to provide maximum quality of the final product? How do we attract the manufacturers' attention in a positive way, to obtain the relevant data and cooperation needed for analysis of the quality of their products?

Better yet, shouldn't we start looking for modules that offer better solutions to edge sealing and protection, similar to that provided by some TFPV manufacturers as Solar Frontiers, who put special effort into ensuring the integrity of the thin film in the module by protecting it with several moisture ingress barriers, including a frame all around the module edges?

Note the carefully designed redundant seal of the module edges in Figure 6-8, where the moisture has to penetrate a thick layer of frame-sealing material, specially designed moisture-resistant MVTR film, and a thick layer of encapsulation, before reaching the active thin film structure.

This module construction is very similar to that of framed c-Si PV modules, with the silicon cells replaced by CIS thin film, thus preserving a solid, impenetrable (glass encapsulation, cell encapsulation, backsheet, frame-all-around) cover.

Based on proven PV industry technology, this module design would provide excellent moisture ingress prevention, and protect the active thin film structure inside much better than the frameless glass/glass PV modules, where only a thin layer of encapsulation separates the thin films from the elements and eventual decomposition.

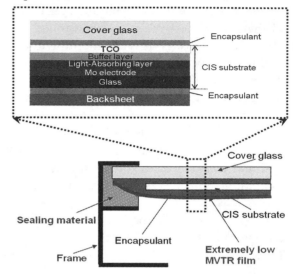

Figure 6-8. TFPV module with moisture barrier

Would these modules with such careful encapsulation operate flawlessly 30+ years in the desert? We don't know the answer to this question, because no module has been field tested for that long, but common sense and years of hands-on experience tell us that framed TFPV modules with triple edge seal barriers will outlast any frame-less module installed in any part of the world by far.

This is even more true for modules operating under extreme desert heat or in humid areas. This is an important consideration that must be taken into account by customers, designers, installers and investors alike, because it might mean the difference between success and failure of our large-scale PV projects in deserts.

Conclusions

Obviously, the PV industry has a lot of work to do before it becomes a proven, mature technology. There are many research groups active in the field of photovoltaics in universities and research institutions around the world.

This research can be divided into three areas:

a)   making solar cells and modules cheaper and/or more efficient to effectively compete with other energy sources;

b) developing new technologies based on new solar cell architectural designs; and

c) developing new materials to serve as light absorbers and charge carriers.

c-Si modules have the advantage of being more efficient, which leads to better space and equipment utilization. Thin-film photovoltaic modules, on the other hand, are advantageous in that they use less than 1% of the expensive raw material (silicon or other light absorbers) compared to silicon wafer-based solar cells, leading to a significant price drop per Watt peak capacity.

Many research groups worldwide actively study different c-Si and thin-film materials and processes; however, it remains to be seen if they can find techniques that can compete with conventional energy sources, which are improving and becoming cheaper.

An interesting aspect of thin-film solar cells is the possibility of depositing cells on all kinds of materials, including flexible substrates (PET for example), which opens a new dimension for applications. The future will surely bring even more good news to the thin film industry.

Practical Applications

Taking a careful look at the field test studies in this text makes us pause and look again. Excessive annual deterioration (much higher than expected) after only 10-15 years in some cases, and other failure mechanisms, are things to think about. How would a 5-10% reduction in power (or total number of failed modules) reflect on the bottom line of our 100 MWp PV power plant? How would we justify the annual power losses and make up for the power expected by, and contracted for, with the utility?

Field failures are another shocking confirmation that many of the unresolved and unaddressed issues presented in this text are real, measurable, and quite significant, so they must be addressed and resolved soon—preferably before we enter into an irreversible large-scale PV nightmare. These test results are frightening even to those of us who have been aware of the problems and issuing warnings for years.

Our long experience with engineering issues and quality control matters cannot process the near 100% defect ratio suggested in the field test results, let alone put it into perspective. The practical advice we have for perspective PV customers or investors is to make sure they know EXACTLY what they are dealing with and getting into:

1. Don't let brand new, shiny PV modules mislead you into thinking that they will perform as well as they look for the long duration—especially when extreme climates like deserts and swamps are in the picture. These modules are not disposable toys, but serious high-tech devices which will be exposed to the most ferocious climate on Earth and endless torture for many years, so they must be made and used properly.

2. Don't let the manufacturers' promises mislead you into believing that they know what their modules will do under excess desert heat and moisture, unless they have long-term test data to show. And most PV module manufacturers today cannot show any long-term tests because they have not been around for very long themselves.

3. Don't let financial gains shown on paper for the first 3-5 years blind you to the fact that what comes during years 10-15 might not be as wonderful.

4. Do your homework thoroughly; do not leave anything to chance or to unconfirmed promises:
   a. Know who the module manufacturer is, and how long they have been making these modules (check if these particular modules are a newly designed product without any desert operation history).
   b. Learn the production process details: equipment, process steps, materials, labor conditions and everything else that goes into the manufacturing process.
   c. Get as much quality-control documentation as you can: initial inspection, in-process QC and test procedures, final test procedures and results, and long-term test procedures and results.
   d. Get information on the supply chain (the quality of materials and chemicals made by a third party).
   e. Learn all you can about the modules: where they were made, where and how they were tested and certified, any long-term test results done—when, by whom, and how.
   f. Become familiar with the details of the technical information available, describing the modules' structure, function and performance.
   g. Discuss and negotiate warranty conditions, in detail.
   h. Parts and labor warranty conditions: What

is considered a failure? By whom, when, and how will the failed modules be replaced? How would the lost time be compensated?

Long-term Performance Conditions
1. Temperature coefficient (Tc) is accepted as 0.5% per degree Celsius measured at STC. But in the deserts we measure over 180 degree Celsius inside the modules. This is a problem that we must acknowledge at the very beginning. We must agree on the modules' performance under desert conditions, where >180°F in-module temperatures during the summer months is daily. If 0.5% per °C is the maximum accepted under the warranty, then what could be done if it gets higher?

2. Annual degradation of 1% per year is a accepted as normal. What if it goes up to 1.5 or even 2% during year 5? How would the replacement be done and lost time compensated?

3. In addition to the anticipated Tc of 0.5% per degree Celsius, and annual power loss of 1%, we need to worry about some exceptional conditions and failures. For example, over 20% total power loss per module should be considered a complete failure and the module(s) must be replaced. How would the replacement be done and lost time compensated?

This is a list of the key basics, but there are a large number of secondary issues to be addressed and questions to be answered to get a complete picture and reduce the risks to a manageable level.

Since some of the issues are quite complex, and since no one person can resolve them all, the intervention of a number of specialists in the different areas of the project is a must.

Conclusions

Crystalline silicon technologies have a long track record of success. The encouraging aspect of their performance is that there are documented tests and long-term installations with flawless records. The overwhelming conclusion from all available documentation, including our own tests through the years, is that the quality, performance and longevity of c-Si PV modules and systems is directly related to the quality of materials, equipment, processes in their construction, and the performance of the people involved in their manufacture.

Basically, well-established, reputable companies do have the ability and experience to produce a world-class product, which functions per the manufacturer's specs. A number of newcomers and low-cost manufacturers are working hard at getting their manufacturing and customer support processes streamlined, and we expect that many of them will succeed.

c-Si PV technology is successfully used in many areas of the world, but we'd like to caution the reader that it is not the ultimate solution to the energy crisis, as some portray it. It is useful, but—in its present shape and form—only in certain geographic areas and in special situations.

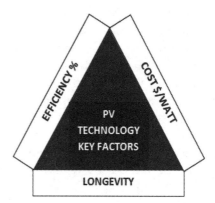

Figure 6-9. Key PV modules parameters

Efficiency, cost, and reliability are the most important factors for the success of long-term deployment of c-Si modules around the world. While efficiency and cost are intensely addressed and resolved daily, *reliability is the wild card*, which most players prefer to keep hidden, or shrouded in vagueness and uncertainty.

In response to that, and due to their large temperature coefficient and humidity dependence, we'd suggest that, at least in the short term, c-Si PV modules could be used without limitations or restrictions in areas with moderate temperatures and humidity. Their use in large-scale power plants in extreme climate areas, such as deserts and humid areas of the world, however, is not well proven and must be approached carefully.

Most of the above facts and conclusions apply to thin film PV modules as well, but since these are relatively new (much newer and unproven than c-Si PV modules), we need more time and data, to make plausible conclusions and predictions.

Efficiency Improvement Techniques

The quest for more efficient solar cells continues. Some of the techniques and design features used to produce the highest possible efficiencies include:

a. Lightly diffused phosphorus to minimize recombination losses and avoid the existence of a "dead layer" at the cell surface;

b. Closely spaced metal lines to minimize emitter lateral resistive power losses;

c. Very fine metal lines, typically less than 20 μm wide, to minimize shading losses;

d. Polished or lapped surfaces to allow top metal grid patterning via photolithography;

e. Small area devices and good metal conductivities to minimize resistive losses in the metal grid;

f. Low metal contact areas and heavy doping at the surface of the silicon, beneath the metal contact to minimize recombination;

g. Use of elaborate metallization schemes, such as titanium/palladium/silver, that would give very low contact resistances;

h. Good rear surface passivation to reduce recombination;

i. Use of anti-reflection coatings which can reduce surface reflection from 10-30%.

Table 6-4 shows a breakdown of the efficiency of cells and modules, around 2011-2012, representing different PV technologies. Obviously c-Si is the most efficient of the cells and modules for everyday use during this time and, in our opinion, for some time to come.

There is a lot of work going on in this area. Some of the directions of these efforts are reflected in Table 6-5.

Improvements in the areas outlined in Table 6-5, such as using less silicon materials in c-Si solar cells, or eliminating the use of toxic materials in thin film PV modules, will bring us much closer to the goal of generating large amounts of efficient, safe solar energy.

The solutions to many of today's energy and *environmental* problems are related to the solutions to many of the *alternative energy* problems. We are confident that the ongoing work is headed in the right direction, and that solutions are on the horizon.

CPV AND HCPV TECHNOLOGIES

The concentrating PV (CPV) and high concentration PV (HCPV) technologies were developed in the 1980s and tested on a small-scale during the 1990s. Though very efficient and promising, these technologies are still unproven for large-scale field operations. Several manufacturers have installed capacities, but they seem to still be mostly in the development stages.

HCPV systems are designed and built for operation under extreme desert conditions. They operate at 50-1000 times the concentration of sunlight with over 42% efficiency at the cell. Efficiency in the grid is lower, of course, but still at least 2-3 times higher than any of the competing PV technologies.

HCPV solar cells are made out of germanium, or GaAs, which are superior semiconductors not affected by temperature extremes. They are also made via sophisticated and precise semiconductor processes, which make them durable and reliable for long-term operation. The optics are made out of glass and silicone, which are also unaffected by sunlight and temperature extremes.

HCPV modules are mounted on trackers which

Table 6-4. Performance of different cells and PV modules

Solar cell material	Cell efficiency η_z (laboratory) (%)	Cell efficiency η_z (production) (%)	Module efficiency η_M (series production) (%)
Monocrystalline silicon	24.7	21.5	16.9
Polycrystalline silicon	20.3	16.5	14.2
Ribbon silicon	19.7	14	13.1
Crystalline thin-film silicon	19.2	9.5	7.9
Amorphous silicon[a]	13.0	10.5	7.5
Micromorphous silicon[a]	12.0	10.7	9.1
CIS	19.5	14.0	11.0
Cadmium telluride	16.5	10.0	9.0
III-V semiconductor	39.0[b]	27.4	27.0
Dye-sensitized cell	12.0	7.0	5.0[c]
Hybrid HIT solar cell	21	18.5	16.8

Table 6-5. Proposed improvements to PV materials, components and processes

TASKS	SURFACE, MATERIALS AND PROCESS OPTIMIZATION AND R&D
Wafers	Thickness reduction, and surfaces cleaning, etching and texturing 100% initial material tests for bulk contaminants and mechanical problems New and/or improved solar materials for c-Si and TFPV cells and modules
Solar Cells	Front surface preparation and passivation for efficiency improvement Front and back metallization improvement via new deposition methods Integration of bypass diodes on the cell level in the module body
Compounds	Optimizing the use of copper, tin, steel, aluminum, etc. structural materials Soldering connections, metal welding operations, glue and bonding use Materials and processes for front and back contacts and integrated connections.
Front Cover	AR coated, self-cleaning and/or textured glass for improved optical performance Synthetic materials (polycarbonate, acrylic etc.) safe for use in some climates
Encapsulants	Optimization of EVA formulations and application methods Developing new autoclave free formulations (PVB, etc.) Optimization of silicones, resins, gels, lamination foils and casting resins Improved polymer systems for encapsulation per specific applications
Back Cover	Optimizing existing and developing new PET-PVF lamination systems Optimizing application of PET materials with vapor barriers New types and configurations of polymer, glass, steel and aluminum back covers
Junction Box	Developing new materials, such as polymers Improving the heat transfer and longevity of the materials and components Developing integrated electronics and automated placement process
Labeling	Introducing new labeling, embossing and embedded marking systems Developing new and more secure electronic marking and labeling methods

have a number of moving parts, control mechanisms and electronics that need regular maintenance and tuning for optimum performance. This requires highly trained personnel and increases O&M expenses.

HCPV cells are mounted in special assemblies on a heat sink, and placed under special (Fresnel) lenses, which concentrate and focus sunlight on the cells at 100 to 1000 times its original intensity. See Figure 6-10.

This is a lot of sunlight falling on a small area, but it allows high efficiency and reliability, and better land utilization, among other benefits.

Cell-lens packages are then assembled into large modules, which are mounted on trackers which track the sun precisely through the day, providing the most power possible.

Efficiencies of 30-32% of AC power, as measured in the grid, are obtainable with these devices.

HCPV systems use very accurate two-axis tracking and this, combined with other sophisticated components, makes HCPV the most efficient, reliable technology to date. Its operation, however, is complex for use in today's energy market, so it will take awhile, but HCPV

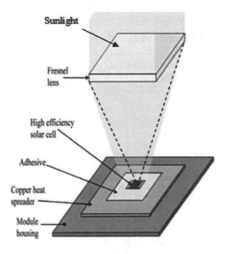

Figure 6-10. HCPV CPV assembly

Figure 6-11. HCPV tracker

technology will eventually find its place.

HCPV is especially efficient, because in addition to over 42% efficient CPV cells, it tracks the sun extremely accurately—with 0.01% accuracy—which allows it to capture every single sunbeam from sunrise to sunset. The problem is that if there is any cloud cover or haze, HCPV optics cannot handle the diffused light going through them, and the efficiency drops quickly and significantly.

Basically, HCPV systems can be used only in bright, hot desert areas, where lots of bright unobstructed sunlight is available all the time, and where other technologies have a hard time surviving the elements.

With time, HCPV's technological issues will be resolved, and we clearly see CPV and HCPV technologies as the primary choice for installation and use in large-scale power plants, especially those in desert regions, during the latter part of the 21st century. At that point, large orders for materials and components will have significantly reduced equipment costs. As the recently formed CPV Consortium states, "... CPV is on the cusp of delivering on its promise of low-cost reliable GW-scale solar energy..."

The recent failure of the HCPV leader, Amonix', which shutting down their new production plant, as well as problems with the HCPV trackers in their new power field in the summer of 2012 were a big step back and gave the industry a black eye that it won't overcome soon. But, HCPV is here to stay, offering power output possible today from no other solar device.

The major problems with the CPV/LCPV/HCPV technologies are:

1. Need for direct and unobstructed sunbeam radiation,
2. Need for complex, precise and expensive tracking mechanisms,
3. Difficulty in efficient and reliable cell-to-heat sink heat transfer,
4. Difficulty with wind gusts, which de-focus the sunbeam, and
5. Difficulty with Fresnel lens materials.

The biggest problem of CPV/HCPV equipment is that it simply stops working every time the sun hides behind even the smallest cloud. Diffused solar radiation is what CPV/HCPV cannot deal with. Power companies cannot tolerated this incapability.

The introduction of energy storage will alleviate this problem and open new opportunities for this and other solar technologies.

The failures of some CPV and HCPV projects and companies during 2010-2012 only show that the CPV/HCPV technologies are not fully developed and that they should not be rushed to market. Advanced materials and manufacturing processes are needed to ensure efficient and reliable long-term operation.

HCPV holds great promise of high efficiency (over 60%), which guarantees power output suitable for large-scale generation of electricity in the deserts during the 21st century.

CPV-T Technology

A different approach to generating power—using the high efficient CPV solar cells—is the introduction of CPV-T hybrid technologies that can produce simultaneously DC power and heat (hot water). This is an efficient and practical approach for a number of applications, so we see its market share rapidly increasing.

Solar DC Power and Hot Water (CPV-T) Generation

Rooftop solar water heating and PV power generation have been successfully used in the US and abroad for several decades. Combining these two energy sources, however, has not been feasible, or cost-effective, until recently.

The newest and most successful development in the field is the use of 43% efficient concentrating photovoltaics (CPV) systems equipped with cooling lines, which allow both cooling of the CPV cells (thus generating the maximum amount of electric energy possible),

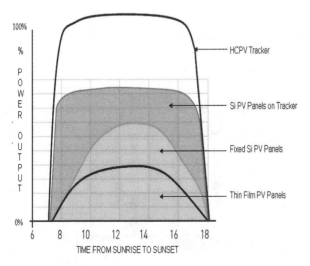

Figure 6-12. Comparing the efficiency of different PV technologies

while at the same time providing hot water (T) for residential or commercial applications (Figure 6-13a).

System Description

The CPV-T hybrid consists of several CPV/DC power generating modules mounted on a two-axis tracking frame which allows the 43% efficient CPV solar cells in the modules to accurately follow the sun constantly, thus providing maximum amounts of power throughout the day.

As an example, a CPV-T system can produce 3.0 kWp from an 8.5 m² area, or 2-3 times more power than today's conventional PV modules plus enough hot water to service a large residential, or small commercial enterprise.

Water running through the heat sinks, on which the CPV cells are mounted, keeps them cool and at the same time heats the water to 190°F, using waste heat energy in the process. This hybrid CPV-T (DC power and hot water) power generator is over 65% efficient, which is the highest in the world.

The "engine" of the CPV-T system is the CPV cells mounted on heat sinks (Figure 6-13b).

Due to the high temperature generated at the CPV cell, the heat sink is critical for the efficient operation of the system and its longevity. Since the solar magnification exceeds 500 suns, the CPV cell and the heat sink get very hot. Air cooling is acceptable for large-scale operations, where the excess heat is dissipated in the surrounding air.

A hybrid system, however, benefits from running a liquid through the heat sink, to provide more efficient cooling, while at the same time providing enough pre-heated water for different applications and maximizing the available solar energy use.

Note: The addition of six regular silicon solar cells peripherally to each CPV cell assembly increases the performance further, bringing the fully equipped CPV-T system to over 80% efficiency of the combined CPV-PV DC power and hot water energy generation system.

In Summary

- The CPV-T hybrid system is the most efficient system ever.

- It can be used (with some modifications) without liquid cooling, using air cooling of the heat sinks only.

- The system can be built in different sizes, from 0.5kWp to 50kWp as needed to fit different roofs and applications—including large-scale ground-mounted installations.

- As an option, 6 regular silicon solar cells are added to each CPV cell assembly in a special arrangement, so that some of the sunlight that is reflected by the CPV cell illuminates the six regular solar cells around it, producing additional DC power.

Note: This configuration increases a 3.0kWp DC power output of a standard 7-modules CPV-T system to 4.5 kWp, without significantly reducing the temperature or amount of hot water.

- Different combinations of these options allow unprecedented efficiency and product flexibility, and offer a number of application choices.

TEST RESULTS

PV modules that pass the pre-test inspection at certification labs are put through a real test and undergo

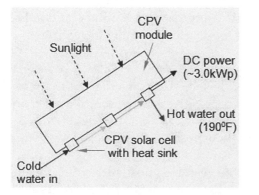

a. CPV-T hybrid system.

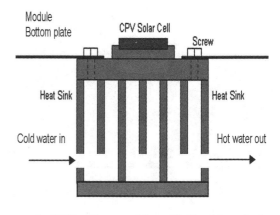

b. CPV cell assembly with liquid cooled heat sink (cross section)

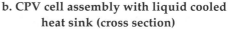

Figure 6-13.

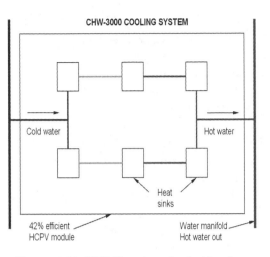

Figure 6-14. CPV-T system, backside view

procedures such as damp heat, UV exposure, heat and freeze, mechanical strength and others that require 3-6 months or more for completion. To make sure the modules will pass the certification tests and to save time and money, manufacturers run similar tests at their facilities before sending the modules for certification; i.e., current standards require approximately 1000 hours of exposure to damp heat, but some manufacturers extend this procedure 2-3 times in their in-house tests, to ensure that the modules perform adequately.

Long test procedures result in the delayed introduction of materials to the PV market; however, so many companies, in an attempt to "accelerate" the aging process, utilize alternative test procedures. For example,

HAST (highly accelerated stress test) has become popular among PV modules and components manufacturers. The HAST procedure provides some useful, albeit superficial test results. They must be standardized and correlated with damp heat and outdoor performance, before they can be relied upon to provide meaningful test results.

So, modules that pass manufacturers' in-house tests are sent to a qualified test and certification lab (usually in the US or EU) for final testing and certification. At the lab they are put through a series of tests to determine their efficiency, overall performance, and longevity.

TUV Rheinland Test Results

From 1997-2009, TUV Rheinland ptl in Tempe, AZ, recorded the results from a series of such tests with several thousand PV modules from several major PV module manufacturers. These results were published in *Photovoltaics International*, May 2010, clearly outlining test failures and related issues as follow:

A. *Crystalline silicon PV modules (extrapolated from Figure 6-15)*

   3% failed at the first wet resistance test (down from 5% in 2007)

   16% failed after 200 thermal cycles* (up from 12% in 2007)

   3% failed UV test

   5% failed after 50 thermal cycles (up from 1% in 2007)

   14% failed after 10 humidity freeze cycles (up from

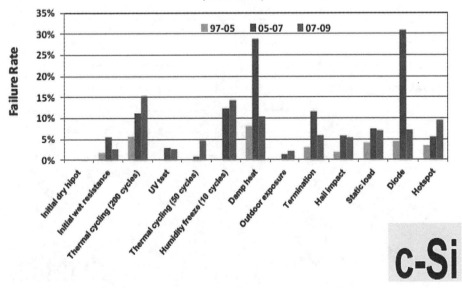

Figure 6-15. c-Si modules tests 1997-2009 (11)

12% in 2007)
11% failed damp test (down from 28% in 2007)
1% failed outdoor exposure test (down from 2% in 2007)
6% failed termination test (down from 12% in 2007)
5% failed hail impact test.
7% failed static load test.
7% failed diode test (down from 31% in 2007)
9% failed hotspot test (up from 6% in 2007)

*NOTE: 200 thermal cycles represent 200 days on-sun operation, while 30 years of thermal cycling = ~10,000 cycles.
**NOTE: 10 humidity freeze cycles represent only part of an Arizona winter season's worth of freeze cycles, while 30 years would consist of over 1,000 cycles, and many more in northern states.

B.   *Thin film PV modules (extrapolated from Figure 6-16)*
1% failed at the first wet resistance test (down from 20% in 2007)
12% failed after 200 thermal cycles* (down from 20% in 2007)
12% failed after 10 humidity freeze cycles** (down from 16 in 2007)
31% failed damp test (down from 70% in 2007).

There is big problem here...
5% failed outdoor exposure test.
12% failed termination test.
6% failed hail impact test.
12% failed static load test.
10% failed hotspot test.

*NOTE: 200 thermal cycles represent 200 days of on-sun operation, while 30 years of thermal cycling = ~10,000 cycles.
**NOTE: 10 humidity freeze cycles represent only part of an Arizona winter's season worth of freeze cycles, while 30 years would consist of over 1,000 cycles, and many more in northern states.

In general, the tested PV modules' quality during 2005-2007 was absolutely unacceptable. Quality has improved in some areas during the past several years, but it is still much lower than the best possible and profitable for use. Even higher degradation and failure rates are possible in extreme climate areas, which might lead to lower power output and profits.

An added uncertainty here is the fact that there were significantly fewer failures (30% less in some cases) during 1997-2005. The number increased during 2005-2007 and decreased again later. This titter-tatter of

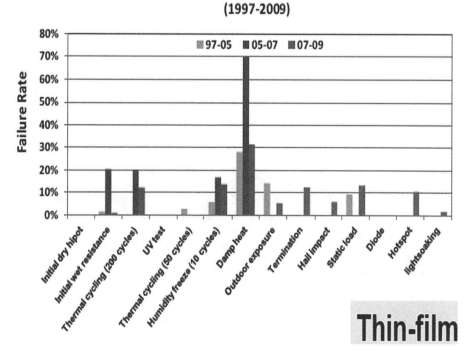

Figure 6-16. Thin film modules tests 1997-2009 (11)

failures can be contributed to factors such as changes in the quality of raw materials and the influx of new solar cells and modules manufacturing facilities around the world. It also might be related to world demand for PV modules, which forces manufacturers to speed up production beyond their ability to control quality.

PI Berlin Test Results

Since 2008, thousands of PV modules have been sent by different manufacturers worldwide for testing at the Photovoltaik Institut (PI), Berlin. The overwhelming fact is that 32% of the tests had to be aborted due to major problems that did not allow completion of the test sequence.

PI reported a summary of the test results in the August 2012 issue of *Photovoltaics International*, with the failures distributed as follows:

30% failed during initial inspection at reception (before any testing was done)
15% failed during light soaking
13% failed during 1000 hr. damp heat testing
10% failed during humidity/freeze testing
8% failed during hot spot testing
8% failed during mechanical load testing, and
5% failed during outdoor exposure testing

These numbers indicate that PV modules are unreliable, and the behavior of different makes and models is unpredictable.

These tests were done within a period of several days or week's time. Guess what would have happened to these modules during 30 months or 30 years of bombardment by the elements in the field.

It is our educated guess that even those modules that did not fail during the certification tests would exhibit symptoms of fatigue, and many would fail, especially in the desert.

How can we justify even a 5% failure rate? Who will pay for the downtime, power loss, and extra labor needed to identify, locate and replace failed modules? How do we know that the new ones we just installed will last any longer? How many times will we go through this?

Basically, where do we start looking for quality? Were the problems due to poor-quality materials or problems with the equipment or labor on the production line? This is a complex set of questions for which we have no answers. Most likely, the problems would be a combination of things, which makes the situation even more complex and hard to manage.

But let's move on. PI Berlin then sorted the failures to those originating in thin film modules, vs. those in c-Si modules. The results are equally shocking:

Thin film PV modules tests

30% failed light soak testing
15% failed 1000 hr. damp heat testing
15% failed humidity/freeze testing
15% failed mechanical load testing
10% failed hot spot testing
10% failed outdoor exposure
5% failed 200x thermal cycle test

c-Si PV modules tests

29% failed 200x thermal cycle test
29% failed 1000 hr. damp heat testing
14% failed hot spot testing
14% failed humidity/freeze testing
14% failed 50x thermal cycle test

Note that this tally does not include modules failed during the reception, where, as we saw above, numbers of modules sent for testing failed during the initial inspection. Imagine that—a reputable company sending samples of their best product, which doesn't even pass first muster.

Author's note: Any good quality control engineer will tell you that the above results are an indication of serious materials, process, equipment, labor, or management abnormalities, and most likely some combination of these. What were those quality control engineers and managers thinking when they allowed defective product to be sent to certification tests?

A 30% failure rate?! Any product failing for any reason fresh from the mass production line at such a rate is absurdly inadequate for use, let alone 30 years of non-stop operation. It is mind-boggling that any manufacturer would allow such a high rejection ratio.

Worse, many of the modules that passed the tests, most likely barely passed. These modules' quality isn't suitable for toys, let alone long-term exposure to desert elements.

Most mind boggling is that anyone would pay for such products, and expect them to work well for 30 years. Amazing!

How do we classify this with other cases in terms of quality control? Is it even worth talking about quality at such high failure rates? Is this product even feasibly useful for anything from a quality point of view?

The above numbers surely confirm that PV modules coming from most manufacturers have problems, but what caused these problems, and can they be eliminated? What do they mean to our 100 MWp power plant in the desert?

PV Modules Yield Data (Simulations and Predictions)

Accuracy of the PV yield prediction process, including meteorological data (direct and diffuse irradiance with its actual spectral composition and spatial distribution), material properties of encapsulation (refractive indices, absorption coefficients, thermal properties), parameters relevant for heat transfer, PV conversion parameters of the cell (temperature coefficients, spectral response, weak light performance, degradation), depends on the quality of the input data applied (derived from literature, data sheets, norms, software tools, or own measurements).

During tests run on PV yield prediction software the following particulars have been found:

Accuracy of the entire yield prediction process—including meteorological data (direct and diffuse irradiance with its actual spectral composition and spatial distribution), material properties of encapsulation (refractive indices, absorption coefficients, thermal properties), PV conversion parameters of the cell (temperature coefficients, spectral response, weak light performance, and degradation—considerably depends on the quality of the input data applied (derived from literature, data sheets, norms, software tools, or own measurements).

Garbage in, garbage out.

So what are we looking for in PV modules' performance, and how can we simulate and predict it? Here are some of the answers, compliments of the engineering team at Photovoltaik Institut in Berlin. (25)

Parameters
- Irradiance could be measured quite precisely (within an accuracy range of 2%) using state-of-the-art pyranometers.
- Usually irradiance measurements are carried out on a horizontal plane and irradiance on tilted surfaces (e.g., the plane of the PV module installation) is extrapolated, causing an inaccuracy of 3% to 5%, depending on the local conditions.
- Irradiance has an almost linear impact on PV yield: while Isc is very linear, Voc and FF are increasing for higher irradiance levels.
- Information on the actual incoming spectrum is quite vague and based more on estimation than on measurements.

- Spectral response of PV modules depends considerably on the cell technology, to a lesser extent on irradiance level, and on temperature. Increasing temperature reduces VOC, but enables slightly higher wavelengths to generate an electron-hole pair, thus expanding the usable wavelength and increasing slightly the ISC.

- Polarization of diffuse irradiance occurs commonly but is rarely taken into account. Not considering polarization leads to an overestimation of PV yield in the vicinity of 0.4-0.7%

- Albedo depends very much on micro-siting of the PV power plant; reflectivity of ground surface may vary significantly and could even change for different weather conditions (e.g., wet vs. dry. Usually the effect of Albedo plays a greater role for locations more distant from the equator, where module elevation angle is more elevated, esp. at PV façade installations. In the tropics the PV module surface is facing towards the sky, thus considerably reducing possible Albedo.

- Shadowing from fixed objects can be determined quite accurately via different tools, however shadowing from living vegetation or smoke (or vapor) from chimneys is hard to determine exactly.

- Outdoor temperature can be measured quite precisely and does not pose a relevant problem for the determination of PV yield.

- Wind speed and wind direction is most often measured at a certain height above ground (typically at 5m or at 10m) and extrapolation of these data to the module front and back surface causes significant errors. Even at a distance of only 1m from the module installation site, the deviation of wind speeds is considerable, however the impact of that inaccuracy on PV power output remains below 3%. Frequency of deviation of measured wind speed on the module surface compared to measurements at a distance of 1m from the module installation site.

- Optical and thermal parameters of the materials of encapsulation such as the refractive index, the absorption and the heat conductivity of glass, EVA, and antireflective coating (cell or module) can be determined accurately in the case of glass, and less accurately for EVA and antireflective coatings, leading to an uncertainty in yield of about 2.7% to 3.7%.

- Thermal parameters at module surface and the ambient cannot be found very precisely; their impact on yield is intermediate (± 3.7%). Instead of using a theoretical approach, convective heat transfer can also be measured.

- Data for temperature coefficients (tc) are often inaccurate (esp. for Isc) and may even change during lifetime (e.g., reduction of tc at a-Si modules after degradation, see [13])

- Cell temperature: Typically, measurements of the module surface temperature are taken for cell temperatures. The error of that method is at least 2 K for backside measurements of glass-foil modules. Also a temperature distribution over the area of the module can be observed, therefore positioning of the temperature sensor is important, while the range of measured temperatures can deviate by 5 K. Front surface temperature can be considerably lower than temperature on the back surface, especially for thick glass-glass modules and for high wind speeds.

- Degradation: Even for the same PV technology degradation may vary: The final loss by degradation can be 16.5% or reach 19.5%—even for the same PV technology (a-Si) from the same manufacturer, thus causing large errors for the prediction of PV yield.

- MPPT-tracking and final power measurement can be performed within an accuracy of 0.5-1% each, leading to the same value in deviation for the determination of PV yield (summarized 1-2%).

Results

A large, less relevant dissimilarity of input data was observed at times, but occasionally with significant effects on the accuracy of PV yield prediction. Consequently, a preliminary screening of inaccuracies has been carried out. The results of that screening are presented in Table 6-6 (preliminary data).

Conclusions

The effects on PV yield range from 0.5% to 20%; consequently, further effort should focus on the improvement of the data precision for those parameters that are responsible for most relevant PV yield inaccuracies:

1. *The degradation effect (in particular at a-Si)*: Its complete understanding and the set-up of a time vari-

Table 6-6. Overview on the parameters that influence PV yield prediction, including qualified guess of values. (25)

Parameter	Variation of data available (±)	Effect on PV yield (±)
Irradiance:		
Horizontal global irradiance	2-4%	2-4%
Tilted global irradiance	3-5%	3-5%
Actual spectral information	10-70%	5-20%*
Polarization of diffuse irradiance	10-20%	0.5-1%
Albedo	9-50%	4-19%
Shadowing	0-20%	0-85%
Meteorological parameters:		
Outdoor temperature	2%	0.5%
Wind speed (at module)	50%	1.5%
Wind direction (at module)	10%	1%
Encapsulation parameters:		
refractive indices	2%	0.5-1%
absorption coefficients	2-10%	0.5%
heat conductivity of module materials	3-10%	0.5-1%
Thermal parameters at module surface relevant for heat exchange with ambient:		
emissivity of module surfaces	2-5%	ca. 0.7%
emissivity of ground surfaces	10-15%	ca. 0.5%
equivalent sky temperature	15-20%	ca. 0.5%
convective heat transfer coeff.	10-30%	ca. 2%
Cell Parameters:		
spectral response	5-10%	5-10%*
temperature coefficients	5-20%	0.5-2%
weak light performance	1-20%	1-10%
degradation (for thin film)	3-20%	3-20%
Electrical yield:		
MPP tracking accuracy	0.5-1%	0.5-1%
Measurement of electrical power output and yield	0.5-1%	0.5-1%

* depending on cell technology and materials

ant model considering the history of the modules including recovery effects due to daily and seasonal changes in temperature and irradiation, possibly a model that is used to describe the lifetime performance of an electrical accumulator, can be used. Some technologies such as CdTe and CIGS do not degrade; they even may improve after light soaking.

2. *Accuracy of spectral meteorological data*: Considering the range of exiting and frequently used models, differences in yield prediction may reach up to 20%. While precise instruments within an accuracy range of 2-5% are available, a significant improvement should be obtainable via an extensive measurement program, focusing on the special needs for the calculation of PV yield. As an ideal result, an accurate database offering all spectra from every direction of the sky sphere, for every moment in time, would be obtainable.

3.   *Albedo & shadowing*: Those parameters depend very much on the actual location of the PV power plant and its installation. An accurate mircositing must be carried out to take Albedo and shadowing into account. In general, those effects are less momentous for locations close to the equator while the modules are facing more towards the zenith, so obstacles close to the ground surface have a reduced impact on PV performance and PV yield.

INVERTERS

Electric power networks are intended to utilize large baseload power plant outputs, but inherently have limited ability to ramp output up or down to keep it at a constant level. The increased variability created by intermittent sources such as photovoltaic (PV) present new challenges, and demands on the system's flexibility increase, which is where DC/AC inverters come into the picture.

They have high conversion efficiency and power factor exceeding 90% for a wide operating range, while maintaining current harmonics THD at less than 5%. Numerous large-scale projects are in operation, with many more planned for the near future. Technical requirements from the a utility power system point of view need to be satisfied, to ensure the efficiency and safety of PV installations and the reliability of the utility grid.

Identifying the technical requirements for grid interconnection and solving interconnect problems such as islanding detection, harmonic distortion requirements and electromagnetic interference are very important issues for widespread application of PV systems.

The control circuit also provides control and protection functions like maximum power tracking, inverter current control and power factor control. Reliability, life span and maintenance needs should be certified through the long-term operation of the PV system.

Further reduction of cost, size and weight is required for more utilization of PV systems. Using PV inverters with a variable power factor at high penetration levels may increase the number of balanced conditions and subsequently increase the probability of islanding.

DC-AC inverters are an integral part of any large-scale PV plant operation, since they receive and handle the output of the PV field and determine the final level and quality of power sent into the grid. They are also a major portion of the capital equipment cost and maintenance expense of the plant, so we will dedicate

this section to their function, operation, and related iss Solar (PV) inverters are a major and critical part of PV electric power generating systems. Inverters are designed to convert the variable DC output from PV modules in the field into a clean sinusoidal 50 or 60 Hz AC electric current at specified voltage, which is then sent directly to the electrical grid, or to an off-grid electrical network.

Power control, data transfer and communications capability are usually included in a good inverter design, for the operators to monitor the system's performance and to take action as needed. Depending on the grid infrastructure, wired (RS-485, CAN, Power Line Communication, Ethernet) or wireless (Bluetooth, Zig-Bee/IEEE802.15.4, 6loWPAN) networking options can be used.

Operational Principles and Components
H-Bridge Circuit

The key component in a simple (single-stage) inverter is the H-bridge circuit. The switches at the left-hand side (S1 and S2) are the power semiconductor switches (MOSFETs or IGBTs). By alternately closing S1 and S4 switches, then S3 and S2 switches, the DC voltage is "chopped" and inverted from a positive (+) to a negative (–) state, which creates a rectangular, and somewhat chaotic, AC waveform.

In a grid-connected inverter, however, the output current needs to be a sine waveform. This requires a more complex operation. The H-bridge puts out a series of on-off cycles to draw an approximated sine wave shape. This is known as pulse width modulation (PWM).

With a 250 Vdc to 600 Vdc input, the H-bridge circuit for a typical 60 Hz transformer-based string inverter will put out an approximated sine wave with an ac voltage of about 180 Vac. The role of the components after the H-bridge is to smooth and change the magnitude of that approximated sine wave.

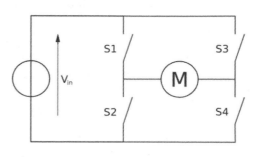

Figure 6-17. H-bridge

MOSFET and IGBT

In most modern grid-tied inverters, DC power from the PV array is inverted to AC power via a set of solid state switches—MOSFETs or IGBTs—that essentially flip the dc power back and forth, creating the characteristic AC power sine wave.

All inverters today use some combination of power semiconductors—MOSFETS (metal-oxide FET), or IGBTs, (isolated gate bipolar transistor), which is a type of a MOSFET, or both of these in some cases—to invert the incoming DC to grid compatible AC power.

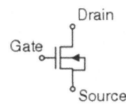

Figure 6-18. MOSFET/IGBT

The insulated gate bipolar transistor or IGBT is a three-terminal power semiconductor device primarily used as an electronic switch and in newer devices is noted for combining high efficiency and fast switching.

It switches electric power in many modern appliances: electric cars, trains, variable speed refrigerators, air-conditioners and even stereo systems with switching amplifiers.

The IGBT combines the simple gate-drive characteristics of the MOSFETs with the high-current and low-saturation voltage capability of bipolar transistors by combining an isolated gate FET for the control input, and a bipolar power transistor as a switch, in a single device.

The IGBT is used in medium- to high-power applications such as switched-mode power supplies, traction motor control and induction heating. Large IGBT modules typically consist of many devices in parallel and can have very high current-handling capabilities in the order of hundreds of amperes with blocking voltages of 6000 V, equating to hundreds of kilowatts.

The IGBT is a fairly recent invention. The first-generation devices of the 1980s and early 1990s were relatively slow in switching, and prone to failure through such modes as latchup (in which the device won't turn off as long as current is flowing) and secondary breakdown (in which a localized hotspot in the device goes into thermal runaway and burns the device

out at high currents). Second-generation devices were much improved, and the current third-generation ones are even better, with speed rivaling MOSFETs, and excellent ruggedness and tolerance of overloads.

Other key components in the main power inversion circuit are inductor(s), capacitors and a transformer, either 60 Hz or high frequency. The latter is used in transformer-based inverters to adjust voltage levels as needed by the topology and to provide galvanic isolation between the solar output to the grid on the other.

Single-stage inverters, like 60 Hz transformer-based string inverters, typically use an H-bridge for inversion from DC to AC.

AC Filters

Inverters contain several pieces of equipment that are referred to as magnetics or magnetic components. These include the inductor and the transformer, shown to the right of the H-bridge.

These magnetic components filter the wave shapes resulting from the PWM switching, smoothing out the sine waves, and bring ac voltages to the correct levels for grid interconnection.

The magnetics also provide isolation between the dc circuits and the ac grid. Note that the ac waveform entering the inductor is raw and triangular; but on leaving the device, it is a clean 180 Vac sine wave.

Because 180 Vac cannot be directly connected to the utility grid, it goes through a 60 Hz transformer. The resulting smooth, sinusoidal 208, 240 or 277 Vac inverter output is connected to the grid. Grid synchronous operation is made possible by grid sensing feedback. Grid voltage information is provided to the inverter's digital signal processor (DSP) or microcontroller, the device that controls the H-bridge.

In conclusion, MOSFETs and IGBTs are manufactured by the world's largest semiconductor fabrication plants, who then sell these devices to the inverter companies. The same is true for most other electronic components (inductors, capacitors, etc.), because very few inverter manufacturers are in the business of fabricating electronic components.

One can deduce that all inverters have similar parts made by the same, limited number of manufacturers. From here on, the difference between the best and mediocre product is in the quality of the design of the inverters, the assembly process, and the rest of the components. Warranty conditions and terms, and maintenance services are other important factors that should be carefully looked at when choosing inverters.

MPPT

Maximum power point tracking is a method an inverter uses to remain on the ever-moving maximum power point (MPP) of a PV array, thus it is called maximum power point tracking (MPPT).

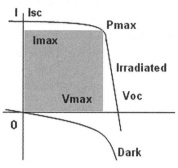

Figure 6-19. I-V curve with MPPT

PV modules have a characteristic I-V curve that includes a short-circuit current value (Isc) at 0 Vdc, an open-circuit voltage (Voc) value at 0 Amps, and a "knee" at the point where the MPP is found—the location on the I-V curve where the voltage multiplied by the current yields the highest value, the maximum power—marked with Pmax here.

MPPT is what grid-tied inverters use to obtain the maximum power possible at each moment of the day. MPPT, therefore, is a variable—a moving target that inverters try to optimize. This is one of the factors that separates good from not-so-good inverters.

The inverter samples the incoming power, and through a number of complex operations, decides what action to take. In simple terms, it applies resistance (load) as needed to obtain the maximum power.

Minimum Input Voltage

As a rule of thumb, the inverter's DC input bus voltage (the total solar power field output voltage) needs to be greater than the peak of the AC voltage on the primary side of the transformer (the power grid side). To maintain this relationship at all times, an additional control and safety margin is required. With a minimum PV input voltage of 250 Vdc, for example, the highest amplitude AC sine wave you can create is about 180 Vac.

The PV input voltage, of course, will greatly exceed 250 Volts DC in many array layouts or temperature and light conditions, with some exceptions, such as cloudy skies, early mornings and late evenings. This is when the solar power field enters into an uncontrolled variability state, and is when special considerations apply.

If 250 Vdc is selected as the inverter's lower dc voltage design limit, then the H-bridge will always create an ac sine wave with a magnitude of 180 Vac. This is true even when the dc voltage present is 300 Vdc, 400 Vdc or higher. This is because the rest of the 60 Hz transformer-based inverter needs to operate on a relatively fixed ac voltage.

The voltage on the utility side of the inverter's transformer—the secondary side—is fixed within a small range of variation. Inverter designers must set the voltage on the primary side of that transformer accordingly.

All this is compounded by other variables, such as:

1. Solar industry immaturity and instability
2. Government regulations (or lack thereof) and/or changes
3. Lack of trained personnel
4. Lack of adequate investment

Key Inverter Functions

Four major functions or features are common to all transformer-based, grid-tied inverters:

- Inversion
- Maximum power point tracking
- Grid disconnection
- Integration and packaging

Inversion

The method by which dc power from the PV array is converted to ac power is known as inversion. Using DC power straight from a PV array is not very practical, except for use in small off-grid systems and small solar gadgets. The power most often used is AC, coming from the National Grid. AC is widely used, because it is easier and cheaper to transport and distribute.

This is mostly due to the fact that AC (voltage) can be stepped up and down and can travel long distances with small losses, using minimal material and resources. DC might replace AC sometime in the future, if more DC energy is produced, stored and consumed.

There are already grandiose plans (albeit still on paper) to generate large amounts of DC power in North African deserts and transport it by underwater cables to Europe. The initial calculations show that transporting and distributing DC power in this case might have a number of benefits. We have the technology to boost DC power to high voltages and transfer it long distances, but the problem is its complexity and expense, so for now we will use AC power from the power plants for our homes and businesses.

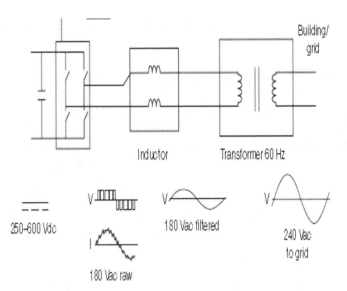

Figure 6-20. A 60 Hz, transformer-based, single-phase inverter circuit

When DC power is generated at a wind or solar power plant, it is run through an inverter, where it is inverted to AC power. The most common inverter methods use a set of solid state switches, named MOSFETs or IGBTs, similar to those used to power and control domestic appliances and industrial equipment.

These switches essentially "flip" the DC power back and forth (or up and down), creating AC power equivalent output. When arranged in a special way (i.e., basic H-bridge) the switches chop the DC power to plus and minus signal.

This raw signal, which retains its maximum ups and downs, is then filtered by another set of electronics (inductors and such), and is finally "conditioned" into a standard sine line form at the set, appropriate voltage levels.

Maximum Power Point Tracking

As discussed above, under ideal conditions, inverters need to receive the maximum amount of DC power coming their way from the PV modules, and convert the majority of it into useful AC power. Producing the maximum AC output possible is done by tracking and controlling the ever-moving voltage level (the maximum power point, MPP) of the PV modules by using a method called maximum power point tracking, MPPT.

PV modules have a characteristic I-V curve that includes a short-circuit current value (Isc) at 0 Vdc, an open-circuit voltage (Voc) value at 0 Adc. When plotting the Vdc and Adc we get a "knee," where the MPP is found. It is actually the location on the I-V curve where the voltage multiplied by the current yields the highest value, which represents the maximum power.

Solar cells generate a lot of electricity when the sun is up, and not so much when the sun goes down. The solar cells' output also decreases with temperature increase.

Sunlight intensity and cell temperature vary substantially throughout the day and the year, so the solar cell (or module) MPP current and voltage also move up and down accordingly. Sunlight intensity varies gradually from zero (at night) to full sun (at noon). It can move fast from full sun to zero and back, when the sun hides. This constant movement greatly affects the inverter design and operation.

Inverters also need to be able to handle the lowest voltages coming from PV modules (early morning, late at night or under cloudy skies), as well as at their highest voltages, which occur at unloaded open circuit array conditions on the coldest days.

Finding the PV modules MPP and "tracking" it (generating output AC power at MPPT) even as the input DC power fluctuates wildly, is the most important function of DC-AC power inverters, especially that of grid-tied solar inverters.

Grid Disconnects

As required by UL 1741 and IEEE 1547, all grid-tied inverters must be disconnect from the grid if the AC line voltage or frequency goes above or below the set limits. The inverter must also shut down if it detects an island (the grid power is shut down).

In either case, the inverter may not interconnect and export power until the inverter records the proper utility voltage and frequency for a period of 3-5 minutes. This eliminates the chance of sending power into disconnected utility wires, which could cause a hazard to utility personnel. If an inverter remained on or came back on before the utility was reliably reconnected, the PV system could back-feed a utility transformer, which could create utility pole or medium-voltage potentials of many thousands of volts.

Three-phase commercial inverters over 30 kW have limits that can be adjusted with the permission of the local utility. This can be very useful in an area with a fluctuating grid, which often results in a significant loss of energy.

Long utility lines, areas with heavy load cycling or an unstable island of power grids can all contribute to grid fluctuation. If a PV system significantly underperforms as a result (beyond nuisance tripping), adjusting the inverter limits can be beneficial.

A specially tuned "resonant load," set up to mirror the utility, tests this inverter function. The resonant load is made up of a very specific inductive, capacitive and resistive network with many settings. Its goal is to attempt to trick the inverter's anti-islanding algorithm into thinking that the utility is really there, at many different prescribed power levels called out in the UL 1741 standard. This load is connected to the inverter operating at full power, and the grid is connected.

The resonant load is set to the exact output power of the inverter. When the whole system is stable, the utility is disconnected while the resonant load maintains voltage and frequency within the inverter's limits. The inverter has a maximum of 2 seconds to successfully recognize that the grid is disconnected and shut off.

Integration and Packaging

In addition to the main power inverting MOSFET and IGBT assemblies, other equipment is built into, or included with, an inverter package, such as:

1. AC and DC disconnects (manual and automatic switches and breakers),

2. EMI and RFI filtering assemblies,

3. Transformer (if the inverter is transformer-based),

4. Cooling system (needed to keep components cool on hot days),

5. GFDI circuit (safety mechanism),

6. LED indicators or LCD display (visual controls),

7. Communication connects and switches for internet access (for data monitoring and system control),

8. The packaging (metal enclosure, doors, locks, wheels, etc.).

Manual AC and DC switches are designed so that the inverter can be disconnected from the grid and the PV array during maintenance or in an emergency. They are used by qualified personnel to turn the system off or to access the components in the inverter enclosure.

Automatic AC switches are used to eliminate nighttime power losses and avoid damage from power surges and lightning strikes. These switches protect and also extend the service life of inverter components.

Inverter enclosure is a convenient way to place components into a single, shippable unit. Large inverters usually ship in several enclosures. The enclosure also provides protection from vandalism and outside elements.

High-quality materials and finishes are needed to meet environmental challenges. Corrosion-resistant materials are important to ensure uninterrupted operation of individual components for the duration under most inhospitable climates.

MPPT Detection and Control Methods

Output power characteristics of the PV system as functions of irradiance and temperature are nonlinear and are influenced by variations of these variables. Under these changing conditions, the MPP of the PV array changes continuously as well; consequently, the PV system's operating point must change to maximize the energy produced.

A number of different MPPT detection and tracking techniques are therefore used to maintain the PV array's operating point at its MPP, thus generating the most power.

There are several MPPT methods available, the most widely used of which we present briefly.

Constant Voltage Method

The constant voltage (CV) algorithm is the simplest MPPT control method. The operating point of the PV array is kept near the MPP by regulating the array voltage and matching it to a fixed reference voltage V_{ref}, which is set close or equal to the actual V_{mpp} of the PV module, or to another calculated best fixed voltage.

The CV method assumes that the solar insolation and temperature variations on the array are insignificant, and that the constant reference voltage is an adequate approximation of the true MPP. Operation is therefore never exactly at the MPP and different data must be collected for different geographical regions.

The CV method does not require any input, but measurement of incoming voltage V_{dc} is necessary to set up the duty-cycle of the DC/DC segment.

The CV method is most useful under low-solar insolation (low-voltage generation) conditions, when it is more effective than any other method. Because of this, the CV method is often combined with other MPPT techniques.

Short-current Pulse Method

The short-current pulse (SC) method achieves the MPP by sending the operating current Iop to a current-controlled power converter. In fact, the optimum operating current Iop for maximum output power is proportional to the short-circuit current Isc under various conditions of irradiance level S as follows:

Iop can be determined instantaneously by detecting Isc, and the relationship between Iop and Isc is proportional (approximately 92%), even when the temperature varies wildly.

This control algorithm, therefore, requires measurements of the current Isc, which is necessary to introduce a static switch in parallel with the PV array, to create the required short-circuit condition.

During the short-circuit, Vpv is 0, so no power is generated by the PV module. As in the CV method, measurement of the PV array voltage Vdc is required to obtain the Vref value able to generate the operating current Iop.

Open Voltage Method

The open voltage (OV) method is based on the fact that the MPP voltage, Vmpp, is always close to a fixed percentage of the open-circuit voltage, Voc. The MPP changes with operating conditions (solar insolation and temperature) within a tolerance band of approximately 2%.

The OV technique uses 76% of the open-circuit voltage, Vov, as the optimum operating voltage, Vop, which is where the maximum output power can be obtained.

This control algorithm requires measurements of the voltage, Vov, and here again it is necessary to introduce a static switch into the PV array, and the switch must be connected in series to open the circuit.

When Ipv is 0, no power is generated by the PV system and if such intervals persist, then the total energy generated by the PV system will be reduced accordingly. In this method, unlike the above two, measurement of the voltage Vpv is required for the PI regulator.

Perturb and Observe Methods

The perturb and observe method (P&O) operates by periodically perturbing (i.e. increasing or decreasing) the PV modules' voltage or current and comparing the output power with that of the previous perturbation cycle.

If the PV array operating voltage changes and power increases, then the control system moves the PV array operating point in that direction; otherwise the operating point is moved in the opposite direction.

In the next perturbation cycle the algorithm continues in the same way.

A common problem in P&O algorithms is that the array terminal voltage is perturbed every MPPT cycle; therefore, when the MPP is reached, the output power oscillates around the maximum, resulting in power loss in the PV system. This is especially true in constant or slowly varying atmospheric conditions, and P&O methods can fail under rapidly changing atmospheric conditions.

In such cases, the actual operating point diverges from the MPP and will keep diverging if the disturbance continues.

a. In the classic P&O technique (P&Oa), the perturbations of the PV operating point have a fixed magnitude.

b. In the optimized P&O technique (P&Ob), an average of several samples of the array power is used to dynamically adjust the perturbation magnitude of the PV operating point.

c. In the three-point weight comparison method (P&Oc), the perturbation direction is decided by comparing the PV output power on three points of the P-V curve. These three points are the current operation point (A), a point B perturbed from point A, and a point C doubly perturbed in the opposite direction from point B.

All three algorithms require two measurements: a measurement of the voltage Vpv and a measurement of the current Ipv.

Incremental Conductance Methods

The incremental conductance (IC) algorithm is based on the observation that the MPP can be tracked by comparing the instantaneous conductance to the incremental conductance, which determines the direction of perturbation leading to the MPP.

Once MPP has been reached, the operation of PV array is maintained and the perturbation is stopped unless a change is noted. In that case, the algorithm decreases or increases Vref to track the new MPP. The increment size determines how fast the MPP is tracked.

Through the IC algorithm it is therefore theoretically possible to know when the MPP has been reached and when the perturbation can be stopped. The IC method offers good performance under rapidly changing atmospheric conditions.

There are two main IC methods available in the literature.

a. The classic IC algorithm (ICa) requires several measurements to determine the perturbation direction: a measurement of the voltage Vpv and a measurement of the current Ipv.

b. The two-model MPPT control (ICb) algorithm combines the CV and the ICa methods: If the irradiation is lower than 30% of the nominal irradiance level, the CV method is used; otherwise, the ICa method is adopted. Therefore, this method requires the additional measurement of solar irradiation.

Temperature Methods

The open-circuit voltage Vdc of a solar cell varies mainly with the cell temperature, whereas the short circuit current is directly proportional to the irradiance level and is relatively steady over cell temperature changes.

There are two temperature methods available:

a. The temperature gradient (TG) algorithm uses the temperature T to determine the open-circuit voltage Vdc. The MPP voltage Vmpp is then determined as in the OV technique, avoiding power losses.
    TG requires the measurement of the temperature T and a measurement of the voltage VPV for the PI regulator.

b. The temperature parametric equation method (TP) determines the MPP voltage instantaneously by measuring the temperature and the solar insolation.

Inverter Types

There are a number of solar power generation DC/AC inverters, but the line commutated is the most widely used for grid-tie operations.

Line Commutated Inverters

These types of inverters use a switching device like a commutating thyristor that can control the turn-on time while it cannot control the turn-off time by itself.

Turn-off is usually performed by reducing circuit current to zero with the help of a supplemental circuit or source.

Self-commutated

These inverters use a switching device that can freely control the on-state and the off-state, such as IGBT and MOSFET. It can freely control the voltage and current (voltage and current type inverter) waveform at the ac side, and adjust the power factor and suppress the harmonic current, and is highly resistant to utility system disturbance.

Most inverters for distributed power sources such as PV power generation now employ a self-commutated

inverter. These are subdivided as follows.

a. Voltage Type

Voltage type is a system in which the dc side is a voltage source and the voltage waveform of the constant amplitude and variable width can be obtained at the ac side. It is employed in PV power generation. It can be operated as both the voltage source and the current source when viewed from the ac side, only by changing the control scheme of the inverter. It produces a sinusoidal voltage output. It is capable of stand-alone operation supplying a local load.

This type of inverter is therefore connected to the grid via an inductance. The inverter voltage may be controlled in magnitude and phase with respect to the grid voltage. The inverter voltage may be controlled by controlling the modulation index, and this controls the VARs. The phase angle of the inverter may be controlled with respect to the grid, and this controls the power.

b. Current Type

Current type is a system in which the DC side is the current source and the current waveform of the constant amplitude and variable width can be obtained at the ac side. It produces a sinusoidal current output. It is only used for injection into the grid, not for stand-alone applications. The output is generated by producing a sinusoidal reference which is phase locked to the grid.

The output stage is switched so that the output current follows the reference waveform. The reference waveform may be varied in amplitude and phase with respect to the grid, and the output current of the inverter follows the reference. The output current waveform is ideally not influenced by the grid voltage waveform quality. It always produces a sinusoidal output current. The current control inverter is inherently current-limited because the output current is tightly controlled even if the output is short circuited.

In the field operation, depending on application, the inverters could be divided as follows.

String Inverters

String inverters are basically smaller inverters that can be mounted in different (and space restricted) areas of the power field. They can handle the high DC voltages and three-phase output required by large-scale installations; and, by having a large number of maximum power point trackers, they can convert power more efficiently. String combiners and external string monitoring systems are not required, saving on wiring expenses and reducing wiring losses.

Compact transformer stations can be used to connect to medium-voltage grid points, and because of their small size the transformers can also be placed at different locations around the field. The 630-kVA transformer station is among the most commonly used and usually has short lead times. As the height of the transformer station is limited, it is possible to place it behind the modules. The opposite substructure of modules is only slightly more shaded if distances between them is unchanged.

Low loss transformers reduce the nightly power consumption to below 0.4% of total production, and short circuit losses in the transformer have little effect on overall yield. Outgoing feeder panels with HH-fuses can be inserted in the medium-voltage area of transformers of this size, instead of the more expensive power switches.

String inverters and transformers are relatively small, easy to install, and have short lead times because they are widely used. Special training is not required to install, maintain or replace string inverters, so service contracts (as needed for central inverters) are not needed. There are additional advantages to using string inverter systems, and they should be considered in some commercial and utility type power plants.

Clustered Inverters

Several inverters are usually used in each commercial and utility power plant. The inverters can be installed and wired in a number of ways, depending on field location and size and power use characteristics. Large-scale PV installations are usually wired in strings, with several strings feeding one inverter.

The strings and inverters can be sized and configured in different ways to fit particular installations, usually feeding centralized inverters at large-scale installations.

One approach for optimizing the efficiency and cost of a large-scale installation is to subdivide the power field into 1.0-MWp clusters, each consisting of a 1.0-MWp string of PV modules fed into one integrated central inverter cluster (1.0 MW total), comprised of 1, 2, or 4 inverters per individual cluster. Larger inverters are available now, so the modular approach could be expanded to 2- to 4-MW and larger clusters.

As an example, a 100-MWp power field could consist of a 100 x 1.0-MWp inverter cluster. If each 1.0-MW cluster consists of 2 pc. 500-kW inverters, then there would be 200 inverters in the field. If each 1.0-MW unit contains 4 pcs. 250-kW inverters, there would be 400 inverters.

Each inverter cluster is fed by several PV module strings—each consisting of several sub-strings. The power generated by each 1.0-MWp PV string is fed into an inverter cluster where the DC power is converted into AC power and is sent into the grid.

The brain of the inverter is a real-time microcontroller, which executes the very precise algorithms required to invert the DC voltage generated by the solar module into clean AC voltage. The controller is programmed to perform the control loops necessary for all the power management functions, including DC/DC and DC/AC.

A good controller has the ability to maximize the power output from the PV system through complex algorithms called maximum power point tracking (MPPT). The PV system's maximum output power is dependent on operating conditions and varies from moment to moment due to temperature, shading, dirt, cloud cover, and time of day, so tracking and adjusting for this maximum power point is a continuous process.

For systems with battery energy storage, the controller manages the charging operations. It also provides uninterrupted power supply to the load by switching over to battery power once the sun sets or when a cloud cover reduces PV output.

The controller contains advanced peripherals like high precision pulse-width-modulated (PWM) outputs and analog-to-digital-converter (ADC) for implementing control loops. The ADC measures variables such as PV output voltage and current, and then adjusts the DC/DC or DC/AC converter by changing the PWM duty cycle. Most units are designed to read the ADC and adjust the PWM within a single clock cycle, so real time control is possible.

Communications on a simple system can be handled by a single processor, while more elaborate systems with complex displays and reporting on consumption and feed-in-tariff payback may require a secondary processor, potentially with ethernet capability. For safety reasons, isolation between the processor and the current and voltage is also required, as well as on the communications bus to the outside world.

In summary, the inverter is an all-encompassing receive-convert-send-control power manipulator, which makes continuous decisions for each situation and adjusts the power variables of the DC to AC transformation for maximum conversion efficiency, power output and safety.

We could conclude that the electrical dynamic performance of a PV plant depends on, and is even largely dominated by, the inverters. That makes them a critical part of the proper, efficient and profitable operation of any PV plant. Also, on one hand there is a strong link between inverters' cost, efficiency, quality and size, and on the other hand the cost of the generated electric power per kWh.

These are the particulars we look at first, when evaluating or designing large-scale power plants—regardless of what PV generating equipment is involved.

The assumption has been that higher quality and larger size inverters ensure a better return on the dollar, so present-day inverters of 1 MW and 2 MW are used in some large PV plants. The trend toward higher quality inverters is justified to a degree, but the price can be exorbitant and does not justify the extra efficiency or longevity from these high-end inverters.

The trend towards larger inverters, however, is more complicated, and its improper application might contribute to higher capital and O&M cost, forcing us to seek alternative ways to use inverters when we design, build and operate large-scale plants.

AC Power Overload

Inverters of larger size (1.0- to 2.0-MW and larger) are recommendable for use in large power fields, but they are the newcomers and have not yet been proven reliable. Since inverters are quite expensive are considered the most expensive part of the power field (when maintenance is included), this is a clear disadvantage, so we advise caution, careful consideration, proper design calculations, and thorough negotiations with manufacturers before making a final decision on using 1- to 2-MW or larger inverters.

One fairly new and beneficial feature offered by some inverter manufacturers is their "overload" capacity. This is an extra design feature that allows a power load of 120% in some cases, or 20% more than the designed 100% power-handling capability. In a 1.0-MW inverter this means that it will be able to handle 20%, or 200 kW power in addition to the designed 1.0-MW capacity.

This feature is exceptionally useful in ensuring that any extraordinary events, forcing higher than design output won't damage the inverters. These events can include solar explosions, field power variations, or any work done at the site that changes the strings' output and forces the power injected into the inverters to go over the design level of 100%. The extra power is efficiently used in the DC-to-AC conversion.

This also allows installation of additional modules to compensate for power loss in the future and under special circumstances. In large-scale power fields, a modular setup (group of 1.0-MWp PV modules and inverters) might be most effective from technical and financial points of view. This modular design allows a greater degree of freedom at the large generating capacity and is recommended for any power farm over 10 MWp DC capacity.

In this case, the PV modules and inverters could be grouped in 250-kW, 500-kW, or 1.0-MWp modules and wired in different ways to optimize conversion efficiency and minimize the cost of installation and operation.

Most inverters cannot operate in deserts without proper cooling because their efficiency drops significantly when the air temperature reaches 110°F. Special enclosures equipped with A/C units are used in these cases. Some power is wasted for the cooling, but this allows the inverter to operate at maximum efficiency and adds years of useful life.

Figure 6-21. 1.0 MW PV field with two 500 kW inverter clusters

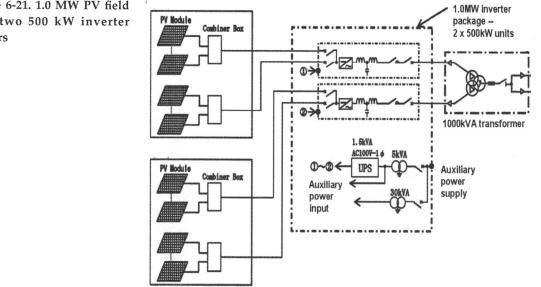

Components Cost Estimates

Although the availability and cost of inverters and related components depend on the manufacturer and type and size of equipment, there are guidelines which we can use for ball-park estimates for a 1.0-MWp modular installation.

Table 6-7. Estimate of 1 MW PV field electric components.*

Qty.	Equipment Description	US$
2 pcs.	500kW Inverters	280,000
2 pcs.	internal transformer (included)	
	Switches & 4-Way Triad	180,000
	1MVA Cooper Transformers	40,000
	Square D low voltage Switchgear	12,000
	Combiner boxes (additional)	?
	Wire, Connectors & Conduit	4,000
3,000ft	1/0 Al Primary cable	6,000
600ft	Copper cable	6,000
	Pole & Feeder upgrades	10,000
	Construction materials and labor	8,000
1.0MW	DC-AC module total cost	$548,000

*Note: This is an estimate only. The actual cost will vary.

Obviously, inverters and related electronic components needed for proper and safe conversion of DC to AC power would add over $0.50/Wp to the cost of the installation. Inverter costs have been going down lately, while their efficiency has been increasing, so this trend will help with reaching a balance in the near future.

The inverters' reliability, however, is not as well defined or predictable. The industry is fairly new (especially larger-size inverters manufacturing) with few years of field operation experience, and it's loosely standardized, so manufacturers have varying quality control criterion and procedures.

For these reasons, some of them have the reputation of "quality" providers, while the rest are either in the lower quality bracket, or are unproven. In any case, due to some bitter lessons from the past, thorough knowledge of the manufacturer's *modus operandi*—including manufacturing and QC/QA procedures and supply chain details—is paramount for the successful choice of reliable equipment.

Inverter Maintenance

Inverters are an integral part of the PV power field structure and are major contributors to its efficiency.

The effect of one inverter's malfunction can reverberate through the entire installation, affecting (usually negatively) the operation of other plant components, and the output will suffer. Inverters come with 5-10 years or more limited warranty (covering parts and components) as well as with a guaranty for a certain level of performance.

More often than not, inverters will malfunction without notice, due to a failed electrical component. To take quick action in such cases, most manufacturers have a stand-by maintenance team to offer diagnostics and repairs by phone or in person.

Figure 6-22 is a schematic of one such support center of a major inverter manufacturer.

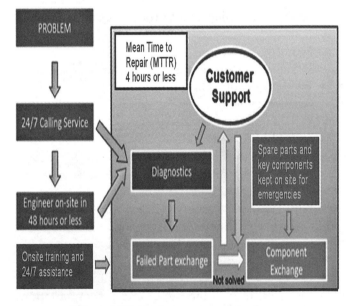

Figure 6-22. Customer support scheme

Provided that a maintenance contract is signed, this manufacturer promises 24/7 technical phone support. If the problem cannot be diagnosed and fixed by phone, then a technical support team will be dispatched. The documented MTTR (mean time to repair) for this manufacturer is 4 hrs. This means that most problems are diagnosed and fixed within 4 hours from notice. This is very good performance, achieved best when the service center is located close to the power plant.

Another major issue is that with time the inverters, just like the PV modules, lose efficiency. Normal wear and tear will reduce the efficiency of a new 95%-efficient inverter to, let's say, 93% by year 5.

Degradation is even more complicated and pronounced in extreme-climate regions. This variable must be well understood and negotiated with the manufacturer at the beginning of the project. It must also be included in the PV power plant design calculations, because it will inevitably affect the installation's power output.

Inverters also have a limited life expectancy, which is reduced further when operated in climates of extreme heat and elevated humidity. Because of that, the plant's O&M budget must allow for periodic overhauls (every 5 or 10 years) or replacement of the inverters. This is a major expense, so its proper estimate and budgeting is of utmost importance to the project's financial success.

Finally, there is a long list of maintenance tasks that must be expected from the manufacturer, or service contractor, in assuring the inverters' proper long-term operation.

The recommended inspection, test, cleaning, repair, tuning and replacement procedures are:

1. Monthly inspection
- Presence of dust and gas in cabinet
- Excess vibrations and noise
- Abnormal heating
- LCD abnormal operation
- LED abnormal operation
- LED failures
- Cleaning of air filters (if necessary)

2. Annual inspection:
- Visual inspection of all components
- Internal parts inspection for corrosion and discoloration
- Check for corrosive gases in the cabinet
- Electrical tests; total output and power variations
- Cleaning of internal parts in the cabinet
- Cleaning of air filters (if necessary)

3. Five-year inspection
- Discharge resistor
- Electrolytic capacitors
- Filter capacitor
- Reactor
- Semiconductors (diodes)
- MCCBs
- Contactors
- PWBs
- Fuses
- Fans
- A/C units
- Wiring
- Bolts, nuts, screws
- Surge protection circuit
- Electrolytic capacitors in control circuit

- Insulation resistance
- Protection circuit check
- Operation time of timers
- Main circuit waveforms
- Timers operation and accuracy
- Output voltage waveforms
- LCD display
- AC output phase check
- Conversion efficiency verification
- Utility interface control parameters
- Full power operation check (if possible)
- Cleaning of the internal parts in the cabinet
- Consumable parts replacement (per contract)

End of Life

If there are no major electro-mechanical accidents or natural disasters, well designed and properly maintained inverters can endure the 20-25 years of non-stop operation even under the harshest climatic conditions.

Realistically speaking, however, inverters have useful lifetimes of 10-15 years, after which the unit(s) must be decommissioned and replaced, or completely refurbished. Both of these tasks are time- and labor-intensive and money-consuming. This issue must be included in the 30-year O&M plan.

We believe that the O&M scheduling and budget plans of some PV projects, hastily thrown together from 2007-2011, will not pass the 10-year line. What will happen to these projects when the inverters have to be replaced or refurbished? With such an unplanned major task, technical, administrative and financial chaos comes to mind. This also highlights the immaturity of the solar industry, so we must remember it as one of the lessons of the present boom/bust cycle.

Inverter Issues and Failures

Inverters are electro-mechanical devices, consisting of complex semiconductor components, transformers, gauges, relays, electronics, connectors and wiring which operate efficiently under "standard" conditions—the accepted, non-controversial STC (25°C). These standard conditions for inverters testing also include incoming DC power with known quality and characteristics.

But things are seldom standard in field operations, and inverters seldom see operation at 25°C and constant power input for long periods of time. They are exposed to a number of environmental attacks and operational mishaps—especially so in harsh regions like deserts where mechanical and chemical attacks are the norm. Inverters, like other electro-mechanical devices, deteriorate and fail under these conditions, and we must be aware of this, to properly design and efficiently operate a PV power plant during its long exposure to the elements.

A number of rules of thumb seem appropriate at this level. Although they might seem incomplete, they do reflect the complexity, and newness, of inverter-related issues. During 30 years of field operation, inverters are exposed to:

1. Daily and hourly solar power fluctuations, which make them work very hard.

2. Temperature extremes in the desert cause power derating.

3. Temperature extremes, dust and humidity can damage critical components.

4. Minimum input voltage causes losses, for all power coming at below 200-250V is wasted, since the inverter switches off.

5. Maintenance is a major issue. An unscheduled 8-hr shutdown of one inverter in a large-scale PV power plant could cost thousands of dollars.

Because of the above issues:

1. Inverters' failures are the most frequent in a PV field. Note the "*most*" qualifier—this is serious!

2. Inverters' life expectancy is 5-10 years.

3. This creates a perceived lack of confidence.

4. Lack of long-term testing contributes to the uncertainty.

5. Manufacturers' inexperience with the design of large size inverters for desert operations is blamed for the failures.

6. New products are rushed to market without proper testing.

7. Components are procured from 3rd parties.

8. Components are procured in small quantities.

9. Large capacitors are not well suited for PV operation.

10. Advanced switching devices need improvements.

11. Advanced controls for low voltage start-up, below 250V, and better MPP tracking of solar insolation fluctuations are needed.

A closer look at potential major inverter issues in field operations follows.

Power Loss

A major problem with inverters is that they tend to have a window of operation—lower and upper voltage limits—so unless the power is kept within these limits something bad will happen.

Because of the fluctuations of the solar insolation and weather conditions, no matter where the fields are (even deserts have partially and fully cloudy days and ferocious monsoon seasons), the inverters' power output fluctuates accordingly. Field design should anticipate and avoid excursions above the upper limit (set at 600 in the US), so the inverters should never "see" voltage exceeding 600V, even if manufacturers provide some over-power flexibility. If this happens, then it is simply a design issue that must be corrected.

The bigger problem in most cases is the lower operating limit, also called "starting voltage," which must be met for the inverter to start operation. Inverters also tend to "drop" power when the voltage of the feeder string approaches their lower limit. So, if the lower limit is 200V, and the string is not providing it, or it goes below that level, then the power output will be interrupted. This can happen when the sun goes behind a cloud and the incoming string voltage drops below 200V.

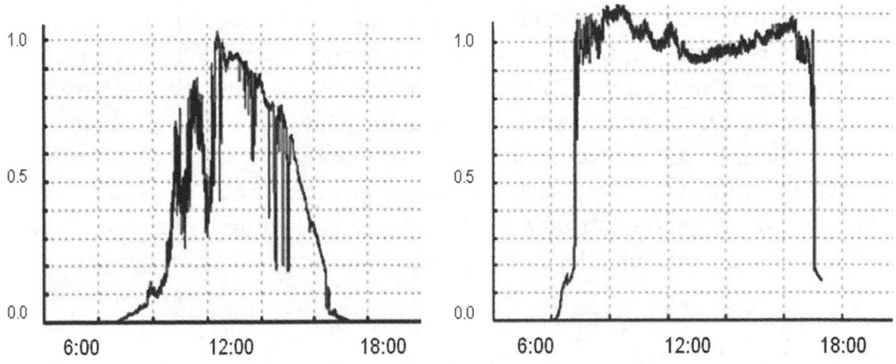

a. Variable hourly solar insolation   b. Stabilized inverter output

Figure 6-23. Optimum inverter operation at variable solar insolation

The power fluctuates during the day with the sun going in and out of the clouds. Optimum (best possible) performance of an inverter operating under variable solar insolation is shown in Figure 6-23b. Note how the voltage is kept almost level while the sun is behind clouds and the feeder string is generating different power output, as that in Figure 6-23a. While the total power (P = V x A) from the inverter is fluctuating, the voltage is kept at a level that allows the inverter to operate uninterrupted, thus converting incoming DC power into AC and feeding it into the grid (albeit at varying power levels). According to the manufacturer, only heavy cloud cover would force the inverter into a stand-by mode.

The good news for large-scale applications is that the power fields, and the inverters in them, are very likely to be installed in desert areas, where the sun is more constant than any other parts of the world. This provides a more consistent sunlight with fewer power interruptions. Even then, there could be some great weather surprises during certain periods of the year, which is where a well designed and sized inverter could make the difference.

Temperature Derating

Another major variable that needs special attention—especially in large-scale installations in the deserts—is the quick and serious temperature derating of most inverters.

Even 98% efficient inverters lose efficiency quickly at elevated temperatures; i.e., most manufacturer specs call for high efficiency rating at temperatures much lower than those in US deserts (which could reach 190°F in the metal enclosures—inverter cabinets—while the units are in operation). Without cooling, these enclosures become ovens in which the inverters bake, and as the temperature climbs, their output drops quickly.

Different inverters have different derating factors, but we dare say that no matter what the derating factor, all inverters operating in desert areas will reduce output and even malfunction without active cooling. A/C units are needed to keep inverters operating at or below 100°F.

If the internal temperature increases enough over 100°C, the internal electronics circuits shut down and inverter stops generating power. Worse, the electronics might fry out, potentially causing major failures and even disasters.

There is no good news in the case of large-scale applications here, because the power fields, and the respective inverters, are most likely to be installed in desert areas, where the heat built up in an operating inverter exceeds the temperature tolerance.

Humidity and Dust Effects

Excess humidity and dust also impedes the inverters' efficient operation. On a monsoon day in the Arizona desert, there could be inverter cabinet temperatures of up to 180°F, humidity over 80%, and a dust cloud as thick as the sand itself, all of which will affect and even damage inverters. Humidity is one of the greatest enemies of all electronic and electric circuits and devices—period. Dust comes next…

Numerous harmful effects that can occur within a circuit when humidity increases above manufacturers' specs. Over time, serious damage to sensitive components occurs from excess humidity, but dust's effect is more indirect (cabinet sand blasting, air filter plugging, etc.).

The inverter's efficiency would be improved, and its life extended if it were housed in an air-conditioned room, where temperature and humidity could be kept within spec, and dust kept out.

This is especially true for utility type projects, which are most likely to be installed in desert areas, where heat and humidity levels exceed inverter design tolerances, and where special precautions must be taken, to ensure long-term reliability.

Inverter failures are the most frequent and most serious incidents in a PV power plant operation. This is often due to manufacturers' lack of experience in early production stages, so newly designed and/or manufactured inverters might have some serious problems. These problems could be compounded by poor power field design, improper installation or negligent operation. Some inverter failures might be caused by weather disturbance, grid glitches, and interconnecting issues, which could cause serious problems and damage, if the inverters are not designed and built to handle such abnormalities.

Large-scale PV power generation is an emerging industry that has the potential to offer improvements in power system efficiency, reliability and diversity, and to help contribute to increasing the PV energy's percentage in the power generation mix.

While a great amount of knowledge has been gained through past experience, the practical implementation of distributed and central PV generation has proved more challenging than originally anticipated—especially for installations in very hot and/or humid areas.

Future Efforts

There are steps that equipment manufacturers and power field design engineers can take to assist in assuring inverter reliability:

1. Research and develop standards and regulation concepts to be embedded in inverters and controllers. Develop dedicated voltage conditioner technologies to integrate with power system voltage regulation, providing fast voltage regulation to mitigate flicker and intense voltage fluctuations caused by local PV fluctuations.

2. Investigate DC power distribution architectures as a future method of improving overall reliability (especially with micro-grids), power quality, local system cost, and very high penetration PV distributed generation.

3. Develop advanced communications and control concepts that are combined with solar energy grid integration systems. These are the key to providing sophisticated microgrid operation that maximizes efficiency, power quality, and reliability.

4. Identify inverter-tied storage systems that will integrate with distributed or multi-string PV generation to allow intentional islanding and system optimization functions (ancillary services) to increase the economic competitiveness of distributed generation.

These and similar initiatives will bring the inverters to the desired level or efficiency and reliability, as demanded by today's PV technologies and utilities power systems.

EQUIPMENT AND PROCESSES STANDARDIZATION

IEC or International Electrotechnical Commission is a standards-making body in the field of electrical and electronics technologies. The IEC works with national committees in different countries in preparing and maintaining standards in this area. IEC is one of the oldest standards making bodies in existence.

The IEC standards making process is handled by various technical committees or TC as they are called. TCs are the key bodies that drive the standardization and are comprised of experts from the national committees whose efforts are completely voluntary. IEC has more than 1,000 technical experts working on standards voluntarily.

This list is intended to detail the various technical committees of IEC, the scope of the committees, their key members and the key relevance and outputs of these committees.

Each technical committee and its standardization efforts are carried out by various working groups within the technical committees.

The most important tasks as far the further development of the solar industry are concerned are addressed by TC 47 and TC 82.

TC 47 deals with standards underway intended to regulate the semiconductor industry materials, equipment design, operation and communications, processing techniques and procedures, and quality control instrumentation and procedures. Many of these are relevant to the equipment and processes in the solar industry—c-Si and thin film PV modules manufacturing in particular.

TC 47 is important to our work in the solar industry simply because many pieces of equipment (especially in the thin film modules manufacturing areas) are taken, or copied with some modifications, from the semiconductor industry.

TC 82 was established in 1981, and its scope is to prepare international standards for systems of photovoltaic conversion of solar energy into electrical energy and for all the elements in the entire photovoltaic energy system. (7)

In this context, the concept "photovoltaic energy system" includes the entire field from light input to a solar cell, and to and including the interface with the electrical system(s) to which energy is supplied.

Note 1: There is some common interest between TC 47 and TC 82, therefore these two committees shall maintain in liaison.

Note 2: Solar cells, except those used in the generation of power, which are specified as components for purposes of direct trade are excluded from the scope of TC 82.

The rapidly growing and highly competitive global marketplace mandates TC 82 coordination and resource leveraging with organizations outside of the IEC. To that end, TC 82 has established liaisons with the European Union's PV Joint Research Centre (JRC) at Ispra (Category A), the International Energy Agency (IEA) PVPS (photovoltaic power system) co-operation program (Category A), and the Global Approval Program for Photovoltaics (PV GAP) (Category A).

TC 82 also maintains loose liaisons with other TC/SCs such as TC 8 Systems aspects for electrical energy supply, TC 21 Secondary cells and batteries, SC 23E Circuit-breakers and similar equipment for household use, SC 32B Low-voltage fuses, TC 34 Lamps and related equipment, SC 37A Low-voltage surge protective devices, SC 48B Connectors, TC 64 Electrical installations and protection against electric shock, TC 57 Power systems management and associated information exchange, and SC 77A Low frequency phenomena.

Business Environment

General

Standards demand—It should be noted that basic standards written within the TC are for design qualification and type approval of flat-plate modules and concentrator PV devices.

These basic standards are used by qualification testing laboratories throughout the world in testing product submitted by manufacturers who wish to enter the PV marketplace. Included in the basic standard realm are safety standards associated with flat-plate modules and concentrator PV (CPV) systems. Obviously, manufacturers in the PV sector are customers who purchase these standards as well as the supporting standards that determine methods of testing, simulator spectrum requirements, and the like. Included in users are teaching and research universities and colleges, government laboratories, and others with an interest in PV technologies.

Standards are also written for balance of systems components—such as inverters and charge controllers—and for grid safety when operating DC to AC inverter systems connected to the utility grid. Customers for these standards fall into the same group as above, i.e., testing laboratories, balance of system manufacturers, universities and colleges, government laboratories, systems integrators, and utilities.

Systems standards are also written for use by systems integrators in commissioning of small and large photovoltaic generating systems.

Technical specifications are also written for use in specifying, commissioning and operating PV and hybrid stand-alone systems or micro-grids in developing countries. Customers here are systems integrators, system owners, utilities, World Bank and governments that provide funding for such systems.

Competing standards organizations include various national standards writing organizations, to include VDE in Germany; in the United States NEC, ASTM, ANSI and especially the IEEE 1547 series of standards that define distributed generation and the use by utilities.

Japan has its own set of national standards for the PV industry as does China.

So, do we need to implement standards in the solar industry? Are we ready for that?

Trends in Technology

Crystalline silicon photovoltaic modules are still the dominant commercial type, but thin film photovoltaic modules have improved considerably and have established a substantial market. Recently, investments into thin-film PV production facilities have begun to increase at growth rates exceeding those of crystalline technologies.

Many types of thin film amorphous silicon and other compounds such as CdTe, CIS and CuInSe2 are now available. Applications for these include building facades, which serve as architectural features as well as power producing elements, and other types of building integrated PV product. Also, large utility-scale installations feature thin-film products more often than before, particularly where the installation area is not a major constraint.

Beyond the PV module, there are hosts of trends involving the other components and overall system design required to translate the power produced from the module into useable power for the application. Some of these include:

- More often than not, the source of problems in fielded systems is either inadequate system design or failure in the power electronics (e.g. balance of systems components). Most crystalline modules have established a respectable track record, and many thin-film technologies prove their reliability also in industrial applications

- Technological trends in power electronics, data monitoring and energy storage all provide opportunities for more efficient, diverse, cost-effective and "dispatchable" PV systems.

- Combined with technological developments and standards, trends in higher quality power requirements, cleaner sources of power and distributed generation also provide opportunities for an increased number of PV based applications.

- Indications of a shift from small individual solar home systems in the developing world to larger "mini" or community power plants with the potential for creating mini-grids that in turn can be networked.

Major companies have recently recognized the industrial potential of PV and are vastly increasing usage and integration of solar products into their product portfolio, and/or in the construction of major production facilities for this purpose. This is complemented by a high degree of commitment for support of photovoltaic technology by continued commercial and government support. This increase in the global PV demand and production capacity indicates an industry poised for sustained, significant growth and contribution to the global economy.

The success of the effort is largely dependent on the industry's ability to sufficiently address key "barriers" (e.g. interconnection, trained workforce) and avoid major pitfalls (e.g. system failures, safety issues, etc.).

Market Trends

Crystalline silicon will continue to be the dominating technology of annual sales. However, thin films will make their marks in overall sales simply because their costs continue to push the crystalline manufacturers out of the picture in many cases.

Other technologies, such as organic and dye sensitized PV cells are just coming out of the research stage and will be developing quite nicely in future years.

Nanotechnology with its entry into the manufacture of modules will assist in increasing efficiencies and lower manufacturing costs—all in the quest to reach grid parity.

Ecological Impact

The operational impact of PV systems on the natural environment is generally minimal and considered benign. PV is one of the most attractive of the renewable energy sources from the environmental point of view as the PV systems have no moving parts and create no waste effluent or residue for disposal in their daily operation.

The processing of the PV devices and the disposal of storage batteries and some PV devices (such as those containing compounds of Cadmium) are the main environmental hazards to be considered in the technology. Recycling processes for current technology batteries, both lead-acid and nickel cadmium are already well established and would appear to require no special consideration. New storage technologies may require special treatment and will be assessed as they are introduced.

The issue of toxicity in the production process and recycling of dangerous products will need to be addressed. In addition, the issues of environmentally acceptable packaging materials, visual pollution, and electromagnetic interference (EMI) are being examined.

ISO standards of environmental management systems will be examined in the future for applicability to PV modules, components and systems. PV standards may be required to comply with the "Performance Evaluation" and/or " Life Cycle Assessment" sections.

System Approach

Standards prepared within TC82 are primarily "product" in nature, with the exception of those written by WG3, (Systems) and the technical specifications of the JCWG.

Various Working Groups (WGs) of TC 82 are Customers of Standards; or where TC 82 contributes to writing standards.

Other TCs suppliers of standards to TC 82 are

TC 21 Battery storage
SC 32B Low-voltage fuses
TC 88 Wind turbines
TC 105 Fuel cells

Customer/Supplier
IEEE
ASTM
UL
CENELEC

Objectives and Strategies

Maintenance of existing standards will be continued by the working groups of the TC. New standards will also be written, especially within the PV concentrator WG to include documents defining trackers and safety. Energy and power rating documents will be refined within both WG2 and WG7.

Building-integrated PV standards will also be explored within WG3. Continued monitoring of advancements in PV technology will be undertaken and standards rewritten to include these technologies once they have been proven to have met the user expectations of long outdoor deployment and safe operation.

Action Plan

All TC 82 working groups plan two meetings per year except for the joint TC21/TC82WG on batteries for PV systems which usually meets only once a year. Project teams within the working groups have been created to facilitate timely completion of the documents.

Due to financial restraints and time limitations for international meetings, it is expected that more homework will be required of the participating experts and that electronic mail communication will play a large role in this work. Initial contact is being made currently between some of the experts with this technique and it is forecast to increase significantly.

Many of the initial IEC PV documents were restricted to crystalline silicon devices for simplicity. These documents will need to be amended to accommodate new thin film devices. Current work anticipates this need and provisions are included to consider improvements or changes in the device technology.

WG 1 will continue to add new terms, symbols, and graphics to IEC TS 61836 as future TC 82 standards are approved for publication. In addition, a more worldwide set of definitions of the terms and symbols used in PV technology will be incorporated in this guide.

The highest priority in WG 2 will be the preparation of standards addressing the safety of PV modules, particularly for high-voltage, grid-connected applications. WG 2 will also develop an energy rating standard to define performance for standard days of high and low ambient temperature and irradiance rather than relying on a single power rating. WG 2 will also revise and correct the translation equations that are used to compare the I-V curves measured under field operating conditions with factory measurements made under Standard Test Conditions (STC).

WG 3's top priority is a standard for defining the performance and test requirements for small, standalone PV systems is nearing final publication, as well as the second edition of its standard for connecting to the electric grid WG 3 will also address the safety issues for all types of PV systems. Standards for individual systems, guidelines and requirements which are applicable to specific performance monitoring, efficiency measurements and acceptance testing at the system level will be receiving higher attention. It will also explore writing Building Integrated PV standards.

WG 6 will prepare standards for the balance-of-system components. Its highest priority is the standard defining safety of grid-connected inverters. As this standard is completed, similar standards dealing with charge controllers will take precedence. The standards from WG 6 will define the requirements and test procedures for determining the component performance, safety and environmental reliability.

Safety guidelines and requirements will also be developed to specify the electrical and mechanical construction characteristics.

An AC module qualification and type approval standard is also being contemplated.

Working Group 7 will prepare standards for concentrator PV systems to define the requirements and test procedures for determining the module and system performance, safety and environmental reliability. Safety guidelines and requirements will also be developed to specify the electrical and mechanical construction characteristics for concentrators.

The JCWG TC82/TC21/TC88/TC105/TC64 will continue to prepare a series of recommendations for small renewable energy and hybrid systems for rural electrification systems employing PV, wind, fuel cells and/or batteries.

The cooperating TCs in the JCG will be responsible for the specific technical requirements of each technology with the JCG providing a coordinating function. The JCWG will also consider topics that apply to all technologies such as project management, project integration, safety and data systems. The group will endeavor to involve other TCs such as TC 4 and TC 64.

To respond to the increasing market and international exchange of PV products, the PV standards being prepared must have a certain flexibility to follow the evolution of the technology. Consideration must be given to the natural environmental impact issues of manufacturing and disposal of PV products.

Other international organizations such as the International Energy Agency (IEA) have initiated PV programs to promote the technology, and closer liaison (Category A) is deemed necessary to maximize the value of the work done under all the programs. Surveys done by others, of PV system failures for instance, could be valuable sources of information to improve qualification and standard requirements.

Future Standardization Work Items Will Include

— System commissioning, maintenance and disposal,
— Updates of existing publications;
— New thin film photovoltaic module technologies such as CdTe, CIS, CuInSe2, etc. characterization and measurement,
— New technology storage systems,
— Applications with special site conditions; i.e., tropical zones, northern latitudes and marine areas,
— Life cycle and disposal analysis for impact on natural environment.

Author's note: Some of these programs have been around for over 30 years (for example in standardizing semiconductor equipment and processes) and some have achieved good results in terms of partially standardizing the solar energy conversion process, including portions of cells and modules manufacturing process and operational procedures.

Certification testing of PV modules is one area that has had significant success, but in our opinion, these standards and guidelines are somewhat weak and outdated; they are too loose and not all-encompassing enough to reflect and predict the performance of PV modules in today's super-large PV power installations, installed in the world's harshest environments.

For example, they do not fully cover the certification of PV modules heated to 180°F day after day during 30 years in desert heat, or in the 100% humidity of a Louisiana swamp, during 30 suffocating summers. Not even close...

Unfortunately these are the exact areas where the largest PV installations are planned for operation today. So, while the good intentions of TC 47 and 82 are quite clear, we feel that they fall short of ensuring proper design, manufacturing, installation and operation of all kinds of PV modules in all kinds of climates.

Also, since there are a number of similar, even competing standards (US, Japan, and China have their own) the confusion in the field will only grow with time.

ONE SINGLE and UNIFORM standard system is needed for the 21st century. This system must cover ALL key areas of the PV cells' and modules' cradle-to-grave lifespan—regardless of where they were manufactured and/or used.

Short of that, we will witness many PV power fields full of low quality, and even unsafe, PV modules, which is not the direction we need to be taking.

PV INSTALLATIONS, SETUP AND PERFORMANCE

Typically, PV modules are not used alone, but in combination with a number of similar modules to create a PV "array." A large array, or several arrays, make a "string." Several strings wired together become part of a PV installation, which we also call PV power field.

A PV array itself is simply a group of carefully selected PV modules, which are equally carefully installed and wired together as a group. The PV arrays come in many forms, shapes and sizes, and can be fixed, sun-tracking with one axis of rotation, sun-tracking with two axes of rotation, or some combination of the above.

Fixed PV array is the cheapest and most common on the PV energy market. It is not as efficient as the tracking array, but it is the simplest, the cheapest, and the one requiring least maintenance.

The function, advantages and disadvantages of fixed vs. tracking arrays is important and must be considered in any PV project. The technical and financial aspects of different types of modules and arrays are discussed in more detail elsewhere in this text.

A fixed array consists of a number of modules that are mounted permanently on a solid frame, usually steel or aluminum angles, which is cemented, or otherwise solidly fixed into the ground or on a structure.

Tracking PV modules are also mounted on a frame, but the frame is heavier and is mounted on a pivoting pedestal, that allows it to rotate, thus following the sun all through the day. The tracking frame moves with the help of one motor-gear assembly for a one-axis system, or two motor-gear assemblies for a two-axis system. A controller, which knows exactly where the sun is, sends a signal to the motors, which then activate the gears to move the frame in position.

Trackers generally have higher comparative efficiency, because they are positioned most advantageously towards the sun (constantly at a 90-degree angle) all day, thus receiving the maximum amount of sunlight and generating the largest possible amount of electric power from sunrise to sundown.

When highly efficient multi-junction solar cells, with 2-3 times higher efficiency than conventional solar cells, are mounted on trackers, the efficiency is greater as well. See Figure 6-25.

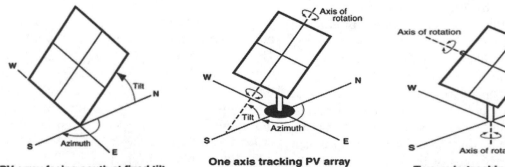

PV array facing south at fixed tilt. **One axis tracking PV array with axis oriented south.** **Two-axis tracking PV array**

Figure 6-24. Types of PV array mounting

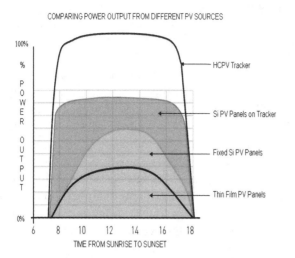

Figure 6-25. Performance of tracking vs. non-tracking PV arrays

Several key considerations must be kept in mind when designing and using PV modules and arrays for serious power generation. Some of these are:

Tilt Angle

For a fixed-mounted PV module, the tilt angle is the angle from horizontal of the inclination of the module top surface (0° = horizontal, 90° = vertical). Or, a PV module is at a 0° tilt angle when lying flat on the ground, or mounted parallel to it. For a sun-tracking PV array with one axis of rotation, the tilt angle is the angle from horizontal of the inclination of the tracker axis. The tilt angle is not applicable for sun-tracking PV arrays with two axes of rotation, because they are always oriented at 0° towards the sun, or are always "looking" straight at it.

The default value of a fixed module is a tilt angle equal to the module's latitude. Installers might consider modifying the azimuth angle +/− several degrees according to the particular location or customer's needs. The overall intent is to maximize annual energy production without changing the tilt angle at any time.

Increasing the tilt angle favors energy production in winter, and decreasing the tilt angle favors energy production in summer. This is due to the low-in-the-sky position of the sun in winter, and high-in-the-sky position in summer. Of course, a compromise which would increase the generated power is the so-called "seasonal adjustment" of the tilt angle. This consists of changing the angle of all modules in the power plant to match closely the seasonal changes of the sun angle. Ideally, this is done 4 times every year at the beginning of each season at the particular location.

Azimuth Angle

For a fixed-mounted PV array, the azimuth angle is the angle clockwise from true north (zero degree) to the south (180 degrees) that the PV array faces. For a sun-tracking PV array with one axis of rotation, the azimuth angle is the angle clockwise from true north of the axis of rotation. The azimuth angle is not applicable for sun-tracking PV arrays with two axes of rotation.

The default value for a fixed module is an azimuth angle of 180° (south-facing) for locations in the northern hemisphere and 0° (north-facing) for locations in the southern hemisphere. This normally maximizes energy production. For the northern hemisphere, increasing the azimuth angle favors afternoon energy production, and decreasing the azimuth angle favors morning energy production. The opposite is true for the southern hemisphere. Installers might consider modifying the azimuth angle +/− several degrees according to the particular location or customer's needs. Seasonal changes of the azimuth angle are also possible, but not recommended, due to little benefit from a fairly major effort.

Power Rating

The size of a PV array is its DC power rating, or nameplate. This is determined by adding the PV module power listed on the PV modules in watts and then dividing the sum by 1,000 to convert it to kilowatts (kW). PV module power ratings are for standard test conditions (STC) of 1,000 W/m² solar irradiance and 25°C PV module temperature.

The maximum power generated by a single cell or module is determined by:

$$Pmax = Voc \times Isc \times FF$$

where
   Voc   is the open-circuit voltage,
   Isc   is the short-circuit current, and
   FF   is the fill factor.

A small commercial type PV system consists of a number of modules, generating 5-500 kW total power. This corresponds to a PV array area (active area) of approximately 30m² to 3,000m², but it depends on the efficiency of the modules, and other factors which determine the overall PV system size; i.e., some thin film modules are only 6-8% efficient, so they will need at least 50% more area for installation plus additional mounting frames, wiring etc.

Large-scale PV systems consist of thousands of PV modules, depending on the nameplate (maximum power generated at SCT, usually as claimed by the modules manufacturer).

DC-to-AC Derate Factor

The overall DC-to-AC derate factor accounts for losses suffered by the DC nameplate (or the originally estimated and calculated power rating), due to losses from the components in the power plant, while generating, transporting, converting and otherwise transferring the generated DC power into the AC grid. It is basically the mathematical sum of the losses (derate factors) triggered or caused by all components of the PV system. So, we need to multiply the nameplate DC power rating by an overall DC-to-AC derate factor of the system to determine its total AC power rating at STC.

Because STC is just a theoretical value provided by the manufacturer and derived basically under lab test conditions, we need to estimate the power rating during everyday operating conditions; i.e., a spring day in the Arizona desert might produce sunlight close to STC and the modules might even operate under STC conditions for a short while. This, however, is not so during a hot day in July, when the PV modules, inverters and other components are boiling hot and when their efficiency drops significantly.

The overall DC-to-AC derate factor of the system is represented by the sum of all components' derate factors (as used by PVWatts), which is the baseline of the system. In the above case, we have a derate factor of 0.77. So 77% of the available power (generated by the modules) will be converted into DC and sent into the grid, if the module or arrays are operating under STC. This is a 23% loss due to the components' resistance operation, and inefficiency.

As mentioned above, the derate factor would be quite different under extreme conditions—extreme desert heat in particular. This difference must be calculated as well, and important decisions must be made on the basis of the difference between STC and in-sun operation under extreme climates; i.e., the efficiency of a PV array or an inverter rating will drop 20-30% from the STC measurements when operated in the summer desert heat. This drop in efficiency in the field must be considered when designing, installing and operating a PV system, because the negative effects under these conditions are usually significant. Ignoring these changes during the design or evaluation stage will result in serious unpleasant surprises in the field.

Operational Characteristics and Issues

PV modules are complex electro-mechanical de-

vices. Cell structure, wiring, electronics and many other factors determine the proper and trouble-free operation of the modules for their long non-stop operation.

Listed below are some of the major issues encountered in PV modules during installation and field operations: (14)

Mismatch or Shading

Serious reduction of output and/or damage to the solar cells in the PV modules might occur if:

a. The cells in the module are mismatched (different output), be it from manufacturing negligence, or subsequent degradation of some cells, and

b. Total or partial shading of the modules might lead to thermal overload and overheating (hot-spots) which are often misunderstood or ignored.

One of the quality characteristics that is verified within UL and IEC qualification testing is the ability of PV modules to withstand the effects of periodic hot-spot heating that occurs when cells are operated under reverse biased conditions. Testing, however, cannot eliminate these effects, and this should be taken into consideration during design, evaluation, and operation of PV systems.

As environmental conditions and the modules' behavior change with time, hot spot issues could appear without warning. Once these problems are present, the affected modules must be inspected, evaluated and removed from operation, or checked periodically to make sure the problem doesn't grow or expand.

Solar Cell Degradation

Overall cell degradation phenomenon ranging from 0.5% to 1.5% of total output loss per year, and the mechanisms thereof, is not well understood and impossible to control. It is assumed that it is primarily due to series resistance (RS) increases over time, most likely due to inadequate metallization, corrosion of the contacts (caused by moisture in the module); and/or deterioration of the adhesion between the metal contacts and the substrate material, due to manufacturing abnormalities, and/or decrease in shunt resistance (RSH) due to metal migration through the p-n junction; and/or antireflection.

(AR) Coating Deterioration or Delamination

There are a number of other reasons for cell degradation, but since we cannot control any of them after the

PV modules have been installed, we must find a way to account for and control them during the long manufacturing sequence—all the way from sand to module.

Short Circuits

Short circuiting at cell interconnections, as illustrated in Figure 6-26, is also a common failure mode, especially for very thin cells, since top and rear contacts are much closer together and stand a greater chance of being shorted together by pin-holes or regions of corroded or damaged cell material.

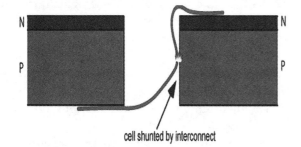

cell shunted by interconnect

Figure 6-26. Severe solar cell shunting mechanism

Temperature Coefficient

The temperature coefficient (Tc) is generic terminology that can be applied to several different PV performance parameters, such as voltage, and current and power fluctuations under temperature variations. Most often, however, it is referred to as the total power output loss upon exposure to excess temperatures. The procedures for measuring the temperature coefficient for modules and arrays are not yet standardized, so it is common to see different Tc expressions and test methods. Also, systematic influences and errors are common in the test methods used, but worse are misconceptions regarding their application under field-operation conditions.

Tc basically is the percent of change (usually decrease) of power output from a solar cell or module per each degree C above standard test conditions, STC (set at 25°C). Tc average of 0.5-0.6% power loss per °C is accepted by the PV industry for c-Si cells and modules. It is slightly lower for some types of thin film PV modules.

Author's note: So, if our c-Si PV module were rated at 100 Wp at STC, it would lose 0.5% output with every degree C above 25°CSTC. When exposed to full sunlight at noon in the Arizona desert, the module would heat up to 85°C (60°C above STC), and its power output would drop 30% (60 x 0.5%). Our 100 Wp PV module would be generating less than 70 W while at this temperature.

This is a significant loss, so Tc is an important parameter in determining PV power system design and

sizing because worst-case operating conditions often dictate the array type and size. In other words, PV modules and arrays are supposed to operate at maximum output around the noon hour, when the sun is highest.

But if the output drops 30% during that time due to extreme heat, we must change the calculation methods in arriving at the peak power output of the array. Or—if our 100 MWp (as measured under STC) power array produces only 70 MWp at high noon in the desert, when we need the power most, we may have to add more PV modules to compensate for the 30% power loss during that critical time.

Because of that anomaly Tc is a most important parameter when considering installations in desert areas where extreme heat conditions will have a great influence on it and overall system performance.

Open Circuits (Cells)

This is a common failure mode, although redundant contact points plus "interconnect-busbars" allow the cell to continue functioning. Cell cracking can be caused by thermal stress; hail; or damage during processing and assembly, resulting in "latent cracks," which are not detectable on manufacturing inspection, but appear later and cause partial or total failure. Interconnect open circuit could be caused by fatigue due to cyclic thermal stress, mechanical damage at a previous stage of the process, or wind loading.

Module Open Circuit

Open circuit failures also occur in the module structure, typically in the bus wiring or junction box.

Module Short Circuit

Although each module is tested before sale, module short circuits are often the result of manufacturing defects. They occur due to insulation degradation with weathering, resulting in delamination, cracking or electrochemical corrosion.

Module Hot-spot Heating

Module hot-spot heating occurs when there is one low-current solar cell in a string of at least several high short-circuit current solar cells. If the operating current of the overall series string approaches the short-circuit current of the "bad" cell, the overall current becomes limited by the bad cell. Essentially the entire generating capacity of all the good cells is reduced by dissipation in the poor cell. The enormous power dissipation occurring in a small area results in local overheating, or "hot-spots," which in turn lead to destructive effects, such as cell or glass cracking, melting of solder, or degradation of the solar cell.

Module Glass Breakage

Shattering of the top glass surface can occur due to vandalism, thermal stress, handling, wind or hail. Once the glass is broken, the interior of the module—including the cells—becomes exposed and vulnerable to the effects of the elements. Moisture penetrating the module could destroy the cells and shunt the entire unit within weeks.

Module Delamination

It is usually caused by reductions in bond strength between the front glass and the EVA or other encapsulant. It caused by environmentally induced moisture or by photo-thermal aging and stress which induced by differential thermal and humidity expansion.

Author's note: This phenomena is especially dangerous for thin film modules, because the active layers are deposited directly on the front glass, where delamination is most often occurring. Once the active layer is separated from the front glass, the fragile thin films would be exposed to the elements and vulnerable to any mechanical (sand blast) or chemical (moisture penetration) damages.

Bypass Diode Failure

By-pass diodes, used to overcome cell mismatching problems, can themselves fail, usually due to overheating, and sometimes due to undersizing, or over-capacity use. The problem is minimized if junction temperatures are kept below 128°C.

Encapsulation Failure

UV absorbers and other encapsulant stabilizers ensure a long life for module encapsulating materials. However, slow depletion, by leaching and diffusion does occur and, once concentrations fall below a critical level, rapid degradation of the encapsulant materials occurs. In particular, browning of the EVA layer, accompanied by a build-up of acetic acid, has caused gradual reductions in the output of some arrays, especially those exposed to higher temperatures, such as in a desert or humid environment.

Author's note: Desert heat, extreme UV radiation and nightly freezing will see to it that all organic materials in the modules (including the encapsulation, lamination and edge sealers) are disintegrated and chemically decomposed. This is what the desert does best, so in all instances, module encapsulation and lamination have

significant roles to play. Their composition and application must be immaculate, to delay the destruction mechanisms. The destruction cannot be prevented, but it can be delayed long enough to complete the useful life of the PV modules.

Optical Losses

Optical losses chiefly effect the power from a solar cell by lowering the short-circuit current. Optical losses consist of light which could have generated an electron-hole pair, but is not because the light is reflected from the front surface, or because it is not absorbed in the solar cell. For the most common semiconductor solar cells, the entire visible spectrum has enough energy to create electron-hole pairs and therefore all visible light would ideally be absorbed.

There are a number of ways to reduce optical losses:

a. Top contact coverage of the cell surface can be minimized (although this may result in increased series resistance).

b. Anti-reflection coatings can be used on the top surface of the cell.

c. Reflection can be reduced by surface texturing.

d. The solar cell can be made thicker to increase absorption (although any light which is absorbed more than a diffusion length away from the junction will not typically contribute to short-circuit current since the carriers recombine).

e. The optical path length in the solar cell may be increased by a combination of surface texturing and light trapping.

Material (Silicon Wafer) Thickness

While the reduction of reflection is an essential part of achieving a high efficiency solar cell, it is also essential to absorb all the light in the silicon solar cell. The amount of light absorbed depends on the optical path length and the absorption coefficient of the material and its thickness.

Thermal Expansion and Thermal Stress

Thermal expansion is another important temperature effect which must be taken into account when modules are designed. Typically, interconnections between cells are looped, to minimize cyclic stress. Double interconnects are used to protect against the probability of fatigue failure caused by such stress. In addition to interconnect stresses, all module interfaces are subject to temperature-related cyclic stress which may eventually lead to delamination.

Mismatch Effects in Arrays

In a larger PV array, individual PV modules are connected in both series and parallel. A series-connected set of solar cells or modules is called a "string." The combination of series and parallel connections may lead to several problems in PV arrays.

One potential problem arises from an open-circuit in one of the series strings. The current from the parallel connected string (often called a "block") will then have a lower current than the remaining blocks in the module. This is electrically identical to the case of one shaded solar cell in series with several good cells, and the power from the entire block of solar cells or modules is lost.

Warranties

PV modules are usually sold with a warranty that covers several aspects of the modules' craftsmanship and performance. The types of standard warranties (different companies have different policies) is as follow:

Materials Warranty (Years)

This is a limited warranty on module materials and quality under normal application, installation, use, and service conditions. For most modules material warranties vary from 1 to 5 years. Most manufacturers offer full replacement or free servicing of defective modules.

Read the warranty conditions carefully prior to purchasing the modules, because they vary from vendor to vendor. The problem that is most often encountered and one that is most disputed is shipping charges and replacement (labor) cost. In other words, the expense of shipping a batch of modules from abroad, uninstalling the defective ones and reinstalling the new batch might be higher than the cost of the modules themselves, thus the vendor might object to absorbing these charges, unless all these conditions and exemptions are clearly identified and addressed in the warranty documents.

Power Warranty (Long-term)

This is a limited warranty for module power output based on the minimum peak power rating (STC rating minus power tolerance percentage) of a given module. The manufacturer guarantees that the module will provide a certain level of power for a period of time—at least 20 years. Most warranties are structured as a per-

centage of minimum peak power output within two different time frames—90% over the first 10 years and 80% for the next 10 years.

For example, a 100 W module with a power tolerance of +/–5% will carry a manufacturer guarantee that the module should produce at least 85.5 W (100 W x 0.95 power tolerance x 0.9) under STC for the first 10 years. For the next 10 years (to year 20), the module should produce at least 76 W (100 W x 0.95 power tolerance x 0.8).

Author's note: This is a significant power loss, which must've been anticipated in the initial contract, as well as calculated into the overall power plant budget. Unforeseen losses (such as weather and maintenance shutdowns), could increase expenses an additional 5-10%, and must also have been used to estimate the final formula.

Module replacement value provided by most power warranties is generally prorated according to how long the module has been in the field. Again, the cost of shipping and replacement of the non-conforming modules could be high, and replacement conditions must be well understood and agreed upon by both parties prior to purchasing the modules.

LARGE-SCALE SOLAR POWER FIELDS

"Large-scale commercial solar PV and central-station utility-scale PV will most likely become the dominant growth market" (EPRI, 2010).

Large-scale PV power generating (commercial and utilities type installations) are the future of our energy independence, and a sure way to clean the environment. They consist of large land areas, usually in the world's deserts, covered with solar energy conversion equipment, such as fixed mount PV modules, trackers etc. With their large size and unlimited power generating capacities, they are on the extreme end of the PV power generation spectrum and represent the best and most efficient way to generate solar electricity.

When properly designed, installed and operated, solar equipment can generate a large quantity of electricity via concentrated solar thermal power (CSP) or photovoltaic (PV) technologies. We will concentrate on the PV side of things in this text, because it is in the area of our expertise and long experience. PV technologies with their great versatility and flexibility will play an ever more important, if not determining, role in our energy independence and clean environment, so they do deserve a closer look.

To be cost effective, CSP technologies are limited to very large installations (over 50 MWp and larger) and also require a large quantity of water for cooling, which is simply not available in the deserts where this technology is most useful. They have also reached, or are close to reaching, their efficiency peak, so little can be done to improve their performance or reduce their cost. A number of cancelled CSP projects in the US, converted into PV installations lately, are a confirmation of the greater flexibility of PV power.

There are several different PV technologies to choose from, most of which can be installed anywhere and in any size. They need no water for cooling and have not yet reached their maximum possible output levels. With continuing research by private and government entities, we expect the efficiencies of some PV technologies to reach 60% before this decade is over and 80% in the near future, which opens great new possibilities for energy generation. By increasing the size of the PV power plants, and adding many GWs of electric power production, the different PV technologies combined could easily surpass the generating capacity of conventional energy sources.

PV technologies are our hope for mass electric power production without the negative effects of global warming and other environmental disasters we are seeing. The transition won't be easy. In the end, however, we will make the right choices and do the right things for our sake and that of future generations.

PV Technologies for Large-scale Installations

The major PV technologies suitable for commercial and utility-scale applications today are solar thermal (CSP), c-Si (mono and poly crystalline silicon) modules, thin film PV (TFPV) modules, and high concentration PV (HCPV) tracking systems, as follow.

CSP Technologies

It goes without saying that CSP is the most understood and broadly applied solar technology—especially for large-scale power generation. Solar water heaters (as part of the solar thermal branch) are also increasing in popularity as some governments are providing generous subsidies to install solar water heaters in new houses and commercial buildings. Others (Spain, for example) are making it mandatory. A similar situation is developing in other EU countries and recently in the US as well.

This is the trend that started the US solar revolution in the 1970s, and it was not very successful then, so why would it be now? Government subsidies sparked an unprecedented increase of solar water heater installations in CA, AZ and other states in the late 1970s and early 1980s. The boom lasted several years and then slowly died, as the subsidies were reduced and finally discontinued.

Will that happen again?

CSP Technologies

For large-scale power generation, CSP technologies are quite adequate, and have been used a long time. Several large CSP power plants are operating successfully around the world, while some are underway. Nevertheless, CSP is also hitting serious barriers lately. Lack of water, the large land requirement, environmental problems, etc. are hindering the full development of CSP power projects in the US and around the world.

CSP, however, is here to stay and we will see many large projects popping-up in the US and world deserts in the future.

Poly-silicon PV Technology

This is a mature technology, which represents the largest segment of the PV industry today, due mostly to lower prices. The materials, equipment, and processes used now are almost the same as those used nearly 40 years ago, when the first 2- and 3-inch silicon solar cells were processed by hand.

Some changes have occurred through the years, of course, where the quality of materials, the sophistication and efficiency of the production equipment and processes have increased significantly. The performance and efficiency of the finished product has increased some too, but is still below the theoretical limits of the respective technologies.

Commercial poly silicon solar cells operate on average at 16-18% efficiency, which means that 1 m^2 PV module packed with the maximum number of polysilicon solar cells will produce a maximum of 160-180W of DC power at STC (1000 W/m^2 solar insolation at 25°C ambient temperature.).

Also, when operating under high desert temperatures in the 160-180°F range (87-95°C) as measured inside the modules, silicon solar cells lose efficiency, which causes significant drop of power output. Output losses due to operation under elevated temperatures range from 0.4 to 0.6% drop per degree C, as measured in US deserts. This loss could amount to 20-30% power loss at noon in summer in the Arizona desert—not a small problem.

In addition, silicon solar cells work well in the beginning of their working life and slowly decrease in efficiency after that, losing ~1.0% per annum. This is widely accepted phenomena, so all PV modules come with a "20 years" performance warranty, which states that the modules will be at least 80% efficient after 20 years of operation.

The long-term loss, combined with the efficiency loss in excess temperatures tells us that c-Si PV modules prefer cooler operating temperatures. Still, well-made poly silicon cells from established manufacturers have proven worthy in long-term field tests, which means that the technology in its present form and at its best quality is capable of handling long-term exposure to the elements. This means that we must choose a reliable manufacturer if we want to use poly Si modules in a large-scale PV power plant.

Mono-silicon PV Technology

This is the most efficient and reliable PV technology today. With its 18-20% efficiency and going higher while prices are going down, it is the technology of choice in many PV projects. It is also the most mature technology, since it has been in use the longest. Its properties and behavior are best known, because mono-silicon (m-S) is widely used in the semiconductor industry too, where it is thoroughly researched and used in mass production.

Yet, TUV Rheinland test center in Tempe, AZ, issued a summary in one of their reports (6) as follows: "Out of 18 (9 mono-Si and 9 poly-Si) modules tested, 9 poly-Si and 7 mono-Si modules passed the hotspot tests of all three standards. Both failures (mono-Si with voltage limited and current limited cells) occurred in the UL method. One failure occurred within the first few minutes (current limited) and the other failure occurred [after] about 70 hours of stress (voltage limited)."

It is undeniable fact that mono-silicon solar cells are the oldest, most stable and reliable PV technology available today, and we are most familiar with their behavior. Still, some of the m-Si modules (20%) failed within 70 hours, while others did not fail for the duration. Why did one of these modules (10%) fail shortly after initiating the test?

The disturbing fact is that modules of this batch would not have lasted very long in the field—or even worse—would have lasted just beyond the product warranty period. Imagine 10-20% of the modules in our large-scale PV power plant failing a month or two after warranty expiration.

Was this anomaly caused by materials, quality or equipment issues, or was it due to manufacturing problems? We will never know, but one thing is for sure: the quality of the most reliable technology coming from different manufacturers varies. This means that the failures are not caused by technology, but by manufacturers.

Conclusions: Mono silicon cells and modules are the oldest, best understood and most reliable solar energy conversion devices available. Mono Si modules, made by reputable and well established manufacturers using proper techniques, have the quality and reliability needed for use in world energy markets. Some m-Si modules made by new and unproven manufacturers, however, have quality issues which we should watch for.

Thin Film PV (TFPV) Technology

This is the new kid on the block, with a very bullish outlook for the solar industry and energy markets in general. And this is good—very good! The energy industry which is dominated by conventional energy interests recently got a good kick in the pants with the quick and efficient deployment of TFPV production capacities, products and large-scale projects.

TFPV manufacturers offer a new approach to solar modules manufacturing—that of elegant, quick, efficient and cheap mass production. Using over a half century of experience with related semiconductor device manufacturing processes, TFPV manufacturers are showing us how PV equipment of the 22nd century will be made.

But efficiency is where TFPV technology stumbles, at 8-9% efficiency at the PV module and 5-6% (we've seen reports of 2%) in the grid, TFPV modules have far to go in competing for efficiency. There are also problems with supplying raw materials for these technologies because some of them are rare, exotic, or toxic, or all of the above. This is exposing the TFPV industry to a serious risk of shortages, price variations, toxicity issues and regulations changes which jeopardize its very existence.

Hi-concentration PV (HCPV) Technology

The HCPV technology was developed in the 1980s and tested in small-scale projects during the 1990s. It was, however, put on the shelf soon thereafter, so it is still waiting to be introduced and proven in field operations. Several manufacturers have installed capacities, but they seem to still be mostly in the development stages.

HCPV systems are designed and built for operation under extreme desert conditions. They operate at 50-1000 times concentration of sunlight with over 42% efficiency at the cell. Efficiency in the grid is lower, of course, but still at least 2-3 times higher than any of the competing PV technologies. CPV solar cells are made out of germanium, or GaAs, which are superior semi-conductors not affected by temperature extremes. They are also made via sophisticated and precise semiconductor processes, which make them durable and reliable for long-term operation. The optics are made of glass and silicone, which are also unaffected by sunlight and temperature extremes.

HCPV modules are mounted on trackers which have a number of moving parts, control mechanisms and electronics that need regular maintenance and tuning for optimum performance. This requires highly trained personnel and increases O&M expenses.

Concentrating PV (CPV), low concentration PV (LCPV) and high concentration PV (HCPV) solar industries are still to demonstrate their feasibility and reliability in large-scale commercial installations, where they are most suitable for operation.

As the recently formed CPV Consortium states, "… CPV is on the cusp of delivering on its promise of low cost reliable GW scale solar energy…" We are still awaiting the actual and successful implementation of this technology in large-scale power fields—an uphill endeavor faced with a number of serious problems.

The shut-down of HCPV leader Amonix's production plant in the summer of 2012 was a big step back in this process, and it gave the industry a black eye that it won't overcome soon.

The major problems with the CPV/LCPV/HCPV technologies are:

1. Need for complex, precise and expensive tracking mechanisms,

2. Need for direct and unobstructed sun-beam radiation,

3. Difficulty in efficient and reliable cell-to-heat sink heat transfer, and

4. Difficulty with Fresnel lens materials.

So, one of the biggest problems of CPV/HCPV equipment is that it simply stops working every time the sun hides behind even the smallest cloud. CPV/HCPV cannot deal with diffused solar radiation. This incapability cannot be tolerated by the power companies, and is one reason the CPV technology lacks wide acceptance.

The failures of some CPV projects and companies during 2012 only show that the CPV/HCPV technologies are not fully developed and that they should not be rushed to market. Advanced materials and manufacturing processes are needed to ensure efficient long-term operation.

We can only hope that with their 40+% efficiency, which is increasing, PV/HCPV technologies will soon become active participants in the energy markets worldwide. HCPV holds great promise of high efficiency (over 60%), which guarantees power output suitable for large-scale generation of electricity in deserts.

Before we build a new solar power plant, we should look at the other components of its design, implementation and long-term life stages. The power grid is one of those key elements, so we need to know as much as possible about its function and issues.

The Power Grid

The DC power produced by most residential and commercial solar power plants is converted to conventional AC power and is sent into the national power grid. Looking at the power grid, we'll start with the conventional power generation and its transmission and distribution around the country.

Because electric power is essential to modern society, a prerequisite for a practical large-scale PV power plant is the ability to transfer generated electricity from the source into the national grid. We have no choice, but to work with the grid, so we need to know all there is to know about it.

In 1940, 10% of energy consumption in America was used to produce electricity. In 1970, that fraction was 25%. Today it is 40%, showing electricity's growing importance as a source of energy supply. Electricity has the unique ability to convey both energy and information, thus yielding an increasing array of products, services, and applications in factories, offices, homes, campuses, complexes, and communities.

The national power grid accomplishes that by its thousands of miles of power lines, substations and other equipment, which convert the power to voltages and frequencies that are adequate for transmission and use.

Looking at the map in Figure 6-27 we see that the national electric grid, just like the human body's circulatory system, carries life-sustaining energy to all the different parts of the country. Without it, most critical activities would simply stop, the economy would crash, and life as we know it would cease. Nothing good can result from a drastic interruption in the national energy supply.

The Basics

The economic significance of generating, distributing and using electricity is staggering. It is one of the largest and most capital-intensive sectors of the econo-

Figure 6-27. Map of the US Power Grid

my. Total asset value is estimated to exceed $800 billion, with approximately 60% invested in power plants, 30% in distribution facilities, and 10% in transmission facilities.

Annual electric revenues—the nation's electric bill—are about $247 billion, paid by America's 131 million electricity customers, which includes nearly every business and household. The average price paid is about 8-15 cents per kilowatt-hour, although prices vary from state to state depending on local regulations, generation costs, and customer mix.

There are more than 3,100 electric utilities in the US:

* 213 stockholder-owned utilities provide power to about 73% of the customers,

* 2,000 public utilities run by state and local government agencies provide power to about 15% of the customers, and

* 930 electric cooperatives provide power to about 12% of the customers.

Additionally, there are nearly 2,100 non-utility power producers, including independent power companies and customer-owned distribution facilities.

The US power grid system consists of three independent networks: Eastern Interconnection, Western Interconnection, and the Texas Interconnection. These networks incorporate international connections with Canada and Mexico as well. Overall reliability planning and coordination is provided by the North

American Electric Reliability Council, a nonprofit organization formed in 1968 in response to the Northeast blackout of 1965.

Power Generation

America operates a fleet of about 10,000 power plants, mostly thermal (coal and diesel) with average efficiency of around 33%. Efficiency has not changed much since 1960 because of slow turnover of the capital stock and the inherent inefficiency of central power generation that cannot recycle heat. Nuclear and hydro plants are more efficient, but much more expensive as well.

Power plants are generally long-lived investments, with the majority of existing US capacity 30 years old or older. They can be divided into:

a.   Baseload power plants, which are run all the time to meet minimum power needs,

b.   Peaking power plants, which are run only to meet power needs at maximum load (known as peak load), and

c.   Intermediate power plants, which fall between the two and are used to meet intermediate and emergency power loads.

The roughly 5,600 distributed energy facilities typically combine heat and power generation and achieve efficiencies of 40% to 55%, accounting for about 6% of US power capacity in 2001 and several times more today.

A shift in ownership is occurring from regulated utilities to competitive suppliers. The share of installed capacity provided by competitive suppliers has increased from about 10% in 1997 to about 35% today. Recent data suggest, however, that this trend is slowing down. Also, cleaner and more fuel-efficient power generation technologies are becoming available. These include combined cycle combustion turbines, wind energy systems, advanced nuclear power plant designs, clean coal power systems, and distributed energy technologies such as photovoltaics and combined heat and power systems.

Because of the expected retirement of many aging plants, and the forecasted growth in electricity demand, America faces a significant need for new electric power generation. Local market conditions will dictate fuel and technology choices for investment decisions, capital markets will provide financing, and federal and state policies will affect siting and permitting. It is an enormous challenge that will require a large commitment of technological, financial and human resources.

PV power generating sources are constantly added to the already complex power generation and distribution system, but due to their location, size and volatile performance, it will take a long time for them to be fully integrated. The utilities are doing their best to accommodate the additional unconventional load, but there are problems related to interconnection, distribution, and control of variability that must be resolved before the new PV sources are truly part of system.

Power Transmission and Distribution

Adequate electric power use in the US is hindered by bottlenecks in the transmission system, which interfere efficient and affordable delivery of electric power to customers. America operates about 157,000 miles of high voltage (>230 kV) electric transmission lines.

Construction of transmission lines and facilities has decreased about 30%, and annual investment in new transmission facilities has declined over the last 25 years. The result is grid congestion, which can mean higher electricity costs because customers cannot get access to lower-cost electricity supplies, and because of higher line losses.

Transmission and distribution losses are related to how heavily the system is loaded. U.S.-wide transmission and distribution losses were about 5% in 1970, grew to 9.5% in 2001, and are even higher today, due to heavier utilization and more frequent congestion.

Congested transmission paths, or "bottlenecks," affect many parts of the grid across the country. In addition, it is estimated that power outages and power quality disturbances cost the economy from $25 to $180 billion annually. These costs could soar if outages or disturbances become more frequent or longer in duration. There are also problems with maintaining voltage levels.

America's electric transmission problems are also affected by the new structure of the increasingly competitive bulk power market. Based on a sample of the nation's transmission grid, the number of transactions has increased substantially. For example, annual transactions on the Tennessee Valley Authority's transmission system numbered less than 20,000 in 1996. They exceed 250,000 today, a volume the system was not originally designed to handle.

Actions by transmission operators to curtail trans-

actions for economic reasons and to maintain reliability (according to procedures developed by the North American Electric Reliability Council) grew from about 300 in 1998 to over 1,000 in 2000 and are much higher today.

Additionally, significant impediments interfere with solving the country's electric transmission problems. These include opposition and litigations by different groups against the construction of new facilities, uncertainty about cost recovery for investors, confusion over whose responsibility it is to build and maintain, and jurisdiction and government agency overlap for siting and permitting. Competing land uses, especially in urban areas, leads to opposition and litigation against new construction facilities.

In Figure 6-28, we get a glimpse into the complexity of the generation-transmission-distribution scheme of electric power transfer. The generator (coal or nuclear power plant) might be miles away from the point of use (POU)—residential or industrial customer.

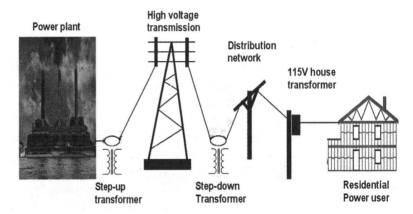

Figure 6-28. The electric power grid (generation and transmission)

The power generated at the power plant is sent to a step-up transformer (substation), where it has to be transformed into higher voltage, as needed for long-distance transfer. Some of this power is used by larger users who have their own sub-stations for power use in their facilities. The rest of the power (most of it) is transported via the national power grid to step-down substations all over the country, where it is converted to lower voltage and is sent by distribution power lines to residential and commercial customers, who also have their own substations or transformers for converting the power to an exact voltage they can use.

Power Distribution

The "handoff" from electric transmission to electric distribution usually occurs at the substation. America's fleet of substations takes power from transmission-level voltages and distributes it to hundreds of thousands of miles of lower voltage distribution lines. The distribution system is generally considered to begin at the substation and end at the customer's meter. Beyond the meter lies the customer's electric system, which consists of wires, equipment, and appliances—an increased number of which involve computerized controls and electronics operating on DC.

There are two types of distribution networks; radial or interconnected.

a. A radial network leaves the power generating station to its final destination with no connection to any other supply in the network. This is typical of long rural lines with isolated load areas.

b. An interconnected network is generally found in more urban areas and will have multiple connections to other points of supply. These points of connection are normally open but allow various configurations by the operating utility by closing and opening switches.

The distribution system supports retail electricity markets. State or local government agencies are heavily involved in the electric distribution business, regulating prices and rates of return for shareholder-owned distribution utilities. Also, in 2,000 localities across the country, state and local government agencies operate their own distribution utilities, as do over 900 rural electric cooperative utilities. Virtually all of the distribution systems operate as franchise monopolies established by state law.

The greatest challenge facing electric distribution is that of responding to rapidly changing customer needs; i.e., increased use of information technologies, computers, and consumer electronics has lowered the tolerance for outages, fluctuations in voltages and frequency levels, and other power quality disturbances. In addition, rising interest in distributed generation and electric storage devices is adding new requirements for interconnection and safe operation of distribution systems.

Finally, a wide array of information technology is entering the market that could revolutionize the electric distribution business. For example, having the ability to monitor and influence each customer's usage, in real time (part of the smart grid solution), could enable distribution operators to better match supply with demand, thus boosting asset utilization, improving service quality, and lowering costs. More complete integration

of distributed energy and demand-side management resources into the distribution system could enable customers to implement their own tailored solutions, thus boosting profitability and quality of life.

The new PV power installations—millions of small residential and hundreds of large-scale plants—add a new dimension to the complexity of the problems that distribution lines are experiencing. They cause power variations and fluctuations that are hard to control, thus new power management schemes and controls must be developed to handle these newcomers. So the new smart grid issues are on the table and discussed daily, with some work underway as we speak.

Power Quality and Reliability Issues

Power quality is a concern for today's power grid and the loads it serves. Computer equipment, in particular, is sensitive to power quality problems, and the ubiquity of computers in today's manufacturing environment means high power quality is very important to most commercial and industrial firms as well as the average homeowner.

Alternating current, which is the predominant way of transporting and delivering electric power, can be illustrated as a sinusoidal wave, as shown in the Figure 6-29. Over time, the voltage oscillates between a positive and negative value that is slightly more than average voltage.

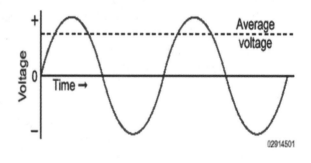

Figure 6-29. AC wave form.

Although alternating current is established as the world's standard, it still has unresolved problems and issues, some of which are important to understanding power quality control and the contribution of PV power generating plants. Below is a list of conditions and characteristics for AC power transmission and use:

1. *Voltage sags and swells.* The amplitude of the wave gets momentarily smaller or larger because of large electrical loads such as motors switching on and off.

2. *Impulse events.* Also called glitches, spikes, surges, or transients, these are events in which the voltage deviates from the curve for a millisecond or two (much shorter than the time for the wave to complete a cycle). Impulse events can be isolated or can occur repeatedly and may or may not have a pattern. The largest voltage glitch, or surge, is caused by a lightning strike.

3. *Decaying oscillatory voltages.* The voltage deviation gradually dampens, like a ringing bell. This is caused by banks of capacitors being switched in by the utility.

4. *Commutation notches.* These appear as notches taken out of the voltage wave. They are caused by momentary short circuits in the circuitry that generates the wave.

5. *Harmonic voltage waveform distortions.* These occur when voltage waves of a different frequency (some multiple of the standard frequency of 60 cycles per second) are present to such an extent that they distort the shape of the voltage waveform.

6. *Harmonic voltages.* These can also be present at very high frequencies to the extent that they cause equipment to overheat and interfere with the performance of sensitive electronic equipment.

Other power quality problems may also be considered reliability problems because they occur when the transmission system is not capable of meeting the load on the system. These are:

1. *Brownouts.* These are persistent lowering of system voltage caused by too many electrical loads on the transmission line.

2. *Blackouts.* These are, of course, a complete loss of power. Unanticipated blackouts are caused by equipment failures, such as a downed power line, a blown transformer, or a failed relay circuit.

3. *"Rolling" blackouts.* These are intentionally imposed upon a transmission grid when the loads exceed the generation capabilities. By blacking out a small sector of the grid for a short time, some of the load on the grid is removed, allowing the grid to continue serving the rest of the customers. To spread the burden among customers, the sector

that is blacked out is changed every 15 minutes or so—hence, the blackouts "roll" through the grid's service area.

How does PV power generation from millions of rooftops and larger PV installations affect all these factors, and what can be done to reduce the negative effects? This important question can be answered by the experts, and they are working on it as we speak. One of the most discussed solutions is the implementation of the upcoming "smart grid."

Variability of PV Power Generation

PV plants produce most power when the sun shines brightly on an unobscured sky. In early morning and late afternoon, the sunlight falls on Earth at an angle, and its power is reduced several fold. When a cloud crosses the sunlight path, it reduces the sunlight power several fold as well. Fog, windy/dusty conditions and other climatic phenomena also contribute to reduction of the solar power coming to the PV modules, thus reducing their output dramatically. This makes them "variable" and unpredictable power generators.

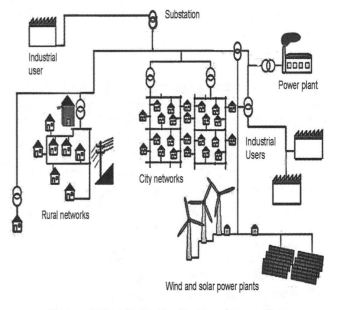

Figure 6-30. Alternative energy integration

Obviously, PV power plant output is not constant because it depends on many factors. It actually varies from season to season, day to day, hour to hour and minute to minute. The variations during a bright sunny day (Figure 6-31a) are insignificant and will not pose any major concerns to grid operators who can predict and handle normal daily production. This includes the early-morning and late-evening low levels of power

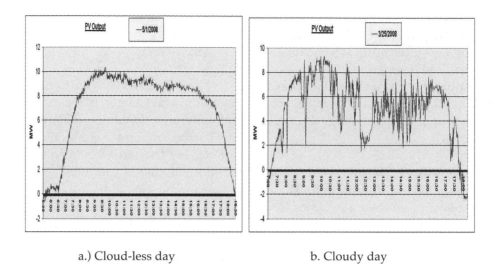

a.) Cloud-less day   b. Cloudy day

Figure 6-31. PV power plant output daily variations

production, before and after full-sun days predominant in desert regions.

In Figure 6-31b however, the intermittent cloud cover forces the solar generation to fluctuate wildly. Thus greatly fluctuating power output is sent into the power grid which is supposed to absorb and distribute it efficiently and without variations. Grid power cannot fluctuate—period!

This is not easily done, especially if we are considering the output from a large-scale solar power plant (over 100 MWp). Worse, several such large plants in different areas of the country, where climate and weather conditions are drastically different, will introduce wild variations of power levels and quality sent into the grid from different locations at different times.

These up and down fluctuations could be significant, with 100 MW power flowing on and off in the grid. This will stress the local grid and challenge the grid operators' ability to maintain steady load conditions. This issue is of great concern today, and since we have no way of making the sun shine consistently, we must concentrate on providing level power output. The smart grid concept and other innovations, such as energy storage and backup, must resolve the variability issues soon, if we are to consider PV a reliable power source. With many large solar power plants coming on line, it is becoming more important and even urgent to do that.

Combined Wind-solar Peak Load Considerations

Solar energy generation could be combined with other energy sources when constant output is the goal. Addition of solar power to conventional power sources (power plants) is one approach. Using energy storage

devices (batteries or water storage) is another. A more natural and most efficient way, however, is combining PV power with wind at locations especially chosen for this purpose. At such locations PV power is complementary to the output of wind generation, since it is usually produced during the peak load hours when wind energy production may not be available. Variability around the average demand values for the individual characteristic wind and solar resources can fluctuate significantly on a daily basis. However, as illustrated by Figure 6-32, solar and wind plant profiles—when considered in aggregate—can be a good match to the load profile and hence improve the resulting composite capacity value for variable generation.

In this example, the average load (upper line) is closely followed during the day by the average output from the combined wind and solar generators (the second from top line) during the same time. This average is created regardless and because of the fluctuations of the individual wind and solar power generators. This is a marriage made in heaven, and this combined power generating combination will work quite well if wind and solar power outputs can be matched as closely as this one.

Although there are areas in the US and abroad that match this wind and solar profile, the combined effort is usually hard to execute, because the best places for wind and solar are at different locations, often miles apart, and also because of lack of infrastructure at the most suitable locations.

Truthfully, it will require a great effort and expense to implement large-scale "wind-solar load matching" schemes anytime soon with existing technologies. Still,

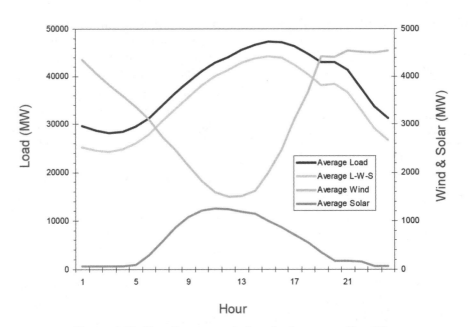

Figure 6-32. Simultaneous wind and solar generation (9)

having as a goal matching wind and solar power outputs will force us to seek the most suitable locations and appropriate technologies for this match.

This won't happen overnight, but if we approach the solution intelligently, we will have a large-scale power output—nationally—that matches the grid power loads.

Wind Power Issues

The implementation and scale-up of a) wind power fields, and b) storage capacities in tandem with solar energy generation, are presently the most important factors for the fast growth of wind power in large-scale deployment, and will even affect its long-term survival.

The wind industry is going through some serious changes, with a number of patterns emerging lately, basically leading to several major manufacturers and developers leaving the U.S. for greener horizons. This is mostly due to Congress' decision to not extend the production tax credit (PTC).

First Wind, based in Boston, borrowed $361 million from Canadian utility Emera Inc. for 49% ownership of Northeast Wind Partners. This partnership would handle First Wind's eight-project, three-state, 385-megawatt northeastern business. In contrast to the US, in Canada renewable energy retains big mandates and incentives, so First Wind expects the new partnership to lead to $3 billion in future investment and 1.2 gigawatts of new wind.

EDP Renewables North America is the second biggest U.S. wind developer after NextEra Energy, and reportedly wants to sell 707 megawatts of operating wind projects and a 1.4-gigawatt development pipeline to Spanish utility Iberdrola Renovables who is looking into U.S. expansion.

The town of Gillett, Wisconsin, population 1,256, lost 45 jobs when Wausaukee Composites, a plastic and fiberglass wind turbine component maker, closed its Gillette factory. Many small businesses in wind's supply chain across the U.S. are following suit.

GE Energy made recently deals in Turkey, Canada and Brazil, and CEO Vic Abate called Europe, Canada, China, Brazil and India "the growth markets of the immediate future." So long, USA…

The situation in China is also critical since the 2011 new wind industry guidelines mandated that Chinese government support for turbine manufacturers would be restricted to those making 2.5 MW or larger machines with up-to-date, transmission-ready technology. The guidelines set off a still-unfolding consolidation in the Chinese wind industry.

Some 80% of China's wind makers may eventually fall or be absorbed. The consolidation is driving heightened competition. Large world class wind power equipment manufacturers (see Table 6-8 below) are trying to penetrate the Chinese energy market, but that won't be easy.

Vestas, burdened by stalled investments in the U.S. while Congress fumbles with the PTC and by European financial crisis-imposed credit limitations, will close its 850-kilowatt turbine manufacturing facility in China's Inner Mongolia Autonomous Region.

Table 6-8. World-wide wind market share

Company	Market share
Vestas, Denmark	12.9%
Goldwind, China	9.4%
GE Wind, USA	8.8%
Gamesa, Spain	8.2%
Enercon, Germany	7.9%
Suzlon, India	7.7%
Sinovel, China	7.3%
United Power, China	7.1%
Siemens Wind Power, Denmark	6.3%
ROW	21.5%
Total sales world wide	40 GW

The increasingly hard push toward bigger and more transmission-friendly turbines was likely one of the factors that drove Sinovel, one of China's wind manufacturing leaders, to pirate software and electronics from American Superconductor (AMSC). It remains to be seen whether AMSC will obtain restitution in China's legal system.

Goldwind, China's second biggest turbine maker, has been as tight-lipped as always about the motives behind its sell-off of 50 percent of its interest in subsidiary Shangdu Tianrun, a development arm. Proceeds appear to have been channeled into a fund that likely will give Goldwind a wider and therefore more hedged investment position.

Subsidiary Goldwind USA has been focusing as much attention in Latin America as in North America over the last few months.   Haizhuang, which owns barely 1.5 percent of China's domestic market, appears to be reacting to the increased competition by taking its efforts to friendlier markets. It just announced a €400 million expansion into eastern Europe in partnership with Chongqing Foreign Trade Group (CTF). Its two-megawatt turbine may eventually be able to compete in the Chinese domestic market. Or Haizhuang, currently in the process of developing a five-megawatt machine, may be biding its time in the incentive-driven eastern European market where slightly smaller turbines are still economic.

The wind industry and its backers are also becoming interested in India's wind market. Researchers at Lawrence Berkeley National Labs recently reported that smaller, lower wind regime turbines make India's undeveloped potential far greater than previously estimated.

Indian wind powerhouse Suzlon sold Chinese subsidiary Suzlon Energy Tianjin to a domestic developer for $60 million. Though there is no clear word on how it will reassign the assets, this obviously opens up the opportunity for more work in India. And Morgan Stanley Infrastructure Partners just spent $210.1 million for a controlling interest in India wind project developer Continuum Wind Energy.

Wind or no wind, solar power generation is plagued by variability issues, which need to be resolved quickly. The world Smart Grid comes to the rescue.

The Smart Grid

The conventional wisdom calls the existing power grid "old, outdated and dumb." It has served us well for almost a century, and it's outdated. But dumb? No. It has enough intelligence to navigate complex up-and-down supply-demand cycles.

Still, billions of dollars are spent in making the grid "smart." What does that mean? Let's take a close look at the proposed smart grid.

One of the most misunderstood concepts in today's energy industry is the new "smart grid." The concept appeared long ago, but first was publically discussed in a 2005 article, "Toward A Smart Grid," by Amin and Wollenberg. The ideas expressed in that article are forward thinking and somewhat revolutionary, but the industry is developing so fast that we might be pleasantly surprised by smart grid technologies winning, or at least leading, the energy revolution.

One major characteristic of this concept is the application of digital processing and communications to the power grid, making data flow and information management essential to smart grid structure and operation. In other words, digital technology would be integrated with power grids, and the new grid's information integrated into utility processes and systems.

Utilities are forced to consider changes in their *modus operandi*, starting with improvement of their infrastructure, adding a digital layer of controls (the foundation of the smart grid), and including business aspects to grid operation, as needed to capitalize on the investments in smart technology.

New, smart substations and distribution automation are part of the smart grid concept, with more possibilities popping up daily.

Generally, the smart grid is a unified response to the challenges of present-day electrical power generation and distribution.

There are no rules or standards covering the new smart grid effort, but a number of unifying factors could be considered.

New technologies applied in the new smart grid will improve grid operation with fault detection and self-healing controls. This will be done without human intervention, ensuring reliable power supply and reduced vulnerability to natural disasters or enemy attacks.

Today's power grid is uni-directional, since the flow of electricity can go only in one direction. If a sub-network generates more power than it is consuming, the reverse flow can raise safety and reliability issues. A smart grid aims to manage these situations, and will be able to handle bi-directional electricity flows. This will allow more flexibility for distributed generation from residential and commercial PV installations and other power generating sources, as well as for use of fuel cells (when charging to/from the batteries of electric cars.)

The smart grid will improve the efficiency of the energy infrastructure mainly by demand-side communications and control. Appliances and machinery will communicate and will be able to determine when is a good time to start operation. They will be able to turn off during spikes in electricity price and/or demand. This means less redundancy in transmission lines, and greater utilization of generators, thus leading to lower power prices.

A smart grid will also reduce a large load temporarily, to allow time to generate more power. It can also continuously control the load as needed to avoid spikes. It will also use prediction algorithms to predict how many standby generators need to be used and control them to provide constant supply/demand ratio.

The smart grid will communicate with smart devices in homes and business, and will track when and how much electricity is used. This will give the utilities a chance to control power generation levels, to prevent system overloads.

This can lead to optimizing peak-level pricing, because smart devices will control when and how much power to use, thus consuming less power during high demand periods. It will also reduce the amount of stand-by generators maintained by the utilities, because the load will be kept level. Consumers will get paid a portion of the peak power saved by turning their devices off.

The smart grid will allow generators and loads to communicate and interact automatically and in real time, thus coordinating the demand level to flatten spikes. If the demand fraction is reduced or eliminated, then the cost of adding reserve generators is reduced or eliminated as well. This in turn, reduces wear and tear and extends the life of the power generating equipment. It also allows users to reduce their energy bills by switching to electricity when it is cheapest.

The smart grid will allow greater use of variable renewable energy sources such as solar and wind power, even without the addition of energy storage—which is impossible at present, since the grid cannot handle many distributed feed-in points and power fluctuations, due to cloudy or gusty weather. Smart grid technology, therefore, is a necessary condition for introducing very large amounts of renewable electricity on the grid.

For utilities, smart grid means taking the risk out of distribution, which is essential and will be key to moving the industry ahead at a pace consistent with consumer expectations. But what's first?

Among industries that have traveled a similar path, the wireless industry is a good example. Wireless companies have improved the technology very quickly and successfully. Starting with a drastic and very complex analog to digital transformation, followed by a business model change from single play (monthly data read) to quad play (data read every second or so)—and all this within the last 15-20 years. Basically, 15-20 years ago we could not even dare think of multi-tasking on hand-held devices and GPS tracking to within an inch of accuracy.

Equally so, we cannot begin to imagine the intricacies and efficiency of power generation, transmission and use under the full control of smart grid technologies. But it is coming—slowly but surely it is part of our lives and we will eventually embrace it whole heartedly, just as we did wireless technologies.

Although there are some differences, such as technological complexity, size, competition and regulations, the lessons to learn from the wireless industry achievements lately are insightful for utility providers, who are now where the wireless industry was 15-20 years back.

Just imagine a national power grid that is smart enough to analyze and control itself. Today's marvel of large-scale technology is getting old—physically and technologically—so an upgrade is needed. The smart grid might provide the solutions sooner than we think possible.

Recently, the Obama administration pledged $3.4 billion toward "smart grid" technology development and implementation. The first step (or as they call it, "the next generation") of infrastructure is geared to stabilize the grid in the event of failures. Then, new green technologies are to be incorporated, the sum of which is thought to vastly improve the overall efficiency.

It's a good first step, but far from bringing the entire national grid into the 21st century, for the efficiency and cost of electrical power to be optimized. A lot of upgrade work is needed for the old power network to meet the new smart grid standards. Adding the smart grid components for total grid automation, optimization and safety, we estimate that over $250 billion, will be needed to bring the outdated national grid into the smart grid era.

But it can be done! The smart grid is here to stay and grow. So what is stopping it?

Coordination

As with any new technologies, the smart grid is now fragmented into different areas, where companies are rushing in and suggesting partial solutions, which may, or may not survive the test of time. A unified approach would be best in this case, so we'd like to suggest that what is needed most right now is to collect and sort the available information, to understand and evaluate the situation best. A practical approach for the short- and long-term design and implementation of different smart grid concepts is needed to complete the planning stage.

Regulations

The next step is to change the regulations. Now is the time for government officials and regulators to create regulatory mechanisms that ensure increased utility investment in cost-effective demand response technologies. Established regulations have been favoring supply-side resources over energy consumption and optimization, and utilities have been encouraged to add new generation in response to demand. And they are willing to do that, because they profit from such investments in their generation, transmission and distribution infrastructure.

Problems

The new smart grid concept, however, creates a conflict of interest for the utilities, so while they show a desire to promote environmentally friendly activities, they must at the same time maximize profits for their shareholders. So, the utilities talk the "green" talk, but walk the established dirty (coal and gas) walk.

Because energy conservation reduces infrastructure investments and consumption, the utilities are not interested. They will always prefer the option to build a power plant that will contribute to profitability, over investing in energy conservation.

Also, the utilities' revenue is directly coupled to electricity usage; the more electricity utility customers use, the more revenue a utility earns, so because energy conservation reduces customer usage, it reduces a utility's revenues.

The most prominent example of this divide between technology and regulation is in France, where the smart grid technology is seriously disadvantaged.

Solutions

We cannot implement a successful smart grid program without the participation of the utilities. Lately, the utilities have taken a "non-resistance" approach, where they show willingness to participate in any "green" program, knowing darn well that, in most cases, they lose money in the process. So they have developed strategies and mass information models to portray themselves as the "good guys," while at the same time continuing to operate in the old ways.

We need to encourage utilities to participate in the smart grid effort, not just with lip-service and misinformation (which some have perfected lately), but with actual deeds:

a. Rewarding utilities for improved efficiency, using some of the savings to reward them for their participation (and potential losses) instead of passing all savings to the customers.

b. Real-time and multi-tariff energy pricing must be encouraged, while at the same time providing adequate consumer protection.

c. The cost and effectiveness of demand response projects and programs must be kept transparent.

d. Information-sharing models must be optimized, so utilities can choose distribution and metering technologies that best fit their needs, again, assuring that consumers are not disadvantaged.

The intelligent transportation sector is evolving due to maturing communications technologies being adopted globally in major cities, along with the ability to send and receive real-time vehicle or infrastructure information.

The market for smart transportation systems, including smart charging for plug-in electric vehicles (EV) and vehicle-to-vehicle systems will continue to grow despite public infrastructure spending flattening out or declining, according to researchers.

The intelligent transportation sector is evolving due to maturing communications technologies being adopted globally in major cities, along with the ability to send and receive real-time vehicle or infrastructure information.

Governments may be looking to reduce their debts, but it appears that funding for Intelligent Transportation Systems will not be sacrificed.

Global investment in four key areas—traffic management systems, public transportation systems, vehicle-to-vehicle systems and smart charging for plug-in electric vehicles—is expected to reach $13.1 billion between 2011 and 2017. Indeed annual spending is expected to surge from $770 million in 2011 to $3 billion in 2017.

The research suggests that the intelligent transportation systems are being implemented with several goals in mind: to enhance mobility, reduce fuel consumption, reduce emissions, improve safety, and increase economic competitiveness.

In general, during the last 2-3 years world governments have provided over $120 billion for smart grid projects to be implemented by 2020. This includes the deploying of 750 million smart meters and 30 million electric vehicles.

Preliminary research shows that by the summer of 2012, there are 267 smart grid projects operating, or underway, worldwide. This is a big business, not unnoticed by industry leaders and governments.

The US government is not far behind in this endeavor. The U.S. Department of Energy through its Distribution Automation in ARRA Funded Smart Grid Investment program has invested close to $8.0 billion in smart grid projects. $3.5 billion of this money was, or is, in the form of direct federal assistance. Grants ranging from $500,000 to $200 million for smart grid technology utility deployments, including smart metering projects were distributed to 99 participants. The program will end in 2015, at which point DOE will decide how to proceed with its involvement in smart grid programs.

DOE's analyses of the smart grid program progress are based on comparing installed assets vs. benefits in $, as follow.

The assets are divided into four categories:
- Advanced Metering Infrastructure
- Customer Systems
- Distribution
- Transmission

The benefits are divided into following categories:
- Economic Benefits
- Peak demand reduction > asset utilization
- Operational efficiencies > reduced O&M
- Energy efficiency > reduced energy needs
- Reliability Benefits
- Environmental Benefits

From this effort (and investment) we get:
- 15.5 million smart meters (commercial and residential)
- 6,500 distribution circuits equipped with automated equipment
- 800 networked phasor measurement units, and of course
- Benefit of more efficient and reliable power grid (the ultimate goal)

A key benefit that can be derived from smart meters is the ability to create residential demand response programs that allow consumers and utilities to better manage electricity usage, as well as to reduce costs and the negative impact on the environment.

For example, Progress Energy was awarded a $200 million grant by the U.S. Department of Energy (DOE) as part of the 2009 federal stimulus package. The funds will be used by the company's two utilities for investments in equipment, systems, technologies and communication infrastructure that will add "intelligence" to the systems used to deliver electricity to more than 3.1 million households and businesses in the Carolinas and Florida.

In addition to providing the utilities real-time information about the state of their electric grids, the smart grid transition will enable customers to better understand and manage their energy use, and will provide for more efficient integration of renewable energy resources.

"The way our customers use electricity is changing," said Bill Johnson, Progress Energy chairman, president and CEO. "At Progress Energy, we are also changing the way we produce and deliver the power they need to run their daily lives. Our investments in the grid will allow customers greater transparency and control over their energy use, and they will allow our utilities to expand our use of alternative technologies, such as renewable energy and advanced transportation. This is just one of the ways we are increasing the quality and reliability of the essential service we provide."

A number of partnerships, based on developing smart grid solutions have been observed lately as well, and are most likely the winning approach of the future development of this technology.

For example, Energate, a leading provider of Demand Response and home energy management solutions for the utility industry, entered into a strategic partnership with Silver Spring Networks, a leading Smart Grid solutions provider, to offer a complete Demand Response solution to the utilities.

Through this partnership, Silver Spring will expand its Demand Response solution to offer elements of Energate's Consumer Connected Demand Response (CCDR) product set, including smart thermostats, load control switches, gateways, and software. The combined solution incorporates Silver Spring's existing Advanced Metering, UtilityIQ and CustomerIQ applications, providing a powerful solution set for utilities of all sizes.

"As utilities develop and deploy their Smart Grid solutions, demand response programs play a critical role, through direct load control and time of use pricing," said Judy Lin, Chief Product Officer, Silver Spring Networks. "We are delighted to formalize a relationship with Energate, a leader in demand response solutions for the Smart Grid, and help more utilities empower more consumers to realize the potential of the Smart Grid."

"We are pleased to strengthen the relationship with Silver Spring Networks. This partnership provides an end-to-end Demand Response solution for utilities," said Niraj Bhargava, Chief Executive Officer of Energate. "Silver Spring has emerged as a Smart Grid leader, and with this agreement, the company is now able to clearly deliver a complete Demand Response to their utility customers."

But there are problems too.

In May 2012, Illinois regulators slapped down big utilities' smart grid spending plans. Ameren utility was affected first, and then Commonwealth Edison (ComEd).

The Illinois Commerce Commission cut $146 million from the rates ComEd expects to collect from customers in 2012. Or, four times more than the $40 million the utility had targeted in cost reductions in the February, 2012 filing.

It is not clear how this will affect ComEd's $2.6 billion in smart meter and distribution automation projects, but the plan is to deploy about 4 million smart meters over the next decade, with Silver Spring Networks as a key partner. This goes along with developing new substations, and distribution grid sensors and controls.

ComEd and Ameren are both trying to prove that their plans for a combined $3.2 billion in smart grid spending are worth it to their customers, as per an Illinois state law passed in December. So far, it's not looking so good for 2012, so ComEd has filed for a 2013 rate increase that would cover losses this year.

Another development during the summer of 2012 points to the most serious vulnerability of solar power generation—its variability—which the smart grid will control in the future, but which is not controlled at pres-

ent. An example of things to come is the recent development in Italy, where the major transmission grid operator and electric power manager, Terna S.p.A., responsible for developing the national electricity transmission grid, has received authorization to shut off distributed generation systems in case of emergency.

So Terna can shut off distributed generation systems up to 100 kW in size if and when they create instability in the grid. Terna will basically disconnect the solar generators anytime it considers it important to ensure the stability of the national electricity grid.

Terna insists that the new disconnection procedure will enable the grid operator to limit the amount of distributed generation on the Italian grid during periods when supply outstrips demand, for example on high irradiation days, during days when the solar power generation is erratic, and/or anytime the grid goes into a fluctuation or other unstable state.

The smart grid will be able to take care of such problems more efficiently than the manually operated Terna power switching network, but that will not happen soon. So the smart grid is needed, but it is still in its design stages, and will be going through serious growing pains for awhile longer.

Author's note: It seems to us that the smart grid efforts in the US are fragmented, without much standardization or central coordination. It is almost as if we are trying to fix an old car with a failing engine by changing the tires and painting the body, but are afraid to open the hood to see the real problem.

We're putting lots of work into a car that needs a total overhaul, not just a face-lift. Similarly, the old US power grid has a failing engine, so we wonder how much good the shiny new gadgets will do to fix that.

Nevertheless, a new smart grid (or better yet, a new power grid with smart components) seems reasonable and inevitable, so only time, effort and money separates us from full deployment in the 21st century. But we need to mention again the large amount of money and effort that are needed for the smart grid (or the new power grid) dream to become reality.

Author's note: A proven benefit, arising from smart grid technologies is the demand response (DR) industry, which provides a predictable, fast-responding, highly reliable and cost-effective solution for energy conservation. DR helps grid operators and utilities improve grid reliability, while at the same time continuing to meet increasing energy demand. Equally importantly, it benefits the environment by reducing the need for new peaking power plants.

As with most new energy initiatives, however, the

utilities are fighting to stop DR's expansion, since it poses a threat to them—it interferes with the peak demand times which drive their profits.

Regulators are working to improve DR, including pro- conservation legislations as Pennsylvania Act 129, which requires utilities to develop plans to reduce electricity consumption by 3%, and reach a 4.5% reduction in peak demand, by 2013.

FERC Order 745 requires that DR be compensated at the same rate as generation in the organized energy markets, while Order 755 finds that there are services and technologies on the market, other than generation assets, that may better regulate frequency. The order also mandates for the organized markets to provide compensation, starting with the value of the frequency regulation for use of these methods.

DR services revenues are expected to increase from less than $1.3 billion in 2011 to more than $6.1 billion by 2016, with the curtailment services segment continuing to be the largest of the three key industry segments. The other two segments are systems integration/consulting services and outsourcing services.

Energy conservation with DR at the helm offers one sure solution to our energy problems. DR delivers real megawatts of power reduction, amounting to tangible decreases in energy consumption, so with some luck and help from the regulators, we just might see DR in the US becoming a prime example of how best to leverage this disruptive technology, to bring efficient utility electricity delivery systems into the 21st century—without increasing power generation levels.

Demand Response Standards

Speaking of standards, the OpenADR (automated demand response) standard is an example of things to come. It was developed by Lawrence Berkley National Lab and promises to bring a level of certainty to the various areas of power use, grid operation stability and better utilities controls across the US and (much later) worldwide.

Customers and systems operators alike would be able to control power use more efficiently than today. Under ideal conditions, automated signals would allow homes and businesses to respond seamlessly to changes in the grid load and the related dynamic pricing of power during different times day.

Demand response is evolving on all fronts, including the markets in which it is authorized to compete and the ways it is delivered between facilities and the grid.

In its present form, Open ADR is comprised of three separate communications standards, OpenADR A, B and C, which specify communication to simple devices (thermostats, residential gateways), building automation systems and communications, back to ISO and aggregators for more complex market interactions, respectively. It does not track whether load is actually dropped, nor does it continuously optimize the load based on weather.

But even with just a communication standard, there will be plenty of progress due to significant incentives from utilities. And PUC will assist building owners to update their controls to allow them to participate in these programs, if they ever see the benefit of it.

Some experts argue that OpenADR will not make much of a difference in the commercial area, because only a small proportion of the structures in the commercial market have a building management system, and those with the largest industrial loads, as well as those with sensitive loads (such as government buildings), tend to want to remain in control. The majority of the power use for DR applications is in industrial and manufacturing sectors, but there is a little chance for these to agree to be fully automated.

Some companies, like Powerit Solutions are already using OpenADR to do just that. Nevertheless, OpenADR has a long way to go before being universally accepted, let alone implemented. Even if all goes well, it most likely won't be the panacea that many would like it to be.

EnerNOC, one of the program supporters, actively questions the extent of the standard's impact in the market. "The major limitation is the customer's willingness to automate their facility," according to them. "Auto DR has its place, but you have to be very realistic about how it's applied and its potential in the market."

Even though most of that load could be dimming lights or tweaking HVAC, if huge markets could be unlocked outside of the commercial load, the peak shaving could be significant. But that is still a big 'if,' as automation is just one piece of unlocking those markets.

The standard is in its infancy (as is the alternative energy industry) so there is much more work to be done, especially in confirming load shed; nevertheless, it could help to expand the conversation with regulators and grid operators about demand response and its role.

Ultimately, although OpenADR would be able to efficiently manage larger capacity networks, its future will depend on how successfully it adapts to the other aspects of the energy markets and especially the technological changes in the energy management of buildings and networks. This case highlights the immaturity and fragmentation of the industry.

Energy Storage

Energy storage is what prevents solar power from getting deployed in large-scale power fields. Presently solar power generation causes grid operators trouble with its huge power fluctuations and up-and-down ramps, so energy storage is badly needed, and is sorely lacking.

Due to its variability, solar (and wind) power generated during some periods of the day or night, must be stored for use at other times, thus leveling their output and reducing the variability.

There are a lot of opportunities for energy storage, but nothing concrete to help that market mature. The question everyone asks is, "Why would one invest in an energy storage facility (battery storage for example), if one can get those services—net metering and balancing—at no cost from the utility?" Why would one?

There are many different answers to the issues, but none is good enough, including the fact that energy storage is still a dream. Large-scale energy storage is especially impractical today, and there is no way to predict when it will become practical… if ever.

So energy storage is the big question, and utilities are seriously concerned with the variability of solar power plants. Years from now the major topic of discussion no doubt will be energy storage.

Solar has a sure advantage during the day, because while wind is most productive at night when not much power is needed, solar has the bulk of its production in or around the middle of the day or a little later, when power is worth more in the daytime. This provides some grid and functional flexibility. For example, San Diego's peak load is moving to later in the day, when solar without storage is available as a lower value resource to the utility.

Solar can be produced close to major load centers, which makes the potential for distributed solar extremely high. It can be installed on rooftops and parking lots and football fields and unused lands—in small or large quantities. It will compete with the conventional energy sources of the future, through its advanced technologies, which can be found in different types, shapes and sizes. It could be incorporated in building materials, road surfaces, clothing and virtually everywhere.

We are expecting a day will come in the not so distant future, when:

1.   Using batteries and other energy storage devices will be cheaper than building peaking plants, and

2.   Using solar power will be cheaper and safer than building a combined-cycle unit.

When this day comes, these innovations will be hugely transformative and disruptive, and will call for redefining the role of the utilities, the grid and the distribution system. Until then, however, we have a lot of work to do on the improvement and the implementation of 1 and 2 above.

Types of Energy Storage

We've seen that variability of PV power plants' output is unavoidable with present-day technologies and energy storage solutions (or lack thereof). PV power might be combined with other energy sources (i.e., wind, or geothermal) to smooth the variations and even match peak load, but even that is not a complete solution.

During cloudy or rainy days, output will be limited and the only way to rectify that is to provide energy storage as a supplemental power source. Stored energy can be used during periods of low energy generation or at night.

There are a number of potential energy storage solutions for use with solar power plants, some of which are applicable for PV power storage too.

See below a complete list of available energy storage technologies, followed by a discussion of the most promising for use today:

Thermal storage
    Steam accumulator
    Molten salt
    Cryogenic liquid air or nitrogen
    Seasonal thermal store
    Solar pond
    Hot bricks
    Fireless locomotive
    Eutectic system
    Ice storage

Electrochemical storage
    Batteries
    Flow batteries
    Fuel cells
    Electrical
    Capacitor
    Supercapacitor
    Superconducting magnetic energy storage (SMES)

Mechanical storage
    Compressed air energy storage (CAES)
    Flywheel energy storage
    Hydraulic accumulator

Hydroelectric energy storage
Spring
Gravitational potential energy (device)

Chemical storage
Hydrogen
Biofuels
Liquid nitrogen
Oxyhydrogen and Hydrogen peroxide

Biological storage
Starch
Glycogen

Electric grid storage

Carbon storage

We will review here only the most suitable for CSP and PV applications storage options.

Thermal Energy Storage

Presently, this is the most widely used method of energy storage in thermal solar (CSP) plants. Heat is transferred to a thermal storage medium in an insulated reservoir during the day, and withdrawn for power generation at night. Thermal storage media include pressurized steam, concrete, a variety of phase change materials, and molten salts such as sodium and potassium nitrate.

The most widely used heat transfer liquids today are:

1. *Pressurized steam energy storage.* Some thermal solar power plants store heat generated during the day in high-pressure tanks as pressurized steam at 50 bar and 285°C. The steam condenses and flashes back to steam, when pressure is lowered. Storage time is short—maximum one hour. Longer storage time is theoretically possible, but has not yet been proven.

2. *Molten salt energy storage.* A variety of fluids can be used as energy storage vehicles, including water, air, oil, and sodium, but molten salt is considered the best, mostly because it is liquid at atmospheric pressure, it provides an efficient, low-cost medium for thermal energy storage, and its operating temperatures are compatible with today's high-pressure and high-temperature steam turbines. It is also non-flammable and nontoxic, and since it is widely used in other industries, its behavior is well understood and the price is cheap.

Molten salt is a mixture of 60% sodium nitrate and 40% potassium nitrate. The mixture melts at 220°C, and is kept liquid at 290°C (550°F) in insulated storage tanks for several hours. It is used in periods of cloudy weather or at night using the stored thermal energy in the molten salt tank to generate steam and turn a turbine, which in turn generates electricity. These turbines are well established technology and are relatively cheap to install and operate.

3. *Pumped heat storage.* Pumped heat storage systems are used in CSP power plants and consist of two tanks (hot and cold) connected by transfer pipes with a heat pump in between performing the cold-to-heat conversion and transfer cycles. Electrical energy generated by the PV power plant is used to drive the heat pump with the working gas flowing from the cold to hot tanks. The gas is heated and pumped into the hot tank (+50°C) for storage and use at a later time. The hot tank is filled with solids (heat absorbing materials), where the contained heat energy can be kept at high temperature for long periods of time.

The heat stored in the hot tank can be converted back to electricity by pumping it through the heat pump and storing it back in the cold tank. The heat pump recovers the stored energy by reversing the process.

Some power (20-30%) is wasted for driving the heat pump and during the transfer and conversion cycles, but the technology can be optimized for use in large-scale PV plants.

In all cases, large heat energy losses accompany the energy storage process. At least one third of the energy is lost during the conversion of stored heat energy into electricity. More is lost during the storage and following cooling cycles.

Battery Energy Storage

No doubt, this is the most direct and efficient way to store a large amount of electricity generated by PV power plants. The generated DC electric energy is stored as DC power in batteries for later use.

There are several types of batteries, the most commonly used as follow:

Lead Acid Batteries

These are the most common type of rechargeable batteries in use today. Each battery consists of several electrolytic cells, where each cell contains electrodes of elemental lead (Pb) and lead oxide (PbO_2) in an electrolyte of approximately 33.5% sulfuric acid (H_2SO_4). In the discharged state both electrodes turn into lead sulfate ($PbSO_4$), while the electrolyte loses its dissolved sulfuric acid and becomes primarily water. During the charging cycle, this process is reversed.

These batteries last a long time, and can go through many charge-discharge cycles, if properly used and maintained. They are affected, however, by high temperatures, when the electrolyte can boil off and destroy the battery. Since there is water in the cells, the electrolyte can freeze during winter weather, which could destroy the battery as well.

Lithium Batteries

These are a mature technology, having been used widely for a long time in consumer electronics. They are actually a family of different batteries, containing many types of cathodes and electrolytes. The most common type of lithium cell used in consumer applications uses metallic lithium as anode and manganese dioxide as cathode, with a salt of lithium dissolved in an organic solvent. A large model of these can be used to store large amounts of electric power generated by a PV power plant, and due to their highest known power density they could be quite efficient—70-85%.

They are suitable for smaller PV installations, too, because scaling up to large PV plants would be a very expensive proposition.

Sodium Sulphur Batteries

These are high temperature, molten metal batteries constructed from sodium (Na) and sulfur (S). They have a high energy density, high efficiency of charge/discharge (89–92%) and long life cycle. They are also usually made of inexpensive materials, and due to the high operating temperatures of 300-350°C they are quite suitable for large-scale, grid energy storage.

During the discharge phase, molten elemental sodium at the core serves as the anode, and donates electrons to the external circuit. The sulfur is absorbed in a carbon sponge around the sodium core and Na+ ions migrate to the sulfur container. These electrons drive an electric current through the molten sodium to the contact, through the electric load, and back to the sulfur container.

During the charging phase, the reverse process takes place. Once running, the heat produced by charging and discharging cycles is sufficient to maintain operating temperatures and usually no external source is required.

There are, however, a number of safety and corrosion problems, due to the sodium reactivity, which need to be resolved before full implementation of this technology takes place.

Vanadium Redox Batteries

These are liquid energy sources, where different chemicals are stored in two tanks and pumped through electrochemical cells. Depending on the voltage supplied, the energy carriers are electrochemically charged or discharged. Charge controllers and inverters are used to control the process and to interface with the electrical source of energy.

Unlike conventional batteries, the redox-flow cell stores energy in the solutions, so that the capacity of the system is determined by the size of the electrolyte tanks, while the system power is determined by the size of the cell stacks. The redox-flow cell is therefore more like a rechargeable fuel cell than a battery. This makes it suitable as an efficient energy storage for PV installations.

A number of additional types of batteries are under development, and some show great potential for use in larger PV installations in the near future. Most batteries have problems with moisture, high temperature, memory effect, and use of scarce and toxic materials, all of which cause longevity problems and abnormally high prices. If and when all these problems are resolved, the energy storage problems of PV power plants will be resolved as well.

Compressed Air Energy Storage

Compressing air into large high-pressure tanks is one of the most discussed and most promising energy storage methods for use with PV power plants today. It is a simple way of energy storage, using a compressor powered by the electricity produced by the PV plant compressing air into the storage tank. A lot of energy is lost by activating the compressor, and heat is wasted during the compression process, so there are several compressing methods that treat generated heat so as to optimize conversion efficiency:

Adiabatic Storage

Adiabatic storage retains the heat produced by compression via special heat exchangers, and returns it to the compressed air when the air is expanded to generate power. Its overall efficiency is in the 70-80% range, with the heat stored in a fluid such as hot oil (300°C) or molten salt solutions (600°C).

Diabatic Storage

Here extra heat is dissipated into the atmosphere as waste, thus losing a significant portion of the generated energy. Upon removal from storage, the air must be re-heated prior to expansion in the turbines, which requires extra energy as well. The lost and added heat cycles lower the efficiency, but simplify the approach, so it is the only one used commercially today. Its overall efficiency is 50-60%.

Isothermal compression and expansion

This method attempts to maintain constant operating temperature by constant heat exchange to the environment. This is only practical for small power plants, which don't require very effective heat exchangers, and although this method is theoretically 100% efficient, this is impossible to achieve in practice, because losses are unavoidable.

There are a large number of other methods using compressed air, such as pumping air into large bags in the depths of lakes and oceans, where the water pressure is used instead of large pressure vessels. Pumping air into large underground caverns is another approach that is receiving a lot of attention lately.

Pumped Hydro Energy Storage

Pumped hydro energy is a variation of the old hydroelectric power generation method used worldwide, and is used quite successfully by some power plants. Energy is stored in the form of water, pumped from a lower elevation reservoir to one at a higher elevation. This way, low-cost off-peak electric power from the PV power plant can be used to run the pumps for elevating the water. Stored water is released through turbines and the generated electric power is sold during periods of high electrical demand. This way the energy losses during the pumping process are recovered by selling more electricity during peak hours at a higher price.

This method provides the largest capacity of grid energy storage—limited only by the available land and size of the storage ponds.

Flywheel Energy Storage

Flywheel energy storage works by using the electricity produced by the PV power plant to power an electric motor, which in turn rotates a flywheel to a high speed, thus converting the electric energy to, and maintaining the energy balance of the system as, rotational energy. Over time, energy is extracted from the system and the flywheel's rotational speed is reduced. In reverse, adding energy to the system results in a corresponding increase in the speed of the flywheel. Most FES systems use electricity to accelerate and decelerate the flywheel, but devices that use mechanical energy directly are being developed as well.

Advanced FES systems have rotors made of high-strength carbon filaments, suspended by magnetic bearings, and spinning at speeds from 20,000 to over 50,000 rpm in a vacuum enclosure. Such flywheels can come up to speed in a matter of minutes—much faster than some other forms of energy storage.

Flywheels are not affected by temperature changes, nor do they suffer from memory effect. By a simple measurement of the rotation speed it is possible to know the exact amount of energy stored. One of the problems with flywheels is the tensile strength of the material used for the rotor. When the tensile strength of a flywheel is exceeded, the flywheel will shatter, which is a big safety problem. Energy storage time is another issue, since flywheels using mechanical bearings can lose 20-50% of their energy in 2 hours.

Flywheels with magnetic bearings and high vacuum, however, can maintain 97% mechanical efficiency, but their price is correspondingly higher.

Electric Grid Storage

Grid energy storage is large-scale storage of electrical energy, using the resources of the national electric grid, which allows energy producers to send excess electricity over the electricity transmission grid to temporary electricity storage sites that become energy producers when electricity demand is greater. Grid energy storage is a very efficient storage method, playing an important role in leveling and matching electric power supply and demand over a 24-hour period.

There are several variations of this method, one of which is the proposed grid energy storage called the vehicle-to-grid, where modern electric vehicles that are plugged into the energy grid can release the stored electrical energy in their batteries back into the grid when needed. Far fetched, yes, but the future will demand many such ingenious approaches, if we are to be energy independent.

Carbon Capture and Storage

Carbon capture and storage technology (CCS), as well as its use and ramifications are (again) the talk of the town. CCS is a new technology, designed to prevent the release of large quantities of CO_2 into the atmosphere. CO_2 generated by fossil fuel burning in power generation and other industries is captured, transported, and pumped into underground geologic formations, or other large vessels.

This gas is securely stored away and kept from being released into the atmosphere, as one way of mitigating the negative contribution of fossil fuel emissions to global warming The CO_2 can be used later on in commercial applications, such as carbonated beverages.

CCS can also be used to describe the scrubbing of

CO_2 from ambient air as a geo-engineering technique. Although CO_2 has been injected into geological formations for various purposes, the long-term storage of large quantities of CO_2 by high pressure pumping is a relatively new and largely unproven concept.

CCS applied to a modern conventional power plant could reduce CO_2 emissions to the atmosphere by approximately 80-90% compared to a plant without CCS. The economic potential of CCS could be between 10% and 55% of the total carbon mitigation effort until the year 2100.

Large amounts of CO_2 can be stored either in deep geological formations, in deep ocean masses, or in the form of mineral carbonates. Deep ocean storage greatly risks increasing the problem of ocean acidification, an issue that also stems from the excess of carbon dioxide already in the atmosphere. Geological formations are currently considered the most promising sequestration sites. According to the National Energy Technology Laboratory (NETL) North America has enough storage capacity for more than 900 years worth of carbon dioxide at current production rates.

A general problem is that long-term predictions about submarine or underground storage security are very difficult and uncertain, and there is still the risk that CO_2 might leak into the atmosphere.

Capturing and compressing CO_2, however, might ease the fuel needs of a coal-fired CCS plant by 30%-40%. These costs will then increase significantly the cost of the energy produced by this method. Retrofitting existing power plants with the CCS technology would be even more expensive especially if they are far from the storage site.

Funding for the CSS technology and the related CCS projects around the world in this economic downturn is not easy. Some efforts have been delayed and some have stalled, according to industry watch dogs.

There are currently 75 large-scale, fully integrated CCS projects in 17 countries at various stages of development, but only eight are operational. And the status of the technology hasn't changed much since 2009.

Funding also remained unchanged at $23.5 billion in 2011 compared to 2010. The U.S. leads the funding of large-scale CCS projects, followed by the European Union and Canada. Although CCS technology has the potential to significantly reduce carbon dioxide emission when used in greenhouse gas-intensive coal plants, developing it to the point that it can make a serious contribution to emissions reduction will require some very hefty investments, which are just not available today.

The future of the CCS technology depends largely on the whim of politicians and regulations. The U.S. Environmental Protection Agency (EPA) recently imposed regulations on CO_2 emissions from power plants, so coal power plants will not be allowed to be built without CCS capabilities. This ruling does not apply to the hundreds of existing coal powered plants.

Therefore, the CCS technology will become increasingly important to comply with regulations, which will lead to a cleaner air in the distant future. For now, we must rely on the good will of coal power operators, politicians and regulators to do something about the old smoke stacks belching out thousands of tons of poisonous gases daily.

Conclusion

There are a number of other energy storage methods such as fuel cells, new types of batteries, superconducting devices, supercapacitors, hydrogen production, and many others under development, so the future looks bright in this area. In practice, however, there are a number of energy storage installations around the world, totaling over 2,100 MW, with the major technologies being:

1. Thermal energy storage over 1,140 MW
2. Batteries energy storage over 450 MW
3. Compressed air energy storage over 440 MW
4. Flywheels energy storage over 80 MW

Energy storage has many advantages, in addition to its unique potential to transform the electric utility industry by improving wind and solar power variability, availability and utilization. It can contribute to overall energy independence and environmental clean-up by avoiding the building of new power plants and transmission and distribution networks. Experts consider energy storage to be the solution to the electric power industry's issues of variability and availability, opening new opportunities for wind and solar power use.

Complexity, safety, price and other restraints, must overcome before any of the energy storage methods become acceptable for large-scale PV installations. The question is when; and don't hold your breath.

California Independent System Operator (ISO) officials say that energy storage is good stuff, but that it is very expensive, so it makes no economic sense for the short term (21-12-2013).

California ISO is in charge of balancing the electrical generation supply to meet the power demands of all California locations. They are familiar with the particular baseloads and the generation sources.

ISO seems enthusiastic about energy storage, but high cost and uncertainty of the regulatory environment will hinder energy storage technology for now. Still, it is not a matter of if, but when... when the technology comes down in price.

Energy storage plus renewables is a marriage made in heaven. Due to the variability of wind and solar we need storage to moderate and control the fluctuations. This is needed to reach the 33% renewable energy mandate set before California, but the economics of energy storage must be sorted out first, according to ISO officials.

ISO is not sure how the growing, but variable, wind and solar power injected into the grid will be balanced. A number of simulations were run for 2015-2020 with a deficit of 4,000 megawatts of ramping capability shortfall, which is partially due to the pending ban on once-through cooling in coastal power plants.

Only about 4 megawatts of energy storage is available on the California power grid now, with 10-14 MW of sodium sulfur and lithium-ion batteries storage scheduled for inclusion in the grid in the near future.

Also, spending lots of money on battery storage devices is questionable, if net metering and power use balancing is doing pretty much the same thing at no cost to the utility or the customers.

There are, however, communities along the coast of California that are considered local constraint pockets, where storage may be a must and worth much more.

So, for now energy storage—even in California, the leader in renewable energies—is only a niche market.

Back to the Future

Now we return to the imaginary new power plant. Upon completion of the preliminary planning stage and final review of (and agreement on) all components of the PV power plant, we now have a good idea what is needed, as far as equipment, location and the related elements of an efficiently functioning solar power plant are concerned, so we can start the actual design process.

This is another extremely important phase of plant setup, which must be thoroughly completed prior to starting construction and installation work. It is an ominous task, that must be done right.

THE PLANNING PHASE

Now that we know what technologies are best suited for large-scale power field installations, we must start planning the entire process of design, installation and operation of the field.

A number of issues must be considered before we start the design and installation of PV modules in the large-scale PV fields. The present-day planning process is complex, and while we see it getting somewhat simpler and faster (on the permitting and regulatory side), the basic outline presented here will not change significantly for a long time to come.

Some of the basic and yet critical topics and issues of the solar plant planning process are discussed below.

Preliminary Tasks

The PV power plant design process is a complex undertaking, which must be evaluated and executed by a team of professionals, specialized in land, environment, government, regulatory, and technical disciplines related to the broad spectrum of the tasks and sub-projects of the design process.

The preliminary work in PV power plant planning is the first assessment of the potential project. It is a high-level review of the main aspects of the project such as the solar resource, grid connection and construction cost to determine if the project is worth taking forward. There are no rules or standards for this procedure, so the remainder of this section represents our long experience in the field.

Pre-feasibility Study

A pre-feasibility study assesses the very basics of the project to see if it is worth undertaking, without committing significant expenditure. This study should include, as a minimum:

- The project site; type, size and boundary area.
- A conceptual design of the project, including estimation of installed capacity.
- Preliminary weather assessment and seasonal sunlight availability
- The approximate costs for development, construction and operation of the project and predicted revenue
- Estimated energy yield, considering the available data
- Grid connection—cost and likelihood of achieving connection
- Permitting requirements and likelihood of achieving those
- Preliminary estimate of long-term operating profit

PV Field Evaluation

a. Solar analysis

b. Land topography and life forms evaluation

c.   PV system planning and design considerations
d.   PV system verifications, calculations and final design

Permits and PPA

a.   Create minimum acceptable proposal for permitting authorities and utilities

b.   Set up and document as needed to proceed with permitting and negotiating a suitable PPA

Planning and pre-design considers major design elements. These steps are submitted as a proposal in response to a Request for Proposal (RFP), or to a private funding (bank, VCs, etc.).

The key steps needed to create a suitable proposal follow.

Report or Proposal Details

Planning, design, construction and operation of a commercial or utility type PV power generating installation is a multiplicity of daunting tasks. It starts with an idea, an opportunity, and writing a report/proposal to a customer, investor, or utility company.

Actually, utility companies have their own way of acquiring the information they need (via RFPs and other such documents), but some of the work needed for writing a proposal applies to the RFP completion as well.

There are no standards, or set procedures for this effort (as is the case in many areas of solar energy), so we will take a shot at it based on our experience, by outlining the key tasks that should be addressed to come up with an accurate, convincing presentation, application, or report of the project in its planning stages.

Below, in no particular order, is a list of such tasks which must be addressed to prepare a thorough presentation for a large-scale PV power generating installation:

Feasibility Assessment

The feasibility phase focuses on the site information as outlined in the pre-feasibility study. It takes into account each of the factors and constraints in more detail. A typical scope for a feasibility study would include:

- Design of a detailed site plan.
- Calculation of solar resource and climate characteristics (sunlight availability, high and low temperature extremes, wind direction and speed, hail and dust events, etc.).
- Assessment of shading (horizon and nearby buildings and objects).
- Outline layout of areas suitable for PV development.

- Assessment of technology options providing cost/benefit for the project locations, including:
  — Module manufacturer, type, size etc.,
  — Mounting system type and size,
  — BOS materials (inverters, junction boxes, wiring, etc.).
- Outline exact power field design.
- Application for outline planning permission.
- Grid connection possibility assessment; likelihood, cost and timing.
- Precise estimate of energy yields.
- Detailed financial modeling.

PV System Design

- Mechanical and construction services
- Electrical service (PV module level)
- Electrical service (field level)
- Final electrical design considerations
- Installation and O&M
- General system considerations

Proposal Submission

A minimum acceptable proposal includes:

- Power generation estimate
- STC DC power rating, derating and estimated AC rating
- Include all conditions and estimates
- Total project cost estimate with rebates, tax incentives, and net cost
- Estimate of permit, interconnection fees, taxes and other costs
- List (and explanation) of variables and unknown costs
- Suggested payment (ROI) schedule
- Incentives paid over time (PBI, FIT, SRECs)
- Construction (installation) timeline and milestones
- List of major equipment
- List of special factors foreseeable issues and complications
  — Module temperature and degradation over time
  — Land issues (erosion, buildings and businesses nearby etc.)
  — System dependency on sun intensity and cloud cover
  — Expected output vs. system capacity
  — Instantaneous power vs. annual energy production
  — Potential cross-over of impact on overall warranty
  — Performance estimation and validation vs. actual output

If all of this seems overwhelming and somewhat chaotic, it is because it is so. The present-day system of "customers-suppliers-installers-investors-regulators-politicians-utilities and everyone else involved in a PV power installation" is a work in progress, for lack of better words.

The lack of proven engineering specifications and standardized procedures for aspects of solar energy components and systems is obvious. While there are on-going efforts in these areas, they are fragmented, and it will take a long time to bring them together. Even when control becomes reality (certainly in the 21st century), the basics outlined here will remain vital parts of the system.

Location, Location, Location

Back to building our large-scale PV plant: Smaller commercial systems are easier to design, install and operate. All we need in most cases is the owner's permission to hang some panels on the roof or in the back yard, get local permits, and install the modules. Installation of large-scale PV power plants is more complex. So where do we start?

The planning of the project is the responsibility of the whole owner/designer/installer/investor team. Team members must be fully aware of land, logistics, regulatory, and financial conditions (and limitations) at the site. Each member has specific responsibilities in their respective areas of expertise.

Every detail should be taken into consideration, analyzed, calculated and entered into the overall formula—with the actual location in mind.

So what's in a location? A large number of variables like sunshine, clouds, fog, dust, snow, rain, hail, population, industrial activities, roads, wild life, and terrain all impact "location."

Are there environmental issues, transmission lines, substations? This is just the beginning. In most cases we don't have the luxury of choosing a location, because the land has been chosen by a customer or an investor, so we must work with what we have.

Predicting Solar Power Generation

Intermittent power output is the worst enemy of the utilities and a major headache to the solar industry. We cannot control sunshine, but it helps when we can predict the variable function of sunlight coming to a large PV power field.

Every passing cloud creates chaos in the utility grid monitoring systems, and as more and more solar is coming onto the grid, utilities are looking into so-phisticated modeling and forecasting tools to help them predict the changes and keep control over the power generation/

There are a number of solar prediction tools that are capable of predicting hour-by-hour and even min-ute-by-minute power fluctuations across entire regions, narrowed down to a square kilometer.

Using a combination of satellite imagery, historical data and available solar installation data from the area of interest, incorporated into patented methods to predict the aggregate of the solar power generators within the area during the moments when hard data just aren't available. The enhanced data packages fill in the gaps between satellite reads by predicting rainstorm development, cloud drifts, and other weather changes.

This way the utilities have at least 5-10 minutes advance warning of changes in the weather—especially changes in the solar insulation over the area and the amount of produced electricity expected in the prediction period.

This is immensely important to utilities that need the greatest possible insight into the generated solar power coming onto their grids. This is exceptionally important for managing the power coming from utility-scale solar projects equipped like conventional power plants, so many solar installers do have and use their own monitoring and predictive analysis hardware and software packages.

Generally, utilities have little information from direct connections to the PV systems, but especially from the residential systems, so any tools that can help the utility anticipate and manage the load will help them maintain the integrity and quality of the power networks.

Several companies are working on developing new and more sophisticated weather prediction tools, some sponsored by various partners, including the California Energy Commission, CAISO (PDF), Pacific Gas & Electric and Sacramento Municipal Utility District. The origins of these packages date from the days when economic valuation tools for solar companies were needed—including power bills, as well as quote tools used by major solar energy providers.

The solar prediction angle was spawned in 2007 as a research project by several companies using satellite irradiance data to gauge how productive solar panels would be when sunlight and weather conditions were changing. The software packages we've seen are accurate as compared with real-world data. The most critical for the weather prediction application, and most advanced features for filling in gaps in time and data are usually covered under patents, with some receiving approval under the fast-track program for cleantech patents launched in 2010.

The best versions of the hardware and software packages not only narrow the focus of the mapping down to 1 square kilometer, but also allow the collection of a number of endpoints into arbitrary groupings. That allows combining sets of solar generation sources into looking at the performance of virtual power plants in terms of predicting how they interact with the grid at large.

We don't know if weather predictive tools could be used successfully for full grid management systems to smooth out all solar power fluctuations, but the work continues and we expect to see some of these packages used soon by the major utilities, grid operators and solar developers.

The ultimate goal is to allow the majority of customers to incorporate the software's predictions into load forecasting, power purchasing and solar development. The 21st century will offer plenty of opportunities for predictive hardware/software packages to help operate the power plants and manage their power output more efficiently.

Bankable Energy Yield

We must be aware during the planning and development stages of the project, that the banks will look at our estimates, plans and drawings to decide the financing options. They will need to see the minimum energy yield that will be required to secure finance, and then figure out the maximum, which would provide profit and successful long-term operation. The energy yield is of utmost importance to the banks, since it is the financial bottom line.

The proper estimate of that yield must be carried out by an independent specialist who works for the bank. This will ensure that confidence can be placed in the results and will help attract investment.

The energy yield estimate should include:

- An assessment of the seasonal variations and yield confidence levels.

- Consideration of site-specific factors, such as soiling or snow, and the required cleaning.

- Full sunlight obstruction and restrictions review of the site, including albedo, etc.

- Detailed losses from all possible climate, natural and unnatural events.

- An official, professional review of the proposed design to ensure that all parameters are feasible and within design tolerances.

Environmental Impact Assessment

An Environmental Impact Assessment (EIA) is a requirement for projects in most locations over a certain size. It is an assessment of the proposed project's estimated impact on the local environment. The EIA should consider the natural, social and economic aspects of a project's construction and operation.

The EIA usually considers the possible environmental effects based on the site's specifics and its surrounding environment. This will determine what specific studies and actions are required. Possible ways of avoiding, reducing, offsetting or mitigating any potentially adverse effects must be included as well. This is a baseline used to determine the impact of the project.

The main factors to consider and present in the EIA are:

- Type, size, and number of potential impacts.
- The magnitude or severity of an impact, reflecting the actual change taking place.
- The importance or value of the affected resource or receptor.
- The duration involved.
- The reversibility of the effect.
- The number and sensitivity of receptors.
- Ability of the land or the environment to absorb change.

In all cases, the EIA assessment and the subsequent written report must is done and/or verified by an experienced and certified environmental impact assessor or similarly qualified person. The report must be submitted together with the other project reports and accompanying documentation.

THE DESIGN PHASE

The preliminary investigative work on planning and designing the PV power plant will follow established principals and methods, starting with looking into the basics, answering preliminary diligence questions, and resolving issues identified during the pre-planning and actual planning processes.

So what do we need to know, and do, if we want to design and develop a solar power plant? There are a number of different answers to this question, depending on location, type of plant, etc., so we would provide a generic outline of the different tasks at hand and suggestions for their proper execution during the design process:

Technical, Administrative and Financial Considerations

The tasks ahead of us, during the power plant's design process (as applicable to California PV projects) can be summarized as follow:

General
a. Verify total useable acreage status and conditions
b. Verify location of solar irradiation and weather pattern characteristics
c. Annual MWh power output estimated for this location
d. PV technology to be used, its efficiency and cost
e. Tracking vs. fixed installation estimates
f. Total MWp of the chosen technology needed vs. MWh/year generated

Electrical
a. Local electrical transmission system condition
b. Ownership and condition of the nearest sub-station
c. Status of the available transmission and distribution lines
d. Electric transmission system interconnection study
e. CAISO/CPUC review and obtaining LGIA or SGIA status
f. Transmission system and interconnect estimate
g. PPA negotiations conditions and barriers
h. CEQA and or NEPA review and downstream transmission upgrades

Land Permitting
a. Permitting authority (CEQA or NEPA) involvement
b. RWQCB, air district, etc. permitting status
c. Biological, water and cultural resources assessment
d. Surveys for listed species and plants
e. Environmental review; ND, MND, EIR, EA/FONSI, EIS status
f. Environmental review and permitting status
g. Environmental impact; issues; bio, water, air, etc?
h. Mitigation efforts and costs
i. 2081 permit/CDFG or Section 10 consultation/USFWS status

Financials
a. Power purchase agreement (PPA) status and conditions
b. Utility ownership and status (muni or investor owned)
c. PPA price point per MW/h
d. CPI escalator, term and investor's estimated IRR
e. Levelized cost of energy (LCOE) estimate
f. Key financing ratios; D/E, DSCR, interest/term on debt assumptions
g. Exit strategy if no PPA; market; assumed off-taker or other

The Design Process

Plant design of large-scale PV projects in California and Arizona consists of the following steps:

Technology Selection and Preparations

Using some of the information and guidelines discussed above, we now need to get all necessary information on the technology we have chosen for this plant. This includes manufacturers' specs, information packages, and customer review information to date.

Our choices are actually limited to:

a. c-Si modules
   i. Monocrystalline silicon
      1. Fixed mount
      2. One axis tracker mounted
      3. Dual-axis tracking (not recommended in most cases)
   ii. Polycrystalline silicon
      1. Fixed mount
      2. One axis tracker mounted
      3. Dual-axis tracking (not recommended in most cases)

b. Thin film modules
   i. a-silicon fixed mounting
   ii. SIGS fixed mounting
   iii. CdTe fixed mounting
   iv. One axis tracking (not recommended in most cases)

c. HCPV modules on two axis trackers

The installation, operation and performance considerations needed for the proper design of a solar power plant are numerous, so we will only provide a list of the most important:

PV Plant Considerations

a. Location, orientation, mounting angle and height, tracking etc., will be carefully considered, and

b. A serious attempt will be made to provide an aesthetically pleasing layout by considering the modules' size and appearance, the shape of the field, and the surrounding area characteristics.

Pre-engineered System Packages

a. Packages from different companies will be reviewed and those with the best options will be paid special attention, and

b. one will be chosen as a model, and then as a final package.

Product and System Warranties

a. Warranty conditions from each supplier will be carefully reviewed and discussed among the team members.

b. Follow-up discussions and negotiations will be conducted to obtain the best conditions possible for the products and services at hand.

Official Listings and Certifications

a. Listings and certifications as required by authorized officials and agencies (e.g. UL 1703, UL 1741, and any applicable evaluation reports from National Evaluation Services (NES) or International Conference of Building Officials (ICBO) Evaluation Services), and any other applicable such will be carefully reviewed and considered.

b. Only products meeting and exceeding these criteria should be considered, and

c. Additional checks and verification should be performed as well, to ensure the quality and integrity of the plant's building blocks.

System Options

a. All system options will be reviewed and considered, making sure the equipment meets the guidelines of local, state and federal programs.

b. State and federal programs will be reviewed to obtain the best conditions for project development.

Local Utility Companies

a. Local power companies will be interviewed and consulted, for the duration.

b. They will be kept posted on all developments and decisions as well.

Documentation

a. Documents will be generated as required by the QC program and reviewed periodically to ensure that the system meets local permitting, interconnection and other requirements.

b. Revisions and additions will be made periodically and/or as needed.

Equipment

a. The type of equipment to be used should be agreed upon, finalized and documented.

b. Equipment purchasing (type quantity and quality) will be done after the documentation package is approved by a consensus.

Buy-down

A power buy-down package should be completed and sent to the appropriate utilities and state authorities and regulators for review and decision.

Local, State and Federal Rules and Regulations

All local, state, and federal rules and regulations, including incentives and subsidies, should be reviewed and incorporated in the design.

Shading and Ground/Sky Obstructions

a. Impact of shading and other obstructions on the PV array layout will be reviewed and considered in the final design.

b. All obstructions will be clearly shown on the final drawings.

Site Drawing

a. The distance between the estimated locations of all system components, including strings and grid interconnect points, will be measured and

b. A complete site drawing will be developed, including one-line diagram of PV system installation as needed for the permit package and for completion of the final field design.

c. The final design package will be certified by state certified local PEs (mechanical and electrical professionals).

Permits

a. A permit package for the local authority having jurisdiction over this project will be assembled and presented at the appropriate time. It should include:

    i. Site drawing showing the location of the main system components; PV arrays, above ground and underground conduit runs, electrical boxes, inverter enclosure (housing), control room, critical load subpanels, utility disconnects, main service panel, utility service entrance, etc.

    ii. One-line diagram showing all significant electrical system components.

    iii. Cut sheets for all significant electrical system components (PV modules, inverter, combiner, DC-rated switches, fuses, etc.).

    iv. Copy of filled out utility contract (PPA agreement).

    v. Mechanical and structural calculations and drawings for the support system and structures.

Quality Control

The project QC manager is responsible for the creation of training manuals, actual training and re-training of all technical personnel. S/he is also in charge of, and fully responsible for, the integrity of the installation and the proper implementation of all QC procedures, including documentation and records keeping, daily and periodic inspections of work sites, and the acceptance and performance evaluation of each step of the planning, design, and installation stages.

With the design phase completed, and the permitting stages under way, we can plan the installation phase of the new PV plant. But first, we need to take one final look, and make a detailed estimate of the performance characteristics of our new power plant.

Performance Factors

Several pieces of information we need at the onset of the planning and design stages of a PV plant are estimates of its availability factor (Af) and its capacity factor (Cf), because these have a great influence on its final design and expected performance.

Plant Availability Factor

The availability factor of a power plant is basically the amount of time it is able to produce electricity over a period of time, divided by the total amount of the time in the period. Partial capacity availability should be considered. The availability factor is not the same as the capacity factor.

So, a plant in maintenance mode is "unavailable" for the duration. Under this definition, most PV plants are over 95% available, although they idle during the night and in periods of no sunlight. Nevertheless, they are "available" only because the plant is "ready and willing" to go into action, weather permitting.

A more realistic figure considers the idle times of the plant during the producing hours of the day only. In this case, the above % Af would be much more meaningful, because it would tell us how available the plant is during the hours we expect it to operate at full capacity—i.e. from 6:00 AM to 8:00 PM in the summer months. Night hours are "dark" hours, so they should not be considered in any of the technical or financial calculations.

Plant Capacity Factor (Cf)

The capacity factor is a more useful and accurate measure of PV plant performance. It is the ratio of actual output of a power plant over a period of time, and its output if it had operated at full uninterrupted power the entire time (or nameplate capacity). So, as in this example of a 1.0M Wp nameplate power plant operating in the Arizona deserts:

$$Cf = \frac{Annual\ MW/h}{(Hours/year * nameplate)} * 100 =$$

$$= \frac{1,800\ MW/hr}{(8760h/y * 1MW)} * 100 = 20.5\%$$

In this case, our PV plant with 1.0 MWp nameplate capacity has operated at full capacity an average of 6 hrs/day during 300 days of the year. The plant produced a total of 1,800 MW/h during that time and when this is divided by the sum of the total hours of the year (8760 hrs.) and the nameplate capacity (1 MW) we get a capacity factor of 20.5, which is about average for the southwestern US.

This can be also understood to mean that the plant was working at 100% capacity 20.5% of the time—or ~1/5 of the 24-hr cycle. PV plants in the Midwest have Cf of 12-15%, while similar plants in Arizona or California deserts would have CF of 19-22%. Wind farms average 40-45%, while nuclear or diesel fuel plants could have Cf of over 90%, simply because they can run unobstructed, non-stop, and at full capacity all the time and for long periods of time.

PV plants cannot run non-stop and we must be very careful in estimating their Af and Cf, when using existing PV technologies to design a power field at a particular location. In estimating the Cf of PV plants we need to take a good look at the Af and all other variables that are expected during a plant's full 30 years of operation.

Historical weather data of the location could be used to approximate the solar availability, and from that the total energy produced. A margin of error must be considered here, based on the variability of the historical data and different trends we observe in it. Historical data of equipment performance (from the manufacturers), along with utility company demand and supply records could provide information for estimating the expected power generation and losses. A margin of error should be considered, based on the available data. Although the margin of error might be significant, these estimates are vitally important, and must be painstakingly worked out and dutifully incorporated in the design and finance process estimates. They could be also used for providing a baseline of the new plant's performance to be used in future O&M and financial analysis.

Solar Resource and Temperature

As part of the plant design's yield calculations and array sizing purposes, the solar irradiance data required by the performance model are of utmost importance. They determine the overall power production, and must be thoroughly researched. Some of the most reliable data are typically obtained from long-term meteorological models providing hourly averaged sunlight values, such as NASA's satellite weather data. These data could be complimented by more detailed weather data for the specific location obtained from local weather stations, which will allow more accurate calculations and predictions. Thus obtained solar irradiance and weather data can be manipulated using different methods to calculate the expected solar irradiance incident on the surface of a photovoltaic module positioned in an orientation that is determined by the power field design. The margin of error in all cases would be significant, because no one can predict the weather exactly—regardless of the accuracy of historical data—and there will always be surprises and errors. We can only hope to minimize these to an acceptable level, which at the end will determine the quality of our work.

The PV module design also has a lot to do with its performance, as far as temperature coefficient, overall efficiency, and longevity are concerned, since these vary from location to location. Does the manufacturer

have such data? Is he willing to share them freely and assist in plant planning and design? In all cases, during long-term performance the solar irradiance in the plane of the module is often a measured value and should be used directly in the performance model. The level of irradiance is very important to the module's performance, especially with the quick expansion of PV plants in the deserts, so it must always be taken into consideration and dealt with carefully.

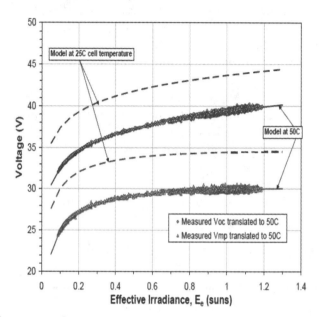

Figure 6-33. Over 3300 measurements recorded on five different days with both clear sky and cloudy operating conditions for 165-Wp mc-Si module (2).

In Figure 6-33, note the 15-20% drop of Voc and Vmp at increasing the temperature from 25°C (room temperature) to 50°C (average—but not highest—desert day temperature). These results could be extrapolated for 85°C, which we measure in the modules during hot summer days in the deserts, and which will show a significant additional drop of output approaching a 30% total drop in power.

Actual cell temperature and back-surface module temperature can be distinctly different for different PV modules, but particularly for concentrator type modules, since their temperatures are much higher—well over 150°C. The temperature of cells inside the module can be related to the module back surface temperature through a simple relationship, but is usually significantly higher. The relationship given in the equation below is based on an assumption of one-dimensional thermal heat conduction through the module materials behind the cell (encapsulant and polymer layers for flat-plate modules, ceramic dielectric and aluminum heat sink for concentrator modules).

Cell temperature inside the module is then calculated using a measured back-surface temperature and a predetermined temperature difference between the back surface and the cell.

$$Tc = Tm + E/Eo \bullet \Delta T$$

where:

Tc = Cell temperature inside module, (°C)

Tm = Measured back-surface module temperature, (°C)

E = Measured solar irradiance on module, (W/m²)

Eo = Reference solar irradiance on module, (1000 W/m²)

ΔT = Temperature difference between the cell and the module back surface at an irradiance level of 1000 W/m²

This temperature difference is typically 2 to 3°C for flat-plate modules in an open-rack mount. For flat-plate modules with a thermally insulated back surface, this temperature difference can be assumed to be zero. For concentrator modules, this temperature difference is typically determined between the cell and the heat sink on the back of the module.

Variations and excesses in module temperature are one of the key reasons for PV module efficiency decrease. They also contribute to annual degradation and excess failure rates of all types of PV technologies. Because of that, PV cells and modules manufacturers must account for these variations by issuing theoretical modeling as in the above example, and actual test data, under different weather conditions—and especially under excess heat.

Short of that, the power plant design team must come up with the data by performing the above calculations and estimates to provide a complete and clear picture of operating conditions in the specific PV field.

PV Modules and Strings Evaluation and Modeling

Executing all design and production steps perfectly means a lot of engineering time and usually results in a higher-cost final product. Nevertheless, some basic requirements must be kept in mind when looking into purchasing those 500,000 PV modules for a 100 MWp power field. Properly modeling the field performance of PV modules is an absolute requirement—during the design, manufacturing, QC, and final test stages.

This should be verified during the design of the power field as well, for it will help estimate the per-formance and longevity of the PV products (modules, inverters etc.).

The string design process must take into consideration that in large PV arrays, individual PV modules are connected in both series and parallel. A series-connected set of solar cells or modules is called a "string." The combination of series and parallel connections may lead to several problems in PV arrays. One potential problem arises from an open-circuit in one of the series strings.

The current from the parallel connected string (often called a "block") will then have a lower current than the remaining blocks in the module. This is electrically identical to the case of one shaded solar cell in series with several good cells, and having the power from the entire block of solar cells lost. Figure 6-34 shows this effect.

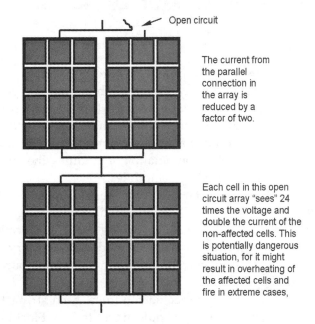

The current from the parallel connection in the array is reduced by a factor of two.

Each cell in this open circuit array "sees" 24 times the voltage and double the current of the non-affected cells. This is potentially dangerous situation, for it might result in overheating of the affected cells and fire in extreme cases,

Figure 6-34. Open circuit effects in larger PV arrays.

Although all modules may be identical and the array does not experience any shading, mismatch and hot spot effects may still occur. Parallel connections in combination with mismatch effects may also lead to problems if the by-pass diodes are not rated to handle the current of the entire parallel connected array.

For example, in parallel strings with series connected modules, the by-pass diodes of the series-connected modules become connected in parallel, as shown in Figure 6-35. A mismatch in the series-connected modules will cause current to flow in a by-pass diode, thereby heating this diode. The affected diode could overheat and eventually fail. In all cases it will also affect the module/string performance negatively.

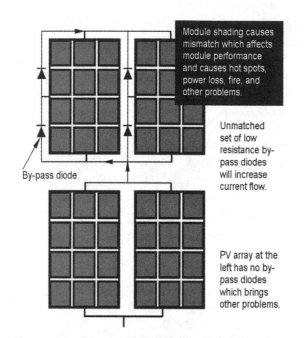

Module shading causes mismatch which affects module performance and causes hot spots, power loss, fire, and other problems.

Unmatched set of low resistance by-pass diodes will increase current flow.

By-pass diode

PV array at the left has no by-pass diodes which brings other problems.

Figure 6-35. Bypass diodes in parallel connections

The current may now flow through the by-pass diodes associated with each module, but must also pass through the one string of by-pass diodes. These by-pass diodes then become even hotter, further increasing current flow. If the diodes are not rated to handle the current from the parallel combination of modules, they will burn out, damaging the PV modules.

In addition to the use of by-pass diodes to prevent mismatch losses, a blocking diode, may be used to minimize mismatch losses. A blocking diode is typically used to prevent the module from loading the battery at night by preventing current flow from the power source through the PV array.

With parallel connected modules, each string to be connected in parallel should have its own blocking diode. This not only reduces the required current-carrying capability of the blocking diode, but also prevents current flowing from one parallel string into a lower-current string and therefore helps to minimize mismatch losses arising in parallel connected arrays.

When modeling, we must start with theoretical suppositions and empirical formulations, to be followed by lab and field tests for final verification. Theoretical estimates are intended to simulate expected operating conditions. We must repeat final tests of a new product, many times, to verify the results and adjust our design and manufacturing operations.

QA/QC Verification and Enforcement

The performance and longevity of a PV power generating installation depend on the quality of its com-

ponents, PV modules and BOS equipment, and the quality of the power field design, installation and operation. PV modules' quality is one of the first discussions taking place when considering a PV installation of any type or size. There are a number of PV module manufacturers with stellar reputations for quality and customer support.

How do we find the best manufacturers of materials, components and services for our PV project?

PV modules manufactured by world-class companies perform perfectly and for a long time without major degradation or failures. Why do these modules do so well, while others fail right out of the box? Installations with proper design and implementation are performing as specified and expected, while others don't fair as well. What is the difference? The simple answer is quality—quality of supply chain materials, equipment, processes and procedures; quality of the quality control; quality of the people and the management team. The quality of the entire cradle-to-grave process—from sand to decommissioning of the power field—is what makes the difference.

Asking manufacturers to provide data on their quality control procedures usually results in silence, or garbled responses. So what is the solution? Of course, buying from reputable companies is the best way to go, but not always possible, so what can we do if we must deal with a new company—especially for the purchase of large quantity PV modules?

Imagine 500,000 PV modules delivered for installation in a 100 MWp power field and having no idea of their quality, other than what the manufacturer provides in the enclosed documentation. What an overpowering feeling of helplessness this must be for any professional who knows the odds of success when using untested products with unverified quality from unproven manufacturers. How can anyone who is in charge of such a project allow installation of such a large quantity of products (which are supposed to last 30 years) without thoroughly verifying the product quality and testing its performance?

Yet, it happens every day. It happens because we are so accustomed to believing what we see on product labels. But a label is the last step in a 6-months, "sand-to-module" (semi-controlled) manufacturing process, done 6,000 miles away by people we don't know, with equipment and procedures that are not thoroughly revealed.

There are several options when seeking assurance of the quality of the PV modules:

1.   Request a copy of the manufacturer's
     a.   Engineering specifications and manufacturing procedures

b. QA/QC system manual and documentation
c. SPC program documentation
d. Quality control inspections and tests documents
e. Supply chain qualification and control
f. Management of change documentation
g. Equipment maintenance and calibration
h. Technicians training/qualification program
i. PM and safety programs
j. Corrective action system
k. Test and performance measurements documents
l. Warranty policies workmanship and long-term performance warranty
m. Customer feedback and suggestions log

2. Request a site visit with access to the production line, people, equipment, process specs, quality control logs, etc. If a visit is allowed, then take a close look at and request documentation of:
    a. Raw silicon supplier qualification, material testing and QC control
    b. Process chemicals (acids, bases etc.) supplier qualification and QC control
    c. Process gasses (Ar, N2, etc.) supplier qualification and control
    d. EVA, front glass and back cover supplier qualification and control
    e. Process specs list and compliance verification
    f. In-process quality control procedures and final QC tests data review
    g. Observe the manufacturing operations

3. If a site visit with a quality verification option is granted, then
    a. Set up a QC station at key steps of the process, including the final test
    b. Verify proper execution of the steps and check the quality if possible
    c. Observe the final test procedure and verify final product quality

Far fetched, maybe, but not when buying 500,000 PV modules to be installed in the Arizona desert, for over 30 years' operation. Not far fetched at all. As a matter of fact, it is our responsibility to find out as much as we can before making a final decision.

This would be the best, if not the only, way to ensure quality and save many long-term problems. Usually, however, it is difficult to even obtain permission for a site visit, let alone performing quality checks there. So

we must figure out other ways to check product quality and accept or reject batches. This requires a lot of experience, money, time—both in the lab and in the field.

Final Design Considerations

We have discussed a number of subjects related to the manufacturing process and the PV products we plan to use, but control of the quality of the power field itself—the planning, design, installation and operation stages—is something we also need to look at.

It's not an easy task, because while some quality control specs and procedures are used in some manufacturing operations, power field quality control is partial at best. Its complete version is only a vision in the minds of some responsible professionals today.

There are no established QC/QA standards for completely integrated control of the quality of all materials, procedures and services during the power field inception and duration. In most cases only certain areas are subject to some quality control and inspection procedures. To our knowledge, no large-scale PV project is under complete and undisputable quality control today.

Quality control in the power field starts at the planning and design stages. It consists of a thorough knowledge and understanding of the major components, their interactions and the forces acting upon them that influence their operation.

Now we need to put the following on paper and use the results in the power field design:

a. Field design considerations and drawings
    i.   Technology choice
    ii.   Land location and preparation
    iii. Support structure choice
    iv.   Mounting choice (fixed tilt, trackers)
    v.   Packing density (row spaces)
    vi.   Equipment positioning (inverters, combiners, etc.)
    vii. Wiring size, pathways and methods
b. Key equipment and control instrumentation selection
c. Installation, QC, and O&M procedures and manuals
d. Mechanical calculations, considerations and drawings
e. Electrical calculations, considerations and drawings
f. Performance charts and calculations
g. Aesthetics and socio-political considerations
h. Permitting points, issues and negotiations
i. NEC, OSHA and EPA considerations

After taking into consideration all technical, regulatory and logistical issues and having chosen the location and the technology to use, we can apply our knowledge to put down the final plant layout on paper. PV modules and arrays will be laid in their respective rows and mounted on the chosen structures. Proper size wires will be run from the rows or strings to the inverters which will then be hooked into the grid. A number of control and measurement systems will be interconnected as well—and the field is ready for operation—at least on paper.

Figure 6-36 is a sketch of our final design for a 1.0 MWp DC power modular unit for use in a large-scale PV power plant in California deserts. This particular design uses 4 ea. 250 MW inverters, but the same concept applies when using different number of inverters. Each of these units is basically a self-contained modular PV power generator. Several of these could be installed and interconnected in any field, using any type of modules, inverters and BOS. This package will work flawlessly, if there is sunshine to generate power and a nearby interconnection point to send the generated power to.

Finally, we need to see how our design looks from the banks' perspective. Actually, we should've considered the "bankability" of the project through the planning and design stages.

Author's note: A project's "bankability" is basically nothing more than a measurement tool the banks use to figure out if a project is feasible and financeable.

THE INSTALLATION PHASE

Installation of components in the large-scale power PV field will start upon agreement by all parties on the technical, administrative and financial conditions, and after thorough evaluation of the design plans, specs, bids, proposals and contracts.

The actual work on the components and systems installation will follow established principals and methods, some of which are outlined below. In the absence of one accepted standard, the text below is only a guideline—a glimpse into the complexity of the PV power plants' design and installation procedures.

The Pre-installation Process

The installation (construction) phase of the power plant starts immediately after approval of the design documentation, obtaining the necessary permits, PPA and financing, all of which is much easier said than done… and requires a great amount of time, resources and money!

But we have gone through all this, and now it is time to start the actual construction of the field and installation of the equipment. First, the project manager and technical coordinators must make sure that all necessary steps of the planning and design processes have been completed successfully, and that the relevant documentation has been properly and thoroughly executed.

The pre-installation process is conducted in parallel with the planning and design stages, and follows the steps outlined below (which are actually part of the

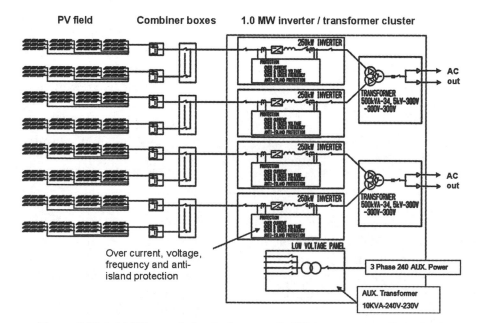

Figure 6-36. 1.0 MWp module of a large-scale PV power generating plant

power plant planning and design segments):

1. Required permit documents and materials will be submitted to the authorities, and preliminary approvals will be issued prior to starting any construction activities.

2. Schedules will be finalized for receiving plant equipment and materials.

3. All equipment will go through a thorough initial inspection to verify quantity received, and to make sure that it has arrived without any modifications or shipping damage.

4. Actual field tests with some equipment (PV modules) should be performed to verify performance and compliance.

5. Installation instructions for each component will be finalized and reviewed, and the responsible personnel will be trained in installation procedures.

6. Length of wire runs from PV modules to combiner and inverter will be verified and documented. Trial runs must be performed to verify the process steps.

7. Ampacity of PV array circuits will be verified to determine the minimum wire size for current flow.

8. Wire runs must be verified based on maximum short circuit current for each circuit and the length of the wire run, and as follow:
   a. The minimum wire ampacity for the wire run from modules to combiner is based on module maximum series fuse rating printed on the label.
   b. The minimum wire ampacity for the wire run from combiner to inverter is based on the number of module series strings times the maximum series fuse rating.

9. The size of the PV array wiring must be such that the maximum voltage drop at full power from the PV modules to the inverter is 3% or less, for
   a. Wire run from modules to combiner
   b. Wire run from combiner to inverter

10. Length of wire run from inverter to main service panel must be established and verified. The goal is 1% voltage drop for AC-side of system (3% maximum).

11. Main service electric panels must be checked to determine if they are adequately sized to receive the PV breakers or whether the panels must be upgraded.

Once these procedures have been completed and signed-off by the responsible personnel—including the QC manager—the installation process can begin.

The Installation (Construction) Process

Installation of PV and BOS equipment should be done by a certified solar energy equipment installer/contractor. Several contractors, specialists in the different areas of the plant's construction, could be employed. The installation process usually follows the sample procedure below:

1. Review instruction manual and train technicians accordingly. The proper execution of this work effort will be supervised by the project manager and verified and documented by the QC manager and his crew.

2. The pre-installation check procedures include:
   a. Check all modules visually. Test the open circuit voltage and short circuit current of each module to verify proper operation, before taking onto the field. See checklist for detailed procedures.
   b. Check plug connectors and connector boxes.
   c. Check the attachment points and methods as indicated in the drawings.

3. The installation procedure includes:
   a. Mount modules on support structure and connect to conduit.
   b. Install PV combiner, inverter, and associated equipment to prepare for system wiring.
   c. Connect properly sized wire to each circuit of modules and
   d. Run properly sized wire (per drawings) for each circuit to the circuit combiners.
   e. Run properly sized wire (per drawings) from circuit combiner to inverter overcurrent/disconnect switch
   f. Run properly sized wire (per drawings) from inverter to utility disconnect switch

g. Run properly sized wire (per drawings) from utility disconnect switch to main service panel and connect circuit to the main utility service.

4. Use the checklist and drawings to ensure proper installation throughout the system by visual inspection. Sign off each step of the installation and verification procedures.

5. Verify that all PV circuits are operating properly and the system is performing as expected. Double check the checklist and drawings prior to executing the system acceptance test.

6. Final inspections will be conducted by the QC manager, the local authorities, and the utilities. Once approval is received, the system is ready for operation

7. The buy-down request form, with all necessary attachments, will be sent to the appropriate authorities to receive buy-down payments per agreed conditions.

8. The quality control manager is fully responsible for the proper training of technical personnel in charge of QC procedures at each step of the design and installation phase. The QC manager also coordinates the proper documentation and records keeping, daily and periodic inspections of work sites, and the actual acceptance and performance evaluation of each step of the installation phase.

The QC manager is ultimately responsible for following proper preparation, installation, inspection, verification and test procedures.

Performance Evaluation

The performance evaluation phase will start immediately after the EPC contractor(s) signals completion of the installation and has performed start-up and initial performance and safety testing of each step of the process

A detailed step-by-step checklist of steps to be taken during the installation and performance evaluation procedures—all of which will be supervised and verified by the QC manager and his QC inspectors—is to be developed ahead of time, to ensure an efficient quality installation.

Upon completion of each step of the installation procedures, the QC manager and his inspectors will conduct performance evaluation of the components and systems. Tests designed to verify and confirm compliance with the design specs and requirements, as well as to test the performance characteristics of the components and systems at each step of the installation process will be executed accordingly. The results of these tests will be verified, approved and signed off by the QC manager.

When all construction stages have been completed, and the related components and systems have been signed off, the final test of the entire power plant complex will begin. With all modules and strings activated and connected in the circuit, a number of tests will be performed with each string, to begin with, and recorded as a baseline for each particular sting under the existing weather conditions.

The total power output—on the DC and AC sides—will be measured, and the appropriate calculations of the plant's efficiency, yield, etc. will be made and documented. The resulting data will be used as a baseline of the power plant performance under existing weather conditions and for comparative purposes in the future.

Final System Certification and Documentation

Thorough documentation and record keeping is the best way to ensure efficient implementation of all phases of this project. A number of documents and record-keeping media will be developed for the different segments of project implementation, such as:

1. General system documentation for installation and operation phases
2. Technical drawings for the same
3. Training documentation
4. QC manuals and inspection documentation
5. Non-conforming materials and procedures documents

These will be needed for inspection and verification during each step of the project by different specialists and responsible parties. The complete package of documentation will be submitted to the inspecting authorities for final inspection, verification and certification of the facility.

At the end of the construction phase, and after all quality checks have been performed and signed off by the QC manager, the appropriate authorities are notified to execute certification checks and tests. These vary from state to state and from project to project, but basically consist of review of all construction and QC documents and verification of the final performance test results.

Again, some operation and quality control policies and standards for the different stages of the PV products manufacturing, installation and operation do exist, but there is no single, all-encompassing, worldwide PV industry standard for these.

Until such a standard is implemented, we will improvise in some cases, and compromise in others. A number of serious gaps still exist in the operating and quality control procedures of sand-to-modules manufacturing and in power plants' design-to-decommissioning processes.

Power Plant Quality Certificate

Another sign of the times, and of the further maturing of the solar industry… how about quality certification for PV power plants? A new concept, yes?

With an increase in the number of utility-scale PV power plants being built, mostly without any standards to speak of, let alone being enforced standards, quality standards across the PV industry are coming into question.

A breakthrough has come in this area, and it's a plausible attempt to provide the highest possible standards from design to operation. Conergy and TÜV Rheinland have partnered to provide an example of how things should be done and establish a "best practices" program that can be used in setting up PV power plants.

The testing of several Conergy power plants will be done by TÜV Rheinland, using their long expertise in the field. Certification of large-scale PV plants is a new, but badly needed, instrument.

The ultimate goal of this undertaking is to provide a level of security, confidence and transparency to the interested parties—investors, customers, etc. This is a new way to introduce and enforce the quality concept to an entire power field, with a focus on ensuring bankability in a manner that would be understood and accepted by everyone.

All future power plants to be built by Conergy will be tested and certified to standards established by TÜV Rheinland. This includes the planning, design, installation and commissioning stages and procedures, including efficiency and safety evaluations and tests.

Power yield estimates in the planning stages, and actual measurements during operation are important factors in this equation, so they will take a special place in the program, with the objective to provide the highest yield possible and to bring confidence in the success of the projects, thus ensuring maximum long-term financial security.

'Critical Design Review' steps will be followed in the pre-construction phase, and repeated quality assurance checks will be made on site during construction.

Safety-related testing of the electrical installation will be done as part of the program, and will follow the standards and specific criteria outlined by TÜV Rheinland.

A number of components and elements will be examined independently, including the quality of the yield forecast, the plausibility of the plant concept, an analysis of the suitability of plant components quality, components and plant safety, long service life and efficient plant performance, and the quality of the installation effort.

This "power plant quality certification" program is a new and welcome development in the solar energy sector. Its humble beginning, with only one participant notwithstanding, shapes the way for great things to follow.

This new quality assurance program is a confirmation that the introduction of quality standards in power plant planning and plant construction work is possible. Everything considered, we are sure that such plants will be set up and operated at much higher levels of quality than are generally accepted today.

The greatest benefit of this new undertaking is that the customers will have peace of mind and be assured that they have a solid deal. This assurance coming from a reliable, single source like TÜV Rheinland is unprecedented, and we are sure that it will find wide acceptance.

We are optimistic that all 21st century solar power plants will have similar quality certificates, and we already see the opposition standing in the way of the new enterprise, especially in the short term. There *will* be struggles.

OPERATION AND MAINTENANCE PHASE

Upon completion of construction and installation stages, the responsibility for the operation of the new facility is transferred immediately to a team of specialists, working for the plant owners, or an independent contractor who is paid a fee to operate the plant. The engagement conditions are carefully discussed and officially agreed upon in advance, and an operation and maintenance (O&M) contract is signed prior to starting the actual O&M work.

The O&M team has a number of important responsibilities, which range from simple janitorial work, to solving the most complex issues. The O&M management team is fully and irrevocably accountable to the project owners for all issues related to the proper and efficient operation of the plant.

Final System Check and Yield Monitoring

As part of our QC/QA plan discussed above, we would test and verify the integrity and longevity of the PV products purchased for our 100 MWp power plant. Once all equipment was installed, the power plant would be switched on, tested, and certified. When all the inauguration champagne had been drunk, we would start the final step—transferring of the plan's long-term operation and management to the O&M team.

The results from the final QC and certification tests would be accepted as a baseline of the plant's performance. Additional tests would be run under different weather conditions, which would tell us if the PV components conform to the manufacturers' specs, and if our planning, design and construction efforts were of the quality we claimed.

Initial O&M system tests consist of thoroughly checking and testing all components, identifying defects and malfunctions, and documenting all events and procedures. A final report of the initial O&M test must be issued by the engineering group conducting the tests. The report should include a thorough description of technical issues and the related financial impact—reflected in equipment malfunctions or yield loss.

System tests include visual examination of the modules and infrared checks of each module at peak performance. These tests, done by qualified personnel,

will spot damaged modules (cracked glass, discoloration, etc.) and will identify thermal malfunctions like hot spots, which could reduce the output power and/or damage the modules.

These tests are especially important, and an absolute must, if no pre-installation tests were conducted. I-V curves for the different strings and sub-strings must be generated at peak power and in different weather conditions as well, to establish a baseline for each of them to use as a future reference. I-V curve tests could reveal inefficient or faulty module performance, wiring and other abnormalities. The tests must be performed by well trained and qualified personnel, to avoid erroneous results and conclusions.

Yield assessment must be performed simultaneously by the same team, ensuring more reliable results. The yield assessment is an even more complex variable, so it suffices to say that unless it is done with utmost care by qualified personnel, the results would be erroneous or unrepeatable at best.

OPERATIONAL CONSIDERATIONS, TASKS AND ISSUES

Once the initial tests are done and the power plant is given a clean bill of health, normal operation starts. O&M procedures of a PV power plant are complex, sometimes expensive, but very important. Daily observations must be documented and periodically analyzed, with appropriate action taken if abnormal conditions or behaviors are noticed. Yield monitoring and assessment is also a daily task—especially in the early days of a plant's operation. The O&M team is in charge of these tasks, and is ultimately responsible for everything going on at the plant, including the prevention of natural or man-made incidents.

Yield monitoring must be carried out properly and precisely too, since yield variations are expected on a minute-by-minute, and hour-by-hour basis. Things can change in an instant or overnight. They surely change monthly and yearly. The dependencies are complex, so their proper relation and documentation are important. Even more important are their interpretations and resulting corrective actions.

The PV power plant is a living entity that can be very dangerous if not treated properly, so only well trained and qualified personnel can run it efficiently, safely and profitably. These three factors are interrelated, and a deep understanding of each component and its function is required.

Figure 6-37. Large-scale PV power plant

O&M team responsibilities include:

1. Materials and labor needed to operate efficiently
2. Solving mechanical and electrical problems
3. Voltage disturbances corrections
4. Frequency disturbances corrections
5. Islanding protection response
6. Power factor support
7. Reconnect after failure or P&M
8. DC injection into AC grid issues
9. Scheduled and unscheduled disconnects
10. Grounding and source circuit abnormalities
11. Ground fault protection failures
12. Overcurrent protection incidents
13. Containment of electromagnetic interferences of any sort
14. Personnel training and management of daily activities
15. Safety and accident prevention programs

PV plants are operated and maintained by an operator/contractor, who is trained and fully responsible for proper and efficient system operation. Some of the duties of the operator/contractor are:

1. Wash PV array at scheduled intervals, or when build-up of soiling deposits is noticed (see author's note below).

2. Periodically inspect the system to make sure all wiring and supports stay intact.

3. Review the output of the system daily to see if performance is close to the specs and previous year's reading.

4. Maintain a log of these readings to identify system problems.

5. Daily monitoring of the system is a key component of its proper operation, since it provides feedback to the operator as to total power and actual energy production metering. Without proper metering the operator will never know if the system is operating appropriately. A properly designed PV power plant will be equipped with the latest in remote monitoring equipment to register and record the power production of each string, as well as the total power pumped into the grid.

6. Weather watch in the form of a basic weather sta-tion is also needed to follow and record weather changes and alert the operators in case of drastic changes. Actual temperature and solar insolation at the site are also important and must be tracked constantly. The data will be analyzed periodically to calculate overall system efficiency, to prevent system problems, and to take corrective actions.

Author's note: A new technology has been developed, where a thin nano-particles type film is deposited on the front surface to repel dust, pollen, water and other particles without hindering the solar panel's ability to absorb sunlight. The coating can maintain this ideal hydrophobic surface for years, they say, reducing overall maintenance.

It is difficult to believe that any coating can repel the massive sand and dust clouds during a desert haboob and stay intact on the front surface for 30 years of desert abuse. For now, we will keep the water bucket handy.

PV Plant Output Variability

The amount of power produced by a PV power plant depends on several factors, with sunlight availability and intensity being the major variables. Sunlight can ramp up and down rapidly, which is important characteristic of large-scale PV power generation, but the utility system operators need to maintain a balance between the aggregate of all generators and loads.

General Principals

These principles must be maintained to ensure the integrity of the grid:

a. National electric system reliability must be maintained, regardless of the generation methods and their variability.

b. Generating sources must contribute to system reliability and should not cause unnecessary complications and interruptions.

c. Power industry standards and operating criteria apply to all generators. They are transparent and based on efficient performance.

Understanding the characteristics of variable PV output over large areas of PV installations and correlation to the load is critical to understanding the potential impacts of large quantities of PV injected into the national energy system.

PV variability can drive localized concerns, which typically manifest themselves as voltage or power quality problems. These issues are distinct from issues of balancing at the grid system level, and should not be confused. Management and remediation options for local power quality problems are generally different from options for maintaining a balance between load and supply at the system level.

The complexity of injecting variable load into a steady power transmitting system is not to be overlooked, or taken lightly. There are a number of things that the utilities could do to control the variability of injected power, but the responsibility of ensuring trouble-free operation falls squarely on the PV power plant owners and operators.

The local utility company can help in most cases, but it will not step in to solve problems at the PV plant. Good understanding of the variability issues and their consequences must be a priority for the owners and managers of the plant.

Large-scale Power Plants Output Variability and Issues

Basically, photovoltaics fall under the broader category of variable generation. Experience with appropriate, unified approaches for managing variable generation will ease integration issues. *The most important fact* is that the dialogue regarding PV variability requires additional time-synchronized data from multiple PV plants and insolation meters over spatial scales ranging from a square kilometer to greater than 10,000 square kilometers.

In all cases, the data will need to cover at least a year of measurements and should be synchronized with comparable load data to understand the net impact on the variability that must be managed by the system operators. Certain questions, particularly those concerning power quality and regulation reserves, will require data with a time resolution as high as multiple seconds.

Analysis of data from multiple time-synchronized PV plants will allow detailed evaluation of the degree to which rapid ramps observed in point measurements will be smoothed by large PV plants and the aggregation of multiple PV plants. Such studies will help remove unwarranted barriers to interconnection and provide the basis for setting appropriate interconnection standards that will allow solar energy from PV plants to reach significant penetration levels.

The output variability observed by a single point of insolation measurement will not directly correspond to the total (daily or monthly) variability of a large PV plant, simply because one end of the plant might be sunny, while the other could be covered by clouds. A single point measurement ignores this, and the sub-minute and sub-hour time scale smoothing that can occur within these multi-kW plants. This difference can be amplified when looking at sub-minute smoothing that can occur within very large-scale plants.

In summary, basically:

a. Extrapolation suggests that further smoothing is expected for short time-scale variability within PV plants that are hundreds of MW, but this needs to be confirmed with field data from large systems.

b. Diversity over longer time scales (10-min to hours) can occur over broad areas encompassed by a power system balancing area. Data from the Great Plains region of the U.S. indicate that the spatial separation between plants required for changes in output to be uncorrelated over time scales of 30-min is on the order of 50 km. The spatial separation required for output to be uncorrelated over time scales of 60-min is on the order of 150 km.

The assumption that variability on a 15-min or shorter time scale is uncorrelated between plants separated by 20 km or more is supported by data from at least one region of the U.S. Additional data are required to examine this assumption in regions with different weather patterns.

c. Multiple methods will be used for forecasting solar resources at differing time scales. Weather variations (clouds, rain, snow, etc.) are the primary influence in the solar forecast. It is important to recognize that clouds (and their rate and direction of movement) are visible to satellites and ground-based sensors, so some successful forecasting can be expected. Over longer time scales, clouds can change shape and grow or dissipate, so numerical weather modeling methods may prove necessary. As with wind forecasting, solar forecasting will benefit from further development of weather models and datasets.

Power output variability is here to stay, and accurate, error-free forecasting and power control are still not available, but error-free forecasting not so far away. We do expect that today's efforts in these areas will bring good results soon, allowing us to predict and manage PV plants' power output variability enough to assure their reliability as grid-connected energy sources. It's another step closer to energy independence.

Energy storage is the Achilles' heel of solar energy. It is absolutely needed for the long-term expansion of solar energy as a 24/7 power source. Without storage it will always be considered a supplementary, uncontrollable variable (and highly unreliable) power source. Adding energy storage to any solar facility is a must for the 21st century.

POWER FIELD PERFORMANCE

Once a field is fully operational, we must assure that it is functioning properly and generating maximum possible power at all times. Along with daily measurement and analysis, we need to analyze overall system efficiency and performance.

The key performance parameters for tracking the operation of a profitable power plant follow.

Final Power Yield

The final yield is a reflection of the bottom line of the power plant performance. It is derived by plugging many variables into a long estimate. It is the net AC energy output of the power field divided by the aggregate nameplate power of the installed PV array at STC ($1000W/m^2$) solar irradiance and 25°C cell temperature.

$$\text{Final Yield} = \frac{\text{kWh AC}}{\text{kW DC}}$$

UL-listed modules have a nameplate number on the back of the module that identifies the STC rated DC power. The total array power can easily be determined by summing the nameplate power ratings for the array.

While the power field nameplate is simply a sum of the modules' output tested at STC in a lab and is easy to calculate, the generated AC energy (kWh AC) in the equation is a multiple of many variables and factors. The final yield, therefore, represents the efficiency, or rather the inefficiency, of the BOS (wiring, inverters, transformers) from the PV modules to the grid. The final yield is a dynamic variable, which varies significantly with the total power output and operating temperatures.

Thorough knowledge of the system variables and operation is needed to arrive at an accurate number during planning and design stages. A team of experts must calculate each variable early on, to protect the bottom line.

Reference Yield

The reference yield is the total in-plane solar insolation (kWh/m²) divided by the array reference irradiance. It represents an equivalent number of hours at the reference irradiance. The reference irradiance is typically equal to $1 kW/m^2$; therefore, the reference yield is the number of peak sun-hours.

$$\text{Reference Yield} = \frac{\text{kWh}/m^2 \text{ (Total insolation)}}{1000W/m^2}$$

Note: This variable is also critical, and here as above, the lower number (STC conditions) is fixed, while the upper number (total insolation) depends totally on a combination of climate and weather. Coming up with a precise number for it is impossible, but a thorough look at historical data and weather patterns could give a good picture of what to guestimate.

Performance Ratio

The performance ratio is the final yield divided by the reference yield and is dimensionless. It represents the system's performance in terms of total losses due to equipment, restrictions, or under-spec conditions. Typical system losses involve DC wiring, cells/modules mismatch, bypass diodes, module temperature effects, annual degrading rate, inverter conversion efficiency degrading (due to temperature or age), transformers issues, and more.

$$\text{Performance Ratio} = \frac{\text{Final Yield}}{\text{Reference Yield}}$$

An accurate estimate of the above variables—major elements of the overall power field yield—is critical for proper design and efficient and profitable operation of a PV power field.

We will take a look below at some additional key variables and factors, which help determine the quality of a power field's design and performance, starting with the solar irradiance and equipment requirements.

Above, we looked at the overall theoretical yield and performance assessment. To complete this effort, we also need to:

1. Verify that the location is appropriate for solar power generation,

2. Verify that the chosen PV technology works as planned by conducting actual field tests and evaluations, and

3. Establish parameters for performance and yield tests during the operation stage.

Field Performance Issues

We will now take a closer look at the practical expression of the PV modules' performance characteristics and related issues.

Optical and Visual Field Degradation

Tests done by the author and his associates with a number of different c-Si PV modules exposed to direct sunshine for more than 10 years under direct desert sunlight show significant visual signs of degradation.

Some of the signs can be easily seen by the naked eye even during the first year or two. These range from discoloration (browning) of central region of cells, to stains (halos and rainbows) on the module's glass surface. The degradation, which causes the change in color of the EVA (yellowing) is demonstrated visually by characteristic yellow or brown stains on its surface.

The EVA discoloration process, according to some researchers, is most likely due to the formation of polyenes and carbonyl-polyenes compounds after long-term exposure to heat and the elements.

Visual inspection of the reflectance from the individual cells shows degradation also of the AR (anti-reflection) film as a discoloration and decrease of intensity of the blue AR film color. Degraded modules also show an overall increase of reflectance, which contributes to the measured electrical losses. The increase of reflectance was attributed at times to optical interference from EVA degradation which is most likely accompanied by production and emitting of chemical byproducts such as lactones, ketones and acetaldehydes.

Once the module encapsulation materials start decomposing, in addition to generating harmful gasses, they will also allow environmental chemicals, gasses and moisture to penetrate into the module's interior. Inside the modules these foreign substances would have ample time and opportunity to attack the solar cells' structure thus causing gradual decrease of power and eventually total failure of the modules.

A disturbing trend lately is the sale and installation of millions of PV modules (TFPV mostly) which lack the much needed edge protection of side frames. These modules consist of two glass panes bonded together with organic thermoplastic materials.

Although the quality of these materials has improved significantly lately, they are no match for the ferocious IR and UV radiation they will be exposed to during 30+ years of non-stop on-sun operation in extreme heat and moisture environments. All organic materials decompose with time under these conditions, and in the case of frameless TFPV modules this opens the door for the elements to penetrate into the module and damage the active PV structure.

The question is how to predict and prevent quick deterioration and premature failure of the protective encapsulation layers and the subsequent destruction of the active module components. What are the variables that we should keep in mind, to ensure proper protection of PV cells and modules in large-scale PV plants?

What kind of quality control is needed to provide maximum quality of the final product? How can we attract the manufacturers' attention in a positive way, so as to obtain relevant data and cooperation needed for the analysis of their products?

Temperature Effect

A PV module exposed to sunlight generates heat as well as electricity. For a typical commercial PV module operating at its maximum power point, only 10 to 15% of the incident sunlight is converted into electricity, with much of the remainder being converted into heat.

The key factors which affect the efficiency and thermal behavior of PV modules are:

a. The degree of reflection from the top surface of the module,

b. The electrical operating point of the module components,

c. Absorption of sunlight by regions not covered by solar cells,

d. Absorption of low energy (infrared) light in the module or solar cells,

e. The packing density of the solar cells, and

f. The overall quality of the solar cells and modules.

Each of these factors is seemingly independent in origin and function, but is inter-related in long-term effect on the module's efficiency and longevity. Any PV cell or module will lose ~0.5-0.6% of its total output per degree C increase of temperature. This is a significant number, which causes a lot of concern about the auditability of PV modules for long-term operation in extreme heat conditions.

Mechanical and Thermal Effects

There are a number of materials in the PV module that act and interact differently with adjacent materials, depending on temperature, pressure and other factors. The major mechanical forces determining the interaction between the different layers in the module are:

a. *Coefficient of friction*—the force that resists the lateral motion of solid surfaces in contact with each other and moving in different directions. This includes the types of materials (substrates and layers of materials stacked on top of each other) of which PV modules are made.

b. *Coefficient of thermal expansion* (linear and volumetric)—the way different materials expand and shrink under different temperature gradients. In other words, all materials tend to change shape and size along the surface and across their volume, with changes in temperature, which ultimately results in mechanical friction if in contact with other materials.

These forces also determine other events and phenomena that occur in the modules over time. Hot and cold days, windy conditions, hail, storms, etc. have effects on the modules, and all these events must considered in the design and operation of PV power plants. The relations and interactions between the module's components could be described as intimate, ongoing, and relentless. In the Arizona deserts, these materials must go through blistering 180°F temperatures during the day in summer and sub-freezing temperatures in winter, not to mention the effects of UV light. Also, moisture, air, environmental chemicals, and gasses penetrating the module will cause quick deterioration and even failure.

Since the coefficient of expansion of these materials is very different, they go through never-ending shifting, sliding and slipping, expanding and shrinking, constantly rubbing against each other—a soup of dissimilar molecules. This is where any poor design, or deviation from the materials or manufacturing specs will be revealed, most likely as decreased performance or failure.

Cracks, chips, pits, crumbling, voids, adhesion failures, delamination, chemical decomposition and reactions within the cells and the modules are expected during the 25-30 years of operation—due to natural elements. Some of these effects, such as yellowing of the EVA, or change of front surface color due to degrading of the AR coating, become visible to the naked eye. Most other internal and some external processes, however, go undetected.

Once the PV modules and BOS are installed, the power field is considered complete, and electricity starts flowing in the grid. There are a number of issues, in addition to the major ones discussed in more detail above (temperature and annual output degradation, etc.). Although these could be considered as less important, the

fact is that we are dealing with thousands of modules and BOS components, and each little "issue" amounts to a big (and often negative) result.

The obvious, sometimes serious, problems observed in the power fields of today, are mentioned below, and are things to watch for.

"Snail Trails"

There are number of reports of visible traces (looking like snail trails) left on the module surface, usually months and years after—but occasionally not long after—installation.

Recently a dozen or so world manufacturers have admitted observing this phenomena in their PV modules. The phenomena is not new, but since it does not cause any significant drop of power, it has been ignored until recently.

Recent studies have not provided a reason for, or solution to, the problem, but numerous explanations have been given for this phenomena: EVA composition, deterioration and/or outgassing, glass contact effects, surface contamination, chemicals contamination, metal grid contamination, moisture in the module, micro-cracks…

While most of the above are considered harmless, and do not significantly affect module performance, micro-cracks are a serious problem.

Micro-cracks in silicon or thin film modules can be created during the manufacturing process—overly stressed wafers coming from the wafer slicing saw; improper loading, unloading and handling; improper temperature regime; assembly into modules; packaging, transport, and installation.

At each step of the process, the fragile wafers and thin films are exposed to some sort of a mechanical and/or thermal stress, each of which (and their combined effect) must be reduced—or eliminated.

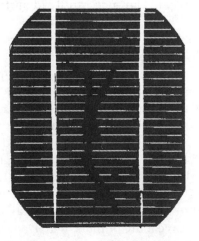

Figure 6-38. "Snail trails" on a solar module (under glass).

Figure 6-39. Potential micro-crack generation during PV modules installation.

Since there are many people handling the wafers, a number of special procedures are needed to ensure cells and modules safety. While some of the process steps are complex and do not allow good control of micro-cracks generation, some are simple and micro-cracks can be eliminated.

Looking at Figure 6-39, it's easy to see how this 150-lb technician stepping on the modules could put immense stress (approximately 75 lbs per square inch) on the glass pane over each cell, which will immediately be transferred to the solar cells directly beneath.

Any slight bend of the glass pane means a slight bend of the solar cells underneath. When a solar cell is bent, the silicon (or thin film) material (especially weak surface areas such as those in the metal contacts' proximity) will be overly stressed and very develop micro-cracks.

Micro-cracks in silicon solar cell or thin film modules are usually identified by a human observer—by a naked-eye inspection or with optical instruments. The electroluminescence (EL) method is most often used for this inspection, and can successfully identify any micro-cracks.

The problem is that once the micro-cracks are discovered, it is too late to do anything about it. The only question left at this point is how the micro-cracks affect the PV module performance.

Analysis of direct impact of micro-cracks on the module power and the consequences after artificial aging show that the initial effect on the module's power generation is small. But this is only the beginning.

When the module is exposed for a long time to the extremes in the field, the presence of micro-cracks is potentially crucial, if not detrimental, with aging. Micro-cracks have a life of their own. Depending on the module's mechanical stress and chemical interaction

(water penetrating into the module), the micro-cracks could grow large enough to affect its performance and eventually make it fail.

There is a real need for quantifying the risk of power loss in PV modules with micro-cracks, and the chance of a failure. This will allow the manufacturers to properly inspect, sort and discard defective modules with high risk of failure, while at the same time keeping modules with uncritical micro-cracks.

With today's lack of procedures (let alone standards) for such elaborate work, we need to design a method to evaluate the micro-cracks and to quantify the risk associated with each type and size.

As to micro-cracks in solar cells, we need to determine the upper limits of the potential power loss that we are willing to live with. This can be done by simulating the impact of inactive solar cell fragments on the power of a common PV module type and PV array. We show that the largest inactive cell area of a double string protected by a bypass diode is most relevant for the power loss of the PV module. A solar cell with micro-cracks, which separate a part of less than 8% of the cell area, results in no power loss in a PV module or a PV module array for all practical cases. Between 12 and 50% of inactive area of a single cell in the PV module, the power loss increases nearly linearly from zero to the power of one double string.

PID

Potential induced degradation (PID) is a new buzz word in the solar industry. Any conventional photovoltaic systems will develop voltage differences between the PV modules' frames and the solar cells assembled inside. This phenomena is especially pronounced under wet conditions (high humidity climates), where it can lead to undesired, and sometimes serious, leakage currents. This leakage can cause a substantial decrease in performance of individual cells, modules, strings of modules and the entire PV power field. Yield loss of 20% or more has been reported in some cases, but some manufacturers claim that it is not so serious a problem, the bad effects of which can be diminished and even prevented.

How this is done is a well kept secret, so it is up to us to figure out if our cells and modules suffer from the negative effects of PID or not.

PID or no PID, solar cells and modules exposed to extreme humidity need to be well encapsulated, sealed and periodically checked. This is very important, because regardless of how PID-safe the modules are, if and when the moisture barriers break down (and they will sooner or later), the moisture will penetrate inside the modules, and the solar cells will soon be damaged.

Fire Hazard

Hot spots and other fire hazard conditions are created in PV modules under different circumstances, but with equally disastrous results. A hot spot, for example, can overheat the module structure and under extreme circumstances can ignite the plastic materials, and melt the glass and metal covers, basically destroying the entire module, the array and/or the field.

A number of mechanical and electrical defects in modules—loose bus bars, poor connections, etc.—can also create a fire.

A number of standards and rules regulating the PV modules' construction and safety are mandatory, while some are equally important, but are not mandated. In all cases, there are ways to ensure the integrity and safety of modules—from their design, through manufacturing, installation and operation stages. Using proper standards and procedures, as well as careful execution is a keys to avoiding poor performance and fires.

Anytime a manufacturer or installer is negligent, or compromises, we have a potential disaster. In the summer of 2012 a large-scale PV installation in California was stopped indefinitely because a state safety inspector found that a large number (millions) of thin film PV modules did not have the Underwriters Laboratory (UL) approval, as required by the project's County-granted Conditional Use Permit (CUP), and as agreed on by all parties prior to initiating the project.

The inspector was reportedly doing a routine site inspection, in service to the CUP, when he discovered there was no UL approval for the panels' electrical connections. State of California building and safety codes require UL approval of electrical connectors in all electrical appliances, and *PV modules are no exception*. But why did it take the inspectors this long to discover the problem?

The project has been in the planning, permitting and design stages for years, and state authorities, including inspectors, have been involved since its inception, yet they had not noticed the discrepancy.

They had also visited the site on several occasions prior to this "discovery," without noticing the lack of the UL stamp, so this is just another confirmation of the immaturity of the solar industry. The standards and procedures were not well defined and/or executed, and that's how this project got an embarrassing and costly delay.

This approval is apparently not required, or its absence was not noticed, in other states such as Nevada where millions of similar panels were installed. It is also a standard to which some companies have not been held at installations done elsewhere in California.

In this particular case, if the panels must be reworked and certified, the effort will cost a lot of time and money, in addition to causing long, expensive delays to the projects. A number of projects can be affected—close to 1.5 GW, to be exact.

And if the inspectors let the projects continue with the "as is" panels—without requiring UL certification of the connections, then we might be looking at a burning field just like the one pictured above some day soon.

Even worse, these particular PV panels contain significant amounts of the toxic and carcinogenic heavy metal cadmium, which in case of fire could contaminate thousands of desert land acres, and harm life in the vicinity as well.

Note: TÜV Rheinland and Currenta have developed a new flammability test for PV modules, based on

Figure 6-40. PV module arcing power

Figure 6-41. PV module on fire

Figure 6-42. PV array on fire

Figure 6-43. PV field, partially burnt by an electric fire

DIN EN ISO 11925-2. The test is developed in response to Germany's plan to include PV modules and solar collectors in the German Building Rules list by the end of 2012.

This test will affect all PV modules and solar collectors and is the first time Deutsches Institut für Bautechnik in Berlin is considering solar products to be subjected to building regulations. In addition to their other tests, manufacturers must verify "normal flammability" of their products via this new test method.

Verification will be done by a conformity certificate from a certification body, and will be a prerequisite for issuing the compliance certificate and building permits required in Germany.

The new flammability test is based on an extensive research project on preventive fire protection in PV systems conducted by TÜV Rheinland, in conjunction with the Fraunhofer Institute for Solar Energy Systems, the German Solar Industry Association, the German Federal Ministry for the Environment and other partners.

As an additional measure, the new tests verify the PV modules' resilience to the effects of external fire in accordance with IEC 61730 and ANSI/UL 1703.

This is great news and a good step towards ensuring the safety of the ever-increasing number of solar installations, some of which contain toxic and flammable materials.

SOLAR POWER FIELDS IN THE 21ST CENTURY

Now we know all there is to know about solar technologies, their materials, and the devices design, manufacturing, and operation. With this information we should be able to predict the future of solar energy in the 21st century… but, the field is changing very fast, with new developments recorded daily.

One thing we know: wind and solar energy power generation will continue to increase. There is simply no other choice, as fossils are being depleted at an amazing speed, and the rising prices associated with their scarcity will force change. Bring it.

CSP Technologies' Future

The dark areas in Figure 6-44 show where the EU community is planning to install a large number of wind, CSP, and PV power plants. CSP technologies will take a major part in this effort which will bring the CSP industry to a new high by 2020—and multiply power output several-fold by 2050.

This, in addition to plans for many additional GWs

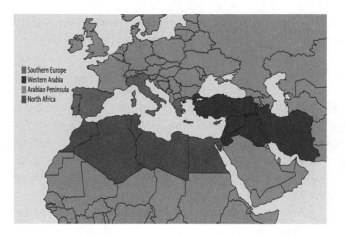

Figure 6-44. CSP technologies: future applications

of CSP installations in the US and Asia, is a very exciting development which paints a bright picture for the future of the CSP technologies as a world-class electric power generator.

We have witnessed the operation of some of the CSP systems over the long term, as well as the quick and successful development new ones of late. We have also seen estimates that place CSP technologies at the top of the list of technologies for use in large-scale power generation.

The only serious issue in need of resolution is the lack of water in the deserts. CSP heat processing equipment—steam generators, turbines, condensers, etc.—need a lot of water to operate efficiently. We're talking about millions of gallons of fresh water lost to heat exchange and evaporation every day.

This is a lot of water to use in the driest places on earth, which is simply not possible, so the technology must be modified to use other cooling methods. Dry cooling is one potential candidate to replace fresh water cooling, but it uses much more expensive equipment and wastes a lot of precious energy. Still, alternative cooling is absolutely necessary for this technology to take its place as a major provider of electricity from the deserts.

The fast pace of the technological developments today, and the quickly changing socio-economic and political climates around the world, however, make it hard to predict what will happen, but we know for sure that CSP is here to stay and that it will grow steadily.

PV Technologies' Future

The future of solar power lies in the full development of the key PV technologies, and their deployment in the different energy markets. Due to their flexibility and ease of use, PV equipment has found a place in many areas of energy generation, and many more are being developed.

Building large-scale PV power plants in the deserts, using efficient and reliable PV technologies, is the best and fastest way to ensuring our energy independence and cleaning the environment.

Important details about PV equipment include:

1. Does not require water for cooling (a major problem with CSP power plants),

2. Operates at relatively low temperature (as compared with CSP technologies,

3. Has very few moving parts (also a problem with CSP equipment),

4. Can be installed in small sizes and on rooftops (none of which is possible with CSP).

We clearly see a lot of PV power plants, small and large—on house roofs, on commercial buildings and in the deserts—popping up in large numbers during the 21st century. The pace and size of these installations will increase with time, simply because future generations will have no choice.

Conclusions

Some of the key issues to be addressed are:

1. Principals must work together. Manufacturers, suppliers, distributors, customers, developers, investors, insurance agents must have the same goals and use the same procedures.

2. Manufacturers must produce quality product, with quality materials delivered by the suppliers. Distributors and developers must provide the best products and services possible.

3. Standardization of these processes will give our solar projects much greater chances of success.

4. Overall solar project implementation (construction and interconnect permitting, PPAs, field design, securing adequate performance insurance, as well as proper installation and O&M procedures) are not well defined today, but new standards are on the horizon.

The issues in 1, 2 and 3 above are adequately understood and well underway toward improvement. The hardest part of the work ahead, as described in point 4, consists of optimizing and coordinating the different steps of the solar projects; permitting, financing, field design, implementation and O&M, all of which are ever changing technically and administratively. They are less understood, and much more difficult to implement.

The critical steps of solar project design, implementation and O&M consist of:

• Professional site and project assessment
• Proper permitting (construction, PPA, ROW, interconnect, etc.)
• Proper budgeting for 30 years non-stop operation under different conditions (weather variability, equipment deterioration, labor issues, socio-economic and political changes, etc.)
• Obtaining adequate financing for the different stages of the project development
• Proper project design and optimization
• Obtaining quality materials and executing proper testing at all steps of the project
• Faultless project installation and commissioning
• Efficient and profitable long-term project operation and maintenance
• Sufficient customer enablement (for the duration)

If and when the steps listed above are well defined, standardized and properly executed, an entire project has a good chance for efficient, profitable operation. Only then that we can call our PV products and power fields successful and profitable.

The work in many of the above outlined areas is not going to be easy, cheap or quick. The solar energy baby is just now beginning to crawl, so it will take a lot effort before it walks and even longer before it can run alongside conventional energy sources.

One last, and most important thing: let's remember the lessons of the 2007-2011 boom-bust cycle. Unlike the ones before it, this one showed us (take a look at the incredible developments in Germany, Spain and the US) that solar energy generation is feasible and practical, that it can compete, and that it holds the promise for energy independence and a clean environment in the 21st century:

Here are some of the lessons to remember, and some of the most serious mistakes that are not to be repeated:

• Government subsidies work, but are not a long-term solution.

Communism was built on government subsidies, but it did not work! Obama threw a lot of taxpayers' money at the solar industry, but it did not work. Most of the subsidies created chaotic situations, and many resulted in embarrassing and wasteful enterprises. Looking back, half of the money went to buying foreign products and services. This should never happen again!

• Political maneuvers, in most cases, cause more harm than good.

The US Department of Energy, BLM, EPA and a number of other political and semi-political entities did not understand their roles and complicated the already complex situation.

• Large companies and projects can fail too.

Solyndra, GE, Applied Materials, to mention a few in the US, along with over 50 other large, medium, and small companies here and abroad failed. Some have shut their doors, while others are still struggling. Poor understanding of the technology, hastily jumping into a crowded market, and unwise use of funding are basic factors contributing to their fall.

• Quick money-making schemes and actions give the industry a bad name.

During the solar boom of 2007-2011, we witnessed thousands of businesses change their name, or add "solar" to it. Anyone who could hold a screwdriver, or had some physics education, jumped at the chance to make a quick buck. It didn't work!.

• Solar installations in cloudy areas have marginal success at best.

Germany's case is an exceptional example of using solar where solar doesn't belong. The Germans spent billions of dollars, making their solar installations 10 to 100 times more expensive than conventional energy sources.

• Solar power is variable, so more efficient ways to handle it are needed.

Utilities and power grid operators had a (bad) taste of the variability of solar installations. New ways of power management, such as adding and using smart grid components (in addition to energy storage) will need to be designed and implemented, in order to provide a more constant output.

• Large-scale desert solar installations are the way to energy independence.

The debate of distributed vs. large-scale solar power generation started during the 2007-2011 solar boom. Our 30 years experience tell us that large-scale power fields—those over 100MWp—are the answer, if we are talking about serious power generation.

• Efficient energy storage is needed to make solar power feasible.

Energy storage will alleviate the problem of solar interference caused by clouds and weather. It is neither easy nor cheap, but is becoming more important by the day. We need to solve this issue.

• All solar technologies do not perform equally.

There are many solar technologies, each with its own advantages and disadvantages. Special consideration must be given to where each technology is used. At present, this fact is not well understood.

• Using cheap materials and labor do not lead to long-term success.

The solar cells and modules manufacturing "explosion" followed by an inevitable "implosion" led by Chinese manufacturers teaches us that only a few manufacturers can be trusted, and even then with a great deal of caution. Hundreds of companies, subsidized by the Chinese government, converted overnight into "solar specialists" quickly producing large quantities of questionable product.

Yes, we must question the quality of these products because PV cells and modules are not toys; they are serious technological tools that are expected to operate hellish conditions for 30 years, and they come with toxic elements.

Yes, we suspect that we will see a lot failures of Chinese made PV modules in the years to come. But at the end, the fault is not theirs for they did not make us buy the untested and unproven modules. We bought them with Obama's money and

Notes and References

1. A Manual for the Economic Evaluation of Energy Efficiency and Renewable Energy Technologies, http://www.nrel.gov/docs/legosti/old/5173.pdf
2. INACCURACIES OF INPUT DATA RELEVANT FOR PV YIELD PREDICTION Stefan Krauter, Paul Grunow, Alexander Preiss, Soeren Rindert, Nicoletta Ferretti Photovoltaik

Institut Berlin AG, Einsteinufer 25, D-10587 Berlin, Germany

3. PV module characterization, Stefan Krauter & Paul Grunow, Photovoltaik Institut Berlin AG, TU-Berlin, Germany

4. PV MODULE LAMINATION DURABILITY, Stefan Krauter, Romain Pénidon, Benjamin Lippke, Matthias Hanusch, Paul Grunow, Photovoltaic Institute Berlin, Berlin, Germany

5. PV module testing—how to ensure quality after PV module certification. Photovoltaic Institute, Berlin. August, 2011

6. Hot Spot Evaluation of Photovoltaic Modules, Govindasamy (Mani) TamizhMani and Samir Sharma, Photovoltaic Testing Laboratory (ASU-PTL), Arizona State University, Mesa, AZ, USA 85212

7. IEC, TC 82, Solar Photovoltaic Energy Systems, SMB/4212/R, 02/2010

8. http://solarbuyer.com/

9. Accommodating high levels of variable generation. NERC, April 2009.

10. Photovoltaic array performance model, D.L. King, W.E. Boyson, J.A. Kratochvil Sandia National Laboratories, Albuquerque, New Mexico 87185-0752

11. Testing the reliability and safety of photovoltaic modules: failure rates and temperature effects," Dr. Govindasamy TamizhMani 2010, "Photovoltaics International, Vol. 8, pp. 146-15

12. Adhesion and Thermomechanical Reliability for PVD and Modules, by Reinhold H. Dauskardt.

13. Hot Spot Evaluation of Photovoltaic Modules," by Govindasamy (Mani) TamizhMani and Samir Sharma

14. PVCDROM, http://www.pveducation.org/pvcdrom/by Christiana Honsberg and Stuart Bowden

15. Integrating Large-Scale Photovoltaic Power Plants into the Grid, Peter Mark Jansson, Richard A. Michelfelder, Victor E. Udo, Gary Sheehan, Sarah Hetznecker and Michael Freeman, Rowan University/Rutgers University/PHI/Suntechnics Energy Services/Exelon Energy

16. Executive Summary, 2009 Top Ten Utility Solar Rankings, SEPA, 2009

17. Modeling of GE Solar Photovoltaic Plants for Grid Studies, GE, 2009 Mesa, AZ, USA 85212

18. Mora Associates, internal report

19. 2009 global PV cell and module production analysis, Shyam Mehta | GTM research

20. Large Solar Plant Woes: Will it Open the Door for Smaller Projects? Greentechsolar, January 2010

21. IREC Net Metering Rules http://www.irecusa.org/fileadmin/user_upload/ConnectDocs/NM_Model.pdf.

22. Decoupling Utility Profits from Sales, SEPA, Feb. 2009

23. Wikipedia

24. Photovoltaic Incentive Programs Survey, SEPA, Nov. 2009

25. PV YIELD PREDICTION FOR THIN FILM TECHNOLOGIES AND THE EFFECT OF INPUT PARAMETERS INACCURACIES. Stefan Krauter, Alexander Preiss, Nicoletta Ferretti, Paul Grunow Photovoltaik Institut Berlin AG (PI-Berlin) Einsteinufer 25, D-10587 Berlin, Germany

26. POLARIZATION EFFECTS AND TESTS FOR CRYSTALLINE SILICON CELLS, Simon Koch, Christian Seidel, Paul Grunow, Stefan Krauter* and Michael Schoppa Photovoltaik Institut, Berlin

27. Photovoltaics for Commercial and Utilities Power Generation, 2011. Anco S. Blazev

Chapter 7

Solar's Importance, Finance, and Regulations

We cannot direct the wind but we can adjust the sails.
—Dolly Parton

Adjusting the sails of the solar industry is absolutely necessary—especially at this critical juncture of strong winds blowing in various directions. Many variables—governments, regulators, utilities, investors, technology gurus, supply chain companies, manufacturers—complicate the picture immensely and determine the overall direction of the solar energy generating industry.

Each variable has an effect on the business. Conflicting opinions and interests have several times put the solar industry into hibernation. These cycles began in the 1970s with the oil embargo. Solar looked *good* when oil was scarce.

Today gas and oil fracking are the current best sellers on the energy market, and solar is again at the bottom. Still, we must keep working at this technology to ensure the progress of the renewable energy industry. This is a team effort; one energy source alone cannot accomplish all the tasks needed for long-term energy independence.

Below we will take a close look at the major drivers that determine the conditions for solar energy development.

FINANCIAL CONSIDERATIONS

The Stakeholders

Who are the key players in today's solar energy business and related markets? Who assures the proper execution of the steps related a PV energy field's environmental issues? Who pays the bills? Who takes the risks?

Government agencies such as DOE, BLM, EPA, CPUC, AZCC, and others are fully engaged in solar projects and are well aware of the existing and potential problems of the different PV technologies. They are tracking the developments in the areas where PV is to be deployed in large quantities, and they are responsible project safety.

They know that the most likely areas for large-scale PV installations are those of higher sunlight intensity,

especially the desert areas of Utah, Colorado, California, Arizona, and Texas, and are working on making those available for solar energy generation. But, these harsh climates provide a serious challenges to all PV technologies, so how are we going to manage the best locations that offer the worst working conditions?

Because there are no official, uniform standards and procedures outlining individual and collective responsibilities a lot of improvisation occurs.

Here is a list of major players in each of the PV installations with varying responsibilities from project to project:

1. Manufacturers, suppliers and dealers of PV modules, and BOS
2. Independent system operators
3. Utilities
4. Public utility commissions
5. Developers and installers
6. Investors and financiers
7. Public and private land owners
8. Mining industry executives/managers
9. Transport companies
10. Recycling facility operators
11. EPA regions/headquarters
12. BLM and other agencies managing land and water
13. Federal and state entities
14. DOE and the National labs and technological institutions
15. Federal and state regulators
16. Tribal governments and communities
17. City governments and chambers of commerce
18. Corporate and local planning commissions
19. Environmental organizations
20. Private parties and neighbors affected by the developments

All these entities are in some way involved in, and responsible for, the solar power plants' planning, finance, design, installation, operation, safety, regulation, and decommissioning.

One of the major problems is that many new players are entering the field only because it promises to be lucrative. Solar "specialists" are arising out of desire more often than experience.

The governing authorities and decision-makers, like their private counterparts, don't have the necessary background and experience to understand the whole PV spectrum well enough to make good decisions. The lack of standards and uniform procedures complicates things further. Things are getting clearer as time goes by, but smooth sailing is still a bit far.

The major issues that we propose for debate and resolution are:

1. The evaluation and implementation of PV technologies coming from new, unproven manufacturers, for efficiency, reliability, and longevity in large-scale PV power fields.

2. The evaluation and implementation of PV technologies containing toxic materials, and their long-term use in large-scale PV power fields in the US deserts.

3. Standardization of the sand-to-grave cycle of PV products.

4. Standardization, permitting, and regulatory procedures must be optimized.

5. The US power grid and its distribution channels must be optimized and properly maintained to support new large-scale solar and wind power generators.

6. Government and state subsidies must be re-evaluated and given to companies, projects and technologies that can guarantee the money won't be wasted.

In Summary

The solar industry's boom-bust cycle of 2007-2012 was the worst on record. What we learned is that there are still a number of unresolved issues which deeply affect solar industry development.

- Government subsidies work, but are not a long-term solution.
- Large companies and projects can fail too.
- Quick money-making schemes give the industry a bad name.

- Solar installations in cloudy areas have marginal success at best.
- Solar power is variable, so efficient installations with energy storage are needed.
- Large-scale desert solar installations are the way to energy independence.
- All solar technologies do not perform equally.
- With materials and labor, we get what we pay for.

Finances and their ramifications are important and must be well understood—be it for residential, commercial or utilities applications. This includes everything from pre-project financial assessment to periodic analysis and optimization of project implementation and plant performance.

According to NREL, the most complete analysis of an investment in a technology or a project, requires the analysis (estimates) of each year of the life of the investment, taking into account relevant direct costs, indirect and overhead costs, taxes, and returns on investment, plus any externalities, such as climate and environmental impacts relevant to the decision to be made.

However, it is important to consider the purpose and scope of a particular analysis at the outset because this will prescribe the course to follow. Also, the ultimate use of the results of an analysis will influence the level of detail undertaken. Decision-making criteria of potential investors must also be considered.

ECONOMIC EVALUATION OF RENEWABLE ENERGY TECHNOLOGIES

Introduction

Below is an overview of procedures involved when developing the key financial elements of a solar product or installation.

1. Net present value (NPV),
2. Total life-cycle cost (TLCC),
3. Revenue requirements (RRs)
4. Levelization (LCOE),
5. Annualized value (AV)
6. Unadjusted and modified internal rate of return (IRR),
7. Simple payback (SPB),
8. Discounted payback (DPB),
9. Benefit/cost (B/C) ratios, and
10. Savings/investment ratios (SIR).

The descriptions and formulas in this section can be used to compare specific alternative investments and/or projects economics on an individual basis. However, the analyst should be aware that when comparing alternatives, different measures may not always provide the same answer.

For example, simple payback requirements reject an alternative that, while having a longer payback period, has strong long-term returns.

Here is a description of the subjects we consider absolutely necessary for a basic economic evaluation and analysis of solar projects, which MUST be done before anything else is even considered (1):

Net Present Value (NPV)

The NPV of a project is one way of examining costs (cash outflows) and revenues (cash inflows) together. A NPV analysis can be composed of many different cost and revenue streams. The analyst must know the form of the different streams (current or constant dollars), so the correct discount rate can be used for the present value analysis. (1)

Alternatively, cash flows can be adjusted to reflect the form of the discount rate.

The formula for NPV can be expressed as:

$$NPV = \sum_{n=0}^{N} \frac{Fn}{(1+d)^n} = F_0 + \frac{F1}{(1+d)^1} + \frac{F2}{(1+d)^2} + \ldots + \frac{FN}{(1+d)^N}$$

Where

NPV	=	net present value
Fn	=	net cash flow in year n
N	=	analysis period
d	=	annual discount rate.

Total life-cycle cost (TLCC)

TLCC analysis is used to evaluate differences in costs and the timing of costs between alternative projects. TLCCs are the costs incurred through the ownership of an asset over the asset's life span or the period of interest to the investor.

Only those costs relevant to the decision should be included in the analysis. TLCC analysis considers all significant dollar costs over the life of the project. These costs are then discounted to a base year using present value analysis.

Any revenue generated from the resale of the investment is also discounted to the base year and subtracted from present value costs.

The formula for calculating TLCC is as follows:

$$TLCC = \sum_{n=0}^{N} \frac{Cn}{(1+d)^n}$$

Where

TLCC	=	present value of the TLCC
Cn	=	cost in period n: investment costs include finance charges as appropriate; expected salvage value; nonfuel O&M and repair costs; replacement costs; and energy costs
N	=	analysis period
d	=	annual discount rate.

Revenue Requirements (RRs)

RRs are often calculated in regulated industries, such as the electric power markets, for a firm (utility) as a whole. The discussion here is directed toward the revenues required to cover the costs associated with a specific project, not a firm.

The RR is the total revenue that must be collected from customers to compensate a firm for all expenditures (including taxes) associated with a project. It is actually no different than the before-tax revenue required of the TLCC outlined earlier.

The RR economic measure is recommended as an appropriate measure when evaluating a regulated investment. The general decision rule for utilities is to choose the alternative for which the present value of the multi-period investment revenue requirement is the lowest (assuming the services provided are identical).

Businesses apply the RR method to project costs over an investment's useful life. Because RR is normally the basis on which regulated investments are defended, it is the recommended measure for evaluating such investments.

RR is not recommended for economic evaluation when deciding whether to accept or reject an investment because RR provides no frame of reference for what are acceptable or unacceptable costs and benefits, and returns are not addressed.

The formulation of the revenue requirements approach is:

$$RR = TLCC = \frac{I - (T*PVDEP) + PVOM\,(I - T)}{(I - T)}$$

Where

RR	=	revenue requirements
TLCC	=	total life-cycle cost
I	=	initial investment
T	=	income tax rate
PVDEP	=	present value of depreciation expenses
PVOM	=	present value of operating and maintenance expenses

The Levelized Cost of Energy (LCOE)

LCOE allows alternative technologies to be compared when different scales of operation, different investment and operating time periods, or both exist. For example, the LCOE could be used to compare the cost of energy generated by a renewable resource with that of a standard fossil fueled generating unit.

The LCOE can be calculated using the following formula:

$$\sum_{n=1}^{N} \frac{Q_n * LCOE}{(1+d)^n} = TLCC$$

$$LCOE = TLCC / \left\{ \left[\sum_{n=1}^{N} \left(Q_n / (1+d) \right)^n \right] \right\}$$

Where

    LCOE = levelized cost of energy
    TLCC = total life-cycle cost
    Q_n = energy output or saved in year n
    d = discount rate
    N = analysis period.

It is important to note that if the system output (Q) or savings remains constant over time, the equation for LCOE can be reduced as follows:

$$LCOE = (TLCC / Q)(UCRF)$$

Where

    TLCC = total life-cycle cost
    Q = annual energy output or saved
    UCRF = the uniform capital recovery factor

$$\frac{d(1+d)^N}{(1+d)^N - 1}$$

The shortcut methodology for estimating the LCOE-required before-tax revenues requires the assumptions that the project not only have constant output, but also constant O&M and no financing. If these assumptions can be made, the shortcut can be used with the application of the following formula:

$$LCOE = \frac{I * FCR}{Q} + \frac{O\&M}{Q}$$

Where

    LCOE = levelized cost of energy
    I = initial investment
    FCR = fixed charge rate, in this case the before-tax

revenues required FCR
    Q = annual output
    O&M = annual O&M and fuel costs for the plant.

Annualized Value (AV)

The annualizing process transforms a swing of cash flows (Fn) into equivalent annual streams. Cash flows are discounted to their net present value and then annualized by multiplying the present value of the cash flow by the uniform capital recovery factor (UCRF); i.e., $[d(1+dy)/[(1+d)'' - I]$.

This is similar to the annualization of required revenues in the Revenue Requirements subsection and can be done with a single equation that combines the NPV and capital recovery factor calculations.

$$NPV = \Sigma^N_{n=1} F_n / (1+d)^n$$

$$AV = UCRF * NPV = UCRF * \Sigma^N_{n=1} F_n / (1+d)^n$$

Where

    NPV = net present value
    AV = annualized value
    Fn = cash flow in period n
    UCRF = uniform capital recovery factor
    d = discount rate.

This formula can be simplified if the cash flow Fn is escalating at a constant rate ΔP:

$$NPV = \sum_{n=1}^{N} [F_0(1+\Delta P)^n] / (1+d)^n$$

$$= F_0 \sum_{n=1}^{N} * k^n$$

$$= F_0[k(1-k^n)/(1-k)]$$

$$AV = UCRF * NPV$$

$$AV = UCRF * F_0[k(1-k^n)/(1-k)]$$

Where

    Fo = cash flow to be annualized
    ΔP = rate of escalation
    k = $[(1+\Delta P)/(1+d)]$.

The Internal Rate of Return (IRR)

IRR for an investment that has a series of future cash flows (Fo, F1, ... Fn) is the rate that sets the NPV of the cash flows equal to zero. The IRR analysis allows for the comparison of a wide variety of investment activities. However, IRR is not recommended when evaluating investments in which further investment after return is required because downstream investments are improperly discounted and multiple positive PRR values can occur.

IRR is commonly used for accept/reject decisions, allowing a quick comparison with a minimum acceptable rate of return (hurdle rate). IRR is not recommended when selecting among mutually exclusive alternatives because the values of differing investment sizes are not considered (an investor will choose to invest more in an alternative with more favorable returns).

This shortcoming may be corrected by applying IRR on an incremental basis. IRR is also not recommended when ranking projects because IRR implicitly assumes reinvestment of returns at the IRR.

The viability of a project is usually assessed with IRR by comparing the IRR of the project with a "hurdle rate" that represents the IRR of the next best alternative (the opportunity cost of capital).

IRR also has the advantage that it can be used to directly compare the after-tax return on readily available financial instruments, such as bonds, thereby providing the investor with a quick qualitative assessment of the project.

IRR equals the rate for which:

$$0 = NPV = \sum_{n=0}^{N} [F_n / (1 + d)^n]$$

Where

NPV = net present value of the capital investment

Fn = cash flows received at time n

d = rate that equates the present value of positive and negative cash flows when used as a discount rate

Modified Internal Rate of Return (MIRR)

For projects of different scale and lives, it is possible for different ranking criteria, such as NPV and IRR to produce conflicting results because of differing reinvestment assumptions.

The NPV method assumes reinvestment at the discount rate, whereas the IRR method assumes reinvestment at the IRR rate.

The modified internal rate of return (MIRR) accounts for varying reinvestment rates and should be used in these circumstances. MIRR is calculated by assuming that all cash inflows received before the end of the analysis period are reinvested at the discount rate until the end of the analysis period.

The terminal (future value) mount at the end of the analysis is then discounted back to the base year. The discount rate that will equate the present value (in the base year) of the terminal amount to the present value of all investment costs is the MIRR.

The formula for calculating MIRR is as follows:

MIRR = solution for "r" in the following equation:

$$MIRR = r, \text{ where } \sum_{n=0}^{N} \frac{F_{n_n}}{(1+d)^2} = \sum_{n=0}^{N} \frac{F_{p_n}(1+d)^{N-n}}{(1+r)^N} =$$

Where

MIRR = r, modified internal rate of return

Fnn = net negative cash flows at time n

Fpn = net positive cash flows a time n

d = rate of return of reinvestment

N = life of investment

Simple Payback (SPB)

SBP is a quick, simple way to compare alternative projects. It is relatively easy to utilize and as such is a very popular financial tool. Simple payback is the number of years necessary to recover the project cost of an investment under consideration.

SPB is commonly used and recommended when risk is an issue (i.e., significant uncertainties are present) because SPB allows for a quick assessment of the duration during which an investor's capital is at risk.

SPB is not recommended when evaluating alternatives involving financing and tax features because their inclusion complicates the analysis and it loses the advantages of simplicity.

SPB is also not recommended for selecting among mutually exclusive alternatives because the values of differing investment sizes are not considered; i.e., an investor will choose to invest more in an alternative having more favorable returns.

This shortcoming may be corrected by applying SPB on an incremental basis. When ranking projects, SPB is not recommended because it ignores returns after payback.

The SPB can be expressed as the first point in time at which:

$$\sum n * \Delta I_n \leq \sum n * \Delta S_n$$

Where

SPB = minimum number of years required for the non-discounted sum of annual cash flows net annual costs to equal or exceed the non-discounted investment costs

ΔI = non-discounted incremental investment costs (including incremental finance charges)

ΔS = non-discounted sum value of the annual cash flows net annual costs.

The Discounted Payback Period (DPP)

DPP is the number of years necessary to recover the project cost of an investment while accounting for the time value of money.

DPB is recommended when risk is an issue (i.e., significant uncertainties are present) because DPB allows for a quick assessment of the duration during which an investor's capital is at risk.

DPB is not recommended when evaluating alternatives involving financing and tax features because their inclusion complicates the analysis and it loses the advantages of simplicity.

DPB is also not recommended when selecting among mutually exclusive alternatives because differing investment sizes are not considered (i.e., an investor will choose to invest more in an alternative with more favorable returns).

This shortcoming may be corrected by applying DPB on an incremental basis. When ranking projects, DPB is not recommended because it ignores returns after payback.

DPB can be expressed as that time in which:

$$\sum n[\Delta I_n / (1 + d)^n] \le \sum n [\Delta S_n / (1 + d)^n]$$

Where
- DPB = minimum number of years required for the discounted sum of annual net savings to equal the discounted incremental investment costs
- ΔI = incremental investment costs
- ΔS = annual savings net of future annual costs (i.e., ΔS equals the incremental energy costs, incremental nonfuel operation, maintenance, and repair costs, incremental repair and replacement costs, minus the incremental salvage costs)
- d = annual nominal discount rate

Benefit-to-cost (BIC)

BIC analysis is conducted and B/C ratios are derived to ascertain whether and to what degree the benefits of a particular project exceed the costs. Such analyses are more frequently performed for projects that involve the public interest. They range from direct government investments in public works (such as infrastructure, parks, or public power projects) to private investments by utilities in which the impacts on the ratepayers, investors, and the environment all may have to be considered to determine if the action is appropriate.

Because B/C ratios are commonly used when evaluating investments from a societal perspective, they are recommended for use when evaluating projects in which social costs prevail.

BIC ratios are not recommended when selecting among mutually exclusive alternatives because the values of differing investment sizes are not considered; i.e., an investor will choose to invest more in an alternative with more favorable returns.

This shortcoming may be corrected by applying B/C ratios to the incremental benefits/costs of the alternatives predominate, from a public policy perspective.

BIC ratios can also be used when ranking projects if benefits predominate, particularly from a public policy perspective.

The mathematical formula for B/C ratios is:

$$B/C = [PV (All\ Benefits)] / [PV(All\ Costs)]$$

Where
- PV (All Benefits) = present value of all positive cash flow equivalents
- PV (All Costs) = present value of all negative cash flow equivalents.

Savings-to-investment Ratio (SIR)

A variation of the B/C ratio, known as the savings-to-investment ratio (SIR), is used when benefits occur primarily as cost reductions. Although SIR is basically a B/C ratio, there is a computational difference. B/C ratios are typically calculated with dl benefits in the numerator and all costs in the denominator.

In the calculation of SIR, only the principle investment costs are included in the denominator. All other costs are subtracted from the benefits in the numerator. This, in essence, makes the numerator a measure of net savings.

The formula for the calculation of the SIR is as follows:

$$SIR = PV (NS) / PV (IRS)$$

Where:
- SIR = savings investment ratio
- PV(NS) = present value of net savings of the activity (i.e., total savings minus all incremental operation and repair costs not directly attributed to incremental investment costs)
- PV(IRS) = present value of the sum of incremental investment costs plus incremental replacement

costs minus incremental salvage values

Now we know the financial factors in play, and/or what is required to fund, install and operate a solar power plant. Assuming that we have done everything right and the plant is up and running properly, let's take a close look at the day-by-day operation from a financial point of view.

Subsidies and Taxes Update, 2012

The status of government subsidies, taxes and investments, related to renewable energy sources and projects for the 2012-2018 time period is as follow:

Production Tax Credit (PTC)

Applicable for wind, geothermal, landfill gas, trash combustion, open-loop biomass, closed-loop biomass, hydropower and wave tide.

- The PTC provides a tax credit for the production of electricity from renewable sources and the sale of that electricity to an unrelated party.

- Credit amount is:
  — 2.2 cents per kWh for wind, closed-loop biomass and geothermal, and
  — 1.1 cents per kWh for other renewable energy resources.

- Available for facilities placed in service before 1 January 2014 (2013 for wind).

- Available for a 10-year period beginning the year the facility is placed in service.

Investment Tax Credit (ITC)

Applicable for solar, geothermal, qualified fuel cell or micro turbine property, combined heat and power systems, small wind, geothermal heat pumps and PTC-eligible facilities placed in service after 2008 and before 2014 (2013 for wind).

- The ITC provides a credit for qualifying energy property.

- The ITC for any taxable year is the energy percentage of the basis of each energy property placed in service during the taxable year.

- Credit amount is:
  — 30% of eligible costs for fuel cell, solar, and small wind property, and
  — 10% of eligible costs for combined heat and power, microturbine property and geothermal heat pumps.

- The ITC is generally available for eligible property placed in service on or before 13 December 2016.

Grant in Lieu of PTC and ITC

Applicable for tangible personal property or other property that is an integral part of a qualified facility (as defined by the PTC and ITC rules).

- The American Recovery and Reinvestment Act (ARRA) enacted a new grant program which provides a cash grant in lieu of the PTC or ITC.

- ARRA permits PTC or ITC projects to elect a grant of up to 30% of costs of construction of PTC or ITC energy property in lieu of tax credits.

- Projects must begin construction before 2012 and submit a grant application no later than 1 October 2012.

- Projects must be placed in service before their PTC or ITC credit expires:
  — PTC expires in 2014 (2013 for wind)
  — ITC expires in 2017.

Renewable Portfolio Standards (RPS)

These standards generally place an obligation on electric supply companies to produce a specified fraction of their electricity from renewable energy sources and numerous mechanisms that are permitted to achieve compliance, such as renewable energy credits (RECs).

Currently no federal RPS legislation has been enacted.

A total of 29 states and the District of Columbia have an RPS. The states include Arizona, California, Colorado, Connecticut, Delaware, Hawaii, Illinois, Indiana, Kansas, Maine, Maryland, Massachusetts, Michigan, Minnesota, Missouri, Montana, Nevada, New Hampshire, New Jersey, New Mexico, New York, North Carolina, Ohio, Oregon, Pennsylvania, Rhode Island, Texas, Washington and Wisconsin.

Armed with the financial basics, we can take a close look at what happens during long-term operation of a solar power plant, and what is needed to ensure its financial success.

PRACTICAL EXAMPLES

Keeping in mind what we discussed thus far, we can apply this knowledge to a practical example of a (virtual) PV power plant, operating 30 years in the Arizona desert.

Pre-construction Considerations

Our new virtual PV plant is 100 MWp and is to be installed in an appropriate location in a US desert. This particular location has 300 days of full sunshine and averages 6 hours per day of full output. The average solar insolation is 900 W/m².

For the duration we estimate, and should expect:

Total solar insolation **per m²** per day = 6 hrs/day x 900 W/m² = 5.4 kWh/m²/day.

Total solar insolation **per m²** per year= 300 days x 6 hrs/day x 900 W/m² = 1,620 kWh/m²/year

This means that we can expect ~1,620 kWh sunlight to fall on each square meter of that location over the entire year. Or, if we had 100% efficient technology we could produce 5.4 kWh from each square meter of land each day, and 1,620 kWh from each square meter during an entire year.

This sounds quite good, BUT:

1.   There is no such thing as 100% efficient PV technologies, so we'll use 15% c-Si PV technology for this example.

2.   We cannot cover 100% of the land with modules, so we need much more land for inter-row spacing, to avoid shading from adjacent modules.

3.   300 days per year and 6 hours per day of full sunlight are averages we use here (Arizona desert), based on average summer-winter sunlight utilization, and related weather variations.

4.   There are other variables, which change noticeably from hour to hour and day to day—even in the deserts—so the numbers could change drastically in one or other direction.

Keeping in mind the above factors, and seeking the best estimate for the operation of this new PV installation, we start with theoretical calculations which go something like this:

A 100-MWp (nameplate) power installation will require 100 MWp of PV modules, fixed-mount or on trackers. Assuming that the modules are 200 Wp each, we'll need 500,000 modules to generate 100 MWp DC electric power.

If we are using trackers, the number of modules to obtain the 100 MWp nameplate rating would be the same, but the daily and annual energy output will be greater. Tracking modules produce more power during the early and late hours of the day simply because by tracking they receive more sunlight than fixed-mount modules all through the day. Around noon, both types have almost the same power output.

Note: The nameplate number of 100 MWp is arrived at by considering the maximum output of the modules at STC (25°C module temperature and 1000 W/m² insolation) as measured by the manufacturer. In reality this power output will be reached only around the noon hour (thus the designation of peak power) every day. Keep this in mind, because it is an important variable, which is sometimes overlooked.

Fifteen percent efficiency means that each square meter covered with these PV modules will generate 150 Wp DC power (at STC). Since we are using STC numbers to design the plant, this means that we need ~667,000m² (165 acres) of active area (area covered by PV modules). Adding 50% space between the rows of fixed-mounted modules, we end up with minimum ~330 acres of land needed to install a 100-MWp power plant.

Usually, however, we need much more than that, to compensate for shade-free operation, adding access roads, abatement, systems, etc., so a typical plant size would be in the 500-600 acres range, which comes to approximately 5-6 acres per 1.0 MWp installed.

And this is when using c-Si, or any 15-16% efficient modules. The land usage increases proportionally when the modules efficiency is lower; i.e., a 100-MWp power plant using 9-10% efficient thin film PV modules will need additional land. Several such power plants have been installed lately on 10-12 acres of land per 1.0 MWp installed.

PV modules mounted on trackers would need less land, depending on their configuration and plant design characteristics (location, installation specifics, etc.). As a rule of thumb, 15% c-Si modules mounted on trackers are expected to generate 15-20% more power in the desert. This difference might be even greater under cloudy conditions, but estimates for such conditions are hard to make due to the great variability of sunlight intensity.

CPV technology using high concentration solar cells with 40% efficiency changes the calculations to where—everything considered—we could get much more power from an acre of desert land. We look at CPV technology in other chapters, so we'll only mention here that although CPV cells need to be mounted on trackers, which are expensive to build and operate, the final financial outcome might still be better.

Most utilities prefer using Watts AC in their calculations, instead of Watts DC, which changes the calculations. To do this, we need to track the power delivered to the grid back to the PV modules. In our case, 100 MW AC power would mean adding at least 20 MWp to the original 100 MWp DC modules to compensate for the losses from the modules to the grid (wires, inverters, transformers etc.), if we are to provide 100 MWp AC to the grid at peak hours. The rest of the estimates would parallel the increased number of modules and power generation capabilities.

With all of this figured out, we find an appropriate land in an acceptable location (sunny climate and cooperative utility) and negotiate a purchase or lease with the owners. Then we finalize the calculations, and obtain pre-construction financing.

Several million dollars would be needed for pre-construction financing, to:

a) design the plant and pay the expenses (engineering services, etc.),
b) generate official electro-mechanical drawings of the installation,
c) secure the land via lease or buy option, and to
d) obtain building and interconnect permits, PPA, etc.

We estimate $4-5 million pre-construction finance would be needed to complete the above tasks on a 100-MWp PV power plant in central California.

After months of meetings, stacks of forms, payments, etc. we get a construction permit, a PPA, and construction finance ($450-500 million to build a 100 MWp, fixed-mount PV power plant in the US during 2011-2012). Then we proceed with the installation of fixed-mounted modules (or on trackers) on the site.

When everything is ready to go, we flip the main switch ON and start measuring the output during different times of the day.

PV Plant Operation, 30 years Scenario

Here is a close, hypothetical, look at the day-by-day operation of a PV plant operating non-stop for 30 years in the SW US deserts.

Note: The operation in any other geographic area would include many more fluctuations and irregularities, due to bad weather (clouds, rain, snow, etc.). So we can think of this example as an "ideal case," which can be used as a foundation for other projects, after extrapolating for the additional variable of each case.

The only exceptions of the "ideal" case scenario in this case are the facts that, a) extreme heat reduces the power output significantly, b) extreme heat tends to shorten the life of the PV modules and BOS, while, c) wild fires, flash floods, sand, rain and hail storms and other desert phenomena can damage the installation.

With all this well understood and kept in mind, we flip the power switch ON, and our new 100 MWp PV power plant starts generating power, sending it into the grid.

Great, but let's take a very close look at what's going on in the field.

Day 1

It is noon on a bright mid-summer day in the desert, the sun bathing our new 100 MWp power plant with unobstructed rays. Solar insolation 900 W/m^2, air temperature 120°F.

Since the modules' output was measured at STC (25°C module temperature and 1000 W/m^2 insolation) we get our first surprises:

a. Because typical solar insolation at high noon in the desert is ~900 W/m^2 instead of the theoretical 1000 W/m^2 (which was used for the nameplate calculations of the modules and the power plant), we see that we have lost nearly 10% of the modules' output. Instead of 100 MWp, we are getting 90 MWp DC power output as a first measurement.

b. Then we notice that the output starts going down quickly. This cannot be. Yes, it can! The modules start heating up under power, and with the help of the hot, noon sun the temperature inside the modules could reaches 150-180°F or higher.

Under these conditions internal resistance increases and other things happen (remember the 0.5% power drop for each degree C increase of temperature), so that we could lose another 10-20% of the power output during the most productive hours of the day.

Now our 100 MWp power system is producing only 80 MWp during the noon hour.

c. While we are busy making these measurements, DC power is running through the wires, combiner boxes, inverters and transformers, and finally flowing into the electric grid. When we measure the actual electric power going into the grid, we find that it has dropped another 8-10% due to conversion and resistive losses in the equipment between the PV modules and the grid.

The higher the ambient temperature, the higher the DC to AC conversion losses.

Now we are down to 70-75 MW AC power generated during the noon hour and sent into the grid.

d.   When the sun starts going down, the power output will go down proportionally until no more power is generated by the system at sundown. As the sun starts going down, the modules will cool some, so the power reduction is not linear.

e.   The cycle starts again the next day with a gradual increase of power output as the sun goes up, until the modules reach their maximum output. In the afternoon hours, output levels drop gradually as the sun goes down, and level at zero during the night.

NOTE: Using trackers will complicate the comparative measurements, but we should assume that the same type PV modules installed on trackers, would generate the same amount of power at noon as fixed-mount modules. During the hours before and after noon, however, tracking modules will generate 10-20% more power than their fixed cousins, due to better orientation towards the sun and better absorption of sunlight.

Temperature extremes and tail-ends have other effects on tracking PV modules, complicating the exercise. Suffice it to say, the end effect is 10-20% higher DC power output from the trackers.

Money Talk

Now let's look at the bottom line of the first day. During the noon hour, we generated 75 MWp AC power, or 75 MWh (75 MW per hour) went into the grid. We actually generated that much power during each hour from 10:00AM until 4:00PM on that summer day in the desert. This was a full 6 hours of maximum power production.

After that (from 4:00PM until 8:00PM) the generated power went down with the sun. The same was true for the hours from 6:00AM until 10:00AM, when the generated power went from zero to maximum as the sun was rising.

So, during that day we had 6 hours of full power generation and 8 hours of partial power generation.

Those 8 hours would average the equivalent to ~3 hours of full power generation. So that day we had a full 9 hours of full power generation (6 hours of full power and 3 hours average from the rest of the day)—or—9 hours x 75 MWh = 675 MWh AC power was sent into the grid.

This is 675.000 kWh, for which the utility company will pay us $0.15/kWh, or a total of $101,250.00 for one day of power generation.

Because we don't have 365 days of full sunshine, and most winter days don't produce even half this amount of energy, we usually average the annual power production to 300 days annually at 6-7 hours per day. This puts our annual gross income in the $20-25 million range.

Note: $0.15/kW payback is another variable, controlled by the utilities. We have seen PPAs from $0.09 to 0.23/kWh, depending on the location and the utility.

The utilities also pay different amounts for power generated during different times of day during different seasons. This complicates things, but basically it means that we can get a higher return during peak demand hours of summer (1-8PM) and much less in winter when the peak hours are 5-9AM and 5-9PM when there is not much sun.

How much of the gross income is profit will depend on the financing package(s), but there is a good chance of ending up with a decent profit the first year of operation.

Year 1

One year has gone by since the first day of operation. At noon of the anniversary, and under the same weather conditions, we notice that the power output is 1-1.5% lower than this time last year. We look at the manufacturer's spec sheet and see that it predicts ~1% drop of power each year. If the loss is less than 1%, we are good and operation continues.

If the output loss is higher than 1%, however, we need to negotiate with the manufacturer and obtain a replacement for the failing modules. But the warranty is often not clear if any replacements can be done during year 1. Also, how do we justify a plant shutdown for the replacement of the failing modules, if an agreement is reached at all?

A day or two of shutdown will result in several hundred MWh loss in production, which is even greater than the power loss of over 1% that we are experiencing. Tough decision…

Money Talk

If we average our power generation (summer and winter months) to 6 hours/day and consider 300 full sunshine days per year, we get 300 days x 6 hours/day x 75 MW AC = 135 MWh/year. This is 135,000 MWh x $0.15 = $20.25 million in gross income expected during year 1. At this rate, the investment would be repaid in ~20 years.

Adding the different subsidies, incentives, carbon credits and perks, the plant might be repaid in 10 years or less—if everything works as planned. IF. Hoping that everything goes this way for the duration is a bit of a stretch, because things usually do not go as planned, and we are yet to discount the expenses from the gross income.

Year 5

The power output from several hundred modules and several inverters has dropped significantly during the last several years and something must be done. The inverters are under warranty for now, so all we have to worry about is the lost power generation during a 2-3 day maintenance shutdown for their upgrade. Unless there is a stipulation in the inverters' warranty agreement, we have to eat the loss.

The failing modules, however, must be removed and replaced, which will take several days and maybe weeks depending on the quantity. This is gross profit that will fall while the operating expense remains constant. Did we include this loss in the initial plant O&M budget?

Using trackers would cause additional headaches for the plant managers at this time, because some of the moving parts, such as bearings, motors, sensors and controllers need to be upgraded or replaced. This is additional expense and more time lost.

This activity and the related expense must have been anticipated during the power plant design stages as well. If proper maintenance and replacement of failing equipment is not done on time, then the trackers just might stop tracking and things will only get worse.

Money Talk

$20.25 million were generated during year 1, dropping by 1% annually each year thereafter, so now we are at least 5% short on power generation and gross income, in addition to the above mentioned loss of time. This means that the 75 MW AC peak power generation we got during year 1, is now down to 71 MW AC power generation at peak hour. That also affects the bottom line, so now we get ~$19.00 million annual gross income.

And some other things have gone wrong; additional maintenance and replacement of hardware and equipment is needed, which will cost at least another $1

million. Basically, our power generation and revenue are now down an additional 6-7%, putting the gross annual income at $18 million. Considering inflation, and that we are 5 years in the future, $18 million might be... even less.

Year 15

Many modules show greater than 20% loss of power output now. Several inverters are failing and their long-term warranty has expired too. This is a critical decision time. What is better, shutting down the plant to replace the failed modules (provided the manufacturer is still around and will honor the warranty), or continuing to operate like this for awhile longer? What if the warranty is no longer an option? Who is going to pay for the replacements?

Do we replace the inverters now or later? This is a major expense so, unless it was provided for in the plant operation budget, it just won't happen. And 15-year-old inverters operating in the desert are a gamble, even with the best maintenance.

Money Talk

Now things get serious. The power output has been dropping 1% every year and we are getting at least 15% less than the first year. Our 100 MWp DC power plant is pumping only 64 MW AC power into the grid. Our gross income for the year is down to ~$17 million at best.

Also, this is the time when capital equipment replacement and major maintenance work must be done (projected or not), the cost varying from $5-$15 million (if a number of inverters are to be replaced).

This might be catastrophic on its own, if proper budget and other measures were not taken in the beginning, because without these the power plant is faced with a major revenue loss.

Year 20

The scenario is even worse now. A proper plant operations budget and management would make the difference between success and failure at this point.

Did we thought of this beforehand? Were we aware of all possibilities, and did we provide for each accordingly? The answers to these questions will determine if the PV power plant remains in operation or goes into bankruptcy, thus converting another 500 acres of desert land into a junkyard.

Money Talk

At this time the power output has dropped at least 20%, so we are down to ~60 MW AC power generation, and $16 million in gross income annually. Is this enough to pay the bills?

Even worse, all manufacturers' warranties have expired and we depend totally on luck and good performance insurance, if any.

Any major failure of, or damage to, large numbers of PV modules or inverters means that we'll have a hard time recovering—even if there is adequate insurance coverage and modules/inverters replacement options (due to lost power generation).

Adequate budget and efficient management will now be the key to success.

Year 30

The power output is down over 30%, so our 100 MWp power plant is now producing less than 50 MWp. All warranties have expired, maintenance costs are increasing, and it is time to shut it down for good.

The plant goes into the final stage of its life cycle—decommissioning. Who will remove the PV modules and the supporting structure? Who will load the metal waste on trucks and pay for its transport, recycling and disposal? Who will bear the extra cost of handling, recycling and disposing of hazardous waste (in case of toxic TFPV modules)? Who will return the land to its original state and decontaminate it as needed?

These are very expensive processes that must be done properly and efficiently. Hopefully we have allowed for this effort in our preliminary planning, design, and budget calculations. If not, there will be a lot of toxic junkyards in the deserts 20-30 years from now.

And speaking of toxic waste management, this is an extremely expensive proposition, which can be taken lightly by manufacturers and project managers. They have some money for the decommissioning of modules, but disassembly, sorting, packing, loading, transporting, recycling, and hauling waste away is very expensive today, and will be much more expensive in later years.

Money Talk

This, in our opinion, is where many PV power plants would be in shambles. With power generation and revenues decreased by 30%, increasing failures and maintenance costs, and pending decommissioning, recycling and land restoration expenses, most managers and owners would prefer the easy way out. Unless, of course, all this has been foreseen and taken care of ahead of time. But there is not much talk about the last stages of PV installations.

At this stage, the winners and the losers will be separated. The surviving technologies and operators

will be proven adequate and safe, and proper initial planning and design will be verified as a standard for the PV industry to follow.

Summary of Our 30-Year Example

We have seen a number of issues, which we need to keep in mind with cells, modules and power field design and use. These issues are partly responsible for the uncertainty in the PV industry, and lack of investments in, and failures of PV installations today.

Some Reasons for Failure

1. PV module manufacturers usually guaranty 80% output by the 20th year, and stop there. This means an average 1% reduction in produced power every year up to year 20, leaving uncertain losses after year 20—which is when PV modules are nearing their end-of-useful-life stage and when they start failing at even faster rates.

2. Some insurance coverage can be obtained to cover damage and performance deficiency of PV power plants, but we are not sure that adequate insurance can be purchased for the period after year 20, which is just around the time many PV plants become profitable. Or, if such insurance exists, can we justify its price?

3. The yearly power decrease represents a minimum 1% annual loss of revenue—1% loss the first year, 2% the year after, 10% by the 10th year, 20% by the 20th year (in best of cases) and a whopping 30% loss or more during the 30th year of operation, if the plant is still operational at all. And this is big IF with no guarantee whatsoever.

    Note: Long-term field test results show that regardless of the manufacturers' claims, most PV modules lose ~1.3-1.5% or even more power annually. There is no way to prove this with new modules, so using the field tests as the most reliable long-term test numbers available, we calculate 13-15% power and revenue loss by year 10, 26-30% loss by year 20, and a staggering 39-45% power loss by year 30.

4. PV modules also lose 0.5% power per degree C above STC (25°C); i.e., PV modules exposed to the Arizona desert sun will heat to over 85°C in summer. This is 60°C over the STC value. The modules, therefore, will lose 30% of their power output during the 4-6 most productive hours of the day. This phenomenon is not well known, and such a sig-

nificant loss should be addressed during the power plant design stages.

5. An equally serious loss, which cannot be easily estimated and is often overlooked, is the loss from various total field failures. There are many reasons for PV modules to fail shortly after installation, and manufacturers usually offer 2-5 years materials and labor warranty, which is good enough in most cases.

    Developers of large-scale PV power plants, however, should also consider negotiating a post-warranty failure warranty with the manufacturers. Such a warranty, or special performance insurance is needed, because there are a number of manufacturing conditions that might cause premature field failures, as discussed in previous chapters.

6. Designers, customers and investors must understand the issues we are discussing herein and obtain assurance from the manufacturers that the PV modules will not degrade excessively, and/or fail prematurely. Purchasing performance insurance might be another way to ensure a plant's survival.

We need to know exactly what would happen in the field during the first year, year 5, 10, 20... and have a clear idea of the measures to be taken in every case. Technical consultants should provide the most realistic theoretical analysis and best/worst-case scenarios based on actual experience and scientific data, backed by field measurements and tests. Much of this work has not yet been started, but it must be done to provide efficient long-term operation.

At the end of the first day we got an unpleasant surprise, which grew even bigger by year 15-20. It is clear that our virtual power plant is not getting the performance we expected, and we are losing money. And around year 30 things start falling apart?

Using highest quality PV modules and BOS components, proper project planning, strong all encompassing budget, and excellent performance insurance are the solutions. But this is not what we are seeing. Front end gains are the priority in most cases. Reliability and longevity are either not well understood, or non-existent.

End-of-life decommissioning and recycling take a back seat—no one wants to talk about a funeral at a wedding.

Hybrid CPV/T System Financial Analysis

We reviewed the design and function of CPV-T hybrid power generating systems in previous chapters,

so here we will present an example of the operation of a small- to medium-size Concentrating PV (CPV) and Thermal (T), or CPV-T hybrid system in US deserts.

This system consists of a CPV tracker equipped with highly efficient solar cells (over 43% DC power at the cell), with cooling water lines, which feed a water heating loop with hot water as a final product.

CPV/T energy systems generate both DC electricity and hot water at the same time at a very high efficiency—40% and 20% efficiency respectively—totaling close to 60% overall utilization of solar power.

The amount of energy produced depends on the size of the system, its actual setup and operation, and—very importantly—the amount of sunlight available for its operation.

Since CPV technology requires direct beam sunlight, it is best suited for use in the U.S. Southwest and a number of sunny locations around the globe.

For the present performance analysis we'll pick a location in the Arizona deserts with the following average performance criteria:

- 300-320 sunny days annually
- Average 900W/m² full solar insolation
- Average full solar insolation of 8 hrs. per day
    — Summertime full solar insolation for CPV use is ~10 hrs,
    — Wintertime solar insolation for CPV use is ~6 hrs.

Under ideal conditions, 900 W/m² x 8 hrs = 7.2 kW/h energy is received daily per m². Each CPV/T unit is over 40% efficient and occupies ~6.0 m² area.

Or 7.2 kW/h x 6.0m² x 40% = ~17.3 kW/h daily generation (vs. 5.5 kW/h for PV panels*) 300 CPV/T units (1800 m²) would generate 1.0 MWp or ~8.0 MW/h daily averaged on annual basis.

At $0.20 avg. PPA (including carbon credits, federal and local subsidies and incentives**) this represents $1,600/day, or $512,000 annual income from DC/AC power generation.

Also, each CPV/T unit generates hot water at ~20% efficiency (180W/m² = 614 Btu/m²) Thus, 614 Btu x 8hrs x 6.0m² x 320 days = 9.43 million Btus are generated annually. Or, our 1.0 MWp plant would generate additional ~$180,000 annually from hot water generation.

---

*Note: CPV is several times more efficient than fixed-mounted PV panels. It also tracks the sun all day, thus getting maximum power possible from morning 'til evening.
**Note: Federal, state and local subsidies might cover 30-40% of the initial cost.

Useful power generation (considering DC/AC and other conversion losses and O&M inefficiencies) would be ~20% less. Therefore, we can expect gross income during the first year of operation from our 1.0 MWp CPV/T plant to be ~$550,000***.

Total capital expense for building the CPV/T power plant is estimated at:

$1.30/Wp for equipment and $2.50W/p for BOS, land and administrative expenses,

or a total of $3.8 million is needed to install and start operation of the 1.0 MWp plant.

Additional expenses of approx. 15% of gross income must be assumed and allocated for annual O&M operations, which brings total net operating income to ~468,000 annually.

Therefore, our 1.0 MWp CPV/T plant will be paid for in approximately 8 years, providing clean, green energy and reducing global CO_2 generation by over 1 million lbs. annually. The number of years will vary as other factors, such as government and local subsidies and incentives, carbon credits, taxes and initial investment repayment, are considered.

The drawback of this type of system, and where the uncertainty lies, is the maintenance of the complex key components—positioning controls, x-y gear assemblies, and the related motors, bearings, etc. In the best of cases, maintenance would be 1-5% higher than fixed-mount modules, but in reality we expect 5-10% higher maintenance costs. We hope technological advances will improve the system and reduce O&M costs.

OPERATIONAL RISKS

There are a number of risks during field operation, associated with the use of PV modules. These risks are augmented in large-scale fields operating under extreme climate conditions (extreme desert heat and/or high humidity.)

Some of the major risks associated with the operation of PV modules, as far as the profitable and safe operation of these is concerned, are as follow:

Low Power Output

Low power condition (lower than expected) could be caused by a number of abnormalities. The most fre-

quent causes of low power condition are defects in the modules, such as improper soldering, micro-cracks in the cells, lamination problems, frame materials and design problems.

Poor installation, bad weather, or a combination of the above factors could also cause lower than expected conditions.

Power Loss (heat)

Excess heat, such as the 120+ days in the desert, causes all PV cells and modules to decrease power output in proportion to the heat levels. An average power loss of 0.5% per degree Celsius is accepted as the norm.

This means that a module operating at high noon on a hot desert day could lose 30% of the power it generated in the morning.

Power Loss (degradation).

All PV modules degrade with time; as they weather, they change in a way that causes them to lose power gradually and consistently. An average 1.0% power loss per year is accepted as the norm.

This means that a 100 Wp PV module will generate 90 Wp during year 10, 80 Wp during year 20, and 70 Wp during year 30.

The actual long-term field tests with c-Si and thin film modules show much greater excursions as time goes by—especially in desert regions, as is to be expected.

Premature Failure

Premature failure of PV modules (the modules failing shortly after installation, and before the expected EOL) could be caused by a number of factors, some of which are outlined in 1 above.

Excess power loss, high annual degradation and other unusual incidents are often early symptoms of premature failure.

Electrical Shock

Electrical shock is caused by short circuits, and/or insulation failure in the modules, overheating, bad frame and connectors design, faulty inverter, poor installation, etc.

Electrical shock is easy to prevent by making periodic electrical measurements of the modules and strings.

Physical Injury

Physical injury (personnel cuts, falls, etc.) are most often caused by bad frame design and manufacturing,

---

***Conventional PV panels lose 1% efficiency annually. CPV/T systems do not. Conventional PV panels lose 0.5% efficiency per degree C. CPV/T systems do not.

where sharp edges and other imperfections could cause serious cuts on personnel.

Faulty tracking systems could also harm people by collapsing, or by other abnormal behavior. Improper handling of heavy objects (inverters, frame elements etc.) is another source of personal injury.

Poisoning

Many PV modules contain toxic materials, Pb, As, Cd, etc. Although these are mostly contained within the modules, trace amounts (from the manufacturing process) could be stuck to the modules' surface. On contact, these toxins could cause skin rashes and internal poisoning.

With the large number of modules installed at each site—millions upon millions in large-scale fields—and with the increased number of CdTe and CIGS modules installed in these, the danger of poisoning has increased proportionally.

Fire

Fire is the most dangerous long-term risk. Individual PV modules could catch on fire on overheating—especially when operating under extreme climate conditions. Once fire has started in one point of a module, it will spread quickly through the entire module by burning the EVA and other plastic components.

When one module starts burning, adjacent modules in the string overheat. This and increase of the internal resistance in the modules might ignite many of them too. In worst cases, the fire can jump from string to string and engulf the entire field.

Solar Storms

Intense solar storms are becoming more frequent and stronger. Sky-watchers in the North get a treat from the aurora borealis, or northern lights, which are caused by solar storms at their peaks. The lights occur across much of the northern tier of the US, as far south as Oregon in the West, Illinois in the Midwest, and the Mid-Atlantic states in the East.

March 10th, 2006, NASA issued a solar storm warning for unusually strong solar activities (solar flares) during 2012. In June of 2010, NASA reinforced their original 2006 warning telling the public to get ready for something new and unexpected—a solar storm.

NASA spokesman said, "We know it's coming but we don't know how bad it's going to be, however, the next solar maximum should be a doozy." NASA expects the sun storm to be in full swing sometime in 2012-2013, but they don't know exactly when, how strong, or long it will be.

March 8, 2012. Massive solar storm (flares) were expected about 7 a.m. ET, according to the Space Weather Prediction Center in Boulder, CO, who predicted the storm to last about a day. This storm, a G3, was considered to be a "strong" one.

July 12, 2012. Another massive solar storm was observed, with strong energy field traveling towards the earth at a high speed. The effects of the high energy stream hitting the earth were expressed in increased intensity of the Northern Lights, which could be observed as far as the central and southern US. No damages were reported.

These storms have the *potential* to trip electrical power grids, and power companies around the world are usually alerted in advance for possible outages. Solar storms can also disrupt GPS systems or make them less accurate. The storms can lead to communication problems and added radiation around the North and South poles, which could force airlines to reroute flights.

NASA reports that the current increase in the number of solar storms is part of the sun's normal 11-year solar cycle, during which activity on the sun ramps up to the solar maximum. This cycle should peak in late 2013, so similar solar storms will occur every month or two for the next few years.

As is always the case with natural phenomena, there are those who wonder "what if." Some speculate that if NASA and other scientists are right, the physical and economic damage to the US from a *super solar storm* would be many times greater than hurricane Katrina. We simply don't know what to expect.

If the earth gets a direct hit by a solar storm producing x-class solar flares, (Coronal Mass Ejection or CME) it could potentially cripple humanity in many ways: cell phones, computers, cars, trucks, planes, trains, ships, computers, GPAs, TV, radio, medical equipment (everything that is battery- or electric-powered) could stop working.

Such flares would damage many essential pieces of equipment, which wouldn't come back online for a long time.

Imagine that our virtual PV power plant has been brought to a halt by a super solar storm. The damage would be great and potentially irreversible. This is the worst-case scenario, but is it possible? And what can we do to protect the field and ourselves from such a tragic event?

There are a number of measures that can be taken to double and triple the safety mechanism of the plant equipment. Such redundancy, however, is expensive, so not many people would even think of it. Insurance, if we have it, would not cover "acts of God" in most cases.

Such monstrous solar storms don't happen every day, but one big one is all it takes to wipe out a great portion of the active PV fields in the affected areas. Gone.

So, planning and designing redundant protection for super solar storm attacks must be on the agenda. All parties involved must be aware that there is—albeit very slight—the possibility of a super solar storm hitting the earth, and badly damaging our power field sometime during its 30 years operation.

In addition to the deteriorating performance of modules, we could also be surprised and badly hurt (financially) by their field misbehavior. What can be done to safeguard from surprise?

How about insurance? The modules are basically insured by the manufacturer to produce the specified power for 20 years. But what about unforeseen events—excess clouds, equipment failures, natural disasters, etc.?

PERFORMANCE INSURANCE

This is a new concept, though discussed in the past, and even partially applied in some cases, but it's never been fully implemented in a single project. Today, however, it is becoming a prerequisite, in our opinion, for starting (and certainly for financing) new solar projects.

We are looking at the newest branch of the US insurance industry—performance insurance of solar power plants.

This type of insurance is too new for us to assign it an exact role in the 21st century, but our long experience shows that it is one (if not the only) way to fill the gaps between quality, performance, and profits.

We know what to expect from the PV manufacturers, installers and operators, so let's take a look at the insurance companies.

The insurance industry is being significantly affected by the current economic downturn, new regulation requirements, and increased losses to policyholders. In the context of these large-scale forces shaping the industry as a whole, the insurance market for PV systems is evolving and maturing rapidly.

The addition of the renewable energy market insurance options to the portfolio of insurance products, has not made the insurance company executives lives any better. Needless to say, investments in PV products and projects are often viewed by underwriters as quite

risky because the technologies are newer (no long history of operational data), and there are fewer installations relative to other insurance avenues such as real estate and automobiles.

Insurers use the "law of large numbers," which says that "the larger the group of units insured, the more accurate the predictions of gain and loss will be." Although solar technologies have been in existence since the 1950s, their deployment at a significant level occurred much later, so the numbers are not that large.

Since 2007 new commercial on-grid installations increased to more than 1,500, which statistically speaking is not a large number. Not enough solar installations have been in place for long enough for underwriters to feel they can accurately predict what the losses associated with them would be. The experiences with some of these fields have not been very reassuring either, so all this fuels the uncertainty of the insurance industry.

On top of that, data of the different projects, products, installation and operation are not readily available, which complicates the process. Insurers have access to data only on projects they insure, so the more insurers there are, the less data each insurer has.

Another complexity of insuring solar installations is the typical involvement of many parties, including installers, developers, investors, lenders, and insurance companies in each project. Typically, each party to the transaction attempts to minimize the risks it assumes while defining the recourse available in case a risk event occurs and leads to actual loss. In many cases, insurance products can be used to mitigate and manage the allocations associated with complex contractual arrangements.

Until the recent growth in the PV market, demand for PV system coverage was not adequate to encourage insurance underwriters to develop PV-specific products. If the upward trend of new PV installations continues, there will be a great demand for its insurance. This growing market of new solar installations, which is being driven by state policies, federal incentives, and corporate responsibility, represents a possible market opportunity for insurance underwriters.

The Issues

We've discussed the major technical, administrative, and financial issues facing the solar industry. We determined that long-term performance and longevity are some of the most important, albeit least discussed, issues in any PV undertaking with any of the existing PV technologies. Looking at the data in the above tests we wonder how anyone could put a positive spin on the results and accept the risks associated with 20-30 years on-sun operation of deserts modules.

Long-term field tests are conclusively assuring us that the risks are real and are warning us of the potential and serious long-term physical deterioration, power degradation, failures and other dangers.

The lack of convincing test data, plus the data in the ASU field test study, are major obstacles to PV projects' financing, because investors are concerned.

A number of insurance companies do offer conditional performance guarantees designed to put risk factors on level ground, making potential customers and investors more comfortable. PV performance insurance is a new field and like the PV industry it is as yet unstandardized and unregulated. In some cases, coverage is available to all participants (manufacturers and distributors of PV modules, integrators, utilities, customers and investors), while in other cases it covers only certain segments of the PV cycle, such as PV modules' efficiency and performance.

Different insurance companies cover different types and percentages of the risks related to PV products and services. That might be just enough to eliminate the majority of risks, but we'd advise customers and investors to take a close look at the guidelines and definitions of the coverage in each policy before making a final decision. This is new territory and should be approached thoughtfully.

Still, this is an encouraging development that might be just what is needed to bridge the gap of financing PV projects today and in the near future. Proper performance insurance fills the gap between quality, longevity and profitability of PV modules and projects, so it might be the key to providing acceptable levels of risk management, thus increasing the customers' and investors' confidence in solar installations.

We hope the insurance companies involved in providing performance insurance to the solar products business will be able to understand and successfully tackle this fascinating but risky business venture.

If everything is properly set, solar project installers, owners and investors would have a choice of insurance protection, which could vary from full, unconditional coverage to a minimal coverage required by law.

1. Manufacturers who use their own PV modules to build PV power plants may not need or want performance insurance, although they still might choose a level of conventional insurance coverage to protect the plant from natural and manmade disasters.

2. Customers using the highest quality PV modules available may also decide to skip insurance coverage altogether for obvious reasons.

3. Other players, however, especially those cutting corners, will be able to choose a level of coverage as they see fit. This way the gap between uncertain quality, deteriorating performance and reduced longevity will be at least partially closed and the PV plant will survive unpredictable events, including excess performance degradation and failures.

Note: There is a difference between "performance" and regular insurance coverage required by state or federal law. Performance insurance in this text refers only to covering the amount of energy lost with time due to poor product performance, or other factors affecting the overall performance and yield.

Nevertheless, there is some overlapping of conditions and issues when talking about any type insurance for power plants, so utmost care must be used when designing or signing insurance contracts.

There are still major challenges in these developing solar and wind energy fields, and so not many (if any) insurance companies would be willing to provide adequate long-term insurance to projects using these technologies.

As an alternative, mid- to large-size renewable energy developers and manufacturers (or projects) concerned about the overall cost of insurance and product warranty conditions and premiums may have the option of establishing a captive insurance company (hereafter referred to as "captive"). Such enterprises are common with fortune 500 companies and have been successfully used for a long time.

A captive subsidiary established with the parent company (or project) explicitly as a risk mitigation arm is intended to maintain control of the operation—including any accidents, underperformance and other issues.

Like traditional insurance companies, the insured (the solar project) pays a premium in exchange for coverage as outlined in the policy. And like any insurance company, the captive should be managed in such a way that it avoids absorbing all the losses of the insured.

This could be an attractive alternative to buying insurance from a traditional underwriter, since insurance premiums for plant construction and operation are limited in scope and are too costly. Insurance premiums of 25% of a system's annual operating expense, and/or 0.25%-0.5% of the total installed cost of the facility, have been reported.

Based on NREL's System Advisor Model (SAM), a 1MW system located in Phoenix, AZ, with a total installation cost of $4.5 million would have an annual insurance cost between $11,500 and $22,500. This means that our 100 MWp at the same location, and insured under the same conditions might pay close to $1.0 million annually in insurance fees alone for incomplete coverage (full performance coverage is not provided). This is a large chunk of the profits no matter how you look at it, so a better alternative is needed.

The higher premiums in most cases are due to the solar industry's turbulent childhood, with its maturing technologies, imperfect business models, and evolving applications—most of which are not fully established and/or standardized.

Also, most insurance companies lack the expertise and investment strategies to manage actuarial capital and the related risks in the solar industry. So the combination of all these factors drives prices unreasonably high, at this stage in the game.

The Challenges

The challenges of captive enterprises are numerous:

a. The captive may lack the actuarial expertise needed to assess risks, especially for the uncertainties associated with innovative technologies, their financial structures, conditions and issues.

b. Setting up a captive includes the costs of transactions, feasibility studies, organization, actuarial services, business planning, permitting, PPA, management, legal counsel, audits, and taxes. These costs could be in the millions for a significant-size projects.

c. Costs to establish a captive range from about $50,000 to over $125,000, depending on the captive's complexity. This means that developers would need to determine the minimum project risk (and optimum operating conditions) that would make a captive worthwhile.

d. Determining where to establish the captive is tricky too, because not all states have regulations conducive to establishing such a venture. Developers must research state and local code to determine possible locations. Vermont is the most common state for captive domiciles, and foreign countries may also be used.

e. A major claim event or series of claims could be costly, so the possibility of such an event must be entered into the equation.

The View from the Top

Insurance executives' view of the solar industry in general is somewhat limited.

1. Manufacturers cannot ensure the long-term efficiency and reliability of their product.

2. Developers do not know exactly what coverage they need.

3. Broker's understanding of solar technology ensures proper coverage.

4. The solar PV market is a maturing industry, but it is not yet mature.

5. Lack of established standardization and regulations is of great concern.

6. The solar PV industry has excellent fundamentals (e.g., strong product demand, declining input costs), but there are still a number of uncertainties and risks.

The View from the Bottom

On the other side of the equation, the solar industry's opinion of insurers is also somewhat skewed.

1. Insurance policies are not affordably priced.

2. Insurance companies sometimes lack the background knowledge of solar PV technologies.

3. Insurance premiums could go down if the insurance industry had better data and understanding about system operation and historical data.

4. Insurance brokers are considered the elephant in the room.

5. If brokers aren't educated on the technologies and risks, then the underwriters won't be either.

6. Some brokers pretend they understand solar technologies and place policies that do not fully cover what needs to be covered.

In a famous case, one insurer asked about the use of molten salt in the system, which is relevant for solar CSP, but not for PV, thus revealing his severe lack of understanding of the business.

Negotiations

Solar project insurance negotiations usually go along the lines of the general insurance business.

1. General Liability—covers death or injury to persons or damage to property owned by third parties (i.e., not the policyholder).

2. Property Risk Insurance—covers damage to or loss of policyholder's property, such as theft, natural disasters, transit of goods, etc.

3. Additional policies, such as Environmental Risk Insurance, Business Interruption, Contractor Bonding etc., AND

4. We will add here Performance Insurance. The long-term (30-year) performance of the equipment is a major issue we are concerned with. We must have some assurance (and insurance) that whatever happens to or with our investment (power generation) is protected against unforeseen events—including and above the manufacturers' warranties.

Summary

Project performance (and any other type) insurance might be the solution to successfully using any type or quality of PV modules and products.

The European and Japanese solar markets are more mature than the U.S. solar market. Most of the growth in PV capacity in Europe has occurred in Germany, which was the world leader in cumulative installed capacity in 2010-2011. Spain and Japan also have robust PV markets.

The maturity of the European PV market gives European insurance companies more experience underwriting renewable energy generation systems than the U.S. market. Because of Europe's greater experience with solar PV installations, some U.S. insurance companies use loss data from Europe to project probabilities of loss and risk in the US.

Provided that the insurance companies are well aware of the risks and accept them (for a price) most PV power plants with adequate performance insurance will survive.

2012 Update

As confirmation of the positive developments in the area of performance insurance coverage, and to free us from our worries and hypothesis about field failures, several companies are actively looking into it, and some are taking immediate action.

LDK Solar and Solarif introduced a new type of insurance package for PV systems in the summer of 2012. It is a full insurance package, including "all risk insurance," with the related warranty and inherent defect insurance. All risk insurance is for all components of PV systems (including BOS), while LDK Solar's power and product warranty covers their PV modules.

LDK Solar's manufacturing process was audited by Solarif and the insurance is covered by German insurer HDI Gehrling. The package consists of full coverage of all products sold on the European continent, including compensation in case of production loss during warranty exchange, transportation and all other costs related to module replacements per warranty conditions.

To our knowledge, this makes LDK the first PV company to introduce such a complete insurance solution.

This is a good start in the right direction. This type of full insurance coverage, we believe, is what is needed today, and is a solution for securing investments in PV projects. Notwithstanding the high price of such insurance, we see it as the missing link that hindered development of solar projects thus far in the US in particular. It is also absolutely the best (if not only) way to ensure power plants' long-term profit. It is how business will be done in the near future, and for sure during the latter part of the 21st century. There is simply no better, way!

In another case, Assurant, Inc. and GCube Insurance Services, Inc., both insurance industry veterans, have partnered and have created a new insurance for commercial-scale solar projects. The insurance package covers solar installations between 100 kW and 3.0 MW in size. In addition to the standard property and liability insurance, the new insurance package offers an unique warranty component at a project-specific level.

The new warranty concept addresses the concern that the manufacturers typically offer 25 years insurance, usually handled by a company that has been in business for at least 4-5 years. The new package brings experience and capital to reassure the financiers that it can stand behind the warranty. Also, the project developer has the option of warranty management coverage as long as the project's property and liability coverage remains with the original insurer.

This new insurance instrument is of huge value, because it is the only one on the market with warranty management available. The warranty management will be done through the manufacturers and the O&M providers, similar to the way auto insurance providers use mechanics and body shops.

As another confirmation of the need and feasibility of performance insurance, ReneSola, a China-based manufacturer of solar wafers and modules, signed a solar products performance insurance agreement with PowerGuard Specialty Insurance Services, who specializes in insurance for solar and other alternative energy companies and products.

A newly coined agreement, "PowerGuard," provides conditional insurance and warranty-related coverage for ReneSola's PV products for 25 years. This insurance will let ReneSola offer its products with a high degree of confidence and increase its market share to include customers with varying risk-management profiles.

This type of insurance addresses the customer needs for reliable long-term operation, and reduces the difficulties that are often encountered by solar project developers, installers and operators. It also allows ReneSola to focus on their area of expertise, thus increasing the quality of their products.

PowerGuard is a good example of things to come. The PowerClip warranty product has been particularly popular, with China Sunergy, Lightway Green New Energy and SunEdison having all adopted the protective cover during 2012.

The Practical Aspects

With the economic and technical hurdles of solar companies and projects of late, investors and customers alike are getting more and more cautious about putting money in solar companies and projects. The fact that the world's PV modules manufacturers are sitting on 50% surplus is one indicator of the realities. The spontaneous failure of solar companies and projects of late has increased the level of uncertainty, threatening the very existence of the solar industry.

Customers and investors are asking PV modules installers to provide insurance from reputable insurance companies. Several such companies are involved in solar insurance. These companies cover warranty claims even if the solar panel maker goes out of business, and this is becoming the accepted practice, if not the only way that solar companies, even large ones, can sell their products on US and EU markets.

Warranty Claims

Warranty claims resulting from manufacturing defects appear to be one of the biggest expenses facing PV modules makers. There were recently a number of news stories of such claims eating into profits at solar companies. Such problems are expected to escalate, as manufacturers down-size with today's economic situation.

Most manufacturers do not use third-party insurance, because most insurers don't have enough confidence in the PV products. They also lack sufficient cash reserves to cover larger claims.

In our opinion, warranty claims will keep coming in ever increasing numbers, but solutions will become fewer and far between. The imbalance will be too severe for some companies and projects, so we won't be surprised if some of them fail.

Manufacturer Quality

What can we expect when buying from different quality manufacturers, while considering adding performance insurance to our large PV project? Let's see:

Poor Quality Manufacturers

Buying poor (or uncertain) quality PV modules and BOS components was widespread in the US and Europe until recently. We are yet to see the full range of consequences from their low performance and prematurely failure.

Medium Quality Manufacturers

In the case of below-average or intermittent quality of PV modules and BOS components, the risks are lower, but still unpredictable. So the risks still need to be evaluated and mitigated, and the problems corrected; be it through warranty, negotiations with the manufacturers or via insurance claims.

Since in most cases the manufacturers are of some reputation, they will agree to cooperate. This makes things much better, but some additional expense (lost power generation revenue, module replacement, modules transport and installation, etc.) is expected to be paid by the customer.

High Quality Manufacturers

The quality of the PV modules and BOS components in this case is good to excellent, which is confirmed via thorough understanding, and or direct inspections and tests, of the manufacturing process. Some of the risks inherent to PV products are still present, but they are monitored and controlled by the customer and/

or a third party, with the full cooperation of the manufacturer.

In all cases, some long-term performance insurance will save a lot of money in the end, giving the insurance industry a decisive word in the world's solar energy game. The evidence points to the fact that solar power plants of the 21st century will be either self-insured, or insured by a third party.

Certification Insurance

Speaking of insurance, here is a new, more elaborate and reliable way to test and certify PV modules. It is a service offered by a few companies, who claim to thoroughly check and test the three key aspects of the modules' quality: overall performance, individual cells' quality, and level of workmanship.

These factors basically do reflect the performance and reliability of PV modules, so these new tests would determine how good a PV module is, and also that a manufacturer is mass producing good PV modules on a large scale.

The tests are done via conventional test methods, but a more thorough check and evaluation is also undertaken. Even more important (and not done in any enterprise thus far) is that 100% of the modules are tested in some way to avoid gaps in individual modules, and/or batch performance.

The results are then sorted and reported as follow.

Performance [weight 34%]

Performance shows mainly the power output of a module, but power alone is not entirely representative of module performance. A module with higher efficiency will reduce the cost of the BOS (balance of system) by reducing the land use and the number of structure elements.

The power output of each module is evaluated, based on the result of a flash test at standard testing conditions (STC) and also at low irradiance, to evaluate the power output on less sunny days.

In addition, the lot tested is evaluated as a whole, as PV modules in the field will be grouped together to produce electricity and the weaker one will affect the power output of all the others.

Also evaluated under this category is the ability of the manufacturer to provide PV modules with positive tolerance on power.

Cells Quality [weight 30%]

The performance is evaluated at the time of testing, but users of PV modules want to get a high power output from their modules for the longest time possible.

A solar cell containing cracks may perform correctly at the moment, but with transportation, mechanical stress and climate constraints, cracks and other cell defects will spread in the cell and result in a cell that does not produce any more electricity.

Under this category, the quality of the cells themselves is evaluated, or more exactly their electroluminescence picture.

Workmanship [weight 36%]

Under this category is grouped the quality related to the making of the PV modules. Assessment includes the visual appearance of the PV modules, safety to the user, the packaging, and the quality of the sealing, construction and other parameters.

The workmanship score indicates the quality of the work that has been performed to produce the PV modules.

Additional Factors

The mark given to each panel is the result of a thorough and detailed testing, during which the following factors are checked and tested step by step:

Positive Tolerance [weight 5%]

A higher score is given to manufacturers providing PV modules with a positive tolerance for their PV modules. This seems easy, but the output will later be evaluated against this positive tolerance.

Output against Nominal Value [weight 12%]

The power output is determined by flash test under STC (25°C, AM=1.5, 1000W/m^2).

To show the ability of the manufacturer to mass produce quality PV modules, we take into account the average power of the lot, the minimum and the consistency (homogeneity in power) of the whole lot.

Efficiency [weight 6%]

To save cost of system and space, the PV modules need to provide maximum of power on the smallest area. A higher score is given to modules with a higher efficiency.

Efficiency at Low Light [weight 6%]

Modules are expected to perform well in sunny weather and on less sunny days. The power output is therefore determined under low irradiance (25°C, AM=1.5, 400W/m^2) to evaluate its performance under these conditions.

Inefficient Area [weight 10%]

Appearing as black areas under EL tests, these areas of the cells do not produce power and reduce power output of the module.

Another risk is heating of these areas, resulting in hotspot (burns on the modules).

I-V Curve Appearance [weight 6%]

When natural sunlight is simulated through a flash test machine, the response from the module in terms of current and voltage is well defined. Bumps and inconsistencies in that curve show some problems in the function of the solar cells or the module itself.

Cracks [weight 10%]

Cracks in solar cells are a threat to the reliability of the PV modules. Cracks are identified under electroluminescence test.

Pollution [weight 5%]

Processing issues during solar cells manufacturing might result in less efficient area which in turn can result in hotspot problems or lower cell output. These defects are identified under electroluminescence test.

Soldering Defects [weight 5%]

Processing issues during manufacturing might result in less efficient area which in turn can results in hotspot problem or lower cell output. These defects are identified under electroluminescence test.

Packing [weight 5%]

Packing is not always considered a major item by manufacturers, but it is crucial to protect the PV modules during long transportation and processing.

Marking [weight 3%]

Required by IEC standards and mandatory to enter European or US market, marking must include specific information and be legible.

Construction [weight 3%]

Under this category are taken into account both the design of the PV modules construction and the quality of the assembly.

Dimensions [weight 5%]

Installers are basing their structure and installation on the dimensions of the PV modules. Dimensions such as length, width, skew and compliance of mounting holes are measured.

Frame Assembly [weight 5%]

Poor docking of the frame elements can result in sharp edges putting the user at risk. The presence of these sharp edges and quality of the work is evaluated.

Sealing [weight 5%]

For the PV module to perform for more than 20 years, the encapsulation of the cells and the sealing must be waterproof to prevent moisture from entering and corroding the inner elements.

Cells Layout [weight 3%]

Spacing between cells and between strings is important to avoid any risk of short circuit inside the modules. A minimum distance between the electrical conductors and the frames must be respected as well, to avoid electrical shock.

Stringing [weight 2%]

To perform correctly, the cells must be correctly connected to each other. The conductors have to be well soldered and properly laid to ensure good conductivity in time.

Component Appearance [weight 3%]

As end users are more and more concerned with the visual appearance of the products they buy, cosmetic defects are taking a higher importance in the choice of PV modules. Components such as glass, frames, junction boxes and cables are evaluated.

Components Usage and Function [weight 2%]

When tested against IEC standards, the PV modules are a combination of different components.

Taken separately, some components might seem equivalent; however, the combination of all these components might not work properly. This is why it is mandatory for manufacturers to respect the use of specified components.

The function of these components such as connecting function of connectors is also evaluated.

In Conclusion

Test labs involved in these types of thorough examinations are looking for a set of product issues related to overall performance and reliability.

One of the major aspects of the solar products and projects is their ability to deliver the expected power for the duration. The maximum power determination is performed according to IEC 60904. Even though IEC 61215 requires the output power determination to be performed on class B sun simulators, AAA Class equipment should be used to obtain the most accurate and reliable results.

The tests are performed in a controlled environment, in conditions as close as possible to the Standard Test Conditions (STC). Not only is the output power controlled, but each electrical characteristic as well as the shape of the IV curve is checked to identify any intermittent malfunction of some solar cells.

One guarantee that modules will perform well over time, is that the solar cells are free from defects such as cracks and pollution. Those defects are invisible to the naked eye, but are detected through electro luminescence. The electro luminescence equipment performs an "x ray " like picture of the module where the dark areas show the less efficient part of the cells.

As a lot of operations remain manual during production, the quality of workmanship can vary considerably from one module to another. Quality of workmanship will not only affect the appearance of the module (which has become a criterion of choice for end users), but it can affect the safety of installers and the product reliability.

A thorough inspection of the construction focuses on the dimensions, the quality of sealing, the components used and the packaging.

Note: The final performance tests (flash tests) are done via solar simulation tests. IEC 60904-9 defines 3 categories of sun simulators A, B, C (A being the best class) based on 3 different criteria:

Temporal Instability

To be accurate, the measure is performed with a stable irradiance (light sent to the module by the simulator) during the whole period of testing (the time to take the entire I-V curve).

Temporal instability refers to the difference between the maximum and the minimum irradiance in percentage. The lower the instability, the better.

Spatial Nonuniformity

Equally, the irradiance provided by the sun simulator shall be the same all over the testing area. Spatial nonuniformity refers to the difference of irradiance between different points of the testing area. The lower the nonuniformity, the better.

Spectral Match

As the natural sunlight is composed of different wavelengths, to provide a good evaluation of the behavior of the module in the field, the sun simulator shall be able to recreate the composites of the natural light. In nature, different wavelengths of light will participate in a certain proportion of the irradiance.

The spectral match is the ability of the sun simulator to provide irradiance similar to the natural light for given wavelength ranges or its deviation from the reference irradiance for different wavelengths.

Summary

In summary, lab tests done by qualified labs using thorough, sophisticated, 100% checks and testing of the product, using the best equipment and procedures available, is a step forward in the battle for efficient and reliable product.

If all PV modules coming to the energy markets were tested this way… Unfortunately, this type of testing is not done by all manufacturers, and even if it were, there is still a chance that many modules would fail with time.

This is because there are perhaps hundreds of materials and process steps with many combinations and permutations of conditions, which affect the performance and quality of modules. Not ALL of these can be checked and tested even with the most vigorous and thorough test methods available today. The key word here is, TODAY.

Today we simply do not have the technology to take a close look at the inner workings of PV modules— especially during their actual under-load field operation (the most relevant part of their function). But we are sure that it will be a routine operation later in this century, when engineers with special equipment will be able to see in real-time exactly what goes on inside each module operating in their power plants. They will be able to correct problems in a timely fashion, avoiding surprises.

All the talk about performance insurance is directly related to ensuring the quality of the final product by strict control of the down-stream materials used in the manufacturing process, as well as the labor quality at all stages of production. The quality battle starts with the materials used in production, most of which come from third-party suppliers we call the supply chain. So let's look at the supply chain.

Supply Chain Quality and Costs

Volume, cost and quality fluctuations of raw materials and consumables in the solar industry is a subject which needs an entire book of its own. We chose to include the supply chain issues under the subject of performance insurance because they are of utmost importance to final quality. The issues are not easy to detect, and some will reveal themselves months and years after installation, so we need to know what we are looking for in supply-chain materials, if we are going to discuss and negotiate performance insurance.

Polysilicon

Polysilicon is a key material needed to manufacture c-Si solar cells, and although it is abundant in nature, it is a bottleneck in the solar photovoltaic energy industry explosion. The high initial capital and technical requirements of polysilicon production plants mean that only large size (over 1,000 tons per annum) production can be economical and competitive. This also means that only large enterprises with government help (as is the case in Asia) can undertake the huge risks in this area.

The gap between demand and supply has been and still is huge in Asia. The total demand in 2005 was 2,825 tons while the supply was only 130 tons. A number of new polysilicon production plants closed the gap somewhat, but PV has been growing about 30% per year lately, and since this growth is faster than anticipated, a temporary shortage in silicon feedstock occurred. With the projected slowdown of the PV industry during 2011-2013, the gap might be closed. In the midst of this, polysilicon prices dropped significantly from $400/kg in the 2008 spot-market to $30/kg in 2010.

The quality of the silicon feedstock material is of major concern. The 130 tons of silicon produced in 2005 has grown to over 30,000 tons today, with plans to reach 90,000 tons in the near future. The volume, however, is going down due to lack of demand, causing shockwaves in the industry, and possibly having a negative effect on product quality.

Metals

Huge quantities of silver, aluminum, copper and other metals, as well as plastics and many chemicals are needed to produce millions of PV modules as demanded by solar markets. Large amounts of silver are used for good ohmic contact between the inside of the cell and the outside circuitry of the PV modules and for reflective backing for mirrors in thermal solar plants.

According to the VM Group in London over 1,000 tons of silver have been projected for making PV modules in 2011. This is over 1.0 million kilograms of silver used today, an amount projected to triple by 2016. Nearly 3,000 metric tons of silver will be used for making PV modules worldwide every year at that time. If we add that much more silver metal for coating heliostat mirrors

and use in the semiconductor and other industries, we end up with some very large numbers.

So the question is, "How much silver will be left in the world in 10, 50 or 100 years, if we use it at this pace?"

Silver is a precious metal, the price having gone sky high during the last 10 years from $4.00/ounce several years ago to nearly $40.00/oz. today. What will these rising prices do to the price of PV modules?

Similarly, the price of copper and aluminum has sky rocketed, and although there are large deposits of these left on Earth, the increased prices will play a significant role in PV modules' use on the world's energy markets.

The quality of these metals is an issue that has no simple answers. The metals are used in different forms—sheet, soldering wire, paste, dust, etc.—and there are issues at every step of their manufacturing process. One cost-saving measure resulting in a contaminated batch of silver paste, for example, could make several thousand PV modules fail immediately, or within a year, in the field.

Encapsulation Materials

A number of materials, mostly of organic nature, are used in the PV modules manufacturing process. Many of these materials were originally developed for application in laminated windscreens and laminated architectural glass.

Encapsulants play a key role in the development of photovoltaic modules. Current and future innovation in formulations will lead to modules with longer useful lives and increased energy conversion efficiency. These advances will result in reduced cost per watt for the module maker and end module user.

EVA is an established material for this purpose, but Polyvinyl Butyral (PVB) is gaining favor in certain segments of the photovoltaic market. PVB is typically provided to a module maker as a resin-based sheet that can be trimmed to fit a given application. While it has been in use for decades in laminated glass applications, recent developments have made it a material of choice for certain PV module designs—including many thin film technologies.

The development of formulas with strong edge stability and chemical compatibility provide module designers with a material that improves manufacturing processes and long-term module durability. This does not answer the question of long-term reliability, because no matter how good the newly developed compounds are, they have not had a chance to be thoroughly tested.

Added to that, the quality of the materials used to make encapsulants is paramount to their performance in the field. A small amount of contamination in a production batch, for example, could make a large number of modules fail during testing or after short time in field operation.

Because manufacturers claim proprietary materials and processes, there is no way to check or enforce the quality of the resin materials. The same is true for other materials and components in the supply chain. So, there are no simple answers.

Thin Film PV Modules

Thin film modules manufacturing requires a number of rare and exotic raw materials, such as cadmium, tellurium, arsenic, indium gallium and selenium. Some of these are already in limited supply and it won't be long before they are depleted altogether.

How and why do we qualify these as "renewable?" They are being depleted into oblivion, their prices will go up with time, according to the supply and demand ratios, and we must do something before we end up on the ugly side of town.

The quality of thin films is less of an issue, due to the sophisticated deposition processes and the related quality control procedures. The problems arise from improper execution of some intermediate steps, such as substrate surface cleaning and preparation, deposition process abnormalities and such.

Having reviewed this complex subject in previous chapters, it suffices to say that product quality of thin film PV modules is harder to measure and of less concern than their overall reliability in long-term operation in the deserts—a condition related mostly to their operational characteristics and not so much to their components or quality.

LONG-TERM FIELD TESTS

What are the consequences of poor quality? What happens if the product contains poor quality materials, or is manufactured with unqualified labor? What happens to our profit margins in such cases?

While the results from laboratory tests done at the manufacturers' and certification labs are important, what happens under actual load conditions in the field is totally different. Very different.

The actual results depend on many factors, so keep in mind that:

1. The results from tests under the heat of the Arizona desert or the high humidity in southern Florida will be drastically different from those under the cloudy skies of Portland, ME, or Berlin, Germany, and that,

2. In all cases, failures and degradations increase exponentially with time, especially when operating in extreme heat and humidity.

The field tests in Table 7-1 show the average annual degradation of different PV modules from different manufacturers operating in moderate Midwest US climate. Most of the results from c-Si PV modules are actually quite encouraging, but notice also that most of the manufacturers whose products show good results are reputable, well-established companies who have been in the solar business for a while now and have figured out how to make a high-quality product. In contrast, most new manufacturers have not had a chance to work out the process bugs and/or conduct long-term field tests, to prove the reliability of their products.

Long-term Desert Field Tests

As a confirmation of the facts and issues discussed in previous chapters, a research study published in the fall of 2010 (26) confirms the presence of major issues in PV modules operating in desert conditions. The study

Table 7-1. Field test results

Manufacturer	Module Type	Exposure Years	Degradation (% per year)
ARCO Solar	ASI 16-2300 (x-Si)	23	-0.4
ARCO Solar	M-75 (x-Si)	11	-0.4
Mfgr. X	X-number (a-Si)	4	-1.5
Eurosolar	M-Si 36 MS (poly-Si)	11	-0.4
AEG	PQ-40 (poly-Si)	12	-5.0
BP Solar	BP-555 (x-Si)	1	+0.2
Siemens Solar	SM50H (x-Si)	1	+0.2
Atersa	A60 (x-Si)	1	-0.8
Isofoton	I110 (x-Si)	1	-0.8
Kyocera	KC-70 (poly-Si)	1	-0.2
Atersa	APX-90 (poly-Si)	1	-0.3
Photowatt	PW-750 (poly-Si)	1	-1.1
BP Solar	MSX-64 (poly-Si)	1	0.0
Shell Solar	RSM-70 (poly-Si)	1	-0.3
Wurth Solar	WS-11007 (CIS)	1	-2.9
USSC	SHR-17 (a-Si)	6	-1.0
Siemens Solar	M-55 (x-Si)	10	-1.2
Mfgr. Y	Y-number (CdTe)	8	-1.3
Siemens Solar	M-10 (x-Si)	5	-0.9
Siemens Solar	Pro 1 JF (x-Si)	5	-0.8
Solarex	MSX-10 (poly-Si)	5	-0.7
Solarex	MSX-20 (poly-Si)	5	-0.5

describes the results from long-term scientific tests done with PV modules operating in an official desert test field for 10 to 17 years.

Note: To our knowledge, this study is the largest-ever evaluation of long-term performance of production-grade PV modules under desert conditions.

The study was done at one of the most prestigious PV test sites in the world, APS's STAR solar test facility in Phoenix, AZ, where most world-class PV modules manufacturers send their modules for long-term testing, so we would not be surprised if the tests include some famous names.

The work was supervised by one of the world's most reputable PV specialists, Dr. Govindasamy Tamizhmani (Mani) from the Arizona College of Technology and Innovation, and President of TUV Rheinland, Tempe, AZ.

Dr. Mani's team thoroughly visually inspected, electrically tested and IR measured each module in the batch of nearly 1,900 PV modules.

The final results from these long-term tests are shown in Table 7-2. Amazingly, almost 100% of the nearly 1,900 PV modules tested showed some visually-apparent deterioration. Some were more impacted than others, but the overall conclusion was that a large percent of PV modules exposed to constant sunlight and heat in the deserts deteriorate beyond the manufacturer's warranty within 10-17 years of operation.

Many of the tested modules average 1.3 to 1.9% annual power degradation, which translates into 23-33% total power loss for the duration (some only 10 years). This, extrapolated for the extended time of use, is well above the manufacturer's warranty of maximum 20% degradation in 20 years.

Interestingly, significant EVA encapsulation discoloration and delamination were observed on many modules, even those exposed to the sun only 10 years. Solar cell discoloring (due to a number of factors) was observed on most cells too. Since the oldest modules in this study were made in the mid-1990s we must assume that encapsulation and cell manufacturing processes have improved some during the last 10-15 years to reduce the defect ratio.

However, we have no proof of such progress, so this becomes another unknown that needs our attention when dealing with PV modules destined for desert applications.

In summary, browning in cell center was almost 100%, encapsulation issues followed as a distant second, followed by frame seal deterioration, hot spots, broken

Table 7-2. Long-term testing in the Arizona desert.

Module Types	Module Count	Modules Affected
A. Mono-Si (17 years; glass/polymer)	**384**	
- Browning in cell center	384	100.0%
- Frame edge seal deterioration	384	100.0%
- Hot spot (IR scan)	4	1.0%
B. Mono-Si (12.3 years; glass/polymer)	**1092**	
- Encapsulant delamination	2	0.2%
- Browning in cell center	1092	100.0%
- Hot Spot (IR scan)	2	0.6%
C. Poly-Si (10.7 years; glass/glass)	**171**	
- Broken cells	47	27.5%
- Encapsulant delamination	55	32.2%
- Hot spot (IR scan)	26	15.2%
- White materialnear edge cells	2	1.2%
- White material browning	56	32.8%
D. Pol-Si (10.7 years; glass/polymer)	**48**	
- Browning in cell center	37	77.1%
- Hot spot (IR scan)	4	8.3%
E. Mono-Si (10.7 years; glass/polymer)	**50**	
- Bubbling substrate	33	66.0%
- Browned substrate near J-box	50	100.0%
F. Pol-Si (10.7 years; glass/polymer)	**120**	
- Discolored cell patches	6	5.0%
- Backsheet bubbling	1	0.8%
- Browned spots on backsheet	2	1.7%
- Metallization discoloration	22	18.3%

cells and glass, etc. Many of these defects are serious enough to cause power loss and eventually lead to total failure due to optical interference, electrical resistance, overheating, mechanical disintegration, or any combination of these.

This is simply not supposed to happen, not in such large numbers at such an important test site, so we must ask, "What caused the problems? Was it the materials quality (silicon, metals, EVA, glass), or the manufacturing process (diffusion, metallization, encapsulation), or was it some combination of these that was primarily responsible for such a large number of defects and failures?"

In any case, this study confirms the unfortunate reality that PV modules of any type and size are prone to rapid deterioration and failure under harsh desert conditions. Although we cannot predict future performance and overall condition of modules during years 20-30, the logical assumption is that the deterioration processes will continue at the same rate, and even faster in some cases. We must conclude that each module type and size must be very well designed and properly manufactured with high quality materials and proper processes to be given a chance to survive desert elements.

Note: The reasons for the above described defects and the related explanations and justifications are too numerous to launch into at this point, but references and explanation to many of them can be found in previous chapters.

Italy Field Test

Another set of long-term field tests, with a reduced number of modules and under different climatic conditions, was conducted by the Institute for Energy, Ispra, Italy. The test field is in northern Italy, so climate conditions are much different from those in Arizona, causing the field inspection and test results to be different as well:

Table 7-3. Long-term field test in Italy

Modules	Power Loss
32%	1-5%
25%	10-20%
24%	5-10%
8%	20-30%
6%	30-50%
3%	75-100%
2%	50-75%

Summary

What can we learn from the above long-term test in a non-desert area?

- 50% of the modules show yellowing after 10 years, and 98% after 20 years of on-sun operation, with 78% showing excessive yellowing.

- 74% of the modules show delamination after 10 years, and 92% after 20 years.

- 76% of the modules show sealant infiltration and 22% have noticeable signs of hot spots.

- The average annual degradation of connected modules is ~1.0%, while that of modules left in open circuit mode was only 0.6%, showing the negative influence of current flow and additional temperature increases.

- Glass-glass modules also had a greater degradation, most likely due to excess temperature levels during operation even at this location.

- High level power losses (>20%) are attributed to series resistance increase, while moderate power losses (<20%) to optical properties degradation.

Note: These results are not as dramatic as some other field tests we've seen (probably due to the moderate climate and good construction—Arco c-Si modules), but are still significant and another confirmation of the fact that PV modules will perform as well as they are made.

Extrapolated to the finances of our 100 MWp power plant, the average power degradation and failures shown in the above long-term field tests represent a major income loss from year 10 on. The question remains, "Can we afford even moderate power and product losses like this?"

The field test results prove that c-Si technology, if properly made and used, can be a viable and reliable energy source, and that some well-established manufacturers have achieved reliability.

Keep in mind, however, that most of the above tests (except one) have been conducted 10-12 years max), so the conclusions reflect only a fraction of the time, stress and abuse all PV modules will experience during 30 years of non-stop exposure. Extreme climate conditions are proven to have detrimental—and usually accelerated—effects.

We must always assume that power degradation after 10-15 years non-stop operation will increase somewhat erratically and perhaps drastically, which could be catastrophic in some cases.

We know by now that the financial health of any solar installation largely depends on the quality of its products. Quality will be priority #1 in the design and implementation of solar power generation in the 21st century.

Where do we start? How about with some solid standardization of the solar industry?

SOLAR INDUSTRY STANDARDIZATION

The best way to ensure high product quality is to establish world-wide standards covering all aspects of the production cycle—from the procurement of the initial raw materials to the last step of manufacturing.

Today's solar industry standards should also cover installation, operation, maintenance and decommissioning.

To provide full insurance coverage in a comprehensive and standardized way, we need standardized products, but we are far from that.

Every beginning is difficult, and the developments in the solar industry today are no exception. There are a large number of manufacturers, installers, regulators, products, processes, regulations and laws operating in a fragmented manner. There is no comprehensive and all-encompassing standardization in any of these areas and activities. Instead, the different parties make their own rules, which have resulted in a patchwork of products and services, some of which are hard to classify and qualify.

Table 7-4 clearly identifies some of the major areas in the PV industry that have limited or no standardization. Full standardization of these key elements must be implemented as we expand solar energy use, but they are complex and expensive to tackle.

A number of fragmented efforts are helping, but an all-encompassing standardization must to be initialized. We believe that it will be difficult to introduce worldwide standards in the solar manufacturing industry, and that it will take decades to report serious progress in this area, but it's a necessary goal.

Until US customers and investors become more educated about the long-term effects of poor-quality PV products and the issues surrounding them, the immature solar industry will continue without standards—see Table 7-4.

Now we know that quality is directly related to financial integrity and profits. The other two aspects of the road to financial stability are cost and reliability of the product. This is so because even if we have the best quality and most durable product, it has to be priced so that we can afford it and eventually make a profit.

The product must also last long enough—which part of the quality cycle, but it is somewhat different, in that it is even harder to control than quality. Here, solid experience with production equipment and processes, as well as understanding how these relate to a long-lasting product are essential.

Looking at Figure 7-2, we can deduce that cost and performance (efficiency) factors have been manufacturers' and users' focus for a long time. Durability and lon-

gevity, however, are the hardest to address because they are a conglomeration of a many complex phenomena which are hard to understand, let alone put in numeric form and solve on paper.

Nevertheless, this is what is needed if we are to install Giga Watts of efficient and reliable PV modules in the deserts of the US and the world. Avoiding the longevity issues, or postponing their resolution (and badly needed standardization) is the wrong thing to do. It is also something that would most likely prove damaging to the industry and the energy markets in the long run.

Standardization of the installation and operation of power fields, as well as that of using the generated power more efficiently is also needed. There are a number of efforts lately to address these issues, and we do take a closer look at some of them in the other chapters, but they are still fragmented and incomplete.

BUSINESS MODEL FOR SOLAR DEPLOYMENT

Residential and Small Commercial

These types of solar installations are getting more popular by the day in the US, but the numbers, while growing, are not spectacular. Even with nodule prices

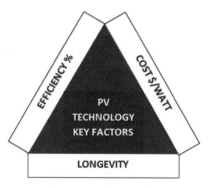

Figure 7-2. Key technology factors

Table 7-4. (a) Key manufacturing, and   (b) Power plant standards needed

Supply Chain	Mfg. Process	QC/QA		PV Products	Permitting	Logistics
Raw materials	Equipment	Initial inspection		PV modules	Federal policies	Land preparation
Process chemicals	Process specs	In-process inspect.		Inverters	State policies	Interconnection
Consumables	O&M procedures	Final inspection		BOX	Local policies	Installation
Waste treatment	Enforcement	Long-term tests		Misc.	Utilities policies	O&M

(a)   b)

falling through the floor, and with the help of subsidies and incentives, we don't see mass drive to install solar on house of office buildings.

Below is an estimate of the costs for installing a 200 kWp ground mounted PV power plant in Arizona during the summer of 2012 using imported c-Si PV modules:

Table 7-5. Small PV installation costs

Item	$/Watt
PV modules	0.75
Racks	0.35
Inverter	0.25
BOS	0.30
Transport	0.10
Labor	0.25
Construction	0.25
Management	0.15
Engineering	0.05
Permits	0.02
Misc.*	0.05
Profit	0.25
Total	$2.77

*Note: The cost or lease of the land, as well as some land-related expenses, are NOT included in this estimate. Adding these costs increases the $/Watt number accordingly.

The above numbers are expected to remain at these levels for the foreseeable future. If so, this and any other plant would be profitable within 8-10 years. If, however, PV modules and/or any other costs go up, and/or the incentives and subsidy are decreased, then it will take longer to get to a profitable state.

Utility-scale PPA Model

Project developers and financiers have traditionally operated under a common and relatively simple project finance model in which a developer invests sponsor equity into a project and a lender provides non-recourse project debt. In the renewable energy sector, a tax equity investor can also provide financing if a developer cannot monetize the renewable tax benefits, including the ITC. With this capital, a developer can build a project and sell electricity into the merchant market, or directly to a utility via a power purchase agreement (PPA). (19)

Host-owned Model

Host-owned projects differ from utility-scale projects under PPAs in several ways:

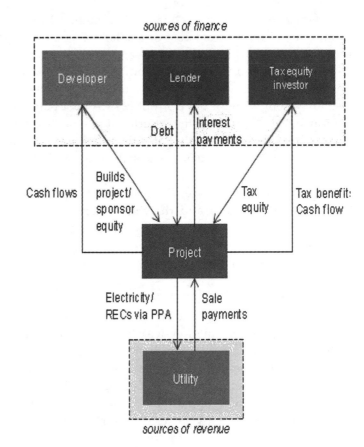

Figure 7-3: Utility-scale PPA Model (19)

• They tend to be small-scale.

• Rather than primarily serving the utility's grid, these projects produce electricity primarily for the host—that is, the owner of the property on which the project sits (e.g., rooftop or adjoining land). Assuming 'net metering' is in place, the system owner receives credit for any excess generation the solar system sends into the grid. Depending on state and utility regulations governing net metering policy, system owners may roll the excess generation over months or years, with the value of excess electricity paid to the homeowner at year end at prevailing wholesale or retail rates.

• These projects are too small to attract tax equity investors such as banks or insurance companies, which tend to want to deploy at least $15-30m at a time. One alternative is for the host to exploit the tax benefits directly. This is possible for some corporate enterprises such as big box retailers or even individual homeowners; either may face sufficiently high income tax obligations to effectively monetize the benefits.

However, many residential and commercial users cannot afford the upfront cost of a solar system, or are unwilling to use their own equity and arrange project debt. This has led to the development of third-party financing models, which offer customers the benefits of a solar system without the upfront cost. A host pays to the third-party financier either a series of payments via a lease ($/month) or PPA payments linked to the system's performance ($/kWh), usually based on a 10–25 year contract.

Effectively, the lease/PPA is a loan agreement between the customer and the third-party financier. Various third-party financiers have developed several different 'flavors' of this business model: the vertical, semi-vertical and financial market structures.

Vertical Model

Under the vertical model, an integrated player handles customer origination, installation, engineering, maintenance and financing services via a lease or a PPA tailored to a customer's location and system size. Such firms essentially serve as both installer and third-party financier to the home or business owner that receives generation from the PV system. SolarCity is today among the largest and best-known firms serving this dual role of installer and third-party financier. Others serve only the third-party role.

Third-party financiers pool multiple leases and PPAs from multiple systems into investment portfolios to attract larger outside project finance lenders and tax equity providers, which invest in the portfolio rather than directly in the third party. The third party and the investor create a fund—which may or may not be leveraged by a lender—to support these projects and build additional ones. As the third party installs more systems, the fund is drawn down. A host's lease/PPA payments, though written to the third party, are typically assigned to the fund.

Semi-vertical Model

This model popularized by Sunrun and Sungevity, is similar to the vertical model. However, rather than having installation and maintenance services in-house, a third-party financier partners with selected installers.

The third-party financier pays the installer a fee, while the host's lease/PPA payments go to the financier.

Financial Market Model

Developed by Clean Power Finance, this model brings in various investors of different types to compete with each other on the basis of yield or other characteristics. The concept is somewhat similar to lending exchanges, in which an intermediary provides an interface to match large lenders with small borrowers.

For example, a tax equity investor could create a financing product for the marketplace, stipulating that the fund requires a certain minimum unlevered yield, a minimum credit score for the host, or other criteria. In-

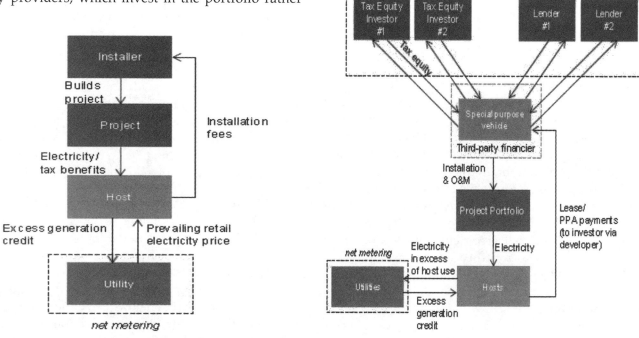

Figure 7-4. Host-owned model (19)   Figure 7-5. Vertical model (19)

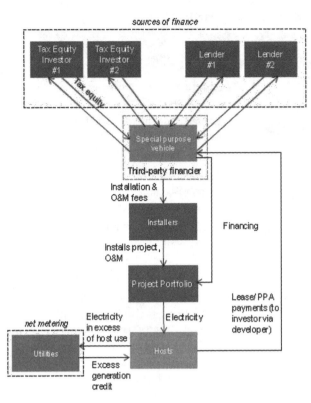

Figure 7-6. Semi-vertical model (19)

stallers with a project meeting those specifications might choose to sell a project to an investor if it provides a 'best fit' and competitive economics. The intermediary (such as Clean Power Finance) underwrites the installers and manages the operational work (e.g., billing, administration) on behalf of the investor. The intermediary collects a fixed fee on each transaction from the installer, and receives a monthly 'risk and asset management fee' from the investor.

Investors' Roles in the Business Models

The various solar business models present different opportunities for investor participation.

Investors and the Evolution of Solar Financing

Five major drivers suggest that there may be an expanded role for new types of investors behind US solar projects:

1. Diminished appetite from the historically most active investors, such as banks and the government.

2. The success of emerging business models for solar deployment.

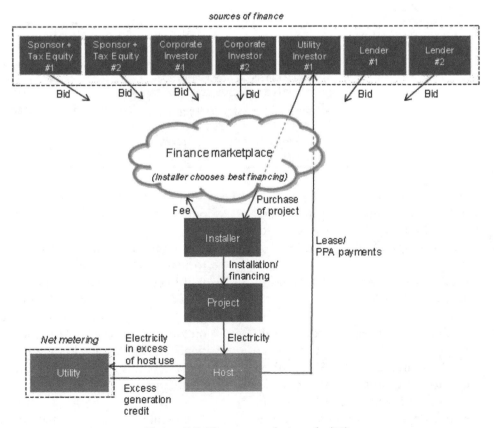

Figure 7-7. Finance market mode (19)

Table 7-6. Investors' Roles in Range of Solar Business Models (19)

Business model	Potential role for investors	Advantages for investors	Disadvantages for investors
Host-owned	• Host, own, and receive electricity from the asset, capture the ITC	• Protects investor from retail electricity costs (since investor is the solar power consumer) • Maximises project value as there are no 'frictional' costs paid to other parties	• Limited to hosts able to afford upfront cost and monetise the ITC • Exposes investor to project-specific risk without the benefit of diversification
Independent power producer (IPP) power purchase agreement (PPA)	• Provide cash equity, tax equity, or debt to the project • Provide cash equity or debt (back leverage) to the developer • Purchase securities (eg, bonds) backed by the project	• Allows investor to enter project development at desired stage to obtain preferred risk/return • Risk evaluation for individual projects is easier than for project portfolios	• Exposes investor to project-specific risks (eg, plant performance, offtaker creditworthiness, manufacturer warranties) • Nascent project bond market currently makes investment via capital markets difficult
Vertical and semi-vertical		• Risk evaluation is easier due to standardised PPAs/leases and contracts arranged by single third-party financier • Contract standardisation can open market up to solar-backed securitisation	• Opacity between the investor and the third-party financier can allow the latter to realise higher returns on the margin between PPA/lease revenue and financing costs
Semi-vertical only	• Provide tax equity to the portfolio (most common) • Provide cash equity or debt to the portfolio (uncommon) • Purchase securities (eg, collateralised loan obligations) backed by the portfolio	• Competition among installers drives down installation costs and reduces upfront cost per lease/PPA, all else equal	• Business model relies on relationships with installers, opening up risk that installers cannot deliver on commitments • Use of different contracted parties for installation and O&M creates uncertainty
Semi-vertical and finance market			• Use of different contracted parties for installation and O&M creates uncertainty
Finance market		• Investors can specify project criteria (eg, minimum host credit score, geography) • Intermediary is an independent third party and will not negotiate down tax equity investor's yield (whereas a project sponsor or third-party financier might)	• Inherent competition brings down overall yields

3. The basic characteristics of solar assets (potential for high returns for early-stage equity investors; likelihood of steady, high single-digit returns for owners and creditors of operating projects).

4. Momentum behind development of new types of high-liquidity financing vehicles.

5. Recognition that the clean energy sector overall, which has surpassed $1 trillion in cumulative investment since 2004 and grew by $260 billion in 2011, will need to increasingly draw from the massive reserves of wealth managed by institutional and other investors to continue to fund this tremendous growth.

The Solar Finance Revolution

Several fundamental changes appear likely:

• Participation from investors who have been historically active (banks and federal government) will likely diminish. European banks, which have for years provided construction debt to US clean energy projects, are grappling with the continent's ongoing credit crisis. At the same time, the set of financial regulations known as Basel III is further causing banks to reconsider their exposure to project finance.

The regulations require that any loans longer than one year be backed by funding of at least one year (e.g., a 20-year project loan must be matched by an asset with a maturity greater than one year, such as a long-term bond). This will result in higher capital requirements and will likely raise the cost and decrease the length of project finance debt.

Though the European banks will feel the squeeze first, Basel III will reduce the ability of many banks throughout the world to provide long-term debt finance. Beyond the banks' diminishing appetite for long-term lending, the US federal government's role as a direct renewable project investor is also decreasing due to the expiration last year of a key loan guarantee program (though the federal government is poised to play a major role as a project host via military projects).

- With many banks scaling back their long-term lending, others will gain market share. While borrowers with well-established lender relationships and solid project development track records may still be able to access long-tenor commercial loans, many borrowers will transition to 5-10 year 'semi-perms' (i.e., shorter tenor loans, which are refinanced in the middle or end term). Banks will continue to be the dominant source of construction debt, as other institutions are not comfortable with the associated risks. Asian banks, which are better capitalized and therefore less affected by Basel III, will also do more term finance in the absence of active European lenders.

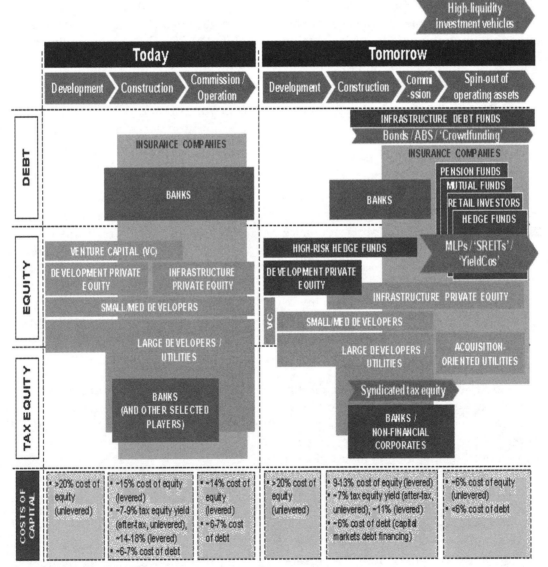

Source: Bloomberg New Energy Finance

Figure 7-8. Potential evolution of US solar financing (19)

- New investors will become comfortable with solar assets' risks/rewards. While PV is a fairly mature technology within clean energy, compared with infrastructure projects such as natural gas pipelines or toll roads, it is new and therefore can be viewed as riskier. As the sector matures, increasing investor interest and decreasing perceived risk will lead to a lower cost of equity, debt and tax equity.

- Institutional investors in particular may become more active. Solar projects are a good potential fit for insurance companies and pension funds, which seek stable, low-risk assets to match the duration of their long-term liabilities. While these investors prefer not to take construction risk, they may become involved in refinancing construction debt, perhaps by 'taking out' short-term lenders. They may also play significant roles on the equity side: some insurance companies have an inclination for buying operating solar assets under all-equity deals and then holding the assets for the remainder of the project lifetime: the reliable cash flows for projects with PPAs nicely match the funds' regular annual obligations.

- High-liquidity vehicles will pave the way for the expanded role of new investors. While some large insurers (e.g., MetLife, John Hancock) have experience investing directly in US solar projects, many institutional investors do not have project finance teams capable of doing the same, or simply prefer investments that are more liquid and more closely resemble other asset classes to which they are accustomed, such as bonds. Other examples of investors who have been on the sidelines to date but who could enthusiastically pursue public markets securities include hedge funds and mutual funds.

- New non-financial corporates will enter the tax equity market. In March 2012, the White House hosted a session to educate new potential corporate investors on tax equity, and earlier this year, Chevron announced its intention to invest in solar projects in the 3-20MW range. Other companies may follow: the yields are attractive, the market is relatively uncompetitive, and some have experience with tax credits via the low-income housing sector. In the long-term, increased supply of tax equity may drive down yields from the current 7-9% range. But in the short-term, one tax equity investor mentioned that yields may trend upwards

as the increased demand—due to the expiration of the cash grant—is met with inelastic supply. Additionally, large utilities with development arms such as Duke, NRG, enXco (affiliate of the European utility, EDF) and others—which typically internalized tax benefits—may not have much tax capacity remaining. Demand for tax equity will further increase as these companies go to the market for financing. However, as will be explained below, the list of new tax equity providers to meet that demand will be short.

- Utilities may become active purchasers of distributed solar portfolios. Residential solar installations reduce their customers' power demand, as system hosts which generate electricity therefore buy less from their local utility. As solar becomes more prevalent, utilities may purchase portfolios of solar PPAs and leases to compensate for revenue lost as a result of those very installations. Two avenues are available. The first is via tax equity, such as PG&E's combined $160m in tax equity funding to SunRun and SolarCity. The second is via the acquisition of portfolios on the tail end of existing tax equity deals, with the utility taking advantage of the cash revenue from leases or PPAs for the remainder of the project lifetime.

- The cost of capital will fall. Sector maturation and the entrance of new investors will lower the cost of financing. In particular, the widespread adoption of high-liquidity vehicles could open up solar to an investable base with a lower required return on capital. Widespread adoption of mortgage-based asset-backed securities drove significant decreases in the cost of capital for the homeowners, and a similar effect could happen in solar.

Barriers to Investor Participation

Bloomberg interviews revealed five major barriers to further participation from the broader investment community. (19)

- *Complexity*: Tax equity, necessary in the solar industry through 2016, is not for the faint of heart. The mechanisms necessary to pass tax and depreciation benefits to investors involves complicated accounting and legal structuring. This has limited participation to a select number of institutions with specialized financial expertise, a strong understanding of the energy industry, or both.

Theoretically, tax equity should be attracting more investors. As one tax equity provider interviewed for this study put it, "For the measure of risk we are taking, the yields have been exceptional."

Thus far, few corporations have been persuaded. Many consider the opportunity, then deem it too complex or far afield strategically from their core business, and too risky. Additionally, the significant time and effort in deal structuring requires either the creation of an internal team or the outsourcing of service providers at significant expense. Finally, it is impossible to start small; investors usually do not consider deals smaller than $15-30m or more because of the 'overhead' involved.

• *Immaturity*: Solar projects—and manufacturers—are not viewed as having a track record of long-term operating success. Investors like certainty. While many active financiers assume that solar panels will operate within a manufacturer's expectations, more risk-averse investors such as pension funds prefer to see evidence first. Panels from newer manufacturers have yet to operate over the 25-year expected lifetime, although the oldest panels, based on the same technology, have been in service for over 35 years now. Another turn-off for potential investors centers on the current wave of consolidations and bankruptcies in the manufacturing sector. To paraphrase one interviewee, "It's difficult to invest in a solar project when you don't know if the manufacturer giving the warranty is going to be in business next year."

• *Illiquidity*: Solar investments are not easily transferrable. Liquid investment vehicles have developed in other capital-intensive industries such as real estate (REITs) or natural gas infrastructure (MLPs), but none has been popularized in the solar industry. Many investors have little interest in assets that cannot be sold at short notice for net asset value. Tax credits are particularly vexing for liquidity-seekers: if a tax equity investor sells a project within five years, the IRS recaptures a portion of the ITC.

• *Lack of Standards*: In the residential and commercial-scale market, there is a lack of standards for payments and services. The numerous third-party financiers have different PPA and lease contracts, and credit quality and maintenance standards. Without standard PPA and lease agreements throughout the sector, it is difficult to rate and is-

sue solar securities. Geographical diversity reduces weather-related risk, but it also brings unique regulatory and market risks, making portfolio evaluation even more difficult. Challenges in bundling assets together may create a shortage of capacity and cash flow required to arrange and issue a solar bond and access capital markets.

• *Availability of Opportunities*: Some investors have expressed considerable interest but have been unable to find suitable investment targets. US solar presents a vast potential market—from utility-scale plants in the Southwest desert, to green-field developments in New Jersey to portfolios of rooftop projects on big-box retail stores across the nation. Yet solar development options inevitably come with assorted challenges: Southwestern utility-scale markets are extremely competitive; Northeast US markets are often beset by SREC pricing volatility; many regions simply do not offer sufficient incentives to make the economics viable; and assembling meaningful-sized portfolios of projects often comes at steep customer acquisition costs. Opportunities abound, but identifying and executing them is rarely easy.

EXTERNAL IMPACT AND COST

The "external" costs of PV plant setup and operation is defined by monetary quantification of the socio-economic and environmental effects and damage they inflict now (and in the future), to a local area, a nation, and the world as a whole during their cradle-to-grave life cycle. These effects could be expressed in $ per kWh generated, for lack of better way, and should account for all materials, procedures, and damages from cradle to grave.

Included in these calculations are environmental effects, human health, materials production, effects of use and disposal (materials, gasses, chemicals) on agriculture, noise, audio and visual pollution, ecosystem effects (acidification, CO_2 damage), and all other effects. Such calculations could be used to provide a scientific basis for legislative and regulatory policies, energy taxes and incentives, global warming policy adjustments, and more.

ExternE Project

The "ExternE" (25) project, addressing these issues in detail, was introduced and funded by the European Commission in order to develop a methodology to calculate external costs caused by energy consumption and production. This is another far-thinking program of the EU community, which, along with their RoHS program, makes them far superior in this area when compared to the near vacuum of such considerations in the US and other developed and developing countries.

The ExternE project defines costs as the monetary quantification of the socio-environmental damage, expressed in eurocents/kWh, with the possibility of providing a scientific basis for policy decisions and legislative proposals such as subsidizing cleaner technologies and energy taxes to internalize the external costs. It looks at all energy-production technologies using a methodology developed for this project that allows the various fuel cycles to be compared. An outline of the initial results for the PV fuel cycle, starting with a very small sample of PV systems, shows that the results are not consistent and require more work.

There has been pressure from the PV industry for the PV fuel cycle to be re-done using a larger sample of more representative systems, but this is still in the making. So, one of the earliest publications from this project by Baumann et al. is still valid. In this paper the author outlines the basic assumptions and requirements of the methodological framework for the quantification of external costs and compares the environmental effects of different energy technologies, including renewables and the conventional technologies of coal, oil, nuclear, and natural gas.

Basically, PV systems must be designed and manufactured with long-term environmental concerns and considerations in mind, which must include:

1. Manufacturing materials and procedures, including the production and use of MG and SG Si, thin-film metals and chemicals, glass panes, metals frames, gasses, chemicals, etc. process supplies.

2. Solar wafers and cells manufacturing processes.

3. Module encapsulation and framing processes.

4. Evaluation of direct and indirect processing energy (including transport, storage, etc.).

5. Gross energy requirements of input materials (supply-chain and internally generated products and byproducts).

6. Allocation schemes used in the calculations.

7. Separate thermal energy, electrical energy, and "material energy" calculations.

8. End-of-life recycling and disposal calculations (including transport and storage).

There are a number of dislocated efforts in the above areas, but no uniform, standardized method presently exists that is capable of capturing all environmental factors into one, all-encompassing methodology. Such a methodology is needed to account for the effects of the above concerns on the PV manufacturing processes and long-term use of the related PV products.

LIFETIME ENERGY BALANCE

Lifetime energy balance (LEB) is the energy used during the PV products manufacturing, transport, installation, and operation, compared with the power they generate during their useful lifetime. LEB is an important factor, which needs to be kept in mind when analyzing and calculating the energy benefits from a PV power plant. LEB basically tells us how much energy we save by using PV energy generating sources.

The solar cells and modules manufacturing process starts with melting and purifying sand in huge, dirty, dangerous, and energy-guzzling furnaces. The produced metallurgical-grade (MG) silicon is crushed in huge mills and further purified in a complex network of chemical and electric equipment, again using enormous amounts of natural resources and energy. The product at this point is solar-grade (SG) silicon of varying purity, the quality of which depends on the raw materials and process quality.

The SG silicon is crushed again and melted again in large furnaces at high temperatures for a very long time. Thus produced ingots of mono or poly silicon are sliced into thin wafers onto which the solar cells will be built. The wafers are cleaned, baked, fired several times, and coated several more times until finally a solar cell emerges at the end of the line. They are then sorted, lined up, and vacuum-bake sealed into the module frames, then transported to the location. The next time these PV modules are transported will be at the end of their life cycle, going to the crusher.

All this requires energy, materials, and resources. Fortunately, today's production equipment and processes are very efficient, but even so, making a solar cell and module is an energy-consuming undertaking.

The LEB Formulas

Consider this equation for lifetime energy balance (LEB):

$$LEB = \frac{E_{prod} + E_{trans} + E_{insst} + E_{use} + E_{decom}}{E_{gen}}$$

Where

E_{prod} = energy used during production of materials, wafers, cells, and modules

E_{trans} = energy used to transport materials, modules, and BoS to PV site

E_{inst} = energy used to assemble and install the PV power plant

E_{use} = energy used to operate the PV power plant

E_{recyc} = energy used to decommission, transport, and recycle the PV field

E_{gen} = energy generated during the life of the PV power plant

In all cases, Egen must be much higher than the sum of the other sources of energy used in order for the system to be an effective energy source.

The term "energy payback" describes this "energy in/energy out" ratio and is what we will have to consider when designing, pricing, and justifying a PV system. In other words, how long do we need to operate a PV system before we recover the energy used, and pollution generated, during its manufacture, transport, and operation before it is decommissioned?

There are energy payback estimates ranging from 1 to 4 years for different technologies; more specifically:

- 3-4 years for systems using current multi-crystalline silicon PV modules

- 2-3 years for current thin-film PV modules

These estimates, however, vary from product to product and from manufacturer to manufacturer. Cost of energy (crude oil in particular) also has a great effect on the estimates, so these numbers must be adjusted periodically. The ever-changing socio-economic situation in different countries is another great factor, and we expect that the present-day economic slowdown and worldwide financial difficulties will drastically reshape these and most other estimates.

Lifetime CO₂ Balance

Similarly, a system's lifetime CO_2 balance (LCB) is a factor that takes into consideration the CO_2 used during the manufacturing and use of PV components vs. the amount of CO_2 saved by using the PV components instead of coal- or oil-fired power generation.

The solar wafers, cells, and modules manufacturing processes generate significant amounts of CO_2 and other harmful gasses, which must be taken into consideration when talking about the advantages of PV technologies over the conventional energy sources.

Consider this equation for lifetime CO_2 balance (LCB):

$$LCB = \frac{C_{prod} = C_{TRANS} + C_{INST} + C_{USE} + C_{DECOM}}{C_{save}}$$

Where:

C_{prod} = energy used during production of materials, wafers, cells, and modules

C_{trans} = energy used to transport materials, modules, and BoS to PV site

C_{inst} = energy used to assemble and install the PV power plant

C_{use} = energy used to operate the PV power plant

C_{recyc} = energy used to decommission, transport, and recycle the PV field

C_{save} = energy generated during the life of the PV power plant

In all cases, C_{save} must be much higher than the sum of the CO_2 generation sources for the system to be an effective energy source. Regardless of the LCB ratio, the PV plant will receive carbon credits for the CO_2-free power generated during its lifetime.

Here again, we have seen estimates that the total quantity of CO_2 generated during the cradle-to-grave cycle of PV components is compensated within 2-4 years of CO_2-free PV power generation, by reducing CO_2 emissions as compared to conventional power generators.

Carbon Tax

A carbon tax is a new taxation imposed on carbon (CO_2 and such gasses) generating facilities. It is usually determined by the carbon content in the fuels used in these facilities.

Carbon is a component in every hydrocarbon fuel (coal, petroleum, and natural gas), which upon burning generates and emits these harmful gasses (CO_2 mostly).

CO_2 is a heat-trapping "greenhouse" gas (GHG), and scientists have proven its potentially negative effects on the climate system upon release in the atmosphere.

Since GHG emissions are related to the carbon content of the respective fuels, a tax on these emissions can be levied by taxing the carbon content of fossil fuels at any point in the product cycle of the fuel.

Other, non-combustion energy sources, such as wind, sunlight, hydropower, and nuclear do not burn, and therefore do not emit CO_2. So, the carbon taxes offer a potentially cost-effective means of reducing GHG.

From an economic perspective, carbon taxes are a type of Pigovian tax. They help to address the problem of emitters of greenhouse gases, not facing the full (social) costs of their actions.

Carbon taxes are also regressive taxes, in that they disproportionately affect low-income groups, because they pay a disproportionately large portion of the tax.

A number of countries have implemented carbon taxes related to carbon content of emitted gasses. The problem is that most GHG-related taxes are levied on energy products and motor vehicles, rather than on CO_2 emissions directly.

Many countries and large users of carbon resources in electricity generation, such as the USA, Russia and China, are resisting carbon taxation, so it remains to be seen if it will survive, and if it does in what form.

INVESTMENT AND FINANCING

The 2008 financial crisis and the lingering economic recovery have significantly affected the development of renewable energy projects, and investors, developers, and consumers are still feeling repercussions. Federal stimulus funding through the American Recovery and Reinvestment Act of 2009 (ARRA) offered short-term relief, but ARRA programs are nearing their end. Renewable development in California will then rely on traditional financial markets and project finance strategies under uncertain economic times.

Consumer education, renewable technology advancement, and generous financial support have enabled market growth during these troubled economic times. However, when compared to other infrastructure-dependent sectors, such as transportation, renewable energy development is still a young asset class.

Renewable development does not enjoy the same technological, financial, and marketplace advantages afforded to traditional infrastructure with more mature supply markets and support structures in place.

Key Financing Issues

Key financing issues facing renewable technologies and efforts to address those issues, include:

- Financing challenges, specifically funding gaps

that occur during the research and development (R&D) and early commercial stages that can affect project development.

- The role of private and public investment in energy-related R&D, including "angel" and venture capital investors, universities' investment in research, and the role of power purchase agreements and financing instruments, like debt and tax equity, in resolving barriers to renewable development.

- Efforts underway to address financing challenges for utility-scale renewable projects including supporting technology innovations and reducing significant capital requirements through tax incentives, accelerated depreciation, and loan and bond financing programs.

- Efforts underway to address financing challenges for distribution-scale renewable projects, including case studies to illustrate technologies and finance programs in use.

- Efforts to finance renewable development in neighboring states and internationally.

- Financing renewable energy—from research and development to project deployment—is dynamic and capital-intensive. Each stage of development requires significant resources, and financing gaps must be addressed for emerging technologies to move to commercial maturity in a timely manner.

Development of Renewable Technologies

Successful development of renewable technologies includes five stages:

- R&D generates ideas and tests intellectual property. If developed successfully, intellectual property can leverage needed funds further along the development continuum. Nevertheless, R&D is a high-risk stage for potential investors given high failure rates.

- Demonstration/proof-of-concept builds the company, designs and tests prototypes, and further develops intellectual property needed to demonstrate the feasibility of an idea or technology.

- Pilot facility moves the technology out of the laboratory to the field where data and results are quantified to improve the prototype and to provide technical information to potential investors. In this phase, companies also begin to market the technology.

- Early commercial enables a company to demonstrate the viability of its technology at scale and prove that manufacturing and/or power generation can be economically scaled.

- Commercial maturity is the widespread adoption of a technology, when it is commercially proven, sold, and distributed at scale.

The inherent uncertainties in R&D can make it difficult to obtain financing, and significant capital is required in the early commercial stage to address technology performance issues and regulatory risk. These two key financing gaps can affect the ultimate development of a renewable project as described below.

Financing Research and Development—Financing Gap 1

California is endowed with abundant natural resources. These assets lend themselves to renewable development but first require research and development, resource assessment, and use impacts. As such, California requires public and private investment in research and development that differs from other states due to its variety of renewable resources.

A recent American Enterprise Institute report indicates that, despite a clear innovation imperative, neither public nor private sectors currently invests the resources required to accelerate clean energy innovation and drive down the cost of clean energy.

Private firms do not invest adequately in new technologies for various reasons, including the higher price of clean energy technologies, knowledge spillover risks from private investment in research, inherent technology and policy risks in energy markets, the scale and long time horizon of many clean energy projects, and a lack of widespread clean energy infrastructure.

R&D externalities and challenges can lead to financing gaps and less than socially optimal technology innovation. The International Energy Agency estimates that globally, solar and wind energy technologies face an annual R&D shortfall of between $2.68 billion and $6.28 billion. To a large extent, knowledge is a public good. In economic terms, it is nonexcludable and nonappropriable because it is difficult for owners to establish enforceable property rights or to dole out usage rights to particular individuals. Because almost anyone can access the knowledge developed in R&D, and a private firm cannot monetize all the public benefits and spillovers of its R&D, private companies tend not to invest in the level of R&D that is most beneficial to society.

Patents help reduce this concern but generally offer limited protection for intellectual property rights. The underinvestment in R&D can be further exacerbated for clean energy technologies because the positive environmental attributes of these technologies, like greenhouse gas emission reductions, may not be adequately valued in the market.

Other R&D challenges increase risk due to the high variance of the distribution of expected returns, the specialized and sunk (and as such not transferable) nature of the asset, and intangibility. Uncertainty and intangibility make financing through capital markets difficult since investors typically want some certainty of return on their investment. Information asymmetry, where one party has better information than another, can exacerbate uncertainty since a technology developer is in a better position to assess the potential of a technology than investors, so investors will require bigger return to address this uncertainty.

These challenges—and given that startup markets are decentralized and involve multiple firms—contribute to a financing gap for R&D, particularly energy-related R&D. Although overall R&D investment in the United States has grown annually by 6%, investment in energy-related R&D is about $1 billion less than a decade ago. The private sector's share of energy R&D investment has also declined to 24% from nearly half in the 1980s and 1990s, with total private sector energy R&D less than the R&D budgets of a few large individual biotech companies.

Corporate R&D spending (reinvestment) can be a significant driver of new technology development. However, according to the National Science Foundation, corporate R&D spending as a percentage of domestic sales in 2008 was 25% for communications, 15% for software, and only 0.3% for energy, which is spent primarily on technology improvements and not on new technology development. This underinvestment in the renewable energy sector affects the development of next generation, lower-cost technologies and illustrates the important role of the public sector in accelerating the demonstration of new clean technologies in the absence of private funding.

Apart from private and public sector investment trends, venture capital investments continue to increase and are still on the same scale as private R&D investments by large companies. Venture capital firms' contribution to innovation is especially important since studies have found that, in general, venture capital investment is three to four times more effective than R&D at stimulating patenting.

The biggest private sector participants in R&D are "angel investors"—an individual or a network of individuals that provide capital to startups in exchange for ownership equity or convertible debt. Angel capital fills the gap between seed funding—usually provided by the startup members and friends and family—and venture capital, and typically ranges from hundreds of thousands of dollars to less than $2 million.

After declines of 26.2% in 2008 and 8.3% in 2009, total angel investments in the United States in 2010 increased 14% over 2009 to $20.1 billion According to a study by the University of New Hampshire's Center for Venture Research, the clean tech sector received about 8% (about $1.6 billion) of those investments. The study also reveals that angel investors have lessened their interest in the startup stage of technological development, with seed or startup capital decreasing in 2010 by 4% from 2009.

The financing gap for energy related R&D may also persist at the roof of concept stage as companies continue to require capital for applied research and pre-commercial growth and activities. The most active private sector participants in this stage are angel investors and venture capital (VC) firms. VC firms invest the financial capital of third-party investors in enterprises that are too risky or too complex for the standard capital markets. Targeted investments are early-stage and high-potential startups that have already received seed investments and have a technology and idea beyond the initial R&D stage.

Venture capital investments in clean tech companies have increased, with $3.98 billion invested nationally in 2010. Investments in the first quarter of 2011 were $1.14 billion, a significant increase from the $743.3 million of venture capital invested during the first quarter of 2010, with California accounting for 56% of total investments.

California perennially tops the list of regions receiving venture capital. Unlike other regions of the United States, California is home to more than one venture hub: Silicon Valley in the north, San Diego in the south, and a burgeoning new corridor in Orange County. In addition, California-based, venture-backed firms and their investors have worked consistently with state policy makers to ensure that young innovative startups and their technologies have the opportunity to grow and succeed within the state's larger business climate. These factors have led to nearly $200 billion in venture investment in California since 1970.

Historically, federal and state governments have played a pivotal role in funding research and demonstrating high-risk technologies through direct procurement. The federal government is the primary source of funding for basic research across all sectors, providing some 60% of funding; the second largest source of basic research funding comes from academic institutions. Universities conduct the majority of basic research in the United States (55% in 2008), with business and industry conducting less than 20%.

The state funds research primarily through state and private universities. The University of California research system received more than $4.3 billion in total research funding in fiscal year 2009-2010, produced more patents than any university in the nation, and secured $8 in federal and private dollars for every $1 in research funding provided. In addition to directly funding energy research, the university system contributes to technology transfer and the overall development of the state's expertise in renewable generation technologies.

In addition to universities, the Energy Commission's Public Interest Energy Research (PIER) Program funds public and private entity research and has a significant role in supporting renewable technologies. The PIER Program has also seed funded incubators such as the Sacramento Area Regional Technology Alliance (SARTA) and the Environmental Business Cluster. Fostering R&D through such organizations as SARTA, CleanTECH San Diego, CleanTechOC, Clean Tech Los Angeles, Silicon Valley Leadership Group, Greenwise Sacramento, and the Bay Area Council accelerates the growth of renewable energy technologies and provides economic benefits that accrue to local communities.

The state's Innovation Hub (iHub) initiative stimulates partnerships, economic development, and job creation around specific research clusters through state designated iHubs. The iHubs leverage assets such as research parks, technology incubators, universities, and federal laboratories to provide an innovation platform for startup companies, economic development organizations, business groups, and venture capitalists.

Financing Early Commercial Development—Financing Gap 2

Another financing gap occurs at the early commercial stage of development. Early commercial can be defined as one of the first three to five deployments at a scale that generates revenue and within the size range consistent with a company's long-term rollout plan. In this stage, companies and technologies face the convergence of high capital needs and scarcity of capital. Significant capital is needed to finance projects to demonstrate the viability of a technology at scale, as well as to prove that the manufacturing and/or power generation can be economically scaled.

There is a significant difference in installed renewable capacity between the residential and commercial sectors for early commercial technologies that can be attributed to: 1) greater federal tax benefits and accelerated tax depreciation provided to commercial systems, 2) the larger commercial project size, and 3) the appropriate project finance structure. Small-scale renewable systems, often found on residential and small commercial properties, have limited financing options and must resort to using established financing tools such as equity loans and cash combined with applicable public programs. Other options include leases, small scale power purchase agreements, and on bill financing.

New financing opportunities, such as third-party models, open up for community scale systems. Partnering with an investor such as project developers and other organizations can also take advantage of tax and depreciation benefits through various partnership structures and power purchase agreement models to reduce costs. Utility or large-scale renewable projects are most commonly financed by a combination of equity and debt. Financing options include corporate balance sheets, use of debt (loan guarantees), equity (federal Section 1603 cash grants), and asset depreciation.

The U.S. Partnership for Renewable Energy Finance indicates where several emerging renewable technologies are currently or are anticipated to be at this early commercial juncture in the next three years. In particular, concentrating solar-power towers, advanced solar manufacturing, and energy storage will encounter Financing Gap 2. These technologies are seeking approval for U.S. Department of Energy Loan Guarantee Program funding, which could be leveraged for additional private capital. Funding for traditional renewable technologies such as solar PV, wind, and geothermal could also advance these technologies in the market.

Traditionally, private equity, debt, and tax equity markets have served as options to firms in the early commercial stage. However, since the financial crisis, these options are either impractical given economic conditions, depend on government incentives to function well, or do not provide sufficient returns for investors.

Finance options and structures have evolved in response to the patchwork of incentives, and the preferred choice depends on several factors such as return and yield rates, presence of a tax investor, cash and financial position (balance sheet financing), exit strategies, and risk-adjusted mitigation measures. Financial innovations are increasingly reducing investment barriers such as high upfront costs, technology and performance risk, and low tax credit appetite. These mechanisms allow investors to extract maximum value from myriad local, state, and federal support programs.

Financing Gap 2 poses a challenge for equity investors such as angel investors, venture capital firms, and other early stage investors accustomed to traditional technology companies' speedy and relatively low-cost paths to commercialization. Capital requirements for renewable energy projects can be tens to hundreds of millions of dollars, depending on size and technology, for a single project. This is beyond the capacity and appetite of the great majority of VC firms.

Renewable projects also rely heavily on project finance, which is not the traditional business model used by VC firms. Project financing relies on the complex management of project risks, mitigation strategies, legal and commercial structuring, significant debt financing, and coordination with a variety of stakeholders, including contractors, large industrial partners and potential customers.

Last, private equity firms, particularly VC firms, anticipate higher equity returns than those expected from renewable energy projects.

Bankability

Bankability, or the perceived ability to show financial stability and marketing ability in order to get bank financing for a solar company or a project, is a new issue, which we feel is totally misunderstood and misused.

A typical financial structure for any solar company or project, including large companies and large-scale projects (over a few MW) is the so-called "non-recourse project financing," and we have seen through the years that banks and other lenders often refuse to finance companies or projects for a number of reasons on the basis of this structure. So how could "bankability" change that?

More often than not, banks refuse financing because they don't understand or like a certain technology or project, or some aspect of it. Sometimes they simply don't like the company, the management, and/or some parts of the information they are given. There are no rules to follow on dodging or changing these sentiments.

This perception, which at times overflows into an established behavior, is what the myth of "bankability" is based on. It is supported and "cemented" by the perceived necessity of the manufacturers to show "bankability," and by so to get on the banks' "bankability" lists.

But, as anything perceived, such lists do not even exist—at least not officially—and there are no defined guidelines for establishing "bankability." Furthermore, most reputable and experienced banks do not want to work with such lists, preferring to work on case-by-case, location-by-location and company-by-company bases.

So "bankability" is a term that we think should be sent packing, as it is not something a company or a product can claim officially. Even if an entity somehow achieved such status, there is no way to maintain it, due to the ever changing character of the solar industry and related markets.

Solar Financing Today and Tomorrow

US solar projects have historically been bankrolled by some combination of energy sector players, banks, and the federal government, but the landscape is rapidly changing. New business models are emerging with an emphasis on third-party financing. New investors, including institutional players, are entering. And new financing vehicles such as project bonds and other securities are being assembled to tap the broader capital markets. (19)

The ongoing evolution of US solar financing, where the market is today, where it is heading, and what's behind this important transition, are outlined below.

- Maintaining US solar deployment growth will require substantially more investment. Asset financing for US photovoltaic (PV) projects has grown by a compound annual growth rate of 58% since 2004 and surged to a record $21.1 billion in 2011, fueled by the one-year extension of the Department of Treasury cash grant program. But funding the next eight years of growth (2012-20) for US PV deployment will require about $6.9 billion annually on average.

- Traditional players are taking a smaller role. Regulations, primarily in the EU, are curtailing banks from providing long-term debt as easily as previously. In the US, the Department of Energy loan guarantee program's expiry has meant less direct federal government support.

- New models are emerging. Distributed generation is driving innovation and creating new models for solar deployment. Few homeowners can afford the upfront cost of a solar system, giving rise to third-party financing models, which allow them to 'go solar' with little or no money down. These models also give investors a diversified opportunity to back solar.

- New investors are taking interest. Institutional players such as insurance companies and pension funds seek stable, long-lived assets to match long-term liabilities; some utilities may seek solar portfolios to offset revenue loss from distributed generation. On the development side, infrastructure funds could achieve targeted returns by bringing these projects to fruition.

- New vehicles are taking shape. Such structures aim to make solar project investments more liquid by allowing developers to tap the capital markets. Options include project bonds, solar real estate investment trusts (S-REITs), public solar ownership funds ('yieldcos'), and others.

- The cost of capital will fall. Solar equipment prices have dropped by more than half since the start of 2011 but financing costs matter too. New financing vehicles and new investors across the solar project lifecycle—development, construction, commissioning, and then long-term operation of assets—will cause the costs of equity, debt, and even tax equity to migrate down.

- This is happening now. Institutional investors have bought solar bonds; publicly listed renewable asset funds exist; solar portfolios are poised for securitization; and pension funds have shown willingness to buy and own renewable assets.

- Policy could accelerate change. Surveyed investors seek stronger SREC programs, new standards, more flexible tax credits, and sanctioned high-liquidity vehicles such as S-REITs.

THE US INVESTMENT AND FINANCE

California is the leader in solar energy development and use in the US, and soon it might become the largest solar energy producer in the world too. California and its neighboring states are the most active in the implementation of solar energy into the energy markets, with a number of different incentives provided to encourage investment in renewable energy generation plants and manufacturing facilities.

Arizona and Oregon have introduced personal and corporate tax incentives, especially geared for solar projects, while Arizona and Nevada have sales tax incentives for renewable energy as well.

Arizona's solar energy generation potential is estimated at 101 million MWh per year. Key efforts include Arizona Renewable Energy Standard and Tariff of 2006, which is similar to California's Renewables Portfolio Standard.

Arizona project development is also bolstered by net metering, state corporate production tax credit, income tax credit, a property tax exemption and incentive, and a sales tax exemption. Most of these efforts do not expire until 2018, providing certainty to the market.

While existing in-state solar generating capacity is relatively low at about 50 MW, more than 2,200 MW of solar projects have been *announced* or *planned*. Renewable energy manufacturing equipment also has a sales and use tax exemption.

Nevada has a large geothermal resource that produces nearly 4% of its electricity, and the state has the potential to generate about 83 million MWh per year of solar energy. Renewable energy development is supported by a renewable portfolio standard, net metering, various rebates and incentives, and laws that prohibit siting restrictions on solar and wind energy systems.

Nevada, like California, has state legislation for a Property Assessed Clean Energy (PACE) Program. The Renewable Energy Producers Property Tax Abatement provides a property tax abatement of up to 55% for up to 20 years for real and personal property used to generate electricity from renewable energy resources including solar, wind, biomass, fuel cells, geothermal, or hydro. Generation facilities must have a capacity of at least 10 MW. The Nevada Energy Renewable Generations Rebate Program provides solar rebates from $2.30 per watt AC; wind rebates from $2.50 per watt; and small hydro rebates from $2.00 per watt.

Oregon's Small Scale Energy Loan Program (SELP) was created as a result of a voter approved constitutional amendment that authorized the sale of bonds to finance small-scale local energy projects that save energy, produce energy from renewable resources, create products from recycled materials, use alternative fuels, and reduce energy consumption during construction or operation of a facility.

Generally, the loans range from $20,000 to $20 million. The Energy Trust of Oregon provides cash incentives and development assistance for renewable energy projects that have a capacity of 20 MW or less. Funding is available for grant writing, feasibility studies, or technical assistance with design, permitting, or utility interconnection.

The Energy Trust will pay up to 50% of eligible project costs for a maximum of $40,000. Incentives are based on the project's costs in comparison to the market value of energy produced.

The Power Grid…

In August 2012, India's power grid failed. It collapsed, actually, leaving half of the country powerless. The electricity has been restored, after 600 million people were left in the dark for several days. Such disaster is unlikely to occur in the US, nevertheless, the old and tired US power grid stability and capacity are of great concern these days.

Figure 7-9. India's power grid

Yes, the picture in the US is quite different than India's, but our old and outdated grid is a reality, which we should not take lightly. Case in point is the fact that in 2003 East Coast outages affected approximately 50 million people across the US and Canada. There were similar outages in California and Arizona during the 2000s as well.

The American Society of Civil Engineers (ASCE) estimates that over $110 billion in additional investment is needed by 2020 to keep the electrical infrastructure whole.

If and when the U.S. economy comes back to life (2015-2016), there may not be enough power to support the new industrial peak demand. We saw new peak records in 2011-2012 set for the first time since 2008, so the demand for electricity is again increasing. Utilities know this, but are hesitating in their decisions.

The events in India, as well as major outages experienced in the US and across the globe, should encourage us to focus on the current power infrastructure and to take the needed measures to implement modern grid technologies, improving the grid's stability.

Basically, the US power grid needs a major face lift.

The Smart Grid

We cannot ignore the importance of the proposed smart grid as a new way to improve energy use worldwide.

The "Smart Grid" concept did not even exist until 5-6 ago, although there were some companies working in this area. Today, however, we see hundreds of millions of dollars of government and VC investment flow into a wide range of smart grid start-ups. This new movement is creating a new market and a large ecosystem. ranging from power generators to grid power distribution and management, commercial and home networks, and a lot of new ideas that go with these.

Some of this is hype, unfortunately, and a lot of it is simply impractical or impossible. Nevertheless, private investment in smart grid projects from 2004 to 2009 was over $1.3 billion, and more than that in the following years. The US government is not far behind. In July 2012, the U.S. Department of Energy issued the Recovery Act Smart Grid Progress Report, which provides background information and the status of the smart grid projects that received funding through the Smart Grid Investment Grant Program. The Smart Grid Investment Grant Program (SGIG) is a $3.4 billion initiative that seeks to accelerate the modernization of the nation's electric grid by deploying smart grid technologies and systems. The government even has a special website, SmartGrid.gov dedicated to the development of the US smart grid.

The "Smart Grid" concept, however, is mostly theoretical, vs. a true engineering project, which in a perfect world could be built quite easily. All we need is to design a self-controlling, self-healing, sensor-operated, robust and secure power network, which allows efficient power generation and distribution, and is accompanied by reasonable pricing.

Because we don't live in a perfect world, we can only attempt to superimpose intelligence and sophisticated technologies over an antiquated and tired power grid that has a mind of its own, many masters and many problems. Utilities are slow to make changes, the regulators and legislators are even slower, and existing standards often compete with proposed innovations.

However, there is a huge momentum in the smart grid field, and funding is increasing. Where there is money to be given away, there are takers, so the field of smart grid designers, manufacturers and users is increasing to the point to where even Fortune 500 companies, like IBM and Intel are starting to show interest. This is immense pressure that might accelerate the smart grid development process, and push the utilities together with the US power grid into the 21st century.

The smart grid idea is a good thing, but the existing grid needs to be fixed before we can launch a truly "smart" grid.

Energy Storage

Energy storage, or the lack of it, is stopping the full implementation of solar energy in the US. Energy storage is badly needed to reduce the variability of solar generation and to provide power during after-sun hours.

The U.S. Department of Energy (DOE) has made US$43 million in funding available to 19 energy storage research projects, looking to improve battery and grid storage capabilities. The funding will go towards the development of energy storage technologies and help support "promising" small businesses.

Twelve projects will receive a total of $30 million to improve existing battery technologies and develop advanced sensing and control technologies. The goal is to both reduce costs and advance battery performance, while $13 million has been awarded to seven projects from small businesses, which will focus on energy storage developments for stationary power and electric vehicles. The goal of these projects is to develop new innovative battery chemistries and battery designs.

Of particular interest is the research proposed by Portland-based Energy Storage Systems, Inc., which was awarded $1.7 million for developing a new flow battery for grid-scale storage. The flow battery will have a target storage cost of less than $100/kWh, which could enable deployment of renewable energy technologies all across the nation's power grid.

ITN Energy Systems, Inc., has proposed to improve Vanadium flow batteries for grid-scale energy storage, and this could also have an impact on small-scale solar generation. It has been awarded $1.7 million towards the integration of a "unique, low-cost membrane with a new flow battery chemistry."

Basically, with these awards, the US DOE is looking to address some of the key challenges in energy storage technologies, the resolution of which could revolutionize solar power generation by storing and using energy on demand. This type of energy storage is also useful as a strategic vehicle to improving the access to energy by the U.S. military at remote areas.

WORLD INVESTMENT AND FINANCE

Solar energy related activities, investments, and finance are once again on the rise. Things have changed drastically since the disproportionate and unsustainable boom of 2007-2011, but in 2Q 2012 the world's renewables investment was $59.6 billion, up 24% from the $27 billion of the 1Q 2012 (the worst figure since 2009). This number, however, is down 18% from the near-record $72.5 billion from the same period in 2011. $33.6 billion were invested in solar and $21.6 billion in wind during 2Q 2012.

Utility-scale projects finance led with $35.9 billion, more than a 50% increase from the previous quarter, but 24% less than the same period in 2011.

Green venture investment during the quarter totaled $1.6 billion, a drop from both the first-quarter of 2012, and 2011 figures. A second-quarter green IPO tally was at $1.8 billion, up from $1 billion in the first quarter but down from $2.2 billion in the same quarter of 2011.

Only small-scale clean power generation—projects of 1 megawatt or less—were worth $21.5 billion in 2Q 2012, up 13% from 2Q, 2011. This attributed to the low prices of PV modules (75% lower than 2009).

Globally, Germany and Italy remain the largest solar markets, but installations in the U.S. are catching up despite their higher cost. California passed 1 gigawatt of customer-owned rooftop solar power capacity in June, 2012.

China's green energy growth was $18.3 billion, a 92% increase from the same quarter last year. The European investment grew only 11% to reach $20 billion, and U.S. investment grew 18% to reach $10.2 billion. Beyond being the world's biggest wind turbine market, China has quadrupled its domestic solar installation targets.

Smart grid and advanced transportation brought in $1.1 billion in the second quarter—a 74% increase from the first quarter, and below the 2011 figures. While smart grid has seen venture capital investment dwindle during 2012, merger and acquisition rates are on the increase. What that means is unclear, but something unusual is going on in that field.

Biomass and waste-to-power technologies brought in $1.4 billion, down 22% from the first quarter. Bio-fuels rose $750 million, down 12% from the previous quarter.

China

China's 2009 total renewable energy investment of $34.6 billion led the world, and has been primarily in the wind energy sector which accounts for 71.1%. Similar to Germany and Spain, China supports renewable energy with a renewable energy standard and clean energy tax incentives. In August 2011, China announced a national FIT for solar power installations. The government offers green energy bonds to encourage development.

The U.S. is a leader in research and development for the solar industry with some $7.1 billion spent in 2009 alone—significantly more than the EU and the rest of the world, including China, according to industry specialists. But solar is a capital intensive industry—about five times more expensive than coal—where mass production requires lots of cash. This is where China has an edge.

Chinese manufacturers entered the market at a time when global demand was high and supply was low, flooding the market with solar products. Unfortunately, time was not taken to develop the techniques necessary to manufacture these products which require great precision, supply-chain management and quality control. Subsequently, a number of solar companies and projects worldwide failed during 2011-2012.

Although China's exports were not the only, or even the main reason for the failures, the flood of inexpensive imports in the hands of would-be experts, funded by unenlightened investors and banks, undermined the basic principles of pricing structures, quality, and everything else the solar industry should be based on.

In the spring of 2012, China's central government issued a new 5-year plan for its solar industry, outlining initiatives such as new financial subsidies, more support in industry, financial and tax policy, and more assistance with the development and manufacturing of equipment that is used to produce solar polysilicon, silicon ingots, wafers, cells and PV modules. The plan includes increased support for the development and commercialization of the budding thin-film PV industry, with emphasis on silicon thin films (a-Si and such) and CIGS technologies. The plan designates solar among seven "strategic emerging industries" that warrant massive support, with subsidies ranging between $1.5 and 2.0 trillion.

China's solar manufacturing capacity is over 30 times greater than its domestic demand, which requires that 90-95% of their solar production be exported.

These events spawned an avalanche of political and industrial upset in the Free World, including legal and legislative maneuvers by governments and businesses alike.

The U.S. International Trade Commission issued a unanimous preliminary ruling on December 2, 2011, declaring that Chinese trade practices are harming the U.S. domestic solar industry.

In a partial preliminary determination in March, 2012, the U.S. Department of Commerce announced that at least 10 categories of Chinese subsidies for its producers of solar cells and panels were illegal. This decision was later on re-affirmed and specifics were provided.

These turbulent events are matters of record and will not be further detailed here.

Mexico

Mexico achieved the biggest increase in renewable energy investment in Latin America, excluding Brazil, due in part to the Renewable Energy Development and Financing for Energy Transition Law enacted in November 2008. Mexico's renewable investment grew 348%, stimulated by major wind projects and geothermal.

Mexico's government raised renewable energy capacity from 3.3% in 2009 to 7.6% by 2012, and wind is intended to make up 4.3% by 2012. Projects are expected to reach $8 billion in 2011 by foreign investment into Mexico with primary interest in the borderland between northern Mexico and the southern US.

The Baja California region has grid connectivity with California and the potential of exporting renewable energy to the state. Historically, most foreign investment in Mexico comes from the United States. As an example, San Diego-based Cannon Power Group is investing $2.5 billion in Mexico to build wind farms to generate more than 300 MW of electricity.

However, until the interior of Mexico has viable power distribution networks—until the infrastructure is upgraded—there is no potential of capturing solar and wind energy. Also, increasing gang violence in US-Mexico border areas stands in the way of significant development of solar energy in the region.

Canada

Canada has invested $CDN 4.9 billion in renewable energy thus far. Canadian incentives for investment include:

1) a 50% accelerated capital cost allowance for clean energy generation;

2) underwriting R&D activities that lead to new, improved, or technologically advanced products or processes;

3) a $CDN 1.5 billion investment to increase clean electricity from renewable sources; and

4) a $CDN 230 million investment in clean energy science and technology that will fund RD&D to support next-generation energy technologies.

The province of Ontario has also successfully implemented a FIT program and a micro FIT program (for projects less than 10 kW) over the past 18 months.

Although Canada is not recognized as a "sunny" country (quite the contrary—most of the country is under constant cloud cover), the government subsidies, high FiT rates, and investment community's enthusiasm have created a sizeable solar industry. We would not be surprised if Canada takes a lead in the solar race of the 21st century.

Europe

Germany and Spain are active participants in expanding renewable energy generation, investing $4.3 billion and $10.4 billion, respectively, in 2009 alone. The bulk of investment in these countries has been in the solar and wind sectors, supported by subsidies, carbon markets, renewable energy standards, clean energy tax incentives, FITs, and encouraged by government procurement requirements.

As of 2009, 29% of German power capacity and 30.1% of Spanish power capacity was generated from renewable sources.

Germany and Spain's FITs have resulted in the installation of thousands of MW of renewable capacity. Germany has already achieved its renewable energy goals and increased generation targets by 5%. Spain's FITs are also highly utilized. Because of higher than expected demand for FITs, regulators were forced to cap annual eligible installations and reduce incentives.

This disruption provided market uncertainty and resulted in a period of excess panels supply, and decreasing prices, as global supply outpaced demand. Spain's incentive policy was not a long-term sustainable design nor was it market responsive. This led to taxpayer backlash and market uncertainty as a boom/bust effect was felt in the growing solar PV market.

Greece has suffered most from the economic downturn, where presently over 25% of adults are unemployed.

Spain is next on the EU's list of most unfortunate. In the summer of 2012, 50% of young adults were reported unemployed.

The other EU countries are in a better shape, but

barely so. This grave economic downturn brought the demise of the solar industry in the EU. As a victim of the situation, solar development has been drastically cut in most EU countries.

The question now is how would the EU solar and wind industry recover? Is it possible? We think so, but that it will take a long time to build rebuild confidence in solar power, and win back government and investor support.

Spain

During the golden solar years 2007-2010 Puertollano, Spain, a small, gritty mining city, enjoyed a gold rush. Puertollano was a coal dust-covered place with not much to show except blistering sunshine most of the year...

Generous incentives from the Spanish government flowed into southern Spain, and soon Puertollano's fields were filled with all kinds of solar power generating equipment. Coal ventures were replaced by solar companies, rich with free government money. A campaign slogan popped up: "The Sun Moves Us." The heck with coal...

By 2010, Puertollano, famous for its Museum of the Mining Industry, had two enormous solar power generating plants, in addition to factories making solar cells and panels. A clean energy research institute was located nearby, as a further proof of the legitimacy and prosperity of the new local solar energy industry.

Miners left the mines and donned factory attire, becoming well paid solar cells and panels manufacturing specialists. They enthusiastically sold land for solar plants. New stores and boutiques opened to support the flux of solar professionals. People from all over the world, seeing business opportunities, moved to the city, which had previously suffered from 20% unemployment and a increasing population exodus.

This where the euphoria ended. Low-quality, poorly designed, and hastily put together solar plants sprang all over Spain's plateaus, including Puertollano. Spanish officials came to realize that many of the failing fields would need indefinite and increasing subsidies; the industry they had created might never produce efficient green energy on its own.

So, in September, 2010, the government abruptly cut payments and lowered the cap on new solar construction. Puertollano and its new and old inhabitants were caught totally unprepared, when the brief boom turned into an ugly bust.

Factories and stores shut down, thousands of workers lost jobs, foreign companies and banks abandoned contracts that had been negotiated. Many farmers who sold or leased their fields to solar developers did not receive the monthly payments they were relying on. Work on new installations stopped overnight. Unemployment went up to 10-12%. People started evacuating the area... again.

"We lost the opportunity to be at the vanguard of renewables—we were not only generating electricity, but also a strong economy," Puertollano's mayor complained. "Why are they limiting solar power, when the sun is unlimited?"

A recent study calculated that Spain has spent approximately €571,138 to create each "green job"—much of which was paid to foreign companies and immigrant labor. This includes subsidies of more than €1 million per each wind industry job. The programs creating those jobs also resulted in the destruction of nearly 110,500 jobs elsewhere in the economy. This means that 2.2 permanent, well-paying jobs were lost for every temporary, low-paid "green job" created.

Each "green" megawatt installed was estimated to destroy additional jobs on average elsewhere in the economy: 8.99 jobs eliminated by photovoltaics, 4.27 by wind energy, and 5.05 by mini-hydro.

The Spanish FiT during 2007-2009 was $0.58 per generated kWh—one of the highest in Europe and absolutely unsustainable under any circumstances. The higher cost of the produced electricity negatively affects production and employment levels in key sectors of the economy, such as metallurgy, non-metallic mining, food processing, beverage and tobacco industries.

These costs and problems are not unique to Spain's approach; they are largely inherent in all countries and with all schemes to promote renewable energy sources via disproportionate and unsustainable government subsidies and incentives.

The wrenching rise and fall of Spain's renewable energy points to the delicate policy calculations needed to stimulate nascent solar industries and create green jobs, and might serve as a cautionary tale for the United States, where similar exercises have been tried.

Germany

This is a good place to note the undeniable fact that Germany has been one EU country that fully embraced the generation of electricity from the sun during this last boom and bust cycle. It presently has almost half of the world's PV installations, but even this amazing progress was not able to avoid economic doom, and now this heartland of solar energy has dark clouds looming over its solar future.

Subsidies are falling quickly. Solar panels and inverters manufacturers are downscaling production, and some have gone bankrupt. Thousands of employees are frustrated and have been holding demonstrations all over the country. One of the largest losses in Germany was the shutdown of First Solar's manufacturing facility, where nearly 2,000 jobs were lost in the spring of 2012. Many other German solar products manufacturers followed this path. See Chapter 5.

The major problem is the recent cut in subsidies by 30%, taking effect in 2013. According to the government, this is because of the great success of the solar scheme. The demand for solar has been great, so the budget allocated for solar has been greatly exceeded. With the subsidy drastically cut, the industry is finding it tougher to continue.

The manufacturers are under enormous pressure, because they have a high over-supply of finished product, making it difficult to sell quickly and at a reasonable price. Now experts say the government made mistakes in how it designed and handled the solar subsidies program.

Imports have been a major headache for the German solar products manufacturers who claim that their government failed to stipulate that subsidies would be used to buy modules from German manufacturers, aiding German workers, not supporting other countries.

The broader question is whether Germany's embrace of solar power was wise at all. Solar works in solar climates—which is not the case in Germany where the solar insolation is close to that of Juno, Alaska.

Because of that, German experts are convinced that it is a mistake to push so much solar under cloudy skies. Solar belongs in Spain, Italy or North Africa, which much more sun and would make solar power much more profitably.

Without solar energy, Germany is looking at total dependence on oil and nuclear power, so we still insist that the energy solution for the entire EU continent is solar energy—solar energy produced in areas where solar belongs—the Sahara desert and surrounding regions. Thus generated electric power can be easily transmitted to Europe via DC cables and used as needed—expensive, but with lasting rewards.

Germany is still the leader in renewable energy. In May, 2012, they set a world record reaching 22 GW of power generation. At that point, 50% of the country's electricity was generated from solar energy sources. The current solar capacity reaches about 28 GW and the goal is to reach 66 GW by 2030. In 2011, Germany had about 22 times more solar power installations per capita than the United States.

In addition to their streamlined regulations and permitting procedures, Germany introduced a three-pronged approach to their solar energy FiT law that has been studied intensively, with a number of conclusions pro and con.

The key German FiT law mandates:

1)   Fairly transparent, long-term pricing of generated electricity that allows repayment of installation costs and reasonable profit, according to location and type of technology;

2)   Officially guaranteed, hassle-free connection to the grid in a timely manner (mostly valid for residential systems); and very importantly,

3)   The utilities are required to pay for all upgrades needed to connect renewable power to the grid for both residential- and utilities-scale solar installations. WOW!

More often than not, these upgrades are pricier than the projects they serve, so while 1 and 2 above will be slowly but surely implemented in the US, #3 is taboo at present. It will be implemented here too, someday, but with heavy modifications, and in the distant future. We predict 2040-2050 (when US gas supplies dwindle and the next solar boom is on the horizon) as a starting date for negotiations on this subject.

POLITICAL CONSIDERATIONS

The 2007-2010 solar boom and bust was remarkable in many ways, but mainly it showed us that:

1.   Solar energy is a viable energy resource, which is waiting to be used;

2.   Solar technologies should be made, deployed, and used properly;

3.   Solar energy is our hope for energy independence and a clean environment;

4.   Solar technologies still have a long way to go in the competition with conventional energy sources;

5.   One-way government intervention is NOT the way to solar prosperity; and

6.   Dealing with complex regulatory procedures is so 20th century. New approaches are needed, if we are

to achieve solar power success in the 21st century.

With all that in mind, we will take a closer look at the factors that played role in the development of solar energy US and around the world during the 2007-2010 boom-bust cycle

US Solar Politics

US solar energy, during the 2007-2010 solar energy boom, followed closely the developments in other countries, albeit at a slower pace. A number of solar projects were built and some are still in construction phase, but the boom is fizzling now, under the pressure of economic and political changes.

Solar in the US is not dead (yet), but we don't expect any great developments (like many GWs of solar energy) to pop up anytime soon in the US.

For more information on failed companies, please refer to Chapter 5, where we examine the failures of different companies and projects, and the mechanisms behind their failures. Our principal goal is to record these as learning tools for the next solar boom—whenever it occurs.

The US Administration

To its credit, in August 2012 the administration announced that seven large solar and wind energy projects will be expedited in Arizona, California, Nevada, and Wyoming. These job-creating infrastructure projects should produce a total of 5,000 MW of clean energy, to power 1.5 million homes.

The solar segment of this undertaking consists of:

Quartzsite Solar Energy, AZ, to be permitted by December 2012

This power plant would be located on approximately 1,675 acres managed by the Bureau of Land Management. It would produce an estimated 100 MW of clean energy—enough to power about 30,000 homes—and help the State of Arizona meet its renewable energy goals.

Desert Harvest Solar Energy, CA, to be permitted by December 2012

This project would utilize photovoltaic technology on approximately 1,200 acres in Riverside County, CA. It would produce an estimated 150 MW of solar energy—enough to power about 45,000 homes.

McCoy Solar Energy, CA, to be permitted by December 2012

This PV project would be situated on 4,893 acres in Riverside County, CA. It would produce an estimated 750 MW of solar energy—enough clean energy to power 225,000 homes—while helping the state of California meet its targets for renewable energy.

Moapa Solar Energy Center, NV, to be permitted by December 2013

This project is being developed in cooperation with the Moapa Band of Paiute Indians on a 2,000-acre site on the Moapa River Indian Reservation and on lands administered by the Bureau of Land Management in Clark County, NV. If approved, the 200-MW project would employ 100 MW of photovoltaic technology and 100 MW of concentrated solar power technology. Once constructed, this proposed project would be one of the first large-scale US solar projects on tribal lands.

Silver State South, NV, to be permitted by March 2013

This project is a solar energy generation plant proposed on 13,043 acres of public land. If approved, it would produce an estimated 350 MW—enough to power approximately 105,000 homes—helping the state of Nevada meet its renewable energy goals. Construction on the 50-MW Silver State North project has been completed, making it the first solar project on public lands to be delivering power to the grid.

The plans are in. The battle among manufacturers, environmentalists, regulators, utilities, local governments, and investors has started, and the results will be announced soon. We only hope that all goes as planned.

Our one reservation is the Silver State South project, because the PV modules planned for this 350-MW power plant contain significant amounts of toxic, carcinogenic, heavy metal cadmium which would be spread over 13,000 acres of desert lands.

Because these modules have not been approved for long-term use in the deserts, we fear, they might create a large-scale environmental debacle.

The US DOE

In the summer of 2012 DOE issued a draft of the Federal Renewable Energy Guide: Developing Large-Scale Renewable Energy Projects at Federal Facilities Using Private Capital.

The Guide's goal is to provide a project development framework for federal agencies, private developers, and financiers in an attempt to coordinate and streamline their work on large-scale renewable energy projects, in order to ensure the achievement of federal renewable energy goals. The Guide describes the efforts of the Energy Department to deploy clean, renewable energy as needed to diversify our nation's energy capabilities, thus ensuring our energy security, while at the same time cleaning the environment.

In addition to providing a development framework to speed the deployment of renewable energy, the Guide is a general resource to federal employees, helping them understand renewable energy projects and the related activities. It is also expected to improve the energy industry's understanding of the related federal processes.

The Guide is also organized to match federal processes in the renewable energy field, focused on private investments required to implement large-scale renewable energy projects.

REGULATORY PROCEDURES

Planning and Permitting

All solar power generating projects in California (the leader in solar and wind power generation) must go through an environmental review and permitting process subject to California Environmental Quality Act (CEQA), and may also be subject to the National Environmental Policy Act (NEPA).

When utility-scale renewable energy generation facilities are proposed on federally owned land in California, both CEQA and NEPA processes are necessary. Permitting is also required at state, federal and local levels, depending on the size, location, and technology of the proposed facility. This effort alone, is extremely time and money consuming.

The additional steps for planning a new large-scale solar project are as follow:

Federal Permits

There are a number of federal permitting processes that one must go through, most of which involve multiple government agencies and timelines that can add a fair level of uncertainty, expense and delay to renewable energy projects licensing process, such as:

a   Biological resource permit processes under the Endangered Species Act (ESA), administered by the U.S. Fish and Wildlife Service (FWS).

When renewable resource projects are proposed on privately owned land with federally listed species found on the site or in the vicinity, the process for obtaining an ESA permit can be extended and difficult due to the lack of a "federal

nexus." Without a federal agency such as the BLM involved in the project, there may not be a ready, timely avenue or "nexus" for the request for consultation, which the FWS needs.

Another federal agency's request for consultation triggers the FWS Determination process regarding a project's impact on a listed species.

b.   Air quality Prevention of Significant Deterioration permits under the Clean Air Act, which require a permit or concurrence by the U.S. Environmental Protection Agency (U.S. EPA).

c.   Water quality permits related to the National Pollutant Discharge Elimination System (NPDES), which adds uncertainty due to changing U.S. EPA regulations and regulatory guidance.

d.   Federal Land Use Entitlements, which involve federal land management agencies like the BLM or the U.S. Forest Service (USFS) and NEPA compliance.

e.   Multidisciplinary permitting of transmission lines, which involve permits from the Western Area Power Administration (WAPA) and/or the California Public Utilities Commission (CPUC), and associated NEPA and CEQA compliance documents.

In the past, federal agencies have been responsive to the state's needs for timely review of power plant licensing applications. However, in 2009 the timing of federal permits became especially important because of specific build and timing conditions for federal stimulus funding under the American Recovery and Reinvestment Act (ARRA).

Challenges in coordinating federal permits in the Energy Commission licensing process include ensuring that necessary information needed by federal agencies is developed and presented early in the process, including the project application pre-filing phase.

Complex biological mitigation measures like desert tortoise relocation plans must be consistent with measures likely to be required by federal agencies. To help identify and address critical issues, federal agencies need sufficient staff to write comments and participate in workshops, scoping, meetings and hearings.

Timing of federal permits can affect the licensing process in a variety of ways, including:

a.   Limiting or delaying information needed to establish project compliance with laws, ordinances, regulations, and standards before licensing.

b. Delaying information needed to establish mitigation measures before licensing decisions, which can result in significant changes to proposed projects, a need for additional analyses, and schedule delays.

c. Subjecting projects to future appeals when federal permit decisions are made after the California Energy Commission has approved a power plant license, appeals over which the Commission may have limited influence.

Permitting at the Local Level

Local governments, primarily larger counties through their planning and redevelopment agencies, also review and permit utility-scale renewable electrical generating facilities (non-thermal projects such as solar PV and wind energy, and thermal projects less than 50 MW).

These permits typically require a multidisciplinary CEQA analysis and a public comment and hearing process. Many local governments do not include utility-scale renewable energy facilities in their general plans or zoning ordinances, so developers may have to apply for general plan amendments or rezoning, which can complicate and delay the review and approval of renewable generation projects under county jurisdiction. Local governments also face resource constraints as planning departments are being downsized following the economic downturn.

As part of its "lessons learned" proceeding, the Energy Commission is reviewing the licensing processes for several PV and wind projects both in California and out of state to compare other jurisdictions' environmental documents with the Energy Commission's siting process.

Local government participation and involvement are also important for projects permitted at the state level. For example, in 2010 the Energy Commission and the BLM permitted numerous large solar thermal projects, for which county fire departments were involved with worker safety and hazardous material management analyses and associated mitigation and conditions of certification.

Local agencies have expressed fiscal and environmental concerns about large renewable energy projects and have sometimes formally participated in the Energy Commission's permitting process. Their input and participation have generally led to conditions of certification addressing some of their resource concerns and requirements for local review and comment on various plans such as construction traffic control.

Challenges for Solar Generation Projects

Renewable distributed generators (DG) of 20 MW or smaller, and some large-scale projects also face challenges associated with environmental impacts and planning and permitting processes.

Environmental Challenges

Depending on the technology and the site location, DG projects can cause a range of environmental impacts similar to those of utility-scale projects. DG projects in California must comply with a number of environmental requirements including permits for air quality, water discharge, building standards (for systems that potentially impact the building environment) and waste discharge permits (for DG facilities that process solid materials or have disposable wastes).

Because of their smaller size, there is greater opportunity to relocate or redesign renewable DG projects to avoid or reduce environmental impacts. In addition, renewable DG technologies like small PV can be located in industrial areas on already disturbed land or on existing residential, industrial, or commercial buildings, which reduces their environmental impact. Similarly, individual or small groups of wind microturbines can be sited and designed to avoid or minimize land use, noise, visual, or habitat impacts.

Biomass DG projects using forest or agricultural waste have smaller footprints than utility-scale facilities and are typically located at or near existing lumber mills or agricultural facilities to maximize fuel access and avoid or minimize land-use conflicts or other environmental impacts. However, these projects can face challenges in securing air permits, particularly in areas with significant air quality issues.

While the impacts from hydroelectric projects can vary, small hydroelectric projects generally cause fewer and less severe impacts than large hydroelectric projects.

Many new small hydroelectric facilities are and will be located in conduits. These projects involve replacing older turbines at existing dams with more efficient equipment; taking advantage of pumped storage opportunities; operating to minimize fish mortality; and making use of run-of-river turbines placed in river, stream, and conduit flowing water. In general, using existing disturbed locations and avoiding construction of new impoundments eliminates impacts or keeps them to acceptable levels.

However, some existing facilities located on natural waterways designated for critical habitat restoration, and new facilities designed with on-stream impoundments adversely affect water resources by changing stream flows, reservoir surface area, the amount of groundwater recharge, water temperature, turbidity (the amount of sediment in the water), and oxygen content.

Potential biological impacts include new lakes that can flood terrestrial habitat and changes in water quality and quantity flowing downstream that may alter fish migration patterns and cause other aquatic life impacts. New reservoirs can also damage or inundate cultural, tribal, archaeological, or historical sites. Scenic or wilderness resources can be lost or degraded, and projects can increase landslides and erosion.

Federal Grid Rules for Solar and Wind Power

Power grid operators have a number of problems, derived from complex power generation sources and distribution channels, and related issues. Until now they had a number of well defined and fixed energy sources, such as nuclear, natural gas and big hydro power. The problems these brought upon the society in large (pollution, radioactive disasters, blocked rivers, etc.) were not their problems, since they were not energy generation or distribution related.

These sources also provided a steady, controllable power, day and night. Scheduling was done a day ahead with hourly real-time updates of dispatch, accounting for, and quickly responding to, unexpected rise and fall in demand. This worked very well for almost a century.

The new "normal" is a minute-by-minute scheduling and dispatch of power generation, based exclusively on detailed forecasts of renewable resources. The variable state of the oldest power plants is also taken into account, a combination which makes the grid operators' job quite hard.

"Many grid operating practices were put in place decades ago to accommodate the particular attributes of fossil, nuclear, and hydroelectric power plants that made up nearly the entire generating fleet at the time," according to the American Wind Energy Association (AWEA) and its allies stated in pre-decision comments on proposed rule changes by the Federal Energy Regulation Commission (FERC).

"Attempting to fit variable renewable energy resources into these operating practices is often like trying to fit a round peg into a square hole."

The secret here is adaptability.

There are many tools and capabilities available to grid operators, offered by smart transmission systems equipped with the newest computing and communications technology. So it is just a matter of incorporating the new (variable and more complex) schedule and forecast changes as needed to integrate wind and solar energy sources.

The new FERC rules:

1. Do require grid operators to offer sub-hourly (15-minute or shorter interval) scheduling;

2. Do not require them to do sub-hourly dispatch of generation; and

3. Do require renewables generators to report their (typically more accurate and detailed) resource forecasts to grid operators.

The greatest omission in the new rule is a requirement to use generation sources in the sub-hourly units in which dispatch can be planned. By continuing to dispatch at hourly intervals, grid operators will need to use a great deal more fossil fuel reserves, but these actions will increase prices and negate the positive impact of renewables.

FERC commissioners' ruling is a good step forward, but it does not go far enough, thus making the process slow and incrementally political. Variable energy resources (wind and solar) increase their share daily, and the new rule is yet to prove whether the cost of implementing the changes will outweigh the benefits. If the rules are not clear enough, and allow grid operators to go back to their old familiar ways, that will cost the ratepayers money, and will cause outages and security breaches.

There is also a question of whether the old transmission systems are ready for the changes. Can they handle the new variable loads? Do the regulators and legislators have enough information about what this will take, and are they ready and willing to facilitate the transfer to the new system?

This dramatic grid makeover will impacting transmission planners, because of the possibility of blackouts. This results in understandable resistance on their part, but change is inevitable.

Farm Bill S-3240

On June 21, 2012, the Senate passed a five-year bipartisan reauthorization of the Farm Bill (S. 3240). The bill provides $800 million for rural energy programs, including loan guarantees for on-farm renewable energy, energy efficiency projects, and research and development for advanced bio-fuels.

Though it does not create any new energy programs, the bill re-authorizes the "Rural Energy for America" program that provides guarantees for renewable energy and energy efficiency projects, as well as other existing programs. This includes the research and development of solar and wind projects, though that is not spelled out in the bill.

There is work to be done in developing and implementing farm-related energy projects. These projects can be used to develop new, or optimize existing, renewable energy sources, to supplement the farmers' energy use in field operations and at home.

Planning and Permitting Process Challenges

Through a review of relevant literature and informal discussions with stakeholders from private and public sectors, Energy Commission staff identified the following permitting challenges related to renewable DG. These are not exhaustive, but they represent the variety of permitting challenges facing renewable DG deployment.

Planning

If a parcel is not zoned for electricity generation, the process to obtain permission to install renewable DG can be lengthy and cumbersome, especially for facilities requiring large amounts of land. The impact of zoning is less of a problem for rooftop PV because the Solar Rights Act requires local land-use authorities to allow roofs to be used for solar electric facilities. Precedent is less clear regarding the application of the Solar Rights Act to ground-mount solar electric facilities. Similarly, Government Code 65893-65899 establishes regulatory limits that local governments must follow when planning for and permitting small wind turbines.

Physical interconnection of DG units to the local distribution network may be complicated, depending on the electricity infrastructure in each community, and upgrades to the distribution system can require local permits.

Zoning

Many cities and counties do not have zoning ordinances that permit and guide the development of renewable DG systems. Without a preferred development pattern for renewable DG projects, developers must request general plan amendments and/or rezoning of developable parcels. General plan amendments and rezones are lengthy and involved processes, and are not preferred choices for expeditious permitting.

Varying Codes, Standards, and Fees

Stakeholders indicate the land-use permitting process for identical renewable DG systems, and it varies significantly from jurisdiction to jurisdiction. This inconsistency makes it difficult for developers to create an efficient process to meet permit requirements, and increases the risk associated with securing project approval. Moreover, permit fees and calculation methods vary statewide.

In 2010 and 2011, the Sierra Club, Loma Prieta Chapter, conducted surveys on permitting costs and processing times for solar PV systems in various jurisdictions throughout California.

The range of fees in the municipalities surveyed in each county varied widely: In Los Angeles, for example, fees for commercial PV projects 131 kW in size ranged from $0 to $46,000. In general, the surveys found that the cost of a PV project does not correlate with the staff hours a municipality must devote to plan review and inspection. In addition, basing a permit fee on the valuation of a PV system tends to generate higher fees than the actual cost to service a permit, since the time involved for review and inspection does not appear to be linear; for example, it does not take 10 times as long to evaluate a 100 kW system as a 10 kW system.

In the counties surveyed, the percentage of municipalities that exceeded maximum estimated cost recovery fees for commercial systems ranged from 17 to 65%.

Williamson Act Issues

Many land owners have Williamson Act contracts on rural parcels that can easily accommodate distributed or utility-scale renewable energy systems. However, it can be cumbersome to complete the process needed to allow nonagricultural use of Williamson Act land. If renewable energy is not considered "a compatible use," land owners must cancel, non-renew, or rescind their contracts to develop renewable energy facilities on Williamson Act land. Senate Bill 618 (Wolk, Chapter 596, Statutes of 2011) provides another way to rescind a Williamson Act contract for the purpose placing a solar use easement on unproductive and nonprime agricultural land.

The process requires local government approval of the rescission as well as approval by the state Department of Conservation in consultation with the state Department of Food and Agriculture. However, challenges remain to balance the need to conserve farmland with the development of renewable energy in California. For example, on October 31, 2011, the California Farm Bureau Federation sued the Fresno County Board of Supervisors, arguing that the county violated the Williamson Act when it approved a partial cancellation pertaining to 90 acres of a 156-acre parcel of prime agricultural soil to allow the land to be opened to development of the 18 MW Westlands solar project.

Vague, Duplicative, and Uncoordinated
Permitting Processes

Permit approvals typically require renewable DG builders to secure approval from local fire departments, building and electrical code officials, and local air districts before receiving a zoning clearance to build a facility. If the proposed site for renewable DG development is not privately held, the public land holding body (for example, the BLM or the Department of Defense) may impose additional permitting requirements.

The permit application process is inefficient, and efforts are duplicated because many agencies are involved in reviewing and approving development applications. According to stakeholders, the lack of coordination among permitting bodies is a barrier for developing all scales of renewable DG, including residential rooftop PV.

Unknown Environmental Review and
Mitigation Requirements

Similar to utility-scale renewable facilities, DG development in California is subject to an environmental review under CEQA and, in some cases, NEPA. On a project-by-project basis, the environmental review requirement adds uncertainties to renewable DG projects. If a lead permitting agency determines that an EIR is necessary to assess potentially significant project impacts, additional time and mitigation requirements may make a project unfeasible. Stakeholders comment that lead permitting agencies have thorough environmental screening and review processes in place for traditional development, but many lead agencies are not prepared to assess environmental impacts associated with renewable DG.

Conclusions

Obviously, there is conflict about energy, worldwide. The principal participants are solar products manufacturers, supply chain manufacturers, customers, developers, power plant installers, power plant operators, investors, lenders, insurers, department of federal and state governments, regulators, and utilities.

It one entity wants is not necessarily the same as the rest, so this becomes a battle on many fronts, and disagreements could become more serious if the world economics worsen.

How secure is our energy future? How clean is our environment? What would it take to take care of the energy and environmental issues? These questions are still to be answered in the 21st century.

Notes and References

1. A Manual for the Economic Evaluation of Energy Efficiency and Renewable Energy Technologies, http://www.nrel.gov/docs/legosti/old/5173.pdf
2. INACCURACIES OF INPUT DATA RELEVANT FOR PV YIELD PREDICTION Stefan Krauter, Paul Grunow, Alexander Preiss, Soeren Rindert, Nicoletta Ferretti Photovoltaik Institut Berlin AG, Einsteinufer 25, D-10587 Berlin, Germany
3. PV module characterization, Stefan Krauter & Paul Grunow, Photovoltaik Institut Berlin AG, TU-Berlin, Germany
4. PV MODULE LAMINATION DURABILITY, Stefan Krauter, Romain Pénidon, Benjamin Lippke, Matthias Hanusch, Paul Grunow, Photovoltaic Institute Berlin, Berlin, Germany
5. PV module testing—how to ensure quality after PV module certification. Photovoltaic Institute, Berlin. August, 2011
6. Module Prices, Systems Returns & Identifying Hot Markets for PV in 2011, AEI, September, 2010
7. Fostering Renewable Electricity Markets in North America, Commission for Environmental Cooperation, Canada, 2007.
8. PC Roadmap, http://photovoltaics.sandia.gov/docs/PVR-MExecutive_Summary.htm
9. European Photovoltaic Technology Platform comment on "Towards a new Energy Strategy for Europe 2011-2020," http://www.eupvplatform.org/
10. Economic Impacts from the Promotion of Renewable Energy Technologies. The German Experience. Manuel Frondel, Nolan Ritter, Christoph M. Schmidt, Colin Vance http://repec.rwi-essen.de/files/REP_09_156.pdf
11. Analysis of UK Wind Power Generation from November 2008 to 2010. Stuart Young Consulting, March 2011
12. Performance Degradation of Grid-Tied Photovoltaic Modules in a Desert Climatic Condition by Adam Alfred Suleske, ASU, Tempe, Arizona
13. CAISO Today's Outlook, http://www.caiso.com/outlook/SystemStatus.html
14. Fracking Problem: Shale Gas may be Worse for Climate than Coal, http://envirols.blogs.wm.edu/2011/04/25/fracking-problem-shale-natural-gas-may-be-worse-for-climate-than-coal/
15. NASA solar data. http://eosweb.larc.nasa.gov/cgi-bin/sse/grid.cgi?uid=3030
16. NREL SAM, https://www.nrel.gov/analysis/sam/download.html
17. http://www.greenpeace.org/international/Global/international/planet-2/report/2009/5/concentrating-solar-power-2009.pdf
18. Weather Durability of PV Modules; Developing a Common Language for Talking About PV Reliability, Kurt Scott, Atlas Material Testing Technology
19. Re-imagining US solar financing, Bloomberg, New Energy Finance, May 2012
20. http://en.wikipedia.org/wiki/Environmental_impact_of_the_Three_Gorges_Dam
21. Clean Energy Patent Growth Index (CEPGI), http://cepgi.typepad.com/files/cepgi-1st-quarter-2011.pdf
22. An Eye on Quality. Intevac, July 2011 http://www.electroiq.com/articles/pvw/print/volume-2011/issue-4/features/an-eye-on-quality.html
23. What is the energy payback for PV? US DOE.
24. Energy Pay-Back Time (EPBT) and CO_2 mitigation potential, Evert Nieuwlaar, Erik Alsema.
25. The ExternE Project. http://www.externe.info/externe_d7/?q=node/6
26. Performance Degradation of Grid-Tied Photovoltaic Modules in a Desert Climatic Condition by Adam Alfred Suleske, ASU, Tempe, Arizona
27. Photovoltaics for Commercial and Utilities Power Generation, 2011. Anco S. Blazev

Chapter 8

Solar Markets in the 21st Century

*For anything worth having one must pay the price; and
the price is always work, patience, love, self-sacrifice"*

—John Burroughs

RENEWABLE ENERGY IN THE U.S.

Figure 8-1. The solar industry, 2012.

We don't know why this guy is pushing such a large ball uphill, but it reminds us of what the solar industry went through several times during the last few decades. It had to roll the ball uphill each time it was dropped by governments, big business, or circumstance.

During 2011-2012 we witnessed another example of the solar ball being dropped. After an exciting 4-5 years of progress and successes, the solar industry is back at the bottom, looking at a steep uphill climb...again.

In recent years, wind and solar power have been among the fastest-growing sources of energy in the world. This was due mostly to government subsidies, incentives and other promises...until they were dropped on short notice, just as during the 1970s, 80s and 90s. The key subsidies, incentives and FiTs were discontinued in mid flight, and promises were broken. The solar industry went into a free fall. Many companies went bankrupt, and others were sold for pennies on the dollar.

So before we start pushing the ball uphill again, there are a number of questions that must be answered:

- What can we expect from government, states and local incentives and subsidies now and in the future? And having learned the lessons of the past, we must also ask:

- How can we ensure government support in key areas, while transferring the responsibility for solar, wind and the entire renewable energy sector, in its entirety, to private enterprise?

For wind power in the US, the situation is especially precarious. The federal production tax credit, which has provided incentives for wind farm operators to produce power since 1992, expires at the end of 2012. Congress extended it in the past, most recently in 2009 as part of the federal stimulus package, but this time things are different, and if the tax is not cut this time, it will be next year. Without the tax credit there would be significant, if not devastating, reduction of new wind projects in the United States. Close to 50,000 jobs will be lost as well.

The solar industry also faces expiration of important tax, subsidy and grant programs in the near future. These were created as part of the stimulus package, and the industry is lobbying for extensions. The expiration of these programs would have a significant compressing effect on the amount of renewable energy that gets financed and installed.

An extension would allow solar business to grow. This includes the 30% federal investment tax credit in place through 2016, which is also threatened. Other technologies, like fuel cells and small wind turbines, have access to similar tax credits through the same time period, and depend heavily on them too.

Historically, the grant program was far more effective than the tax credit because it provides incentives for a broader range of private investors to help finance projects, as opposed to merely those with high tax obligations, which the credit helps offset. The grant program also applies to wind power, though wind insiders consider the tax credit more important for their industry's future development.

Another major federal program, the provision of loan guarantees to aid large renewable energy projects, ended in 2011. That program became controversial after Solyndra, the first solar recipient, filed for bankruptcy, leaving taxpayers potentially liable for more than $530 million.

While the loan guarantee program helped finance emerging technologies and "higher-risk projects," the grant program aided "extremely low-risk projects" using off-the-shelf and proven technology.

The continuation of federal incentives is important and like all forms of energy—including fossil fuels—relies on a complex web of state and federal credits and aid conditions. More established technologies including oil and gas, coal and nuclear power are still taking advantage of incentives that were established in the early 20th century. One important federal incentive for oil and gas drilling, for example, has been in place since the 1910s, at a cost of over $6 billion annually, and although there is a lot of talk about it, it will be with us for a long time, it seems.

It is clear that, all things considered, it is time to make large investments in renewables, just as we've invested heavily in oil and gas for a long time now. For the new technologies, trying to establish themselves and reduce their costs, government incentives may be a make-or-break situation, just as it was for gas, oil, coal and nuclear in the beginning of the 20th century.

Unlike established oil and gas companies today, who make money even if they didn't have some of the credits they have, the wind and solar industries won't develop properly, and some may even die, without this help.

On top of that, the perpetual threat that "the incentives will expire soon" makes financing of projects and long-term planning difficult to impossible. Lack of certainty is a serious issue, so companies hope to be able to rely on the federal grant programs, but are forced to plan for life without any subsidies and grants.

For renewable energy developers, the threatened expiration of incentives may have a few benefit. For example, the wind deadline at the end of 2012 caused customers to accelerate decisions. This is a good, albeit one-time, temporary solution to a larger, long-term problem.

For that reason, 2011-2012 was a banner year for

wind installations, and maybe the largest thus far in the U.S., in terms of numbers of installations. But the threat of a dramatic fall-off in 2013 looms over the entire alternative energy industry.

Provided that Congress renews the production tax credit for another four years, wind power will probably reach 4-5% of the country's electricity supply by 2016, according to industry insiders. This is up from 3%. Short of that, a sudden halt of activities in 2013-2014 could.

The costs of solar and wind have dropped substantially in recent years, but still not enough to compete with conventional energy sources, and this includes present-day tax credits and grants.

Low natural gas prices and the renewed interest in gas fracking—as dangerous as it might sound from an environmental point of view—have made it tough for wind and solar to compete. It would be nearly impossible to do so if and when the production tax credit ends.

This is exactly what happened to the solar industry in the late 70s and several times since. We are accustomed to ups and downs in the industry and will probably survive this one too, but its demise would be very bad news for our energy independence and for the well-being of future generations.

With all this said, the questions remain: Is it practical, or even possible, to continue the dependence on government, state and local subsidies and other third-party incentives? Aren't we exposing the US solar industry to a great risk—like that experienced in Spain and other countries—where the subsidies and incentive programs were cancelled abruptly due to lack of money? What will happen to the dozens of US solar companies and projects who depend on the government?

Author's note: Between 2002 and 2008, renewable energy received $12.2 billion in government support, with $6 billion in direct spending and $6.2 billion in tax breaks, according to industry sources. At the same time fossil fuel industries got $70.2 billion. $16.3 billion directly and $53.9 billion in tax breaks. And oil companies have recorded massive record profits all through this period.

Corn ethanol, for example, is alive only thanks to federal subsidies. It had benefited from $11 billion in tax breaks and $5 billion in direct spending during this time.

From 1918 to date, the oil and gas industries have been getting $4.9 billion annually, and the nuclear industry $3.5 billion. Subsidies for renewable energy started in 1994 at a $0.35 billion annually—or exactly 10 times less and 14 times less than the nuclear and fossil industries respectively.

Still, we cannot and should not expect governments to prop up and support the solar and wind industries. It might help in the short term, but it won't work in the long run. Private enterprise is the only key to successful solar industry in the 21st century.

THE US SOLAR MARKET, 2010-2011

2010-2011 were remarkably successful years for the US solar market. There were approximately 900 MWp of solar power installations in the US in 2010, and almost 2.0 GWp in 2011. While the residential installations remained at a constant rate, and the commercial rates fluctuated, the utility scale installations rose remarkably fast during that time, and especially at the end of that period.

Table 8-1. U.S. installed solar power, 2010-2011

Residential installations	2010	2011
Q1	70.0 MWp	70.0 MWp
Q2	70.0	70.0
Q3	70.0	70.0
Q4	70.0	70.0

Commercial installations	2010	2011
Q1	70.0 MWp	175.0 MWp
Q2	70.0	200.0
Q3	110.0	175.0
Q4	110.0	250.0

Utility scale installations	2010	2011
Q1	10.0 MWp	35.0 MWp
Q2	45.0	60.0
Q3	10.0	230.0
Q4	180.0	450.0

PV installations during 2011 totaled approximately 1.85 GWp, with 280 MWp in residential, 800 MWp in commercial, and 775 MWp in utilities PV installations— good progress during one of the worst financial crises the world has seen.

The large PV modules manufacturers who were most active in the US market during that time—mostly Chinese and Asian—sold a lot of product.

Asian PV modules manufacturers dominated the markets in the US and Europe. The quality of their product is still unknown, but will be revealed during the decade, when the installations—especially those in deserts—start showing signs of fatigue.

After more than 5.7 GWp of total solar installations (3.2 GW are PV installations), and additional 3.5 GWp in

Table 8-2 Major PV modules suppliers in the US, 2010-2011.

Residential market %	
SunPower	22.0
SunTech	12.0
Kayocera	6.0
Sharp	8.0
Yingli	10.0
Other	42.0

Commercial market %	
SunTech	19.0
SunPower	17.0
Yingli	10.0
Trina	12.0
SolarWorld	5.0
Other	37.0

planned or under construction, the US solar industry is still in reactive, instead of a proactive mode. A lot progress has been made, but this a good time to look at the long-term solution to our energy problems by creating a vertically integrated, 100% US-made solar and wind industries.

The solar industry has enough internal problems, shown by the introduction of bills in Congress to reduce and eliminate federal investment tax credit (ITC). Wind's production tax credit (PTC) is also on the chopping block, so the alternative energy sector as a whole is at a crossroad.

Rooftop solar installations might suffer too. Homeowners are making large upfront investments, the equivalent of paying for 30-40 years of electricity with uncertain payback and contracts with slowly rising annual lease payments. Installers can easily have gone out of business by the time a homeowner needs maintenance or repairs.

"The future is not an inheritance," Clinton said, "it is an opportunity and an obligation." It is obligation for manufacturers, contractors and owners to be well informed and do the right thing. We are not there yet, and much work remains to be done, covering all aspects of its manufacturing and use, before solar becomes a truly perfect energy source.

Potential Estimates

The estimates of U.S. technical potential for renewable energy generation and capacity in the summer of 2012 were reported by the National Renewable Energy Laboratory (NREL).

Six technologies were evaluated, and the total annual generation potential of these was estimated at 481,800 TWh, compared with roughly 3,754 TWh of total generation in all 50 states in 2010.

The total cumulative capacity potential of the country was estimated at 212,224 GWp—everything included.

The US has excellent solar resources in the midst of large population areas, like Texas and California, which could be considered leaders in estimated urban utility-scale PV potential.

Texas has approximately 14% of the annual U.S. generation potential for rural utility-scale PV, and 20% of the estimated yearly U.S. generation potential for CSP generation.

California could count on the highest annual generation potential for rooftop PV, in addition to great potential for large-scale power fields.

There is a sharp distinction between states that have solar potential and those that don't, because of the relatively high solar resource minimum threshold of 5 kilowatt-hours per meter-squared per day. The breakdown is shown in Table 8-3.

Table 8-3. Total estimated US renewable power generation and capacity potentials.

TECHNOLOGY	Capacity GWp
Utility-scale PV	154,200
Rooftop PV	664
CSP	38,000
Wind power	15,200
Geothermal power	4,000
Hydro power	98
Bio power *	62

* Biofuels are not included

The US total renewable energy generating capacity is enormous—over 200,000 GWp. This is a lot of power to be generated from free and renewable energy sources—keeping in mind that the electric power generating capacity of the entire US in 2010-2011 was less than 1,500 GW.

Add another 500 GW for other power generating sources (heat, etc.), and this is still 100 times less than our total renewable energy generating potential. What are we waiting for?

THE STATE (THE SOLAR COUNTRY)
OF CALIFORNIA

This is the brightest spot for the solar markets in the US, and maybe the entire world…at least for now. Or maybe especially now!

California is one of the largest states in the U.S., and is the largest consumer of electricity. It also has the nation's most successful and most efficient energy policies, especially when it comes to solar and wind power. The number and quality of such installations of late confirm that.

In 2012 there were over 2.5 GWp in residential, commercial and utility type solar installations in California—almost half of the total US solar installations—with a significant number under construction. And almost that much is planned for installation by 2015.

The reasons for this success are numerous. On the utility side, the local utilities—SCE, SDW&P, and PG&E—are very proactive in achieving the 33% mandate by 2020. They launched a number of programs and initiatives, which led to high numbers of utility type installations.

On the residential side, initially, state utility rebates made solar very affordable. Although, the generous rebates of $2,500 per kilowatt in 2006 have steadily been decreasing down to just over $600 per kilowatt in 2012. This reduction has been balanced by the federal 30% tax credit which kicked in during the same period, keeping after-incentive costs at about the same level.

Another factor in the continuing solar boom in California is reduced PV panel prices, dropping close to $0.50/Watt, which was prompted by EU countries dropping their overly generous FiTs. This caused the oversupply to explode, so prices dropped precipitously. The lowest PV module prices are offered by Asian manufacturers, like Suntech Power, Sharp, Yingli, Trina Solar, and Kyocera.

At the same time, several solar energy bills were introduced and passed in California. They basically extended, confirmed and/or improved existing policies designed to encourage wide-scale adoption of solar energy technology in the state. The main theme of the bills is lowering the cost of solar energy by removing initial costs and investments and facilitating the entire process, since those are the major barriers to the solar power market expansion in the state.

For example, Senate Bill 2249 amends the Solar Water Heating and Efficiency Act of 2007, which created a 10-year $250 million program aimed at installing 200,000 solar water heaters in homes and businesses across California. It extends incentives to commercial pools and removes cost burdens that have caused towns and schools to stay away from purchasing solar heating devices.

Another bill, SB-1222 reforms the solar PV permitting system, where, PV permitting fees would be capped at $500 plus $15 per kilowatt for each kW above 15 kW of residential rooftop power. Fees for commercial rooftop systems would be capped at $1000 plus $5 for every kilowatt exceeding 250 kW.

Solar net metering, addressed by SB-594 targets net metering customers with multiple electric meters. It allows aggregation of the electrical loads across the property and surrounding the solar generation, if the customer is the sole owner of the property. This would be particularly beneficial to farmers with multiple meters, or people who have separately metered rental units.

So, California is well on its way to a bright solar future. Even though there are clouds on the horizon, we believe the California lawmakers and utilities will find a way to achieve the 33% mandate by 2020. Unfortunately, we cannot say the same about most of the remaining states.

Author's note: The California case needs some clarification from a technical point of view. Things are still moving in the right direction, since it increased the renewable portfolio standard (RPS) from 20% to 33% by 2020. This means that 33% of all electric power in California must be generated by alternative power generators. Is it feasible?

1. Assuming that California's entire electric power generation capacity is 70 GW AC, and that

2. 14% (~10 GW AC or 12 GW DC) of the energy mix today comes from renewables (biomass, geothermal, small hydro, wind and solar), then

3. By 2020 California must add at least another 15 GW AC of alternative energy power generators, to meet the 33% RPS mandate by 2020.

Looking at the solar and wind power contribution (which is the largest by far), and considering the part-time operation and hourly variability of solar and wind generators, adding energy storage capacities, and losses in converting DC to AC increases the total needed to meet the mandate to ~25 GW DC. Or approximately 25 MWp solar and wind power generation must be added to the California energy mix by 2020, which will cost ~$150 billion.

Replacement of old and outdated technologies from existing power fields (bio and hydro generators, steam turbines, coolers, CSP trackers, liquid transfer lines, pumps, heliostats, inverters, PV modules, etc.), and addition of transmission lines, substations etc. would cost, let's say, another $50 billion. So a minimum of $200 billion (in our estimate) is needed for this under-

taking to be completed by 2020.

This is equivalent to adding 3.0 GWp, and spending nearly $30 billion, every year from 2013-2020.

These are staggering figures under any circumstances, but especially during the current financial slump, the end of which is not in sight. Granted, much of this money would come from private and institutional investors, government loan guarantees, and grants. And another, not insignificant part, will be paid by the consumers; but (and here is the catch), even if all the money were available, how would this huge undertaking be accomplished?

Is this even possible, technically and logistically? What technologies will be used? Where will the hardware come from? Where would the labor come from? Would we cover the deserts with unreliable PV modules and the hills with low-quality wind generators?

What about the complicated land ownership, permitting, and rights-of-way issues? Currently, it takes years to go through those processes for each installation. And let's not forget the interconnection points, transmission lines, and substation upgrades which are big show-stoppers. How fast can we build or upgrade substations and run transmission lines across the deserts?

Our energy future depends on finding answers. How much of this added energy is going to be generated by solar, and how much of the solar component is going to be PV? We have seen plans for adding ~2.0 GW of PV power generation by 2015, plus several GW of other alternative energy technologies (CSP and wind mostly), but the total is still far from the 25 GW DC, or even 15 GW needed to meet the RPS goals.

Of a serious nature is the fact that concentrated solar power (CSP) or thermal solar electric power generation has been going through some major changes lately, so we are not sure how this will affect solar power development in general, and PV in particular, during the next few years.

There is still time, though, and we should not discount American ingenuity or the resolute Californians, who are especially productive under duress.

The California Solar Initiative

The California Solar Initiative (CSI) was made into law by a 2006 Senate Bill, the combined design of California Public Utilities Commission (CPUC) work, and Governor Arnold Schwarzenegger's "million solar roofs" vision. It's initial goals were to:

1. Build 1.94 GW of solar in California in supported system allotments of one kilowatt to one mega-

watt. A General Market Program of 1,750 megawatts was aimed at residential and non-residential settings and another 190 megawatts targeted low-income settings, and

2. Transform the solar market and make solar "sustainable, vibrant and even mainstream" in the state and the country.

The California Energy Commission (CEC) was to oversee the $400 million New Solar Homes Partnership (NSHP), which was intended to increase installations of new-home solar systems in the territories of the three IOU utilities, PG&E, SCE, and SDG&E IOU.

A voluntary program for POU was budgeted at almost $800 million.

The CPUC was allotted the balance of the funding, approximately $2.2 billion, to oversee new and retrofit non-residential solar and residential retrofits in the IOU territories.

The program has six segments, one residential (up to ten kilowatts) and one non-residential (ten kilowatts to one megawatt) for each of the IOUs, with the SDG&E segments administered by the California Center for Sustainable Energy (CCSE).

Unlike programs before it, that lowered rebate rates when the money began to become depleted, the CSI program built in step-downs in accordance with a megawatts-achieved schedule, not a time schedule.

Rebates are paid in two ways; a) payment on installation of a system, and b) second payment made monthly that is based on the customer's meter reading.

The progress of the different programs varies, but one can get a daily update at Solar Initiative Rebates webpage, http://www.gosolarcalifornia.ca.gov/csi/rebates.php

A long-term concern is that there is a very close relationship between the third-party-owner model and the ITC, so when the ITC goes down from 30% to 10% percent after 2016, the impact is unpredictable.

Some speculate that neither the change in the ITC, the consolidation in solar manufacturing, nor the import tariffs on foreign modules will drive the price up, because there is enough competition to compensate, so prices might even continue moving down.

Short-term Direction

The world solar industry is going through serious changes, and although we cannot foresee what will happen in 2050, we can clearly see what will happen by 2015.

Below is an attempt to summarize the present conditions and project them into the near-term future of solar energy in the US and the world—and especially in California.

Regulatory Changes

CEC (California Energy Commission) and CPUC (California Public Utilities Commission) are looking into the definition of "energy generation," taking into consideration the addition of solar power, and its negative effects. Solar power variability is a great problem, which causes abnormalities in the proper and trouble-free grid operation. This disadvantage gives priority to companies and projects with energy storage that provides the stability and continuity of power generation that utilities need.

This simply means that projects with energy storage (CSP and such) will benefit immensely and immediately. PV companies will be forced to look for ways around energy storage (or lack of), or insist that CEC and CPUC reverse the decision as unfair, or untimely.

Since California leads the way in many areas—including solar energy development—we expect the energy storage issue to escalate, which will cause some confusion and changes in the solar field in the US, and perhaps abroad too. It is a serious issue in need of a final resolution, to execute projects that function efficiently in the grid.

Utilities Participation

a. Utilities are already very concerned, and are voicing their complaints about the significant negative impact of solar power plants' "variability" on grid management. Cloudy days and hourly solar insolation changes introduce uncontrollable fluctuations and instability in the grid, which in turn makes it difficult for grid operators to maintain quality of power. It also confuses the "peaker" power plants, which are turned on and off frequently, at great expense, and causes additional confusion for grid operators.

    The operation of the peaker power plants is of great concern here, because idling them, or not being able to build new ones as original business plans call for, cuts right into the utilities' bottom line. As a matter of fact, these and other utilities concerns brought up the proposed regulatory changes discussed above. A number of similar changes and adjustments are expected in California in the near future as well.

b. The decision of the major California utilities and CPUC in 2011 to give priority (permitting, PPA etc.) to smaller PV projects (10-20 MWp) is still being studied and we are not sure how it will affect the overall development of solar projects in the US.

There are a number of pros and cons to be considered here, with the overwhelming conclusion being that developing large-scale fields (and large profits) is no longer feasible, which caused a number of companies to change their strategy or to give up their plans all together.

In other states solar is not realistic. For example, Georgia Power and Solar Design & Development (SD&D) proudly announced the development of a series of solar projects totaling 19 MW, which is Georgia Power's first retail utility-scale solar power development. This will more than double Georgia's solar production if and when complete, but it is a far cry from making Georgia even close to a "green" state.

This series of solar projects is part of Georgia Power's 50 MW large-scale solar initiative approved last year by the Georgia Public Service Commission. Georgia Power has contracted to purchase the output for the next 20 years.

Adding 50 MW solar power in 20 years means that Georgia will replace one of its mid-size coal-fired plants with solar within…400 years. Georgia has a lot of sunshine, but needs hundreds of these developments before the presence of solar is even felt in the state.

We must mention here the example of the Arizona-based SRP utility, the second largest in the state. It is a POU, therefore they don't feel directly obligated to be part of the solar game. Instead, they have a semi-solar program which they offer to customers. It offers power generated by others (in other states), which "sort of" puts SRP in the solar game…but indirectly so, and without the hassles.

Chinese Imports

The majority of PV installations in the US contain Chinese component—PV modules and BOS—which contributed to the price reductions during 2011-2102 and the relative success of solar in general here.

The new un-proportional taxation imposed by the US DOC on Chinese PV modules imports this month means that:

a. At least temporarily (2012-2013), the prices of PV installations in the US will increase by 15-20%, and some planned PV projects might be cancelled because of the price increases. The lower prices might contribute to significant reduction of present-day estimates as well.

b. Replacing cheap, unproven imported components will have the benefit of building long-term quality and responsibility into US-made products and projects.

Customers' Sentiment

The successful (and sometimes not-so-successful) development of new solar programs and initiatives from 2008 to date have been paid in part by the customers. Be it in form of increased utility bills, or taxes, this fact was not obvious to the ordinary customer until now, but it is becoming more evident and well known with every new project and program.

This will likely become even more of an issue as the public becomes better educated, and as government and state subsidies and incentives are reduced. Changes in regulations, billing, and taxation will need to be introduced to keep things moving, and the customers happy, which will change operating conditions again.

Federal and State Subsidies and Incentives

A number of government programs have already been cut, and several are under revision. These are too many and far too complex to list here, so the outcome and impact of these programs, cannot be properly estimated today.

There will be changes in government and state subsidies, incentives and taxation, and they will vary, but the major power sources the US is going to rely on and develop are coal, gas and oil.

Long-term Direction

We have seen solar energy at its finest when oil prices are high. During these times, it has been treated as an urgent solution to our drive towards energy independence and environmental cleanliness.

During this last solar cycle, governments and VCs allocated billions of dollars, in the form of loan guarantees, grants and investments, to the solar industry. Many companies and projects failed because the technologies weren't ready, and too much money was available with too little administration.

In the short-term, the US solar industry faces hurdles (cheap coal, gas, and oil) and political barriers. But solar energy is the cleanest, most abundant energy source available. And it is the only free energy source on earth! All we need to do is capture and use it properly and efficiently. As Winston Churchill said, "Americans can always be counted on to do the right thing...after they have exhausted all other possibilities." How true this is now. Unfortunately, the course for the near future (next 4-5 years) is set with fossils taking the lead...again.

The development of the energy markets overseas and their influence on US energy markets cannot be overemphasized. There will be great changes around the world during the 21st century, with the most important developments happening in the US, China and Western Europe.

a. Solar will continue to grow in the US, no doubt. It will be limited to location—concentrated in the Southwest US. States like California are committed to solar, and we will see great things happening there.

b. China is positioning solar and wind as major power producers by 2020. China has the money, and the demand for large amounts of power to sustain its economic development, so many are predicting that they will be number one in alternative energy generation in the world within the next several years. This is already attracting many foreign companies to China.

c. The other "hot spot" for solar development is the plan for providing power for Europe from solar and wind power fields in North Africa, a la Desertec model. Europe has no better energy solution, so we predict great developments in that part of the world as soon as it emerges from its present economic slump. This will also attract many solar companies from around the world.

What the solar markets need is quite simple—a level playing field, and technology that is efficient and reliable enough to compete. Meeting these needs is not so simple.

The situation depicted in Figure 8-3 is quite clear. The PV energy markets rely on many factors which have been, or are in the process of being addressed by different entities. The state of PV technology cannot be resolved by a regulatory change or political decree alone. Looking at the figure we can see clearly how the markets are caught in the balancing act of the solar technologies and their unresolved, or partially resolved issues—cost, quality, and efficiency.

While cost and efficiency are on the drawing boards of most companies and institutions, quality is

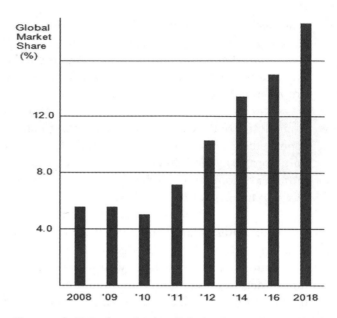

Figure 8-2. U.S. share in the global solar energy market.

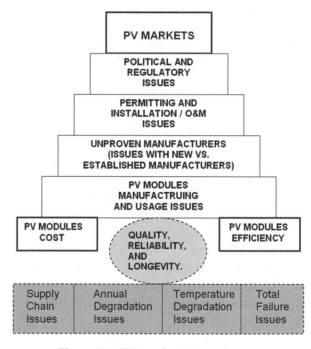

Figure 8-3. PV markets dependence.

still veiled in secrets and promises. The fact that most companies are too young to have a product with a good track record is too obvious, and yet it is often swept under the rug. This uncertainty is growing with the increase in mega-fields of unproven foreign PV products.

We will not be able to claim having reliable, long-term solar technology until these products show successful long-term operation in all their useful climates. Major issues to be resolved are the quality of the PV modules' manufacturing supply chain, the annual and temperature degradation of PV modules, and the related field failures—complex, interwoven, and expensive issues.

Available data from long-term field testing thus far show a number of imperfections, and a high percentage of failures. For example, right now we have no long-term data for any PV product and/or manufacturer that show 100% reliable and safe long-term (30 years) field operation in the world's deserts and high-humidity areas.

This simply shows the immaturity of solar technology, and emphasizes the need for additional work on developing new materials and processes. This is where the government subsidies might be most helpful.

A number of federal and state solar and wind energy generation programs and projects are in the planning stages, and we will look at the most significant ones under development, or planned as of the fall of 2012.

Government PV Projects

As a part the Administration's "We Can't Wait" initiative, seven nationally and regionally significant solar and wind energy projects will be expedited, including projects in Arizona, California, Nevada, and Wyoming. These job-creating infrastructure projects would produce nearly 5,000 megawatts (MW) of clean energy—enough to power approximately 1.5 million homes, and support the Administration's strategy to expand American-made energy.

The Office of Management and Budget is charged with overseeing a government-wide effort to make the permitting and review process for infrastructure projects more efficient and effective, saving time while driving better outcomes for the environment and local communities.

Mohave Wind Energy (BP Wind), AZ

Target date for completing federal permit and review decisions: January 2013.

The proposed Mohave County Wind Farm is a wind-powered electrical generation facility that would be located on approximately 38,099 acres of public land managed by the Bureau of Land Management and 8,960 acres of land managed by the Bureau of Reclamation in Mohave County, Arizona. If approved, it would produce up to 425 MW of wind energy and help the state of Arizona meet its targets for renewable energy.

Quartzsite Solar Energy (Solar Reserve), AZ

Target date for completing federal permit and review decisions: December 2012.

The proposed concentrating solar power plant would be located on approximately 1,675 acres of land managed by the Bureau of Land Management. It would produce an estimated 100 MW of clean energy—enough to power about 30,000 homes—and help the State of Arizona meet its renewable energy goals.

Desert Harvest Solar Energy (enXco), CA

Target date for completing federal permit and review decisions: December 2012.

The proposed Desert Harvest Solar Energy project would utilize photovoltaic technology on approximately 1,200 acres in Riverside County, CA. The project would produce an estimated 150 MW of solar energy, enough to power about 45,000 homes.

McCoy Solar Energy (NextEra), CA

Target date for completing federal permit and review decisions: December 2012.

This proposed solar photovoltaic array would be situated on 4,893 acres in Riverside County, CA. It would produce an estimated 750 MW of solar energy—enough clean energy to power 225,000 homes—while helping the state meet its targets for renewable energy.

Moapa Solar Energy Center (RES Americas), NV

Target date for completing federal permit and review decisions: December 2013.

This solar project is being developed in cooperation with the Moapa Band of Paiute Indians on a 2,000-acre site on the Moapa River Indian Reservation and on lands administered by the Bureau of Land Management in Clark County, NV. If approved, the 200 MW project would employ 100 MW of photovoltaic technology and 100 MW of concentrated solar power technology. Once constructed, this proposed project would be one of the first large-scale solar projects on tribal lands in the U.S.

Silver State South (First Solar), NV

Target date for completing federal permit and review decisions: March 2013.

The Silver State South Solar Energy project is a solar energy generation plant proposed on 13,043 acres of public land. If approved, it would produce an estimated 350 MW of clean energy utilizing photovoltaic technology—enough to power approximately 105,000 homes—and help the state meet its renewable energy goals. Construction on the 50 MW Silver State North project has been completed, making it the first solar project on public lands to be delivering power to the grid.

*Chokecherry/Sierra Madre Wind Energy
(Power Company of WY), WY*

Target date for completing federal permit and review decisions: October 2014.

The proposed Chokecherry and Sierra Madre project, located on approximately 230,000 acres in Carbon County, WY, could produce up to 3,000 MW of wind energy—enough to power over 1 million homes. The Chokecherry and Sierra Madre Wind Farm Project is the largest proposed wind farm in North America. The project, as currently configured, avoids critical sage-grouse habitat identified as "Sage-Grouse Core Areas." Chokecherry is a multi-tiered decision process that includes a land use plan decision anticipated in October 2012, followed by review of a series of right-of-way applications.

After extensive environmental analysis and stakeholder collaboration, U.S. Department of the Interior Secretary Ken Salazar signed the Record of Decision approving the Chokecherry and Sierra Madre Wind Energy site located in Carbon County, WY, as suitable for wind energy development.

This act, and the willingness of the US government to allocate areas suitable for solar energy development is a good first step in the road to developing and operating successful solar technologies. Overcoming the other political and regulatory issues (permitting, interconnects, distribution, etc.) is the next step in this process.

US Army Contracts

The US Army and the US Army Corps of Engineers have been active participants in the development and implementation of solar power generation since day one. In August 2012, they proposed a new program:

The U.S. Army Corps of Engineers, through its Engineering and Support Center, Huntsville, AL, issued a Multiple-Award Task Order Contract (MATOC) Request for Proposal (RFP) for $7 billion in total contract capacity to procure reliable, locally generated, renewable and alternative energy through power purchase agreements. This money will be used for the purchase of alternative energy over a period of 30 years or less from renewable energy plants that are constructed and operated by contractors using private sector financing.

The intent of this action is only to purchase energy that is produced somewhere else and by someone else, and does not include building or acquiring any energy-generation assets. The private contractors will finance, design, build, operate, own and maintain the energy plants, and the Army will purchase the power for up to 30 years. The terms and conditions will be driven by site or project-specific agreements, resulting from task orders awarded under multiple Indefinite Delivery (ID)/Indefinite Quantity (IQ) contracts.

The project could be located on any federal property within the U.S. including Alaska, Hawaii, territories, provinces or other property under the control of the U.S. government for the duration of contract performance. The contracts will be awarded to large and small businesses according to four different renewable energy technologies—solar, wind, geothermal and biomass. Task orders will be executed against the basic ID/IQ contracts (mentioned above), using fair opportunity procedures established in the Federal Acquisition Regulation Part 16.

The Army's development of large-scale renewable energy projects is critical to achieving installation energy security, mission effectiveness and resilience objectives while supporting the DoD's goals to enhance installation energy security and reduce installation energy costs. Energy security and sustainability are operationally necessary, financially prudent and mission critical.

By awarding these contracts, the Army will have a streamlined process that uses private sector financing to develop large-scale renewable energy projects. This approach will help speed overall project development timelines to ensure the best value to the Army and private sector.

This will also ensure the propagation of solar energy around the US and will give it the necessary boost in proving its long-term operational reliability.

Future Large-scale Solar Installations in the US

The US Department of Energy (DOE) and the Bureau of Land Management (BLM) (the agencies) have completed a Draft Programmatic Environmental Impact Statement for Solar Energy Development in Six Southwestern States (Solar PEIS), which evaluates utility-scale solar energy development in Arizona, California, Colorado, Nevada, New Mexico, and Utah. The agencies are preparing the Solar PEIS to reach goals established by Congress to make suitable BLM-administered lands available for solar energy development. These are BLM-administered lands that would be available for right-of-way (ROW) application under each of the alternatives evaluated by the BLM in the Draft Solar PEIS.

The preliminary list goes something like this:

1. In *Arizona*, approximately 9,218,009 acres (37,304 km2) of land would be available for ROW applica-

tion under the no action alternative, and 4,485,944 acres (18,154 km2) would be available under the solar energy development program alternative.

Three special energy zones (SEZs) would be identified: Brenda (3,878 acres [16 km2]), Bullard Wash (7,239 acres [29 km2]), and Gillespie (2,618 acres [11 km2]).

2. In *California*, approximately 11,067,366 acres (44,788 km2) of land would be available for ROW application under the no action alternative, and 1,766,543 acres (7,149 km2) would be available under the solar energy development program alternative.

Four SEZs would be identified: Imperial East (5,722 acres [23 km2]), Iron Mountain (106,522 acres [431 km2]), Pisgah (23,950 acres [97 km2]), and Riverside East (202,896 acres [821 km2]).

3. In *Colorado*, approximately 7,282,061 acres (29,469 km2) of land would be available for ROW application under the no action alternative, and 148,072 acres (599 km2) would be available under the solar energy development program alternative.

Four SEZs would be identified: Antonito Southeast (9,729 acres [39 km2]), DeTilla Gulch (1,522 acres [6 km2]), Fourmile East (3,882 acres [16 km2]), and Los Mogotes East (5,918 acres [24 km2]).

4. In *Nevada*, approximately 40,794,055 acres (165,088 km2) of land would be available for ROW application under the no action alternative, and 9,587,828 acres (38,801 km2) would be available under the solar energy development program alternative.

Seven SEZs would be identified: Amargosa Valley (31,625 acres [128 km2]), Delamar Valley (16,552 acres [67 km2]), Dry Lake (15,649 acres [63 km2]), Dry Lake Valley North (76,874 acres [311 km2]), East Mormon Mountain (8,968 acres [36 km2]), Gold Point (4,810 acres [19 km2]), and Millers (16,787 acres [68 km2]).

5. In *New Mexico*, approximately 12,188,361 acres (49,325 km2) of land would be available for ROW application under the no action alternative, and 4,068,324 acres (16,464 km2) would be available under the solar energy development program alternative.

Three SEZs would be identified: Afton (77,623 acres [314 km2]), Mason Draw (12,909 acres [52 km2]), and Red Sands (22,520 acres [91 km2], and Millers (16,787 acres [68 km2]).

6. In *Utah*, approximately 18,182,368 acres (73,581 km2) of land would be available for ROW application under the no action alternative, and 2,028,222 acres (8,208 km2) would be available under the solar energy development program alternative.

Three SEZs would be identified: Escalante Valley (6,614 acres [27 km2]), Milford Flats South (6,480 acres [26 km2]), and Wah Wah Valley (6,097 acres [25 km2]).

This land mass could power the US many times over, and provide energy to the entire world. Only time will tell how much of this land will actually be used for solar power generation, but the US government is making a giant step in the right direction. It is now up to the states, the regulators, the utilities and the US solar industry to take advantage of the situation.

Most importantly, we must find a way to reduce and eventually eliminate our dependence on imported energy—whether oil or PV modules. Importing mass quantities of strategic energy products (which PV modules are) is not a solution.

The US is set to increase its global market share, though slower than anticipated due to the newly developed financial and energy situations at home and abroad. Still, the solar energy sector is now a major employer, where the total number of steadily growing green jobs is outnumbering those in oil and gas.

Unfortunately, the US solar market is still concentrated in a half dozen states which have more than 85% of all installations (a 5% increase from 2010), and this situation will not change soon.

The CSP market is falling behind at present, with no CSP projects completed in Q1 2011, although over 1.0 GW CSP plants are under construction, some built by foreign companies benefiting from generous US government loan guarantees.

Overall, the US solar market, as fragmented and uncoordinated as it might be, has the best chance to survive the test of time. California is leading the pack, and is expected to continue doing so for the foreseeable future.

Hopefully, California's solar principals, the US government, the regulators, and the scientific and business communities will work together on expanding the progress of solar energy, and focus on the implementation American-made components. We have the know-how, the resources and the need.

THE U.S. PV INDUSTRY

Solar energy, and PV particularly, continues to grow in the US and has a significant traction, building on the successes of the past. During 2011, installed PV capacity reached the 2 GW level. 880 MW of that were residential and commercial installations, while 760 MW were utility power plants.

Serious challenges in the beginning were addressed, identified and resolved over time. There were lengthy negotiations, before any project could be started. The issues often began and ended with what technology to use and why, where and how to install the PV arrays, and what size.

Land issues—buy or lease, EPA and construction permits, etc.—were on top of the list of issues to tackle. There were complex—even mysterious—interconnection permits, distribution lines, and PPA negotiations with the utilities. Who will pay for the substation upgrade, where will the lines go, who will purchase the electricity, and most importantly, how much will they pay for the generated power?

The most important issue—the one that made or broke projects—was financing. PV projects were products of their financing or lack thereof. Government and state subsidies and associated incentives included many different types of technical and financial assistance.

The type of technology to be used in different areas is still in dispute simply because there are no track records of long-term operation.

Soon, utility-scale PV installations in the U.S. will be growing substantially, with over 24 GWp PV projects currently in the planning stages. Some have already signed contracts, while others are awaiting PPA and/or project financing.

Author's note: PPA vs. financing is one of the greatest complexities of PV project development today. In most cases, one must have a PPA, because it will determine if the utility is willing to buy the power and how much it will pay for it.

Obtaining PPA, however, is an expensive and complex process, usually requiring a number of specialists and several million dollars. Initial financing is hard to get, because of the risks involved. What that means is that only financially healthy, or well-connected, companies can get to a PPA stage without initial financing.

As for permitting, federal and state policies dictate which locations are targeted by PV developers. BLM has a list of such locations (see Future Large-scale Solar Installations in the US above), so those are likely where most of the PV projects will be located.

Several U.S. states have some level of commercial or utility PV capacity, but California continues to dominate the PV installations in the U.S., followed by New Jersey. New Jersey, however, has less than half of the sunlight of the California deserts, so we don't see it as a serious contender—especially after government subsidies are eliminated.

California's planned and installed PV projects reached 80% of national demand recently, but when all PV projects in the pipeline from 2010 to 2015 are considered, California's share will decline to around 25%.

Table 8-4. US solar projects, 2010-2012

State	%
CA	25.0
NJ	16.0
MA	12.0
PA	6.0
AZ	5.0
NC	4.0
HI	3.0
TX	3.0
NY	3.0
NM	2.0
Other	22.0

A change made by some utilities established a trend towards smaller power fields, so that now nearly 30% of all planned or installed PV projects in the U.S. are in the 1.0 to 5.0 MWp range.

There are still problems with developing megawatt-scale PV fields, for they are not supported by the public. Local residents are often the most vocal against solar in their proximity, so many planned solar installations have encountered serious opposition from them, or from environmental groups. Sometimes projects end in court. Some of the most disputed are lands with historical or cultural significance, and agricultural lands.

Table 8-5. US solar projects per type, 2010-2012

Installation Type	%
Roof mount	47
Ground mount	33
Ground tracking	6
Carports	9
Other	5

US PV MARKETS

The US PV markets are segmented, and vary from place to place and from application to application. In most European countries today we see restrictions on large megawatt-scale installations in favor of roof-mounted PV. This has not happened, and most likely will not happen in the US.

So, in the US we have many viable options as to where to install PV systems. For now, ground-mounted PV installations account for most of our capacity, while rooftop installations make up almost half of the U.S. PV projects.

A number of large companies (retailers, universities, and others) are installing significant quantities of PV on their roofs mostly. This is a good way to use otherwise-wasted space and to bring free electricity to the owners. It also brings new employment opportunities to local solar installers.

These large customers are more likely to obtain bank financing for their PV installations than small business owners. Most small businesses (such as schools and municipalities) prefer third-party ownership of their PV systems, so they basically just lease their roofs.

Suitable areas for PV projects may be more difficult to find for some businesses, so new options seem to be appearing, such as floating PV arrays. These are simply PV arrays suspended (floating) in a lake or a pond. This is particularly practical for use in regions where land space is highly valued, such as community parks or agricultural properties.

Building-integrated options are another example of limited space utilization, where the solar arrays are incorporated into the building itself.

PV carport installations can be quite large, and some are over 1 MWp in size. These are constructed in municipal or university parking garages, and other structures suitable for solar integration.

ADDRESSING TRADE ISSUES

We witnessed a number of failures in the solar industry in the US during 2007-2012. Many of the companies forced to close their doors claimed their demise was due to unfair Chinese practices of flooding the world markets with low-cost PV modules. Benefiting from large government subsidies, Chinese manufacturers have been able to continue production and build large inventories. Now, these inventories must go, and at a price far below the manufacturing cost.

Because of that, a complaint was filed in 2011 by a number of US solar products manufacturers against this practice. One of the most significant cases is the shutdown of the grandfather of the US solar industry—SolarWorld in Camarillo, CA. That plant was the oldest and one of the most sophisticated in equipment and procedures facilities in the country.

The U.S. Department of Labor eventually determined that all manufacturing employees laid off from SolarWorld Industries America Inc. as a result of the company's shutdown of their 35-year-old manufacturing facility—are eligible for federal trade-adjustment assistance, including grants for education to retrain them for new work.

The determination that Chinese imports helped cause the shutdown resulted from an investigation by the department's Office of Trade Adjustment Assistance. This means that many of the 186 laid-off SolarWorld employees can tap federal assistance with job placement: expenses for job searches, relocation and retraining; income support during full-time retraining; and a tax credit on health-insurance premiums.

According to U.S. law, the Labor Department may certify workers for trade-adjustment assistance only if it finds that an increase in competing imports "contributed importantly" to the decline in sales or production of a firm and to the cause for worker layoffs.

Though SolarWorld invested tens of millions of dollars automating the Camarillo plant after purchasing it in 2006, the company determined it needed to consolidate its U.S. manufacturing in Hillsboro, OR, where it operates the Western Hemisphere's largest solar plant. Consolidation was required, according to SolarWorld, to contend with the illegally subsidized and dumped solar products of China's government-backed export drive.

U.S. Department of Energy researchers have concluded that without state sponsorship, Chinese manufacturers would face a 5 percent cost disadvantage in producing and delivering solar products into the U.S. market.

The closure of the Camarillo factory was one of many within the U.S. crystalline silicon solar manufacturing industry since 2010.

"We welcome the federal help that might ease the plight of our former factory workers in Camarillo," said Gordon Brinser, president of SolarWorld Industries America Inc. "But it will neither make them whole nor offset the loss of their pioneering know-how to the world solar industry.

"How many more U.S. manufacturing jobs must the United States lose in this most promising renewable-energy industry, which Americans pioneered, before adequate remedies are put in place to offset the illegal practices of Big China Solar?" Brinser asked.

In the first major ruling in the trade cases, the U.S. International Trade Commission issued a unanimous preliminary ruling on Dec. 2, 2011, that Chinese trade practices were injuring the domestic manufacturing industry. So far, in its ongoing investigation, the Department of Commerce has determined 10 categories of Chinese subsidy programs to be illegal under U.S. and world trade law.

SolarWorld is one of the founding members of the Coalition for American Solar Manufacturing, a coalition of more than 190 U.S. employers of over 16,000 American workers that is advancing trade cases to seek anti-subsidy and anti-dumping duties against the government-underwritten Chinese manufacturers.

POLITICAL EVOLUTION

In the 2012 Blueprint for A Secure Energy Future Progress Report, the Obama administration clearly abandoned its support for renewable energies, to focus on the conventional energy sources.

Key Segments of the Blueprint

In 2011, U.S. crude oil production reached its highest level since 2003, increasing by an estimated 120,000 barrels per day over 2010 levels to 5.6 million barrels per day.

Since 2009, the United States has been the world's leading producer of natural gas. In 2011, U.S. natural gas production easily eclipsed the previous all-time production record set in 1973.

Overall, oil imports have been falling since 2005, and net imports as a share of total consumption declined from 57 percent in 2008 to 45 percent in 2011—the lowest level since 1995.

Currently, the United States has a record number of oil and gas rigs operating—more than the rest of the world combined.

In March 2011, the President set a bold but achievable goal of reducing oil imports by a third in a little over a decade. He directed DOI to determine the acreage of public lands and waters that had been leased to oil and gas companies but remained undeveloped. DOI's report reached several important conclusions: First, the Department has offered substantial acreage for leasing and resource development, but much of this acreage has not been leased by industry. Second, tens of

millions of acres that are currently under lease remain idle. Soon after the release of the report, the President released the Blueprint for a Secure Energy Future, which called for a number of reforms to incentivize efficient oil and gas development.

Developing Region-specific Strategies to Facilitate Responsible Development of Energy Resources

Currently, the Gulf of Mexico supplies more than a quarter of the nation's oil production, and the Central and Western Gulf remain the two offshore areas of highest resource potential and industry interest—and the areas where the infrastructure supporting the oil and gas industry, including the resources to support an oil spill response, are the most mature and well developed. Implementing the most comprehensive reforms to oversight of offshore oil and gas activity in U.S. history following the Deepwater Horizon oil spill, DOI has approved plans and permits for exploration activities throughout the Gulf of Mexico.

These new safety measures include heightened drilling safety standards to reduce the chances that a loss of well control will occur, as well as a new focus on containment and response capabilities in the event of an oil spill. New workplace safety rules, including significant ones recommended by the National Commission on the BP Deepwater Horizon Oil Spill and Offshore Drilling, have also been implemented and an enhanced proposal is expected to be finalized before the end of 2012 following a public comment period which closed late last year.

The Blueprint directed the Secretary of Energy Advisory Board (SEAB) to establish a subcommittee and identify measures that can be taken to reduce the environmental impact and improve the safety of shale gas production. The final SEAB report was issued in November 2011. The Department of Energy, the Environmental Protection Agency, and the Department of the Interior are acting on many of the recommendations.

Protecting Consumers by Strengthening Oversight of Energy Markets

Combating manipulation and fraud, The Administration has pursued unprecedented coordination through the oil and gas prices fraud working group and active monitoring of gasoline and diesel projects in 360 cities across the nation. The CFTC has also taken steps to address loopholes that allowed financial trades to evade oversight by trading in unregulated or overseas markets.

Author's note: Clearly, oil and gas are the focus for now, with coal and nuclear following closely. Solar and wind are mentioned later on in the Blueprint, but only as an alternative, after gas, oil, nuclear, and coal.

We will see a lot of oil drilling, gas fracking, coal digging, and nuclear development during the coming years, while solar and wind take a back seat—again.

LCOE OR NOT LCOE?

The levelized cost of energy (LCOE) measures the total generation cost in dollars per kilowatt-hour over the lifetime of a PV system. Looking at energy cost, as opposed to capital cost, it allows us to assess the competitiveness of solar, versus conventional power generation.

For years, solar companies and projects have been benefiting from a 30% solar investment tax credit (ITC). But what if it is slashed or reduced, as it was done in many other countries?

The ITC has been the driver of renewable energy development in the US. Introduced in 2006, this federal policy was designed to support deployment of solar installations in the U.S. It is to be credited with the development of the majority of solar companies and projects.

But the ITC will be stepped down to 10% after 2016. A number of analysts have undertaken preliminary analysis of what the U.S. solar market would look like in a post-ITC world, and it is a mixed picture.

For example:

1. An average 5 kW residential system installed in 2013, operating 25 years with 0.5% annual degradation (power loss), 28% federal tax rate, 100% debt financing, and installed cost of $3.25 would be at or below LCOE in the states of AZ, CA, CT, HI, NV, NJ, NM, and NY, but not in most other states.

    With an ITC reduced to 10%, only the state of Hawaii would be able to keep LCOE below grid parity. In the rest of the states, residential solar installations would break even, and even lose money in the long run. Basically, the residential market looks much weaker, and we expect to see more problems in this area.

2. On the other hand, evaluating an average 500 kW commercial system installed in 2013, operating 25 years with 0.5% annual degradation (power loss), 35% federal tax rate, split financing, and installed cost of $2.75 would be at, or below, LCOE in most states.

Dropping the ITC to 10% reduces the LCOE advantage, but the commercial solar installations in most states that are presently active in solar energy would still remain below LCOE.

The final numbers in both cases will depend on the amount of state and local incentives, utilities regulations, etc., but we expect solar (and PV in particular) to be technologically and economically viable options for commercial customers in the near future—even after a step-down of the ITC from 30 to 10%.

What If?

Assume for a moment that ITC is dropped. What will happen then? In our opinion, short of state and local governments stepping in to fill the gap, only a very few states would be able to stay below LCOE.

The margins are so slim—even with 30% ITC—that there would be no profit to speak of, in which case many more solar companies and projects would fail.

Finally, the profit/loss scenario in all cases will be determined primarily by how well the system operates. A poor quality system that breaks down often, or one that operates under cloudy skies, would not get even near LCOE—ITC or no ITC. As we have very little control over the quality of the system components (PV modules, inverters, etc.) and no control over weather patterns, surprises are expected.

Other Factors

This is the dark side of the equation, because when discussing LCOE we need to take into consideration a number of other factors, which are usually ignored, or underestimated. Keep in mind that we are talking about competition here—the established and well supported conventional energy sources vs. the new and still neglected renewables.

Billions of dollars of direct and indirect financing are allocated to conventional energy sources, while the solar and wind industries are left struggling.

Billions of dollars are spent on remediation of the immediate physical damages caused by the use of conventional energy sources. We must calculate the serious long-term environmental damage done to soil, water and air by those energy sources, including global warming, climate changes, air and water qualities.

These costs are "hidden," but are real and must be considered in LCOE calculations.

So there are a number of serious unknowns and shifting factors in the U.S. energy industry, which make for a very complex situation. The solar industry is the new kid on the block, undergoing growing pains, and with a great number of combinations and permutations at play.

THE WORLD

The solar boom cycle started in 2007 with amazing news of new installations in Germany, Spain, Italy, France, the US and other countries. The number and size of the new installations was staggering:

World solar energy has gone through several boom and bust cycles since the 1970s, but this latest one is unprecedented in its size and speed of development. The generous, albeit disproportionate and unsustainable government financial, technical and logistics support contributed to a large extent to this incredible rise of solar power capacities worldwide, with Germany as the leader, followed by other EU countries, Japan, China and the US. A large number of liberal, and in retrospect hasty, VC investments fueled some of the companies and projects that were not ready for the action. We have seen such rises before, but this time the effects were much greater and longer lasting.

The solar boom-bust cycle of 2007-2012 is definitely different from those in the past. It is absolutely different from anything we should expect in the future. The incredible progress made in a few short years at the beginning, and the comparably great number of failures at the end taught us many valuable, embarrassing, and painful lessons.

This book is a summary of the events—good and bad—during this period and we hope that those who

Table 8-7. Global PV markets, 2011

Country	%
Italy	28
Germany	25
USA	10
China	8
Japan	4
France	4
Australia	2
UK	2
India	2
Canada	2
ROW	12

are involved in solar business will take a look at what happened and understand why it happened. This alone could save a lot of headaches.

As for the long run, some indicators—unlike those of the 1970s, 80s and 90s—are that solar energy is maturing, and given a chance will grow as expected.

PV Modules Manufacturing Surplus

After several years of sustained profits and strong (if not always stable) growth, the PV manufacturing sector has found itself in the midst of a protracted downturn. A combination of overly aggressive capacity build-up in 2010 and 2011, along with severely curtailed subsidies in major feed-in tariff markets have resulted in a massive supply-demand imbalance that manifested itself in early 2011, and is not expected to abate until at least 2014.

Table 8-6. Worldwide PV power capacity.

Country	INSTALLED PV CAPACITY (MWp)							
	2007	2008	2009	2010	2011	2012	2015	2020
Germany	1,100	1,500	3,800	7,200	7,500	6,000	4,000	6,500
Spain	560	2,500	75	850	500	650	1,200	2,000
Italy	75	420	550	4,200	9,200	4,500	4,500	5,000
France	55	150	250	700	1,500	1,500	3,750	4,200
Japan	201	230	480	975	1,500	2,500	2,200	4,500
China	70	110	175	580	2,800	4,500	3,000	5,000
India	10	15	30	80	1,500	300	900	4,500
USA	207	342	480	975	2,100	4,000	5,500	7,500
World	Balance	Balance	Balance	Balance	Balance	Balance	Balance	Balance
Annually	~2,800	~5,600	~7,500	~16,250	~27,500	~32,000	~28,000	~30,000
Total	~9,550	~15,150	~22,650	~38,900	~65,000	~75,000	~180,000	~300,000

Table 8-8. PV modules manufacturing and installations supply and demand discrepancy (GWp)

Country	2008	2010	2012	2015e
China		2.5	15.0	48.0 52.0
Japan	2.0	3.0	5.0	5.0
Asia	1.0	2.0	4.0	4.0
Europe	2.0	6.0	8.0	4.0
USA	1.5	2.5	2.5	2.5
ROW	0.5	1.5	2.5	4.0
Total inventory	9.5	30.0	70.0	72.0
Total installations	5.6	16.2	32.0	28.0
Total surplus	3.9	13.8	38.0	44.0

PV modules inventories have remained high across the board, which in turn has contributed to severely decreased gross profit margins for PV module suppliers. During the 2008-2009 price peaks, it was normal to see PV modules sold for $3.00/Watt, while in the fall of 2012 the price hovered around $0.50/Watt.

Adding insult to injury, a significant factor is the perception in the minds of the average buyer that quality and reliability are the same regardless of the manufacturer. As one manufacturer put it, "My customers look at the modules, and if they are shiny enough they are happy." This behavior has allowed inferior and unproven, but very shiny, Asian PV modules to flood the world's markets.

The rampant overcapacity and commoditization of PV modules, combined with lower prices and slimmer profit margins, resulted in an avalanche of plant closures, market exits and insolvencies during the 2010-2012 time period. The victims thus far have mostly been smaller producers in high-cost regions, but this fact has done very little to alleviate the industry's troubles. Larger manufacturers have been affected too, but they have more staying power by shutting down excess capacity. This is what Solar Frontiers did in the fall of 2012 by shutting down one of its 3 Japanese manufacturing plants.

The overcapacity is likely to persist through much of 2013 and even 2014, while the balance sheets of most PV components manufacturers are under pressure.

Consolidation is another way to stay afloat, and we foresee these activities increasing as well. In all cases, the global PV module landscape is headed for a significant transformation.

The Chinese module suppliers have the largest number of production plants with the highest through-

put. The Chinese PV modules industry is dominated by large-scale crystalline silicon manufacturers, representing over 65% of the Chinese manufacturing capacity. A number of diversified c-Si firms play an important role, with over 30% of the domestic capacity share.

One amazing thing to note here is the fact that most of these are privately owned, and that the large state-owned enterprises have not yet jumped into PV manufacturing in a big way.

A number of smaller manufacturers are also looking into thin-film silicon (a-Si) technologies, but their numbers and output are still limited. Even fewer of small firms are interested in the other thin film PV technologies, such as CdTe, CIGS, etc.

Table 8-9. Worldwide solar energy markets, 2020-30 (GWp)

Area	2020	2030
S. America	15	50
N. America	15	100
Europe	80	200*
Asia	50	250
Other	10	20

*The 200 MWp solar power estimate for Europe is based on the plans of the EU community to build several very large-scale solar and wind power fields in the North African deserts. This generated DC power will then be send to Europe through the Mediterranean Sea, via underwater cables. The so-called Desertec initiative is part of these plans. If all goes well, Europe may be able to generate 2,000 GWp in the N. African deserts during the 21st century.

Solar energy markets will continue to grow, once the dust of the 2007-2012 debacle settles. Asia (and conditionally Europe) are where most of the growth is expected, but with the unpredictability of this market, anything is possible.

The EU Experience

At the end of 2012 Europe had 51 GWp combined solar energy producing capacity, or over 70% of the entire world's installed capacity, which was estimated at approximately 70 GWp.

More than $140 million in government subsidies have already reached China (via purchases of Chinese solar products) from Germany alone. This happened along routes designed to bolster the solar industry in the country that has already become the dominant global market player in some arenas.

The Chinese got much more than that from the US and the other EU countries too. How did that happen?

The German development bank of the government-owned KfW group of banks put a plan in action, designed to support China's green industry with low-interest loans. The German Investment and Development Company (DEG), also a subsidiary of KfW, was/is one of the major financial backers of the Chinese industry giant Yingli Solar. That was done with good intentions of helping Chinese businesses and the budding green energy industry in particular (and profiting in the process).

Those German development bank loans are still active. According to KfW officials, it was their goal to help Chinese solar manufacturers to promote the environmentally friendly technology internationally.

As a matter of fact, the German government's climate protection program for China is a huge success, with over $12 billion spent in seed capital, using funds generated from the sale of CO_2 pollution certificates to the German industry.

The same program also consists of low-interest loans issued by the KfW Development Bank, with the first loan of €75 million going to the state-owned Export-Import Bank of China. This is a low-interest loan, expiring in December 2013. China's banks then added more funds and distributed the total to China's industry giants like Yingli, Sunergy and JA Solar.

LDK Solar took the money back to Germany, but then it bought part of German solar cell producer Sunways, based in Konstanz, Southwest Germany. Subsidizing solar may be part of the German government's climate protection policy, but for China it is part of an aggressive industry development, with export objectives.

Yingli, which is still one of the world's largest solar producers, can also take its time to repay the $25 million it borrowed from German banks under very favorable terms. Its last payment is due in September 2013, but until then Yingli can use the money as it wants. So Yingli is doing very well; production is up, and the number of employees is growing. Germany is STILL Yingli's best customer, since it sells almost half of its systems to German homeowners and solar farm operators.

Believe it or not, this is precisely what German development officials had hoped for. In 2008 Winfried Polte, then the representative of DEG, a subsidiary of KfW, said that his organization was proud of helping Yingli with its expansion strategy. After the first loan agreement he said, "The expansion will create thousands of new jobs in China."

It did, and it *cost* thousands of jobs in Germany and the US. As saying goes, "It is not crazy who eats two cakes, but he who lets him."

Now while the German solar industry is suffering from the flood of inferior Chinese energy products, the German banks are doing well.

Anti-dumping

Following the example of the US companies, who filed an anti-dumping (AD) suit against China, European companies are doing the same thing with a twist. In the summer of 2012, Germany's SolarWorld filed an anti-dumping complaint against Chinese rivals with the European Commission.

This will hurt Chinese companies, but will benefit Taiwanese solar cell suppliers and other non-silicon solar cells and modules makers, since their products won't be taxed…for now.

The European Union penalties will be at least a third lower than the US imposed, and it is quite likely that the US AD case will have a certain level of influence during the complaint phase. Once the EU Commission initiates the investigation, there will be an independent assessment of dumping and the US case will not have any direct influence.

If there is sufficient evidence, and at least 25% of the total production capacity to support the complaint, a preliminary determination would be reached within 9 months at the latest.

The size of the European PV market would make it hard for Chinese cell suppliers to circumvent any duties using cells made outside China given capacity constraints in Taiwan. The issue of community interest will likely prove the key point.

The key differences between the EU and US AD cases are:

1.   Injury margin
     Instead of just looking at the dumping margin like the US, the EU also calculates an injury margin, and any final duty imposed is the lower of dumping margin and injury margin. The US used a supply chain buildup to determine the cost of a cell in Thailand, while in the EU they look at pricing of competing products outside of China. In the EU, non-market economy prices are generally based on a constructed price in a third country. Companies like First Solar may not be part of the discussion if the scope is limited only to cells.

2.   Community interest
     Instead of just looking at the injury to a defined industry as in the US, EU looks beyond that to see if the decision would be contrary to community interest as a whole. This allows upstream/downstream companies and consumers to have greater influence in the EU than in the US. Of recent cases that have been dismissed, most of the industries were not present in the EU or the EU would not otherwise be able to serve the market.

3.   Market economy treatment

While the US will treat China as a non-market economy, the EU allows Chinese companies to request market economy treatment (MET) based on five criteria. Between 2005-2010, roughly 20% of the requests received the market economy status. However, in recent years, there has been a tendency to grant MET less frequently.

4.   Single authority

The US has a bifurcated system with both Department of Commerce (DoC) and International Trade Commission (ITC) in charge. The EU has one investigating authority, the EU Commission.

5.   Transparency: Unlike the US, lawyers for participants do not have access to the non-confidential data of the industry and other respondents.

In 2012, German environment minister Peter Altmaier offered the government's support to German solar companies' efforts to launch AD proceedings in the EU against Chinese PV manufacturers. Many EU cell producers joined this action, and so far using EU-made modules still allows positive return for project owners.

Suntech's response: "Suntech rejects SolarWorld´s allegations that it has received illegal subsidies and is dumping solar products in Europe and will cooperate fully with any investigation. As the market leader with a global presence and customers in 80 countries, Suntech will continue to demonstrate its adherence to fair international trade practices.

As a NYSE-listed company, we are transparent with regards to our cost of production and cost of capital. Suntech's growth is due to its efficient manufacturing operations and long-term investments in R&D to create high performance solar products. We hope that the European Commission will recognize that any protectionist measures would harm the entire European solar industry and that a misguided trade war would undermine years of progress.

Protectionist measures would increase the cost of solar energy in Europe and delay the transition from fossil fuels to renewable energy. Tariffs would also destroy thousands of jobs in the European solar industry," added Stokes. "The EU solar industry provides employment for around 300,000 people and more than 80% are employed in upstream and downstream industries such as raw material suppliers, equipment manufacturing, system design, installation and project financing, and not in cell production.

In addition, the global supply chain for solar panels is complex and interconnected. Most solar systems installed in Europe are made up of components and services from manufacturers around the world. Suntech, for instance, sources a significant portion of its manufacturing equipment and raw materials from Europe. In 2010 and 2011 we procured a total of approximately €600 million of equipment and materials from European suppliers. We are concerned that any tariff to support European cell and module manufacturers would damage all other parts of the value chain. Suntech, along with the vast majority of European and global companies in the solar industry stand together in our support for free trade and our determination to see a trade war averted," concluded Stokes.

Germany

Germany became a front runner in the solar energy industry overnight, with power installations and large power plants popping up all over the place. At the end of 2012 Germany had almost 70% of the EU (70 GWp) capacity, or over 30 GWp.

The average German PV system price in 2012 was estimated at $2.24/Wp. Residential and commercial rooftop systems are predominant in Germany, representing approximately 72% of all PV installations in 2011, so this is an impressively low number.

2012 installations in Germany are estimated at around 6.5 GWp—a slight decrease, as compared to 7.5 GWp installed in 2011. Residential system prices fell by 4.8% from Q4 2011 to Q1 2012.

Module prices of around $0.90/Wp at the time, implies an average balance of system cost of $1.34/Wp. This seems to be one of the major reasons why the German PV market is still alive and well.

At the same time, the U.S. average PV system price was $4.44/Wp in 2011 and slightly lower in 2012. Why is there such an enormous discrepancy in the average price per watt in Germany versus the U.S.?

Since the PV components pricing is roughly the same for both countries, the culprit for the high prices in the U.S. might be the soft costs of permitting and financing, in addition to higher costs of the engineering, procurement, and construction (EPC) process.

If we add to that the potential increase in module pricing due to the China trade tariffs, then the U.S. market faces some real headwinds in driving down the cost of solar.

The soft costs are the main goal for the DOE's "Sunshot" program, which is designed to lower the cost of financing and other soft costs. Also, programs like the "ten-day permitting" process in Vermont bring some promise for reducing soft costs in the U.S.

Germany's 2012 FiT legislation introduces monthly tariff degressions, with the amount of the FiT cut to be variable with a maximum annual degression fixed at 29%.

In May 2012 some 8,500 new PV installations with a total capacity of ~2.5 GWp were reported, accounting for approximately 10% of Germany's energy supply during that month, according to the Federal Association of Energy and Water (BDEW). This capacity was then added to more than 1.2 GW installed in March and capacity from the rest of the period. Annual installations of approximately 7 GW to 8 GW have been the norm since 2010, which is when the German solar market was considered to have reached maturity, following six years of development and growth. That trend, however, will decline in 2013, only to rise in 2014 and beyond if and when investments pay off, regardless of the FiT rates.

Nevertheless, Germany's solar glory days went with the withdrawal of subsidies and the rising competition. The German government slashed the subsidies for electricity from solar energy in 2011-2012, at which point the share prices of solar-panel makers plunged and the entire solar game changed. Then, after several successive rainy days, the outlook for Germany's solar economic miracle started looking worse than ever.

When the German solar program began in 2000, it offered index-linked payments of €0.51 for every kWh of electricity produced by solar PV, guaranteed for 20 years. This is similar to the UK's initial subsidy, of 41 p. As in the UK, the solar subsidy was, and remains, massively greater than the payments for other forms of renewable technology.

The net cost of all solar PV installations in Germany during 2000-2008 was €35 billion. A further cost of €18 billion for the 2009-2010 period was incurred, for a total of €53 billion spent in ten years. These investments make sense for the residential owners who could afford to install the panels, but the lucrative returns are guaranteed by taxing the rest of Germany's electricity users—some of which cannot afford it.

The Results

What has been achieved by the generous government programs and lavish spending that followed? By the end of 2008, PV installations were producing a grand total of 0.6% of Germany's electricity. 0.6% at a cost of €35 billion. This is a lot of money for a fraction of a percent.

These incredible payments were supposed to also stimulate innovations, to achieve cost reductions, but the specialists' estimates show that saving one ton of CO_2 through use of solar energy in Germany still costs €716. And according to the International Energy Agency, it will increase to €1,000 per ton in 2013. Due to the cloudy German skies, we believe, there are many better ways to save carbon for a fraction of the cost of solar in that country.

Ruhr University issued a paper on the subject, that comes out against using feed-in tariffs to stimulate wind power as well, but in this case it shows that large-scale wind power in Germany is likely to become cheaper than conventional power by 2022, at which point subsidies will become redundant.

The paper, however, makes no such prediction for solar. It reinforces the point that while Germany, like the UK, belongs to the European emissions trading scheme, any carbon savings made by feed-in tariffs merely allow polluting industries to raise their emissions. The net saving is zero.

The paper also suggests that a far more cost-effective mechanism would be to crank down the emissions cap under the trading scheme—then let renewable technologies fight it out to offer the biggest carbon saving per euro.

As for stimulating innovation, the report shows that Germany's feed-in tariffs have done just the opposite. Like the UK's scheme, Germany's is digressive—it goes down in steps over time. What this means is that the earlier you adopt the technology, the higher the tariff you receive. If you waited until 2009 to install your solar panel, you'll be paid €0.43/kWh (or its inflation-proofed equivalent) for 20 years, rather than the €0.51 you get if you installed in 2000.

This encourages people to buy existing technology and deploy it right away, rather than to hold out for something better. In fact, the paper shows the scheme has stimulated massive demand for old, clunky solar cells at the expense of better models beginning to come onto the market. It argues that a far swifter means of stimulating innovation is for governments to invest in research and development.

But the money has gone in the wrong direction. While Germany has spent some €53 billion on deploying old technologies over ten years, in 2007 the government spent only €211 million on renewables R&D.

In principle, tens of thousands of jobs have been created in the German PV industry, but this is gross jobs, not net jobs. Had the money been used for other purposes, it could have employed far more people.

The Ruhr paper estimates that the subsidy for every solar PV job in Germany is €175,000. In other words the subsidy is far higher than the money the workers are likely to earn. This is a wildly perverse outcome. Moreover, most of these people are medium or highly skilled workers, who are in short supply in Germany. They have simply been drawn out of other, more viable industries.

The Ruhr researchers say that: "Any result other than a negative net employment balance of the German PV promotion would be surprising. In contrast, we would expect massive employment effects in export countries such as China." And sure enough, right they were.

The National Grid

With a steep growth of power generation from photovoltaic (PV) and wind power and with an 8 GW base load capacity suddenly taken out of service, the situation in Germany has developed into a nightmare for system operators.

The peak demand in Germany is about 80 GW. The variations of wind and PV generation (especially PV modules operating under cloudy skies) create fluctuations in the regulation and transport of power. The grid capacity is far from sufficient to handle the variations. The result is a remarkably large number of curtailments of RES (renewable energy sources).

Reports from the European Network of Transmission System Operators for Electricity (ENTSO-E) and the German Grid Agency reflect concern for the operational security of the power system. The risk of a prolonged, widespread blackout was earlier recognized by the German Bundestag and discussed in an official report.

Since January 2012, all four German system operators have published estimated photovoltaic generation based on representative samples. The data will give research environments a new opportunity to analyze the impact of RES in Germany.

Some observations are possible from the evidence:

- Wind power peaks seem not to be simultaneous with PV peaks. This means that PV does not add its full peak capacity to the grid problems during high wind periods.

- Most German wind power is installed in the northern part of the country, while most PV capacity is installed in Bavaria. The nuclear moratorium has created the most serious supply problems in the southern part of Germany. This observation suggests additional PV generation is needed to relieve the supply problems.

- PV generation cannot reduce the need for peak capacity. The reason is that there is no PV generation during the evening peak load.

- The regulating work which must be made by controllable power sources grows considerably with the growth of wind power and PV. TenneT is one of Germany's four main grid operators. In the TenneT area, a calculation for April 2011 has shown that wind power alone would extend the regulating range by more than 50 percent, while the actual combination of wind power and PV has doubled the regulating range.

Although PV may be able to give some relief to the grids, PV is also a variable power source, and cannot reduce the need for peak capacity. Additional PV might even cause a considerable growth in the need for additional regulating capacity—another headache and another huge expense.

The German Grid is Europe's Backbone

On November 4, 2006, a German 380-kilovolt line had to be temporarily disconnected. Due to insufficient coordination of protection systems, a circuit tripped and started cascading outages. The result was that the continental grid in Europe was divided into three islands and about 17 gigawatts of load was shed. The case demonstrates how a local event in Germany can turn into a widespread European disturbance.

In April 2012, the president of ENTSO-E, Daniel Dobbeni, stated his concern about security of power system operation in Europe in a letter to the European Commissioner for Energy, Günther Oettinger.

ENTSO-E: "As long as RES generation in certain regions expands faster—partly as a function of national support schemes—than the transmission network can accommodate, the risk of insecure system operation coupled with costly generation curtailments will rise significantly."

A recent briefing paper gives an overview of the current situation. The rapid increase of wind power and other renewable energy sources (RES) without a corresponding reinforcement of the electric grids has caused the problems.

The paper explains: "Heavy 'unplanned' transit flows added to scheduled flows cause severe loading on southern interconnectors (PL/CZ, PL/SK, DE/CZ, and also SK/HU and SK/UA) and lead to non-compliance with fundamental network security criteria. The high level of flows on the interconnectors leads to overloading of the network in Germany and neighboring countries Poland, Czech Republic, Slovakia and Hungary.

Among the countermeasures of the transmission system operators (TSOs) is the use of the HVDC links across the Baltic Sea for a redistribution of power flows. A common procedure has been developed by German and Polish TSOs and two Nordic TSOs (Energinet.dk and Svenska Kraftnät). However, the remedial actions cannot be guaranteed, as they depend on prevailing system conditions.

The countermeasures have cost implications and cannot be implemented without cost sharing agreements.

ENTSO-E makes reference to its Ten-Year Network Development Plans (TYNDP). The timely implementation of the projects will require the active support of European policy makers.

The paper estimates the necessary investment for reinforcement of the western and the eastern transport corridors in Germany to be 30 billion euros for the next decade. The German reinforcements must be coordinated with investments in neighboring countries.

Efficient market arrangements are important for efficient congestion management, secure grid operation and overall market efficiency. Therefore, the organization of more consistent markets and redefinition of bidding areas deserve consideration.

The ENTSO-E paper concludes: "If this infrastructure does not materialize in due time, then the rate of RES increase should be examined under a more pragmatic prism."

A German Performance Report for 2011-2012

The federal German Grid Agency has confirmed the assumption of a strained grid in a 120-page report on the supply situation for electricity and gas in Germany during the winter season of 2011-2012.

It is useful for the general understanding of the significance of the infrastructure when an authority evaluates actual system conditions and publishes annual reports for better or for worse. Unfortunately, that sort of report is rare in the electricity business.

The 10 points of the report in a summary:

1. The situation of the power grid was very strained during the winter 2011-2012.

2. Besides the scenarios described in the Grid Agency report of August 31, 2011 the shortage of natural gas in February 2012 was followed by an unexpected event which added to the load on the

electric grids and required additional measures from the transmissions system operators for maintaining system security.

3. In addition to that, an unusually large number of forecast errors caused an exhaustion of the regulating reserves. Therefore, the transmissions system operators had to resort to additional measures. The Grid Agency will create incentives for improvements of the forecasts by adaptation of the price system for balancing power.

4. The synchronous compensator Biblis was commissioned in February 2012 and provided the expected relief of the voltage problems.

5. German and Austrian power plant reserves were used in several cases for the relief of power lines and as a supplement to already exhausted regulating capacity. About the same magnitude of power reserves will be needed next winter.

6. The power plant capacity has developed unfavorably. Planned extensions have been delayed. Further decommissioning of conventional (coal and nuclear mainly) power plants cannot be defended in Germany for the time being. The prevention of decommissioning of power plants for conventional production will require regulating and legal measures. If more power stations should be decommissioned in southern Germany, the needed reserve capacity would increase correspondingly. Besides, the need for capacity mechanisms should be intensively investigated in the medium term.

7. The supply of more power from renewable sources than can actually be transferred by the grid would add to overloading of the grid, because the price signals would displace conventional power plants in the merit order and the electricity export from Germany in the internal market would increase. It is the understanding of the grid agency that the existing legal framework allows the transmission system operators to use measures which can reduce the supply to a level that can be transferred by the grid. Nevertheless, a normative clarification seems to be expedient.

8. The cooperation between grid operators for electricity and gas must be improved in order to take

account of the growing significance of gas power plants and gas supply to the security of supply of the electric grids. Even here, changes of the legal framework are recommended.

9. No technical valid measures can replace grid extensions. A consistent use of the established instruments for acceleration of the reinforcement of the grids is required.

10. The reduced supply of gas in February 2012 has revealed the weak points of the gas grids. Action is needed for the gas grids. Fortunately, this need is clearly inferior to the need for action in the electricity grids.

The general view seems to be the concern for the future capacity of power plants, and regulating power and reserves, and other concerns of the grid operators. A strong grid is important, and several integration and other measures deserve careful consideration, including control of the renewable power generation.

The increasing trend in the use of §13.1 of the German Energy Industry Act (EnWG) for re-dispatch and in the use of §11 of the RES Act (EEG) and §13.2 of EnWG for reduction of feed-in of power is demonstrated in the report. The data are valid for the transmission grid.

Re-dispatch is used for the relief of highly loaded grid components. For both years, most re-dispatch concerned the Remptendorf-Redwitz line between Germany and Austria.

Feed-in reduction was initiated 197 times during the winter season of 2011-2012, compared to 39 times the previous year. In 184 cases, wind power caused high feed-in from distribution grids into the transmission grids. Five cases were remarkable and affected the entire grid:

This information confirms that the German electricity supply had narrow margins during the winter of 2011-2012 without much room for additional political intervention. Hopefully, the messages of the Grid Agency will be understood, so a better harmony between the transition of the production facilities toward green solutions and the necessary adaptation of the infrastructure can be achieved.

The US power grid, although much larger, and somewhat more resilient to interferences, is still vulnerable to similar problems, including power fluctuation from large-scale solar and wind power plants. Is the U.S. going to have the problems like those described above?

Germany's Energy Future

Germany has a goal to discontinue all nuclear power by the year 2020 and replace it with renewable resources. This goal prompted a government policy called feed-in-tariffs (FITs). A feed-in-tariff is a policy designed to encourage the adoption of renewable energy of all kinds to accelerate the cost of renewables down to grid parity.

FITs typically include three provisions:

1. guaranteed grid access,

2. long-term contracts for the electricity produced, and

3. prices that are based on the cost of renewable energy generation with a downward trend towards grid parity. Besides PV, they include other renewable technologies, such as concentrated solar power, wind, and geothermal.

In almost all of Africa, Pakistan, Hawaii, Italy and large portions of Japan, the price of electricity from fossil fuels is already in excess of the cost of electricity from solar. There is a ready market for today's solar electricity without any subsidies. As the price of solar comes down every year, more and more locations will benefit from making the switch to solar when new capacity is added.

Other countries with major PV feed-in-tariff programs are Italy, Japan and China. These countries will help take up the growth slack from Germany who has achieved the initial goals and will be reducing incentives in the future.

As can be seen above, the US is way behind in solar installations. The US has considered a feed-in-tariff, but has yet to form a federal consensus to pass legislation. However, there are 14 US states and the District of Columbia that regulate retail electricity markets in which customers may choose "alternative" power suppliers.

In addition, some states such as California and Arizona, have implemented their own aggressive incentive programs to encourage alternative power. The US, especially in California and Arizona, has several very large solar installations in progress, over 1.5 GWp, which when finished will catapult the US into a major position internationally.

Spain

Spain's situation is somewhat different. Although there is no problem with sunshine in most of southern Spain, the solar industry fell on hard times. The problems were somewhat different from those in Germany. The final results were different as well, caused not because of lack of sunshine, but for lack of good planning and execution. This, combined with the economic downturn, proved too much for the Spanish solar industry.

The example below is only one of the Spanish solar industry failures. Please read carefully—it's a lesson well worth learning.

Case Study

Puertollano, Spain, is a small, gritty, mining city in south Spain, which has a lot of sunshine and a lot of unemployed people. The federal and local governments saw an opportunity in the beginning of the solar boom-bust cycle and decided to enter the 21st century with a bang.

For a long time, the primary occupation of Puertollano's blue collar workers has been mining coal, so Puertollanos grabbed the opportunity to get out of the coal mines, and get deeply into another energy source—their sunshine. Puertollano has bright, relentless, scorching sunlight almost every day.

Too, there were so many attractive and generous incentives from the Spanish government that it was hard to ignore the opportunity to jump-start the national solar energy industry. This was a lifetime opportunity to make lots of money, and lead the nation in solar energy.

So, the city made ambitious plans to replace its failing, dirty, coal economy by offering its sun-bathed fields to anyone who wanted to make use of them by generating solar energy. They even invented a way to attract solar companies with a new campaign slogan that read convincingly, "The Sun Moves Us." Puertollano enjoys more sunlight than most other Spanish regions, and many times that measured in Germany, UK and Scandinavia.

Within months, Puertollano's coal industry was locked in the new Museum of the Mining Industry, and the city became a host to two enormous solar manufacturing plants. Hundreds of people were employed, making solar panels and installing them in the Spanish countryside. A new clean energy research institute was built and started operation in the area.

Over half the total solar power installed globally in 2008 was installed in Spain, and Puertollano played a big part in that. Bliss was smiling upon the small sun-lit city in the midst of its new-found prosperity. Coal miners became installers of new, shiny solar panels in the new PV power fields. Farmers sold the land that had fed them for generations, as needed to accommodate the new, and very promising, solar plants. New businesses grew overnight. People from all over the world saw the newly created business opportunities and moved to the city. This influx of new business eliminated quickly the previous 20%+ unemployment, and reversed the population exodus.

Life was good in Puertollano…until the problems started. And the problems were many—while it is easy to take the miner from the mine and call him a "solar specialist," it is not easy to quickly train him properly, to understand and follow quality standards.

Similarly, the design engineers and the regulators were too green. It took them awhile to figure out that just because a PV module is new and shiny, it won't necessarily perform as specified.

Due to materials and procedural issues, a number of low-quality, poorly designed solar plants sprang up all over Spain—including several in Puertollano.

The regulators, who were also to blame for these problems, figured out, somewhat late, that the government would have to subsidize many of these shoddy operations indefinitely, and that the industry they had created might never produce efficient and profitable green energy on its own.

Then the ax fell on the unsuspecting Puertollano citizens. In the fall of 2009 the Spanish government abruptly changed its solar policies. The generous payments were cut to the bone, and solar construction was limited and capped to very low levels.

Puertollano's short-lived boom turned bust overnight and life in the city became a nightmare. The new factories and stores were closed and thousands of solar-related workers lost their jobs. But the worst was yet to come: foreign companies and banks abandoned contracts that had already been negotiated and projects that had been started.

Soon Puertollano's citizens went back to the dreary existence they had before—back to the coal mines and agricultural fields they abandoned not that long ago. Worse, the mines were becoming depleted, and many of the agricultural fields were now junkyards, filled with rusting solar panels.

Puertollano is only one small, though bright example, of boom gone bust within a short time, but it is also a very good example of the importance of government insight into technology. We hope this example serves others, because we saw similar oversights in our own backyard.

On a broader scale, the collapse of the Spanish solar market in 2009-2010 was commonly believed to have been a solar bubble. However, a closer look reveals what Spain's real problems were, and how to avoid them.

The extra money coming from Spain's FiTs, as required to make renewables profitable, does not come from the government's budgets. The additional costs are actually paid by the consumers in the form of higher electricity rates.

Here is the catch: the Spanish electricity market did not allow these extra costs to be passed to the customers. Instead, at the beginning of each year, the Spanish government sets retail power rates, depending on outside factors, such as price of natural gas and gasoline. It then charges (or reimburses) the power providers and grid operators according to the difference, thus keeping electricity rates artificially low. Of course, part of the cost is covered by taxpayers according to government calculations.

Then, the EU got into a budget deficit, and with this scheme, the Spaniards are in fact transferring the cost of presently subsidized electricity consumption to future generations. In an attempt to fix the untenable practice, Spain started to phase out the entire renewable energy system. They estimate that by 2013 the entire system will be dismantled and replaced by a more rational and attainable one.

But now (2012-2013) Spain is getting deeper and deeper into the economic debacle that engulfed the continent in 2011. The "energy deficit" has risen to over €14 billion since 2000 (around €300 per Spaniard), and growing fast. The renewables alone cannot be blamed because energy prices in general have increased lately.

Nevertheless, the sharp, poorly planned and executed rise in solar installations in 2007 and 2008 by over 100% each year, took Spanish budget planners and regulators by surprise. So they had to act. The decree that revised the solar rates states, "Energy sources under this special regime constitute a risk for the system's sustainability because of their effects on power prices."

But the risk in the system was created by the system creators—the unreal subsidies and other creative financial instruments that encouraged fast development of solar installations without a long-term vision, let alone an exit plan.

The new changes include a ceiling on the amount of solar installations around the country, and the addition of a registry which will also cover installed wind capacity.

New projects must be entered in the registry and the government will then decide whether a project can be connected to the grid, when, and at what rate.

Italy

Italy's solar industry has been the leader in new installations during the last several years. Then things slowed down, and the uncertainty grew with the political changes, but it was lifted recently by the Gestore dei Servici Energetici (GSE). The authorities and the public were notified that the Italian solar industry has reached the €6 billion benchmark, or 14,3 MW of installed capacity, or the equivalent of 400,000 PV systems a year.

The government was then free to launch the contentious Conto Energia V, which it did in August, 2012. The old Conto Energia IV bill remains in place for awhile, with a corresponding bi-annual reduction of the FiT. The new Conto Energia V bill allocates €500 million per year to renewables, with €200 for PV installations. Once again, roof-top systems are favored, with a special advantage to those systems replacing asbestos roofs. PV products containing materials produced in Europe are also on top of the list of special favors and benefits.

Registration of new PV systems now costs €3/kW, and €2/kW for larger systems. There will be a charge of €0.05 per kW/h generated. Additional costs can be imposed on any system at any time, as needed to ensure grid stability.

PV systems over 12 kW also benefit from simplified registration procedures. Owners of smaller PV systems who are willing to forego the FiT for 20% of the power they generate don't even have to register. Installations on public buildings are another exemption.

There are also four other system-type groups that do not need to register: systems of up to 50 kW that are carried out in combination with the removal of asbestos, systems using concentrator technologies, building integrated installations with innovative characteristics, and systems built by public entities. However, the funding of each of these groups is capped at an additional annual amount of €50 million.

In September 2012, about €256 million of the €700 million reserved for the Conto Energia V had already been reached, so Conto Energia V will most likely phase out by the end of 2012. After the end of the tariff-based funding, some PV investors will still be able to benefit from the possibility of a tax deduction for investments into renewables and from "Scambio sul Posto," a net-metering type of scheme for systems up to 200 kW.

Even with this somewhat good news, it is unlikely that they will save the Italian PV market from going through a further 'shakeout' phase. Many PV installers in Italy have exited the PV industry already or are expected to do so during 2013-2014. This is the case for smaller companies, and those for whom PV had represented the main source of revenues.

Italy had been widely considered for some time as perhaps one of the first large PV markets that could survive, independent of incentives. But any post-incentive market in Italy is likely to be more complex and less efficient than before.

The reliance on rooftop systems, involving self-consumption (in combination with storage), other renewable types, and overall energy management will require companies to develop new business models.

Managing the integration of PV into an overall energy system is another barrier, which makes any move towards full transition to a self-sustaining PV market difficult and unpredictable.

Because in 2011 the Italian PV market generated €2 billion in taxes, the Italian government is trying to support the trend, which helps to revive the economy and improve the budget. The government, however, cannot commit to any additional financing, so it must reduce the bureaucratic procedures that contribute to the increase in residential PV system prices.

The imposition of the register, the quota of available resources and the inadequate time management surrounding the introduction of the decree rule the market, making it accessible to only a few. More red tape and uncertainty could cause a boomerang effect.

The premature announcement of legislative change, for example, has generated a rush to install new PV systems, which increases the amount of incentives paid, and which effectively nullifies the intentions of the fifth energy bill.

It is quite possible that solar power in Italy will continue to grow, but there are still the ongoing economic problems in the EU, which could derail some of the plans.

France

In retrospect, France is the best example and a confirmation of volatile government support. Wind and solar power were at the heart of a big new push by the French government to increase the renewable share of the country's total energy consumption at the beginning of the century, from 6.7% in 2004 to 20% by 2020.

The government set the target of raising its installed capacity for wind power from 810 megawatts (MW) in 2006 to 25,000 MW by 2020. Also, installed capacity for photovoltaic (PV) power is to increase from 32.7 MW in 2006—about 100 times less than Germany—to 3,000 MW by 2020.

In addition, 5 million solar thermal units were to be installed in buildings by 2020, 80% of these in homes.

Biomass accounts for two thirds of all the renewables used in France and hydro power for another third, while solar and wind power still play a marginal role.

"These targets mark a new era in the development of wind and solar power in France, and though they are ambitious, they can be achieved," Jean-Michel Parroufe then-head of the renewable energy division at the French Environment and Energy Management Agency.

The plan would change the structure of France's primary energy consumption—275 million TOE (tons of oil equivalent) in 2006—so that 20% would come from renewables, 25% from nuclear and 55% from fossil fuels by 2020, saving 20 million tons of crude oil.

"From now on, a bigger range of renewable energies, and not just biomass, will help meet the challenge of fighting global warming in France," Parroufe said at the time. "In spite of the growth in the wind and solar sectors, biomass will continue to provide the lion's share of renewables in France even in 2020."

The government has already laid solid foundations for growth in renewables by introducing more favorable feed-in tariffs for electricity from wind and solar power in 2006 as well as tax breaks. As a result of the tax breaks, solar thermal systems grew by 80% in 2006, to reach 210 MW of installed capacity.

Growth in PV installed capacity was 150% in 2006, boosted by a base feed-in tariff (FiT) of 30 cents per kW/h for PV electricity in cities.

With 9.3 million tons of oil equivalent (TOE) in 2006, France was the biggest consumer of fuel wood in Europe after Sweden and Finland; more than 40% of all domestic heating systems in the country used wood as fuel—and the number was growing. However, expanding the use of biomass would require setting up a better network for collecting wood from the country's forests.

Other measures that the French government has announced on the renewable front include huge new investments in renewable energy research, like developing second generation biofuels. To boost the use of biogas, in 2006 the government increased the price by 50% as an incentive for drivers to use cleaner cars, such as electric and hybrid models.

Also, energy performance certificates recording the carbon emissions of new cars became obligatory in May 2006 and financial incentives were introduced to make cars with low carbon emissions more attractive.

President Nicolas Sarkozy announced the new push for more renewables and more energy-efficiency to fight climate change, following a three-month consultation period with representatives from environmental, business and social groups. Cutting carbon emissions would be factored in to all government decisions in the future, including the construction of new buildings and the handling of waste.

In 2008 the FiT was raised to $.65/kW/h and solar installations were on everyone's agenda. This was one of the highest returns for electricity generated by renewables. It put renewables on the profit side of the equation and the excitement was palpable. Alas, the euphoria did not last too long...

Fast forwarding to 2011, France had over 2.8 GW of photovoltaics at the end of the year, more than half, or more than 1.6 GW, installed that same year. This was up from the total of 32.7 MW France had in 2006.

The electricity generated in France by photovoltaics was 1.8 GWh in 2011. The largest solar parks were the 115 MW Toul-Rosieres, and the 67.2 MW Gabardan Solar Park.

Of the 242,295 installations completed by the end of 2011, 0.2% were over 250 kW, and made up 38.6% of the total. 89.1% were 3 kW or less, and made up 20.3% of the total.

February 2012

Effective January 1, 2012, the government changed the feed-in tariff. For residential PV systems, it is down by 4.5% and for all other types of facilities it is down by 9.5%.

April 2012

There were further cuts to the FiT. Residential BIPV applications amounted to 37.4 MW and non-residential applications were 102.4 MW, resulting in a decrease to the FiT of 4.5% and 9.5% respectively. Ground-mounted installations fell to €0.1079 and residential BIPV arrays have been set to a maximum of €0.1934.

In time, the French government released a draft of the decree administering a 10% bonus when manufacturers use modules that are at least 60% of European origin. Following submission to the Supreme Council of Energy, the Union of Renewable Energy and the Liaison Committee for Renewable Energy published its response in the Official Journal before the first round of presidential elections. The changes apply to facilities whose connection request is filed on or after January 1, 2013.

October 2012

The French Energy Regulation Commission has fixed the FiT down to $0.045, and $0.097 for the fourth quarter of 2012, having fallen from $0.058, and $0.123 residential and commercial installations respectively. In some cases this is over twelve-fold decrease in FiT, compared with the Bonanza of 2008-2009. Not 2, or 3 times, but 12 times drop in profits.

A close look at the development in France during the last solar boom-bust cycle, reveals some of the weak points of government subsidized and regulated renewable energy generation. The solar industry is too immature, and is vulnerable to the whims of its benefactors. It simply cannot proceed without outside help, and since the help it receives is equally immature, the solar industry will have to find a better way to survive, if it is ever to step on its own feet.

The United Kingdom

The UK has been trying different things during the latest solar boom-bust cycle. To begin with, its FiT plan is nearly identical to Germany's; yes, that same one that produced woeful amounts of energy, jobs and innovation and which caused the bust in Germany.

Solar power in the UK, and the FiT designed to encourage it, as well as recent results, show unequivocally that this is not the solution for the UK. There are many reasons, but one that is head and shoulders above the rest, is the fact that the UK simply does not have any significant sunlight.

Preventing runaway climate change is a noble task, but it must be done in a way to get the biggest available bang for the buck, instead of going broke. Money spent on ineffective solutions is not just a waste; it's also a lost opportunity.

The new energy law draft, published in May, 2012, aims to restructure the power market and spur $177 billion of investment, which is needed to support aging power stations and to upgrade the grid by 2020.

The energy draft suggests a guarantee for prices for low-carbon electricity through so-called contracts for difference. It would also set up a capacity market where producers are paid for providing back-up supplies when wind, solar and marine power fall short.

Several oil and coal-fired power plants with 14% of the nation's capacity are due to retire by 2015, as part of the plan to retire older plants, and as tight pollution rules force other power plants to close. The government is also seeking to encourage new nuclear stations to replace the current plants due to close by 2030.

Davey said he has "taken on board" investor concerns that they want a "robust" counterparty to guarantee the long-term power prices secured in the contracts for difference. He said the counterparty will probably be a company owned by government.

He also said the department identified "shovel ready" projects that ministers can clear quickly, and that the government intends to remove a "legal ambiguity" in the 1986 laws about natural gas, which could bring

£160 million of investment over eight years.

No doubt the government sees the energy issue as increasingly pressing. The electricity grid regulator is warning that supplies will tighten in the coming years, and many utilities are expected to continue increasing power prices.

Ultimately, faster expansion of solar energy sources doesn't seem to have a high priority in the UK, and most politicians support gas and shale production, which are expected to get some tax breaks and other perks. The bottom line is that solar power cannot be drawn from a constantly cloudy location.

We don't see any plans for developing solar and wind power fields in the North African deserts, which, in our opinion, is the only power source the Londoners will have available to them by the end of the 21st century. There is enough sunlight falling in those deserts to burn to a crisp the entire European continent, and it's there every day. There is enough wind in the wind corridors of those areas to blow Britain's roofs away too.

Summary of EU Solar Developments

Is the sun setting on European solar subsidies? Europe once accounted for about 80% of the world's demand for solar power equipment, but governments from Germany to Portugal have started to remove their guarantees of long-term—artificially high and unsustainable—rewards for anybody who puts solar panels on their roof or invests in an industrial-scale solar energy farm.

Industry experts say those feed-in tariffs, which are usually guaranteed for 20-25 years, have succeeded in their primary aim of kick-starting the solar energy industry by encouraging enough investment and economies of scale to bring down the prices of solar power equipment. In some parts of Europe, such as southern Italy, solar power is almost able to compete in price with fossil-burning electricity for retail consumers, and broad areas of the continent should reach the same point by 2015.

But a sudden policy shift by the UK government clouded plans for grabbing a slice of Europe's solar energy boom, as the UK joined other EU countries in rolling back the consumer-paid subsidies that have fostered the world's most dramatic growth in solar power installations.

The problem for the UK's fledgling solar industry is that the coalition government has backtracked on its commitment to the FITs. A "rethink" of the program was announced just 10 months after the scheme was introduced as a tool to encourage home owners to install solar panels on their roofs.

Under the old solar scheme, anybody who installs solar panels is given a 25-year guaranty to receive up to $0.66 per kW generated, regardless of who uses the power. If, however, the electricity is fed into the system, then the generator gets an extra $0.04 on top of the base pay. These rates are almost 10 times higher than the market price.

Now, the promise is gone, and the government's about-face is exasperating for companies and investors that have ploughed money into preparing for the scheme. This is hardest on companies who have already received planning permission for the development of solar energy parks, especially in the Cornwall province, which has the best sunshine and lowest average income in Britain.

It is hard to believe that government specialists didn't realize how much demand the large FITs would generate, and calculate from there the potential incremental (or in this case exponential) increase of solar installations. So, this is a case of the FITs having been too successful for their own good.

Much more investment was attracted than government experts expected, and suddenly they were face with potentially billions of euros in higher costs that consumers would have to pay. The financial crisis is making the government uncomfortable about raising utility charges.

Reducing the FITs is the only control they have over the size and speed of developing solar plants and with that the amount of the subsidies.

Germany pioneered the scheme with spectacular results with 6.5 GWp of solar energy capacity installed in 2009 alone. This is more than the total capacity of the next two largest solar nations, Spain and the US, put together. This at a cost of $7.6 billion in higher electricity charges for German consumers. This means that 5.6% of the nation's combined electricity bill, or 0.2 percent of its total economic turnover, goes to subsidize solar producers.

One positive development for Germany's expensive solar investment is that it is now well positioned to reach its climate change commitment to generate 18% of its energy from renewable sources by 2020. To date, Germany produces over 10%, while Britain, which has promised to reach 15 percent, has only achieved 2.5%.

France, on the other hand, unexpectedly suspended its FiT scheme for three months in December 2010, and imposed a limit on the solar installations it will subsidize each year. It also introduced a tender system for permitting and developing large-scale solar plants.

The subsidies are actually highest in the Czech Republic, where 7.4% of electricity payments go to subsidize solar power, meaning that 0.37% of GDP is devoted to producing a world-high 3.3% of the nation's electricity. This scheme is also unsustainable.

Industry insiders claim that solar power will on average reach price parity with other forms of electricity in Europe by 2015, and the FITs will fade away shortly thereafter. The problem is that if the tariffs are too low, the market would be small, and if they are too high then a boom and bust follows, as we've seen, in which case the crucial thing is not IF, but WHEN and HOW to cut the tariffs. Germany and Italy have the most reasonable approach to FIT reductions by the implementation of scheduled, gradual cuts, tied to the industry's annual growth. This is also correlated with lowering the price of solar panels and other costs. Carefully planned and executed actions have basically avoided surprises and retrospective changes for long-term investments.

Spain, on the other hand, stands out as an example of how NOT to do it. High demand kept prices up until 2008, but then Spain hit a bump and the prices crashed. But they actually fell to a reasonable and sustainable level, where they should have been to begin with. The government then over-reacted by implementing overnight retroactive changes, and not by reducing the actual tariff rates, but by imposing a cap on the number of hours output you can claim for each year. This move cannot be justified by falling solar module prices, as some sources insist.

While many solar plant operators and owners are now faced with avoiding financial losses on their projects for the next 25 years, many have gone bankrupt due to poor planning and sudden changes in FiTs.

The changing policies have upset many investors, but that will not derail the long-term viability of the industry, according to Europe's leading analysts. Europe's solar industry is going through a certain amount of adjustment, but there is still good growth in the future.

One good development is that in some parts of southern Italy, where retail electricity prices are high and sunlight is plentiful, unsubsidized solar power is already as cheap as carbon-burning electricity— IF solar installation costs are the same as those in Germany.

Although there is no real reason for the costs to be higher than Germany's, pricing in Italy has been value-based instead of cost-based. This means that Italian solar plant owners have been charging high prices simply because they can. These unnaturally high FiTs mean that Italy's solar boom could also hit a sudden downturn.

Even if Italy's FITs were significantly reduced, it already has a viable solar industry, which will continue to thrive, and some other EU countries are not far behind. This excludes parts of Europe (most of northern Europe) where lack of sunshine and cheap electricity means that solar power is not going to reach grid parity soon, if ever.

According to EU industry specialists, since retail electricity prices are much higher than commercial prices, household solar power units will reach price parity first, IF module prices keep coming down. If this trend continues long enough, a point will be reached where solar power simply won't need any subsidies. This is the goal for 21st century Europe.

In all cases, the experts predict that by 2015 there will be parity across a lot of Europe.

Figure 8-4. By 2050 Desertec (and similar projects) in North Africa could provide a large part of the electricity needed to power Europe.

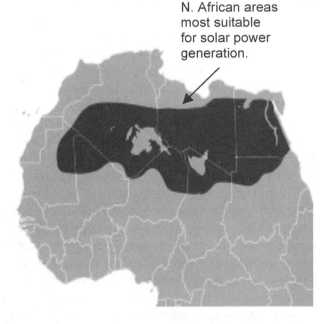

N. African areas most suitable for solar power generation.

We conclude this section by mentioning that, in our expert opinion, Europe just doesn't have enough sunshine and enough open spaces to implement the large-scale solar plants needed to generate large amounts of electricity from the sun. Solar power from roofs and other such distributed energy generators helps, of course, but large-scale power generation—in desert locations—is the key to energy independence. Period. Anything else is a temporary bandage.

Also, growth in the European renewable energy sector, brought on by government policies such as capital grants, subsidies, tax credits and energy production payments, is also a temporary gain that may be damaging the long economic stability of the region. While effective in the short run, the level of subsidization available on renewable power generation cannot be sustained by the current financial environment.

Because of that, we are looking—again—at the planned Desertec and similar projects, where many solar and wind farms installed in the North African deserts will provide electric power to EU countries. The projects will generate many jobs in the area, and will improve the stability in the region. Desertec alone will create 240,000 jobs in Europe, while potentially generating over $2.5 trillion worth of electricity by 2050. This is a significant goal, which cannot be achieved under cloudy skies, or by rooftop installations. It is also the only feasible alternative energy source for the 21st century EU population.

Author's note: The "Desert Industry Initiative" (DII) is a project which plans to use the abundance of sun and wind in the MENA region, thus enabling the creation of a joint power network that will entail more than 90% renewables. Most of the generated power will be transported to the European members. This benefits all parties involved; it provides jobs and cheap power for the locals, while developing an export industry from their excess power with an annual volume worth more than €60 billion.

By importing up to 20 percent of its power from the deserts, Europe could save 30 euros per each MW/h of desert power, according to the experts. This would be an economical strategy for successfully achieving the EU's energy independence and climate protection goals.

With its constant supply of wind and solar energy throughout the year, the MENA region can cover a significant portion of Europe's energy needs. A further benefit of the power network is the enhanced security of supply to all nations concerned. A renewables-based network would lead to mutual reliance among the countries involved, complemented by inexpensive imports from the south and the north.

Desert energy could be a stimulus for economic growth and bring peace to the region as well…if and when all 16 countries and a number of interested parties reach an agreement. Thus far the negotiations have stalled and a number of issues are to be resolved. We do believe that the DII will be a successful undertaking, but it will take much longer to implement, due to technical and political problems.

China

The US is not the only country that has provided extensive government financial support to the solar industry. China has poured billions of dollars into solar companies and projects in an attempt to build a large, world-class solar industry.

The result has been great progress, and that caused rapid increase of manufacturing capability, which led to oversupply followed by ridiculously lower PV modules costs. This also made many heavily subsidized solar companies insolvent, or barely solvent for now, with a grim outlook for the future.

Now there is a vast oversupply of PV materials and products. Thousands of tons of polysilicon, and millions of solar modules are sitting in storage areas. As the supply and demand ratios grow, and the cash reserves dwindle, it is inevitable for China's government-run banks to let some of its solar companies fail.

The government cannot support the current level of losses indefinitely, and the losses are expected to grow in the next 2-3 years. So, the Chinese authorities must decide which manufacturers will fail, and most likely, the victims will be smaller companies with the most short-term debt and the worst financial performance.

Chinese solar manufacturers have varying degrees of short- and long-term debt as well as cash levels, and these vary from day to day, so it is hard to get a good picture of the situation. But at a glance, the financial results of 2011 show: LDK Solar and JA Solar have $1.2 billion and $1.0 billion long-term debt, and $2.2 billion and 7.2 billion short-term debt, respectively.

At the same time, Trina Solar and Yingli Green Energy have $383 million and $0 in long-term debt, and $343 million and $1.1 billion in short-term debt, respectively.

As of the summer of 2012, approximately half of China's PV manufacturing firms have failed. Industry reports show that while 50% of the manufacturers have stopped production altogether, 30% are working at drastically reduced capacity, while the remaining 20% are struggling to maintain their current levels of production. Only some tier-1 solar companies have been able to keep capacity utilization rates at or above 75% during 2011.

Also, Chinese PV manufacturers are facing high charges as a result of environmental contamination (i.e., JinkoSolar's massive fluoride spill and a number of such environmental disasters in China). Steep additional charges have been implemented in an effort to prevent, and/or mitigate, other such accidental environmental damages caused by the fledgling PV manufacturers.

Is the end near for Chinese PV manufacturers, or does this signal a new phase of their development? As US, EU and other world markets are slowly closing to

Chinese PV products, will the Chinese open new markets in Asia, South America, or Africa, or gear up for marketing their products in China proper? We believe the latter is most likely.

As confirmation, the city of Xinyu has paid all $80.0 million of LDK Solar's debt. This is the first time any Chinese government entity has openly paid off the debt of a private solar company.

As a continuation of the government bailout trend, the city of Wuxi, where Suntech is based, has made an emergency $32 million loan to allow the company to keep operating. The local government has also taken additional measures with subsidies and loans to Suntech and other solar cell makers in the area.

The China Syndrome

During the last decade, China's government has made an effort to exercise full control over the national economy—clearly changing the population's focus to material goods.

This led to the implementation of massive dumping policies of cheaply made goods in the US and worldwide. A struggle began in the 1990s, which has led to ruinous competition in the global markets for all kinds of mostly disposable consumer products. This includes the export of wind machines and solar modules, which are made the same way as all other goods, though they are not considered disposable. In many countries the struggle was escalated by pursuing predatory pricing policies, designed to displace local companies even in their own local markets.

Chinese government subsidies provide unlimited capital to competing manufacturers, something most companies—even developed Western countries—do not have access to. So the prices of all energy products first rose and then started falling, going into a free-fall in 2011-2012. The Chinese strategy is quite simple, and consists of 3 key steps:

a. Flood the markets, usually at a loss in the short term,
b. Put the competition out of business in the mid term,
c. Dominate the markets in the long run.

The strategy is working. See chapter 5 of this text for a long list of failed companies in the US and abroad—many due to the Chinese aggressive dumping strategy.

How did this happen, and how did the West allow it? The answer is surprising, and reveals a number of abnormalities, which are leading the solar industry down the wrong path:

Investment

China plans to invest around $39.5 billion in domestic solar power generation over a 5-year period between 2011 and 2015, according to a development plan released by the Chinese National Energy Administration.

As part of China's recently revised 12th Five-Year Plan (2011-15), China will install 10 GW of large-scale PV capacity over the five-year period. It will also add 10 GW of distributed PV power generation as well as 1 GW of solar light and heat power generation systems, helping to create up to 500,000 jobs in the industry.

China offers great potential for large-scale PV systems with a lot of suitable land for PV development. The Qinghai-Tibet Plateau, the Loess Plateau and the Inner Mongolia Plateau have been identified as suitable areas for the development of large-scale PV projects.

The ongoing trade disputes between China, the EU, and the US are complicating the plans, however, shifting the focus to the domestic PV markets. China may see some negative effects at first, but eventually it may benefit by the desire of overseas solar companies to enter the large Chinese solar market.

China is a major producer of PV products, designed mostly for export. In 2010-2011 China accounted for approximately 60% of the global PV cells and modules production capacity. The majority of the installed PV capacity during that time was in Germany, Italy, Spain and Japan.

Power Transmission

China's population is growing in size and wealth, while the old power grid is barely covering 1/3 of the country. In an attempt to catch up with its development, China is building new power transmission lines, some of which are of the world's highest size and capacity. For example, construction on an ultra-high voltage power transmission line designed with the world's largest capacity started in May 2012 in China's far western Xinjiang region.

This is an 800 kV ultra-high voltage, using direct current (UHVDC) transmission lines, which will connect the energy base of Hami prefecture in eastern Xinjiang with the central city of Zhengzhou, Henan in western China. The project is developed by the State Grid Corporation of China, as the main contractor, with many local sub-contractors.

The 1,400-miles-long UHVDC line will transverse the vast areas of Xinjiang, Gansu, Ningxia, Shaanxi, Shanxi, and Henan provinces at a cost of $3.7 billion. It is designed to have a transmission capacity of 8 million kW, when completed sometime in 2014. This is a grand undertaking, which is setting a new world record, and clearly shows that China is capable of being a major player in the world' energy struggle.

Figure 8-5. China power grid upgrade world record

When completed, the new UHVDC line will be capable of transmitting the average of 37.0 billion kWh annually, according to the State Grid Corporation of China spokesman. By using the new and most efficient line, 320,000 tons of SO_2 and 260,000 tons of NO_2 will be prevented from escaping into the atmosphere.

Another, to be built in parallel, 750-kV high voltage DC (HVDC) transmission line will link Xinjiang province with the main network of northwest China was also started in May, 2012. At a cost of $1.5 billion, the new 1,300-miles-long HVDC line will serve as a major route for the transmission of wind and solar power generated in Gansu province to the rest of the country.

The construction of this infrastructure is quite timely, as China is planning to increase the solar and wind generation in the western and north provinces, since it is where the large energy bases are located. The new HVDC and UHVDC transmission lines will be able to transmit the electricity from the energy-rich west to the booming economies of the central and eastern provinces.

In addition to solar and wind power capacities, the western provinces have major coal reserves. There are an estimated 2 trillion tons of coal in the Xinjiang province, some of which is in Hami. Hami is also one of the country's major wind power generation centers.

In any case, China is in need of power generation

and transmission upgrades, because many provinces are suffering from prolonged and worsening power shortages; i.e., in the spring of 2011 a large power shortage swept through most southern and eastern provinces, due to a power gap of approximately 30 million kilowatts of electricity.

To alleviate this unacceptable condition, there are plans (in addition to the above major upgrades) to construct four AC and three DC ultra-high voltage power transmission lines across the country, with an investment exceeding $50 billion during 2012-2013.

China is seriously and decisively entering the 21st century by meeting its energy needs with these major and very expensive projects.

Author's note: Ultra-high voltage power transmission (UHVPT) is a new, unproven, technology, so we must reserve judgment.

The China Connection

In September 2012, China Technology Development Group Corporation (CTDC), a growing clean energy group that provides solar energy products and solutions, entered into cooperation framework agreements with several PV industry players (mostly Chinese) regarding cooperation on PV projects, with a total of 2.123 GW capacity to be deployed in the next three years.

CTDC has signed a total of four cooperation framework agreements as detailed below:

• 973 MW PV projects cooperation framework agreement with GCL-Poly Energy Holdings Limited and China Merchants New Energy Holdings Limited,

• 200 MW PV projects joint development agreement with Hareon Solar Technology Company Limited and China Merchants New Energy Holdings Limited,

• 450 MW PV projects cooperation framework agreement with Znshine PV-Tech Company Limited and China Merchants New Energy Holdings Limited, and

• 500 MW PV projects cooperation framework agreement with Astronergy Company Limited and China Merchants New Energy Holdings Limited.

Within the next 3 years, CTDC will take an intermediary role as a coordinator among different parties for the upcoming PV projects and will be remunerated with a commission fee based on services provided. "These activities are expected to bring more diversified business opportunities to the Company, which is expected to play an important role in global solar project development," according to CTDC spokesman.

As with any international trading arrangement, it is important to ensure that trading and environmental laws are followed, such that the price of solar cells cannot be unfairly manipulated. For example, one country could temporarily "flood" the market with artificially inexpensive products in an attempt to capture market share from local producers, who may be following the rules and thereby making products that are relatively more expensive. This has happened before with China as well as other countries.

China's Solar Water Heating Market

China's "Big Four" solar thermal firms, with their vast distribution networks, are racing ahead of a fast-consolidating solar water heating market. Energy savings and reduced energy costs are the goal, because heating showers with these is far cheaper than using gas or electricity.

Direct marketing is successfully used in this market, where it provides a vital edge in an increasingly fierce tussle for sales. As an example, there are six thermal solar shops in Jinan, a city of six million in the coastal province of Shandong, where each shop sells about 1,600 systems.

A nationwide chain, Linuo Paradigma has 20,000 franchise shops, who are financially independent but must use the manufacturer's logos and marketing material. Solar shops generally sell systems at prices set by the system supplier, ranging from $390 to $940 per heating system. Discounts are set by the supplier's central sales department and further rebates are forbidden. Installation is included in the fixed gross system price and must be covered by the shop owner's margin.

Solar water heating has a compelling sales pitch; solar water heaters are the cheapest way to heat domestic water. According to calculations from the Chinese Solar Thermal Industry Federation (CSTIF), over its lifetime, a solar water heater costs a family 3.5 times less than an electric water heater and 2.6 times less than a gas one.

This straightforward selling technique has enabled the Chinese solar water heater industry to grow to an enormous size. There are estimates that 2,800 companies (1,600 assemblers and 1,200 suppliers) had a combined turnover of $11.5 billion in 2010.

Rapid consolidation is also underway. Recently, the market share for the top 100 brands has rocketed from 40% to 70%. One estimate is that 1,000 solar thermal companies have closed their doors in the last two years, mostly because they did not have access to the fast growing solar project market, and because the sales networks of the larger companies are getting tighter and tighter.

The "Big Four" thermal solar firms are the Sunrain Group, the Linuo Group, Himin Solar and Sangle Solar. Each has a solar thermal business valued above $313 million, and is heavily committed to sales and marketing. Often salespeople outnumber production workers, and in one firm there are over 5,000 sales staff and less than 2,000 in production.

The Chinese solar industry aims to raise its export business more than 12-fold from $20 million to $250 million by 2015. Yet $250 million would be only 2 percent of the industry's estimated overall turnover.

Chinese water heater manufacturers are also exporting their products, which include a wide range of non-pressurized thermosiphon systems for water heating to complex large-scale solar thermal systems generating process heat. The key markets for now are Vietnam, Malaysia, South Korea, Australia and Germany.

The Big Four are faced with cutthroat competition for exports from their compatriots; low-cost Chinese solar thermal system suppliers with lower-quality products. The other obstacle to exports is the rising strength of the Chinese currency on international currency markets.

The Chinese answer to the tighter national and international solar energy markets is to become a service provider for complete micro-emission solutions. To that purpose they are increasing the product range to incorporate various clean energy technologies for customer groups such as hotels, schools and office buildings.

The Chinese are quickly developing their water heating market, so we see it growing very fast in the future, locally and internationally.

China's 5-year Plan

China launched its 1st Five-Year Plan (FYP) in 1953, and five-year plans have been both a blueprint for the immediate future and a showcase of the political economy of the day ever since. Although the gains have been huge, some critics voice concern over the unsustainability of the Chinese model of rapid industrialization and the government's emphasis on GDP growth above every other factor.

The government's prime concern for the latest

12th FYP (covering 2011-2015) is the related question of re-balancing a society that has suffered from a severely uneven distribution of the benefits of growth.

Addressing the root causes of discontent in the distribution of benefits, land rights, health and social security should pay off in security savings. One aspect of the plan that could have profound effects is the ambition to move small farmers into rural conurbations of up to 250,000 people, allowing better delivery of services and more efficient farming.

The new 2011-2015 plan's targets include clear strategies on resource management and environmental protection.

Some details of the 2011-2015 Plan are as follow:

Energy. 16% cut in energy consumed per unit of GDP, 17% cut in carbon emitted per unit of GDP, and a boost in non-fossil fuel energy sources to 11.4% of primary energy consumption (it is currently 8.3%). 40 GW of nuclear and some hydro power will be added, 70 GW of wind power and 5 GW of solar plants are in the plan too.

Total annual energy use in the country will be limited to 4 billion tons coal equivalent by 2015.

Note: Both the nuclear and solar estimates were revised in 2012. The nuclear plant permitting was suspended after the Fukushima disaster until further notice. Solar power will be increased logarithmically to offset some of the nuclear energy losses, and to accommodate the large PV modules surplus of Chinese manufacturers.

Pollution. There is an 8% reduction target for sulphur dioxide and chemical oxygen demand and a 10% reduction target for ammonia nitrogen and nitrogen oxides, the latter of which come mainly from China's dominant coal sector. There will also be a focus on cutting heavy-metal pollution from industry.

Water. Water intensity (water consumed per unit of value added industrial output will be cut by 30% by 2015.

Forestry. Forestry cover will be boosted to 600 million m^3 to cover 21.66 forested lands.

Climate. Carbon taxes and carbon trading "may" be introduced, but no definite decisions have been made yet.

Government investment. The present-day generous subsidies, grants, and loans are projected to increase to 3 trillion yuan. Most of it will go into pollution control.

35,000 km of high-speed rail will be build to connect cities of over 500,000 inhabitants.

Overall, the plan's goal is to achieve average GDP growth of 7% (this is less than the present-day 9-10%. It also will provide for equal distribution of benefits to alleviate the social inequality and protect the environment.

36 million low-income housing units will be constructed, or renovated.

45 million jobs will be created in urban areas, keeping unemployment below 5%. Pension schemes will cover all rural and urban residents.

Population growth will be limited to 1.39 billion people and life expectancy will be increased by one full year.

Seven strategic emerging industries will be supported via tax breaks, and other benefits; biotechnology, new energy, high-end equipment manufacturing, energy conservation and environmental protection, clean-energy vehicles, new materials and next-generation information technologies. These are expected to contribute 8% to the GDP (from 5% today).

So, China is planning to enter the green (r)evolution and get closer to capitalism. We must welcome this development, but keep in mind the lessons we've learned.

Japan

The much-anticipated feed-in tariff (FIT) was launched in Japan in July, 2012, with the goal of accelerating the non-residential power generating segment that previously represented just 10% of the Japanese PV market.

All told, Japan is expected to spend $10 billion, adding over 3 GWp of new solar power installations during 2012-2013. This number is expected to grow by another 5 GWp by 2015-2016.

Keep in mind that the price of new PV installations in Japan is still over $6/Wp, which is twice the rate in other western countries.

We believe this somewhat abnormal situation reflects a combination of the country's desire for quality and its desperation for solar energy.

The new FIT rates are $0.53/kWh (with estimated IRR of 3.2%) and cover 10 years for residential PV installations. They are similarly high, $0.51-0.53/kWh, or estimated 6% IRR for commercial installations, although that covers 20 years of their operation.

Japan's residential power is purchased on "excess generated electricity" basis only, while all of the power produced by the commercial installations will be grid connected and credited to the power source.

The residential IRR is lower because the residential

installations still qualify for up-front 'capacity-based' federal incentives, which prompted the past growth of their residential solar markets.

These attractive rates, however, are implemented on a trial basis and cover only the first three years of the program. This will provide a good ROI for the duration, and will encourage adequate investments in the domestic solar market.

In the spring of 2012, the program consisted of one rate, based on an up-front, capacity-based rebate, providing 48 yen/watt for residential PV systems, if the total installed system cost was less than 600/watt. The program now offers two rates to further stimulate system cost reductions: 35 yen/watt if the installed system cost is less than 475 yen/watt, and 30 yen/watt for systems with installed cost below 550 yen/watt, thus the lower the installed system cost, the higher the rebate rate provided.

Efficiency Preference

The FIT portion (production based incentive) has prompted investors and potential PV owners to consider PV systems that optimize and maximize electricity production. Therefore, domestic module providers have started to offer more efficient modules, such as mono-silicon PV modules. These ultimately produce more electricity than 'standard' rated PV modules and require less space for installation.

Several companies started offering mono cell based c-Si modules, specifically targeted at the domestic 'FiT'-specific solar energy generating Japanese market.

Sharp started mass production of high efficiency mono c-Si modules during 2010 and now promotes this type of c-Si module for residential applications.

Mitsubishi Electric recently announced an intention to produce only mono c-Si modules and to terminate production of multi cell based c-Si modules.

Canadian Solar has also unveiled a higher-efficiency module offering to the Japanese market from July. This module is based on their (metal wrap-through) 'ELPS' c-Si cell technology.

Canadian Solar also announced plans to build a manufacturing plant in Japan as early as FY'2013, potentially becoming the first foreign manufacturer to produce solar panels domestically.

Panasonic Corp. (formerly Sanyo-brand) manufactures PV modules that are acknowledged as among the most efficient mono-based module types today. The "HIT Double" heterojunction-based modules are the most efficient for the non-residential energy markets. They are already sold into the European and US mar-

kets, and are capable of generating electricity simultaneously on the front and back surfaces.

The shift to mono-Si based PV modules is clearly seen when comparing Q1'10, when multi-Si modules accounted for over 70% of production, while in Q1'12 this ratio had effectively been 'reversed', with mono c-Si modules making up almost 70% of the overall c-Si demand.

Process-cost reduction has been the key driver for c-Si manufacturers globally, but now Japan is changing the game, to where new and more efficient technology and product innovation is the priority.

Japan has limited ground space and high electricity consumption (it is #4 globally), so high-efficiency modules will be the energy markets' favorites, this overshadowing the traditional multi-Si based PV technologies.

This preference also indicates the tendency towards higher quality, more reliable, and safer PV modules. This coincides with the deductions made in this text, that conclude that mono-silicon PV modules are the most efficient, reliable, and safe of all PV modules available today.

So Japan is now shaping up as a test bed for the quality, efficiency and safety of the 21st century PV technologies, and it is a good place for such a test, because Japan has:

1. The technical expertise to build, certify, install, and use PV modules,

2. Limited space and higher standards, and

3. The need for lots of reliable and safe energy (after shutting down all nuclear stations).

The combination of these factors, if properly implemented, will make the new programs a win-win for all parties involved.

The only problem we see, in addition to serious lack of ground for solar installations, is their location. Japan is not known for its bright unobstructed sunlight. On the contrary, the coastal and mountain areas, as well as most rural locales are severely deprived of sunlight most of the year. So, how much power can be generated in Japan remains to be seen.

Also, the tenuous weather over most of the country—rain, drizzle, snow, earthquakes, storms—will bring a lot of additional variability and fluctuations in the national power grid. These will worsen with time (as the solar power input increases).

The Japanese don't have much choice in the energy matter, so we are going to see many great things (as is usually the case of people acting under duress), including advancements in the solar technologies manufacturing and applications, coming out of that small island during the 21st century.

Status Quo

Until recently, the Japanese solar PV market was dominated by the 'Big Four' module suppliers: Kyocera, Sharp, Sanyo (now Panasonic), and Mitsubishi. These giants shaped the PV markets in Japan while serving their primary suppliers.

Now, Chinese competitors like Suntech, Canadian Solar, Yingli, Trina, and JA Solar, as well as some prominent Japanese newcomers such as Solar Frontier, are putting up a good fight. Another fight in Japan is that for the residential solar market, since it is almost 5 times larger than the commercial. The country's new feed-in tariff (FIT) and other factors are in its favor, so the competing companies are gearing up to increase their residential sales and services.

Residential systems in 2011 averaged \$6.93/W for retrofits and (\$6.02/Watt for new homes. The major factors contributing to these high prices are due to significant subsidy programs, lack of downstream competition, use of higher-cost, domestically produced panels and BOS materials, preferential, complicated distribution networks, relatively small system sizes, sales of products in kits that include monitoring and LCD screens, the value of the Japanese yen, and finally, high labor costs relative to other countries.

Commercial installations also abide by similar rules, but there are so few such installations, mostly due to lack of suitable large areas in this small country where every square foot is precious. The climate in 80% of country is unsuitable for large solar field installations, where clouds, rain and snow would make solar generation excessively unreliable and variable beyond the control of grid operators.

The Japanese Smart Grid Market

Japan has ten utility companies, all of which have been paying close attention to demand-side management in recent years. The government is also looking into it, and is driving the energy management systems for residential (HEMS) and commercial (BEMS) customers, into the product development and deployment stages.

There are 50 million homes in Japan, so the HEMS market alone would be worth well over \$2.3 billion by 2015, and the BEMS market will be significantly larger. There are, however, a number of factors influencing the rapid evolution of HEMS and BEMS technologies in Japan. Some of these are: Japan's mature transmission grid, the pressing need to reduce energy demand, and an opportunity to become a world leader in a new and very large technology market.

After the Fukushima Daiichi nuclear disaster in March 2011, demand-side management technologies became a need. The lukewarm smart grid market changed overnight into a national priority in the drive to reduce energy demand as much and as quickly as possible. Many major Japanese companies have announced some sort of participation in the HEMS and BEMS products and related pilot studies. Toyota Housing Corp, DENSO, Fujitsu, Panasonic, Sharp, Toshiba, TEPCO, Tokyo Gas, Osaka Gas, Omron, Hitachi, NTT DoCoMo, and many more are fully submerged in the new field and are working hard to win the battle for this new and promising opportunity.

A clear business case exists for reducing load for all customers, including faster return on investments, higher profit margins, larger customers, and a more reliable market. HEMS is a promising market that includes home automation technologies, and Japanese firms are actively pursuing both markets. An AMI meter connects a HEMS or BEMS to the grid; therefore Japan has mandated a rapid smart meter rollout. For example, TEPCO will install 17 million AMI meters in homes over the next five years, and another 10 million AMI meters in buildings over the next 10 years.

The communication infrastructure in Japan is highly developed and is driving the move to widespread use of HEMS and BEMS.

To facilitate productive competition, the Japan Smart Grid community has formed domestic and international standards in HEMS and BEMS technologies, such as ECHONET, which is one example of a communications standard for HEMS.

To speed up HEMS and BEMS development, Japan initiated a series of "smart city" projects. Private companies and government are planning to spend \$1.64 billion on four smart city pilot projects, with HEMS, BEMS, and community EMS (CEMS). These pilot projects are expected to expand the smart city concept across Japan, making it a world leader in HEMS, BEMS, and CEMS.

Toshiba and Panasonic are planning to offer "smart city" packages to cities globally, by incorporating various HEMS and BEMS solutions into their product lines. These plans are then expanded to include exports of these technologies into Asia and the rest of the world.

We will watch and learn from the solar power and smart grid developments in Japan.

Australia

Australia has an estimated 1.5 MW of installed PV power, contributing an estimated 2.5% of total electricity production in the country. This represents a 10-fold increase in the amount of installed PV capacity in Australia between 2009 and 2011, but is still insignificant, compared with the immense solar energy potential of the country. Favorable FiTs and mandatory renewable energy targets, designed to assist renewable energy commercialization in Australia are mostly responsible for the rapid increase.

The first commercial-scale PV power plant opened in October 2012 at Greenough River Solar Farm with a capacity of 10 MW. It's not much, as far as powering the country goes, but a good first step in the right direction.

Australia's recent power issues have an interesting history of events, some of which should not have happened, and some that should have but didn't. Recently, some parts of New South Wales experienced large-scale bushfires—some even a month before the "normal" start of the bushfire season. This climate chaos forced the Australian government to find a way to reduce the emissions of greenhouse gas pollution...at least on paper.

In reality, the opposite is happening. The construction of Australia's first large-scale solar power plant has been put in doubt after the EPC, Solar Systems, was put under voluntary administration. In addition, Solar Systems' manufacturing facility in Melbourne, which had the capacity to build 500 MWp of PV a year has ceased production.

The $420 million, 154 MWp solar power plant that Solar Systems was to construct near the town of Mildura, and which would have been the largest in the world, was also mothballed. The Mildura solar power plant project fell victim to the capitalist system's driving force—the profit motivation—since its solar dish technology wasn't "competitive enough," thus it was not going to bring enough profit to the owners. Instead they decided to cut their losses and get out of the solar game.

Then, the federal government cut the rebate for solar hot water systems, marking the end of another attempt to incentivize households to save on carbon emissions. How much carbon would have been saved, and at what cost, by the renewable energy bonus scheme which paid a $1000 rebate to householders who bought a solar hot water system and $600 for a heat pump, was never satisfactorily established.

The government went ahead with the rebates, raising the expectations of the solar companies that manufacture these systems. The government then pulled the rug out from under them by cancelling the subsidies prematurely and without notice, leaving those companies with considerable unsold stock and unfinished projects.

The government did a similar thing with the home insulation program, which ceased in February 2010, costing taxpayers $1.5 billion. The NSW government also abruptly stopped a solar bonus scheme in April in 2011 that cost another $1.9 billion, citing "policy failure and an extraordinary level of mismanagement."

The list goes on... The Australian Capital Territory (ACT) Labor government's minister for the environment and sustainable development, Simon Corbell announced in May, 2012 that the large-scale solar auction is in its final stage and that the FiT support scheme has shortlisted 22 proposals.

Under the two-stage process, prequalified proponents have been invited to submit final proposals and set out the value of the FiT they need. There were 49 proposals, 15 of which were prequalified to submit offers in June 2012 in the fast-track stream, while those remaining have until early 2013 to complete their bids. After that, the projects will move to stage 2.

In July 2012 the Australian government announced plans to establish AUD$10 billion clean energy agency to assist green energy financing and projects. Moving forward, the Clean Energy Finance Corp. will support large projects, while smaller projects are on their own.

The federal government's mandatory renewable energy target (MRET) is designed to ensure that renewable energy sources obtain at least a 20% share of the electricity supply in Australia by 2020. The MRET will increase from 9,500 gigawatt-hours to 45,000 gigawatt-hours by 2020, and will last until 2030.

The MRET requires wholesale purchasers of electricity (such as electricity retailers or industrial operations) to purchase Renewable Energy Certificates (RECs), created through the generation of electricity from renewable sources. These sources include Wind, Hydro, Landfill Gas and Geothermal, as well as Solar PV and Solar Thermal, providing a stimulus and additional revenue for these technologies.

Australia's dry climate and favorable latitude give it a high potential for solar energy production. Most of the Australian continent receives in excess of 4 kWh m^2 per day of solar insolation during winter months, with a region in the north exceeding 6 kWh/day. This greatly exceeds the average values in Europe, Russia, and most of North America. Only the desert areas of northern and southern Africa, the south western United States and the adjacent area of Mexico, and regions on the Pacific coast of South America can be compared to the Australian deserts.

The problem, however, is that the areas with highest insolation are in the interior, away from population centers and with inadequate infrastructure.

While the future of solar energy in Australia looks bright, there is understandable hesitation in the deployment of solar installations lately.

Saudi Arabia

Solar power in Saudi Arabia is becoming more important by the day for a number of reasons, not the least of which is the fact that in 2011, over 50% of electricity in the country was produced by burning oil.

So, the Saudis announced in May 2012 that the nation would install 41 gigawatts of solar capacity by 2032. It is projected to be composed of 25 gigawatts of solar thermal, and 16 gigawatts of photovoltaics.

At that time there was a measly 3 MWp solar installed in the entire Kingdom. The richest and sunniest place on Earth has less solar power than some of the poorest and cloudiest countries.

Saudi Arabia's first solar power plant was commissioned on October 2, 2011, on Farasan Island. It is a 500 kW fixed-tilt photovoltaic plant, that is expected to generate 864,000 kWh/year. A 200 kW rooftop installation is planned for Riyadh, and is expected to generate 330 MWh/year. 1,100 megawatts of photovoltaics and 900 megawatts of solar thermal (CSP) is expected to be completed by early 2013.

Early in 2012, representatives from the King Abdullah City for Atomic and Renewable Energy (KA-CARE), the government body directing alternative energy development, announced the country's ambitious long-term goals for solar power. There are plans to install 25 GW of concentrated solar power (CSP), and 16 GW PV projects by 2030.

This is in addition to new geothermal, biomass, wind and nuclear plants that are also in the planning stages. These efforts would cost many billions of dollars, with the ultimate goal to produce almost 25% of the Kingdom's electricity from solar power installations in the deserts.

The grand plan is for solar PV projects to supply daytime electricity, while high-capacity CSP plants will provide the rest (the majority) of solar power, including thermal storage facilities for nighttime use. If only a part of this happens, Saudi Arabia will be one of the world's largest solar power producers and users.

Saudi Arabia will be conducting two bidding rounds for the construction of over 5 GW solar power plants, with the first round of bidding to be held in the first quarter of 2013. Another 1.1 GW worth of PV project and 0.9 GW of CSP will be open for bidding as well.

The second round of bidding will take place later on in 2014, with 1.3 GW worth of PV and 1.2 GW of CSP projects. The minimum project size is expected to be around 5 MW. The project offers will be then assessed on qualitative measures such as experience in the development of solar projects as well as suggested price per kWh of produced electricity.

This effort will diversify Saudi Arabia's energy mix, will create new jobs, and will generate green power. As an added bonus, the new solar power fields will make use of useless desert land parcels, and will reduce hydrocarbon emissions in the Kingdom.

All this is part of Saudi Arabia's National Renewable Energy Law to be finalized in 2013, and many world companies are eyeing the Kingdom as an opportunity for the solar market to open up...again.

In 2011 Saudi Electric Company (SEC) inaugurated Saudi Arabia's first solar power plant on Farasan Island. This is a 500 kW Solar Power Plant project that is expandable to 8 MW. It uses CIS Solar Thin Film Technology from Solar Frontier of Japan.

Plant construction started in July 2011 and was completed in August. Showa Shell Sekiyu (SSSKK) built the plant on a Build Own and Transfer (BOT) basis and will transfer it to SEC after 15 years or less. SEC will operate and maintain the project. The plant is directly connected to SEC's distribution system.

This project gives SEC a better understanding of the difference between planned and actual kWh output of the solar power plant which will be used as reference for future projects. The variability is attributed to a number of things, including the type of PV modules used and weather variations. With this experience, the Saudi government will be in a better position to negotiate on the terms and conditions with any solar project proponents for future projects.

However, there are issues that need to be carefully studied before shifting to the use of large-scale solar power which requires connection with the transmission grid. One major hesitation is the reliability (and variability) of solar plants. Although there are new developments that would help resolve (or at least alleviate) some of the problems, the available technologies are still too complex, expensive and unreliable for now, so there are still many questions to be answered. Nevertheless, the government is hopeful that it might be possible to implement some of the most promising technologies in the near future.

An important issue here—even for a rich country like Saudi Arabia—is the financial support needed to construct large-scale solar projects, as well as their operation and management. The Saudis did not get rich by wasting money, so we don't expect them to throw good money after bad product or services. We are confident they would not allow poor quality, unproven PV modules, or shoddy services to flood their energy fields.

Arab Renewable Energy Congress

This growing interest in integrating solar power projects is something that is being seen across the entire MENA region. In 2011 Saudi Electric Company joined CEOs and top level management of other MENA utilities including Egypt's EEHC, Tunisia's STEG, Algeria's Sonelgaz and Jordan's NEPCO. They met at the Arab Renewable Energy Congress in Jordan.

Leading CSP, CPV and PV companies including SunEdison, Belectric, Abengoa, Skyfuel, Amonix and Siemens joined the discussions to further guide the region's energy policies and iron out concerns on how solar power can be successfully integrated into the region's power mix.

The official conference of the Arab Renewable Energy Commission was held Jordan in 2011 by HRH Prince Asem Bin Nayef, H.E. Adnan Badran, former Prime Minister of Jordan & Board Member of the Masdar Institute of Science & Technology. Designed as a business-to-business forum with the ears of the region's governments, the conference developed the key points for officials and stakeholders to act upon to drive forward the sector's renewable energy market.

The present-day economic slowdown and the world's solar industry shakedown are slowing the progress of these programs as well, but we believe they will be renewed in the near future. This will allow Saudi Arabia and the MENA countries to take their rightful place as the world's largest solar energy producers in the 21st century.

Sun is what Saudi Arabia and MENA countries have—even more than sand and oil. As an added incentive, the sun doesn't get depleted, doesn't pollute, and doesn't go up in price, so logic dictates that this is where their future lies.

Only time separates these countries from jumping in and becoming leaders in solar power generation.

India

In 2009 grid-connected solar power in India was a grand total of 10 MW. Yes, the entire country of India had just 10 MW of power. Another 25.1 MW solar power was added in 2010, and an additional 468.3 MW in 2011.

By the end of 2012, installed grid-connected PV installations had increased to over 1 GWp. Now India expects to install an additional 10 GW by 2017, and a total of 20 GW by 2020 at a cost of $19 billion. This represents less than a $1/Wp installed, which is a very difficult, close to impossible, target.

Under this plan, the use of solar-powered equipment and applications would be made compulsory in all government buildings, hospitals and hotels.

Due to its location, India is ranked number one in terms of solar energy production per watt installed, with an insolation of 1,700 to 1,900 kWh/kWp, so we cannot discount India's plans as impossible.

Some large solar projects have been proposed recently, and a 35,000 km^2 area of the Thar Desert has been set aside for solar power projects. This area, if properly developed, is sufficient to generate over 2,100 GW solar power.

According to a 2011 expert report, India is facing a perfect storm of factors that will drive solar photovoltaic (PV) adoption at a furious pace during the next several years, as follows:

- The falling prices of PV panels, mostly from China but also from the U.S. has coincided with the growing cost of grid power in India,

- Government support and ample solar resources have also helped to increase solar adoption, and

- Perhaps the biggest factor has been the need for additional clean energy in India. As a growing economy with a surging middle class, India is facing a severe electricity deficit that often runs between 10 to 13% lower than the average daily need.

Another immediate problem is that Indian banks restrict loans to Indian solar projects, after the recent debacle where the government fined 13 projects for delays and other discrepancies, including one owned by Indian Oil Corp. (IOCL). Too, 13 of 20 additional solar projects in Rajasthan state missed their deadlines, according to the Ministry of New and Renewable Energy.

The delays and penalties are a setback for lenders and may make it more difficult and costly, especially for new and inexperienced solar companies to raise finance.

A shortfall in commercial project finance could stall the solar industry in India, one of the biggest growth markets. And since India will need at least $3.2

billion of debt finance in the next three years to complete already-announced solar projects, this might cause additional, even more serious, delays.

The 13 delayed projects face a total penalty of about $8 million.

According to the rules, the companies posted bank guarantees when they won contracts in 2010, some of which were forfeited in installments during the first three months of delays. Next, a daily fine of 100,000 rupees is levied for another three months. After that, the project is canceled and the company loses its contract.

Recently four solar companies lost their entire deposit and are being fined daily, according to bank officials. Four more companies have gone to court to protest the fines, according to the list. As if this were not enough, there were murky deals in the mix too, so the Rajasthan state suspended some officials who earlier issued false completion certificates to delayed projects.

This is only one of the problems in India. As a matter of fact, the July 2012 issue of *Photovoltaics International* had an article about the solar capital costs in India printed upside down (most likely unintentionally), but that pretty much summarizes the situation there.

It is one of extremes with politics and regulations which are simply not conducive to large-scale solar development. The country's incredible power grid anomaly is only one contributing factor here. Figure 8-6 is a real picture of a cosmopolitan center in a large city that reflects the deficiency—for lack of better words—of India's national power grid.

Figure 8-6. India's power grid

As a confirmation of the serious grid anomalies, in August 2012, over 680 million people in India went for several days without power, due to total power grid failure.

That major power grid failure is just one event of many. The most amazing results from the analysis of this event is that most of the 680,000 people who were left without power didn't even know that there was a big problem.

Most did not notice the massive failure, because they are so accustomed to flipping the switch and having nothing happen. Power is turned off and on several times every day in most areas, and people have learned to rely very little on power coming from the grid.

In fact, most well-to-do people, businesses, hospitals, shopping malls and other enterprises that cannot afford the interruptions, have installed and operate on daily basis diesel or gas power generators. Keep in mind that recent statistics show that India spends over $8 billion every year on diesel purchases, most of which is spent on diesel generators powering residential and commercial enterprises.

So when we speak about new technologies and building new renewable power sources, designed to show the progress of India, we need to first take a close look and figure out the hidden problems—those that are not readily visible nor openly discussed, and those that are obvious. For example, plugging a significant amount of variable solar or wind power sources into the existing mess of wires called a power grid might even make things worse.

Case Study

Uttar Pradesh is one of the six states in India with a solar policy, and just in time, for the state's economic growth is restricted by inadequate power supply. As of June 2012, the state had a monthly peak power deficit of 9.7% and an energy deficit of 15.3%. The renewable energy share in the state is just 5.14% of its total installed capacity.

The additional solar capacity that is to be installed by end of 2017 is a positive step, and it will narrow the supply and demand power gap. Under the Jawaharlal Nehru National Solar Mission (JNNSM), Uttar Pradesh got just one 5 MW project in Phase 1 of the program. This is a drop in the bucket, and will not help the state's problems much. Going forward, Uttar Pradesh's new solar policy is positioned to tap its untouched solar potential, and will hopefully bring much more solar development to the state.

Qualifications

Qualification criteria are quite stringent in the state, letting only the most experienced developers take part in significant projects. Unlike other state policies, Uttar Pradesh has clearly defined targets for the near future, though no capacity has been allocated to FY13. Project size is limited to small (2 to 10 MW), medium (10

MW to 25 MW), and large (25 MW and larger). There is no clarity on proportion between thermal and PV technologies for which the commissioning deadline is set for 12 and 18 months respectively.

No action has been taken in the rooftop category either, which Uttar Pradesh plans to implement through the Ministry of New and Renewable Energy only. Projects will be allocated through tariff-based competitive bidding where power purchase agreements will be signed with a distribution licensee/utility and/or an REC mechanism where a third-party sale will also be allowed.

Lessons Learned

As fragmented as this program might seem, Uttar Pradesh has learned from the mistakes made under JNNSM, and one thing will be done right: it will avoid the possibility of default from employing novice project developers.

Qualification criteria are quite stringent, letting only the most experienced developers take part. Project developers will be required to have successfully commissioned aggregate capacity of 1 MW, 3 MW and 5 MW in last three years (assuming 1 MW costs around INR 10 million and cost incurred is proportional to capacity in MW) when bidding for small, medium and large size projects respectively.

Net worth of INR 30 million per MW (additional INR 20 million per MW beyond 25 MW) and annual turnover of INR 50 million per MW is necessary from any of the last three annual accounts.

The required infrastructure for building solar power capacity in Uttar Pradesh is in the planning stages. To make projects bankable, a Renewable Energy Fund will be created (from fees levied on conventional power generators) for promoting renewable energy such that INR 1000 million from the fund will be available under a "Solar Incentive Scheme" for payment to developers on PPAs signed.

Solar parks will be created by the government for project development (by providing land on a nominal lease basis) and associated manufacturing facilities. Also, necessary common infrastructure will be provided. Uttar Pradesh New and Renewable Energy Development Agency will act as a single clearinghouse for projects making the process hassle free.

Other attractive incentives for developers:

- Exemption from transmission/wheeling and open access charges for third-party sale and captive units,

- Relevant incentives under industrial policy of state will be available for power plants,

- Exemption from electricity duty on energy used by developer for own use,

- Timely provision of evacuation infra by transmission and distribution utility.

Uttar Pradesh's state policy is well planned as far as required incentives, infrastructure and administrative support for solar power development. The policy takes into consideration the mistakes made in other solar policies, such that only experienced and serious players will be invited to participate, prevent project default.

Investments by developers are made risk-free by devising the necessary payment security system, greatly increasing the bankability of the projects. The only thing missing is the rooftop segment, where significant capacity could have been allocated and new aspirants could also take part in solar power development in Uttar Pradesh.

Outside Interferences

India plans to build 22 GW of solar generation capacity by 2022, thus it kick-started the nation's solar manufacturing industry in 2010. So today, India's 50 module makers have a 2 GW manufacturing capacity and its 19 solar cell makers have a 900 MW production capability—a good combination of need and capacity. On paper. Instead, there is bankruptcy, loan restructuring and pleas to the government for support against international competition (from China and the U.S.).

For example, one production line at Indosolar, the country's biggest polysilicon-based solar cell manufacturer, stopped making cells in January 2011. The other was shut down in September. Soon, 80% of India's solar manufacturing capacity was off-line. This was despite all the benefits, plans and promises discussed above. In a power-starved country, and in the midst of implementing its most ambitious solar energy mission, we see most of the national solar energy shutting down plants and sending its workers home.

In 2012 India's solar manufacturers were hit by the bust cycle, and many face bankruptcy and loan restructuring while turning for help to the government.

Part of India's problem—similar to that of all other world manufacturers—is the global solar panel oversupply, which is nearly twice the global demand. But another factor has created a loophole in the India's plans. In its first phase, India's solar mission required silicon-based PV modules to be sourced in India. In the second phase, the requirement was extended to both modules and cells. But they did not institute a domestic sourcing requirement for thin film PV modules, because Moser Baer was India's only thin-film module producer.

So… thin-film PV modules made in the U.S. are much cheaper, and although they are much less efficient, nearly 60% of India's PV projects have chosen thin film PV modules for their projects. In comparison, thin film PV modules have been the choice in only 14% of worldwide PV projects—and nearly 5 times more thin film PV modules are to be used in India? Price? Most likely.

Developers can also get very low-cost thin film modules from the U.S., in addition to very low interest loans from U.S. banks, none of which could be matched in India. The production cost of solar cells and modules is around $1.00 per watt-peak, but U.S. and Chinese firms sell PV modules in India between $0.55 and $0.65 per watt-peak. This competition would be hard to beat.

For example, MEMC, Suntech Power, and CSun modules from the U.S. and China were used at the 214-megawatt Charanka Solar Park in Gujarat. The Moser Baer plants use LDK, Trina and other Chinese panels, and the 40-megawatt Reliance Power project in Rajasthan chose U.S.-based First Solar modules.

This is causing Indosolar, Maharishi Solar, Tata BP and other Indian plants to reduce output, lay off workers and even shut down, threatening the well-being of the entire Indian solar manufacturing industry.

Indian developers are importing from China primarily because of extremely low prices, while importing from First Solar and other U.S. companies, is done because the U.S. Export-Import Bank and the Overseas Private Investment Corporation are providing low-interest loans, as part of a $30 billion Fast Start Financing program to stimulate U.S. manufacturers' exports. India developers got $248.3 million in loans through this program during 2010-2011 and $57.3 million in 2012.

And… we in the US complain about unfair Chinese government subsidies…

The low-interest loans are attached to the mandatory condition that the equipment, solar panels and cells are purchased from U.S. companies. This has distorted the market completely in favor of U.S. companies.

The U.S. ExIm bank's 3.18% loans are extremely seductive to Indian developers, who have to pay at least 14% or more to their own banks.

The Solar Manufacturer Association of India recently filed a federal anti-dumping case alleging that China, Taiwan, Malaysia and the U.S. dump solar equipment in the country; the group is seeking anti-dumping duties on imports. It is also pushing for new domestic sourcing requirements for thin film materials; but, of course, Indian project developers, who benefit from the low prices and cheap loans, are not part of this movement.

India's Future

India's solar industry is well on its way to the 21st century, but there are problems that must be resolved before we can see great progress. Industry experts recommend the following for successful implementation of solar projects in India:

- Public incentive programs as well as private players need independent (experienced) advisors and contracted third-party service providers for support on key decisions,

- Project developers can improve their bankability through verification by an independent professional services partner,

- Collaborating with experienced and strong partners can lower the project risk and make project financing terms more favorable,

- Solar is an increasingly attractive option for decentralized power supply to increase supply security and reduce power costs in India, so more studies should be done to detail the success factors,

- India's power sector supplies inadequate and unstable power to the country, while the power demand is continuously on the rise, so a daily balance must be sought,

- India's spotty track record in building large-scale infrastructure calls into question the entire concept of a centralized power infrastructure, so it must be rebuilt,

- India needs to incorporate measures—perhaps even implement a system of penalties—to improve Renewable Purchase Obligations (RPO) enforcement,

- India should incentivize RPO compliance through REC and increase the REC window,

- India needs to heighten the price stability and increase the bankability of RECs.

India has enormous solar potential, but to get its benefits the country needs more action, less talk, and even less hesitation. There are a number of challenges which must be faced directly and honestly, which can be done only by the introduction of courageous and assertive political decisions.

This is especially the case, since Indian politicians need to allow for more free markets and less political interference in order for the economy to get back on a higher growth path.

India is another country with favorable sunlight, where solar power could be used to supply most of the domestic power needs. The high solar insolation is ideal for using solar power in combination with wind power generation, where India is already a world leader.

TECHNOLOGIES OF THE 21ST CENTURY

In James Bond's movie, "Die Another Day," the bad guys develop a powerful piece of equipment, named appropriately "Icarus," that can capture and concentrate enormous amounts of energy from the sun onto the earth. With the Icarus, they can light dark areas on the earth's surface, or by focusing it on a specific area, they can burn and devastate whatever is at the focal point. The Icarus technology worked amazingly well… like everything else in James Bond movies.

Some of the infamous developments during the last solar energy boom-bust cycle remind us of the bad guys in that movie, and their Icarus solar technology.

Intentionally, or unintentionally, a number of present-day not-so-bad guys came up with plausible (so they seemed) schemes for solar technology that would miraculously solve all the problems and provide incredible benefits to society.

They received millions of dollars (from gullible investors and government bureaucrats) and promised much magic, but none of the claims materialized. Now, those folks are no longer in the solar business, but looking for a new "Icarus" and other unsuspecting victims. There is no panacea here.

We foresee only incremental improvement of the existing technologies. We do not see a new, disruptive solar technology replacing all existing technologies any time soon, if ever.

Instead, we must take a realistic look at the technologies we have, and their capabilities, and find an appropriate place for each, according to its abilities.

Supply and Demand

Supply and demand is a tricky variable, especially where immature business such as PV power generation (as we know it today) is concerned. W can say confidently that the present-day solar business (and PV in particular) is as immature as they come. We insist that it is in its embryonic state, and although it is showing some signs of maturity, these are not a sure indication that a growth trend has even started.

The supply and demand picture is a reflection of the PV industry's immaturity, but represents only one facet of the complex picture of technical, political and financial issues it faces at present. It is exposed to the growing trend of uncertainty, lack of consistency and any type of market stability or controls in the solar industry.

Note the higher levels of (past and future) global supply, vs. demand in Figure 8-7.

We believe that the ongoing world's economic crisis will bring even more serious surprises in this area during 2011-2012, with surplus hitting record levels—unless China and other Asian countries find a way to use the surplus in their own territories.

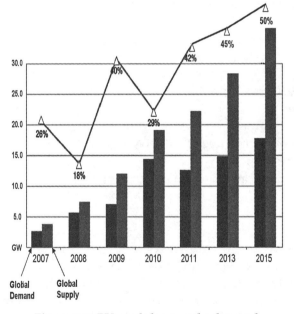

Figure 8-7. PV modules supply-demand

This is a complex picture to be sure, but overwhelmingly with supply consistently much higher than demand. This, and the related price drops, allowed serious growth of PV installations in 2010, where global PV cell and module production peaked at ~24 GW and 20 GW respectively. This represented an annual increase of 110% over the previous year. Thus world PV installations reached 16.2 vs. 7.5 GWp in 2009, or a staggering 146% increase.

This is due to several factors:

1. The strong European FiT markets until recently, mostly in Germany and Italy,

2. Lower PV module prices that have kept pace with subsidy cuts; i.e., Chinese c-Si modules dropped from $2.35/Wp in 2009 to less than $1.50/Wp in 2011, and then to less than $0.50/Wp in 2012, and finally but very importantly,

3. Surplus of polysilicon material, wafers, cells, and modules due to unprecedented expansion of manufacturing capacities, with China and Taiwan leading the pack of low-cost manufacturers which is bringing polysilicon prices down to $20/kg, vs. more than $500 in 2008.

The demand during several periods (especially 2nd and 4th quarters) of 2010, however, challenged the ability of suppliers to keep up despite the significant increase of production capacities in that part of the world. Based on this, we believe that PV installations could have exceeded the 20 GW mark in 2010 if there were no temporary and unpredictable shortages.

The breakdown of PV technologies manufactured in 2010 was approximately as follows:

c-Si modules 19,800 MWp
CdTe modules 1,450 MWp
a-Si modules 1,350 MWp
Super m-Si modules 950 MWp
SIGS/CIS modules 450 MWp

The situation was similar in 2011, but in 2012 the worldwide slow-down reduced the volume, so now the PV module manufacturers have warehouses full of inventory, the size of which is hard to estimate.

Silicon-based technologies were well ahead of the rest, volume-wise, and we believe they will remain so until new, much more efficient technologies make their way onto the energy markets.

China was responsible for nearly 50% of the PV modules manufacturing capacity in 2010 (mostly silicon based), and is planning to increase its share on the world PV market in the years to come.

The picture of significant surplus of PV modules observed in the past will most likely dominate the solar industry in the future. There are several reasons for the unexpectedly slow growth of PV projects in the US, with land permitting, regulatory changes, interconnection and transmission issues contributing heavily.

Another very important factor affecting large-scale PV projects is lack of confidence in the quality, reliability and longevity of the PV modules—especially those coming from newly hatched manufacturers. Potential customers and investors are not convinced of the reliability of these modules and inverters (in particular) for long-term use in large-scale PV installations in the desert.

Lack of confidence in cheaply made PV modules is a main subject of every meeting of PV professionals, installers and investors. Many books have been written on the matter and entire conferences dedicated to it. The results thus far, however, are spotty.

While the numbers of manufactured modules are impressive, the quality of the modules from the various Asian manufacturers is not. How, when, and where did these manufacturers test their products for long-term operation? How do they calculate the reliability and longevity of their modules? Annual power loss, temperature degradation, and failure rates of PV products during 30 years of non-stop operation in the deserts are the best kept secrets of the PV industry.

But not everyone can be misled. A major solar modules and services provider, using high-quality, US-made PV modules issues the following long-term power output warranty: "The System's electrical output during the first ten (10) years of the Lease Term shall not decrease by more than fifteen percent (15%)." This is an average 1.5% annual power loss for US-made modules in 2011.

Keep this in mind as a "best case" practical scenario to be compared when designing or purchasing power-generating systems. Many other, low-cost, manufacturers are making promises (some that sound much better than the above) that cannot be verified, so we can close our eyes and taking the plunge, or we can look closely at what we are buying.

Supply is expected to be high, demand low, and many of the large-scale PV projects which could solve the demand problem are at a standstill. Due to the complexity and high cost of large-scale PV installations, their large size and long duration, there is an obvious need to make sure that the quality of all components in these large fields is acceptable. We need to ensure that modules, inverters, etc. do not degrade excessively or fail during their long exposure to harsh elements.

A standardized and well coordinated team effort is needed to measure these requirements, but it's not readily available, nor easy to do, presently. Still, developers and investors should open the communication channels and start serious discussions with the manufacturers on reliability issues. Plant visits, QC system and documentation inspections, test and other data collection would help to get an idea of a manufacturer's readiness to provide quality product and stand behind it for the duration. But *the best approach would be for manufacturers to participate as part owners in projects using their products.* This would best ensure the long-term success of each project.

Polysilicon Production

The ongoing price wars and never ending expansion of polysilicon production capacity will likely continue to generate an oversupply for the next several years. This in turn will drive polysilicon manufacturers to further innovate processes and reduce operating costs in preparation for an impending wave of new demand.

Subsidies and tax credits have helped spur expansion of production capacity around the world, but most of the expansion was and still is in China. The solar energy industry with its fast production ramp-ups has helped these developments, which have brought prices significantly below the previous record level.

The expansion in production capacity means a silicon and PV modules oversupply for the next few years, with the only way to change the imbalance being potentially spectacular solar success and worldwide expansion, which is unlikely from this vantage point.

Silicon producers are presently using two chemical vapor deposition (CVD) methods; the well established silicon production approach known as the Siemens process, and a fairly new development based on fluidized bed reactors.

For the time being, production-capacity increases appear to have overshot demand, which combined with the entry of new companies into the silicon production industry, will ensure over-capacity of silicon at least for the next several years.

The production capacity increases were centered in Asia, including both major players in the polysilicon production area and smaller companies alike—all rushing to fill the need for more and cheaper silicon.

In the 2009 expansion, the leading silicon producer Hemlock Semiconductor Corp., started operations of a $1.0 billion expansion at its Hemlock, MI, headquarters. That raised its production capacity to over 36,000 metric tons per year (MT/yr). Hemlock was also constructing a new polysilicon facility in Tennessee, where the construction was scheduled to begin in 2012.

Wacker Chemie AG, Germany, was also planning a new polysilicon production site in Tennessee within the next 2-3 years.

The Renewable Energy Corp. (REC), Norway initiated operation of its newly expanded polysilicon plant in Moses Lake, WA.

LDK Solar, China, began operating a 15,000 MT/yr polysilicon production plant in Xinyu, China.

Yes, these were the plans, but most of these companies reduced production, and some stopped it all together during the 2011-2012 solar bust.

Government subsidies and tax credits helped the expansion of production capacity of the Chinese silicon producers, where the overall production capacity was multiplied several fold within a matter of 2-3 years.

The US government was not much behind the race. During the 2009 solar frenzy, the U.S. government handed out $1.0 billion in tax credits to the solar industry (some to foreign companies) supposedly for job creation investments. Hemlock and its majority owner, Dow Chemical Co. received a total of $169 million in tax credits for expansion projects, while Germany's Wacker Chemie, AG received $128.4 million for its new production facility in TN.

The U.S. arm of REC received tax credits worth $155 million for its recent expansion project at Moses Lake, WA.

Prices for purified silicon rose sharply in 2007 and 2008 due to a shortage of the material—in 2008, prices for polysilicon peaked above $500/kg on the spot market. Prices plummeted throughout 2009, however, falling to about $50-55/kg on average by year's end. Prices kept dropping over the next years, although not as precipitously as in 2009, but by 2012 prices hovered slightly over $20/kg.

Although the wild expansion in production capacity was meant as a long-term strategy and was therefore essential to move solar energy toward "grid parity," it actually hurt most manufacturers. There appears little doubt that polysilicon production will be dominated by the solar PV market in years to come. While the need for high-purity silicon in its traditional market—the semiconductor industry—remains, most silicon produced today is destined for the solar photovoltaic (PV) market. By 2008, the size of the market for silicon for the solar energy industry had overtaken that of the electronics industry. In 2010, 70% of the purified polysilicon produced entered the solar market, versus 30% for semiconductors. Now, however, the situation has changed, and maybe even reversed in favor of semiconductor polysilicon.

Polysilicon Production Processes

To meet the needs of a solar-dominated future, high-purity silicon companies are exploring process improvements mainly for two chemical vapor deposition (CVD) approaches—an established production approach known as the Siemens process, and a manufacturing scheme based on fluidized bed (FB) reactors.

It appears likely that improved versions of the two types of processes will be the workhorses of the polysilicon production industry for the near future.

There are presently several ways to produce solar grade polysilicon:

Siemens Process

The Siemens reactor was developed in the late 1950s and has been the dominant production route historically. In 2009, about 80% of the total polysilicon manufactured was made through a Siemens type process. The Siemens approach involves deposition of silicon from a mixture of purified trichlorosilane or silane gas, plus excess hydrogen, onto hairpin-shaped filaments of high-purity polysilicon crystals. Silicon growth occurs inside an insulated "bell jar," which contains the gases. The filaments, which are assembled as electric circuits in series, are heated to the vapor deposition temperature by an external direct current. As the gases enter the bell jar, the high temperature (1,100-1,175 °C) on the surface of the silicon seed filaments, with the help of the hydrogen, causes trichlorosilane to reduce to elemental silicon and deposit as a thin-layer film onto the hot seed filaments. HCl is formed as a by-product.

Temperature control is critical to the process—the temperature of the gas and filaments must be high enough for silicon from the gas to deposit onto the solid surface of the filament, but the temperature cannot be so high that the filament starts to melt. Further, the deposition rate must not be too rapid, or the silicon will not deposit in a uniform, polycrystalline manner, rendering the material of little use for semiconductor and solar applications.

Hemlock Semiconductor has a highly proprietary Siemens-type process that is capable of producing silicon of "11-nines" purity for the semiconductor industry and "9-nines" purity for the solar-PV market.

Fluidized Bed Process

Several companies are developing polysilicon production processes based on fluidized bed (FB) reactors. The FB approach to polysilicon production has its origins in a 1980s-era program sponsored by the U.S. Department of Energy whose goal was to devise less energy-intensive methods for making silicon. FB approaches to polysilicon production offer the ability for continuous production, as opposed to the batch production of the Siemens route. In addition, FB polysilicon reactors consume less energy. REC Solar, for example, says their fluidized-bed polysilicon process consumes only around 10% of the energy required to run a Siemens-type process.

In an FB process, tetrahydrosilane or trichlorosilane and hydrogen gases are continuously introduced to the bottom of the reactor at moderately elevated temperatures and pressures. High-purity silicon particles are inserted from the top and are suspended by the upward flow of gases. When the reactor operates at high temperatures (750°C), the silane gas reduces to elemental silicon which deposits on the surface of the seed particles. As the seed crystals grow, they fall to the bottom of the reactor, where they are removed continuously. To compensate for the removal of silicon granules, fresh seed crystals are injected into the top of the reactor.

MEMC Electronic Materials (St. Peters, Mo.; www.memc.com), a silicon wafer manufacturer, has been producing granular silicon from silane feedstock using a fluidized bed approach for over a decade. Several new facilities will also feature variations of the FB. Wacker Chemie is expected to announce the operation of a fluidized bed reactor facility using trichlorosilane as the working fluid in mid-2010. The plant, located in Burghausen, Germany, is designed specifically to make solar-PV material. Several major players in the polysilicon space, including Wacker and Hemlock, are developing FB processes, while at the same time continuing to produce silicon using the Siemens process as well.

Polysilicon Production 2012

China's polysilicon industry, the biggest supplier to solar-panel manufacturers worldwide, with about 45% of global production capacity of solar grade (SG) purify polysilicon, has idled almost one-third of its production. The plants will be kept idling, or closed down, until prices recover from a recent over 60% plunge.

The price tumble spurred the smallest producers including units of Baoding Tianwei Baobian Electric Co and Dongfang Electric Corp, to halt production and aborted plans for new plant development. The freeze in production will last until polysilicon prices rise to about $47/kg, which is almost twice the current price of SG polysilicon. Below that level, the polysilicon production—especially for smaller producers—is unprofitable.

Polysilicon is expected to average about $25-30

during 2012, and the rebound from polysilicon's price slump could be alleviated (only) by China's increased interest in installing photovoltaic power plants on its own soil. If that happens, then the Chinese producers will double the PV modules production volumes. That would also absorb some of the industry's excess inventory, which initially led to the drop in prices and profits.

The expectation that China will increase installations this year has led some solar companies to keep plants running; i.e., GCL-Poly, LDK Solar Co and Asia Silicon (Qinghai) Co have continued to operate their plants, although some at very low volume.

It is the smaller companies that have struggled the most. Leshan Ledian Tianwei Silicon Science & Technology Co and Xinguang Silicon, units of Baoding Tianwei, halted production in 2011-2012 to reduce losses and operating costs. Dongfang Electric Emei Semiconductor Material Co, a unit of Dongfang Electric Corp, also stopped manufacturing and most of these are on the brink of bankruptcy. Zhejiang Xiecheng Silicon Industry Co filed for bankruptcy in December, 2011, signaling the beginning of the Chinese solar bust.

Prices may be stuck near $30 a kilogram for a year or two, but this may be enough for the bigger companies like LDK and GCL to continue production, while the smaller companies may not be able to wait that long. It is also quite unlikely that the larger companies will seek to acquire the struggling smaller ones. Smaller companies can't operate efficiently, so acquiring them is costlier than expanding the existing large operations.

Future Plans

Polysilicon production is a big business. During 2010-2011 there were grandiose plans to expand and build new facilities, most of which have been curtailed or abandoned. Some of these plans included:

- South Korean chemical firm OCI and Germany's Wacker Chemie are each investing about $1.5 billion to increase production of polysilicon for solar cells. The combined additional annual capacity of 42,000 metric tons will help to meet increasing worldwide demand, experts say.

- OCI plans to invest $1.6 billion in a 20,000-metric-ton plant at its site in Kunsan, Jeollabuk-do, South Korea. In addition, OCI will increase its output at an existing polysilicon plant by 7,000 metric tons. The company forecasts that it will be the world's largest polysilicon supplier in 2013 with a total capacity of 62,000 metric tons.

- Wacker has disclosed details about a planned facility near Cleveland, TN, that the firm first announced in February 2009. To cost $1.5 billion, up from about $1 billion originally, the plant will have capacity for 15,000 metric tons and create 650 jobs when it is completed.

- Tokuyama (Tokyo) has decided to build an additional polysilicon plant at the Samalaju Industrial Park in the state of Sarawak, Malaysia. The second plant will have the capacity to produce 13,800 m.t./ year of polysilicon for solar cells. The project is expected to involve an investment of about Yen 100 billion ($1.24 billion). The plant is expected to become operational in January 2015. Tokuyama's Tokuyama Malaysia (Kuching) subsidiary is currently building a polysilicon facility at the Samalaju Industrial Park with an investment of about Yen 80 billion including infrastructure and utilities. The plant will produce polysilicon for solar cells, and will have a production capacity of 6,200 m.t./year. Operations are due to begin in September, 2013.

- LG Chem will invest $455 million to build the company's first polysilicon plant at Yeosu, Korea. Construction of the plant, which will have a production capacity of 5,000 m.t./year of polysilicon for solar cells, will begin next month, and it is expected to be completed by the end of 2013. LG Chem plans to enter the polysilicon market with its own technology.

- Other Korean companies are also building polysilicon capacity. OCI (Seoul), formerly DC Chemical, announced last month it will invest won1.8 trillion to build a fifth polysilicon plant with a production capacity of 24,000 m.t./year at the Saemangeum Industrial Complex in North Jeolla province, Korea. The plant is expected to be completed by December 2013. The completion of the fifth polysilicon plant in December 2013 will give OCI total polysilicon manufacturing capacity of 86,000 m.t./ year, to become the largest polysilicon producer in the world.

- Hanwha Chemical (Seoul) is moving ahead with its plans to build a polysilicon plant in Korea. The plant will be located at Yeosu and it is due for completion by July 2013. The plant will involve an investment of over won 1.0 trillion and will have a production capacity of 10,000 m.t./year.

Trillions of dollars are slated for these facilities, but since the solar bust, most of these are on hold. We don't know how many of the planned expansions and new construction will resume when the dust settles, but it is obvious that the money and the will are there...if and when the markets allow.

PV Modules Manufacturers' Saga

2011-2012 saw amazing changes in the PV modules manufacturing circles too. A number of companies failed, but most—especially the large ones—kept going, albeit at a slower pace. Free falling panel prices, and never-ending economic havoc pushed solar manufacturing into survival mode, creating a spiral of chaotic and abnormal behavior with some manufacturers.

Industry insiders think that the oversupply of silicon wafers, cells and modules will be here for quite awhile, which will bring the demise of some of the present-day players, since they cannot simply wait for demand to increase.

Manufacturers' capacity is more than twice the actual world demand, so big companies can afford to stick around and even get stronger, while the smaller ones cannot.

A disruptive situation, consisting of several energy disasters that increase the cost of energy is one way to make solar the best, if not only, value proposition. And while waiting for demand to increase, manufacturers are busy improving the quality of their products. That will make the large companies even stronger, simply because they can afford to play this game, and will further accelerate the demise of the weaker companies. Consolidations are most likely one way out for many of them.

Another disruptive way out of the situation is for enterprises and governments to begin efficiency measures with very short paybacks. One of these is the plan to retire power generating assets, to be replaced by natural energy resources. This trend, according to the experts, could start in the middle of 2013 and accelerate thereafter.

The manufacturers' dilemma is quite complex, with the following key factors:

1. Government support, which was the key to success of many manufacturers, is winding down in some areas and is non-existent in others.

2. FiT rates which were high—some exuberantly—are down to below profit-making margins.

3. Supply and demand balance is becoming more im-

balanced, forcing some companies out of business

Note: Everyone blames the Chinese for this anomaly, but bad planning is also a big factor, that some people just won't admit.

4. Wild card markets, such as Germany, Spain and France, where solar activities were going through the roof during the 2006-2010 boom cycle, are dormant now, and with no prospect of revival anytime soon.

5. Gross profit margin for PV manufacturers during the solar boom of 2006-2010 was over 20%. Now it is non-existent, and most companies are forced to sell their product at cost, or at a loss.

6. The latest and final PV products cost reduction cycle is now over, and prices are stabilized at the lowest possible level. The cycle was driven by the price of polysilicon, which is now stable at near-manufacturing costs. Because of that, we don't foresee any further decrease in PV products prices.

What we should blame most, when asking why this happened, is the lack of foresight and planning on the part of the manufacturers. As an example, REC, a large European player, is still alive because of some insightful decisions made in 2008. The management then (in the boom era, when everybody thought that it would last forever) made a not very popular decision to cut costs by 50%. It also invested in its FBR deposition technology, which allows manufacturing of silicon by using 80-90% less energy. The management also made a very tough decision to close some Norwegian facilities. All this was done during the boom cycle, and many people thought this was the wrong direction to take, but now, thanks to these decisions, REC is able to hang in there longer than those who expanded during the boom, only to shut down when the market collapsed.

Another example, JA Solar, diversified during the boom by moving from pure sales of cells to module manufacturing, shipping 2 MW of modules in 2010 and 180 MW in 2011, thus providing a more balanced revenue from cells and modules, with the module business model following closely that of the cell business.

And finally, with another approach, Canadian Solar is moving toward development and vertical integration. Since Canadian Solar has no ability to vertically integrate, they partnered with other companies to achieve that goal. This way they can provide diverse, higher efficiency, cost-effective, and quality products that will meet the needs of future projects.

As a result of the present-day solar stand-by, companies and organizations are re-focusing their efforts and energy on "non-production," but equally important issues that might bring the solar industry to a higher level with time.

Standardization of all segments of the manufacturing process is sorely lacking, so they may have more time and incentive to work on it now. This would ensure better quality and bring costs down, without compromising product differentiation.

End markets are shifting away from Europe and moving to new markets such as China, India, Japan, the U.S., Latin America, Australia, Southeast Asia and elsewhere. It is our prediction that no single business model, or single market, will dominate solar industry development anytime soon.

Adjusting to local demands is a must, and looking at innovative approaches, including new technologies and new financing schemes is key. No paradigm shifts are expected soon, so manufacturers are focused on incrementally improved efficiencies at an increasingly rapid pace, as the key to solar industry survival in the 21st century.

Top Brand PV

This is a new concept, which was thought ridiculous just 2-3 years back. As a sign of the times—and as a new way to separate good from bad product—the EU community has established a new approach to recognizing solar companies, via its "Top Brand PV" award, which is used to describe and distinguish the systematic development of a PV brand (or company) within a target group and its differentiation from the competition. When this is done well, brand management can allow a brand to become functional, relevant and emotionally "charged," and thus to become a true corporate value.

Based on years of experience, EuPD Research has developed a model in which over 30 individual factors are analyzed to quantify and evaluate brand management. With the seal "TOP BRAND PV," the leading market research and opinion research firm EuPD Research has expanded their expertise in the field of certification into the PV sector.

Alongside the leading award in the field of sustainable management—the Corporate Health Award—EuPD Research now offers manufacturers in the PV branch new benefits.

For years, EuPD Research provided companies and market-listed corporations with certificates in the fields of brand management, sustainable management and production standards.

Together with business partners and test institutes, such as the TÜV Süd Life Service, EuPD Research is currently leading the market in the certification of corporate sustainability. With the seal "TOP BRAND PV," manufacturers can distinguish themselves from the competition, appear united in the eyes of end users and strengthen their own brand.

In a competitive market, seals of approval such as the "TOP BRAND PV" seal offer not only benefits for manufacturers, but intermediaries such as installers can also differentiate themselves once they have proven that they are offering high quality brands. This generates extra marketing benefits without additional costs. For end users, the quality seal offers a point of orientation in a highly intricate market. The seal is both easy to understand and recognizable.

As an example, REC received the independent seal of approval "TOP BRAND PV" recently. The Norwegian company was awarded the seal in the three most important European sales markets: Germany, France and Italy.

The basis for the award is built on independent surveys of installers and end users, conducted by the Bonn-based market and opinion research company EuPD Research. REC achieved the best performance in Germany where the company scored the best ranking as a European, non-German brand.

Unfortunately the award did not help REC much, and the solar bust took its toll on them too.

Nonetheless, only a select few leading manufacturers who have undergone a rigorous quality check are allowed to use this seal. Therefore, they belong to a small group of elite brands worldwide who are permitted to display the seal "TOP BRAND PV."

The Ongoing TFPV Industry Saga

The US thin film PV (TFPV) industry went through the recent boom-bust cycle, taking the deadly curves with mind-boggling speed and aptitude. It started with the polysilicon bottleneck during 2004-2009, when thin-film PV seemed to be the only alternative. Shipments grew from 68 MW to 2 GW in 3 to 4 short years.

By the end of 2009, thin film PV modules production reached nearly 10% percent of the total solar modules market and was growing by the day.

Then polysilicon prices rose to $450 per kilogram, which gave new wings to the TFPV industry and thin-film solar startups popped up like mushrooms after a pouring rain.

Thin film manufacturers quickly and smartly consumed $3 billion of private investment, with $1.25 billion during the peak of the TFPV frenzy in 2008 alone.

Silicon Valley CIGS startups were the winners. Solyndra, Nanosolar, and MiaSolé were the first to benefit from the chaos the frenzy brought upon unsuspecting investors.

Seeking to generate jobs, the Administration opened the flood gates to public debt financing for solar manufacturers, via the hastily put together Department of Energy (DOE) Loan Guarantee Program.

Solyndra wasted $535 million in constructing new buildings, which they did not need, and buying custom-designed equipment.

DOE threw $1.3 billion at solar manufacturers during the frenzy years, most of which (nearly $1.1 billion) was given to thin-film solar PV cells and modules manufacturers.

While TFPV production grew to 3.7 GW in 2011, the c-Si prices went way down and dominated the markets. Then, as China's c-Si module prices kept falling, the oversupply increased and the European markets shut down one by one; the TFPV industry saw itself on the wrong end of things. Low efficiency and high prices were working against it.

Solyndra made the splash, declaring bankruptcy and flushing $535 million in tax payers' money down the drain. Abound Solar ceased production and retooled for a recovering demand market. First Solar started shaving expenses by eliminating its CIGS research division, ceasing construction of the new Arizona facility, scrapping plans for production in Vietnam, closing its Germany facility, and idling Malaysian plants.

The bad news for First Solar didn't stop there. They had a $246 million loss in returns, rejects and recalls, mostly due to faulty production and the fact that their PV modules cannot withstand the harsh desert climate. Then couple of their large projects (Antelope Valley and Desert Sunlight) were stopped due to improper certification of their PV modules.

And there are a number of law suits filed against First Solar presently as well, all of which brought their stock shares from $300 a few years ago to $14 in the summer of 2012.

Falling c-Si module prices from Chinese module manufacturers were blamed for the bad times, and there is a chance they will go even lower. So the question is how will the TFPV industry respond in the short term, and will it be able to survive in the long run?

We see the game changing with the GE's acquisition of Primestar in the spring of 2012, and Tokyo Electron purchasing Oerlikon Solar. We also see Asian investors looking into thin-film PV as the only substitute to c-Si which they don't have.

Hyundai Heavy Industries JV'd with Saint-Gobain's Avancis division in Korea in a 100-megawatt TFPV production plant.

And there are AVACO and TSMC's plans to use Stion's technology. SK invested $50 million into HelioVolt, TFG Radiant's bought 39% ownership in the potential $275 million Ascent Solar production, and Mitsubishi Heavy Industries' entering a MOU with Taiwan Oerlikon Micromorph® customer Auria Solar.

Additional TFPV (CIGS mostly) research and development efforts are reported at LG, AUO and Samsung (who is planning a move in the space applications with MiaSolé as a potential partner.

Thin Films Production Levels

Thin film companies include First Solar, Stion, NuvoSun, SoloPower, ISET, HelioVolt, Ascent, and AQT. The price wars are turning the industry upside down, with prices nearing $0.50 per Watt, and most of the discussions and negotiations leading to LCOE.

First Solar leads both in volume and manufacturing cost, so most of its thin film competitors still must prove they can compete with the likes of First Solar and Chinese manufacturers like Yingli. The competition is on the performance metrics, while making a sustainable margin as panel prices continue to drop.

Bankability is another issue, and is a difficult case to make for new companies with only a few short years in existence.

The thin film PV modules production numbers in 2011 are shown in Table 8-10.

Table 8-10. Thin film producers 2011-2012

Company	2011	2012
First Solar	2000 MW	1,500 MW
Solar Frontier	577	650
Solibro	95	50
MiaSolé	60	75
Solyndra	40	—
Avancis,	25	100
Global Solar	19	60
Soltecture	14	—
Nanosolar	10	15

Note: As a reference point, the Chinese giants Suntech and Yingli produced respectively 1,866 MW and 1,554 MW PV modules during 2011.

Table 8-11. TF Markets 2010-2020 (estimate)

MWp	010	2012	2014	2016	2020
a-Si	1000	600	600	1000	2500
SIGS	500	500	1000	2500	5000
CdTe	1000	1500	1500	1500	1000

The bad news is that most, if not all, thin film PV modules manufacturing firms are in trouble to one degree or another. Solibro changed ownership; Soltecture declared insolvency. Solyndra... Nanosolar is in a similar state as MiaSolé, boasting some technology advances, but unable to sustain full-scale manufacturing operations without outside funding or a new partner.

Additional indicators point to further erosion in the thin film PV modules manufacturing sector, so we would not be surprised if some of these companies follow Solyndra's path into oblivion.

Manufacturing Cost and Materials Availability

The equipment manufacturing cost for CSP and wind power generation is going up and down according to world commodities values, but the excursions are not drastic, so with that in mind we can safely conclude that their prices won't change much in the future.

The situation is more complex and totally unpredictable with the c-Si and thin film PV technologies. Cost (c-Si raw material) and availability (thin film exotic materials such as cadmium, tellurium, indium, arsenic, etc.) are variables that can change overnight, but for now, and up to 2015, we see the scenario in Table 8-12 unfolding:

CdTe

CdTe PV modules manufacturers and projects have been in the news since 2008, and recently CdTe power installations and large-scale power fields are covering many acres of desert land in the US.

The sharp drop in c-Si PV module prices lately has forced some CdTe manufacturers to change strategy, but they are holding tightly to their plans to develop many utility-scale power plants, because they see it as the only way to survive these downtimes.

But there are problems with the plans for large-scale deployment of CdTe PV modules, especially large-scale deployment of these in the US deserts:

1. Significant amounts of the rare and exotic cadmium and tellurium metals are used in the CdTe modules. Their prices will go up with time, which

Table 8-12. Projected manufacturing cost and materials availability, 2013-2015

Company	Projected cost $/W		Materials availability
Yinghli Solar	0.60 and up		Unlimited
Trina Solar	0.65	"	"
Renesola	0.65	"	"
Jinko Solar	0.65	"	"
JA Solar0.65		"	"
Suntech Power	0.65	"	"
Candian Solar	0.65	"	"
Aide Solar	0.65	"	"
TBEA SunOasis	0.65	"	"
CdTePV mfgrs	0.75	"	Limited
CIGS PV mfgrs	0.85	"	Limited
Organic PV cells*	0.45	"	Unlimited
Nano PV cells*	0.35	"	Unlimited

*Note: The questions of low efficiency and unproven reliability of these technologies remain and need to be solved before they can compete with their more mature cousins. With good reason, however, we do expect their market share to increase substantially in the very near future.

makes the manufacturing of CdTe solar modules unsustainable.

2. Cadmium and tellurium are dug out of the world's dirtiest and most dangerous mines, where many people—including children—get sick and die every day. As many such cases, this is a well kept secret.

3. Cadmium is one of the 6 most toxic materials on earth and a carcinogenic heavy metal, which causes serious damage, including cancer, to any life form.

4. Each CdTe PV module contains 8-9 grams of cadmium poison, and since millions of CdTe PV modules are used on thousands of acres of desert land, CdTe technology must be classified as "toxic" and extremely dangerous.

5. CdTe PV modules are not proven efficient in desert operation, where the intended large-scale fields are located. The modules' efficiency drops significantly with temperature increase, and the overall output drops with time too, due to internal module changes.

6. Since CdTe PV technology is relatively new, the CdTe PV modules have not been tested in long-

term desert trials, so we might be looking at huge CdTe waste fields, 10-15 years down the road, with millions of modules, covering thousands of acres of public land rendered useless and/or toxic.

7. CdTe PV modules are not proven safe for long-term desert operation, where the intended large-scale fields are located. The 2-pane glass modules are "glued" together with plastic compounds, which would disintegrate in time—like anything left in the desert.

When that happens, air and moisture easily penetrate into the modules and attack the fragile CdTe/CdS thin films. These then disintegrate mechanically and decompose chemically, causing solid or liquid leakage and/or outgassing of toxic cadmium materials, which could contaminate the air, soil and water table on thousands of acres.

8. A number of cases of defective production have been reported, followed by several field failures and product returns and recalls in 2010 and 2011. This on/off reliability of the CdTe manufacturing process puts the integrity of the CdTe PV modules in doubt, especially these destined for long-term desert use.

9. It was discovered recently that a large CdTe manufacturer has been installing CdTe PV modules without proper UL certification of their electric connectors. That is unacceptable when dealing with anything electrical in the US.

Note: The abnormality was discovered by a lowly safety inspector. That forced LA county officials to stop one of the largest CdTe projects in the world. The project was on hold for several months, but county officials allowed the installation to proceed without certification. Now, millions of CdTe modules are operating without proper UL certification, so they would be responsible for any problems, such as fire, arising from the uncertified modules. Fire in a large-scale power CdTe field, accompanied by toxic cadmium fumes or leaching into the soil, means long-term devastation of the environment in the affected locale.

10. US CdTe PV modules manufacturers obtained over $5.0 billion in US government loan guarantees, grants, subsidies and other incentives in the recent past. This puts them in the different category of being accountable to US taxpayers.

In our professional opinion, CdTe thin film PV technology is NOT renewable (limited supply of Cd and Te), is not clean (a lot of dirty work goes in Cd and Te mining), and is not safe (due to toxic and carcinogenic Cd and Te materials content). Too, its efficiency and reliability during 30 years of desert operation are still unproven.

We believe that the installation and operation of CdTe thin film PV modules must be limited to moderate climate regions—until proven reliable and safe for use in the deserts. It is too late for thousands of acres of U.S. desert lands that are covered with millions CdTe PV modules, so we must monitor these fields carefully. If problems occur, we must hold the principles responsible for their remediation.

PV MANUFACTURERS' FUTURE

At least half of the world's existing photovoltaic manufacturers have been living under the threat of complete solar industry collapse. The fear is that if things don't improve significantly, many of them will go bankrupt, or be taken over by competitors.

There are three key trends in the industry: financing innovations; residential grid parity; and the trading of large project portfolios, that are keeping some of the affected manufacturers and other related companies alive. A number of companies have been forced to announce insolvency, including some of the large players, such as Germany's Q.Cells.

The renewable energy country attractiveness indices, issued by industry watchdogs state that although PV module prices are way down, and have stabilized since 2011, current costs of multicrystalline silicon modules manufacturing are still above US$1.0/Watt. The funny thing here is that the actual sales quotes have been as low as $0.55-0.65/Watt. This means that somebody is losing money, which is an indication of a temporary, unsustainable situation, which has already led to a number of insolvencies.

This is shaping the way of the near-term future, 2013-2014, when at least half of the world's manufacturers will go bankrupt, or consolidate, either through bankruptcy or acquisition.

China is leading this trend, with over half of its solar products manufacturers going out of business, or switching to other products. This is simply because they are too deep into it and have no way out.

One way out of it for some of them, being forced by the U.S.-China trade war, is to acquire smaller Western

Table 8-13. Number of PV module manufacturers and vendors worldwide, 2006-2020

		manufacturers worldwide
2006	~50	"
2008	~80	"
2010	~175	"
2011	~210	"
2012	~160	"
2014e	~100	"
2016e	~80	"
2020e	~50	"

companies, which will allow them access to the US and Western energy markets.

We are also expecting the creation of new and innovative financing models, as well as increase in the buying, bundling and de-risking of large portfolios of solar assets. These could be floated as traded bonds, and access pools of capital that can only make liquid investments, due to increased investor understanding (or is it?).

Warren Buffett's MidAmerican Holdings sold $850 million worth of bonds to finance the Topaz Solar Farm in California, which it bought from First Solar for $2.4 billion.

The 1.0 GW Blythe project, under development by the now insolvent Solar Trust of America (Solar Millenium/Solarhybrid), was also sold to the highest bidder. This project also uses potentially toxic CdTe PV modules, so we just have to see how it goes.

One good bit of news here is that in some countries PV cost is getting close to residential grid parity—German PV prices of systems below 100 kW averaged €2.43/W in Q4, 2011. Denmark, Italy, Spain, Hawaii and parts of Australia can make over 6% profit when investing in solar systems, if they can use every kWh generated to replace the one they would have had to buy. This is real grid parity at the residential level, or 'socket parity', and it is happening today. Good news!

Large-scale PV projects, however, cannot work without subsidies, at least for now; although, there are places (Spain and Mexico) where FiT-less system is under trial, some showing good results. It might work in the long run, but the economics here are tight and will require excellent product and execution from the get go. Especially in the beginning stages.

We do, however, see this as the opportunity of the 21st century, so we are absolutely sure that this type of project trial will increase with time. As the need for cheap renewable energy and clean environment grows, so will the importance of using large-scale solar power without government subsidies and/or FiTs. Under capitalism, if there is a need there is a way, so we know that solar energy will find the way sooner or later.

Today, while the long-term prospects for the solar energy industry look bleak, the short-term future looks quite positive. Industry insiders claim that clean energy investment levels in 2012 were the weakest since the depths of the financial crisis in 2008-2009.

One of the most disturbing happenings today is the ever-widening gap between developed and emerging markets. While the West is suffering from policy abnormalities, developing nations are looking into implementing new strategies and incentive schemes. And the gap between these two situations is growing.

The recent financial challenges reduced European and US policy support, while Asian competition, decreasing carbon prices, and the U.S. oil and shale gas "boom" created a bleak short- to medium-term outlook for the solar energy sector.

Nevertheless, we, and future generations, have no choice. The wind and solar technologies will mature and move steadily towards grid parity. We have no doubt that the renewable energy sector will grow and win the battle with the conventional fuels in the long run.

The Unlimited Warranty

The unlimited PV modules warranty is one of the new solutions to the PV markets woes at least in the long run. It was also thought impossible and ridiculous some 2-3 years back. An example of it is SunPower's launch of the industry's first "combined warranty" that covers both power and product for 25 years.

Please note that this is complete coverage of the product's workmanship and performance, efficiency, reliability and total power generation. This is in addition to a "linear power warranty" that offers the promise of the lowest power degradation over the life of the system.

This is the only straightforward and complete 25-year total, global product warranty, which is 15 to 20 years longer than conventional solar warranties, offered by most solar companies in 2012. The PV modules are guaranteed to produce at least 95% power for the first five years, followed by a maximum degradation rate of 0.4% each year thereafter. This results in an unprecedented 87% power level at the end of 25 years, warranting a total of 9.1% more energy than the industry standard over the first 25 years of the solar panel life.

The warranty, therefore, covers all aspects of the

proper performance of the PV modules, with no hidden angles or catches—including a guarantee to produce at least 87% of the initial power during year #25. If a solar panel needs to be repaired or replaced for the duration of the warranty, an identical or functionally equivalent panel will be provided, or the cost of the panel will be fully refunded. The related labor and transportation costs for customers whose solar panels were installed by SunPower dealer/partners are covered as well. If the panels were installed by someone else, then different conditions apply. In such a case, the panels will be replaced, but the cost of labor and transport may or may not be covered—depending on additional contractual conditions.

The new warranty is also written in simple terms that detail the coverage, and the company stands behind its promises. According to company spokesmen: "SunPower's 25-year combined power and product warranty is possible due to our proven, differentiated Maxeon solar cell technology. These cells make our high-efficiency panels incredibly durable, allowing SunPower to offer our customers this first-of-its-kind warranty increasing performance with more energy delivery. SunPower is standing behind its products with this new easy-to-understand warranty, which serves our customers well for any issues that may arise."

The new warranty went into effect on July 1, 2012 and is available to all residential, commercial and utility-scale power plant customers installing SunPower solar panels. This is the direction the solar industry is headed in. Like a long-term unconditional warranty comes with buying a reliable car, so it will with solar products.

In our opinion, this new unprecedented and very much needed warranty paves the road to success for the world solar industry.

To Consolidate, or Not to Consolidate?

A number of industry specialists estimate global PV supply to hover in excess of demand by 30-40 GWp throughout 2015. This means more closures, more acquisitions, with an estimate of over 150 additional PV modules manufacturers to either expire or succumb to acquisition during the same time.

The largest number of firms exiting the markets, will be those operating in the high-cost manufacturing markets in the U.S., Europe, Japan, and Canada. In our estimate, over 75 firms will become victims of the new situation in the near future (2013-2014).

The manufacturing costs in these countries is about $0.80/Watt, while the actual cost for most Chinese made modules is between 45 cents and 65 cents per watt.

The writing is on the wall for the high-cost companies: reduce price, submit to acquisition and get what you can, or get out of the market.

One third of the doomed-to-fail (estimated 150) companies are small Chinese manufacturers. Most of them are so-called "solar zombies," with manufacturing capacities less than 300 MWp, who have come to a hasty existence with the help of government support, were never successful, and continue operating at a loss.

The number of these unfortunate enterprises could go much higher, if the Chinese government doesn't step in and start building large PV power fields on its territory. If it does, however, this will surely prop up many domestic suppliers, small and large, and might even result in the creation of a very strong PV industry. Strong not only in numbers and cost, but also in quality.

China announced in the summer of 2012 that it plans to increase its cumulative 2015 solar installations from 15 GW to 21 GW, which will provide captive demand for many firms that are struggling, thus buying them more time to recover and establish a reputation.

The Chinese government will continue to provide financial support to some of the more established, large-work-force, manufacturers, which move is justified by the need to cover near-term debt obligations. This possibility was evidenced by the municipal loan to LDK Solar in July 2012, where the local government put several hundred million dollars in LDK to save it from insolvency. The China Development Bank has also renewed its pledge to support 12 selected domestic suppliers, so it seems like China is not giving up yet, and is moving the solar game on the local turf.

The large Chinese manufacturers, Trina Solar, Yingli Green Energy, Suntech Power, JA Solar, Jinko Solar and Renesola, who make up more than 20% of existing global module capacity, are the most likely beneficiaries of China's new politics. The financial support might encourage some of the large solar companies, as well as non-solar Chinese industrial conglomerates, to get into solar by acquiring some of the smaller companies.

Nevertheless, the consolidation in the PV manufacturing sector of late has not relieved the ongoing problem of overcapacity. This is because, although the PV module manufacturers have some control over the supply side, they have very little control over the demand side, thus the great supply and demand discrepancy exists.

There are no indications that the world economic situation will improve soon, so unless China creates an internal solar market, the supply and demand dilemma will continue at least until 2014. We foresee a more stable balance between supply and demand after that, which will help a select group of manufacturers to generate profits and stay in the game longer.

The overwhelming indications, however, point to a number of uncertainties, which could derail expectations and predictions. One thing remains certain: unless something drastic happens, over 50% of the existing PV components manufacturers will be forced to exit the market one way or another.

Table 8-14. The survivors…2015 and beyond.

Company	Country
Canadian Solar	China
First Solar	U.S.A.
Hanwha Solar	China
JA Solar	China
Jinko Solar	China
SunPower	China
Talesun	China
Trina Solar	China
Yingli Green Energy	China

Table 8-15. c-Si manufacturers who might be forced into consolidation or…by 2015.

Company	Country	MW/y
SolarWorld	US, Germany	950
Siliken	Spain	280
Conergy	Germany	250
Solaria Energia	Spain	250
Bosch Solar	Germany	190
Isofoton	Spain	140
Photowatt	Canada, France	130
Hevel Solar	Russia	130
Solar Photonics	U.S.	100
Helios	U.S., Italy	90
Martifer Solar	Portugal	50
OpSun	Canada	40
Mage Solar	U.S.	40
Motech	U.S.	40
1SolTech	U.S.	35

The companies listed in Table 8-15 are, in our opinion the PV manufacturers that would be able to survive the downturn of the latest solar cycle. Some of those might close, or be forced into consolidation, or some other exit strategy. But the uncertainty is great.

Amazingly, almost everywhere we look in the US, we see Chinese products. Solar is no exception. It is on our roofs, in our gardens, and in our large-scale power plants. In search of energy independence from Arab oil, we've jumped into Chinese solar products…

The situation is rather dangerous, so several US and EU firms are taking the matter to a higher level.

Anti-dumping Action, 2012

One of the most significant events of the last solar boom-bust cycle was the action taken by several US and EU companies against the low-cost Chinese imports. A complaint filed by several US companies in 2011, was decided in their favor by the US Commerce Department. The commotion generated by the decision is quite significant, and it will be a long time before the dust settles.

Chinese manufacturers, and by extension their government (with its huge subsidies), are the greatest losers, because they now have to pay additional taxes, which jeopardizes their profit base. This new tax is actually on top of the existing 5-6% countervailing duties and taxes they are already paying on their exports.

After more than six months' investigation and deliberation, the US Department of Commerce (DOC) issued a final ruling in October, 2012 that there will be no change of the scope, as previously determined, except that small consumer goods will now be excluded.

The final Department of Commerce ruling in October, 2012 was:

- There will be no change of scope (some suggested changes were denied).

- DoC did not agree to SolarWorld's petition to include China module assembly.

- Small consumer goods are now excluded.

- The effective rate on Suntech has gone up a few percentage points, down 4% for Yingli and Canadian Solar, and with Trina down 12% from the preliminary numbers.

- Commerce recommended anti-subsidy duty percentages of 14.78% for imports made by Suntech, 15.97% by Trina Solar, and 15.24% for all other Chinese manufacturers.

- Critical circumstances are in effect, save for Suntech's anti-dumping tariffs.

The final DOC decision did not significantly increase the tariff from its preliminary decision in May, 2012. It basically determined that crystalline silicon solar (PV) cells produced in China, whether or not assembled into modules, would be subject to anti-dumping

Table 8-16. The new DOC import duty on Chinese solar cells.

	AD Prelim	CVD Prelim	Total Prelim		AD Final	Less Export Subsidy	AD Final Adjusted	CVD Final	Total Final
Trina	31.14	4.73	35.87		18.32	10.54	7.78	15.97	23.75
Suntech	31.22	2.9	34.12		31.73	10.54	21.19	14.78	35.97
Separate Rate	31.18	3.61	34.79		25.96	10.54	15.42	15.24	30.66
Countrywide Rate	249.96	3.61	253.57		249.96	10.54	239.42	15.24	254.66

and countervailing duties.

So, here is the loophole: PV modules assembled in China from cells produced in third countries do not fall within the scope of the duties. As an example, and based on the new ruling, PV cells produced in China by JA Solar will be subject to an anti-dumping duty of 25.96%, and countervailing duty of 15.24%.

Please note that this duty affects only solar cell made in China, so this means that solar cells made anywhere else and assembled into PV modules in China will be exempt. So, there is a big loophole in this ruling.

But the case is not yet closed. The International Trade Commission will make its decision known later on, and a new wave of discussions and negotiations will most likely ensue. More time will pass before a final decision and ruling is made, which then might affect the US DOC ruling, and a catch 22 situation could propagate from there as well.

With all this going on, concerns about the growing global trade war are growing too. The new US DOC ruling, as well as the pending one from ITC might hurt American and other solar industry jobs, growth, and consumers as well. Success in the long run will be determined by American companies, making solar products for America.

The tax margins are high, but not as high as in many previous U.S.-China anti-dumping cases, as with electric blankets, steel grating, tires, and other products flooding US markets in the past.

Nevertheless, in today's race for lower prices, the new taxes are certainly much higher than the Chinese manufacturers were expecting. This brings their solar products' prices on par with those of US modules manufacturers.

Still, PV module prices in the U.S. will increase in the short term, and—at least for awhile—the demand and installation growth of the solar industry will be dampened even further.

The US DoC ruling, however, is a preliminary decision, so Chinese manufacturers and CASE representatives will surely contest the findings in the days ahead—

especially since the decision was driven in part by the fact that China is considered a "non-market economy."

The Chinese have two choices:

1. Manufacture the solar products outside China, or

2. Use other Asia-based suppliers to make solar cells (increasing the cost by approximately 10%), and assemble the modules in China.

In both cases, Chinese manufacturers can bypass import tariffs.

Basically, the anti-dumping decision is positive, creating a temporary boost in the US PV industry. The comments of some industry leaders in the wake of the DoC decision are quite revealing:

Suntech, Andrew Beebe, Chief Commercial Officer, "These duties do not reflect the reality of a highly-competitive global solar industry. Suntech has consistently maintained a positive gross margin as revenues are higher than our cost of production. We will work closely with the Department of Commerce prior to their final decision to demonstrate why these duties are not justified by fact. As a global company with global supply chains and manufacturing facilities in three countries, including the United States, we are providing our U.S. customers with hundreds of megawatts of quality solar products that are not subject to these tariffs. Despite these harmful trade barriers, we hope that the U.S., China and all countries will engage in constructive dialogue to avert a deepening solar trade war. Suntech opposes trade barriers at any point in the global solar supply chain. All leading companies in the global solar industry want to see a trade war averted. We need more competition and innovation, not litigation."

Yingli, Robert Petrina, Managing Director, "We felt validated after the Department of Commerce's preliminary CVD decision in March, which determined that we are not being substantially subsidized as the petitioners claim. Today's preliminary anti-dumping tariff recommendation was not unexpected given the historical tariff levels in these types of cases. We will continue to aggressively defend ourselves and remain optimistic that we will persevere in the final determination. The overwhelming majority of the U.S. solar industry supports access to affordable solar energy and fair market trade. We are grateful to the tens of thousands of U.S. solar installers, developers, manufacturers, and suppliers who stand behind us today."

SolarWorld, Gordon Brinser, President, "The verdict is in. In addition to its preliminary finding that Chinese solar companies were on the receiving end of at least 10 WTO-illegal subsidies, Commerce has now confirmed that Chinese manufacturers are guilty of illegally dumping solar cells and panels in the U.S. market. We appreciate the Commerce staff's hard work on this matter. Commerce today put importers and purchasers on notice about the consequences of importing illegally subsidized and dumped products from China. We understand U.S. Customs and other federal agencies are already aggressively enforcing the countervailing tariffs in order to prevent circumvention, and we expect they will be equally vigilant with the anti-dumping tariffs."

SCHOTT Solar, Tom Hecht, President, "The solar industry has been awaiting today's decision from the U.S. Department of Commerce—and for good reason. U.S. project developers and investors need clarity and confidence to make critical supply decisions. Today's decision brings clarity—but creates another issue for U.S. developers. As they look to keep projects on track over the next three to four months, many will be trying to close on sources for PV panels not subject to the new tariff structure. SCHOTT has over 50 years' experience in solar. With factory sites in the U.S., Europe and Asia, SCHOTT Solar has been preparing to supply modules to our customers without interruption, regardless of the government decision. Longer term, the U.S. solar industry and government must focus on energy policies that will provide long-term certainty to the market and continue to encourage investment. Now is the time to support the industry in its efforts to create energy security for our nation and create additional jobs in manufacturing and services."

Solar Energy Industries Association (SEIA), Rhone Resch, President and CEO, "The solar industry calls upon the U.S. and Chinese governments to immediately work together towards a mutually-satisfactory resolu-

tion of the growing trade conflict within the solar industry. While trade remedy proceedings are basic principles of the rules-based global trading system, so too are collaboration and negotiations. Importantly, disputes within one segment of the industry affect the entire solar supply chain—and these broad implications must be recognized. In addition, the U.S. solar manufacturing base goes well beyond solar cell and module production and includes billions of dollars of recent investments into the production of polysilicon, polymers, and solar manufacturing equipment, products which are largely destined for export. If the U.S.-China solar trade disputes continue to escalate, it will jeopardize these U.S. investments. Given these broader implications, it is imperative that the U.S., China, and other players in the dynamic global marketplace work constructively to avert or resolve trade disputes that will ultimately hurt consumers and businesses throughout the solar value chain."

Canadian Solar, Shawn Qu, CEO: "Canadian Solar is disappointed by today's decision from the DOC. Imposing an obligation to post large bonds on solar imports at this preliminary phase of the antidumping investigation is unwarranted and will inflict losses on the entire solar industry. Limiting trade in solar products will cause panel prices to increase, defeating America's goal of driving down costs and hindering its move toward a clean energy future. Our first priority should be to support the health of the industry as a whole through the financing and installation of solar, which is the key driver to expanding jobs in the US solar market."

CASE, Jigar Shah, President, "Today SolarWorld received one of its biggest subsidies yet—an average 31 percent tax on its competitors. What's worse, it will ultimately come right out of the paychecks of American solar workers. Fortunately, these duties are much lower than the 250 percent tax that SolarWorld originally requested. This decision will increase solar electricity prices in the U.S. precisely at the moment solar power is becoming competitive with fossil fuel generated electricity. At the same time, CASE recognizes that today's decision is 'preliminary.' Between now and a final decision before the end of the year, there are many issues that will be addressed and whose resolution would lead to a significantly lower tariff. CASE will continue to fight SolarWorld's anti-consumer and anti-jobs efforts to ensure a better result for America's solar industry."

SunEdison, Kevin Lapidus, Sr. VP of Legal and Government Affairs, "The U.S. solar industry has been growing, adding new solar electric systems, creating jobs and investing billions of dollars in the U.S. energy infrastructure. By increasing the price of modules and therefore the price of solar energy, these tariffs will undermine the success of the U.S. solar industry and reduce the ability of solar energy to compete with electricity generated from fossil fuel."

Verengo Solar, Ken Button, President, "As the second largest residential solar company in the country, Verengo has helped thousands of middle class families save money during tough economic times by installing solar. Because our customers are very price sensitive, today's decision to increase costs for solar cells and panels will make it harder for American families to access solar."

REC Silicon, Tore Torvund, CEO, "This decision is short-sighted in the extreme and a severe setback for President Obama's clean energy program with its goal of expanding the use of solar and other renewables. Further, we are very concerned about the increased likelihood that China will retaliate with their own tariffs on polysilicon exports from U.S. producers such as REC Silicon. Triggering a solar trade war is not in the best interests of the U.S. solar industry or its customers."

GT Advanced Technologies, Tom Gutierrez, CEO, "Today's Department of Commerce decision subsidizes a German-owned company to the tune of an average 31% tax on its competitors and potentially harms U.S.-headquartered companies like GT Advanced Technologies, Dow Chemical, REC Silicon and MEMC. Ultimately, protectionism fosters dependence and high-cost business models, rather than the innovation and agile approaches required for companies to succeed in the global marketplace. Now is the time for the U.S. solar industry to move forward with the development of advanced technologies that create jobs and enhance our energy security—in spite of this new barrier. American solar manufacturing can compete without special protections."

Jeffries Group Inc., Jesse Pichel, "Environmentalists and the unemployed should be equally disappointed with this decision because lower cost solar panels make solar more competitive with dirty fossil fuels. It should be clear by now that there are more U.S. jobs on the installation side of the solar business than on manufacturing. These cases have a chilling effect on business and it will linger for a long time. It's unfortunate that

SolarWorld has taken this scorched Earth approach and that they are distracting from the growth of U.S. jobs and affordable solar energy."

Author's note: A close look at the above statements shows a clear division along country lines, increased polarization on the subject, and points to dangerous developments haunting the U.S. world solar energy industry in the near future.

Worldwide cooperation and voluntary standardization of the industry are still on the back burner. Marketing tricks and other unconventional strategies for gaining market share are in.

On the bright side, America was built by people who faced problems head on. It might take some time, but this US solar industry baby will grow to be a healthy, self-sustaining, competitor and contributor to 21st century energy.

What Happened?

This is not the first time we've seen the solar energy taken out of the closet and dusted up, only to be thrown back in, but this time was a bit different. The speed and size of the new developments were mind boggling.

Following is a preliminary recounting of events and the lessons we might be able to derive from them for future use:

Basically, politics and regulators brought the industry out of its decade-long slumber in 2007-2008. Propped up by billions of dollars, generous FiTs, and other government perks, the solar industry jumped on its feet very quickly. But when the props were removed, it was not able to stay up on its feet—a sure sign of immaturity.

- Solar is a serious political (partisan) issue. At the intersection of big money and big government, solar became a politically divisive topic and the target of powerful negative scrutiny. Taking the politics out of the agenda and replacing them with economic objectivity and reason is the key to stability.

- Dependence on decreasing and uncertain solar incentives, such as FiTs and other perks, made many solar companies projects more difficult to finance. This put unnecessary pressure on project developers to lower capital costs of PV plants. Since price alone is not a factor to success, many of these companies and projects failed.

- Shifting solar incentives and massive competition caused companies to shift priorities, forcing them to serve multiple and moving end-markets, which is expensive and difficult to execute. Very few were successful in this complex multi-tasking environment.

- Establishing value proposition of large-scale solar energy with utilities, regulators and politicians is a big issue. Stakeholders were not convinced that solar energy is competitive with fossil fuels on a price basis. They also did not consider the benefits of energy security, diversity, long-term price stability and the environmental advantages of solar in the overall price evaluations.

- The speed of change of the solar industry was ignored by the principals. Instead, they counted on a certain planned growth of their primary technology; i.e., CSP manufacturers focused on utilities, while PV start-ups focused on getting "high cost" silicon out of their product. Market changes threw their plans off balance and many of these companies and projects failed.

- Sharp imbalance of supply and demand drove many companies to their demise. Thin film companies' (CIS, CIGS and CdTe) technologies were the most affected. Building large-scale thin film PV power plants with this technology, as done presently in the US, is an expensive experiment, with a large degree of uncertainty.

- Challenges that the solar industry was not able to resolve include a) companies that scaled up manufacturing and drove prices down, b) continued global subsidizing of fossil and nuclear fuels, and c) PV industry immaturity that set up a precedent of bankruptcy, mergers and acquisitions.

- Public relations, in the midst of a flood of media and political attention focused on the struggles of some companies, and DOE loan-guarantee recipients. This focus has derailed efforts to point to the good side of the solar industry, including record numbers of successes in the U.S., Japan and China.

- Another stumbling block is the public's perception of the cost and effectiveness of solar. The general public falsely thinks solar is much more expensive than it actually is, and that it is not an effective technology. In fact, solar is much more flexible, complex, and generally less understood than fossil fuels, so the industry must be present the differences in clear and believable ways.

- Over emphasis on pricing dominated the markets, instead of diverting the focus to quality and reliability. Solar electric power plants, with their 25-30 year life expectancy must be high quality, reliable technologies with minimal O&M. Clarification of the need for maintenance is needed, because no energy technology is maintenance-free. That point was missed by the solar proponents.

- The high-growth opportunity of solar energy attracted many new entrants who raced to gain cost advantage by scaling capacity, which created overages through much of the supply chain. Many were not specialists in the field and their miscalculations contributed to the large number of failures.

- The most serious issue in this cycle is the looming global trade war started in the U.S. in the summer of 2012 by SolarWorld. When the solar industry was emerging as a cost effective solution for over 20% of global electricity sales, we were forced to deal with peripheral, very complex and expensive issues. Disruptions in the free flow of solar goods can raise prices, and eliminate jobs in the global solar industry.

- "Grid parity," and lowest "levelized cost of energy" (LCOE) dominated discussions during the last few years. Most people don't know what these words mean, but when they hear them from the experts it's like pushing a panic button. "No way, solar can reach grid parity or LCOE comparable to fossils," the common thinking goes.

- As to financial performance, the more critical trend and phrase for the duration was, "the race to the bottom." The ever-accelerating push to reduce costs in photovoltaics has put many companies on life support and has others facing margin compression and profitability evaporation. This, at a time when compared to most industries, the PV industry growth is spectacular.
  — The solar industry is still struggling with tight credit lines, which drove many of them out of business and will continue to do so for the next few years. To be competitive under these conditions, solar companies need a strong project finance capability, attracting project equity and debt at low cost. There are not many companies left in this class.

- As vendor bankability improves, financing risk is reduced. This leads to reduced financing costs and the viability of a greater range of projects. With the increase in the size of projects, vendor bankability designed to support larger utility installations becomes even more important. Yet, with 2007-2011 financial conditions—accompanied by large government financing offers—company bankability was a confusing subject, added to an already confusing situation.

- Large-scale PV projects grew in size, from a few hundred kilowatts, to utility-scale applications of hundreds of megawatts. There are millions of dollars invested in these projects, so one key to success is the participation of key component suppliers in the project's 25-30-year operation and end-of-life decommissioning. The reliability of vendors to support those projects and the availability of replacement parts is crucial to project success. The best warranty is worthless without a manufacturer to back it up.

- With so many issues on the agenda, the pricing and quality of the balance of system (BOS) components (support structures, inverters, wiring, etc.) was put on a back burner. Yet, with the proliferation of multi-billion-dollar, utility-scale solar projects, BOS should've been at the top of the agenda. Instead, commercial-scale hardware and electronic solutions (which were more advanced in their development) were shoe-horned to fit utility-scale applications. This added installation time and equipment costs, and uncertainties.

- The variability of solar systems became quite obvious, and is quickly becoming a serious stumbling block in their future deployment. Grid operators cannot easily control power fluctuations coming from PV power fields on bad weather days. Utilities complain that their costs increase with the increase of solar installations, because their peaker power plants are used more often. Variability will actually become an even larger issue as PV plants sizes increase. Energy storage is the solution, but we are far from storing enough power for nighttime or bad-weather power generation.

- The biggest issue for the global solar industry is market uncertainty. Government support is inconsistent, with an atmosphere of uncertainty leading to project delays, price drops, and a decline in hiring and expansion plans.

The absence of strong federal energy programs and the lack of standardization in state-to-state programs in the US are issues that have been around since the 1970s. The lack of a federal energy policy is the most important deficiency hindering development of the solar industry in the country.

- Geographic inequality of solar installations was emphasized during this boom-bust cycle too. Most PV installations, and all CSP such, were in developed countries. The developing world—especially Africa—saw very little of the progress solar made for the duration. This inequality gave the world's cleaner and safer technology the reputation of something "only the rich can afford."

Forecasts for 2013 and Beyond

PV module manufacturers (if China expands the national PV fields development) will return to an acceptable level of profitability sometime in 2013, in our opinion, and slowly increase the profit margins. Revenue, however, will continue to decrease during the first half of 2013, before returning to normal.

During 2013-2016, global PV installations are expected to grow 10% annually. Asian markets, and particularly—China and Japan—are expected to compensate the declining demand from the US and Europe.

PV module manufacturing is expected to stabilize at around 50 GWp for the duration. This way, overcapacity built up by the massive investments in 2009-2011 will no longer effect prices. Average c-Si PV module prices will reach and stabilize at $0.55/Wp at the end of 2013, and will most likely start going up slowly after that. This way, most solar products manufacturers will be able to schedule their activities without the threat of major spikes in demand and supply.

However, the solar energy industry is still in a state of shock, following the recent boom-bust cycle, and a number of uncertainties hang over it. Anything can happen in such fragile situation, so we just hope and the worst is over and look toward better planned, smoother sailing.

NICHE AND SPECIALTY ENERGY MARKETS IN THE 21ST CENTURY

The future will bring us many new developments, accompanied by even greater surprises. Here is a short list of some which we believe will make the biggest difference in the 21st century.

Carbon Market (Trading)

The European Union Emissions Trading System (EU ETS) was the first large emissions (carbon) trading scheme in the world. Launched in 2005, it was designed to combat and reduce climate change. It is now a major pillar of the EU climate policy.

The first ETS trading period lasted from January 2005 to December 2007. The second trading period began in January 2008 and spans a period of five years, until December 2012. The third trading period is from January 2013 to December 2020.

The EU ETS now covers more than 11,000 factories, power stations, and other installations with a net heat excess of 20 MW in 30 countries. All 27 EU member states, plus Iceland, Norway, and Liechtenstein are part of it. This represents the installations regulated by the EU ETS, who are collectively responsible for almost 50% of the EU's emissions of CO_2, and 40% of its total greenhouse gas emissions.

These installations monitor and report their CO_2 emissions, and every year they return leftover emission allowances to the government. In order to neutralize irregularities in CO_2 emission levels due to extreme weather events (harsh winters or very hot summers), emission credits are doled out for a period of several years called a "trading period."

Installations currently receive trading credits from their national allowance plans (NAPs), administered by the governments of participating countries. If a factory's carbon emissions exceed what is permitted by its credits, it can purchase trading credits from other installations or countries.

At the same time, if an installation has performed well at reducing its carbon emissions, it can sell its leftover credits for money. This allows the system to be more self-contained without necessitating too much government intervention.

Like any other financial instrument, trading consists of matching buyers and sellers between members of the exchange and then settling by depositing a valid allowance in exchange for the agreed financial consideration. Much like a stock market, companies and private individuals can trade through brokers who are listed on the exchange, and need not be regulated operators.

When each change of ownership of an allowance is proposed, the national registry and the European Commission are informed in order for them to validate the transaction. During Phase II of the EU ETS, the UNFCCC also validates the allowance and any change that alters the distribution within each national allocation plan.

The EU Allowance Unit of one ton of CO_2, or "EUA," was designed to be identical ("fungible") with the equivalent "Assigned Amount Unit" (AAU) of CO_2 defined under the Kyoto Protocol. Hence, because of the EU's decision to accept Kyoto-CERs as equivalent to EU-EUA's, it is possible to trade EUA's and UNFCCC-validated CERs on a one-to-one basis within the same system.

During Phase II of the EU ETS, the operators within each Member State must surrender their Phase I allowances for inspection by the EU before they can be "retired" by the UNFCCC. And so on.

This is a good way to at least monetarily balance pollution responsibility and suffrage. A number of issues remain to be resolved when optimizing the carbon trading scheme. The most important of these is location inequality, where the environment and people close to the largest pollution generators suffer most, but are not equally rewarded.

So, this is perhaps the best way to bring attention to the issues related to carbon pollution, and to assign responsibility and accountability to the generators.

Environmental Economy

At the Copenhagen climate conference in 2009, 167 countries met to discuss the global environmental changes and to look for ways to remediate them. A number of these countries are responsible for almost 90% of global greenhouse gas emissions, and most of them agreed that the power platforms of their economies are raising the global temperature.

An increase of over 2°C, they agreed, is unacceptable. To avoid catastrophic climate change, and to stay below that level, their aggregate economies must not put more than an additional 565 gigatons (GT) of carbon dioxide (or its equivalent in other greenhouse gases) into the air before 2050.

Ultimately, and in very general terms, the nations agreed, they must switch to using power generating technologies that do not generate more greenhouse gases than oceans and plants can absorb without causing catastrophic instability in the climate system.

The global economy put into the atmosphere a record level of emissions in 2011, producing 31.6 GT (gigatons) of greenhouse gases. This means that the Copenhagen agreement intends to slash the emissions by over 60%. Is this possible? And if it is possible, who will enforce it?

During the same time, Chinese emissions rose 9.3%, while American emissions fell 1.7% (most likely due to decrease of industrial activities as a result of the economic crisis). The developing nations are burning massive amounts of coal to produce electricity, and wasting a lot of the produced energy in the process as well. Many nations rise from poverty only to enter a tragic paradoxical effect of threatening to destroy the very environment they live in. In the streets of Beijing and other large Asian cities, the smog and smoke are so dense that many people wear face masks.

Developed nations are not far behind in their damaging activities. Per capita, Americans consume about 250 kilowatt-hours of electricity (kWh) each day, in the form of heating, lighting, air conditioning, transportation, and consumption of products made with electricity. Europeans and the Japanese consume about half as much, mostly via conservation efforts.

This means that investment in efficiency and infrastructure can reduce the amount of energy that we here in the U.S. use too. The problem is that regulatory, legal, and financing obstacles block hundreds of billions of private sector investment in efficiency measures.

The United States consumed 4,326 TWh of electricity in 2011, roughly as much as China. Excluding hydropower and nuclear power, renewable energy in the U.S. constituted only 4% of electric consumption, but in China, it was merely 1.3%. Fossil fuels (gas, coal and oil) generated 70% and 80%, respectively. So, these two economies are roughly the same in terms of the energy solutions they use and the problems they face.

China consumed 17% hydropower; the United States consumed 20% nuclear power; which is the most they can produce in those segments to address economic needs or climate change requirements. It makes sense for both countries to have similar plans for moving to power generating platforms. But, as of the fall of 2012, power generating options for the US were in favor of the status quo, with the discovery of large natural gas and oil reserves.

China is taking a serious look at solar and wind, so we expect it to make some progress in these areas. Still, coal and gas exploration are on the increase.

China and the U.S. are also looking into converting to electric transportation. China has already announced a plan to have a market of 5 million electric vehicles by 2020, and 500,000 by 2015. If China succeeds in achieving this goal, it will have 43% of the expected global electric vehicle market, and 0.4 percent of the global passenger fleet will be electric.

China also will have created about 20,000 direct jobs in this electric car market. The U.S. is dabbling in it also, but there are no definite goals.

The government must continue to be the driving force in developing clean energies. Just as it helped organize, finance and build the U.S. highway and railroad infrastructures, it must lead the building of the renewable energy infrastructure.

There are a number of ways the government can do this:

1.   Lower the cost of initial capital investment, so the industry can step on its own feet,
2.   Modify the tax laws to attract private investors,
3.   Enforce renewable energy standards at the state level,
4.   Reform utility function and regulation as needed to attract private investment
5.   Promote awareness and provide solutions to the world's climate crisis.

DC Power Transmission

High-voltage direct current (HVDC) electric power transmission lines use direct current (DC) for bulk transmission of electric power, while the presently used alternating current (AC) power transmission lines use AC power. DC power has different characteristics than AC power, and it so happens that for long-distance transmission, HVDC systems may be more suitable, less expensive, and suffer lower electrical losses.

HVDC is much better and more economical for uninterrupted large distance transport, as well as for underwater power lines, since it avoids the heavy currents required by the AC line capacitance.

HVDC also allows power transmission between unsynchronized AC distribution systems, and can increase system stability and reliability by preventing cascading failures from propagating from one part of a wider power transmission grid to another.

The controllability of current flow through HVDC rectifiers and inverters, their application in connecting unsynchronized networks, and their applications in efficient submarine cables mean that HVDC cables are often used at national boundaries for the exchange of power. As an example, HVDC connections divide much of Canada and the United States into several electrical regions that cross national borders, although the purpose of these connections is still to connect unsynchronized AC grids to each other.

AC transmission lines can interconnect only synchronized AC networks (those that oscillate at the same frequency and in the same phase). Many areas, however, share power but have unsynchronized networks. For example, the UK, Northern Europe and continental Europe are not synchronized, while Japan uses dual (50 Hz and 60 Hz) power transmission networks. Continental North America is operating at 60 Hz throughout, but is subdivided into regions which are unsynchronized: East, West, Texas, Quebec, and Alaska. South America's Brazil and Paraguay share the enormous Itaipu Dam hydroelectric plant, but operate at different frequencies (60 Hz and 50 Hz respectively).

Lately the growing number of offshore wind farms require undersea cables, but their turbines are unsynchronized, so they require complex and expensive equipment to transmit and synchronize the produced power before plugging it into the national grid.

By using HVDC transmission, all these different networks and systems can be uniformly interconnected, and as an added bonus, the resulting voltage and reactive power flow can be precisely controlled and made much more stable than using AC power lines.

With the increased power generation by alternative power sources, HVDC may become the preferred transmission current, and could provide the needed stability and efficiency to the grid of the 21st century. The best way to do this is to start with the conversion of local networks, and work up to national and international power grids.

In the near future, the installation of DC transmission lines could be easily converted from a niche market into a large-scale effort on a national level—converting the grid to DC power. This undertaking could bring thousands of jobs, and more efficient power transport and distribution.

Electric Cars

Electric cars have been surrounded by mystery since the very beginning, and have a special meaning and designation in the US and Europe. The person driving an electric car is thought to be avant-guard. Lately, however, the mystery is being uncovered as less than forward-looking.

While these cars run "clean," the electric energy they use is far from it. It was generated somewhere else by burning dirty coal. And while, using electricity to charge the car's batteries does not contaminate the local environment, it did contaminate the environment around the power generating plant. So the net gain is reduced to transferring responsibility from one place to another.

What would happen if most of us bought an electric car? Around 5:00 PM, as everybody got home, they would plug the electric car battery charger in the wall socket. The sun goes down, and there is no wind to speak of, so solar and wind generators are down.

With millions of electric cars plugged in at the same time, the peaker power plants would kick in, and our electric bills and the probability of blackouts would increase significantly.

While no one can argue that electric cars have the potential to do a lot of good for the air and the planet, they also push utilities toward a reckoning with the so-called smart grid. Our national electric grid at its present state is already stretched to the limit. It is simply not designed to handle the load.

As more governments encourage the use of electric cars through mandates, tax breaks, and other mechanisms, utilities will need to figure out how to deliver power to those cars efficiently, without cutting further into already razor-thin profit margins.

This is not an easy task, considering that providing more power at night (which is not the case today) is only part of the problem. The utilities also must create and service an electric version of the gas station network that has been in place for over 100 years.

The grid will be forced to get smarter...at a small, several billion dollars worth, charge. Overseas the situation is even more complex. For example, in Germany the government has set a 2022 deadline for shutting down all nuclear power plants, which until recently provided nearly 25% of the country's power generating needs. In Japan, lingering concerns about the safety of nuclear power following the tsunami-related meltdown at the Fukushima Daiichi facility, have resulted in the ongoing shutdown of almost all the country's nuclear power sector.

The likely replacements for nuclear (and coal, oil, and natural gas), solar and wind power, are as unreliable and anxiety-inducing to the current electric grid as are electric vehicles. Neither provides what the grid is designed for—predictable flow.

With wind power generation, utilities may have only a day or even just a couple of hours' notice to confirm whether generating capacity will meet or exceed power needs. Some of the time, the wind may blow when it is not needed, so utilities will need storage solutions for excess energy that they can feed back into the grid when the wind dies down and power demand picks up. At the same time, nobody can predict where and how long the sun will shine, so the utilities are totally relying on an unreliable power source.

While we exaggerated the situation to a worst-case scenario, let's not forget that renewables and electric vehicles are in their infancy. They still represent great opportunities for new revenue, new markets, and even for new business models for the respective industries, which today are still learning to walk.

Although it is unlikely EVs to take over the world, we do believe that they will eventually account for 10-15% of vehicle sales volume in developed countries. Some governments (i.e., Germany and Ireland) have already set these goals. For the utilities, this new activity represents an opportunity to open a new and very large market.

The most profitable new business opportunity may come from building service stations and operating the charge points that EV owners will need to recharge their batteries and otherwise service their cars. As another case of immature technology, these charge points are a challenge, and they are few and between. In Europe, it costs over $15,500 just to install a single public charging station. Then there are the costs of maintaining and operating these charge points.

Energy companies are looking at a variety of strategies for defraying charge point installation costs or adding revenue streams:

- Collaboration with businesses where drivers don't mind waiting. Restaurants, retailers, and shopping malls are prime candidates.

- Managing electric car fleets to provide car-sharing. Utilities could take on the management of corporate and rental electric car fleets, just as some of them provide other solutions for power management.

- Partnering with technology companies. This will speed up the efficient installation and operation of networks of charging stations.

- Adding electric car charge and service to conventional gas stations. This is the easiest solution, since the infrastructure is already in place and adding charging components would increase business.

- Using the charge station for advertising displays could bring additional income.

A decisive factor in this development of the EV industry would be the development of the smart grid. Only with a well developed smart grid could we cope with the demands of millions of consumers plugging their cars into the grid at the same time, either in their homes or at public charging points. In any event, the utilities will need to find a way to manage the extra demand with smart grid systems that can intelligently allow EVs to draw from the grid during periods of low energy demand, such as during the overnight hours.

Utilities could allow customers to recharge their EVs at any time, but provide financial incentives to recharge a car (or run any major electric appliance) during low-demand periods.

This immature technology is very attractive, and EVs are here to stay—getting more reliable and cheaper by the day; and, as governments boost incentives and mandates for EV development, more utilities are getting interested.

The 21st century is a time for bold experimentation, when EVs, smart grids, and renewable energy technologies are growing in importance and promise to bring us, among other things, a new and exciting utility-transportation system that is environmentally safe.

Internet Data Center Cooling

Cooling for servers and their hardware is an important part of ensuring the efficiency and reliability of internet providers. This is the internet's hidden cost. A data center with efficient cooling design costs 50% less energy to run than an inefficient one on average, depending on location and operational specifics.

Today power and cooling costs outpace the cost of the actual equipment in some places, like Phoenix, AZ, where the summer temperatures are in the mid 110s. This is because companies are paying not only for the power that is necessary to run equipment, but also for power to cool it during hot days. A lot of power is needed to keep computers and super-computers cool and within operating specs.

Imagine a 110°F day in Phoenix in a room full of computers belonging to a local internet provider. The ambient temperature by 2:00 PM approaches the outside air temperature, and the operating equipment adds even more heat. If the equipment in the room were left unattended and with no cooling, by 6:00 PM it would reach critical temperature levels.

Most CPUs are equipped with over-heating protection interlocks, so it is very likely that the computers in that uncooled room in Phoenix, AZ, would start shutting down. Those without overheating protection would continue operation until something inside melts. In either case, around 6:15 PM that day, the local internet provider would shut down. So coolers and air conditioners are absolutely necessary there, and in fact, they are quite busy.

Countries in cold climates have seized the opportunity and have been marketing their environmentally friendly (for internet servers) lands as a cheap solution to this cooling issue, which would also bring jobs to their citizens and reduce the energy bottom line of web-focused businesses. As a matter of fact, Facebook and other major players have been contemplating this idea, and some are planning to move their data centers abroad to cooler places.

Environmentalists have been criticizing Facebook for using energy derived from coal, and this new solution is one more step in the company's campaign to change a perception of social irresponsibility in energy.

The best way to keep your data center cool is to use the cool, ambient air of your natural surroundings. But if your company doesn't have the luxury of locating the data center near the moderate Oregon coast, like Facebook, or near the fjords of Norway, the cost of cooling can be significant.

Data center efficiency is big business today, so startups like Power Assure are getting into the business along with power engineering giants IBM, General Electric and HP. SM Group International (SMi), an international engineering firm based in Montreal, is tackling the cooling methods with claimed savings of up to 50%.

SMi's patent-pending technology isn't actually new, but rather a paradigm shift in how to cool computers. The key is to pressurize the space and blow air directly over the servers before the heat builds up. Using traditional pre-cooling of the equipment helps too. But in hot areas like Phoenix, AZ, pre-cooling doesn't help much during the day, and only a little at night.

Amazingly, data centers account for about 2% of total electricity use in the U.S., and that figure is increasing with the increase of internet use.

While SMi was working on optimizing the pre-cooling process, they also started looking into putting the computers in a wind tunnel, pushing cold air in a systematic way, instead of pumping it into a chaotically organized square room, where cooling cannot be efficiently controlled. The basic direction of moving air efficiently, and in a controlled way in a pressurized environment could theoretically cut energy use by up to 50%.

The average data center has a power use effectiveness (PUE) ratio of about 2 to 3. Facebook's latest Oregon data center sports a PUE of 1.07 and consumes 38% less power than the average setup, thanks to a new server design. The new GE's ultra-efficient data center rates a score of 1.63.

SMi' cooling approach could cut the PUE to as low as 1.1 for a newly designed facility. Retrofits results would vary, but if a data center had a PUE of 2, it could be brought down to about 1.3.

Cooling, of course, is just one aspect of data center power use. Other ideas for curbing power usage have included weather mapping inside data centers' computer rooms, application shifting, improved AC controls, switching from AC to DC power, swapping disks for flash memory, better power conversion, and smaller, more energy efficient servers.

All this helps reduce power use in the ever-increasing use of the internet. And since the internet is one thing that will continue growing disproportionately during the 21st century, and as we depend more and more on it, we need to improve the way data centers operate—including switching to power efficient equipment and cost-saving cooling methods.

This, somewhat hidden and obscure niche market, promises to grow to great proportions. What we learn from these efforts could be then applied in other industries as well.

Specialized Cleaning Services

New areas for specialized services will be opening in the near future in the solar mega fields. One such service of great importance might be the periodic cleaning of PV modules in utility solar power plants. Contrary to the claims of some companies that their PV modules do not need cleaning, experience shows that the millions of PV modules, parabolic troughs and mirrors installed in large-scale solar power plants do need cleaning.

Some locations are much worse than others, with deserts being the dirtiest places on Earth. Sand storms, monsoons, and other weather phenomena, in addition to scorching days and freezing nights, dominate the desert landscape. PV modules stuck in the middle bare the brunt of it all. With every monsoon storm, the shiny front glass of the PV modules and the reflective surfaces of parabolic troughs and mirrors are covered with dust and mud spots. These must be cleaned immediately, or there would be much less power generated. If the dust and mud are not cleaned quickly, they get hard and baked in with time, thus becoming more difficult to remove.

Here is where we envision specialized cleaning crews jumping into action, and attacking the dirt and mud in an organized and efficient way. New cleaning chemicals, processes and technologies would be developed to solve these problems quickly, efficiently and cost-effectively. A niche market would be born, and entire industries might develop around this service as well.

Author's note: The claim of some reputable companies that their PV modules "need no water," or "need cleaning once or twice a year," is further confirmation of the ignorance surrounding the complex issue of operating solar power equipment in the deserts. Such misinformation, coming from large and reputable companies, is often interpreted as fact based on scientific research and long-term experience—but far from the truth.

Materials

Since all solar devices are made of one or another physical material, the type and quality of these determines the performance and reliability of the solar cells and modules made from them. Garbage in/garbage out is the most appropriate qualifier in this case. We cannot possibly expect great performance of a device made of poor-quality materials.

The situation is not that extreme in most cases, so what we most often get is something in between—something that may or may not work well. Because of that, the research on existing materials and the development of new ones is an ongoing and very large business.

The work in some of the areas of materials R&D includes:

Hi Purity Polysilicon

Metallurgical-grade (MG) silicon, which is usually produced from sand, is about 99% pure. According to generally accepted guidelines regarding silicon purity, electronic grade (EG) material suitable for the semiconductor industry requires at least 99.9999999% pure material (known as "9-nines" purity), which represents impurity concentrations of about 0.0005 ppm.

The rising needs for faster and better operating semiconductor devices, however, have pushed the polysilicon producers of EG materials to routinely achieve 11-nines purity. At present, the purity requirement for the solar grade (SG) polysilicon for the PV market is somewhat less stringent. The prevailing opinion is that silicon with a purity of 99.9999% (6-nines) is the lower limit for a viable solar cell.   Research in the area suggests that the efficiency with which solar cells convert sunlight into electricity is correlated with material purity, therefore a solar cell made out of 6-nines SG would be less efficient, and less reliable, than one made of 7 or 8-nines. The price goes up according to the level of purity, so a balance must be struck by the manufacturers in terms of efficiency and reliability vs. purity and quality of polysilicon materials.

More and more crystalline photovoltaics companies are pushing for higher-quality material, and substantial R&D throughout the solar industry is underway, intended to validate the correlation between polysilicon purity and PV cell performance. In Japan, for example, the limited space available for solar installations is forcing the use of the most efficient cells and PV modules. This trend will grow with time, as customers start finding out that higher quality materials provide not only higher efficiency, but a greater long-term reliability.

Going beyond the established and accepted purity limits of the polysilicon materials will play and increasingly important role in the solar installations of the 21st century. Finding new ways to achieve and ensure high purity, thus increasing the efficiency and reliability of the final products, will be on the agenda of all solar cells and PV modules manufacturers.

New Feedstock for FB Process

A development-stage company called Peak Sun Silicon Corp. (Albany, OR) is working on an FB process using tribromosilane (TBS) as the feedstock, rather than either silane, or trichlorosilane. While TBS costs more than its cousins, Peak Sun's process uses less, so the feedstock costs are almost even.

The inclusion of massive bromine atoms in the feedstock compound also offers several key advantages, such as a much wider tolerance of temperatures and pressures, which is an advantage over the tight control that must be exerted in other processes.

TBS processes operate at lower temperatures, which enables energy savings and allows additional quality control. It also helps avoid formation of submicron-diameter amorphous silicon dust through a homogeneous nucleation route, which are significant challenges encountered by silicon reactor operations that use silane or trichlorosilane as the feed gas.

Homogeneous nucleation can occur when materials move from vapor to solid phase and reach a "critical nucleus" of solid-phase material. At certain values of "re" atoms, the thermodynamics of the system favors self-nucleation, rather than deposition on the surface of a seed particle. The phase-change energy barrier is overcome more readily at higher values of n, since surface area-to-volume ratio decreases as particle diameter increases. Silane and trichlorosilane have a natural propensity for self-nucleation when overcoming the phase change energy barrier.

By virtue of its relatively large molecular size, TBS requires a greater "n" value to achieve critical nucleus, and molecules prefer to transition to solid phase by nucleating on an existing silicon surface. The result is that TBS deposition favors growth on the surface of crystalline seed particles over formation of low value amorphous silicon dust.

Other advantages to the Peak Sun TBS-FB process include the formation of dense metallic beads with a narrow particle size distribution. Silicon beads formed by the TBS-FB process tend to have less surface oxidation and less gas molecule inclusions, which gives the silicon better melt properties.

When completed, the Peak Sun project will be the first demonstration of an FB silicon unit using TBS. Peak Sun claims that its process will have lower capital costs and 25% lower operating costs than a Siemens process unit.

Crystallization Technologies

Another key process in the manufacture of high-purity silicon involves a re-crystallization step that converts polycrystalline material to monocrystalline silicon. The process is more important at present for semiconductor-grade silicon, since polycrystalline silicon is suitable for PV cells.

Manufacturers typically use some variation of the so-called Czochralski process, in which a seed crystal is introduced to a silicon melt and slowly withdrawn to generate a long mass of monocrystalline silicon.

An alternative is a modified float-zone process, in which impurities are segregated during a transition that occurs as a mass of polysilicon passes through a radio-frequency heating coil, which creates a localized molten zone from which the pure crystal grows.

This way the purity and the final quality of the product can be tightly controlled and significantly improved.

UMG

One technology approach to polysilicon production that appears to be under consideration for low-cost manufacturing of solar cells, is upgraded metallurgical grade (UMG) silicon. UMG is produced by melting metallurgical grade silicon and slowly and directionally recrystallizing it. The approach would offer a less expensive route to material, but the 5-nines or 6-nines purity that UMG can achieve wouldn't be viable if the higher-purity methods are cost competitive.

The drive today is for achieving high levels of purity—over and above what UMG can offer—but UMG is a viable alternative for some specialized applications, where low-cost silicon can be used. It also might be considered for applications where the PV modules would be used temporarily.

Nevertheless, UMG offers a niche market worth exploring, for use in special cases, and if and when polysilicon prices reach the $500 mark again.

Military Micro-grids

Traditionally, if the power goes out at a military base, each building within the base will switch to a backup energy source, most often provided by a diesel generator. The military isn't crazy about this setup, in part because this is an expensive solution, and also because generators can fail to start—especially bad news for base hospitals and other critical operations. And if a building's backup power system doesn't start, there is no way to use power from another building's generator.

Also, most generators are oversized and use dirty, increasingly expensive fuel. So the military is working on ways to change that by connecting clean energy sources like solar and wind in micro-grids that can function when commercial power service is interrupted.

A new $30 million initiative could transform the way U.S. military bases deal with power failures. Termed SPIDERS—short for Smart Power Infrastructure Demonstration for Energy Reliability and Security—the three-phase project will focus on building smarter, more secure micro-grids on military bases and other facilities that make use of renewable energy sources. Sandia Labs has been chosen as the lead designer and technical support for the SPIDERS project.

Diesel generators are used all the time to provide emergency power to buildings, but they are usually not interconnected with alternative energy sources like solar, hydrogen fuel cells, etc., as needed for a stable and significant energy source. It's a real integration challenge, according to Sandia's engineers working on the SPIDERS program. So, Sandia will work to set up a smart, cyber-secure micro-grid that will allow renewable energy sources to stay connected and run in coordination with diesel generators, which can all be brought online as needed. The new system is expected to not only make the military's power more reliable, it will also lessen the need for diesel fuel and reduce its carbon footprint.

The project is being funded and managed through the Defense Department's Joint Capability Technology Demonstration with the support of the U.S. DOE.

Someday soon, the departments hope to use the SPIDERS plan for civilian facilities like hospitals.

Power Plants for Sale

PV power installations "for sale" is a growing business. There are different reasons for the projects—in different stages of development —to be for sale. Some are operational, which makes their "for sale" status even more questionable.

The list in Table 8-17 is an example of the variety of PV projects for sale worldwide in the fall of 2012.

Table 8-17. Partial list of PV projects for sale, fall of 2012.

EU:	86.5 MW portfolio of operational PV plants in Germany, Italy, Spain & Czech Republic.
	Note: Some of these have a substantial long-term debt in place
Bulgaria:	2.0 MW operational solar farm
Bulgaria:	1,7 MW operational solar farm
Bulgaria:	1.0 MW operational solar farm
Bulgaria:	2.0 MW under construction
Ecuador:	20.0 MW solar farm project rights
France:	1.86 MW portfolio of operational roof projects with high legacy FIT rates
France:	1.7 MW portfolio of 100 kW new-build hangar projects @ 0.23€/kWh, developer turnkey offer
France:	1.6 MW portfolio of 100 kW new-build hangar projects @ 0.23€/kWh, developer turnkey offer
France:	1.3 MW portfolio of 100 kW new-build hangar projects @ 0.26€/kWh & 0,23€/kWh, developer turnkey offer
France:	1.3 MW portfolio of 100 kW new-build hangar projects @ 0.21€/kWh, developer turnkey offer
France:	912 kW portfolio of operational roof projects with high legacy FIT rates
France:	874 kW operational roof projects @ 0,42€/kWh
France:	500 kW: 2 operational new-build hangars on single site @ 0,60€/kWh
France:	500 kW: 5 PTFs @ 0.21€/kWh
Italy:	3,9 MW portfolio of operational solar farms. Substantial long-term debt
Italy:	1.0 MW operational solar farm. Substantial long-term debt
Romania:	2.5 MW under construction
Romania:	58.9 MW solar farm—developer turnkey offer
Romania:	28.0 MW RTB project rights
Romania:	8.2 MW 3 solar farms—developer turnkey offer
UK:	10.75 MW operational domestic roofs
UK:	10.0 MW solar farm project rights
UK:	2.1 MW solar farms, under construction, developer turnkey offer
UK:	1.4 MW roofs, under construction, developer turnkey offer
USA:	380 kW ground mount PV in Arizona, permits and PPA in place.
USA:	560 kW roof mount PV in California, permits and PPA in place.
USA:	10.0 MW utility power plant in permitting stages.

Why would anyone who had gone through the hassles and expense of permitting and constructing a PV plant decide to sell an operational one? What's wrong with it? Do the P and L calculations for any of these plants would show low power generation, and/or low profits?

Most likely, many of these plants were built on the basis of the old FiT rates, at which time they showed profit, so investors financed them. Today, the FiT in all these countries have been reduced drastically, and other unfavorable conditions have been added too, so the profitability has been removed from the equations.

So, we have a surplus of PV plants under construction or built recently, which due to the above (or whatever) reasons are for sale.

Recycling of PV Modules

PV companies and government bodies worldwide are actively looking into the process of recycling and disposal of PV products, both as manufacturing waste and as end-of-life (EOL) modules.

EU companies have proposed a voluntary take-back system capable of meeting the future waste recycling demands. The new recycling process lines must be capable of processing crystalline silicon and thin-film modules alike. The recycling issues are now at the pilot stage, and reuse of PV material will be timely for the industry.

US companies have been reclaiming solar cells and semiconductor process wafers for many years, so the expertise and equipment are available. Scaling up to accommodate the large demand of the future is a key to success here.

The forecast is that 40,000 MT of PV components will be ready for decommissioning and recycling by 2020. This number will double or triple during the following decade. This includes silicon and thin-film based PV components recycling. The thin-film component will be ~20% by then. 8,000 MT CdTe thin-film PV modules will be the majority of the recycled products in this category.

8,000 MT CdTe is a lot of PV modules which, with an average of 8-9 grams of cadmium in each module, could create a serious hazmat debacle if not handled properly.

As can be clearly seen in Table 8-18, the manufacturers are legally responsible for the safety and proper execution of all steps from the beginning to the end of the useful life of their products. Retailers, installers, and plant operators are also responsible in some stages of the product life.

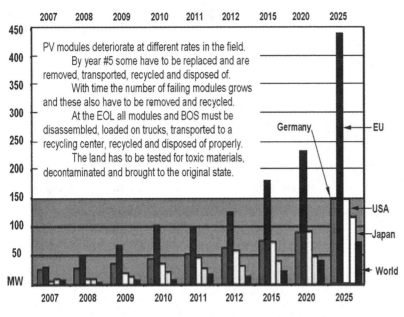

PV modules deteriorate at different rates in the field.
By year #5 some have to be replaced and are removed, transported, recycled and disposed of.
With time the number of failing modules grows and these also have to be removed and recycled.
At the EOL all modules and BOS must be disassembled, loaded on trucks, transported to a recycling center, recycled and disposed of properly.
The land has to be tested for toxic materials, decontaminated and brought to the original state.

Figure 8-8. Forecast of PV waste recycling in the decades to come

Cadmium compounds are toxic and are regulated in many countries because of their toxicity to all life forms and humans. In some countries the sale of CdTe-containing PV modules is prohibited because of the policies regulating cadmium in photoelectric semiconductor devices.

Cadmium is associated with numerous human illnesses, such as lung, kidney and bone damage. It is especially dangerous due to its absorption in the body, where it is cumulatively collected and can remain for decades.

Cadmium sometimes leaches from landfills to be collected in ground and surface water. It can also contaminate the atmosphere while escaping from incinerator smokestack emissions.

Using air pollution control equipment at incinerators could trap cadmium gasses in the

Table 8-18. End-of-life outline of the participants' responsibilities

EOL TREATMENT OF TFPV MODULES	DECISION AND EOL* NOTICE ⇨	REMOVAL ⇨	TRANSPORT ⇨	RECOVERY & DISPOSAL
GENERAL REQUIREMENTS	Notify proper authorities of EOL	Proper and safe procedures use	Authorized Hazmat carrier use	Authorized facilities use
ORGANIZATIONAL RESPONSIBILITY	Plant operator	Manufacturer and operator	Manufacturer	Manufacturer
TECHNICAL RESPONSIBILITY	Plant operator	Manufacturer and operator	Manufacturer	Manufacturer
FINANCIAL RESPONSIBILITY	Plant operator	Manufacturer and operator	Manufacturer	Manufacturer
LEGAL RESPONSIBILITY	Plant operator	Manufacturer and operator	Manufacturer	Manufacturer

Overall, all entities who have been involved in the power plant's planning, design, installation, and operation are responsible to one degree or another for the proper execution of the respective steps, including the final decommissioning, recycling, and disposal of plant components, and land clean-up.

Used PV materials and waste are different, because of the long lag time between production and decommissioning. It is more than a generation, and many things might change in that time.

The amount of waste will vary dramatically after 2030-2040, because different modules and power fields have different life spans and require different handling. The situation is further complicated by the fact that many PV modules contain hazardous materials such as cadmium, tellurium, lead and selenium.

ash, which can cause problems later by leaching from the containment sites.

So, it is of utmost importance to realize that a lot of solar waste is, and will be, created in large quantities during the 21st century, and some of this waste contains hazardous material. It is, therefore, the responsibility of the regulators to anticipate the developments and ensure the safe processing and disposal of these materials.

The major types of PV modules worth recycling are c-Si, a-Si, CdTe and CIGS. A number of other upcoming technologies will change the landscape, but most have far to go.

c-Silicon Modules

Silicon based modules consist of glass, aluminum, silicon solar cells and plastic encapsulants. Recycling

Table 8-19. PV modules recycling steps and methods

RECYCLING OF PV MODULES AT EOL

c-Si MODULES			TFPV MODULES		
RECYCLABLE	kg/m²	%	RECYCLABLE	kg/m²	%
Glass	10.0	90	Glass	15.0	90
Aluminum	1.4	100	Aluminum	0	0
Solar cells	0.5	90	Solar cells	0.1	90
EVA, Tedlar	1.4	0	EVA, Tedlar	0	0
Ribbons	0.1	95	Ribbons	0	0
Adhesives	0.2	0	Adhesives	0.2	0

RECYCLING AND DISPOSAL METHODS

Mechanical	Crushing, attrition, density separation, flotation, adsorption, radiation, laser beam, metal separation, and other methods.
Chemical	Acid / base treatment, extraction with solvents, dissolving, precipitation, slagging, and other methods.
Thermal	Incineration, burn-out, pyrolysis, melting, slagging, and other similar methods.
EOL disposal	Recycling into the same product, recycling into another product, recovery of energy from thermal treatment of organic layers, utilization of mineral fractions (e.g., concrete, road material, etc.), and landfill disposal.

these materials is not easy and, due to their low market value, not profitable, except for removing and reprocessing the glass.

Since the toxicity level in these modules is very low (some trace amounts of Pb and Sn), dumping the majority of the modules' materials in waste disposal sites does not seem to pose much danger, so very little effort is dedicated today to recycling the materials in silicon based PV modules.

In the future, however, when millions upon millions of silicon modules reach their EOL, their "as is" disposal might become a problem, and recycling might be necessary. In any case, recycling would be un-economical, so plans must be made even now as to when, who, and how the silicon modules will be handled at EOL.

a-Silicon Modules

The semiconductor material in a-Si PV modules is composed of a very thin layer of silicon material, which has a very low market value and is not considered toxic.

Because of that, there is currently no widely accepted drive to recycle these modules at EOL. It is assumed, therefore, that if needed, the a-Si based PV modules recycling and disposal can be done by any of the commonly accepted techniques.

It is clear from work done with a-Si, that the economic benefit from recycling these modules would be negative, even if a good way were found to recycle the top glass with the pre-deposited transparent conducting oxide intact. Finding a good way to do this would significantly improve the economics for all thin film modules, because the glass is a major part of the overall cost of the finished product.

CdTe Modules

CdTe PV modules fall into a totally different category. As much as the manufacturers and their supporters try to paint them as green and environmentally friendly, the presence of significant quantities of toxic, carcinogenic heavy metals necessitates using expensive HazMat procedures during their EOL disassembly, transport, recycling and disposal process.

The recycling process involves chemical stripping of the metals and other materials that have been in contact with them. This is followed by electro-deposition, precipitation and evaporation of the different compounds, as needed to separate and recover the base metals cadmium and tellurium. The plastic solute can be skimmed from the chemical solution, decontaminated and possibly re-used. The glass and frame can be recovered, decontaminated and reused as well.

The cost of recycling and disposal of CdTe modules is much higher than any of the other types of modules, simply because they contain a substantial amount of cadmium which is a toxic heavy metal and considered a hazardous material.

Recycling CdTe PV modules is required, and although it might be beneficial for the recycling mandate, the total profitability of recycling CdTe modules is heavily negative. Recycling CdTe modules is not a profitable process, but it is part of the environmental responsibility of the manufacturers and related parties.

Handling, processing and disposing of hazardous materials is a complex, dangerous and very expensive process, and as chemicals and labor costs increase, in addition to tightening regulations, so will the cost of the recycling and waste disposal processes.

This is something the investors and owners of large-scale PV power fields using CdTe PV modules must be well aware of. A good understanding of the situation is a must, and a contract detailing each of the above mentioned EOL procedures must be signed. This will ensure the proper handling of the hazardous materials in case the manufacturer is no longer in business, or if something happens to the modules in the field.

CIGS Modules

CIGS modules contain a number of metals, such as copper, selenium, indium, and gallium. Some of these are also toxic, albeit in much smaller quantities. Nevertheless, the disassembly, transport, recycling and waste disposal process of CIGS PV modules is complex and expensive too, due to the presence of toxic materials.

The recycling process involves smeltering of the materials, or processing them in acid baths as needed to dissolve and recover the different metals. These are dirty, dangerous, expensive and heavily regulated processes. With increased chemicals and labor costs and tightening regulations, they are getting more and more expensive too.

The top glass is usually processed through thermal decomposition, solvent or acid dissolution to remove any remaining PV layers, and could be recovered.

It has been estimated by researchers that the profit from re-using or re-selling the recovered semiconductor material, metals and glass from CIGS PV modules exceeds the cost of recycling. However, the presence of cadmium, tellurium and other toxic materials in most CIGS modules, increases the cost of recycling significantly, which makes re-selling the recovered materials uneconomical.

There is the need to recycle the toxic materials, and the combined economic benefit of not dumping toxics in landfills.

Basically, there is no economic benefit derived from recycling any type of PV modules. So, recycling is presently looked at as a necessary evil in the case of modules containing toxic materials. In this case, recycling and proper disposal are critical and mandatory. Because of that, there is a growing need to modify existing policies to ensure recycling, and to consider the growing environmental and social benefits not well addressed at present.

Nevertheless, PV modules recycling will grow from a niche market, which it is now, into a big business by mid-21st century.

Non-market based approaches to handling the recycling are also needed to ensure the environmental sustainability of the EOL dilemma. This could be done via corporate and social responsibility, or through governmental regulations, banning the disposal of any PV modules without proper treatment.

RECs

Renewable Energy Certificates (RECs) are a fairly new branch of the energy markets, and basically represent the desire of people and companies to participate, albeit indirectly, in the drive towards energy independence and environmental cleanup.

Most people and organizations are environmentally conscious and are willing to do whatever is in their power to protect the environment. Most, however, haven't the opportunity or funds to install solar panels or wind mills on their property, so they are willing to pay for electricity that is produced on their behalf by someone else and at some distant location, using clean, renewable sources of power generation.

Renewable electricity is attractive for its environmental and greenhouse gas reduction benefits, when compared to conventional fossil fuel-based electricity generation. To help its successful growth in the US and the world, individuals and organizations have been given a chance to participate in several green power product options. These options include buying renewable energy certificates (RECs).

RECs represent the environmental and other non-power attributes of renewable electricity generation and are a component of all renewable electricity products. They are usually measured in single megawatt-hour increments and are created at the point of electric generation.

Buyers can select RECs based on the generation resource (e.g., wind, solar, geothermal), when the generation occurred, and the location of the renewable generator.

RECs provide key information about the generation of renewable electricity delivered to the utility grid. Since RECs represent only the environmental or non-power attributes of renewable electricity generation, they are not subject to electricity delivery constraints.

The information conveyed by a REC allows buyers to make specific environmental claims about how their electricity is produced.

RECs usually include the following primary attributes and information:

- The type of renewable resource producing the electricity (i.e., wind, solar, etc.)

- The vintage of the REC (i.e., the date when it was created)

- The vintage of the renewable generator, or the date when the generator was built

- The renewable generator's exact location

- The RECs eligibility for certification or renewable portfolio compliance

- The renewable generation's associated greenhouse gas emissions (if any)

RECs are increasingly seen as the "currency" of renewable electricity and green power markets. They can be bought and sold between multiple parties, and they allow their owners to claim that renewable electricity was produced to meet the electricity demand they create.

RECs and the attributes they represent are an ingredient of all green power products. REC providers—including utilities, REC marketers, and other third-party entities—may sell RECs alone or bundled with electricity. As of 2007, more than 50 percent of utility customers have access to green power bundled products, whereas all customers have access to buying renewable energy certificates.

Potential buyers and interested parties can get more information, and identify green power suppliers using EPA's Green Power Locator tool: www.epa.gov/greenpower/pubs/gplocator.htm.

The new RECs market identifies the problems and offers some solutions, but it will not solve our energy and environmental problems by itself. It is just another of the energy and environmental instruments at our disposal today.

Third-party Residential Solar

Third-party residential solar installations and financing is a new market niche that is growing exponentially in the US. Third-party financiers eliminate the upfront cost of solar panels to customers. Customers can install solar panels for no money down and pay just for the solar electricity they produce at prices below utility rates. The third parties (or their partners) manage the process of permitting, installation and long-term O&M under a set of strict conditions.

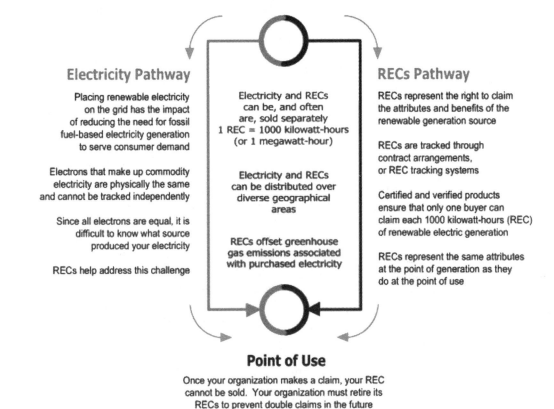

Figure 8-9. RECs chart, (according to EPA, 2012)

Third-party financing is being entertained for other energy services like solar hot water, and solar on commercial buildings, in addition to a number of start-ups that are looking into this new way to finance and operate.

The major players and their market share in the summer of 2012 are shown in Table 8-20.

Table 8-20. Third-party residential solar providers

Company	Share
SunRun	37%
SolarCity	20
SunPower	13
CPF	11
Sungevity	8
American Solar	5
ROW	6

This market's growth is ensured, at least for now, by the risk-takers like Citi bank, Credit Suisse and others, who will collect the profits from the generated power during the next 20 years. The only unknown is how much power will be generated, since the multi-components variables are hard to calculate, and there is no precedent.

Roof Installations

Solar rooftop installations impose a number of challenges that could slow their widespread adoption. The issues include quality of PV devices, maintenance of roof water-tightness, roof warranty issues, continuous vs. variable operation, aesthetics, reliability and stability, durability of materials, and high costs of labor for installation and maintenance.

The benefits to utilities are huge, since this is a low-risk investment that encourages energy efficiency and conservation, and reduces dependence on the electricity grid.

To successfully address these issues, the roofing system should be designed and installed to serve as a permanent platform for the continuous operation, service and maintenance of the PV system. In terms of investment, the quality of the roofing system is especially critical. Not only must the remaining service life of the existing roof system be adequate to meet or exceed the investment horizon of the PV system, but there should be a detailed review of the compatibility of the existing *roof* to accommodate the PV system both initially as well as during ongoing operations.

Advances in affixing the PV system to the roof using state-of-the-art roof racking systems not only address the above challenges, but drive down the costs of PV installation.

Residential. Roofs that are usually steep-sloped present certain issues for PV module racking. Several manufacturers now offer solar shingles that greatly reduce the negative aesthetic impact of PV solar panels.

However, most residential PV systems are elevated above the roof, but still retain a low-profile appearance. The racking structures in these designs use stand-offs and fasteners that penetrate the roof to keep the assembly from sliding off, and provide mechanical stability in high winds. The integrity of the penetration points is crucial, as each must be properly sealed to prevent water leaks. Additionally, most residential roofs have surfaces facing different directions, which can impact area utilization for electrical generation.

Commercial. Roofs that are low-sloped (standard for commercial buildings) are geometrically simpler, walkable, and therefore lend themselves to ballasted PV module technology. A ballasted system relies on the weight of the ballasted and racking structures to hold the PV modules in place. A nearly horizontal roof presents good exposure to mid-day sunshine when daily solar production reaches its peak.

Materials. An issue common to all rooftop PV systems is the durability of the materials constantly exposed to the sun and weather. Fortunately, recent material innovations increase service life. The newest generation of PV systems entails Fibrex, a lightweight composite material with excellent weathering characteristics and insect resistance. Such materials do not corrode or degrade quickly and can provide a serviceable lifetime consistent with the PV cells.

Labor Costs. The cost of PV solar panels has declined considerably lately, but still represents approximately 40% of the total system cost, down from 80% just a few years ago. Economy of scale has similarly led to cost reductions in inverters and other system components.

Despite the cost reduction of PV modules, and other "balance of system" items, widespread adoption in the U.S. has been stymied, in part, by issues related to roof racking products that involve complex, labor-intensive assembly and installation procedures requiring skilled labor and special tools.

Labor costs for solar roof projects involve installation, as well as design and engineering. Therefore, as component costs of PV systems decrease, labor has become a significant cost driver.

In all cases, rooftop solar installations (already a large business), also offer a number of opportunities for niche market improvements, new designs of materials, components and methods. These and other improvements are expected to drive the rooftop solar installations during the 21st century.

Solar Power Leasing

The successful deployment of solar power generation is critical for America's future energy supply, but its financing has been a problem since day one. Recently, a lease financing mechanism was introduced as one of the most powerful drivers of solar power deployment in the US. Though solar leases have helped grow the industry, they come at an inflated and higher than intended cost to US customers, if compared to cash purchases.

In October 2011 leased systems surpassed cash purchases in California, according to the California Solar Initiative (CSI). In 2008 leased residential PV systems represented ~6% of the number of installations in the market tracked by the CSI. As of April 2012 that number was ~60% and growing. Like many other financial products that proliferate without being matched by a change in an underlying hard asset, the solar lease warrants further scrutiny.

Counter to normal business logic, profit in the solar leasing model is driven in part by higher costs (and not higher production), as far as permitted by regulation. The solar lessors can exaggerate costs to inflate tax benefits and to minimize the risks. They can also profit by syndicating these tax benefits to large corporations or banks to decrease their tax liabilities.

Accounting Principles. As a result, the US taxpayer may over time pay for the entire system in lost tax revenue and dubious tax credits. For bookkeeping purposes under Fair Accounting Standards Board (FASB) and Generally Accepted Accounting Principles (GAAP) rules, a solar lease must be structured as an "operating lease," and must avoid characteristics that would trigger a "capital lease" designation.

The FASB/GAAP treatment primarily applies to businesses seeking off-balance sheet financing, a controversial accounting method. In order to qualify as an operating lease and not a capital lease (also known as a financing lease, which would appear as a liability), the following requirements apply:

1. Lease payments under contract must not exceed 90% of the fair market value of the equipment,

2. The term must be less than 75% of the useful life of the PV system,

3. Ownership must not default to the lessee (typically the host of the PV system) at the end of the lease term, and

4. The buyout price must be about fair market value—no US$1 basement buyout price. As the referenced patent reads: "…customers are given fair market value buyout options, which are priced so that they would not fail one of the capital lease tests."

The manipulation of these rules is a controversial tax tactic used in other capital intensive industries, such as airlines. Simply put, an operating lease appears in a company's income statement as an operating expense. This lets the lessee/host take on the liability of the lease without the lease affecting its balance sheet, which would reduce its book value and leveragable assets. Solar leasing companies use IRS and Treasury rules to make every installation as valuable a tax asset as possible.

A key method for achieving this goal is to structure leases as a true lease rather than a tax-oriented lease. As a true lease, SolarLease Co. and SunFund, our hypothetical lease originators and finance backers, are owners of the financial attributes of the system, and are able to collect depreciation benefits that a residential system purchased with cash is ineligible to collect.

The financial attributes of a system are created through several key levers:

1. For the customer: paying for the lease via a tax deductible home equity line of credit. The tax deduction is an effective subsidization.

2. For SolarLease Co. and SunFund, inflating the total installed cost of the project generates
   a. The largest investment tax credit possible
   b. A higher depreciable tax basis

To a homeowner paying for a system in cash, the installed cost dictates the long-term net present values and rates of return, which are the fundamental metrics used to determine whether to purchase a PV system. On the other hand, if a homeowner receives a flat monthly lease rate, they probably don't care what the cost benchmarks and rules are for tax treatment. This is part of the appeal.

One of the largest leasing company's average reported cost under the California Solar Initiative for residential projects installed in 2011 was US$7.91/Wp.

Another large leasing provider in California reported an average cost of US$8.28/Wp. This compares to the first company's average cash, non-third-party owned system, cost of US$6.40/Wp. Meanwhile, many firms install residential projects on a cash basis for a cost of US$4-5/Wp. Property owners who have purchased and installed their own PV systems reported an average installed cost of US$4.56/Wp in 2011.

In other words, homeowners and wanna-be solar operators have choices, but must understand what's behind them, and make wise decisions. Solar is good, as long as it doesn't break, and/or as long as it doesn't break the bank.

Solar power leasing represents a niche market that has grown out of proportion, and some of its components need to be balanced. This opens a number of smaller—niche—markets for consultants, accountants, installers and other professionals involved in the solar field. The market will grow proportional to the ease of access to the products and projects, so an army of these professionals will be needed to guide it in its future development.

Solar Thermal (Water Heating)

The California Solar Initiative (CSI) promises to support the solar hot water industry, and to put it on a footing with natural gas. Although the working conditions are favorable for manufacturers and contractors, customer resistance and confusion is driving the market into uncertainty, exacerbated by a few inferior products and installations that give the industry a bad name.

After decades of diminished sales and low expectations, solar thermal hot water is hoping again (again is a key word here) to capture a sliver of the enormous opportunity. The US DOE's own, Energy Information Administration estimates the current total revenue for solar thermal collector shipments at only $100 million annually. In 2010 forecasting business decisions was difficult because of uncertainty in the economy, incentive ambiguities and the unpredictability of fossil fuel markets—especially natural gas. Despite these concerns, the quality and number of sales leads arriving in the second half of 2010 put the industry back on track, and even show some growth.

Expectations are great, and industry executives expect the next five years to bring a lot of hot water business. Commercial and industrial markets also offer some optimism. Commercial customers today are more affected by, and more aware of, solar thermal hot water technology and are much more willing to install it, provided the cost is right.

To compete with taxpayer-funded oil and gas exploration, however, solar thermal needs government incentives to generate new sales. The typical incentive program of late included an upfront payment of approximately $1,500 for a single family residential hot water system, while commercial incentives are based on the square footage of installed collectors. In addition, public benefit funds may pay for thermal incentives too, which are funded by surcharges on electric bills while replacing natural gas.

For any incentives to be most effective, lawmakers must enact well-designed, consistent programs. Poorly designed programs often turn out to be difficult to implement. On the other hand, consistent, long-term programs have the maximum impact, create the best opportunities and ultimately transform the market. But extended programs cost more than short-term programs, so legislators are motivated to enact programs incrementally.

California Solar Initiative

The solar thermal industry is watching closely the developments of the ambitious multi-billion California Solar Initiative (CSI) program, which was championed by the California Solar Energy Industry Association (CALSEIA), with the help of Environment California, which provided the necessary advocacy support.

CSI is the solar rebate program for California consumers, customers of Pacific Gas and Electric (PG&E), Southern California Edison (SCE), and San Diego Gas & Electric (SDG&E). It funds solar installations on homes, commercial, agricultural, government and non-profit buildings. This program funds both solar photovoltaics (PV), as well as other solar thermal generating technologies. It is sometimes referred to as the CSI general market program, and it consists of:

1. A solar hot water rebate program for customers in PG&E, SCE, and SDG&E territories. This program funds solar hot water (solar thermal systems) on homes and businesses. This program is called the CSI-Thermal program.

2. A solar rebate program for low-income residents that own their own single-family home and meet a variety of income and housing eligibility criteria. This program is called the Single-family Affordable Solar Homes (SASH) program.

3. A solar rebate program for multifamily affordable housing. This program is called the Multifamily Affordable Solar Housing (MASH) program.

4.   A solar grant program to fund grants for research, development, demonstration and deployment (RD&D) of solar technologies is the CSI RD&D program.

The CSI offers solar customers different incentive levels based on the performance of their solar panels, including such factors as installation angle, tilt, and location rather than system capacity alone. This performance framework ensures that California is generating clean solar energy and rewarding systems that can provide maximum solar generation.

The CSI program has a total budget of $2.167 billion between 2007 and 2016 and a goal to install approximately 1,940 MW of new solar generation capacity.

The CSI-Thermal portion of the program has a total budget of $250 million between 2010 and 2017, and a goal to install 200,000 new solar hot water systems.

The CSI program is funded by electric ratepayers and the CSI-Thermal portion of the program is funded by gas ratepayers. It is overseen by the California Public Utilities Commission and rebates are offered through the program administrators.

The residential incentives program was introduced in May, 2010, and its commercial counterpart was launched in October, 2010. Unlike the commercial program, participants in the residential program cannot reserve its funding prior to installation which means it's too early to gauge level of participation.

Sun Light & Power installed the first commercial application to receive a CSI rebate check for solar hot water in California. In November of 2010, the company installed 2 Heliodyne Gobi 410 collectors on the roof of the Taco Bell restaurant in Albany, CA. The combined rebate of $3,654 on top of the 30% investment tax credit (ITC), paid for half of the total project cost.

The planned $25 million marketing program to broadcast the CSI has not yet been implemented, but the marketing of the CSI may be as important as the structure of the program itself.

Cogenra, a Khosla Ventures-funded firm, combines solar hot water with photovoltaics and seems to be gaining some traction.

Solar Space Heating And Cooling

A number of manufacturers are working on developing solar thermal space heating systems, which may finally develop into a viable market. The DOE has encouraged this development with a new 222,000 square-foot research support facility for the National Renewable Energy Laboratory. This building demonstrated solar space heating with a 9,000 square-foot Conserval's SolarWall system integrated on the south wall to preheat the fresh ventilation air.

Space cooling is an even more attractive market, because it works best when demand is high (mid-day) and because air conditioning, which is the largest load of a commercial building, works harder during these hours. Several equipment manufacturers have taken the lead by demonstrating the viability of their solar thermal cooling systems. Absorption chillers allow the building to use the thermal collectors to power its air conditioning. The water heated by solar energy in the collectors is used to initiate a thermal dynamic process involving low-pressure chambers that chill water to around 44°F.

Another new technology, using the sunlight as a power generator is successfully used to dehumidify ambient air by using a reusable desiccant—not unlike the small packets that keep packaged products dry. It uses solar thermal heat to evaporate the moisture from the desiccant and keep the air dry in a number of applications.

This dehumidification process, however, handles only the latent load or the heat responsible for humidity. Additional solar capacity could be used to power absorption chillers or air conditioning units for large space cooling.

Process water or air heating and cooling is of interest to agricultural livestock businesses, especially poultry farming and domestic pigs.

Hospital laundry services, food processing and dairy farms are ideal candidates for solar thermal process heating. Even though process heating is eligible for an ITC, it is the accumulated cash flow over the 30-year life of the system that is especially attractive.

The Solar Rating and Certification Corp (SRCC), which tests solar thermal collectors and other equipment, has continued to restructure its operations in the face of challenges caused by testing delays. Over the last several years, demand on the only two accredited testing labs grew significantly, causing costly certification delays for manufacturers.

To address this, SRCC instituted an interim certification process—offering manufacturers a temporary listing for their collectors to get collectors into the market faster. More significantly, SRCC recently accredited 15 new testing labs worldwide, so the load is now evenly spread, and the certification process was significantly accelerated.

Geo-thermal Power

Geo-thermal is widely used in a number of countries, and the US is getting involved in it as well. As an example, a new geo-thermal power plant, the 50 MW Hudson Ranch 1, was built on the shores of Salton Sea in California by EnergySource. Construction of the project, everything included, is estimated at $400-450 million.

The power will be bought for 30 years by Salt River Project (SRP), a municipal power and irrigation district in Tempe, AZ. And SRP has in fact already agreed to a second power purchase agreement from EnergySource's next 49.9 MW geothermal plant, Hudson Ranch II.

But… in order to pay for the purchased power produced by someone else, and to avoid building its own renewable energy plants, SRP is increasing its rates by a flat $6.00 per month. Each household serviced by SRP in Arizona will pay $72 a year more than they presently pay. Half of the those serviced by SRP can't pay their monthly bills as is.

Looking at the broader picture, this is what we will see in the future—increasing cost of power produced by expensive additions to the renewable energy sources.

Solar Water Heating

Solar hot water generation was heralded as a major energy saver in the 1980s. Many federal and state subsidies went into developing and installing different types of solar water heaters. Entire industries popped up as a result of this effort—before the bust.

According to specialists, 100 million water heating systems of different types are operating in the U.S., with half using electricity, fuel oil or propane. This constitutes a market opening of more than 50-million homes and commercial enterprises that don't use natural gas to heat water.

While there is a well defined and accessible market niche for solar water heating, it is not well supported by the government. So the market might stay dormant until the government removes the barriers and creates adequate, stable, long-term policies for solar thermal— including economic incentives.

As we learned from past experience, manufacturing, installation and operation standards are needed, as is well-funded research and development of appropriate technologies for different geographic and economic locations.

Any advance in solar thermal energy generation must be accompanied by advances in thermal storage, for this is how the market will be expanded into the small- and medium-size commercial enterprises.

The U.S. military is also warming up to solar hot water. The Eneref Institute, founded in the wake of the September 11 attacks to help shift the nation's energy use, has recently taken up the cause of solar thermal

and brought it to the Pentagon where, due to budget constraints, it has been well received by a military eager to save money and energy.

Third-party financing has completely changed the solar photovoltaic landscape—opening up new customer sectors, new investment areas and huge entrepreneurial opportunities. Could third-party financing do the same for solar hot water?

In any case, solar water heating is a huge niche market, which seems to be headed to expansion, even under today's weary socio-economic conditions.

Solar Inverters

Solar inverters are electro-mechanical devices that convert the DC power generated by PV modules into AC power. Thus converted AC power is conditioned to match the grid frequency and is sent into the distribution networks.

While there are mixed signals and business environment scenarios for the solar sector with over 150 active suppliers, inverter manufacture revenues have been steadily increasing in 2012. There is a reported 3% revenue increase, and a 25% increase in shipments…?

Projections are that inverter suppliers will continue to see high shipment growth, but will be struggling with the equilibrium of top- and bottom-line growth. Inverter prices decreased by double-digits in 2012, partly driven by product mix change and shifts in demand to lower-cost countries. The main driver, however, is the price erosion forced by the stagnation of major markets. The good news is that although the market shrank in terms of revenues, shipments increased by more than 12% from 2011, which is a bonus for inverter manufacturers, freeing them from excess inventory from the prior year.

The PV inverter market in Europe is expected to continue to lose market dominance as key markets such as Germany and Italy experience PV system installations decline from 2012 onwards, due to serious FiT reductions and even more serious financial crunch.

Revenues will not to return to 2011 levels until 2016-2018, which in itself presents a huge challenge to suppliers—most of whom are European, with the majority of their facilities and customers located in Europe.

Nevertheless, the inverter market is still growing around the world, albeit in a more hesitant and fragmented manner. So, even in bad times, shipments of inverters are forecast to increase over the next five years, with revenues projected to exceed $9.0 billion by 2016-2018.

Market share rankings are unchanged, though SMA Solar Technology, which is/was the market leader is showing weakness and continues to see its share erode. The market share and revenues of the other major players, Power-One, Kaco, Fronius, and RefuSOL are expected to follow the above expectations without any major changes.

Surprisingly, the biggest market share gainers in 2011-2012 were the second-tier inverter manufacturers, like the start-ups Enphase Energy and SolarEdge. These smaller manufacturers, show guts in moving onto their larger cousins' turf and gaining significant dominance of the alternative energy markets. This also shows that they are not going to give up, and we expect other entrants in the field. Although consolidation of the inverter industry was previously predicted, we now think that it won't happen in a significant way.

Smart Home

Energy is an important pillar of the connected home market, and big box retailers are capitalizing on emerging trends to converge on the energy market. The proliferation of smart appliances, smart meters, and devices such as smart thermostats as part of the connected home will be further driven by growth in smart phone and tablet applications that allow users to more easily monitor, manage and control these technologies while away.

Energy management and the ability to lower energy bills using dynamic or real-time pricing programs can be attractive to customers. Such services, particularly energy management, increase the stickiness of one's offerings and, once the market moves beyond the first generation of offerings, lead to new recurring revenues. 65% of the U.S. broadband households rate at least one energy-related value-added service as highly appealing, with demand for energy management broader than the demand for security.

"Unified communications" is increasingly being looked at by businesses as a way to contain costs, expand reach and simplify communications—both internally and externally. But there are stumbling blocks. Interoperability, proof of ROI and a lack of understanding are all contributing to what has been fitful progress for the technology.

Development of smart grid in general and home energy management systems in particular is creating a need for a range of new niche service providers. Utilities are increasingly enlisting the help of retail stores to get the message out about energy efficiency, specifically, smart home energy management, including smart thermostats, lighting controls, and alarms.

Retail establishments are partnering with energy

management service providers to deliver the same message. For example, Sears, Home Depot and Best Buy all have programs that promote the benefits of smart home energy monitoring. Instead of focusing just on energy, the stores are offering products for the fully connected home. These appealing sales propositions are more about control and comfort than kilowatt per hour savings.

SRECs

Solar Renewable Energy Certificates (SRECs) or Solar Renewable Energy Credits are a form of Renewable Energy Certificate or "Green tag." SRECs exist in states that have Renewable Portfolio Standard (RPS) legislation with specific requirements for solar energy, usually referred to as a "solar carve-out."

The states with viable and functioning SREC systems are California, Delaware, District of Columbia. Maryland, Massachusetts, Missouri, New Hampshire, North Carolina, Ohio, and Pennsylvania. The other states either do not have an SREC program, or it is not used.

SRECs represent the environmental attributes from a solar facility, and are produced each time a solar system produces electricity. For every 1000 kW/h of electricity produced by an eligible solar facility, one SREC is awarded. In order for a solar facility to be credited with that SREC, the system must be certified and registered.

The additional income received from selling SRECs increases the economic value of a solar investment and assists with the financing of solar technology. In conjunction with state and federal incentives, solar system owners can recover some of their investment in solar by selling their SRECs through spot market sales or long-term sales, both described below.

In order to produce SRECs, a solar system must first be certified by state regulatory agencies, usually public service commissions or public utility commissions, and then registered with the registry authorized by the state to create and track SRECs. Once a solar system is certified with the state agency and registered with a registry such as PJM-GATS or NEPOOL-GIS, SRECs can be issued using either estimates or actual meter readings depending upon state regulations.

Solar RPS requirements are meant to create a marketplace for SRECs and a dynamic incentive for the solar industry. Solar RPS requirements demand that energy suppliers or utilities procure a certain percentage of electricity from qualified solar renewable energy resources in a state.

These energy suppliers and/or utilities can meet

solar RPS requirements by purchasing SRECs from homeowners and businesses who own solar systems and produce SRECs. Homeowners and businesses can then utilize the sale of the SRECs they generate to help finance their solar systems. SRECs can be sold in a variety of ways, such as on the spot market, at auction, or by negotiating long-term contracts.

SREC supply in a particular state is determined by the number of solar installations qualified to produce SRECs and actually selling SRECs in that state. As more solar systems are built, SREC supply will increase. SREC demand is determined by a state's RPS solar requirement, typically a requirement that a certain percentage of the energy supplied into a state originate from qualified solar energy resources. Load-serving entities or organizations that supply electricity into the state must meet these requirements. RPS solar requirements in many states are set to increase in the coming decade.

Typically, there is no assigned monetary value to an SREC. SREC prices are ultimately determined by market forces within the parameters set forth by the state. If there is a shortage in SREC supply, pricing will rise, resulting in an increase in the value of the incentive for solar systems and an intended acceleration in solar installations. As SREC supply catches up to SREC demand, pricing will likely decrease, resulting in an intended deceleration in solar installations.

Spot prices for SRECs are generally higher than prices found in long-term contracts since the system owner is taking on market risk. If increases in supply outpace the growing demand, spot prices could fall. SRECs have traded as high as $680 in New Jersey. Meanwhile other state SREC market prices range from $45 in Delaware to $271.05 in Massachusetts.

In addition to providing cash flow security and stability, long-term SREC contracts are often required by banks or other lending institutions unwilling to accept market and legislative risk associated with SREC markets. However, SREC contracts longer than 3 years can be difficult to secure in some SREC markets because in deregulated electricity markets, energy suppliers rarely have electricity supply contracts longer than three years.

In general, SRECs have contributed to the rapid development of solar energy in a number of states— some of which don't even have much sunshine to justify solar installations. SRECs served the purpose of kick-starting the solar power generation in places where it didn't have much of a chance.

While SRECs, together with government subsidies and incentives, were able to do some good things in the US solar economy, it is not clear now how much good they can do alone. Without the subsidies and incentives—especially in areas with restricted solar insulation—SRECs alone would not be able to support the fast growth of solar installations. We might be surprised, but common sense dictates that the solar energy of the 21st century will be generated in areas with the most sunshine…SRECs or no SRECs.

THE 21ST CENTURY DIRECTION

No doubt, California is the leader of alternative energy generation in the US, and soon it will be first in the world as well. Things have gone fairly smoothly in California, but now a wave of change is hitting the shores of the sunshine state.

Actually it has been coming for awhile now, where CSP projects have been cancelled and later converted into PV projects, and other unexpected anomalies within the renewable industry have been noticed.

Could it be that wind and PV are turning against each other? The three major California utilities are dedicated to providing 30% of their power from renewable sources by 2020. Because of that, they and the regulators are no longer concerned about the quantity of the renewables. That point has been decided and agreed upon!

As a matter of fact, by May 2012 PG&E has renewables capacity equal to 20% of its 2011 electric power generation. SDG&E has procured approximately 21% percent, and SCE has over 21%. Renewable energy investors confirm that the IOUs have as much as three-quarters of their 2020 obligations under contract.

Now the game is switching from quantity to quality (long-term profitability). So, the utilities are much more carefully considering the quality (in terms of efficiency and longevity) of the solar and wind projects and calculating the best ways to fit them into their portfolios.

Net Market Value

In an attempt to determine the best economic choices for the implementation of the remainder of the California renewables, the CPUC is considering a new formula. It comes as a redefinition of the 2004 "least cost, best fit" formula for capturing the full range of costs and benefits of renewables selected to meet the RPS.

Net market value measures all the benefits and costs of delivering energy into the system. It is a critical measure of the type of system to use, where and when. As confirmation of our claim that the world solar energy is an embryo undergoing growing pains, CPUC, and major California utilities are looking into a new way to value the performance and costs of solar power plants.

Accurate assessment and effective management of the costs of integrating the variability from large wind and solar power generation into the existing transmission systems is the big issue. Some solar companies are questioning the way it is done. The valuation methods are outdated, because they use the established—albeit irrelevant to wind and solar—standards of fossil and nuclear generation.

Case in point, CPUC has been considering a new formula for a cost assessment, based on the "least cost, best fit" formula, designed to capture the full range of costs and benefits of renewables, which are to be selected to meet the state's 33% renewables by 2020 renewable portfolio standard (RPS).

The Net Market Value of a generation source is defined as:

$$R = (E + C) - (P + T + G + I)$$

Where
    R   is NetMarketValueofapowergenerationsource,
    E   is Energy Value
    C   is Capacity Value
    P   is Post-Time-of-Delivery Adjusted Power Purchase Agreement Price
    T   is Transmission Network Upgrade Costs
    G   is Congestion Costs, and
    I   is Integration Costs.

The proposed formula for Adjusted Net Market Value is,

$$A = R + S$$

Where:
    A   is Adjusted Net Market Value,
    R   is Net Market Value, and
    S   is Ancillary Services

These terms are still to be defined for the point of use. CPUC is working with researchers at NREL and LLNL to finalize the definition of these terms and quantify the factors that are used in the decision-making processes. Still a lot of work is to be done in this area until then by all involved—NREL, LLNL, Cal ISO, etc.—whose emphasis is on the need to look at the full cost of operating a system when comparing different power generating sources.

The new approach would drastically change the way utilities purchase renewable energy in California, since it proposes to substantially increase the role of transmission planning in renewable procurement. And with the addition of ancillary services into the equation, it includes a generation source's potential positive impact on grid operation.

In contrast, until now, solar and wind power developers have sited their projects where the wind blows and the sun shines. They relied on the old and proven, "if you build it, transmission will come" paradigm. The new value formulation would promote development of "cost-effective and location-appropriate transmission" first. It could save ratepayers billions in the RPS procurement process.

The new, comprehensive method of valuation proposed by the CPUC shows the concentrating solar power (CSP) plants with energy storage will be highly valuable to utilities in their pursuit of meeting the RPS without compromising reliability.

According to the industry principals, utilities are obligated to balance reliability and affordability and are committed to environmental stewardship. A utility has different ways of meeting the energy requirements of their customers, as needed to comply with the required reliability standards, which include constant and reliable service.

In contrast, wind and PV (and CSP without storage) power generators will also provide variable output, which utilities do not like quite as much. Instead, they require balancing generation as a prerequisite to maintaining reliability and low cost of their services.

Output variability impacts grid operations and adds costs for ancillary services, spinning reserves, non-spinning reserves, and other types of market specific products used extensively under variable load.

Wind, for instance, may be most productive during off-peak hours, which is not in alignment with the system's maximum demand. During the 2006 California "heat storm," for example, the California ISO reached its all-time maximum demand. At that moment it had about 3,000 megawatts of wind available, but the amount of wind delivering electricity into the system when it hit its peak demand was only 1%.

What it comes down to is that the actual time of delivery has an intrinsic value, and energy delivered at different times of the day has different values, with the on-peak availability having the highest value.

CSP power generator with energy storage has the advantage of delivering power whenever demand is peaking and that is usually worth from two to three times what it is compared to off-peak. How much storage is needed is a another question, and makes sense as an economic optimization variable with the focus on the total value of the energy being delivered to the market when needed.

For example, one project in the Mojave Desert near Las Vegas has a 32% energy storage capacity factor, which in the CPUC formula plays a major role in the value of a generation resource. With added storage, the capacity factor increases. Four to six hours of energy storage will bring the capacity factor above 50%, and the beauty of it is that the cost of operation is not increasing for the duration. So the fixed costs are spread out over more hours of the day or night, which ultimately drives costs down.

This way, the operation of peaker power plants can be reduced, and plans for building new plants can be cancelled, saving a lot of money in and adding more value to renewables power generation.

This is a new angle from which to look at solar energy. It represents economic optimization, which provides a new definition to the variability and overall cost of solar energy.

So, CPUC is basically trying to formalize the "least cost, best fit" methodology used for procurement, and is also looking at the individual components using the best approach for their quantification. These components include capacity value, energy value, transmission upgrade costs, and integration of renewables.

"Capacity value," for example, is a new approach to quantifying the long-term potential of a renewable power resource to reduce, delay or avoid entirely the need to build additional coal or gas peaker plants.

Since peaker plants are built with only the goal of increasing power generation as needed during peak hours, they are expensive to operate and maintain. Reducing the need to use existing or build new fossil-fueled peaker plants is considered a capacity benefit and to therefore have capacity value used in the formula. The only question is how to assign an exact numeric value to the C factor in the formula.

Here is the rub: solar and wind are variable power sources. They have the potential of providing power during peak hours—different amounts during different hours at different locations—but their contribution is seriously hampered by the availability (or lack thereof) of wind and sunshine at certain times of day—including peak hours. Lack of readily available energy storage, or its high cost if and when available, complicates the situation. This makes the new formula CPUC is working on an educated guess at best.

For example, during the summer of 2006 "heat storm" in California, ISO reached all-time maximum demand, while sitting on over 3.0 GW of available wind energy generating capacity. Nevertheless, the amount of electric power delivered into the grid at that particular time of extreme peak demand was only 1%.

At the same time CSP and PV plants around the state were working full blast, producing a lot of power, which helped the emergency situation. This makes them more valuable during summer peak hours, according to industry specialists.

For the sake of the formula, since wind produced about 30 MW of power during that emergency, and since peaker plants are about 50 MW in size, the wind contribution in this case avoided 60 percent of the cost of using an existing peaker, and contributed to saving money for building a new one.

Nevertheless, most specialists agree CSP with energy storage is more likely to provide a higher capacity value during summer noon peak hours than wind. But, capacity and, more broadly, the energy value refers to power generation during the entire year. It then relates to the total value of fuel saved during these special moments when renewables' generation supplements power plant generation. Wind might have a higher value in such cases, simply because it is operational, albeit unpredictable and variable, day and night.

In conclusion, the work is complex and must be focused not only on quantifying the particular values of power generation and use, but also on understanding how they change with increased penetrations into the national energy portfolios.

The immediate challenge is making procurement decisions, which must include the benefit of renewable resources as defined in the "least cost, best fit" methodology. It must also consider the LCOE for the involved resources (which by the way is most commonly, and somewhat unjustly, used). In reality, it is the combination of these two factors that gives the best picture of a particular situation, so both must be used for best results.

As an example, at 0% penetration of a PV source with a hypothetical value of $90 per megawatt-hour and a cost of $100 per megawatt-hour, the renewable premium, according to CPUC definition of "Net Market Value" is in the range of $10 per megawatt-hour.

A CSP plant, on the other hand, might provide the same value of $90 per megawatt-hour, but a hypothetical cost might be more like $200 per megawatt-hour. Then the renewable premium is $110 per megawatt-hour.

So it looks like a competition between CSP, PV and wind is brewing on all levels. Until more data are available, we must not look at the renewables as "either/or" solutions, but instead we should consider each for what it is worth and what it can do.

We will have to wait for the completion of the work on these useful definitions and important components in the procurement equation: energy value, capacity value and transmission costs.

Entering the 21st century, we should not consider "wind only," or "solar only" as immediate solutions, but use the combined powers of these technologies for longer-term benefits. With the help of the utilities and regulators, this will be how solar power generation and the entire alternative energy business is be done in the 21st century.

Net Metering

To encourage small electrical customers to install environmentally friendly sources of electrical power, the state of California has passed laws to make the process of interconnecting solar or wind powered generation systems with a total installed capacity of 1 MW or less as simple and economical as possible.

This legislation also provides for "Eligible Customers" to receive the additional benefits provided by a billing process referred to as Net Energy Metering (NEM). SCE supports these objectives and has tailored its tariff rates and interconnection requirements to minimize the "red tape" while ensuring the safe operation of the equipment its customers may install.

NEM allows a customer to net the energy produced by its generating facility against the energy received from its electric service provider. The "netting period" is established as the 12 months following the date NEM service is initiated for an eligible customer.

During this period, if the customer has produced more energy than it has consumed, it will be charged only for the "non-energy" related components of the electric service it has received. If the customer's generation has not met all of its energy needs, the customer will also be billed for the shortfall, or net energy supplied by its energy service provider. Residential and small commercial customers may elect either a monthly or annual billing option.

All other NEM customers will continue to be billed monthly with surplus energy credits carried forward from month to month, although the state code does not provide for surplus energy credits to be carried forward past the 12-month netting period, and surplus energy is not purchased by SCE. In all cases, customers are to be provided with monthly statements to allow them to monitor the status of their account.

The only issue left was that of how to define the NEM cap, and CPUC addressed that on in May 2012. The definition of the cap had become a battleground over NEM pitting Investor Owned Utilities (IOUs) against renewables advocates.

The CPUC decision effectively lifts the ceiling from 2.4 GW of small-scale residential and commercial solar to 5.2 GW. Commissioners went much further than before when they blocked San Diego Gas & Electric's general rate case, which sought to scrap the "silent subsidy."

NEM was capped at 5% of the utility's "aggregate customer peak demand." But how many actual MWs that meant varied, because California's three main IOUs were doing their own separate calculations. So the unknown was whether they were using the correct formula to reach their own numbers in calculating the NEM cap.

The commissioners voted unanimously against the IOUs and in favor of a definition of the NEM cap that will allow for much more distributed generation (DG) going forward. Like 43 other states, California has a NEM program that allows owners of DG systems of up to one megawatt in capacity, like small wind turbines, combined heat and power systems and rooftop solar systems, to reduce their electricity bills. For the kilowatt-hours they send to the grid, system owners' meters turn backwards as they are credited at the same retail rate they pay for the kilowatt-hours they consume.

So the fire is officially, partially, and temporarily (until 2014) put out by CPUC, but the battle is far from over. The utilities are looking into ways to level the field for more freedom of maneuvering in the future, fearing that the variability of renewables power generation will bring them more surprises and more expenses.

The Smart Grid

"Smart grid" is a fairly new concept, with growing importance. It is a common denominator for a wide range of developments that make power generation and the related medium- and low-voltage grids, as well as the use of electric power, much more intelligent and flexible than they are presently.

Communications among the key elements, and control of them, will allow them to be managed more efficiently and much easier. The main motive for smart grid initiatives is that such developments improve reliability of supply and/or support the trend towards a more sustainable energy supply.

Presently, most medium- and low-voltage networks cannot be remotely observed and controlled, so, when fully developed, smart grid components will eventually solve that problem. At least this is the goal of industry specialists and the companies involved.

Various companies are already developing technologies aiming at creating and supporting different smart grid network segments. However, some of these developments are either futuristic, or are based on technological possibilities, rather than in a sound problem analysis and a coordinated, structured smart grid approach.

Recently, a great variety of sensors, protocols, communication equipment and such has been designed to support the move toward smart grids. However, because of differences in opinion and lack of standardization, most of them have not found wide application. This can also be at least partly attributed to the fact that they simply do not provide significant solutions to the existing problems.

In other words, there is too much technology push and too little market pull. The fact that some manufacturers of unsuccessful technologies even blame network operators as conservative instead of improving the price performance ratio of their products, reflects the anomaly and further hampers a real take-off of smart grid concepts.

In the longer term, smart grid technologies will play an important role in maintaining reliability of supply and improving sustainability. The complexity of electricity distribution increases, as new PV plants are connected to the grid, and the number of wind turbines and solar power plants increases as well. This also applies to small generators, such as rooftop PV installations, where smart grids support these developments by continuously monitoring and controlling the grid and the generators.

The smart grid vision is becoming clearer over time, and ever greater efforts will be spent on developing smart grid technologies in cooperation with commercial energy companies, other grid operators, and suppliers. These efforts will also help to increasingly focus discussions between regulators and the government on the future energy supply and the role of smart grids.

The Smart Grid Markets
Europe

In May 2012, members of the Smart Energy Collective in Europe approved the second phase of the smart grid initiative, which involves the development of five large-scale smart grid demonstration projects in the Netherlands.

Schiphol Airport, ABB and Siemens offices, and several residential districts were the chosen sites for smart grid implementation and tests.

"In order to make this possible, an intelligent energy system is required that uses a combination of innovative technologies and services, which will enable us to keep the costs of our energy supply at a reasonable level with a comparably high level of reliability," said the principals.

This effort would be an important step towards standardization, in addition to testing the smart grid operation in actual use. A survey during the first phase showed that there are more than 6,000 relevant standards playing a role in the new technologies, which according to the group, will be introduced to the market in the coming years.

Working groups have been established for standardization, market mechanisms, services and business cases, smart grids, privacy and security, and ICT infrastructure, in order to establish a solid foundation for the design of the five demonstration projects, as well as for future projects.

The Smart Energy Collective is a sector-transcending cooperation involving a wide range of companies working for smart energy and smart grid implementation. Members include: ABB, Alliander, APX Endex, BAM,DELTA, DNV KEMAEnergy & Sustainability, Efficient Home Energy, Eaton, Eneco, Enexis, Essent, GEN, Gemalto, Heijmans, IBM, ICT Automatisering, Imtech, KPN, Nedap, NXP Semiconductors, Philips, Priva, Siemens, Smart Dutch, Stedin and TenneT.

Asia

Asia is quickly becoming a major player, and the center of global smart grid activity. The combined smart grid market in China, Japan and South Korea is estimated at over $10.0 billion, with an estimated increase to over $30.0 billion by 2020, according to industry specialists.

As Asian countries are irreversibly becoming the predominant smart grid markets, the competition and the positioning of different vendors is increasing as well. Lack of clear understanding of the energy scenarios (present and future) in the major Asian countries (China, Japan, and South Korea) is an obstacle that needs to be resolved first.

It is widely expected that the smart grid markets in Asia will move forward at a breakneck pace during the next decade or two. The developed countries are already positioned for the race with over $45 billion in funding, earmarked by the respective governments and utilities across China, Japan and South Korea. The majority of funding and related opportunities are located in China.

A level of uncertainty is still distorting the smart grid vision, so determining the trends and establishing a clear path of energy policies and currents will allow much faster implementation of smart grid technologies in these countries. Once the uncertainties are lifted, a meaningful entry in the Asian smart grid markets will proceed quickly.

China

The growth of these markets will be characterized by the special needs of each country's energy demands, as well as the local utilities and existing grids specifics. For example, the smart grid investment in China is focused on transmission and distribution automation, to support the plans for a new power grid (planned to be developed) and robust renewable energy (planned to be built.)

China is aiming to become a world leader in smart grid technologies in the next decade. The "Strong and Smart Grid," an 11-year plan revealed in 2009, outlines the ambitious steps to get there. It involves all aspects of the power grid, including increase of power generation capacity, implementation of smart meter programs, emphasis on large-scale renewable energy, and a large transmission lines and substation build-out.

Today, the plan is alive and well. The State Grid Corporation of China, which is actually one of the largest utilities in the world, and the executor of China's smart grid plans, is already in phase two of the three-phase program—the construction phase, which is scheduled for completion in 2015.

New transmission lines are a major focus for State Grid in the construction phase, which is struggling to meet the growing energy demands of the rising middle class in the east and south. Most coal, hydro, wind, and solar load sources are over 1,000 kilometers away from the populous east and south. High voltage (HV, under 300 kilovolts), extra-high voltage (EHV, 300 kilovolts to 765 kilovolts), and ultra-high voltage (UHV, 765 kilovolts and up) lines are being installed currently, with at least one 1,000-kilovolt UHV AC or DC line installed annually until 2015. Overall transmission line investments for 2015 are approximately $269 billion, equivalent to the combined market cap of ABB, GE, and Schneider Electric as of May 21, 2012.

China is adding so much new transmission capacity and so many power lines that it could build three quarters the length of a new American transmission grid in just five years. When the dust settles, there will be over 200,000 kilometers of new 330-kilovolts-and-up transmission lines built, for a total of 900,000 kilometers of transmission lines, compared to 257,500 kilometers of transmission lines presently in the U.S.

At a cost of $1.05 million per mile for UHV transmission line and equipment, each UHV line requires billions of dollars to build, and State Grid put in a staggering $80 billion investment into 40,000 kilometers of UHV lines for the 2011-2015 construction phase. The business case is readily apparent: a 2,000-kilometer, 800-kilovolt UHV DC line has an incredibly low 3.5 percent line loss rate per 1,000 kilometer and a high 6.4-gigawatt transmission capacity, all the while being 30 percent cheaper than a 500-kilovolt EHV DC or 800-kilovolt UHV AC line of the same length. By 2020, UHV lines will have 300 gigawatts of transmission capacity, roughly split 60 percent AC and 40 percent DC.

The competitive business environment seen in the transmission grid build-out is indicative of the rest of the smart grid market in China—high-quality goods, competitive costs, and a well-built relationship with State Grid all go a long way toward winning a contract. Fierce vendor competition exists, due in part to State Grid's competitive construction procurement process. All projects costing over $300,000 to build are required to go through an open bidding process that aims to enforce fairness and transparency, but State Grid still holds the reigns tightly on choosing project developers. In the process, State Grid has the final say and does a rough 45/45/10 split when evaluating meters, based on quality, cost, and bankability of the company.

With the promise of power shortages disappearing and a stable energy supply base, the build-out of the transmission grid is ushering in the next era of smart grid opportunities in China. Smart meters and renewable integration are already big businesses, and new substation infrastructure has brought with it a vibrant and growing substation automation market. The need for better monitoring equipment has risen as China is keen on decreasing its system average interruption duration index (SAIDI) and improving power quality to its customers.

State Grid has earmarked over $40 billion toward these smart grid technologies between 2011 and 2016, with smart meters alone being a $2.5-$3 billion annual market. State Grid has paid special attention to substation automation technologies, and plans to install 74 new digital substations for 63 kilovolts to 500 kilovolts by 2015.

While this number is small compared to the existing 40,000+ substation base, State Grid has stated it intends to include digital technology in all new substations built. Companies such as BPL Global have been expanding their substation operations in China, which has been met with stiff domestic competition. The substation market offers promising growth over the next ten years.

The transmission grid build-out also has an impact on technologies at the distribution level and downward. China is building 36 million new urban homes between 2011 and 2015, and modern building automation and smart meter technologies will be utilized. The coming years promise to create a new and vibrant building automation market, but for the time being, the market continues to focus on meeting demand shortfalls and other key infrastructure challenges.

Expect to see an exciting shift toward technologies at the distribution level and downward in the next five to ten years, as China's grid solidifies its transmission grid and generation sources. If the past three years are any indication of future progress, expect to see China become a leading smart grid market for the next five to ten years. The distribution grid build-out and digitization will be the next major indicator of China's smart grid prowess.

In Japan, the shutdown of many nuclear plants created a need for demand response, energy management and smart meter deployments. Smart grid is on the agenda. Money is available, the need is there, the customers are willing, so only time separates Japan from the full implementation of efficient smart grid concepts.

We would not be surprised if Japan leads the world in this area in the near future, and even becomes the first country to claim complete smart grid deployment.

The South Korean market is quite different. As the country with the most reliable grid in the world, South Korea is looking into developing the next-gen smart grid technologies and components (hardware and software) across all segments, but primarily for global export.

The major players in the US and Asia are ABB, Accenture, BPL Global, Echelon, Freescale, GE, Holley Metering, Moxa, RuggedCom, Siemens, State Grid Corporation of China, Wasion, XD Electric, and XJ Group.

Many other companies worldwide are working feverishly on developing new components for the smart grid technologies, which is leading to new inventions and concepts that will prove helpful in the future.

In summary, the global smart grid future is bright. It has the potential to develop into one of the most important parts of the world's energy markets, thus bringing us closer to the overall goals of achieving energy independence and a clean environment.

The 21st century will bring a lot of developments

in this area, and the Asian countries will be the first to claim smart grid implementation and use.

The U.S.

In December 2007, Congress passed Title XIII of the Energy Independence and Security Act of 2007 (EISA). EISA provides legislative support for DOE's smart grid activities and reinforces its role in leading and coordinating national grid modernization efforts. Key provisions of Title XIII include:

- Section 1303 establishes at DOE the Smart Grid Advisory Committee and federal Smart Grid Task Force.

- Section 1304 authorizes DOE to develop a "Smart Grid Regional Demonstration Initiative."

- Section 1305 directs the National Institute of Standards and Technology (NIST), with DOE and others, to develop a Smart Grid Interoperability Framework.

- Section 1306 authorizes DOE to develop a "Federal Matching Fund for Smart Grid Investment Costs."

The Office of Electricity (OE) is the national leader, partnered with key stakeholders from industry, academia, and state governments to modernize the nation's electricity delivery system. OE and its partners identify research and development (R&D) priorities that address challenges and accelerate transformation to a smarter grid, supporting demonstration of not only smart grid technologies but also new business models, policies, and societal benefits.

OE has demonstrated leadership in advancing this transformation through cooperative efforts with the National Science and Technology Council (NSTC) Subcommittee on Smart Grid and the federal Smart Grid Task Force.

The Department of Energy's Smart Grid Investment Grant Program is one example of an ill-conceived (well intentioned) intervention. Grants ranging from $500,000 to $200 million are to be issued for deployment of smart grid technologies, most of the recipients of which seem to be working on smart metering solutions (since it is the easiest, and most accepted by the public, smart grid related technology).

In addition, IRS issued a guidance in 2010 providing a safe harbor, under which the $3.4 billion of federal Smart Grid Investment Grants (SGIGs) issued under the American Recovery and Reinvestment Act of 2009 (ARRA) will not be taxable to corporate recipients.

The catch is that energy conservation, derived from the full implementation of smart meters (as encouraged by the grant and tax programs), means lost revenue to utilities at a time when the utilities are in the midst of expensive changes and are not ready for additional revenue losses. This lost revenue is untimely and is not aligned with the objectives of most utilities and their shareholders.

In addition, long-established regulations have favored supply-side resources over energy conversation, so utilities have been encouraged to add new generation because they earn a rate of return on investments on their assets—mainly the power generation, transmission and distribution infrastructure. Now, without warning, the utilities must abandon the profitable business model and gear for energy conservation, which simply translates to loss of revenue.

This misalignment might cause delays and even failure of some energy-efficient technologies and services in the U.S. and reflects the disconnect and fragmentation in the sector. Some of the disconnect is fueled by the ignorance of the regulators as well, which completes the circle of incompetency, and which will make the smart grid implementation that much harder.

The technologies that promote energy conservation have evolved much faster than the regulations that govern utilities and other suppliers. This creates an imbalance, which will continue until new regulations, promoting the implementation of demand response and other energy conservation programs are implemented. To cross the divide seamlessly, appropriate and timely regulations and other policies should be properly structured to justify and encourage full utility participation and investments in demand response.

The work has started, but it is full of gaps, fragmentation and misdirection. To coordinate and encourage the activities, new legislation must be approved, and regulation must be implemented to create incentives for utilities to reconsider investing in generation, transmission and distribution assets and promote energy conservation and demand response instead. There must be a balance between the approaches.

There should also be some type of shared-savings program that allows utilities to participate in the savings customers receive from reducing their energy usage. This will persuade them to promote energy conservation because they, too, will benefit. In addition, there should be penalties if utilities do not encourage customers to reduce their energy usage.

Another way to encourage utility participation would be to compensate a utility for a portion of its avoided supply costs obtained through demand response and other related programs.

So, the future of the smart grid in the US is not clear. Solar was the hope for energy independence in the early 1970s, and it still is…albeit in an embryonic state. After 40 years, we know what solar can do, but we still don't know what to expect from it.

Smart grid technologies are headed in the same direction, changing at an exponential pace. Ten years ago we didn't even have smart phones, and the term 'social media' was still in the making.

To imagine where, or what, the "smart" might be in 2015, 2020, 2050 and beyond is not possible, but we can say with certainty that there will be progress, with a heavy component of customer (partially virtual) service.

The utilities are well aware of the demographic changes and the pending changes. They see the new generation of informed and proactive young customers, as well as the large retiring workforce. By 2020, half of the utility workforce will be retired—and they know too much…

Customers are accepting new technologies quickly, as evidenced by the fact that, according to some statistics, the volume of text messages now exceeds that of phone calls. Ten years ago text messaging did not even exist as we know it. So, information is the foundation, customer acceptance is the route, and integration is the vehicle.

The existing data are enormous, and growing exponentially. All participating systems must be integrated in parallel, to receive and process the data for maximum efficiency and lowest cost. The smart grid and its smarter virtual applications are here to stay. Whatever new is coming in the area of smart grids will have a large virtual component with a high level of self sufficiency and reliability. The utilities must be given a chance to get ready for the new order of things.

Grid Parity

Grid parity is a sound bite that we hear every time we talk about solar or wind energy. It is complicating the solar industry's struggle to compete with conventional energy sources. It has signaled the end of many new careers, accompanied by the loss of talent—the very expertise the solar industry badly needs to continue innovating and growing.

Grid parity is on the agenda of discussions, conferences and contracts, most of which project a positive attitude about the solar industry's achievement of late.

Most praising the industry for its achievements, and predict that it is "almost there," or at least where it has learned to compete with conventional energy without help or subsidies.

Conventional energy sources are also beneficiary of enormous financial aid, and direct and indirect subsidies, and have been since the early 1900s. Billions of taxpayer dollars have gone into nuclear, coal, gas and oil power generation, research, programs and projects.

But low manufacturing costs, high quality and efficiency require decades of research and development, and a lot of money. Conventional energy sources are the result of long hours in the lab, and years of development followed years of field operations and proving product reliability.

It almost seems—with all this talk of "grid parity" success—that we are trying to bypass the long road toward stabilizing the fragile solar industry. Instead of focusing on supporting incremental improvements and successes, we are claiming champion achievements, such as higher cell efficiency.

Triumphantly waiving partial results, and loudly announcing achievements and roadmaps as actual data, we have placed solar on the slippery slope of empty promises, wishful thinking, and unmet expectations. Grid parity rules, and has been accepted as a measuring stick for all that the solar industry stands for.

The grid parity motto has been even used as a reason to sacrifice profits, but to move forward we must accept the fact that the solar value chain needs some profit to stay afloat, and a lot of profit to prosper. Ignoring these simple principles of successful capitalist enterprises leads to unsustainable development, as we've seen in the long list of failures.

For manufacturers, realizing reasonable profit margins requires depth of understanding and acceptance of expenses over which they have no control. These costs include the price and quality of raw materials, consumables, equipment and labor.

On the other hand, installers, operators and owners of solar installations (especially those of cost-intensive utility-scale installations), must know well that performance is the most important metric, but that they have no control over the most important factor—local weather conditions. So, the cost of unexpected weather events such as excess cloudy days, snow, hail, and sand storms, must be thoroughly understood and taken into account by investors and system owners—during the design stages and well before the system is installed.

Recently some industry supporters have tried to cover up the misdeeds of low-quality solar products manufacturers and their spectacular failures, blaming lower prices of PV modules as the reasons for industry failures. These same people tout the lower prices of PV modules as proof that the industry has achieved grid parity (or is on the brink of it) and will continue to grow despite these losses and failures.

One has only take a quick look at financial statements from the leading manufacturers (even those in China) to see the unsustainably low prices. One needn't be a specialist to see that this is a temporary situation which won't end well for all involved.

Other specialists have claimed lately that the system price is the most important metric when discussing grid parity—not the cost of the module. But... you'd be hard pressed to build a PV power field without modules, and without the artificially low module prices there would be no talk of grid parity.

While the dropping cost of PV modules has been blamed for the downward price pressure lately, there are significant improvements in their efficiency and durability, which needs to be plugged into the formula. There are also cost and efficiency improvements at the balance-of-system (BOS) level.

In addition, real opportunities for efficiency improvements are yet to be explored at the permitting and installation level. As a matter of fact, while the cost of solar modules and other components fluctuates, it is a variable that we are familiar with and can follow. In contrast, the complex and unclear costs and uncertainties of land permitting, environmental assessment, obtaining PPA and connection rights, as well as other administrative costs artificially raise the system cost and constrain margins for demand-side participants. Solar and wind power grid parity is far from a reality, when we consider all the variables.

Combined Wind and Solar

The future of alternative energies is not clear, but we see a winner in combining wind and solar. One example is the wind and solar plant by Element Power in New Mexico.

According to insiders, it there were plans for two different projects—wind and PV—and they were, as a matter of fact, developed separately. The demand for renewables in that part of the country is immense, and it just so happened that it is a place where both solar and wind can be operated at maximum capacity. And the synergy is not just in sharing land and resources, but in conveniently spread out peak time power generation, which provides much more power than wind or solar can produce by themselves.

The wind power plant is composed of 28 Vestas V100 turbines, 1.8 MW each, which were installed with labor, services and materials from the local community. This cooperation always brings prosperity to the locals, which helps build the trust between the parties—an important elements in the success of any project.

The solar plant is a 50-megawatt PV installation, using conventional solar modules. It also has a PPA with the local utilities, so the power produced by both plants will be pumped into the local power grid, via a nearby substation, which was built for this purpose.

The fact that there is excellent solar radiation, and good winds at the right times, enables both wind and solar to be cost-competitive sources of energy for the local utilities. Combining these two power generating sources is a big plus as well, which brings more profit and contributes further to the success of the project.

Transmission system operators are always concerned with managing wind and solar variability, but since these two plants are near each other, it makes the task somewhat simpler. Since wind and solar generate at different times, there is alternating of the generated power; i.e., wind feeds the transmission lines 30-40% percent of the time, but is usually weak during the mid-day hours. Solar kicks in exactly during those hours, thus generating power when is most needed—during the noon peak hours.

In general, the combination of wind and solar power generation reduces the variability, these power sources experience when operating separately, which is good news and welcomed by grid operators.

Solar energy generation could be combined with other energy sources when constant output is the goal. Addition of solar power to conventional power sources (power plants) is one approach. Using energy storage devices (batteries or water storage) is another. A more natural and most efficient way, however, is combining PV power with wind at certain locations especially chosen for this purpose. At such locations PV power is complementary to the output of wind generation, as with the New Mexico site.

As illustrated by Figure 8-10, solar and wind plant profiles—when considered in aggregate—can be a good match to the load profile and hence improve the resulting composite capacity value for variable generation.

In this example, the average load (upper line) is closely followed during the day by the average output from the combined wind and solar generators (the second from top line) during the same time. This average is created regardless of the fluctuations of the individual wind and solar power generators—a marriage made in heaven for sources matched as closely as these.

Although there are areas in the US and abroad that match this wind and solar profile, the combined effort is usually hard to execute, because the best places for wind and solar are often miles apart, and also due to lack of infrastructure at the most suitable locations. Implement large-scale "wind-solar load matching" schemes with existing technologies would require great effort and enormous investment.

Still, having matching wind and solar power outputs as a goal will force us to find the most suitable locations and appropriate technologies for this match.

Solar and wind combinations are not uncommon in residential and small commercial applications. The combo solar panels and small wind turbine can maximize local resources, provide power for remote locations, pump water, run appliances. etc.

On a larger scale, China's biggest power network operator, State Grid Corp, installed and is operating a 140-megawatt combined wind-solar hybrid project, where 100 megawatts of wind and 40 megawatts of PV solar are installed and operating side-by-side since January, 2012.

This is the largest on-line, utility-scale, solar-wind hybrid project in the world. It also has a 20-megawatt battery storage capability, which provides even more stability to the hybrid power generation.

A smaller example is the Western Wind's fully integrated 10.5-megawatt hybrid system in Arizona, which consists of five 2.0 MW Gamesa turbines and a 500 kWp

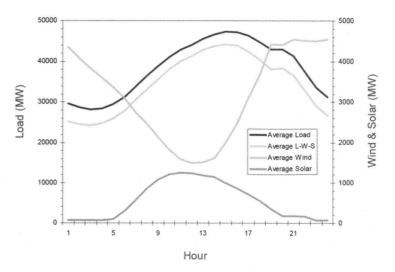

Figure 8-10. Simultaneous wind and solar generation

c-Si PV array. There are a number of plans to develop similar projects—either new installations or retrofitting existing solar with wind power generation capabilities, as needed to reduce the harmful variability issue.

Presently enXco's Pacific Wind/Catalina Solar undertaking is the largest in the US. It has 70 REPower turbines, 2 MW each, combined with the largest Solar Frontier CIGS solar field in the world. It will be used to test the effectiveness of feeding the grid with both resources.

We foresee the wind-solar hybrid generation as one of the most reliable and widely spread alternative energy generators in the 21st century. The combined action of these two power sources, with added energy storage, will eliminate (or reduce significantly) the generated power variability, which is one of the major barriers to the development of large-scale power fields today. The wind-solar hybrid is also the surest known way to successfully competing with conventional energy sources.

FiT-less Solar Power

Here is the new trend of solar power development in the 21st century:

Solaria Energía y Medio Ambiente, S.A., leads the way into a new era of solar power development and use—without FiT—head-to-head with the big guys.

The new 60 megawatt (MW) photovoltaic plant in Spain will not receive a FiT, due to the Spanish government's renewable energy moratorium on subsidies imposed in 2011 The new plant will feed its generated electricity into the Spanish grid at the going market rate. It will be competing directly with the local energy sources.

The company will manage the design and the EPC services, using its own PV modules to build the project at a cost of around $65 million.

Solaria is the first solar company in the world who puts its neck on the line. This shows real commitment to, and confidence in, the Spanish energy market. And sure enough, grid parity is becoming a reality in Spain, and PV generation is a mature technology which dares to compete with conventional energy sources.

Another company, Gehrlicher Solar, is planning a 250 MW photovoltaic plant in Spain on the same principle—without any FIT and/or any economic incentives. If all goes well, the new plant will start construction in 2013, to be completed in 2015.

Author's note: In 2011 the Spanish government suspended all subsidies and incentives for renewable energy, due mostly to its poor financial health. There are no indications as to when the moratorium will end, if

ever. This, of course, had an immediate negative impact on the photovoltaics industry, ending with the cancellation of several PV projects there.

Industry insiders estimate the new policy will halt many more installations for the duration. Still, an increase of up to 4,177 MW is possible by 2013, compared with 3,985 MW in 2011. The solar industry will stop growing at least until 2015-2016, unless the above discussed FiT-less projects succeed, at which point the Spanish solar industry will take on a life of its own. That would signal the beginning of a new era of grid parity, and the Spaniards would lead the world again in new installations.

Presently, solar electricity can be produced for $0.06-0.07 per kWh. If the PV modules are purchased at $0.55-0.65/Wp, then this becomes an attractive ROI of approximately 5-6% percent.

But there are risks.

The electricity market price in Spain is fluctuating wildly, which makes it impossible to predict what will happen next year, let alone 20-25 years down the line.

Green electricity certificates, sold for clean energy credits, are one of the hopes for additional income, but the market for these is just developing and its future is not secure.

The biggest problem we see in Spain's solar future is the fact that there is not much sun in most of the country, and the desert-like areas are of limited size. Also, these areas are far away from population and industrial centers, and improvements such as new transmission lines and substations would be required.

Still, the pioneering developments toward grid parity prices in Spain are incredibly positive. It has captured the attention of investors, developers, and potential customers alike. A number of planned PV projects are looking into the FiT-less system, as they have no other choice. Maybe this is what is needed to push the solar industry over the barriers.

The FiT-less model appears to be the best, if not the only, way to implement large numbers of large-scale PV power fields in the 21st century, so we wish the Spaniards good luck in this undertaking.

As confirmation of the legitimacy of the FiT-less concept, the Asian Development Bank (ADB) has been reviewing options to support solar projects without the need for feed-in tariffs. Discussions within the bank's renewable energy group are currently centered on supporting solar development programs specifically without subsidies or FIT.

ADB has penciled an initial $100-million credit facility to be considered by interested solar power developers and owners, and expects "fierce competition" on capital access for clean energy projects, including those on renewable energy (RE) developments.

The Manila, Philippines, bank touts its own solar rooftop-installed power system, showing that the technology can already be utilized viably without subsidies. Where did ADB learn about solar, and how long did it take to come to these conclusions?

Presently the Energy Regulatory Commission (ERC) prepares a ruling on the FITs for various renewable energy developments, and regulators are cautioned on subsidy impositions for solar, especially with reports that the technology is already at grid parity in various jurisdictions.

In fact, sources from the National Renewable Energy Board's (NREB) technical working group have announced that the initial high FiT calculations were not supported by the data presented by the prospective developers. So ADB is willing to extend technical assistance to the Department of Energy (DoE) for a study to assess the viability of solar installations in the Philippines, primarily for rooftops. This option might later be extended to industries and businesses.

The bank is convinced that rooftop-installed solar facilities are now economically feasible for higher-demand customers—even without subsidies and FiTs. But the poverty level in the Philippines is above 26%, so bringing down the cost to the level of the low-to-medium class residential end-user is key, and it will take a long time.

At the same time, the bank's Asia Solar Energy Initiative (ASEI) is targeting solar capacity installations of up to 3,000 megawatts within Asia and the Pacific over three years, ending in 2013. The ADB is looking into financing up to $2.25 billion worth of projects under its ASEI program, and will also leverage an additional $6.75 billion in solar investments, using instruments such as London interbank offered rate (LIBOR)-based loans, donor contributions, grant funds, innovative risk mitigation mechanisms, carbon market support measures, and direct support for its solar lending program.

This initiative might be more doable, but we question the availability of money, appropriate land, and infrastructure needed to undertake such a large program. While plans are good for building investor confidence, building large-scale PV plants in that part of the world, where the sun is a precious commodity, is not that easy.

1603 Grant

A new Office of Management and Budget (OMB) report indicates that the full value of the 1603 Treasury

grant may be in jeopardy, and developers expecting a 30% cash grant for their renewable energy projects may be in for a surprise in 2013. The report indicates that the 1603 program may be subject to budget cuts, also known as sequestration, which could result in cash grant values closer to 27.72% rather than the expected 30%. This is a 7.6% reduction of the cash grant value which could significantly impact the economics of a project.

OMB issued this report in response to the Sequestration Transparency Act of 2012 which asks OMB to outline automatic spending cuts if Congress does not enact a plan to reduce the nation's deficit by $1.2 trillion as required under the Budget Control Act of 2011. Anticipating that Congress might not meet that deadline, OMB set out to outline how the $1.2 trillion reduction would take place. OMB's proposed a 7.6% cut to 1603 would result in a total spending reduction of $279 million for the government for the 2013 fiscal year. This news came as a shock to many in the renewable energy industry, as the Treasury Department had informally indicated only weeks prior that they did not believe the Treasury grant was within the scope of sequestration.

A number of unknowns cloud the actual implementation of sequestration, regarding to the Treasury Grant. For example, it is unclear whether the 7.6% reduction would be enforced per grant as noted above, or whether the reduction would be allocated in a different manner. It is also unclear on the timing of the sequestration and what projects would be subject to the reduced cash grant.

Would projects placed in service in 2012 be subject to the reduced cash grants if they did not complete the final grant application until 2013, or would they be exempt from the budget cut as the systems went operational in 2012? Details such as these are important for understanding the full ramifications of sequestration. In any case, change is coming, if not in 2013, then in 2014, or 2015, but inevitably 1603 will undergo change.

Mandatory Time-of-use Program

PG&E, one of California's largest power generators, including solar and wind power, announced a new energy saving program, designed to help its small and medium sized customers to prepare for a mandatory shift to time-of-use pricing of the power they use.

The CPUC-mandated switch will not affect residential customers, but is designed to help customers save money while easing the strain on the power provider. Large commercial and industrial customers have already been moved to time-of-use pricing, so only smaller businesses and enterprises will be affected by the new program, and PG&E is trying to make the process as transparent and painless as possible.

Sensitive to customers' needs and wants, PG&E is in the process of asking state regulators to consider an opt-out option for customers who wish to remain under the current pricing model for awhile longer.

According to PG&E analysis, 60% of the customers would benefit from the new pricing structure, since the rates would increase only about 2% for the affected customers.

PG&E thinks that conserving even a small amount of energy during afternoon peak demand can lower costs to everyone. It also helps the environment by reducing the need for utilities to buy power from fossil-fueled power plants and increases the reliability of the state's electric grid.

SCE and SDG&E are working on similar programs and estimate switching to time-of-use pricing by March 2013.

This is one way to save energy without adding energy generating sources.

Utilities Surcharges

Maryland regulators may undo a provision that allows state utilities to bill customers for the first 24 hours of an electricity outage. The Maryland Public Service Commission conducted a hearing Monday to consider the merits of its Bill Stabilization Adjustment mechanism, which was introduced in 2007 as a method to allow utilities to charge customers a nominal fee to recoup revenue losses caused by outages. The regulation covers Maryland's utilities, including Baltimore Gas & Electric, Pepco, Delmarva Power & Light Co., and the Southern Maryland Electric Co-op. Its impact was reduced in January, with the Commission deeming it applicable only for the first 24 hours of an outage.

The fee kicks in for any storm with more than 100,000 outages, or 10 percent of a utility's Maryland service territory, whichever is less, according to the PSC.

The fee came under further scrutiny in June, 2012, after a severe storm stuck numerous Maryland electric customers with long-lasting outages. Paying a supplemental fee while going without power is tough to explain to customers—even if it rings in at around $0.50 a head—but utilities counter that, since they cannot control the weather, they deserve to be protected against unexpected revenue losses.

Following the hearing, the PSC did not give any indication of whether or not it would further restrict the provision, saying only that it would rule as soon as possible, according to a *Baltimore Sun* report.

At meetings throughout the state, the commission heard from utility customers who think that paying the utilities an energy-distribution fee on top of the expenses incurred during an outage—the cost of lost food, for instance—adds insult to injury.

After the quick Derecho thunderstorm in June 2012 caused an eight-day outage for some customers in the Baltimore region, the commission proposed eliminating the "bill stabilization adjustment" that Baltimore Gas and Electric Co. and the state's other major utilities bill customers for the first 24 hours of a major outage.

In 2007, the commission began allowing BGE to collect the adjustment to compensate the utility for money it might lose because of programs to reduce energy consumption. Potomac Electric Power Co., Delmarva Power & Light Co. and the Southern Maryland Electric Cooperative also have been allowed to institute bill stabilization programs.

In 2011, the commission decided to investigate whether allowing the adjustment fee to be charged during outages "eliminated a critical incentive to restore service quickly," according to letters the commission sent to the utilities in February 2011.

The commission decided in January that the billing stabilization adjustment did not align the utilities' "financial incentives with reliability goals." It amended the rule to prohibit utilities from charging the fee from the second day of an outage until all customers had power restored. The prohibition applies to any weather event that causes "more than 10 percent or 100,000, whichever is less, of the electric utility's Maryland customers" to lose power for more than a day.

Following complaints from utility customers that they were charged the fee during the first day of the outage, the commission decided to take up the issue again and decide whether the fee should be withheld during the first 24 hours of a storm as well.

If the additional prohibition were adopted, it would reflect the rules regulating the adjustment fee in Washington, where Pepco is prohibited from charging customers the fee during an outage that lasts more than a day.

Consumer advocates believe that the fee prohibition should be expanded into the first day of an outage. "The additional experience provided by the Derecho storm now weighs in favor of the Commission applying the same standard to the 24 hours after commencement of a Major Outage Event," said Ronald Herzfeld, from the Office of the People's Counsel.

The principle of paying for a service that is not working should be enough to prohibit the fee from being assessed during an outage, according to a representative of the residents.

PV Industry's Large-scale Project Strategies

While on the subject of best approaches, we must also consider the way the PV industry operates today. Manufacturers, installers, operators and everyone else involved are part of the capitalist system, so they must consider the bottom line first. However, the ultimate success of the projects in which their products are used is often left to chance.

As for the PV industry's involvement in large-scale power generation and future trends in this area, remember, these are the embryo years of the industry. Manufacturers are just now realizing that involvement in large-scale projects is the ticket to success.

At this time PV products manufacturers of all types and sizes have one basic *modus operandi*: sell what you have. But, they are becoming more aware of the fact that due to technological, logistical, and financial considerations, they must get involved hands-on in the projects where their products are used.

Sell What You Have

This option is what manufacturers are best at— selling their own products (modules, inverters, frames, etc.) to customers and developers. This is what most present-day suppliers are doing, to gain a share in the industry. They are constantly creating and introducing new products and services, tailored especially for large-scale projects (expecting large sale volumes). These strategies fall into one of two categories:

a. *Products geared for large-scale installation (modules, inverters, etc.).* At a minimum, most suppliers have introduced a larger series of modules and/ or inverters especially for the utility market. The larger size or capacity of these special products is intended to drive down installation time and materials costs. Larger inverter products also often contain additional features that are attractive to utilities and grid operators such as control of reactive power, variability, and voltage ride-through.

b. *Pre-designed (modular type) systems* have been developed by a number of suppliers as integrated systems, intended for sale and ready to go as a single package.

Develop and Control the Entire Project

This is the second and still novel approach used by a few PV products manufacturers. It consists of the design and development of their own power plants from scratch, or in partial participation, in the projects at hand. These new approaches are going through constant revisions and changes and are under serious evaluation by a number of US, EU, and Asian manufacturers. They fall into two basic categories:

a. *Full project integration.* Here the manufacturer is playing the role of a designer, supplier, developer, installer, and operator. In the best of cases, products and services are owned and managed by the manufacturer/installer/operator entity. There are several examples of this type of operation, and we feel that there will be many more. This is the best way for manufacturers to control their volume, price and quality.

b. *Semi-integration* is the most likely scenario simply because there are presently no manufacturers who make all products needed for a large-scale PV installation. Semi-integration means that the manufacturers maintain their own brands but have the option of sourcing third-party components. Parent entities' products have inherent advantage, of course. Lately, many manufacturers are looking into the second approach because it is the surest way of placing their excess inventory and ensuring increase in sales.

It is the author's opinion, that manufacturers should get fully involved in project development without losing focus on their primary activity—providing quality product. Full involvement in PV projects, as in "a" and "b" above, will provide confidence in the efficiency and longevity of the products. This will eliminate a number of technological, logistical, and financial barriers standing before the PV industry.

In the best of cases, however, a number of issues still need to be resolved, including the availability of land suitable for PV applications (including interconnect-ability), and utilities willing to purchase the generated power.

There is a lot of land available, but it is not ready for PV, and the utilities have a limited appetite for new PV power, so they pick and choose from the flood of proposed PV projects. Clearly the ball is in the utilities' court, but most of them are not in a hurry, and many large-scale PV power projects will be delayed or cancelled because of that.

The Future of Utility-scale Projects

According to industry specialists, utility-scale PV solar projects are getting harder to come by and even tougher to execute. In 2011 utilities issued several RFPs. More than 70 GWp of applications were submitted a short time after that, and they shortlisted about 2.5 GWp.

At the same time, the ISO utility queue has dropped from 80 GWp to 50 GWp, and utilities are cautiously combing through the contracts, looking for problems and show stoppers. Competition for PPAs is extremely tough, so many developers are giving up and monetizing or cancelling their portfolios.

This points to a U-turn in the US solar market, where in the next several years it will morph into over 50% distributed generation and 50% utility scale projects. Over 85% of the 3.5 GWp of utility projects are already in the queues and reserved for different companies, which severely limits opportunities in the utility segment. This forces developers to reevaluate the growing sector of distributed generation, since the solar markets are getting so much harder to scale up.

Many larger developers are switching to the small generation interconnection queue of 10- and 20-MWp projects, which have been named the "doable renewable" projects. However, the smaller projects are not getting through any faster, and high transactional costs really hurt small projects; but, if a company has a few of the small projects and one or two larger ones, then they have enough economies of scale in procurement to be able to justify the higher costs.

There are some glaring discrepancies, where some developers market projects below 20 MWp that are just documents that they are trying to sell. There is also disconnect between what smaller developers think projects are worth and what they are actually worth. In projects above 20 MWp, most developers greenfield, because of transmission interconnections, getting the right location, and ensuring entitlement.

Most California projects that will be built by 2016 are already in the utilities' queues. Their numbers are high enough to meet the 33% mandate by 2020 and to replace contracts that default, so the future is not bleak... at least for those who can get in the ques.
CSP technologies are suffering the most, because a major part of the U.S. is in a severe drought, so using 5,000 gallons of water per minute for cooling a steam turbine power generating plant is a challenge that has no solution at present.

Nevertheless, the conditions are still favorable. Rising retail rates, and falling PV module costs, point to a rush for solar projects before the federal investment tax credit drops from 30% to 10% in 2016.

In the long run, for utility-scale solar to become a reality, costs need to come down to less than $1.50/Watt, gas needs to go up to about $6 per Btu, and developers need to get ~$20 per MW/h credit for capacity, to compete with $0.06 or $0.07 wholesale power.

Until then utility-scale solar will be going through the growing pains with a lot of uncertainties. The situation will most likely become more dramatic after 2016, when the ITC will be reduced and most other subsidies and incentives will be gone too.

Analytics

Analytics is a new field in the solar industry, brought on by the fact that future utilities will depend on an increasingly instrumented and automated power grid. That grid will benefit from the application of predictive analytics, optimization and advanced computing to drive better business and operational results.

This will mean better response to weather-induced power outages, such as the recent New York disaster caused by the super-storm Sandy. Through data analysis and computer modeling, utilities will be able to predict the impact a storm will have on the electrical grid, days in advance. By following that prediction, a utility could foresee the most probable damage and pre-position repair crews to begin restoration as quickly as possible.

Integrating renewable and distributed energy resources into the grid can also benefit. These energy sources are prone to a lot of variability (wind may or may not blow, clouds stop solar panels from producing, and consumers may choose to plug in electric vehicles at different times of the day). The ability to predict these variables accurately will help utilities balance supply and demand.

This is a drastic change from the old model, where ratepayer data were gathered monthly for little more than billing purposes. In light of this changing landscape, analytics are vitally important to improving utility operations; but, with technology rapidly changing, it's not enough for utilities to prepare themselves for a future built on analytics. They must also adjust quickly to today's market while laying the groundwork for success, anticipating the continued evolution of the analytics market.

Grid analytics can be broken down into two categories: 1) grid optimization (the more popular offshoot of analytics at the moment), and 2) asset optimization, which is still an up-and-coming area of utility focus. Merging these paths is key to analytics' success.

Utilities are gradually beginning to get the message and readying themselves for the changes. A Utility Analytics Institute survey found that 27% of utilities treat energy-efficiency analytics as their primary focus, 73% have a customer analytics program underway, and 40% of these focused on meter data analytics. The focus on meter data analytics makes sense given the developing state of the smart grid and the relatively new uptick in smart meters. But that is all changing as smart technology expands beyond the customer home and out to the rest of the grid.

The involvement of the utilities executive level staff into these programs shows progress. Executives, however, can only provide the initial spark, while all employees (whether judging or advocating) are the key to the success of new operations.

Part of promoting an understanding of the benefits of smart grid technology is a continual tracking and sharing of data in order to defend return on investment projects and justify expenses. It's important to track the benefits actively throughout the deployment and report them, in order to estimate the improvements and the impacts these systems have on the bottom line and on operational efficiencies.

Whatever path utilities take, and however they decide to manager their data, it's clear the current interest in the role of analytics is growing and should enjoy sustained momentum. The Utility Analytics Institute projects that grid analytics spending will grow about 33% by 2016, and that overall analytics spending in North America will balloon from $552 million in 2011 to nearly $2 billion in the same time.

Case Study

Some time ago, IBM started thinking of all the data being generated from increasingly sophisticated grid components, such as smart meters and sensors on the grid, and how that data can be maximized for operations. To study and commercialize applications, IBM just launched a Smarter Energy Research Institute (SERI), which will be a collaborative model between different utilities to use data for analytics that are both "predictive and prescriptive." The institute uses predictive analytics, optimization and advanced computation. The effort will combine IBM Research's capabilities in mathematical sciences, computer science and high-performance computing with the extensive power and engineering knowledge of the participating utilities.

Today's energy and utility companies face a changing landscape as technology transforms both the delivery and requirements of energy. One of these changes is the increasing participation of consumers, who want to monitor their energy use in real time. To meet this and other related challenges, the Smarter Energy Research Institute has defined five innovation tracks. Members can choose to focus their joint-research efforts on outage planning optimization, asset management optimization, integration of renewables and distributed energy resources, wide-area situational awareness and the participatory network. Within this framework, members of the Institute can work on specific research projects, collaborating with IBM researchers. Each member obtains usage rights within their enterprises for all the innovations (algorithms, code, patents) created by all the participants of the Institute working with IBM Research.

The new members of SERI are: Hydro Quebec, which is Canada's largest electric utility and exporter of electricity to the U.S. It is also one of only seven large electric utilities in the world to operate its own research center and has a portfolio of intellectual property of more 1,000 patents in the portfolio field energy.

Alliander is a major energy distributor and specialist in renewable energy, sustainability, technical innovations, and complex power systems in the Netherlands, serving more 3 million customers.

DTE Energy is involved in the development and management of energy-related businesses and services across the United States. Its largest operating subsidiaries are Detroit Edison, an electric utility serving 2.1 million customers in southeastern Michigan, and Michigan Consolidated Gas Co. (MichCon), a natural gas utility serving 1.2 million customers in Michigan.

So, analytics is becoming an integral part of the utilities operations, and with it we are looking toward more a reliable and more automated power grid. Analytics, with the application of predictive analytics, optimization and advanced computing will ensure better business and operational results in the US electrical net.

The Carbon Markets

At the United Nations climate change conference in Durban in December 2011, the European Union graciously agreed to keep the faltering Kyoto Protocol alive by committing to continue operating its European Trading Scheme (ETS) carbon market.

The EU also agreed to impose higher energy prices on its citizens in the hope that somehow this would encourage the rest of the world to do the same by 2020. The principle here seems to be: "If I bash my head with a baseball bat now, maybe you'll bash yours in a couple of years."

One of the chief goals of boosting the price of energy produced by burning fossil fuels is to encourage a shift toward energy generated by "cleaner" solar and wind power. That's not working out so well, according to industry insiders.

Later on, an executive at Eon, the German energy group that is one of Europe's largest, pronounced the carbon market broken. "Let's talk real," he said. "The ETS carbon market is bust, it's dead."

Upon its launch seven years ago, the worldwide carbon market was supposed to work on a simple premise; proponents hoped that by putting a price on carbon (contained in power generating and other emissions) and forcing companies to pay for the emissions, it would prod Eon and others to pour money into green technologies and greater efficiency. But, as a result of a subsequent recession and poor management, the market is saturated—and could be for years to come—with permits that give companies the right to emit carbon without penalty.

That has led to a prolonged slump in the carbon price. At roughly €7 per ton, compared with a peak of nearly €30 in July 2008, it is a fraction of what policymakers and analysts had forecast it would have reached by now—and well below the levels necessary to justify the desired investments.

The European Union, already facing criticism after expanding carbon trading this year to airlines, has opened five battlegrounds in its bid to resuscitate its flagging market, according to Societe Generale SA.

"It's very clear their primary goal is to increase prices," industry analysts said. EU permits are at their highest level, after the Climate Commissioner said the bloc may propose a delay in supply at carbon auctions starting in 2012. Lawmakers would also continue to consider another plan to temporarily cut supply, known as a set-aside.

There are three other policy initiatives that could also boost prices. The bloc is considering installing a 2030 target for the first time, tightening its 2020 target, and even expanding the market to include cars and trucks.

The bloc should consider expanding its emissions trading system to cover the region's transport industry and shouldn't judge the effectiveness of the program by a decline in carbon-permit prices, Poland said in 2012.

Permits for delivery in December 2012 are in the $9.80 per metric ton on the ICE Futures Europe exchange in London.

The commission's plan is the first step in a process to strengthen the ETS, and the EU will continue to explore "longer-term structural issues," including the set-aside of carbon allowances, or withholding a number of permits from the market.

The inclusion of flights to and from EU airports in the European carbon cap-and-trade program is another battle with opposition from China, the U.S. and Russia. The group insists that Europe should let the United Nations' International Civil Aviation Organization decide on greenhouse-gas limits for the industry.

It is easy to see that the carbon market is a new phenomena, and like the immature solar industry, is looking for its legs. Nevertheless, they both are here to stay, so we must learn to deal with them for the benefit of all.

Future Directions

A number of factors have shaped the path of the solar industry, and many more will determine its future direction. Some of these are the technology types, proper manufacturing and use. Since it all starts with the selection and use of materials, we'll look at these first.

The starting materials for c-Si PV represent 75% of the total cost structure, with polysilicon as the single largest contributor to the total costs.

Thin film modules, such as CdTe modules (containing cadmium and tellurium heavy metals), and CIGS PV modules (containing In, As, Ga and other metals) account for only 10% of the total cost of the finished PV modules. On the surface, this means that thin film PV technologies are less dependent on a potentially volatile commodity metals markets.

The problem, however, is that most of these metals (cadmium, tellurium, indium, arsenic, etc.) are exotic, rare and some very toxic. Most of these are also produced under extremely dirty and dangerous working conditions in Third World countries.

In other words, silicon is a large part of silicon PV manufacturing, but it is a well understood commodity, and its future quite clearly points towards increased efficiency, reliability and reduced price.

At the same time, thin films PV modules' future (due to metals' availability and toxicity), in addition to unproven long-term reliability, is veiled in uncertainty.

Author's note: The uncertainty surrounding rare, exotic and toxic metals, used in thin film PV modules, is due to changing regulations and other conditions, which affect the availability and price of these commodities. We expect that, as they become more popular and more attention is given them, the problems will increase.

Also, the materials' availability and pricing discrepancies can be a double-edge sword; the price pressure can provide a large incentive for c-Si manufacturers to reduce silicon material costs and usage, whereas thin film suppliers are locked into the dependency on commodity materials that are uncontrollable by virtue of their nature, location, toxicity, availability and price.

Since cost and the future of the raw materials availability of thin film's PV technologies are uncertain at best, thin film manufacturers must find other ways to reduce costs; i.e., increasing the efficiency of their modules. Although this makes sense, it will be hard (and very expensive) to do, because most thin film technologies are already close to their maximum efficiencies (for mass production).

Thus, reducing material costs on a per-watt basis and reducing the balance-of-systems penalty suffered by low-efficiency products is a goal, which all thin film manufacturers are chasing, but it is like hitting a moving target. Some manufacturers will find the right balance, and others will fail due to hastiness or choosing the wrong direction.

During 2011, record efficiencies were announced, such as CdTe record cell at 17.3%, 17.8% CIGS cell aperture-area efficiency, flexible module aperture-area efficiency of 13.4 percent, etc. But these are lab trials that have very little to do with mass production and even less with long-term field performance. For now, the commercially available technologies suffer from much lower efficiencies and unproven long-term reliability. Nevertheless, the record lab-test efficiencies are a promising indication that high-efficiency thin film products could be commercially available some time in the distant future.

What can be done to bridge the gap between thin film and c-Si PV manufacturing costs? Each manufacturing process and technology is different, so the method and size of the gap would vary.

Several years ago a 0.5% increase in module efficiency resulted in approximately 5-6% savings in total module manufacturing costs. The cost savings decrease logarithmically as the efficiency increases and get closer to the maximum possible. The higher the efficiencies go, the more thin films cost to achieve quality in mass production.

So this is not the cheapest or fastest way to catch up with c-Si technologies, because their efficiencies can be increased faster and cheaper. One great advantage of thin film manufacturers is that their equipment is much more sophisticated and flexible, and readily lends itself to automation. This simply means that thin film manufacturers must focus on increasing process yields,

Table 8-21. Comparing the c-Si and thin film technologies of today and tomorrow

Condition / Task	c-Si PV	Thin films PV
Materials Availability	Unlimited	Limited
Process equipment	Simple	Complex
Production process	Simple	Complex
Mfg. plant setup / scale up	Cheap	Expensive
Final PV module price	Decreasing	Going up
Future PV module cost	Decreasing	Leveled
Present PV module efficiency	15-16%	9-11%
Future PV module efficiency	20-22%	12-13%
BOS equipment (quantity)	100%	125-150%
Power field land use	4-6 acres	10-12 acres
Module frame	Wrap-around	Frame-less
Esthetics	Normal	Better
PV modules reliability	Proven	Unproven
PV modules toxicity	Slight to none	Considerable
Recycling	Not needed	Mandated (toxic)
Restrictions	None	Increasing
Surprises	None expected	Many expected
Future success	Clear, upward	Complex...???

throughput/uptime, and the always-important manufacturing scale.

Scale-up of the production process is another tool that, if properly used, could result in cost reduction. Several years back, doubling of total production capacity of a thin film manufacturing facility could result in 15-20% drop in production costs, while the same production ramp-up could bring only 10% (or less) decrease in c-Si wafer, cell, and module costs production.

Another key aspect of cost reduction is production capacity utilization. Production expansion (especially today) is not a guaranteed success move—regardless of any and all other factors and conditions. Production utilization rates have been dropping across the board—in all countries and all types of technologies. A number of factors must be kept in mind when considering levels of utilization. Growing too quickly is one of them, and will often result in significant capacity underutilization—which is exactly what we saw with a number of Chinese companies who were forced to down-scale or even shut down completely as a result of the untimely manufacturing underutilization.

This usually greatly affects the fully loaded costs (which also include depreciation) of PV modules and other components.

Finally, capital equipment initial cost and yearly depreciation are of utmost importance to the bottom line as well. Since c-Si capital manufacturing equipment is much simpler, readily available and cheaper, it costs less and its annual depreciation amounts to only 5-8% of the module costs.

In the thin film modules manufacturing process, however, equipment depreciation usually accounts for 20-25% of the operational cost (provided that it is operated at full capacity). So expansion of scale (to reduce costs) may increase costs (via capex), while at the same time c-Si manufacturers continue to successfully produce and sell PV modules at very thin and (recently) even at negative margins.

Industry analysis forecasts over 20 GW of PV module manufacturing capacity is to come offline by 2015. This is partially due to the global financial slump, accompanied by energy markets' supply-demand imbalance, lack of energy storage, reliability issues, etc.

PV module manufacturing capacities will stabilize by 2015 (see Figure 8-11), with the costs for Chinese Tier-1 suppliers projected to fall as low as $0.45 per Watt, with an average selling price as low as $0.61 per Watt by 2015. At the same time, the average c-Si module efficiency is expected to hit 20% by 2015, compared to 14.5-15% in 2012.

			Modules	
Year	Wafers	Cells	c-Si	Thin film
2010	1.6	0.25	0.5	0.0
2011	1.8	1.0	0.75	0.2
2012	3.2	4.2	4.0	1.25
2013*	9.5	8.0	8.0	1.5
2014*	4.0	4.5	3.5	3.0
2015*	0.5	1.0	1.0	1.0

*Estimates

Figure 8-11. PV manufacturing capacity fluctuations (in GW)

Most current PV manufacturers will have to make critical decisions in the near future. Their choices are to either exit while they can, or continue fighting in the component markets, while looking for marginal advantages and finding ways to leverage them to compensate for the risks which could be as unpredictable as gambling, and could eventually also end in a failure.

The Future

The future is looking bright for many PV technologies, but some significant changes are expected and needed to bring them to LCOE, and full utilization. Some causes of disruptive change will be:

1. Direct solidification of silicon material provides the cheapest wafers. Direct solidification of molten Si offers true kerfless wafering (which eliminates losses from sawing). It is the Holy Grail for the solar industry, which has been the goal of many companies since the 1970s. This technology has a potential market size of up to $600 million, and is expected to be the first to reach full commercialization by 2015.

2. Alternatives to increased cell efficiency, such as optimized anti-reflective and light-trapping coatings. These are also 2nd-tier technologies, looking at a market size of over $600 million. By providing active competition, they will pave the way to cost-effective efficiency gains. Commercialization of the new technologies is expected most likely around 2015 and beyond.

3. New active thin film layers. New processes and materials are expected to dominate the 21st century PV technological gains, such as:
   a.) Copper-zinc-tin-sulfide (CZTS) cell technology to replace some of the competing thin film technologies (possibly CIGS and CdTe) and gain significant market share through use of cheaper, safer materials, with the major advantage of eliminating the use of exotic, expensive and toxic materials, such as indium, gallium, cadmium.
   b Epitaxial Si (epi-Si) technology, and variations of, have the potential to replace amorphous silicon (a-Si) infrastructure and reach higher efficiencies at lower cost than present-day a-Si modules.

Basically, the game changers for the PV modules markets will be emerging PV technologies (or modifications of existing ones). Success belongs especially to those technologies that are reliable, easy to scale-up, provide the highest efficiency, and ensure reduced final costs per generated Watt.

The game will intensify, as government subsidies dwindle, and as the needs and requirements become better understood and enforced. Standardization of PV products manufacturing and related power plant processes will contribute to the successful deployment of solar energy and its ability to compete with conventional sources.

The Desert Phenomena

The deserts, where everyone thinks the solar revolution will take place in the 21st century, are where the sun shines most brightly. So it is logical for solar manufacturers and developers to look at them as the best possible opportunity. What many of them don't even suspect, however, is the traitorous nature of the climate in there.

We have quoted in this text a number of examples of premature deterioration and failures of PV modules operating under desert conditions. In many cases it takes only weeks or months for PV modules to show signs of physical changes and performance problems.

The latest such test of great significance—maybe the greatest and most important thus far—is the test done in 2012 at the planned site of Masdar City in Abu Dhabi. This test is extremely important because it is an initial test of the feasibility of using different PV modules for deployment in the multi-billion solar power generation projects, planned by UAE countries, Saudi Arabia and the European Community. Billions of dollars are planned to be used in these projects, to provide renewable solar energy to these countries in the near future.

For the purposes of the Masdar City project, in 2011-2012 the Abu Dhabi Future Energy Company assembled 41 solar panel systems from 33 firms, for testing in a desert site. The test was scheduled to last 10 months in the harsh desert climate near Masdar City—real world conditions, not a lab.

This is first time a test on this scale has been performed in the deserts of the UAE countries. We all know that manufacturers test their PV modules in labs, under controlled conditions. Field operation—especially in the deserts is quite different.

At the Masdar City test site, the conditions are much less than ideal. There is sun every day, but haze and humidity partially block the sunlight. Sand-filled wind deposits dust particles on the exposed PV modules every day, significantly reducing daily power production.

Too, when there is enough sunlight, it is so hot that on the hottest days of summer, the surfaces of some panels measure over 180°F, which further reduces the efficiency of silicon photovoltaic (PV) cells.

This test is not a science experiment. It is a test of which PV modules Masdar will purchase for its first 230 MWp solar project. The overall plan calls for solar power to provide 80% of the power for the planned zero-carbon Masdar City. A contract of this nature would be worth billions of dollars.

PV modules are planned to cover every roof and every square inch of available surface area in the city of 50,000, so it is extremely important to test the different modules, to see how each type and model would actually perform in the field.

The city is doing its homework up front, according to officials. The difference between the different models and manufacturers has never been tested in this area of the world, so Masdar is the real-life test site for the world's manufacturers and their best products.

The test site has a control room, where all PV modules are monitored and documented at all times. This is the place where the PV modules demonstrate the potential payoff of the undertaking.

There are 41 PV systems of 1 kWp each, a mix of c-Si and thin film PV modules, made by different world-class manufacturers. Each system is wired and equipped with identical electrical inverters. The actual performance of the systems is displayed as the total amount of electricity produced by each unit.

The results one late afternoon were not impressive. As a matter of fact, they might be characterized as "far below expectations." Remember that each system is rated (and tested in the lab) to generate exactly 1.0 kWp DC power. This means that between 10:00 AM and 3:00 PM Masdar time, each system should generate 1,000 Watts. The readings that afternoon, however, were much lower.

The best system was generating DC power at just above 400 Watts. The worst measured less than 200 Watts. The lower readings were due to overheating the PV modules which creates power degradation of 0.5% per degree C above 25°C, or in Madar this is 80 − 25 = 55°C. 55 x 0.5 = 27.5% less power during the most productive periods of the day. So, a 1,000-Watt system could produce only about 700 Watts at that time. Additionally, dust collected on the modules and haze in the sky would drop the power down to 400-500 Watts per each 1 kW system.

Why some of the systems read 200 Watts, however, is unclear. We assume that this abnormality is due to manufacturing defects, where the modules are underperforming under the harsh conditions. This also might be a case of premature total failure of some of the PV systems.

There is nothing Masdar City can do about the modules failing due to high temperatures, but they can clean the dust from the surface, so PV modules at the Masdar City test site were cleaned at least every other day. When the PV systems were left untouched for a month, the collected dust alone lead to significant reduction of generated power. This is forcing Masdar to look at developing automated systems to clean the PV modules in the field.

The Masdar City tests run 18 months, at which point Masdar plans to release a report on the results. The winner, if there is a winner, will be awarded a contract to supply the PV modules for the first large Masdar City' solar project.

This test offers another set of important lessons about the use of PV modules in desert conditions. The foremost and immediate challenge is the dust that settles on the glass coating, blocking the solar exposure of the cells, since dust cover on the top surface of the PV modules has a great negative impact on the overall performance of the systems.

Masdar City is also planning a second solar competition involving concentrated photovoltaic (CPV) and solar thermal (CSP) systems during 2013-2014. With these tests, Masdar is taking steps to ensure that when solar technologies enter the mainstream energy market, they will have been vetted by their harshest challenger, the Arabian Desert.

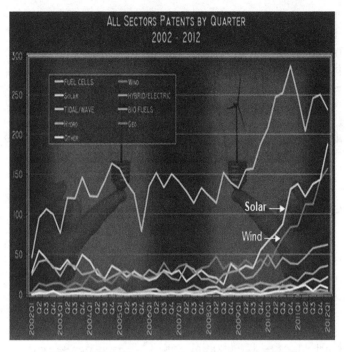

Figure 8-12. Clean energy patents issued (3)

Patents

Patents are a clear indicator of the level of effort and expenditures in the area. The Clean Energy Patent Growth Index, presented by the Cleantech Group at Heslin Rothenberg Farley & Mesiti P.C. shows the patent activities in the renewable energy field.

During the 1st quarter of 2012, granted solar patents (188) once again topped the remaining components and its closest competitor, wind (157), by 31. Solar and wind were tied the previous quarter at 143. This quarter, wind was up by 14 (to 157) and solar up by 45. Both also greatly exceeded the results of the first quarter of 2011 with wind topping the previous year by 71 and solar up 50. Hybrid/electric vehicle patents (62) were up two, relative to the 4th quarter and up 24 as compared to a year prior. Biomass/biofuel patents (36) were up 2 from the 4th quarter, and more than doubled relative to the 1st quarter of 2011. Hydroelectric patents (5) were up four compared to the quarter of a year prior and down one as compared to the 4th quarter. Tidal patents were up six at 22 from the 4th quarter and up 13 over the year before.

Toyota emerged to take the quarterly Clean Energy Patent crown for the first time since 2009 in the first quarter of 2011 with 49 patents. Toyota's patents were primarily in fuel cells at 35 with an assist from hybrid/electric vehicle patents at 14 and a biofuel patent. The leader for 2011, GE, followed with 33 patents (30 in wind, 2 Solar, and 1 each in hybrid/electric vehicles and hydroelectric.

Vestas Wind Systems moved into third with 30 patents—all in wind. General Motors slipped to fourth with 28 patents—all in fuel cells except four hybrid/electric vehicle and one solar patent. Electronics giant Samsung was in fifth with 17 fuel cells patents and five more in solar. One-time leader Honda had 21 patents—again all in fuel cells except a single hybrid/electric vehicle patent. Siemens took the seventh spot on the strength of 13 wind patents and a pair of patents for fuel cells.

Ford and Mitsubishi tied with 11 patents. Ford's patents were in hybrid/electric vehicles (7), fuel cells (3), and biofuels (1). Mitsubishi picked up its patents in wind (5), solar (5), and hybrid/electric vehicles. Hyundai rounded out the top ten with 10 patents of its own (6 for hybrid/electric vehicles, 5 for fuel cells, and 1 for biofuels).

Geographically, Japan was the first quarter leader among non-U.S. holders of U.S. clean energy patents and the individual U.S. states with 150, up 17 over the fourth quarter, and up 19 over the same quarter a year ago, to again claim the quarterly geographical clean energy patent crown.

California was in second place for the second consecutive quarter at 70 clean energy patents, down 9 from the fourth quarter and up 11 compared to a year prior, leading Germany with 51 patents, which has seen large increases over the last year—up 14 from last quarter and up 37 against the same period in 2011.

Michigan followed with 49 patents (down 1 over the fourth quarter and down 3 over the year before) trailed by New York with 47, down 23 and up 12 compared to last quarter and the year prior, respectively. Korea and Denmark had 45 and 32 clean energy patents, respectively. Taiwan had 28 while New Jersey (16), Texas (15) Massachusetts (14), Connecticut (12), France (11) and Delaware (10) all had clean energy patents in the teens.

CONCLUSIONS

To get a better understanding of the present, and make an educated guess as to the future of the world energy markets, one could reduce the global energy demand from all sources of energy to liquid form.

In a quick calculation—converting all energy to crude oil—we come up with global energy usage as the equivalent of 125,000 gallons of oil used per second, 7.5 million per minute, 450 million per hour, 10.8 billion per day, 324 billion per month, and 118.26 trillion per year.

Multiplying this by 100 (the number of years we have used crude oil thus far) we come up with a staggering equivalent of energy used reaching 11.8 quadrillion gallons of oil. This is an entire ocean of liquid—with oil, gas, and coal making up for 80% of it.

This is an ocean of energy pumped and dug out of the ground and used—burned, or converted into other forms—today and in the past, with renewables representing a drop in it—approximately 2% of the entire global energy ocean.

Amazingly, renewables have been around for 50-60 years, but their development has been hindered by a number of forces—most of them boiling down to cost and reliability.

More amazingly, even in the 21st century, firewood and animal dung generate much more energy worldwide than all renewables combined.

One conclusion we must make, and keep in mind at all times when discussing and analyzing energy, is that it takes all forms of energy to run this world. They all have a specific role to play, which is changing, but at a very slow pace.

Changing energy use systems and methods is a difficult process, and given the enormous scale it will take time. Making overnight changes (as attempted recently in Germany, Spain and other countries) is not the way to a sustainable energy future.

Natural gas is emerging as the energy solution of the 2015-2050 period, and will help to bridge the energy gap, while other energy sources (solar, wind, biomass, etc.) are finding their way into the mix.

1. The natural gas supply in the U.S. is far bigger than was thought even a few years ago. It is now estimated that we have more than a 100-year energy supply in the form of natural gas. US crude oil and coal resources are also much larger than previously thought, all of which means that we have enough energy resources to ensure our mid-term energy future.

2. Still, the U.S. is now and will be in the near future, dependent on oil imports at least for the next 25-30 years, unless a significant breakthroughs in alternative energies, or transportation are found.

3. All energy sources are rising in cost, and will continue to.

4. Most conventional energy sources (including natural gas) are polluters that pose a serious threat to our environment and human health.

Assessing points 1-4 above one can easily conclude that we should not be overly worried about our short-term energy supplies—they are here and we know how to use them—but at the same time we must look for new, more efficient, cheaper and less-polluting energy sources. Then, we should use all of them as reasonably and smartly as possible until the best permanent combination is found and implemented.

We have time, so let's not jump into untested and unproven technologies (ridiculously impractical Solyndra and dangerously toxic CdTe types come to mind) that will fail and even further damage the environment. Let's take a scientific and rational approach to solving the energy problem by slowly and deliberately developing all kinds of alternative and renewable energies—solar (PV and CSP, wind, bio-fuels etc.) and deploying them at a sustainable pace.

There is no question that this will be done by the future generations—they simply will have no choice. But we must set the pace and point the direction.

The Future...

The International Finance Corporation, World Bank Group, conclusions and recommendations for solar fields were published in the *Utility Scale Solar Power Plants: A Guide For Developers and Investors, 2012*, as follow:

Solar PV Technology Conclusions (23)

Photovoltaic (PV) cell technologies are broadly categorized as either crystalline or thin film. Crystalline wafers provide high efficiency solar cells but are relatively costly to manufacture; they are sub-divided into mono-crystalline or multi-crystalline silicon.

Mono-crystalline silicon cells are generally the most efficient, but are also more costly than multi-crystalline.

Thin film cells provide a cheaper alternative but are less efficient. There are three main types of thin film cells:

- Amorphous Silicon—The low cost of a-Si makes it suitable for many applications where low cost is more important than high efficiency.

- Cadmium Telluride—Modules based on CdTe produce a high energy output across a wide range of climatic conditions with good low light response and temperature response coefficients.

- Copper Indium (Gallium) Di-Selenide (CIGS/CIS)—Commercial production of CIGS modules is in the early stages of development. However, it has the potential to offer the highest conversion efficiency of all the thin film PV module technologies.

The performance of a PV module will decrease over time due to a process known as degradation. Typically, the degradation rate is highest in the first year of operation and then it stabilizes. PV modules may have a long-term degradation rate of between 0.3% and 1% per annum. Banks often assume a flat rate of degradation rate of 0.5% per annum.

Modules are either mounted on fixed angle frames or on sun-tracking frames. Fixed frames are simpler to install, cheaper and require less maintenance. However, tracking systems can increase yield by up to 34%.

Tracking, particularly for areas with a high direct/diffuse irradiation ratio, also enables a smoother power output.

Inverters convert DC electricity generated by the PV modules into AC electricity, ideally conforming to the local grid requirements. They are arranged either in string or central configurations. Central configuration inverters are considered to be more suitable for multi-megawatt plants. String inverters enable individual string MPPT and require less specialized maintenance skills. String configurations are becoming increasingly popular as they offer more design flexibility.

PV modules and inverters are all subject to certification, predominantly by the IEC. However, one major absence in the standards is performance and energy rating testing other than at standard testing conditions (STC). A standard is being prepared for this, which should enable easier comparison of manufacturers.

The performance ratio (PR) of a well-designed PV power plant will typically be in the region of 75% to 85%, degrading over the lifetime of the plant.

The capacity factor should typically be in the region of 16%. In general, good quality PV modules may be expected to have a useful life of 25 to 30 years.

On a world-scale, we foresee the following scenario unfolding during the next several years and decades:

1. The US installed 1.2 GWp in 2010, 2 GWp in 2011, 2.5-3.0 GWp in 2012 and the number will grow steadily as time goes on—regardless of the political and regulatory winds of change. We do, therefore, foresee the US as the leader in world solar power installations during 2013-2020.

2. China installed 500 MWp in 2010, 1.0 GWp in 2011, 2.0-2.5 GWp in 2012, and is projecting over 2.2 GWp solar power to be installed by 2020. China has deserts and needs electricity badly, and they have decided that solar power is IT. And so we foresee the number and size of solar power installations growing quickly in north and southwest China.

3. Germany installed 8 GWp in 2010, 5-6 GWp in 2011, and 4-5 GWp in 2012, but the pace clearly

Figure 8-13. 21st century large-scale PV power plant (23)

points toward a gradual slowing down—especially in the utility scale area, which is close to saturation. Germany, however, doesn't have many large sun-bathed fields, let alone deserts, which, in addition to the ever cloudy skies, is simply not conducive to reliance on large-scale solar power generation. Also, the "German experiment" showed that solar power installations in Germany are too expensive and simply not a good investment.

4. Italy installed 4 GWp in 2010, less than 3 GWp in 2011, and 2-3 GWp in 2012. There is a lot of talk about Italy becoming the next "solar energy" capital of Europe, and maybe the world, but Italy is deep in the present financial situation, so expenditures must wait.

5. Japan installed 800 MWp in 2010, 1 GWp in 2011, and 1.5-2.0 GWp in 2012, most of which (80%) residential, with some small commercial additions. Here again, Japan has no large desert areas, so it is not expected to jump into large-scale solar installations.

6. India installed 50 MWp in 2010, 400-500 MWp in 2011, and 800-900 MWp in 2012, most of which was utility scale installations. India has poor infrastructure and other serious socio-economic issues that will not allow solar power development to rival that of the US or China.

Note. Australia has tremendous solar potential, but did not make our humble list. It has mostly a small residential market thus far, and since most of the subsidies were eliminated, the solar boom is fizzling = quickly. Several large-scale solar projects presently underway will be completed, but not many new ones are planned.

Near-term Consolidation Enables Evolution

The energy revolution that has been described throughout this text is one that we expect to unfold over the next several years. For this to happen, however, the make-up of the US solar industry will itself need to evolve in the near term. Specifically, sector consolidation will enable these innovations in solar financing.

The industry will likely transition from one with many small players—particularly in residential and commercial markets—to one that has a small number of large, well-capitalized players. The expiration of the cash grant will accelerate industry consolidation. Small-

er developers who thrived on the grant will not be able to raise tax equity because providers have minimum investment thresholds of $15-30m or more.

Additionally, smaller developers are typically unable to provide indemnification for tax equity providers who are looking to protect themselves against the risks of ITC recapture (recapture may be triggered by circumstances such as foreclosure on the property on which the solar project sits). Sponsor size matters, and smaller developers unable to raise tax equity may not be able to stay in business.

Consolidation will facilitate the emergence of investment options that require scale such as asset-backed securities or 'yieldco' issuances. It will lower risks for investors, and lead to a decrease in cost of capital. Consolidation may particularly benefit the commercial market. Large third-party financiers such as SolarCity are well-versed in obtaining tax equity finance for the residential market, and large developers such as NRG are similarly equipped for utility-scale projects.

However, the commercial space has been largely neglected by tax equity investors, as project developers today are too small to provide standardized commercial PPA contracts that are more complex than residential PPAs and generally too small to offer indemnification to tax equity investors. In a consolidated industry, developers may have the scale to raise tax equity for commercial-scale projects.

Even with consolidation, it is unclear if the tax equity 'majors' will be willing to finance small distributed solar projects. Many providers have experience only with utility-scale projects and may be turned off by the due diligence involved with small projects or portfolios. In the near-term this may require new entrants to finance these deals as well as potentially tax equity guarantors to protect investors against the risk of ITC recapture.

Policy Options

Under the status quo, US solar will experience significant growth in the coming decade, enabled by rapidly falling costs and by the drivers identified in this report—new business models for deployment, new investors, and new vehicles for investment. With adjustments to policy, however, this growth could potentially be accelerated, and these drivers could be reinforced.

Solar investors, developers, and industry advocates have articulated five major themes as high priority. These cover policies which (a) have either been proposed or are already in place and could be further strengthened, (b) could be folded relatively easily into existing market structures or existing policies, and (c) have been cited by investors as examples of policies which could pave the way for significant increases to investment and solar penetration.

- Strengthen existing Solar Renewable Energy Credit (SREC) programs: Solar REC markets, pervasive in the northeast US, are often touted by observers who have only a superficial understanding of those markets.

- Experienced players operating there know the challenges of those markets—the scarcity of long-term SREC contracts, which makes project investments a highly risky merchant bet, and the boom-bust cycles whereby build rates hurry ahead of regulators' expectations, creating cases of massive SREC oversupply.

Investment in these markets could be supported by strengthening existing standards, which can be achieved through various means, the most straightforward being lifting or accelerating the targets for solar carve-outs. New Jersey and Pennsylvania are examples of markets that are currently flooded with oversupply and whose legislatures are evaluating proposals to accelerate SREC targets.

A further policy push could focus on mandating availability of long-term contracts. These are sometimes not easily compatible with load-serving arrangements in deregulated markets but in some cases have been implemented through creative workarounds. Connecticut, for example, recently introduced a carve-out, expected to be filled mostly through solar, which requires electric distribution companies to allocate funds towards the procurement of SRECs, which will be procured in standard, 15-year contracts.

- Introduce standards: Adoption of standardized programs propels investments—the California Solar Initiative and the state's Renewable Auction Mechanism allow developers to efficiently submit and process projects. Standards around solar leases and PPAs, furthermore, would spark the growth of solar-backed securitization.

- Enable incorporation of solar into the grid: Relatively slight changes to regulation could facilitate deeper penetration of distributed solar. Two prominent examples are wider adoption of net metering rules and adjustments to PURPA rules.

Under net metering, system owners are compensated for electricity that the system feeds back into the grid (i.e., the portion of generation not consumed by the owner). Most US states allow net metering up to a certain percent of peak load (typically 1-5%); advocacy efforts could focus on promoting net metering rules at the federal level or lifting those limits.

The 1978 Public Utility Regulatory Policies Act (PURPA) mandates that an electric utility buy power from a qualified generator at the utility's avoided-cost rate. The law was the first major step toward unbundling of electricity service, prompted the emergence of IPPs, and opened the door to competitive renewable projects. Yet, for the time being, with solar still pricier than fossil fuels, the technology does not usually meet the avoided-cost metric. Introduction of a solar-specific but still competitive avoided-cost rate could ensure the uptake of solar generation by utilities that would otherwise purchase fossil-based generation.

- Make ITC use more flexible and amenable to new parties: Previous Bloomberg New Energy Finance analysis has shown that the ITC works for investors: tax equity investors in a typical solar project could achieve IRRs of 14-48%, depending on the structure. Yet many potential investors are discouraged from undertaking investments, often due to the illiquidity of the renewable energy tax credits, especially compared with tax credits in other sectors. Solar financing professionals have identified three adjustments to the ITC rules for solar which could make them more aligned with other sectors and increase their appeal to new investors.

- Sanction liquid, tax-advantaged vehicles: Whether they take the form of renewable MLPs or S-REITs, vehicles that allow investors to buy shares in public markets, offer tax-efficient financing to the sector, and are without an illiquidity premium have the potential to expand the base of investors and lower the cost of capital. Of the two, S-REITs seem more politically viable, as they can be implemented without legislative change and do not entail butting heads with deficit hawks.

As an initial approach, usages of the S-REIT structure may simply feature the bundling solar systems as part of an existing REIT-eligible real estate property. The longer-term play, though, is to enable the creation of pure play S-REITs—funds composed entirely of solar assets.

Realizing these kinds of structures would require that PPA revenues earn qualification as rental income and likely also that solar equipment earn qualification as part of property.

A final wrinkle to be smoothed to further pave the way for S-REITs, would be to allow for production-based tax credits (PTC) in lieu of ITCs for solar. According to advocates of this idea, the steady long-term drip of PTCs would more suitably fit the cash flow profile of a dividend-yielding REIT than would a large lump-sum payment of the ITC. However, this would also expose an investor to performance risk.

In conclusion, solar power generation is here to stay…and compete. It is a much needed commodity that will continue growing worldwide, slowly and in the shadow of gas, oil, coal, wind etc.

As world finances improve and the need for affordable and clean energy increases, the growth of solar energy will immerge with another solar boom.

We now know what works and what does not, so the expectations are high that with that knowledge and clearly defined objectives we will make solar power generation the primary energy source of the 21st century.

The Deserts…Again

We'd like to complete this chapter by over-emphasizing the fact that deserts are where the solar energy of the 21st century will be generated and where it will make the greatest difference. Deserts have bright, direct sunlight that can generate the most power per area.

Deserts, on the other hand, have a harsh climate and unforgiving sunlight. Strong UV and IR radiation during the long summer days (14-16 hours long) demolish anything in their way.

The desert has no friends, so very few materials known to man can survive the blistering sunlight, sand storms, monsoons, and other maladies the deserts throw at them.

Most plastics decompose quickly in the desert, glass becomes semitransparent, steel rusts, wood disintegrates. Remembering that PV modules, inverters, wires and other components used in the solar power fields contain some of these materials (especially plastics), we must be very careful of what products we use in the deserts, and how we use them. When one material in a system is damaged, the rest suffer too, usually resulting in decreased performance and eventually total failure of the PV modules and systems.

This, combined with the fact that the PV modules installed today will operate day after day, exposed to the desert's whims for 25-30 years, means that only the best will endure the test of time. Still, we fear that many will start losing power or even fail after the first few months of operation.

Most designers and manufacturers of PV products have never seen the deserts, let alone know of their mighty destructive powers. This leads us to believe there will be a lot of efficiency and reliability problems in the PV power fields in the deserts in the near future.

There are few animals who live in the desserts, and they are especially adapted for the harsh environment. Camels are the best example. We need camels for work in the desert, because any other animal won't make the long journey and will give up, or die in the process.

Similarly, we need PV modules with camel-like performance and endurance for operation in the deserts. Unfortunately, what we see instead are expensive-looking racing horses (PV modules unproven for desert operation). They look good and run fast for awhile, but then they slow down and even die, simply because they are not designed to endure in such a harsh climate.

Many of the PV technologies used today were not properly tested, not proven efficient, reliable and safe for long-term desert operation. Some even contain toxic materials, increasing the long-term dangers, where in addition to failing modules, we might even end up with contaminated desert lands.

The 21st century will surely provide the answers to most questions raised here. We believe that by mid-century there will be a fully grown solar industry, which will have resolved the issues—technical, political, regulatory, and financial. Solar power generation will be a major factor in ensuring energy independence for developed countries and improving the life of the people in the underdeveloped parts of the world.

A major portion of the environmental and climate change problems will be resolved by future 21st century generations. They won't have many other choices, and will be forced to find alternatives—solar power being a bright part of their achievements.

Notes and References

1. *Re-imagining US solar financing*, Bloomberg, May 2012
2. PVCDROM
3. Patents 2012, Heslin Rothenberg Farley & Mesiti P.C., http://www.hrfmlaw.com
4. Module Prices, Systems Returns & Identifying Hot Markets for PV in 2011, *AEI*, September, 2010
5. Fostering Renewable Electricity Markets in North America, Commission for Environmental Cooperation, Canada, 2007. http://www.cec.org/Storage/60/5230_Fostering-RE-MarketsinNA_en.pdf
6. 6 Key Drivers of Emerging Markets, by: Frank Holmes, CEO of U.S. Global Investors, April 25, 2010
7. PC Roadmap, http://photovoltaics.sandia.gov/docs/PVR-

MExecutive_Summary.htm

8. European Photovoltaic Technology Platform comment on "Towards a new Energy Strategy for Europe 2011-2020," http://www.eupvplatform.org/

9. Economic Impacts from the Promotion of Renewable Energy Technologies. The German Experience. Manuel Frondel, Nolan Ritter, Christoph M. Schmidt, Colin Vance http://repec.rwi-essen.de/files/REP_09_156.pdf

10. Analysis of UK Wind Power Generation from November 2008 to 2010. Stuart Young Consulting, March 2011

11. Performance Degradation of Grid-Tied Photovoltaic Modules in a Desert Climatic Condition by Adam Alfred Suleske, ASU, Tempe, Arizona

12. DOE database

13. CAISO Today's Outlook, http://www.caiso.com/outlook/SystemStatus.html

14. Fracking Problem: Shale Gas may be Worse for Climate than Coal, http://envirols.blogs.wm.edu/2011/04/25/fracking-problem-shale-natural-gas-may-be-worse-for-climate-than-coal/

15. NASA solar data. http://eosweb.larc.nasa.gov/cgi-bin/sse/grid.cgi?uid=3030

16. NREL SAM, https://www.nrel.gov/analysis/sam/download.html

17. http://www.greenpeace.org/international/Global/international/planet-2/report/2009/5/concentrating-solar-power-2009.pdf

18. Weather Durability of PV Modules; Developing a Common Language for Talking About PV Reliability, Kurt Scott, Atlas Material Testing Technology

19. An Eye on Quality. *Intevac*, July 2011 http://www.electroiq.com/articles/pvw/print/volume-2011/issue-4/features/an-eye-on-quality.html

20. http://en.wikipedia.org/wiki/Environmental_impact_of_the_Three_Gorges_Dam

21. http://www1.american.edu/ted/itaipu.htm

22. Clean Energy Patent Growth Index (CEPGI), http://cepgi.typepad.com/files/cepgi-1st-quarter-2011.pdf

23. *Utility Scale Solar Power Plants. A Guide For Developers and Investors*, 2012. International Finance Corporation, World Bank Group.

24. *Photovoltaics for Commercial and Utilities Power Generation*, 2011. Anco S. Blazev

Chapter 9

Energy Generation and the Environment In the 21ˢᵗ Century

The environment is everything that isn't me.
—Albert Einstein

ENVIRONMENTAL CONSIDERATIONS

Albert Einstein is not the first man to officially declare that the environment is everything that surrounds us, but he is surely the first on record to put it that way. He is also perhaps the first man to fully realize the influence of the environment on everything that "IS me," so I must care about it.

He continues, "The world is a dangerous place to live; not because of the people who are evil, but because of the people who don't do anything about it." Now, just look around. You will see that most of us are concerned about the environment, but few have a full understanding of the issues at hand, and even fewer are responding.

We have hurt the environment horribly during the last 100 years; it remains to be determined to what degree the harm affects us. We need to know how much of this damage is man-made, and what part of it is irreversible.

We need to understand the problems, their sources, related issues and design solutions. Nothing should stop us from being responsible people in the 21st century.

Will disasters like the BP's oil spill in the Gulf in 2010 (man-made event), or the Dust Bowl phenomena of 1934-1940 (natural event) happen again? Yes, very likely, but regardless of their origin and type, we are more prepared to protect the affected life and property than previous generations.

We are also more aware of the difference between man-made disasters and those which are completely natural. This gives us more control over the environment, our actions, our own health and destiny. But how much control do we have? To answer this question, we must fully understand what is involved.

Author's note: Environmental damage in a broad sense is a misnomer, because the environment only changes from one state into another. What seems like

Figure 9-1. BP Deepwater Horizon, 2011

a "environmental damage" to us, is no more than just another form of its evolution. An ice age might kill most life forms on earth, but the environment will survive and recover.

The recovery might bring an environment much different than what we know, but it will keep changing,

Figure 9-2. The Dust Bowl, 1936

and that's its natural state. In all cases, the environment will survive in one form or another no matter what.

The environment recovers from change, but change is always at the expense of the living things in it, since they (we) are much less tolerant to change and adaptation.

We humans have a very narrow window of tolerance (40 to 100 degree F air temperature, and minimum 10% oxygen in the air), and even narrower comfort zone (60-80 degree F air temperature, and 15-18% oxygen in the air). This makes us very susceptible to environmental changes, and they affect us in measurable ways.

We may not like the present changes, and we are afraid of upcoming changes. We know that if the current pattern continues, we may not survive the next changes, but the environment will. In the worst of cases, the environment will survive and recover—even if humanity disappears altogether.

Since today's serious environmental changes started several decades back, we have seen devastating effects on the human race, so we need to consider the preservation of the environment (in its present form and shape) as a prerequisite for sustaining life on Earth, and a vehicle to our own wellbeing.

Human activities have been blamed for recent negative environmental changes and, because there are no human activities that are absolutely pollution-free and totally environmentally friendly, there will always be some environmental effects and damages brought about by them. Being aware of the consequences of our activities and taking measures to balance the good and bad is a major issue we must tackle during the 21st century.

Since most of the environmental changes and damages are being blamed on the energy generation process, we will focus on its effects on the environment and human health.

The goal of this text is to make a detailed analysis of the cradle-to-grave processes of energy generation—and how it affects the environment during the different stages. We will allow ourselves to make some conclusions and provide recommendations, but it is you, the reader, who will make the final call. And all of us need to agree on a plan to take the necessary actions.

Lessons of the Past

There are a number of cases that illustrate our relation with the environment and its changes through the centuries. Some of the changes, are undoubtedly related to our activities, while others have occurred naturally.

A glaring case of human impact on the environment and human health is found in the 1800s in London.

At exactly 10:00 PM every night, bells would ring signaling the time when the chamber pots could be emptied directly on the streets. The bells were intended also to warn unsuspecting pedestrians to watch their step, and to be wary of stuff falling on them from open windows.

Imagine the noise of thousands of windows and doors opening at exactly 10:00 PM. Imagine the stench. With no place to go, the sewage sat there, building up nightly. Specialists of the day believed there was no alternative. They could not even imagine central canalization and waste treatment, as we know them.

That was a man-made localized environmental disaster for which the knowledge of the day had no answers. The overall effect on human health of this lifestyle was not small, or positive. Amazingly enough, there are places in the world where people live in similar—or worse—conditions even now, in the 21st century.

From our early-London case, we can draw a parallel with what is happening today, and we can imagine what the people of the 22nd century will think of us— ignorant—not unlike the Londoners of 1800. But we are guilty on a much larger scale; while Londoners were killing only themselves, we are killing the environment as we know it. We are dumping sewage, garbage, and many pollutants indiscriminately and in large amounts, in the soil, air and the oceans worldwide.

Among our errors are serious crude oil spills on land and at sea, strip-mine scars on the earth's surface, coal miners suffering and dying from their vocation, and thousands of tons of CO_2, SO_2, NO_2 and other poisons sent into the air daily. Would our 22nd-century descendants add, "There are large areas of desert land covered with inferior solar technologies, rusting in the sun, some containing poisons like lead, arsenic, cadmium?"

China and India will account for 25% of the world's energy use by 2025, and doubling during the next decades. Pollution will increase proportionally. Waste will increase rapidly as the billions in develop-

Figure 9-3. The future is here…today.

ing countries begin driving cars. The need for energy increases proportionately with population growth, so we will pump more oil, and gas fracking will dominate environmental discussions.

History of the World's Environment

Following is a list of acts and policies set forth during the 1970s. Such efforts continue, but now we wonder if our slow call to arms is responsible for the great rise in natural disasters. That list, too, is shown below, along with recent manmade disasters.

1970—National Environmental Policy Act signed, creating the Council on Environmental Quality (CEQ), which gives the President advice on environmental issues.
— General Motors president Edward Cole promises "pollution free" cars by 1980 and urges the elimination of lead additives from gasoline to allow the use of catalytic converters. (Catalytic converters are in, but pollution free cars are still in the queue and still a long way from reality.)
— Earth Day celebration in San Francisco organized by John McConnell.
— First nationwide Earth Day organized by Sen. Gaylord Nelson and Dennis Hayes.
— Clean Air Act passed.
— Natural Resources Defense Council created.
— Friends of the Everglades founded by Marjory Stoneman Douglas.
— Lake Michigan Federation founded.
— Environmental Protection Agency signed into law.
— Occupational Health and Safety Administration (OSHA) bill signed into law.

1971—Chamber of Commerce warns of dangers arising from enforcing pollution regulations.
— Passage of Animal Welfare Act and Wild and Free Ranging Horse and Burro Protection Act.
— President's CEQ acknowledges racial discrimination negatively affects urban environment.
— Greenpeace founded in Victoria, B.C., to oppose atomic testing in Alaska.

1972—W. Eugene Smith completes his essay on the crippling effects of mercury pollution.
— First regional treaty to regulate dumping of radioactive wastes in Europe.
— EPA announces all gasoline stations required

to carry nonleaded gasoline.
— Buffalo Creek disaster occurs in West Virginia, where strip mining kills 125 people.
— Congress passes Federal Water Pollution Control Act, Coastal Zone Management Act, Ocean Dumping Act, and the Marine Mammal Protection Act.
— Toxic Substances Control Act (TSCA) law passed.
— First bottle recycling bill passed in Oregon.
— Supreme Court supports Sierra Club over Disney Inc. in battle over development.
— United Nations Conference on the Human Environment convenes in Stockholm, Sweden.
— UN Environment Program (UNEP) acts on the recommendations of Stockholm meeting.

1973—Eighty nations sign the Convention on International Trade in Endangered Species (CITES).
— Arab oil embargo panics US and European consumers; prices quadruple.
— Congress approves Alaska Oil pipeline.
— A group of Himalayan villagers stop loggers from cutting down a stand of hornbeam trees.
— Endangered Species Act passed by Congress.
— Tellico Dam controversy; Endangered Species Act blamed for stopping project.

1974—F.S. Rowland and M.J. Molina blame CFCs for breaking up ozone in a catalytic cycle.
— Congress passes Safe Drinking Water Act to be administered by EPA.
— K. Silkwood dies in a suspicious accident, involving Kerr-McGee nuclear weapons facility.
— Worldwatch Institute founded.

1975—Atlantic salmon return to Connecticut River after 100-year absence.
— Congress passes Hazardous Waste Transportation Act.
— Greenpeace leads the Great Whale Conspiracy battle.
— Standoff over logging in Brazil's Amazon region by local rubber tappers.
— Federal court says EPA has authority to regulate leaded gasoline.
— Catastrophic failure of Grand Teton Dam in Idaho causes 14 deaths and much damage.
— Chemical explosion in Milan, Italy, spreads dioxin, causing chloracne in 300 school children.
— National Academy of Science report on CFCs

gasses warns of damage to ozone layer.
— Congress passes Resource Conservation and Recovery Act (RCRA), Federal Land Policy Management Act, and the Whale Conservation and Protective Study Act.
— Urquiola oil spill, La Coruna, Spain.
— Liberian tanker Argo Merchant crashes by Nantucket Island, leaks 9 million gallons of oil.
— The Land Institute founded in Salinas, Kansas.
— The International Primate Protection League formed in Thailand.
— American Museum of Natural History forced to halt cat experiments.

1977—U.S. Department of Energy is created by President Jimmy Carter.
— Congress passes Soil and Water Conservation Act, and the Surface Mining Control and Land Reclamation Act.
— Ecofisk oil well blowout occurs in the North Sea.
— U.S. Supreme Court upholds the 1973 Endangered Species Act and stops construction of Tellico Dam.
— Allied Chemical Company and state of Virginia settle lawsuit over extensive contamination of James River.
— Federal Clean Air Act amendments require review of all National Ambient Air Quality Standards by 1980.
— Congress adds additional protection for Class I National Park and Wilderness air quality.

1978—Propylene gas explosion occurs in Tarragona, Spain.
— The Amoco Cadiz wrecks off the coast of France and loses 68 million gallons crude oil.
— Energy Tax Act creates federal ethanol tax incentive of 5 cents per gallon.
— Lois Gibbs and her neighbors form the Love Canal Homeowners Association.
— Robert Bullard begins investigating Triana, AL, where DDT contaminated a stream. Environmental justice movement is born as a result.
— US Congress passes National Energy Act, Endangered American Wilderness Act, and the Antarctic Conservation Act.

1979—Three Mile Island nuclear power plant loses coolant and partially melts down.

— IXTOC I oil well blowout occurs in Bay of Campeche, Mexico; large area contaminated.
— Earth First! organized by Dave Foreman, Howie Wolke, and Mike Roselle.
— Bean v. SWM lawsuit filed, challenging the siting of a waste facility.
— EPA suspends and later bans domestic use of 2,4,5 T, Agent Orange component.
— Appropriate Community Technology demonstration-one of the first alternative energy exhibitions on the national and international level is held in Washington, DC, mall.
— Greenpeace vessel rams the Portuguese pirate whaler Sierra on the high seas.
— J.J. LaFalce and D.P. Moynihan propose "superfund" legislation.

——————————

1986—Chernobyl nuclear plant reactor number four suffers a series of explosions. The environment in local and adjacent areas is inhabitable for the next century or more. Thousands died, many more thousands became ill, and the disaster is still unfolding.

1989—The Exxon Valdez ran aground, resulting in the second largest oil spill in US history, estimated at 500,000-750,000 barrels and listed as the 54th largest spill in history.

2003—Summer heat wave in Europe takes 35,000 lives.
— Earthquake in Iran kills 40,000.

2004—Hurricane Jeannine kills 3,037 people.
— Asian tsunami kills 250,000 people.

2005—Hurricane Katrina devastates New Orleans and kills 1,836 people.
— Earthquake in Pakistan kills 75,000 people.

2008—Myanmar cyclone kills 146,000 people.
— China earthquake kills 70,000 people.

2009—Global swine flu kills 11,800 people.

2010—BP's Deepwater Horizon oil rig spills millions of gallons of crude oil into the Gulf of Mexico. (This disaster will damage the environment and threaten life forms and humans in the Gulf for years to come.)

— Haiti earthquake kills over 300,000 people.
— Iceland volcanic eruption paralyzes European air traffic for several days.
— Mining accident in Chile buried 30 miners and riveted the world's attention to the saga of their survival and eventual rescue.

2011—Japan earthquake and tsunami in March devastates several hundred miles of populated coastal area. Thousands of people were killed and many are still missing. Four nuclear reactors were damaged and still leak radiation. Japan is confronted with long-term uncertainty and large recovery expense.
— Record floods in Thailand, Cambodia, Myanmar, Vietnam and Laos, affected over 3 million people and killed 2,828. World-wide supply-chain disruptions occurred in technology sector, and billion-dollar losses and severe parts shortages rippled to corporations of developed nations. The assumption of safety from floods is now in question in many nations thought or assumed to be prepared.

2012—Extreme heat and draught conditions in the several US regions caused heavy damage to crops and livestock; food prices are rising daily and are not expected to drop until the new harvest, provided that the draught does not continue through 2013.
— Strong earthquake in Iran kills several hundred and thousands are injured.
— Floods in North Korea and Philippines create a desperate situation, where thousands are homeless and entire areas are threatened by hunger and disease.

The greatest and most dangerous natural events—earthquakes, floods, fires, draughts, extreme temperatures, and killer storms—seem to be increasing in frequency and magnitude. Why? Can anything be done to prevent or lessen some of these?

A BRIEF OVERVIEW OF
US ENVIRONMENTAL LEGISLATION

Environmental awareness in the US grew in the 1960s, and in 1970 President Nixon signed the National Environmental Policy Act (NEPA), initiating the "environmental decade" in the US. NEPA created the Council on Environmental Quality to oversee the envi-

ronmental impact of events caused by federal actions. Then the Environmental Protection Agency (EPA) was created, consolidating efforts into a single authoritative entity.

During the 1970s effort was focused on estimating and controlling the generation of pollutants in the air, surface water and groundwater, and solid-waste disposal. Pollutants such as particulates, sulfur dioxide, nitrogen dioxide, carbon monoxide, ozone, as well as the issues of acid rain, visibility, and global warming were put on the agenda. Some preliminary measures were taken and some pollution limits were set.

Dissolved oxygen, bacteria, suspended and dissolved solids, nutrients, and toxic substances such as metals in surface water, as well as biological contaminants, inorganic and organic substances, and radionuclides in ground water were regulated. Solid waste contaminants from agriculture, industry, mining, municipalities, and others were put on the agenda too.

Several amendments of the Federal Water Pollution Control Act and the Clean Air Act emphasized the environmental concerns and moved the field into uncharted territory. The limits and standards prescribed by acts, to be enforced by the individual states, were without merit, and some were unattainable. The state-of-the-art was simply not able to achieve these goals.

The confusion was augmented by the fact that each state was responsible for the preparation and implementation of the environmental plans, which still had to be approved by the EPA. This included the provision of obtaining permits from the EPA to emit pollution into any and all surface waters.

Congress enacted a massive public works program, designed to assist in the construction of water and waste treatment plants for municipalities. There were also deadlines and penalties for automobile emission standards in new cars, which eventually resulted in the development and adoption of catalytic converters, greatly reducing automobile pollution.

During the following administration, heads of departments were appointed, and the design and operations of environmental protection measures were changed to almost voluntary cooperative regulation.

Environmental laws were written and interpreted more favorably for industry interests. The Office of Management and Budget (OMB) was given power to require a favorable cost-benefit analysis of any environmental regulation before it could be implemented.

Subsequent administrations introduced a mixture of innovation and restriction in the environmental regulations, then a total freeze was put on new regulations. Industry-favorable rulings, such as the redefinition of wetlands and the allowance of untreated toxic chemicals in local landfills, were enacted.

Still later, regulatory authority was returned to the respective agencies. The EPA's budget was increased, and much of the country's natural resources were put under greater protection, such as the restoration of the Everglades, and more notably the increase in size of the Everglades National Park.

But the Clean Air Act had provisions to control air pollution on a national level, in which a program called New Source Review (NSR)) required power plants to add anti-pollution technologies before they could expand.

The renewable agenda got another hard push forward, but no follow-through, so the solar boom-bust cycle ended in 2011-2012 and the tide turned against renewables. The US is again focused on coal, oil, gas and nuclear as the most promising energy sources.

While renewables are on hold now, the environmental issues are still with us. Fracking for oil and gas, and mining for coal is on the increase.

Regulation Fog

There are so many regulations that it is nearly impossible to keep track of them all, so we live in a regulation fog. We see clearly only when we are directly affected by what is being regulated.

Environmental regulations are a huge part of business life, at least in the US. Starting in the late 1960s, the US government established regulations toward protection of our environment. By 1998, following a number of environmental disasters, the Environmental Law became one of the governing factors in business life in the US. While it might have contributed to cleaner environment, it also was the driving factor behind the mass exodus of many US companies.

A number of agencies administer different aspects of the Environmental Law. Table 9-1 gives a list of the most important entities and their responsibilities.

The sheer number of agencies and the wide diversity of their responsibilities suggests inefficiency and fragmentation. Still it's a necessary step towards protecting the environment, and we are confident that it will be more unified in time.

These agencies have introduced a number of environmental legislations since the beginning of the 20th century, as shown in Table 9-2.

Table 9-1. List of government agencies involved in environmental work, and their responsibilities.

Federal Agency	Environmental Responsibilities
White House Office	Overall policy, Agency coordination
Office of Management and Budget	Budget, Agency coordination and management
Council on Environmental Quality	Environmental policy, Agency coordination, Environmental impact statements
Department of Health and Human Services	Health
Environmental Protection Agency	Air and water pollution, Solid waste, Radiation, Pesticides, Noise, Toxic substances
Department of Justice	Environmental litigation
Department of the Interior	Public lands, Energy, Minerals, National parks
Department of Agriculture	Forestry, Soil, Conservation
Department of Defense	Civil works construction, Dredge and fill permits, Pollution control from defense facilities
Nuclear Regulatory Commission	License and regulate nuclear power
Department of State	International environment
Department of Commerce	Oceanic and atmospheric monitoring and research
Department of Labor	Occupational health
Department of Housing and Urban Development	Housing, Urban parks, Urban planning
Department of Transportation	Mass transit, Roads, Aircraft noise, Oil pollution
Department of Energy	Energy policy coordination, Petroleum allocation research and development
Tennessee Valley Authority	Electric power generation
Department of Homeland Security / United States Coast Guard	Maritime and environmental stewardship, National Pollution Funds Center (NPFC)

Table 9-2. The US Environmental Laws

Note: See Appendix A for a detailed list of
US environmental laws and regulations.

Year	Law	Year	Law
1899	Refuse Act	1974	Safe Drinking Water Act
1918	Migratory Bird Treaty Act of 1918	1975	Hazardous Materials Transportation Act
1948	Federal Water Pollution Control Act	1976	Resource Conservation and Recovery Act
1955	Air Pollution Control Act	1976	Solid Waste Disposal Act
1963	Clean Air Act (1963)	1976	Toxic Substances Control Act
1965	Solid Waste Disposal Act	1977	Clean Air Act Amendments
1965	Water Quality Act	1977	Clean Water Act Amendments
1967	Air Quality Act	1980	CERCLA (Superfund)
1969	National Environmental Policy Act	1984	Resource Conservation and Recovery Act Amendments
1970	Clean Air Act (1970)	1986	Safe Drinking Water Act Amendments
1970	Occupational Safety and Health Act	1986	Superfund Reauthorization
1972	Consumer Product Safety Act	1986	Emergency Wetlands Resources Act
1972	Federal Insecticide, Fungicide, and Rodenticide Act	1987	Clean Water Act Reauthorization
1972	Clean Water Act	1990	Oil Pollution Act
1972	Noise Control Act	1990	Clean Air Act (1990)
1973	Endangered Species Act	1993	North American Free Trade Agreement
2003	Healthy Forests Initiative		Forest services

PRESENT-DAY ENVIRONMENTAL STATUS REPORT

"The sun's rays are the ultimate source of almost every motion which takes place on the surface of the earth." —Source unknown.

How true! There are many miracles around us, but the biggest is the way the sun rises every day to warm the Earth, revive the vegetation and provide abundant energy, most of which is simply wasted. We barely notice the sun on its way up in the sky, but imagine for a moment what would happen if the sun never shone again.

The Earth is positioned at the right place in the universe, with the perfect equilibrium between the sun's energy coming down to Earth, the amount of energy used, and that radiated back into space—an amazing phenomenon that has kept life on Earth going for many millennia.

If the Earth or its atmosphere were slightly out of place, or if the atmosphere were not composed of the proper amount and type of gasses and particles, life on Earth would be quite different, or non-existent. This is a marvelous and yet fragile combination, which needs to remain this way to preserve human life.

The radiation reflected back into space is regulated by the amount and types of gases and particles in the Earth's atmosphere. In the absence of atmosphere, the temperature on Earth would be about –16°C, and life would cease. Carbon dioxide (CO_2) in the atmosphere absorbs the reflected radiation and keeps some of its energy in the atmosphere, thus warming the Earth enough to keep life at its present levels. This equilibrium maintains the Earth's temperature at ~15-16°C on average.

CO_2 is absorbed mostly in the 13-19 μm wavelength band, while water vapor (also abundant in the atmosphere) is absorbed in the 4-7 μm wavelength band. Therefore, most outgoing radiation (70%) that escapes into space is in the "window" between 7-13 μm. If that window were filled with other gasses, then the escaping energy would be trapped. And if that window were packed with these harmful gasses, temperatures would rise and life on Earth would change drastically, as we have heard many times lately.

Since our "comfort zone" lies in a very narrow temperature window, somewhere between 60 and 80 degrees F, and our minimum-maximum temperature tolerance is only slightly wider, we depend on the earth's atmosphere to keep us alive and comfortable. We also need the right concentration of oxygen and clean water to survive, and that makes us even more dependable on Mother Nature.

Basically, we live in a borrowed place and on borrowed time, and are still learning how to live in it safely and yet productively. We are learning that many natural events could modify and even drastically change life as we know it. Just imagine a large meteor hitting Earth and pushing it slightly out of orbit. What would that do to gravity (if we survived the impact)? What would happen to the already narrow inner energy exchange window, gravity, ocean tides, and even more importantly to the sunlight falling on Earth and its effect on living things?

Human activities are another complex unknown of great importance. During the last century, humans have contributed significantly to changing the environment. Increase of "anthropogenic" gases in the atmosphere is one example. These gasses are the most dangerous for the environment because they absorb in the 7-13 μm wavelength range. They are the particularly harmful gasses—carbon dioxide, methane, ozone, nitrous oxides, and chlorofluorocarbons (CFCs). These gases disturb and even prevent the normal exchange and escape of energy, which leads to an increase in the temperature of the Earth, and its rivers and oceans, thus changing climate and weather patterns.

There is now scientific evidence that CO_2 levels will double by 2030, which will most likely cause significant global warming and increasing air temperatures by 2~4°C on average. This increase might accelerate with time, taking the global warming to unimaginable levels.

This accelerated warming trend could change wind patterns and rainfalls, causing the interiors of continents to dry out, while ocean levels increase. The increasing release of anthropogenic gases must be interrupted. Technologies with low environmental impact and no greenhouse gas emissions will become increasingly important over the coming decades.

Because the energy sector is the major producer of greenhouse gases via the combustion of fossil fuels, technologies such as wind and solar that can substitute for fossil fuels must be considered seriously and used broadly.

Do We Know Why There is Global Warming?

Rapid increase in CO_2 levels has been blamed for recent environmental changes and particularly for global warming. We don't know how much of this is true, and how much is just media hype, but the scientific facts accurately represent the events of the last century and a half.

On the other hand, there is evidence that these events are a transitional part of the Earth's natural evolution. Just as there was an Ice Age, there might be come a "Torrid Age." In fact, there are numerous arguments against the man-made global warming theory, some of which make sense and deserve a second look.

Note: The author supports neither theory, nor is he involved in any related debates or activities. We believe the glass is neither half full nor half empty because it has been clearly observed that all work is in progress and that things can go either way, depending on human activities and Mother Nature's plans.

There is a natural balance that man has not been able to decipher, or to maintain. Still, environmental issues are serious matters, with many serious people on both sides of the debate who have plausible pros and cons, so they all deserve equal time and respect.

The views and facts supporting environmental decay are overwhelming, while the counter-views are not as well-advertised and are sometimes even laughed at. These counter-views, however, are supported by 17,000 serious and respected scientists, who have signed a petition circulated by the Oregon Institute of Science and Medicine, claiming that there is no convincing scientific evidence that human release of carbon dioxide, methane, or other greenhouse gases is causing catastrophic heating of the Earth's atmosphere.

Either way, we all agree that the climate is changing somewhat and that this is a serious problem if it continues uncontrollably. We just don't agree entirely as to how fast it is changing or why.

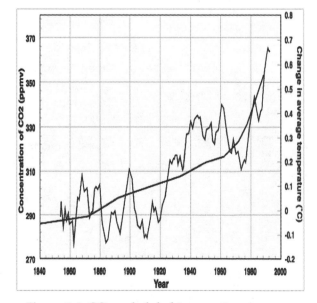

Figure 9-4. CO_2 and global temperature increase

We must agree that as responsible citizens of Mother Earth, we must do our best to contribute to the well-being of the environment, on which our long-term survival as a species depends. The short-term solution is to consider the environment in everything we do. Installing efficient light bulbs and turning off the AC and heaters when not at home are good first steps.

Replacing fossil fuel use with solar and wind energy (efficiently and safely) is the best way to provide energy while reducing pollution. Because energy generation is a big business, and environmental protection is not so big, these ideas will take time to gain popularity.

Energy-related Environmental Facts

Different fuels, and materials used as fuels, produce different amounts of energy and emit different amounts of pollutants. Some examples follow.

1. One MT (metric ton) of coal emits 745 kg. carbon as CO_2 and similar gases. One ton is 1,000 kg, so the coal-to-pollutant ratio is 1.0:0.75. Basically, only ¼ of the coal is converted into energy, while the rest goes up in the atmosphere as CO_2 or is hauled away (as hazardous chemicals containing ash) to a county dump.

2. A typical coal-burning plant (1 GW nameplate) generates:
   a. 3,500,000 tons of CO_2 per annum, which, combined with the waste from thousands of similar plants around the world, goes up in the atmosphere and contributes to the greenhouse effect, feeding global warming trends.
   b. 10,000 tons of sulfur dioxide (SO_2), causing acid rain that damages forests, lakes, and buildings, and forms small airborne particles that can penetrate deep into the lungs.
   c. 500 tons of small airborne particles, which can cause chronic bronchitis, aggravated asthma, and premature death, as well as haze-obstructed visibility.
   d. 10,000 tons of nitrogen oxide (NO_x), as much as would be emitted by half a million late-model cars. NO_x leads to formation of ozone (smog), which inflames the lungs, burning through lung tissue and making people more susceptible to respiratory illness.
   e. 700 tons of carbon monoxide (CO), which causes headaches and places additional stress on people with heart disease.
   f. 200 tons of hydrocarbons, volatile organic compounds (VOC) which form ozone.
   g. 1/70th of a teaspoon mercury which deposited in a 25-acre lake can make the fish unsafe to eat.
   h. 200 pounds of arsenic, which will cause cancer in one out of 100 people who drink water containing 50 parts per billion.
   i. 110 pounds of lead, 4 pounds of cadmium, other toxic heavy metals, and trace amounts of uranium. These are extremely toxic and deadly in very small quantities.

3. One MT (310 gal) of crude oil emits 584.5 kg. of carbon as CO_2 and similar gases. Oil burning power plants—tens of thousands around the world—have similar problems to those described for coal plants. The emitted gas types and quantities are different, but the end effects are the same—heavy environmental pollution and serious threat to human health.

4. One gal. of gasoline emits ~2.77 kg. of carbon as CO_2 and similar gases harmful to human life and the environment.

5. One cubic meter of natural gas emits 0.5 kg. of carbon as CO_2, as well as similar gases.

6. Biofuels (trees and bushes) emit 50% of their weight in carbon-based gases during burning. The majority of the African population still uses these biofuels for their daily needs, so we need to find a way to replace them with renewables; solar is the fastest, cheapest, and cleanest way to do this.

7. One ton of natural gas floating free in the atmosphere traps as much global warming-causing radiation as 20 tons of CO_2 in the same place.

8. Fine-particle pollution from US power plants cuts short the lives of over 30,000 people each year. Hundreds of thousands of Americans suffer from asthma attacks, cardiac problems, and upper and lower respiratory problems associated with fine particles from power plants.

   The elderly, children, and those with respiratory disease are most severely impacted by fine-particle pollution from power plants. Metropolitan areas with large populations near coal-fired power plants feel their impacts most acutely; their attributable death rates are much higher than in areas with few or no coal-fired power plants.

9.   Power plants outstrip all other polluters as the largest source of sulfates, the major component of fine-particle pollution in the US. Approximately two-thirds (over 18,000) of the deaths due to fine-particle pollution from power plants could be avoided by implementing policies that cut power plant pollution containing sulfur dioxide and nitrogen oxide.

Some progress in this area has been made, and more activities and regulations are planned for the near future in the US, but this large-scale pollution will continue for the foreseeable future, especially in developing countries.

The only good news here is that environmental awareness in the US and abroad is increasing, so that events and accidents get more publicity, and average folks are more involved in the debate. This attention helps foster more rapid resolutions as with the BP oil spill, and Hurricane Katrina disasters.

The Major Issues

Since the environmental movement of the 1970s, the nature of environmental issues has changed. While the initial emphasis was on conventional air and water pollutants, which were the most obvious and easily measurable problems, newer issues are long-term problems that are not easily discernible and can be surrounded by controversy.

The major environmental issues, in different stages of understanding and mitigation are as follow.

CO₂ Generation

Present-day consensus is that of all environmental culprits, CO_2 is the most notorious. It is the evil of all evils, and the reason for everything bad happening to the environment. As global warming's most urgent concern, we need to examine CO_2 because PV installations are measured in terms of tons of CO_2 emissions saved, as compared with those generated by fossil fuel plants of the same size during a certain period of time.

A gallon of gasoline, which weighs about 6.3 pounds, could produce 20 pounds of carbon dioxide (CO_2) when burned. The $C + O_2$ combination, expressed in CO_2 units, is a wicked one, with some very special properties. When gasoline or other carbon-containing fuels burn, the carbon and hydrogen separate, the hydrogen combines with oxygen to form water (H_2O), and carbon combines with oxygen from the surrounding air to form carbon dioxide (CO_2).

The reaction is $C + O_2 = CO_2$, where a carbon atom

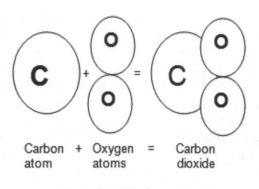

Carbon + Oxygen = Carbon
atom   atoms   dioxide

Figure 9-5 CO₂ formation

has an atomic weight of 12, and each oxygen atom has an atomic weight of 16, giving each molecule of CO_2 an atomic weight of 44.

To calculate the amount of CO_2 produced from a gallon of gasoline, the weight of the carbon in the gasoline is multiplied by 3.7 (44/12). Since gasoline is about 87% carbon and 13% hydrogen by weight, the carbon in a gallon of gasoline weighs 5.5 pounds (6.3 lbs. x .87). Then we multiply the weight of the carbon (5.5 pounds) by 3.7, which equals approx. 20 pounds of CO_2 produced by a gallon of gasoline.

Amazing! A gallon of gas creates 20 lbs. of poisonous CO_2. Who knew? This means that an average car leaves approximately 1 lb. of poison with every traveled mile. This is doubled for SUVs, tripled and quadrupled for larger trucks and RVs, and multiplied many times over for jet planes, boats and other large vehicles.

So, consider the case of the family van that burns one gallon of gasoline every 10-15 miles (or every 5 miles if we drive a Hummer, or an RV). Every 10-15 miles, we leave 20 lbs. of CO_2 behind us, which goes up in the atmosphere to accelerate global warming.

Every 100 miles driven in the family van generates 400 lbs. of CO_2—not a commendable footprint. A summer vacation trip of 1,000 miles will leave 4,000 lbs. (over 2 tons) of CO_2 footprint behind. A hundred thousand families taking similar vacations will load the atmosphere with an additional 400,000,000 lbs. (200,000 tons) of CO_2. A million families…

How about a 2 GW coal-burning power plant bellowing dense clouds of smoke all day long, or a cruise ship navigating aimlessly around the Gulf of Mexico burning thousands of gallons of diesel? Multiply that by thousands of such activities and you'll see that millions of pounds of harmful CO_2, and other gasses are emitted into the atmosphere every day.

Acid Rain

Acid rain and dry deposition (particles) of contaminated matter onto populated centers and other critical areas of human activities is primarily the result of SO_2 and NO_2 being emitted into the air. These gasses are generated in large quantity by fossil burning power plants. They travel freely, whichever direction the wind blows, and land in different places. Upon contact with different surfaces, the chemicals in these gasses could change the acidity of the water or the soil they fall on.

Acid rain conditions in the northeastern US from the burning of coal, and in the western US from gasses generated by utilities and motor vehicles, have been causing problems since the beginning of the 20th century. The situation was partially exacerbated by the Clean Air Act, which forced coal power plants to use taller smoke stacks, resulting in transmission of acid rain gasses for much longer distances, resulting in the contamination of larger land areas.

This is a good example of the complexity of the environmental issues, and how sometimes the best ideas, geared to solve problems, produce disastrous results.

During the Carter administration, a risk-averse policy was undertaken through the EPA and Council on Environmental Quality (CEQ), aimed to research and control the pollutants suspected of causing acid rain. The next administration believed that the scientific uncertainties surrounding exposure levels did not justify the necessary expenditures, and serious efforts in that direction might curtail energy security and economic growth.

In 1990 new Clean Air Act amendments would curtail SO_2 and NO_2 emissions by over 12 million tons per year. A market-like system of emissions trading was implemented, and a cap on emissions was set during 2000, which was partially achieved by the installation of industrial scrubbers on the large emitters.

According to environmentalists, the initial costs in cutting emissions levels, to be paid by the utilities, was expected to be over $4.6 billion, resulting in a 40% rise in electricity costs. Instead, the total cost impact was only about $1 billion, which resulted in only a 2-4% rise in electricity costs. This "discrepancy" can be attributed to the fact that low-sulfur coal was used extensively in coal-fired plants, so no major facilities upgrades were needed to reduce the emission levels.

Ozone Depletion

There has been talk about ozone depletion for decades, and know that means reduced concentration of ozone in the Earth's stratosphere (the ever-shrinking ozone layer). We also know that ozone is needed to block part of the sun's harmful UV radiation. But we still don't know the size of the problem, its actual effect on the earth's environment, or exactly what causes ozone layer problems.

Chlorofluorocarbons (CFCs), which have been used extensively in the 20th century, were blamed for much of the depletion of the ozone layer, so EPA and FDA banned CFCs in aerosol cans in the late 1970s. Did that solve the problem? Not sure...

In the 1980s we learned that the problem was much worse than before, and a massive (albeit controversial) hole in the ozone layer over Antarctica was identified. International agreements were made to reduce ozone-damaging substances, i.e., the Vienna Convention, the 1987 Montreal Protocol, and a third agreement in 1990 in London. The 1990 Clean Air Act Amendments phased out production of CFCs in the US and required recycling of CFC products.

So, the phase-out of CFCs and similar policies are seen as a success, and a crisis seems to have been averted, but the ozone layer is still depleted. According to the proponents, the longevity of CFC particles in the atmosphere will prevent our seeing signs of recovery until 2024-2025.

A lot of time, effort and money was spent on fixing the hole in the ozone layer. Most of these efforts were beneficial, but did we act in time, and will the ozone layer ever be as intact as it was in the 19th century?

Hazardous Wastes in the United States

With the advance of American industry and agriculture during the 20th century, hazardous wastes became a serious problem. There was increased contamination of air, soil and water in all sectors of American commerce. In the mid-1970s, cradle-to-grave regulations were introduced (the Resource Conservation and Recovery Act—RCRA), to govern hazardous waste from its initial generation to final disposition. At the same time, the Toxic Substances Control Act (TSCA) was designed to anticipate possible hazards from chemicals.

In the early part of century, someone had an idea to build a "dream community" on a three-block tract of land on the eastern edge of Niagara Falls, NY. A short canal was proposed to connect the upper and lower Niagara Rivers. Cheap electric power generated by the rushing water would fuel the industry and homes of this model city. It was the infamous Love Canal project.

Bad economic and political circumstances crushed the dream, so by 1910 it was abandoned. All that was left was a partially dug ditch where construction of the canal had begun.

For ten years the canal was used as a municipal and industrial chemical dumpsite. Tons of unidentified chemicals and hazardous materials were dumped in it with no supervision. As such, Love Canal will remain a perfect model for *how not to.*

Then the property owner covered the canal with earth and sold it to the city for one dollar. Like the ignorant Londoners who dumped their chamber pots in the street every night, this chemical company dumped its hazardous waste in the lap of the city. Blindly, city officials were happy to buy it so cheaply.

In the late '50s, the city built about 100 homes and a school on top of the covered waste. Twenty-five years later someone dug into the contaminated soil of Love Canal and found a bona fide industrial dump. There were 82 different chemical compounds identified, 11 of which were suspected carcinogens. These chemicals were not dormant underground, but were working their way upward through the soil, as their metal containers rotted and leached the contents into the backyards and basements of homeowners onto the school ground.

And then... Love Canal "exploded." Triggered by record rainfall, the heavily leaching toxic chemicals débuted. Corroding drums broke ground in backyards; trees and gardens turned black and died; one swimming pool popped up from its foundation, afloat in a sea of chemicals. Puddles of noxious substances filled yards, basements, and the school grounds.

A faint, choking smell filled the air. People coming in contact with the chemicals had burns on their hands and faces. Then birth defects and general ailments became so obvious that the citizens of Love Canal were evacuated from their homes.

After the events at Love Canal, the Comprehensive Environmental Response, Compensation, and Liability Act (CERCLA, or Superfund) was enacted in 1980 to assist in the cleanup of abandoned hazardous waste disposal sites. In the mid-1980s, the Hazardous and Solid Waste Amendments (1984) and the Superfund Amendments and Reauthorization Act (1986) were passed.

The aim of hazardous waste regulation is to prevent harm from occurring due to hazardous waste and to pass the burden of cleanup on to the original producers of the waste. Some of the problems of hazardous waste regulation are that its negative effects can be controversial and difficult to detect, and that due to the amount of hazardous waste generated (214 million tons in 1995), regulation can be difficult and costly.

In 1986, Congress passed the Superfund Amendments and Reauthorization Act, increasing funding and providing for studies and new technologies. By 1995, Superfund cleanup still took an average of twelve years per site, and costs for one site can range in the billions of dollars.

There are hundreds of "Love Canal-like" chemical dumpsites across this nation. Unlike Love Canal, however, very few are situated so close to human settlements. Because of that, we might never hear about them, but without a doubt, their contents are leaching out, and the next target could be a water supply, or a sensitive wetland.

A key problem with these sites is that their ownership frequently shifts, making liability difficult to determine, once it is suspected; no secure mechanisms are in effect for determining such liability.

Firewood Pollution

Finally, we must point to the amazing fact that most of the world's energy (per capita and in daily Btus) is generated by firewood. Trees, brush, branches, twigs and stumps are burned by millions of people in developing countries.

Figure 9-6. Daily supply of firewood.

Over 600 million people is Sub-Saharan Africa, 800 million in India, 500 million in China and 100 million in South America are using firewood for their daily needs. These people have no other choice.

Consumption of firewood is estimated at approximately 800 kg. per person per year. This comes to only 2-3 lbs. per person per day, but it is a large amount of wood—especially in areas that have been cleared from trees and vegetation for years. This creates great problems, including deforestation and tensions, especially in the more densely populated areas.

Let's assume for the purposes of this rough calculation that 500 kg. firewood are used per person annually by approximately 2 billion people worldwide.

Firewood's energy content is equivalent to 0.35 tons of oil per ton of firewood. Assuming that firewood stoves and fire pits burn at maximum 15% efficiency, we can conclude that to replace the firewood used worldwide, we would need over 150 million tons of oil a year.

This is nearly a quarter of US oil consumption—OR—300 million US inhabitants use 4 times the amount of oil used by *two billion* people in the Third World—OR—the US uses over 20 times more energy per person than the entire developing world.

Making matters worse, firewood is a big business in many countries. Large forests and wooded areas are cut to replenish supplies for firewood markets in the cities. Large quantities of wood are also used to make charcoal, which is also a hot commodity in the populated centers.

Though most countries have anti-deforestation laws, many people are involved in wood gathering, preparation and sales of firewood and charcoal. Some are not aware of the laws; others don't care or simply have no choice. So, the large-scale deforestation of the poorest countries in the world continues.

Providing oil to the Third World is not feasible, nor is any of the conventional energy sources (coal, gas, nuclear). Only renewables could alleviate the worsening worldwide firewood situation.

Firewood is also used in large quantities in rural areas of developing countries, and in the US, the 1973 oil crisis brought out an entire wood-burning industry, complete with the most efficient wood-burning stoves.

Generally, though, firewood use in developed countries is not of necessity; it is a type of luxury. Cuddling with a book by a roaring wood fire is part of the American dream, and firewood use has increased to a point where there are many winter days with local wood-burning bans, due to increased CO_2 pollution.

Of course, with every wood fire there is smoke. Wood burning is one of the largest energy generators and polluters in the world. It is the least discussed subject in the energy field, and one that many energy principles tend to ignore because most of the world's wood-burning people are the poorest segment of the world's population.

It is hard to say if wood burning classifies as a non-renewable energy, but we hesitate to call it renewable because it leads to uncontrolled and extensive destruction of forestation in entire regions. We should find ways to replace it with a more sustainable and less damaging energy systems.

Author's note: Flying over some areas of Asia or Africa at night, one can easily see hundreds of fires below. Even large cities, like Taipei and Beijing are marked by the fires and plums of smoke coming from their suburbs at night. In daylight, large areas of wasteland can be seen for miles too.

Wood burning cannot be eliminated completely, for it has some useful purposes, but providing alternative energy sources for people in places such as India and sub-Saharan Africa, would contribute significantly to cleaning the environment from harmful gasses and the related effects.

Forest Fires

Forest fires are one of the largest sources of particulate matter pollution. They can have significant impacts on local air quality, visibility and human health. Emissions from forest fires can travel large distances, affecting air quality far from their origins. Their emissions include particulate matter, carbon monoxide, atmospheric mercury, ozone-forming chemicals, and volatile organic compounds.

Industry specialists estimate that an acre of primarily coniferous forest that burns completely will emit approximately 4.81 metric tons of carbon into the atmosphere. Eighty to ninety percent of the released gasses are in the form of carbon dioxide (CO_2). The rest is carbon monoxide (CO) and methane (CH_4).

In the United States, record-setting 96,385 wildfires in 1996 destroyed about 9.87 million acres of forest. These fires accounted for 47.47 million metric tons of carbon emissions in the affected areas. The nation's total annual carbon dioxide emissions are estimated at around 6.049 billion metric tons.

In 1997 several large-area fires burned through thin-trunked tropical trees in Indonesia. The devastated forests were covered in carbon-rich peat, measuring up to 20 meters thick in places. Those fires were estimated to have released approximately 2.0 Giga tons (2,000 billion tons) of carbon—almost one third of the world's annual emissions at the time—or more than 300 times the total amount of CO_2 released in the US that year.

Planned and controlled burning is practiced to manage large forest fires, and is more desirable than uncontrolled wildfires. Other types of wood burning are industrial, slash and wood-residue, land clearing, agricultural, yard leaves and debris, outdoor wood-fired boilers, campfires and beach fires, garbage, and construction debris. The combined negative effects from air pollutants emitted by these sources should not be overlooked.

The overall effect on the environment, from our daily activities, is negative, and increasingly so.

Risk Control Policy

Risk control, in the framework of environmental protection and remediation, is a new concept. Underlying the policy decisions made by the United States, the concept of risk control consists of two parts: risk assessment and risk management. The science behind risk assessment varies greatly in uncertainty and tends to be the focus of political controversy. For example, animal testing is often used to determine the toxicity of various substances for humans. But assumptions made about expected dosage and exposure to chemicals are often disputed, and the dosage given to animals is typically much larger than what humans normally consume. While industry groups tend to take a risk-tolerant position, environmentalists take a risk-averse position, following the precautionary principle.

Another issue is the effect of chemicals relative to lifestyle choices. Cancer, for example, typically surface decades after first exposure to a carcinogen, and lifestyle choices are frequently more important in causing cancer than exposure to chemicals. While a governmental role in mitigating lifestyle-choice risks is controversial, chemical exposure through lifestyle choices can also occur involuntarily if the public is not properly educated.

Finally, the way that threats are presented to the public plays a large role in how they are addressed. The threat of nuclear power and the environmental effects of pesticides are overstated, some have claimed, while many high-priority threats go unpublicized. To combat this discrepancy, the EPA published a Relative Risk Report in 1987, and a follow-up report published by the Relative Risk Reduction Strategies Committee in 1990 suggested that the EPA should adopt a more pro-active posture, educating the public and assigning budgetary priorities for objectively assessed high-risk threats.

Overall Impact

Since the major environmental legislations of the 20th century were enacted, great progress has been made in some areas, but environmental protection has come at a high price. Between 1970 and 1996, air pollutants dropped 32% while the population grew by 29%. Other pollutants have been more difficult to track, especially water pollutants. While air and water standards have been slowly improving, over 60 million people still live in US counties that don't meet EPA ozone standards. Also 30% of rivers and 40% of US lakes don't meet minimum standards for all uses (swimming, fishing, drinking, supporting aquatic life). Nearly 50% of the 960 endangered species are still far from recovery.

The overall cost of environmental regulation in the United States is estimated to be about 2% of the gross domestic product—similar to many other countries, but calculating the cost is challenging, both conceptually (deciding what costs are included) and practically (with data from a broad range of sources). Almost $122 billion was spent in 1994 alone on pollution abatement and control—$35 billion in direct government spending, $65 billion spent by business, and $22 billion spent by individuals.

Critics of environmental legislation argue that the gains made in environmental protection come at too great a cost. The cost of meeting Occupational Safety and Health Administration workplace exposure standards, for example, can be as high as $3 million per life-year for benzene protection in coke and coal factories, or $51 million per life-year for arsenic protection in glass manufacturing plants.

Due to these high costs, non-compliance with environmental rules is rampant in some areas, such as water pollution. Contaminating the water table has long-term effects that cannot and should not be ignored, because it affects not only people in the immediate locale, but all life forms downstream.

The overall effect of the above listed issues is global environmental changes, resulting in warmer climate, and another set of problems.

Global Warming Issues

Pollution from energy generation accounts for over 50% of all air, soil, and water pollution today. Electric power production from coal and oil, the largest contributor to global warming, is blamed for environmental damages. EPA estimates that fossil fuel-based power generation has an environmental health cost of 10.5 cents per kilowatt hour—almost as much as it's actual cost. This means that its actual cost is double its present value, half of which is adverse health effects.

Environmental buzzwords are all around us: *anoxic waters, ocean deoxygenation, climate change, global warming, global dimming, fossil fuels, rise in sea level, greenhouse gas, ocean acidification, shutdown of thermohaline circulation, conservation, species extinction...* it's quite an array of words, conditions, and situations. (See Appendix B for a complete, alphabetized list.)

So how do we know what is good and what is bad behind each of these terms? Where do we start, once we have determined that there is a problem? What are the most important issues, and how are these interconnected? What should we do to help the environment, ourselves, our way of life, and that of future generations?

Examples of Environmental Damage

On December 5, 1952, the residents of London, England, awoke to the dawn of a five-day reign of death. A temperature inversion had trapped the coal smoke from the city's furnaces, fireplaces, and industrial smokestacks, creating a "killer fog" that hovered near the ground. People began to die from respiratory and cardiopulmonary failure. Not until that weather system finally loosened its grip, and the soot-filled air cleared out, did death rates return to normal. The end of the episode saw more than 3,000 dead. It is not hard to imagine such a scenario developing again, and over more densely populated areas—something we are just not prepared for.

Driving north from Hong Kong into mainland China, you'll notice dozens, if not hundreds, of large factories belching black smoke into the air and discharging fizzling, green-brownish liquids into the ground. On the streets of major Asian cities, you'll notice most people wearing masks over their mouths and noses. This unprecedented environmental pollution on such a large scale started some 20-30 years ago, and is getting worse. It coincides with the tremendous growth of the Asian industrial complex at the expense of local and world environments and the immediate health of those who live in the region.

But you don't need to go to Asia to see examples of man-made hazmat dumps and generators of poison gasses. Drive by any large paper, cement, or chemical factory, or any coal-burning power plant in the US or Europe, and you'll see the signs of pollution: heavy smoke in the air, dead trees, dusty fields, and lagoons of foul water.

Other stark examples are Chernobyl, the Exxon Valdez, Deepwater Horizon and Fukushima Daiichi nuclear plants, the 2010 collapse of the coal mine in Chile, and the 2011 Japanese nuclear disaster. Each event in itself is profound enough, but connected they have an enormous negative effect on the environment and human life. Making this connection, is a complex undertaking, with many variables and unknowns, thus the split among scientists.

Solar cells and modules manufacturing, transport, installation, and recycling processes also emit large amounts of CO_2 and other gasses into the atmosphere. They fill county dumps with rusting and harmful materials, so we will take a closer look at these later on in this text.

There are failed solar power plants in the US, that were supposed to bring us a bright energy future back in the 1980s and 1990s, but instead created "solar junk yards." Why did these high-visibility installations fail? Was it the technology, the installation, the operation, poor planning and management? Did the failed equipment cause any damage to the surrounding environment beyond the visual pollution and the rusting metal parts?

All industries and energy-generating technologies use natural resources and generate waste by-products. Renewable energies (wind, solar, etc.) are no exception. The question is how to estimate, control, and reduce the negative effects. We don't have the answers to most of these questions, but hopefully this text will bring some of the issues out for discussion.

Author's note: Environmental deterioration is all around us and increasing daily. The recent oil drilling and gas fracking energy-boom in the US is quite evident in parts of Texas, Pennsylvania, Ohio, Colorado and other states. Hundreds of new holes for gas and oil fracking and geothermal power are drilled daily. 1.4 million geo-thermal drill holes have been recorded in Texas alone. Leaking chemicals and gasses can be found in the nearby air, soil and water, backyards, rivers and lakes.

The impacts are compounded. When one hole is drilled, its negative effects are combined with those of adjacent holes—underground and above—so we cover large land mass with *ad-continuum* dangers and unknowns (from chemical and geo-mechanical points of view), the long-term effects of which are, well, unknown...

ENERGY GENERATION AND THE ENVIRONMENT

Pollution from energy generation accounts for over 50% of all air, soil, and water pollution today. This is a complex subject, the surface of which we can barely scratch here.

Electric power production from coal and oil, is the largest contributor to global warming, and is blamed as a major contributor to environmental damage. To repeat, EPA estimates that fossil fuel-based power generation has an environmental health cost of 10.5 cents per kilowatt hour—almost as much as it's actual cost.

For every kWh produced in the US, the environmental damage on *human health* (and the cost to remediate it) is averaged at 10.5 cents. We're talking about hospital bills, lawsuits, sickness and death under different circumstances—all in the process of energy production.

When we add the cost in health damages of one kWh to the base cost of one kWh, the actual cost of electric energy is double. As consumers, we are paying twice for our energy—once for the electric power, and again for medical expenses caused by its production.

Pollution from conventional power generators (coal, oil, gas and nuclear) is a well known problem, and a lot of discussion is underway on the subject. Environmental damage from renewable power generators—solar, wind, biomass—is a new issue, and much less understood, but it is there, and we will take a closer look at it below as well.

In any case, we separate the power generators, and their respective environmental issues—those created by conventional methods (coal, oil, gas and nuclear), and renewable energy sources (solar, wind bio-fuels, etc.).

CONVENTIONAL ENERGY GENERATORS

Comparing conventional energy sources with today's solar technologies is not a straightforward process, but there are clear differences. Our objective is to provide a detailed technical discussion and analysis of the different technologies from an environmental point of view, discuss their characteristics, and draw some conclusions to open an honest discussion.

Activities related to electric power production via conventional energy sources (coal mining, oil drilling, gas fracking and nuclear power) with the related digging, crushing, pumping, transport, recycling and disposal processes no doubt contribute to environmental pollution, global warming, acid rain, and disease.

Increased production and use of fossil fuels, especially coal, has severe local and regional impacts. Locally, air pollution takes a significant toll on human health. Regional acid rain precipitation and other forms of air pollution degrade downwind habitats—especially lakes, streams, and forests—and they, in turn, are damaging crops, buildings and outdoor equipment.

One recent study warns that in the absence of sulfur abatement measures, acid depositions in parts of southern Asia could eventually exceed the critical toxicity load for major agricultural crops by a factor of 10.

Without the use of the best available technology and practices, coal mining leads to land degradation and water pollution, as does the disposal of hazardous coal ash. And while the effects of natural gas fracking and nuclear power use and waste disposal are less obvious, they are estimated to be even greater in the long run than those of coal mining and use.

Following are detailed technical analyses of the cradle-to-grave processes in the attainment of conventional energy sources, use, recycling, and disposal.

1. *Sourcing process*—Exploring areas suspected of having deposits of coal, oil, natural gas, etc. Locating, testing, probe drilling.

2. *Regulatory process*—Permits, purchase agreements and other private, state and federal negotiations and regulatory authorizations as needed before the extraction of the goods.

3. *Extraction preparation*—Activities related to the preparation for extraction: construction of the main and auxiliary production facilities, surface layer removal for surface mines, digging deep holes for underground mines, hole drilling for oil pumping and gas fracking, etc.

4. *Extraction process*—Extracting the energy sources begins and is executed for many years. Coal and uranium are dug up and trucked out of surface mines, or railed out of underground mines. Deep holes are drilled underground for oil pumping, and toxic fracking chemicals are pumped underground, under high pressure, for gas extraction.

5. *Transport*—Usually thus obtained materials cannot be used on site, so they must be railed, piped or trucked to dedicated facilities, for additional processing and/or conversion to electricity.

6. *Conversion process*—Most of thus extracted and transported materials (coal, oil, gas, uranium) are burned, or otherwise converted into energy. The process usually consists of conversion into thermal energy in the form of steam, which then drives turbines to generate electric energy.

   Some of these energy sources are used for heating homes and business, cooking and other needs, where they are usually burned for their calorific value.

7. *Distribution*—The electric energy that has been generated is now sent into the national or local power grid, for use by homes and businesses. Local and national utility companies provide the equipment and services to accomplish this task.

   Natural gas that is destined for use in homes and businesses is sent to its destination via pipes. The same utilities that manage the electrical supply also support the gas lines and control equipment in most cases.

8. *End use equipment*—The electricity in the national grid is used by numerous appliances and equipment, for lighting, heating, air conditioning, and many other domestic and industrial needs.

9. *Electric and heating services*—The electrical and gas heating appliances are manufactured, serviced and disposed of by specialized personnel and companies.

Please note that a lot of energy (electricity, heat, etc.) is used during each and every step of the cradle-to-grave process. A lot of waste gasses, liquid and solids, are generated during each step of the process, as well as during the production and use of each energy type.

This is exactly what we will be focusing on—the environmental impact of generating and using energy, and the variables that accompany each step of the energy generation and use process.

Coal

Coal is one of the most controversial fuels. It is both needed and unwanted. It is both useful and damaging. We can't live with it or without it. It is absolutely needed to ensure our daily comfort, but it is killing us in the process. Coal generates 54% of our electricity, and at the same time it is the single biggest air polluter in the U.S., emitting close to half of the CO_2 and other gasses.

Coal pollutes all through its cradle-to-grave energy generation cycle. It pollutes when it is mined, transported to the power plant, stored, burned, and when its waste is disposed of. It causes all kinds of environmental damage in the process.

Coal levels mountains, creates huge scars in the earth and pollutes the land, water, and air. Burning coal causes smog, soot, acid rain, global warming, and an amazing variety of toxic emissions.

Other waste products generated are ash, sludge,

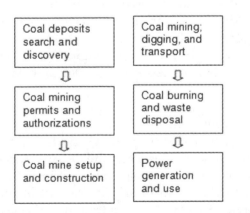

Figure 9-7. Cradle-to-grave coal power generation

toxic chemicals, and waste heat; all of which create more environmental problems. On top of that coal plants need billions of gallons of fresh water for process cooling. All of these activities and products harm the environment, humans and wildlife in the local areas and far beyond.

Coal Mining

60% of US coal is scraped and ripped out from the earth in surface mines. The rest is dug out from underground mines. Surface coal mining is especially damaging to the landscape and its life forms. Coal companies often remove entire mountaintops to expose the coal below. The wastes are generally dumped in valleys and streams. In other cases they dig enormous holes that are often abandoned as great scars in the earth' surface, causing irreparable damage to the area's environment.

For example, in West Virginia, more than 300,000 acres of hardwood forests (half the size of Rhode Island) and 1,000 miles of streams have been destroyed by surface mining.

Underground mining is one of the most hazardous occupations. Coal mining also causes chronic health problems.

Coal Transportation

A typical coal plant requires approximately 40 railroad cars daily to transport the coal to the power plants. That's 14,600 railroad cars a year. Railroad locomotives and trucks used to transport the coal to the power plants burn large quantities of diesel fuel. Their engines emit nearly 1.0 million tons of nitrogen oxide (NO_x) and 52,000 tons of coarse and small particles in the United States during their way from the mines to the power plants. Coal dust blowing from coal trains contributes significant particulate matter to the air in the affected areas.

Coal Storage

Massive piles of coal to be burned by power plants are typically stored nearby in uncovered piles. Dust blown from the open coal piles harms humans who come in contact by irritating the lungs and eyes. It also often settles on nearby houses, yards and agricultural fields. Rainfalls create runoff from the huge coal piles, leaving pollutants to leach into and contaminate land and water in the locale.

Solid Waste

Coal burning power plants generate enormous amounts of solid waste before, during, and after the burning process. Waste created by a typical 500-megawatt coal plant includes more than 125,000 tons of ash and 193,000 tons of sludge from the smokestack scrubber each year. Nationally, more than 75% of this waste is disposed of in unlined, unmonitored onsite landfills and surface impoundments.

Toxic substances in the waste—including arsenic, mercury, chromium and cadmium—can contaminate drinking water supplies and damage vital human organs and the nervous system.

In one study, one out of every 100 children who drink groundwater contaminated with arsenic from coal power plant wastes were at risk of developing cancer.

Local ecosystems too have been damaged—sometimes severely or permanently—by the careless disposal of coal plant waste.

Cooling Water

A typical 500-megawatt coal-fired power plant draws about 2.2 billion gallons of water each year from nearby lakes, rivers, or oceans, to create steam for turning its turbines. This is enough water to support a city of approximately 250,000 people.

When this water is drawn into the power plant, 21 million fish eggs, fish larvae, and juvenile fish may also come along with it—and that's the average for a single species in just one year. In addition, EPA estimates that up to 1.5 million adult fish per year may become trapped against the intake structures. Many of these fish are injured or die in the process.

This water is used to keep the plant operating properly, by cooling the post-burning steam and steam equipment. Then the cooling water is released back into the lake, river, or ocean, but is now 20-25°F hotter than the water into which it is dumped.

This "thermal pollution" affects life in the area as well. Its effects are numerous, starting with decreased fertility and increased heart rates in fish. Then, power plants add chlorine or other toxic chemicals to their cooling water to decrease algae growth in the area. These chemicals are also discharged back into the environment and negatively affect life forms in it.

Waste Heat

Much of the heat produced from burning coal is wasted. A typical coal power plant uses only 33-35% of the coal's heat to produce electricity. The majority of the heat is released into the atmosphere or absorbed by the cooling water.

Gas Emissions

A typical (500 megawatt) coal plant burns 1.4 million tons of coal each year. There are about 600 US coal plants, so the total is over 800 million tons of coal—dug out, transported, burned, and its waste discarded—every year in the US alone. The world total is several times this number. Billions of tons worldwide. Every year.

In an average year, a typical coal-fired power plant generates:

- 3,700,000 tons of carbon dioxide (CO_2).

- 10,000 tons of sulfur dioxide (SO_2).

- 500 tons of small airborne particles.

- 10,200 tons of nitrogen oxide (NO_x).

- 720 tons of carbon monoxide (CO).

- 220 tons of hydrocarbons, volatile organic compounds (VOC), which form ozone.

- 170 pounds of mercury.

- 225 pounds of arsenic.

- 114 pounds of lead, 4 pounds of cadmium, other toxic heavy metals, and trace amounts of uranium.

Multiply these toxins, carcinogens and other poisons by 600 and you will get a picture of the intense gaseous, particulate and liquid pollution we send into our air, soil and water every single day. Then multiply the above numbers by several thousand—the number of coal burning plants around the world—and we can begin to see why we have environmental problems.

On a global level, increased burning of fossil fuels means an accompanying rise in greenhouse gas emissions, along with the potential adverse impacts of global warming and other climate changes. Nuclear fuel, too, has obvious environmental costs associated with materials production and disposal, although nuclear power produces virtually none of the air pollution and carbon dioxide discharges of fossil fuels during power generation. Its drawback is the thousands of tons of waste materials stored at all nuclear plants worldwide.

How long can we keep pumping such large quantities of deadly materials in our air, soil and water?

The Hidden Costs of Coal

A recent estimate of the hidden costs of burning coal to generate electricity in the US alone set the number at $62 billion a year. These are the costs related to environmental damage, which does not include expenses related to global warming. This is part of the annual $120 billion in total damages from use of energy in the past.

These are the external expenses and benefits associated with production, distribution and use of energy not already reflected in market prices. Nuclear power plants and renewable energy generators are also excluded, because of their quite low effect as compared with those of fossil-fuel-based power plants.

These are basically the costs related to air pollutants on human health, grain crop and timber yields, building materials, recreation and visibility of outdoor vistas.

Aggregate damages associated with sulfur dioxide, nitrogen oxides and particulate matter emitted by the facilities amounted to $156 million on average per each plant. Multiplied by 600 plants, the costs are $90 billion in the US alone.

Cars, trucks, trains and the burning of transportation fuels generate $56 billion in health and other non-climate related damages annually.

Oil Processes and Impacts

Drilling, fracking and transportation of crude oil and oil products cause problems, as detailed in the coal section above. At the same time, burning oil (and its products) creates another set of environmental and health problems.

Crude oil is the most sought after energy source. It is also a commodity that has found a place in the commodity markets, where it is manipulated by traders, creating chaos in the world's energy markets.

The cradle-to-grave oil production and energy generation process is similar to that used in coal production and power generation.

Fracking for oil is a recent development, and is the worst of all methods of oil production. Like natural gas fracking, oil fracking consists of pumping huge quantities of water under very high pressure into the ground. The water contains acids and other chemicals, which re-

act with the soil and rock formations, assisting the high pressure solution to break the rocks and release the oil deposits. This crude oil is pumped to the surface, stored and transported to refining facilities or to oil-burning power plants.

The consequence of the large-scale application of this method in some states, in addition to the poisons released during oil burning, is serious air, soil and water contamination during the fracking process. The powerful mechanical and chemical forces cause leakage of toxic chemicals in streams, lakes and backyards, as well as land mass collapses.

Other serious oil-burning pollution comes from the millions of vehicles which emit toxic gasses every day. In large cities, especially, we can sense the extremely large quantities of these gases.

Acid rain is another effect from burning large quantities of oil products. Sulfur dioxides and nitrogen oxides are pumped in the air during oil burning, and when they merge with moisture, acid rain is created. This acid penetrates our water systems and dependent flora and fauna. It eats away at buildings, statues and other structures.

Global warming is another concern surrounding the burning of oil. Carbon dioxide, an oil-burning by-product, traps heat in the atmosphere, and the planet becomes hotter.

It is hard to quantify the amount of oil being burned in the US, and even harder when looking at the world level.

Natural Gas Impacts

Getting to natural gas deep in the earth involves drilling vertical, horizontal or multi-lateral wells to the target gas deposits. A number of different techniques, with hydraulic fracturing as the most promising, are used today to create an effective connection between the well and the targeted natural gas formation.

Before drilling, geologists complete a full analysis of the geology using proprietary and public data. They assess results from other wells drilled in the vicinity, including water wells, producing oil and gas wells and nonproducing wells (dry wells). A plan is developed for drilling and completing the well that must be approved by state regulators. Key stakeholders, including communities, officials, government agencies and regulators are involved in the planning process.

The actual drilling process consists of leveling the ground, setting up a drilling rig, protecting the fresh-water aquifer by means of multiple barriers of cement and steel casting, the drilling thousands of feet into the

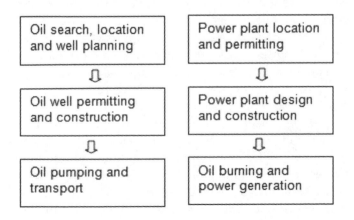

Figure 9-8. Oil sourcing and power generation

formation as perforated casing is installed throughout the length of the potential gas-producing section.

Then the fracking operation commences, and when a sufficient gas flow is obtained, the surface processing facilities (pumps, storage tanks, etc.) are installed and gas production starts. Some of the affected area, that is not needed for gas production, is restored, but the underground activities continue for the next 20-30 years.

While natural gas has been hailed as a "clean" energy source, its production creates a much larger quantity of harmful gases (mostly due to raw methane) than any other industry. Raw methane gas released in the atmosphere is worse than CO_2 in accelerating the green house effect.

Officially, America's oil and natural gas industries have a long-standing commitment to safety and protecting the environment. Their environmental investments represent a crucial aspect of today's energy exploration and production process. Since 1990, they have invested $250 billion toward improving the environmental performance of products, facilities and operations. This represents approximately $800 for every man, woman and child in the United States.

Between 2000 and 2010, the US oil and natural gas industries invested $71 billion in technologies that reduce greenhouse gas emissions, far more than the federal government ($43 billion).

This is all very good, but what are the long-term effects of the actual daily activities of mining, oil drilling and gas fracking?

Shale Gas

Shale gas is also raising our hopes for cheap energy independence. However, it too brings several dangers and creates a serious environmental impact. The drilling and high pressure water pumping (fracking) are so extensive that some geologists fear that these activities might destabilize the ground in affected areas.

When drills penetrate through aquifers, the boreholes must be well sealed, so water doesn't leak in and waste doesn't leak out. If this isn't done right, there's trouble. Also, some of the fracking water injected into the newly drilled well gets absorbed by the shale, but some "burps" back out, contaminated with toxic chemicals and must be disposed of as hazardous waste.

These are complex and dangerous activities, even when done right. When not done right, they simply contaminate and devastate the area.

As our energy demands grow and related activities increase, we find ourselves dealing with new phenomena, such as gas and oil fracking, but we don't know what to watch for—what to safeguard against—or what to expect…

Nuclear Energy

Since the 1960s nuclear energy has been hailed as the safest and most economical energy source. In fact, it has proven to be so…with only few exceptions, but that's where we draw the line, because these exceptions have been too costly.

The health risks and greenhouse gas emissions from nuclear fission power are small relative to those associated with coal, but there are "catastrophic risks" such as the possibility of over-heated fuel releasing massive quantities of fission products into the environment. The 1979 Three Mile Island accident and 1986 Chernobyl disaster, along with high construction costs, ended the rapid growth of global nuclear power capacity.

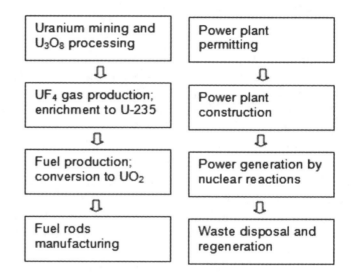

Figure 9-9. Uranium processing and Nuclear power generation

The 2011 Japanese nuclear disaster brings another set of questions, which affect human life in a most direct, serious, and permanent way. The fact that it might be 100 years before the plant can be even approached, to remove the melting fuel rods safely is a frightening reminder of the power and negative effects of nuclear power.

And the serious issue of long-term storage of nuclear waste materials is still unresolved.

Still, nuclear energy is still a driving force in the world's energy mix, and it will remain so for a long time—regardless of the efforts of some governments to reduce or eliminate it. A major EU funded research study known as ExternE, undertaken between 1995 and 2005 found that the environmental and health costs of nuclear power, per unit of energy delivered, was

€0.0019/kWh. This is lower than that of many renewable sources including the environmental impact caused by biomass use and the manufacture of photovoltaic solar panels, and it was more than 30 times lower than coal's impact of €0.06/kWh, or 6 cents/kWh.

The energy source of the lowest external costs was found to be wind power at €0.0009/kWh, which is an environmental and health impact just under half the price of nuclear power.

According to the US Department of Energy, America needs 20,000 MW (or 20 GW) of new power generation every year for the next 20 years to meet projected demand. An additional 2,000,000 MW are needed around the world over the same period. This represents the need for adding 15-20 conventional power plants per year in the US and 1,500-2,000 additional power plants worldwide.

Natural resources will be depleted by the end of this century if we continue digging and pumping at this rate, and our health will deteriorate even faster as well if we don't find and implement cleaner power sources in the very near future. So, the overwhelming conclusion is that we need wind and solar. Yes, but solar, although called "green" and "clean" energy, has its own problems.

Renewable Energy Generators

Solar, and the other renewable energy generation sources (wind, bio-fuels, and hydro) are hailed as green and clean. There is no question that most of them are renewable and safe during operation; as long as the sun continues to shine, the wind blows, the rivers run, and the earth turns. The degree of their "green" and "clean" nature, however, varies from technology to technology and from use to use.

Focusing on solar power generating products, which are mostly green and clean during operation, we see a number of disturbing problems with the materials in their supply chain, and during the different manufacturing steps. There are also serious environmental issues to be expected during their transport, installation, and even the operation and recycling processes.

Here is a brief review of the hidden environmental problems created during the manufacture of PV modules.

Environmental and Safety Concerns of
PV Manufacturing and Use

The manufacturing of PV cells and modules, their transport, installation, operation, end-of-life decommissioning, and recycling all generate pollution. Safety and environmental concerns start from the moment we dig a shovel full of sand to make the silicon material from which we will make the solar cells and modules. Having a better understanding of the processes and their related environmental damages will allow us to find ways to control them.

The actual impact is only "suspected" in some cases, without proven qualitative value, so we can only summarize the environmental impact and health issues due to solar energy activities (equipment manufacturing and use).

1. Mining of raw materials: sand, silver, aluminum, lead, cadmium, telluride, copper, etc. is a dirty and extremely polluting business accompanied by air pollution, soil erosion, and water contamination, to mention a few. Some of these materials are heavily toxic and have contributed to workers' illness and death as well.

2. Large numbers of chemicals in gaseous, liquid and solid form are used all through the silicon material, solar wafers, solar cells and PV modules manufacturing process. Some of these are toxic, others are unstable, but all lead to pollution, contamination and negative health impacts on all life forms they come in contact with.

Some of the chemicals and materials used during the long cradle-to-grave cycle of solar cells and modules are:

a. Heavy metals, benzene compounds, diesel fumes and radionuclides, contribute to reduction in life expectancy and diseases such as lung and other cancers, osteroporosia, ataxia, and renal dysfunction.

b. O_3, SO_2, PM10, PM25, CO, and O_3 are responsible for respiratory illnesses, congestive heart failure, chronic bronchitis, restricted activity days, asthma attacks, and frequent hospital admissions.

c. Mercury is proven responsible for loss of IQ in young children.

d. NO_2, NO_x, SO_2 and O_3 cause yield reduction in wheat, rye, barley, oats, potatoes, sugar beets and sunflowers.

e. CO_2, CH_4 and N_2O influence worldwide morbidity and mortality rates, coastal and agricultural impact, and economic impacts due to temperature change and sea level rise.

f. Acid rain, SO_2, NO_x, and NH_3 cause elevated acidity, eutrophication, and 'PDF' of species.

g. Fatalities from traffic, and workplace related accidents (mining and refining are prime suspects).

h. Long-time exposure to noise can be blamed for life expectancy reduction, and some operations are so noisy that one must shout in another's ear to be heard.

These chemicals and activities are found at different stages of solar equipment manufacturing and operations. Remember that the solar industry consists of a myriad of operations, executed in different production plants and factories worldwide. The manufacturing process of c-Si and TFPV modules includes some very large industrial operations—metals mining and refining, sand melting and silicon refining are huge (in size and volume) enterprises that use a lot of energy and chemicals, to make and transport tons of materials and products. These activities sometimes cause significant damage to air, soil, and the water table in an area, including damage to life forms in the vicinity.

The facts herein are related mostly to the manufacturing process, because the operation of PV power plants is a much cleaner undertaking. There is a small amount of outgassing measured during field tests of c-Si PV modules, but the amounts are too small to cause any concern at this point.

Thin Films

Thin film PV modules are somewhat different in that many contain toxic chemicals (Cd, Te, As, Se and such), and they are a relatively new product not yet proven safe for large-scale, long term, desert use.

We must be aware of potential dangers because of the rapid escalation of TFPV modules deployment in the deserts, with the PV power plants getting very large in size.

We should not close our eyes to the fact that many (thousands) of thin film modules containing toxic and carcinogenic chemicals, such as the heavy metal cadmium are spread over very large areas (thousands of acres) desert land.

This concern might grow into a much larger issue in the near future, if and when the cadmium (in CdTe thin film modules) and other poisonous chemicals (in CIGS thin film modules) start leaking and outgassing with time. Who will be held responsible for the damages and the necessary reparation and remediation—the manufacturers, the regulators, the government?

Now let's look at the environmental problems created by solar energy equipment manufacturing and use.

Table 9-3. Environmental and health impact of solar industry activities

Raw Materials	Serious air pollution, soil and water contamination during mining
Energy and CO2 payback	2-4 years are needed to replace the energy used, and compensate for the CO_2 emitted during the materials, cells and modules manufacturing processes
Resource depletion	The availability of some metals used in c-Si and TFPV modules is limited
Land use	Environmental impact varies between PV technologies and world's locations
Manufacturing safety	Safer materials and processes must be considered and developed
	Emphasis must be on prevention of accident-initiating events
	Capturing accidental releases and preventing human exposure is a priority #1
Field operation safety	Thorough measurements of contaminating species must be implemented
End-of-life recycling	The EOL recycling process must be well designed, executed and enforced
Waste handling	Safe waste handling during manufacturing and at EOL is an absolute must
	Flammable/explosive gases like silane, phosphine, germane, and toxic metals like cadmium in TFPV and lead in c-Si PV need to be thoroughly controlled
	Efficient and controlled recycling facilities and landfills must be implemented
	The PV industry should learn HSE mgmt. from the semiconductor industry
Climate change	PV technologies provide clean energy and reduce greenhouse gas emissions, which is their greatest contribution to improving quality of life.

RENEWABLE, GREEN AND SAFE?

We all agree that conventional energy generators (coal, oil, gas, and nuclear) are not renewable, green, and/or safe. Comparing conventional energy sources with today's solar technologies is not a straightforward process, and no line can be drawn between them, but there are definite differences.

While our objective is not to compare the technologies from an environmental point of view, we can discuss the characteristics and draw some conclusions. So how renewable, green, and safe are the new alternative energy sources? Solar energy proponents and manufacturers claim their PV products are "renewable," "green," and "safe."

This means that from an environmental and safety point of view the related technologies and their components are nearly perfectly clean. We, see it differently, however, and will show in this text that the cradle-to-grave process of the manufacture and use of solar products and projects is not harmless, and that some solar technologies are much worse than others. We'll start with the manufacture and use of silicon solar cells and modules.

Silicon Production

Silicon solar cells are made of silicon—one of the most abundant materials on Earth—sand. But not all sand is suitable for solar cells manufacturing, and a lot of pollution is generated during the mining, transport, and processing of silicon. In one estimate, over 7.0 tons of CO_2 are generated in producing just one ton of SG silicon. Then that much, or more, is produced during the subsequent steps of refining it and converting it into PV cells and modules.

Around 25% of the Earth's crust consists of silicon, which is not pure, but is mostly in the form of "clay" and other alumino-silicate materials. Pure silicon dioxide (SiO_2), a.k.a. silica, from the Latin "silex," is a mineral best suited for making solar grade (SG) silicon. Silica is also quite abundant, making up ~12% of the Earth's crust. So the best type of sand for our purpose is "silica sand," formed from the weathering of silicate minerals and rocks, as part of a natural cycle.

Naturally occurring silicate materials in contact with CO_2 and water are eroded over time into Silica and $CaCO_3$. We have only to find a deposit, dig out the silica sand, and extract it from the $CaCO_3$ and other ingredients with which it is mixed. The purer the sand, the less energy it takes to convert it into useful SG silicon. There are a number of "pure" silica sand mines around the world, but their purity varies significantly.

Desert "sand" is often misunderstood to be the kind of Silica sand we use for solar cells, but it is not even close; it is nothing more than dried earth (clay). One way to tell is to spit into the palm of your hand, add a small quantity of the "sand" and then rub it. If it is clay, it will turn into brown mud, but if it is silica, it will just become damp and will clump together. Thus, as the Sahara Desert advances because of lack of rainfall, more sand appears, but really it is just earth turning into dust.

We've seen a number of plans to convert the Sahara's desert sand into solar cells, so we wonder what these engineers and scientists are thinking, or if they know something we don't and will be having the last laugh. The few places on Earth where pure silica sand can be found are often isolated, meaning that unless the silicon foundries are built nearby, the sand has to be transported—another great expense.

Silica and silica sands are also widely used for the manufacturing of many everyday products such as glass, optical fibers, diatomaceous earth, cement, and ceramics. They are used as additives in foods, not to mention the use of silica sand in making millions of semiconductor-type silicon wafers. There is a lot of competition for it, yet we don't foresee major shortages any time soon, although prices will certainly fluctuate with overall demand and energy costs.

Metallurgical grade (MG) silicon's production starts with a dozen or so very large bulldozers digging a large hole in a sand pit and digging enormous amounts from it. In the midst of the noise and dust generated in the area, this special type of sand is loaded on trucks or train cars and transported to a melting facility for processing.

Here, a huge furnace, burning coal and oil is heated to thousands of degrees with the sand piled inside. After baking it for many hours, and using tons of additives, the resulting MG silicon melt is taken out, cooled down and crushed into chunks. The place is full of dust, noise and stink.

The MG silicon chunks are then loaded on trucks or train cars to be transported to another production plant that will refine them to solar grade (SG) silicon.

Production of Polysilicon material (the raw material needed for producing Silicon solar cells) is on an extremely large scale. Mountains of sand are dug out and moved for conversion into solar grade silicon. As a matter of fact the quantity and quality of thus produced polysilicon raw material is a good indicator of the solar industry development.

China is again one of the largest producers and

Table 9-4. Polysilicon production in China and the world

China Polysilicon Production

Year	2005	2006	2007	2008	2009	2010	2011	2012	2013
Production(Ton)	80	230	1000	2000	7000	13000	18000	23000	28000
Demand(Ton)	1151	3688	9194	16724	24143	28864	34465	25687	27265
Shortage(Ton)	1071	3458	8194	14724	17143	15864	16465	0	0

Global Polysilicon Production

Year	2005	2006	2007	2008	2009	2010	2011	2012	2013
Production(Ton)	30680	33390	37500	51000	73500	96500	115200	142000	168000
Demand(Ton)	33850	39520	46900	62940	81340	103440	121560	102150	122000
Shortage(Ton)	3170	6130	9400	11940	7840	6940	6360	0	0

consumers of Polysilicon, as follow:

The by-products of digging, transporting and processing large quantities of sand are hard to assess. Most of these processes release tons of dust, liquids and gasses into the soil, water table and atmosphere. The enormous size of these undertakings is a good indication of their impact (measurably negative) on the environment, and the health of those in the area.

So how renewable, green, and safe is this material?

Renewable? If the mountains of sand dug out to be melted are eventually replaced, so that the surrounding environment and life forms are not damaged, perhaps it's renewable.

Green? Not so much, due to the extremely large size and quantity of the materials and the energy used in production. Also, excessive air, soil and water contamination occur during these processes.

Safe? Maybe, if properly manufactured and used. The concern here is with the mining and refining operations, in which many people are exploited and exposed to unsafe working conditions.

Silver, Copper and Other Metals

Huge quantities of silver, aluminum, copper and other metals, as well as plastics and many chemicals are needed to produce millions of PV modules. These metals are also dug out of mines on the earth's surface or deep into it. Here again, massive amounts of dirt are moved and processed, to get the pure metals out.

Again, the dirt is dug out and processed via heat and chemicals, which processes emit great amounts of poisonous gasses and liquids, contaminating the air, soil and water table.

Large amounts of silver metal are used to provide

good ohmic contact between the metal grid on top of the cell and the interconnecting wires. Silver is also used for reflective backing for mirrors in thermal solar plants. According to the VM Group in London, over 1,000 tons of silver were projected for making PV modules in 2011. This was more than the 1.0 million kilograms of silver used previously, and the amount is projected to triple by 2016, to nearly 3,000 metric tons of silver used for making PV modules worldwide every year. If we add that much more silver metal for coating heliostat mirrors, we arrive at some very large numbers. So the question is, "How much silver *is there* in the world?"

As a precious metal, the price of silver has gone from $4.00/ounce several years ago to over $40.00 today. Prices will go higher. What will that do to PV module prices?

Similarly, prices of copper and aluminum metals have sky rocketed lately, and although there are large deposits of these left on Earth, the increased prices will play a significant role in PV manufacturing operations and cost.

Though there are significant amounts of these metals around the world, they cannot be considered "renewable." Though they are non-toxic and could be considered "green," mining and refining operations are far from green. Safe? Yes, if we ignore the hazards of mining.

Silicon Wafers and Solar Cells Production

MG Si produced at the mining and refining operations is delivered to poly silicon production plants to be purified and converted into solar grade silicon (SG Si). SG Si refining plants are actually huge chemical factories where a number of solid, liquid, and gaseous chemicals (mostly hazardous, toxic, and poisonous) require special handling, processing, transport, and disposal.

Here, all these materials, chemicals, and gasses are mixed, boiled, baked, sifted, crushed, liquefied, gasified, and solidified. A number of liquid chemicals in this process are quite expensive, so they are recycled via complex distillation, filtration, and other processes, all of which use huge amounts of electricity, cooling water, and additives.

Millions of cubic tons of CO_2 and other toxic liquids and gasses are the by-products of these processes. Some are difficult to recycle or capture, so they are just freely exhausted or disposed of into the environment without any treatment.

Organic chemicals such as silane, dichlorosilane (DCS), trichlorosilane (TCS), silicon tetrachloride (STC), and many others are mixed, heated, evaporated, condensed, and transported, along with inorganic liquids and gasses such as HCL, HF, HNO_3, and H_2O_2. Some of these are dumped into the soil or vented into the air. The resulting mixture of organic and inorganic compounds sometimes stagnates over population centers, becoming part of the atmosphere and accelerating global warming.

Hazardous liquid by-products, some dangerously corrosive and even pyrophoric (self-igniting), are created, transported, and processed along the complex process sequence. All of these require special handling, placing chemical and fire safety at the top of management's priorities in most facilities. Personnel are trained in the proper handling of these chemicals, including emergency procedures. Proper equipment and building designs/procedures are used throughout the facilities, to include ventilation, electrical system safety, static electricity control, control of all ignition sources, personal self-contained breathing apparatus use, and, of course, a no-smoking ban.

At the end of the SG Si purification process, the silicon material is crushed again for ease of transport and sent to different facilities for the manufacture of solar wafers, cells, and modules. At these facilities the SG Si chunks are melted in special, high-temperature furnaces in the presence of different gasses, where they are shaped into long cylindrical ingots (single crystal silicon) or square blocks (poly crystalline silicon). The rods and blocks are sliced into thin wafers on special saws, after which they are ready for processing into solar cells.

Wafers go through an elaborate manufacturing process to convert them into solar cells. After that they are tested and either shipped to another facility for processing into PV modules, or are assembled and encapsulated into PV modules at the spot.

Large amounts of electric power, chemical additives, slurries, liquids, and gasses are used during the melting, shaping, cutting, slicing, cleaning, etching, baking, and assembly operations. The gasses, waste chemicals and slurries, and cleaning liquids must be processed and disposed of eventually—another hazardous undertaking. In many cases the gasses are released in the atmosphere, while the chemicals are dumped into a nearby lagoon.

The thermal and chemical operations used in the sand-to-PV modules manufacturing process use a lot of electric energy and thousands of gallons of chemicals, liquids, and gases. These expenses must also be considered in the manufacturing budget.

Work safety in these facilities is executed in clean areas where, in most cases, safety and environmental standards are the norm, but the disposal of gases and chemicals is often problematic, and sometimes the local environment is badly damaged.

Though over their lifetime crystalline silicon solar panels generate 10-20 times the energy required to produce them, the environmental damage during their production is sizable and in many cases un-repairable. So how renewable, green and safe is the solar wafers and cells manufacturing process? Somewhat, when compared with conventional energy sources, but much work remains to be done before their manufacturing processes are environmentally neutral.

Thin-film PV Materials

Thin-film PV (TFPV) technologies leave their signature mark on the environment as well. The TFPV manufacturing process also starts in a mine somewhere in Asia or Africa. It's low-paid work in deplorable conditions, digging, loading, processing, and transporting the precious and often toxic metals. Health problems and even death are common occurrences among these workers, but the actual facts are not widely publicized.

The freshly mined raw materials are transported to refining facilities to be refined into useful form and shape. They are exposed to a number of complex mechanical, chemical, and electro-chemical operations to separate and purify the different metals. Dangerous gasses and chemicals are used in these processes too, and some of the resulting equally dangerous and toxic gases and chemicals are vented into the atmosphere or dumped nearby. Some of the worst-case mining and refining operations look, feel, and smell like they have just jumped from the pages of *Dante's Inferno*.

Thus refined metals are shipped to TFPV module manufacturing facilities for processing into modules. In sharp contrast with their mining and refining operations, TFPV manufacturing facilities are sparkling clean, modern, semiconductor-type fabrication plants, furnished with state-of-the-art equipment. Engineers and technicians follow well-defined processes executed in clean rooms under strict controls. The safety, efficiency, and productivity of these facilities is the envy of the PV industry, and we have only praise for their setup, operation, and quality control.

The refined and still toxic materials are placed in special chambers, where they are deposited in the form of very thin films onto glass substrates (panes) under near-ideal conditions. The deposited films on the first glass pane are then covered with a second pane and encapsulated between them. The resulting TFPV modules are flash-tested and packed for shipping. This is a most efficient, cost effective, and high-volume production process that is leading the PV industry.

Nevertheless, although the metals used in TFPV modules manufacturing are by-products of the mining and production of other metals, they are still scarce, in addition to being toxic. These are serious issues that are not discussed freely, so we hope they will soon be brought to the table.

Another issue of environmental concern here is the fact that most TFPV modules are frameless. The active thin film structure (CdTe or CIGS) is deposited on a glass pane, covered by an encapsulant and another glass pane. So, the thin film structure is protected from environmental attacks by a thin film of plastic which is exposed to the elements at the open edges of the modules. No plastic material can last very long under harsh desert conditions, so thin edge seals deteriorate and allow moisture and harmful environmental gasses to destroy the thin films. This shortens the useful life of the TFPV modules, and can result in environmental contamination.

While this might not be a serious problem in small size installations, the millions of TFPV modules installed on thousands of acres of desert land in the newly proposed large-scale PV power plants pose a potential danger to the environment and life in the affected areas. More preliminary work must be done before a full proliferation of these unproven PV technologies in the US deserts occurs.

It was discussed in previous chapters that frameless modules are more prone to degradation due to ingress of moisture and environmental gases into the glass/glass modules (front and rear glass covers) which also retain excess heat more than modules with plastic or metal back covers. The excess heat could result in a number of unwanted effects (delamination, overheating, hot spots etc.).

Some TFPV modules are both frameless and of glass/glass construction; therefore, the active thin films inside are more prone to degradation and power loss. The frameless structure also facilitates the escape of toxic materials from the module in case of mechanical disintegration or chemical decomposition.

As discussed in the previous chapter, there are ways to improve the edge sealing of the TFPV modules. Some manufacturers actually wrap up the thin film structure in a sheet of protective material, similar to that used in c-Si PV modules. This completely seals the thin film structure and the module edges in one continuous envelope.

Additionally, some manufacturers add a metal frame around the module edges which is "glued" to the glass panes by edge sealer, providing a solid, several-layer barrier to protect thin films for a long time regardless of environmental attacks.

Most TFPV manufacturers, however, continue making frameless all-glass modules to keep costs down. Price vs. environment? This doesn't sound very safe or green.

Looking to the future, a number of issues need to be resolved by the thin film PV industry. Making the cheapest product is one thing, but making it truly reliable and safe is another. Thin film PV modules manufacturers have not proven yet that their products are reliable and safe for 30 years operation—especially under desert conditions.

Materials used in making thin film PV modules:

Cadmium

Cadmium (Cd) is a toxic, carcinogenic heavy metal, generally recovered as a byproduct of zinc concentrates. Zinc-to-cadmium ratios in typical zinc ores range from 200:1 to 400:1. It is used for making NiCd batteries, electroplating, lasers, electronics, paints, and most recently in thin film PV modules.

In January 2010 USGS estimated ~600,000 tons of cadmium reserves worldwide (calculated as a percentage of available zinc reserves), of which the world mines and uses ~19,000 tons annually. If USGS is correct, in 32 years there will be no more cadmium.

Cadmium displaces zinc in many metallo-enzymes in the body, so cadmium toxicity can be traced to a cadmium-induced zinc deficiency. It concentrates in the kidneys, liver and other organs, and is 10 times more toxic than lead or mercury. Inhaling cadmium-laden dust leads to respiratory tract and kidney problems which can be fatal. Ingestion of significant amounts of cadmium causes immediate and irreversible damage to the liver and kidneys. Japanese agricultural communities consuming Cd-contaminated rice developed itai-itai disease and renal abnormalities, including proteinuria and glucosuria.

Cadmium is one of several substances listed by the European Union's Restriction on Hazardous Substances (RoHS) directive, which bans certain hazardous substances in electrical and electronic equipment but allows certain exemptions and exclusions from the scope of the law. In February 2010 cadmium was found in an entire line of Wal-Mart jewelry, which was subsequently removed from the shelves. In June 2010 cadmium was detected in paint used on McDonald's tumblers, resulting in a recall of 12 million glasses.

Compared to other serious toxins, cadmium is more dangerous because it accumulates in the human body and is not dissipated over time. Even a negligible dose of cadmium in the air or water, inhaled or ingested daily, will eventually accumulate to toxic levels, causing a variety of illnesses, cancer and/or organ failure.

USGS says, "Concern over cadmium's toxicity has spurred various recent legislative efforts, especially in the European Union, to restrict its use in most end-use applications. If recent legislation involving cadmium dramatically reduces long-term demand, a situation could arise, such as has been recently seen with mercury, where an accumulating oversupply of by-product cadmium will need to be permanently stockpiled."

So how "renewable, green and safe" are the cadmium-based PV technologies? From the above facts we see that cadmium is not a renewable commodity *per se*, but we don't foresee a shortage during the next 3 decades. It is, however, far from green or safe! This raises a number of serious questions, which manufacturers, investors and owners of cadmium-based solar products will need to answer eventually.

Tellurium

Tellurium (Te) metal is produced by refining blister copper from deposits that contain recoverable amounts of tellurium. Relatively large quantities of tellurium are also found in some gold, lead, coal, and lower-grade copper deposits, but the recovery cost from these deposits is too high to be worthwhile.

Tellurium is used mostly in making steel and copper alloys to improve machinability, in the petroleum and rubber industries, and for making catalysts and some chemicals. One of the rarest elements in the Earth's crust, it is found in considerable quantities as a secondary metal in mining operations.

The world produces 100-200 tons of tellurium annually, and while its total availability is uncertain, we do not foresee a shortage anytime soon. Tellurium is used as cadmium telluride in manufacturing thin film PV modules. It is a mildly toxic material, but utmost precautions must be taken when handling its pure form or its basic compounds as contained in the PV modules.

Tellurium and its compounds are known to cause sterility in men working with tellurium-containing materials, even under strict monitoring conditions such as in semiconductor fabs and hard disk manufacturing operations. Although there is significant amount of tel-

lurium around the world and it cannot be considered renewable, we do not foresee a shortage anytime soon. Due to its toxic properties, however, it is not green, or safe.

CdTe PV modules contain significant amount of both cadmium and tellurium as CdTe compounds in their active thin films. Millions upon millions CdTe PV modules have been installed in US deserts, and more are planned for installation in the near future. We cannot help it but ask why. Why are products containing toxic materials allowed for such large-scale deployment in the most inhospitable regions of the world, where they have the greatest chance of failing with time and spreading their poisons in the environment?

Past experience tells us that deploying dangerous materials in large quantities eventually leads to negative effects on the environment and human life. Evidently, we have not yet learned that lesson.

Selenium

Selenium (Se) is a non-metal that is chemically related to sulfur and tellurium. It is obtained by mining sulfide ores, and is used in glassmaking, metallurgy, and pigments. While toxic in large amounts, trace amounts of selenium are needed for cellular function in most animals.

Selenium toxicity was noticed first by doctors who found increased sickness among people working with it. A dose as small as 5 mg per day can be lethal, causing selenosis. Symptoms include a garlic odor on the breath, gastrointestinal disorders, hair loss, sloughing of nails, fatigue, irritability, and neurological damage. A number of cases of selenium poisoning of water systems were attributed to agricultural runoff through normally dry lands.

Selenium quantity and price depend on mining operations of other metals and minerals, but we don't foresee a shortage anytime soon. Its use is estimated at 1,500-2,000 tons annually. Though there is a significant amount of selenium worldwide, it cannot be considered renewable, green, or safe.

Arsenic

There are over 1 million tons of arsenic (As) worldwide, of which 54,000 tons are extracted annually, mostly in the form of arsenic sulfur compounds. Another 11 million tons might be recovered from copper and gold ores. The main use of metallic arsenic is for strengthening alloys of copper and especially lead used in automotive batteries. Arsenic is toxic and poisonous. Although there is a significant amount of arsenic worldwide, it cannot be considered renewable. Due to its toxic properties, it is not green, or safe.

CIGS PV modules contain significant amounts of As, Se, Cd and other toxic metals. Although the quantities in each module are not that high, the fact that many millions CIGS modules are deployed world-wide, makes us think that we would should carefully monitor all large-scale PV plants that contain any toxic metals.

Gallium and Indium

These and other hard-to-find-and-isolate mildly toxic metals are also presently used in significant amounts in thin-film PV modules manufacturing processes. These elements are rare, so they cannot be considered renewable. They are mildly toxic, but could be qualified as green, or safe, if properly produced and used.

EVA and Other Plastics

A number of organic materials (plastics) are used throughout the entire PV module manufacturing process. They are too varied and complex to qualify or quantify in this text. Since their production is based mostly on fossils (extracts from crude oil, coal and such) they are not renewable, but we don't foresee shortage anytime soon. Because they are manufactured with the help of poisonous solvents and other toxic materials, we cannot classifying them as green or safe. They are, however, safe enough to work with, following basic precautions like wearing gloves and masks. Due to outgassing, their safety in long-term operations has been questioned and needs more research—especially in light of the large-scale PV installations in the deserts, where organics are most vulnerable and unpredictable.

The complications here are endless, and so interwoven that we could write an entire book on the properties and effects of the different combinations and permutations of the availability, mining dangers, toxicity, adverse short- and long-term environmental and health effects, improper field use, etc. of the materials and components used in making and using PV cells and modules.

Many questions remain to be answered in these areas, so we must insist that these issues are thoroughly discussion by all parties involved, including manufacturers and proponents of TFPV technologies.

How do we justify the claim that PV technologies are renewable, green, and safe, if some of the materials used to manufacture them are not? Should we not qualify and use the different PV technologies according to the type and amount of non-renewable, not green and unsafe materials used during their manufacture and use?

Note: At the time of this writing, several manufacturers have applied for "non-toxic" certification of their PV modules. The trend will continue until it becomes an industry-wide standard, and we envision different types and levels of "toxicity" assigned to different types of PV modules in the not-so-distant future.

This process will take awhile, but it is unavoidable, because consumers must be fully aware of how renewable, green, and safe the products they use are and what to expect in the long run.

Aluminum, Steel and Copper Metals

Solar power plants—PV or CSP—contain large quantities of aluminum, steel, and copper metal products. Aluminum is used as a back cover for most c-Si, and some thin film, based PV modules. It is also used for making the supporting frames onto which the PV modules are mounted. Aluminum, steel and copper are also used to make the support structures, inverters, transformers, distribution boxes and other support components that are critical for the operation of solar power plants.

Copper is the primary metal used in making wires and cables, which are found in PV modules and support equipment, connectors, transformers, and inverters. Thousands of feet of copper wire are used to interconnect the modules into strings and the strings into blocks. Each block is then connected to the next one, or to an inverter.

One estimate is that one single 100 MWp plant uses over 2,000 miles of copper wires and cables of different sizes.

Note: All PV modules are interconnected with AWG #10 metal clad copper wires. At the end of each module row (string), the wires are spliced into AWG #2 cables, which run to the marshalling and disconnect panels, or the inverters. These cables are then spliced into AWG 4/0 cables to connect the inverters to the isolation transformers. From there they lead into the grid connect point (substation), which could be thousands of feet, or even miles, away.

CSP power plants also use copper in the fabrication of the receiver tubes and some portions of the hydraulic and liquid transfer lines. This is also a lot of copper.

Aluminum and steel are used to support the PV modules during operation. For example, a 500 MWp solar project would use between 4 and 8 million PV modules (depending on their make and efficiency). Each of these modules is bolted onto an aluminum or steel frame, so several pounds of metals are used to support each individual module. Multiplied by 4 or 8 million, this amounts to a lot of aluminum and steel metal used in a single installation (several hundred tons, depending on the style of the frame and its use, fixed or tracking).

The same is true for CSP power plants. The troughs in the parabolic trough solar plants are made of aluminum sheets bolted onto heavy-duty aluminum or steel support frames. The frames and driving mechanisms are bolted onto support pillars which are cemented into the ground.

The reflective mirrors of the power tower CSP plants are also mounted on and supported by heavy-duty aluminum or steel frames. The frames and the driving mechanisms are also bolted on the support pillars, which are cemented in the ground.

Many tons of aluminum, steel and copper are used for this purpose around the world. So, where does all this aluminum and steel come from?

These metals are also dug out of the ground in the form of the corresponding ores, at the expense of leveling mountains and digging large, deep holes in the ground, damaging the environment—at times beyond repair. A lot of fuel and electricity are used in the process, and there are connected morbidity and mortality rates.

The ore is transported to a refining facility, where, again with the use of megawatts of electric power, tons of diesel fuel and much human suffering it is crushed and refined into metal blocks. The metal blocks are sent to another plant, where they are melted and formed into the corresponding shapes—wires, beams, pillars, gears, bolts, etc. More electricity is used and a lot of toxic liquids and gases are generated by these facilities, causing additional environmental damage.

So, the right answer here is that these metals are not renewable, nor can they be considered green or safe.

Smart Meters Health Problems

Here is the newest environmental and health risk phenomena, as a result of the energy revolution of late. This was an unexpected blow for the budding smart grid technology. We need to look closely, because it defines the moment and its problems.

As a result of a number of federal, state and utilities programs, thousands of smart meters were installed in some states lately, replacing the old analog meters. It appears now, however, that radio frequency (RF) radiation emitted by some smart meters can have ill effects, causing sleep loss, heart palpitations, dizziness and other problems.

PUC has a duty to look into all health concerns before installing the new devices. Now PUC is providing an option to opt out of the new meters, but the opt-out provision doesn't assure safety for those who keep smart meters and is also meaningless to people who live in congested neighborhoods where they're surrounded by smart meters.

Maine's highest court ruled in July 2012 that state regulators have failed to properly address the safety concerns about the new smart meters installed in many homes. The ruling, however, had no immediate impact on more than 600,000 smart meters already installed statewide.

Human health and energy generation go hand-in-hand. The questionable smart meters might be getting our smart grid technologies' development and implementation off to a shaky start.

Author's note: The author was involved in a number of projects that use RF radiation (RF plasma deposition and etch in the semiconductor and solar industries), and has personally experienced its effects. While the effects of RF radiation on human health are not well known or understood, in light of extremely large-scale deployment of these radiation sources, we'd like to bring the issue up for discussion by the scientific community and the public.

ENVIRONMENTAL IMPACT OF SOLAR INSTALLATIONS

Above we looked at the environmental impact of the materials used in the manufacturing of different "renewable" energy-generating technologies. We saw that some of them are NOT renewable, while others are not green or safe—as used in those manufacturing processes.

Now we will look at the impact of related technologies during their installation and long-term use. Like anyplace where people and equipment operate, PV installations have some impact on the local environment—and quite negative in some cases.

Energy Balance

Environmental impact studies and analyses show that large amounts of CO_2 and other environmentally unfriendly gasses and toxic liquid by-products are generated during the manufacturing, transport, and recycling stages of PV modules and other components. Usually this is much less than the amounts generated by fossil-fuel power production of similar size, during the 20-30 years of operation. Even if we assume the least advantageous 1:6 ratio used by some experts for energy payback comparison, PV systems will still prevent many tons of pollution from entering the atmosphere by using solar energy instead of fossil fuels. The larger the PV plant, the few the harmful gasses released.

Various sources use different calculation methods arriving at different estimates, so we will use middle-of-the-road estimates, assuming that each kilowatt hour (kWh generated by burning coal produces ~1.5 lbs. CO_2 and other harmful gasses (CO, SO_2, NO_2 etc.). This means that a small, commercial PV system of 10 kWp operating in Arizona will produce 20,000 kWh electric power annually, or a total of 500,000 kWh (considering losses) during its 30-year lifetime. Our small 10 kW commercial PV system will save the environment from 750,000 lbs. of harmful gasses during its lifetime, *minus* the CO_2 that was generated during the production of its components. The manufacture, transport, installation, operation, and recycling of PV systems are estimated to create an average of 0.25 lbs. of CO_2 per kWh produced, or, 125,000 lbs. of CO_2. In the end, our small 10 kWp system will still save 625,000 lbs. (750,000 – 125,000) of poisonous gasses from entering the atmosphere during 30 years of operation.

Looking at our 100 kWp system, using the same logic, we see that it will prevent 6,250,000 lbs. of harmful gases from entering the environment every year. How about a 1.0 MWp system? ...62,500,000 lbs. A very large-scale 100 MWp system will keep a total of 6,250,000,000 lbs. from changing the climate and causing global warming during its 30 years of operation. A 500 MWp plant will reduce the CO_2 emissions by 5 times this amount...

Of course these are only estimates, with a significant margin of error, but the numbers leave no doubt that we are talking about major forces at play—forces that must be understood and controlled if humans are to inhabit Earth for another millennia, or more.

Note: No large quantity of CO_2 gas is produced during the actual on-sun operation of PV modules, but there is some measurable outgassing of other gasses and substances from them and the peripheral components during long-term field operation.

Also, as new technologies (thin-film modules containing poisonous elements as Cd, Te, As, etc.) are being installed in ever-increasing numbers on thousands of acres of desert, we need to consider and reconsider where we install and how to use them.

Emitted gas amounts are very small, but as power

fields get larger, so we need to pay more attention to the outgassings; some of the new PV technologies have no history, no precedent, and no data to show their safety for the duration.

CSP Power Generation

In general, concentrating solar power (CSP) technology is considered environmentally friendly, except for the environmental impact of manufacturing the components (mirrors, support structures, steam turbine, cooling towers etc.), when a lot of energy is used. Tons of metals and other materials are consumed too, and their production which requires a lot of energy, and generates large amounts of harmful liquids and gasses.

Mirror production, compared to the overall PV cells and modules manufacturing process, is less energy-intensive and more environmentally friendly, yet uses lots and lots of energy and toxic materials, which must be entered into the equation.

Like all power plants, CSP power generation has some impact on the local area:

- CSP equipment uses a lot of water which is not abundant in desert areas where most CSP plants are located.

- Since the technology is based on mirror use, the terrain needs to be leveled, seriously disturbing thousands of acres of virgin desert land. The impact of this intrusion is significant, but not yet measured.

- Mirrors are low maintenance, but must be cleaned at a small cost (water and workforce) and a lot of mess—muddy terrain and run-off—which causes concerns with water table contamination.

- The plants are almost neutral for landscape except for the solar power towers, which rise high above the ground and are at times considered a nuisance or visual pollution.

- The noisiest parts of CSP systems are the steam turbines and the Stirling engines (in dish Stirling case), but the plants as a whole are quiet. To our knowledge, there are no documented reports of noise complaints.

- CSP equipment may have occasional spilling of oil or coolant, but this is negligible and preventable.

Water usage in CSP energy generation process—steam generation and cooling—is one of the major issues plaguing this technology. Water is the most precious commodity in the desert.

Table 9-5. Cooling and process water use per MW/hour generated by the different energy sources.

Technology Used	Water consumption Gal/MWh generated
Hydroelectric	4500*
Geothermal	1400
Solar Trough	850
Solar Tower	850
Nuclear	600
Fossil Thermal	450
Biomass	450
Coal, IGCC	350
Natural Gas	180
PV	5
Wind	0

* Due to river/lake evaporation

Surprise! Hydroelectric wastes more water than any other energy generator.

- 100 MW hydroelectric plant operating 24 hrs. a day, 365 days a year, wastes nearly 3.9 billion gallons of water every year.

- 100 MW geothermal power plant operating 24 hrs. a day, 365 days a year, wastes nearly 1.2 billion gallons of water every year.

- 100 MW solar trough, or solar tower plant operating 24 hrs. a day, 365 days a year, wastes nearly 370 million gallons every year.

- 100 MWp photovoltaic plant operating 24 hrs. a day, 365 days a year, wastes only 1-2 million gallons of water every year, mostly used for washing the modules.

Or the ratio of 3900:1200:370:2=Hydro:Geo:CSP:PV.

In all cases this is a large waste of a precious resource, especially in the deserts. The striking fact is that the major CSP technologies—solar trough and solar tower—are always installed in high deserts, simply because they operate most efficiently there.

Two or three billion gallons of fresh water wasted every year in each of these installations is something that deserts are not capable of providing. We have seen drop in the water reservoirs and water tables in a number of locations in the Arizona and California deserts, so we see this power generation method as unsustainable, impractical and even harmful to the environment, because it is robbing the deserts of their most precious resource—water.

PV Power Generation

PV power plants are basically environmentally friendly, but general environmental conditions apply here too. As PV technology was developing, it was not until the 1980s that researchers began to consider its environmental implications. As its use expands, pressure is applied by the scientific community and administrations that all costs of the creation and use of energy systems be taken into account.

The first German-American workshop was organized in Ladenburg, Germany, in October 1990 to discuss the "External Environmental Costs of Electric Power: Analysis and Internalization." Papers at this workshop considered environmental damage and ways in which environmental costs may be internalized. PV was not discussed specifically at this workshop; however, the impacts of external costs on wind energy were covered and can be easily related to those of the PV industry.

Ottinger's group at Pace University, Centre for Environmental and Legal Studies, published an extensive review of environmental costs/risks and covered all electricity generation technologies, including photovoltaic installations. Various environmental costing models were also discussed, forming the basis of many studies on external/environmental costs that have been published.

The summary of a 1997 report, "Environmental Aspects of PV Power Systems" (28) indicates that the immediate risks to human health and the environment from production and operation of PV modules seem to be relatively small and manageable.

The methodology developed was applied to multicrystalline and amorphous silicon cells and gives a comprehensive breakdown of the technologies involved and how the life cycle analysis is structured. The results are divided into environmental, social, economic, and other impacts, showing a significant impact on its respective area of influence. Impact type and size vary from case to case, but the overall conclusion is that the environmental impact of all PV technologies must be analyzed and considered in the energy life cycle calculations.

Table 9-6. Immediate and long-term environmental and health issues considered.

TASK	IMMEDIATE CONCERNS	LONG TERM CONCERNS
SUPPLY CHAIN OPTIMIZATION	~Investigate the availability and short term constant supply of rare and exotic materials such as Cd, Te, In, Ga, As, Ge and Ag now and in the future ~Optimize the methods and improve the efficiency and safety of rare, scarce and exotic materials mining, manufacturing operations and field use.	~Investigate solutions for supply chain shortages ~Develop thinner active film layers in TFPV modules ~Optimize efficiency of rare and toxic materials use ~Optimize EOL module decommissioning procedures ~Design new materials and products for complete recycling and safe waste disposal systems
EFFICIENT ENERGY USE	~Develop methods to reduce energy use during silicon material, cells and modules production ~Optimize energy use for module frames and BOS	~Optimize energy consumption of Si processes ~Optimize energy consumption of recycling processes ~Optimize energy-efficient frame and BOS designs
IMPROVEMENT OF CLIMATE CHANGE CONDITIONS	~Optimize CO_2 mitigation potential of PV technologies ~Investigate release of fluorinated (FFCs) and other toxic and unsafe compounds during manufacturing ~Investigate air and land contamination from PV modules operating in extreme climates ~Design & suggest hybrid energy generation options	~Investigate the CO_2 mitigation potential of autonomous PV systems ~Develop FFC alternatives for use in PV production ~Optimize gas and liquid release methods ~Investigate the role and impact of dynamic assessment methods
HEALTH AND SAFETY CONCERNS	~Optimize safe use of compressed and explosive gases in the manufacturing processes ~Investigate and develop procedures for safe use of "black list" materials such as Cd, Te and As in manufacturing and long term field operations	~Develop safer materials and safer alternatives ~Investigate using thinner active cell layers ~Develop more efficient material utilization methods ~Investigate the long-term risks from (low-level) releases of "black list" materials in large scale fields
RECYCLING AND WASTE CONCERNS	~Investigate leaching of heavy metals from modules in long term landfills ~Optimize module & BOS waste management options	~Investigate the environmental aspects of relevant recycling and waste management methods ~ ~Investigate and optimize long term landfill safety

PV power fields are growing in size. From the 5-10 MWp maximum size in the 1990s, we are now witnessing 500 MWp PV power plants under development. This represents millions of PV modules installed on thousands of acres, so the expansion in size means larger quantities of harmful effects which need to be seriously considered.

The largest PV power plants are to be installed in the US deserts, where failure rates are many times higher than in moderate climates. The negative consequences of this new phenomena could be too great over time, and should not be ignored.

Worst of all, the largest fields in the world have been installed in US deserts, where many technologies (thin film PV included) have not been previously used, so we have no data to confirm their reliability and safety.

Because of that, below we will take a closer look at TFPV modules from an environmental and safety point of view.

Environmental Impact of Solar Power Plants

Regardless of the manufacturers' claims, and media buzz on the subject, all solar power plants do have some effect—usually negative—on the environment.

Disturbing the land and expelling indigenous life forms (including humans) from the plant's territory is how every project starts. Traffic around the site, digging holes in the ground, bringing millions of tons of cement and metals, with the accompanying dust and exhaust fumes do leave scars on the locale.

And things happen during 30 years of desert elements. PV modules are not everlasting; they are actually quite fragile. With fragile organic materials used for encapsulating and sealing the equally fragile active layers, it's just a matter of time until they are no longer viable.

c-Si PV Power Plant

Among all other types of solar generators, silicon-based PV modules are considered the most durable and least damaging to the environment.

Looking at Figure 9-10 one can see that the solar cells are well protected by encapsulation and sealers, so it would take the desert elements some time before penetrating the module and damaging the solar cells.

Silicon-based cells pose minimal risks to human health or the environment according to reviews conducted by a number of national labs and other entities. This is due mostly to the fact that the solar cells are en-

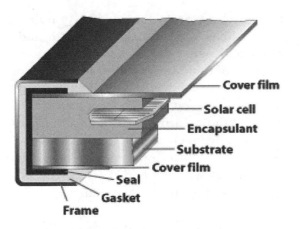

Figure 9-10. c-Si PV module

cased in heavy-duty glass and plastic materials, so there is little risk that any of the contents can be released into the environment in the form of gas, liquid or particle contamination.

In the event of fire, it is theoretically possible for hazardous fumes to be released, posing a risk to humans, but these risks are not substantial given the short-duration of fires and the relatively high melting point of the materials in the solar modules.

Also, the risk of fire at ground-mounted solar installations is remote because of the precautions taken during site preparation and the removal of flammable and burnable materials.

A much greater risk in PV power fields is the potential for shock or electrocution during normal operation, or in an emergency situation.

Overall, the strength of electromagnetic fields produced by photovoltaic systems does not approach levels considered harmful to human health established by the International Commission on Non-Ionizing Radiation Protection. Also, the occasional small electromagnetic fields produced by photovoltaic systems rapidly diminish with distance and would be indistinguishable from normal background levels within several yards.

Thin Film PV Power Plants

Thin-film PV cells and modules are the "new kid on the block," and have not been thoroughly tested and proven safe, especially when used in large-scale power fields in the deserts. There are a number of issues which we will address only briefly herein, but we would like to encourage the reader to take a closer look at the available literature and make an independent decision on the subject.

CdTe and CIGS thin-film PV modules have seen a quick rise and have been deployed successfully in large

numbers around the world lately. The low efficiency and other issues are obviously not hindering the TFPV technology deployment (most likely due to favorable incentives and subsidies), so they are growing faster than the other PV technologies currently available. Amazingly enough, they are the first PV technologies planned to be deployed in gigantic 250 MW and 550 MW power fields in US deserts.

The key issues here are:

1. Some thin film PV modules contain compounds of cadmium (Cd), tellurium (Te), selenium (Se), arsenic (As), and other metals and chemical.

2. The active thin film layers in the TFPV modules are mechanically fragile structures, which could easily be disintegrated by heat-freeze action, and/or mechanical friction, bending or impact on the modules.

3. Thin films are also, under the right conditions, chemically active compounds. They can easily be decomposed, or affected by moisture such as rain water with its weak acids and other chemicals, as well as by most anything else that enters the modules.

4. Presently most thin film modules consist of two glass panes within which the thin films are encapsulated and sealed within a plastic layer. This is not enough protection, so damage to the active structure during long-term operation is expected.

5. The active thin films in some TFPV modules are encapsulated by layers of organic polymers, which could easily break down, disintegrate and decompose under intense IR and UV radiation. Once that happens, thin film structures are vulnerable to the elements.

6. Toxic materials-based thin film PV modules used where desert temperatures within the modules can reach 180°F, and where storms, rain and hail would aggravate the above mentioned degradation could be expected to suffer major changes with time:
   a. The never ending heat-freeze action of desert temperatures will stress and eventually damage the thin film structure.
   b. The encapsulating layers will eventually break down, allowing the elements to penetrate the module, react with the toxic thin films and decompose them.

c. From the above actions, resulting particulates, liquid and gaseous contamination might pose threats to human health and all life forms in the area.

Note in Figure 9-11 that the oxide and active materials of the thin film PV modules are deposited directly onto the front glass. There is no encapsulation at that level, so there is only a thin layer of sealer at the edge of the module to stop environmental moisture and gasses them from penetrating and disintegrating the active layers.

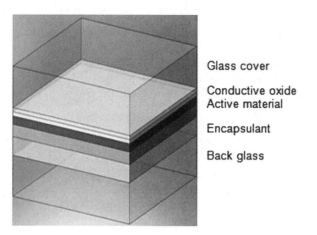

Glass cover

Conductive oxide
Active material

Encapsulant

Back glass

Figure 9-11. Thin film PV module

How long would it take the desert to wreak havoc with these modules? What will happen if the toxic layer is attacked? Mixing toxicity with solar power—especially in the desert—is an uncomfortable subject, and most people are shying away from it. But we must not wait until the unthinkable happens, before addressing the issues.

The increasing number of solar installations using PV modules containing toxic materials has encouraged a number of scientists and scientific organizations to publish papers on the subject:

1. Scientists at Brookhaven National Laboratory (BNL) in the USA, have been publishing health and environmental information of interest to the PV industry since the 1970s; mostly funded by the U.S. Department of Energy. The work was started by Moskowitz and then continued until the present day by Fthenakis. It was the early studies by Moskowitz that first alerted the PV industry and environmental scientists to the possible hazards associated with the manufacture, use and disposal of PV modules containing toxic materials.

2. In 1995, BNL and other scientists published a comprehensive study on the health and environmental issues of the manufacturing, use and disposal of thin film modules. They examined the hazards associated with producing and using thin film modules, focusing on the potential of workers in manufacturing facilities to be exposed to chronic, low levels of Cd. They also review regulations and control options that may minimize the risks to workers and discuss recycling and disposal options for spent modules.

3. Hynes, Baumann and co-workers in the UK have published several papers on environmental aspects of many thin film deposition processes. These include environmental risk assessment and hazard assessment of the manufacture of CIGS-based thin film PV cells, the chemical bath deposition of CdS and the deposition of alternative window materials to CdS for use in thin film modules. Steinberger from the Fraunhofer Institute in Munich has investigated such risks of Cd-containing thin film modules in the operation phase, considering hazards that may occur due to fire, weather, or damage from mishandling.

4. Steinberger first identifies the amount of material in the modules and then investigates the concentrations of selenium, cadmium and tellurium in the air due to a fire lasting about 1 hour. His results show that there is no acute danger posed by such fires, and that releases of cadmium or selenium from a burning PV module are less than those of a coal-fired power station in normal operation.

5. Dr. Fthenakis has considered the cadmium emissions from cadmium telluride thin film cells also. He has taken a cradle-to-grave approach and comes to some interesting conclusions. The Cd present in a NiCd battery is elemental Cd and not CdTe, a much more stable and insoluble form. Coal- and oil-burning power plants routinely produce Cd emissions, while a PV cell does not produce any Cd emissions during its operation. Cd can either be used or discharged into the environment, where it is normally "cemented" and buried or land-filled as hazardous waste.

In summary, the above works have several things in common:

1. They all recognize the serious implications of using cadmium and other toxic metals in PV products but offer no concrete solutions to avoid or minimize the dangers.

2. They are all old, outdated and incomplete, but due to the seriousness of toxic exposure (and to avoid liability and further responsibility), they always recommend more studies and further investigations harmful effects.

3. They usually refer to tests done under "standard," or "normal" conditions, which generally means dry air and 25°C operating temperature. These conditions, however, are far from the extreme heat and humidity in the areas where the majority of these products are planned for installation.

4. The actual lab tests, although properly executed for lab tests, offer no real solutions. Their conclusions and recommendations lack scientific proof of the point at hand, and fail in the attempt to somehow extrapolate the results of a small lab bench test to the behavior of the millions of modules in a mega PV power field in the desert.

5. They offer no actual test data, nor any kind of scientific proof for long-term exposure of potentially toxic metals containing PV modules to a harsh desert environment. We have seen no long-term tests to date, done under actual extreme-desert or excessively humid climatic conditions. That is our main concern.

What will happen to the toxic and carcinogenic materials in these PV modules (that are untested and not unproven for extreme climates), after 30 years in the desert? How would we explain such large-scale environmental contamination? Who will be responsible for the damages? Who will pay for the remediation?

Some of these issues have been superficially addressed by present manufacturers and others, but they are far from fully addressed, let alone resolved. There is a lot of money to be made before the window of opportunity closes, so the race is on.

Manufacturers and responsible parties are closing their eyes to the dangers, due to ignorance or negligence. We hope this changes, because if something goes wrong in those mega-fields, we could have the greatest environmental disaster ever known to man.

Safety and Environmental Issues in the Desert

This is a good time and place to emphasize the fact that deserts are where the maximum amount of sun energy is concentrated, and where we see the maximum MW of PV modules installed recently. It is also where we expect most of the efficient and profitable future large-scale power generating plants to be located.

This brings into play, a number of issues which are critical to the future of the PV and energy industries.

To start with, most PV products manufacturers have never seen a desert and have little idea what it is or what it can do to their fragile products that are no match to the extreme desert forces.

Deserts are the harshest and most unforgiving areas in the world. Anything left on the desert floor (metals, glass, ceramics, plastics and every other element and substance known to man), will eventually be changed or destroyed by the desert.

Work and operation in remote desert areas is very difficult, and at times impossible. Even in civilized countries and developed areas, paved roads and other infrastructure are few and far between, which makes regular maintenance also quite difficult. PV modules and people alike must endure the climate extremes—heat and freezing, dust, storms, etc. IR and UV radiation in desert attack PV modules, causing serious changes, deterioration, and failure. Damage must be expected in these areas—foreseen and unforeseen—more than in any locations in the world.

A number of creatures live in desert areas, and since there is no abundance of food, some of these desert dwellers have learned to eat anything they find—including copper wire and cables. Gnawing across cable bundles, some of these creatures go through the shielding and cause disruptions, short circuits and even fires. A fire in a field filled with PV modules containing toxic materials is never good news, and some large-scale fires might have serious environmental consequences.

Other serious and unavoidable dangers in the desert are unpredictable sand storms and lightning strikes, which can damage solar structures mechanically or electrically. They can also cause additional damage by sand-blasting the PV modules' optical elements. A sand storm usually causes surface soiling, thus reducing the PV modules' efficiency and if proper measures are not taken, it will eventually render them useless. Of course, even under normal circumstances, frequent and serious surface soiling is unavoidable, and is just another expense we must face in the desert.

There is no known method of preventing wind and sand damage, but some precautions can be taken during the design and installation stages:

1. Gravel-type deserts are best suited for large-scale PV installations, because they contain less free sand.

2. Build protective wind barriers around the edges of the power plants to reduce wind speed and minimize sand-blasting effects.

3. Incorporate in the overall system design an efficient way to stow trackers, or to turn them in the direction of the wind, thus protecting their optical side from sand-blasting damage.

We need to remember that work in the desert is always difficult. Large-scale PV installations are a new phenomena, and we don't have enough experience to know all the precautions needed for safe and efficient operation. No doubt the first large-scale installations will be the guinea pigs of utility-scale PV power generation in the US.

Let's hope they survive the test of time, and even more importantly, that they behave well, and don't make us victims of our haste, negligence and ignorance.

The US Desert Phenomena

US deserts, especially those in southern Arizona and southeast California, are as ferocious as any in the Sahara. As a matter of fact, the highest temperatures ever recorded in a desert are those in Death Valley in California.

US deserts, including Death Valley, are unique in that elaborate, extensive and well maintained roads and water irrigation systems were developed in their midst during the last 50-60 years. This converted some parts into productive and very busy man-made agricultural and commercial areas. Many small and large population centers are thriving there as well. Places where just a short time back nothing could survive, are now equipped with excellent infrastructure (irrigation canals, railroads, highways, electrical transmission lines, substations etc.). Most of these developed areas are now well established and functioning efficiently in the middle of the desert as if it was not even there.

Solar power plants in these areas could be an ideal solution to solar energy generation, and the established infrastructure should greatly simplify adding solar capacity there.

The fact that energy costs are increasing, supplies are decreasing, and the world is choking on pollution should add to the ease and desire of implementing solar energy in US deserts.

When energy independence and environmental cleanup are priorities, we do see a few large-scale PV plants implemented in the deserts. More would be... more... and the deserts are always there, reminding us that abundant energy is right under our noses. Getting it will not be easy, or quick, but we need to persist in our goal of using our desert sun as the energy source we so desperately need.

Here is a detailed list of US deserts suitable for large-scale PV power plant installation:

- Red Desert is in Wyoming
- Alvord Desert is in eastern Oregon
- Owyhee Desert is in northern Nevada, SW Idaho and SE Oregon (Yp Desert is a portion of the Owyhee Desert in Idaho)
- Black Rock Desert is a dry lake bed in northwestern Nevada
  (Smoke Creek Desert is an extension of the Black Rock Desert)
- Great Salt Lake Desert is in Utah
- The Great Basin Desert is in Nevada
- Tule Desert is in Nevada
- Amargosa Desert is in western Nevada
- Painted Desert is in Arizona
- Mojave Desert is in California (Death Valley is in California (also considered part of the Mojave Desert and part of the Great Basin.
- Chihuahuan Desert is in Arizona, Texas, New Mexico, and Mexico
- Trans-Pecos Desert is in west Texas
- White Sands is in New Mexico
- Sonoran Desert is in US and Mexico
- Lower Colorado Desert is in California and Arizona
- Low Desert of Southern California is in California, USA
- Yuha Desert is in Imperial Valley, California
- Lechuguilla Desert is in southwest Arizona
- Tule Desert is in Arizona and Mexico
- Yuma Desert is in southwest Arizona

The most suitable for large-scale PV power generation deserts are located mostly in the Southwest, and occupy significant portions of Arizona, California, New Mexico, Nevada and Texas. There are also some areas in Utah and Colorado with desert, or desert-like climate suitable for solar installations.

The US deserts vary in climate, flora and fauna,

and are mostly poor in natural resources, but are all blessed with one efficient, constant and unending energy resource—a lot of hot, bright sunshine—almost every day of the year, nearly all day long—precious sunlight, waiting to be harvested.

The downside is, again, those desert extremes which are so hard on PV equipment. These problems are being addressed as we speak, and it is encouraging to know that even small portions of the US deserts are capable of providing a major portion of the electric power to the US and neighboring countries in the future. The climate and behavior of the deserts are well understood, so with the necessary precautions and proper solar technology, deserts should become the hub of solar energy generation for the US, Mexico and Canada in the not-so-distant future.

World Desert Power

The world produces and uses ~15 terawatts of electricity annually, which would require ~75 million acres of land to produce, assuming 5 acres per each MW generated at 15% efficiency, and provided that the PV modules last long enough to make a difference. This means ~120,000 square miles of desert are needed to power the *world* with this type of equipment.

The world's land surface area is ~55 million square miles. The Sahara Desert covers over 5 million square miles (almost 10% of the world's land area), so we would need only a small sliver of its (otherwise useless) land to power the entire world. This isn't possible at this time, for technological, logistical and financial reasons, so we need to find ways to use the deserts closest to the population centers for which we could provide enough power to significantly reduce their use of conventional fuels.

The US generates ~1.1 terawatts of electricity, which would require ~8,000 square miles of 15% ef-

ficient PV modules installed in the desert to generate. This is an area of 80 miles by 100 miles in the Arizona or Nevada deserts. Imagine driving from Phoenix, AZ, to Las Vegas, NV, among a forest of solar equipment and installations feeding the national power grid. Far fetched, yes, but this is all the future generations will have left, so we must dare to dream.

As mentioned above, and worth repeating, the most practical use of PV power today would be to locate smaller solar power generating plants near populated and industrial centers. This could provide enough electricity to reduce our dependence on foreign oil and partially eliminate the pollution associated with its production, transportation, use.

Imagine large-scale PV fields where 40 or 50% efficient PV equipment is operating in Arizona desert wastelands. This automatically reduces the land use by a factor of 3 or 4. Imagine that. Suddenly, things look more doable. Instead of 120,000 square miles, we are now talking about 30,000-40,000 square miles to power the entire world, and only 2,000-3,000 square miles to power the entire US.

This is a huge difference but still far removed from today's reality, although some advances in the technology, design, manufacturing, and use might bring us close to the target soon. One ray of hope is the high concentration PV (HCPV) technology which is 42% efficient today. When all the bugs are worked out, HCPV just might take the lead in desert power generation.

PV Modules Recycling

While the solar cell is the heart of a photovoltaic system, on a mass basis it accounts for only a small fraction of the total materials required to produce a solar panel. The same is true, and even more so, in thin film PV modules. The outer glass cover, in all cases, constitutes the largest share of the total mass of a finished crystalline photovoltaic module (approximately 65%), followed by the aluminum frame (~20%), the ethylene vinyl acetate (EVA) encapsulant (~7.5%), the polyvinyl fluoride substrate (~2.5%), and the junction box (1%). The solar cells themselves only represent about four percent (4%) of the mass of a finished module, while thin films active layer is much less than 1%.

Proper decommissioning and recycling of solar panels both ensures that potentially harmful materials are not released into the environment and reduces the need for virgin raw materials. In recognition of these facts, the photovoltaic industry is acting voluntarily to implement product take-back and recycling programs at the manufacturing level. Collectively, the industry

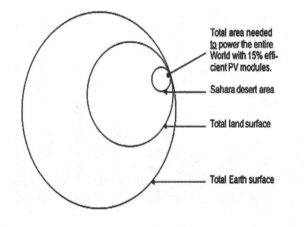

Figure 9-12. Land area needed to power the entire world

recently launched PV Cycle, a trade association to develop an industry-wide take-back program in Europe. In the United States, product take-back and recycling programs vary by manufacturer.

While recycling methods and take-back policies vary by manufacturer, the most frequently recycled components are the cover glass, aluminum frame, and solar cells. Small quantities of valuable metals including copper and steel are also recoverable. The ethylene vinyl acetate encapsulant and polyvinyl fluoride substrate are typically not recoverable and are removed through a thermal process with strict emission controls and the by-product ash land-filled. Following this process, the glass and aluminum frame are separated and typically sold to industrial recyclers.

The solar cells are then reprocessed into silicon wafers with valuable metals recovered, reused or sold. Depending on the condition, the wafer can then either be remade into a functioning cell or granulated to serve as feedstock for new polysilicon.

If not properly decommissioned, the greatest EOL health risk from crystalline solar modules arises from lead-containing soldering compounds. Under the right conditions it is possible for the lead to leach into landfill soils and eventually into water bodies.

There is a much greater risk in recycling of thin film based PV modules, due to the high toxic content in some of them. Proper, much more complex and expensive procedures must be followed as needed to ensure the safety of the decommissioned land, the labor during decommissioning, the transport and the actual recycling and disposal processes.

PV companies worldwide are actively moving towards the recycling of their products, both for manufacturing waste and end-of-life modules. EU companies have proposed a voluntary take-back system capable of meeting the future waste recycling demands. The new recycling process lines must be capable of processing crystalline silicon and thin-film modules alike. The silicon recycling is now at the pilot stage, and reuse of silicon material will be timely for the industry. US companies have been reclaiming solar cells and semiconductor process wafers for many years, so the expertise and equipment are available. Scaling up to accommodate the large demand of the future is a key to success here.

Recycling and reclaiming reduce the energy payback time by a factor of 4 in the best case scenario, but this depends on the

available insolation at the PV power site and the particular PV technology used.

The forecast is that 40,000 MT of PV components will be ready for decommissioning and recycling by 2020. This number will double and triple during the following decade. This includes silicon and thin-film based PV components recycling. The thin-film component will be ~20% by then. CdTe thin-film PV modules, or 8,000 MT, will be the majority of the recycled products in this category. 8,000 MT CdTe PV modules is a formidable amount. With an average of 8-9 grams of cadmium in each module, 8,000 could create a serious hazmat debacle if not handled and processed properly.

Note: Cadmium is one of the six most toxic, carcinogenic heavy metals on Earth, which together with tellurium, selenium, arsenic and other toxic metals and their compounds must be handled with utmost caution during uninstall, transport, crushing, and disposal. Disposal of some components as hazardous waste is another major issue to be addressed and fully resolved in the years to come, because special handling, transport, processing, and containment would be required.

The EU already has directives for voluntary and extended manufacturer responsibility, where the decommissioning, transport, storage, processing, and disposal of the modules are the ultimate responsibility of the original manufacturer. The directives have been integrated into the legal system of several member states. Different paragraphs outline the registration, packaging, transport, waste disposal, documentation, legal responsibilities, etc. These components of the overall effort to protect the environment and life during all stages of the manufacture, use, and recycling of PV products clearly place the responsibility on the manufacturers' shoulders. Customers and users are required

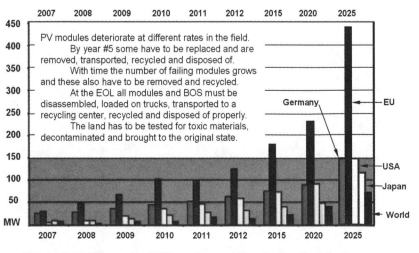

Figure 9-13. Forecast of PV waste recycling in the decades to come

to be aware of the issues at hand and to observe and complete their obligations as well.

This means that whoever made the PV products is legally responsible for them and obligated to dispose of them at the end of their useful life. The manufacturer is obliged to assure the proper execution of this process in all cases—even if no longer in business. Special arrangements, such as insurance or delegation of responsibility must be provided at the time of installation and be well-documented prior to starting the actual work on each project. The cost of the decommissioning process is to be agreed upon and reflected in the installed cost of the power plant.

Every year around the world, a large quantity of PV modules will reach their end-of-life state, become obsolete, or fail. From all these old modules, structural steel, aluminum, inverters, wiring, and other hardware will have to be disassembled, sorted, loaded on trucks, and transported long distances for recycling and/or disposal. This is a great effort, which in many cases is not taken into consideration during the PV plants' design and financing stages.

Table 9-7 contains a number of operations that must be properly executed during recycling and/or disposal procedures. Some of these operations are accompanied by toxic gas and liquids generation, so special caution must be exercised during their execution as well.

Also, the proper final disposal of waste materials and related chemicals and gases is absolutely necessary. That too is the responsibility of the project owners and products manufacturers.

Recognizing and embracing cradle-to-grave responsibility for products and land safety, as well as proper handling and disposal, must be one of the conditions to participate in any PV project in the US. We have no reason to ignore this critical aspect of PV industry development, regardless of what other countries do or do not do. The future generations' well being depends on our decisions.

The PV plants' designers, installers, investors, and operators, in addition to being responsibility

for safe plant operation, are also fully responsible for proper decommissioning and disposal of the plant's components, as well as for final cleaning and returning the land to its original state.

As can be clearly seen in Table 9-8, the manufacturer is legally responsible for the safety and proper execution of all steps from the beginning to the end of the useful life of their products. The retailers, installers, and plant operators are also responsible in some stages of the product life. Overall, all entities who have been involved in the power plant's planning, design, installation, and operation are responsible to one degree or another for the proper execution of the respective steps, including the final decommissioning, recycling, and disposal of the plant components and land clean-up.

Environmental Cost of PV Power Plants

The environmental costs of PV plants' setup and operation is defined by monetary quantification of the socio-economic and environmental effects and damage they do (and will) as they inflict a local area, a nation, and the world as a whole during their cradle-to-grave life cycle. These effects and the related damages could

Table 9-7. PV modules recycling steps and methods

RECYCLING OF PV MODULES AT EOL

c-Si MODULES			TFPV MODULES		
RECYCLABLE	kg/m²	%	RECYCLABLE	kg/m²	%
Glass	10.0	90	Glass	15.0	90
Aluminum	1.4	100	Aluminum	0	0
Solar cells	0.5	90	Solar cells	0.1	90
EVA, Tedlar	1.4	0	EVA, Tedlar	0	0
Ribbons	0.1	95	Ribbons	0	0
Adhesives	0.2	0	Adhesives	0.2	0

RECYCLING AND DISPOSAL METHODS

Mechanical	Crushing, attrition, density separation, flotation, adsorption, radiation, laser beam, metal separation, and other methods.
Chemical	Acid / base treatment, extraction with solvents, dissolving, precipitation, slagging, and other methods.
Thermal	Incineration, burn-out, pyrolysis, melting, slagging, and other similar methods.
EOL disposal	Recycling into the same product, recycling into another product, recovery of energy from thermal treatment of organic layers, utilization of mineral fractions (e.g., concrete, road material, etc.), and landfill disposal.

be expressed in $ per kWh generated, for lack of better way, and should account for all materials, procedures, and events from cradle to grave.

Included in these calculations are environmental effects, human health, materials production, effects of use and disposal (materials, gasses, chemicals) on agriculture, noise, audio and visual pollution, ecosystem effects (acidification, CO_2 damage), and all other effects. Thusly obtained numbers could be used to provide a scientific basis for legislative and regulatory policies, energy taxes and incentives, global warming policy adjustments, etc.

Costs are defined as the monetary quantification of the socio-environmental damage, expressed in eurocents/kWh, with the possibility of providing a scientific basis for policy decisions and legislative proposals such as subsidizing cleaner technologies and energy taxes to internalize the external costs. It looks at all energy-production technologies using a methodology developed for this project that allows the various fuel cycles to be compared. An outline of the initial results for the PV fuel cycle, starting with a very small sample of PV systems, shows that the results are not consistent and require more work.

There has been pressure from the PV industry for the PV fuel cycle to be re-done using a larger sample of more representative systems, but this is still in the making. So, one of the earliest publications from this project by Baumann et al. is still valid. In this paper the author outlines the basic assumptions and requirements of the methodological framework for the quantification of external costs and compares the environmental effects of different energy technologies, including renewables and the conventional technologies of coal, oil, nuclear, and natural gas.

Basically, PV systems must be designed and manufactured with long-term environmental concerns and considerations in mind, to include:

1. Manufacturing materials and procedures, including the production and use of MG and SG Si, thin-film metals and chemicals, glass panes, metals frames, gasses, chemicals, etc. process supplies

Table 9-8. End-of-life outline of the participants' responsibilities

EOL TREATMENT OF TFPV MODULES	DECISION AND EOL* NOTICE	REMOVAL ⇨	TRANSPORT ⇨	RECOVERY & DISPOSAL ⇨
GENERAL REQUIREMENTS	Notify proper authorities of EOL	Proper and safe procedures use	Authorized Hazmat carrier use	Authorized facilities use
ORGANIZATIONAL RESPONSIBILITY	Plant operator	Manufacturer and operator	Manufacturer	Manufacturer
TECHNICAL RESPONSIBILITY	Plant operator	Manufacturer and operator	Manufacturer	Manufacturer
FINANCIAL RESPONSIBILITY	Plant operator	Manufacturer and operator	Manufacturer	Manufacturer
LEGAL RESPONSIBILITY	Plant operator	Manufacturer and operator	Manufacturer	Manufacturer

2. Solar wafers and cells manufacturing processes

3. Module encapsulation and framing processes

4. Evaluation of direct and indirect processing energy (including transport, storage, etc.)

5. Gross energy requirements of input materials (supply chain and internally generated products and byproducts)

6. Allocation schemes used in the calculations

7. Separate thermal energy, electrical energy, and "material energy" calculations

8. End-of-life recycling and disposal calculations (including transport and storage)

There are a number of dislocated efforts in the above areas, but no uniform, standardized method presently exists that is capable of capturing all environmental factors into one, all-encompassing methodology. Such a methodology is needed to account for the effects of the above concerns on the PV manufacturing processes and long-term use of the products.

Lifetime Energy Balance

Life time energy balance (LEB) is the energy used during the PV products manufacturing, transport, installation, and operation, compared with the power they generate during their useful lifetime. This is also related to the environmental benefits of using solar energy, as related to the total amount of CO_2 that is not emitted in the cradle-to-grave process, as compared to conventional energy sources.

LEB is an important factor, when analyzing and calculating energy benefits from a PV power plant. LEB basically tells us how much energy we save by using PV energy generating sources.

As outlined numerous times herein, the solar cells and modules manufacturing process uses enormous amounts of natural resources and energy, and generates a lot of toxic gasses and liquids. At the end of their life cycle, more resources are used in breaking them down for disposal or recycling.

Even though the production equipment and processes are very efficient these days, making a solar cell and module is an energy-consuming and environmentally harmful undertaking, so we hope that their 30 years of operation will justify the expense and the pollution. It all depends on how efficient and durable the PV modules are.

Consider this equation for lifetime energy balance (LEB):

$$LEB = \frac{Eprod + Etrans + Einst + Euse + Edecom}{Egen}$$

Where
Eprod = energy used during production of materials, wafers, cells, and modules
Etrans = energy used to transport materials, modules, and BoS to PV site
Einst = energy used to assemble and install the PV power plant
Euse = energy used to operate the PV power plant
Erecyc = energy used to decommission, transport, and recycle the PV field
Egen = energy generated during the life of the PV power plant

In all cases, Egen must be much higher than the sum of the other sources of energy used, for the system to be an effective energy source.

The term "energy payback" describes this "energy in-energy out" ratio and is what we must consider when designing, pricing, and justifying a PV system. How long must we operate a PV system before we recover the energy used, and justify the pollution generated during its manufacture, transport, and operation?

Energy payback estimates ranging from 1 to 4 years for different technologies indicate that 3-4 years is reasonable for systems using current multi-crystalline silicon PV modules, and 2-3 years for those using current thin-film PV modules.

These estimates vary from product to product and from manufacturer to manufacturer. Cost of energy (crude oil in particular) also has a great effect on the estimates, so these numbers must be adjusted periodically.

The ever changing socio-economic situation in different countries is another great factor and we expect that the present-day economic slow-down and worldwide financial difficulties will drastically reshape these and most other estimates as well.

Lifetime CO_2 Balance

Similarly, a system's lifetime CO_2 balance (LCB) is a factor that takes into consideration the CO_2 used during the manufacturing and use of PV components vs. the amount of CO_2 saved by using the PV components instead of coal- or oil-fired power generation.

The solar wafers, cells, and modules manufacturing processes generate significant amounts of CO_2 and other harmful gasses, which must be taken into consideration when talking about the advantages of PV technologies over the conventional energy sources.

Consider this equation for lifetime CO_2 balance (LCB):

$$LCB = \frac{Cprod + Ctrans + Cinst + Cuse + Cdecom}{Csave}$$

Where
Cprod = energy used during production of materials, wafers, cells, and modules
Ctrans = energy used to transport materials, modules, and BoS to PV site
Cinst = energy used to assemble and install the PV power plant
Cuse = energy used to operate the PV power plant
Crecyc = energy used to decommission, transport, and recycle the PV field
Csave = energy generated during the life of the PV power plant

In all cases, Csave must be much higher than the sum of the CO_2 generation sources for the system to be an effective energy source. Regardless of the LCB ratio, the PV plant will receive carbon credits for the CO_2-free power generated during its lifetime.

We have seen estimates that the total quantity of CO_2 generated during the cradle-to-grave cycle of some PV components is compensated within 2-4 years of CO_2-free PV power generation, by reducing CO_2 emissions, as compared to those generated conventionally.

These numbers, however, seem low, considering the intensity of energy consumption and CO_2 emissions during manufacture of PV modules and all related components involved in a PV power installation.

These numbers will always be case-specific, depending on the type and location of the installation.

ENVIRONMENTAL CHALLENGES

All renewable energy facilities located in desert and non-desert locations face environmental issues, depending on the technology, location, and size. The desert locations are more affected, because they are usually very large—thousands of acres containing millions upon millions of PV modules and BOS equipment, working day and night under extremely adverse conditions.

All power generating facilities have some adverse effect. Solar PV facilities can cause problems including potential glare and glint hazards to aircraft, trains, and highway traffic, and can impact agricultural lands and open habitat areas.

Depending on their location, technology, and site design, wind farms have the potential for killing migrating or foraging raptors and bats, adversely affecting the visual landscape, creating aviation hazards, and causing noise problems if located near urban areas.

Geothermal facilities can affect rare and endangered plant and animal species, cultural resources, the quantity and quality of local water supplies, and visual landscapes. Geothermal projects can also cause or contribute to air quality problems through emission of moderate amounts of regulated air pollutants, though they are required to use best available emissions control technology and provide emission offsets to comply with local air district regulations.

Biomass plants can cause regional increases in criteria pollutants and particulate matter, post ash disposal and local land use concerns, and increase water use (if a facility uses water cooling towers or employs "wet scrubbers" to reduce hydrochloric acid emissions).

For renewable facilities in the desert, the primary environmental concerns are biological and cultural resources, water supply, visual impacts, transportation related visual hazards, and land use, as discussed in the following sections. Depending on the project, there may also be air quality, hazardous materials, noise, public safety, and local community concerns.

So, following is a brief list of the negative impacts most frequently associated with renewable energy installations.

Biological Resource Impacts

Proposed desert locations for utility scale solar and those for wind energy projects often provide habitat for sensitive species like raptors, bats, tortoises, kit foxes, various reptiles and amphibians, ground squirrels, and sensitive plants. Solar PV and solar thermal parabolic trough projects generally cause greater habitat loss than wind farms and solar thermal heliostat and power tower projects, because sites often need to be leveled to accommodate a linear design, which typically cannot be altered to avoid sensitive areas.

Heliostat and power tower projects do not necessarily require that a site be leveled, so habitat impacts can be less. The site topography can be maintained and some vegetation left intact since that technology has far greater flexibility regarding where the mirrors are located.

Still, solar installation hardware covers the ground surface, causing sudden environmental changes. Shaded terrain usually undergoes a slower, but more drastic change in flora and fauna. Planning to minimize long-term impact before, during and after installation is important.

Wind energy projects located in key migration routes or foraging areas can affect bird and bat species through collisions with turbine blades and through barotraumas (tissue damage and lung failure) caused by rapid air-pressure reduction when bats and some birds get too close to moving turbine blades.

Wind farms generally cause less absolute habitat loss within a project footprint than utility-scale solar facilities because habitat for plant and wildlife species remains between turbines. Indirect impacts to wildlife outside the turbine footprint can result from roads, vehicles, and noise—possibly rendering a site generally unusable by wildlife depending on usage on the site and density of turbines. It may be easier to protect rare plant populations on a wind energy site. In addition to habitat loss, most large renewable generation projects (both solar and wind) can potentially affect wildlife movement patterns, particularly if they are proposed in or near migration corridors or impede the connections between sensitive species populations, which can be critical to their survival.

Water Supply Impacts

Water is limited in the desert, and groundwater basins are often already in an overdraft condition. Fresh water is an increasingly critical resource, not only in desert regions, but throughout California. More and more, power plants may be competing with other local users for diminishing water supplies. As California's population and water demand continue to grow, the Department of Water Resources anticipates that the state will experience water supply shortfalls of more than several million acre feet within the next 10 years.

In the desert region, the majority of proposed utility-scale renewable energy facilities use either wind or solar technologies. The solar technologies are further categorized as either solar thermal or PV. Wind and solar facilities require large areas of open desert to take advantage of higher wind speeds or to maximize the collection of solar radiation, but their water use can vary significantly.

Solar thermal facilities, which use steam turbine generators, must dissipate waste heat. The preferred technology for heat dissipation (cooling) is evaporative cooling. Use of this technology requires a sizable volume of water during operation.

Wind technologies and PV do not require thermal cooling equipment, resulting in significantly less water use. Activities such as grading and dust control for construction of both PV and solar thermal projects may require significant water use.

Mirror washing throughout the life of a project may also contribute to significant water use. Water needs for PV panel washing are estimated as one-tenth of the requirements for solar thermal power mirror washing values.

Renewable energy facilities can take advantage of different strategies to reduce their water consumption. Solar thermal facilities can also use alternative approaches, such as dry cooling (air cooled condensers) and hybrid cooling, which are available and commercially viable. This can reduce a project's water demand by up to 90 percent, and simplify the analysis involved in the permitting process. Rather than using fresh water, renewable projects can use degraded water, also known as non potable water, which can be treated and reused for power plant process water.

Surface Water Impacts

Federal and state regulations protect many of the ephemeral and intermittent streams in the desert region because they are important sources of sediment, water, nutrients, seeds, and organic matter for downstream ecosystems and provide habitat for many species. Unlike other streams, those in the desert typically have relatively long periods where no flow occurs, punctuated by episodic flows of relatively short duration and high intensity.

Site design must also be modified where important biological resources are identified and site drainage would have an impact. Diversion of high velocity flows through and around a site can be difficult where potential impacts up and downstream of a project site must be decreased and there is a need to mimic natural conditions. Temporary erosion and sediment protection measures should be installed to control soils disturbed by construction.

Visual Impacts

Utility-scale solar thermal power plants or wind farms can cover many square miles, including the power block facilities, access roads and transmission lines, and cause major visual changes in non-industrialized desert or mountainous landscapes with scenic values. The steam plumes produced by the wet cooling towers of solar thermal power plants may also change the view of the landscape.

Geothermal power plants, including well pads, steam pipelines, power generation facilities, access roads and transmission lines, may occupy as much as 350 acres, and power plant wet cooling towers can produce steam plumes—all potentially causing similar visual impacts on undeveloped desert or mountainous terrain.

Cultural Resources Impacts

State laws define cultural resources as buildings, sites, structures, objects, and historic districts. There are three kinds of cultural resources: prehistoric (related to prehistoric human occupation and use of an area); historical (associated with Euro-American exploration and settlement of an area); and ethnographic (materials important to the heritage of a particular ethnic or cultural group, such as Native Americans).

A major challenge to the development of lands especially in the southern desert, is the lack of comprehensive information regarding the locations and significance of cultural resources. While some archaeological sites are small and well defined, historic and prehistoric landscapes can stretch for miles.

Many of the elements within these and other areas of historical significance have not been identified or evaluated. Information on both historical and archaeological sites is scattered among city, county, state, and private archives, multiple information centers, and state and federal agencies, such as the California Office of Historic Preservation, National Register of Historic Places, and the California Register of Historic Resources.

Scarce and fragmented information, along with confidentiality requirements limiting access to cultural resource information, can make it difficult for developers to select sites that will avoid significant cultural resources. This can cause delays or inaccuracies in the resource analysis during the licensing process and create the need for more extensive site surveys, especially in remote desert areas.

Much of the land under consideration for solar and wind development includes Native American ancestral lands that are centuries old and contain artifacts, burial and historic village sites, trails, plants, animals, landscapes, and vistas with cultural and spiritual significance. The spiritual value of these areas and artifacts is separate from the archaeological and historical value of these or other cultural resources and, from a Native American perspective, the loss of the use of these lands or their spiritual context within the landscape cannot be mitigated. Information regarding landmarks and other areas of significance to Native Americans is often known only to tribal elders or tribal historic preservation officers.

Another significant challenge to avoiding or mitigating cultural resource impacts, especially under the California Environmental Quality Act (CEQA), is the lack of flexibility in site location and design once a project reaches the application phase. Developers frequently fail to adequately consider the potential cultural sensitivity of a site through appropriate resource studies and discussions with knowledgeable technical specialists and Native American tribal representatives before settling on a final location. To avoid cultural resources, staff and tribal representatives must have the opportunity to identify resources before site finalization.

Land Use Impacts

Most desert lands are owned by the federal government and managed for multiple uses by the BLM and National Park Service. The Department of Defense also owns and manages large tracts of desert land for military purposes. Siting renewable energy facilities on BLM and National Park Service lands can affect existing and future multiple uses such as recreation, wildlife habitat, livestock grazing, and open space.

For example, solar thermal facilities within the BLM's California Desert Conservation Area Plan have significantly impacted or restricted other uses of the land. Similarly, siting renewable energy facilities on or near Department of Defense lands may affect military operations and related programs.

If located in productive agricultural areas, solar thermal or PV projects that require grading of many acres may result in the permanent loss of crop and grazing lands. In contrast, wind energy projects are generally compatible with agricultural land uses and may even help farmers preserve their farms with supplemental income received from leasing land to wind developers.

The average wind farm requires 5.5 acres of land to produce 1 MW of electricity, allowing land outside the turbine footprint to remain available for planting and grazing. However, wind projects can affect agricultural resources through soil disturbance during construction and the loss of agricultural land from installing access roads, wind turbine towers, and transmission lines.

Geothermal facilities are usually land intensive and can result in permanent loss of productive agricultural land due to geothermal steam well field development, steam pipeline installation, construction of the power plant facilities and transmission lines, and permanent access roads to develop and maintain all of the steam field and power plant facilities.

Transportation-related Visual Hazards

Solar thermal, PV, wind, and geothermal technologies present significant hazards to general aviation and military flight activities, as well as to motorists, and railroad crews on their normal runs.

Solar thermal and PV plants, with their huge numbers of mirrors and collectors, can emit glint and glare that can pose a nuisance to pilots of any plane flying in the vicinity. They can even cause flash blindness, which is especially dangerous during take-off and landing, so solar plants located near airports could be considered especially dangerous.

Wind turbines cause turbulence, which can affect low-flying aircraft (light planes and helicopters) due to upward airflow disruption. Wind turbines are also being blamed for interrupting birds' flight paths and even killing them due to collision with the turning blades and/or disorienting them by means of the heavy airflow disruption in the field.

The evaporative and dry cooling towers of some solar thermal plants and cooling towers of geothermal plants may emit high velocity, hot air plumes, disrupting airflow and potentially causing severe turbulence to low-flying aircraft as well.

The Department of Defense has also raised concerns about thermal plumes from power plants located near military flight areas and the effect on radar operations. All very tall structures, including wind turbines and solar power towers, have the potential to interfere with low-flying aircraft and with military flight zones that have structure height restrictions.

Efforts to Address Environmental, Planning, and Permitting Challenges

The importance of streamlining renewable permitting processes is widely recognized; therefore a number of efforts are either completed or underway to help promote the development of utility-scale renewable electricity generating facilities, and especially those in the state of California. Some of these efforts and initiatives follow.

Renewable Energy Transmission Initiative

In 2007, the Renewable Energy Transmission Initiative (RETI) was initiated as a joint statewide effort combining land use and transmission planning factors among the CPUC, the Energy Commission, the California ISO, and investor-owned and publicly owned utilities. The primary goals of RETI were to:

1. Help identify the transmission projects needed to accommodate California's renewable energy goals;

2. Ease the designation of corridors for future transmission line development; and

3. Facilitate transmission line and renewable generation siting and permitting.

The stakeholder-driven process identified Competitive Renewable Energy Zones throughout the state with the greatest potential for cost-effective and environmentally responsible renewable energy development.

Renewable Energy Action Team

To address challenges with permitting renewable projects in sensitive California desert regions, the Renewable Energy Action Team (REAT) was formed in 2008 to streamline and expedite the permitting processes for renewable energy projects, while conserving endangered species and natural communities at the ecosystem scale. Based in part on recommendations from the RETI process, the REAT is developing a Desert Renewable Energy Conservation Plan (DRECP) for the Mojave and Colorado Desert regions.

The REAT also published the multidisciplinary Best Management Practices and Guidance Manual: Desert Renewable Energy Projects in December 2010 to help project developers design projects that minimize environmental impacts for desert renewable projects. The manual provides guidance on initiating permitting processes, conducting land-use assessments and surveys, decisions on water use and quality, roadway planning, avoiding conflicts with aviation, and grid interconnection issues.

Desert Renewable Energy Conservation Plan

In conjunction with other federal, state, and local agencies and stakeholder groups, the REAT is developing the DRECP to identify areas in the Mojave and Colorado Desert regions suitable for renewable energy project development and areas that will contribute to the conservation of sensitive species and natural communities. The DRECP encompasses about 22 million acres in Kern, Inyo, Los Angeles, San Bernardino, Riverside, San Diego, and Imperial Counties. It will promote development of solar thermal, utility-scale solar PV, wind, and other forms of renewable energy along with associated infrastructure like transmission lines.

The DRECP will be a Natural Community Conservation Plan (NCCP) and will serve as the basis for one or more Habitat Conservation Plans (HCP). As required by state and federal law, the environmental impact of the DRECP will be analyzed in a joint environmental impact report and statement anticipated to be completed by December 2012, along with the NCCP.

Solar Energy Development Programmatic Environmental Impact Statement

At the federal level, the US Department of Energy, Energy Efficiency and Renewable Energy Program, and the US Department of the Interior, Bureau of Land Management, are preparing a Solar Energy Development Programmatic Environmental Impact Statement (PEIS) to assess environmental impacts from programs intended to promote environmentally responsible utility-scale solar energy in six Western states.

The draft PEIS was issued December 16, 2010, and in May 2011 the Energy Commission and the California Department of Fish and Game (DFG) submitted joint comments on the draft with the following recommendations:

a. Abandon further consideration of the Iron Mountain Solar Energy Zone (SEZ).

b. Consider designating and studying additional SEZs on previously disturbed lands in the western Mojave Desert.

c. Delay final PEIS-triggered amendments to the affected Land Use Management Plans until the DRECP process is complete in 2012.

d.   Fully consider and address all Energy Commission and DFG comments made previous to and in response to the publication of the draft PEIS. A supplement to the draft PEIS was released in the fall of 2011.

Cross-agency Coordination

State and federal agencies are working to streamline the permitting of renewable energy projects in California by increasing cross-agency cooperation and coordination, with several multi-agency agreements already in place:

In 2007, the Energy Commission, the US Department of the Interior, and the BLM signed a memorandum of understanding (MOU) on agency roles, responsibilities, and procedures for joint environmental review of solar thermal projects proposed on federal land.

In 2009, the Energy Commission entered into an MOU with the California State Lands Commission to ensure timely and effective coordination during the Energy Commission's thermal power plant review process.

In 2010, the State of California and FERC signed an MOU to coordinate and share information for reviewing offshore wave and tidal energy projects.

In 2010, the Energy Commission and the Departments of General Services, Corrections and Rehabilitation, Transportation, Water Resources, and Fish and Game signed an MOU to promote the development of renewable energy projects on state buildings, properties, and rights-of-way. The State Lands Commission and University of California subsequently joined the MOU.

Transmission Infrastructure Issues

In addition to the environmental, planning, and permitting issues identified herein, states can face challenges to planning and permitting power lines and other transmission infrastructure needed to bring electricity generated by large-scale renewable facilities to consumers. Permitting procedures alone can take as long as eight years.

Carbon Tax

Carbon tax is a new phenomena, used to measure and control the emission of gasses such as CO_2, generated by fossil fuels. This tax is levied according to the carbon content of the fuels, and the amounts used.

Carbon is a significant part of every fossil fuel (coal, oil, and natural gas) and most of it is released as CO_2, when fuels are burned. CO_2 gasses contribute to the greenhouse effect, so to compensate for those damages, fossil fuels users (power plant operators) are taxed on emissions at any point in the production and use cycles of the fuels.

Carbon tax is basically a way of encouraging the reduction of greenhouse gas emissions by penalizing those who emit. In reality, most greenhouse gas emissions in developed countries are levied on energy products and on motor vehicles, rather than directly on CO_2 emissions. This system places a disproportionate portion of the tax on the end user.

A number of countries have implemented carbon taxes or energy taxes that are related to carbon content. The tax was designed and implemented in some US states (there is no national carbon tax in the US) to address the urgent issue of environmental damage from fossil fuel use.

Collecting carbon tax money and distributing it is another issue. In Canada, for example, 1.6 million families got $325 million in compensation from the carbon tax. Benefits varied from $69 to $100 per family.

There is talk of a $20 carbon tax per ton generated GHGs, which might amount to $1.2 trillion in 10 years. It was suggested that it be applied to reducing the national debt, which would do very little to reduce air pollution.

The carbon tax issue is complicated by the fact that harmful gasses are generated in one area, the power is used in another (usually far away), while environmental damage is done in yet another area.

This creates an imbalance between fossil fuel power generators and residential and commercial power users, with environmental damage as one of the major negative results.

It is unlikely that the carbon tax will make any difference in reducing the greenhouse effect as a whole, simply because the CO_2 gas emissions cannot be re-

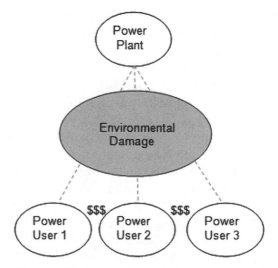

Figure 9-15. Carbon tax and the environment

duced by monetary means. Shifting the cost of fossil fuels usage to the largest greenhouse gas generators (power plants operators) will not make them reduce the use of carbon, nor will it hurt them, because in most cases they can charge the customers for the difference.

So the carbon tax is a good idea, but it is attempting to address a very complex issue, which cannot be solved by monetary means alone.

PROBLEMS

Presently, coal-fired power plants emit approximately 1,500 lbs. CO_2/MWh. A 500 MW coal power plant emits 500,000 lbs of CO_2 in the atmosphere every hour, and 12,000,000 lbs CO_2 every day.

A 500 MW coal-fired power plant emits 6.6 billion lbs. CO_2 per annum. Guidelines for newly built plants will reduce this amount to 4.4 billion lbs. CO_2 per annum.

There are 594 coal-fired power plants in the US (2009 count) with approximately 330,000 MW of combined power generation. Several dozen new coal-fired power plants are in the making, so the number will grow significantly with time. This is 2,890,800,000,000 lbs. (2.9 trillion lbs.), or 1.3 billion tons of CO_2 pumped into the atmosphere every year by US power plants alone.

China's 2,000 coal-fired power plants alone generate 2 billion MWh/annum (almost 10 times the US numbers), and the country also burns that much coal for other domestic and industrial purposes. By 2025-2030, China is expected to emit as much poisonous gasses as the entire rest of the world.

Using these statistics on coal as one example, we can cut to the chase and admit that *all* sources of energy have downsides. The upside is that we can improve how we do things.

SOLUTION(S)

Coal Sequestration

Clean coal technology is here, and it is just a matter of implementing it. A good example of this technology is the new GreenGen power station, in Beijing, China. It is a 400 MW coal-burning plant that is using a coal-to-gas conversion and gas-burning process, which is much cleaner than solid-coal burning.

The plant is also equipped with efficient gas sequestration equipment, which ensures that the gasses that are released in the atmosphere are 90-95% clean. Thus generated CO_2 (the major component in the waste gasses) is trapped, segregated, purified and stored for later use. It is then sold to be used in bottling of soft drinks and other commercial needs.

GreenGen demonstrates multiple Integrated Gasification Combined Cycle (IGCC) technologies that can be scaled to simultaneously address several environmental challenges in response to climate change and energy security concerns.

The $1.0 billion GreenGen plant features pre-combustion technology that will strip pollutants such as SO_2 and particulates from the coal-burning process.

The second phase will implement fuel cell power generation and carbon capture and sequestration (CCS) technology for nearly zero-emission power generation.

The third phase of the project is planned for completion by 2016, when the plant would produce a total of 650 MW and 3,500 tons of syngas per day.

When completed and optimized, this model will be duplicated in many such IGCC-CCS plants. With China continuing to build about 30 power plants a year, these technologies could have a significant impact on air quality and greenhouse gas emissions not only in China but worldwide.

Grain Alcohol

Another great example of energy generation diversification is the massive use of alcohol to power cars and commercial enterprises in Brazil. Starting in the mid-1970s, Brazil initiated a large modification of its energy infrastructure by cultivating millions of acres of sugar cane for the production of alcohol. Cars in Brazil can run on both gas or alcohol, and any mixture of those.

The American alcohol experiment in 2000 didn't catch on. While grain alcohol will not be a primary fuel source here anytime soon, it is still used for mixing in gasoline during certain times of the year, and we foresee it making a comeback, eventually.

Conservation

Here at home, by just implementing energy-saving measures in homes and commercial buildings, the US could save 1/4 of its energy demand immediately. It's food for thought.

PV Installations on Contaminated Land Sites

Identifying and using land located in areas with high quality, renewable energy resources will be an essential component of developing electricity from renewable energy sources. As the number of large PV projects increases, the demand for large pieces of land will increase. Land that has been previously used for activities such as mining and waste disposal will become increasingly important for this application.

The US Environmental Protection Agency (EPA) estimates that there are approximately 490,000 sites and almost 15 million acres of potentially contaminated (or previously contaminated) properties across the United States. Most of these are tracked by EPA for potential future use in a number of projects, including solar and wind power generation.

This estimate includes Superfund, Resource Conservation and Recovery Act (RCRA), Brownfields, abandoned mine lands, reconstituted municipal garbage dumps, and more.

Cleanup goals have been achieved and controls put in place on lands, to ensure long-term protection for more than 917,000 acres. Through coordination and partnerships among federal, state, tribal, and other government agencies, as well as utilities, communities, and the private sector, many new renewable energy facilities can be developed on these contaminated properties.

The EPA Office of Solid Waste and Emergency Response (OSWER) Center for Program Analysis (CPA) is looking into opportunities to facilitate the reuse of contaminated properties and active and abandoned mine sites for renewable energy generation.

These lands are environmentally and economically beneficial for siting renewable energy facilities because they:

1. Offer thousands of acres of land with few site owners.

2. Often have critical infrastructure in place, including electric transmission lines, roads, and on-site water, and are adequately zoned for such development.

3. Provide an economically viable reuse for sites with significant cleanup costs or low real estate development demand.

4. Take the stress off undeveloped lands for construction of new energy facilities, preserving the land carbon sink.

5. Provide job opportunities in urban and rural communities.

Figure 9-14. EPA tracked potential PV sites

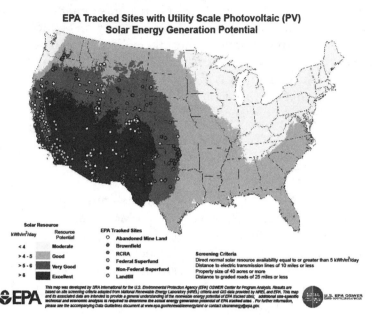

6. Advance cleaner and more cost effective energy technologies, reducing the environmental impacts of energy systems, like greenhouse gas emissions.

What are the advantages and disadvantages of installing PV power plants on particular contaminated land site? Are there risks associated with such a move? How do we separate the past, present, and future issues, be they environmental, health, etc.? Who is ultimately responsible for the environmental and health safety of these potentially dangerous fields, used in an untested, unproven, and unprecedented manner? What are US regulators saying on the matter? These are important questions that demonstrate again how young and immature the PV industry is.

We have a number of options to address the energy crisis. As with any new thing, we must protect it from abuse and misuse. The solutions we choose will determine our energy future, and that of our descendants.

Notes and References

1. A Manual for the Economic Evaluation of Energy Efficiency and Renewable Energy Technologies, http://www.nrel.gov/docs/legosti/old/5173.pdf
2. INACCURACIES OF INPUT DATA RELEVANT FOR PV YIELD PREDICTION Stefan Krauter, Paul Grunow, Alexander Preiss, Soeren Rindert, Nicoletta Ferretti Photovoltaik Institut Berlin AG, Einsteinufer 25, D-10587 Berlin, Germany
3. PV module characterization, Stefan Krauter & Paul Grunow, Photovoltaik Institut Berlin AG, TU-Berlin, Germany
4. PV MODULE LAMINATION DURABILITY, Stefan Krauter, Romain Pénidon, Benjamin Lippke, Matthias Hanusch, Paul Grunow, Photovoltaic Institute Berlin, Berlin, Germany
5. PV module testing—how to ensure quality after PV module

certification. Photovoltaic Institute, Berlin. August, 2011

6. Module Prices, Systems Returns & Identifying Hot Markets for PV in 2011, AEI, September, 2010

7. Fostering Renewable Electricity Markets in North America, Commission for Environmental Cooperation, Canada, 2007.

8. PC Roadmap, http://photovoltaics.sandia.gov/docs/PVR-MExecutive_Summary.htm

9. European Photovoltaic Technology Platform comment on "Towards a new Energy Strategy for Europe 2011-2020," http://www.eupvplatform.org/

10. Economic Impacts from the Promotion of Renewable Energy Technologies. The German Experience. Manuel Frondel, Nolan Ritter, Christoph M. Schmidt, Colin Vance http://repec.rwi-essen.de/files/REP_09_156.pdf

11. Analysis of UK Wind Power Generation from November 2008 to 2010. Stuart Young Consulting, March 2011

12. Performance Degradation of Grid-Tied Photovoltaic Modules in a Desert Climatic Condition by Adam Alfred Suleske, ASU, Tempe, Arizona

13. CAISO Today's Outlook, http://www.caiso.com/outlook/SystemStatus.html

14. Fracking Problem: Shale Gas may be Worse for Climate than Coal, http://envirols.blogs.wm.edu/2011/04/25/fracking-problem-shale-natural-gas-may-be-worse-for-climate-than-coal/

15. NASA solar data. http://eosweb.larc.nasa.gov/cgi-bin/sse/grid.cgi?uid=3030

16. NREL SAM, https://www.nrel.gov/analysis/sam/download. html

17. http://www.greenpeace.org/international/Global/international/planet-2/report/2009/5/concentrating-solar-power-2009.pdf

18. Weather Durability of PV Modules; Developing a Common Language for Talking About PV Reliability, Kurt Scott, Atlas Material Testing Technology

19. Re-imagining US solar financing, Bloomberg, New Energy Finance, May 2012

20. http://en.wikipedia.org/wiki/Environmental_impact_of_the_ Three_Gorges_Dam

21. Clean Energy Patent Growth Index (CEPGI), http://cepgi.typepad.com/files/cepgi-1st-quarter-2011.pdf

22. An Eye on Quality. Intevac, July 2011 http://www.electroiq.com/articles/pvw/print/volume-2011/issue-4/features/an-eye-on-quality.html

23. http://www.aproundtable.org/tps30info/globalwarmup.html against global warming.

24. What is the energy payback for PV? US DOE.

25. RE-Powering America's Land: Renewable Energy on Potentially Contaminated Land and Mine Sites, EPA, 2008.

26. Energy Pay-Back Time (EPBT) and CO_2 2 mitigation potential, Evert Nieuwlaar, Erik Alsema.

27. The World's Worst Environmental Disasters Caused by Companies. http://www.businesspundit.com/the-worlds-worst-environmental-disasters-caused-by-companies/

28. Environmental Aspects of PV Power Systems, http://www.energycrisis.org/apollo2/pvenv1997.pdf

29. ExterneE, Impact pathways of health and environmental effects, http://www.externe.info/

30. USGS: http://minerals.usgs.gov/minerals/pubs/commodity/cadmium/mcs-2010-cadmi.pdf

31. http://abebooks.com/servlet/BookDetailsPL?bi=1108560583&s earchurl-ds%3D30%26isbn%3D9783540541844%26sortby%3D17

32. http://www.fas.org/ota/reports/9344.pdf

33. http://www.iiasa.ac.at/Admin/PUB/Documents/IR-02-073

34. http://www1.eere.energy.gov/femp/pdfs/largereguide.pdf

35. Hazards and Hazardous Materials—SLO County Planning Commission http://www.google.com/search?q=C.9.3.4+Impact+Assessment.+Topaz+Solar+Farm+Project&hl=en&sourceid=gd&rlz=1Q1GGLD_enUS473US474

36. Photovoltaics for Commercial and Utilities Power Generation, 2011. Anco S. Blazev

Appendix A

A. Clean Air Act and Amendments

1955 Pollution Control Act

1959 Reauthorization

1960 Motor vehicle exhaust study 1963 Clean Air Act Amendments

1965 Motor Vehicle Air Pollution Control Act 1966 Clean Air Act Amendments of 1966

1967 Air Quality Act of 1967 National Air Emission Standards Act

1970 Clean Air Act Amendments of 1970

1973 Reauthorization

1974 Energy Supply and Environmental Coordination Act of 1974

1977 Clean Air Act Amendments of 1977 1980 Acid Precipitation Act of 1980

1981 Steel Industry Compliance Extension Act of 1981

1987 Clean Air Act 8-month Extension 1990 Clean Air Act Amendments of 1990

1991 Technical correction to list of hazardous air pollutants

1995-96 Relatively minor laws amending the Act

1998 Amended Section 604 re methyl bromide 1998 Border Smog Reduction Act of 1998

1999 Chemical Safety Information, Site Security and Fuels Regulatory Relief Act

2004 Amendments to §209 re small engines

2005 Energy Policy Act of 2005 (amended §211 re fuels) 2007 Energy Independence and Security Act of 2007 (amended §211, re. fuels)

P.L. 84-159

P.L. 86-353

P.L. 86-493

P.L. 88-206

P.L. 89-272, Title I P.L. 89-675

P.L. 90-148

P.L. 91-604

P.L. 93-15

P.L. 93-319

P.L. 95-95

P.L. 96-294, Title VII

P.L. 97-23

P.L. 100-202

P.L. 101-549
P.L. 102-187
P.L. 104-6,
P.L. 104-59,
P.L. 104-70,
P.L. 104-260
P.L. 105-277, Section 764
P.L. 105-286
P.L. 106-40
P.L. 108-199, Division G, Title IV, Section 428
P.L. 109-58 P.L. 110-140

B. Clean Water Act and Amendments
1948 Federal Water Pollution Control Act
1956 Water Pollution Control Act of 1956
1961 Federal Water Pollution Control Act Amendments
1965 Water Quality Act of 1965
1966 Clean Water Restoration Act
1970 Water Quality Improvement Act of 1970
1972 Federal Water Pollution Control Act Amendments
1977 Clean Water Act of 1977
1981 Municipal Wastewater Treatment Construction Grants Amendments
1987 Water Quality Act of 1987
P.L. 80-845 (Act of June 30, 1948)
P.L. 84-660 (Act of July 9, 1956)
P.L. 87-88
P.L. 89-234
P.L. 89-753
P.L. 91-224, Part I P.L. 92-500
P.L. 95-217

C. Ocean Dumping Act and Amendments
1972 Marine Protection, Research, and Sanctuaries Act
1974 London Dumping Convention Implementation
1977 Authorization of Appropriations 1980 Authorization of Appropriations
1980 Authorization of Appropriations 1982 Surface Transportation Assistance Act
1986 Budget Reconciliation 1986 Water Resources Development Act
1987 Water Quality Act of 1987 1988 Ocean dumping research amendments
1988 Ocean Dumping Ban Act
1988 US Public Vessel Medical Waste Anti-Dumping Act of 1988
1990 Regional marine research centers 1992 National Coastal Monitoring Act
1992 Water Resources Development Act
P.L. 92-532
P.L. 93-254

P.L. 95-153
P.L. 96-381
P.L. 96-572
P.L. 97-424
P.L. 99-272, §§6061-6065
P.L. 99-662, §§211, 728, 1172
P.L. 100-4, §508
P.L. 100-627, Title I
P.L. 100-688, Title I
P.L. 100-688, Title III
P.L. 101-593, Title III
P.L. 102-567, Title V P.L. 102-580, §§504-510

D. Safe Drinking Act and Amendments
1974 Safe Drinking Water Act of 1974 P.L. 93-523
1977 Safe Drinking Water Act Amendments of 1977 P.L. 95-190
1979 Safe Drinking Water Act Amendments P.L. 96-63
1980 Safe Drinking Water Act Amendments P.L. 96-502
1986 Safe Drinking Water Act Amendments of 1986 P.L. 99-339
1988 Lead Contamination Control Act of 1988 P.L. 100-572
1996 Safe Drinking Water Act Amendments of 1996 P.L. 104-182
2002 Public Health Security and Bioterrorism Preparedness P.L. 107-188 and Response Act of 2002

E. Solid Waste Disposal Act, Resource Conservation and Recover Act (RCRA)
1976 Law to Control Hazardous Wastes, End Open Dumping, and Promote Conservation of Resources
1980 Hazardous Waste Regulations
1981 First Storage Permit Under RCRA
1984 Hazardous and Solid Waste Amendments of 1984,
1988 Standards For Underground Storage Tanks
1990 Land Disposal Restrictions for Hazardous Wastes
1999 Revision of Standards for Air Emissions from Hazardous Waste Combustors

F. Insecticide, Fungicide, and Rodenticide Act and Amendments
1938 Federal Food, Drug, and Cosmetic Act
1947 Federal Insecticide, Fungicide, and Rodenticide Act P.L. 80-104
1954 Federal Food, Drug, and Cosmetic Act Amendments
1958 Food Additive Amendments of 1958 (including the Delaney Clause)
1964 Federal Insecticide, Fungicide, and Rodenticide Act Amendments P.L. 88-305

1970 Transfer of pesticide residue responsibility to EPA

1972 Federal Environmental Pesticide Control Act P.L. 92-516

1975 Federal Insecticide, Fungicide, and Rodenticide Act Extension P.L. 94-140

1978 Federal Pesticide Act of 1978 P.L. 95-396

1980 Federal Insecticide, Fungicide and Rodenticide Act Amendments P.L. 96-539

1988 Federal Insecticide, Fungicide, and Rodenticide Amendments of

1990 P.L. 100-532 1990 Food, Agriculture, Conservation,

and Trade Act of 1990 P.L. 101-624

1991 Food, Agriculture, Conservation and Trade Amendments of 1991 P.L. 102-237

1996 Food Quality Protection Act (FQPA) of 1996 P.L. 104-170

1996 Food Quality Protection Act of 1996

2004 Pesticide Registration Improvement Act of 2003 P.L. 108-199, Division G, Title V

2007 Pesticide Registration Improvement Renewal Act P.L. 110-94

Appendix B

Key Environmental Terms

acid mine drainage
acid rain
air pollution
air quality
algal bloom
anoxic waters
arsenic poisoning
asbestos poisoning
asthma
bioaccumulation
biomagnification
birth defects
blast fishing
bottom trawling
burial problems
bycatch
cadmium poisoning
carbon monoxide
carbon oxide
chlorofluorocarbons, CFC
clear-cutting
climate change
coal mining
conservation
consumer capitalism
consumerism
conventional energy sources
coral bleaching
coral reefs extinction
cyanide fishing
dam impacts
DDT
deforestation
desertification
dioxin

earth quakes
efficient energy use
electromagnetic fields
electromagnetic radiation
endangered species
endocrine disruptors
energy conservation
environmental degradation
environmental effects of meat production
environmental health dependency
eonoculture
eutrophication
e-waste
exploitation of natural resources
fish kill
floods
forest fires
forest depletion
fossil fuels
fracking soil pollution
genetic engineering
genetic pollution
genetically modified crops
genetically modified foods
ghost nets
global dimming
global warming
Great Pacific Garbage Patch
greenhouse gasses
habitat destruction
habitat modification
habitat fragmentation
heat pollution
herbicides
holocene extinction
hurricanes
Illegal fishing

Illegal logging
indoor air quality
intensive farming
invasive species
irrigation-water pollution
island flooding
land degradation
land misuse
land pollution
landfill
leach rate
lead poisoning
light pollution
littering
logging
marine debris
marine pollution
medical waste
mercury in fish
mining
mountaintop removal mining
nanopollution
nanotechnology
nanotoxicology
natural gas fracking
nitrogen oxides
noise pollution
nonpoint source pollution
nuclear fallout
nuclear issues
nuclear meltdown
nuclear power
nuclear waste
obesity
ocean acidification
ocean level increase
ocean warming
ocean deoxygenation

Appendix C

The Periodic Table of the Elements

Glossary

A

Absorber is a PV device, the material that readily absorbs photons to generate charge carriers (free electrons or holes).

Absorption is the energy of an incident photon large enough to excite an electron into the conduction band from the valence band (energy ≥ band gap). Depending on the wavelength of the photon, and the type of material, it will occur more readily, and at a certain depth into the material.

Absorption coefficient is the distance into a material that a particular light wavelength can penetrate before it is absorbed. Depends on the wavelength of the photon (light) and the type of material the light is incident on. In units of cm-1.

AC is alternating current.

AC PV building block is a complete, environmentally protected PV modular system consisting of a PV module, a complete integrated inverter enclosed with a housing, eliminating exposure of any dangerous voltage and generally doubling as the module frame or mounting structure that also encloses all of the necessary AC bus work, interconnects, communication, surge protection and terminations.

Acceptor is a dopant material, such as boron, which has fewer outer shell electrons than required in an otherwise balanced crystal structure, providing a hole, which can accept a free electron.

Air mass is the cosine of the zenith angle—that angle from directly overhead to a line intersecting the sun. The air mass is an indication of the length of the path solar radiation travels through the atmosphere.

Alternating Current (AC) is a type of electrical current, the direction of which is reversed at regular intervals or cycles. In the United States, the standard is 120 reversals or 60 cycles per second.

Ambient Temperature is the temperature of the surrounding area.

Amorphous Silicon is a thin-film, silicon PV cell having no crystalline structure.

Amortize is to recover or pay back, on a regular schedule, a sum of money.

Ampere (amp) is a unit of electrical current or rate of flow of electrons.

Ampere Hour (Ah) Meter is an instrument that monitors current with time.

Ampere-Hour (Ah/AH) is a measure of the flow of current (in amperes) over one hour.

Analysis, Breakeven—a determination of the conditions under which the economic values of two alternatives are equal.

Analysis Period is the amount of time or the period an analysis covers.

Angle of Incidence is the angle that a ray of sun makes with a line perpendicular to the surface.

Anode is the positive pole of a device where electrons leave and current enters the system.

Antireflection Coating (AR) is a thin film coating on a solar cell surface, intended to reduce light reflection and increase light transmission.

Annualization is the conversion of a series of transactions to an equivalent annuity.

Annuity is a series of equal annual payments occurring over a period of time.

Array Current is the electrical current produced by a PV array.

Array Operating Voltage is the voltage produced by a PV array connected to a load.

Application specific integrated circuit (ASIC) is a highly integrated circuit package containing hundreds of logic functions that is modified by burning away internal paths to produce application-specific circuit functions.

Autonomous System is a stand-alone system.

Availability is the availability of a PV system to provide power to a load.

Azimuth Angle is the angle between true south and a point on the horizon directly below the sun.

B

Balance of System (BOS) are all components other than PV modules.

Band Gap is the energy difference between the conduction and valence bands.

Band Gap Energy (Eg) is the amount of energy (in eV) as required to transfer an electron from the valence to the conduction band.

Base Load is the average amount of electric power that a utility must supply in any period.

Base Year is the year in which all cash flows are converted.

Benefit Cost Ratio—BIC is the ratio of the SUM of all discounted benefits accrued from an investment to the sum of the associated discounted costs.

Bi-directional inverter is an inverter that can be operated in all four quadrants of the voltage/current regime. It may function as an inverter or as a rectifier by applying the proper drive signals. Power flow may be in either direction.

Building-integrated Photovoltaics (BIPV) is design and integration of PV technology into buildings.

Black body radiation is when a black body absorbs all radiation incident on its surface, and emits radiation based on its temperature. The sun approximates a black body. The sun emits light in the visible region, and thus can be harnessed by PV.

Blocking Diode is a one-way valve that allows electrons to flow forwards, but not backwards.

Book is a general term that connotes the financial records of a company, usually related to rate making and investment miters.

Book Life is the period of time over which an investment amount is recovered through book depreciation.

Boron (B) is a chemical element used as the dopant in solar cells manufacturing.

Btu (British thermal unit) is the heat required to heat 1 lb. of water 1 degree F.

Burden is the impedance (load) of the circuit connected to the secondary winding of an instrumentation transformer. For voltage transformers it is expressed in terms of the equivalent volt-amperes and power factor at a specified voltage and frequency.

Bypass Diode is a device connected across a solar cell to protect it from overcurrent.

C

Cadmium (Cd) is a chemical element used in making certain types of solar cells and batteries.

Cadmium Telluride (CdTe) is a polycrystalline thin-film PV material.

Capitalize is to place an investment on the books of a company.

Capitalization is the total of all debt and equity in a company.

Capacity Factor is the ratio of the power output to the capacity rating over a period of time.

Capacity, Rated is the maximum capacity that a generating unit can sustain over a specified period of time.

Capital Investment is a major investment in equipment, or infrastructure, or other project-related necessities.

Carrying Charges—the revenue needed to support an investment.

Carrying Charge Factor (Rate) is the amount of revenue per dollar of investment that must be collected from customers in order to pay the carrying charges on that investment.

Cash Flow—F is the net income plus amount charged off for depreciation, depletion, amortization, and extraordinary charges to reserves.

Cathode is the negative pole of a device where electrons enter and current leaves the system.

Cell Barrier is a very thin region of static electric charge between the positive and negative layers in a solar cell.

Cell Junction is the area between the positive and negative layers of a solar cell.

Charge Carrier is a free and mobile conduction electron or hole in a semiconductor.

Charge Controller controls the flow of current to and from a PV system.

Chemical Vapor Deposition (CVD) is thin film deposition method using heat and reactive gasses.

CHP is combined heat and power.

Cleavage of Lateral Epitaxial Films for Transfer (CLEFT) is a GaAs cells manufacturing process.

Cloud Enhancement is the increase in solar intensity caused by nearby clouds.

Combined Collector is a PV module that provides heat energy in addition to electricity.

Common Stock is a type of equity security in which the holder is a part owner in the company.

Concentrator is a PV module, using optics to concentrate sunlight onto the solar cells.

Concentration gradient represents when the concentration of carriers varies from one area/region in the solar cell to another.

Conduction is movement of the free valence electrons in a conductor material. In a conductor material, electrons are not tightly bound to the atom, and as a result, are free to participate in conduction.

Conduction Band is the energy level (band) in which electrons can move freely. It is at a higher energy level than the valence band.

Conductor is the material through which electricity is easily transmitted.

Constant Dollar Analysis is an analysis made without including the effect of inflation, although real escalation is included.

Consumer Surplus is the difference between the total amount of money consumers are willing to pay for a given quantity of a commodity minus the total expenditures that are necessary to acquire that quantity of the commodity.

Contact Resistance is the resistance between metallic contacts and a semiconductor.

Converter is a unit that converts certain DC voltage to higher or lower voltage level.

Converter is a device used for changing direct current power to alternating current power or vise versa, or from one frequency to another.

Copper Indium Diselenide ($CuInSe_2$, or CIS) is thin-film PV material.

Copper Indium Gallium Diselenide ($Cu(In,Ga)Se_2$ or CIGS) are the improved CIS devices.

Cost of Capital is the return required by investors to attract investment capital for a utility.

Cost of Debt is the amount paid to the holders of debt securities for the use of their money.

Cost of Equity is the earnings expected by an investor when purchasing equity shares in a company.

Covalent bond is a chemical bond whereby pairs of electrons are shared by two atoms.

Crystalline Silicon is type of material silicon (c-Si) solar cells are made from.

Current is flow of electric charge carried out through the moving electrons in a conductor material.

Current at Maximum Power (Imp) is maximum power generated by a PV cell or module.

Current-controlled inverter is an inverter that converts DC power to AC power where the output current is controlled and unaffected by output voltage fluctuations.

Current Dollar Analysis is an analysis that includes the effect of inflation and real escalation.

Cutoff Voltage is the voltage level when charge controller disconnects the PV array from the load.

Czochralski process is a method of growing large size, high quality single crystal silicon ingots.

D

Dangling Bond is a disjointed bond hanging at the surface layer of a crystal.

Data Acquisition System (DAS) is a system that receives data from one or more locations. (from IEEE Std. 100-1996).

Days of Storage is the time a solar system will generate power without solar energy input.

Debt Ratio is the ratio of debt money to total capitalization.

Decision Analysis is the evaluation of decision options and the estimation of the value of additional information or testing, using the time and risk preferences of the decision maker.

Declination angle (δ) is determined by the fact that the earth is tilted at a 23.45 degree angle, which causes the angle of the suns rays incident on the earth to vary throughout the year. On the spring and autumnal equinoxes, the angle is 0 degrees, whereas on the winter and summer solstice, the angle is 23.45 degrees.

Deferred Income Tax is the portion of the total income tax liability that may be deferred to a future date.

Dendrite is a slender threadlike spike of pure crystalline material, such as silicon.

Dendritic Web Technique is a method for making sheets of polycrystalline silicon.

Depletion Recovery of an investor's economic interest in mineral (including oil and gas) reserves through federal tax deductions related to removal of the mineral over the economic life of the property.

Depletion Zone is a thin region, depleted of charge carriers (free electrons and holes).

Depreciation is the accounting mechanism for the reduction in value of a capitalized item.

Depreciation Period is the amount of time required for the original capital investment to be fully recovered.

Depreciation, Accelerated is any schedule of depreciation that reduces a sum of money more rapidly than would be done with straight line depreciation.

Depreciation, Book is a component of the carrying charge, it is the revenue required to repay the original investment. In the utility industry, it is usually calculated on a straight-line basis.

Derate factor is the DC power produced by a PV system vs. AC power delivered into the grid.

Design Month is the insolation and load requiring maximum energy from the PV array.

Diffuse Insolation is sunlight received as a result of scattering due to clouds, fog, haze and dust.

Diffuse Radiation is sunlight received after ground reflection and atmospheric scattering.

Diffusion is the flow of free carriers from regions of high concentration to regions of low concentration, leading to a net movement of free carriers to regions of low concentration. Over time, the free carrier concentration will become uniform throughout the material.

Diffusion length is the average distance a carrier can move from point of generation until it recombines.

Diffusivity is the rate at which diffusion occurs. It depends on the velocity at which carriers move and on the distance between scattering events.

Diffusion Furnace is used to create p-n junctions in semiconductor devices and solar cells.

Diffusion Length is the mean distance a free electron or hole moves before recombining.

Diode is an electronic device that allows current to flow in one direction only.

Direct Beam Radiation is sunlight received directly from the solar rays.

Direct Current (DC) is electricity that flows in one direction through the conductor.

Direct Insolation is sunlight falling directly upon a collector.

Disconnect switch is a switch gear used to connect or disconnect components in a PV system.

Discount Rate is the rate used for computing present values, which reflects the fact that the value of a cash flow depends on the time in which the flow occurs.

Distributed Energy Resources (DER) is a variety of small modular power-generators used to add to or improve the operation of the power delivery system.

Distributed Generation is localized or on-site power generation.

Distributed Power is any power supply located near the point where the power is used.

Distributed Systems are installed at or near the location where the electricity is used.

Dollar Year is the year in which constant dollar results of an analysis are reported.

Donor is n-type dopant (Phosphorus) that puts an additional electron into an energy level very near the conduction band and is easily exited into the conduction band by incoming sunlight.

Donor Level is the level that donates conduction electrons to the system.

Dopant is a chemical element added a pure semiconductor the electrical properties.

Doping is the addition of dopant type chemicals to a semiconductor.

Downtime is the time when the PV system cannot provide power for the load.

Drift is type of transport that occurs as a result of an electric field being superimposed on the semiconductor (also occurs in metals). The result is a net movement of electrons in the direction opposite of the electric field and of holes in the direction of the electric field.

Duty Cycle is the ratio of active time to total time, or operating regime of PV system loads.

Duty Rating is the amount of time an inverter can generate full rated power.

E

e-beam is electron beam (e-beam ray, or deposition via electron beam).

Earnings is that portion of revenue that remains after all charges, including interest, have been satisfied.

Economic Dispatch is commitment and operation of electric generating units or load control activities so as to meet demand with minimum total system operating costs.

Edge-Defined Film-Fed Growth (EFG) is a method for making sheets of silicon material for PV devices.

Electric Circuit is the path followed by electrons through an electrical system.

Electric Current is the flow of electricity in a conductor (wires), measured in amperes.

Electric Power System (EPS) is a facility that delivers electric power to a load. (from IEEE Std 1547-2003).

Electrical grid is an integrated system of electricity transmission and distribution.

Electricity is the energy resulting from the flow of electrons or ions.

Electricity Demand is the rate at which electric energy is delivered, expressed in units of power, such as kilowatts, at a given instant (indicated) or averaged (integrated) over a designated period of time.

Electrochemical Cell is a device containing two conducting electrodes of dissimilar materials immersed in a chemical solution that transmits positive ions from the negative to the positive electrode.

Electrode is a conductor used to make contact with the non-conductive part of a circuit.

Electro-deposition is a process where metal is deposited from a solution of its ions.

Electrolyte is a liquid used to carry ions to be deposited at the electrodes of electrochemical cells.

Electromagnetic interference or compatibility (EMI/EMC) is electromagnetic interference (radio frequencies) produced by a device and electromagnetic compatibility (EMC) of the device. EMI may be radiated as a radio wave or conducted on the AC and DC lines.

Electron is an atomic particle with a negative charge and a mass of $1/1837$ of a proton.

Electron Volt (eV) is the amount of kinetic energy gained by an electron when accelerated through an electric potential difference of 1 V, equivalent to 1.603×10^{-19}.

Elevation angle is the angular height of the sun in the sky measured from the horizontal, at $0°$ at sunrise and $90°$ when the sun is directly overhead.

Embedded Cost is the average or fully distributed cost.

Energy is the capability of bodies or systems to do work or convert energy into other forms.

Energy Audit is a survey that shows how much energy is used in a place.

Energy Contribution Potential is the recombination occurring in the emitter region of a PV cell.

Energy Density is the ratio of available energy per pound in storage batteries.

Energy efficiency is the ratio of output (useful) energy to input (used) energy during an identified period.

Energy Level is the energy represented by an electron in the band model of a substance.

Energy of a photon is denoted by: $E = hc/\lambda$: where h = Planks' constant, c = speed of light. Light with high energy photons has a short wavelength, and vice-versa.

Epitaxial Growth is the growth of one crystal on the surface of another crystal.

Equilibrium is a system in balance with no net flow of energy. In semiconductors, equilibrium is where there is no external bias, no illumination and no transient changes.

Equilibrium carrier concentration is equal to the number of carriers in the conduction or the valence band.

Equinox is the two times of the year when the sun crosses the equator and night and day are of equal length. These events occur on March 21st (spring equinox) and September 23 (fall equinox).

Equity is that portion of a company's total capitalization resulting from the sale of common and preferred stock and retained equity earnings.

Equity Ratio is the ratio of equity money to total capitalization.

Equivalent series inductance (ESL) is the inductance associated with the construction and leads of capacitors.

Equivalent series resistance (ESR) refers to the power losses of a capacitor.

Emitter-turn-off thyristor (ETO) is a solid-state switch consisting of a thyristor device under development that is configured to facilitate device turn-off via emitter signals and generally switches faster than the commercial GTOs and can handle more power than IGBTs.

Escalation, Apparent is the total annual rate of increase in cost.

Escalation, Real is the annual rate of increase of an expenditure that is due to factors such as resource depletion, increased demand, and improvements in design or manufacturing (negative rate).

Expected Value is the mean or average value of a variable.

Expense is a cost for goods and services that are normally utilized.

Extrinsic Semiconductor is the product of doping a pure semiconductor.

F

Fermi Level is the energy level at which the probability of finding an electron is one-half.

Fill Factor is the ratio of a PV cell's actual power to its maximum current and voltage levels.

Financing is the sources of funds raised from external sources in order for a firm to invest and to conduct organizational operations.

Finance Period is the period of time for which an investment's financing is structured (e.g., a loan is amortized over 30 years).

Fixed Charge Rate is the factor by which the present value of capital investment is multiplied to obtain the annual cost attributable to the capital investment.

Fixed Tilt Array is a PV array set in at a fixed angle with respect to horizontal.

Flat-Plate Array is a PV array that consists of non-concentrating PV modules.

Flat-Plate Module is an arrangement of PV cells or material mounted on a rigid flat surface with the cells exposed freely to incoming sunlight.

Float-Zone is a method of growing high-quality silicon ingots.

Flow Through Accounting is an accounting practice used by regulated utilities in which deferred income taxes are passed on immediately either to the rate payers through a decrease in rates or to the stockholders through an increase in earnings (return of equity). It is the opposite of normalization accounting.

Free carriers is the carriers (electrons in the conduction band or the holes in the valence band) that are free to move about the semiconductor lattice, and therefore participate in conduction. This can only occur when an electron in the valance band is excited into the conduction band, therefore making it free to participate in conduction.

Frequency is the number of repetitions per unit time of a complete waveform, expressed in Hertz.

Frequency Regulation is the variability in the output frequency.

Fresnel Lens is an optical device that focuses light like a magnifying glass.

Full Sun is the sunlight received on Earth's surface at noon on a clear day (about 1,000 W/m^2).

G

Gallium (Ga) is a metal used in making solar cells and semiconductor devices.

Gallium Arsenide (GaAs) is a compound used to make solar cells and semiconductor material.

Gassing (or outgassing) is the evolution of gas from different materials.

Generation occurs when electron-hole pairs are generated due to the absorption of photons. It also denotes age or order, such as 1st and 2nd generation devices.

Gigawatt (GW) is a unit of power equal to 1 billion Watts; 1 million kilowatts, or 1,000 megawatts.

Grid Lines are metal contacts on the solar cell surface.

Grid-connected Systems act like central generating plants by supplying power to the grid.

Grid-interactive System is similar to grid-connected system, but may be more flexible in use.

H

Harmonic Content is the number of frequencies in the output waveform in addition to the primary frequency (50 or 60 Hz.).

Heat Rate is the amount of energy expressed in Btu required to produce a kWh of electric energy.

Heterojunction is a region of electrical contact between two different materials.

High Voltage Disconnect is the voltage at which a charge controller will disconnect the PV array.

High Voltage Disconnect Hysteresis is the voltage difference between the high voltage disconnect setpoint and the voltage at which the full PV array current will be reapplied.

Hole is the vacancy where an electron would normally exist, behaving like a positive particle.

Homo-junction is the region between an n-layer and a p-layer in a single material, PV cell.

Hurdle Rate is the minimum acceptable rate of return on a project.

Hybrid System is a PV system that includes other sources of electricity generation, such as wind.

Hydrogenated Amorphous Silicon is a-Si with small amount of incorporated hydrogen.

I

Ideality factor is a measure of how closely the current voltage characteristics of a diode follow the ideal diode equation.

Incident Light is light that shines onto the face of a solar cell or module.

Indium Oxide is a semiconductor, used as a front contact or a component of a solar cell.

Infrared Radiation is electromagnetic radiation from 0.75 to 1000 micrometers wavelength.

Inflation is the rise in price levels caused by an increase in available currency and credit without a proportionate increase in available goods and services of equal quality. Inflation does not include real escalation.

Input Voltage is determined by the total power required by the alternating current loads and the voltage of any direct current loads.

Insolation is the solar power density incident on a surface of certain area and orientation.

Insulator is a class of materials that do not participate in conduction because the valence electrons are tightly bonded to their atoms.

Integrated Resource Planning is a set of analytic tools used to determine whether benefits are greater than costs for various resource planning options.

Interconnect is a conductor (wire) that connects the solar cells electrically.

Interconnection (system)—equipment and procedures necessary to connect a power generator to the utility grid.

Internal Rate of Return is the discount rate required to equate the net present value of a cash flow stream to zero.

Internal Rate of Return, Modified is the discount rate required to equate the future value of all returns to the present value of all investments.

Intrinsic carrier concentration is the concentration of carriers in undoped material. This concentration is typically significantly lower than the concentration of the majority carrier concentration in doped materials. In intrinsic material, the concentration of electrons in the conduction band is equal to the concentration of holes in the valence band.

Intrinsic Layer is pure, undoped, semiconductor material in the solar cell structure.

Intrinsic Semiconductor is a pure, undoped semiconductor.

Inverter is a device that converts DC to AC electricity.

Investment is an expenditure for which returns are expected to extend beyond a 1-year time frame.

Investment Useful Lifetime is the estimated useful life of a capital investment.

Investment Tax Credit is an immediate reduction in income taxes equal to a percentage of the installed cost of a new investment.

Investment Year is the year in which a capital or equipment investment is fully constructed or installed and placed into service.

Ion is an electrically charged atom that has lost or gained electrons.

Irradiance is the direct, diffused, and reflected solar radiation that strikes a surface.

Islanding is an unwanted condition where a portion of the grid is energized by a local generator, while that portion of the grid is supposed to be disconnected.

ISPRA Guidelines are guidelines for the assessment of PV power plants, published by the Joint Research Centre of the Commission of the European Communities, Ispra, Italy.

I-Type Semiconductor is a material that is left intrinsic.

I-V Curve is a graphical presentation of the current versus the voltage from a PV device as the load is increased from the short circuit (no load) condition to the open circuit (maximum voltage).

J

Joule is a metric unit of energy or work; 1 joule per second equals 1 watt or 0.737 foot-pounds.

Junction is a region of transition between semiconductor layers, such as a p-n junction.

Junction Box is an enclosure on the module for connectors, wires and protection devices.

Junction Diode is a semiconductor device passing current in one direction better than the other.

K

Kilowatt (kW) is a standard unit of electrical power equal to 1000 watts.

Kilowatt-Hour (kWh) is an energy measure of 1,000 watts acting over a period of 1 hour.

L

Langley (L) is unit of solar irradiance equal to 1 gram calorie/cm^2, or 85.93 kWh/m^2.

Lattice is the regular periodic arrangement of atoms in a crystal of semiconductor material.

Levelization is the conversion of a series of transactions to an equivalent value per unit of output.

Levelization Period is the time over which a series of transactions is converted to an equivalent value per unit of output.

Levelized Capacity Factor is a constant annual capacity factor for an electric generating unit such that the present value of the energy produced during the analysis period using constant annual capacity factors is the same as the present value of the energy produced by the individual annual capac-

ity factors.

Levelized Cost of Energy (LCOE) is the cost per unit of energy that, if held constant through the analysis period, would provide the same net present revenue value as the net present value cost of the system.

Life Cycle Cost (LCC) is the sum of all costs and expenses from the design to decommissioning stages of the project. Future costs must be calculated by considering the changing value of money.

Lifetime is the traditional definition of operational lifetime of the device. In photovoltaics the system lifetime may be 30 years or more. Lifetime also frequently refers to the minority carrier lifetime.

Light Trapping is the trapping of light inside a semiconductor material by refracting and reflecting the light at critical angles; trapped light will travel further in the material, greatly increasing the probability of absorption and hence of producing charge carriers.

Light-Induced Defects are defects, such as dangling bonds, induced in an amorphous silicon semiconductor upon initial exposure to light.

Line-Commutated Inverter is an inverter that is tied to a power grid or line.

Listed Equipment is equipment, components or materials included in a list published by an organization acceptable to the authority having jurisdiction and concerned with product evaluation, that maintains periodic inspection of production of listed equipment or materials, and whose listing states either that the equipment or materials meets appropriate standards or has been tested and found suitable for use in a specified manner. (from the National Electrical Code; Article 100.)

Load is the demand on an energy producing system; the energy consumption or requirement of a piece or group of equipment, expressed in terms of amperes or watts in reference to electricity.

Load Circuit is the wire, switches, fuses, etc. that connect the load to the power source.

Load Current (A) is the current required by the electrical device.

Load Duration Curve is a chart showing electric demand in decreasing magnitude plotted against total duration of occurrence over a specified period of time

Load Factor is the ratio of the actual energy consumed during a designated period to the energy that would have been consumed if the peak load were to exist throughout the designated period.

Load Management is the application of measures to influence customers' use of electricity so as to modify the demand and load factor.

Load Profile Curve is a chart showing chronological electric demand plotted against time of occurrence. It illustrates the varying magnitude of the load during the period covered.

Load Resistance is the resistance presented by the external electrical load.

Localized Surface Plasmon Resonance (LSPR) is represented by collective electron charge oscillations in metallic nanoparticles that are excited by light. It is specific for nanometer-size structures.

Loss of Load Probability (LOLP) is the proportion of time that the available generation is expected to be unable to meet the system load.

Low Voltage Cutoff (LVC) is the level at which a charge controller will disconnect the load.

Low Voltage Disconnect is the voltage at which a charge controller will disconnect the load.

Low Voltage Disconnect Hysteresis is the voltage difference between the low voltage disconnect setpoint and the voltage at which the load will be reconnected.

Low Voltage Warning is a warning signal indicating that a low voltage setpoint has been reached.

M

Majority Carrier is the current carriers (either free electrons or holes) that are in excess in a specific layer of a semiconductor material.

Marginal Cost of capital is the cost of additional (perhaps unplanned) funds.

Maximum Power Point (MPP) is the point on the current-voltage (I-V) curve of a module under illumination, where the product of current and voltage is at a maximum.

Maximum Power Point Tracker (MPPT) is the means of a power conditioning unit that automatically operates the PV generator at its maximum power point under all conditions.

Maximum Power Tracking (Peak Power Tracking) is the operation a PV array at the peak power point of the array's I-V curve where maximum power is obtained.

Megawatt (MW) is 1,000 kilowatts, or 1 million watts is a measure of PV array generating capacity.

Megawatt-hour is 1,000 kilowatt-hours or 1 million watt-hours is a measure of generated energy.

Metalorganic is a crystalline compound, consisting of metal ions or clusters coordinated to often rigid organic molecules to form one-, two-, or three-dimensional structures.

Micro CHP is micro-combined heat and power.

Microgroove is a small groove scribed in the solar cell's surface, filled with metal for contacts.

Minority Carrier is a current carrier, either an electron or a hole, which is in the minority in a specific layer of a semiconductor material.

Minority Carrier Lifetime is the average time a minority carrier exists before recombination.

Modified Sine Wave is a waveform that has at least three states (i.e., positive, off, and negative).

Modularity is the use of multiple inverters connected in parallel to service different loads.

Module Derate Factor accounts for lower PV module output due for field operating conditions such as dirt accumulation on the module.

Modular inverter is compatible with the paralleling or summing with one or more inverters of the same or similar design.

Monolithic means that it was fabricated as a single structure.

Movistor is a metal oxide varistor, used to protect electronic circuits from surge currents.

Multicrystalline is a material composed of variously oriented, small, individual crystals. It is also referred to as poly-crystalline or semi-crystalline.

Multijunction (MJ) Device is a high-efficiency PV device containing two or more cell junctions.

Multi-level inverter uses circuit topology that switches segments of the energy source in and out of the output circuit in order to synthesize a current sourced low frequency (typically 50 or 60 Hz) sine waveform.

Multi-Stage Controller is a charging controller unit that allows different level charging currents.

N

National Electrical Code (NEC) contains guidelines for all types of electrical installations. The Article 690, "Solar Photovoltaic Systems" should be followed when installing a PV system.

National Electrical Manufacturers Association (NEMA) sets standards for some non-electronic products like junction boxes, frames and such.

Net Present Value is the value in the base year (usually the present) of all cash flows associated with a project.

Nickel Cadmium Battery contains nickel and cadmium plates and an alkaline electrolyte.

Nominal Voltage is a reference voltage, describing modules and systems (i.e., 12-volt system).

Non-islanding inverter shuts off in 10 cycles or less when subjected to islanded loads that are >±50% mismatch to inverter real-power output and power factor is less than 0.95 [45]. It also shuts off within 2s if the load to inverter match is <50%, the power factor is >0.95, and the quality factor is 2.5 or less.

Non-radiative recombination processes are where the energy of recombination is given to a second electron which then relaxes back to its original energy by emitting phonons.

Normal Operating Cell Temperature (NOCT) is the estimated temperature of a PV module when operating under 800 w/m² irradiance, 20°C ambient temperature and wind speed of 1 meter per second.

Normalization Accounting is an accounting practice used by regulated utilities in which deferred income taxes are accumulated in a reserve account and effectively used to purchase new investments. The rate base is reduced by the accumulated reserve.

N-Type is a negative semiconductor material in which there are more electrons than holes.

N-Type Semiconductor is a semiconductor produced by doping an intrinsic semiconductor with an electron-donor impurity (e.g., Phosphorus in silicon).

N-Type Silicon is a silicon substrate that has been doped with a material that has more electrons in its atomic structure than silicon.

O

Ohm is a the electrical resistance of a material equal to the resistance of a circuit in which the potential difference of 1 volt produces a current of 1 ampere.

One-Axis Tracking is a system capable of rotating about one axis.

Open-Circuit Voltage (Voc) is the maximum possible voltage across a PV cell.

Operating Point is the current and voltage that a PV module produces when connected to a load.

Opportunity Cost is the rate of return on the best alternative investment available.

Optical path length is the distance that an unabsorbed *photon* may travel within the device before it escapes out of the device.

Orientation is placement of PV modules with respect to

N, S, E, W direction.

Outgas is the emission of gas from some materials under certain conditions.

Overnight Construction Cost is the value of total plant investment if construction had occurred overnight and all expenditures were made instantaneously.

P

Packing Factor is the ratio of array area to actual land area of the PV system.

Parallel Connection is connecting positive leads together and negative leads together.

Parallel/Paralleling: The act of synchronizing two independent power generators (i.e. the utility and a photovoltaic power plant) and connecting or "paralleling" them onto the same bus. In practice, it is used interchangeably with the term interconnection. IEEE 100 Def.: "The process by which a generator is adjusted and connected to run in parallel with another generator or system."

Passivation is a chemical reaction that eliminates the detrimental effect of electrically reactive atoms on a solar cell's surface.

Payback Period is the time required for net revenues associated with an investment to return the cost of the investment.

Payback Period, Simple is the payback period computed without accounting for the time value of money.

Payback Period, Discounted is the payback period computed that accounts for the time value of money.

Peak Demand/Load is the maximum energy demand or load in a specified time period.

Peak Power Current is the amperes produced by a PV module or array operating at the voltage of the I-V curve that will produce maximum power from the module.

Peak Power Point is the operating point of the I-V curve for a solar cell or PV module where the product of the current value times the voltage value is at a maximum.

Peak Sun Hours is the number of hours per day when the solar irradiance is at maximum.

Peak Watt is the maximum nominal output of a PV device, in watts (Wp) under STC.

Performance test conditions (PTC) is a fixed set of ambient conditions that constitute the dry-bulb temperature (20°C), the in-plane irradiance (1000 W/m² global for flat-plate modules, 850 W/m² for concentrators), and wind speed (1 m/s) at which electrical performance of the PV system is reported.

Phonon is encountered when at a given temperature the crystal lattice vibrates. Vibrations in the crystal lattice are described by phonons the same way that electromagnetic vibrations are described by photons. Phonons modify the way that electrons and photons interact and they explain absorption and emission in greater detail.

p-n junction is a structure formed between a p- and n-type layer.

Phosphorous (P) is a chemical element used as a dopant in making n-type semiconductor layers.

Photocurrent is an electric current induced by radiant energy.

Photoelectric Cell is a device for measuring light intensity.

Photoelectrochemical Cell is a type of PV device in which the electricity induced in the cell is used immediately within the cell to produce a chemical, such as hydrogen, which can then be used.

Photon is a particle of light that acts as an individual unit of energy.

Photon flux is a term used in determining the number of electrons which are generated, and hence the current produced from a solar cell. Defined as the number of photons per second per unit area.

Photovoltaic Array is an interconnected system of PV modules that function as a single electricity-producing unit.

Photovoltaic Cell (solar cell) is the smallest semiconductor element within a PV module to perform the immediate conversion of light into electrical energy.

Photovoltaic Conversion Efficiency is the ratio of the electric power produced by a PV device to the power of the sunlight incident on the device.

Photovoltaic Device is a solid-state electrical device that converts light directly into direct current.

Photovoltaic Effect is the phenomenon that occurs when photons in a beam of sunlight knock electrons loose from the atoms they strike, thus creating useful electric current.

Photovoltaic Generator is a PV array, or system, which is electrically interconnected.

Photovoltaic Module is the smallest environmentally protected, essentially planar assembly of solar cells and ancillary parts, such as interconnections, terminals, [and protective devices such as diodes] intended to generate direct current power under non-concentrated sunlight.

Photovoltaic Panel is often used interchangeably with "PV module" (especially in one-module systems), but more accurately used to refer to a physically connected collection of modules (i.e., a laminate string of modules used to achieve a required voltage and current).

Photovoltaic System is a complete set of components for converting sunlight into electricity by the PV process, including the array and BOS components.

Photovoltaic(s) (PV) pertains to the direct conversion of light into electricity via the photovoltaic effect.

Photovoltaic-Thermal (PV/T) System is a PV system that, in addition to converting sunlight into electricity, collects the residual heat energy and delivers both heat and electricity in usable form—physical.

P-I-N is a semiconductor PV device structure that layers an intrinsic semiconductor between a p-type semiconductor and an n-type semiconductor, used mostly with amorphous silicon PV devices.

Point-contact Cell is a high efficiency silicon PV concentrator cell that employs light trapping techniques and point-diffused contacts on the rear surface for current collection.

Point of common coupling is the point at which the electric utility and the customer interface occurs.

Polycrystalline Silicon is a material used to make PV cells, which consist of many twisted and intersecting crystals unlike the uniform single-crystal silicon structure.

Power Conditioning is the process of modifying the characteristics of electrical power.

Power Conditioning Equipment is used to convert power from a PV array into a form suitable for subsequent use. A collective term for inverter, converter, battery charge regulator, and blocking diode.

Power conditioning unit (PCU) is a device that converts the DC output of a PV array into utility-compatible AC power (inverter). It may include the array maximum power tracker, protection equipment, transformer, and switchgear.

Power Conditioning Subsystem (PCS): The subsystem (inverter) that converts the DC power from the array subsystem to AC power that is compatible with system requirements. (From 100-1996) See also Inverter.

Power Conversion Efficiency is the ratio of output power to input power of the inverter.

Power Density is the ratio of available power from a source to its mass (W/kg) or volume (W/l) and is calculated by multiplying photon flux by the photon energy.

Power Factor (PF) is the ratio of actual power being used in an electric circuit, expressed in watts or kilowatts, to the power that is actually drawn from a power source in volt-amperes or kilovolt-amperes.

Power foldback is an operational function whereby the unit reduces its output power in response to high temperature, excessive input power, or other conditions.

Preferred Stock is a type of security in which the holder is a part owner in the company.

Present Value is the value in the base year (usually the present) of a cash flow adjusted for the time-value differences in those cash flows between the time of the actual flow and the base year.

Producer Surplus is the amount of money paid to producers by consumers for a given quantity of a commodity over and above the amount the producers would have been willing to accept for the given quantity of the commodity.

Projected Area is the net south-facing glazing area projected on a vertical plane.

P-Type Semiconductor is a semiconductor in which holes carry the current; produced by doping an intrinsic semiconductor with an electron acceptor impurity (e.g., boron in silicon).

Pulse-width-modulated (PWM) is a wave inverter that produces a high quality (nearly sinusoidal) voltage, at minimum current harmonics.

Pyranometer is an instrument used for measuring global solar irradiance.

Pyrheliometer measures direct beam solar irradiance with aperture of 5.7° to transcribe the solar disc.

Q

Quad is one quadrillion (1,000,000,000,000,000) Btu.

Qualification Test is a procedure applied to commercial PV modules involving the application of defined electrical, mechanical, or thermal stress tests in a prescribed manner and amount.

R

Radiative recombination processes are where the energy of recombination results in the emission of a photon.

Rate Base is that portion of total assets (principally investments in plant and equipment) for regulated utilities, as defined by a rate regulatory body, upon which a utility is allowed to earn a return.

Rated Module Current (A) is the current output of a PV module measured at STC.

Rated Power of the inverter is the full, per spec, power an inverter can generate.

Reactive Power is the sine of the phase angle between the current and voltage waveforms in an alternating current system.

Recombination is the action of a free electron falling back into a hole.

Rectifier is a device that converts AC into DC current.

Regulators prevent excess power by controlling charge cycle to conform to specific needs.

Reinvestment Rate is the rate of return at which cash flows from an investment are reinvested.

Remote Systems are same as stand-alone (non-grid connected) systems. **Reserve Capacity** is the amount of generating capacity a central power system must maintain to meet peak loads as and when needed.

Resistance (R) is the electromotive force needed for a unit current flow, or the property of a conductor to oppose the flow of an electric current resulting in the generation of heat in the conductor.

Resistive Voltage Drop is the voltage developed across a cell by the current flow through the resistance of the cell.

Return on Debt is a component of the carrying charge, return on debt is the revenue required to pay for the use of debt money.

Return on Equity is a component of the carrying charge, return on equity is the revenue required to pay for the use of equity money.

Revenue Requirement is the amount of money that must be collected from customers to compensate a utility for all expenditures associated with an investment.

Reverse Current Protection is any method of preventing unwanted current flow from the load to the PV array (usually at night).

Ribbon (PV) Cells is a type of PV device made in a continuous process of pulling material from a molten bath of PV material, such as silicon, to form a thin sheet of material.

Risk Analysis is a method of quantifying and evaluating uncertainty.

Root Mean Square (RMS) is the square root of the average square of the instantaneous values of an AC output. For a sine wave the RMS value is 0.707 times the peak value. The equivalent value of alternating current, I, that will produce the same heating in a conductor with resistance, R, as a DC current of value I.

S

Sacrificial Anode is a piece of metal buried near a structure to protect it from corrosion.

Salvage Value is the value of a capital asset at the end of a specified period.

Satellite Power System (SPS) is a concept for providing large amounts of electricity to Earth via one or more satellites in geosynchronous Earth orbit.

Savings Investment Ratio is the ratio of discounted net savings accrued from an investment to the discounted capital costs (plus replacement costs minus salvage value).

Scenario Analysis is the evaluation of a set of conditional relationships between variables.

Schottky Barrier is a cell barrier established as the interface between a semiconductor, such as silicon, and a sheet of metal.

Scribing is the cutting of a grid pattern of grooves in a semiconductor material, generally for the purpose of making interconnections.

Self-commutated inverters use switches and controls that may be turned "on" or "off" at any time, using a PWM method to generate a synthesized waveform. Semiconductor is any material that has a limited capacity for conducting an electric current.

Semi-crystalline is the same as multi-crystalline.

Sensitivity Analysis is the evaluation of a project under a number of different assumptions on the values of one or more uncertain variables.

Series Connection is a way of joining PV cells by connecting positive leads to negative leads.

Series Resistance is parasitic resistance to current flow in a cell due to mechanisms such as resistance from the bulk of the semiconductor material, metallic contacts, and interconnections.

Short-Circuit Current (Isc) is the current flowing freely through an external circuit that has no load or resistance. It is the maximum current possible.

Shunt Controller is a charge controller that redirects or shunts the charging current away from the battery. The controller requires a large heat sink to dissipate the current from the short-circuited PV array.

Shunt resistance is parasitic resistance in parallel with the solar cell. A high efficiency solar cell requires a high shunt resistance.

Siemens Process is a commercial method of making purified silicon.

Silicon (Si) is a semi-metallic chemical element that makes an excellent semiconductor material for PV devices. It crystallizes in face-centered cubic lattice like a diamond.

Silicon controlled rectifier (SCR) is a thyristor that cannot be switched from "on" to "off" with gate controls unless current through it passes below a holding threshold (typically through zero).

Simulated Utility: an assembly of voltage and frequency test equipment replicating a utility power source. Where appropriate, the actual Area EPS can be used as the Simulated Utility. (From IEEE P1547.1).

Sine Wave is a waveform corresponding to a single-frequency periodic oscillation that can be mathematically represented as a function of amplitude versus angle in which the value of the curve at any point is equal to the sine of that angle.

Sine Wave Inverter is an inverter that produces utility-quality, sine wave power forms.

Single Junction (SJ) Device has one single p-n junction, vs. multi-junction (MJ) PV devices.

Single-crystal Material is a material that is composed of a single crystal structure.

Single-crystal Silicon is material with a single crystalline formation.

Single-stage Controller is a charge controller that redirects all charging current as the load dictates.

Solar Cell is same as photovoltaic (PV) cell.

Solar Constant is the average amount of solar radiation that reaches the earth's upper atmosphere on a surface perpendicular to the sun's rays equal to 1353 W/m², or 492 Btu/ft².

Solar Cooling is the use of solar thermal energy or solar electricity to power a cooling appliance.

Solar-grade Silicon or intermediate-grade silicon is used in the manufacture of solar cells.

Solar Energy is the electromagnetic energy transmitted from the sun (solar radiation). The amount that reaches the earth is equal to one billionth of total solar energy generated, or the equivalent of about 420 trillion kilowatt-hours.

Solar Noon is the time of the day, at a specific location, when the sun reaches its highest, apparent point in the sky; equal to true or due, geographic south.

Solar Radiance is the sun's instantaneous power density in units of kW/m². It varies throughout the day from 0 (at night) to 1 (max during the day). It can be measured either globally, or directly. This is also known as solar radiation. It can go over 1 kW/m².

Solar Resource is the amount of solar insolation a site receives, usually measured in kWh/m²/day, which is equivalent to the number of peak sun hours.

Solar Spectrum is the total distribution of electromagnetic radiation emanating from the sun.

Solar Thermal Electric Systems are solar energy conversion technologies that convert solar energy to electricity by heating a working fluid to power a turbine that drives a generator.

Solar time or the *local solar time* (LST) is when the sun is the highest in the sky, local time (LT) = varies from LST because of time zones, eccentricity of earth's orbit, etc.

Specific Gravity is the ratio of the weight of the solution to the weight of an equal volume of water at a specified temperature.

Spinning Reserve is the electric power plant or utility capacity on-line and running at low power in excess of actual load.

Split-spectrum Cell is a compound PV device in which sunlight is first divided into spectral regions by optical means. Each region is then directed to a different PV cell optimized for converting that portion of the spectrum into electricity, thus achieving significantly greater overall conversion efficiency.

Sputtering is a process used to apply PV semiconductor material to a substrate by a physical vapor deposition process where high-energy ions are used to bombard the source material, ejecting vapors of atoms that are then deposited in thin layers on a substrate.

Square Wave is a waveform that has only two states, (i.e., positive or negative) and contains a large number of harmonics.

Square Wave Inverter is a type of inverter that produces square wave output.

Staebler-Wronski Effect is the tendency of amorphous Silicon PV devices to degrade (drop) efficiency upon initial exposure to light.

Stand-alone System is an autonomous or remote PV system, not connected to a grid.

Stand-alone inverter (S-A) operates with the loads connected directly to its output and independent of any other AC power source.

Stand-by loss is the active and reactive power drawn from the utility grid when the power conditioner is in stand-by mode.

Standard Deviation is a statistical term that measures the variability of a set of observations from the mean of the distribution.

Standard Reporting Conditions (SRC) is a fixed set of conditions (including meteorological) to which the electrical performance data of a PV module are translated from the set of actual test conditions.

Standard Test Conditions (STC) are the conditions under which a module is tested in a laboratory, set at 1,000W/m² illumination and 25°C temperature.

Standby Current is the amount of current used by the inverter when no input power is available.

Stand-off Mounting is a technique for mounting a PV array on a sloped roof, which involves mounting the modules slightly above the pitched roof and tilting them to the optimum angle.

Static power converter (SPC) is a device used in some standards for any static power converter with control, protection and filtering functions used to interface an electric energy source with an electric utility system.

Storage Battery is a device capable of transforming energy from electric to chemical form and vice versa. These are used to store energy for night use, or in case of cloudy conditions.

String is a number of PV modules or panels interconnected electrically in series to produce the operating voltage required by the load.

String inverter is used in a single PV string of modules for its input.

Subjective Uncertainty is a probabilistic distribution representing a person's beliefs about the possible values of a variable, based on that person's experience and knowledge.

Substrate is the physical material upon which a PV cell is applied. Silicon, GaAs, glass, etc.

Subsystem is any one of several components in a PV system (i.e., array, controller, batteries, inverter, load, etc.).

Sunk Cost is any cost incurred by a prior decision that cannot be affected by the current course of action.

Superconducting Magnetic Energy Storage (SMES) is a technology that uses superconducting characteristics of low-temperature materials to produce intense magnetic fields to store energy. It has been proposed as a storage option to support large-scale use of photovoltaics as a means to smooth out fluctuations in power generation.

Superconductivity is the abrupt and large increase in electrical conductivity exhibited by some metals as the temperature approaches absolute zero.

Superstrate is the cover (glass usually) on the top side of a PV module, providing protection for the solar cells from impact and environmental degradation while allowing maximum transmission of the incoming sunlight.

Supervisory control and data acquisition (SCADA) is equipment used to monitor and control power generation, transmission, and distribution equipment.

Surface Plasmon Resonance (SPR) is the resonant (collective) oscillation of the valence electrons in the surface of the material, caused by stimulation from energy due to incident light. The resonance condition is observed when the frequency of the incoming light photons matches the natural frequency of the material's surface electrons, which are oscillating while opposed by the restoring force of positive nuclei.

Surface Recombination is an area of high recombination rates because the lattice structure is disrupted, leaving dangling bonds.

Surge Capacity is the maximum power, usually 3-5 times the rated power that can be provided over a short time.

System Availability is the percentage of time (usually expressed in hours per year) when a PV system will be able to fully meet the load demand.

System Operating Voltage is the PV array output voltage under load.

T

Taxable Income is that portion of revenue remaining after all deductions permitted under the Internal Revenue Code or a State Revenue Code have been taken.

Tax Recovery Class is one of several classes that determines the period of time over which the cost recovery tax deduction occurs.

Tax Preferences are the incentives designed to encourage investment as a stimulus to the overall economy.

Tax Rate is the rate applied to taxable income to determine federal and state income taxes.

Temperature Coefficient is used to determine decrease in PV module output at high temperature.

Temperature Factor is used to determine decrease in the current carrying capability of wire at high temperature.

Texturing is a process, usually an etch, to structure the surface of a solar cell to reduce reflectivity.

Thermo-photovoltaic Cell (TPV) is a device where sunlight concentrated onto an absorber heats it to a high temperature, and the emitted energy is used by a PV cell designed for that purpose.

Thick-crystalline Materials are semiconductor materials, typically measuring from 200-400 microns thick, cut from ingots or ribbons.

Thin Film is a layer of semiconductor material, such as copper indium diselenide or gallium arsenide, a few microns or less in thickness, used to make PV cells.

Thin Film PV Module is a PV module constructed with sequential layers of thin film semiconductor materials.

Tilt is the way the modules are inclined in respect to the horizon. Higher values of tilt angle usually increase the power production in winter and decrease it in summer.

Tilt Angle is the actual angle at which a PV array is set to face the sun relative to a horizontal position.

Tin Oxide is a wide band-gap semiconductor similar to indium oxide; used in heterojunction solar cells or to make a transparent conductive film, called NESA glass when deposited on glass.

Total AC Load Demand is the sum of the alternating current loads used in selecting an inverter.

Total Harmonic Distortion is the measure of closeness in shape between a waveform and its fundamental component.

Total Internal Reflection is the trapping of light by refraction and reflection at critical angles inside a semiconductor device so that it cannot escape the device and must be eventually absorbed by the semiconductor.

Total Plant Investment is the total plant cost as modified by escalation and interest during construction.

Tracking Array is a PV array that follows the path of the sun to maximize the solar radiation incident on the PV surface. One axis tracker is where the array tracks the sun east to west, while two-axis tracking is where the array points directly at the sun at all times. Tracking arrays use both the direct and diffuse sunlight. Two-axis tracking arrays capture the maximum possible daily energy.

Transformer is an electromagnetic device that changes the voltage of alternating current electricity.

Transient surge device is used to suppress transient power surges.

Tray Cable (TC) is used for interconnecting BOS components.

Tunneling is a Quantum mechanics concept whereby an electron is found on the opposite side of an insulating barrier without having passed through or around the barrier.

Thyristors are a family of semiconductor switching devices characterized by bi-stable switching (either "on" or "off") through internal regenerative feedback.

Two-axis Tracking is a PV array tracking system capable of rotating independently about its two x and y axes (e.g., vertical and horizontal).

U

Ultraviolet light is electromagnetic radiation in the wavelength range of 4 to 400 nanometers.

Uncertainty is the range of possible values a variable may have in the future.

Underground Feeder (UF) is used for PV array wiring if sunlight resistant coating is specified.

Underground Service Entrance (USE) is used for interconnecting BOS components.

Uninterruptible Power Supply (UPS) is the designation of a power supply providing continuous uninterruptible service.

Uniform Capital Recovery Factor is the uniform periodic payment, as a fraction of the original investment cost, that will fully repay a loan, including all interest, over the term of the loan.

Utilization (Array Utilization) is the ratio of the energy (or power) that is actually extracted from the module or array to the maximum energy (power) potentially available from the array. Array utilization less than 1.0 is a result of inaccurate Maximum Power Point Tracking.

Utility is the organization having jurisdiction over the interconnection of the PV system and with whom the owner would enter into an interconnection agreement.

Utility-interactive inverter (U-I) is connected to the utility grid or other stable AC source and usually incorporates an MPPT to maximize power delivered to the grid.

Utility-interactive Inverter is an inverter that can function only when tied to the utility grid, and uses the prevailing line-voltage frequency on the utility line as a control parameter to ensure that.

V

Vacuum Evaporation is the deposition of thin films of semiconductor material by the evaporation of elemental sources in a vacuum.

Vacuum Zero is the energy of an electron at rest in empty space; used as a reference level in energy band diagrams.

Valence Band is the highest energy band in a semiconductor that can be filled with electrons.

Valence Electrons are electrons that can participate in the formation of bonds with other atoms.

Valence Level Energy/Valence State is the energy content of an electron in orbit about an atomic nucleus. It is also called bound state.

Varistor is a voltage-dependent variable resistor, used to protect sensitive equipment from power spikes or lightning strikes by shunting the energy to ground.

Vapor Deposition is a method of depositing thin films by evaporating the film materials.

Vertical-junction field-effect transistor (VJFET) is a field-effect (SiC) device.

Vertical Multijunction (VMJ) Cell is a compound cell made of different semiconductor materials in layers, one above the other. Sunlight entering the top passes through successive cell barriers, each of which converts a separate portion of the spectrum into electricity, thus achieving greater total conversion efficiency of the incident light. It is also called a multiple junction cell.

Volt (V) is a unit of electrical force equal to that amount of electromotive force that will cause a steady current of one ampere to flow through a resistance of one ohm.

Voltage is the amount of electromotive force, measured in volts, that exists between two points.

Voltage at Maximum Power (Vmp) is the voltage at which maximum power is available from a PV module.

Voltage-controlled inverter converts DC to AC power where the output voltage is controlled. Typically used in stand-alone applications since the output voltage must be regulated within the inverter.

Voltage Protection is used where many inverters have sensing circuits that will disconnect the unit from the load if input voltage limits are exceeded.

Voltage Regulation indicates the variability in the output voltage. Some loads will not tolerate voltage variations greater than a few percent.

W

Wafer is a thin sheet of semiconductor (PV material) made by cutting it from an ingot.

Watt is the rate of energy transfer equivalent to one ampere under an electrical pressure of one volt.

One watt equals 1/746 horsepower, or one joule per second.

Waveform is the shape of the phase power at a certain frequency and amplitude.

Wave Packet is a collection of waves which may interact in such a way that the wave-packet may either appear spatially localized (act as a particle) or a wave. This concept is known as "wave-particle duality."

Weighted Average Cost of Capital is the weighted average of the component costs of debt, preferred stock, and common equity.

Window is a wide band gap material chosen for its transparency to light. Generally used as the top layer of a PV device, the window allows almost all of the light to reach the semiconductor layers beneath.

Wire Types are reviewed in detail in Article 300 of National Electric Code.

Work Function is the energy difference between the Fermi level and vacuum zero, or the minimum amount of energy it takes to remove an electron from a substance into the vacuum.

X

x-y Drive is the gears and controls of a two-axis tracker, which determine the path the tracker follows during the day in tracking the sun.

Z

Zenith Angle is the angle between the direction of interest (of the sun, for example) and the zenith (directly overhead).

List of Abbreviations

α	Maximum elevation at solar noon
	Absorption coefficient.
	Angle (rad)
Δ	Difference (delta)
Δn	Excess electron concentration.
ΔGa	Free enthalpy of anodic decomposition (kJ mol^{-1})
ΔGc	Free enthalpy of cathodic decomposition (kJ mol^{-1})
ΔGloss	Free energy losses related with anodic and cathodic over-potentials
$\Delta G(H_2O)$	Free energy of H_2O formation
λ	Wavelength of light in µm or nm.
λ_{mp}	Temperature coefficients of maximum power voltage of the PV modules
σ	Electrical conductivity (Ω^{-1} cm^{-1})
µi	Mobility of ionic charge carriers (cm^2 V^{-1} s^{-1})
µn	Mobility of electrons (cm^2 V^{-1} s^{-1})
µp	Mobility of electron holes (cm^2 V^{-1} s^{-1})
η	Viscosity, efficiency
η_e	Electricity generating efficiency, dimensionless
ηg	Fraction of efficient solar irradiance
ηch	Chemical efficiency of irradiation
ηQE	Quantum efficiency
λ	Wavelength (nm)
λi	Threshold wavelength
φ	Work function (eV)
φa	Work function of photo-anode (eV)
φel	Work function of electrolyte (eV)
Φ	Work function
Ø	Diameter
3D	Three dimensional
a	Anode/photo-anode
A	Anode
	Ampere
	Normally open contacts
	irradiated area (m^2)
AA	Atmospheric air,
	Atomic absorption
	Automatic adjust
ABC	Automatic background calibration
AC	Alternating current
ACB	Air circuit breaker
ACDC	Alternating current, direct current
A/D	Analog to digital

ADIO	Analog/digital input/output
AFD	active frequency drift (frequency bias)
AFDPF	AFD with positive feedback (aka SFS)
AFM	Atomic force microscopy
Ag paste	Silver-(epoxy) paste
AH	Ampere-hours
AI	Analog input
Al	Aluminum
Alq_3	Tris (8-hydroxyquinolinato) aluminum
AM	Air mass, amount (length) of atmosphere light passes through
ARC	Anti-reflective coating
As	Arsenic
ASIC	Highly integrated circuit package containing hundreds of logic functions, modified by burning-away internal paths to produce application specific Circuit functions (application specific module).
ASTM	American Standards for Testing Materials
ATE	Automatic test equipment
ATM	Atmosphere (measurement)
ATRP	Atom transfer radical polymerization
AWG	American wire gauge (measurement)
B	Boron
Ba	Barium
BCP	Bathocuproin
BE	Binding energy
BHJ	Bulk hetero-junction
BIPV	Building-Integrated Photovoltaics.
BEOL	Back end of line
BIF	Barrier improvement factor
BOM	Bill of materials
BOS	Balance of system components
c	Speed of light in vacuum (2.99793×10^8 m/s)
	Cathode/photo-cathode
C	Carbon
	Capacitor
c^2	Speed of light
Ca	Calcium
CAE	Computer aided engineering
CAFM	Conductive atomic force microscopy
CB	Carbon black
	Circuit breaker
	Conduction band
CdTe	Cadmium telluride
CEA	Commissariat à l'Énergie Atomique (EU)
CFM	Cubic feet per minute

CHA	Concentric hemispherical analyzer	Eg	Band gap (eV)
Cl	Chlorine	$E(H^+/H_2)$	Energy of the redox couple H^+/H_2 (eV)
CMP	Chemical mechanical polishing (wafers process)	Ei	Threshold energy (eV)
COC	Cost of consumables	$E(O_2/H_2O)$	Energy of the redox couple O_2/H_2O (eV)
COO	Cost of ownership	E_{loss}	Energy loss (eV)
CP	Process (performance) capability	E_1	Eectrolyte (or electrolyte state)
	Conductive polymer	$En_{,d}$	Free enthalpy of electrochemical oxidation (per one electron hole) (eV)
	Cathodic protection		
CPV	Concentrated photovoltaics	$Ep_{,d}$	Free enthalpy of electrochemical reduction (per one electron) (eV)
CPV/T	Concentrated PV plus thermal output (hybrid systems)		
		E_v	Energy of the top of the valence band (eV)
CPU	Central processing unit	EMF	Electromotive force (open circuit voltage) (V)
C_T	Empirical constant relating to the impact of cell temperature on output		
		EBPVD	Electron beam physical vapor deposition
Cu(In,Ga)Se$_2$	Copper indium-gallium di-selenide	ECN	Energy Research Centre of the Netherlands
CuPc	Copper phthalocyanine	EDG	Electron donating group
CV	Cyclic voltametry	EDX	Energy-dispersive x-ray analysis
CVD	Chemical vapor deposition	EOL	End-of-life
CZ	Czochralski process (silicon mfg.)	epi	Epitaxial
DA	Dry air	E_{PV}	Energy to grid (kWh)
DAC	Digital to analog converter	EQE	External quantum efficiency
DC	Direct current	ERDA	Elastic recoil detection analysis
DCS	Di-chloro silane	ES	Engineering specification
DER	Distributed Energy Resource		Expert system
DF	Diode factor, dimensionless	ESCA	Electron spectroscopy for chemical analysis
DG	Distributed Generation	ESL	Equivalent series inductance.
DLARC	Double layer anti-reflection coating	ESR	Equivalent series resistance.
dppp	1,3-bis (diphenylphosphino) propane	F	Faraday constant (F = eN_A) (9.648 x 10^4 C mol^{-1})
DSC	Differential scanning calorimetry		
DSP	Digital signal processor	ETO	Emitter-turn-off thyristor
DI	De-ionized (water, or air)	EVA	Ethyl vinyl acetate
DOE	Design of experiment	EWG	Electron withdrawing group
DTR	Diffusion transfer reversal	F	Frequency
DUT	Device under test	F_8BT	Poly (9,9-dioctylfluorene-co-benzothiadiazole)
Dye	Photo-sensitizer at ground state		
Dye$^+$	Dye at excited state	FA	Failure analysis
Dye$^+$	Dye at charged state	FACS	Flexible AC Transmission Systems
e	Electron	FCC	federal communications commission
	Electric charge on an electron, 1.60×10^{-19} C	FEOL	Front end of line
		FET	Field-effect transistor
	Elementary charge (1.602×10^{-19} C)	FF	Fill factor
e′	Quasi-free electron	FWHM	Full width at half maximum
E	Energy (eV)	FZ	Float zone (silicon mfg.)
	Energy (potential, non-dimensional)	G	Gibbs energy (free enthalpy) (kJ mol^{-1})
Ea	Activation energy	G^0	Standard Gibbs energy (standard free enthalpy) (kJ mol^{-1})
E_A	Array output energy (kWh)		
E_B	Potential energy related to the bias ($E_B = eV_{bias}$)	GaAs	Gallium arsenide
		GB	Gigabit
E_c	Energy of the bottom of the conduction band (eV)	GEN	Generator
		JFET	Junction field-effect transistor
E_F	Fermi energy (eV)	GFCI	Ground fault circuit interrupter

GHz	Gigahertz		Integrated control panel
GND	Ground	IGBT	Insulated gate bi-polar transistor
gr	Gram	Impp	Current at maximum power point (I_{mp})
G_{STC}	Reference irradiance at STC (1 kW/m^2)	InGaAsP	Indium Gallium Arsenide Phosphide
GTO	Gate turn off device	I/O	Input/output
GW	Gigawatt = 10^9 W	IPCE	Incident photon-to-current efficiency
h	Planck constant (6.626 x 10^{-34} J s)	IPR	Intellectual property rights
h·	Quasi-free electron hole	IR	Infrared
h_c	Convective heat transfer coefficient (W/(m^2 K)	ITN	Iso-thianaphthene
		ITO	Indium tin oxide
h_r	Radiative heat transfer coefficient (W/(m^2 K)	IV	Current-voltage
		IV-curve	Current-voltage diode characteristics
H	Henry	I-V-L	Current-voltage-luminance
	Hybrid	J	Joule
H$^+$	hydrogen ion (can be considered as hydro-nion ion H$_3$O$^+$)		Flux density (flowing across a given area—unit area perpendicular to the flow—per unit time, e.g. number of particles) (m^{-2} s^{-1})
	Concentration of hydrogen ions (M)		
H$_2$	Hydrogen	Jg	Flux density of absorbed photons (m^{-2} s^{-1})
$H_{(t,\beta)}$	Incident irradiance in the plane of the PV generator	K	Potassium
			Acceleration factor
H$_2$O	Water	k	Boltzmann constant, or kB. 1.38 × 10^{-23} J/(K mol)
HALT	Highly accelerated life tests to reveal component weakness related to premature failure mechanisms and mean-time-to-first-failure (MTBF).		
		K_a	Thermal conductivity of air (W/(m K))
		$k$deg	Degradation constant
		KE	Kinetic energy
HAZCOM	Hazard Communication Standard	kg	Kilogram
HC-PEDOT	Highly conductive PEDOT	KOH	Potassium hydroxide
HCPV	High concentration photovoltaics	kV	Kilovolt
HCPV/T	High concentration PV plus thermal output (hybrid system)	kW	Kilowatt (1000 Watts)
		l	Length
HF	High frequency	L	Luminance
	Hydrofluoric acid		Distance from entry point (m)
HOMO	Highest occupied molecular orbital		Inductor
HPE	hybrid photo-electrode	LC	Liquid chromatography
HPLC	High-performance liquid chromatography		Liquid crystal
H_T	Mean daily irradiance in array plane (kWh/m^2 d)		Inductance / capacitance
			Limit cycle
HTO	Tritium-containing water	LCD	Liquid crystalline display
HV	High voltage	LED	Light-emitting diode
HWE	Horner-Wadsworth-Emmons	LiF	Lithium fluoride
i	Concentration of ionic charge carriers (cm^{-3})	LIP	Localized irradiation probe
I	Current	LPCVD	Low pressure chemical vapor deposition
	Intensity	LUMO	Lowest unoccupied molecular orbital
	Input	m	Mass, meter, micron
I_L	Light generated current (A)		Number of parallel connected cells (dimensionless)
I_o	Diode current (A)		
I_r	Incidence of solar irradiance (W m^{-2})	mA	Miliampere (0.001A)
Isc	Short-circuit current	mfp	Mean free path
IC	Integrated circuit	M	Metal
	Ion chromatography	MOx	Metal oxide (x corresponds to oxygen stoichiometry)
ICP	Inductively coupled plasma		

MB	Megabyte	O_2	Oxygen
MALDI-TOF	Matrix-assisted laser desorption/ionization–time-of-flight	ODCB	Ortho-dichlorobenzene or 1,2-dichlorobenzene
MCBI	Main cycles between interrupts	OEM	Original equipment manufacturer
MDMO-PPV	Poly (2-methoxy-5-(3,7-dimethyloctyloxy)-1,4-phenylenevinylene)		Original engineers model
		OFR	Over frequency relay
MEH-PPV	Poly [2-methoxy-5-(2-ethylhexyloxy)-1,4-phenylenevinylene]	OLED	Organic light-emitting device or organic light emitting diode
MeOH	Methanol	OLTC	On Load Tap Changers of transformers
MeV	Megaelectron volt	OPRF	Optimal Reactive Power Flow
Mg	Magnesium	OPV	Organic photovoltaic
MOCVD	Metal organic CVD	OPVC	Organic photovoltaic cell
Mol wt	Molecular weight	OTR	Oxygen transmission rate
MOS	Metal on silicon	p	Concentration of electron holes (cm^{-3})
MOSFET	Metal oxide field-effect transistor.	pc	Polycrystalline
MOV	Metal oxide varistor	pH	$-\log [H^+]$ (degree of acidity or alkalinity of a solution)
MSD	Monitoring units with allocated all-pole switching connected in series.		
		Pin	Incoming solar power
MT	Metric ton	Pmax	Maximum power
MTBF	Mean time between failure	P_{mp}	DC power supplied by the PV generator at maximum power point
MTTF	Mean time to failure		
MTTR	Mean time to repair	P_O	Peak power (W_P) is the maximum power of the PV generator
mV	Millivolt (0.001V)		
MW	Megawatt	Pout	Output electrical power
n	Concentration of electrons (cm^{-3})	P	Phosphorous
	Number of series connected cells (dimensionless)		Pico-
			Potential
ηe	Energy conversion efficiency	P_3CT	Poly(3-carboxythiophene-co-thiophene)
ηn	Viscosity in a spin-coating process	PCB	Printed circuit board
N_A	Avogadro number (6.022×10^{23} mol^{-1})	PCBM	[6,6]-phenyl C61-butyric acid methyl ester
N_{eff}	Efficient number of incidents	PCC	Point of common coupling
N_{tot}	Total number of incidents	PCE	Power conversion efficiency
N_2	Nitrogen	PCS	Power conditioning system (aka inverter)
N	Negative (connection)	PCT	Patent cooperation treaty
	Neutral	PD	Poly-dispersity
	Nano	PE	Poly(ethylene)
	Number of panels in surface (dimensionless)	PEC	Photo-electrochemical cell
	Number of photons		Positive (connection)
Na	Sodium		Power output (W)
NBS	N-bromosuccinimide	PECVD	Plasma enhanced chemical vapor deposition
NEC	National Electrical Code	PEDOT	Poly(ethylendioxythiophene)
N(E)	Distribution of photons with respect to energy (s^{-1} m^{-2} eV^{-1})	PEN	Poly(ethylenenaphthalate)
		PEOPT	Poly[3-(4_-(1__,4__),7__-trioxaoctyl) phenylthiophene]
NFPA	National Fire Protection Association		
NHE	Normal hydrogen electrode	PERL	Passivated Emitter Rear Locally diffused
NIR	Near infrared	PET	Poly (ethylene terephthalat)
NMR	Nuclear magnetic resonance	PFB	Poly (9,9-dioctylfluorene-co-bis-N,N-(4-butylphenyl)-bis-N,N-phenyl-1,4-phenylenediamine)
NRA	Nuclear reaction analysis		
NREL	National Renewable Energy Laboratory		
OH$-$	Hydroxyl ion	PFO	Poly(9,9-dioctyl-fluorene)
O	Output	PITN	Poly(isothianaphthene)

PJD	Phase jump detection
PLD	Pulsed laser deposition
PLED	Polymer light-emitting device or polymer light emitting diode
PLL	Phase locked loop
POCC	Point of common coupling
POMeOPT	Poly[3-(2-methoxy-5-octylphenyl)-thio-phene]
PPV	Poly(phenylenevinylene)
PSS	Poly(styrene sulfonic acid)
(PTC)	Performance test conditions
PTCA	Perylene tetracerboxylic acid
PTCBI	Perylene tetracarboxylic acid bisimide
PTOPT	Poly[3-(4-octylphenyl)-2,2_-bithiophene]
PTV	Poly(thienylenevinylene)
PV	Photovoltaic
PVC	Poly vinyl chloride
PVD	Physical vapor deposition
PVUSA	PVs for utility scale applications.PWM Pulse width modulated PWM.
Q	Solar insolation (W/m^2)
QA/QC	Quality assurance/quality control
Q_{ref}	Reference insolation (usually 1000 W/m^2) (W/m^2)
R	Universal gas constant (8.3144 J mol^{-1} K^{-1})
R	Resistance (Ω)
	Reliability
	Redundant
Rs	Series resistance
R(H$_2$)	Rate of hydrogen generation (mol s^{-1})
R2R	Roll-to-roll
RBS	Rutherford backscattering
RCMU	Residual current monitoring unit.
RF	Radio frequency
RMS	Root mean square
ROI	Return of investment
RPC	Reactive Power Controller
RR	Rectification ratio
sc	Single crystal
S	Surface area (m^2)
S-A	Stand alone inverter
SAD	Silicon Avalanche device, a transient surge suppression device.
SBIR	Small business innovative research program
SCADA	Equipment used to monitor and control power generation, transmission, and distribution equipment
SCR	Thyristor that cannot be switched from "on" to "off" with gate controls, unless current through it passes below a holding threshold (usually zero)

SD	Sputter deposition	
SEC	Size exclusion chromatography	
SEM	Scanning electron microscope	
SFS	Sandia frequency shift	
SMS	Slide-mode frequency shift	
SRC	Standard reporting conditions	
SOV	Silicon oxide varistor, a transient surge suppression device	
(SPC)	Static power converter	
STC	Standard test conditions	
Si	Silicon	
	a-Si	amorphous silicon
	c-Si	crystalline silicon
	mc-Si	multi-crystalline silicon
	MG-Si	metallurgical grade silicon
	p-Si	poly-crystalline silicon
	sc-Si	single crystal silicon
	SG-Si	solar grade silicon
SiO$_2$	Silicon Dioxide (sand)	
SiC	Silicon carbide	
SiN	Silicon Nitride	
SIMS	Secondary-ion-mass spectrometry	
SMU	Source measure unit	
SofA	State of the art	
SOI	Silicon on insulator	
STC	Standard test conditions	
SVC	Static VAR Compensator	
t	Time	
T	Temperature	
	C Centigrade	
	F Fahrenheit	
Ta	Annealing temperature	
T_c	Temperature in the rear part of the cell or PV module (K)	
TCO	Transparent conductive oxide	
TCR	Thyristor Controlled Reactor	
TE	Thermal evaporation	
TEM	Transmission electron microscopy	
TEMPO	2,2,6,6,-tetramethylpiperidin-1-oxyl	
Tg	Glass transition temperature	
THF	Tetra-hydro-furane	
TF	Thin film	
TFPV	Thin film photovoltaics	
TMY	Typical Meteorological Year	
TOF	Time-of-flight	
TOF-SIMS	Time-of-flight secondary-ion-mass-spectrometry	
T_{ref}	Reference temperature (usually at 298 K)	
TSC	Thyristor Switched Capacitor	
TSD	Transient surge device	
TW	Terawatt = 10^{12} W	

x	Number (related to nonstoichiometry in chemical formulas)	V_h	Potential drop across the hybrid photo-electrode (V)
X	Anion in salts, such as Cl^- or SO_4^{2-}	$V_{ph}(Si)$	Photo-voltage across the Si cell (V)
z	Number of electrons (electron holes)	$V_{ph}(TiO_2)$	Photo-voltage across the oxide photo-electrode (V)
U_a	Anodic over-potential (V)	VD	Vapor deposition
U_c	Cathodic over-potential (V)	VE	Vacuum evaporation
U_{fb}	Flat band potential (V)	VJFET	Vertical-junction field-effect transistor
UFR	Under frequency relay	Vmpp	Voltage at maximum power, or V_{mp}
UHV	Ultrahigh vacuum	V_{oc}	Open circuit voltage at reference values (V)
U-I	Utility-interactive inverter	Voc	Open-circuit voltage
UIPV	Utility-interactive photovoltaic (system)	VTE	Vacuum thermal evaporation
UPS	Ultraviolet photoelectron spectroscopy	Vth	Theoretical paste volume
UV	Ultraviolet radiation	W	Watt
UV-vis	Ultraviolet-visible radiation	W/m^2	Watt per square meter (solar insolation)
v	Frequency (Hz)	Wp	Watt peak
V	Volume	WVTR	Water vapor transmission rate
V_{bias}	Bias voltage (V)	XPS	X-ray photoelectron spectroscopy
V_B	Surface potential (corresponding to band curvature) (V)	XRD	Grazing-incidence x-ray diffraction
$Vn_{,d}$	Cathodic decomposition potential (V)	XRD	X-ray diffraction
$Vp_{,d}$	Anodic decomposition potential (V)	XRF	X-ray fluorescence
V_H	Potential drop across the Helmholtz layer (V)	ZJ	Zetta joule = 1021 J

Index